Advances in Intelligent Systems and Computing

Volume 1115

The series "Advances in Intelligent Systems and Computing" contains publications on theory, applications, and design methods of Intelligent Systems and Intelligent Computing. Virtually all disciplines such as engineering, natural sciences, computer and information science, ICT, economics, business, e-commerce, environment, healthcare, life science are covered. The list of topics spans all the areas of modern intelligent systems and computing such as: computational intelligence, soft computing including neural networks, fuzzy systems, evolutionary computing and the fusion of these paradigms, social intelligence, ambient intelligence, computational neuroscience, artificial life, virtual worlds and society, cognitive science and systems, Perception and Vision, DNA and immune based systems, self-organizing and adaptive systems, e-Learning and teaching, human-centered and human-centric computing, recommender systems, intelligent control, robotics and mechatronics including human-machine teaming, knowledge-based paradigms, learning paradigms, machine ethics, intelligent data analysis, knowledge management, intelligent agents, intelligent decision making and support, intelligent network security, trust management, interactive entertainment, Web intelligence and multimedia.

The publications within "Advances in Intelligent Systems and Computing" are primarily proceedings of important conferences, symposia and congresses. They cover significant recent developments in the field, both of a foundational and applicable character. An important characteristic feature of the series is the short publication time and world-wide distribution. This permits a rapid and broad dissemination of research results.

**** Indexing: The books of this series are submitted to ISI Proceedings, EI-Compendex, DBLP, SCOPUS, Google Scholar and Springerlink ****

More information about this series at http://www.springer.com/series/11156

Zdenka Popovic · Aleksey Manakov ·
Vera Breskich
Editors

VIII International Scientific Siberian Transport Forum

TransSiberia 2019, Volume 1

Springer

Editors
Zdenka Popovic
Department of Roads,
Airports and Railways
University of Belgrade
Belgrade, Serbia

Aleksey Manakov
Siberian Transport University
Novosibirsk, Russia

Vera Breskich
Peter the Great St. Petersburg
Polytechnic University
St. Petersburg, Russia

ISSN 2194-5357 ISSN 2194-5365 (electronic)
Advances in Intelligent Systems and Computing
ISBN 978-3-030-37915-5 ISBN 978-3-030-37916-2 (eBook)
https://doi.org/10.1007/978-3-030-37916-2

This Springer imprint is published by the registered company Springer Nature Switzerland AG
The registered company address is: Gewerbestrasse 11, 6330 Cham, Switzerland

International Siberian Transport Forum—TransSiberia 2019

Preface

International Scientific Siberian Transport Forum—TransSiberia 2019—took place in Novosibirsk, Russia, on May 22–27, 2019.

The forum was organized by the Government of the Novosibirsk Region, the Ministry of Transport of Russia, the State Duma of the Russian Federation, the Legislative Assembly of Novosibirsk Region, and Siberian Transport University.

Leading experts in the field of transport from more than 15 countries met in Novosibirsk to exchange the latest scientific achievements, to strengthen the academic relations between the leading scientists of the European Union, Russia, and the World, and to create favorable conditions for collaborative research and implement collaborative projects in the fields of transport.

The authors from several countries submitted 575 qualified papers to TransSiberia 2019 conference. Of these, 214 papers were accepted. All papers passed a strict scientific, technical, and grammatical review.

Only papers of original research-type performing results of original studies are accepted.

The recommended size of a paper is 9–15 pages. The obligate condition for a scientific paper to submit is the accordance with IMRAD structure.

Within the framework of technical review, all papers are thoroughly checked for the following attributes:

(1) For compliance with the subject of the conference.
(2) For plagiarism, acceptable minimum of originality is 90%.

(3) For acceptable English language, all papers are reviewed by a native speaker.
(4) At the same time, papers are checked by a technical proofreader (quality of images, absence of Cyrillic, etc.).

Scientific review of each paper is made by at least three reviewers. If the opinions of the reviewers are radically different, additional reviewers are appointed. «Potential reviewers» recommended by authors are not used. Authors have a right to answer the remarks of reviewers and submit revised versions of their papers.

Live participation in the conference is an indispensable condition for the publication of the paper.

The conference included workshops and plenary sessions dedicated to the issues of road transportation, railroad transport, road engineering complex, air transport, public transport, transport and logistics complex, road safety, and passenger taxi.

Within the conference, the participants discussed a wide range of issues concerning digitalization and innovative development of transport, road network and road facilities, application of federal legislation in the field of passenger traffic, comfortable urban environment for low-mobility passengers, energy efficiency in transport, and other topics. The key agenda of discussions was the implementation of the national project "safe and high-quality roads" and the comprehensive plan for the expansion and modernization of infrastructure, including in the Siberian Federal District.

The members of our organizing committee express their deep gratitude to the crew of your journal. We appreciate your help in preparation of our TransSiberia 2019 conference volume.

The members of our organizing committee express their deep gratitude to the crew of Advances in Intelligent Systems and Computing journal for the publication of selected papers of the Siberian Transport Forum conference!

The Steering Committee of the VIII International Siberian Transport Forum—TransSiberia/TransLogistica 2019

Andrey Travnikov
Governor of the Novosibirsk Region, Chairman of the Steering Committee

Vladimir Znatkov
First Deputy Chairman of the Government of the Novosibirsk Region, Deputy Chairman of the Steering Committee

Anatoly Kostylevsky
Minister of Transport and Road Facilities of the Novosibirsk Region, Secretary of the Steering Committee

Vladimir Burovtsev
Deputy Head of the Siberian Territorial Administration of the Federal Agency for Rail Transport

Ivan Goncharov
Head of the Department for Investment Policy and Territorial Development of the Office of the Plenipotentiary Representative of the President of the Russian Federation in the Siberian Federal District

Alexander Gritsai
Head of the West Siberian Railway—a branch of the Open Joint Stock Company "Russian Railways"

Natalia Donskaya
Deputy Chairman—Head of the Office of the Chairman of the Executive Committee of the International Association for Economic Cooperation of the Subjects of the Russian Federation "Siberian Agreement"

Valery Ilyenko
Deputy Chairman of the Legislative Assembly of the Novosibirsk Region

Igor Leontyev
Head of Commercial Services of the Novosibirsk International Airport (Tolmachevo) of the Joint Stock Company "Tolmachevo Airport"

Alexey Manakov
Rector of the Federal State Budgetary Educational Institution of Higher Education "Siberian State Transport University"

Fedor Nikolaev
Chairman of the Committee of the Legislative Assembly of the Novosibirsk Region on Transport, Industrial and Information Policy

Sergey Pavlushkin
Head of the Federal Budget Institution "Administration of the Ob Basin Inland Waterways"

Daniyar Safiullin
Deputy Mayor of the city of Novosibirsk

Vyacheslav Sorogovets
Acting Head of the West Siberian Interregional Territorial Administration of Air Transport of the Federal Agency of Air Transport

Alexandr Starovoytov
Member of the Committee on Transport and Construction of the State Duma of the Federal Assembly of the Russian Federation

Elena Tyrina
Deputy Chairman of the Committee of the Legislative Assembly of the Novosibirsk Region on Transport, Industrial and Information Policy

Alexey Fursov
Deputy Chairman of the Executive Committee of the Interregional Association for Economic Cooperation of the Subjects of the Russian Federation "Siberian Agreement" for project activities

Evgeny Chernyshev
General Representative of S7 Airlines in the countries of Central Asia and at the Tolmachevo Airport

The Program Committee of the VIII International Siberian Transport Forum—TransSiberia/TransLogistica 2019

Aleksey Manakov
Rector of Siberian State Transport University, Russia

Jacek Szołtysek
Head of Social Logistics Department, University of Economics in Katowice, Poland

Andrey Abramov
Vice-Rector of Siberian State Transport University, Russia

Zdenka Popovic
Department of Roads, Airports, and Railways, University of Belgrade, Serbia

Miomir Miljković
Faculty of Civil Engineering and Architecture, University of Niš, Serbia

Olli-Pekka Hilmola
Lappeenranta University of Technology, Kouvola University, Finland

Sergei Kudriavtcev
Vice-Rector of Far Eastern State Transport University, Russia

Askar Zhusupbekov
President of Kazakhstan Geotechnical Society, L.N. Gumilyov Eurasian National University, Kazakhstan

Miroslav Živković
Faculty of Engineering, University of Kragujevac, Serbia

Andrii Bieliatynskyi
National Aviation University, Ukraine

Mareks Mezitis
Head of Institute, Institute of Transport, Riga Technical University, Latvia

Dragan Rakić
Faculty of Engineering, University of Kragujevac, Serbia

Ivan Belošević
Faculty of Transport and Traffic Engineering, University of Belgrade, Serbia

Paulo Cachim
Department of Civil Engineering, University of Aveiro, Portugal

Milorad Jovanovski
Faculty of Civil Engineering, Ss. Cyril and Methodius University in Skopje, Skopje, Nord-Macedonia

Sergey Baryshnikov
Rector of Admiral Makarov State University of Maritime and Inland Shipping, Russia

Valerii N. Li
Far Eastern State Transport University, Russia

Slobodan Ognjenovic
Faculty of Civil Engineering, Ss. Cyril and Methodius University in Skopje, Skopje, Nord-Macedonia

Luka Lazarević
Faculty of Civil Engineering, University of Belgrade, Serbia

Jaroslav Matuška
Faculty of Transport Engineering, University of Pardubice, Czech Republic

Ivana Barišić
Faculty of Civil Engineering, University of Osijek, Croatia

Mirjana Vukićević
Faculty of Civil Engineering, University of Belgrade, Serbia

Tomasz Kozlowski
Kielce University of Technology, Kielce, Poland

V. N. Paramonov
St. Petersburg State Transport University, St. Petersburg, Russia

Andrey Benin
Head of Research Department, Emperor Alexander I St. Petersburg State Transport University, Russia

Valerii Pershakov
National Aviation University, Ukraine

Yu. A. Davydov
Rector of Far Eastern State Transport University (FESTU), Khabarovsk, Russia

Jozef Melcer
Faculty of Civil Engineering, University of Zilina, Slovak Republic

Stanislav Vlasevskii
Far Eastern State Transport University, Russia

Izolda Li
Admiral Makarov State University of Maritime and Inland Shipping, Russia

Yurii Ezhov
Admiral Makarov State University of Maritime and Inland Shipping, Russia

Contents

Transportation Engineering

Estimation of Digital Substation Reliability Indices

Pavel Pinchukov(✉) and Svetlana Makasheva

Far Eastern State Transport University, Seryshev st. 47, Khabarovsk 680021, Russia
pinchukov-pavel@mail.ru

Abstract. The paper deals with the problem of improving the reliability of digital power supply facilities, namely digital substations. The goal of the study is to identify factors of a negative impact on the digital substation reliability indices. Based on the influencing factors analysis, technical measures for increasing reliability indexes are proposed. Based on the basic principles of the classical theory of reliability a numerical estimate of the probability of failure-free operation for the main equipment digital substation was made taken as an example the real digital substation switchgear 220 kV. Detailed equivalent circuits are given to ensure the reliability of the main units of digital substations. The numerical values of the reliability indices of the main elements are given in tabular form. The digital substation blocks with low reliability indices, as well as the most vulnerable elements of equivalent circuitry of its nodes, are identified. It was revealed that a large number of secondary circuits used in digital measuring instruments and interface devices negatively affect the reliability of digital substation units. Recommendations on maintaining the required level of reliability of power supply are given.

Keywords: Digital substation · Theory of reliability · Fault rate · Reliability index · Digital current transformer · Mean Time Between Failures · Survival function · Merging Unit · Standalone Merging Unit

1 Introduction

It is well known that the digital technologies have firmly entered into the modern man life, affecting almost all the main areas of his activity. The widespread adoption of digitalization technologies at electric power facilities, including the world's railways, is no exception.

According to the information contained in the research papers [1–3], digitalization of electric power facilities and systems allows to reduce the duration of power outages, the frequency of technological violations and the accident rate. These factors are directly related to reliability indicators of power supply for both individual consumers (or power industry objects) in particular, and power supply systems in general. Regarding digital technologies in power supply systems, it should be noted that one of the key elements in the power supply system is an substation. Electrical substation is designed for receiving and converting voltage in an AC network and then for

Z. Popovic et al. (Eds.): TransSiberia 2019, AISC 1115, pp. 3–14, 2020.
https://doi.org/10.1007/978-3-030-37916-2_1

distributing the electricity in power supply systems for various consumers needs. For currently widely used in the Russian Federation and in the other worlds countries the traditional substations (also named analoge) the basis for the exchange of information between units and elements of substations is the results of analog measurements. This type of measurements of electrical quantities is done by means of current and voltage measuring transformers.

The reliability issues of traditional substations are well studied and adequately represented in the Russian and foreign educational and scientific literature [3–8], but the same cannot be said about the reliability issues of new, digital substations. Nowadays there are no examples of a numerical assessment of reliability indicators (or indexes) in the available information sources. Also, in modern Russian and world literary sources there are no results of a comparative analysis for reliability indexes in cases of consumers power supply from ordinary, analog electric power equipment (or system) and digital equipment, etc. Such a situation, in our opinion, is a rather serious deterrent for the digitalization possibilities wide and comprehensive engineering assessment and its intensive distribution in electric power systems worldwide. Comprehensive engineering assessment. In other words, a comprehensive engineering assessment of the reliability indexes for a digital substation is extremely relevant and practically significant.

The main goal of this research is an increasing the reliability indexes of a digital substation. To achieve the goal, it is necessary to solve the following tasks:

- creating the reliability equivalent circuitry for digital substation's main blocks;
- calculation and evaluation the main reliability indexes;
- identification the "weaknesses" of digital substations based on the results of a comparative analysis of reliability indexes;
- developing the necessary measures to maintain the required level of reliability of power supply to consumers.

2 Materials and Methods

The solution of these tasks is achieved through comprehensive analytical studies that rely on the basic, traditional principles of reliability theory [3–5]. The analysis of consumers power supplying reliability is based on the compilation and subsequent calculation of equivalent circuits for reliability indexes by presenting the topology of the considered electric network in the form of series-parallel connections [3–6].

At the same time, the provisions of the Markov and Poisson theory of random processes [3–8] are applicable to the tasks to be solved for calculating the reliability indexes of restored objects at power supply systems. The object of study is a real substation 220 kV, located at the Far East of Russia, in particular - its 220 kV switchgear. The substation is under reconstruction and modernization to the digital type at present.

The result data were presented in the form of graphs, descriptive and analytical tables. This method has many attractive practical applications as understanding of the "weaknesses" blocks (or objects) at a particular substation and determination the numerical volume of reliability indexes. The benefits can help to find a solution to eliminate the malfunction and make timely adjustments to the substation digital blocks as much as possible for better operation and widespread all over the world.

3 The Main Reliability Indices Term and Equations

The term "reliability" means the complex property of an object to retain in time the ability to perform the required functions in the specified modes and conditions of use, maintenance, storage and transportation. Reliability is a property that may include availability, failure-free operation, longevity, maintainability and storability or a certain combination of these properties depending on the purpose of the particular object under consideration and the conditions of its stay. According to the classical theory of reliability [1–3], the quantitative indicators (named the reliability measures) of availability, failure-free operation, longevity, maintainability and storability, as well as complex indexes are used to assess reliability. They are characterizing the readiness and efficiency of using technical objects (in particular, electrical objects or parts of power supply systems).

The probability of failure-free operation, R, is the most common criterion for assessing the reliability of an object in the world engineering practice [3, 9–11]. This criterion characterizes the possibility the situation, within the specified operating time, an object failure will not occur. If we take into account the assumption that the time distribution of the number of equipment failures included in electric power systems occurs according to the exponential law [3–6], then the probability of failure-free operation of the facility, R, can be determined by the equation

$$R = e^{-\lambda \cdot t} \tag{1}$$

In the Eq. (1), the λ symbol means failure rate, 1/hour, and the symbol t denotes a specified operating time interval in hour.

The probability of failure, Q, is the probability that the object will fail at least once at a given time interval t. The probability of failure Q is determined by equation

$$Q = 1 - R \tag{2}$$

Mean time between failures (MTBF), determinate as T, is the amount of time counts from the first objects using (or from its restoration) until the failure time. MTBF is the predicted time in hours between inherent failures during normal object operation. MTBF can be calculated as the arithmetic mean (average) time between failures of a system. The term is used for repairable systems, while mean time to failure (MTTF) denotes the expected time to failure for a non-repairable system. This index can be determined by equation

$$T = \frac{1}{\lambda} \tag{3}$$

The probability of failure-free object operation, written as R (t) at a given time interval t, corresponding to the characteristics of the manufacturer, is determined by formula

$$R(t) = e^{-\lambda t} \tag{4}$$

If the reliability equivalent circuit contains a block consisting of two parallel-connected elements marked 1 and 2, the probability of this block's failure-free operation Rblock_par will be calculated by equation

$$R_{block\_par} = 1 - Q_1 \cdot Q_2 = 1 - (1 - e^{-\lambda_1 \cdot t}) \cdot (1 - e^{-\lambda_2 \cdot t}) \tag{5}$$

The probability of failure-free operation for the block, consisting of two series-connected elements marked 1 and 2, R_{block_ser}, will be calculated by equation

$$R_{block\_ser} = R_1 \cdot R_2 = e^{-\lambda_1 \cdot t} \cdot e^{-\lambda_2 \cdot t} \tag{6}$$

Thus, in the classical statement of the theory of reliability [3–6], in order to calculate the probability of failure-free operation R of an object for the time interval from the moment of switching on to any given time t, it is necessary to provide some operations step-by-step:

1. to draw up an equivalent circuits for reliability indexes based on the source information on the topology of the object in question;
2. to determine the average uptime or constant failure rate from the reference (or experimental) data) for each element of the equivalent circuits for reliability indexes;
3. to determine the final probability of failure-free operation R of the object by using the basic formulas (1–6) and the phased transformation of the equivalent circuits for reliability indexes.

4 Digital Substation Infrastructure

A typical line diagram of a 220/110/6 kV substation is presented at Fig. 1.

As shown at Fig. 1, there are all substation devices connected by an information network called the Local Area Network (LAN) for transfer sampled values of current and voltage by the Standalone Merging Unit (SAMU). SAMU is a device that enables the implementation of an IEC61850 process bus by converting the analog signals from the conventional current transformers and voltage transformers into IEC61850 Sampled Values [7, 9]. Through this information network, all signals, including instantaneous currents and voltages values at control points, are digitized and transmitted as a digital stream. The processes occurring at the digital substation are controlled remotely. For example, the switching apparatus is controlled either from the automated workplace of the operational personnel or from the dispatch control point, which is in charge of one or another switching devices.

Thus, the unified telecommunication infrastructure of a digital substation, made on the basis of modern technologies through digital exchange between intelligent electronic devices, has several advantages over old-type, analog substations.

So, the digital substation's infrastructure allows:

1. to do monitoring all processes in close proximity to information sources;
2. to transfer significant amounts of data to all subsystems using fiber-optic communication lines;
3. to make virtual most of the functions performed at the substation.

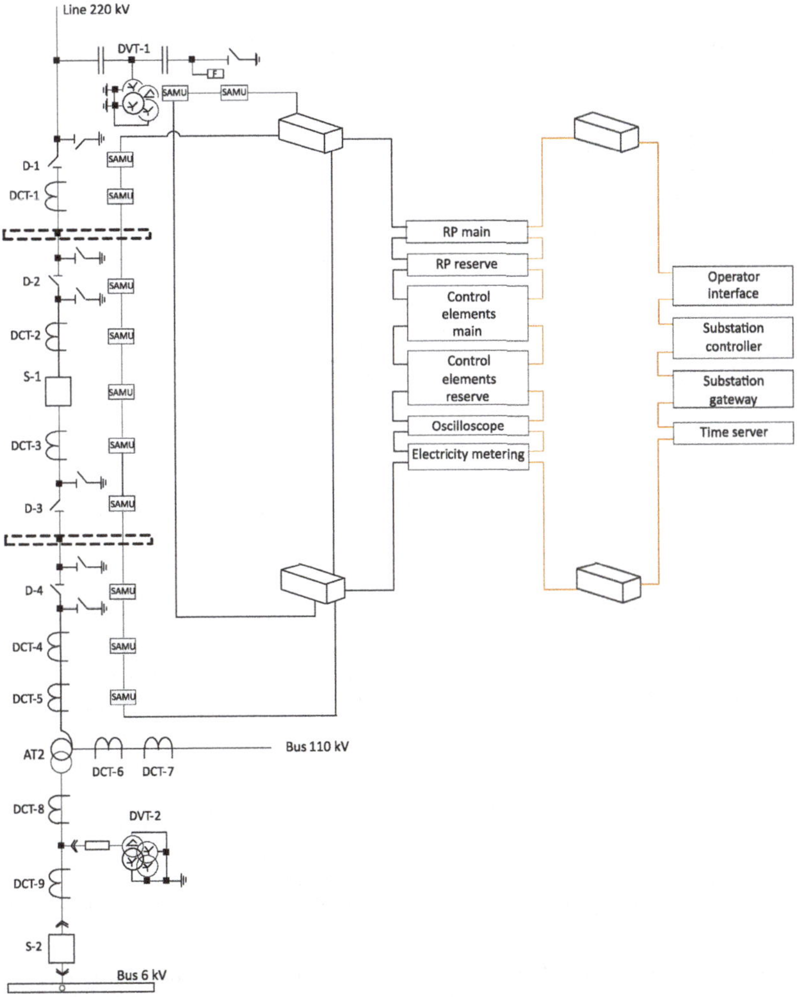

Fig. 1. Line diagram of typical 220/110/6 kV substation

With such a digital substation infrastructure, all measuring devices become sources of information, and all embedded intelligent electronic devices become consumers of this information. When the equivalent circuits for reliability indexes of a digital substation is constructing, it is necessary to take into account a significantly larger number of conditions. For example, the reliability of a digital substation circuit breaker will depend not only on its factory characteristics, but also on the characteristics of the interface to the process bus, the parameters of the optical fiber line through which data is exchanged, as well as the reliability of switches and relay protection terminals. In addition, it will be necessary to take into account the correct operation of the operator interface [11–16].

5 Calculating the Probability of Failure-Free Operation of Digital Substation Main Blocks

5.1 220 kV Disconnector Block

The 220 kV disconnector of the digital substation is indicated in Fig. 1 as D – 1 220 kV. The equivalent circuit for the reliability indexes for disconnector block is shown at Fig. 2, a in general, and at Fig. 2b in details.

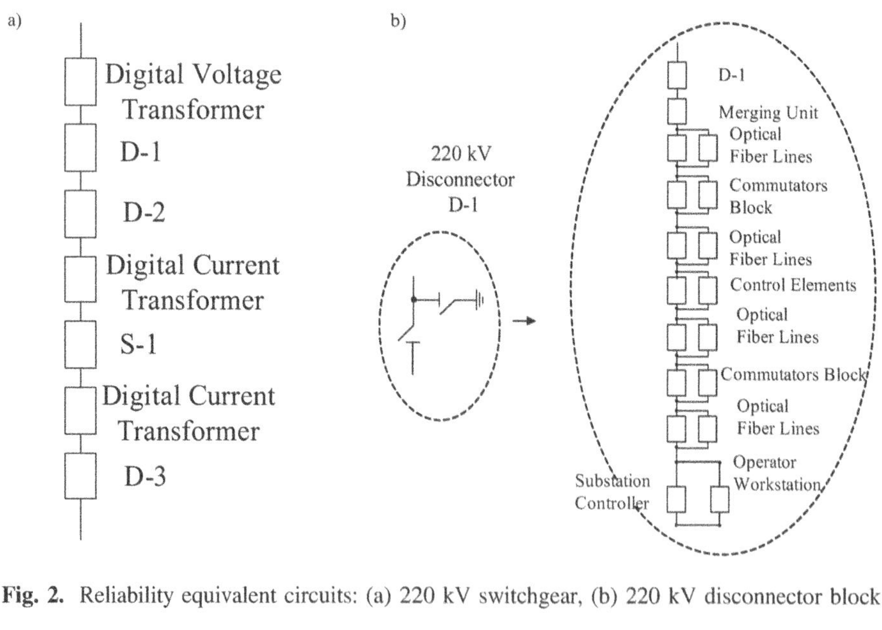

Fig. 2. Reliability equivalent circuits: (a) 220 kV switchgear, (b) 220 kV disconnector block

The disconnector reliability depends not only on the factory characteristics, but also on the combination of various control actions on it. So, for example, a disconnector failure can occur due to SAMU failure, therefore, in the circuit of Fig. 1, redundancy with a loaded duplication of the disconnector control circuit is provided, which increases the reliability of the entire circuit.

We set the estimated time period equal t = 2190 h. This duration is due to the frequency of maintenance of IT equipment at the substation according the Russian Standards. According to the digital substation's design documentation, maintenance for digital devices at the substation should be performed once a quarter, i.e. once every three months, which corresponds to duration of 2190 h.

Next, using data from electrical equipment manufacturers, we can calculate the probability of failure-free operation of the D−1 disconnector:

$$R_{D-1} = e^{-0.76 \cdot 10^{-5} \cdot 2190} = 0.983$$

Similarly, calculations were made for other elements of Fig. 2 according to formulas (1–4) and results are presented in the Table 1.

Table 1. Reliability indexes of the disconnector block D−1 220 kV

Equipment (model, manufacturer)	MTBF, T, hour	Failure rate, λ, 1/hour	Probability of failure-free operation, R
Disconnector (SDF245pII* + 2E/3MD50, ABB)	131 400	$7.6 \cdot 10^{-5}$	$R_{D-1} = e^{-0.76 \cdot 10^{-5} \cdot 2190} = 0.983$
Merging Unit (BA2704v700, EKRA)	125 000	$0.8 \cdot 10^{-5}$	$R_{MU} = e^{-0.8 \cdot 10^{-5} \cdot 2190} = 0.983$
Commutators (RedBox RED25, Hirschmann)	345 000	$2.9 \cdot 10^{-6}$	$R_{block\_com} = 1 - (1 - e^{-2.9 \cdot 10^{-6} \cdot 2190}) \times (1 - e^{-2.9 \cdot 10^{-6} \cdot 2190}) = 0.999$
Optical fiber line per 1 km (S.I. Tech) 1. Line 1 = 15 m 2. Line 2 = 20 m	–	$3.88 \cdot 10^{-7}$	$\lambda_{OF1} = 0.15 \cdot 3.8 \cdot 10^{-7} = 0.582 \cdot 10^{-7}$ $\lambda_{OF2} = 0.2 \cdot 3.88 \cdot 10^{-7} = 0.776 \cdot 10^{-7}$ $R_{block\_OF} = 1 - (1 - e^{-0.582 \cdot 10^{-7} \cdot 2190}) \times (1 - e^{-0.776 \cdot 10^{-7} \cdot 2190}) = 0.999$
Control Elements: 1. Operator Workstation 2. Substation Controller (SPRECON-E-C, Sprecher Automation GmbH)	100 000 100 000	10^{-5} 10^{-5}	$R_{block\_control} = 1 - (1 - e^{-10^{-5} \cdot 2190}) \times (1 - e^{-10^{-5} \cdot 2190}) = 0.999$

From the calculation results shown in Table 1 and at the Fig. 2, b, it follows, that two elements in the reliability equivalent circuit have the probability of failure-free operation R equal to 0.983: the D–1 disconnector and the merging unit. The remaining 8 elements in the circuit of series-connected elements have a probability of failure-free operation equal to 0.999. Thus, minimizing the serial connection circuit step-by-step, we can get the resulting probability of failure-free operation for the disconnector block equal to:

$$R_{D-1\_Block} = 0,983 \cdot 0,983 \cdot 0,999^{8} = 0,959$$

Since the interaction pattern between the disconnector and the secondary circuits is the same for all three disconnectors, indicated in Figs. 1 and 2, a, as D–1, D–2 and D–3, the probability of failure-free operation for all these disconnectors equal to the value calculated for D– 1, i.e. $R_{D-1} = R_{D-2} = R_{D-3} = 0.959$.

5.2 220 kV Circuit Switch Block

The calculation of the probability of failure-free operation of the 220 kV switch indicated in Figs. 1 and 2, a, as S–1, is carried out according to the above-mentioned algorithm. The elements, where influence the total probability of failure of the circuit switch, are shown In the equivalent circuit for reliability at Fig. 3.

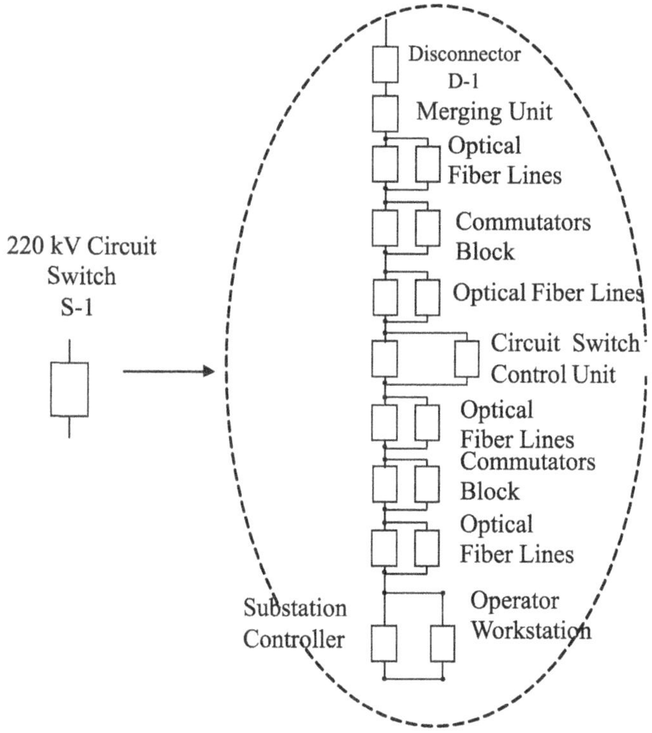

Fig. 3. Reliability equivalent circuit of 220 kV circuit switch block

As can be seen from Figs. 2b, and 3, reliability equivalent circuits for the disconnector and for the circuit breaker are similar: there are also 10 series-connected units, 8 of which consist of two elements connected in parallel. When calculating the probability of failure-free operation of the 3AP1 DT-245 circuit breaker block (manufactured by ABB), the MTBF is $T_{S-1} = 131\ 400$ h, and the failure rate is $\lambda_{S-1} = 0.76 \cdot 10^{-5}$ 1/hour [13].

The probability of failure-free operation of the circuit switch for a 2190 h's time interval is calculated by formula (1):

$$R_{S-1} = e^{-0.76 \cdot 10^{-5} \cdot 2190} = 0.983$$

The structure of the circuit breaker control unit of the S-1 switch, indicated in Fig. 3, includes the terminal for line distance relay protection marked SHE2607 09 (ECRA manufacturer) and the relay protection terminal SHE2607 016 (the same manufacturer). For each of these terminals, the mean time between failures is T = 125000 h, and failure rate $\lambda = 0.8 \cdot 10^{-5}$ 1/hour. The probability of failure-free operation of the circuit breaker control unit for the 2190 h interval can be calculated as:

$$R_{S-1\_Control\_Unit} = 1 - \left(1 - e^{-0.8 \cdot 10^{-5} \cdot 2190}\right) \cdot \left(1 - e^{-0.8 \cdot 10^{-5} \cdot 2190}\right) = 0.999$$

Thus, the probability of failure-free operation for the circuit-breaker block S – 1 of the digital substation, by analogy with the disconnector D – 1 can be calculated as:

$$R_{S-1\_Block} = 0.983 \cdot 0.983 \cdot 0.999^{8} = 0.959$$

5.3 Digital Current Transformer and Digital Voltage Transformer

The calculating the probability of failure-free operation was made for TTEO –220 digital current transformer and DNEE digital voltage transformer (both manufactured by Profotec, Russia). The reliability indexes for each device are equal: MTBF is $T_{DCT} = T_{DVT} = 120\,000$ h, and the failure rate is $\lambda_{DCT} = \lambda_{DVT} = 0.83 \cdot 10^{-5}$ 1/hour.

Thus, the probability of failure-free operation for the digital current transformers and digital voltage transformers of the digital substation, by analogy with the disconnector D – 1 can be calculated as:

$$R_{DCT} = R_{DVT} = e^{-0.83 \cdot 10^{-5} \cdot 2190} = 0.981$$

After determination parameters of each element and blocks shown at Fig. 2, a, we can calculate the resulting probability of failure-free operation for the 220 kV switchgear of digital substation.

5.4 220 kV Switchgear

Similarly, calculations were made for each block of digital substation's 220 kV switchgear according to formulas (1–4) and results are presented at the Table 2.

Table 2. Reliability indexes of digital substation's 220 kV switchgear

Probability of failure-free operation, R_i,	Probability of failure-free operation for block		
	Disconnector block	Circuit switch block	Digital current (voltage) transformer block
Main element	0.983	0.983	0.981
Merging unit	0.983	0.983	–
Commutators block	0.999	0.999	–
Optical fiber lines block	0.999	0.999	–
Control unit block	0.999	0.999	–
Summary	0.959	0.959	0.981

In the reliability equivalent circuit shown at Fig. 2(a), seven elements follow each other and form a multi-element chain, where each element contributes its share of failures to reduce the overall system's reliability. For such scheme type failure of one element leads to failure of the entire system as a whole.

The probability of failure-free operation of digital substation's 220 kV switchgear for series-connected elements is calculated by the formula (6)

$$R_{Switchgear220kV} = R_{DVT} \cdot R_{D-1} \cdot R_{D-2} \cdot R_{DCT} \cdot R_{S-1} \cdot R_{DCT} \cdot R_{D-3} = 0.959^4 \cdot 0.981^3 = 0.798512$$

6 Discussion

The resulting value of the probability of failure-free operation R for electro energy objects should tend to a value equal to 0.996 in accordance with Russian Federation Federal Grid Company requirements [15]. In this case, the reliability of the object is considered quite high. As the calculation result shows, the probability of failure-free operation of the 220 kV switchgear of digital substation, equal to 0.798512, is significantly less than the values recommended by Russian Federation Federal Grid Company. Moreover, the calculated indicator is significantly lower than the values recommended even for ordinary, analog substations [9, 15].

The calculated low value of the probability of failure-free operation of the digital substation's 220 kV switchgear can be explained by the following factors:

- a large number of elements of the digital substation secondary circuits;
- the presence of circuit elements with relatively low reliability characteristics: digital measuring transformers, merging unit for process bus couplers;
- there are a number of non-redundant elements in the reliability equivalent circuit (a device for interfacing with the process bus, power equipment, digital current transformers and digital voltage transformers).

Improving the digital substation reliability a can be achieved by eliminating weaknesses in the digital substation circuit, that is, by reserving elements at risk of failure. Also, the improvement of reliability in the future can result in improved quality indicators of electrical equipment manufacturers. This is due to the active development of digitalization of electric power facilities, which gradually increases the experience of operating digital equipment at power supply objects. This experience and failure statistics of new equipment will dictate to manufacturers higher requirements for the reliability of the equipment being manufactured.

At the same time, the equipment maintenance strategy based on its actual technical condition also contributes to a certain decrease in the probability of failure, since contributing to an increase in the overhaul interval and the elimination of unnecessary manipulation of equipment [1, 3, 12, 14–16].

7 Conclusion

To achieve the intended purpose of the study a topological diagram for a modernized to digital type substation were compiled in accordance with the digital substation concept. Based on data from electrical equipment manufacturers a reliability equivalent circuit has been compiled in the form of series-parallel connection of elements. A preliminary analysis of the compiled reliability equivalent circuit revealed a negative effect on the reliability indices numerous elements of substation secondary circuits.

The calculation of the digital substation main block's probability of failure-free operation as the example of 220 kV switchgear were made. The calculated value of the probability of failure-free operation for the 220 kV switchgear of digital substation is equal $R_{Switchgear220kV} = 0.798512$. This value is significantly low then recommended level by the Russian Federation Federal Grid Company even for equipment of analog type substation.

Obtained results leads to the following conclusions:

1. It was revealed that for a digital substation, backup of communication elements is performed, for example, an optical data transmission ring for SV and GOOSE protocols, but, at the same time, a number of elements in the risk zone do not have a reserve. For example, there is no redundancy at the interface to the process bus. Such blocks in a digital substation are treated as vulnerable, weak points of digital type substations.
2. Relatively low digital substation's reliability indices, identified as a result of studies, can become a negative factor that hinders the development of digital substation technology, therefore, it is necessary to take them into account at the stage of system design.
3. In order to increase the digital substations reliability, it is proposed to use redundancy for single elements at risk of failure, as well as to tighten the requirements for reliability indices of equipment (mean time between failures, failure rate, etc.), which is laid down by manufacturers of electrical and electronic equipment.
4. Improving the reliability of digital substations and eliminating barriers will allow to fully experience the benefits of digital technologies development in the electric power industry: increase the level of automation, reduce the cost of design, commissioning, operation and maintenance of energy facilities.

References

1. Hunt, R., Zapella, M., Pimentel, C., Silvano, G.: Tutorial on networking for digital substations. In: 72nd Conference for Protective Relay Engineers (CPRE), College Station, TX, USA (2019). https://doi.org/10.1109/cpre.2019.8765874
2. Barbosa, J.D., Santos, R., Romero, J.F., Asano, P.T., Neto, A.V., Junior, J.R., Junior, J.B., Cugnasca, P.S.: Methodology for determining reliability indices of electric power substations. IEEE Lat. Am. Trans. **16**, 1959–1968 (2018). https://doi.org/10.1109/TLA.2018.8447363
3. Allan, R.N., Billinton, R.: Reliability Evaluation of Power Systems. Plenum Press, New York/London (1996)
4. Iqbal, F., Siddiqui, A.S., Deb, T., Khan, M.T., Rana, A.S.: Power systems reliability – a bibliographical survey. Smart Sci. **6**(1), 80–93 (2018). https://doi.org/10.1080/23080477.2017.1407985
5. Ali, A.: Failure Analysis and Prevention. Books on Demand (2017). http://dx.doi.org/10.5772/65149
6. Liu, N., Panteli, M., Crossley, P.A.: Reliability evaluation of a substation automation system communication network based on IEC 61850. In: 12th IET International Conference on Developments in Power System Protection (DPSP) (2014). https://doi.org/10.1049/cp.2014.0057
7. Zhang, Y., Sprintson, A., Singh, C.: An integrative approach to reliability analysis of an IEC 61850 digital substation. In: IEEE Power and Energy Society General Meeting, San Diego (2012). https://doi.org/10.1109/pesgm.2012.6345699
8. Sangno, R., Singh, P.D., Pudu, R.: Int. J. Curr. Sci. Technol. **04** (2016) https://www.researchgate.net/publication/324013357_Reliability_Evaluation_of_a_13233_kV_HV_Substation
9. IEEE Guide for Electric Power Distribution Reliability Indices – Redline: In IEEE Std. 1366-2012 (Revision of IEEE Std 1366-2003), pp. 1–92, Redline (2012). http://ieeexplore.ieee.org/stamp/stamp.jsp?tp=&arnumber=6329910&isnumber=6329909
10. Webstore International Electrotechnical Commission IEC 61850:2019 SER Series Communication networks and systems for power utility automation - ALL PARTS. https://webstore.iec.ch/publication/6028
11. Voloshin, A.A., Voloshin, E.A., Gracheva, N.P., Blagorazumov, D.O.: Application of multi-agent system for calculation of reliability indexes of PACS for digital substation. In: International Youth Scientific and Technical Conference Relay Protection and Automation (RPA) (2018). https://doi.org/10.1109/rpa.2018.8537224
12. Efron, B., Hastie, T.: Computer Age Statistical Inference: Algorithms, Evidence, and Data Science. Cambridge University Press, Cambridge (2016)
13. ABB Asia Brown Boveri Ltd. CH. https://new.abb.com/
14. Pinchukov, P.: Experimental evaluation of sustainable functioning of overhead wire relay protection at traction substations of the Far Eastern Railway. Russ. Electr. Eng. **87**(62), 6–8 (2016). https://doi.org/10.3103/S1068371216020152
15. Pinchukov, P., Makasheva, S.: Improving methods for reliability assessment of electric power systems. In: Advances in Intelligent Systems and Computing, vol. 692, pp. 162–169 (2018). https://doi.org/10.1007/978-3-319-70987-1_17
16. Makasheva, S., Pinchukov, P.: Expanding the functionality of automated monitoring systems for traction substations. In: MATEC Web Conference, vol. 239, p. 01014 (2018). https://doi.org/10.1051/matecconf/201823901014

Simulation of Devices for Voltage Regulation in 25 kV AC Electric Traction Network

Yuliya Konstantinova, Valerij Li, and Andrey Konstantinov(✉)

Far Eastern State Transport University, Seryshev st. 47, Khabarovsk 680021, Russia
kamfes@mail.ru

Abstract. This paper proposes a MATLAB/Simulink simulation model aimed at energy efficiency estimation of different devices for voltage regulation in the AC traction network. The electrified railroad section with the length of 40 km is modelled. The model contains blocks of high-voltage power supply, a high-voltage overhead line, symmetrical three-phase load, power transformers with on-load tap changer (OLTC), an overhead line, rails, and electric rolling stock and device with booster transformer (BT). The model allows to estimate a technical loss in all elements of the power supply system. The operating regime is simulated with the three-phase and single-phase load combination and different devices for voltage regulation. The speed of trains is 50 km/h, the train interval is 10 min. The total technical loss when applying of transformer OLTC is 12.08% more than for the option without adjustment, due to an increase in losses in the overhead line. The use of a device with BT will reduce annual losses by 2.06%.

Keywords: Simulation model · AC electric traction network · Technical losses · On-load tap changer · Booster transformer

1 Introduction

The traction power supply system should provide power to traction loads, as high-priority consumer, from two independent sources and provide, as a rule, two-way power supply to the contact network. In addition to the requirements of reliability of power supply, two-way power supply ensures uniform load of the contact network and significantly reduces the loss of electrical energy and wear of the insulation of transformers of traction substations. All these theses are valid only with equal voltage on the busbars of adjacent substations. However, the voltages on the busbars of adjacent AC substations, as a rule, differ in phase and modulus.

The voltage phase difference of adjacent traction substations is 3–5 electrical degrees. Normative documents in Russia that regulate the acceptable voltage profile on substation busbars allow voltage deviations within 10%.

The main reasons for the voltage differences on the 27.5 kV busbars of adjacent traction substations are presented in Fig. 1.

Z. Popovic et al. (Eds.): TransSiberia 2019, AISC 1115, pp. 15–24, 2020.
https://doi.org/10.1007/978-3-030-37916-2_2

The reasons (Fig. 1) cause the voltage difference on the 27.5 kV busbars of adjacent traction substations, both jointly and independently from each other.

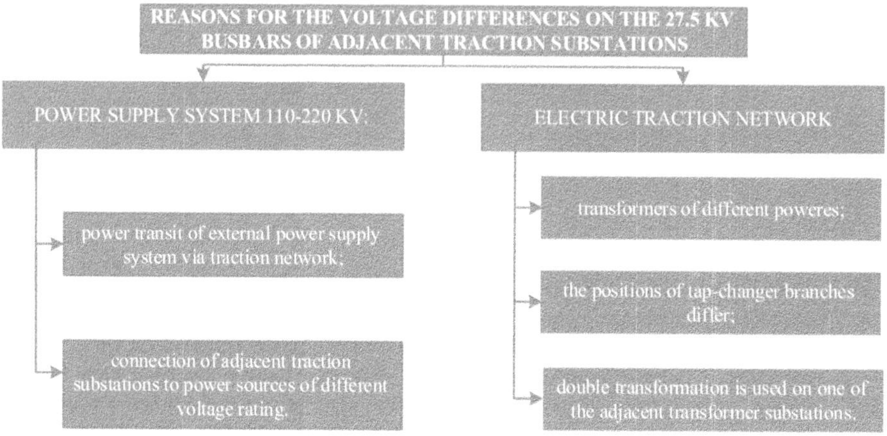

Fig. 1. Reasons for the voltage differences on the 27.5 kV busbars of adjacent traction substations

If there is a voltage difference along the traction network, a balancing current begins to flow. The path for the balancing current flow consists of secondary windings of substations transformers, an overhead line, rails.

The balancing current caused by the voltage difference during two-way power supply is a regular occurrence during the operation of the traction power supply system, but at the same time, the average electric energy losses from the power supply for each inter-substation area are 450 MWh/year.

Thus, for the Far Eastern Railway of Russia, which includes 30 inter-substation areas, with an electricity cost of 3.2 rubles. per kWh, the damage from equalizing current is about 43.2 million rubles per year.

This article presents a simulation model that allows you to calculate the loss of electricity that occurs during the transit of electricity through the traction network from an external network, as well as when using voltage regulation devices. The purpose of this article is to compare the energy efficiency of devices for voltage regulation in the traction network in the application of on-load tap changer (OLTC) which is placed in transformer (is the most common method for voltage regulation in distribution network) [1–6] and the proposed device with booster transformer (BT) [6–8].

2 Materials and Methods

The subject of research is the traction power supply system (Fig. 2), which consists of a power supply system 110 kV and an electric traction network. The power supply system 110 kV contains a three-phase source (source), a symmetrical three-phase load

(load) and sections of overhead power transmission line (line) [9]. The electric traction network includes three-phase transformers (T1, T2), sections of the overhead system (OS1 … OS10) and electric rolling stock (ERS) (1 … 10) [10]. The distance between substations is 40 km, there is a sectioning point (SP) in the middle of the inter-substation area. The speed of trains is 50 km/h, the train interval is 10 min.

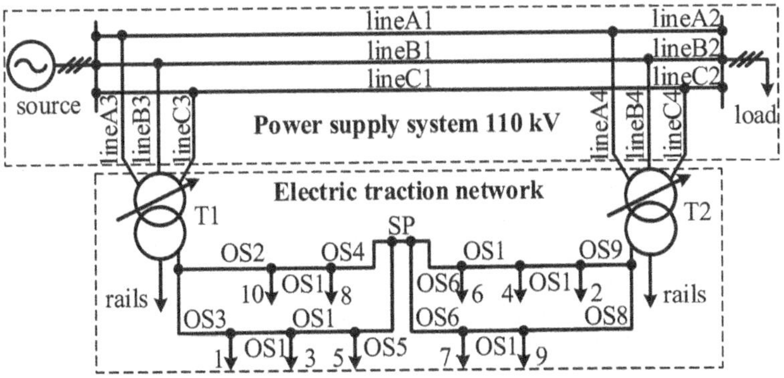

Fig. 2. Schematic diagram for power supply

The calculation is made by the method of instantaneous circuits. Given the given parameters, if at the initial moment of time ERS 1 locates at transformer T1, and ERS 2 places at transformer T2. Five ERSs are located on each track, the distance between adjacent ERSs is 8.3333 km. These ESRs move towards the opposite transformer through $\Delta t = 6.6 \cdot 10^{-3}$ h to 333.33 m.

Thus, it is enough to calculate 26 instantaneous circuits with a step of 333.33 m with different positions of the ERS, then the process will continue cyclically. Three models with different arrangement of trains are enough to calculate. The arrangement of ERS to the right and to the left of SP is presented for the models presented in the Table 1.

Table 1. The parameters of the model elements

Model no.	Instantaneous circuits	ERSs located T1–SP		ERSs located SP–T2	
		Down train	Up train	Down train	Up train
1	1 … 11	1, 3, 5	8, 10	7, 9	2, 4, 6
2	12 … 21	1, 3	6, 8, 10	5, 7, 9	2, 4
3	22 … 26	1, 3	6, 8	5, 7	2, 4

A simulation model of the studied system was created in the MATLAB/Simulink program (Fig. 3). Elements of the model and their parameters are presented in Table 2 [9, 10]. The three-phase load power is selected based on the condition of the maximum allowable load of wires of the power line in the area between source and transformer T1.

Parameters of simulation during simulation: Simulation type - Discrete; Sample time (s): 2e−5; Discrete solver - Tustin/Backward Euler (TBE); Start simulation with initial electrical states from: blocks.

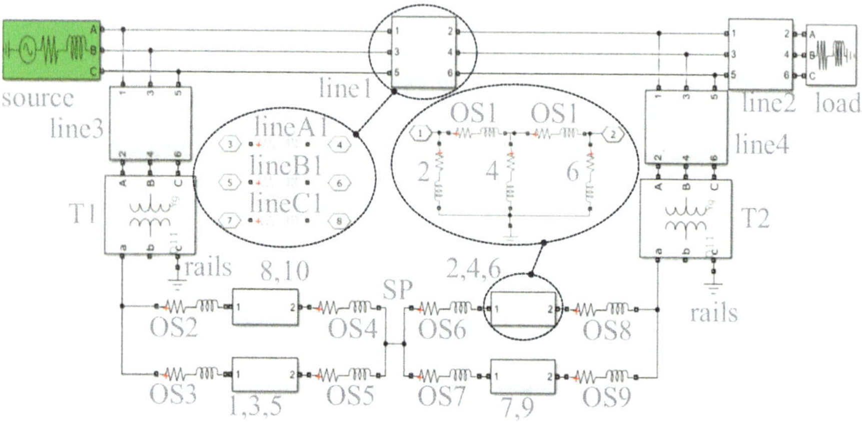

Fig. 3. Simulation model

Table 2. Parameters of model elements

System element	Model element	Parameters
source	Three- phase voltage source	U = 115 kV, f = 50 Hz
load	Three-Phase Series RLC Load	U = 115 kV, f = 50 Hz, S = 20.5 MVA, cosφ = 0.85
lineA1 … lineC4	Series RL Branch	R = 0.244·l_m Ω, X = 0.427·l_m Ω
T1, T2	Three-Phase Transformer	S = 40 MVA, f = 50 Hz, $U_{\omega 1}$ = 115 kV, $U_{\omega 2}$ = 27.5 kV, $R_{\omega 1}$ = $R_{\omega 2}$ = 0.9 Ω, $X_{\omega 1}$ = 35.5 Ω, $X_{\omega 2}$ = 20.7 Ω, Q_x = 320 kVAR
OS1 … OS10	Series RL Branch	R = 0.249·l_i Ω, X = 0.421·l_i Ω
ERS 1 … 10	Series RL Load	U = 25 kV, f = 50 Hz, P = 6560 kW, cosφ = 0.9

Three options were modelled.

Option 1. There is the traction power supply system without voltage regulation (WVR). Option 2. There is change of position of the OLTC, which is located on the primary winding of the transformer T2. Voltage regulation is performed by changing the voltage on the primary winding T2 is given as follows:

$$U_{TCUL\omega 1T1} = \frac{U_{2i}}{U_{1i}} \cdot U_{\omega 1T1}, \tag{1}$$

where $U_{\omega 1T1}$ is the catalogue value of the voltage on the primary winding of the transformer T1, $U_{\omega 1T1} = 115$ kV; U_{1i} is the measured value of the voltage on the secondary winding of the transformer T1 without voltage regulation, kV; U_{2i} is the measured value of the voltage on the secondary winding of the transformer T2 without voltage regulation, kV.

Option 3. A device for circulating current decreasing [8] consists of a control system (CS), an input transformer, a rectifier (R) [11, 12], a controlled inverter (CI) [12–14], a filter [15] and a booster transformer (BT). The primary winding of the input transformer is connected to OS and the rail, the secondary winding is connected to the input terminals of the rectifier, the output terminals of the rectifier are closed to the inverter input (the snubber is placed between output R and input CI), output CI through the filter is connected to the primary winding of the BT, the outputs of the secondary winding of the BT are connected between the SP buses. The control system consists of a measuring units (MUs), the evaluation unit (EU) and unit the shaping of boosting voltage (USBV). The inputs of the MUs are connected to the secondary windings of transformers T1, T2, the outputs of the MUs are connected to the inputs of the MU, the output of the MU is connected to the input of the USBV, the output of the USBV is connected to the inverter.

The device operates as follows. The transformer's voltages ($\mathbf{U}_1$, $\mathbf{U}_2$) are measured by MUs and subtracted in BV.

The calculated voltage is converted into control signals for the inverter in the USBV. Meanwhile, the input transformer lowers the voltage of the contact network, which is rectified and smoothed.

From the rectified voltage, an inverter generates a sinusoidal signal, which is increased by the BT and supplied to the traction network. Thus, boosting voltage is generated in the contact network of the SP, which is equal in amplitude and opposite in phase to the geometric voltage difference of the secondary windings of the transformers $\mathbf{U}_1 - \mathbf{U}_2$.

The criterion for comparing options is the technical losses of electricity in the traction network and the power supply system 110 kV, are given by:

$$\Delta W = \sum_{i=1}^{26} \left\{ \left[l_{OS1} \cdot \sum_{j=1}^{n} I_{ji}^2 + \sum_{k=2}^{9} \left(l_{OSki} \cdot I_{ki}^2 \right) \right] \cdot R_{OS0} + R_{\omega 1} \cdot \sum_{d=1}^{6} I_{T1id}^2 + R_{\omega 2} \cdot \sum_{d=1}^{6} I_{T2id}^2 \right\} \cdot \Delta t, \tag{2}$$

$$\Delta W = \sum_{i=1}^{26} \sum_{h=1}^{3} \sum_{m=1}^{4} \left(I_{line\_ihm}^2 \cdot l_{line\_hm} \right) \cdot R_{line0} \cdot \Delta t, \tag{3}$$

where l_{OS1} is the length of the OS section between adjacent trains, which is equal to the distance travelled by ERS at a speed of 50 km/h for 10 min, $l_{OS1} = 8.3333$ km; I_{ji} is current flowing through OS section with a length of 8.3333 km, A; n is the number of

OS sections with a length of 8.3333 km, $n = 8$ at i from 1 to 21, $n = 6$ at i from 22 to 26; l_{OSki} is length of OS section connecting the circuit node (transformers or SP) with the nearest ERS, km; I_{ki} is current flowing through OS section from circuit node to the nearest ERS, A; R_{OS0} is electrical resistivity of the OS wires, $R_{OS0} = 0.249$ Ω/km; $R_{\omega 1}$ is the resistance of the transformer primary winding, $R_{\omega 1} = 0.9$ Ω; I_{T1id} is current flowing through the transformer primary winding, A; $R_{\omega 2}$ is the resistance of the transformer secondary winding, $R_{\omega 2} = 0.9$ Ω; I_{T1id} is current flowing through the transformer secondary winding, A; Δt – time for which the train travels a distance 333.33 m, hour; l_{line_hm} is the length of the section m of the overhead power transmission line phase h, km; I_{line_ihm} is current flowing through the section m of the overhead power transmission line phase h, A; R_{line0} is electrical resistivity of the overhead power transmission line wires, $R_{line0} = 0.244$ Ω/km.

3 Results

The results of calculating the technical losses in the traction network and the power supply system 110 kV are presented in Figs. 4 and 5. The maximum energy losses of 14.42 kWh (for WVR) in the traction network occur in instantaneous circuit 11 when the trains are located at the SP, that is, at the maximum distance from the transformers, which is 4.26% more than for instantaneous circuit 1, when the trains are as close to the transformers as possible. The decrease losses in instantaneous circuit 22 by 9.8% (relative to instantaneous circuit 1) and further is due to the departure of the inter-substation area by one train on each track and the decrease in the number of trains from 5 to 4 train pairs.

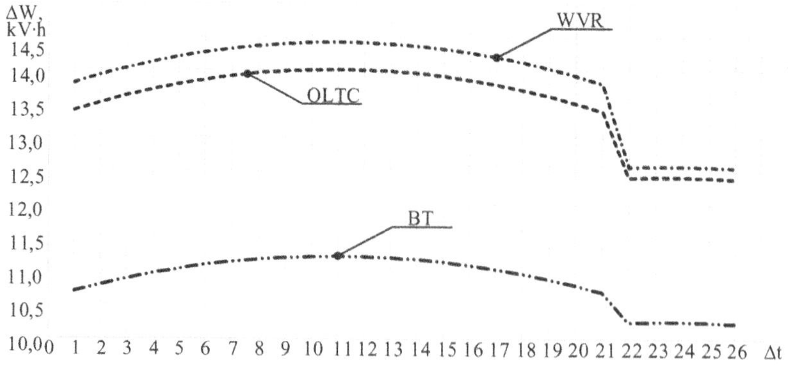

Fig. 4. Graph changes technical losses in the traction network

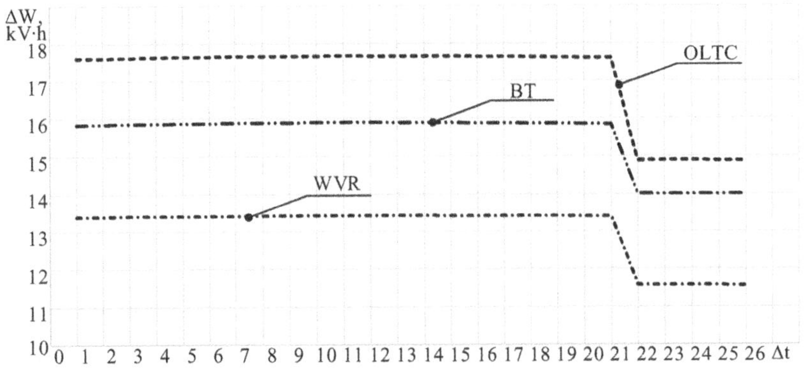

Fig. 5. Graph changes technical losses in the traction network

Technical losses in the power supply system 110 kV also depend on the number of trains in the inter-substation area, losses are 16.1% less with 4 pairs of trains compared to 5 pairs of trains. But the location of trains on these losses affects only 0.36%.

As can be seen from the graph in Fig. 4, the use of voltage control devices reduces power losses with using transformer OLTC from 1.27 to 2.98%, with devices with BT - from 18.46 to 22.52%. However, the maximum energy losses in the external power supply system occur when using transformer OLTC is 28.73–31.71% more than WVR, and when using a device with BT it is 18.28–21.1% more.

4 Discussion

The decrease in technical losses in the electric traction network ΔW_{etn} (Fig. 4) is explained that the application of transformer OLTC lowers the algebraic voltage difference of the secondary windings of transformers $|\mathbf{U}_1| - |\mathbf{U}_2|$, and a device with BT eliminates the geometric difference $|\mathbf{U}_1 - \mathbf{U}_2|$ in the traction network. This means that both methods reduce the transit of electricity from a power supply system 110 kV through the traction network. As can be seen from the graph of changes in voltage differences on the secondary windings of transformers (Fig. 6), although the shapes of the graphs of the algebraic and geometric differences coincide, but the algebraic difference varies from 2647.5 V to 2842.9 V, and the value of the geometric difference is from 3574.4 V to 3769.97 V. That is, the presence of a difference in the angle between voltages of the transformers from 5.91° up to 6.31° increases the voltage difference modulus by 32–39%.

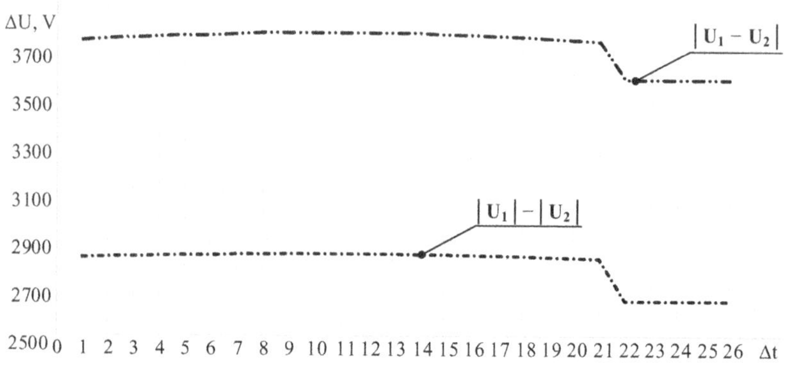

Fig. 6. Graph change module geometric and algebraic of the voltage difference

The methods considered, eliminating (in whole or in part) the voltage difference on the secondary windings of transformers, significantly reduce the transit of electricity from a power supply system 110 kV to a traction network. Since transit electricity currently flows mainly through the power supply system 110 kV, in the system, these technical losses ΔW_{pss} increase (Fig. 5).

Technical losses when applying of transformer OLTC are higher than those of the other options, since in addition to returning transit energy to the power supply system 110 kV, due to the increase in voltage on the secondary winding of the second transformer, the load of this transformer and the overhead transmission lines are also increased by the traction load.

Table 3 shows the technical losses from the considered options per year and the relative (∂_W) and absolute (Δ_W) changes in energy losses during voltage regulation in the traction network.

Table 3. Annual technical losses

Option	ΔWpss, MW•h/year	ΔWetn, MW•h/year	ΔWpss + ΔWetn, MW•h/year	Δ_W, MW•h/year	∂_W, %
1	18427.76	17328.29	35756.05	–	–
2	17946.43	22722.09	40668.52	4912.47	12.07929
3	14442.64	20592.41	35035.05	−720.996	−2.05793

As can be seen from Table 3, the total technical loss when applying of transformer OLTC is 12.08% more than for the option without adjustment, due to an increase in losses in the power supply system 110 kV. The use of a device with BT will reduce annual losses by 2.06%, which is 721 MWh/year.

5 Conclusions

The model of traction power supply system in Matlab/Simulink allows to estimate energy efficiency of devices and methods of voltage regulation in traction network. Application of transformer OLTC at elimination of transit of the electric power on a traction network is inefficient and promotes increase in total technical losses of the electric power. The proposed device with BT increases the energy efficiency of the traction power supply system by eliminating the transit of electricity through the traction network from the external network. In the future of research, it is possible to consider the use of a device with BT as an element of Smart Grid.

References

1. Zhang, X., Kang, R., McCulloch, M., Papachristodoulou, A.: Real-time active and reactive power regulation in power systems with tap-changing transformers and controllable loads. Sustain. Energy, Grids Netw. **5**, 27–38 (2016). https://doi.org/10.1016/j.segan.2015.10.006
2. Feizifar, B., Usta, O.: A new arc-based model and condition monitoring algorithm for on-load tap-changers. Electr. Power Syst. Res. **167**, 58–70 (2019). https://doi.org/10.1016/j.epsr.2018.10.024
3. Hu, J., Marinelli, M., Coppo, M., Zecchino, A., Bindner, H.W.: Coordinated voltage control of a decoupled three-phase on load tap changer transformer and photovoltaic inverters for managing unbalanced networks. Electr. Power Syst. Res. **131**, 264–274 (2016). https://doi.org/10.1016/j.epsr.2015.10.025
4. Iria, J.P., Heleno, M.L., Cardoso, G.: Optimal sizing and placement of energy storage systems and on-load tap changer transformers in distribution networks. Appl. Energy **250**, 1147–1157 (2019). https://doi.org/10.1016/j.apenergy.2019.04.120
5. Moolamkunnam, S.S., Mathew, J., Joseph, S.: Power electronic on-load tap changer for HVDC converter transformer. Int. J. Innovative Res. Electr. Electron. Instrum. Control Eng. **7**(4), 50–58 (2019). https://doi.org/10.17148/IJIREEICE.2019.7408
6. Plakhova, M., Mohamed, B., Arboleya, P.: Static model of a 2 × 25 kV AC traction system. In: 6th International Conference on Power Electronics Systems and Applications (PESA), pp. 1–5. IEEE Press, Hong Kong (2015). https://doi.org/10.1109/pesa.2015.7398917
7. Kosarev, A.B., Kosarev, B.I.: Determination of the parameters of a compensating unit in a traction power supply system with a booster transformer. Russ. Electr. Eng. **89**(9), 531–535 (2018). https://doi.org/10.3103/S1068371218090092
8. Konstantinova, Y.A., Li, V.N., Tryapkin, E.Y.: Enhancing energy efficiency of 25 kV traction power system due to balancing current decrease. In: 2018 International Multi-Conference on Industrial Engineering and Modern Technologies (FarEastCon), pp. 1–5. IEEE Press, Vladivostok (2018). https://doi.org/10.1109/fareastcon.2018.8602632
9. Shenoy, U.J., Sheshadri, K.G., Parthasarathy, K., Khincha, H.P., Thukaram, D.: MATLAB/PSB based modeling and simulation of 25 kV AC railway traction system - a particular reference to loading and fault conditions. In: 2004 IEEE Region 10 Conference TENCON, pp 508–511. IEEE Press, Chiang Mai, Thailand (2004). https://doi.org/10.1109/tencon.2004.1414819
10. Worku, G.B., Kebede, A.B.: Autotransformer fed traction power supply system: analysis, modeling and simulation. Glob. Energy Interconnection **1**(2), 187–196 (2018). https://doi.org/10.14171/j.2096-5117.gei.2018.02.011

11. Samsudin, N.A., Ishak, D., Ahmad, A.B.: Design and experimental evaluation of a single-stage AC/DC converter with PFC and hybrid full-bridge refctifier. Eng. Sci. Technol. Int. J. **21**(2), 189–200 (2018). https://doi.org/10.1016/j.jestch.2018.03.003
12. Açikgöz, H., Keçecioğlu, Ö.F., Gani, A., Yıldız, C., Şekkeli, M.: Optimal control and analysis of three phase electronic power transformers. Proc. Soc. Behav. Sci. **195**, 2412–2420 (2015). https://doi.org/10.1016/j.sbspro.2015.06.240
13. Somkun, S.: Unbalanced synchronous reference frame control of singe-phase stand-alone inverter. Electr. Power Energy Syst. **107**, 332–343 (2019)
14. https://doi.org/10.1016/j.ijepes.2018.12.011
15. Ibrahim, S., Cramer, A.M., Liao, Y.: Integrated control of voltage regulators and distributed generation inverters. Electr. Power Syst. Res. **169**, 45–52 (2019). https://doi.org/10.1016/j.epsr.2018.12.003
16. Pomalis, M., Leborgne, R.C., Herrera-Orozco, A.R., Bretas, A.S.: NSGAII optimization for single phase passive filter allocation in distribution systems. Electr. Power Syst. Res. **176**, 1–9 (2019). https://doi.org/10.1016/j.epsr.2019.105923

Method for Increasing Power Factor of Multi-range Converter

Yuri Kulinich, Sergei Shukharev, and Denis Drogolov(✉)

Far Eastern State Transport University, Serysheva str., 47, 680021 Khabarovsk, Russia
chesser@yandex.ru

Abstract. It is proposed to use a new method to generate control pulses for a multi-range voltage rectifier, which is based upon changing of minimal unregulated angle of opening of thyristors depending on regulation range number. The comparison of energetic indicators of standard and modified converter circuit has been carried out. The essential advantage of the modified circuit is the simplicity of implementation of new algorithm for multi-range voltage rectifier control. Quoted results of numerical modelling in MATLAB/Simulink environment as well as the results of experimental research of the model confirm that it is possible to change the control algorithm for multi-range converter thyristors.

Keywords: Multi-range converter · Power factor · Mathematical model · Commutation · Experimental model

1 Introduction

Successes in the development and production of new types of power semiconductor transistors served as a basis for creation a new type of electric drive – three-phase induction motor with square-cage rotor. Converters with an intermediate DC voltage link, which consist of voltage source inverter (VSI) creating three-phase symmetric voltage system to feed up the motors and also of a rectifier, switched on from the mains, which supplies power to VSI using direct current voltage, are used in order to feed up motors. These links are interconnected with an intermediate DC voltage link representing a high-value capacitor. In order to limit rectified voltage ripple, a resonant *LC*-circuit is to be integrated into in parallel to the capacitor.

Rectifiers with range and phase regulation of rectified voltage U_d have become the most commonly encountered among rectifiers with current self-switching, which are used in asynchronous drive. The regulation of voltage U_d is carried out through changing the angle of opening of the rectifier thyristors.

There are two approaches when choosing the rectifier structure: in the first case (Fig. 1a), the rectifier consists of two series-connected circuits of controlled (*V5–V8*) and straight (*V1–V4*) bridge rectifiers, each of which is connected to corresponding transformer winding. The regulation of rectified voltage average value U_d is performed using controlled rectifier *V5–V8*.

Z. Popovic et al. (Eds.): TransSiberia 2019, AISC 1115, pp. 25–32, 2020.
https://doi.org/10.1007/978-3-030-37916-2_3

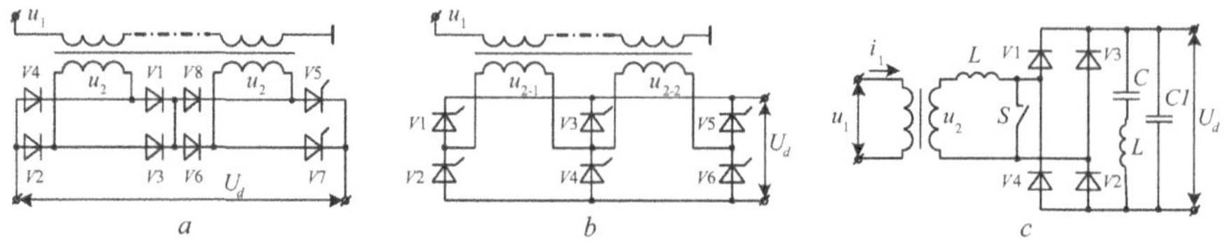

Fig. 1. Circuits of rectifiers

The basis of the second approach is the partitioning of a transformer secondary winding in combination with joining adjacent valve arms of the rectifier sections (Fig. 1b). In case of the equality of voltages $u_{2-1} = u_{2-2}$ of the transformer secondary windings such circuit ensures the two-range regulation of the rectified voltage U_d. In the first regulation range, the voltage U_d is generated from one section of the transformer winding u_{2-1} using thyristors *V1–V4*. In the second regulation range, the voltage u_{2-2} of the second transformer winding is added to a load circuit, connected to it using thyristors *V5, V6*.

Major drawback of rectifiers with current self-switching is low power factor value, especially peculiar to the low values region of the rectified voltage U_d. In order to increase rectifiers power factor, current forced commutation nodes are integrated into them. Simplified circuit of a power circuit of a rectifier, designed by *BBC* company, is shown in Fig. 1c [1, 2]. Converter includes a modulator S, connected through throttle *L* to the second winding of a transformer, rectifier *V1–V4*, and also *LC*- and *C1*-filters. Switching on and off of the modulator *S* is carried out so that to ensure the absence of phase angle of shear φ between feeding voltage u_1 and voltage consumed from the mains with current i_1. In this case, the highest values of power factor due to the increase in one of its components cosφ are attained.

The modulator S represents the configuration of bridge circuit, performed based upon power *IGBT*-transistors [3], controlled via pulse width modulation method [4] according to harmonical law. The practical implementation of such rectifier is connected to significant sophistication of its circuit. The paper discusses the method of increasing power factor of multi-range rectifier without changing its structure.

2 Proposal for Increasing Power Factor

Power factor of an electric power consumer is normally defined by an equation [5]:

$$PF = \cos\varphi \cdot \nu \tag{1}$$

where φ is the angle of shear between feeding voltage and the first harmonic of consumed current, ν is the consumed current shape distortion ratio.

Angle φ for multi-range converter is to be calculated by the approximation:

$$\varphi \approx \alpha_0 + \gamma/2 \quad (2)$$

where α_0 is the minimum unregulated angle required to open thyristors of multi-range rectifier, γ is the angle characterizing the duration of current commutation process in a thyristor's rectifier.

Equations (1) and (2) show that the increase in value of power factor *PF* may be reached for due to the decrease of minimum angle of thyristors α_0 and the reduction of the duration of commutation γ as well as via the improvement of consumed current waveform. Let's consider the possibility of the increase in PF for the account of the decrease in minimum angle of opening of thyristors α_0.

Currently fixed value of minimum angle of opening of thyristors $\alpha_0 = 9°$ is taken for power converters [6] in all four regulation ranges, i.e. control pulses come on a thyristor's rectifier with a delay to angle α_0 in relation to the beginning of half-cycle of mains voltage. It is necessary for direct voltage U_F between the anode and the cathode to reach the required level ensuring its reliable switching-on by the time of thyristors opening.

In accordance with the Eq. (2), switching of a rectifier thyristors at lower value of angle α_0^{mod} promotes the decrease of phase angle of shear φ and, accordingly, the increase of power factor *PF* of a converter. Besides, the decrease of minimum delay angle to α_0^{mod} allows increasing the average value of rectified voltage U_d at converter output at the same level of mains voltage. It is conditioned by the fact that with the decrease of minimum delay angle from α_0 to α_0^{mod}, the duration of negative polarity intervals in rectified voltage curve u_d also decreases.

The decrease of minimum angle of opening of thyristors α_0 may be performed in the highest regulation ranges with the preservation of potential conditions on thyristors at the time of their switching-on. The value of minimum angle $\alpha 0 = 9°$ in all regulation ranges is taken as equal on standard converters [7], at that anode voltage on thyristors coming into operation at the moment of their opening in the first regulation range is:

$$U_{F.max} = \sqrt{2}U \sin 9° = \sqrt{2} \cdot 315 \cdot 0.156 \approx 70\,\mathrm{V} \quad (3)$$

This means that such voltage on a thyristor ensures its guaranteed opening in the beginning of a half-cycle with all possible distortions of shape of mains voltage.

Let's establish at which value of minimum angle of opening of thyristors their reliable opening in standard circuit of multi-range converter in the first and the highest regulation ranges is ensured. In Fig. 2, shapes of positive half-wave of alternating voltage with effective values of 315–1260 V, coming to a thyristor's converter from transformer secondary windings in 1–4 regulation ranges are shown.

The figure analysis shows that in the first regulation range for thyristors arms coming into operation in the beginning of a half-cycle, the anode voltage $U_{open} = 70$ V required for their opening is reached with the delay angle $\alpha_0 = 9°$. In the second regulation range, the total voltage of 630 V of two sections of a transformer secondary winding is applied to therefore it is possible to perform their switching-on with less value of minimum delay angle $\alpha_0 = 4.5°$. In the third regulation range, thyristor arms coming into operation, carry the voltage of 945 *V* of a transformer secondary windings,

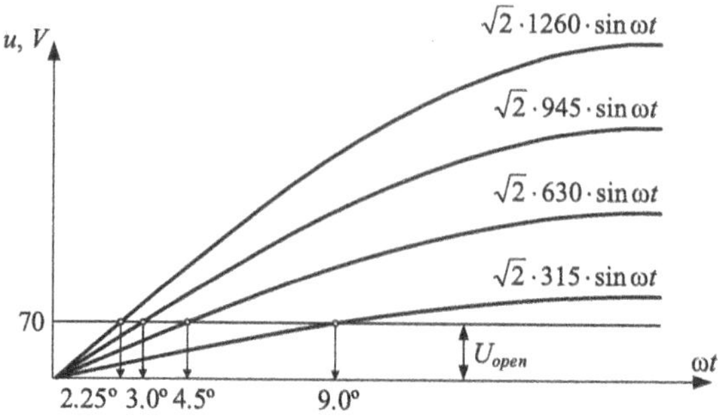

Fig. 2. Definition of minimum value of delay angle α_0

at that the voltage U_{open} = 70 V required for their opening is achieved at α_0 = 3.0°. In the fourth regulation range, switching-on of thyristor arms is possible to carry out with a delay α_0 = 2.25° at effective value of anode voltage 1260 V.

Thus, the anode voltage required for thyristors switching-on my be generated at various values of minimum delay angle α_0 depending on the regulation range. This allows for the possibility to increase power factor of multi-range rectifier due to the decrease in minimum unregulated angle of thyristors opening and, accordingly, to increase cosφ of one of the components.

3 Mathematical Modelling

In order to confirm the functional capability of a converter as well as to define its energetic indicators in MATLAB [8] software package, mathematical model given in Fig. 3 has been developed.

The model consists of the source of AC voltage source *AC*, multi-winding transformer (*Multi-Winding Transformer*), multi-range rectifier (*Rectifier*), rectifier control unit (*Control system*), and also load bank (*Load*).

Measurement of energetic indicators of the rectifier is performed using measurement unit (*Measurement_1T*) and indicator unit (*Indicators_2*). The model operating mode is set using regulation range controller (*zone*), delay angle αr (*Alpha_r*) and minimum angle α_0 (*Alpha_0*), unregulated in phase. The shape of rectified voltage and current is controlled using the unit (*Indicators_1*). Commutating angle sensors *I2* and *I3* allow realizing the method of progressive switching in large and small circuits of the multi-range converter commutation. As an example, Fig. 4 shows the process of progressive network current switching obtained by means of the mathematical model.

The process of progressive network switching [9] γ consists of two intervals γ_1 and γ_2. In the beginning of the process of progressive switching with the duration of γ_1 commutation current $i_к$ takes place in the large circuit between thyristors *V*1 and *V*7. At that the current of a valve *V*7, coming into operation, is defined by commutation current $i_к$, and current of a valve *V*1, being closed, is defined as the difference of currents I_d and $i_к$. The process of commutation in the small circuit is completed for the time γ_2 in the course of transfer of load current from thyristor *V*8 to thyristor *V*4. Since the

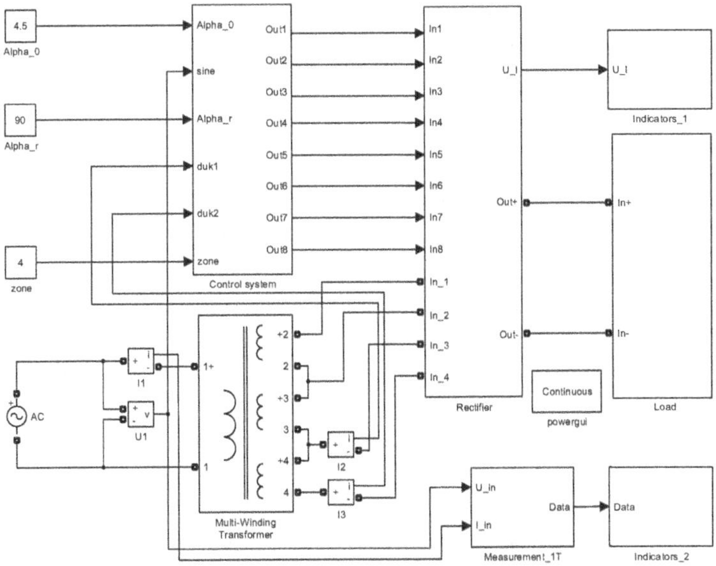

Fig. 3. Mathematical model

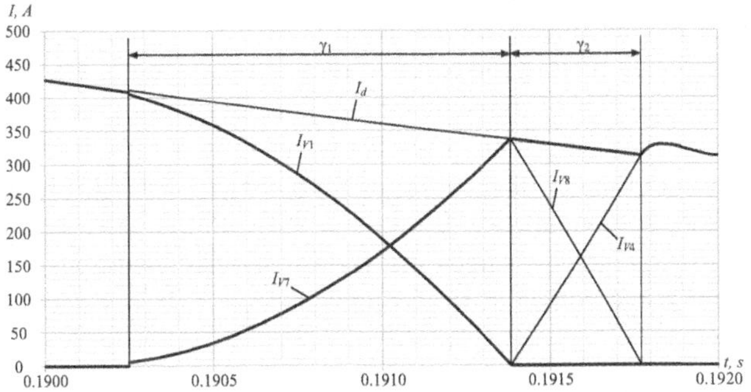

Fig. 4. The process of progressive network current commutation in thyristors of the multi-range converter

commutation in the small circuit occurs with greater voltage of the secondary transformer winding, its duration $\gamma_2 < \gamma_1$, at that the waveform of the current of commutated valves *V*4 and *V*8 approaches to straightforward shape.

4 Mathematical Modelling Results

Using mathematical modelling as the main method of analytical studies, the analysis of three optional of converter operation has been performed: in a standard circuit, and modified circuit of a rectifier with two values of minimum delay angles $\alpha_0 = 4.5°$ and

$\alpha_0 = 2.25°$. In each option of the circuit, the values of power factor PF, average value of rectified voltage Ud as well as the duration of network commutation $\gamma = \gamma_1 + \gamma_2$ have been calculated. The results of modelling are given in Fig. 5.

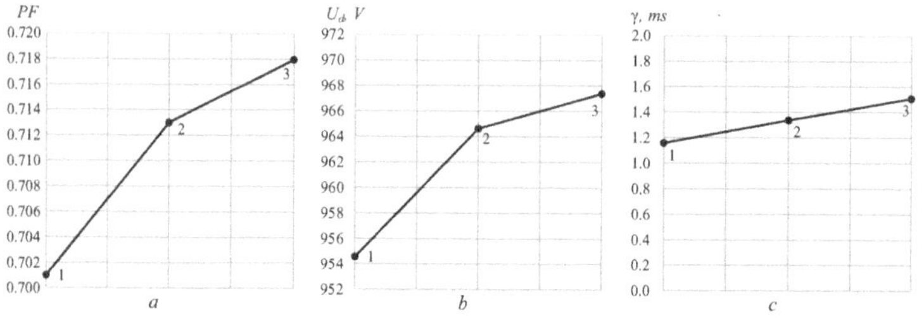

Fig. 5. Mathematical modelling results: *a* – power factor value, *b* – rectified voltage value, *c* – the duration of network commutation

It is seen from the figure analysis that the transfer from the first to the third option of a converter operation is accompanied by the increase in all the said indicators. Thus, the transfer from the standard circuit (1) to the second option of the circuit (2) (Fig. 5a) is accompanied by the increase of power factor by 1.7%. The most visible increase PF of a converter to 2.42% is reached in the course of the transfer from the standard (1) to modified circuit (3), controlled with the angle $\alpha_0 = 2.25°$ (3rd option). At that, due to the decrease in negative lines of the curve of rectified voltage U_d, the increase from the middle value from 954.6 V of the standard circuit to 967.4 V of the modified circuit in the third option of the calculation (Fig. 5b) occurs. As it appears from Fig. 5c, the reduction of minimum angle of opening of thyristors α_0 leads to the increase in the duration of network commutation γ. It is connected with the decrease of instantaneous value of rectified voltage U_d, under the action of which currents commutation in thyristors of multi-range converter is accomplished.

5 Experimental Results

In order to confirm the results of mathematical modelling, a multi-range converter (given in Fig. 6) model experimental study has been performed.

The experimental model allows to accomplish four-range regulation of voltage on DC motor 4 with the capacity of 250 W. Resistive torque valve on the motor shaft is set with DC load generator with independent excitation. Excitation winding is fed from laboratory power supply 5. Control system 2 is designed for automatic double-circuit motor control. Set values of current and motor speed enter the control system 2 from mode selector 3. Control system 2 is designed based upon micro-controller PIC18F452. Control over electromagnetic processes is accomplished by data acquisition device *USB*-6009 [10] by *National Instruments*. The conversion of the received information is made based upon PC 6 using a program created in graphical programming language in *LabVIEW* [11] environment.

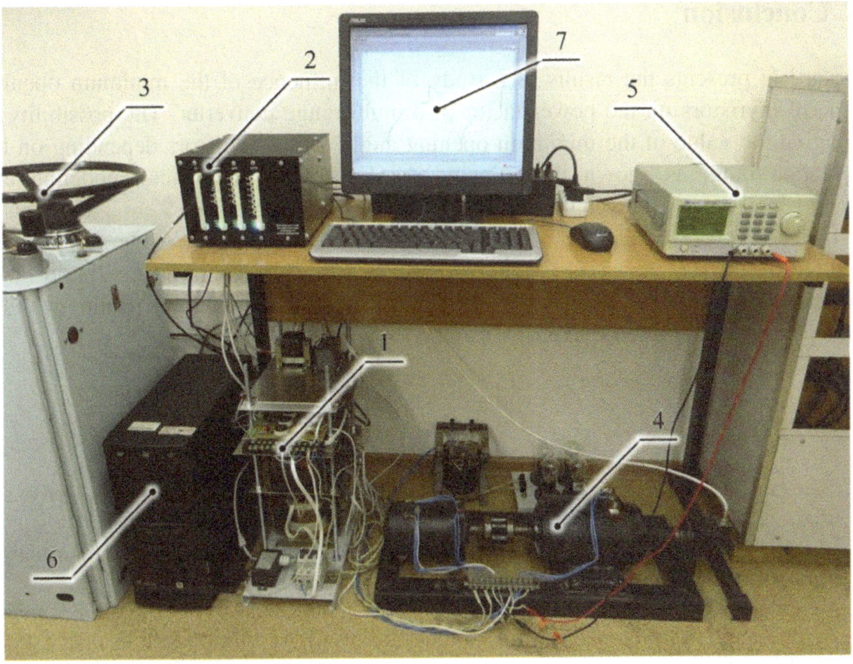

Fig. 6. Exterior of the experimental model

As an example in Fig. 7(a, b) there are curves of rectified voltage U_d received using experimental model with various values of minimum angle of opening of thyristors α_0.

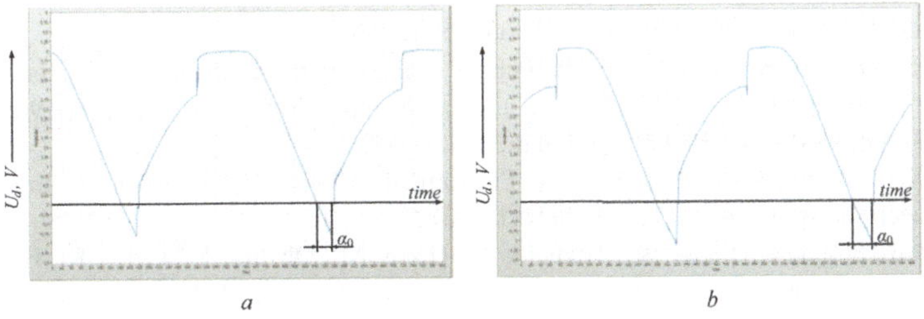

Fig. 7. Rectified voltage diagrams

The results of experimental modelling show that the decrease in delay angle twice lead to the increase in power factor of a converter from 0.66 to 0.71. The result received using the experimental model confirms the effect of increasing power factor due to the decrease of minimum angle of opening of multi-range converter thyristors.

6 Conclusion

The article presents the results of a study of the influence of the minimum opening angle of thyristors on the power factor of a multi-range converter. The possibility of changing the value of the minimum opening angle of the thyristors depending on the regulation range is shown theoretically. Analysis of mathematical modeling and experimental results showed that reducing the minimum angle from 9° to 2.25° leads to an increase in power factor *PF*, as well as an increase in the average value of the rectified voltage U_d by 1.34%. The proposed method of controlling a multi-range converter may be useful to developers of power conversion equipment, dealing with the issues of improving their energy indicators.

References

1. Salmeron, P., Litran, S.P.: Improvement of the electric power quality using series active and shunt passive filters. IEEE Trans. Power Deliv. **25**(2), 1058–1067 (2009)
2. Hosny, W., Park, H.E., Song, J.H.: Investigation of shunt active power filters in railway systems, substation installation. J. Energy Power Eng. **7**(10), 1974–1979 (2013)
3. Choi, U.M., Blaabjerg, F., Lee, K.B.: Study and handling methods of power IGBT module failures in power electronic converter systems. IEEE Trans. Power Electron. **30**(5), 2517–2533 (2014)
4. Youssef, A.B., El Khil, S.K., Slama-Belkhodja, I.: State observer-based sensor fault detection and isolation, and fault tolerant control of a single-phase PWM rectifier for electric railway traction. IEEE Trans. Power Electron. **28**(12), 5842–5853 (2013)
5. Kulinich, Y., Shukharev, S.: Increase in the power factor of the electric ac loco-motive due to a new structure of compensator operation. Proc. Eng. **165**, 1512–1519 (2016)
6. Khairudin, M., Efendi, Purwatiningsih, N., Irawan, W.: The implementation of silicon controlled rectifiers for DC motor control. In: AIP Conference Proceedings, vol. 1977, no. 1, p. 030035 (2018). https://doi.org/10.1063/1.5042955
7. Baimel, D., Kuperman, A.: Full theoretical analysis with simulation-based verification of thyristors-bridge-type SFCL operation modes. Simul. Model. Pract. Theory, **94**, 349–366 (2019). https://doi.org/10.1016/j.simpat.2019.03.009
8. Katsikis, V. (ed.): MATLAB - A Fundamental Tool for Scientific Computing and Engineering Applications, vol. 1. InTech (2012)
9. Perez, J., Sugai, T., Jiang, W., Tokuchi, A., Horie, M., Ohshio, Y., Ueno, K.: High current pulse forming network switched by static induction thyristor. Matter Radiat. Extremes **3**(5), 261 (2018). https://doi.org/10.1016/j.mre.2018.04.001
10. Han, W.H., Park, T.Y., Kang, C.J., Lee, S.I., Choi, Y.J.: Development of scanning field emission current microscopy. Curr. Appl. Phys. **9**(2), 29–32 (2009)
11. Blume, P.A.: The LabVIEW Style Book. Pearson Education (2007)

Mathematical Modeling of Transient Heating Processes in the System of Three Single-Pole Cables

Svetlana Kolomiitseva[1(✉)], Donat Sulyandziga[2], and Elena Sulyandziga[1]

[1] Far Eastern State Transport University, Khabarovsk, Russia
skolomy@gmail.com
[2] Moscow State University, Moscow, Russia

Abstract. In this paper, we calculate a mathematical model of the transition process of forming the temperature field of a cable line with insulation from cross-linked polyethylene at various currents close to the maximum allowed. The model of underground cable laying with a triangle in standard climatic conditions is considered. When the current is applied, the temperature difference between the medium and the cable is a solution to the ordinary linear differential equation, which implies that the thermal balance of the system occurs 30 min after the start of operation. The model of formation of the thermal field of a cable using an equivalent screen shows proportionally close heating of the core and the screen. The temperature change of all cable elements describes a two-dimensional parabolic type equation with zero Neumann boundary conditions. The simulation is done in PDETool Matlab. Empirically, the heating time of the cable core to temperatures of 27, 40, 50° was determined. Numerical modeling takes into account the change in the thermophysical parameters of the cable with increasing temperature. For this, the simulation time is divided into three iterations with updating of the initial conditions and parameters of the equation. The proposed model makes it possible to evaluate the limiting currents of the cable load, possible changes in the materials used, laying methods, and determine the interdependence of the model parameters. Thermal analysis is used to solve operational problems, such as determining the values of the cross-sectional area of the cable and screen, methods of connecting and grounding the shielding shells.

Keywords: Mathematical modelling · Thermal fields · Single-phase XLPE cable

1 Introduction

Power cable heating by current is one of the important factors affecting cable performance. In this case, the formation of the temperature distribution depends on the physical characteristics of the materials, laying methods, design features, cable current load, etc.

Z. Popovic et al. (Eds.): TransSiberia 2019, AISC 1115, pp. 33–41, 2020.
https://doi.org/10.1007/978-3-030-37916-2_4

Currently, there are many works that discuss heat transfer problems in different cable management systems [1–4], but the purpose of the work is to analyze and evaluate the thermophysical parameters of soil and backfill, their effect on the heat transfer of the cable system, optimization of material costs during installation cable [2, 5–9]. Note that in all these works more complex models and various methods of numerical simulation are used [10–12], but the calculation results lead to the conclusions presented in this paper [13, 14].

The main idea of this work is to calculate the temperature field of the cable core at close to the limiting currents according to the mathematical model of the transition mode of heating the cable line with one copper core, with a copper screen, with wire armor and an outer shell made of cross-linked polyethylene, built in PDEToolbox Matlab.

2 Methods

Consider a block of three single-phase cables with a copper core, with a copper screen, wire armor and an outer polyethylene sheath with a voltage of 64/110 kV. The cable is laid by a "triangle" in an earthen trench 0.8 m wide at a depth of 1 m, the screen is grounded on both sides. Then the currents in the screens and the associated electrical power losses affect the thermal mode of the cable and its throughput: you can't load the cable as much as it would be possible in the absence of parasitic screen currents [6, 7]. Soil temperature 20 °C. A necessary condition for observing the temperature regime of the cable is when the surface temperature of the core, at a given current value, does not exceed the maximum permissible value of +90 °C. The structural diagram of one core of the power cable is shown in Fig. 1.

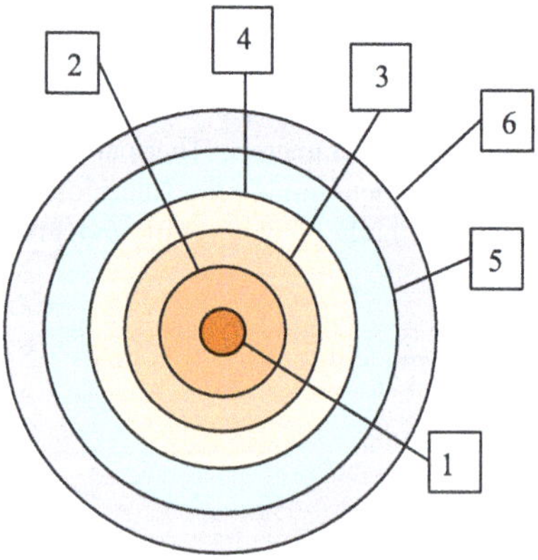

Fig. 1. Block diagram of one core of a power cable 64/110 kV. (1) Copper (aluminium) core; (2) Screen section (copper); (3) Armor section (aluminum); (4) Insulation (polyethylene); (5) Waterproofing layer; (6) Outer sheath (polyethylene).

We assume some approximations:

1. The radius of the cable is much less than its length, (where is the cross-sectional area and is the cable length.
2. The system has axial symmetry.

The characteristic dimensions of the cable elements and its necessary physical characteristics are given in Tables 1 and 2. The physical parameters are temperature dependent, approximated using the Lagrange interpolation polynomial from the tabular values [8] taken in the interval [300; 800] K.

Table 1. Metric cable parameters

Structural elements	
Core section	185 mm^2
Screen section	150 mm^2
Armor section	538 mm^2
Equivalent screen	130 mm^2
Copper core current	$\leq$425A

Table 2. Thermophysical cable parameters

	Coefficient of thermal conductivity k, $\frac{W}{m \cdot K}$	Material density ρ, $\frac{kg}{m^3}$	Specific Heat c, $\frac{J}{kg \cdot K}$	Resistivity ρ_c, $Ohm \cdot m$
Copper 20 °C	393	8930	381	$1.68 \cdot 10^{-8}$
Copper 27 °C	401	8933	385	$1.725 \cdot 10^{-8}$
Copper 40 °C	399.7	8920.6	386.85	$1.813 \cdot 10^{-8}$
Copper 50 °C	398.8	8912	388.2	$1.818 \cdot 10^{-8}$
Copper 60 °C	397.9	8905	389.5	$1.948 \cdot 10^{-8}$
Copper 100 °C	392	8900	399	$2.34 \cdot 10^{-8}$
Aluminum (armor)	237	2697	903	$2.73 \cdot 10^{-8}$
Polyethylene (insulation)	0.42–0.44	950	2300	10^{14}
Soil	0.833	1750	800	
Gravel with sand	2.26	2300	845	

If there is current, the temperature difference between the cable and the medium will change until the heat balance sets in (equality to zero the algebraic sum of the heat that the cable gives off to the environment and the heat that is released due to the current). This task corresponds to the Cauchy problem:

$$\begin{cases} cM\theta' + kS_{sur}\theta = I^2R, \\ \theta(0) = 0, \end{cases} \tag{1}$$

where k – is the coefficient of thermal conductivity, S_{sur} – is the area of the side surface of the cable, M – is the mass of the cable, c – is the specific heat of the conductor material.

The solution of task (1) is:

$$\theta(t) = \frac{I^2R}{kS_{sur}}\left(1 - e^{-\frac{kS_{sur}}{cM}}\right). \tag{2}$$

The function $\theta(t)$ has an exact upper bound $\theta_e(t) = \frac{I^2R}{kS_{sur}}$. We introduce the characteristic heating time $\tau = \frac{cM}{kS_{sur}}$, the stationary mode sets in 4τ, for the current task parameters this time is 30 min.

Transient thermal processes are described by the heat equation, and when the steady state is reached, the Poisson equation.

We calculate the temperature field of the cable when heated by current without taking into account the separation of elements into their component conductors at a load factor $k_l = 1$. The density of heat sources is described by the Joule-Lenz law

$$f(x, y, t) = \frac{I^2\rho_c}{S^2}, \tag{3}$$

where I – is the current, ρ_c – is the resistivity, S – is the cross section of the conductor.

The calculated densities of the heat sources of the copper core at different currents and temperatures are given in Table 3.

Table 3. Thermal parameters of a copper core

Current, A	Density of heat sources copper core ($\frac{W}{m^3}$) at different temperatures				
	20 °C	27 °C	40 °C	50 °C	60 °C
350	60131	61742	64892	67326	69724
375	69028	70877	74493	77287	80040
400	78539	80643	84757	87936	91068
420	86589	88909	93444	96949	100402
430	90762	93193	97947	101621	105240

To calculate the density of the heat sources of the copper screen and aluminum-aluminum armor, we use the equivalent screen method [9], the heat release density of which is proportional to the density of the heat sources of the core

$$\frac{f_s}{f_c} = \frac{S_c}{S_s\left(1 + \left(\frac{R_s}{X}\right)^2\right)}, \quad (4)$$

where f_S, f_C – heat density of the screen and core, S_S, S_C – screen and core sections, R_S^o – screen resistance, X – core and shield mutual inductance. The corresponding values of the proportionality coefficient are presented in Table 4.

Table 4. The coefficient of proportionality of heat loss

	The ratio of heat loss of copper screen and core at different temperatures				
	20 °C	27 °C	40 °C	50 °C	60 °C
$\frac{f_s}{f_c}$	0.91	0.89	0.86	0.83	0.81

We use the obtained values of in our model both for the copper screen and for the armor.

3 Results

The simulation of cable heating is described by an inhomogeneous two-dimensional heat equation of parabolic type

$$\rho c \frac{\partial u}{\partial t} = k\Delta u + f(x, y, t). \quad (5)$$

We will consider the problem in the space-time domain $G \times [0, T]$, where $G = \{(x, y) : 0 < x < 0, 8; 0 < y < 1\}$ describes the characteristic dimensions of the cable laying trench, [0, *T*] – the time interval corresponding to the onset of the steady state (for our task T = 1800 s), *u(x, y)* – is the temperature.

The physical characteristics of the object describe piecewise constant functions with the following parameters: k – thermal conductivity coefficient, ρ – material density, c – specific heat, $f(x, y, t)$ – density of heat sources.

At the boundary of the domain we set the zero Neumann boundary conditions

$$\left.\frac{\partial u}{\partial n}\right|_{\partial G} = 0 \quad (6)$$

Initial conditions $u(x, y, 0) = 20$ °C.

So, the two-dimensional initial-boundary problem of heat conduction is posed

$$\begin{aligned} \rho c \frac{\partial u}{\partial t} &= k\Delta u + f(x, y, t), \\ \left.\frac{\partial u}{\partial n}\right|_{\partial G} &= 0, \\ u(x, y, 0) &= 20\,^{\circ}\mathrm{C} \end{aligned} \tag{7}$$

Note that this model can only be used to describe the transient process in the time interval [0, T] sec. Also, this model is applicable for estimating the maximum value of the supplied current at known maximum temperature values.

Simulation is carried out in two stages, given the invariability of the characteristics of the armor:

3. at constant physical parameters (taken at a temperature of 20 °C),
4. in several iterations, determined by the reference points for changing the physical parameters from temperature (27, 40, 50°). During each iteration, we update the initial condition, the physical parameter, and the values of the heat sources. During each iteration, we consider the physical parameters to be constant.

To solve the problem of finding the temperature field of the cable system taking into account their laying, the PDEToolbox Matlab software environment is used, which uses the finite element method to solve it [15] (Figs. 2, 3 and 4).

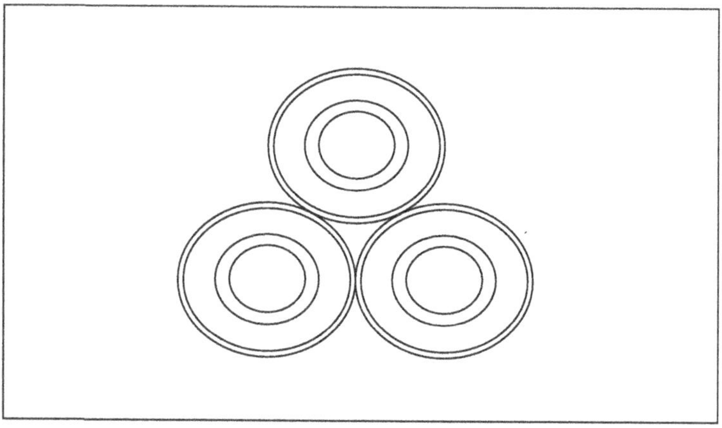

Fig. 2. General view of the calculation model when laying in the ground by a triangle

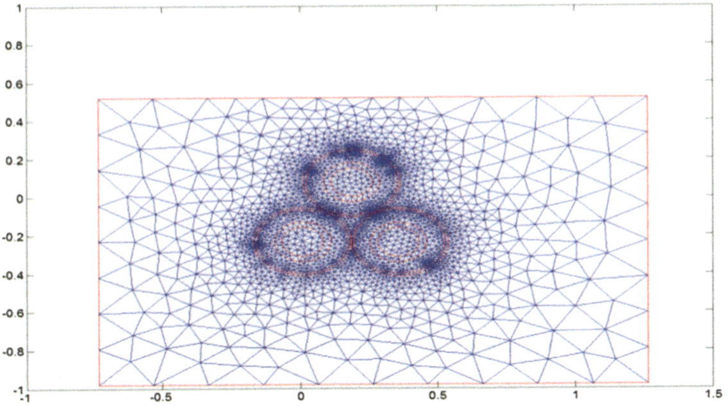

Fig. 3. Butt-laying with triangle triangulation of the computational domain

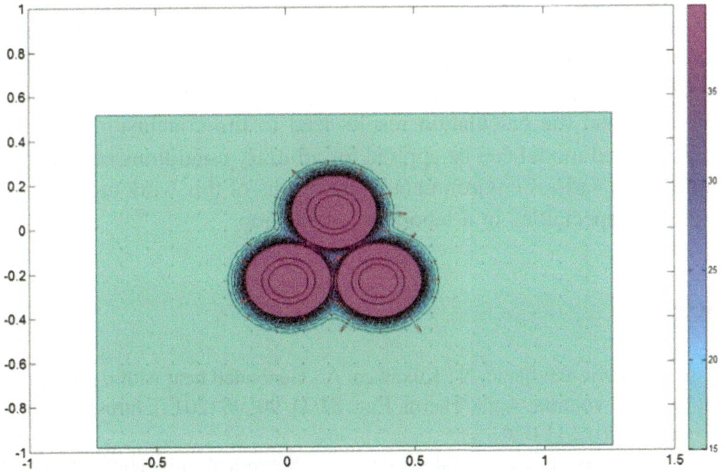

Fig. 4. The color distribution of the thermal field in the section of the cable system 185 mm^2 with a copper shield of 150 mm^2 and the cross-section of the armor 538 mm^2 when laying in the ground with an end-to-end triangle, the armor and the shield are grounded from both sides.

The results of numerical simulation are presented in Tables 5 and 6.

Table 5. The steady temperature for different structural elements without specifying physical parameters

Current, A	Core temperature, $^\circ C$	Armor temperature, $^\circ C$	Insulation temperature, $^\circ C$
430	62.3	62.3	54.0

Table 6. The steady-state temperature of various structural elements with a phased refinement of physical parameters

Current, A	Core temperature, $^\circ C$	Armor temperature, $^\circ C$	Insulation temperature, $^\circ C$
430	65.24	65.2	59.1

4 Discussion and Conclusions

As a result of simulation using the equivalent screen method, we obtained the temperature distribution in a system of three single-phase cables laid in a triangle at a current close to the limiting one for the transition mode with constant coefficients in the heat equation. Also, modeling was carried out with an iterative change in the physical parameters of the core and the equivalent screen, when it is easy to notice the refinement of the steady-state heating temperatures of the cable elements. The ratio of heat release between the screen and the core, calculated using the equivalent screen method, is close to unity.

Numerical simulation confirms the high heating of the screen, which leads to a decrease in possible operational characteristics and determines the further direction of technological development. In [11], in solving a similar problem, a more complex model was used, and the calculation results lead to the conclusions obtained in our work. The developed model can be applied for arbitrary conditions of cable laying with other parameters in various environments. The results of this work are used by students of energy-electric specialties in a laboratory workshop.

References

1. Kovac, N., Grulovic-Pavljanic, N., Kukavica, A.: Generated heat within power cable sheaths per unit time and volume. Appl. Therm. Eng. **52**(1), 90–96 (2013). https://doi.org/10.1016/j.applthermaleng.2012.11.020
2. Oklon, P., Cisek, P., Pilarczyk, M., Taler, D.:. Numerical simulation of heat dissipation processes in underground power cable system situated in thermal backfill and buried in a multilayered soil. Energy Convers. Manag. **95**, 352–370 (2015). https://doi.org/10.1016/j.enconman.2015.01.092
3. Oklon, P., Bittelli, M., Cisek, P., Kroener, E., Pilarczyk, M., Taler, D., Rao, R.V., Vallati, A.: The performance analysis of a new thermal backfill material for underground power cable system. Appl. Therm. Eng. **108**, 233–250 (2016). https://doi.org/10.1016/j.applthermaleng.2016.07.102
4. Quan, L., Fu, C., Si, W., Yang, J., Wang, Q.: Numerical study of heat transfer in underground power cable system. Energy Proc. **158**, 5317–5322 (2019). https://doi.org/10.1016/j.egypro.2019.01.636
5. Oklon, P., Cisek, P., Rerak, M., Taler, D., Rao, R.V., Vallati, A., Pilarczyk, M.: Thermal performance optimization of the underground power cable system by using a modified Jaya algorithm. Int. J. Therm. Sci. **123**, 162–180 (2018). https://doi.org/10.1016/j.ijthermalsci.2017.09.015

6. Salata, F., Nardecchia, F., Gugliermetti, F., Vollaro, A.: How thermal conductivity of excavation materials affects the behavior of underground power cables. Appl. Therm. Eng. **100**, 528–537 (2016). https://doi.org/10.1016/j.applthermaleng.2016.01.168
7. Salata, F., Nardecchia, F., Vollaro, A., Gugliermetti, F.: Underground electric cables a correct evaluation of the soil thermal resistance. Appl. Therm. Eng. **78**, 268–277 (2015). https://doi.org/10.1016/j.applthermaleng.2014.12.059
8. Ahmad, S., Rizvi, Z., Khan, M.A., Ahmad, J., Wuttke, F.: Experimental study of thermal performance of the backfill material around underground power cable under steady and cyclic thermal loading. Mater. Today **17**, 85–95 (2019). https://doi.org/10.1016/j.matpr.2019.06.404
9. Metwally, I.A., Al-Badi, A.H., Al Farsi, A.S.: Factors influencing ampacity and temperature of underground power cables. Electr. Eng. **95**(4), 393 (2013). https://link.springer.com/article/10.1007/s00202–012-0271-5. Accessed 19 Aug 2019
10. Gołębiowski, J., Zareba, M.: The simplified method for transient thermal field analysis in a polymeric DC cable. Electric. Eng. **93**(4), 209–216 (2011). https://link.springer.com/article/10.1007/s00202-011-0207-5. Accessed 19 Aug 2019
11. Bure, I.G., Khevsuriani, I.M., Bystrov, A.V.: Influence of the shield grounding system on choice of section. Russ. Electr. Eng. **87**(11), 641–646 (2016). https://doi.org/10.3103/S1068371216110031
12. Zarubin, V.S., Kuvyrkin, G.N., Savelyeva, I.Y.: Temperature state of electrical insulation of a superconducting DC cable. Cryogenics **95**, 1–4 (2018). https://doi.org/10.1016/j.cryogenics.2018.08.003
13. Brito Filho, J.P.: Heat transfer in bare and insulated electrical wires with linear temperature-dependent resistivity. Appl. Therm. Eng. **112**, 881–887 (2017). https://doi.org/10.1016/j.applthermaleng.2016.10.172
14. De Lieto, R., Fontana, L., Vallati, A.: Experimental study of thermal field deriving from an underground electrical power cable buried in non-homogeneous soils. Appl. Therm. Eng. **62** (2), 390–397 (2014). https://doi.org/10.1016/j.applthermaleng.2013.09.002
15. Trufanova, N.M., Navalikhina, E.Y., Gataulin, T.V.: Mathematical modeling of nonstationary processes of heat and mass transfer in a rectangular cable channel. Russ. Electric. Eng. **86**(11), 556–560 (2015). https://doi.org/10.3103/S1068371215110139

Electric Energy Storage Units Applicability Assessment of Different Kinds in the Conditions of Moscow Central Ring

Vladislav Nezevak(✉), Vasily Cheremisin, and Alexander Shatokhin

Omsk State Transport University, Omsk 644046, Russia
VladislavNezevak@gmail.com

Abstract. The article deals with the issues of improving the energy efficiency of the traction power supply system of the Moscow Central ring through the use of energy storage systems. The features of work of system of accumulation of electricity in the context of the predominance of passenger traffic to select the kind of storage of electricity. The paper deals with three common types: electrochemical, supercapacitor and electromechanical storage, identified the most preferred, as which it is proposed to consider hybrid systems of energy storage.

Keywords: Traction power supply system · Moscow central ring · Improving energy efficiency · Level of charge

1 Introduction

Electric energy storage units' development led to appearance of powerful and и capacious systems, which cost falls during the last years. Specific cost reduction forecast of electric energy storage units' different kinds until 2025 year focuses on cost reduction of most known storage units up to a level of 35–40 th. r./kWh in current prices [1]. At present by national and foreign industry there was mastered a production of different electric energy storage units [2]. Prevailing by established capacity among common storage units' there is hydro accumulating power stations, a part of which in capacity structure exceeds 90%. Among other kinds the most dissemination got the electrochemical accumulators, supercapacitor, kinetic storage units and storage units, which use energy of compressed air. The use questions problematic of storage units system, which is bound with electricity consumption issues solving, peak load cover, voltage stabilization, energy efficiency increase is topical in electro energetics and on transport, in particular, in the electric traction system (further – ETS) of Moscow central ring (further – MCR).

Besides energy intensity, one of the main characteristics is electric energy storage units' capacity. On subway lines with possibility exception of capacity and energy loss in addition to electric energy storage units' there are set in the inverter converters. Used in the world on subway lines electric energy storage units of different kinds (electrochemical, kinetic and supercapacitors) are characterized with high enough power. The most common capacities range of electric energy storage units in this case is 1.0–2.0 mW.

Z. Popovic et al. (Eds.): TransSiberia 2019, AISC 1115, pp. 42–51, 2020.
https://doi.org/10.1007/978-3-030-37916-2_5

Existence of electric energy storage units system of different kinds determines a question formulation about kind choice, which is the most appropriate for the electric traction system specific, and necessary parameters determination.

2 Problem Statement

At present in the electric traction system of national railway transport electric energy storage units didn't find a use. In Russia in transport there was performed an experimental operation of electrochemical storage units in Moscow subway [3]. Foreign practice demonstrates successful experience of electric energy storage units operation in the subway (voltage – 750 V), for example, in the subway of the city Osaka (Japan), where a storage unit of electrochemical type is used with capacity about 500 kWh for voltage 750 V. Storage unit, which was established by Tesla on st. Osaka in 2019 has capacity about 7000 kWh [4]. External characteristic of this storage unit is close to characteristic of uncontrollable rectifier with statism coefficient $k = -0.05$.

Researches in the field of inertial energy storage unit creation (IESU), intended for usage on MCR, led to appearance of experimental sample with capacity 5–10 mW and energy intensity 100–150 mJ (28–42 kWh). In Table 1 there are given the main parameters of projected inertial storage unit for MCR [5].

Table 1. The main energy parameters of inertial storage unit

№ in order	Parameter	Value
1	Electric energy exchange stock	42 kWh
2	Voltage at the input of IESU by charge	3550–4000 V
3	Voltage at the output of IESU by discharge	3300–3550 V
4	Maximum current of charge-discharge I_{max}	3000 A
5	Duration of charge-discharge peak	15–30 s
6	Medium efficiency current value during an hour	650 A
7	Medium efficiency current value during overcharge	1600 A
8	Peak capacity of IESU	10 mW

Supercapacitors constitute capacitors of high capacity with nominal voltage up to 1000 V. By their characteristics a supercapacitor takes an intermediate position among capacitors and chemical current sources. As an example there are supercapacitors of «Technokor» company production with nominal voltage from 14 up to 700 V and capacity from 95 up to 0.15 F respectively (Table 2).

Storage unit test results in the electric traction system of DC 3.3 kV of Polish transport company allowed estimating of potential electric energy saving on passenger trains traction, which was 1 490 th. kWh per year [6].

Table 2. Base capacitors characteristics of «Technokor» company

Type, kJ/V	Voltage, V	Current, A	Capacity, F	Internal resistance, Ohm	Intencity, kJ	Capacity, kW
9/14	14	670	95.0	0.0060	9.3	8.2
40/28	28	4000	104.0	0.0055	40.8	35.6
40/64	64	2125	18.0	0.0300	36.9	34.1
40/96	96	1300	9.0	0.0600	41.5	38.4
20/150	150	750	1.8	0.2000	20.3	28.1
15/175	175	700	1.0	0.2500	15.3	30.6
60/200	200	1000	3.0	0.2000	60.0	50.0
27/220	220	1100	1.1	0.2000	27.1	60.5
18/350	350	700	0.3	0.4000	18.4	76.6
64/400	400	1000	0.8	0.4000	64.0	100.0
36/700	700	1000	0.2	0.7000	36.8	175.0

Comparison of lead-acid, nickel metal hydride and lithium ion accumulators, supercapacitors and kinematic storage units for the conditions of Polish railways showed, that on ten year interval of operation the most profitable is the use of super capacitor (estimated economic effect of electric energy saving is \$1.9 mln), further in decreasing order of profitability: lead-acid accumulator (\$2.5 mln), kinetic storage unit (\$3.5 mln), lithium ion accumulator (\$4.8 mln) and nickel metal hydride (\$5.1 mln). Characteristics of researched storage systems are given in Table 3 based on date for lead-acid accumulators [7], nickel metal hydride [8], lithium ion [9], supercapacitors [10] and kinetic flywheels [11].

Existing advantages and disadvantages of electrochemical and capacitive electric energy storage unit against each other [12], led to appearance of hybrid storage systems, which allow using storage units advantages of different kinds.

3 The Simulation Modeling

Work specific of the electric rolling stock of MCR is characterized by frequent stops and appropriate modes changes of traction and regenerative braking. The received by results of recorders date processing traffic parameters of electric trains of series ES2G we can get voltage change characteristic on current collector (Fig. 1a) and electric train current (Fig. 1b).

Received distributions cannot be described with classic distribution lows, as well as load characteristics of traction substations, for example, as it is shown in [13]. In this regard for description of these processes it is necessary to use modified distribution lows or empirical distributions.

Received distributions allow estimate the voltage range of current collector as wide for areas with plain profile of MCR in the range from 3000 V up to 4000 V, current from −600 A up to 1000 A.

Table 3. Electric energy storage units' main characteristics (DOD – depth of accumulator discharge; C – nominal accumulator capacity, Ah)

Parameter	Battery Hitachi LL1500-WS (Pb)	Kawasaki Gigacell (NiMh)	Hitachi CH-75-6 (Li-Ion)	Maxwell BMOD0063 P125 b08 (CK)	Vycon VDC (flywheel)
Number of elements/voltage of an element, V	1650/2	91/36	148/22.2	27/125	6/600
Energy intensity, Ah/kWh	1500/3	1800/5.4	75/5.0	–/39	–/0.52
Maximum current of one chain, A	900	750	225 (3C)	100 (100 × 10)	150 (150 × 6)
Maximum capacity, mW	3.0	2.5	4.9	4.9	3.0
Number of cycles for DOD	70%–4500	1.8%–367000	75%–4000	106	–
Cost, $/1 kWh	300	1412	900	5100/on 1 CK	6000/on 1 flywheel
Approximate cost of control system, $	500000	500000	500000	500000	500000
Total cost, $mln	2.5	5.1	4.8	1.9	3.5

Interaction simulation modeling of electric rolling stock with electric traction system, which was performed for conditions of MCR (based on electric trains' traction characteristics, executed traffic graph and electric traction system characteristics in the program complex CORTES), allows receiving daily load graphs and voltage changes on traction substations bus bars and linear units.

Electric energy volumes of energy return on linear units' bus bars by modeling results allow determining a proportion, which is typical for charge and discharge modes. In borders of MCR ETS there are located four posts of sectionalization, for which there are received coefficients of energy volumes correlation, which are determined by formula (1):

$$k = \frac{W_3}{W_p}, \tag{1}$$

where W_3, W_p – electric energy volume, which appropriate the charge and discharge mode.

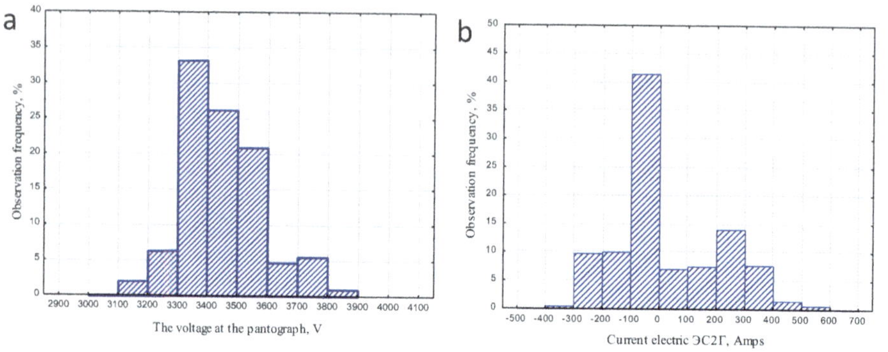

Fig. 1. Frequency histogram for voltages on current collector (a) and traction load (b) of electric trains of series ES2G

For the conditions of posts of sectionalization of MCR PSK 270, PSK 445, PSK-1 and PSK Cherkizovo these coefficients are: 1.34; 0.42; 0.17 and 1.19 respectively. Only for two variants of storage units placement, for which $k > 0$, energy exchange processes are balanced, for the other two variants for energy volume balancing there is a need to perform a storage units charging from the contact network, what reduces the usage efficiency for the considered conditions. The given results were received for setpoints by voltage for charge and discharge mode 3600 V and 3550 B respectively. Setpoints values change by voltage leads to significant picture change of electric energy volumes distribution by operating modes and influences required level of useful energy intensity.

The feature of the electric energy storage systems operating is the short duration of charge/discharge modes and the observed absence of work episodes alternation [14] that leads to characteristic graph of electric energy storage unit charge level in the conditions of MCR. The average episodes number of for storage unit charge and discharge modes, which were received by simulation modeling results, is about 500 for both modes (Table 4). For specified storage unit work conditions it was received, that work episodes duration in different modes doesn't exceed 5 min, in particular, for PSK 270 of MCR doesn't exceed 2 min, and electric energy volume by episodes doesn't exceed 50 kWh.

Electric energy storage units charge level is determined by a number of factors: initial voltage, internal resistance, discharge current, electricity number, ecapacity, discharge current (charge). In calculations of charge level we use equations of Shepherd, Haskina-Danilenko, Romanova [15], another empirical dependences, equations, which were received with the help of regressive analysis device or artificial neuronal networks. One of the estimate ways of charge level is the way based on counter Ah or electricity number. Modification of this way is the way based on counter WH for charge and discharge modes, which satisfies the receiving of storage unit work assessment. Change graph building of energy cumulative volume allows receiving required level of useful energy intensity WEESU as a difference between maximum W_{max} and minimum cumulative value W_{min} per day. Specified value W_{EESU} allows estimating storage unit charge level on the k-th calculations step according to the following formula:

$$SoC_k = \frac{\sum_{t_0}^{t_k} u_k \cdot i_k \cdot \Delta t_k}{W_{\text{НЭЭ}}} \cdot 100, \tag{2}$$

where u_k, i_k – voltage and current values for k-th time period; Δt_k – time step.

Table 4. Quantitative characteristic of storage units work episodic nature on posts of sectionalization

Placement	Operating mode	
	Discharge	Charge
PSK 270	561	537
PSK 445	495	451
PSK-1	604	205
PSK Cherkizovo	509	424
Average value	542	404

Processing of simulation modeling date allows with the help of formula (2) receiving charge level and estimating required level of useful energy intensity. The simulation modeling results show that by restrictions absence of useful storage unit energy intensity values in relation to post of sectionalization PSK 270 the required useful energy intensity W_{EESU} will be 1185 kWh. At restrictions introduction of useful storage unit energy intensity in the amount of 500 kWh the change graph shows that SOC doesn't change the cyclicity during the considered day. The total electric energy volume per day in discharge mode in the second case falls on the first from 7054 kWh up to 5920 kWh, that is by 16.1%, in charge mode it falls from 7166 kWh up to 6168 kWh, that is by 13.9%.

4 Comparison of Types of Drives to These Conditions

Lithium-ion accumulators, which are produces with the use of lithium iron phosphate technology (as cathodic material there is used $LiFePO_4$) have the following characteristics. Working life of considered accumulators is no more than 8 years [16]. By discharge depth of 80% resource is no more than 2000 cycles of charge/discharge. For the conditions of MCR the electric energy storage unit work envisages cyclicity at level of 2–3 cycles per day. Discharge depth in the given conditions will lead to resource exhaustion of accumulators work in 2–3 years. By discharge depth from 15 up to 25% resource is determined by exploitation conditions and is 15–20 years.

Manufactures recommend charge/discharge currents by its exploitation at level 0.2 C_n that is 20% of nominal capacity. In relation to conditions of MCR with uneven electric traction load it requires current stabilization in charge/discharge mode at the specified level, herewith current in charge/discharge modes will be less than nominal one hour current five times.

Operating mode of rechargeable batteries envisages work in buffer mode, which is unclaimed for the conditions of MCR. Discharge mode duration from level 1.0 C_{ch} up to 0.6 C_{ch} current 0.2 C_n is about 2 h, similar duration is typical for charge mode. Electric energy storage unit work conditions on MCR, which differ from long charge and discharge (about 200–300 min or 3–4 h), which consist of some episodes, following in a row, create an opportunity for long charge/discharge. Work feature in the conditions of MCR is charge/discharge episodes short duration, which duration doesn't exceed 5 min. This circumstance needs influence estimate on charge/discharge processes of lithium-ion accumulator, accumulators' manufacturer doesn't give such information. Capacity and energy intensity of storage unit lithium-ion accumulators changes by means of parallel consecutive modules connection.

The main advantages and disadvantages of lithium-ion batteries usage on MCR can be formulated in that way:

1. insufficient work resource. Resource increase of lithium-ion accumulators is possible up to 15–20 years by realization of discharge depth up to 25%. In this case the nominal capacity of accumulator battery will exceed the useful used capacity four times. So, by usage of accumulators useful capacity for the conditions of MCR at the level of 500 kWh, the nominal capacity will be 2000 kWh, for the level of 1000 kWh it will be 4000 kWh;
2. necessity of stabilization by current in charge/discharge modes at the level of 0.2 C. In the case of storage units use of such type it will lead to increase necessity of nominal capacity for covering of peak traction load of MCR or traction load reduction regarding nominal capacity five times;
3. necessity absence of buffer work mode from behind the electric traction load specific of MCR;
4. enough duration of charge/discharge processes in the conditions of MCR. The influence of short alternating continuous storage unit charge/discharge episodes requires the additional estimate in the work conditions on MCR on resource and storage unit charge level.

Supercapacitors (ultra capacitors) have high energy characteristics and long working life. Supercapacitors BOOSTCAP of Maxwell company (dealer in RF – LCC «Titan Power Solution») work efficiently in a range of nominal voltage up to the level ½ of nominal voltage that allows using about 75% of available energy. These supercapacitors are electrochemical double layer capacitors (EDLC) [17].

Supercapacitors are characterized by typical time constant, which is equal about 1 s. This time shows supercapacitor charge duration from 0% up to 63.2% or discharge from the level of full charge up to 36.8% of discharge level.

Supercapacitors are susceptible to capacity degradation, which is expressed in the fact that nominal capacity reduction by 30% for a supercapacitor, supported by voltage 2.7 V, will happen after 5500 h at temperature 65 °C, after 11 000 h at temperature 55 °C, after 22000 h at temperature 45 °C, after 44000 h at temperature 35 °C, after 88000 h at temperature 25 °C.

The charge/discharge duration of supercapacitors doesn't exceed values, which are in a range of 1–5 min. Work resource is over 500 000 charge/discharge cycles. These characteristics agree well with electric traction load features in the conditions of MCR: by work resource, energy intensity use and charge/discharge cycles duration. Storage unit capacity and energy intensity on supercapacitors change in wide range by means of modules consistently parallel connection.

The main advantages and disadvantages of supercapacitors use on MCR can be formulated in this way:

1. announced of manufacturer supercapacitors work resource (>500 000 cycles) allows estimating device work duration in the conditions of MCR at the level of capacity and performance degradation, and not by number of charge/discharge cycles, which is not critical. In this case supercapacitor heating level will be defining;
2. supercapacitors useful energy intensity is about 75% from the nominal, what compared with lithium-ion accumulators allows reducing substantially the nominal storage unit energy intensity;
3. supercapacitors charge/discharge processes duration agree well with continuous episodes duration of charge/discharge storage unit by work in the conditions of MCR and allows realizing energy exchange processes frequency, which are specified by traffic schedule.

Known kinetic storage units are intended for work in buffer mode. Increase of charge and discharge duration leads to increase necessity of weight and dimensional parameters or modules number in the system of electromechanical electric energy storage unit. There are known flywheel kinetic energy storage units for urban electric transport of series NKE-3G [18], which working life is 20 years. Storage unit of specified brand has the next characteristics: capacity – 180 kW, energy intensity – 1.95 kWh. These characteristics are insufficient in order to satisfy work conditions in MCR. In this case there is a need of quantitative increase of storage units date in the system. Presented device brand is characterized with enough low energy intensity specific weight of 0.55 kWh/t, expected weight for energy intensity of 100 kWh will be about 181 t. Bearing resource is estimated at the level of 5 work years. Storage units' maximum voltage of specified series is 825 V. Manufacturer doesn't indicate date of time duration, which is necessary for output to nominal energy intensity and reset of kinetic energy. The other samples of kinetic storage units have similar characteristics [19]: acceleration time is 420 s, EF is up to 97%, and maintenance time of voltage failures is up to 7 s. Соотношение acceleration/braking duration correlation is 60/1.

The main braking advantages and disadvantages of kinetic storage units' use on MCR can be formulated in this way:

1. resource sufficiency of storage unit work for the conditions of MCR, which is defined by bearing working life and is independent of device work cyclicity, the total working life is 20 years;
2. storage unit characteristics discrepancy in acceleration and braking mode of episodes work duration and working time in charge/discharge modes in the conditions

of MCR, discharge value is substantially less than acceleration time of electric rolling stock on MCR;
3. main operating mode discrepancy (buffer) to work conditions on MCR. By this criterion storage unit doesn't satisfy on MCR, where the cyclical operating mode is necessary;
4. known national samples have low specific characteristics, that for necessary of MCR energy intensity values leads to significant increase of device weight.

5 Conclusion

Thus we can make the following conclusions:

1. the use of lithium-ion or similar accumulators in the conditions of MCR is bond with a necessity of nominal capacity increase, which is more than calculated values, what bond with a necessity of working life and charge/discharge conditions increase;
2. storage units as supercapacitors are the most appropriate to electric traction character on MCR, charge/discharge modes duration, allow reducing energy intensity in comparison with lithium-ion accumulators kinds, but which have a row of disadvantages, which are bond with capacity degradation and cost parameters;
3. kinetic electric energy storage units, which are intended for work in buffer mode, don't satisfy the traction load conditions on MCR, calculated graphs of devices charge, charge/discharge durations and low specific characteristics, and need development and modification;
4. the given characteristics of electric traction load and the main storage units kinds, calculated operating modes for the conditions of MCR make expedient the use of hybrid technologies, which include modules based on rechargeable batteries and supercapacitors. Hybrid electric energy storage unit, which combines the advantages of rechargeable battery and supercapacitor, is closer electric traction load characteristics of MCR and allows reaching the maximum efficiency of device work. Specified direction sets a row of tasks on a parity of energy intensity of specified modules in device structure, their control algorithm subject to traction load on adjoining traction substations and is the object of separate consideration.

Acknowledgements. The researches are performed with the support of the Russian Foundation for fundamental initiatives under the grant № 17-20-01148 ofi_m_RZD/18.

References

1. Radcliffe, P., Wallace, J.S., Shu, L.H.: Stationary applications of energy storage technologies for transit systems. In: 2010 IEEE Electrical Power & Energy Conference, pp. 1–7 (2010)
2. Takahashi, H., Aoki, K., Maoka, A., Kim, Y.I.: Current and future applications for regenerative energy storage system. Hitachi Rev. **61**(7), 336 (2012)

3. Kobayashi, H., Saito, T., Kondo, K.: A study on a method to design energy capacity of wayside energy storage devices in DC-electrified railway systems. In: 2015 17th European Conference on Power Electronics and Applications, pp. 1–9 (2015)
4. Tesla Powerpack. https://twitter.com/Tesla/status/1110712884286226432
5. Yu, J.G., Schroeder, M.P., Teumim, D.: Utilizing wayside energy storage substations in rail transit systems–some modelling and simulation results. In: 2010 Rail ConferenceAmerican Public Transportation Association, pp. 1–8 (2010)
6. Radu, P.V., Drazek, Z.: Analysis of wayside energy storage devices for DC heavy rail transport. In: MATEC Web of Conferences, vol. 180, p. 04001 (2018)
7. Ribeiro, P.F., Johnson, B.K., Crow, M.L., Arsoy, A., Liu, Y.: Energy storage systems for advanced power applications. Proc. IEEE **89**(12), 1744–1756 (2001)
8. Szeląg, A.: The influence of the voltage in the 3 kV DC traction network on the energy and traction parameters of supplied vehicles. The Institute of Scientific Publishing Spatium, Radom, vol. 158 (2013)
9. Divya, K.C., Østergaard, J.: Battery energy storage technology for power systems—an overview. Electr. Power Syst. Res. **79**(4), 511–520 (2009)
10. Datasheet. 125 V Heavy Transportation Module. http://www.maxwell.com/images/documents/125vmodule_ds_1014696-7.pdf
11. VYCON REGEN. The Proven Flywheel Energy Storage System for Rail. https://vyconenergy.com///wp-content/uploads/2018/06/vycon-regen-kinetic-brochure_4_page_new_web-single-pagesnewbrand.pdf
12. Larminie, J., Lowry, J.: Electric Vehicle Technology Explained, 2nd edn. Wiley, Chichester (2003)
13. Gatelyuk, O.V., Nesevac, V.L., Erbes, V.V.: Analysis of measurement data in a direct current traction power-supply system with uncontrolled rectifiers. Russ. Electr. Eng. **90**(2), 180–186 (2019)
14. Nezevak, V., Cheremisin, V., Shatokhin, A.: Assessment of energy intensity of the drive for traction power supply system. In: Energy Management of Municipal Transportation Facilities and Transport, pp. 524–538 (2018)
15. Galushkin, N.E., Yazvinskaya, N.N., Galushkin, D.N.: Nonlinear structural model of the battery. Int. J. Electrochem. Sci. **9**(11), 6305–6327 (2014)
16. Martyanov, A.S., Korobatov, D.V., Sirotkin, E.A.: Modeling of battery charging algorithms. In: 2016 2nd International Conference on Industrial Engineering, Applications and Manufacturing, pp. 1–4 (2016)
17. Review of Supercapacitors BOOSTCAP. LCC «Titan Power Solution». https://www.titanps.ru/files/obzor-superkondensatorov-maxwell.pdf
18. Flywheel Kinetic Energy Storage Units for Urban Transport KESU-3G. LCC «Kinemac». http://www.kinemak.ru/wp-content/uploads/2014/07/List1a.pdf
19. UPS Based on Kinetic Energy Storage Unit. LCC «Russian Superconductor». http://rhsc.ru

Extending the Life of Power Transformers of Traction Substations of Alternating Current of Electric Railways

Anton Voprikov(✉), Nikolay Grigoriev, and Arseny Parfianovich

Far Eastern State Transport University, Serysheva Street 47, Khabarovsk 680000, Russia
voprikov.anton@mail.ru

Abstract. The article proves the relevance of extending the life of power transformers of traction substations of alternating current of electric railways. The service life of the transformers of many substations on the Far East Railway exceeds the standard service life. Replacing existing transformers with new ones will require huge investments. Every fourth case of transformer failure is associated with damage to the winding insulation. According to the results of the analysis of the operation of transformers, it was proved that the wear of winding insulation is uneven. To assess the wear of winding insulation of existing transformers, a method has been developed that allows determining the residual insulation resource. To reduce the wear rate of winding insulation, which limits the life of existing transformers, new connection schemes are proposed. Extending the life of the winding with the greatest insulation wear increases the life of the transformers. Determination of the predicted life of the corresponding transformer by thermal wear of winding insulation ensures their monitoring in operation. The determination of the relative wear of winding insulation in retrospect is based on the analysis of the executed train schedules. For each substation, a transformer connection diagram is determined, which ensures a decrease in the wear rate of winding insulation, which limits the life of the transformer in the future. The work proves the technical and economic efficiency of extending the life of transformers of traction substations of alternating current of electric railways.

Keywords: Traction substation · Power transformers · Insulation wear of windings · Connection diagrams

1 Introduction

The development strategy of the railway transport of the Russian Federation until 2030 [1] provides for increasing the efficiency of the traction power supply system through modernization and reconstruction, extending the service life of expensive equipment, reducing unit costs for its maintenance, etc. Modernization and reconstruction of traction power supply system are a long-term development program and require significant capital investments.

Z. Popovic et al. (Eds.): TransSiberia 2019, AISC 1115, pp. 52–59, 2020.
https://doi.org/10.1007/978-3-030-37916-2_6

The main and most expensive equipment of railway traction substations (TS) are power transformers. Due to the existing funding shortage, the rate of renewal of power transformers is much lower than the rate of their aging. Therefore, today, the development and improvement of measures aimed at extending the service life of power transformers is relevant and in demand for the railway transport infrastructure.

Based on the data on the actual state of winding insulation, the reliability of operation over the past period, the technical level, the operating life in relation to the full service life, and growth of loads, it is customary to leave the existing traction transformers, mainly because of their high cost. In addition, the electrical parameters of the new traction transformers, which characterize the operating economy, idling and short circuit losses, remain unchanged.

For example, a power transformer of the TDTNZh-40000/220 (three-phase three-winding power transformer with ONAF cooling system) type manufactured in 1981 by its main electrical characteristics ΔP_{id} = 51 kW and ΔP_{sc} = 200 kW (passport data for the operating traction transformer) corresponds to a transformer of the TDTNZh-40000/220 type manufactured in 2018: ΔP_{id} = 51 kW and ΔP_{sc} = 200 kW (data from the stock catalog of the Togliatti Transformer LLC).

Failure of power transformers of general-purpose substations in 16% of cases [2] is caused by failure of the windings. According to the data presented in [3], 80% of cases of winding damage are internal faults, the cause of which in 17% of cases is wear of winding insulation.

2 Materials and Methods

With the generally accepted diagram for connecting traction transformers (Fig. 1), the distribution of the current of the supply legs in the high voltage windings (i_{AX}, i_{BY}, i_{CZ}) is usually determined by the formulas:

$$\dot{I}_{AX} = \frac{2}{3}\cdot\dot{I}_r - \frac{1}{3}\cdot\dot{I}_l,\ \dot{I}_{BY} = -\frac{1}{3}\cdot\dot{I}_r - \frac{1}{3}\cdot\dot{I}_l,\ \dot{I}_{CZ} = -\frac{1}{3}\cdot\dot{I}_r + \frac{2}{3}\cdot\dot{I}_l, \tag{1}$$

where i_l and i_r – currents of the left and right supply leg of the substation, respectively, reduced to a higher voltage.

The current distribution over the transformer windings (see formulas 1) leads to uneven wear of their insulation [4]. If the transformer fails due to wear of insulation of one of the windings (for example, the AX winding), the residual insulation resource of the CZ(ΔL_{CZ}) and BY(ΔL_{BY}) windings is significant, ΔL_{CZ} and ΔL_{BY} will be determined by the formulas:

$$\Delta L_{CZ} = [L] - L_{CZ_{[T_{l.t}]}},\ \Delta L_{BY} = [L] - L_{BY_{[T_{l.t}]}}. \tag{2}$$

According to Fig. 2, the service life of the traction transformer due to insulation wear is limited by the AX winding at time $[T_{t.l}]$.

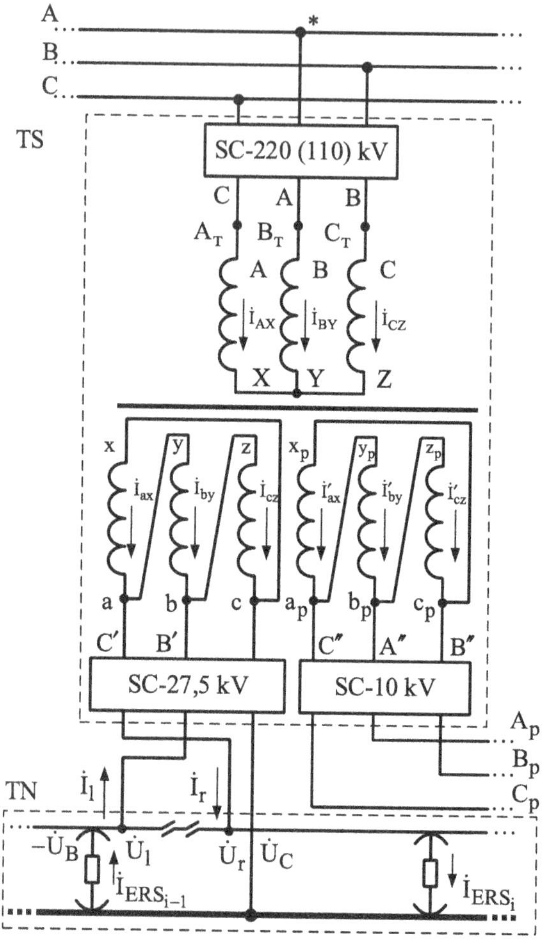

Fig. 1. Design scheme: * – the least loaded phase of the transmission line; ERS – electric rolling stock; TS – traction substation; TN – traction network; U_l and U_r – voltage of the left and right leg of the power substation

3 Results

In order to reduce the wear rate L′ of winding insulation, which limit the service life of the traction transformer, new connection diagrams have been developed with the following names: a and b. The names of the connection diagrams are taken according to the designation of the input winding of the transformer with the voltage of 27.5 kV, connected to the rail network [5]. At the same time, it was proposed to determine the intensity of insulation wear of each winding of the traction transformer using the developed system of phase-by-phase online monitoring of the state of winding insulation [6]. The system allows determining the insulation wear of each winding by the sum of wear for a separate i-th quantization interval of the measurement with a

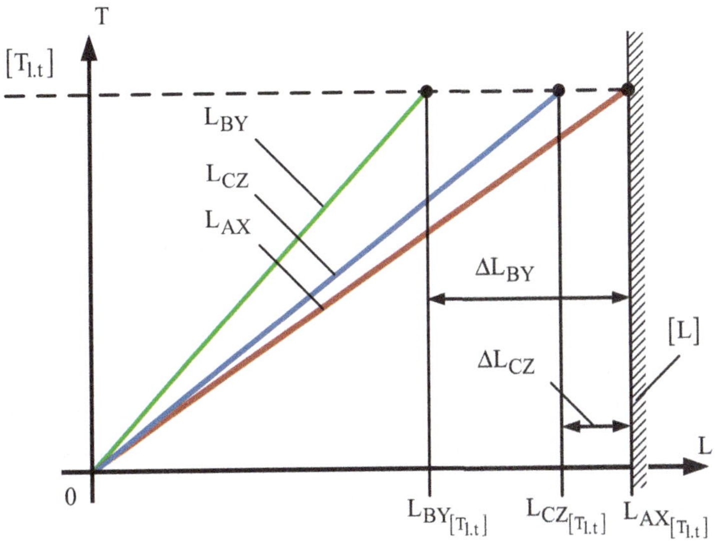

Fig. 2. Dependence of the service life on the wear of winding insulation with the conventional connection diagram

duration Δt, at which the most heated point ($\Theta_{h.p}$) of windings, moisture content of insulation w, acid number $K_{a.n.}$, and the oxygen concentration C_{O2} of the transformer oil can be taken constant. The equations for determining the wear of winding insulation are as follows:

$$\left.\begin{array}{l} L_{AX} = \sum\limits_{i=1}^{m} \exp\left[\ln 2 \cdot (\Theta_{AX_i} - 98)/\Delta + \alpha \cdot \ln(w_i/w_{b.f.}) + \beta \cdot \ln(K_{a.n_i}/K_{b.f.}) + \gamma \cdot \ln\left(C_{O_{2_i}}/C_{O_2 b.f.}\right)\right] \cdot \Delta t \\ L_{BY} = \sum\limits_{i=1}^{m} \exp\left[\ln 2 \cdot (\Theta_{BY_i} - 98)/\Delta + \alpha \cdot \ln(w_i/w_{b.f.}) + \beta \cdot \ln(K_{a.n_i}/K_{b.f.}) + \gamma \cdot \ln\left(C_{O_{2_i}}/C_{O_2 b.f.}\right)\right] \cdot \Delta t \\ L_{CZ} = \sum\limits_{i=1}^{m} \exp\left[\ln 2 \cdot (\Theta_{CZ_i} - 98)/\Delta + \alpha \cdot \ln(w_i/w_{b.f.}) + \beta \cdot \ln(K_{a.n_i}/K_{b.f.}) + \gamma \cdot \ln\left(C_{O_{2_i}}/C_{O_2 b.f.}\right)\right] \cdot \Delta t \end{array}\right\}, \tag{3}$$

where L_{AX}, L_{BY} and L_{CZ} – the wear of the insulation of the windings AX, BY and CZ of higher voltage of the traction transformer; Θ_{AXi}, Θ_{BYi}, and Θ_{CZi} - the temperatures of the most heated points of the winding AX, BY and CZ of higher voltage when measured with serial number i; i - serial number of the measurement, i = 1, 2… m; m - the number of measurements of the wear of winding insulation, m = T/Δt; T - the estimated operating time of the transformer; $w_{b.f.}$, $K_{b.f.}$, $C_{O2b.f.}$- the basic factors of the moisture content of insulation, acid number and oxygen concentration, respectively.

The phase-by-phase accounting of the wear rate of winding insulation of higher voltage of the traction transformer during the time T is performed according to the formulas:

$$\left.\begin{aligned} L'_{AX_T} &= L_{AX_T}/(m \cdot \Delta t) \\ L'_{BY_T} &= L_{BY_T}/(m \cdot \Delta t) \\ L'_{CZ_T} &= L_{CZ_T}/(m \cdot \Delta t) \end{aligned}\right\}. \quad (4)$$

Let's consider the section of the Far East Railway, consisting of 3 traction substations. The operating life of traction transformers at the substations under consideration exceeds the standard service life (25 years) and is: for transformers of substation No. 1-33 years, substations No. 2; 3-35 years. The connection diagram of the power transformers of the substations under consideration to the substation switchgears (SC) is shown in Fig. 3.

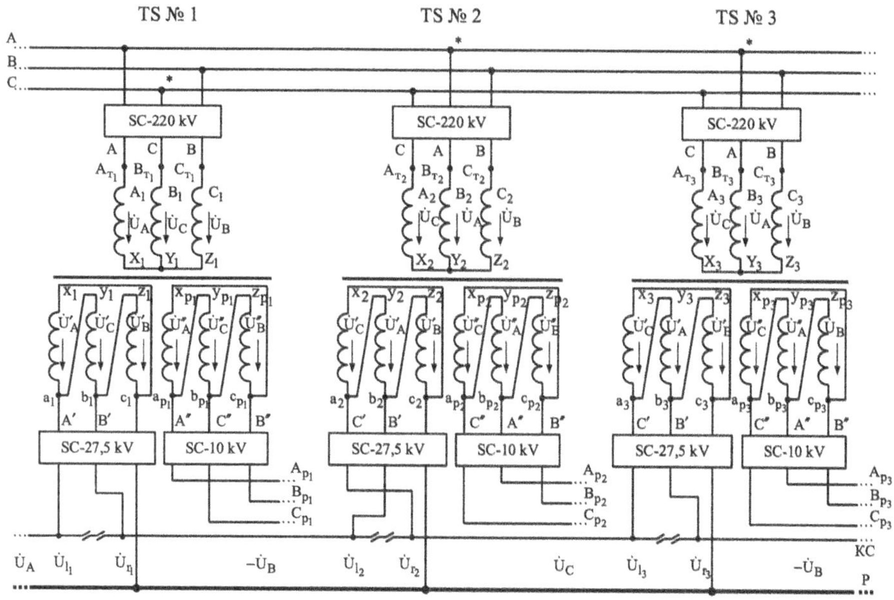

Fig. 3. The existing connection diagram for traction transformers

Based on the simulation results of the considered section, it is determined that $\Theta_{h.p}$ of windings CZ (cz) of traction transformers at substations 1 and 3 are higher than $\Theta_{h.p}$ of windings AX (ax) and BY (by). At substation 2, $\Theta_{h.p}$ of windings AX (ax) is higher than $\Theta_{h.p}$ of windings BY (by) and CZ (cz). The life of the transformers in accordance with wear of winding insulation is limited by the insulation of the winding with a larger $\Theta_{h.p}$. To extend the service life, let's make the choice of connection diagrams for existing traction transformers of substations according to the theory described in [7].

To extend the operating life of traction transformers of substations 1, 2, and 3, they should be connected as follows: at substation 1 - according to scheme "a", substation 2 - according to scheme "b", and substation 3 - according to scheme "a".

The connection of bushings of traction transformers of the substations under consideration to the live parts of the switchgear according to a new diagram is shown in Fig. 4.

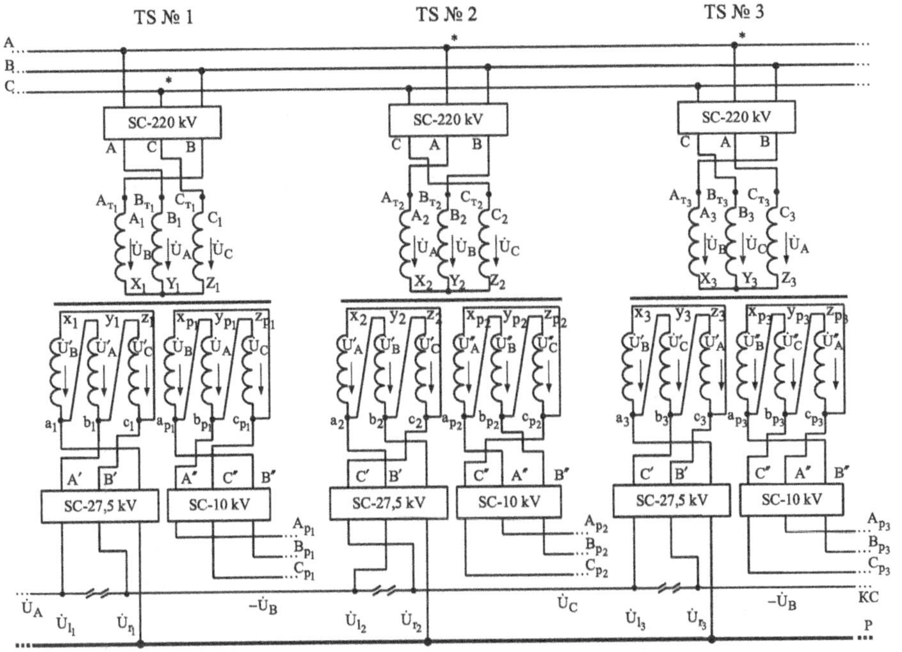

Fig. 4. Connection of traction transformers according to new diagrams

To assess the effectiveness of the application of new diagrams for connecting traction transformers of the substations under consideration, we determine the increase in their service life.

4 Discussion

The wear rate of the insulation of the transformer windings with the maximum, average and minimum insulation wear (L'_1, L'_2 and L'_3) during the operation of the transformer T_1 according to the existing diagram 1 (see Fig. 3) is determined by the formulas:

$$L_1' = L_1/T_1,\ L_2' = L_2/T_1,\ L_3' = L_3/T_1, \tag{5}$$

where L_1, L_2 and L_3 – the wear of insulation of windings 1, 2 and 3, respectively, with maximum, average and minimum wear of the windings in diagram 1.

The maximum permissible operating life of the transformer $\left[T_{1.t}^{(1)}\right]$ with the existing connection diagram and the intensity of the insulation wear of the most loaded windings is determined by the formula:

$$\left[T_{1.t}^{(1)}\right] = [L]/L_1'. \tag{6}$$

The residual life of insulation of windings 1 (ΔL_1), 2 (ΔL_2) and 3 (ΔL_3) at T_1 is determined by the difference between the maximum permissible and actual wear

$$\Delta L_1 = [L] - L_1,\ \Delta L_2 = [L] - L_2,\ \Delta L_3 = [L] - L_3. \quad (7)$$

The maximum permissible operating life of the windings according to the new connection diagram (diagram 2) (see Fig. 4) and the average rate of insulation wear according to the existing diagram (diagram 1) is determined by the formulas:

$$T_{1_2} = \Delta L_1/L'_3,\ T_{2_2} = \Delta L_2/L'_1,\ T_{3_2} = \Delta L_3/L'_2. \quad (8)$$

The extension of the operating life of the traction transformer when it is connected according to new diagrams and when the operating life of the transformer is limited by the insulation wear of the winding 1 is determined by the formula

$$\Delta T_1 = ([L] - L_1)/L'_3 - ([L] - L_1)/L'_1. \quad (9)$$

The extension of the operating life of the traction transformer according to diagrams 1 and 2, and the winding 2, which is limited by the insulation wear, is determined by the formula

$$\Delta T_2 = (L_{1_1} - L_{2_1})/L'_1. \quad (10)$$

The results of determining the insulation wear of windings of traction transformers with existing (see Fig. 3) and proposed (see Fig. 4) diagrams of connection to substation switchgears are presented in Fig. 5.

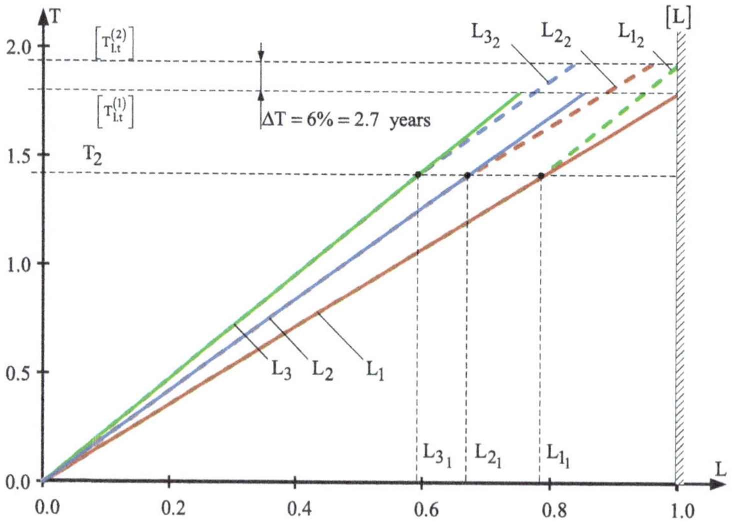

Fig. 5. Dependence of the operating life of the transformer on insulation wear with existing and proposed connection diagrams

5 Conclusions

Thus, the connection of traction transformers of the substations under consideration according to new diagrams will extend the life in accordance with the wear of the insulation of their windings by more than 6% of the standard one, which will be more than 2.7 years.

In addition, it was determined that when connecting one traction transformer according to the new diagram, the savings of the reduced annual costs will be more than 15 000 thousand rubles/year under the given conditions. In accordance with the failure of transformers due to the wear of winding insulation, the extension of their service life gives a conditional annual savings of the reduced costs in size of 340 thousand rubles/year for one transformer.

References

1. Strategy for the development of railway transport in the Russian Federation until 2030, approved by order of the Government of the Russian Federation of June 17, 2008 No. 877-r
2. Vanin, B.V., Lvov, Yu.N., Lvov, M.Yu., Neklepaev, B.N.: On damage of power transformers with voltage of 110–500 kV in operation. Electr. Stations **9**, 53–58 (2001)
3. Lvov, M.Yu.: Development and improvement of methods and criteria for assessing the technical condition of power transformers and autotransformers with voltage of 110 kV and above. Thesis of Doctor of Engineering, p. 253, Moscow (2009)
4. Voprikov, A.V.: Increasing the efficiency of operation of power transformers of traction substations of alternating current railways. Thesis of Ph.D. in Engineering, p. 122. Khabarovsk (2017)
5. Grigoriev, N.P., Voprikov, A.V.: The method of connecting traction transformers in an alternating current system of 25 kV. Pat. 2540685 of the Russian Federation (2013)
6. Grigoriev, N.P., Voprikov, A.V., Klykov, M.S.: System of electrified railways of alternating current of 25 kV. Pat. 2550582 of the Russian Federation (2013)
7. Grigoriev, N.P., Voprikov, A.V.: Choice of connection diagrams for power transformers of traction substations of electrified railways of alternating current of 25 kV. Vestnik Rostovskogo Gosudarstvennogo Universiteta Putey Soobshcheniya **4**, 90–97 (2014)

The Use of Technology 'Big Data' and 'Predictive Analytics' in the Power Supply System of Railways

Evgeniy Tryapkin(✉) and Natalia Shurova

Far Eastern State Transport University, Serysheva Street 47, Khabarovsk, Russia
evgenmehanik@mail.ru

Abstract. To date, the infrastructure of the RZhD OJSC (Russian Railways) is aimed at digitization and informational support. Today, it has the considerable number of digital information systems in place and the figure still grows. Continuous digitalization generates the huge amount of data. This tremendous data size needs to be collected and processed, including, by using a new technology named "big data". This article describes the major challenges associated with big data analytics. Thus, the better part of the data should be processed on a real-time basis to save the resources of data storage and transportation devices. Besides, the presented tools allow a rich toolkit to solve the transport problems related to identification of the equipment operating modes and the deployment of the predictive analytics tool to assess the abnormal equipment status. The article gives specific examples of using the Big Data technologies to assess the existing infrastructure conditions. Various information systems, automated monitoring systems and microprocessor systems installed at infrastructural facilities were used as the sources of information.

Keywords: Big data · Predictive analytics · ANN · Digitized railways

1 Introduction

To date, one of the major priority growth area of the Russian Railways Company is the transition to the digitized railways. Principally, this transition should allow the data accessing on operating modes and health of infrastructural facilities. In particular, traction substations and auxiliary facilities shall serve as such that supply the power to the electric stock.

The project of the traction substation upgrading with its transition to digitization is in process. As per the target concept, traction substations should be allowed to run without the permanent staff. The substation is equipped with the intelligent systems of the equipment operation mode control using the Industrial Internet of Things (IIoT) technology [1].

Remote control and monitoring of the equipment status using the remote control and remote measurement have already been implemented. These are the tasks of the microprocessor protection, automation and telemechanic tools. Moreover, the automated data measuring system of power revenue metering (AIIS KUE) has been

Z. Popovic et al. (Eds.): TransSiberia 2019, AISC 1115, pp. 60–68, 2020.
https://doi.org/10.1007/978-3-030-37916-2_7

implemented to allow the data acquisition on power consumption at traction substations. The automated monitoring system (AMS) [2] has been implemented at the Anisimovka and Sibirtsevo traction substations to collect and transmit mode parameters of traction substations, such as the effective current and voltage, amplitude value and phase difference between them.

Thus, the data size of the power supply system that can be obtained today is quite great. There are many measurement and control systems and complexes in place at substations to constantly collect and process the data for the business process performance. The volume of such information is constantly increasing. And it’s good time to discuss the quality of its application. The question is the new information that appears from existing data flows. The new information will ensure, for example, to improve the efficiency of control activities towards the power supply system. New data acquisition may be associated with the use of big data technology.

The “Big data” is usually a set of data that is large and compound enough to be processed by the conventional methods of data management [3].

There is a range of papers on big data analytics in terms of railways and energy on whole. For example, [4] shows the algorithm for the train delay prediction using the “big data” technology. The [5, 6] describes approaches to ensuring safety in the railway transport using the technology of predictive analytics. The work [7] offers the technique to predict the power industry equipment status and malfunction as per the data of power consumption. The paper [8] is devoted to big data and application thereof to solve the urgent problems in the electric power sector, including to create smart greed.

Normally, the “big data” may be presented in different forms: structured, unstructured, computer-based, in natural language, graph-based, multimedia files, thread-specific data [3]. Structured and unstructured data are of particular interest in terms of the use of big data methodology to study the operation mode of the equipment for the electrified railways.

The structured data are well arranged and can be normally stored in common databases, tables, Excel files. Unstructured data include the data size that is difficult to represent using standard templates.

The data quality is also a problem. Technical deficiency, interference, noises that may occur in process of data collection, have the pernicious influence on the quality of information. According to [3], the most common are the following types of errors: extra spaces, impossible values, null values, outliers. In addition, errors may occur when aggregating data from different sources using different units of measurement (for example, volts and kilovolts) for the same measurements.

Data errors need to be addressed at the earliest stages of analysis. Extra spacing removal is available by the default in most data analysis programs. Impossible values are deleted by the so-called “rationality check”. Outliers are easy to identify during the analysis of the data dependency curve. Outliers considerably affect the data modeling, so they should be excluded from the data selection. There are several methods to handle the null data. For example, the exclusion of missing values, or replacement by the null parameter. You can assign the static value, such as zero, a previous or a subsequent value, or an arithmetic mean. The missing data may be also calculated based on the estimated distribution or modeling of the missing data.

A range of the specialized software systems has been developed to process the big data: MapReduce, Apache Hadoop, Apache Spark [9]. Such programming languages such as R, Java, Python [3] may be used.

Yet, the well-known programs may be used. For example, Excel is the part of the Microsoft Office Suite or LibreOffice package. The big data problems may be addressed using this program: clustering, classification, and regression. Trends and regression analysis are used for forecasting. This program is limited in terms of the volume of input data (1,048.,576 rows and 16,384.00 columns) and does not apply to all statistical problems [10].

In our study, we will use the Matlab software package that uses the same programming language [11]. The tools for data visualization and analysis, forecasting and classification are in-built in this program. Therewith, there is the option to integrate the program with others, including those above.

This article studies the possibility of using the "big data" technology to forecast the voltage level in the middle of the inter-substation zone on the evidence from the specific site.

2 Materials and Methods

The current, voltage and phase difference values will be used as input data during this study. Power consumption values are obtained by sensor connecting to the microprocessor protection devices located at the sectioning point. The sectioning point contains 4 feeders. A specially designed module was connected to the microprocessor terminals for the data reading and writing from the internal storage of safety devices. During the normal operation of the microprocessor unit, these data are erased from memory, but they were recorded for further use for purposes of the study. Measurements were made on each of them continuously for 24 h.

In addition, the train schedule from the Gibral Software was derived. The train schedule data are presented not only in graphics but in tables, either. The table shows the trains in separate columns that go in even and odd directions of their mass and the time of the checkpoint passage.

The voltage level on the current shoe of the electric stock is the important parameter. The voltage value directly affects the speed and safety. And in some cases, it affects the motion possibility [12]. In fact, the level of voltage immediately affects the quality of services delivered by the Russian Railways JSC, as well as economic performance: speed, cargo turnover, volume of power consumed and its losses.

The data are not adequate to identify the voltage level in the contact network, since there are no data on the current consumption by each electric stock. The voltage level may be calculated by the train schedule in the CORTES software package. However, such a calculation is difficult to apply to the situational train environment, given the invariance, as well as facilitation of the CORTES software package.

It was decided to take an attempt to predict the voltage level depending on the train schedule. To solve this problem, the train schedule presented in the table is not suitable. The table shows the mass values of trains pointwise at one time at one point. While the train affects the voltage level during the entire time of travel on the inter-substation zone. This impact is ambiguous and depends not only on the current consumption, but the distance of the electric stock from the reference point, either. Therefore, it was decided to transform the train schedule. To achieve this, the weight of the train was distributed depending on the time of the train travel on runs. In addition, the duration of train stops on the run was considered.

It was decided to use the impact of the train on the voltage level as the equivalent mass of trains. It can be determined by the formula at each specific time:

$$M_i^k = M^k \cdot \frac{t_i}{t}, \tag{1}$$

where M^k – mass of the k-train; t_i – the number of time interval remaining prior to or after the train is at the reference point; t - the time of the train travel on runs in seconds.

During the standing time, the calculation is suspended and the equivalent mass is zero. As the train moves, the calculation goes on from the last non-zero value. This technique allowed the conversion of the train schedule into the structured data.

A diagram explaining the process of converting train schedules into a numerical series suitable for machine learning is presented in Fig. 1.

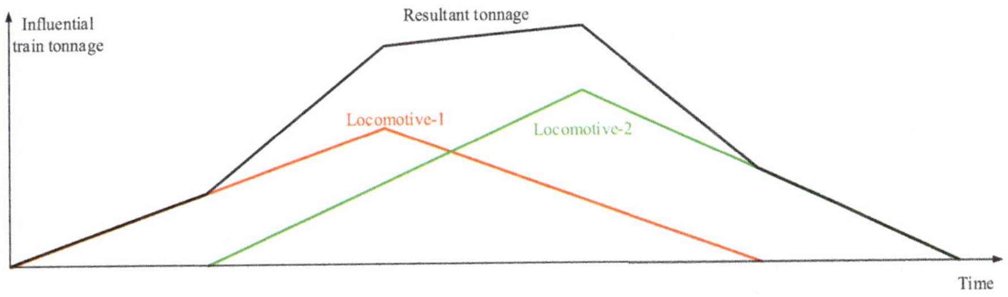

Fig. 1. A diagram explaining the process of converting train schedules

2.1 Data Collection and Cleaning

Data analysis proceeds from the dependence curve plotting resulting from measurements of data in time. Of the total completed measurements, a series of 1,500 values was sampled to form the basis for further analysis. Let us present the recorded data in tables with the time code of the measurement and immediate values of measured parameters – 'timetable' [11].

The Fig. 2 shows the dependence curves of voltage, current and current phase on time. As shown in Fig. 2, the data contains outliers that distort the overall data pattern. The volume of outliers is less than 1% of the total series.

When outliers are deleted, null values (NaN) appear in the series that make the further analysis impossible. To find and replace the outliers in the data, we use the

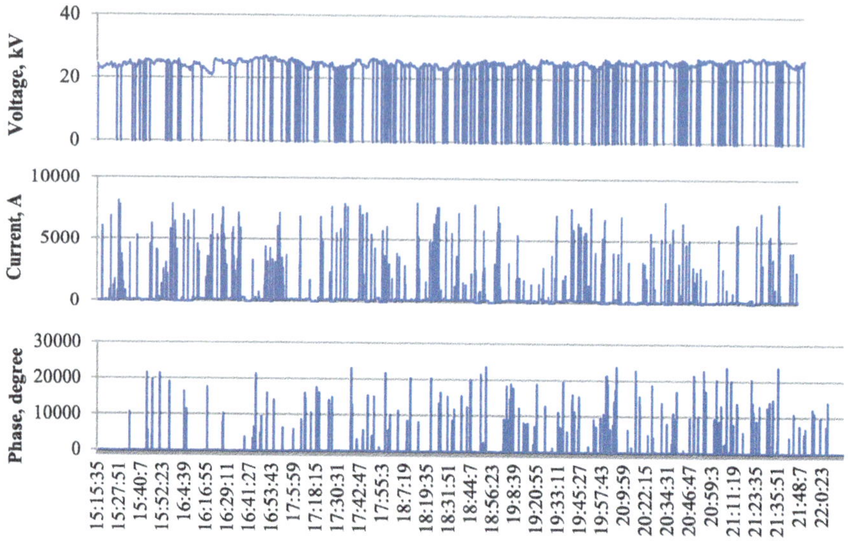

Fig. 2. Dependence of voltage, current and phase on time

'filloutliers' function that finds and replaces outliers by the selected replacement method. Methods such as [11] are also proposed:

- filling with scalar value;
- substitution of the arithmetic mean;
- substitution of the previous or next value other than an outlier;
- linear, piecewise and cubic interpolation;
- modified cubic interpolation.

In this paper, the outliers are replaced at the simplest algorithm that is linear. The dependence curves of voltage, current and current phase on time after removal of the outliers are shown in Fig. 3.

2.2 MatLab-Based Prediction

To predict the time series using the MatLab, you can use the shallow neural networks [11].

1. Nonlinear autoregression (NAR).

Allows the forecast of the series considering its past values. As the initial data to train the artificial neural network (ANN), the values of the time series are used, in our case they are the values of voltage in the middle of the inter substation zone.

2. Nonlinear auto-regressive with external (exogenous) input (NARX).

To predict the series y(t) considering its previous values y(t) and other series x(t). As a series x(t), we will use the values of the trains mass on the section, divided by the motion direction and the series y(t) – the voltage values.

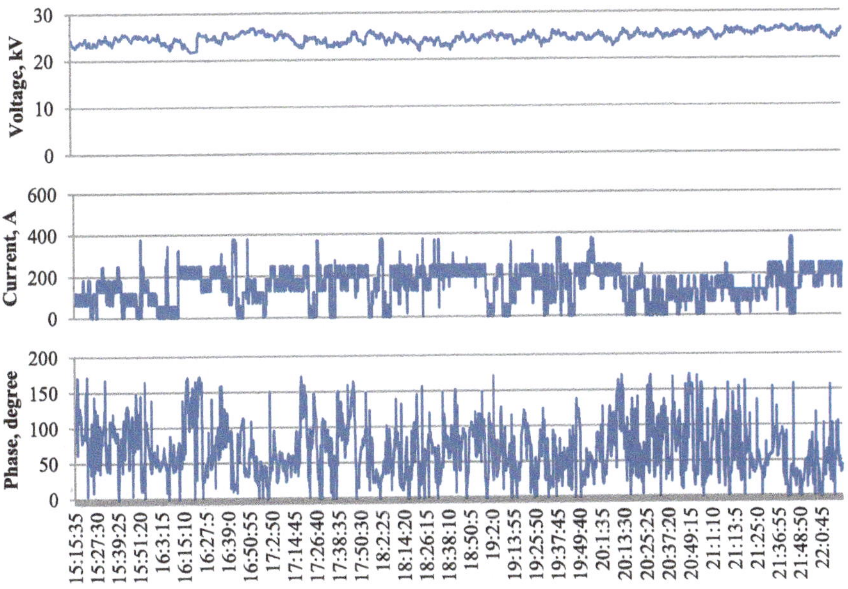

Fig. 3. Dependence of voltage, current and phase on time after removal of the outliers

3. I/o and curve approximation.

In this case, the set of output data y(t) is approximated depending on the change of input parameters x(t).

All three networks have input, one output layer, and one hidden layer of type NAR and NARX, that is, networks with back propagation of the error. The hidden layer activation function is the sigmoid:

$$\Psi = F(\psi) = \frac{1}{1 + e^{-\psi}} \tag{2}$$

Output layer activation function is linear.

3 Results

The results of ANN training of NAR type are shown in Fig. 4.

The curve obtained as a result of forecast is practically a straight line, so that it is not possible to obtain the required forecast of the stress value with high accuracy.

Despite the rather high regression during the training, the resulting forecast has a high MAPE error. In fact, the data in the forecast changes only at the beginning, within the first 100 points, and after the change they are absent in values.

To improve the accuracy of forecasting, predictors were added to the neural network model – the values of train masses that run in even and odd directions.

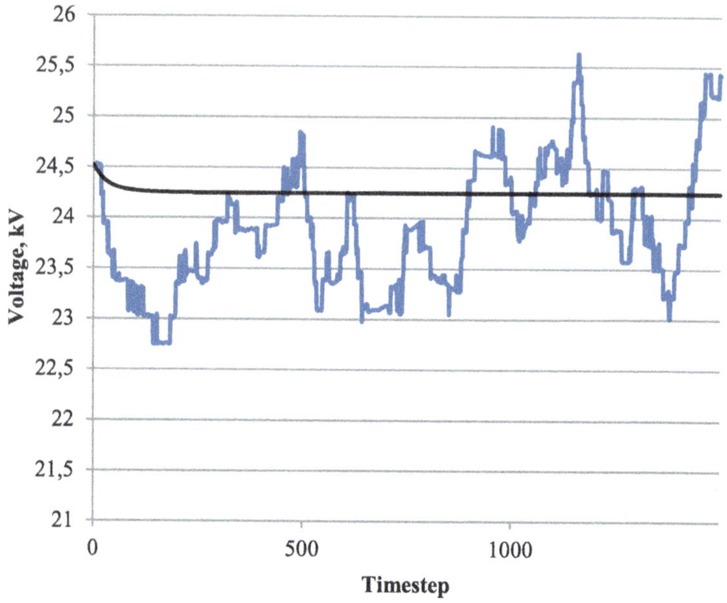

Fig. 4. NAR-induced prediction results

The result of the NARX neural network training with the reverse error propagation is shown in Fig. 5. This method describes the law of voltage variation more accurately than the NAR network.

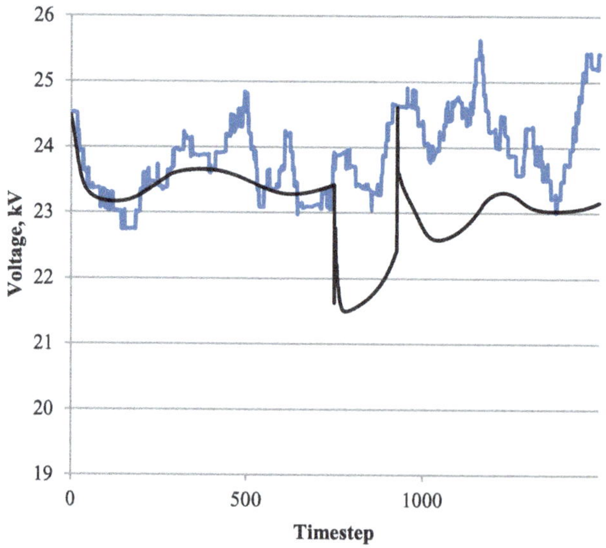

Fig. 5. The forecast results using NARX

Figure 6 shows the prediction result by the curve approximation method. This method shows a more accurate trend of voltage changes.

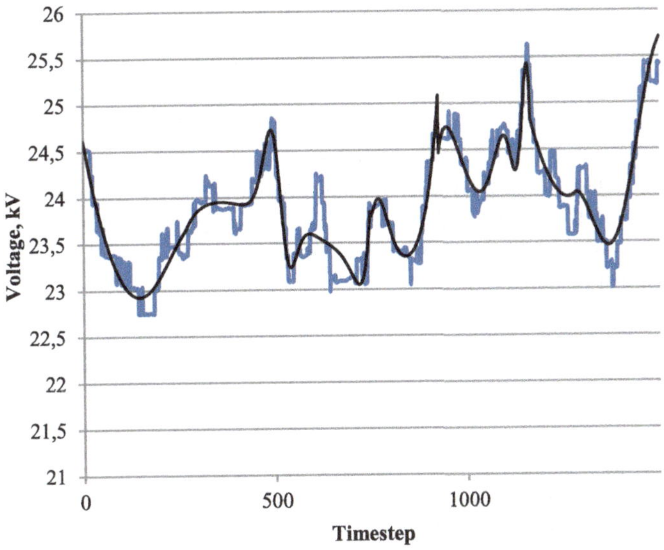

Fig. 6. Forecasting results by the method of curve approximation

4 Discussion

The level of voltage in the contact network is the important index that defines the speed and, in certain cases, the possibility of the train movement. Timely measures taken to maintain the permissible voltage level in the contact network will help to avoid losses caused by train delays. The study showed that the prediction of the voltage value using the artificial neural networks in view of the train mass is the good tool to forecast the expected voltage level. Further researches should target on improvement of forecast accuracy of the variable time series through the profound training and expansion of the list of predictors by adding trains in the speed series that are on the inter-substation area.

5 Conclusions

This paper describes the study aimed at the forecast of the voltage level in the middle of the inter-substation zone depending on the mass of trains running along the section.

Forecasting was made using the MatLab Software. Three types of neural networks were considered to achieve the stated objectives: nonlinear autoregression, nonlinear autoregression with external (exogenous) inputs and input-output and curve approximation.

Of the three neural networks under the study, the most accurate adjustment of the level change curve was demonstrated by the curve approximation method. It also had the lowest average absolute error (MAPE = 0.7%).

Further topic expansion requires several areas. The first is to improve the accuracy of forecast assessments. This objective may be achieved by introducing additional predictors into the predictive model, or by replacing the forecasting method with, for example, the profound training. The second is the transition to online forecasting as per the incoming data from ASM or other automated system. It requires the introduction of industrial computing devices to process data, save data in the convenient format for use before being deleted as, for example, in microprocessor protection devices. The third is the integration of the developed model into the complex problem of the abnormal operation prevention of the traction power supply system, as well as the synthesis of intelligent control of substations.

References

1. Tryapkin, E.Y., Ignatenko, I.V., Keino M.Y.: Development of "digital substation" technology for power supply of railways. In: MATEC Web Conference, vol. 239, p. 01051 (2018). https://doi.org/10.1051/matecconf/201823901051
2. Tryapkin, E.Y., Keino, M.Y., Protasov, F.A.: Synchronous phase measurements in the automated monitoring system of railway power supply facilities. Russ. Electr. Eng. **87**, 110 (2016). https://doi.org/10.3103/S1068371216020176
3. Sien, D., Meysman, A.D.B., Ali, M.: Introdusing Data Science Big Data, machine learning, and more, using Python Tools. Manning Publications, New York (2016)
4. Octay, T., Sayar, A.: Analyzing big security logs in cluster with apache spark. In: Advanses in Big Data, vol. 529, pp. 131–138 (2017). https://doi.org/10.1007/978-3-319-47898-214
5. Parkinson, H.J., Bamford, G.J.: Big data and the virtuous circle of railway digitization. In: Advanses in Big Data, vol. 529, pp. 314–322 (2017). https://doi.org/10.1007/978-3-319-47898-2_32
6. Parkinson, H.J., Bamford, G., Kandola, B.: The development of an enhanced bowtie railway safety assessment tool using a big data analytics approach. In: ICRE (2016). https://doi.org/10.1049/cp.2016.0510
7. Seo, C.K., Kim, J.H., Kwon, S.Y.: A study on modeling using big data and deep learning method for failure diagnosis of system. In: 2018 IEEE International Conference on Big Data (Big Data), pp. 4747–4751 (2018). 978-1-5386-5035-6/18/$31.00
8. Krylov, V.V., Krylov, S.V.: Big Data and Their Electricity Applications. Nobel Press, Moscow (2016)
9. Karau, H., Konwinski, A., Wendell, P., Zaharia, M.: Learning Spark. O'Reilly Media Inc., Sebastopol (2015)
10. Foreman, J.: Data Smart: Using Data Science to Transform Information into Insight. John Wiley, Indianapolice (2013)
11. MatLab: MathWorks website. https://www.mathworks.com/help/matlab/. Accessed 19 Aug 2019
12. Shurova, N.K., Li, V.N.: Increase of energy efficiency of electric power distribution networks by adjustment of reactive energy consumption in a traction network. In: MATEC Web Conference, vol. 239, p. 01049 (2018). https://doi.org/10.1051/matecconf/201823901049

Diagnostics of Electrical Connections of Electric Traction Network

Ivan Ignatenko(✉) and Sergey Vlasenko

Far Eastern State Transport University,
47, Seryshev Street, Khabarovsk 680021, Russia
eao79@yandex.ru

Abstract. Full-scale structural measurements of new and existing electric traction networks are becoming increasingly important because of the ever-increasing train speeds and associated effects. Higher speeds lead to increased loads and greater structural dynamic responses, increasing mechanical and electrical loads. Electrical contact connections are an integral part of the electric traction network: under operating conditions, the electrical installation circuit in many cases should allow to separate the elements from each other. At the same time, electrical contacts are a weak component in energy distribution systems, since the occurrence of large transient resistances in the contact area is one of the most common cases leading to heating of the connection and breakage of the catenary. The purpose of the work is to improve the performance reliability and prevent damage to the electric traction network by reducing the complexity of servicing bolted electrical connections through the development and mass implementation of visual diagnostics. The main parameter characterizing the bolted electrical connection is its resistance, the value of which depends on two interrelated factors: parameters of the contacting surfaces and the contact force. Research of thermal processes in the contact connection and determination of its performance characteristics allow to create a method of diagnosis of the state of the electrical connection on the basis of new methods using modern equipment.

Keywords: Electrical connection · Mathematical model · Transition resistance · Diagnostics

1 Introduction

Any electrical system consists of connections through which electricity is transmitted from the source to the end consumer. In the electric traction network, energy is transmitted from electrical substations to the electric rolling stock. And in case of failure of one of the electrical connections (EC), there may be significant economic damage due to the occurrence of interruptions in the movement of trains [1].

The problem with EC in the electric traction network arises due to degradation processes associated with the deterioration of the transient resistance [2]. When its value increases, the entire EC overheats and, as a consequence, the temperature rises to critical values.

Z. Popovic et al. (Eds.): TransSiberia 2019, AISC 1115, pp. 69–78, 2020.
https://doi.org/10.1007/978-3-030-37916-2_8

Therefore, the determination and prediction of the state of EC is an important problem to improve the operational reliability of the main electric railways [3].

2 Problem Definition

There are distinguished two states of electrical connection: satisfactory and unsatisfactory. The state depends on the value of transient resistance (RT), resulting from contiguity of contact surfaces of a wire and clamp dies. Under service conditions the value RT increases as tightening torque of clamp dies decreases (loosening) and oxide films are formed in places of a wire and clamp contact. For satisfactory state the value of transient resistance should not exceed the value equal to the ohmic resistance of a conductor. The main indicator of the state of the electrical connection is the defect rate and the contact transition resistance [4].

The heating defect rate is not a constant value and is a function of the time of flow of current through the electrical connection. Moreover, it can be assumed that the value of the defect rate itself does not depend on the current value, but on the ratio of the physical (mass, electrical resistance) and thermophysical parameters (specific heat capacity and heat removal) of the clamp and the wire that form the EC.

3 Solution

In work [4] it is proved that in the electric current supply mode indirect estimate of value RT by means of defectiveness coefficient (Kθ), with heating considered, is possible and, as a result, the same holds good for estimate of electrical connection state. However, the diagnostic data on current-carrying clamps obtained with the help of a thermal imaging camera have shown that the lack of the similar data analysis of cooling mode leads to doubtful estimate of current state of bolted-type electrical connection. Therefore, there arises a need to investigate not only heating, but also cooling of an electrical connection, since only joint consideration of both modes can offer profound qualitative evaluation of the existing indicators influencing the electrical connection state estimate.

At present there is no mathematical model of electrical connection estimate within cooling mode. The authors of the given paper offer to transform the universal equation of heat balance of electrical connection [5], without current being taken into account, and receive an expression:

$$(C_j\, m_j + C_w\, m_w) \cdot \frac{d\theta}{dt} = -[\,0.01312 \cdot \theta^{\,1.49} + \varepsilon\, S \cdot 5.67 \cdot 10^{-8}[(T_j + 273)^4 - (T_a + 273)^4]] \tag{1}$$

where C_w is the specific heat capacity of a wire material $\left(\mathrm{W \cdot s\, kg^{-1} K^{-1}}\right)$; C_j is the specific heat capacity of a clamp material $\left(\mathrm{W \cdot s\, kg^{-1} K^{-1}}\right)$; m_w is the mass of a wire clamped between dies (kg); m_j is the mass of a clamp (kg); S is the area of an exterior surface of a clamp (m^2); T_a is the air temperature (°C); T_j is the temperature of a clamp

(°C); $5{,}67 \cdot 10^{-8}$ $\left(\mathrm{W\,m^{-2}K^{-4}}\right)$ is the Stefan-Boltzmann constant; ε is the relative emittance for bronze (ε = 0.6); θ is the excess of temperature of electrical connection over ambient temperature (°C).

To determine temperature of electrical connection we will transform the above Eq. (1) and receive the dependence defining temperature change on time in the course of cooling [6]:

$$\varDelta\theta_C = \left(0.0108Pt\,(C_j\,m_j + C_w\,m_w)^{-1} + \theta_0^{-0.49}\right)^{-2.03} \tag{2}$$

where P is the perimeter of a wire cross section (m); t is the time period of current flow (s).

The received expression (2) describes temperature variation of clamp dies in the course of time when traction current is not flowing through the electrical connection. It gives a possibility to investigate process of cooling and to estimate its influence on K_θ, which represents the ratio of temperature variations of electrical connection to temperature variations of a wire [6]. As mathematical models of heating and cooling for a wire and an electrical connection are known (Fig. 1,a), it is possible to investigate K_θ temporal variation in the course of the "heating - cooling" cycle and consider the additional factors influencing the estimate of an electrical connection [7].

In Fig. 1,b, showing K_θ variation in "heating-cooling" cycle, it is possible to conditionally single out four areas of a curve which describes the thermalphysic processes occurring in the feeding clamp and branching wires. In the Inertial Component area when temperature of connection is much lower than that of wire, the function of the defectiveness coefficient, with heating considered, is of the minimum value, as for the defectiveness coefficient value, it depends on connection mass and the material it is made of. A point of an extremum, as regards to thermal characteristics, is explained by inertial properties of a clamp [8].

Then there is a process of determining a coefficient value due to balancing of a gradient of thermal fields of a clamp and the connected wires [9]. The third area represents a steady-set coefficient value in the heating mode. The coefficient value will depend on two components which influence heat balance of the system under discussion. The first component is a ratio of connection resistances and a wire (defectiveness coefficient with resistance considered). The second is a ratio of heat emission coefficients of a connection and a wire.

The fourth area shows sharp increase in K_θ value with a distinctive point of a maximum. Such a behavior of the coefficient is explained by smaller cooling of an electrical connection, than that of a wire. It should be noted that at this moment K_θ value considerably exceeds the rated value and is equal to 1.5 that has testified to unsatisfactory state of an electrical connection. At the same time the value of transient resistance of 25 μΩ adopted in modeling is considered to be rated. Thus, it is proved that process of cooling cannot be neglected on estimating an electrical connection state, otherwise research results may turn to be erroneous and misleading.

The maximum value of coefficient being reached, its value decreases up to its steady-state value.

On the basis of the offered mathematical model the Program for calculations in Mathcad computer environment was developed. The Program allows a user not only to consider cooling process but also to change electric current parameters by substituting its value and time period of current flow [7] and [8].

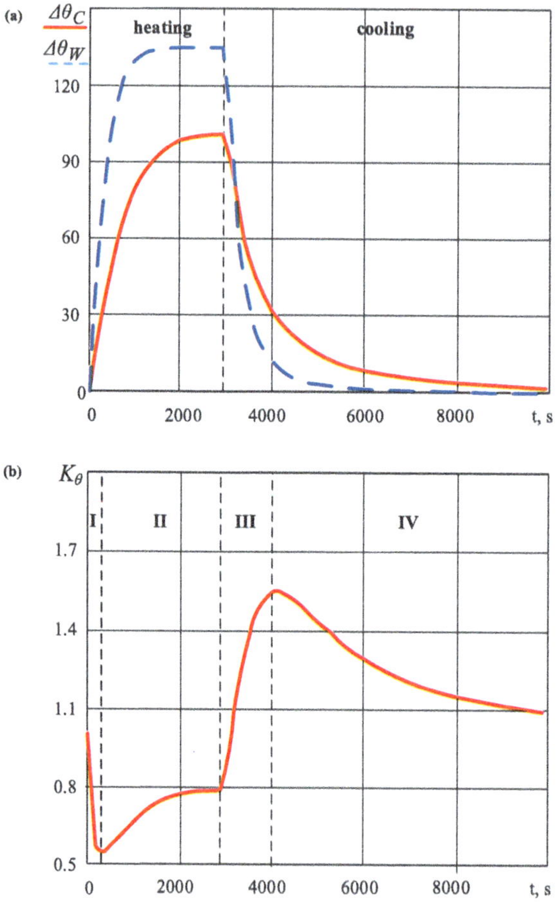

Fig. 1. "Heating-cooling" cycle of a clamp: a – temperature variation of a wire $\Delta\theta_W$ and an electrical connection $\Delta\theta_C$ related to the time period of conventional nominal current flow through them $I = 600$ A, at $T_a = 24$ °C; b – change of defectiveness coefficient K_θ with heating considered at $R_T = 25$ μΩ

The Program performs calculations of differential equations concerning wire heating and electrical connections [3]. Applying the Program operation algorithm, the value of K_θ is determined.

The Program includes the following data: 6 values of current which correspond to variables $I_1(t)$, $I_2(t)$, $I_3(t)$, $I_4(t)$, $I_5(t)$, $I_6(t)$. The current values are chosen one after another, proceeding from time intervals. The chosen time intervals are designated by a variable – t_m, t_n, t_k, t_d t_c – covering the maximum and minimum values of changes of the graph [11]. The value of transient resistance (r, μΩ) is set as well.

Plotting the graphs of dependence of coefficient K_θ on time, the Program retains the previous data on an electrical connection and a wire and considers the when plots new graphs.

To obtain actual data of the algorithm performance, calculations of transient resistance of different value were carried out. In addition, values of current flowing through an electrical connection during heating and cooling modes were changed too.

Reasoning from the calculations made one can conclude that, in case of an electrical connection cooling, value K_θ, can exceed the rated parameter used to estimate satisfactory and unsatisfactory states of an electric connection.

Being aware of the value of the inclination angle of a change curve K_θ, and knowing that the angle shows the speed at which defectiveness coefficient increases during heating mode, one can predict current-carrying clamp state under the impact of electric load within the whole time interval of [4].

To confirm the above hypothesis, models of heating and cooling of the wire and EC are constructed. The ratio of these models makes it possible to investigate the defect rate over time. And in this case, it is possible to estimate the change in the defect rate at different values of the transient resistance, talking about the state of EC.

The study [9] showed that each transient resistance corresponds to a characteristic change in the defect rate. Thus, there is a theoretical assumption that during operation, it is possible to indirectly determine the transient resistance of EC by changing the defect rate, that is, to determine its current state.

To confirm the theoretical developments and approve new criteria for assessing the state of EC, on a specially developed bench, tests were carried out with real currents of 200, 400, 600, 800 A and with EC transition resistances of 20, 40, 60 μOhm. Operational tests confirmed mathematical models of new evaluation criteria.

Based on these data, an algorithm for diagnosing EC was created, which allows to make an accurate assessment of the state of EC [5].

The measurement algorithm procedure (Fig. 2):

1. Constant monitoring of the defect rate value. To do this, the temperature is measured at two points with a time interval Δt.
2. The diagnostic algorithm is selected by the value of the defect rate.
   - 2.1 In the "cooling" mode, the state is evaluated by two criteria after the time has elapsed.
   - 2.2 In the "heating" mode, the minimum point is measured after the time has elapsed.
     - 2.2.1 Constant temperature measurement at two points with a time interval of 5 min + Δt.
     - 2.2.2 At the end of the measurement time, the slope of curve of the change in the defect rate value is determined.
3. The result of measurements of the electrical connection state parameters.

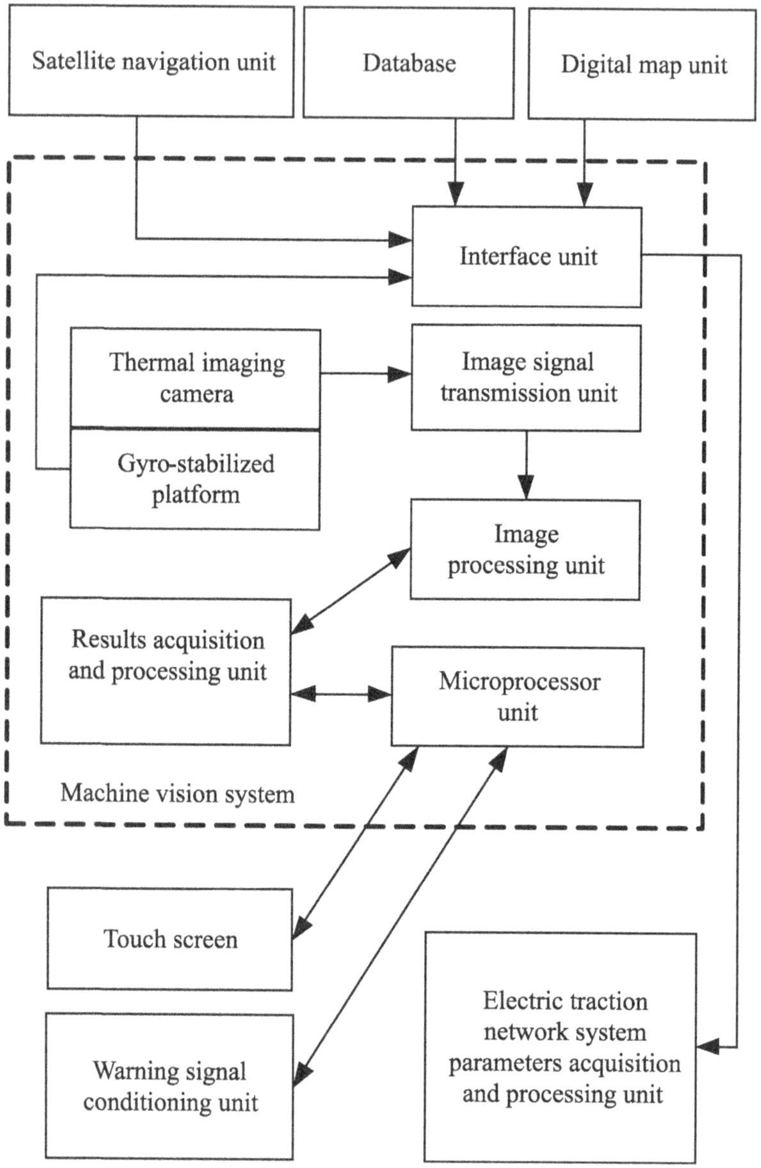

Fig. 2. The algorithm of the method for remote determination of the state of EC

Based on the obtained values of the slope and the value of the defect rate, a conclusion is drawn about the current state of the electrical connection. Next, activities are proposed depending on the current and forecasted state.

It is also necessary to pay attention to the different times required to obtain an opinion on the state of EC. In the "heating" mode, the measurement time is longer than in the "cooling" mode. However, in the "cooling" mode there is no possibility to make a forecast.

From the described algorithm it can be seen that the account of both heating and cooling in the technique makes it possible to more accurately assess the state of the electrical connection. A failure to take into account one of the modes can lead to a false assessment. The assessment of the use of special methods and means of diagnosing EC is carried out [6].

To automate this algorithm, a measurement method is proposed, which is based on the acquisition and processing of images from a thermal imaging camera. It can be used to determine the temperature at the connection point and on the wire of the electric traction network. These two points are processed according to the presented models and the result of the current state of EC is given.

On the basis of the received technique it is possible to make operation algorithm of a remote method of monitoring of the EC state.

The method is based on the acquisition and processing of images from a thermal imaging camera, which can be used to determine the state of EC, as well as to determine other parameters of the traction power supply system (TPS) using measuring equipment [7].

The place of installation of the measuring equipment can be a car roof, or an embankment if the equipment needs to be moved by crew.

The process of obtaining and converting information (Fig. 3) starts with image acquisition by thermal imager. Then the image gets into the image signal transmission unit, and then into the interface unit, where the streaming video (or a series of frames) is generated. In turn, the interface unit receives geographical coordinates (when the position of the measured equipment changes, or the rolling stock moves), date, time, etc. from the satellite navigation system (GPS) unit [8]. The interface unit also receives terrain information from the digital terrain map unit. In turn, the streaming video of the interface unit is sent to the database and to the unit for acquisition and processing of TPS parameters, which is available at the Department of Power Supply Systems. This unit allows to process such parameters as I, U, φ, etc., obtained both in research of catenary system, and in research of traction substations [9].

Next, the image from the image signal transmission unit enters the interface unit, in addition, it enters the image processing unit (Fig. 3), which interacts with the unit for acquisition and processing of the obtained results (Fig. 4). With the help of a microprocessor, the information from this unit is transmitted to the warning signal conditioning unit and directly to the touch screen.

The key information provided by modern mobile laser scanning systems is information about the laser pulse reflection points. Each reflection point has its own coordinates in the global coordinate system and a certain set of characteristics (intensity, color, accurate time stamp, and others). The set of laser reflection points forms a point cloud.

In the future, it is proposed to create a database of the desired objects. Each database item contains information about the object. In the future, the database element will be called the object model. Since the railway infrastructure facilities are sufficiently standardized, their geometrical parameters can be defined in advance and contained in the model itself. Thus, the problem of automatically determining the geometric

Fig. 3. View from the thermal imager

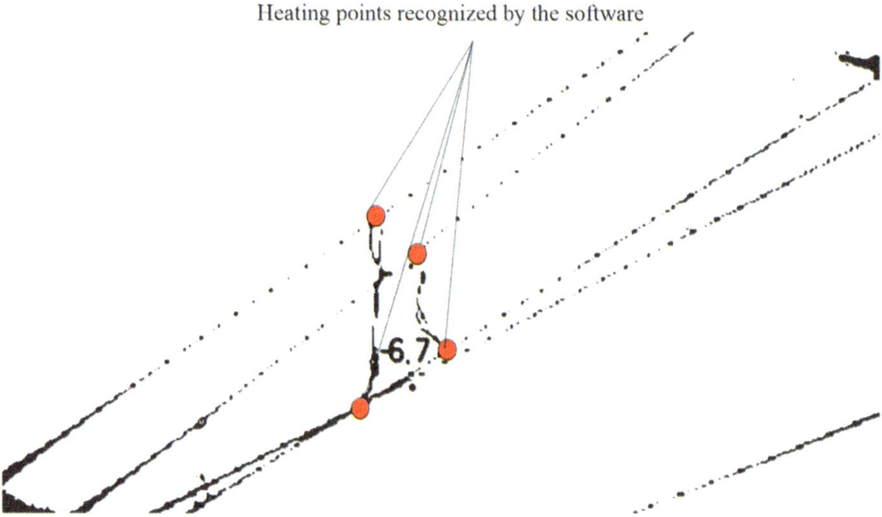

Fig. 4. View from the processing unit

parameters of objects is reduced to the problem of detecting objects corresponding to the database models in the point cloud, and then determining the orientation and position of the found objects in a certain global coordinate system.

The image processing algorithm is based on machine vision technology. To construct a plane, two parallel or intersecting lines are required. Characteristic lines by which we determine the location of the electric traction network wires and determine the location of the EC: always present in the analyzed image; it is possible to construct two characteristic lines that are parallel; have a known size of the segment of the distance between the characteristic lines, have their own linear parameters. Highlighting these parameters, as well as areas with high temperature will allow to compile data on the heating of the electric traction network wires and EC [10].

Currently, the units for processing received information can be compact [11], which allows them to be placed in a mobile control device or permanently on the rolling stock.

4 Conclusions

Thus, the characteristic features of determining the state of EC in the operating mode are revealed. This will make it possible in the future to create stationary and mobile measuring complexes for diagnosing the connecting elements of the electric traction network, which will significantly reduce the damage caused by interruptions in the movement of trains.

There is no doubt that this technology of diagnosing EC can be in demand in other sectors of the economy in terms of safety of processes and human life [12]. For example, according to the statistics of emerging fires, about 30% is due to EC overheating as a result of emergency operation of the electric network due to high transient resistances [13].

At facilities with their own electrical system (buildings, ships, aircrafts, production, etc.), devices for monitoring the EC quality according to the proposed diagnostic technique can be created, which can save people's lives and reduce economic damage due to process flow disruptions.

References

1. Grigor'ev, N., Voprikov, A., Parfianovich, A.: Increasing the operating life of power transformer winding insulation when upgrading 25-kV alternating current traction substations. Russ. Electr. Eng. **87**(2), 100–103 (2016). https://doi.org/10.3103/S1068371216020048
2. Li, V., Shnurova, N.: On the method of determining optimal electricity consumption from an electric traction network. Russ. Electr. Eng. **87**(2), 97–99 (2016). https://doi.org/10.3103/S1068371216020127
3. Li, V., Klimenko, S.: Determination of thermal softening of overhead wire exposed to an electric arc effect. Russ. Electr. Eng. **87**(2), 77–79 (2016). https://doi.org/10.3103/S1068371216020115
4. Xin, Z., Schoepf, T.: Detection and formation process of overheated electrical joints due to faulty connections. In: 26th International Conference on Electrical Contacts (ICEC), p. 12909251 (2012). https://doi.org/10.1049/cp.2012.0663
5. Ignatenko, I., Vlasenko, S.: Determination of criteria to find the quality of an electrical connection. Russ. Electr. Eng. **87**(2), 65–67 (2016). https://doi.org/10.3103/S1068371216020085

6. Wilson, C., McIntosh, G.; Timsit, R.S.: Contact spot temperature and the temperature of external surfaces in an electrical connection. In: 26th International Conference on Electrical Contacts (ICEC), p. 12909202 (2012). https://doi.org/10.1049/cp.2012.0614
7. Bukhanov, S., Doykina, L., Tsirkunenko, A.: Evaluation of the electromagnetic situation at open switchgears of power stations and substations. In: IEEE Conference of Russian Young Researchers in Electrical and Electronic Engineering (EIConRus), p. 18508901 (2019). https://doi.org/10.1109/eiconrus.2019.8656642
8. Tryapkin, E., Keino, M., Protasov, F.: Synchronous phase measurements in the automated monitoring system of railway power supply facilities. Russ. Electr. Eng. **87**(2), 110–112 (2016). https://doi.org/10.3103/S1068371216020176
9. Grigor'ev, N., Klykov, M., Parfianovich, A., Shestukhina, V.: Increasing the power efficiency of a 25-kv alternating current traction power supply system. Russ. Electr. Eng. **87** (2), 113–115 (2016). https://doi.org/10.3103/S106837121602005X
10. Konstantinova, Y., Li, V., Tryapkin, E.: Enhancing Energy Efficiency of 25 kV Traction power system due to balancing current decrease. In: International Multi-conference on Industrial Engineering and Modern Technologies (FarEastCon), p. 18393188 (2018). https://doi.org/10.1109/fareastcon.2018.8602632
11. Galkin, A., Kovalev, A., Shayukhov, T.: Integrated approach to the planning of energy consumption by non-traction railway consumers. In: MATEC, vol. 239, p. 01056 (2018). https://doi.org/10.1051/matecconf/201823901056
12. Tumanov, A., Gumenyuk, V., Tumanov, V.: Development of advanced mathematical predictive models for assessing damage avoided accidents on potentially-dangerous sea-based energy facility. In: IOP Conference Series: Earth and Environmental Science (EMMFT 2017), vol. 90, p. 012027 (2017). https://doi.org/10.1088/1755-1315/90/1/012027
13. Bukhanov, S., Zhuravleva, L., Kornienko, M.: Improving the reliability of the relay automation system for power facilities. In: IOP Conference Series: Materials Science Engineering, vol. 450, p. 072008 (2018). https://doi.org/10.1088/1757-899x/450/7/072008

Regulated Single-Phase Rectifier Circuit Solutions and Their Impact on Power Coefficient

Olga Malysheva[1](✉), Stanislav Vlasevskii[1], Igor Barinov[2], Vitaly Skorik[1], and Ekaterina Buniaeva[1]

[1] Far Eastern State Transport University, Seryshev Street, 47, Khabarovsk 680021, Russia
zavkaf@festu.khv.ru
[2] Irkutsk State Transport University, Chernyshevskogo Street, 15, Irkutsk 664074, Russia

Abstract. Locomotive electric drives with DC commutator motors and smooth voltage regulation systems usually apply reversible power rectifiers based on partially or fully controllable semiconductor devices. The paper presents theoretical analysis of working processes of different single-phase bridge circuits. Reviewed circuits include: initial (or type) half-period thyristor-based circuit; asymmetrical circuit with two thyristor arms and two diode arms; half-period thyristor circuit with a shunting diode arm in parallel with the rectified sub-circuit; contemporary transistor-based bridge circuit. Detailed time diagrams of working processes of the circuits demonstrate that application of different schematic solutions and new control algorithms allows for a significant decrease or complete elimination of phase shift between current and voltage curves on AC side of the electric drive (i.e. power transformer primary winding), making it possible for power coefficient of the rectifier to increase up to a maximum of 1.

Keywords: Electric drive · Power rectifier · Shunting diode · Thyristor-based circuit · Transistor-based bridge circuit · Power coefficient

1 Introduction

Domestic power electronics converter industry possesses multiple regulated single-phase rectifier circuit solutions based on partially or fully controllable power semiconductor devices (thyristors or transistors). These solutions include a zero-point circuit; a zero-point, zero diode circuit and a bridge circuit [1–4].

The initial (or type) regulated rectifier circuit is represented by a thyristor-based bridge circuit working within an assumption that $XT > 0$ and $Xd = \infty$, where XT is the inductive reactance of the transformer windings and Xd is the inductive reactance of the rectified sub-circuit (i.e. the section of the electrical drive past the rectifier that includes a smoothing reactor SR and a motor M), Fig. 1.

Z. Popovic et al. (Eds.): TransSiberia 2019, AISC 1115, pp. 79–86, 2020.
https://doi.org/10.1007/978-3-030-37916-2_9

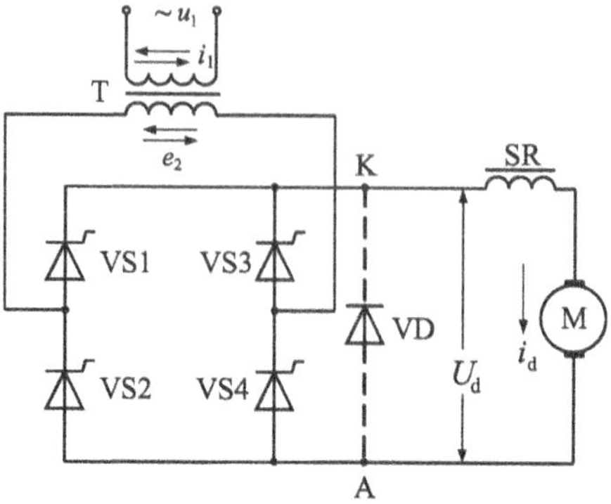

Fig. 1. Bridge circuit of a half-period thyristor-based regulated rectifier

For this type of rectifier, the working algorithm of its thyristor arms suggests firing pairs of thyristor arms (e.g. VS1, VS4 or VS2, VS3) each half-period of the supply grid voltage u_1 by applying control signals that are regulated by their phase, α_{reg}. The regulation range of the control signals can theoretically span the entire half-period of the voltage, with the beginning point being π (or 180 electrical degrees) and the ending point being 0.

Working processes of such a rectifier are presented in Fig. 2.

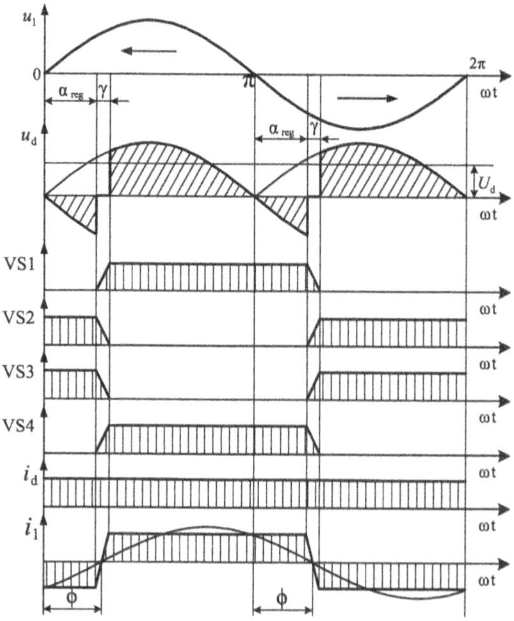

Fig. 2. Diagrams of working processes of the regulated rectifier

The main advantage of this rectifier is simplicity of its control algorithm. However, despite there being a smoothing reactor SR, the necessity of further pulsation limitation of the rectified current i_d decreases the apparent range of rectified voltage U_d regulation to only about a half of 180 electrical degrees. Another drawback is the drastic decrease in locomotive power coefficient χ as the regulation angle α_{reg} proceeds to widen [5, 6].

Figure 7 demonstrates the χ value of this rectifier with curve 1.

2 Materials and Methods

Another method of thyristor-based regulated rectifier bridge circuit organization proposes one pair of transverse thyristor arms being fired by unregulated control signals with a phase of $\alpha_0 = 10°$, and another pair of arms being fired by regulated pulses with a phase of α_{reg}. The same result in terms of regulation would be achieved if, instead of thyristor arms being fired by unregulated signals, diode arms were applied. This proposition simplifies the overall rectifier circuit. Figure 3 demonstrates an electric circuit that implements this method, and Fig. 4 presents diagrams of main working processes of this rectifier. Such a circuit is traditionally referred to as an asymmetrical bridge circuit regulated rectifier.

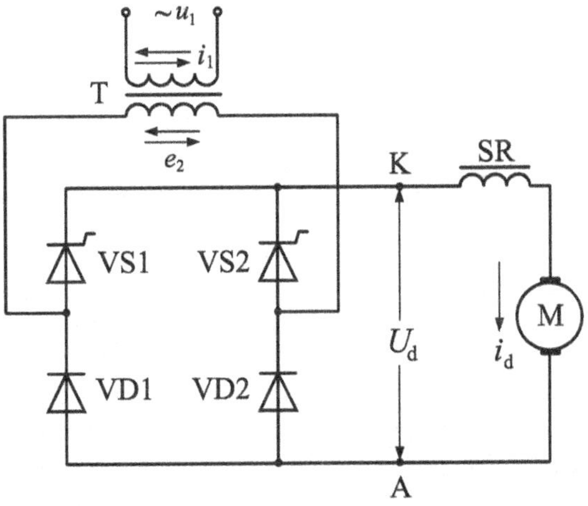

Fig. 3. Regulated rectifier asymmetrical bridge circuit

These results can also be achieved on the basis of type rectifier if an additional diode arm is introduced and attached in parallel with the rectified sub-circuit of the electric drive (see hatch line AK and diode VD in Fig. 1).

The advantage of the circuit shown in Fig. 3 is fewer number of thyristor arms (two instead of four), which simplifies both the structure and the control algorithm of the rectifier. In addition, apparent range regulation of the voltage applied to the load (motor

M) is increased, while power coefficient of the circuit is also slightly higher than that of the type bridge circuit shown in Fig. 1.

Figure 7 shows power coefficient of this rectifier with curve 2.

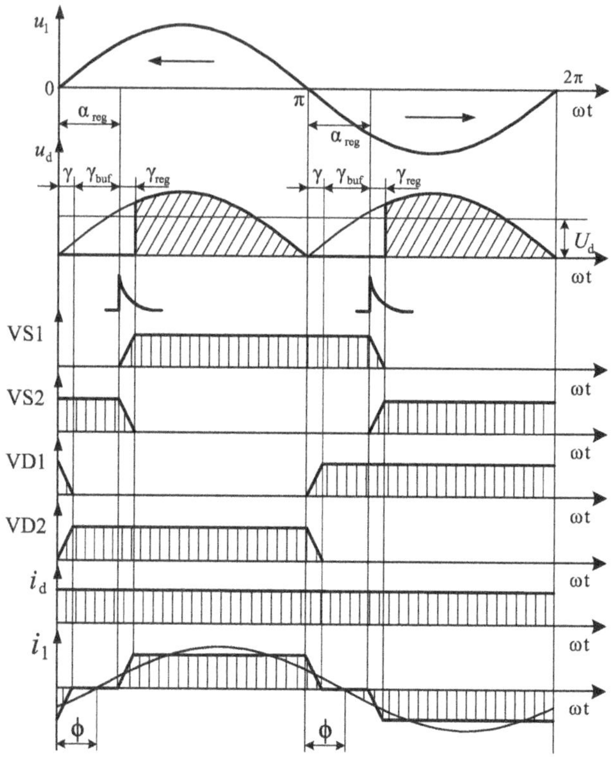

Fig. 4. Diagrams of working processes of the asymmetrical bridge circuit regulated rectifier

From power coefficient standpoint, the most preferable regulated rectifier circuit is the one based fully on transistors, as comprehensively shown in [7, 8]. The method presented in this paper also suggests an inclusion of a shunting diode VD between cathode K and anode A bus-bars of the rectified current sub-circuit. Figure 5 demonstrates such a circuit.

The main distinguishing feature of this control algorithm is the fact that both opening and closing of each transistor arm is executed within a single voltage half-period. Table 1 presents this algorithm.

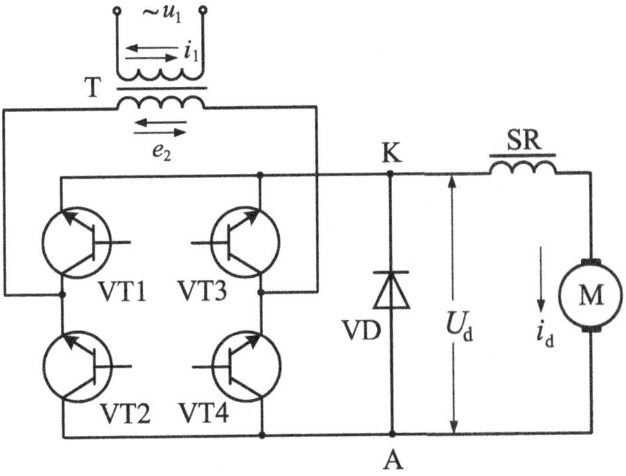

Fig. 5. Transistor-based rectifier bridge circuit

Table 1. Transistor-based rectifier control algorithm

Half-period	Transistor arm			
	VT1	VT2	VT3	VT4
←	α_{reg} (opening) β_{reg} (closing)	-	-	α_{reg} (opening) β_{reg} (closing)
→	-	α_{reg} (opening) β_{reg} (closing)	α_{reg} (opening) β_{reg} (closing)	-

Figure 6 shows working processes of a transistor-based regulated rectifier with the control algorithm from Table 1.

The process of rectified voltage U_d regulation consists of the following. Firing of transistor arms executed by applying control signals starts slightly before π/2 point (approximately 80 electrical degrees), and their work lasts until slightly after π/2 (approximately 100 electrical degrees). To increase rectified voltage U_d value, application of regulated control signals α_{reg} responsible for transistors' opening shift to the beginning of the half-period (i.e. 0 or π depending on the half-period), while their ending points referred to as β_{reg} shift to the end of the half-period (π or 2π respectively). Shifting α_{reg} and β_{reg} points is executed synchronously and with equal steps, which allows rectified voltage increase to proceed smoothly and symmetrically in relation to half-period starting and ending points.

After each pair of transistor arms switches off, the diode shunt VD enters working mode and discharges electromagnetic energy previously accumulated in inductive elements of the circuit. As the next pair of transistor arms is fired by control signals, the diode ceases to work via current self-switching.

This type of control algorithm ensures that the curve of the current i_l in power transformer primary winding is symmetrical in relation to the overhead line voltage curve u_l, thus practically eliminating the phase shift angle φ between u_l and i_l.

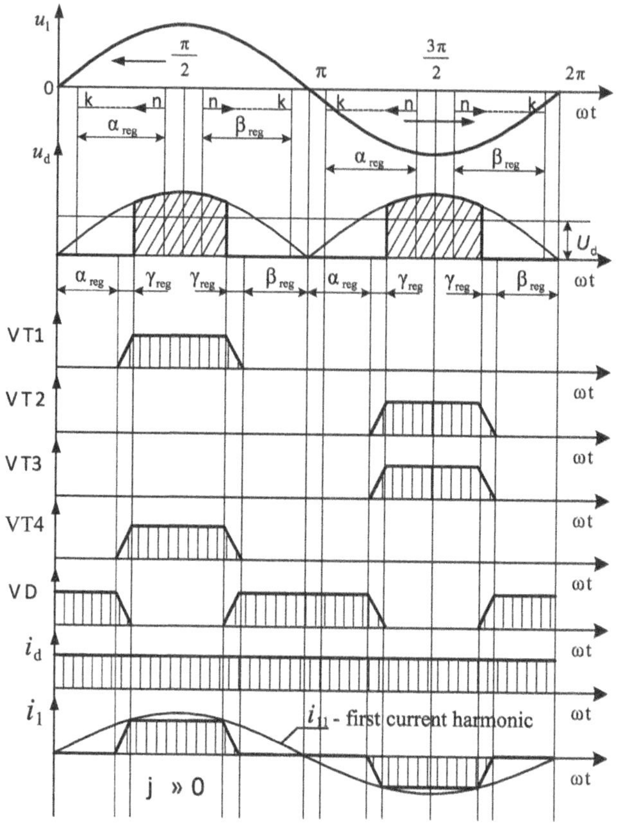

Fig. 6. Diagrams of working processes of the transistor-based bridge circuit regulated rectifier

Power coefficient χ of the transistor rectifier is represented by curve 3 in Fig. 7. Due to the presence of higher harmonics in i_l curve power coefficient of the rectifier is slightly less than 1.

3 Results

Figure 7 presents a relation between power coefficient χ of a rectifier and rectified voltage applied to the load; value U_{d0} is the idling voltage of the rectifier at maximum rectified voltage value, $U_{d0} = \frac{2\sqrt{2}}{\pi} U_2$; U_d is the rectified voltage applied to the load with $\alpha_{reg} = var$ and $\beta_{reg} = var$.

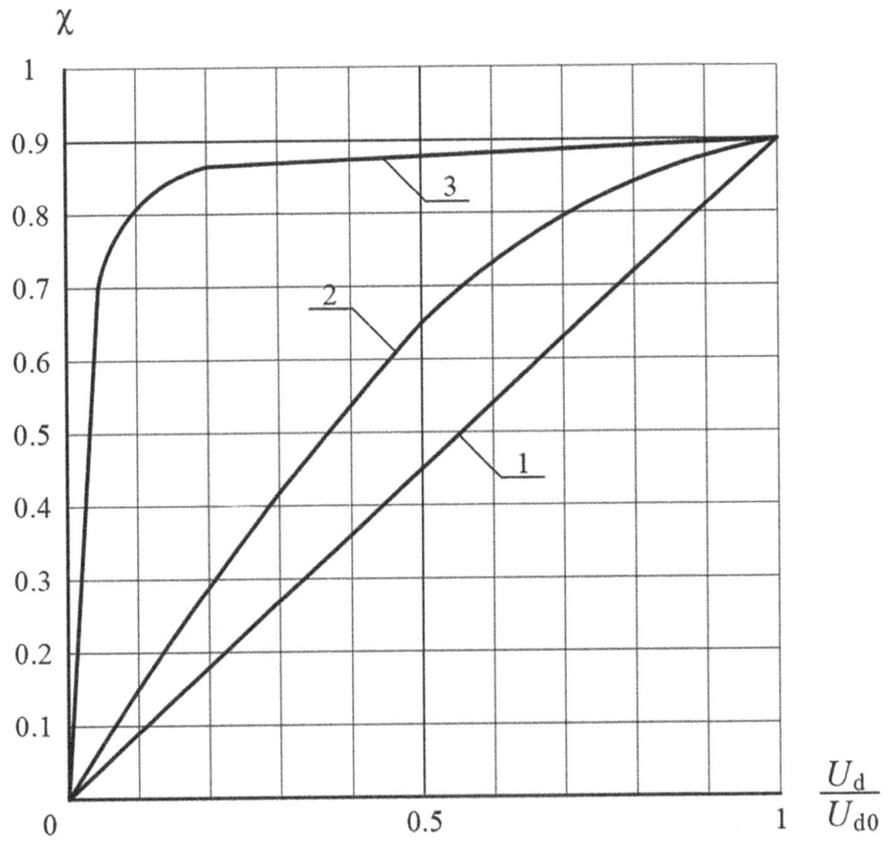

Fig. 7. Graph representing $\chi = f\left(\frac{U_d}{U_{d0}}\right)$ relation

4 Discussion

If χ value is to be examined on the alternating current side (transformer primary winding), its increase is inversely proportional to coefficient of voltage distortion. The latter being $\frac{2\sqrt{2}}{\pi} = 0.9$ gives χ an increase in 1.11. As active power P is tied to power coefficient by $\chi = \frac{P}{S}$ equation, this means the increase can be demonstrated as $P = 1.11P_d$, where P_d is the useful power consummated by the load, $P_d = U_d I_d$.

Previously examined circuit solutions of the regulated rectifier (Figs. 1 and 3) show that decreasing rectified voltage by extending regulation angle α_{reg} leads to a drastic decrease in power coefficient. At α_{reg} being close to π/2 power coefficient can fall as low as 0.45, while further shift in α_{reg} towards the end of a half-period makes power coefficient converge practically towards zero.

5 Conclusions

Thanks to application of fully controllable semiconductor devices and an original control algorithm, the transistor-based regulated rectifier shown in Fig. 5 manages to sustain χ value close to a maximum of 1 along the entire length of rectified voltage regulation. This creates a very economically sound rectifier circuit solution applicable for all modern domestic AC-wire DC-motor locomotives.

References

1. Villablanca, M.E.: Harmonic-free line-commutated ac/dc rectifiers. Electr. Power Syst. Res. **79**(11), 1531–1537 (2009). https://doi.org/10.1016/j.epsr.2009.05.006
2. Singh, S.P., Narasimharaju, B.L., Kumar, N.R.: Performance analysis of AC-DC power converter using PWM techniques. Energy Proc. **14**, 880–886 (2012). https://doi.org/10.1016/j.egypro.2011.12.1027
3. Yang, S., Bryant, A., Mawby, P., Xiang, D., Ran, L., Tavner, P.: An industry-based survey of reliability in power electronic converters. IEEE Trans. Ind. Appl. **47**(3), 1441–1451 (2011). https://doi.org/10.1109/TIA.2011.2124436
4. Zhou, L., Wu, J., Sun, P., Du, X.: Junction temperature management of IGBT module in power electronic converters. Microelectron. Reliab. **54**(12), 2788–2795 (2014). https://doi.org/10.1016/j.microrel.2014.07.002
5. Vlas'evskii, S.V., Mel'nichenko, O.V., Malysheva, O.A.: Emergency operation processes in the thyristor rectifier of an alernating current electric locomotive. Russ. Electr. Eng. **87**(2), 68–72 (2016). https://doi.org/10.3103/S1068371216020188
6. Bunyaeva, E.V., Skorik, V.G., Vlas'evskii, S.V., Fokin, D.S.: A method for improving the energy efficiency of an alternating current electric locomotive in the regenerative braking mode. Russ. Electr. Eng. **87**(2), 73–76 (2016). https://doi.org/10.3103/S1068371216020024
7. Perpina, X., Jorda, X., Vellvehi, M., Rebollo, J., Mermet-Guyennet, M.: Long-term reliability of railway power inverters cooled by heat-pipe-based systems. IEEE Transp. Ind. Electr. **58**(7), 2662–2672 (2011). https://doi.org/10.1109/TIE.2010.2087298
8. Schulze, H.J., Niedernostheide, F.J., Pfirsch, F., Baburske, R.: Limiting factors of the safe operating area for power devices. IEEE Transp. Electr. Devices **60**(2), 551–562 (2013). https://doi.org/10.1109/TED.2012.2225148

Energy Efficiency Electrified Section with Automatic Voltage Regulation

Boris Arzhannikov, Irina Baeva(✉), and Timofey Tarasovskiy

Ural State University of Railway Transport, Kolmogorova Street, 66, 620034 Yekaterinburg, Russia
irina.baeva.01@yandex.ru

Abstract. The paper considers an increase in carrier and traffic capacity of sections (lines) by means of improving 3 kV DC traction power supply system As a method of improvement the authors propose a system of contactless automatic voltage regulation with the application of various regulators: reactor voltage regulation, thyristor-reactor and thyristor. The paper presents simulation of the 3 kV DC traction power supply system with the use of KORTES program (complex for calculating traction power supply) that was developed by the JSC "VNIIZhT" (Railway Research Institute) and that is being used in the JSC «Russian Railways» . The authors used for calculation a real railway section of the Sverdlovsk Railway with a heavy gradient and a locomotive 2ES10 in case of passing single freight train with increased weight. During calculation a developed clarifying method with the use of corrective current coefficients was applied. Based on the simulation results, the authors carried out the study of technical and energy indicators of the traction power supply system that allowed assessing the carrier and traffic capacity of sections (lines). The results obtained proved the efficiency of a thyristor switching device in comparison to a reactor and a thyristor-reactor switching devices.

Keywords: Capacity · Electrified sections · Direct current · Voltage regulation · Calculation complex of traction power supply "KORTES" · Train speed · Energy losses and energy consumption

1 Introduction

With the enhance in traffic on the railways there is a need to increase their carrier and traffic capacity. Train speeds and load limits are being increased, new more powerful electric locomotives are being created. In the traction power supply system for the assessment of traffic capacity and planning measures to strengthen the existing technical equipment the determining factors are train weight, number of trains in the feeder belt and their routing, train-to-train interval. On turn-round sections of trains with increased weight traction power supply system must have the appropriate loading capacity. In case of passing trains with weight more than 6000 tons current load significantly increases in system and, consequently, equipment heating occurs more intensively, catenary voltage level decreases, energy losses increase and operating conditions of protection devices against short-circuit currents are getting more complicated [1–7].

Z. Popovic et al. (Eds.): TransSiberia 2019, AISC 1115, pp. 87–97, 2020.
https://doi.org/10.1007/978-3-030-37916-2_10

One of the options for strengthening the traction power supply system is the application of transformers with regulation under load [8–16]. The paper considers adjustable converter units with the system of contactless automatic voltage regulation that provide stabilization of rectified voltage in the range from 3500 V to 3700 V with a regulation error of plus or minus 0.6%.

Functional scheme of the contactless automatic voltage regulation system is shown on Fig. 1. The object of regulation is the converter unit (converter transformer and rectifying unit). Controlled reactors (CR) connected to the taps of converter transformer are used as the regulation parts. The system of contactless automatic voltage regulation uses automatic voltage control box (ShAUN 5) as the automatic regulator.

The ShAUN 5 box receives a signal from a voltage sensor about the voltage level on converter unit buses and compares it with the specified level of stabilization. Depending on results of the comparison, the box controller sends a control signal to the thyristor bridges that magnetize the controlled reactors thereby ensuring the stabilization of the rectified converter unit voltage at a specified level.

At the present stage of switching devices development one of the controlled reactors is replaced by an uncontrolled reactor (UR) as shown on Fig. 1b.

The main disadvantages of the existing system of contactless automatic voltage regulation are the significant losses in electrical steel and copper of reactors as well as their weight and size parameters. According to the Sverdlovsk Railway data, about 30 million roubles are spent annually on reactor control.

Figure 2 presents the external characteristics of the converter unit with the system of contactless automatic voltage regulation.

The Transport Power Supply Department of the Ural State University of Railway Transport carries out research works aimed at development of the contactless automatic voltage regulation system with a thyristor-reactor switching device (Fig. 3) and a thyristor switching device.

According to preliminary calculations, in comparison with reactor regulation devices, the replacement of a controlled reactor with a thyristor key will reduce losses in a switching device by 5 times, with the replacement of an uncontrolled reactor – by 10 times. Also, the replacement of reactors with thyristor keys will significantly reduce the cost of switching devices.

The system of contactless automatic voltage regulation with a thyristor switching device uses thyristor keys TK1 and TK2 connected to the taps of a converter transformer primary winding as a regulation part and uses the ShAUN 6 box as an automatic regulator. Like the ShAUN 5 box, it compares the voltage on the converter unit buses with the specified value and, according to the result, sends control signals to the conditioners. The 1-st and the 2-nd conditioners have noninverting inputs and are designed for thyristor control of TK1 key, the 3-rd and the 4-th have inverting inputs and are designed for TK2 key control (Fig. 4).

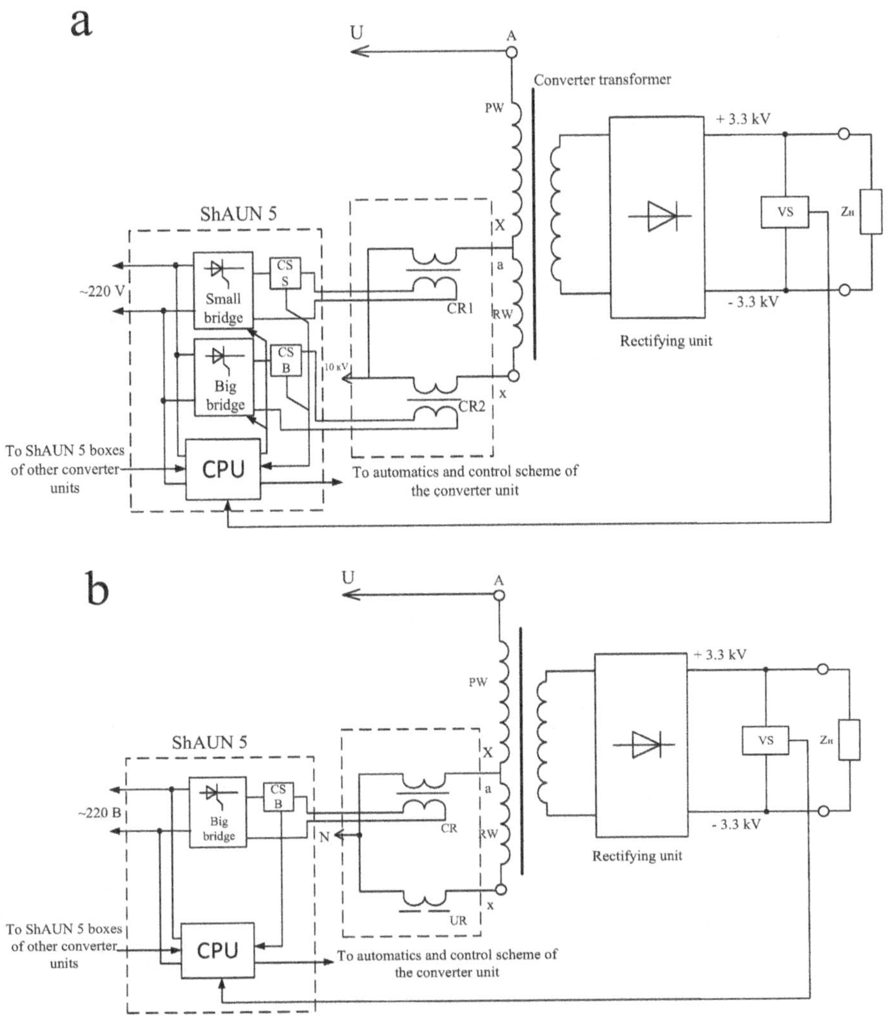

Fig. 1. Functional scheme of the contactless automatic voltage regulation system with: (a) two controlled reactors CR1-CR2; (b) controlled and uncontrolled reactors CR-UR; VS – voltage sensor; CS – thyristor bridges current sensor; PW, RW – power and regulating winding of converter transformer

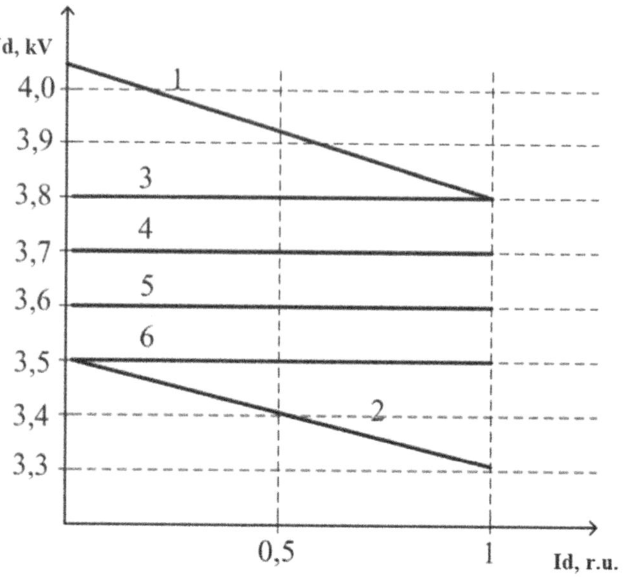

Fig. 2. External characteristics of the converter unit with the system of contactless automatic voltage regulation. 1–2 – natural; 3–6 – automatic

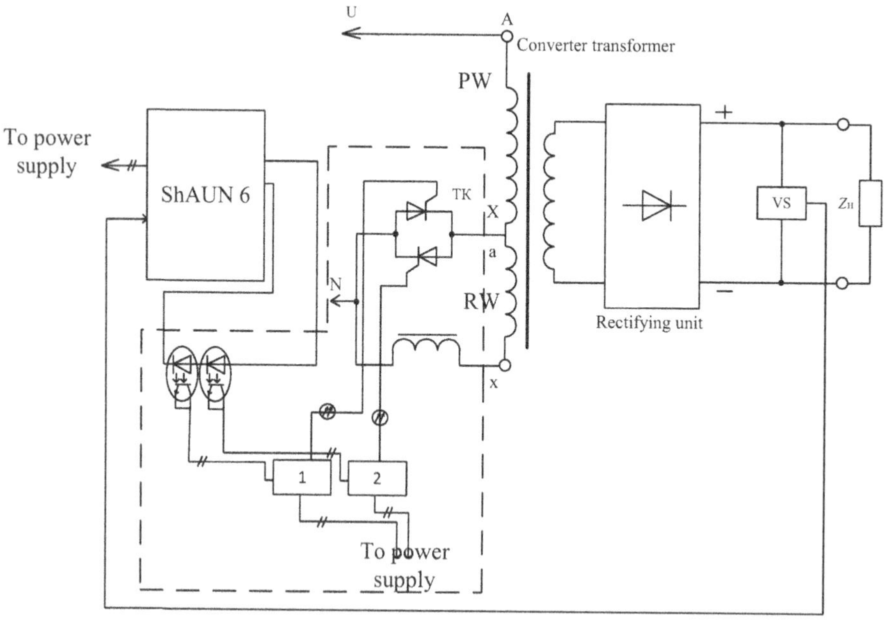

Fig. 3. Functional scheme of the contactless automatic voltage regulation system with a thyristor-reactor switching device (TRSD): 1, 2 – conditioners of light signal to the photothyristor control electrodes of TK key; Zl – full load resistance

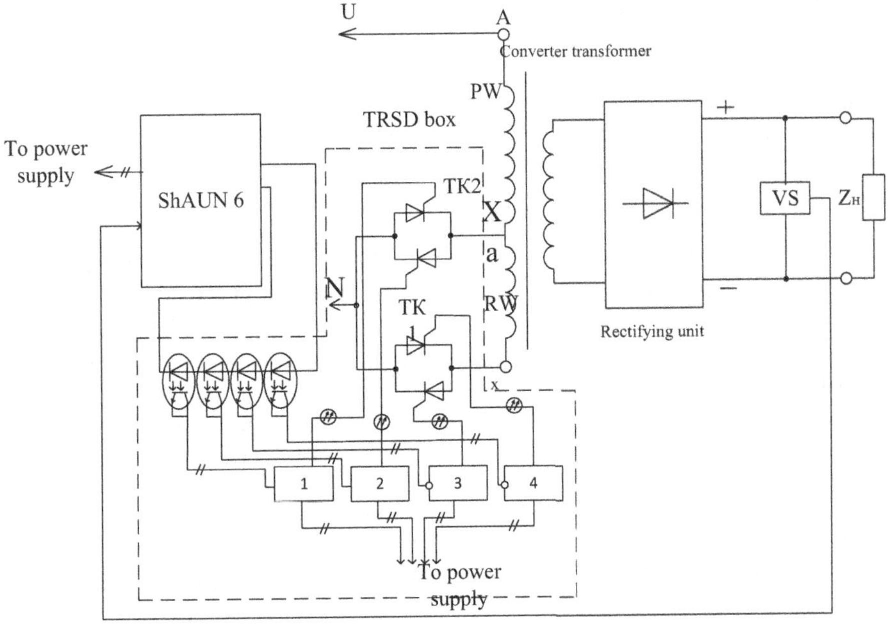

Fig. 4. Functional scheme of the contactless automatic voltage regulation system with a thyristor switching device: 1, 2, 3, 4 – conditioners of light signal to the photothyristor control electrodes of TK1 and TK2 keys

Figure 5 presents the external characteristics of the converter unit equipped with the system of contactless automatic voltage regulation with a thyristor switching device.

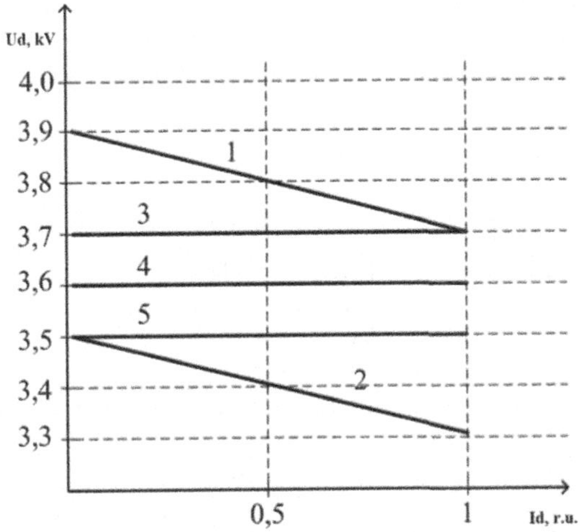

Fig. 5. External characteristics of the converter unit equipped with the system of contactless automatic voltage regulation with a thyristor switching device: 1–2 – natural; 3–5 – automatic

From the above we can state the effectiveness of the contactless automatic voltage regulation system as the method of strengthening. The system of contactless automatic voltage regulation will increase the carrier and traffic capacities of sections (lines). For the further research of the contactless automatic voltage regulation system with various switching devices it is necessary to carry out electric calculations of the traction power supply system on a real railway section.

2 Technical and Energy Analysis of Traction Power Supply System Operation with Various Switching Devices

The most effective method of investigating traction power supply system operation is simulation modelling [17–30]. The JSC "VNIIZhT" (Railway Research Institute) has developed and introduced on railway sections electrified with a direct current a program complex for calculation of traction power supply "KORTES" based on simulation modelling.

Tables 1 and 2 contain the input data for carrying out electric calculations of the regulated traction power supply system. Figure 6 presents track grading, power equipment for the real Shalya – Khrustalnaya section of the Sverdlovsk Railway.

Table 1. Conditions for calculation of the regulated traction power supply system

Train operation section	Shalya – Khrustalnaya
Locomotive type	2ES10
Train set weight, t	6000
Speed, km/h	50
Catenary voltage for grade computations, V	3000
Voltage on buses of traction substations	U_{st} = 3700 V (Stabilized characteristics with voltage regulation devices)
Number of transformers included into operation at each substation	One main step-down transformer and two converter transformers

Table 2. Parameters of the converter transformer with various regulators

Type of converter transformer and regulator	Rated power of power winding, kVA	Rated voltage, kV	Short-circuit voltage, %	Losses, kW		No-load current, %
				No-load	Short-circuit	
TRDP-16000/10 (CR-UR)	13430	10	9.6	19	118	5.0
TRDP-16000/10 (TRSD)	13430	10	9.6	13	93	5.0
TRDP-16000/10 (TSD)	13430	10	9.6	14	87	0.8

Table 3 presents the results of electric calculations of the regulated traction power supply system with various switching devices. The table contains energy consumption on traction At, energy losses in catenary Δ Ac, energy consumption according to electric motive power counters Aemp, energy losses in main step-down and converter transformers Atr, the total energy losses at traction substation Ats with the inclusion of At, Δ Ac, Aemp, Δ Atr and specific energy consumption on motive power att. The paper shows the electrical calculations of the traction power supply system with the application of new more realistic program addition development for clarification of electric locomotive Ie currents in case of the introduction of voltage regulation devices [29, 30] based on the "iteration method" [31] – the approximation of calculated energy consumption on motive power of a locomotive Aemp to the basic Aempb.

Table 3. Results of the traction power supply system modeling

Parameter	Contactless automatic voltage regulation with controlled and uncontrolled reactors CR-UR	Contactless automatic voltage regulation with thyristor-reactor switching device TRSD	Contactless automatic voltage regulation with thyristor switching device TSD
Energy consumption on traction $A_t = A_{emp} + \Delta A_c$, kW*h	9520	9520	9520
Energy losses in catenary Δ Ac, kW*h	519	519	519
Energy losses in electric motive power Aemp, kW*h	9001	9001	9001
Energy losses in transformers Atr load/xx, kW*h	40/1458	36/1231	35/1156
Energy consumption on motive power at traction substation $A_{ts} = A_t + \Delta A_{tr}$, kW*h	10499	10268	10192
Specific energy consumption on motive power a_{tt}, kW*h/104 t*km	152.16	148.81	147.71
Reduction of a_{tt}, % after the introduction of new switching device		2.25	2.92

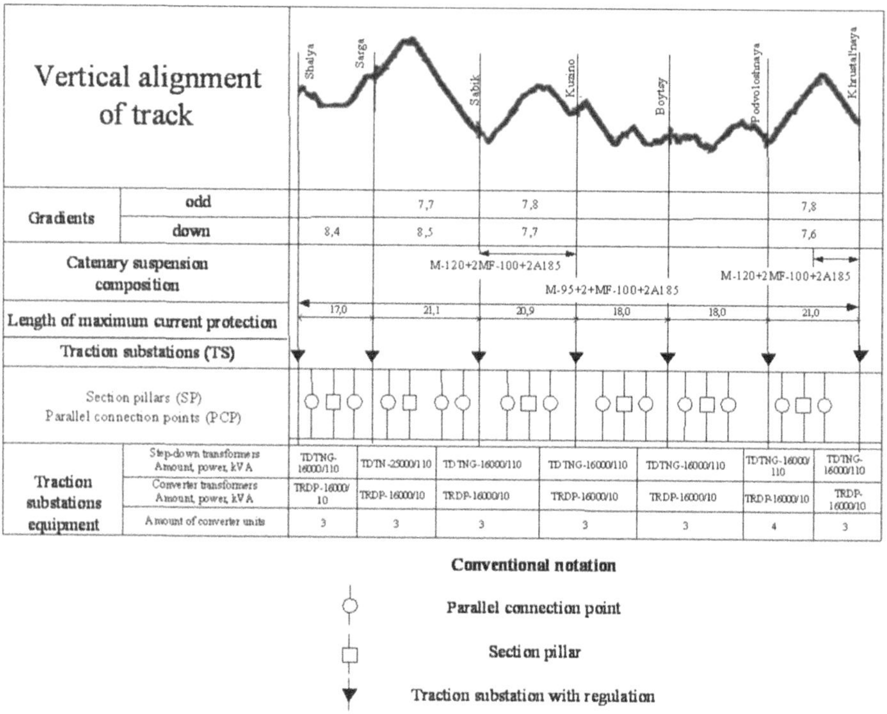

Fig. 6. Track grading on the Khrustalnaya – Shalya section

3 Results

Table 3 shows that energy consumption on motive power including energy losses in catenary and energy consumption according to counters of electric motive power is equal in all variants of electric calculations. The level of catenary voltage in all variants of calculations is above the minimum level of 2700 V (Engineering instructions). Energy losses in trans-formers are the lowest with the use of contactless automatic voltage regulation with a thyristor switching device. Reduction of specific energy consumption on traction after the introduction of contactless automatic voltage regulation with a thyristor-reactor switching device is 2.25% and with a thyristor switching device is 2.92%. Energy losses of no-load mode in transformers are higher than losses of load mode because the calculation was carried out for the section with seven traction substations and one traction load that is why transformers operated in no-load mode.

4 Conclusion

The research presented in the paper shows and justifies the possibility of increasing the carrier and traffic capacity of sections (lines) due to the application of contactless automatic voltage regulation system with a thyristor switching device. The electric calculations of traction power supply system showed the reduction of specific energy consumption after the application of contactless automatic voltage regulation with a thyristor switching device compared with the contactless automatic voltage regulation with CR-UR. The replacement of CR-UR with a thyristor switching device will reduce losses in the switching device by 10 times and will significantly decrease the switching devices cost.

Nowadays the Transport Power Supply Department of the Ural State University of Railway Transport carries out research works aimed at development of contactless automatic voltage regulation system with a thyristor switching device TSD. The physical model is developed and successfully passed tests.

References

1. Esslingera, P., Witzmann, R.: Improving grid transmission capacity and voltage quality in low-voltage grids with a high proportion of distributed power plants. In: ICSGCE 2011, Energy Procedia, Chengdu, China, vol. 12, pp. 294–302 (2011). https://doi.org/10.1016/j.egypro.2011.10.040
2. Kilter, J., Sarnet, T., Kangro, T.: Assessment of transmission network voltage unbalance in connection of high speed electrical railway connection. In: Electric Power Quality and Supply Reliability Conference (PQ), pp. 329–334 (2014). https://doi.org/10.1109/pq.2014.6866835
3. Ahmadi, S., Dastfan, A.: Energy saving in urban railway using speed profile optimization. In: 24th Iranian Conference on Electrical Engineering, Shiraz, pp. 1076–1081 (2016). https://doi.org/10.1109/iraniancee.2016.7585681
4. Vojtek, M., Kendra, M., Stoilova, S.: Optimization of railway vehicles circulation in passenger transport. Transp. Res. Proc. **40**, 586–593 (2019)
5. Sirong, Yi.: Strengthening of the railway transport capacity. In: Principles of Railway Location and Design, vol. 7, pp. 473–534 (2018). https://doi.org/10.1016/j.trpro.2019.07.084
6. Jiajian, L., Ullrich, M., Yong, C.: Increasing performance of railway systems by exploitation of the relationship between capacity and operation quality. J. Rail Transp. Plann. Manag. **7** (3), 127–140 (2017). https://doi.org/10.1016/j.jrtpm.2017.08.002
7. Arzhannikov, B., Baeva, I.: Influence of the organization of the movement of the united freight trains on the increasing of the capacity of the plots electrified on a dc. J. Transsib Railway Stud. **2**(34), 50–64 (2018)
8. Hua, J., Marinelli, M., Coppob, M., Zecchinoa, A., Bindner, H.: Coordinated voltage control of a decoupled three-phase on-load tap changer transformer and photovoltaic inverters for managing unbalanced networks. Electric Power Syst. Res. **131**, 264–274 (2016). https://doi.org/10.1016/j.epsr.2015.10.025/
9. Santiyanona, D., Hongesombutb, K., Srisonphanc, S.: Simulation on voltage unbalance reduction in railway electrification system by different special transformers. In: International Electrical Engineering Congress iEECON2016, Chiang Mai, Thailand, pp. 373–376 (2016). https://doi.org/10.1016/j.procs.2016.05.104
10. Kalantari, M., Sadeghi, M.J., Fazel, S.S., Farshad. S.: Investigation of Power Factor Behavior in AC Railway System Based on Special Traction Transformers. https://doi.org/10.4236/jemaa.2010.211081

11. Chovančíková, N., Dvořák, Z.: Effect of a power failure on rail transport. In: 13th International Scientific Conference on Sustainable, Modern and Safe Transport (TRANS-COM 2019). Transportation Research Procedia, Novy Smokovec, Slovak Republic, vol. 40, pp. 1289–1296. (2019). https://doi.org/10.1016/j.trpro.2019.07.179
12. Hammerschmidt, T., Gwisdorf, B., Borchard, T., Rehtanz, C.: Technical and economic evaluation of voltage regulation strategies for distribution grids with a high amount of fluctuating dispersed generation units. In: 2010 IEEE Conference on Innovative Technologies for an Efficient and Reliable Electricity Supply (CITRES), pp. 8–14. IEEE, Waltham (2010)
13. Coppo, M., Turri, R., Marinelli, M., Han, X.: Voltage management in unbalanced low voltage networks using a decoupled phase-tap-changer transformer. In: Power Engineering Conference (UPEC). 49th International Universities, pp. 1–6. IEEE, Cluj-Napoca (2014). https://doi.org/10.1109/upec.2014.6934809
14. Esslinger, P., Witzmann, R.: Regulated distribution transformers in low-voltage networks with a high degree of distributed generation. In: 2012 3rd IEEE PES Innovative Smart Grid Technologies Europe (ISGT Europe), pp. 1–7. IEEE, München (2012)
15. Dib, M., Ramzi, M., Nejmi, A.: Voltage regulation in the medium voltage distribution grid in the presence of renewable energy sources. In: Materials Today: Proceedings, vol. 13, pp. 739–745 (2019). https://doi.org/10.1109/ISGTEurope.2012.6465833
16. Coto, M., Arboleya, P., Gonzalez-Moran, C.: Optimization approach to unified AC/DC power flow applied to traction systems with catenary voltage constraints. Int. J. Electr. Power Energy Syst. **53**(1), 434–441 (2014). https://doi.org/10.1016/j.ijepes.2013.04.012
17. Che, Y., Wang, X., Lv, X., Hu, Y., Teng, Y.: Study on probability distribution of electrified railway traction loads based on kernel density estimator via diffusion. Int. J. Electr. Power Energy Syst. **106**, 383–391 (2019). https://doi.org/10.1016/j.ijepes.2018.10.021
18. Rao, Z.: Traction Calculation of Trains. China Railway Press (2016)
19. Yang, S., Li, X., Song, K., Wu, M.: A novel modeling approach of negative-sequence current for electrified railway traction substation. Int. J. Electr. Power Energy Syst. **107**, 462–471 (2019). https://doi.org/10.1016/j.ijepes.2018.12.004
20. Zhang, G., Tian, Z., Tricoli, P., Hillmansen, S., Liu, Z.: A new hybrid simulation integrating transient-state and steady-state models for the analysis of reversible DC traction power systems. Int. J. Electr. Power Energy Syst. **109**, 9–19 (2019). https://doi.org/10.1016/j.ijepes.2019.01.033
21. Mellitt, B., Mouneimne, Z.S., Goodman, C.J.: Simulation study of DC transit systems with inverting substations. IEE Proc. - B Electr. Power Appl. **131**(3), 38–50 (1984). https://doi.org/10.1049/ip-b.1984.0008
22. Arboleya, P., Mohamed, B., El-Sayed, I.: DC railway simulation including controllable power electronic and energy storage devices. IEEE Trans. Power Syst. **33**(5), 5319–5329 (2017). https://doi.org/10.1109/TPWRS.2018.2801023
23. Dincel, E., Nak, H., Akkaya, Ş., Canevi, M., Mutlu, İ., Söylemez, M.T.: Robust control of railway traction system. IFAC-PapersOnLine **51**(25), 171–177 (2018). https://doi.org/10.1016/j.ifacol.2018.11.100
24. Dullinger, C., Struckl, W., Kozek, M.: Simulation-based multi-objective system optimization of train traction systems. Simul. Model. Pract. Theory **72**, 104–117 (2017). https://doi.org/10.1016/j.simpat.2016.12.008
25. Feng, D., Lin, Sh, Sun, X., He, Zh: Reliability assessment for traction power supply system based on quantification of margins and uncertainties. Microelectron. Reliab. **88–90**, 1195–1200 (2018). https://doi.org/10.1016/j.microrel.2018.06.032

26. Fernandez, P.M., Sanchís, I.V., Yepes, V., Franco, I.R.: A review of modelling and optimisation methods applied to railways energy consumption. J. Cleaner Prod. **222**, 153–162 (2019). https://doi.org/10.1016/j.jclepro.2019.03.037
27. Açikbas, S., Soylemez, M.: Coasting point optimisation for mass rail transit lines using artificial neural networks and genetic algorithms. IET Electr. Power Appl. **2**, 172–182 (2008). https://doi.org/10.1049/iet-epa:20070381
28. Chen, D., Han, X., Cheng, R., Yang, L.: Position calculation models by neural computing and online learning methods for high-speed train. Neural Comput. Appl. **27**, 1617–1628 (2016). https://doi.org/10.1007/s00521-015-1960-6
29. Arzhannikov, B., Baeva, I.: The effect of voltage regulation on capacity of electrified DC sections and on the consumption of electric energy on the traction of trains. Transp. Urals **4** (55), 71–75 (2017). udc: 621.331.3.024
30. Baeva, I.: Calculating procedure for the system of traction power supply of dc 3,0 kV in case of applying voltage regulating devices. In: Proceedings of Petersburg Transport University, vol. 1, pp. 51–58 (2019). udc: 621.331.3.024
31. Vilenkin, N.Ya.: Approximation method. Popular mathematics lectures 35. Science, Moscow (1968)

Development of Thermally Conductive Compound Based on a Colloid Nanosuspension

Polina Vinoogradova[1,2(✉)], Tatyana Kozhevnikova[2], and Ilya Manzhula[2]

[1] Far Eastern State Transport University, Seryshev Street, 47, Khabarovsk 680021, Russia
vpolina17@hotmail.com
[2] Computing Center, Far Eastern Branch of the Russian Academy of Sciences, Kim Yu Chen Street, 65, Khabarovsk 680000, Russia

Abstract. The most important condition for the formation of modern transport infrastructure is its digitalization. In general, the processes of informatization of domestic transport enterprises and the introduction of intelligent transport systems on Russian roads are quite intensive. Digital technology has affected virtually all areas of the transport enterprise. However, their successful implementation and operation requires the improvement of automation devices, remote control and communication. Modern powerful devices of automation, telemechanics and communications (electronic devices) are characterized by high heat release. For their normal operation is required to ensure heat dissipation. To solve this problem, cooling systems are used, a component of which is a thermal interface, the common elements of which are heat-conducting compounds (thermal paste). In this paper, a mathematical model was developed by means of a mathematical apparatus. This model describes the processes of heat and mass transfer in the considered liquid-phase medium with nanoparticles, used as a thermally conductive composition to optimize the heat exchange processes that occur during the cooling of components in electronic devices. In the future, based on the obtained mathematical model, it is planned to identify the main chemical components that make up the thermally conductive composition for further development. The developed methods and the presented results of the work can be useful to specialists in the field of heating, as well as other researchers of similar subjects.

Keywords: Digital technologies · Electronic devices · Heat and mass transfer · Thermal interface · Thermal paste · Mathematical modeling

1 Introduction

Digital technologies have affected almost all areas of the transport enterprise, from providing information on timetables, routes and tariffs to the operation and repair of rolling stock, maintenance and diagnostics of transport infrastructure facilities. Digital technologies make it possible to use personal and public transport more efficiently, increase citizens' mobility, increase the level of safety and comfort of city trips, and generally optimize traffic flow management.

Z. Popovic et al. (Eds.): TransSiberia 2019, AISC 1115, pp. 98–108, 2020.
https://doi.org/10.1007/978-3-030-37916-2_11

In our country, the information field of passenger services is being actively formed, including navigation, ticket solutions, choosing a travel route, providing Internet access and related services. The most relevant areas of digital technology in the transport sector today are:

- the management of urban traffic flows, using Artificial Intelligence;
- the development of road transport infrastructure based on Big data and predictive analytics;
- the preparation of public transport routes, location of transport hubs, road construction, planning of parking space based on the results of the analysis of Big Data;
- the application of IoT solutions in transport, the development of V2X (vehicle-to-everything) communications, the interaction between vehicles and the exchange of data between a vehicle and road infrastructure;
- the application of autonomous unmanned vehicles in various areas of urban economy;
- the introduction of an online monitoring system for the level of satisfaction of passengers of public transport, the adjustment based on the monitoring results will ensure greater personalization of transport services;
- the Intelligent transport system for maximum efficiency of traffic management and safety of traffic, control of traffic lights and route optimization.

Solving such problems through digital technology complicates electronic devices. An analysis of the development of electronic devices shows that over 10 years the complexity of electronic devices has increased 10 times. During the same time, the speed of electronic devices has increased significantly.

Modern powerful semiconductor devices (transistors, diodes, thyristors, chipsets and high-speed processors) have a high heat output. For their normal operation, it is necessary to ensure heat withdrawal. This task is provided by cooling systems. An integral part of cooling systems is the thermal interface. The following types of thermal interfaces are distinguished:

- heat-conducting paste compounds;
- polymerizable heat-conducting compounds;
- silver thermally conductive adhesives compounds;
- heat-conducting gaskets;
- solders and liquid metals.

In this paper, all attention is paid specifically to the consideration of heat-conducting paste compounds (hereinafter referred to as thermal grease, thermal paste). Therefore, it makes no sense to describe the distinguishing characteristics and applications of each type of thermal interface.

Thermal paste is a multicomponent plastic substance with high thermal conductivity, used to reduce thermal resistance between two contacting surfaces. Typical and most common thermal greases of domestic production are KPT-8, AlSil-3, as well as a series of heat-conducting pastes Steel Frost, Cooler Master, Zalman, etc.

The basic requirements for thermal paste are:

1. the smallest thermal resistance;
2. the stability of properties over time and storage;
3. the stability of properties in the operating temperature range;
4. the ease of application.

The fillers with high thermal conductivity in the form of micro- and nanodispersed powders and their mixtures are used as heat-conducting components in the manufacture of thermal paste:

- metals (tungsten, copper, silver);
- microcrystals (diamond);
- metal oxides (zinc, aluminum, etc.);
- nitrides (boron, aluminum);
- graphite/graphene.

Mineral or synthetic oils, liquids and their mixtures having low volatility are used as binders. There are thermal pastes with a polymerizable in the air binder. Sometimes, in order to increase the density, volatile components are added to their composition, which make it possible to have a sufficiently liquid thermal grease during application and a highly dense thermal interface with high thermal conductivity. Such thermal greases usually reach maximum thermal conductivity within 5–100 h of normal operation. There are thermal pastes based on liquid at 20–25 °C metals, consisting of pure indium and gallium and alloys based on them.

The best (and most expensive) thermal greases have silver or aluminium oxide base. Both bases are having the lowest thermal resistance. The cheapest (and least effective) thermal pastes have a ceramic base. Some thermal greases have electrical conductivity. Therefore, it is necessary to avoid getting such a thermal grease on the terminals of electronic components located near the place of thermal grease application.

It is worth paying attention that thermal grease should not be too liquid and not very thick. Since in the first, it simply will not create the desired contact, and in the second it is laid unevenly on the surface.

Thermal grease is used in electronic devices as a thermal interface between heat-generating elements and devices for removing heat from them (for example, between a processor and a radiator). The main requirement when using heat-conducting paste is the minimum thickness of its layer. Therefore, it is necessary to follow the manufacturer's recommendations when applying heat-conducting pastes. A small amount of paste applied to the thermal contact area is crushed when the surfaces are pressed against each other. In this case, the paste fills the smallest recesses in the surfaces and contributes to the emergence of a homogeneous environment for the spread of heat.

Thermal grease is used to cool electronic components that have a heat dissipation greater than that permissible for this type of enclosure: power transistors and power microcircuits (switches) in switching power supplies, in horizontal scan units of televisions with a kinescope, and transistors for output stages of powerful amplifiers.

The works on the development of modern thermally conductive compounds is carried out everywhere. You can see thermal pastes are produced in the USA, China, Norway, Taiwan and Russia on the shelves of retail stores.

Along with the study of thermal processes, today the physical properties and characteristics of colloidal suspensions, which are both in a static state and under various loads, including thermal loads, are being actively studied. This topic is relevant, and any advances in research of various kinds of colloidal suspensions are characterized not only by new theoretical knowledge, but also often carry applied values.

Nanofluids are participated in the creation of efficient cooling or systems of control large heat fluxes [1–3]. They are also characterized by special transport properties. Unlike large dispersed particles, nanoparticles practically do not sediment, they do not erode the channels through which they move.

The rapid progress in the field of nanotechnology has provided researchers the arsenal of new materials with unique optical and physicochemical properties. With the increase in the productivity of electronic devices and the development of high-energy technologies, the need arises to create efficient cooling systems and control large heat fluxes. In addition, nanofluids find their application in other areas of human activity. For example, positive results have been obtained on the use of gold, silver, silicon oxides, or titanium nanoparticles in radiation protection, in biosensorics, immunoanalysis, genomics, and the treatment of oncological and infectious diseases in recent years [4–7].

Nanofluids are also effective for realizing a number of low threshold nonlinear optical phenomena in the visible range. Their important advantage is the possibility of varying nonlinear optical properties by means of a matrix from a liquid-phase medium, as well as by selecting the concentration, size, and shape of nanoparticles [8, 9].

In these media, unlike homogeneous ones, the nonlinear optical response arises due to a change in the refractive index and absorption coefficient induced by the light wave due to the effects of thermal diffusion and electrostriction of particles. Thus, in microheterogeneous multicomponent media, specific mechanisms of nonlinearity exist due to redistribution of component concentrations (thermophoresis in gases, suspensions, Soret effect in liquid-phase media).

Along with the applied aspect, the study of transport processes and nonlinear optical phenomena in nanofluids is an important fundamental component, because there is no consistent physical theory of these processes and phenomena. Despite the large accumulated experimental and theoretical results, systematic data are not available, and experimental results are often contradictory. It is difficult enough to determine how the properties of the transfer of nanofluids will change when varying one or another parameter of the system: the material of the nanoparticles, their shape, size, volume concentration, etc., to predict their properties. This requires an understanding of the transport mechanisms in nanofluids. At the same time, it should be noted that in recent years both experimental and theoretical results have been obtained regarding the thermophysical constants of the medium.

There are phenomenological approaches to the description of transport processes in open systems such as nanoparticles + liquid + laser radiation, based on the first principles: the laws of conservation of energy and mass. The solution of such a system of balance equations containing specific thermophysical and optical parameters makes it possible to study the dynamics of the concentration of nanoparticles, determine the temperature fields and study the nonlinear optical properties of the nanofluid. This system of equations was poorly studied analytically and numerically. The effects of the self-action of light fields in a nanofluid against the background of the dependence of the radiation absorption coefficient on the concentration of particles, as well as the contribution of thermal diffusion and electrostrictive mechanisms of optical nonlinearity, have not been adequately studied. If we disregard the article [10], then the problems of the formation of dissipative structures and processes of propagation of concentration switching waves in relation to the open systems described above were not considered.

Obviously, the systematic application of nanofluids become possible only if their properties can be predicted in advance.

The aim of the scientific research of this article is to develop a mathematical model that describes the phenomena of heat and mass transfer in nanofluids.

We are taking into account various physical processes that occur when using the investigated nanofluid as thermal paste to protect the components of electronic devices from overheating.

The object of the study is a liquid-phase medium with nanoparticles, which is under the influence of an electromagnetic field.

2 Development of a Mathematical Model

To achieve this goal we resort to methods of mathematical modeling. We consider a mathematical model that describes the phenomena of heat and mass transfer in a nanofluid, taking into account various physical processes that occur when using the developed nanofluid as thermal paste to protect the components of electronic devices from overheating. We assume that the particle sizes satisfy the following condition: $a_0 \ll \hat{\lambda}$, where a_0 is its linear dimension, and $\hat{\lambda}-$ is the wavelength of the light wave. Thus, we do not consider the processes of diffraction and light scattering, we also exclude the processes of particle sedimentation.

It is known that temperature and concentration gradients arise in the medium as a result of the action of a light field. These phenomena are described by a system of balance equations for the temperature and concentration of particles [16]:

$$\rho_{ef} \cdot C_p \frac{\partial T}{\partial t} = div\,(\lambda \cdot gradT) + \alpha \cdot \tilde{I}(r), \tag{1}$$

$$\frac{\partial C}{\partial t} = div\,[D\, gradC + D_T\, C(1-C)gradT - \gamma\, C\, grad\, I]. \tag{2}$$

We note that in Eq. (1) there is no addend responsible for the Dufour effect, in view of its smallness. Here we have adopted the following notation: T - is the temperature of the medium, $C = C(r,t) = m_0/m$ - is the mass concentration of the particles (m_0 - is the mass of particles contained in a unit volume, m - is the mass of a unit volume of liquid), C_p, ρ - are thermophysical constants liquids, $\lambda(C)$ - is the thermal conductivity, $I(r)$ - is the light intensity, α_0 - is medium absorption coefficient, D, D_T - are diffusion and thermal diffusion coefficients, respectively. Further, λ is the coefficient of thermal conductivity, $\gamma = \frac{4\pi\beta D}{\bar{c}n_{ef}kT}$, $\beta = \varepsilon_l \frac{\varepsilon_p - \varepsilon_p}{\varepsilon_p + 2\varepsilon_l} a_0^3$, ε_p, ε_l are the dielectric constants of the particle and liquid, respectively, β is the polarizability of particles, k is the Boltzmann constant,

$\bar{c}$ is the speed of light in a vacuum, n_{ef} is the effective refractive index of the medium. We note that in Eq. (2) the last addend expresses the contribution from the particle flux. This addend is related to the action of the gradient force on the side of the light field (electrostriction of particles). We shall assume that the profile of the light beam is Gaussian.

Next we use the following approximations:

$$div\,(\lambda \cdot gradT) \approx \lambda \nabla^2 T,$$
$$div\,(C(1-C)gradT) \approx C(1-C)\nabla^2 T,$$
$$div\,(C \cdot gradI) \approx C\,\nabla^2 I.$$

Further, we will investigate the dynamics of nanoparticles, assuming that the processes of mass transfer occur against a background of steady temperature. In this case, we set $\tau_T \ll \tau_D$. Thus, in the heat equation we set $(\partial T/\partial t = 0)$.

Let us assume that the concentration dependence of the thermal conductivity can be represented in the form

$$\lambda(C) = \lambda_0 + \beta C = \lambda_0(1 + pC),$$

here $p = \frac{\beta}{\lambda_0}$ and $pC < 1$, λ_0 is the coefficient of thermal conductivity of a fluid.

Similar dependence was theoretically found in a paper [7] and was experimentally confirmed in publications [11–14].

Using the approximation $(\lambda(C))^{-1} = (\lambda_0 + \beta C)^{-1} = (\lambda_0)^{-1}(1 - pC)$, and the stationarity condition we write the Eq. (2) in the form (we consider the one-dimensional case)

$$\frac{\partial C}{\partial t} = D\frac{\partial^2 C}{\partial x^2} - \left(D_T\frac{\alpha_0}{\lambda_0}\right)\tilde{I}_0\, e^{-\frac{x^2}{x_0^2}}C(1-C)(1-pC) + \delta\left(1 - \frac{x^2}{x_0^2}\right)e^{-\frac{x^2}{x_0^2}}C. \quad (3)$$

Here $\tilde{I}\left(\frac{x}{x_0}\right) = \tilde{I}_0\, e^{-\frac{x^2}{x_0^2}}$, $\delta = \frac{4\,\tilde{I}_0\gamma}{x_0^2}$.

In the paper [7] the model problem on the conditions of bistable behavior of such a nanofluid was studied. The problem was reduced to a parabolic equation of the type (3) with constant coefficients and a similar cubical nonlinearity.

We introduce dimensionless variables and carry out the parametrization of Eq. (3). Then

$$\frac{\partial C}{\partial t} = b\frac{\partial^2 C}{\partial \rho^2} - e^{-\rho^2}C(1-C)(1-pC) + m\,(1-\rho^2)\,e^{-\rho^2}C, \quad (4)$$

where $\tau = D_T\frac{\alpha_0}{\lambda_0}\tilde{I}_0 t$; $\frac{x}{x_0} = \rho$; $b = \frac{\lambda_0}{S_T\,\alpha_0\tilde{I}_0 x_0^2}$; $m = \frac{4\lambda_0\gamma}{S_T\,\alpha_0 D x_0^2}$; $S_T = \frac{D_T}{D}$ - is the Soret coefficient.

Consider the following initial boundary conditions for Eq. (4):

$$C(\rho,0) = C_0\, e^{-\rho^2}\left(\rho - \frac{L}{x_0}\right)^2 \rho^2 + C_1, \left.\frac{\partial C}{\partial \rho}\right|_{\rho=0} = 0 \ , 0 \le \rho \le L/x_0, \tag{5}$$

$$\left.\frac{\partial C}{\partial \rho}\right|_{\rho=\frac{L}{x_0}} + S_T C(1-C)\left.\frac{\partial T}{\partial \rho}\right|_{\rho=\frac{L}{x_0}} - \frac{\gamma}{D} C \left.\frac{\partial I}{\partial \rho}\right|_{\rho=\frac{L}{x_0}} = 0. \tag{6}$$

The parameter L expresses the characteristic length, for example, the length of the cell in the experiment, the positive constants $C_0, C_1 < 1$ correspond to the distribution of particles at the origin.

These boundary conditions are obtained from the condition that all flows on the boundary are equal to zero. As can be seen, the second boundary condition contains the temperature gradient. An explicit expression for it can be obtained from the heat equation, assuming in it $\partial T/\partial t = 0$.

It can be shown that the second addend in the Eq. (6) is much less than the third addend. Therefore, neglecting it, we obtain the following boundary condition on the right boundary:

$$\left.\frac{\partial C}{\partial \rho}\right|_{\rho=\frac{L}{x_0}} + \frac{2\gamma L}{D x_0} S_T C|_{\rho=\frac{L}{x_0}} e^{-\frac{L^2}{x_0^2}} = 0. \tag{7}$$

The paper [8] also studied the dynamics of particles in the nanosuspension, a linearized equation with the dominance of the contribution from particle convection was considered. The results of the solutions indicated the existence of concentration waves under these conditions.

We use numerical methods for solving the problem (4), (5), (7). In this case, we use the following values of variables and parameters:

$$b \approx 100 \div 1000; m \approx 0.8 \div 4; p \approx 1.5; L \approx 10^{-2};\ x_0 \approx 10^{-4};\ 0 \le \rho \le \frac{10^{-2}}{10^{-4}} \approx 10^2 \div 10.$$

To solve nonlinear evolution equations, finite difference methods are often used [12–15]. The nonlinearity of the original equation imposes stringent requirements on difference methods, so the problem of constructing effective and economic algorithms remains relevant today. The use of explicit difference schemes for the numerical solution of problem (4), (5), (7) is not effective due to strict limitations on their stability. Therefore, it is preferable to use implicit schemes with relatively weak constraints. We use the predictor-corrector method (see [13–15]) to construct a numerical algorithm for solving the problem (4), (5), (7).

Let $h = \frac{L/x_0}{N}, \tau = \frac{1000}{M}$ are the steps on ρ and t, where N, M are natural numbers, $\rho_i = ih$, $t_k = k\tau$. We define the grid function $y_i^k = y(\rho_i, t_k)$.. We introduce the uniform grid $\omega_h = (\rho_i, t_k) : i = 0, \ldots, N, k = 0, \ldots, M$. To simplify the further notation, we set

$$f(\rho, C) = -e^{-\rho^2} C(1 - C)(1 - pC) + m(1 - \rho^2) e^{-\rho^2} C.$$

When calculating the intermediate values of the desired function, at the predictor stage, the Eq. (4) is approximated by the following difference equation

$$\frac{y_i^{k+\frac{1}{2}} - y_i^k}{0.5\tau} = \frac{y_{i+1}^{k+\frac{1}{2}} - 2y_i^{k+\frac{1}{2}} + y_{i-1}^{k+\frac{1}{2}}}{h^2} + f(\rho_i, y_i^k), \; t_{k+\frac{1}{2}} = \left(k + \tfrac{1}{2}\right)\tau,$$
$$y_i^0 = C_0 e^{-\rho_i^2} \left(\rho_i - \tfrac{L}{x_0}\right)^2 \rho_i^2 + C_1, \; i = 0, 1, \ldots, N,$$
$$y_1^{k+\frac{1}{2}} - y_0^{k+\frac{1}{2}} = 0, \; k = 0, 1, \ldots, M - 1,$$
$$\frac{y_N^{k+\frac{1}{2}} - y_{N-1}^{k+\frac{1}{2}}}{h} + \frac{2\gamma L}{D x_0} S_T e^{-\frac{L^2}{x_0^2}} y_N^{k+\frac{1}{2}} = 0.$$

Then, at the stage of the corrector, a following symmetrical six-point scheme is used:

$$\frac{y_i^{k+1} - y_i^k}{\tau} = \frac{1}{2}\left(\frac{y_{i+1}^{k+1} - 2y_i^{k+1} + y_{i-1}^{k+1}}{h^2} + \frac{y_{i+1}^k - 2y_i^k + y_{i-1}^k}{h^2}\right) + f\left(\rho_i, y_i^{k+\frac{1}{2}}\right),$$
$$y_i^0 = C_0 e^{-\rho_i^2} \left(\rho_i - \tfrac{L}{x_0}\right)^2 \rho_i^2 + C_1, \; i = 0, 1, \ldots, N,$$
$$-3y_0^{k+1} + 4y_1^{k+1} - y_2^{k+1} = 0, \; k = 0, 1, \ldots, M - 1,$$
$$\frac{y_{N-2}^{k+1} - 4y_{N-1}^{k+1} + 3y_N^{k+1}}{2h} + \frac{2\gamma L}{D x_0} S_T e^{-\frac{L^2}{x_0^2}} y_N^{k+1} = 0.$$

The considered scheme is unconditionally stable and has the accuracy of order $O(\tau^2 + h^2)$.

3 The Results of Mathematical Modelling

Further, software written in the C++ programming language was developed to calculate the mathematical model. To confirm the adequacy of the mathematical model, we have compared the heat transfer coefficient of a medium without nanoparticles with the heat transfer coefficient of a medium with nanoparticles. As a medium we use distilled water (thermal conductivity coefficient 0.613 W/(m · K)), and graphene particles (thermal conductivity coefficient ~5000 W/(m · K)) are used as nanoparticles. The volume concentration of nanoparticles with respect to the liquid was 0.1%.

Figure 1 are showed the time dependence of the solution for a fixed spatial coordinate. Distilled water without nanoparticles is a solid line, distilled water with nanoparticles is a dashed line at m = 0.8 (m - is the indicator responsible for the contribution to the concentration flow from electrostriction).

The shape of the curves are showed that the nature of the transition processes strongly depends on the origin: near the origin, the corresponding curve has a pronounced maximum. As the value of the time variable increases, both curves correspond to different concentration values.

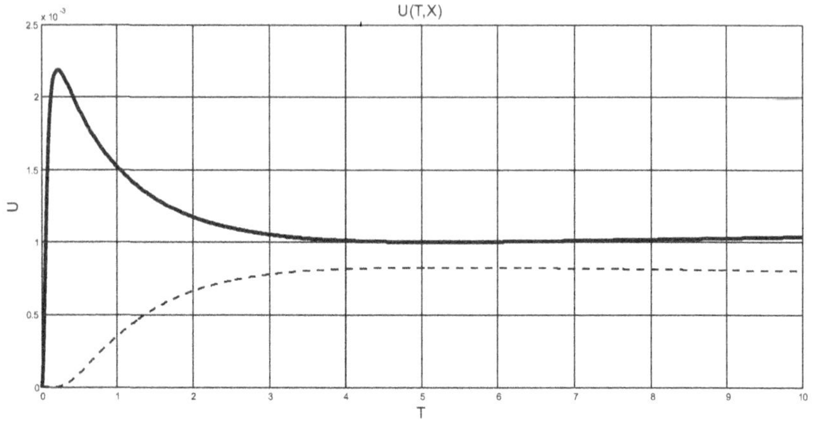

Fig. 1. Time dependences of the solution to problem (4), (5), (7)

Figure 2 are showed that in time, the "spreading" of the initial Gaussian distribution of particle concentration occurs. At the same time, portions of space far from the origin are involved in the transfer processes.

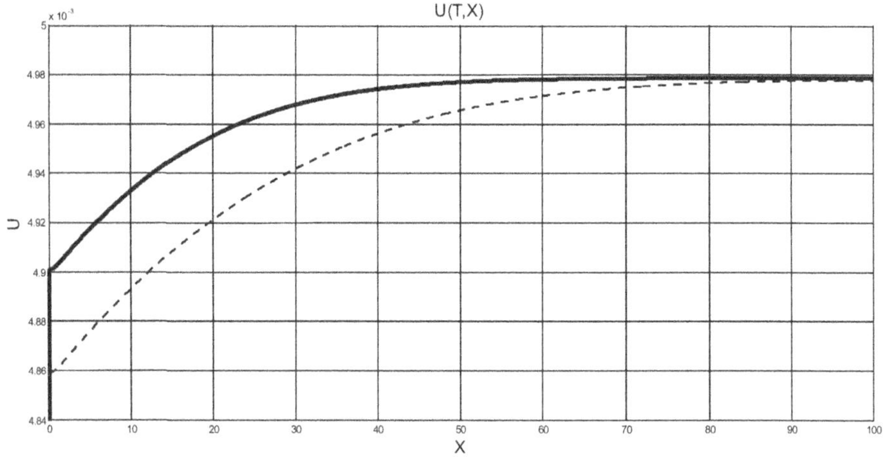

Fig. 2. Coordinate dependences of the solution to problem (4), (5), (7).

The numerical experiment are showed that the application of nanofluids makes it possible to intensify the local heat transfer over the entire length of the cell by more than 10%. Similar dependences are obtained for other nanofluids, however, the effect decreases with a decrease in the concentration of nanoparticles. This decrease is due to a change in the density and viscosity coefficient of the nanofluid compared with the corresponding values for water.

4 Conclusion

A scientific study is showed that the application of colloidal nanosuspensions (nanofluids) as a thermal interface to protect the components of electronic devices from overheating has large prospects. However, the lack of systematic experimental data regarding the heat transfer coefficient of nanofluids requires the use of mathematical modeling methods in this area. This methodology eliminates the need for complex field experiments, the acquisition of expensive laboratory equipment for these experiments.

The difficulty is in the fact that the study of the heat transfer coefficient is a complex task, taking into account the viscosity of nanofluids. Thus, for a more complete understanding of the processes of heat and mass transfer in nanofluids, taking into account various physical phenomena that can occur when using nanofluids as thermal grease to protect components of electronic devices from overheating, it is necessary to complicate the mathematical model, which is described in this paper.

Acknowledgement. The studies were supported by the Government of the Khabarovsk Krai (grant No. 24C/2019 of July 28, 2019).

References

1. Terekhov, V.I., Kalinina, S.V., Lemanov, V.V.: The mechanism of heat transfer in nanofluids: state of the art (review). Part 1. Synthesis and properties of nanofuids. Thermophys. Aeromech. **1**, 1–15 (2010)
2. Krishtop, V., Doronin, I., Okishev, K.N.: Improvement of photon correlation spectroscopy method for measuring nanoparticle size by using attenuated total reactance. Opt. Express **20**, 25693–25699 (2012). https://doi.org/10.1364/OE.20.025693
3. Kulchin, Y.N., Shcherbakov, A.V., Dzyuba, V.P., Voznesenskiy, S.S., Mikaelyan, G.T.: Nonlinear-optical properties of heterogeneous liquid nanophase composites based on highenergy-gap Al2O3 nanoparticles. Quantum Electron. **38**, 154–158 (2008). https://doi.org/10.1070/QE2008v038n02ABEH013529
4. Livashvili, A.I., Krishtop, V.V., Yakunina, M.I.: Electrostrictive self-action mechanism of radiation in nanofluidsю. Adv. Condens. Matter Phys. Article 1–5 (2013). https://doi.org/10.1155/2013/591087
5. Livashvili, A.I., Krishtop, V.V., Bryukhanova, T.N., Kostina, G.V.: Concentration dynamics of nanoparticles under a periodic light field. Phys. Proc. **73**, 156–158 (2015). https://doi.org/10.1016/j.phpro.2015.09.141
6. El-Ganainy, E., Christodoulides, D.N., Rotschild, C., Segev, M.: Soliton dynamics and self-induced transparency in nonlinear nanosuspensions. Opt. Express **15**, 12207–12218 (2007)
7. Livashvili, A.I., Krishtop, V.V., Karpets, Y.M., Bryuhanova, T.N., Kireeva, N.M.: Laser beam-induced bistability of concentration in nanofluids. J. Phys: Conf. Ser. **737**, 012011 (2016). https://doi.org/10.1088/1742-6596/737/1/012011
8. Krishtop, V.V., Livashvili, A.I., Vinogradova, P.V., Kostina, G.V., Bryukhanova, T.N.: Dynamics of nanoparticle concentration in nanofluids under laser light field. IOP Conf. Ser.: J. Phys.: Conf. Ser. **936**, 012079 (2017). https://doi.org/10.1088/1742-6596/936/1/012079
9. Ivanov, V.I., Livashvili, A.I.: Self-action of a Gaussian radiation beam in a layer of a liquid-phase microheterogeneous medium. Atmos. Oceanic Opt. **23**(1), 7–8 (2010). https://doi.org/10.1134/S1024856010010021

10. Livashvili, A.I., Kostina, G.V., Yakunina, M.Y.: Temperature dynamics of a transparent nanoliquid acted on by a periodic light field. J. Opt. Technol. **80**(2), 124–126 (2013). https://doi.org/10.1364/JOT.80.000124
11. Olafsen, J.: Non-equilibrium thermodynamics. In: Sturge's Statistical and Thermal Physics, pp. 297–306 (2019). https://doi.org/10.1201/9781315156958-17
12. Samarskii, A.A., Vabishchevich, P.N.: Computational Heat Transfer 1. Mathematical Modelling. Wiley, Chichester (1995)
13. Richtmyer, R.D., Morton, K.W.: Difference Methods for Initial-Value Problems. Interscience Publishers, Geneva (1994)
14. Matus, P.P., Hieu, L.M., Vulkov, L.G.: Analysis of second order difference schemes on non-uniform grids for quasilinear parabolic equations. J. Comput. Appl. Math. **310**, 186–199 (2016). https://doi.org/10.1016/j.cam.2016.04.006
15. Shi, D.: Numerical Methods in Problems of Heat Exchange, Moscow (1988)

Aspects of Railway Vehicles Vibrations with Nonlinear Spring Suspension Characteristics

Anatoly Savoskin[1(✉)] and Stanislav Vlasevskii[2]

[1] Moscow State University of Railway Engineering, Obraztsova Street 9, 127994 Moscow, Russia
zavkaf@festu.khv.ru

[2] Far Eastern State Transport University, Serysheva Street 47, 680000 Khabarovsk, Russia

Abstract. Random vibrations in railway vehicles with nonlinear characteristics of the spring suspension are considered to be nonstationary. The paper examines a method that allows to account for nonlinear vibrational system different aspects by the example of research of railway vehicle random vibrations with nonlinear characteristics of transverse forces that are close to a cubic parabola. Probabilistic characteristics of such oscillations are determined on the basis of the procedure of averaging over many realizations. In this case the correlation function and the spectral density is three-dimensional. In the spectral densities, in addition to the main maximum, which lies in the main plane on the diagonal, there are also side peaks corresponding to ultraharmonic vibrations of nonlinear systems. Taking into account that the amplitudes of the side peaks are smaller than those of the main maximums it can be said that in this nonlinear system ultraharmonic vibrations appear on frequencies that are about twice as high as the frequencies of the main maximums.

Keywords: Nonlinear system · Non-stationary random oscillation · Averaging over many realizations · Ultraharmonics oscillations

1 Introduction

Spring suspensions of railway vehicles usually have elements with nonlinear characteristics of proportional action (rubber parts), piecewise linear (characteristics of gaps between band rims and rails, vibration restrictors etc.), polysemantic (frictional vibration damper) and serpentine (leaf springs). This means any vibration research of railway vehicles should consider nonlinear nature of spring suspension characteristics and peculiar vibrations inherent in such nonlinear systems. Particularly, when it comes to nonlinear systems the amplitudes of forced vibrations are only ambiguously tied with frequencies and amplitudes of the disturbances. There is also a possibility for subharmonic and ultraharmonic vibrations to appear, as well as self-excited vibrations etc. [1].

Today, studies of dynamic properties of the "vehicle-track" system usually revolve around researching virtual models with finite number of outputs and a single input that transmits random stationary disturbance to the system. In doing so, researches consider

Z. Popovic et al. (Eds.): TransSiberia 2019, AISC 1115, pp. 109–116, 2020.
https://doi.org/10.1007/978-3-030-37916-2_12

nonlinearities of the system, but do not study their influence on vehicle movement aspects deeply enough and do not explore nonlinear effects inherent in these systems. This results from the fact that vibration research of nonlinear systems tends to set the disturbance as a single execution of a random process of rail geometrical imperfection, as if there was an implied suggestion that random vibrations of generalized coordinates are stationary or even ergodic.

2 Materials and Methods

The paper examines a method that allows to account for nonlinear vibrational system different aspects by the example of research of railway vehicle random vibrations with nonlinear characteristics of transverse forces that are close to a cubic parabola. Such characteristics are typical for rubber-metal "silent blocks" connecting wheelset axle-bearings with bogie chassis, as well as for interaction forces between wheels and rails. The latter are defined by these interaction forces when the disturbance amplitudes are smaller than the railway gauge gap; if the amplitudes are larger than the gap, the rail reactance also adds to the interaction forces.

Since, as was previously mentioned, forced vibrations amplitudes only ambiguously depend on frequency and amplitude of the disturbance, the examined random vibrations are nonstationary. Therefore, to determine probability characteristics of such nonstationary vibrations the averaging by a set of realizations of the random vibration process must be executed [2, 3]. And to acquire the set, the computational generation of disturbance process realizations must be done.

For this work, the railway vehicle was represented by a four-input system with disturbances defined by vertical and horizontal imperfections of the left and right rails of the track. These disturbances were treated as probability-connected stationary random processes. To generate said disturbances, the computer environment was introduced to a special algorithm suggested by authors in [4–6] and based on application of generating filters' pulse characteristics and folding integral.

During determination of a random disturbance process generation parameters the choice of the required random process realization duration t_r and its discretization step T was executed by approximate, frequently applied equations

$$t_r \geq \frac{5\ldots10}{f_n} \text{ and } T \leq \frac{1}{(5\ldots10)f_v}. \tag{1}$$

These equations lead to the number of points in random process generation being determined by

$$N = \frac{t_r}{T} = (25\ldots100)\frac{f_v}{f_n}. \tag{2}$$

For assumed values of f_n = 0.1…0.2 Hz and f_v = 10 Hz, t_r = 50…100 s, T = 0.01 s and the number of points N = 2500…1000. As the spectral analysis of random processes applied fast Fourier transformation, its realization required the number of points to conform to $N = 2^n$. The acquired N range conforms to $N = 2^{12} = 4096$ points number, which, together with the discretization step $T = 0.0093$ s gives the realization duration value of $t_r = 38.1$ s. Programming of the problem of carriage random vibration calculation was accomplished in MATLAB Simulink environment, where the largest integration step is 0.0031 s, so the number of realization points for $t_r = 38.1$ s appeared to be equal to $N_r = 12288$.

For every movement speed $N = 4096$ realizations of a four-dimensional random disturbance process were generated, $N_r = 12288$ points in each realization. This way, the total number of points in all realizations of a single imperfection amounted to 50331648; for every point a numerical integration of the nonlinear differential equation system was executed, defining vehicle vibration.

As an example, further follow the results of calculation of sideway vibrations in a form of random vibration processes realization for a carbody lateral displacement [7, 8] $y_k\ (t)$ with a predetermined multidimensional random disturbance process, Fig. 1. The number and length of $y_k\ (t)$ realizations are equal to the number $N = 4096$ and length $N_r = 12288$ of disturbance realizations. As Fig. 1 shows, individual realizations $y_k\ (t)$ substantially differ from each other both in random vibrations span and frequency composition, which confirms nonstationary nature of output random vibration processes caused by nonlinear nature of spring suspension.

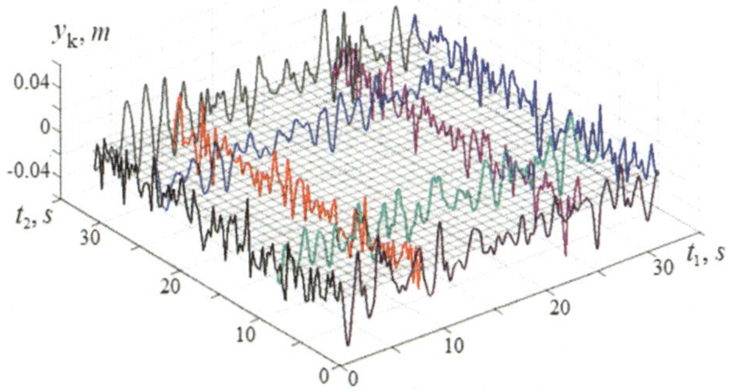

Fig. 1. Graphs of random vibration processes realization for an electric multiple unit carbody lateral displacement at 20 m/s speed

To accomplish probability analysis of nonstationary random vibration process, the matrix of calculation results must be preliminarily reduced to a square form. Because the number of values $y_k\ (t_1)$ was assumed to be equal to $N_r = 12288$, and the number of realizations was equal to $N = 4096$, out of all acquired values of $y_k\ (t_1)$ every third was read and so the discretization step for t_1 amounted to $\Delta t_1 = 3T = 0.0093$ s. Furthermore, for the second axis t_2 the number of realizations 0, 1, 2, …, n, …, N_r was replaced with time $t_2 = n\Delta t_2$. Discretization step Δt_2 was also assumed to be equal to

$\Delta t_2 = 3T = 0.0093$ s. This way, the total size of the value matrix $y_k(t_1, t_2)$ amounted to 4096×4096, or in $t_1 \times t_2 = 38.1 \times 38.1$ s terms of time.

3 Results

Using this matrix, a probability analysis of the nonstationary random process $y_k(t_1, t_2)$ was executed. As known from [2, 3], to analyze such a process first a two-dimensional density of random values distribution $f[y_k(t_1), y_k(t_2)]$ of the examined process must be acquired, then its correlated function $R_y(t_1, t_2)$ and spectral density $G_y(f_1, f_2)$ using

$$R_y(t_1, t_2) = \int_{-\infty}^{\infty}\int_{-\infty}^{\infty} y_k(t_1)y_k(t_2)f[y_k(t_1), y_k(t_2), t_1, t_2]dy_k(t_1)dy_k(t_2) \tag{3}$$

$$\begin{aligned} G_y(f_1, f_2) &= \int_0^{\frac{1}{4}N}\int_0^{\frac{1}{4}N_r} R_y(t_1, t_2)e^{-j2\pi f_1 t_1}e^{-j2\pi f_2 t_2}dt_1 dt_2 \\ &= \int_0^{\frac{1}{4}N}\int_0^{\frac{1}{4}N_r} R_y(t_1, t_2)(\cos 2\pi f_1 t_1)(\cos 2\pi f_2 t_2)dt_1 dt_2. \end{aligned} \tag{4}$$

First, a research of the two-dimensional statistic distribution $f[y_k(t_1), y_k(t_2)]$ of momentary values of the process $y_k(t_1, t_2)$ was carried out. Verification by the power criterion showed that smoothing of the two-dimensional statistical distribution for researched processes can be accomplished by the two-dimensional Gaussian law.

Using the acquired two-dimensional statistical distribution and (3), the self-correlated function was calculated. The graph of this function, Fig. 2, is a surface symmetrical to the level [0, $R(0)$, $R(t)$, t]. This surface is represented in a form of damping cosine curves and its negative amplitudes are significantly smaller than the positive ones. At the origin of coordinates the value of $R_y(t_1, t_2)$ is, as expected, equal to dispersion of $S_y^2(0) = 3 \cdot 10^{-5}$ m^2. Sectioning of this surface by a diagonal level [0, $R(0)$, $R(t)$, t] reveals that ordinates $R_y(t = t_1 = t_2)$ decrease practically under an inclined line and by the end of realization the dispersion value amounts to $S_y^2(8s) = 0.5 \cdot 10^{-5}$ m^2, which indicates a substantial decrease in dispersion as time passes (almost by 6 times), thus showing the random process $y_k(t_1, t_2)$ to be nonstationary.

Using the $R_y(t_1, t_2)$ graph and (4), the spectral density of a carbody lateral displacement vibrations $G_y(f_1, f_2)$ was calculated in $f_n = 0.1$ Hz to $f_v = 10$ Hz frequency range. This characteristic has a nonnegative surface form symmetrical to the diagonal level *0GGf* with a row of maximums. Its form is evidential of the fact that all the energy of nonstationary random vibrations process for a carbody lateral displacement at 20 m/s speed is distributed within the 0.2…2.25 Hz frequency range.

The highest maximums corresponding to the main vibration frequencies of the linearized system lie on the main diagonal and are related to frequencies: $f_{1-1} = 0.24$ Hz, $f_{4-4} = 0.375$ Hz and $f_{7-7} = 0.8$ Hz.

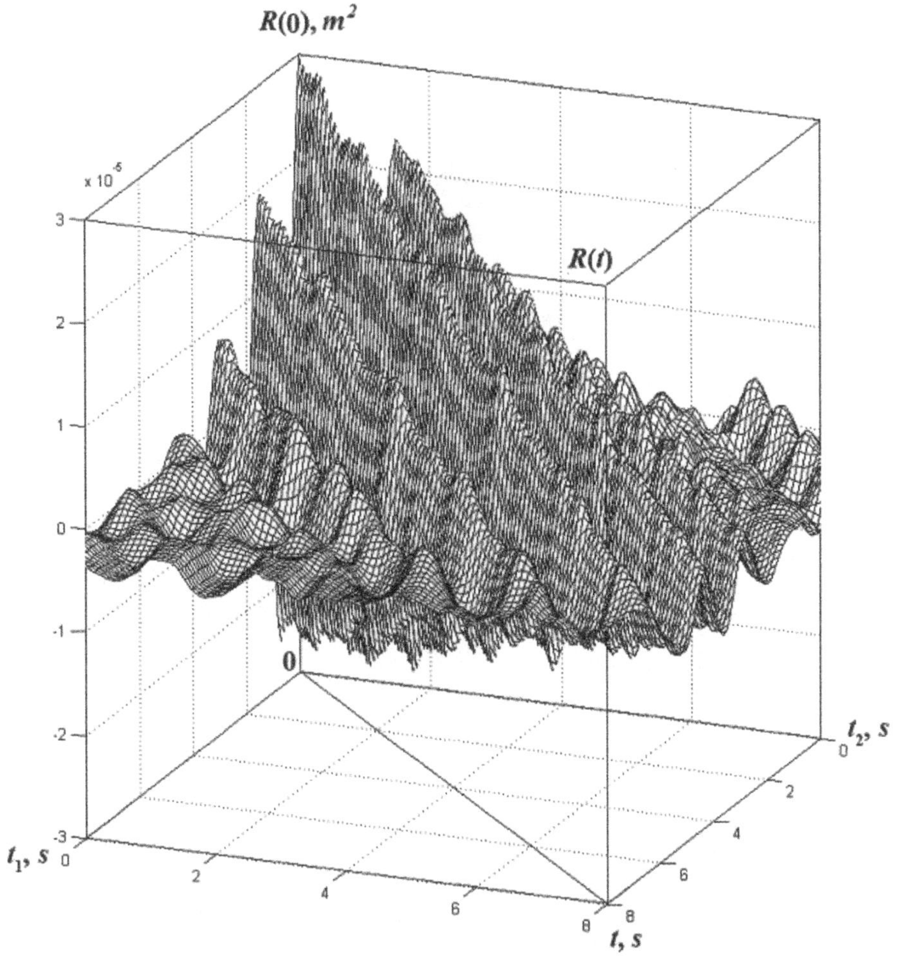

Fig. 2. Self-correlated function of nonstationary random vibrations for a carbody lateral displacement at 20 m/s speed

Frequencies of the side peaks have the following approximate ratios: $\frac{f_{2-1}}{f_{1-1}} \approx \frac{0.59}{0.24}$; $\frac{f_{5-2}}{f_{4-4}} \approx \frac{0.75}{0.375}$; $\frac{f_{10-1}}{f_{7-7}} \approx \frac{1.5}{0.8}$ (here the first number of an index correlates to the point number on the diagonal on Fig. 3, bottom, and the second number of an index correlates to the frequency value on f_1 or f_2 axes of Fig. 3). Generally, these frequency ratios are close to the 2:1 ratio.

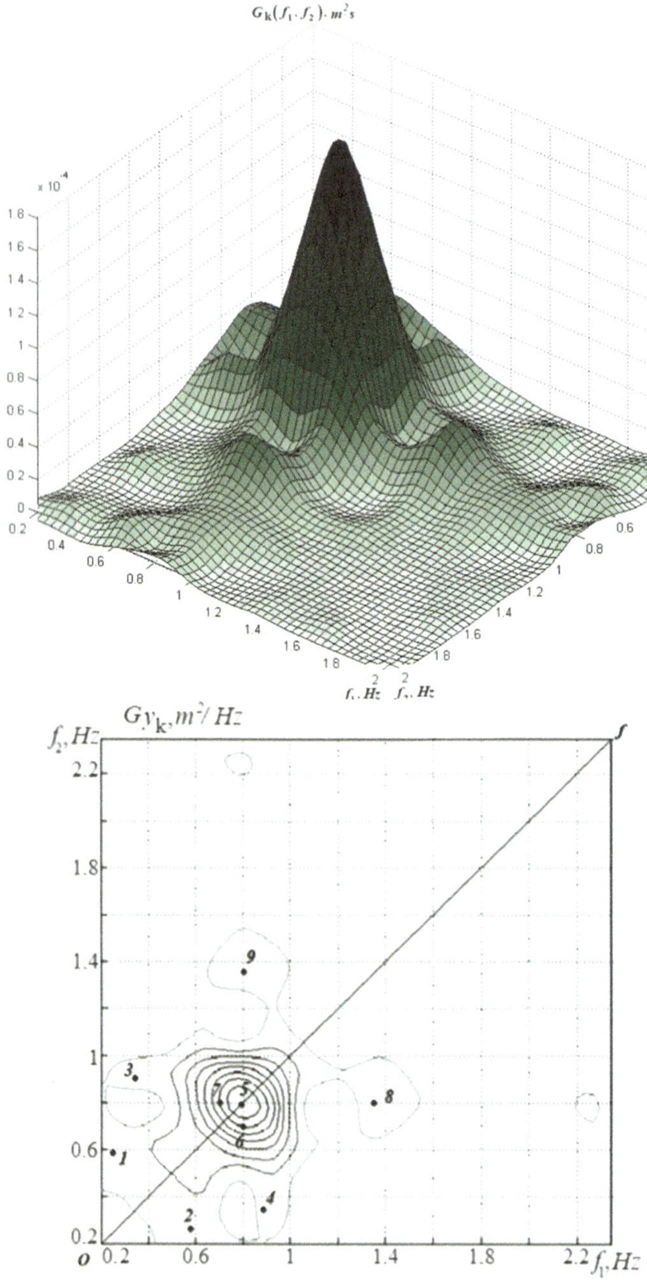

Fig. 3. Spectral density of nonstationary random vibrations for a carbody lateral displacement at 20 m/s speed: top – frequency axis starts view; bottom – vertical view.

4 Discussion

Spectral densities of the lateral displacement of bogies and wheelsets have a more complex form with more maximums. But even for these G_y (f_1, f_2) the frequencies of the side peaks are also about twice as high as the frequencies of the main maximums. Taking into account that the amplitudes of the side peaks are smaller than those of the main maximums it can be said that in this nonlinear system ultraharmonic vibrations appear on frequencies that are about twice as high as the frequencies of the main maximums.

Thus, to achieve more reliable calculation results of probability characteristics for railway vehicles nonstationary random vibrations, being a system with nonlinear elements, it is essential to apply the principle of averaging by a set of realizations of the generalized coordinates. This allows to reveal peculiar aspects of nonlinear system vibrations and leads to results that are substantially different from the usually applied approach where there is only one realization of nonlinear system random vibration process being calculated.

5 Conclusions

Since random vibrations of railway vehicles with nonlinear characteristics of their spring suspensions are nonstationary, their probability analysis requires a procedure of averaging by a set of realizations, using no less than 4096 realizations of the disturbance with the discretization step defined by the highest frequency that appears in vibration processes of the output coordinates of the system.

Spectral densities of the random processes of the system's generalized coordinates are functions of two frequencies and are symmetrical to the main diagonal that the maximums, which correspond to the main frequencies of the nonlinear vibration system, lie on.

Besides the maximums on main frequencies, spectral densities of main processes of the system's generalized coordinates also contain the side peaks that lie on frequencies that are 2, 3 etc. times higher than the maximums of the main vibrations, which is evidential of the presence of ultraharmonic vibrations in the system and displays peculiar aspects of nonlinear systems vibrations.

References

1. Grumondz, V.T.: Investigation of oscillations in nonlinear mechanical systems of arbitrary order. J. Appl. Math. Mech. **42**(1), 65–74 (1978). https://doi.org/10.1016/0021-8928(78)90089-8
2. He, J., Gong, J.: Estimate of small first passage probabilities of nonlinear random vibration systems by using tail approximation of extreme distributions. Struct. Saf. **60**, 28–36 (2016). https://doi.org/10.1016/j.strusafe.2016.02.003

3. Xu, L., Zhai, W., Gao, J.: A probabilistic model for track random irregularities in vehicle/track coupled dynamics. Appl. Math. Model. **51**, 145–158 (2017). https://doi.org/10.1016/j.apm.2017.06.027
4. Shariyat, M.: A generalized global–local high-order theory for bending and vibration analyses of sandwich plates subjected to thermo-mechanical loads. Int. J. Mech. Sci. **52**(3), 495–514 (2010). https://doi.org/10.1016/j.ijmecsci.2009.11.010
5. Wang, W., Zhang, Y., Ouyang, H.: An iterative method for solving the dynamic response of railway vehicle-track coupled systems based on prediction of wheel-rail forces. Eng. Struct. **151**, 297–311 (2017). https://doi.org/10.1016/j.engstruct.2017.08.017
6. Sirong, Y.: Dynamic Analysis of High-Speed Railway Alignment, Theory and Practice, pp. 25–71 (2018). https://doi.org/10.1016/978-0-12-812887-9.00002-7
7. Sheng, X., Thompson, D.J., Jones, C.J.C., Xie, G., Iwnicki, S.D., Allen, P., Hsu, S.S.: Simulations of roughness initiation and growth on railway rails. J. Sound Vibr. **293**(3–5), 819–829 (2006). https://doi.org/10.1016/j.jsv.2005.08.050
8. Jin, X.S., Wen, Z.F., Wang, K.Y., Zhou, Z.R., Liu, Q.Y., Li, C.H.: Three-dimensional train–track model for study of rail corrugation. J. Sound Vibr. **293**(3–5), 830–855 (2006). https://doi.org/10.1016/j.jsv.2005.12.013

Impact of Return Traction Current Harmonics on the Value of the Potential of the Rail Ground for the AC Power Supply System

Ivan Ignatenko, Evgeniy Tryapkin(✉), Sergey Vlasenko, Alexander Onischenko, and Vladimir Kovalev

Far Eastern State Transport University, Serysheva Street 47, Khabarovsk 680021, Russia
Evgenmehanik@mail.ru

Abstract. The problem of the increased potential of the rail-ground with the flow of electric trains reverse traction current operating in the area of the AC feeder is considered. The studies were caused by the need to ensure the electrical safety of the staff and the infrastructure facilities reliability when passing trains of increased weight. In the process of experimental studies of the high potentials causes in rail circuits, data were obtained indicating the complex nature of the electromagnetic interaction of reverse current harmonics with rail circuits. For studies of electromagnetic compatibility between railway subsystems, it requires the use of simulation programs that help to reveal most critical conditions for electromagnetic compatibility conditions, and allow the evaluation of electromagnetic interference from rolling stock in rails in the worst conditions. A developed the traction current harmonics distribution model in the rails is considered in this paper. The traction current harmonics distribution in the rails was calculated for a direct AC power supply of 25 kV depending on the distance from the power supply substation, the conductivity of the railway and the ground and the number of trains in the feeder zone. The results of the harmonics distribution simulation are in satisfactory agreement with the experimental data and will allow the technical measures development to reduce the rail-ground potential.

Keywords: Electromagnetic compatibility · Railway · Traction · Potential

1 Introduction

Railway transport is the most important part of the country's transport system. Railway transport accounts for about 70% of domestic freight turnover and almost 60% of passenger traffic. More than 80% of cargo transportation is carried out by electric traction.

In recent years, the government of Russian Federation has set tasks related to the increase in cargo turnover. In this regard, programs have been developed to increase the weight and length of freight trains. The achievement of the indicators set by the government is inevitably associated with a significant increase in the power of electric

Z. Popovic et al. (Eds.): TransSiberia 2019, AISC 1115, pp. 117–127, 2020.
https://doi.org/10.1007/978-3-030-37916-2_13

rolling stock (ERS), a reduction in inter-train intervals and, as a consequence, an increase traction current in the traction power supply system (TPSS).

The increase in traction current already today causes problems associated, in particular, with the normal functioning of the traction power supply system of 25 kV alternating current (AC). One of the negative consequences of the increase in traction currents (especially in the pass sections with heavy traffic) is the increased value of the potential "rail-ground", which has already caused a number of accidents and related traffic interruptions in the Far Eastern and other sections of Railways electrified by AC.

For all types of traction power supply systems, rails are used as the return wire. This circumstance requires the implementation of an additional set of measures aimed at improving the quality of the rail chains used, in addition, in the alarm system, centralization and blocking (SCB). In addition, the requirements of electromagnetic compatibility (EMC) of traction power supply devices and SCB and communication devices, the main objectives of which are to ensure the safety and regulation of train traffic, must be observed [1–3].

The value of the potential "rail-ground", in addition to traction current, is largely influenced by the value of the transient resistance [4–7]. This geoelectric parameter depends on resistance of the sleepers, ballast prism and earth resistance, which varies seasonally (in winter increases). The results of experimental studies (for different types of soil) found that in winter the value of the transient resistance may increase by 40–50 times in comparison with summer period. In addition, the increase of transition resistance and promote measures for protection from frost (cryogenic) heaving of soils, which include laying body of the subgrade polymer and foam polystyrene coatings.

This article calculates potentials of "rail-ground" for area of Far Eastern Railway "Rozengartovka – Bikin", assesses effectiveness of such measures to reduce the potentials as the transition to a system with shielding and reinforcing wire (SRW) and introduction of suction transformers with an additional wire, calculates the economic efficiency of the proposed measures.

2 Materials and Methods

Consider distribution of traction current in reverse traction network (RTN) [8–10] for the scheme shown in Fig. 1, assuming the absence of inductive coupling between contact and rail networks.

Traction substations TS1 and TS2 are connected by supply lines to the contact network. Electric locomotive current I_{IRS} represents the sum of the currents from the left I_{01} and right I_{02} substations. At the exit I_{IRS} of the electric current as is divided into two parts I_{03} and I_{04}. In turn, the rails currents I_{03} and и I_{04} Are also divided into components I'_{03}, I''_{03} and I'_{04}, I''_{04}, respectively.

The system of equations current distribution in RTN at location area of electric locomotive is as follows:

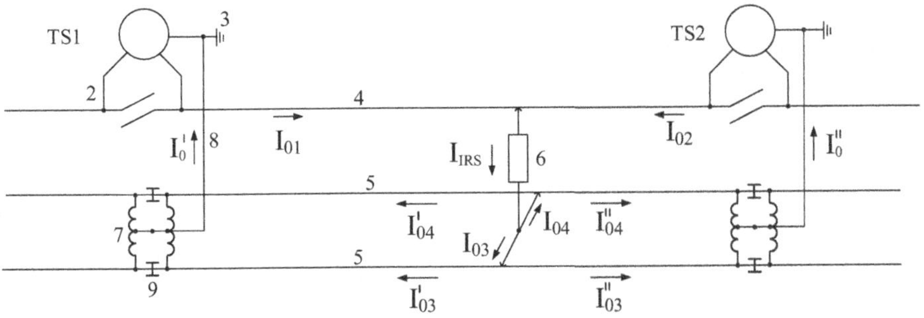

Fig. 1. Scheme construction traction power supply system of 25 kV. 1 – supply line; 2 – substation ground loop; 3 – contact network; 4 – rails; 5 – electric rolling stock; 6 – choke-transformer; 7 – reverse current feeder.

$$\begin{cases} I_{IRS} = I_{03} + I_{04}; \\ I_{03} = I'_{03} + I''_{03}; \\ I_{04} = I'_{04} + I''_{04}. \end{cases} \quad (1)$$

When making assumptions about the absence of leakage currents from the rails to ground, the currents I'_{03} and I'_{04} completely flow on rails. In the place of incorporation choke-transformer, these currents flow along half-winding a choke-transformer, flocking to the middle point and total current I'_0 by feeder reverse current returns to the substation by. The feeder reverse current is connected to the ground loop of the substation, to which, on the other hand, the input "C" of the power transformer is connected. Similar processes occur with currents I''_{03} and I''_{04}.

From 2016 to 2018 the Far Eastern railway recorded at least 12 cases of train delays due to spark gaps, and in 4 cases overlap of signal points due to the asymmetry of traction current. A total of 66 freight and 2 passenger trains were delayed for 94.7 h. The greatest number of traffic delays occurred at the Rozengartovka – Bikin section. The longest break traffic occurred in the area, "Rozengartovka – Boytsovo" 30.12.17 due to violation of the interval of passing dual heavy freight train weighing 10964 tons and the train weighing 6634 tons. The value of the potential "rail-ground" exceeded the breakdown voltage of the spark gap so that there was an electric arc between the grounded clamp and the reinforcement support №710 with the burnout of the steel clamp with a diameter of 18 mm. This incident led to a delay of 53 trains at 83.22 h. According to the results of measurements of the potentials "rail-ground", produced at the site "Rozengartovka – Bikin" after incident, it was found that the greatest value of potential in measurement at time of passage heavy trains exceeded 1000 V.

The increase in potential of "rail-ground" entails problems related to provision of electrical safety maintenance personnel. The main risks faced by personnel performing work in conditions of possibility high potential "rail-ground" should include:

1. Risk of electric shock when servicing of signaling arrangements associated with rail (relay cabinets, choke-transformers, mast lights, relays, racks on posts of electric centralization);
2. Risk of electric shock during maintenance of power supply devices (grounding of contact network supports, spark gaps, cable boxes);
3. Risk of electric shock during maintenance of grounding devices of artificial structures, dimensional gates;
4. Risk of electric shock when servicing communication devices (loudspeaker speakers, repeaters).

2.1 Influencing Factors

In normal operation mode of TPSS law of potential distribution "rail-ground" is probabilistic and statistical in nature, because it depends on number, mass and coordinates of location trains within inter-substation zone.

To ensure normal operation of motion control devices and TPSS elements, it is necessary that the currents and potentials do not exceed the permissible values set for these elements. Numerous studies have found that at alternating current, value of the potential rails can reach 2500 V, which is unacceptable both under conditions of normal operation of TPSS, SCB system, traffic safety systems, and under conditions of electrical safety for service personnel and passengers.

If distance between the traction substations is constant, when mass of the compositions increases, average value of the "rail-ground" potentials in normal operation mode of TPSS increases in proportion to current loads. When representing RTN as a line with distributed parameters, the potential "rail-ground" at any point "x" is determined by expression (2):

$$\varphi_{p-3}(x) = 0,5 \cdot \dot{I}_0 \cdot \underline{z}_w \cdot e^{-\gamma \cdot x} \tag{2}$$

Where $\underline{z}_w$ – wave resistance of a long line, Ohms; γ – coefficient of propagation of a long line, 1/km.

According to [11], wave resistance of rail network is calculated by formula (3):

$$\underline{z}_{wr} = \sqrt{\frac{\underline{z}_{r,\,m} \cdot r_{t,\,1}}{m_r}} \tag{3}$$

Where $\underline{z}_{r,\,m}$ – resistance of contours "rails of m ways-ground", Ohm/km; r_t – transition resistance "rails of one way-ground", Ohm•km; m_r – number of tracks;

The coefficient of distribution rail network is calculated by formula (4):

$$\dot{\gamma}_{wr} = \sqrt{\frac{\underline{z}_{r,\,m} \cdot m_r}{r_{t,\,1}}} \tag{4}$$

As can be set from expression (2) maximum potential "rail-ground" occurs at x = 0, that is, at the location of ERS. Then value of the potential depends on value of current electric locomotive and wave resistance of line, which in turn depends on parameters included in formula (3). Let's take a closer look at these parameters.

Current ERS varies widely and depends mainly on the mass composition, speed of movement and path profile. In relation to conditions of heavy traffic it is established that at dual traction electric locomotives consume current more than 600 A, and at triple – about 900 A.

The difficulty of determining resistance of rails is due to fact that the magnetic permeability of rail material depends on magnitude of current flowing through them. The current in rails along their entire length is not same, because at the same time within inter-substation zone there are several electric locomotives, and, moreover, part of reverse traction current flows through ground. In addition, if there is only one concept of electrical resistance in DC networks, when on AC networks introduced the concepts of active, reactive and total electrical resistance of the conductor.

Active is calculated resistance of conductor, equal to ratio of power losses in this conductor to square of current flowing through this conductor. When alternating current flows through wire, a surface ("skin") effect is observed, which additionally leads to an increase in power losses in wire.

The surface effect is called uneven distribution of electric current over the conductor cross section. This phenomenon manifests stronger, than greater frequency of current and cross-section of conductor, smaller its surface and, finally, greater magnetic permeability of conductor material.

The active and internal inductive resistance of AC conductors can be determined using the Umov – Poynting theorem. The formulas given in [12] take into account the dependence of magnetic permeability conductor material "μ" on magnetic field strength in conductor H. Complexity of determining the resistance of conductor by this method is due to need pre-calculate module and argument of the Bessel functions of zero and first order of first and second kind, which is difficult without the use of special software packages.

The exact results calculation of resistance rails can be obtained in presence of graphs dependence μ = f(H) constructed for a given grade of steel on basis of specially conducted experiments [13]. The following graphs for railway rails are given (Fig. 2). The main parameters of railway rails are given in Table 1.

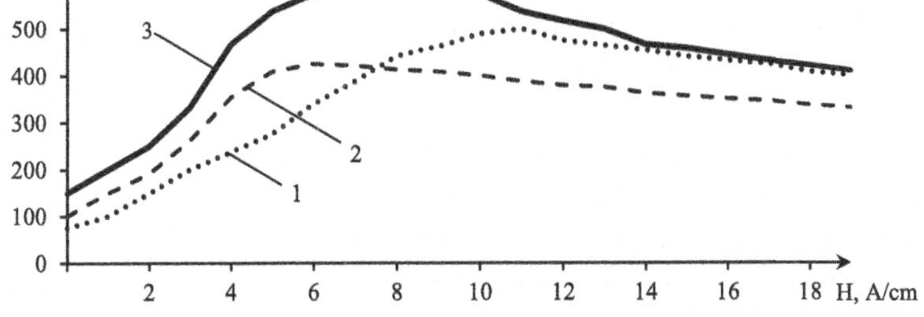

Fig. 2. Dependences $\mu = f(H)$ for rail steel

Table 1. The main parameters of railway rails

Type of rails	Cross-sectional area, S, cm^2	Perimeter P, cm	$r^S_{эк}$, on cross section, cm	$r^P_{эк}$, on perimeter, cm	Rail DC resistance at 20 °C, Om•km
P–43	55.7	56	4.21	8.92	0.042
P–50	64.5	62	4.54	9.87	0.036
P–65	82.9	70	5.14	11.15	0.028
P–75	95.1	74.5	5.5	11.86	0.024

These curves (especially 1 and 2) correlate most satisfactorily with the range μ in which the curve for steel with a carbon content of 0.82% (rail steel according to [8]) should be located.

The value μ is calculated as the arithmetic mean of the three μ values found from the curves shown in Fig. 2. The magnetic field strength on the surface of the rail, calculated by formula (1.5), A/cm:

$$H = 0.4 \cdot \pi \cdot \frac{I_r}{P} \tag{5}$$

Where I_r – current flowing on the rail, A; P – perimeter of cross section rail, cm.

2.2 Carrying Out Measurements on Railway Section

From 04.10.18 to 05.10.18 on railway section "Rozengartovka – Boytsovo" conducted research aimed at identifying causes of high potentials in rail chain at the station "Boytsovo". To ensure requirements of electrical safety during measurements at traction substations and at the "Rozengartovka – Bikin" section, specialists of repair and revision section (RRS), relay protection and automation (RPA) of power supply distance (ECH-2) were involved. During the tests following tasks were set:

1. Determination of nature reverse traction current flow at traction substation "Bikin";
2. Determination of current distribution in traction network on section "Rozengartovka – Bikin";
3. Determination of total and differential resistance rail circuit to flow of reverse traction current in the selected area;
4. Determination of "rail-ground" transition resistances and conditions affecting them;
5. Measuring the resistance of ballast track circuit.

The data of rolling stock within place of measurement "Bikin – Rozengartovka" during the tests (from 8 to 8:32 h local time) are given in Table 2.

It should be noted that similar tests were carried out on section by staff of signalization, centralization and blocking distance and power supply distance in the winter (February 2018). Then, according to test results, when passing signal point "H" composition, weighing 7000 t rail potential on choke-transformer of signal point "H" was about 1200 V.

Table 2. The data of rolling stock within place of measurement

Rolling stock	Weight, t	Coordinate, km (time 8:00 h)	Coordinate, km (time 8:30 h)	Direction of movement
9232 СППМД	12300		17.1	Even
1636 ПМ СР	6143	17.1		Even
8801	1500		11	Odd
2843 Д	1731		22	Odd
2943	1690		33.5	Odd
2260 ПМ	6290	335		Even

As a result of tests in the autumn period, values of potentials and currents in rail circuits were recorded in section with a length of 2.9 km (at a distance of 17.1 km from the Rozengartovka station) when passing on an even path of trains weighing 6143 t and 12300 t. When passing composition weighing 6143 t, maximum potential of rail circuit (at middle point of choke-transformer relative to external grounding) was 593 V, maximum leakage current was 5 A. When passing composition weighing 12300 t, maximum potential in rail circuit could not be fixed, as it was above measuring limit of monitor choke-transformer (more than 650 V). The expected value of potential – 1000 V, leakage current was 11 A. Value of transient resistance "rail-ground" according to results of measurements was 25 Ω km. In addition, results of studies were based on dependence of resistivity rail track on value of current flowing along the rails (Fig. 3).

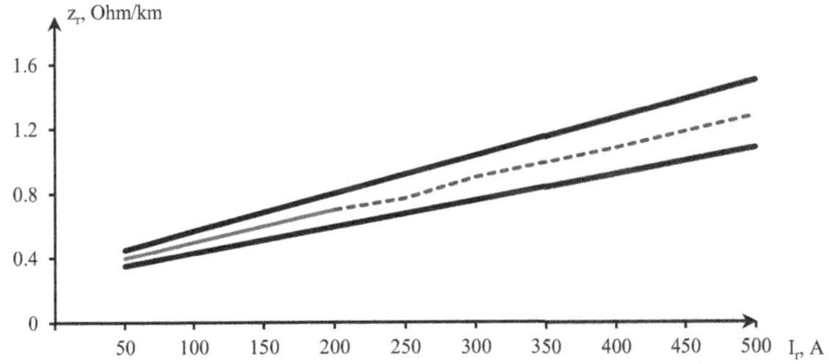

Fig. 3. Dependence of specific resistance rail track from current according to results of experiment

In Fig. 3, a solid thin line shows the average value of resistivity rail track, obtained from results of measurements, dotted – result of approximation, thickened solid lines is limit to expected area of change in the value. However, this dependence is in contradiction with the values of resistances obtained in literature and calculated analytical methods. For a more in-depth consideration of electromagnetic processes influence of traction current on rail circuit was simulated.

2.3 Modeling

Traction current is composed of high harmonic components. To assess their impact on electrical processes in the rail circuit can be used to quantify harmonics of traction current. However, to consider breakdown of spark arresters in grounding circuits of supports, it is necessary to consider electromagnetic processes and factors affecting them. The source of higher harmonics of current in traction network 25 kV AC is rectifier-inverter Converter (RIC). The power circuit of RIC provides four zones of regulation rectified voltage. The order of opening RIC shoulders in rectifying mode is determined by algorithm of control system electric locomotive converters, which in turn forms and distributes control pulses of required parameters with a given phase and in given by algorithm sequence. Thus, as a result of alternate opening thyristor arms on IV control zone, several sharp voltages jumps on traction motor are created, leading to current surges for the half-period of sinusoid supply voltage. To assess the effect of current curve on rail circuits, a simulation of traction current source was performed [15–17].

The simulation was performed in MATLAB Simulink software environment. The following parameters will be taken into account in modeling process:

- leakage current through ballast and grounding structures;
- longitudinal asymmetry of traction current;
- presence of harmonics of traction current;
- stress distribution down the track circuits.

The simulation is performed in MATLAB SIMULINK using Simscape package. To be able simulate area with most approximate conditions, simulated chain was divided into a significant number of separate circuits similar to rail track with a length of 100 meters. The general scheme of simulated area was 10 km. The result of voltage curve at location of electric locomotive is shown in Fig. 4.

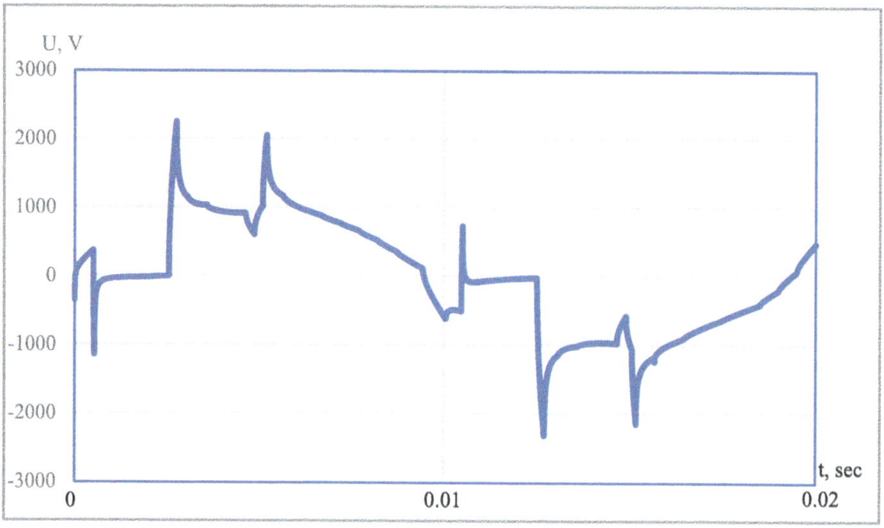

Fig. 4. The curve of rail potential change in the location of locomotive

3 Discussion

The simulation results show that in rail circuit under influence of traction currents there are overvoltages of a pulsed nature. The overvoltage data allow to explain the data obtained during tests carried out at landfill, taking into account specifics of measuring equipment used. Also, the obtained form of voltage curve allows us to put forward a theory about participation of rail circuit in formation of high-frequency electromagnetic interference in train radio communication devices, which takes place in areas with metal supports of the contact network. The rails in traction network 25 kV AC represent a significant inductive resistance to alternating current of industrial frequency, mainly due to the surface effect, as well as due to the appearance of additional resistance due to magnetic coupling (mutual induction resistance) both between rails themselves and between rails and the contact suspension. Therefore, complete modeling of all factors becomes a very time-consuming task, solution of which is selection and identification of the most strongly influencing processes. The data obtained allow to develop technical means to reduce potential in track circuits, taking into account influence of nonsinosuidality traction current. In course of further work, different ways of reducing the "rail-ground" potential were tested.

4 Findings from the Research

Thus, on basis of theoretical and practical research, following conclusions can be drawn:

1. The electromagnetic properties of rail track (as a conductor of reverse traction current), as well as the high resistance of ballast prism and significant currents in rail networks contribute to growth of potential in reverse traction network, in addition to train current;
2. Non-sinusoidal traction current can create significant overvoltage pulse character.
3. A significant increase in potential of rail is additionally caused by anti-beam measures used in accordance with [9], as well use of reinforced concrete sleepers, since there is almost complete isolation of rail from ground;
4. The simplest to implement is reduce the potential by reducing current in rail networks, but their principle of operation and technical performance must be consistent with the operation of adjacent infrastructure devices, mainly with operation of signal rail circuits.
5. According to results of research, it is planned to design and manufacture prototypes of technical means to reduce the potentials in role chains, followed by a test to confirm planned technical effect [18].

Acknowledgment. We express our deep gratitude to all our colleagues, authors of the works, familiarization with which greatly helped us in conducting research and obtaining relevant results, although they may not agree with all our interpretations and conclusions.

References

1. Shamanov, V.: The magnetic properties of rail lines and level of interferences for the apparatus of automatic control and telemechanics. Russ. Electr. Engin. **86**, 548–549 (2015). https://doi.org/10.3103/S1068371215090102
2. Colella, P., Pons, E., Tortora, A.: Rail potential calculation: impact of the chosen model on the safety analysis. In: AEIT International Annual Conference, 18346905 (2018). https://doi.org/10.23919/aeit.2018.8577295
3. Mariscotti, A.: Induced voltage calculation in electric traction systems: simplified methods, screening factors, and accuracy. IEEE Trans. Intell. Transp. Syst. **12**, 201–210 (2011). https://doi.org/10.1109/TITS.2010.2076327
4. Godyaev, A., Onischenko, A.: Automated monitoring of level distortion coded signals transmitted on rail line. In: International Multi-Conference on Industrial Engineering and Modern Technologies (FarEastCon), 18393174 (2018). https://doi.org/10.1109/fareastcon.2018.8602696
5. Rui, P., Mingli, W., Shaobing, Y.: Performance of the integrated grounding system of Hefei-Nanjing passenger dedicated railway. In: International Conference on Sustainable Power Generation and Supply, SUPERGEN 2009, 11001154 (2009). https://doi.org/10.1109/supergen.2009.5348003
6. Ogunsola, A., Mariscotti, A.: Electromagnetic Compatibility in Railways, p. 528 (2013). https://doi.org/10.1007/978-3-642-30281-7
7. Matta, V., Kumar, G.: Unbalance and voltage fluctuation study on AC traction system. In: Electric Power Quality and Supply Reliability Conference (PQ), pp. 303–308 (2014). https://doi.org/10.1109/pq.2014.6866831
8. Shamanov, V.: Formation of Interference from Power Circuits to Apparatus of Automation and Remote Control. IEEE East-West Design & Test Symposium (EWDTS), 18248723 (2018). https://doi.org/10.1109/ewdts.2018.8524676
9. Stell, R.W.B.: Maximum permissible rail potential. IEEE Veh. Technol. Mag. **6**, 12206592 (2011). https://doi.org/10.1109/MVT.2011.942537
10. Tetiana, S., Vitaliy, K., Yevheniia, K.: About electromagnetic compatibility of rail circuits with the traction supply system of railway. In: IEEE 3rd International Conference on Intelligent Energy and Power Systems (IEPS), 18323445 (2018). https://doi.org/10.1109/ieps.2018.8559541
11. Bellan, D., Spadacini, G., Grassi, F., Fedeli, E., Pignari, S.A.: Modeling strategies for conducted and radiated emissions in high-speed railway lines. In: Asia-Pacific Symposium on Electromagnetic Compatibility (APEMC), 15670514 (2013). https://doi.org/10.1109/apemc.2013.7360635
12. Mirzaei, M., Ripka, P.: Analysis of material effect on rail impedance. In: 53rd International Universities Power Engineering Conference (UPEC), 18319950 (2018). https://doi.org/10.1109/upec.2018.8541860
13. Havryliuk, V.: Modelling of the return traction current harmonics distribution in rails for AC electric railway system. In: International Symposium on Electromagnetic Compatibility (EMC EUROPE), pp. 251–254 (2018). https://doi.org/10.1109/emceurope.2018.8485160
14. Serdiuk, T.: Modeling of influence of traction power supply system on railway automatics devices. In: International Symposium on Electromagnetic Compatibility - EMC EUROPE, 17337406 (2017). https://doi.org/10.1109/emceurope.2017.8094637

15. Serdiuk, T., Feliziani, M., Serdiuk, K.: About electromagnetic compatibility of track circuits with the traction supply system of railway. In: International Symposium on Electromagnetic Compatibility (EMC EUROPE), 18149852 (2018). https://doi.org/10.1109/EMCEurope.2018.8485034
16. Baranovskii, A.: Computer simulation and experimental investigations of noise levels and spectra for railway transport. Autom. Control Comput. Sci. **45**, 293–294 (2011). https://doi.org/10.3103/S0146411611050026
17. Charalambous, C., Cotton, I., Aylott, P.: Modeling for preliminary stray current design assessments: the effect of crosstrack regeneration supply. IEEE Trans. Power Deliv. **28**, 13596015 (2013). https://doi.org/10.1109/TPWRD.2013.2259849
18. Tryapkin, E., Ignatenko, I., Keino, M.: Development of "digital substation" technology for power supply of railways. Siberian Transport Forum – TransSiberia. In: MATEC Web Conferences, vol. 239, p. 01051 (2018). https://doi.org/10.1051/matecconf/201823901051

Improvement of Automatic Pneumatic Car Brakes

Evgeny Drozdov and Evgeny Kuzmichev(✉)

Far Eastern State Transport University, Seryshev Street, 47, Khabarovsk 680021, Russia
accord@festu.khv.ru

Abstract. In article the design of an automatic pneumatic brake of the car with the braking accelerator is offered. Use of the developed accelerator of braking of an automatic pneumatic brake by freight cars allows to reduce a brake way of the train and to reduce longitudinally dynamic forces in the train when braking.

Keywords: Pneumatic brake · Highway charging · Braking accelerator · Brake way · Time preparation of brakes for action

1 Introduction

The formation and driving of heavy freight trains is one of the effective measures to reduce the operating costs of Russian Railways, as it reduces the need for locomotives and increases the carrying capacity of roads.

Due to the rapid economic growth in neighboring China and Japan, freight turnover on the Far East Railway is continuously increasing. Besides, in summer time it is necessary to have time to carry out repair works of the way. It is necessary to optimize the transport process.

In order to increase the capacity of the Far East branch of the railway, trains now run along the Baikal-Amur Mainline.

Due to formation of trains of increased weight and length, capacity of road sections is increased, capacity of stations is released and excess fleet of cars is reduced.

In recent years Russian Railways pays great attention to the creation of systems allowing to drive trains of increased weight and length of 9000 tons and more than 100 cars respectively, and in the future 18000 tons.

The most important indicator when driving long-length heavy trains affecting traffic safety is the dynamics of the train as a whole, especially during braking. This allows to preserve rolling stock and increase service life of wagons, and as a result to reduce the cost of transportation.

At the same time the main parameter affecting these processes is the speed of brake wave propagation. The slow speed of the brake wave when the pneumatic brake is applied results in significant longitudinal dynamic forces along the length of the train, as well as increases the time of preparation of the brakes for operation and, accordingly, increases the overall braking distance. A significant increase in the speed of the brake wave propagation can be achieved with the use of an electro pneumatic brake (EPT).

Z. Popovic et al. (Eds.): TransSiberia 2019, AISC 1115, pp. 128–137, 2020.
https://doi.org/10.1007/978-3-030-37916-2_14

It is well-known that the full brake way of St consists of the brake way passed by train during preparation of brakes of Sp, and the valid brake way of Sd, St = Sp Sd [2, 3]. Significant reduction of a full brake way is reached only due to reduction of the brake way passed by train during preparation of brakes of Sp which is defined by time from the moment of the beginning of a discharge of the brake highway the driver's crane on the locomotive until creation of brake force on the last car of the train tp.

Time of preparation of brakes when braking the cargo train long is 1000…1200 m (n = 70 eight-wheel cars) 10…12 s. and the brake way passed by train during preparation of brakes of Sp at an initial speed of braking 70…75 km/h is equal 200 … 250 m. At the same time the valid brake way makes 240…260 m [4, 5].

Reduction of time of preparation of brakes of tp happens generally for the account increase in speed of distribution of a brake wave on train length.

2 Materials and Methods

Over the past decades, Russian Railways has been trying to develop radio-controlled braking systems. However, no suitable technical solutions have been found to date.

It should be noted that in 2013 the development and testing of the train brake control system was completed [9]. A feature of the system is the installation of special units for discharge of the brake line in different places of the long-length train.

Figure 1 shows the formation of a long-length freight train with car braking units installed in different places. The units are controlled via a radio channel from the locomotive and allow to control both all blocks in the train at the same time and selectively only the car braking blocks.

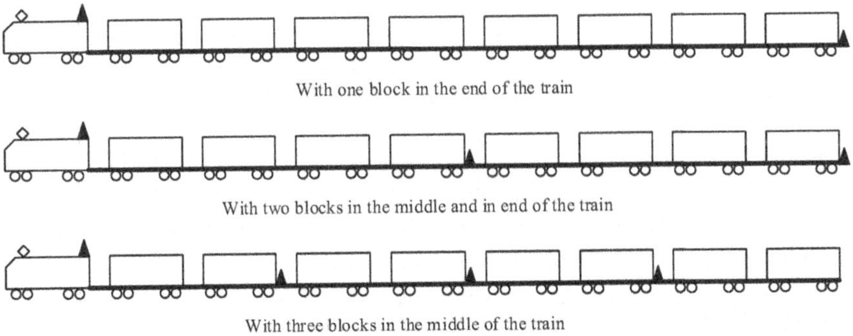

Fig. 1. Versions of braking blocks arrangement along train length

Tests of the system have shown that when braking only from the head of the train (from the locomotive and without braking blocks), the time of pressure reduction per stage of 0.08 MPa along the length of the braking line is 6.6 s.

When braking from the head of the train (from the locomotive) and the presence of a braking blocks, added at the end of the train, the time of pressure reduction at the stage of 0.08 MPa – under the brake main line is 3.7 s.

When braking from the head of the train (from the locomotive) and there are two braking blocks, which are located at the end of the train and its middle, the time of pressure reduction at the stage of 0.08 MPa along the length of the braking line is 2 s.

If more brake units are used along the length of the train, the pressure reduction time per stage of 0.8 MPa along the length of the brake line will be reduced accordingly.

However, the application of the distributed braking control system does not increase the speed of propagation of the braking wave, which preserves the longitudinal dynamic forces during braking, which are not evenly distributed along the length of the composition. As the experience of this system shows, the reliability of its operation was not high enough, and due to failures in transmission of control signals over the radio channel.

For increase in speed of distribution of a brake wave, reduction of time of preparation of brakes of the train and a brake way in modern brake systems of automatic pneumatic brakes of cars braking accelerators are used. The accelerator of braking is located in the air distributor and provides when braking an additional discharge of the brake highway of the car [6].

However the accelerator makes a discharge of the brake highway on the air distributor remote from the main pipeline by means of the bringing tube. The cross-sectional area of the bringing tube is twice less than the cross-sectional area of the main pipe. In result operation of the accelerator on the air distributor increases the speed of distribution of a brake wave slightly. The best modern systems of automatic pneumatic brakes of cars provide the speed of distribution of brake wave no more than 250 … 270 m/s [3].

At such speed of distribution of a brake wave from the car to the car, when braking, in the train significant longitudinally dynamic forces are created. Thus, the existing brake system of an automatic pneumatic brake of the car limits possibilities of driving of trains of big length and weight with big speeds [7].

The brake pneumatic wave extends from the train head on the brake highway of the considered car, then on connecting sleeves of the brake highway between cars and further on the brake highway of the following car. Thus, the brake pneumatic wave consistently extends on the brake highway of all train [8].

The accelerator of braking of an automatic pneumatic brake which is intended for operation only at emergency brake application [10] is known.

The accelerator of emergency brake application of an automatic pneumatic brake is the accelerator of a discharge of the brake highway which contains building 1, a mobile partition 2 sprung output valve 3 (Fig. 2).

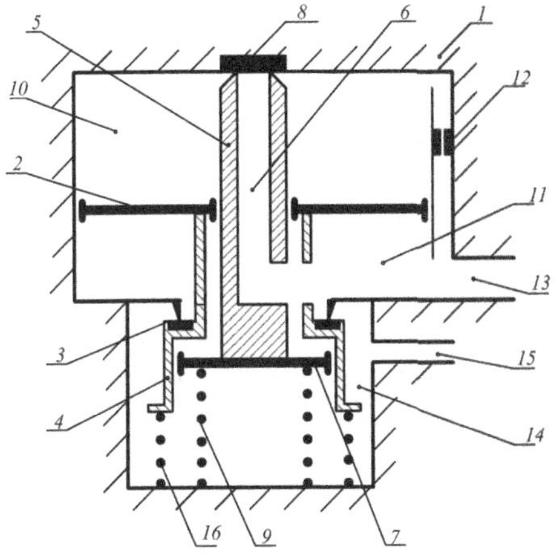

Fig. 2. Scheme of the accelerator of emergency brake application: 1 – the case of the accelerator, 2 – a mobile partition, 3 – output the valve, 4 – the case of the output valve, 5 – the hollow rod, 6 – the axial channel, 7 – the piston, 8 – the valve with a sealing element, 9 – an adjustable spring, 10 – the accelerating camera, 11 – the main camera, 12 – a throttle opening, 13 – the brake highway, 14 – an atmospheric cavity, 15 – the atmospheric channel, 16 – the spring of the output valve

The case of accelerator 1 is fixed on the brake highway of an automatic pneumatic brake of the car.

The mobile partition 2 intended for movement of output valve 3 is located in the case of the accelerator and forms accelerating camera 10 with its walls.

The saddle of the output valve with a mobile partition forms the main camera 11 in the case of the accelerator. Output valve 3 divides the main camera 11 and an atmospheric cavity 14 reported with the atmosphere channel 15. The valve is intended by output for production of air from brake highway 13 in the atmosphere, sprung by a spring 16. In building 4 of output valve 3 piston 7 of the blocking body which is rigidly connected with the hollow rod 5 mounted with a possibility of movement along a mobile partition 2 is installed. The adjustable spring 9 blocking body is also placed in building 4 of output valve 3 which lean in piston 7. Sealing element 8 of the valve of the blocking body is placed in the case of 1 accelerator with a possibility of overlapping of axial channel 6 of a hollow rod 5.

The main camera 11 pneumatically is connected with brake highway 13. Accelerating 10 and main the 11th cameras pneumatically are connected among themselves through a throttle opening 12.

At office braking air pressure in the brake highway goes down the rate of an office discharge. In accelerators of braking of an automatic pneumatic brake air pressure in the main camera also goes down the rate of an office discharge. At the rate of an office discharge air through the calibrated opening manages to flow from the accelerating camera in the main camera, keeping in them equal pressure of air. The mobile partition of the accelerator remains motionless and output the valve is in a closed position pressed to the saddle. The accelerator of braking of an automatic pneumatic brake does not make an additional discharge of the brake air route, keeping invariable the speed of distribution of a brake wave and time of its distribution on the brake highway of the train.

At emergency brake application air pressure in the brake highway and the main camera of the accelerator of braking goes down the rate of the emergency discharge which approximately by 3 times exceeds the rate of an office discharge. At the same time the accelerator begins to work at once as soon as air pressure in the brake highway begins to decrease the rate of the emergency discharge without expenses of time for preparation it to work. At the rate of the emergency discharge of the main camera air through the calibrated opening does not manage to flow from the accelerating camera in the main camera, keeping in the accelerating camera pressure of air is higher, than in the main camera.

At the expense of the difference of pressure in the main camera and the accelerating camera the piston of the accelerator is displaced towards the main camera and opens output the valve. Output valve reports the main camera and the brake highway with the atmosphere, providing additional fast pressure decrease of air in them up to the atmospheric pressure. Opening of the output valve leads to increase in rate of pressure decrease of air in the main camera of the accelerator of braking and the brake highway and to acceleration of a brake wave.

Pressure decrease of air in the brake highway leads to operation of a brake on braking on this car. The brake pneumatic wave extends to a train tail on the brake highway of the considered car, then on the brake highway between cars and further on the brake highway of the following car to its accelerator of braking which comes into action. Thus, the brake pneumatic wave consistently extends on the brake highway of all train.

3 Results

We solved the problem consisting in development of the accelerator of braking of the automatic pneumatic brake providing decrease in time for preparation of brakes at the expense of a discharge of the brake highway of each car along with two of its ends. For the solution of an objective on the car two interconnected accelerators of a discharge of the brake highway located on the opposite ends of the car forming the uniform accelerator of braking of an automatic pneumatic brake are installed.

At installation electrically of the connected two accelerators of a discharge of the brake highway on the opposite ends of n-go of the car along with the main brake wave on one end of n-go of the car on the opposite end of the car the additional brake wave extending on the brake highway as car n-go and on the brake highway n + 1 car is created.

Creation of an additional brake wave on n-ohm the car leads to turning on of the brake system n + 1 car through time of distribution of a pneumatic brake wave on connecting between car sleeves. Time of distribution of a pneumatic brake wave on the brake highway between these cars is insignificant.

Almost simultaneous leads turning on of the brake system on all cars in process of braking to reduction of time for preparation of brakes, and, as a result, to reduction of a brake way.

In Fig. 3 the scheme of the accelerator of braking of an automatic pneumatic brake is submitted [11].

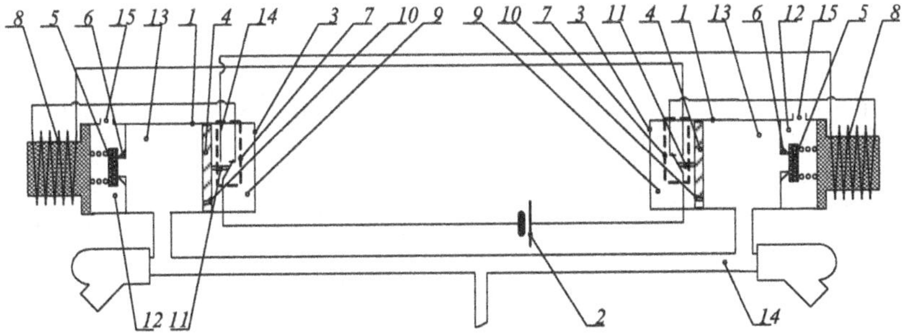

Fig. 3. Scheme of the accelerator of braking of an automatic pneumatic brake: 1 – the device of a discharge of the brake highway, 2 – the DC power source, 3 – the brake highway of the car

The accelerator of braking of an automatic pneumatic brake contains two same devices of a discharge of the brake highway and a source of direct electric current. The accelerator of braking of an automatic pneumatic brake is installed on each car of the vehicle.

And the first and second device of a discharge of the brake highway of each accelerator of braking are located on the opposite ends of the car, and the source of direct electric current can be located in any its place.

Each device of a discharge of the brake highway (Fig. 4) is building 1 in which are located piston 2, output valve 3, a saddle of output valve 4 and on-off electric switch 6. Outside of the device of a discharge of the brake highway electromagnetic coil 7 is located.

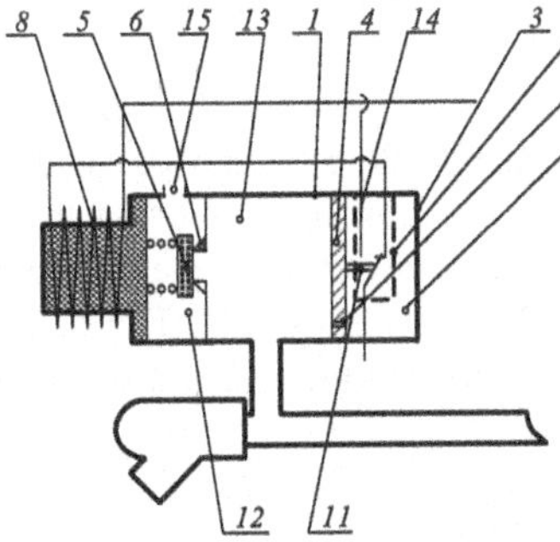

Fig. 4. Device of a discharge of the brake highway: 1 – case; 2 – piston; 3 – output valve; 4 – saddle of the output valve; 5 – a spring of the output valve, 6 – the on-off electric switch; 7 – electromagnetic coil; 8 – accelerating camera; 9 – the calibrated opening in the piston; 10 – piston rod; 11 – atmospheric camera; 12 – the main camera, 13 – the atmospheric channel

Piston 2 with the calibrated opening is intended for short circuit or disconnection of contacts of electric on-off switch 6 and forms accelerating camera 8 with case walls.

The on-off switch has one closing and one disconnecting contact and is placed in the accelerating camera of the case with arrangement of the closing contact near the piston. A mobile part of contact is connected with a rod of the 10th piston.

The saddle of output valve 4 is rigidly connected with walls of the case and forms with them atmospheric camera 11.

At the same time piston 2 and a saddle of output valve 4 with walls of the case form the main camera 12.

Output valve 3 is intended for production of air from the brake highway in the atmosphere. It is made of ferromagnetic metal, located in the atmospheric camera and sprung to a saddle of the output valve.

Electromagnetic coil 7 is intended for opening - closing of the output valve and rigidly fixed coaxially with it on the case.

The main camera pneumatically is connected with the brake highway of the train and through the calibrated opening 9 in the piston is reported with the accelerating camera.

The atmospheric camera is connected with the atmosphere channel 13.

In each accelerator of braking of an automatic pneumatic brake both of its devices of a discharge of the brake highway and a source of direct electric current are connected among themselves electrically.

At the same time a mobile part of the electric switch of each device of a discharge of the brake air route electrically is connected with a braking accelerator DC power source.

One end of a winding of the electromagnetic coil of the first device of a discharge of the brake air route is connected to a motionless part of the disconnecting contact of the electric switch of this device, and other its end – to a motionless part of the closing contact of the electric switch of the second device of the accelerator of braking. The winding of the electromagnetic coil of the second device of a discharge of the brake highway is connected similar to connection of a winding of the electromagnetic coil of the first device of a discharge.

Operation of the accelerator of braking of an automatic pneumatic brake is shown on the example of operation of the accelerator of the braking established on the car, the first from the locomotive.

In an initial condition of a brake in the absence of air pressure in the brake highway in each accelerator of braking pistons of each device of a discharge of the brake highway are in extreme situation. The closing contact of the electric switch of each device of a discharge of the brake highway at the same time is in the opened state and the electrical circuit electromagnetic coils is opened. The output valve is pressed to a saddle by a spring, blocking pneumatic communication of the brake highway with the atmosphere.

Control of a brake of all train is exercised of the driver of the locomotive by change of pressure of air in the brake highway.

At brake charging air pressure in the brake highway increases that leads to increase in pressure of air in the main camera and through the calibrated opening in the accelerating camera of each device of a discharge of the brake highway of the accelerator of braking. At the same time pistons of each device of a discharge remain in extreme situation, and the electrical circuit electromagnetic coils remains opened. Output the valve keeps the situation, leaving to the brake highway blocked pneumatic communication with the atmosphere.

At office braking at the beginning of the car from the locomotive pressure decrease of air in the brake highway begins the rate of an office discharge, creating the main pneumatic brake wave which extends on the brake highway by the end of the car and further to the following car.

At achievement of a brake wave of the highway of the first car the accelerator of braking of an automatic pneumatic brake of this car works.

At the same time in the main camera of the first device of a discharge of the brake highway pressure of air goes down the rate of an office discharge. It leads to emergence of a difference of pressure of air in main and in accelerating cameras. Air through the calibrated opening slowly flows from the accelerating camera in the main camera, reducing air pressure in it speed smaller, than the rate of an office discharge. In the accelerating camera pressure of air remains bigger, than air pressure in the main camera. As a result the piston is displaced towards the main camera, its rod closes the closing contact of the switch of the first device of a discharge of the brake highway. The electrical circuit which is turning on the power supply closing contact of the first device of a discharge of the brake highway, the electromagnetic coil of the second device of a discharge and disconnecting contact of the second device of a discharge is as a result created.

At the same time the electromagnetic coil of the first device of a discharge of the brake highway remains is deenergized.

Under the influence of the electromagnetic field created by electromagnetic coil 7 of the second device of a discharge of the brake highway valve 3 of the mentioned device opens output, and its main camera 12 through channel 13 pneumatically connects to the atmosphere. It leads to pressure decrease of air in the main camera 12 and as the investigation to pressure decrease in the brake highway on the end of the car from a train tail the rate of an office discharge, and also to creation of an additional pneumatic brake wave.

4 Discussion

If including time of distribution of electric current on car length equal to zero, then the discharge of the brake highway on the car occurs along with two ends of the car: at the beginning from the locomotive and at the end from a train tail. On one end of the car from the locomotive the discharge of the brake highway is caused by arrival of the main brake wave. On the opposite end of the car the discharge of the brake highway is caused by operation of the accelerator of braking of an automatic pneumatic brake, in particular – the second device of a discharge of the brake highway of the first car.

Additional and main pneumatic brake waves, extending in the brake highway of the car, at a meeting create in it the lowered air pressure that t/2 leads to operation on this car of a brake on braking through time equal to time of distribution of a pneumatic wave on the brake highway from the end of the car to its middle.

Extending in the direction of the second car, the additional brake wave passes across the brake highway between cars during t1, and leads the accelerator of braking of an automatic pneumatic brake of the second car which works similar to operation of the accelerator of braking of the first car to work.

After operation of a brake of the first car and the accelerator of braking of an automatic pneumatic brake of the second car the accelerator of braking of the first car is switched-off. At the same time in the second device of a discharge of the first car the piston is displaced towards the main camera and the rod disconnects an electrical circuit of power supply of the electromagnetic coil of the mentioned discharge device. The electromagnetic field of the electromagnetic coil on output valve is terminated, and it returns to the place, on a saddle of the output valve, separating the main camera and the brake highway with the atmosphere.

The accelerator of braking of an automatic pneumatic brake of the first car is switched off from work.

The automatic pneumatic brake of the second car works on braking through time equal to time of distribution of a pneumatic wave on the brake highway between the first and second t1 cars, and time of its advance to the middle of the mentioned t/2 car.

Thus, brake response time on the second T2 car from the moment of the beginning of braking by the driver of the locomotive is determined by a formula:

$$T2 = 2t1 + t/2 \tag{1}$$

where t1 is time of distribution of a pneumatic brake wave on the brake highway between cars or between the car and the locomotive;

t/2 is time of distribution of a pneumatic brake wave on the brake highway of the car from its beginning to the middle.

Time before operation of a brake of Tn on the last car at n-s the number of cars is equal in the train:

$$Tn = nt1 + t/2. \tag{2}$$

At emergency brake application the accelerator of braking of an automatic pneumatic brake works just as at office braking.

Use of the developed accelerator of braking of an automatic pneumatic brake will allow to reduce a brake way of the train due to reduction of time for preparation of brakes.

5 Conclusions

Thus, use of the offered accelerator of braking of an automatic pneumatic brake by freight cars allows to reduce a brake way of the train and to reduce longitudinally dynamic forces in the train when braking. As a result the brake of freight cars on characteristics will practically be equal to an electric air brake.

The braking equipment presented in the article will make it possible to create an effective brain system for the freight train, free of disadvantages of systems, both electro-pneumatic brake and braking systems based on transmission of signals over the radio channel.

References

1. Rules of maintenance of the brake equipment and management of brakes of the rolling stock, 224 p. LLC Tekhinform, Moscow (2014)
2. Asadchenko, V.R.: Calculation of pneumatic brakes of railway mobile structure. Marshrut, Moscow (2004)
3. Rules of traction calculations for train work. Are approved by the order of JSC "Russian Railways" No. 867 of 12.05.2016, Moscow, 514 p (2016)
4. Afonin, G.S., Barshchenkov, V.N., Kondratyev, N.V.: Automatic brakes of the rolling stock. Academy, Moscow (2010)
5. Asadchenko, V.R.: Automatic brakes of the rolling stock. Marshrut, Moscow (2006)
6. Galay, E.I.: Increase in efficiency of electric air brakes of the train. BSUT, Gomel (2002)
7. Kravchuk, V.V., Verkhoturov, V.K., Nikulin, Yu.V.: Management of traffic safety of trains: monograph. DVGUPS, Khabarovsk (2010)
8. Krylov, V.I., Efremov, V.N., Demushkin, P.T.: Brake equipment of railway rolling stock. Transport, Moscow (1989)
9. Yegorenkov, N.A., Athaniev, S.I., Chuev, S.G., Populovsky, S.A.: System for distributed control of brakes of high-weight and long trains. Patent utility model description 144186 of the Russian Federation, MPK B60T17/22
10. Kozyulin, L.V., Egorenkov, A.A., Inozemtsev, V.G., Smelov, V.N., Krylov, V.V.: Accelerator of emergency brake application of a brake of the railway vehicle. Patent 2141417 of the Russian Federation, MPK B60T15/30
11. Drozdov, E.A.: Accelerator of braking of an automatic pneumatic brake. Patent 2623790 of the Russian Federation, MPK B60T15/30

Methodology for Assessing the Condition of the Contact Wire by the Value of Thermal Softening

Sergey Klimenko(✉) and Valeriy Li

Far Eastern State Transport University,
Seryshev Street, 47, Khabarovsk 680021, Russia
sergu_n@mail.ru

Abstract. At this time the problem of monitoring the state of the contact wire is relevant. The condition of the contact wire is estimated by the values of mechanical and electrical wear, while the thermal wear of the contact wire is not determined. The paper considers the methodology for assessing the condition of the contact wire of electrified railways by the value of thermal softening. The presented methodology is based on the criterial approach. The criteria for softening the material of the wire and the transition of the material to the liquid state are introduced. By comparing the amount of heat entering the wire in the event of an electric arc, with these criteria, the contact wire is diagnosed in terms of thermal wear. Calculations have been made using the proposed methodology which makes an assessment of a condition of a contact wire on thermal softening depending on the initial data received in the operating of source data. When the calculation results are combined into databases, it is possible to estimate the residual life of the contact wire.

Keywords: Contact wire · Thermal softening · Electric arc

1 Introduction

Today, railway electric traction is the most promising way to increase the carrying capacity of railway sections [1, 2]. In this case, the system for current pick-up is the main method to transmit power energy to electric rolling stock; and contact wires play a crucial role in it. Various damages to contact wires may result in railway-related incidents up to traffic stop. The main task today is to maintain the operating condition of contact wires, conduct diagnostics throughout the railway electrified sections and forecast their service life.

During operation, a contact wire is impacted by various climatic factors, thermal processes occurred during the traction current flow in electric network components, mechanical impact by current collectors, electric corrosive processes, and tension forces. Thus, a contact wire has to operate reliably while experiencing enormous loads [3].

The interaction process between a current collector and a contact wire is not continuous [4, 5]. Sparks and electric arcs are formed in the gap between them. As a result of this interaction, the wire is heated. Researches of Far Eastern State Transport

Z. Popovic et al. (Eds.): TransSiberia 2019, AISC 1115, pp. 138–145, 2020.
https://doi.org/10.1007/978-3-030-37916-2_15

University (FESTU) have found out that the wire material may heat up to recrystallization temperature due to the impact of an electric arc with sufficient power and impact time. Thermal strength degradation (softening) of a contact wire due to the electric arc is called thermal wear [6]. Currently, thermal wear of a contact wire is not measured; and its condition is estimated by mechanical and electrical wear. However, contact wires often break without any visible signs of damage, which may indicate their thermal softening, when the wire material structure changes and its mechanical characteristics deteriorate.

Ultimate tensile strength and hardness are the most reliable characteristics when defining the contact wire wear extent. FESTU researchers found out that the initial step of the contact wire softening occurs at 170 °C to 220 °C. At the same time, material hardness and ultimate tensile strength drop. All transformations accumulate in the wire internal structure and negative after effects take place within a short time. When a current collector passes this site next time, repeated heating will occur that will gradually lead to further softening.

Currently, the problem of the monitoring over the wire condition remains relevant. Non-destructive testing of the contact wire throughout its whole length requires heavy expenses, so it is proposed to firstly find areas, where local softening sites are more likely to appear [7, 8].

Figure 1 illustrates the wear behavior of contact wire.

Thus, implementation of the term "thermal wear" requires solution of several tasks:

1. Searching the ways to assess the extent of the contact wire thermal softening;
2. Creation of the hardware and software suite to calculate the wire softening extent without its withdrawal from operation;
3. Working out recommendations to maintenance staff regarding the further operation of the wire subject to thermal wear.

2 Materials and Methods

To calculate the contact wire softening extent, it is necessary to know electric arc current, its travel rate, and impact time. It is also important to register the arcing point [9].

To simplify the understanding of the wire heating process, let's assume that the whole exposed material is heated (Fig. 2) to the melting point. Then, energy is spent to transform the whole wire mass from solid to liquid phase. Then, the liquid phase is heated to boiling point. The remaining energy is spent to evaporate liquid metal. During phase transitions, energy is spent to transform the material from one phase to another but not for further heating of individual fractions.

Let's take the following main assumptions when considering the solution of the task at hand:

1. The heating area shape is considered as hemisphere;
2. The arc existing time is less than the temperature release time; and heat transfer into the environment may be neglected (all the energy released by the electric arc is absorbed by the wire material and is spent for its heating);

3. Thermophysical properties of the material change a little depending on the temperature; their average values are taken as estimated (T = 20 °C).

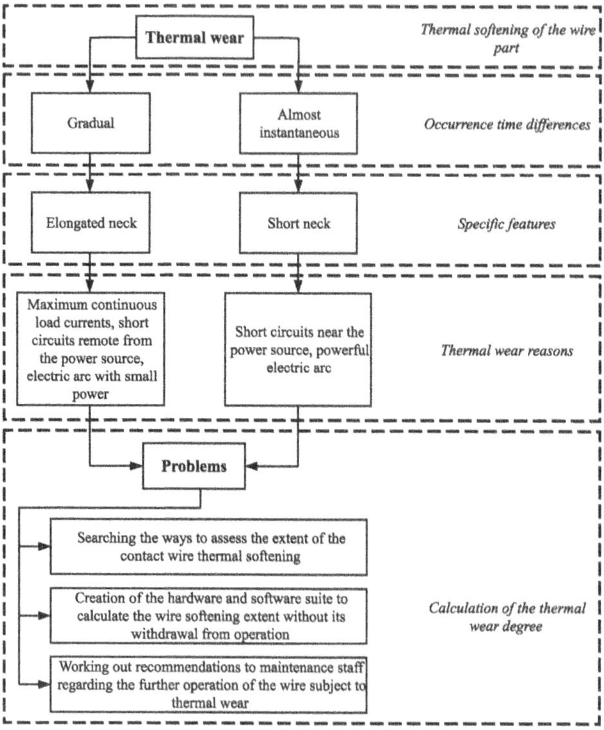

Fig. 1. Thermal wear behavior for a contact wire

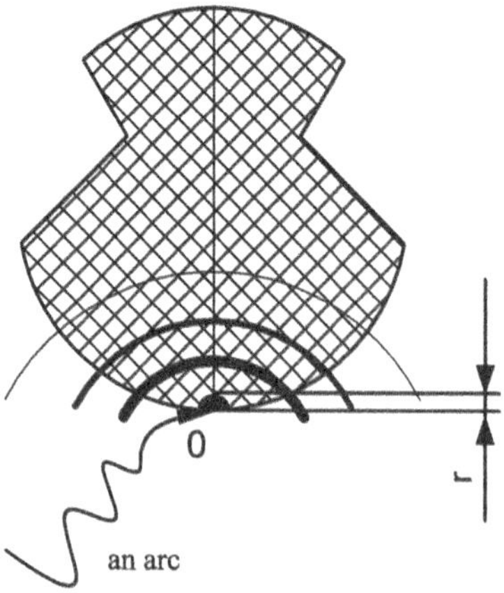

Fig. 2. Cross section of the wire impacted by electric arc

The criterion approach was taken as a basis of the assessment procedure for the contact wire condition by the thermal softening value [9, 10]. This approach allows comparing the value of the electric arc impact on the contact wire with the proposed criteria. Criteria permit to assess the extent of the damage caused by an electric arc to contact wire locally. Criteria of the wire material softening (Q_{CRIT1}) and material transfer into a liquid state (Q_{CRIT2}) were introduced.

$$Q_{CRIT\,1} = m \cdot c \cdot (180 - T_0) \tag{1}$$

where m – is the contact wire material weight, g;

c – is the representative value of specific heat capacity, cal/g × degree;
T_0 – is the initial wire temperature, °C.

Softening criterion indicates what heat amount is necessary to heat the contact wire until its softening at a certain scale.

This criterion is directly proportional to the heated material weight, its specific heat capacity and the heating initial temperature. At the same time, the final temperature value is set to 180 °C. With this temperature, wire mechanical characteristics drop. Once lower temperatures have been achieved, the wire material can restore its properties.

It is also proposed to decrease the contact wire tension during its thermal wear due to the loss of mechanical properties in a certain part of the wire cross section. If tension is not decreased, a neck may form in the softening site due to excessive tension loads that may eventually result in the wire break [10].

$$Q_{CRIT\,2} = m \cdot c \cdot (T_{MEL} - T_0) + m \cdot \lambda \tag{2}$$

where λ – is the specific melting heat, cal/g;

T_{MEL} – is the melting temperature, degree.

The criterion of the material transfer into liquid state illustrates what heat amount is necessary to melt the contact wire at a certain scale. The heat amount less than this criterion means wire softening without its melting. Criterion value is defined as a sum of two thermal processes. At first, the wire material is heated to its boiling point; then, the amount of heat necessary for melting is determined.

The proposed dividing allows to distinguish the processes occurred in the wire material at a local scale; so that it is be possible to obtain more detailed information about the wire damage.

The criteria introduced are directly proportional to the weight of the wire material heated during arcing. Contact wires are made of copper and bronze; so, it is possible to calculate their weight using the density of these materials. The motion speed of the electric locomotive current collector also plays an important role: the higher the speed, the bigger wire section is subject to arcing.

To define the amount of heat entering the wire material, it is necessary to define the current collected during arcing as well as arc duration.

$$Q_{ENT} = q \cdot t_{\text{heat}} \tag{3}$$

where q – is the arc heat power absorbed during contact, W;

t_{heat} – is the arc impact time, s.

Power q absorbed during contact depends proportionally upon the electric locomotive during arcing.

A visual flowchart of the procedure proposed is shown in Fig. 3.

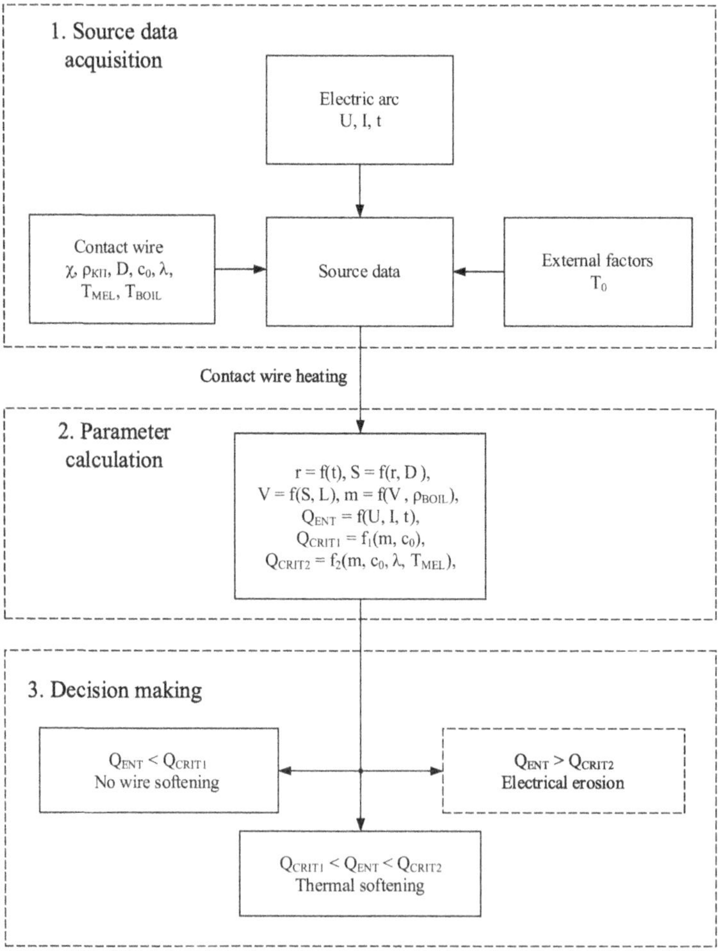

Fig. 3. Algorithm for calculating the softening extent of a contact wire due to the electric arc impact

3 Results

To make calculations by the assessment procedure for the contact wire condition using the thermal softening value, it is necessary to take into account a lot of factors including wire metal characteristics, environmental conditions and electric arc parameters. Since it is necessary to process large data arrays, the need appears to use application software to calculate the contact wire softening. Microsoft Excel was used for calculations. Figure 4 illustrates a calculation example carried out by this procedure.

Source data										
Electric arc			Contact wire							External factor
Voltage U, V	Current strength I, A	Burning time t, s	Diffusivity χ	Material density ρ	Specific heat c	Specific heat of fusion λ	The melting point of the material T_{MEL}, °C	The boiling point of the material T_{BOIL}, °C	The typical size of the section of the wire D, cm	Ambient temperature T0, °C
70	300	0.01	1.14	8.94	0.092	48.94	1083	2560	1	20
Calculation of the degree of softening of the contact wire										
radius, cm										
0.01	0.02	0.03	0.04	0.05	0.06	0.07	0.08	0.09	0.1	0.11
cos										
0.01	0.02	0.03	0.04	0.05	0.06	0.07	0.08	0.09	0.1	0.11
corner φ										
1.561	1.551	1.541	1.531	1.521	1.511	1.501	1.491	1.481	1.471	1.461
89.439	88.8658	88.293	87.720	87.147	86.574	86.001	85.428	84.855	84.282	83.709
Radial cross-sectional area of the heating area										
0.000156	0.000622	0.001391	0.002460	0.003823	0.005475	0.007412	0.009627	0.012117	0.014875	0.017898
Volume of material										
0.0005	0.0021	0.0046	0.0082	0.0127	0.0183	0.0247	0.0321	0.0404	0.0496	0.0597
Mass of material										
0.0047	0.0185	0.0415	0.0733	0.1139	0.1632	0.2209	0.2869	0.3611	0.4433	0.5334
Arc power										
525.0000	525.0000	525.0000	525.0000	525.0000	525.0000	525.0000	525.0000	525.0000	525.0000	525.0000
Amount of heat entering the wire										
5.2500	5.2500	5.2500	5.2500	5.2500	5.2500	5.2500	5.2500	5.2500	5.2500	5.2500
1.2540	1.2540	1.2540	1.2540	1.2540	1.2540	1.2540	1.2540	1.2540	1.2540	1.2540
Softening criterion										
0.0685	0.2727	0.6103	1.0791	1.6770	2.4017	3.2511	4.2229	5.3150	6.5251	7.8511
Melting criterion										
0.6833	2.7185	6.0838	10.7574	16.7173	23.9416	32.4086	42.0962	52.9826	65.0458	78.2639
The result of the influence of the arc										
M	S	S	S	N	N	N	N	N	N	N

Fig. 4. Sample calculating for the procedure of the contact wire condition assessment by the thermal softening value

The possibility to get different dependencies due to the source data change appears at all calculation steps. As an example, Fig. 5 plots the dependency of the heat amount entering the considered point, which is required to soft or to melt the wire material, upon the arc motion speed.

It can be seen that at a high speed, the threshold value increases for material softening and its transition into liquid state

For the sake of convenience, source data and calculation results are displayed in a separate software window; and a relation is established with the basic calculation (Fig. 6).

The result output window contains options of the contact wire condition after arcing. Depending on the calculation results, a corresponding option is selected with color and is marked with a "+" sign. At the same time, the maximum possible damaged area is displayed for the more detailed assessment, and the damaged section is indicated.

Figure 6 shows the calculation procedure example. Thus, for the electric arc having current 300 A and moving with a speed of 60 km/h, heat amount Q_{ENT} = 1.254 cal enters the section of the 100 mm^2 contact wire within arcing time t = 0.01, which means softening of the part of material and its melting within a radius of 0.01 cm.

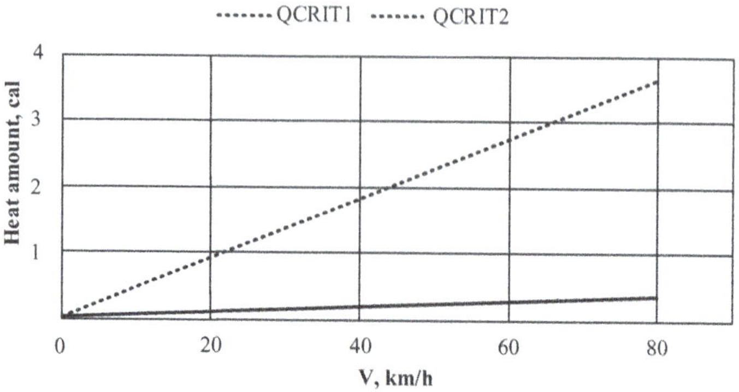

Fig. 5. Criteria Q_{CRIT1} and Q_{CRIT2} vs. the arc motion speed

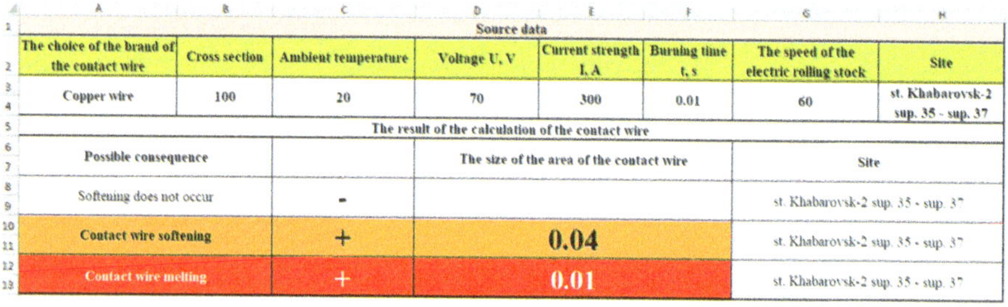

Source data							
The choice of the brand of the contact wire	Cross section	Ambient temperature	Voltage U, V	Current strength I, A	Burning time t, s	The speed of the electric rolling stock	Site
Copper wire	100	20	70	300	0.01	60	st. Khabarovsk-2 sup. 35 - sup. 37
The result of the calculation of the contact wire							
Possible consequence		The size of the area of the contact wire				Site	
Softening does not occur	-					st. Khabarovsk-2 sup. 35 - sup. 37	
Contact wire softening	+	0.04				st. Khabarovsk-2 sup. 35 - sup. 37	
Contact wire melting	+	0.01				st. Khabarovsk-2 sup. 35 - sup. 37	

Fig. 6. Source data input and result output windows

Maximum possible softening zone is 0.04 cm under these conditions. If the arc motion speed decreases to 30 km/h, maximum radius of the area being softened increases to 0.06 cm.

4 Conclusions

Today, monitoring over the contact wire condition is an important task in railway power engineering [11]. The study revealed that a contact wire undergoes thermal wear due to the electric arc impacts. It is required to detect the wire thermal wear during its operation for further diagnostics of its condition and defining its residual service life. A procedure has been developed to estimate thermal wear, which is based on the criteria allowing assessment of the wear effect on a contact wire. A calculation example is given for the procedure that evaluates the contact wire condition by its thermal softening depending on the source data acquired during operation. When collecting calculation results into databases, it becomes possible to estimate the residual service life of contact wires.

References

1. Li, V.N., Shurova, N.K.: On the method of determining optimal electricity consumption from an electric traction network. Russ. Electr. Eng. **89**(11), 679 (2018). https://doi.org/10.3103/s1068371218110147
2. Shurova, N.K., Li, V.N.: Increase of energy efficiency of electric power distribution networks by adjustment of reactive energy consumption in a traction network. In: MATEC Web of Conferences, vol. 239, p. 01049 (2018). https://doi.org/10.1051/matecconf/201823901049
3. Ignatenko, I.V., Vlasenko, S.A.: Determination of criteria to find the quality of an electrical connection. Russ. Electr. Eng. **87**(2), 65–67. https://doi.org/10.3103/s1068371216020085
4. Bucca, G., Collina, A., Tanzi, E.: Experimental analysis of the influence of the electrical arc on the wear rate of contact strip and contact wire in a.c. system. In: Advances in Italian Mechanism Science, pp. 449–456 (2016). https://doi.org/10.1007/978-3-319-48375-7_48
5. Kim, Y., et al.: Fatigue life prediction method for contact wire using maximum local stress. J. Mech. Sci. Technol. **29**(1), 67–70 (2015). https://doi.org/10.1007/s12206-014-1210-3
6. Li, V.N., et al.: Mechanisms of weakening and destruction of aerial contact wires. Russ. J. Nondestr. Test. **39**(12), 913–918 (2003). https://doi.org/10.1023/b:runt.0000026012.71709.cf
7. Styskala, V., et al.: The operation principles and implementation of automated system for arc disturbances control for AC and DC electrified railway lines. In: 2015 16th International Scientific Conference on Electric Power Engineering (EPE). https://doi.org/10.1109/epe.2015.7161060
8. Zharkov, Y.I., et al.: The principals of operation of the automated current collection diagnostic system for electrified railways. In: 2015 16th International Scientific Conference on Electric Power Engineering (EPE). https://doi.org/10.1109/epe.2015.7161202
9. Li, V.N., Klimenko, S.V.: Determination of thermal softening of overhead wire exposed to an electric arc effect. Russ. Electr. Eng. **87**(2), 77–79 (2016). https://doi.org/10.3103/s1068371216020115
10. Klimenko, S.V., Li, V.N.: Possibilities of diagnosing the condition of contact wire in terms of thermal wear. In: MATEC Web of Conferences, vol. 239, p. 01050 (2018). https://doi.org/10.1051/matecconf/201823901050
11. Tryapkin, E.Y., Keino, M.Y., Protasov, F.A.: Synchronous phase measurements in the automated monitoring system of railway power supply facilities. Russ. Electr. Eng. **87**(2), 110–112 (2016). https://doi.org/10.3103/s1068371216020176

Principles of Creation of an Optimal Train Motion Trajectory

Vladimir Anisimov(✉)

Far Eastern State Transport University, Serysheva Street 47, Khabarovsk 680021, Russia
avv@avis.khv.ru

Abstract. The main criteria and types of problems of constructing the optimal trajectory of the train are presented. The reasons for the emergence of time reserves for passing track sections by train related to the peculiarities of developing a train schedule are given. The energy efficiency of train speed reducing by decreasing the used running positions of the controller of the driver or switching to idle when there are time reserves in the train schedule is substantiated. The concept of "equivalent energy/fuel consumption" and a methodology for its determination are proposed. The proposed concepts allow to evaluate the reduction in energy consumption when passing the same section with different speed and time. The basic techniques of driving a train, due to which an effective speed reduction is achieved, are considered. The restrictions and particularities of driving trains, which should be taken into account when implementing the methodology and algorithm for determining the optimal trajectory of a train, are given.

Keywords: Traction calculations · Train schedule · Performance charts · Equivalent fuel/electricity consumption · Rational train driving modes · Optimal train motion trajectory

1 Introduction

Traditionally there are the following criteria for optimizing the train movement [1–6]:

- minimum train travel time;
- minimum fuel or electricity costs;
- minimum operating costs for the train movement;
- minimum mechanical work.

In compliance with these criteria, the statement of the problem of determining the optimal motion trajectory of the train can be formulated as follows.

The First Task. Determine the optimal train motion trajectory by the criterion of the minimum train travel time due to the maximum use of traction effort with known parameters of the train (locomotives, wagons), its movement conditions (weather conditions, permissible speeds, stops at operation points, brake test areas, etc..), as well as the characteristics of the devices and structures of the area of calculations (plan, profile, neutral sections, etc.). Such a statement of the problem is regulated by the Rules

Z. Popovic et al. (Eds.): TransSiberia 2019, AISC 1115, pp. 146–154, 2020.
https://doi.org/10.1007/978-3-030-37916-2_16

of Traction Calculations [5] and is used in the development of train schedules, projects for new railway lines or reconstruction of existing ones.

The Second Task. Determine the optimal train motion trajectory according to the criterion of the minimum fuel consumption (electricity, operating costs for the train movement, mechanical work) at known travel times, train parameters, conditions of its movement, as well as characteristics of devices and structures of the area of calculations. It is used in the development of regime maps.

The Third Task. Determine the optimal travel time and the corresponding optimal train motion trajectory knowing the parameters of the train, the conditions of its movement, as well as the characteristics of the devices and structures of the area of calculations. It is used in the development of regime maps or the selection of optimal travel times built into train schedule.

The calculation of the train movement with the shortest possible travel time is basic in the study of the theory of traction calculations and, in one form or another, is implemented in each program of traction calculations. At the same time, the practical implementation of such a movement leads to maximum energy costs and, as a consequence, operating costs. The increase of the travel time and the use of rational train driving modes, in the vast majority of cases, can reduce costs and expenses.

When a real train moves in a plot, it usually has a time reserve in comparison with movement based on the maximum use of traction effort (locomotive power). The reasons for the emergence of this margin of time are associated with the features of the development of train schedule:

- station-to-station time calculations are performed for trains with a unified mass. In practice, a significant proportion of trains (often more than half) have a smaller mass;
- when performing the calculations, temporary warnings concerning the limitation of permissible speeds (repairs, construction works, single replacement of railroad ties, etc.), which may be absent at the time of the train passage, are taken into account;
- before being accepted to the train schedule, the calculated times are rounded up to a minute, usually upwards (for freight and passenger trains the decimal part for more than 0.2 min is rounded up upwards);
- when laying the threads of the schedule, additional time reserves arise due to the need to link the reception and departure of trains.

The opposite situation is also possible when the train cannot travel along the section with the time specified in the train schedule(for example, severe weather conditions, the appearance of unintended warnings about speed limits, the mass of the train exceeds the unified mass, etc.). But such situations are extremely rare.

Thus, when managing a train, the locomotive driver faces a second task - the rational use of the available time reserve, subject to the indisputable observance of the regulated (by schedule or developing train situation) travel times. To solve it, in theoretical and practical works of scientists, various optimization criteria were proposed.

The criterion of minimum energy resources for train traction is understandable for railway specialists, but the share of energy costs in operating costs for train movement is, averagely, 40–60%.

The criterion of minimum operating costs is comprehensive in terms of technical and economic assessment of train traffic. At the same time, its use leads to the solution of a multi-criteria problem, because the structure of operating costs includes components that depend on:

- the travel time (locomotive-hours, wagon-hours, operating hours of locomotive brigades, etc.);
- the distance traveled (locomotive-kilometers, wagon-kilometers and ton-kilometers);
- the train driving modes (mechanical operation of the locomotive, resistance and braking forces, fuel or electricity consumption).

2 Methods

Because of the fact that in the problem taken under consideration, the distance traveled and the travel time are fixed values, then the variable parameter that affects the value of operating costs is the train driving modes (traction, idle motion or braking) depending on the path. For movement in traction mode (electric braking), it is also necessary to determine the position of the locomotive driver's controller or the degree of use of traction effort (braking) for locomotives with smooth regulation of traction effort (allowing the realization of any value of effort up to limited in adhesion capacity, current, etc.).

The mechanical work of all the forces acting on the train is expressed by the formula:

$$R_m = \int (F(V) + W(V) + B(V))dS \tag{1}$$

here F – full traction effort, kN;
W – full resistance to movement, kN;
B – full breaking force, kN.

Let's consider the components of formula (1) in terms of optimizing train movement.

Braking Force. The usage of braking leads to a decrease in the kinetic energy of the train previously gained due to movement in traction mode (energy costs). The more intense is the accumulation of kinetic energy in traction mode, the more intense will be its loss in braking mode. Even in cases where electricity recuperation is possible on an electric locomotive, the amount of electricity returned to the network does not cover its costs for an equivalent increase in kinetic energy due to the use of locomotive traction.

Traction Effort. With the increase of speed, the cost efficiency of energy resources for the implementation of traction decreases. By energy efficiency, we mean the amount of traction per unit of current consumption (1 A) or fuel (1 kg). Figure 1 shows a graph of the cost-effectiveness of the current of a 2ES4K DC electric locomotive.

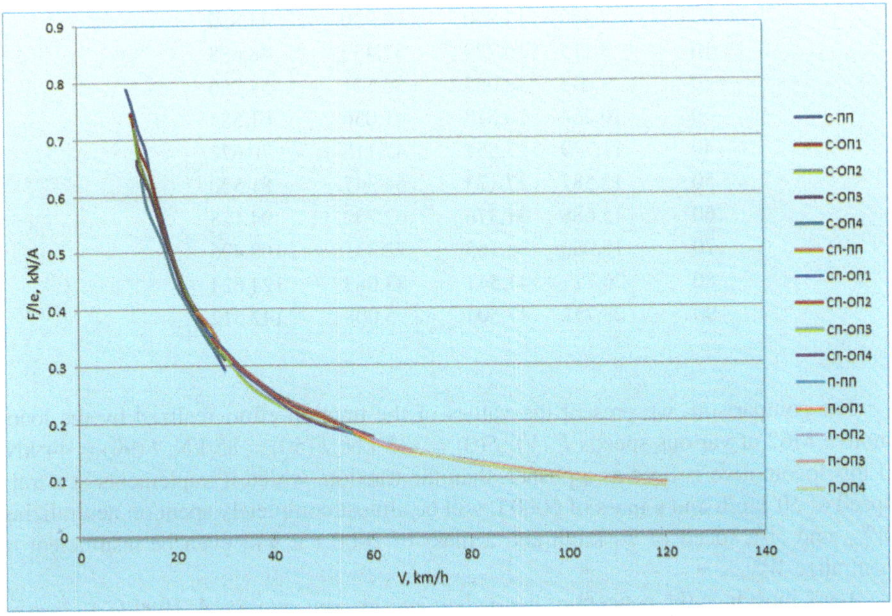

Fig. 1. The effectiveness of expenditures of current for 2ES4K electric locomotive

The nature of the dependence of effectiveness of expenditures indicates that it decreases with the increase of speed, and for the same speed values it practically does not depend on the position of the locomotive driver's controller. A similar situation is also characteristic of AC electric locomotives and diesel locomotives.

Resistance to Movement. One of the significant components of the resistance to the movement of the train W is the main resistance W_m, calculated, in the general case, by the formula:

$$W_m = \left(A + B \cdot V + C \cdot V^2\right) \cdot (P + Q) \tag{2}$$

where A, B and C – coefficients of the formula for the basic resistivity w_m;

P – estimated locomotive mass, t;
Q – train set mass, t.

The main resistance of the train set W''_m for four-axle open wagons with an axial load q_a = 15 t/axle on a continuous welded rails is shown in Table 1.

Table 1. The main resistance of the train set

V, km/h	w''_m, N/t	W''_m, kN		
		Q = 2000 t	Q = 4000 t	Q = 6000 t
0	7.480	14.960	29.920	44.880
10	8.115	16.229	32.459	48.688
20	9.043	18.085	36.171	54.256
30	10.264	20.528	41.056	61.584
40	11.779	23.557	47.115	70.672
50	13.587	27.173	54.347	81.520
60	15.688	31.376	62.752	94.128
70	18.083	36.165	72.331	108.496
80	20.771	41.541	83.083	124.624
90	23.752	47.504	95.008	142.512

For comparison, we present the values of the traction effort realized by the locomotive M62 at various speeds $F(V)$: $F(0)$ = 338 kN, $F(50)$ = 85 kN, $F(90)$ = 44 kN. If this locomotive is used as a pusher, then the traction, which it implements at a train speed of 50 km/h and a mass of 6000 t, will be almost completely spent on neutralizing W''_m, and at a speed of 90 km/h and a mass of 2000 t it will even be insufficient to neutralize W''_m.

Let's introduce the concept of *equivalent energy consumption* A_e (fuel G_e) - energy consumption at speed V_1 relative to speed V_2 for performing of the same "useful" work on train moving, excluding the expenses on overcoming the main resistance of the rolling stock and additional resistances (from curves, slopes, etc..). For example, when driving in traction mode at a constant speed of 60 km/h, the train will follow a section of 1 km in 1 min. In this case, some work $R_{mu}(60)$ will be accomplished. Then the equivalent energy (fuel) consumption at a speed of V relative to a speed of 60 km/h can be calculated by the formulas:

$$A_e(V) = A(V)\frac{R_{mu}(60)}{R_{mu}(V)} \quad (3)$$

$$G_e(V) = G(V)\frac{R_{mu}(60)}{R_{mu}(V)} \quad (4)$$

where $A(V)$, $G(V)$ – energy (fuel) consumption for the passage of a section of 1 km at a speed of V, kWh/km (kg/km);

$R_{mu}(V)$ – "useful" mechanical work on moving a train at a speed of V, kN km.

3 Results

Tables 2 and 3 show the results of calculating the equivalent energy consumption for a DC electric locomotive 2ES4K (Donchak) and a diesel locomotive 2TE25K (Peresvet).

Table 2. Equivalent power consumption 2ES4K electric locomotive, Q = 4000 t, four-axle open wagons with q_a = 15 t/axle

V, km/h	F, kN	W''_m, kN	I_e, A	Per 1 km			
				Δt, min	$R_{mu} = (F - W''_m)$ 1, kN km	$A = \frac{U \cdot I_e \cdot \Delta t}{60 \cdot 1000}$", kWh	A_e, kWh
50	464.9	54.3	2419	1.20	410.6	145	137
60	451.4	62.8	2658	1.00	388.6	133	133
70	413.0	72.3	2840	0.86	340.7	122	139
80	296.2	83.1	2320	0.75	213.1	87	159
90	222.5	95.0	1970	0.67	127.5	66	200

Table 3. Equivalent power consumption 2TE25K diesel locomotive, Q = 4000t, four-axle open wagons with q_a = 15 t/axle

V, km/h	F, kN	W''_m, kN	g, kg/min	Per 1 km			
				Δt, min	$R_{mu} = (F - W''_m) \cdot 1$, kN·km	$G = g \cdot \Delta t$, kg	G_e, kg
50	287.0	54.3	18.25	1.20	232.7	21.90	16.87
60	242.0	62.8	18.25	1.00	179.2	18.25	18.25
70	211.0	72.3	18.25	0.86	138.7	15.64	20.22
80	186.0	83.1	18.25	0.75	102.9	13.69	23.84
90	164.0	95.0	18.25	0.67	69.0	12.17	31.61

Calculations show that at a speed of 90 km/h it is required to spend 50–75% more energy for the same amount of "useful" work (but not for passing a Sect. 1 km long) than at a speed of 60 km/h. For locomotives of old construction, this parameter reaches 100%.

The calculations did not take into account the main resistance to the movement of locomotives, additional resistance from low temperature and wind, which also increase with increasing speed. This suggests that at low speeds, energy costs will be even more efficient.

Thus, minimization of the mechanical work of the locomotive $\int F(S)\,dS$ directly leads to a decrease in the cost of energy resources (operating costs) aimed at moving the trains, while minimization of the mechanical work of the resistance forces $\int W(S)\,dS$ and the braking forces $\int B(S)\,dS$ leads to a more rational use of kinetic energy accumulated during movement in traction mode, and, as a result, to reduction of energy costs (operating costs).

4 Discussion

The effectiveness of speed reduction (in the sense of influencing the value of the optimization criterion) largely depends on the conditions of train movement, the parameters of the plan and the profile of the area of calculations. Among the main methods of driving a train, due to which an effective speed reduction is achieved, three types of techniques can be distinguished (see Table 4).

Table 4. Effective speed reduction techniques

№	Characterization	Illustration
1	Increase in coasting (idling) during a speed reset before speed limitation or braking at station	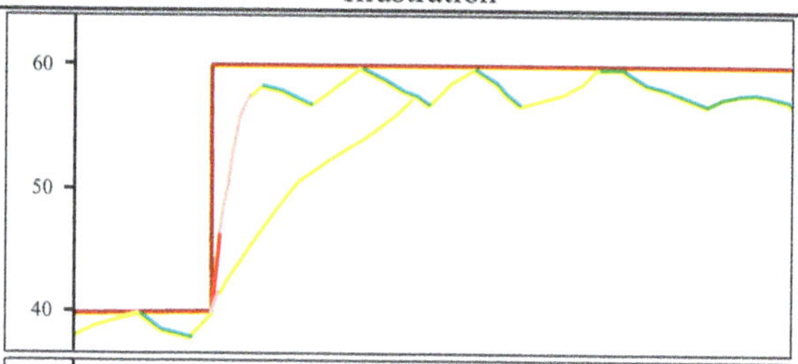
2	Early transition to idle before the section with a constant allowable speed, where to maintain the speed you need an adjusting braking	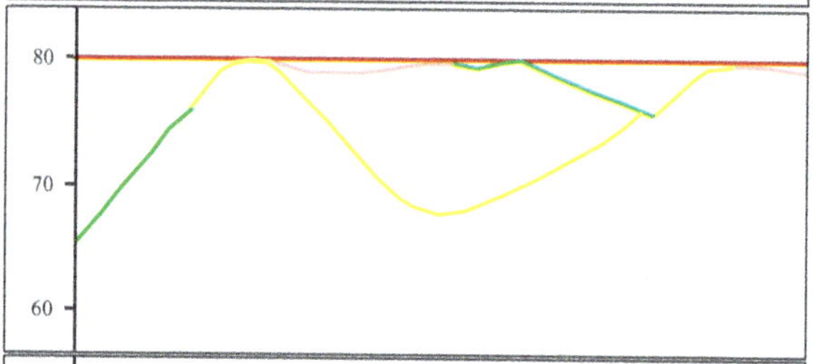
3	Decrease in speed on subsections traversed in traction mode due to lowering the running position (up to idle speed)	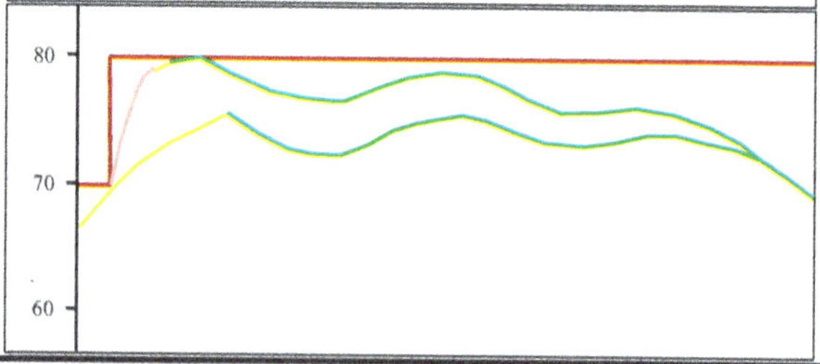

Note. Movement modes: green - traction, yellow - idle, red - braking, pink - adjusting braking or braking during filling of the brake cylinders.

It should be noted that experienced locomotive drivers successfully use type 1 and type 2 techniques when driving due to their obvious effectiveness and relative ease of implementation in practice.

Other effective techniques include the use of electrical braking, especially in areas where traction substations can receive recuperated current.

When implementing the methodology and algorithm of determination of the optimal train motion trajectory, a decrease in speed should be achieved taking into account a number of restrictions and particularities of train driving:

- there are restrictions on the maximum travel time for individual station-to-station blocks or for the entire section;
- the speed in certain sections of the path should not be lower than the specified one, for example:

  – approaching to the areas of testing the brakes with a regulated speed;
  – passage of sections of the path in traction mode at a speed not lower than the calculated minimum (with the exception of sections with a permissible speed below the calculated minimum speed of the locomotive);

- the implementation of traction effort or electric braking for most locomotives depends on the selected position of the locomotive driver's controller;
- there are restrictions on the use of the maximum driving position on individual subsections in order to prevent overheating of the engines or exceeding the allowable longitudinal forces in the train;
- the number of transitions between modes and position switching should be minimal.

The above mentioned principles, limitations and particularities of determining the optimal train motion trajectory were implemented in the module «OMM» (Optimal Modes of Motion) of the «ERA» software and technology complex (Expertise, Calculation, Analysis) [7, 8].

References

1. Ke, B.-R., Lin, C.-L., Lai, C.-W.: Optimization of train-speed trajectory and control for mass rapid transit systems. Control Eng. Pract. **19**(7), 675–687 (2011). https://doi.org/10.1016/j.conengprac.2011.03.003
2. Bigharaz, M.H., Afshar, A., Suratgar, A., Safaei, F.: Simultaneous optimization of energy consumption and train performances in electric railway systems. IFAC Proc. Volumes **47**(3), 6270–6275 (2014). https://doi.org/10.3182/20140824-6-ZA-1003.02368
3. Yan, X., Cai, B., Ning, B., ShangGuan, W.: Online distributed cooperative model predictive control of energy-saving trajectory planning for multiple high-speed train movements. Transp. Res. Part C Emerg. Technol. **69**, 60–78 (2016). https://doi.org/10.1016/j.trc.2016.05.019
4. Albrecht, A., Howlett, P., Pudney, P., Vu, X., Zhou, P.: The key principles of optimal train control - Part 1: formulation of the model, strategies of optimal type, evolutionary lines, location of optimal switching points. Transp. Res. Part B Methodol. **94**, 482–508 (2016). https://doi.org/10.1016/j.trb.2015.07.023
5. Albrecht, A., Howlett, P., Pudney, P., Vu, X., Zhou, P.: The key principles of optimal train control - Part 2: existence of an optimal strategy, the local energy minimization principle, uniqueness, computational techniques. Transp. Res. Part B Methodol. **94**, 509–538 (2016). https://doi.org/10.1016/j.trb.2015.07.024

6. Wang, P., Goverde, R.M.P.: Multi-train trajectory optimization for energy efficiency and delay recovery on single-track railway lines. Transp. Res. Part B Methodol. **105**, 340–361 (2017). https://doi.org/10.1016/j.trb.2017.09.012
7. Information Systems for the Design, Operation and Maintenance of Railways [Electronic Resource]. http://sites.google.com/site/isystemgdt. Accessed 06 Jan 2019
8. Anisimov, V.A., Anisimov, V.V.: ERA system: purpose, composition and capabilities - designing the development of a regional railway network. Sat. Scientific tr/ed. V.S. Schwarzfeld, no. 4, pp. 127–141. VGUPS Publishing House, Khabarovsk (2016)

Increasing the Functional Stability of Distance Relay Protection for Various Types of Catenary Support Grounding

Svetlana Makasheva(✉), Pavel Pinchukov, and Alexey Kostin

Far Eastern State Transport University, Seryshev Street 47,
680021 Khabarovsk, Russia
jap_svet@mail.ru

Abstract. The paper deals with the problem of incorrect operation of the feeder distance relay protection during short circuits in the AC contact network. The goal of the study is to eliminate number of cases of incorrect feeder distance relay protection operation. To achieve the goal, the existing catenary support grounding strategies are analysed in detail. Taking into account grounding strategy, the catenary support resistance components while a short circuit are considered in detail. The complex impedance characteristics measured by distance relay protective devices are considered and given at graphics. In the considered railway section, an analysis of proportional distribution of the catenary support resistance based at full-scale measurement is analysed. Factors affected the short circuit impedance value and making the feeder distance protection incorrect operation are classified. In addition to the impedance measured by the feeder relay protection, the necessity of using another parameter for determination the short circuit mode is justified.

Keywords: Railways · Power supply · Relaying protection system · Catenary support grounding · Distance relay protection · Short circuit · R-X characteristics · Transition resistance · Input impedance · Death grounding

1 Introduction

It is well known that the railways power supply system is a complex of engineering structures with a large number of elements, included traction substations, traction power supply system, railway traction rolling stock, overhead catenary, supporting structures, grounding, etc. In order to carry out the train movement on the rail, a train locomotive needs energy. This required energy or electric power is transmitted via a contact wire from the traction substation busbars to the electric locomotive over a long distance.

In addition to contact wire, a complex supporting structure, including a contact network supports, fittings, etc., called railway overhead catenary, is needed to transfer electric energy from the traction substation to the locomotive. However, despite the fact of importance of the electrical energy transfer, the electrified railways overhead catenary does not have a redundant. This fact increases the requirements for reliability and uninterrupted power supply of consumers, namely, railway traction rolling stock.

Z. Popovic et al. (Eds.): TransSiberia 2019, AISC 1115, pp. 155–166, 2020.
https://doi.org/10.1007/978-3-030-37916-2_17

In such conditions, the correct and reliable operation of relay protection devices is especially important.

The feeder relay protection system is an important part to ensure the safe operation of traction power supply systems. The function of the relay protection devices is to operate during a short circuit in the protected area and timely shutdown of the damaged area [1].

It is common for protective relaying systems for railway feeders to use distance relaying principles [2, 3]. The main parameter measured by distance relay protection is the impedance of the short circuit. The short circuit impedance includes the impedance of the generation power supply system, traction substation and traction network. The traction network impedance forms the overhead catenary impedance and the contact line support impedance, which depends on catenary support grounding type. Moreover, all metal structures of the electrified railways, such as catenary pole, masts, viaducts, bridges, etc., located closer than 5 m from live parts must be grounded to the rail [4, 5].

The direct connection of the contact line support to the rail is called dead earth. Modern relay protection types of the overhead catenary ensure its reliable shutdown in emergency operation of the overhead catenary with a dead earth of the contact line support, when the transition impedance between the contact line support and the ground is quite high. However, integrity violation of the contact line support grounding system changes the impedance value measured by the relay protection devices. In this case, the relay protection devices begin to work incorrectly, that is, they do not respond to a short circuit and do not turn off the damaged section [6, 7]. This case is interpreted as a relay protection failure.

The integrity violation of the contact line support grounding system can occur for various reasons. First of all is the collisions of equipment during work on the track maintenance or breakage of grounding conductors when someone working on communication devices. In addition, recently, cases of integrity violation of the contact line support grounding system have become more frequent on account of the increasing of the train weights and their traffic intensity called heavy haul trains movement [6–9]. This is due to the operating currents large values, which are comparable in value with short-circuit currents. The electric current (or the electric power) in the contact wire of the overhead catenary network is proportional to the train weight, that the locomotive drives.

The flow of such large currents along the arresters or spark gaps included in the ground circuit of the supports of the contact network leads to their burning off [10]. The flow of large quantities of operating currents is perceived by the protection system as a short circuit. In this case, the undamaged section of the contact network is disabled by relay protection, which is also interpreted as the faulty operation of relay protection devices. Such a situation with the relay protection devices failure can lead to serious accidents. This fact significantly reduces the safety of train traffic and the whole power supply system reliability. Therefore, this situation should be eliminated. Since the catenary support grounding play an important role for the described situations, it is necessary to study the degree of influence of various types of catenary support grounding on the correct and no correct relay protection operation.

2 Materials and Methods

In this study the experimental research methods, based on full-scaled measurement of controlled parameters and subsequent calculation and analysis are used. In order to identify a connection between the catenary support resistance value and the number of incorrect relay protection trips, the statistics data from relay protection terminal protocols and data logs recording information were collected. It is monitored on the 200 km in length railway section located at the Far Eastern Region of Russia. Then, the percentage of catenary supports with different resistance values as a fraction of the total number of all supports was calculated and classified. The result data were presented in the form of graphs, descriptive and analytical tables. This method has many attractive practical applications as understanding of the measure of the faulty operation of the relay protection devices at a particular railway section and the numerical assessment of resistance values in a percentage of the catenary support total number. The benefits can help to find a solution to eliminate the malfunction right solution for no correct relay protection operation and make timely adjustments to the relay protection settings as much as possible for better operation.

3 Catenary Support Grounding

3.1 Catenary Support Grounding Strategies

There are some different grounding strategies for AC railways catenary support and any infrastructure objects located near the railway track in the world: ungrounded, direct grounded (or dead earth), and diode grounded [2, 4, 11]. The grounding method involves the use of special grounding devices. Grounding devices can be either individual (for each catenary pole) or group (for several catenary poles). Grounding method using an individual or group grounding conductors connected to the traction rails directly or connected to the midpoint of throttle-transformer. Diode grounded using special protective devices such as arresters, spark gaps, etc. The protective devices such as spark gaps should be included in the catenary supports grounding system if the resistance of individually grounded catenary pole is less than 100 Ω in accordance with the existing Russian Standards.

The choice of grounding method can be determined by a number of requirements based on the real resistance value of each of the grounded devices, as well as their other qualitative characteristics. In any case, the main thing is to ensure the safety conditions for working personnel near the railway track. The procuring of short-circuit current flows only thru the connection circuit of any grounding and protective devices is also important. Moreover, during a short-circuit, the current flowing through the damaged catenary support did not cause its electro thermal overheating and damage, because this provides the possibility of further reliable operation of the power supply system [11–13].

3.2 The Short-Circuit Current Path Through the Catenary Support Grounding

The short-circuit current path depends on the catenary support grounding strategy. In case of isolated catenary support, if the pole does not have grounding or for some reason grounded system has been damaged, a short-circuit current flows directly through the catenary support body. In a case of the catenary support is grounded, a short-circuit current flows through the system of grounding conductors. The short-circuit current path through the catenary support grounding via the spark gap is shown at Fig. 1 as an example for the diode grounding strategy.

As follows from Fig. 1, when the short circuit occurs in AC catenary, the short-circuit current I_{SC} enters to the individual grounding system of a reinforced concrete catenary support (or catenary pole). This short-circuit current flows through the grounding conductor (called down conductor) attached to the sole of one of the rails with a hook bolt. The down conductor is insulated from the ground with a wooden impregnated half-sleeper and insulated from the catenary pole by means of wooden gaskets.

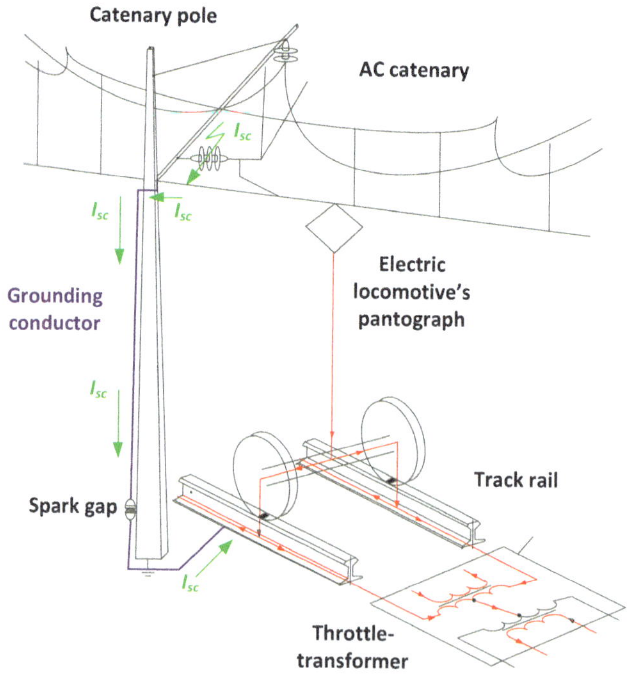

Fig. 1. Short-circuit current path in case of catenary support diode grounding

When the console insulator is broken, a short-circuit current flows through the circuit "traction substation – catenary feeder – catenary – down conductor on a catenary support with damaged insulation – track rails – reverse current feeder – traction substation". In this case, the current and resistance values in the short-circuit path play an important role for the correct operation of distance relay protection. For instance, the traction load current value while heavy haul traffic is often close to the short circuit current value, that why the catenary support resistance value as a level mark is very important for correct relay operating.

3.3 Catenary Support Resistance Value

Reinforced concrete catenary supports of the overhead contact network are most common on the world's railways. The consoles of these supports and their other metal parts, where live parts are suspended that are not electrically connected to the support reinforcement and insulated from concrete, must be grounded to the rail. Grounding is carried out using the grounding down conductor, located on the railway track internal side.

Catenary Support Resistance Structure. The resistance value of the contact network support in the event of a short circuit will depend on the path of the short circuit current. This path, in turn, depends on the adopted catenary support grounding strategy. Figure 2 shows the equivalent circuit for the catenary support with pole resistance components.

Neglecting the resistance value of the electric circuit elements at the point of attachment of the grounding conductor to the rail and the resistance of the element at the point of its attachment to the grounding support elements, we can obtain the equivalent circuit of the reinforced concrete support with the external location of the grounding conductor, shown at Fig. 2.

Figure 2 shows the following components of the reinforced concrete catenary support: 1 - transition resistance between the details of the reinforcement of the support and concrete; 2 - concrete resistance between parts of reinforcement and reinforcement itself; 3 - transition resistance between concrete and reinforcement; 4 - transition resistance between the reinforcement of the base part and concrete; 5 - concrete resistance in the foundation part; 6 - resistance to current spreading from the foundation into the ground; 7 - transition resistance of the soil - rail.

In case of a short circuit and the high voltage appearance in the catenary support upper part, the short-circuit current flows from the fault point to the earth. However, the resulting resistance of catenary support will be determined by the adopted catenary support grounding strategy.

Resistance of Grounded Catenary Support. In case of a short circuit and the high voltage appearance in the catenary support upper part, the short-circuit current flows along the down conductor to the track rail. Thus, a "metallic" fault occurs and it is switched off by feeder relay protection. The short-circuit current flows along the path: fault point in overhead contact network – point A – point B – track rail, shown at Fig. 2 as path number 1. The down conductor has a very small resistance (compared to other elements in this circuit, especially catenary support resistance. Thus, down conductor cannot limit the value of the short-circuit current, so the significant value of the short-circuit current identified by relay protection devices and switched off.

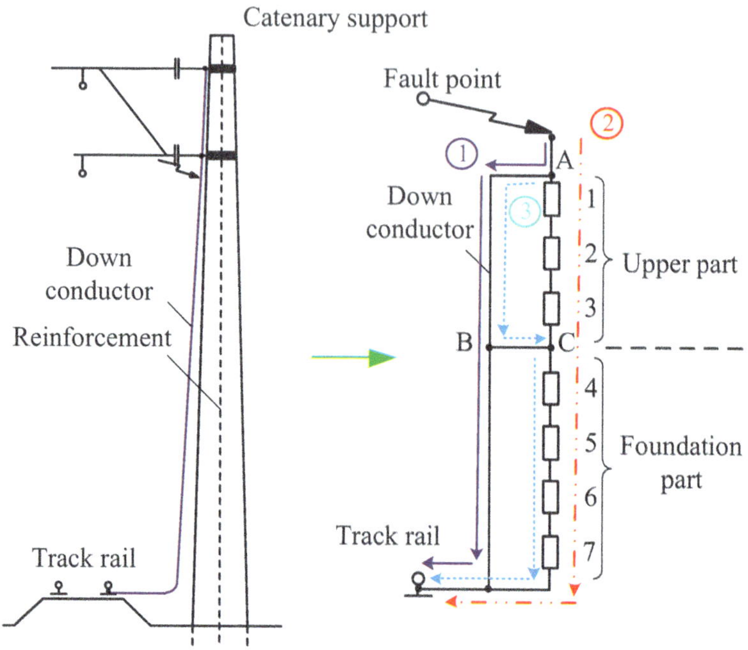

Fig. 2. Catenary support equivalent circuit

Resistance of Isolated or Grounded-Damaged Catenary Support. If there was no grounding conductor at the pole, then all the short-circuit current would flow through the catenary pole body. In case of the down conductor's absence a short circuit process will develop in the catenary support body and will be accompanied by concrete heating, losing its insulating properties and the occurrence of an arc discharge in the concrete thickness. The short-circuit current flows along the path: fault point in overhead contact network – point A – point C (thru all resistances 1–7) – track rail, shown at Fig. 2 as path number 2.

However, it is possible that an electric arc breaks through a concrete layer between embedded parts and reinforcement, in which case a short-circuit current flows along the support armature along the path showed in the Fig. 2 as path number 3: fault point in overhead contact network – point A – point B – point C – resistance 4–7 – track rail. This path excludes all components from the equivalent circuit corresponding to the upper part of the catenary support. The same thing happens when the down conductor is located in the body of the catenary support.

The catenary pole has a big resistance, so the value of the short circuit current would sharply decrease and the relay protection devices could not feel and detect this mode as a short circuit. In the absence of operation or in case of incorrect operation of the relay protection devices, an electric arc near the damage pole can burned out wires and supporting structures. Also, the step voltage of a big value appears near the catenary support, which is extremely dangerous for people's life [13–15].

Next, the numerical value of catenary support at the real railway section was measured and considered as percentage distribution in the total volume.

4 Catenary Support's Ohmic Resistance Percentage

Based on the ohmic resistance measuring protocols for the catenary supports taken place at Far Eastern railway sections during 2018 year, the ohmic resistance diagrams was created and shown at Fig. 3. Wide resistance range analysis was made regardless of grounding type as percentage of all catenary supports at the object of study.

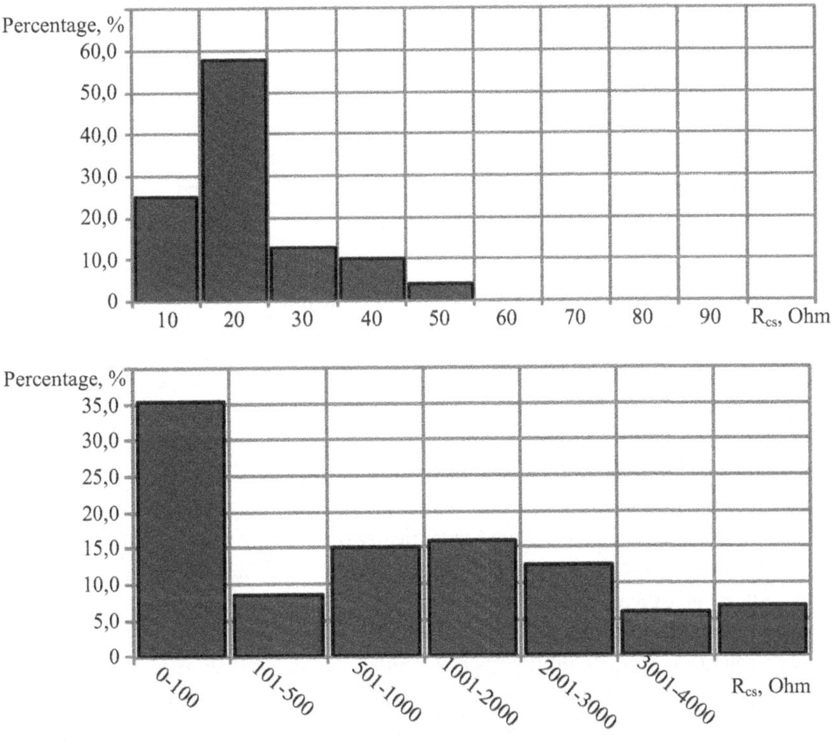

Fig. 3. Catenary support ohmic resistance percentage

It was revealed that the average value (mathematical expectation) of the catenary support ohmic resistance at the considered section was R_{CS} = 10.03 kΩ. About 47% of the catenary supports have ohmic resistance up to 5 kΩ, and about 75% – up to 20 kΩ. According to the full-scale measurements, the ohmic resistance of the catenary supports in some cases has a value exceeding 80 kΩ, which indicates the absence of a metallic ground connection with the reinforcement of the support.

Obviously, in this case we are talking about the unintentional grounding of the catenary support, which causes an increase in the active ohmic resistance of the support. Consider the effect of an increase in the catenary support resistance while a short circuit in case of accidental, unintentional grounding of the catenary supports on the existing relay protective device operation.

5 Functioning Conditions of Relay Protection

Currently, a typical scheme for overhead contact system is a nodal scheme with one section pillar, which is located approximately in the middle of the feeder zone, as shown in Fig. 4. Another common variant of a typical circuit is a parallel connection scheme with one sectioning pillar and two parallel connection points [2, 4, 14].

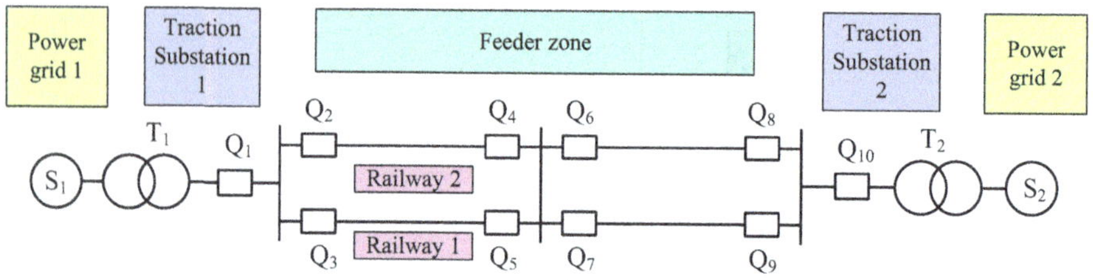

Fig. 4. Feed zone structure

At the same time, relay protection devices in the 25 kV traction power system are set on the power circuit-breakers of the contact network feeders at the traction substations and the sectioning station $Q_2 - Q_9$ and on the traction transformers inputs – Q_1 and Q_{10}, as shown at Fig. 4. The main and backup feeder relay protection at traction substations and sectioning pillars should form, as a rule, a directional multi-stage system (relay protection set). In this case, a short-circuit on the feeder at any point in the inter-substation zone must be detected by at least two relay protections (or two steps of the relay protection set) at the nearest switch and at least one protection (one step of the protection kit) on the switch of the adjacent element. For example, in case of short-circuit at railway 2, shown at Fig. 4, the relay protection devices on the Q_2 or Q_4 power circuit-breakers should trip.

To accelerate shutdowns of short-circuits switch off, a special protection level can be provided at the beginning of the feeder line, as well as additional protection in the form of a current cut-off without time delay [1, 3, 14].

Distance relay protection of the traction network based on a impedance relay that responds to the supplied current and voltage ratio. In the short-circuit mode, this ratio is proportional to the impedance Zcs of the protected line from the point where the relay protection is set to the short circuit point. The 25 kV traction network can be represented as an electric circuit with active R and inductive X reactance components. The capacitive component is neglected due to its insignificant value. Thus, in load mode the resistance r_L is greater than the reactance x_L, as shown at Fig. 5. The phase angle of the load current main harmonic of the electric locomotive varies from 25 to 40°, in other words, the load character of the electric locomotive is defined as an active-inductive character [6, 7, 11].

The most common type of short circuit on the contact network is the insulators overlap. In this case, the impedance of the short circuit is a sum of the power grid impedance, traction substation impedance, the impedance of the contact network and the rail circuit, the impedance of the down conductor (if it present) and the arc resistance. In this case, the end of the impedance vector Z_1, measured at the impedance relay terminals, falls into the short circuit region, as shown at Fig. 5.

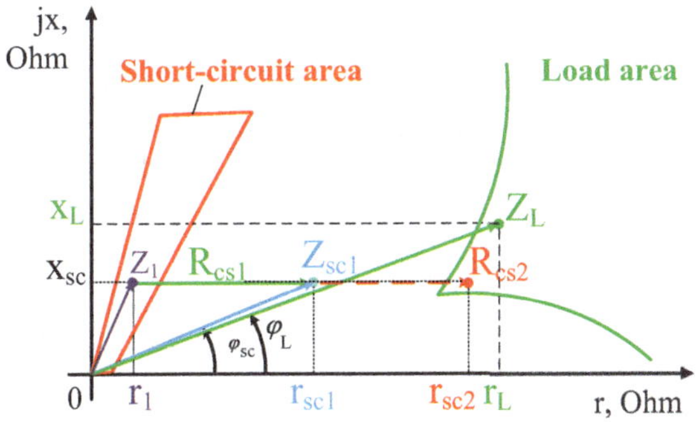

Fig. 5. R-X characteristics

Accidental violation catenary supports grounding leads to an ohmic resistance increasing in short-circuit mode. The resulting vector Z_{cs1} increases by the ohmic resistance dimension R_{cs1} and shifts to the right from Z_1 to Z_{sc1} point, shown at Fig. 5. The Z_{sc1} vector goes beyond the response of the angular characteristics of the distance protection three zones and distance protection does not switch-off the short circuit. In such cases, the fourth zone of distance protection is used, having an angular response characteristic located along the axis of resistance r, which responds to short circuits characterized by high ohmic resistance through a large transition resistance.

In case of a big catenary support resistance R_{cs2}, the short circuit active resistance became the value r_{cs2}. The resulting impedance vector shifts to the right along the ohmic resistance axis, takes the value Z_{sc2} and so falls into the load zone at Fig. 5. In this case, none of the four zone of distance protection able the short-circuit switch-off, because it cannot identify it.

6 Results of Modeling and Long-Term Observations of the AC Traction Network's Relay Protection Functioning

At one of the traction substations located at the Far East of Russia, chosen as the object of study, field measurements of the current and voltage parameters in load and emergency modes were carried out for more than 8 years [1, 9]. During the last few years, measurements are carried out by an automated monitoring system, which has ample opportunities for measurement, protection, control, etc. [15]. The feeders currents and voltages values during the observations are fixed by the automatic monitoring system, and then the impedance values were automatically calculated. These values are plotted as points on the load area as shown at Fig. 6. In the same figure, the short-circuit zone and the response zones of all 4 zones of distance relay protection are marked. The x and y axes at Fig. 6 indicate the numerical values of the reactance and active ohmic resistance for object of study. In addition to the values obtained from the results of field measurements, Fig. 6 shows the simulation results of distance protection parameters and its calculations results.

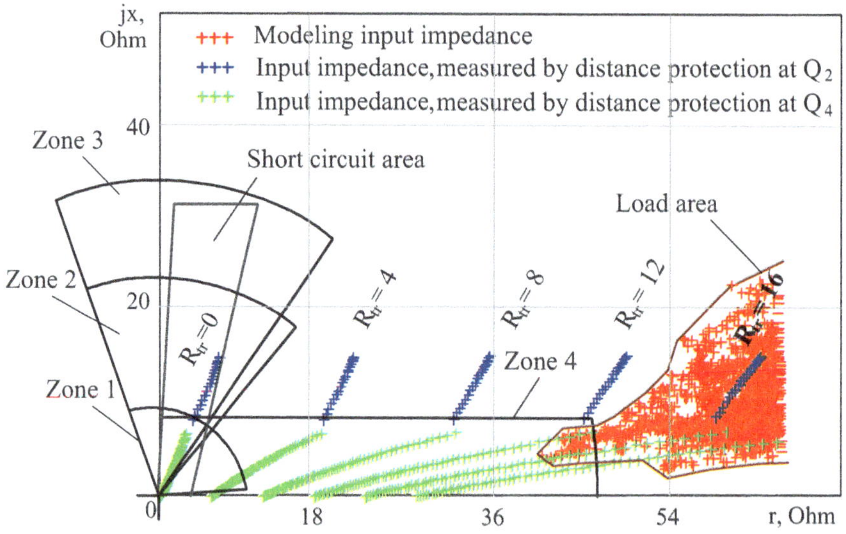

Fig. 6. R-X characteristics while short circuits through transition resistance R_{tr}

Also, for the considerated inter-substation zone, a simulation model of the distance protection devices was created. This model was created at MATLAB® programming platform and it allows to calculate the short-circuit parameters taking into account the various transition resistance values. By means of this model, the impedance values measured by the impedance relays of the distance protection devices installed on the switches Q_2 and Q_4 according to Fig. 4 were calculated. Thus, images of simulated and measured parameters were plotted on the R-X characteristics plane for the same inter-substation zone. Input impedances, measured by distance protection devises, located at Q_2 and Q_4 power circuit-breakers are grouped for each value of transient resistance R_{tr}.

7 Discussion

As we can see at Fig. 6, with an increase in the transition resistance R_{tr}, the input impedances measured by the distance protections are more and more shifted towards the load zone as was previously described. Already at values of transition resistance R_{tr}

equal to or greater than 12 Ω, the input resistance distance protection leaves the 4th zone of distance protection, therefore, both the main and backup relay protection do not disconnect the short circuit. In case of a short circuit at the catenary support with "death grounding", when the value of transition resistance R_{tr} is small, the distance relay protection successfully disconnects the damaged area with the selectivity and sensitivity requirements.

8 Conclusion

Obtained results leads to the following conclusions:

1. In the event of a short circuit at the non-grounded catenary support, a transition resistance appears which disrupts the stable operation of the distance relay protection.
2. The big value of the transition resistance at the short-circuit point increases the input impedance vector and changes its angle and argument, measured by distance relay protection. Thus, the impedance vector does not fall into the angular response characteristic of the distance relay protection that why the short-circuit cannot switched-off.
3. The cause for the relay protection failure is the input impedance vector displacement into the load zone due to the big value of the transition resistance in the short circuit.
4. In this case, the distance relay protection cannot distinguish short circuit from the normal operating mode of the traction network, guided only by the impedance value.
5. If the settings for the operation of the relay protection increase, the distance protection will turn off both the emergency and normal operation of the traction network. Therefore, an additional feature is required by which it will be possible to identify the short circuit mode and distinguish it from the load mode.
6. For identifying the short circuit mode and switching it off, an additional feature, sign or parameter is required in addition to impedance. For example, the spectra of current harmonics during the short circuit mode and load mode can be considered as an additional parameter for short circuit mode detecting.
7. Finding and using an additional parameter for identification short circuit can increases the functional stability of distance relay protection for various types of catenary support grounding.

Acknowledgments. This research was supported by the Foundation for Assistance to Small Innovative Enterprises in Science and Technology (FASIE) in the framework of the Program "UMNIK" (the Russian abbreviation for "Member of the Youth Research and Innovation Competition").

References

1. Pinchukov, P.S.: Experimental evaluation of sustainable functioning of overhead wire relay protection at traction substations of the far eastern railway. Russ. Electr. Eng. **87**(62), 6–8 (2016). https://doi.org/10.3103/S1068371216020152
2. Novo, J.M., Teston, L.S.M., Calero, F., Arroyo, I.: Protective relaying for railway feeders. In: 41st Annual Western Protective Relay Conference on Spokane, Washington (2014)
3. Ma, J., Wang, Z.: Basic theories of power system relay protection. In: Hierarchical Protection for Smart Grids. IEEE 2017 (2018). https://doi.org/10.1002/9781119304814
4. The Railway Technical Website. http://www.railway-technical.com
5. Pinchukov, P.S., Makasheva, S.I.: Research of AC traction network's relay protection operating under heavy haul traffic conditions. In: IOP Conference Series: Earth and Environmental Science, vol. 272, p. 022069 (2019). https://doi.org/10.1088/1755-1315/272/2/022069
6. Pinchukov, P.S., Makasheva, S.I.: The railway transport magazine, vol. 8, pp. 40–42 (2018). http://www.zdt-magazine.ru/%D0%B0%D0%BD%D0%BD%D0%BE%D1%82%D0%B0%D1%86%D0%B8%D0%B8-%E2%84%968-2018/
7. Abramov, E.Y., Schurov, N.I., Rozhkova, M.V.: Electric transport traction power supply system with distributed energy sources. In: IOP Conference Series: Materials Science and Engineering, vol. 127, p. 7 (2016)
8. Guo, J., Yuan, M., Liu, B., Li, L.: Research on feeder protection for a traction network based on adaptive parameter calculation. Int. J. Comput. Consum. Control (IJ3C) **5**(4) (2016)
9. Makasheva, S.: Simulation of a return current system for AC power traction network. In: IOP Conference Series: Earth and Environmental Science, vol. 272, p. 022069. https://doi.org/10.1088/1755-1315/272/2/022071
10. Han, Z., Dong, Z., Gao, S., Bo, Z.: Protection scheme for out-of-phase short-circuit fault of traction feeding network. In: 11th IET International Conference on Developments in Power Systems Protection, DPSP, pp. 1–4. Birmingham (2012). https://doi.org/10.1049/cp.2012.0133
11. White, R.D.: AC DC railway electrification and protection. In: IET Railway Electrification Infrastructure Systems Conference, London (2014). https://doi.org/10.1049/cp.2014.1439
12. Li, W., Hu, P.: Overview of analysis on state evaluation of relaying protection system. J. Power Energy Eng. **07**(05), 1–7 (2019). https://doi.org/10.4236/jpee.2019.75001
13. Chen, G., Wang, D., Qiu, Y., Wang, S., Qi, X.: Challenges and development prospects of relay protection technology. Autom. Electr. Power Syst. **41**(16), 1–11 (2017). https://doi.org/10.7500/AEPS20170219004
14. Romanov, Y.V., Voronov, P.I.: Assessing the sensitivity of relay protection. Power Technol. Eng. **51**, 1–4 (2018). https://doi.org/10.1007/s10749-018-0900-x
15. Makasheva, S.I., Pinchukov, P.S.: Expanding the functionality of automated monitoring systems for traction substations. In: MATEC Web of Conferences, vol. 239, p. 01014 (2018). https://doi.org/10.1051/matecconf/201823901014

Dependence of Power Losses in an Overhead Wiring on the Current Waveform Consumed by an Electric Locomotive

Iurii Kabalyk(✉)

Far Eastern State Transport University, Serysheva Street, 47, Khabarovsk 680021, Russia
kabalyk@festu.khv.ru

Abstract. Overhead wiring was considered as a line with non-linear parameters, where the AC locomotive operates. The parameters of the locomotive were different for obtaining different quality of electrical energy consumption. Depending on the parameters of the locomotive, it consumes a different form of current, which affects the power loss in the overhead wiring. To research this phenomenon was carried out computer simulation. The results of the research show different power losses in the overhead wiring with different locomotive parameters.

Keywords: Electric locomotive · Reactive power · Voltage losses · Overhead wiring · Reactive power compensator

1 Introduction

The railroad electricity system is a rather complex system. This complexity lies not only in the large number of system elements, but also in the fact that the parameters of the overhead wiring change as the locomotive moves along the way. The total electrical resistance of the overhead wiring is determined by the distances between the electric locomotive and the traction substations along the edges of the traffic area. So, when the electric locomotive is near the traction substation, the electrical resistance of the overhead wiring is practically absent, and when the electric locomotive is in the middle of the inter-substation zone, the overhead wiring electrical resistance is have maximum value. The length of the railway section between traction substations is an average of 40–50 km. According with [1, 2], the overhead wiring is a line with distributed non-linear parameters depending on the current frequency. Despite the fact that the AC overhead wiring is powered by a voltage of one frequency (50 Hz), the current consumed by the electric locomotive contains higher harmonic components, the frequency of which reaches 2000 Hz or more [2, 3]. The presence of these factors leads to the fact that the exact determination of power loss in the overhead wiring with an analytical way is a difficult task.

Most AC electric locomotives operating at the Russian railroads contain multi-zone power electric converters based on thyristors. The main energy drawback of this converters is their low power factor value. Moreover, the value of this factor is not

Z. Popovic et al. (Eds.): TransSiberia 2019, AISC 1115, pp. 167–176, 2020.
https://doi.org/10.1007/978-3-030-37916-2_18

constant and changes with regulating the realized power. Changing in the power factor value of these electric locomotives is from 0.65 to 0.85 in the traction mode and from 0.5 to 0.75 in the regenerative braking mode. The low power factor of electric locomotives leads to increased values of reactive energy in the railway electric power supply [4]. The presence of reactive energy additionally reduces the voltage level in the overhead wiring, which negatively affects the operation of electric locomotives (voltage losses) [5, 6]. Also reactive power leads to the appearance of additional power losses in the energy system. To increase the power factor, and as a result of reducing reactive energy, electric locomotives are equipped reactive power compensators. There are many varieties of such compensators, but the most common and already repeatedly tested are passive compensators made in the form of series-connected capacitances and inductances forming an oscillatory circuit [7–9]. The oscillatory circuits of passive compensators are tuned to a frequency close to the third harmonic, since this harmonic is one of the most pronounced in the current spectrum of an electric locomotive. The use of passive compensators can reduce the phase-angle displacement between current and voltage to almost zero, and thereby increase the power factor of an electric locomotive to values close to one.

The use of electric locomotives with reactive power compensators further complicates the system of "traction sub-stations – overhead wiring – electric locomotive". Due to the large number of elements, analytical calculations of electromagnetic processes are necessary. For this reason, when making calculations in the traction power supply system, many assumptions are made, including linearization of circuit elements [10]. In reality, a distorted current of an electric locomotive flows through the overhead wiring, causing additional losses in nonlinear elements. The simplest way to solve this problem is to computer simulate the operation of an electric locomotive using nonlinear elements of the overhead wiring.

2 Materials and Methods

To calculate the electromagnetic processes, computer simulation was carried out with various layout options for the electric power circuit. By computer simulation tools created a system of "traction substations – overhead wiring – electric locomotive". As a model of the overhead wiring, the scheme proposed by the authors of [1] was taken. An electric locomotive model was performed taking into account the components described in [11, 12]. In all experiments, the model of the 2ES5K electric locomotive, located in the middle of the overhead wiring section 50 km long, along the edges of which traction substations were installed, was studied. Also, for all the experiments, the realized power of the electric locomotive was the same, which was taken as the electric power of all the traction motors. This was done with the consideration that the realized power is decisive in the operation of an electric locomotive. In other words, in all experiments, an electric locomotive realizes the same power, but consumes electric energy differently. A value of 5600 kW was taken as the realized power, which is 92% of the power of the continuous mode 2ES5K electric locomotive. When reactive power is compensated, the voltage in the overhead wiring rises, as a result of which the voltage on the traction motors increases and their power increases. In this case, we can increase the thyristors opening angle of the converter, which will reduce the voltage at its output. However, a change in the opening angle will lead to a change in the harmonic composition of the consumed current, which will not allow us to equally evaluate different methods of reactive power compensation. For this reason, to align the realized power when changing the voltage in the overhead wiring, a change in the locomotive speed was used: an increase in speed leads to an increase in the counter-emf of the motors, which reduces their current and electric power. In the auxiliary circuits of the electric locomotive, powerful converters are not used, so the auxiliary load was modeled by an active resistance of 1.44 Ω, which at a voltage of 380 V gives a load of 100 kW. Since the attention in the experiments was focused on electromagnetic processes in the overhead wiring, the model of the traction substation was maximally simplified and consisted of a voltage source, active and inductive resistance connected in series.

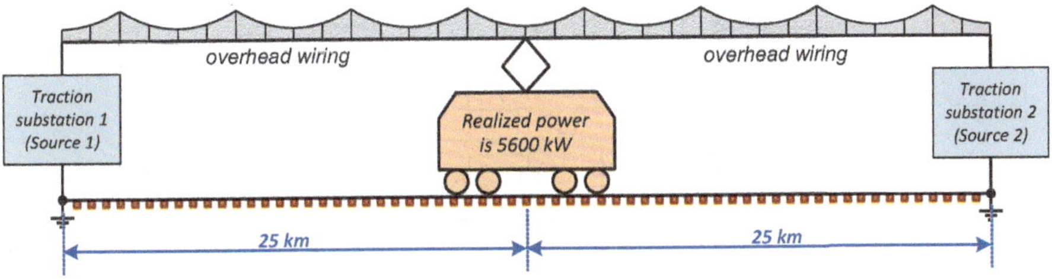

Fig. 1. General view of the simulated system

Simulating was realizing with the following circuit versions:

1. The standard scheme of an electric locomotive (Fig. 1, elements *a* & *b*). In this version, the scheme of the electric locomotive was similar to the standard one, power converters operated in the middle of the fourth regulation zone. This version allows us to evaluate the impact of the typical operation of an electric locomotive on losses in the overhead wiring;
2. Regular electric locomotive consuming undistorted current (Fig. 1, elements *a* & *c*). In this case, instead of an electric locomotive circuit, a sinusoidal current source was installed, which generates a current equal to the first harmonic from the current spectrum of the electric locomotive of the first circuit version. This version allows us to evaluate the impact of undistorted current on the losses in the overhead wiring;
3. A standard electric locomotive with an ideal current phase displacement compensator (Fig. 1, elements *a, b* & *c*). The scheme of the electric locomotive was similar to the first version, however, a sinusoidal current source was installed in parallel with the electric locomotive, the phase of which was offset by 90 electrical degrees relative to the voltage. The amplitude of the current in the source was calculated from the results of the first version and amounted to 316.6 A. This version allows us to evaluate the impact of a distorting factor on losses in the overhead wiring;
4. Electric locomotive with passive reactive power compensator (Fig. 1, elements *a, b* & *d*). As a compensator, the scheme developed by the employees of VELNII and

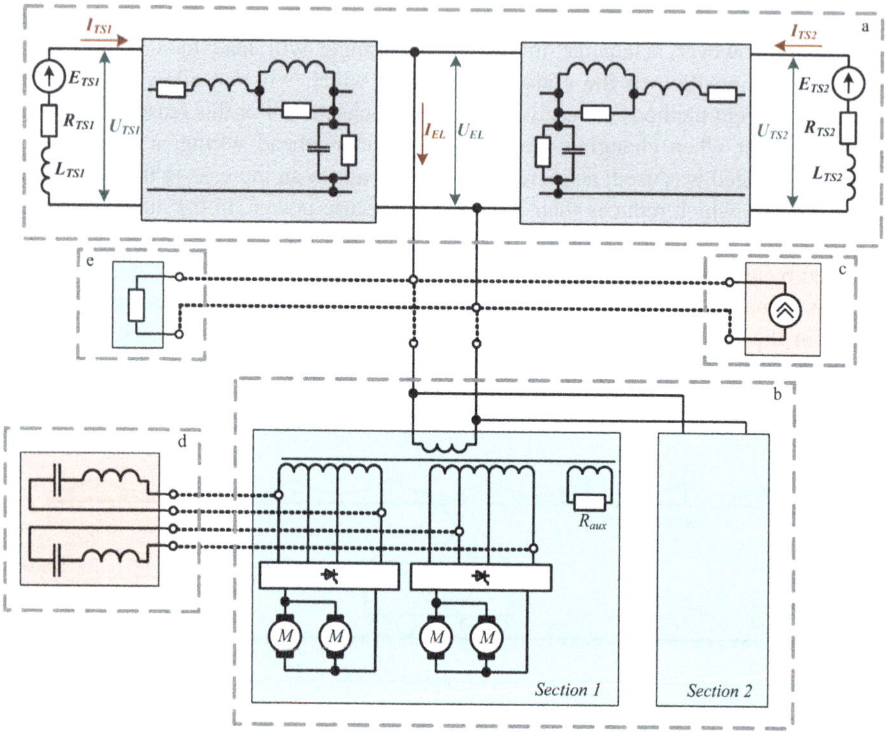

Fig. 2. The elements layout of the simulated system: a – overhead wiring line with traction substations at edges; b – electric locomotive; c – sinusoidal current source; d – passive reactive power compensator; e – resistor simulating the ideal electric locomotive

presented in [7, 8] was taken. The total power of the compensator was selected so as to maximize the power factor of the electric locomotive. The compensator is installed in the circuit between the secondary winding of the transformer and power converters (the compensator for the second section is not shown in Fig. 1). This version allows us to evaluate the impact of passive compensation on the losses in the overhead wiring;

5. An electric locomotive with a power factor that is equal to one (Fig. 1, elements *a* & *e*). Instead of an electric locomotive circuit, a resistor of 122.4 Ω was taken, which at a voltage of 26.6 kV has a power of 5800 kW. This power value was taken as the sum of the electric power of all motors, auxiliary power needs and power losses in the elements of the electric locomotive circuit. We can say that with this version, an "ideal" electric locomotive is modeled from the point of view of the quality of electricity consumption, for which the power factor is equal to one. This version allows us to evaluate the impact of a purely active current on the losses in the overhead wiring.

The research calculated the following parameters: active power of traction substations and electric locomotive; apparent power of traction substations and electric locomotive; effective values of current and voltage on an electric locomotive and traction substations. These parameters were calculated by software based on the instantaneous values of current and voltage.

The parameters are calculated for the following points of the circuit: output of the first traction substation (TS1), output of the second traction substation (TS2) and input of the electric locomotive (EL). The calculation was carried out according to the following formulas:

$$I_x = \sqrt{\frac{1}{T}\int_0^T i_x dt} \tag{1}$$

$$U_x = \sqrt{\frac{1}{T}\int_0^T u_x dt} \tag{2}$$

$$P_x = \frac{1}{T}\int_0^T u_x i_x dt \tag{3}$$

$$S_x = \frac{1}{T}\sqrt{\int_0^T u_x^2 dt \int_0^T i_x^2 dt} \tag{4}$$

where T – period of supply voltage, T = 0.02 s.; x – parameter belonging to a point in the circuit («*TS1*», «*TS2*» or «*EL*»); u – instantaneous voltage value; i – instantaneous current value.

3 Results

Figure 3 shows the oscillograms of the electric locomotive current for one half-period of the supply voltage for the first four circuit versions. To compare the scales and determine the efficiency of compensation, the current waveform of the "ideal" electric locomotive from the fifth circuit version using a dashed line is shown. Figure 2(a) shows that the current waveform of the electric locomotive has significant distortion and phase shift. The first current harmonic of a standard electric locomotive, shown in Fig. 2(b), has a phase displacement about 45 electrical degrees. A comparison of Figs. 2(c) and 2(d) shows that when using a passive compensator, the full distortion of the current waveform is reduced and the current waveform becomes closer to the sine wave.

Table 1 shows the main parameters of the simulated system with various circuit version. From these parameters it can be seen that the active power of an electric locomotive with passive compensator is less than the active power of a standard electric

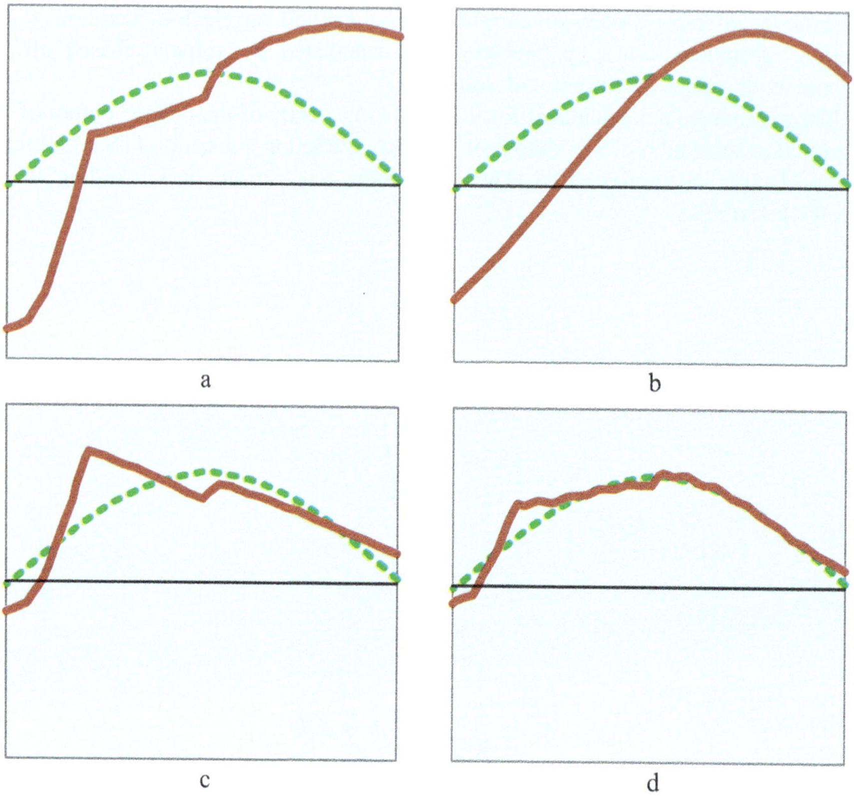

Fig. 3. Oscillograms of the half-period of the current waveform of the electric locomotive for the first four circuit versions (solid line) compared with the purely active current waveform obtained with the fifth circuit version (dashed line): a – first circuit version; b – second circuit version; c – third circuit version; d – fourth circuit version

locomotive by 69.7 kW or 1.1%. This is due to the fact that the total current of the electric locomotive has decreased, which means that losses in the wires from its current have decreased. Also, the table shows a tendency to reduce the apparent power of the electric locomotive with the decrease of distortive power (versions 2 and 5) and with the use of compensators (versions 3 and 4).

4 Discussion

Table 2 presents the calculation of energy parameters based on the results of computer simulation. The table shows that with the accepted parameters of the electric locomotive, the use of a passive compensator reduces the reactive power by more than 6 times. At the same time, the power factor rises to 98.7%, which is acceptable for modern energy-efficient consumers. The smallest voltage loss in the overhead wiring was obtained using an current phase displacement compensator (third circuit version).

Table 1. Research results obtained on a simulated system

Parameter	Circuit version				
	1	2	3	4	5
Active power of two traction substations $\Sigma P_{TS} = (P_{TS1} + P_{TS2})$, kW	6134.4	5859.4	6046.5	6064.7	5934.5
Apparent power of two traction substations $\Sigma S_{TS} = (S_{TS1} + S_{TS2})$, kVA	8637.6	8018.2	6415.9	6158.9	5937.9
Effective value of voltage on traction substation $U_{TS1} = U_{TS2}$, kV	26.46	26.4	27.09	27.04	27
Effective value of current on traction substation $I_{TS1} = I_{TS2}$, A	163.2	151.8	118.4	113.9	110
Active power of the electric locomotive P_{EL}, kw	5886.1	5640.9	5899.3	5926.5	5804.1
Apparent power of the electric locomotive S_{EL}, kVA	8462.1	7883.9	6236.3	6002.8	5804.1
Effective value of voltage on the electric locomotive U_{EL}, kV	25.52	25.36	26.82	26.67	26.58
Effective value of current on the electric locomotive I_{EL}, A	331.6	310.9	232.5	225.1	218.4
Electric power of all motors ΣP_M, kW	5599.7	—	5597.8	5601.1	—
Active power of two traction substations $\Sigma P_{TS} = (P_{TS1} + P_{TS2})$, kW	6134.4	5859.4	6046.5	6064.7	5934.5

When using a passive compensator (fourth circuit version) and with an "ideal" electric locomotive (fifth circuit version), the voltage loss is approximately the same and 2.5 times less than in a standard electric locomotive. In general, in the presence of current distortion (1, 3 and 4 versions), lower voltage losses in the overhead wiring are observed than with a sinusoidal current (2 and 5 versions). Such an unobvious result is most likely associated with the frequency characteristics of the overhead wiring model and with a changing in the effective voltage value with the presence of harmonic distortions. The author additionally conducted several test experiments with the simulated system, however, the values turned out to be the same. A simulation error is unlikely, since the overhead wiring model and the subsystem for determining the parameters did not change during the transition from one circuit version to another.

It is also seen from the calculated results that the power losses in the overhead wiring are reduced from the circuit version to the fifth. With a standard electric locomotive (first circuit version), the power loss in the overhead wiring is 4.2% of the active power of the locomotive, and with a purely active load (fifth circuit version) - 2.2%. A comparison of the first and second circuit versions suggests that current distortion causes an additional power loss of 29.8 kW. The third circuit version differs from the fifth only in the presence of current distortion and a slightly higher active

Table 2. The results of energy parameters calculations in the simulated system

Parameter	Circuit version				
	1	2	3	4	5
Reactive power of an electric locomotive $Q_{EL} = \sqrt{S_{EL}^2 - P_{EL}^2}$, kVA	6079.5	5507.8	2022.3	954.0	0
Power factor of an electric locomotive $PF = (P_{EL}/S_{EL}) \times 100$, %	69.6	71.5	94.6	98.7	100
Voltage losses in the overhead wiring $\Delta U_O = (U_{TS} - U_{EL})$, kV	0.94	1.04	0.27	0.37	0.42
Power losses in the overhead wiring $\Delta P_{OW} = (\Sigma P_{TS} - P_{EL})$, kW	248.3	218.5	147.2	138.2	130.4

power of the electric locomotive. In this regard, we can find the difference in power losses in the overhead wiring between these options (which will also indicate the influence of a distorting factor), which will be 16.8 kW. This difference is significantly less than the difference between the first and second circuit versions. This result can be explained by the fact that the power losses in the active resistances are determined by the square of the current, and in the first two cases the effective current value is greater. At the same time, there is a clear relationship between the current value of the current of the electric locomotive and power losses in the overhead wiring.

It should also be noted that the use of a passive compensator reduces power losses in the overhead wiring more than the use of an ideal current phase displacement compensator (option 3). This means that the passive compensator not only reduces the current phase displacement, but also reduces the magnitude of the higher harmonics of the current of the electric locomotive.

To analyze the mutual influence of the parameters, we studied the dependences of the electrical efficiency of the system and the voltage losses in the overhead wiring on the power factor of an electric locomotive (Fig. 4).

From Fig. 4(a) direct relationship between the electrical efficiency of a system and power factor of an electric locomotive is visible. Each increase in the first parameter leads to an increase in the second. Therefore, this once again proves that the use of reactive power compensators can reduce the consumption of electric energy on rolling stock. Despite this, it can be seen from Fig. 4(b) that the voltage losses in the overhead wiring are not directly dependent on the power factor of the electric locomotive. We can talk about a general inverse relationship: the higher power factor, the less voltage losses. However, if you look at some indicators separately (circuit versions 3–5 or 1–2), then a direct dependence is observed and an increase in power factor leads to an increase in voltage losses.

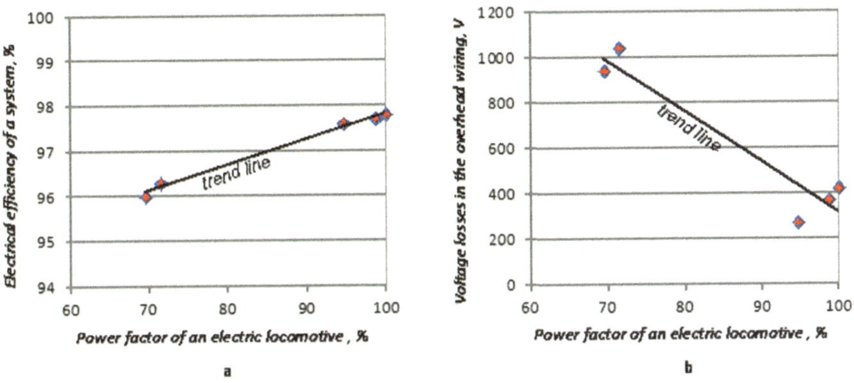

Fig. 4. Dependences of the system parameters on the power factor of an electric locomotive

5 Conclusions

1. Power losses in the overhead wiring are from 2.2% to 4.2% of the active power of an electric locomotive, depending on the magnitude of its power factor
2. The use of a passive compensator is an effective and, to some extent, sufficient means of compensating the reactive power of an electric locomotive. This compensator not only reduces the current phase displacement, but also reduces the harmonic distortion of the current curve
3. Voltage losses in the overhead wiring depend not only on the magnitude of the current flowing through it, but also on the level of distortion of this current.

References

1. Savos'kin, A.N., Kulinich, Y.M., Alekseev, A.S.: Mathematical modeling of electromagnetic transients in a dynamic system: contact network - electric locomotive. Electricity ("Elecktrichestvo" – Russian) **2**, 29–35 (2002)
2. Kulinich, Y.M.: Adaptive Automatic Control System for Hybrid Compensator of Reactive Power of an Electric Locomotive with Stepless Voltage Regulation: Monograph. FESTU, Khabarovsk (2001)
3. Makashova, S.I.: Monitoring the Quality of Electric Energy in the Traction Power Supply System of Alternating Current: Monograph. FESTU, Khabarovsk (2009). (in Russian)
4. Savos'kin, A.N., Garbuzov, I.I., Kulinich, Y.M.: Simulation of energy data during ac locomotive motion with an onboard reactive power compensator on a standard section of an electric railway. Russ. Electr. Eng. **86**(9), 540–547 (2015). (in Russian)
5. Grinberg, R.P., Savos'kin, A.N., Kulinich, Y.M.: Increasing the power factor of ac electric locomotives. Russ. Electr. Eng. **73**(5), 15–20 (2002). (in Russian)
6. Savoskin, A.N., Pudovikov, O.E., Garbuzov, I.I.: Analysis of electromagnetic processes in the traction network under the interaction of electric locomotives in the regeneration and traction modes. Russ. Electr. Eng. **85**(5), 305–311 (2014). (in Russian)

7. Kopanev, A.S., Khripkov, P.A., Volkov, V.M., Volt, P.S.: Tests of reactive power compensation system on electric locomotive 3es5k. Bull. All-Russ. Res. Des. Inst. Electr. Locomot. **2**, 14–36 (2010). (In Russian)
8. Foshkina, N.V., Malyshev, V.M., Kozlovsky, I.L.: Use of reactive power compensator on a 3ES5 K electric locomotive. Bull. All-Russ. Res. Des. Inst. Electr. Locomot. **2**, 37–42 (2010). (In Russian)
9. Kulinich, Y., Shukharev, S.: Increase in the power factor of the electric ac loco-motive due to a new structure of compensator operation. Procedia Eng. **165**, 1512–1519 (2016)
10. Marquadt, K.G. (ed.) Railway Power Handbook, vol. 1, Moscow (1980)
11. Vlasyevsky, S.V.: Mathematical Modeling of Switching Processes in Rectifier-Inverter Converters of Single-Phase to Direct Current Electric Locomotives: Monograph. FESTU, Khabarovsk (2001)
12. Kabalyk, I.S.: Reducing of Higher Voltage Harmonics in the Traction Network of Electrified Railways Through Active Filtration: Monograph. FESTU, Khabarovsk (2015)

Budgeting Direct Costs of Track Complex of JSC "Russian Railways" in the Light of Modern Classification of Railway Lines

Nikolay Kovalenko(✉), Boris Volkov, Aleksandr Kovalenko, and Nina Kovalenko

Russian University of Transport (MIIT), Obraztsova Street 9/9, Moscow 127994, Russia
kni50@mail.ru

Abstract. Conceptual bases of planning of expenses on the current maintenance of objects of infrastructure at linear, regional and Central levels taking into account production and normative budgeting of expenses on production operations are considered. To assess the risks of additional costs and reduce revenues to minimize losses, the article considers the mathematical apparatus of the risk management algorithm. The paper discusses the development of tools for quantitative analysis of the investment project. Operation specification of works and volumes in accordance with the techniques of track work required to reconcile the revealed with the execution of the KOSP of the types of works adopted their nomenclature in accordance with Regulatory and technical documentation.

Keywords: Production and regulatory budgeting · The cost of the current maintenance of infrastructure · Linear · Regional and central levels of budgeting · Risk assessment

1 Introduction

The existing method of determining direct costs for the current maintenance of the track, approved by the order of JSC "Russian Railways" from 30.12.2008 № 2897r, is based on the principles of normative-target budgeting. This approach involves the planning of direct costs on the basis of the distribution of the share of income generated in a comparable pre-plan period, based on its results.

This approach of direct cost planning does not take into account the modern features of the classification of railway lines, their specialization and classification of tracks. In addition, the planning of direct costs on the basis of the distribution of the share of income received does not have stimulating factors for the growth of interest and productivity of both ordinary workers and managers at all levels of management: linear, regional and Central.

The disadvantage of the existing methodology, approved by the order of JSC "Russian Railways" from 30.12.2008 № 2897r, is also the lack of a number of

Z. Popovic et al. (Eds.): TransSiberia 2019, AISC 1115, pp. 177–183, 2020.
https://doi.org/10.1007/978-3-030-37916-2_19

professions as ordinary workers and managers primarily at the linear level. For example, route supervisors, distributors of the work, signalised and others.

2 Materials and Methods

The aim of the study is to develop a conceptual framework for planning the costs of the current maintenance of infrastructure, depending on the classification of railway lines at the linear, regional and Central levels, taking into account the production and regulatory budgeting of costs for production operations (Fig. 1).

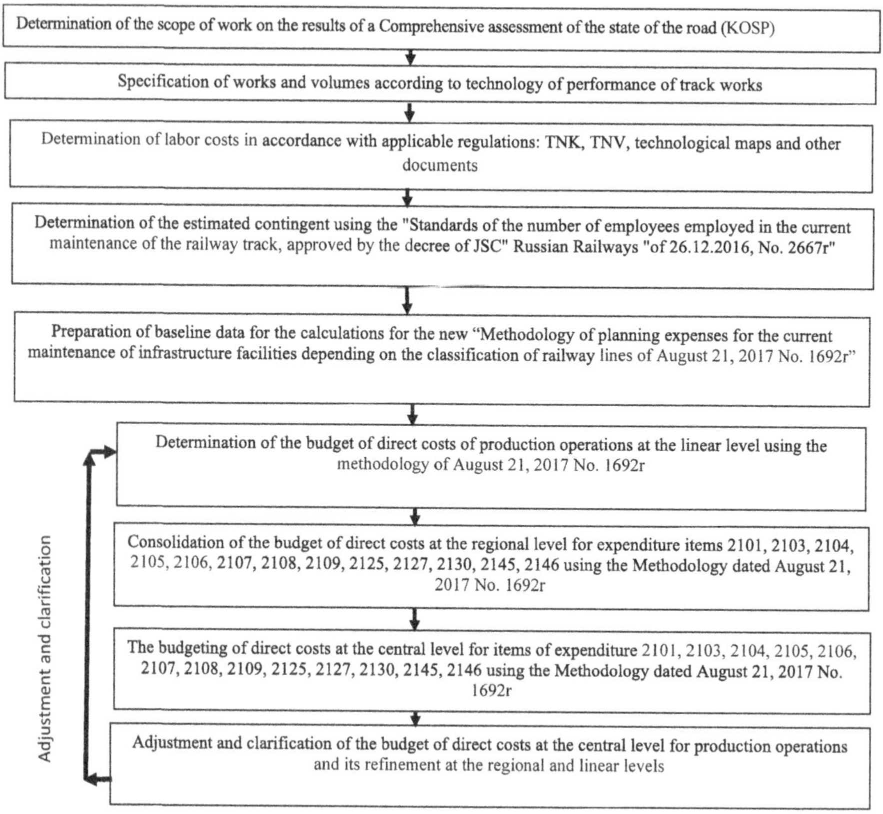

Fig. 1. Block diagram of the production and regulatory budgeting of direct costs for production operations of the track complex

In this block diagram, a comprehensive assessment of the state of the track (KOSP) [1] includes an assessment of the geometry of the track, the state of the elements of the upper structure of the track [2], the roadbed and artificial structures and is designed to improve the quality of the track content [3], determine the effectiveness of its maintenance.

Operation specification of works and volumes in accordance with the techniques of track work required to reconcile the revealed with the execution of the MICS of the types of works adopted their nomenclature in accordance with Regulatory and technical documentation [4].

This is followed by the stage of determining the labor costs in accordance with the current standards of TNK, TNV, technological maps and other documents [4].

A contingent of staff is determined by using "standards of the number of employees on the current contents of Railways, approved by the order of JSC "RZD" from 26.12.2016, No. 2667p" with the changes, clarifications and additions.

The next stage, the preparation of initial data for calculations on the new "Methodology for planning the costs of the current operation of Railways and facilities, depending on the classification of Railways, approved by the order of JSC "Russian Railways" dated August 21, 2017 № 1692r" [4], is characterized by the preparation of information from the relevant statistical reports of the EK ASUI, EK ASUFR, KASANT and other reporting forms [4].

3 Results

This is followed by the stage of performing calculations to determine the budget of direct costs for production operations at the linear level using the "Methodology for planning costs for the current operation of Railways and facilities, depending on the classification of Railways, approved by the order of JSC "Russian Railways" dated August 21, 2017 № 1692r". Features of the budget definition take into account the rules of formation and control of execution of normative-target budgets, classification and specialization of railway lines of JSC "RZD", TNK, local norms, norms of consumption of materials and products for the current content, operational norms of consumption of materials and other normative, technical and technological documents.

For the analysis of information uncertainty, an information model is proposed, for which the risk is considered as a possibility of losses (P) at a linear enterprise (L), arising from the adoption of investment decisions under uncertainty.

In addition, it is recommended that risk assessment take into account individual risk tolerance (J), which is described by indifference or utility curves. Thus, to describe the risk (P_{isc}), you can use any of these three parameters, the dependence (1)

$$P_{isc} = \{P; L; J\} \tag{1}$$

When carrying out a comparative analysis of all the above risk criteria, the advantages and disadvantages of their practical application are revealed. On the basis of the analysis, a generalized criterion is proposed – "risk price" (C_{risk}), which characterizes the amount of conditional losses possible in the implementation of the investment decision:

$$C_{risk} = \{Z; P\} \tag{2}$$

where Z - is defined as the sum of direct losses from the investment decision.

To determine the price of risk, such indicators are used that take into account the coordinates of the "vector": variance, standard deviation, coefficient of variation and so on. The refined definition of risk of the investigated project (further, SP) is offered.

IP risk (RIP) is a system of factors manifested in the form of a set of risks, individual for each participant of IP, both in quantitative and qualitative terms, dependence (3):

$$R_{IP} = \begin{bmatrix} R_{11}, & R_{12}, & R_{13}, & R_{14}, & \ldots & R_{1n} \\ R_{21}, & R_{22}, & R_{23}, & R_{24}, & \ldots & R_{2n} \\ \ldots & \ldots & \ldots & \ldots & \ldots & \ldots \\ R_{m1}, & R_{m2}, & R_{m3}, & R_{m4}, & \ldots & R_{mn} \end{bmatrix} \tag{3}$$

where n is the number of risks SP; m - number of project participants.

The emphasis in the risk definition is that the risk of IP is a complex system with numerous relationships that are manifested for each of the IP participants in the form of an individual combination - complex, that is, the risk of the i-th participant of the project (R_i) will be described in the form of dependence (4):

$$R_i = \{R_{i1}, R_{i2}, R_{i3}; R_{i4}, \ldots, R_{in}\} \tag{4}$$

The matrix column (3) shows that the value of any risk for each project participant is also shown individually.

This approach is the basis of the risk management algorithm. It allows us to consider the development of tools for quantitative analysis of IP. To solve the problems of this stage, an adequate, improved tools have been developed.

In particular, the tools of portfolio analysis in investment design, where it is proposed to use the theory of the portfolio to solve the problems of investment design.

For example, the calculation of the discount rate in determining the criteria for the effectiveness of IP. To calculate the discount rate, we use a model that is a synthesis of the model (CAPM - Capital Asset Pricing Model) and the cumulative approach:

$$r = r_c + b\,(r_n - r_c), \tag{5}$$

where: r_c - no risk free rate of return; r_n - market rate; b - risk factor, determined by the formula (6).

The advantage of the proposed method is that it combines the advantages of both models. A feature of the method is the calculation of the risk coefficient - b:

$$b = \frac{Cov(F_1, r)}{Var(r) \times b_1} + \frac{Cov(F_2, r)}{Var(r) \times b_2} + \ldots\ldots + \frac{Cov(F_n, r)}{Var(r) \times b_n} \tag{6}$$

In addition, for a comprehensive assessment of innovative projects in railway transport, the need to take into account the spatial relationship, which in turn requires the use of methods of Geoinformatics and digitalization.

4 Discussion

The conceptual framework for planning direct costs for the current maintenance of infrastructure facilities for different levels of management: linear, regional and Central have similar methodological aspects for the main calculation elements. Information for planning is formed in the structural units of the linear level with its subsequent consolidation at the regional and Central levels. As an example, the methodological aspects of planning direct costs for the current maintenance of infrastructure at the linear level (for the distance of the path) are considered.

Planning the amount of work on the maintenance of the superstructure (article 2101) is carried out on the basis of data from the Integrated Assessment of the Condition of the Track (hereinafter referred to as COSP). The COSP materials for the autumn complex commission control of the track and infrastructure are taken as the basis for labor planning.

Planning of direct production costs for the elements at the linear level under article 2101, is carried out according to the method in accordance with the dependence (7):

$$Z_p = Z_{pl} + B_{np} + O_{cn} + M_{tcn} + T_{tcp} + E_{tcp} + M_{np} + Z_{np} \quad (7)$$

where, Z_P – direct costs under article 2101; Z_{Pl} – labor costs of employees engaged in the execution of works on the current maintenance of infrastructure facilities; B_{np} – the amount of contributions under non-state pension agreements concluded in favor of employees. O_{cn} - website social contributions, including contributions to the unified social fund (ESF) and the amount of insurance premiums for compulsory social insurance against industrial accidents and occupational diseases for the planning period; M_{tcn} - the cost of materials on the current maintenance of infrastructure; T_{tcp} - fuel costs for the planning period (Article 2101); E_{tcp} - electricity costs attributed to article 2101; M_{np} - other material costs, including the costs of paying bills to outsourcing companies and third-party organizations that perform work on the current maintenance of the permanent structure under article 2101 under contracts; Z_{np} - other expenses, including the costs of paying bills for certification and inventory, payment of fees for registration, installation and operation of radio stations in accordance with the agreements concluded with third parties for the implementation of the current maintenance of infrastructure.

Payment for work for the link (group, brigade, and so on) of the normalized number, according to the techno-normalization card (hereinafter referred to as TNK), at the stage of the current maintenance of infrastructure facilities, is performed according to (8):

$$Z_{pl} = Z_m + Z_{pm} + Z_{br} + Z_c + Z_r + Z_k \quad (8)$$

where, Z_m - the cost of labor reckoning of road workers who are engaged in the performance of work on the current maintenance of infrastructure facilities per plan period; Z_{pm} - the cost of labor of workers engaged in the maintenance of traveling machines and mechanisms when performing work on the current maintenance of infrastructure facilities; Z_br - the cost of labor brigadiers (exempt), employed in the performance of work on the current maintenance of infrastructure; Z_c - the cost of labor signalists (not included in the standards of TNK), the performance of work relating to the article 2101; Z_r - the cost of labor distributors works in accordance with the technology of their implementation; Zk - is the tariff wage of railway condition controllers in accordance with the current infrastructure maintenance technology.

The definition of B_{np}, O_{cn}, M_{tcn}, T_{tcp}, E_{tcp}, M_{np}, Z_{np} Z_m, Z_{pm}, Z_{br}, Z_c, Z_r, Z_k, etc. is given in the MIIT developed under the guidance of Dr. Sc. prof. Volkova B. A. "Methods of planning expenses for the current operation of railways and facilities, depending on the classification of railway lines, approved by order of JSC "Russian Railways" of 21 August 2017. No. 1692r".

Define Z_p, Z_{pl}, and so on for expense items 2103, 2104, 2105, 2106, 2107, 2108, 2109, 2125, 2127, 2130, 2145, 2146 listed in the above Methodology, approved by the order of JSC "Russian Railways" of August 21 2017. №1692r.

The formation of the production and regulatory budget of direct costs for the current maintenance of infrastructure at the regional and Central levels, is carried out for the above items of expenditure in the form of their consolidation. At the regional level – by the number of line units, at the Central level – by the number of regional associations. This methodology is given in the above Methodology approved by the order of JSC "Russian Railways" of August 21, 2017. №1692r.

The consolidated budget transferred to the central level is adjusted and adjusted in accordance with the parameters of the income and expenses of the master budget of the whole branch of Russian Railways. After that, its adjusted parameters are sent first to the regional level, where its characteristics are refined for line enterprises and transferred to the enterprises.

The economic effect of cost planning, taking into account the classification and specialization of railway lines, compared with the previous Method is determined in accordance with the following dependence (9):

$$E_g = Z_{pc} - Z_{pn}, \tag{9}$$

where, E_g – annual economic effect of the introduction of the new Methodology; Z_{pc} – the main direct costs for the current maintenance of the way for the year, determined by the previous Method; Z_{pn} – the main direct costs for the current maintenance of the path for the year, calculated by the new Method.

5 Conclusions

The efficiency is determined by estimating the direct costs of the real distance of the path (hereinafter, if).

Direct costs (Z_{pc} and Z_{pn}) take into account costs that vary depending on the methods used.

These estimates do not include costs for the implementation of the activities of PCH that does not depend on the classification and specialization of railway lines.

To determine the direct costs for maintenance of infrastructure in the present drive by the new Method calculated the number of regulatory fitters way employed on the current contents of the path, depending on the class of train paths using the "standards of the number of employees on the current contents of the railway track", approved by order of JSC "RZD" from 26.12.2016, No. 2667p.

The economic effect of the calculations performed on the new "Method of planning costs for the current operation of Railways and structures, depending on the classification of Railways from August 21, 2017 № 1692r", taking into account the class and specialization of lines, in comparison with the previous "Method of planning costs for the current operation of the track from 30.12.2008 № 2897r", is characterized by a decrease in the wage Fund of workers engaged in the maintenance of the railway track by 7.8%, as well as a decrease in the cost of materials for the maintenance of the track by 7.75%.

References

1. Volkov, B.A., Kovalenko, N.I., Dobrin, A.Y., Kovalenko, A.N.: Reduction of costs for the current operation of the track depending on the classification of railway lines. Path Track Facil. **6**, 18–21 (2019)
2. Volkov, B.A., Kovalenko, N.I., Dobrin, A.Y., Kovalenko, A.N.: On the method of planning costs for the current operation of the track. Path Track Facil. **5**, 23–26 (2018)
3. Romanov, I.A.: Principles of evaluating the effectiveness of innovative projects in the field of transport. Adv. Mod. Nat. Sci. **5**, 135–136 (2013)
4. Tsvetkov, V.Y.: Conceptual model of the innovative projects efficiency estimation. Eur. J. Econ. Stud. **1**(1), 45–50 (2012)
5. Markelov, V.M., Romanov, I.A.: Innovation and geoinformatics. Distance Virtual Learn. **12**, 53–57 (2012)

Assessment of the Mitigation of Consequences Resulting from Incidents at the Railway

Nikolay Kovalenko(✉), Valentin Ponomarev, Nina Kovalenko, Nadezhda Fomina, and Aleksandr Kovalenko

Russian University of Transport (MIIT), Obraztsova Street 9/9, Moscow 127994, Russia
kni50@mail.ru

Abstract. The article presents materials on the development of a method for determining the estimated time of the train start and minimization the time required to mitigation of consequences resulting from incidents (crash, accidents, rolling stock derailment, equipment failure) on the railways of the JSCo "RZD". The method is aimed at establishing the minimal level of the reduced time spent on repairs of the railway track and infrastructure in order to determine the estimated time when the train circulation is restored and minimizing the time required to mitigation of consequences resulting from incidents under operation management.

Keywords: Railway incident · Emergency train · Emergency response and recovery operations · Accident · Reduced costs · Time spent for the track repair · Train schedule (TS) · "Window" in the train schedule · Level of risk · Risks of additional costs · Minimum of reduced costs of repairs · Stop in the train circulation

1 Introduction

The incidents at the railway include crashes, accidents, rolling stock derailment, equipment failures [1, 2].

Before the derailed rolling stock can be lifted and removed from the track, the supervisor of the emergency train (hereinafter referred to as ET) checks whether the rolling stock is properly fixed on the track and inspects the point of derailment [1].

The works on recovery includes [3, 4]:

- identification of the scope of work, the necessary resources and manpower;
- selection of a procedure and order of the work performance, development of recovery plans;
- security of work sites and fencing around them;
- provision of persons involved in the work performance with personal protection means;
- required fire safety measures;
- assessment of the decisions made from the point of view of the damage they cause to the environment.

Z. Popovic et al. (Eds.): TransSiberia 2019, AISC 1115, pp. 184–194, 2020.
https://doi.org/10.1007/978-3-030-37916-2_20

The works must be performed in a certain order in compliance with the approved plan. After the derailed rolling stock is removed from the roadbed, the workers of the emergency trains go on to repair the track, adjust the damaged overhead line supporting structures, restore communication facilities and signaling-arrangement systems [5].

The EP supervisor bears responsibility for the strict observance of the operational recovery plan in terms of lifting the derailed rolling stock. Any additional damage to the rolling stock and containers with hazardous cargo is prohibited [1].

The crane (cranes) is (are) operated under the supervision of the person responsible for safe crane operation, who was certified and selected for appointment from among the engineers and technicians by the order of the enterprise or the railway division [6].

The rolling stock containing explosive materials, flammable gases, and highly flammable and combustible cargos shall not be lifted or repaired without non-sparking tools and engineered equipment (based on copper fouling or aluminum alloys) [7].

Explosion-proof lighting devices shall be used to illuminate the work site, if necessary [7, 8].

In the work plan, the supervisor must anticipate a possible inspection of the incident scene by the competent persons (video filming of the area, photographing, charting, sampling, etc.) [9].

The scope of work on mitigation of consequences resulting from traffic accidents and other disruptions of the traffic safety on the railway as well as recovery of tracks shall be aimed at improving the strength, bearing stability, durability and other reliability indicators for the railway as a whole, and its constituents and elements to ensure the safe passage of trains through the site [10, 11].

The recovery of the railway track covers works on the plan and profile of the line, superstructure, roadbed, small and medium bridges (over-bridges), pipe-culverts considered to be reconstructive in nature according to the Registry (List) of the works involved in reconstruction and repairs of the railway track and its permanent facilities, as enacted by the order of JSCo "RZD" dated January 30, 2009 No. 182p, as well as in accordance with the Guidelines on the classification of works for the recovery of engineering structures of JSCo "RZD", enacted by the order of the JSCo "RZD" dated December 30, 2010 No. 2795p.

Recovery of the roadbed and its facilities shall proceed from engineering processes developed for certain works which require the available mechanization means.

If a message on a disruption in the normal functioning of the power supply units is received, the electric power dispatcher establishes the location, nature, scope and specifics of the damage, takes measures to disable the damaged section, issues the necessary prohibitions or warnings for the train handling, arranges the visit of workers and the collection of emergency repair tools of the power supply division, monitors the crew passage to the damaged section. Jointly with the supervisor, he determines the order of restoring the damaged power supply units, taking into account the current situation with trains [8].

When the facility is serviced by the field operation and maintenance crew in case of an emergency, the electric power dispatcher arranges the works engaging the emergency crews supervised by the head of the overhead line area of the power supply division, head of the network area of the power supply division, head of the railway substation of the power supply division.

2 Description of the Problem

Depending on the extent of damage to the power supply units, the electric power dispatcher sends the emergency response and recovery equipment, motor-rail locomotives, rail motor trolleys and breakdown lorries with crews or arranges the arrival of the staff with trains passing by, sends the train dispatcher a request and monitors the timely dispatch of emergency response tools and, if necessary, the emergency trains. In addition, jointly with the dispatcher of the regional communications center (RCC), works on the telephone connection at the damaged site for the reconstruction manager to contact the train dispatcher, the electric power dispatcher, the electricity supply managers, the electricity supply service of the Railway Infrastructure Administration and any other employees involved in works on recovery [12].

Before the arrival of the crew at the site of damage of the overhead line, the electric power dispatcher should, if possible, remove the voltage from the damaged section by switching-off the disconnectors, switches and disassembly their circuit. Works should be done, as a rule, by the supervisor, senior electrical engineer, electrical engineer or electrician of the overhead line area of 5th qualification level who has the right to act as the works supervisor. In case of works with power supply units of power supply areas up to 1000 V, this should be an electrician of at least 4th qualification level.

Upon completion of the works on recovery of the power supply units, the electric power dispatcher notifies the train dispatcher about the lifting of restriction or prohibition on the handling of trains, about the supply of voltage and about the completion of works with making a record in the "Dispatching orders book" of the DU-58 form. If the train dispatcher is located at a distance from the electric power dispatcher, then the notification is transmitted via selective telephony, equipped with a communications recorder, or via telephone message by fax indicating the surname of dispatcher, date and time of transmission [13].

The electric power dispatcher records all the disruptions in the normal operation of power supply units, disruption in the operation of telecontrol devices, power dispatch communications with a corresponding record in the daily sheet of the EU-89 form or in the Inspection and Failure Book of the EU-83 form indicating the time, reason, nature and extent of the damage, time of calling the crew and their arrival at the site of emergency response and recovery facilities and crews, temporary and full recovery, as well as train delays.

The power dispatcher shall immediately inform the train dispatcher on duty, the head of the relevant power supply unit, the power supply division engineer, the electrification and power supply service managers of the Railway Infrastructure Administration, the territorial center for infrastructure status management of the Management Center for Infrastructure Maintenance (Management Center For Infrastructure Maintenance of the Infrastructure Administration).

In case of an emergency power supply disconnection of the overhead line, the signaling system in terms of external power supply and impossibility to supply voltage, the electric power dispatcher must immediately inform the train dispatcher specifying the disconnection time and recording the notification number and the appropriate time in the operational register [14].

In case of an emergency disconnection of the main and spare automatic blocking lines and impossibility to supply voltage to one of the lines, the electric power dispatcher must notify the division engineer and the regional communication center dispatcher.

The timely performance of all works on the infrastructure aimed at mitigating the consequences of the emergency situations at the territorial level of management is controlled by the Management Center For Infrastructure Maintenance of the Infrastructure Administration [12].

3 The Goal of Developing a Methodology

The development of efficient alternatives for the track repairs (recovery after an incident) and establishment of the necessary degree of reliability and safety must be based on the economic rationale influencing the duration of the "windows".

The methods for assessing the efficiency with which the consequences of the incident are mitigated provide for determination of the estimated time when the circulation is restored and minimization of the time spent (hereinafter, Methods) are aimed at establishing the minimum of the reduced time spent on repairs of the track and the infrastructure under operational management.

4 Materials and Methods

1. Determination of the minimum reduced costs of repairs, taking into account the additional costs spent on the infrastructure and transportation-related losses, when choosing the duration of "windows" to perform the works on the track.
2. Taking account of the risks associated with additional costs and decreased profit when choosing the duration of "windows" to perform the works on the track, ensuring the lowest possible losses.
3. Calculation of the costs incurred due to stopped circulation of trains caused by "windows" as well as losses on their circulation through the section (risk of a reduced profit to be made by JSCo "RZD").
4. Determination of costs related to the track facilities when choosing the duration of "windows" for the works to be performed on the track, which allows for minimization of time loss.

5 Results

Under the current recommendations, the reduced costs associated with the "window" duration are calculated as the sum of the time spent within a single period to repair the railway track and infrastructure as well as the time spent in the course of transportation.

The reduced time spent (R_{zan}, hours) is expressed as the following ratio:

$$\mathrm{R_{zan}} = \mathrm{T_{tr}} + \mathrm{Z_{px}} + \alpha_{\mathrm{pot}} \cdot \mathrm{P_d}, \tag{1}$$

where: T_{tr} – the time spent on the emergency train passage from the dislocation base to the incident scene and their return to the connection station, hours;

Z_{px} – time spent within a single period of the "window" for repairs of the track and the related infrastructure in the course of mitigating the consequences of an incident, hours;
P_d – loss of time resulting from train delays (additional stops and standing time), changes in the schedule and routes during the determination of the "window" duration and after its end, hours;
α_{pot} – empirical parameter resulting from approximation of the generalized time loss caused by delays, idle hours, changing routes of the trains, etc. during the "window" and after its end, showing their changes when the "window" duration is growing.

The time spent by the machines and working trains to get to the work site and back (T_{tr}) is calculated according to the following formula:

$$T_{tr} = \sum_{n=1}^{N} n_{lok} \cdot t_{lok}, \text{ hours} \tag{2}$$

where n_{lok} is the number of locomotives involved in servicing the working trains according to the operational project, items;

$n \ldots N$ – number of operational trains involved in mitigation of the consequences resulting from the incident;
t_{lok} – the time during which the locomotive is used, hours.

$$t_{lok} = 2\left(\frac{l_1}{v_1} + \frac{l_2}{v_2} + t_{baz} + t_{ct}\right), \text{ hours} \tag{3}$$

where l_1 - average distance from the base to the incident scene and back, km;

l_2 – average distance from the base to the assigned depot of the locomotive and back, km;
v_1 – average speed of the operational trains circulation on the section (approximately v_1 = 30–40) km/h;
v_2 – the same for a single locomotive ($v_2 = 40 \div 50$ km/h), km/hour;
t_{baz} – duration of the shunting service at the base and holding time for the route (about 0,5 h), hours;
t_{ct} – time for switching and waiting for the "window" duration at the station adjacent to the area under management, hours.

The instruction on the procedure for the provision and use of "windows" for repairs and construction and assembly on the railways of the JSCo "RZD" [4, 6] recommends the following: "The required period of time for the planned scope of work with the running line closure (t_{okna}) is to be determined by the limiting machine":

$$t_{\text{okna}} = \frac{L_{\text{f}}}{V_{\text{ved}}}, \text{ hours} \tag{4}$$

where L_f – scope of work to be performed within the "window", km;

V_{ved} – the speed of performing the leading operation for the track repairs or technological performance of the leading machine, km/hour.

Technological performance refers to production per a unit of time, taking into account the anticipated loss of time (to shift the trains to the related tracks, to lay polystyrene plates, to replace geotextile coils, to reload the contaminants, to tighten the sets of links, etc.) which does not depend on the "window" duration.

Performance within a "window" will correspondingly reach:

$$L_f = t_{okna} \cdot V_{ved}, \text{ km} \tag{5}$$

The analysis of practical "window" arrangement shows that repairs and construction comply with the engineering processes that require a fixed duration of "windows" (running meters/hour, km/day).

The time spent on the repairs of the tracks and infrastructure recovery ($Z_{voc.infr}$) carried out in the course of mitigating the consequences of the incident is established either by estimating the actual losses for a specific event or proceeding from the costs of the production technology, the corresponding typical engineering process per a unit of length, for example, dependence on 1 km (5).

$$Z_{\text{voc.infr}} = Z_{\text{px}} + Z_{\text{infr}}, \text{ hours} \tag{6}$$

Where: Z_{infr} – costs necessary to repair the infrastructure damaged in the incident.

The time spent on the track works is based on the costs of production technology, the corresponding typical engineering process per a unit of time (Z_{px}, hours) proceeds from the following ratio (6):

$$Z_{px} = \frac{(T_{px} + \Delta T_{nac} + \Delta T_{vyem} + \Delta T_{platf} + \Delta T_{pereezd} + \Delta T_{ctr.perevod} + \Delta T_{izoct} + \Delta T_{blok-poct}) \cdot Kp_c}{L_{p.np}} + \\ + \rho_{otk} \cdot C_{OTK}, \tag{7}$$

where: T_{px} – average time spent on the track works aimed at mitigating the consequences of the incident equal to the corresponding set of technological operations inherent in a typical engineering process (or according to the actual labor costs), hours;

$L_{p.np}$ – reduced length of the estimated area, km;

K_{pc} – factor of the time adjustment;

ρ_{otk} – average statistical probability of the equipment failure at the track facilities (average statistical probability of risks) for the considered section of the route;

C_{OTK} – average time of the equipment failure at the track facilities for the considered section of the track, hours.

The factor of the time adjustment in relation to the operation conditions (K_{pc}) is based on the following Eq. (8):

$$K_{pc} = K_{pl} \times K_{cg} \times K_{\text{‰}} \times K_{gi}, \quad (8)$$

where: K_{pl} – factor of adjustment for the increasing time (or labor costs) while performing works at the curved track;

K_{cg} – factor of adjustment for the time (or labor costs) depending on the distance from the base of the emergency trains;
$K_{\text{‰}}$ – factor of adjustment for the time (or labor costs) depending on the incline of the physical profile of the line;
K_{gi} – factor of adjustment for the time (or labor costs) depending on the working capacity of the section under consideration;
ΔT_{nac} – additional time for work performance (labor costs) within a high bank;
ΔT_{vyem} – additional time for work performance (labor costs) within a heavy cut;
ΔT_{platf} – additional time for work performance (labor costs) within a high or low platform;
$\Delta T_{pereezd}$ – additional time for work performance (labor costs) within a crossing;
$\Delta T_{ctr.perevod}$ – additional time for work performance (labor costs) in the course of disassembling and laying track switches;
$\Delta C_{izoctyk}$ – additional time for work performance (labor costs) in the course of disassembling the existing insulated joint and installing a new one;
$\Delta T_{blok\text{-}poct}$ – additional time for work performance (labor costs) in the course of installing a new block station at the running line within the area under repairs and disassembling the existing one.

Time spent on the recovery of infrastructure (Z_{inf}) damaged by the incident when performing additional works, for instance, rearrangement of the overhead line ($Z_{kont.ceti}$), signaling and communication means ($Z_{czb.cvyzi}$) and other associated works is provided for by individual engineering processes aimed at repair and construction (dependence 9):

$$Z_{inf} = Z_{kont.ceti} + Z_{czb.cvyzi}, \text{ hours} \quad (9)$$

Certain financial documents provide for the costs of repairing the damaged cars and locomotives, insurance payments resulting from the damage or loss of the transported cargo, injuries or death of passengers.

When calculating the optimal duration of the "windows", the maximum delays of the trains for the entire period of repairs associated with mitigating the consequences of the incident and the maximum performance within the "window" are taken into account, covering the guaranteed scope of work to be performed [12].

The operating time of transportation facilities (T_{pr}, hours) as associated with the arrangement of the trains to pass through within a "window" (t_{okna}) is determined on the basis of the Eq. (10):

$$T_{pr} = (1+\xi)\cdot\left(\sum_{m=1}^{m} N_{gr}\cdot t_{gr} + \sum_{p=1}^{p} N_{pac}\cdot t_{\text{pac}} + n_{\text{oct}}^{\text{gr}} * t_{oct}^{gr} + n_{\text{oct}}^{\text{pac}} * t_{oct}^{pac} + n_{oct}^{prig} * t_{oct}^{prig}\right) \tag{10}$$

where $\sum_{m=1}^{m} N_{\text{gr}}\cdot t_{\text{gr}}$; $\sum_{p=1}^{p} N_{pac}\cdot t_{\text{pac}}$ – total period of delays of freight and passenger trains, respectively, including commuter trains, train-hours;

$n_{\text{oct}}^{\text{gr}}, n_{\text{oct}}^{\text{pac}}, n_{\text{oct}}^{\text{prig}}$ – total number of additional stops for freight and passenger trains, respectively, including commuter trains, items;

$t_{oct}^{gr}, t_{oct}^{pac}, t_{oct}^{prig}$ – time (duration) of one additional stop of a freight, passenger and commuter trains, respectively, hours.

ξ – the probability of taking into account an undesirable event or situation caused, for example, by exceeding the "window" period or other factors classified as the risk of additional time to be spent and the costs to be incurred [7, 8].

Additional train-hours are determined as the sum of the products of the delayed trains' number by the duration of the "window". Data on the number of trains and the time of the delay for each of them are taken in accordance with the schedule alternative.

The turnaround of a freight car ($\Delta\vartheta_{ок}$) due to the "windows" proceeds from the following equation:

$$\Delta\vartheta_{ok} = \frac{\Delta\sum_{n=1}^{n} N_{gr}\cdot t_{gr}\cdot m_{vag}}{24\cdot U_{\text{vag}}},\ \text{hours (days)} \tag{11}$$

where, m_{vag} – the number of cars in a freight train, cars; U_{vag} – the work of the freight car fleet (loading, unloading, acceptance of loaded cars), cars.

The growth of the freight car turnaround due to the "window" increases the need for additional resources to ensure a given scope of transportation:

$$\Delta n_{\text{vag}} = \Delta\vartheta_{ok}\cdot U_{vag} = \frac{\Delta\sum_{n=1}^{n} N_{gr}\cdot t_{gr}\cdot m_{vag}}{24},\ \text{cars.} \tag{12}$$

Additional need for locomotives within a "window" is determined in the following manner:

$$\Delta M_{lok} = \frac{\gamma_{lok}\cdot\Delta n_{\text{vag}}}{m_{vag}},\ \text{locomotives} \tag{13}$$

where, γ_{lok} – factor of the need for locomotives per one pair of freight trains, taking into account the additional need for a locomotive fleet to ensure uninterrupted circulation of trains.

Decreased performance of the locomotives ($\Delta\omega_{lok}$) within a "window" is determined on the basis of the following Eq. (14):

$$\Delta\omega_{lok} = \omega_{l\ pl} - \frac{\sum_{l=1}^{L} P_{lok\ netto} \cdot L_{lok}}{\gamma_{netto} \cdot (M_{n\ lok} + \Delta M_{lok})}, \text{ thous. t} \cdot \text{km, gross weight} \quad (14)$$

where, $\omega_{l\ pl}$ – planned average performance of a locomotive on a daily basis, thous. t · km, gross weight;

$\sum_{l=1}^{L} P_{lok\ netto} \cdot L_{lok}$ - locomotive transportation flow, thous. t · km, net weight;

γ_{netto} – factor for converting thous. t · km, gross weight, to thous. t · km, net weight, (approximately, γ_{netto} = 0,67);

$\Delta M_{n\ lok}$ – planned scope of the locomotive fleet, locomotives;

The general ratio for determining the losses of transportation (P_d, rubles) is established as follows (15):

$$P_d = E_{pr} + (1 + \psi) \cdot T_R, \quad (15)$$

where, T_R – time (duration) of possible losses in transportation when estimating the costs resulting from an undesirable event or situation (risk), hours (days);

ψ – the probability of an undesirable event or situation (risk).

6 Debate

As an example, the figure shows the calculations of the optimal "window" duration when restoring the path after the incident (overhaul of the track with new materials), taking into account the costs of related infrastructure facilities (Fig. 1).

Determination of the current costs for railway infrastructure uses the rates depending on the spent crew-hours of the locomotive crews, locomotive kilometers, locomotive hours, cost of electricity, etc.

Analysis of the results produced in the course of determining the optimal duration of the "window" within the recovery of the track after the incident (overhaul of the track with new materials) taking into account the costs of related infrastructure facilities (see the figure) established the following:

- if the "window" duration increases, the costs per unit of work on the track in one hour of the "window" are reduced in a decreasing, virtually, parabolic fashion;
- transportation losses and costs necessary for related infrastructure facilities grow with the duration of work.

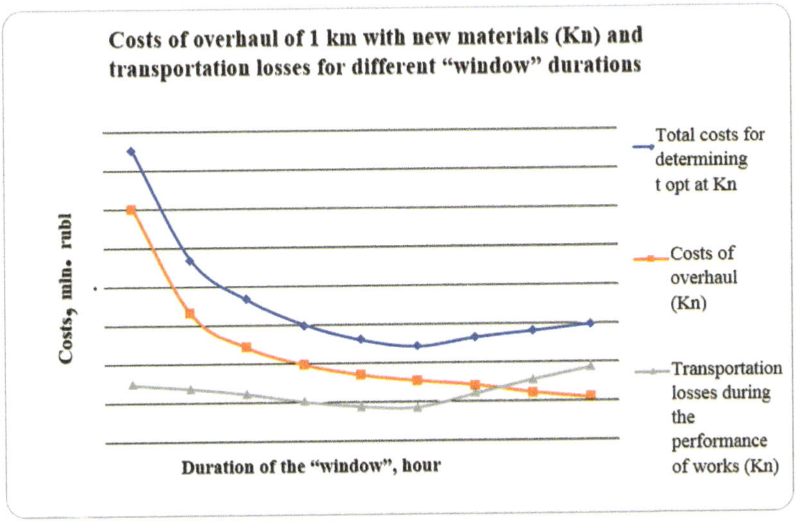

Fig. 1. Determination of the optimal "window" duration when restoring the path after the incident (overhaul of the track with new materials), taking into account the costs of related infrastructure facilities

7 Conclusions

The increase is taking place under the polynomial dependence. Approximately in the first half (first third) of the "window", losses decrease to a certain degree. Then they are increasing virtually along a parabolic curve.

Such changes in transportation losses and costs necessary for related infrastructure facilities are caused by the production reserves in the train schedule and the operation of related infrastructure facilities (the first half (first third) of the "window"). A further increase in losses which is considered significant results from both the production and technological nature of the transportation and related infrastructure facilities.

In the example shown in the figure, the duration of the "window" which is spent to restore the track after the incident is 12 h. The estimated duration of the optimal "window" for the entire complex of works aimed at restoring the track after the incident (overhaul with new materials) is 9 h.

References

1. Santos-Reyes, J., Beard, A.N.: A systemic analysis of the Edge Hill railway accident. Accid. Anal. Prev. **41**(6), 1133–1144 (2009)
2. Debois, S., Hildebrandt, T., Sandberg, L.: Experience report: constraint-based modelling and simulation of railway emergency response plans. Proc. Comput. Sci. **83**, 1295–1300 (2016)
3. Cacciabue, P.C.: Human error risk management methodology for safety audit of a large railway organisation. Appl. Ergon. **36**(6), 709–718 (2005)

4. Schipper, D., Gerrits, L.: Differences and similarities in European railway disruption management practices. J. Rail Transp. Plann. Manag. **8**(1), 42–55 (2018)
5. Cheng, Y.H., Liang, Z.X.: A strategic planning model for the railway system accident rescue problem. Transp. Res. Part E: Logist. Transp. Rev. **69**, 75–96 (2014)
6. Lam, C.Y., Tai, K.: Network topological approach to modeling accident causations and characteristics. Analysis of railway incidents in Japan. Reliab. Eng. Syst. Saf. **193** (2020). Article 106626
7. Xu, J., Tang, Z., Yuan, X.: A VR-based the emergency rescue training system of railway accident. Entertain. Comput. **27**, 23–31 (2018). https://doi.org/10.1016/j.entcom.2018.03.002
8. Wang, D., Zhang, T.: Engineering accidents in society: a comparison of Chinese and American railway accident investigation. Technol. Soc. **43**, 69–74 (2015)
9. Ouyang, M., Hong, L., Yu, M.H., Fe, Q.: STAMP-based analysis on the railway accident and accident spreading: taking the China-Jiaoji railway accident for example. Saf. Sci. **48**(5), 544–555 (2010)
10. Bababeik, M., Khademi, N., Chen, A.: Increasing the resilience level of a vulnerable rail network: the strategy of location and allocation of emergency relief trains. Transp. Res. Part E: Logist. Transp. Rev. **119**, 110–128 (2018)
11. Li, C., Tang, T., Chatzimichailidou, M.M., Jun, G.T., Waterson, P.: A hybrid human and organisational analysis method for railway accidents based on STAMP-HFACS and human information processing. Appl. Ergon. **79**, 122–142 (2019)
12. Khademi, N., Babaei, M., Schmöcker, J.D., Fani, A.: Analysis of incident costs in a vulnerable sparse rail network – description and Iran case study. Res. Transp. Econ. **70**, 9–27 (2018)
13. Li, K., Wang, S.: A network accident causation model for monitoring railway safety. Saf. Sci. **109**, 398–402 (2018)
14. Sañudo, R., Moura, J.L., Ibeas, A.: Decision making system for stopping high speed trains during emergency situations. Proc. – Soc. Behav. Sci. **162**, 429–438 (2014)

Improving of the Electrothermal Characteristics of the Contact Line

Alexander Galkin, Alexander Buynosov, Alexander Paranin, and Andrey Batrashov(✉)

Ural State University of Railway Transport, Kolmogorova Street, 66, 620034 Yekaterinburg, Russia
cool45@mail.ru

Abstract. The currently used methods of electrothermal calculation of the DC contact network have a number of assumptions that do not take into account the contact network clamps, link strings, current collection by electric rolling stock, the interrelated calculation of the current distribution and temperature. The aim of the study is to improve the design of the contact network, which allows to eliminate the places of increased current and heat load in its elements. The methods used in the work allowed to obtain a high degree of conformity of the mathematical model to the real physical process of current and temperature distribution. The results of comparison of calculated and experimental data are presented. The places of increased current and thermal loads are identified, as well as constructive solutions for their elimination are proposed. The proposed solutions make it possible to equalize the distribution of current and temperature in the elements of the contact network, which in turn contributes to an increase in the load capacity of the contact network by 6%. The developed technique is implemented in an application for modeling using COMSOL in the form of an executable file. It can be used by catenary engineers and does not require knowledge of mathematical modeling.

Keywords: Contact line · Direct current · Current carrying capacity · Electrothermal calculation · Finite element method · Current distribution · Temperature distribution · Validation

1 Introduction

A significant part of Russia's Railways is electrified with direct current. The transport strategy of the Russian Federation provides for the development of heavy, high-speed and high-speed traffic. A significant part of the development sites is electrified on direct current. First of all, these are areas near agglomerations-millionaires around cities such as Moscow, St. Petersburg, Yekaterinburg, Novosibirsk, etc.-nodal points in the period of intensive electrification of the 50–60 – ies of the last century. Due to the relatively low voltage level of 3 kV DC system has high current loads, which in turn entails significant thermal loads. In addition, in the areas of heavy trains, there is an increased number of failures of those elements in the contact network for which no justifying calculations were made in the design. Thus, the need to increase the volume of traffic

Z. Popovic et al. (Eds.): TransSiberia 2019, AISC 1115, pp. 195–205, 2020.
https://doi.org/10.1007/978-3-030-37916-2_21

requires a qualitative increase in the approach to the electrical and thermal calculations of the DC contact network.

For electrothermal calculation it is necessary to solve two interrelated equations: electric potential distribution and temperature distribution. In this case, the problem of electrothermal calculation is nonlinear, since the physical properties of the materials of the elements of the contact network, such as electrical conductivity, heat capacity, thermal conductivity are temperature dependent. Numerical methods are widely used to model nonlinear processes. In particular, the finite element method (FEM) has proven itself well [1–6]. This method was used as a basis for improving the methods of electrothermal calculation of the DC contact network.

The aim of the study is to improve the thermal calculations and characteristics of the DC contact network. The scientific novelty of the work lies in the development of methods of electrothermal calculation of the DC contact network taking into account the contact network clamps, temperature dependences of materials and contact of the contact wire with the current-collecting plates.

2 Mathematical Model and Technique of Electrothermal Calculation

The contact network is an extended object. The diameters of the wires are much smaller than their length, which allows us to consider the wires of the contact network in the form of lines, i.e. one-dimensional objects. In the process of calculation, the basic required values are the electric potential φ and temperature T. the Contact network contains many structural elements that affect the redistribution of current near them. These include spring wire, the average contact wire anchors, transverse connectors, non-insulated console and strings, parts of the mate anchor sites. To account for all elements that affect current distribution, the exact geometry of the catenary must be drawn. Clamps and other fittings of the contact network are also represented as one-dimensional objects. In the design geometry, they are located at the joints and fastening wires of the contact network. At the intersection of wires and clamps are set the conditions of thermal and electrical contact between the elements. An example of the calculated anchor section of the contact network is shown in Fig. 1.

Fig. 1. The calculated geometry of the contact network

The mathematical model is based on the equations of continuity of electric current and the equation of heat balance, which are reduced to a unit of length, i.e. the linear characteristics of the simulated elements of the contact network and the current load are given. The system of differential equations (1) is as follows:

$$\begin{cases} g_{wire}^{lin} \cdot \nabla^2 \varphi_{wire} = i_{clamp} - i_{pant}; \\ C_{wire}^{lin} \frac{\partial T_{wire}}{\partial t} - k_{wire}^{lin} \cdot \nabla^2 T_{wire} = q_{wire}^{res} + q_{wire}^{sh} - q_{wire}^{conv} - q_{wire}^{ST} + w_{pant} - w_{clamp}; \\ g_{clamp}^{lin} \cdot \nabla^2 \varphi_{clamp} = -i_{clamp}; \\ C_{clamp}^{lin} \frac{\partial T_{clamp}}{\partial t} - k_{clamp}^{lin} \cdot \nabla^2 T_{clamp} = q_{clamp}^{res} + q_{clamp}^{sh} - q_{clamp}^{conv} - q_{clamp}^{ST} + w_{clamp}; \\ i_{clamp} = \frac{\varphi_{wire} - \varphi_{clamp}}{R_{w-c} \cdot l_{clamp}}; \\ w_{clamp} = \frac{h_{w-c} \cdot S_{w-c} \cdot \left(T_{wire} - T_{clamp}\right)}{R_{w-c} \cdot l_{clamp}}; \\ C_{pant} \frac{\partial T_{pant}}{\partial t} = \left(Q_{cw-p}^{res} - Q_{cw-p}^{fr}\right) \cdot p_{pant} + Q_{pant}^{res} + Q_{pant}^{sh} - Q_{pant}^{conv} - Q_{pant}^{ST} - Q_{pant}^{cw}; \\ w_{pant} = \frac{h_{pant} \cdot S_{cw-p} - \left(T_{pant} - T_{cw}\right)}{b_{pant}}, \end{cases} \tag{1}$$

where *wire*, *clamp*, *pant* – indices related to the wires of the contact network, the terminals, and current collecting plates; g_i^{lin} – linear conductivity, S · m; φ_i – calculated value of electric potential, V; i_{clamp} – linear current source characterizing the current flowing into the wire from the clamp, A/m; i_{pant} – linear current source characterizing current collection by electric rolling stock, A/m; C_i^{lin} – linear heat capacity, J/(m · K); T_i – calculated temperature value, K; q_i^{res} – linear heat flow from Joule heating, W/m; q_i^{sh} – linear heat flow from solar radiation heating, W/m; q_i^{conv} – linear heat flow from convection cooling, W/m; qi^{ST} – linear heat flow from cooling by radiation, W/m; w_{pant} – heat source in contact with current collector, W/m; w_{clamp} – heat flow from wire to clamp, W/m; R_{w-c} – transition resistance in contact of clip and wire, Om; l_{clamp} – clip length, m; h_{w-c} – thermal conductivity of the contact in the connection of the clamp and the wire, W/(m^2 · K); S_{w-c} – contact area in connection of clamp and wire, m^2; Q_{cw-p}^{res} – Joule heating in contact with current collector, W; Q_{cw-p}^{fr} – heating about friction in contact with the pantograph, W; p_{pant} – the proportion of heat released in contact relating to current-collecting plates; Q_{pant}^{res} – Joule heating of current-collecting plates, W; Q_{pant}^{sh} – heating of current-collecting plates by solar radiation, W; Q_{pant}^{conv} – convective cooling of current-collecting plates, W; Q_{pant}^{ST} – cooling of current-collecting plates by radiation, W; Q_{pant}^{cw} – heat removal from current-collecting plates to the contact wire, W; h_{pant} – thermal conductivity of contact with current-collecting plates, W/(m^2 · K); S_{cw-p} – contact area in connection with current-collecting plates, m^2; T_{cw} – contact wire temperature, K; b_{pant} – total width of current-collecting plates, m.

The first two equations of the system (1) refer to the wires of the contact network, the next pair of equations refers to the terminals of the contact network. The fifth and sixth equations describe the conditions of thermal and electrical contact. The calculation of the contact thermal conductivity h_{w-c} and the transient electrical resistance R_{w-c} is carried out on the basis of contact interaction models presented in [7, 8]. The seventh and eighth equations of the system relate to the calculation of the average temperature of current-collecting plates and the condition of their contact with the contact wire with a similar model of contact interaction, as well as according to [9]. In Fig. 2 clearly shows the calculated area of wires, clamps and their intersections.

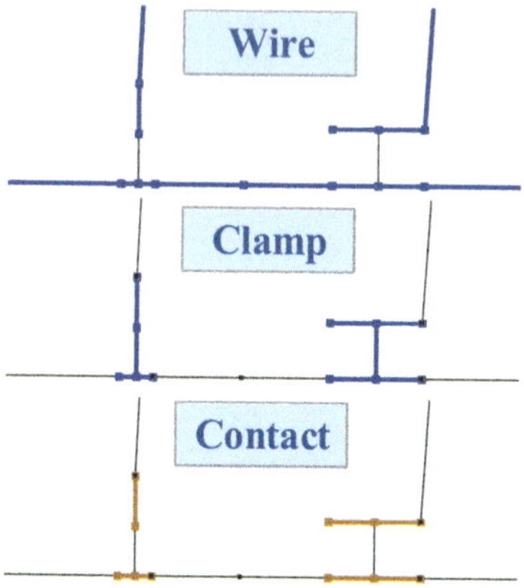

Fig. 2. Computational domain

The calculation uses two types of representation of the contact network clamps – T- and H-shaped. T-a figurative representation of the characteristic of supporting clips, such as stranovye clamps, latch clamps to the contact wire, saddle suspension cable, etc. H-figurative representation of the clips is typical for the connection and supply of clamps, clips the average embedment, etc., i.e. most often it is current-carrying clamps. Different performance results from different functional and structural design of the clips. The dimensions of the linear elements describing the clamp are defined as projections of the corresponding dimensions of the contact network clamp, i.e. length and height. The linear elements are also given equivalent parameters reduced to a unit of length.

Conditions of convective cooling for wires is defined according to [10–12] as for elements of round section, and for clamps-rectangular.

3 Validation of the Developed Analysis Model

Validation of the developed mathematical model [13] for different variants of the design of the contact network based on the scale model is given in [14].

To confirm the correspondence of the current and temperature distribution in the real elements of the contact network, studies were carried out on the real elements of the contact network on the suspension section of the PBSM-70 + MF-100 with link strings made of wire BSM-4 (Fig. 3).

In Fig. 4 the comparison of calculated and experimental temperature values near the seat of the carrier cable KS-008 is presented.

In Fig. 5 the comparison of calculated and experimental temperature values near the fixing clamp of the contact wire KS-049 is presented.

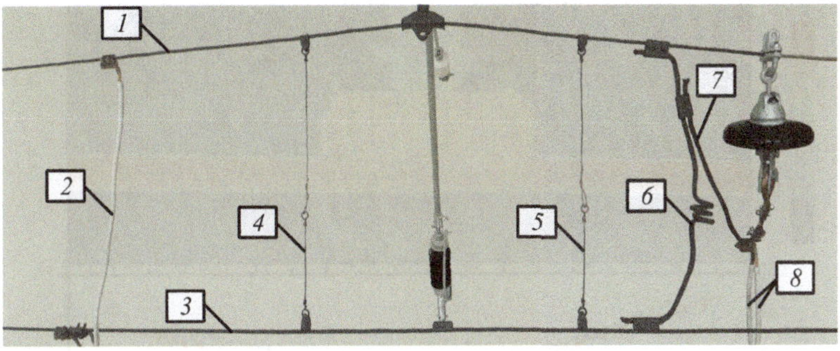

Fig. 3. The monitoring of the contact network: *1* – suspension cable PBSM-70; *2* – the supply wire I, wire M-70; *3* – contact wire MF-100; *4* – isolated string I wire BSM-4; *5* – isolated string II wire BSM-4; *6* – transverse electrochemical I wire MG-95; *7* – transverse electromedical II of the wire MG-95; *8* – the lead wires of the wires M-70

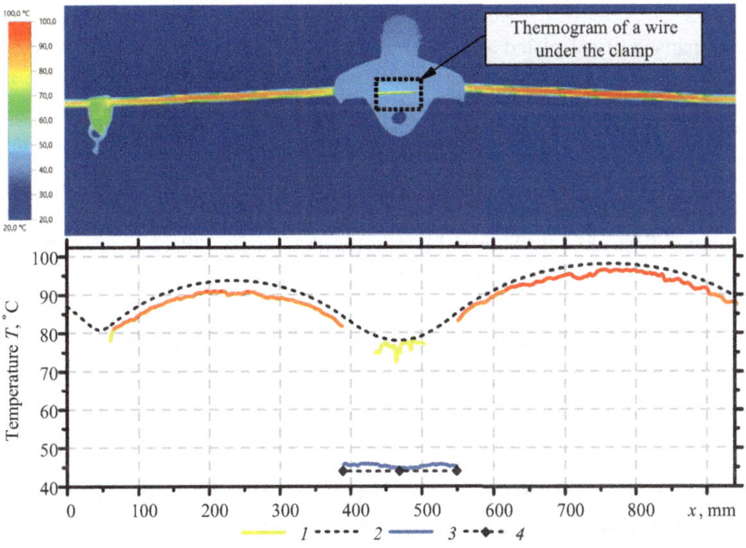

Fig. 4. Comparison of calculated and experimental data for the saddle suspension cable KS-008 and bimetallic copper-steel wire PBSM-70: *1* – the results of experimental measurements PBSM-70; *2* – results of calculation of the temperature PBSM-70; *3* – the results of experimental measurements of the seat KS-008; *4* – results of calculation of the temperature of the seat KS-008

From Figs. 4 and 5 it can be seen that the calculated and experimental values have a high degree of convergence. According to the results of processing the data of deviations of measured and calculated values for the entire study area, the law on the exponential distribution of the deviation value on the basis of the Pearson criterion was adopted. When the number of degrees of freedom $k = 4$, the observed value of 2,16 $\chi^2_{observed}$ equal to the critical value of 9.49 $\chi^2_{critical}$ equal. The average deviation for the

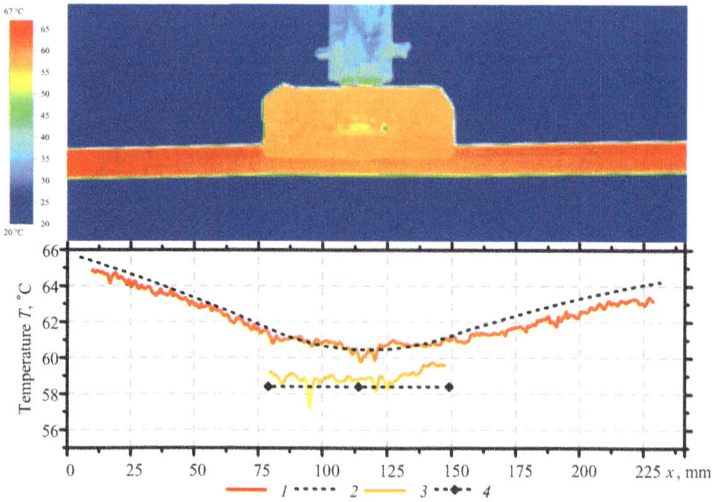

Fig. 5. Comparison of calculated and experimental data for the clamp KS-049 and the contact wire MF-100: *1* – the results of experimental measurements of MF-100; *2* – the results of calculating the temperature of the MF-100; *3* – the results of experimental measurements of the clamp KS-049; *4* – results of calculating the temperature of the clamp KS-049

entire section was 0.68 °C, which is within the instrumental error of the device used-thermal imager Testo 881-1. Thermal coating with a surface blackness coefficient equal to 0.85 was applied to all the studied elements.

On the basis of the conducted experimental researches the conclusion is drawn about compliance of the developed mathematical model to the real physical process within the intended use, namely electrothermal calculation of a contact network of a direct current.

4 Development of Solutions to Improve the Electrical Characteristics of the Contact Network

Structural study of the contact system in the presence of excessive current and heat load showed the following bottlenecks on the existing hangers: sectional non-insulated wire sections mate anchor sections, place the connection of the supply wires, the node anchoring the middle of the contact wire, at the connection points of the transverse electrical connectors to the reinforcing wires. In Fig. 6 the dependence of the current in the strings in the conjugation spans of the anchor sections is presented. The figure shows that the current in the strings of the conjugation spans is 40%, and in the spans closest to the conjugation is 20% higher than in the spans far from the conjugation. Increased current load in these areas entails increased Electromechanical wear of the strings, which is confirmed by the data from the operation.

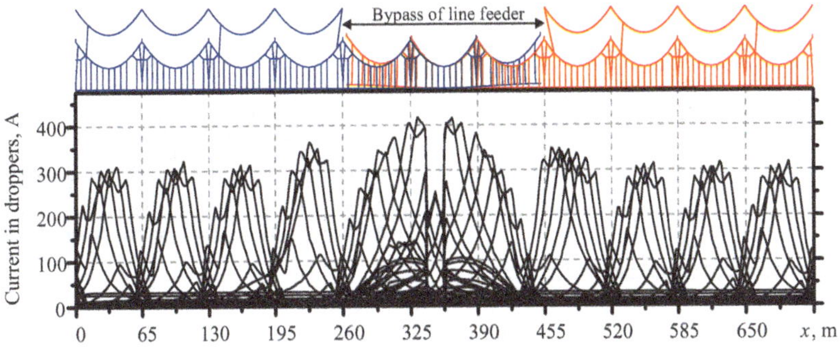

Fig. 6. The current in the strings on the mates of the anchor sections of the contact network during the passage of the electric rolling stock with a current of 4400 A

Studies have shown that the installation of transverse electrical connectors unloads no more than two nearby strings, which is not enough to completely eliminate the increased Electromechanical wear of the strings. According to the results of the study, it is concluded that the reduction of string failures due to Electromechanical wear when handling heavy trains is possible only when the link strings are isolated or when conductive strings are used. At the same time, it should be noted that during the operation of conductive strings on high-speed lines, increased wear is observed in the places of crimping [5].

In Fig. 7 presents structural changes in the contact network, eliminating the identified places of increased current and heat load in the contact network. The connection node of the supply wires Fig. 7a is upgraded by installing an electrical connection between the reinforcing wires and the transverse electrical connectors of the carrier cable and contact wires. Thus, the current and thermal loads in the longitudinal wires of the contact network are equalized from the connection points of the supply wires to the first transverse electrical connector. In the bypass node of the reinforcing wires Fig. 7b it is proposed to install an electrical connection between the supporting cable of the anchored branch and the bypass wires. This connection allows to evenly distribute the load between the longitudinal wires of contact suspensions on the interface, as well as to reduce the load on the strings in the spans of the interface of anchor sections. Modernization of the middle anchorage unit Fig. 7c includes transverse bypass tether anchoring the middle with the two sides of the main and additional runs made by extending the middle rope anchoring. Studies [15] show that the alignment of current and heat load near the middle anchorage node is most important on suspensions with isolated strings, especially with a bronze cable of the middle anchorage. A similar solution is used for semi-compensated contact suspension.

In Fig. 8 shows a temperature dependency of the longitudinal wires along the length of the before and after measures to modernize the connection node of the supply wires, as well as a host medium for anchoring catenary M-95 + 2MF-100 + 2A-185 with sungevity strings, the most common operation. The colored lines refer to the temperature of the wires after the upgrade, and the shaded areas reflect the change in temperature compared to the suspension without the upgrade.

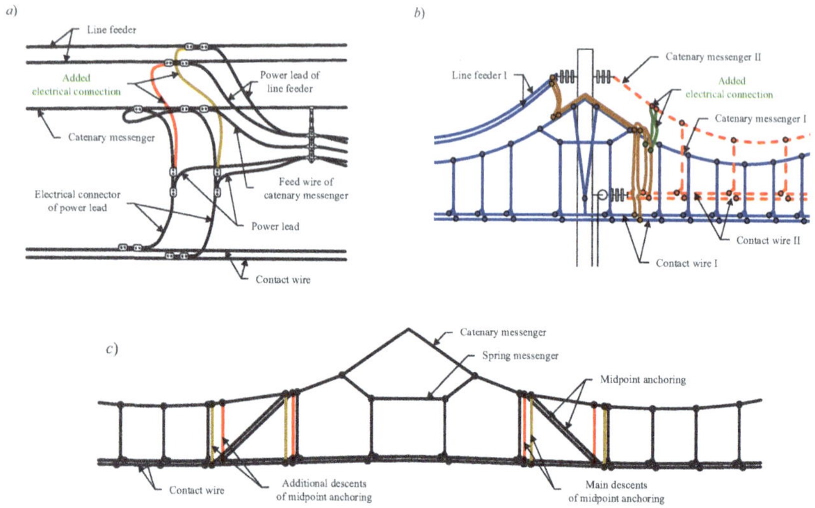

Fig. 7. Structural additions to the nodes of the contact network: *a* – node connecting the supply wires; *b* – node removal of the contact wire; *c* – node of the middle anchorage.

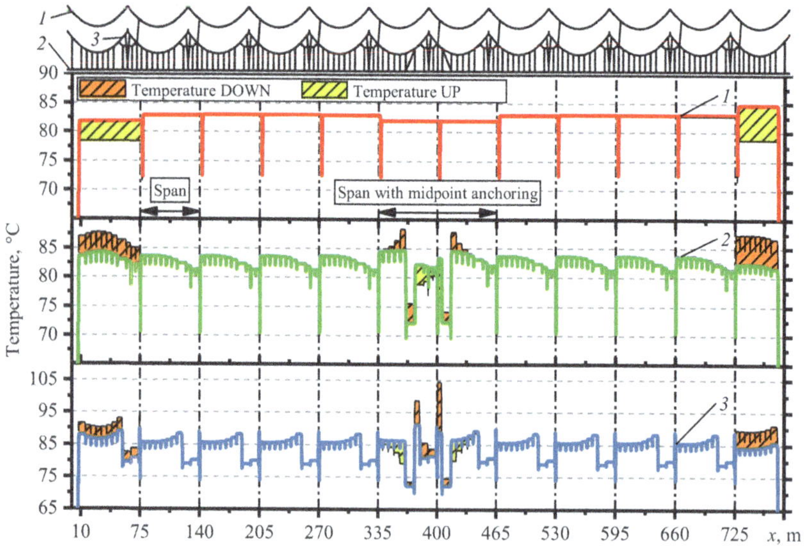

Fig. 8. Temperature profile of the longitudinal wires of the contact network with the improvement of the connection nodes of the supply wires and the middle anchorage: *1* – reinforcing wire; *2* – pin wire; *3* – carrying cable

From Fig. 8 it can be seen that from the connection points of the supply wires to the first transverse electrical connector, the current and, accordingly, the thermal load are equalized. Temperature equalization in close proximity to the site an average of embedment by transverse bypass surgery basic and advanced slopes and levels the temperature in the carrier and contact wires.

Consider the change in the temperature of the carrier cable near the middle anchorage node when passing a package of trains 6300 – 9000 – 6300 tons at intervals of 10 min, shown in Fig. 9.

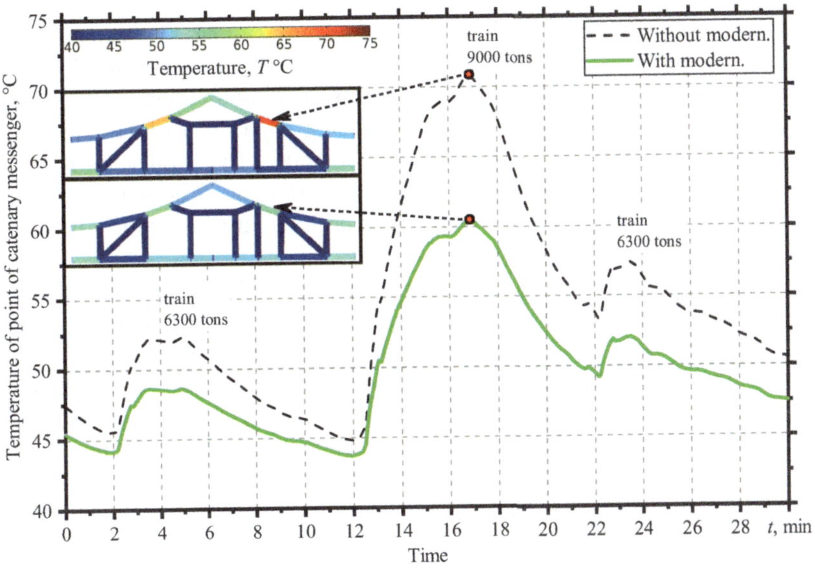

Fig. 9. The temperature of the carrier cable near the middle anchorage before and after the modernization of the DC contact network under operating loads

From Fig. 9 one can see a significant decrease in the temperature of the load-bearing cable and an increase in its uniformity along the length near the middle anchorage node, especially noticeable when passing a train weighing 9000 tons. In addition to the presented design changes in the strengthening and modernization of the DC contact network, it is proposed to duplicate the clamps connecting the transverse electrical connectors with the reinforcing wires. Due to the fact that between the bearing cable and the contact wire includes a significantly larger number of electrical connections is string, the console, the more frequent location of the power connectors between the main cable and the contact wire, the greater the contact resistance of the clamps transverse electric connectors installed on cable or contact wire does not cause significant heating of these compounds is due to the redistribution of parallel current paths. Studies have shown that defective clamps of transverse electrical connectors on the contact wire and the carrier cable do not lead to significant heating, which can lead to failure, since the temperature of the defective clamps does not exceed the temperature of the wires on which they are installed. The situation is different with the connections of transverse electrical connectors and reinforcing wires. Duplicate current pathways for such connections are only adjacent transverse electrical connectors.

Defective connections of transverse electrical connectors with reinforcing wires are subject to an order of magnitude greater thermal loads than connections with the carrier and contact wires. In this regard, it is proposed to duplicate these connections.

It is worth noting that the mathematical modeling was carried out in the software environment COMSOL Multiphysics. The basic module of the program was used in the mode of solving equations specified by the user through the physical interface Coefficient Form Edge PDE. The use of such software requires certain skills, which is not always possible for personnel who maintain, diagnose or design a contact network. To solve the problem by expanding the possibilities of use of mathematical simulation of contact network, and consequently and effectiveness of solutions from the use of such modelling tool have been developed modeling application ELTECAT AB, that Deals with the use of the product COMSOL Compiler. The application is designed directly for use by a wide class of specialists in the field of design, operation and maintenance of the contact network, with no experience in mathematical modeling in specialized software.

The application has the ability to set the parameters of the contact suspension, such as materials and sections of its wires, geometric dimensions. The environmental conditions and operating loads are specified in such design modes as current collection by electric rolling stock (one or two current collectors), stationary transit of traction current, transit current flow changing in the time domain. As a result of the calculation, the user can see the results of the current and temperature distribution in the elements of the contact network in the considered design modes.

5 Conclusions

As a result of the conducted researches constructive offers on strengthening of a contact network of a direct current eliminating places of the increased current and thermal loading were developed:

- in the node of the middle anchorage of the contact wire;
- in the node connecting the supply wires;
- node bypass reinforcing wires on the mating anchor parts;
- in places of connection of transverse electrical connectors to the reinforcing wire.

Justified the decision to eliminate high Electromechanical wear link string by insulation, or the use of conductive wires.

The application of the developed design proposals allow to increase the load capacity of the contact network by 6% compared to the option without modernization.

The obtained conclusions became possible due to computer modeling using the finite element method, which allows to solve the problem of electrothermal calculation of the DC contact network, implemented in the software environment COMSOL Multiphysics. The developed method of calculation is confirmed on a site of a real contact network that allows to consider reliable the presented results of calculation on improvement of electrothermal characteristics of a contact network of a direct current.

To expand the opportunities to use simulation tools for staff directly servicing, diagnosing or designing a network designed for the simulation ELTECAT AB USURT implementation, which is implemented at the Sverdlovsk railway. It is proposed that further development and use of modeling application ELTECAT AB USURT when performing electro-thermal calculations of the contact network.

References

1. Shimanovsky, A., Yakubovich, V., Kapliuk, I.: Modeling of the pantograph-catenary wire contact interaction. Procedia Eng. **134**, 284–290 (2016). https://doi.org/10.1016/j.proeng.2016.01.009
2. Cho, Y.H.: Numerical simulation of the dynamic responses of railway overhead contact lines to a moving pantograph, considering a nonlinear dropper. J. Sound Vib. **3**(315), 433–454 (2008). https://doi.org/10.1016/j.jsv.2008.02.024
3. Bruni, S., Bucca, G., Collina, A., Facchinetti, A.: Numerical and hardware-in-the-loop tools for the design of very high speed pantograph-catenary systems. J. Comput. Nonlinear Dyn. **7** (4), 1–8 (2012). https://doi.org/10.1115/1.4006834
4. He, D., Gao, Q., Zhong, W.: A numerical method based on the parametric variational principle for simulating the dynamic behavior of the pantograph-catenary system. Shock Vibr. **2018**, 1–21 (2018). https://doi.org/10.1155/2018/7208045
5. Liu, X.Y., Peng, J.F., Tan, D.Q., Xu, Z.B., Liu, J.H., Mo, J.-L., Zhu, M.H.: Failure analysis and optimization of integral droppers used in high speed railway catenary system. Eng. Fail. Anal. **91**, 496–506 (2018). https://doi.org/10.1016/j.engfailanal.2018.02.003
6. Collina, A., Bruni, S.: Numerical simulation of pantograph-overhead equipment interaction. Veh. Syst. Dyn. **4**(38), 261–291 (2002). https://doi.org/10.1076/vesd.38.4.261.8286
7. Mikic, B.B.: Thermal contact conductance; theoretical considerations. Int. J. Heat Mass Transf. **2**(17), 205–214 (1974). https://doi.org/10.1016/0017-9310(74)90082-9
8. Cooper, M.G., Mikic, B.B., Yovanovich, M.M.: Thermal contact conductance. Int. J. Heat Mass Transf. **3**(12), 279–300 (1969). https://doi.org/10.1016/0017-9310(69)90011-8
9. Wu, J.: Chapter 6 – Electric Contact Properties of Pantograph and Contact Line. Pantograph and Contact Line System, pp. 193–238 (2018). https://doi.org/10.1016/b978-0-12-812886-2.00006-9
10. Incropera, F.P., Dewitt, D.P.: Fundamentals of Heat and Mass Transfer, 7th edn. Wiley, Hoboken (2011)
11. Bejan, A., Kraus, A.D.: Heat Transfer Handbook. Wiley, Hoboken (2003)
12. Kiessling, F., Puschmann, R., Schmieder, A., Schneider, E.: Contact Lines for Electric Railways, 2nd edn. Publicis Publishing, Erlangen (2009)
13. Batrashov, A.B., Paranin, A.V., Shumakov, K.G.: Improvement of mathematical model for calculation of electrothermal processes in a DC catenary suspension. Transp. Urals **1**(60), 90–96 (2019). https://doi.org/10.20291/1815-9400-2019-1-90-96
14. Paranin, A.V., Batrashov, A.B.: Validation of mathematical model of electrothermal calculation of DC catenary on the basis of scale model. Vestn. Railw. Res. Inst. **4**(77), 222–229 (2018). https://doi.org/10.21780/2223-9731-2018-77-4-222-229
15. Batrashov, A.B.: Constructive decisions on levelling of current and thermal load of DC catenary wires. Her. Ural. State Univ. Railw. Transp. **1**(41), 79–88 (2019). https://doi.org/10.20291/2079-0392-2019-1-79-88

Organization of Operation, Maintenance and Repair of Gas-Turbine Installations in the Far East Railway Section

Alexander Buynosov(✉), Alexander Pavlyukov, Aleksandr Mironov, and Sergey Laptev

Ural State University of Railway Transport, Kolmogorova Street, 66, Yekaterinburg 620034, Russia
byinosov@mail.ru

Abstract. Currently widely used methods for determining the fleet of locomotives, determining the system of maintenance and repair consider the locomotive as a whole, without taking into account the technical condition of its individual components that have a small life cycle. In addition, the methods currently used do not take into account the time of relocation from the operating locomotive depot to the service depot and Vice versa. The aim of the study is to develop methods for calculating the number of locomotives based on the volume of expected work for the far Eastern region of Russia, the necessary technical services and repairs, the optimized structure of the repair cycle on the far Eastern railway. Determined the location of the objects of the equipment for locomotives, designed operated a fleet of turbine-electric locomotives for use on the far Eastern Russian Railways. Calculated the number of services and repairs of gas turbine locomotives, as well as, according to the annual program of repair of locomotives, taking into account the downtime for maintenance and current repairs, the number of required repairs.

Keywords: Railway · Landfill · Gas turbine locomotive · Repair · Maintenance · Operation · Organization

1 Introduction

The development of the Baikal-Amur railway is an economically important task for the Russian Federation. Near the Baikal-Amur mainline (BAM) there are both active mineral deposits and well-studied ones where the economic evaluation of the development efficiency was carried out.

This railway will help the further development of the Far East, will become one of the points of growth in production in the district, increase production of raw materials, construction of new industrial enterprises along the entire route of BAM, as well as increase the capacity of the ports of Vanino and Sovetskaya Gavan.

The Baikal-Amur mainline expansion project is one of the latest infrastructure initiatives of the Russian government and the leadership of Russian Railways aimed at

Z. Popovic et al. (Eds.): TransSiberia 2019, AISC 1115, pp. 206–217, 2020.
https://doi.org/10.1007/978-3-030-37916-2_22

increasing Russia's transport connectivity with the countries of the Asia-Pacific region and improving conditions for industrial development in the Russian far East [1, 2].

According to the order of the government of the Russian Federation of October 24, 2014 No. 2116-R the investment project "Modernization of railway infrastructure of the Baikal-Amur and TRANS-Siberian railway lines with development of throughput and carrying capacities" is realized [3], this investment project is directed on achievement of the purposes defined within "Strategy of development of railway transport in the Russian Federation till 2030", approved by the government of the Russian Federation reconstructed to increase capacity and increase the weight limit for freight trains train participants-Ki Sovetskaya Gavan – Komsomolsk-Sortirovochny (475 km) and Komsomolsk-sorting – Volochaevka (327 km) and Komsomolsk-sorting – Postyshevo (198 km), Postyshevo – New Urgal (276 km), New Urgal – February (283 km), February – Verhnezejsk (339 km), Verhnezejsk – Tynda (329 km), Tynda – Hani (484 km).

The Baikal-Amur mainline on the territory of the far Eastern railway from the station Khani to the station Sovetskaya Gavan is not electrified, the mainline runs along a very complex and diverse path from a large number of intersections of deep rivers to zones with high seismicity. To improve capacity in such circumstances promising to use rolling stock will be locomotives with gas turbine power plant (LGS), liquefied natural gas (LNG) – gas turbine-electric locomotives (GT1h) made at the Lyudinovo diesel locomotive plant and Sarco-mengawasi yourself in trial operation on the Sverdlovsk railway. The gas consumption of this locomotive, in value terms, is almost a quarter lower than that of diesel fuel.

On this basis arises the task of organizing the operation, maintenance and repair of the main turbine-electric locomotives GT1h at the site of the far Eastern railway. For the organization of maintenance of gas turbine locomotives (GT) on the BAM sections, the appropriate infrastructure is required. Figure 1 shows the existing infrastructure.

Fig. 1. The existing infrastructure of BAM: ▲ – rest house of locomotive crews; ■ – operational locomotive depot; ▬ – PTOL and venue TO-2; ● – working capital operational locomotive depot

2 Method of Research

Our proposed placement of gas supply facilities (locomotive equipment) (Fig. 2) uses existing facilities because of this it is more cost-effective to upgrade these facilities than to build new ones.

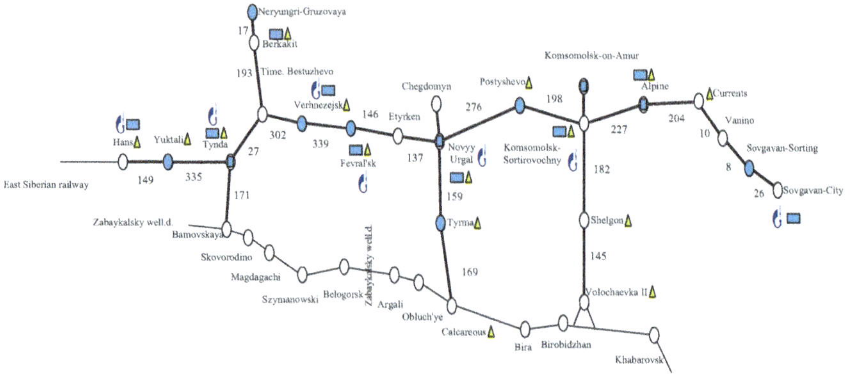

Fig. 2. Placement of gas supply facilities: ▵ – rest house of locomotive crews; ▫ – operational locomotive depot; ▭ – PTOL and venue TO-2; ● – working capital operational locomotive depot; ⛽ – required LNG production facilities

Such placement takes into account the peculiarities of working with cryogenic liquids, such as LNG, which are subject to serious requirements for their minimum reserve in storage tanks, necessary to maintain the required temperature of the walls of the container [4].

In case of unscheduled repairs and during idle periods when the gas turbine locomotive is set aside, the pressure in the cryogenic LNG tank increases by about 0.5 atmospheres per day. Storage of LNG in a cryogenic container is allowed for 5 days [5]. During storage, maintenance personnel should periodically monitor the pressure in the cryogenic tank, do not exceed the critical pressure of 6 atmospheres. If necessary, the excess gas pressure must be relieved to prevent the safety valves from actuating. Discharge of natural gas into the atmosphere is strictly prohibited by regulations, in addition it leads to the loss of an expensive product. To avoid LNG losses when gas turbine locomotives settle, Egorshino depot developed a technology for fueling LNG locomotives from onboard cryogenic tanks of gas treatment systems by the method of compression. At the same time there is a decrease in pressure in the BKE and there is no need to discharge LNG into the atmosphere. For the application of this method of refueling in the operational locomotive depot Egorshino developed and approved the regulations of refueling by the method of displacement, made the original flange [6, 7].

For repairs and maintenance of gas-turbine at the site of the far Eastern glands Noi road, you need to calculate the rate of repair and maintenance on parts of the Soviet Harbour – Komsomolsk-Sortirovochny, Komsomolsk-Sortirovochny – Volochaevka,

Komsomolsk-sorting – By-stalevo, Postyshevo – New Urgal, Urgal-New – February, February – Verhnezejsk, Verhnezejsk – Tynda, Tynda – Hani [8, 9].

Consider the section Sovetskaya Gavan-Komsomolsk-Sortirovochny (475 km), the number of pairs of trains – 6, to calculate the operating fleet of gas turbine engines, we take the technical speed of movement along the section in the even and odd direction of 50 km/h [10].

The local speed is determined by a given coefficient of local speed [11]:

$$V_s = 0.8 \cdot V_t \tag{1}$$

where 0.8 is the service speed coefficient.

Full turnover of the gas turbine:

$$T_1 = \frac{2L}{V_s} + t_m + t_{to} \tag{2}$$

where L is the length of the operational section, km; t_m is downtime for technical inspection for one turnover cycle, hr; t_{to} is locomotive downtime in turnover point, hr.

Factor the needs of GT [11]:

$$K_r = \frac{T_1}{24} \tag{3}$$

where T_1 is locomotive turnover in the operational section, locomotive-km.

The requirement factor shows the number of GT required to service one pair of trains per day. Then the operated fleet of locomotives on the site of the Soviet Harbor – Komsomolsk-Sortirovochny:

$$N_o = K_r \cdot n \tag{4}$$

where n is the number of train pairs.

In accordance with the found number of gas turbine fleet (7 locomotives) on the site Sovetskaya Gavan – Komsomolsk-Sortirovochny, we will determine the repair program and the need for operational resources.

Find the daily mileage of GT by the expression [12]:

$$S_d = 2 \cdot L \cdot n \tag{5}$$

where L is section length, km; n is the number of train pairs per day.

Annual and monthly mileage GT:

$$S_{year} = S_d \cdot 365 \tag{6}$$

$$S_m = S_d \cdot 30.4. \tag{7}$$

The average daily mileage of one GT is determined by the expression:

$$S_{\text{av.d}} = 2 \cdot L \cdot \frac{n}{N_{\text{o}}} \quad (8)$$

where N_{o} is the operated park of GTL; L is section length, km; n is the number of train pairs.

3 Experimental Data and Results

The results are summarized in Table 1. Similar results were obtained for the sites of Komsomolsk-Sortirovochny – Volochaevka, Komsomolsk-Sortirovochny – Postyshevo, Postyshevo – New Urgal, Urgal-New – February, February – Verhnezejsk, Verhnezejsk – Tynda, Tynda – Hani. All calculations made on the considered sites, as well as by types of maintenance and repair are presented in Table 2.

Table 1. Technical and operational indicators of the use of a gas turbine locomotive on the site Sovetskaya Gavan-Komsomolsk-Sortirovochny

Parameter	Signification
Local speed, km/h	40
Average daily mileage of the locomotive, km	814.2
Operating fleet of locomotives, PCs	7

Table 2. Annual program of technical services and repairs of GT, units

Location	Types of maintenance/repair						
	TO-2	TO-3	TR-1	TR-2	TR-3	SR	KR
Sovetskaya Gavan – Komsomolsk-Sortirovochny	511	166	21	10	5	2	1
Komsomolsk-Sortirovochny – Volochayevka II	438	115	14	7	4	1	1
Komsomolsk-Sortirovochny – Novyy Urgal	511	166	21	10	5	2	1
Novyy Urgal – Lime	438	115	14	7	4	1	1
Novyy Urgal – February	365	99	12	6	3	1	1
February – Verhnezejsk	438	119	15	7	4	1	1
Verhnezejsk – Tynda	438	115	14	7	4	1	1
Tynda – Hani	584	170	21	11	5	2	1
In total	3723	1065	132	65	34	11	8

The required number of GT fleet operated on the section of the Baikal-Amur mainline is shown in Table 3.

Table 3. The operated Park of gas turbine locomotives on the section of the Baikal-Amur mainline

Location	Number of GT, pieces
Sovetskaya Gavan – Komsomolsk-Sortirovochny	7
Komsomolsk-Sortirovochny – Volochayevka II	6
Komsomolsk-Sortirovochny – Novyy Urgal	7
Novyy Urgal – Lime	6
Novyy Urgal – February	5
February – Verhnezejsk	6
Verhnezejsk – Tynda	6
Tynda – Hani	8
In total	51

To solve the problem of the best loading of each SLD to perform gas turbine locomotives TR, it is possible to build a model with a rather complex topology of interrelations between TCE and SLD; In the integrated development environment of the software for Microsoft Windows (Embarcadero Delphi), the executable file NW.exe was created by launching the “Project Wizard” window to the user. Information on the characteristics of the project (model) is entered in steps in order to eliminate erroneous data entry.

In the first step, the “project name” is entered, the number of both TCE and SLD, the cost of 1 h of gas turbine carrier idle when carrying out the TR and 1 h of transportation GT1h, the number of repair items in SLD.

As a result of the distribution of electric locomotives according to SLD, on the basis of the above algorithm, it is possible to determine the optimal parameters of the model and construct a unit cost diagram (Fig. 3). The duration of the simulation period should be as long as possible (several decades) in order to accumulate the necessary statistical information. Since modern PCs with high computing power allow modeling in a few minutes, let us take the duration of the simulation period TM = 1.000.000 h = 114.16 years. As can be seen from the obtained diagrams of specific costs for repair of gas turbine carriers, before optimization, the diagram has a falling form with high costs, and the diagram after optimization is characterized by a more stable form with minimal and stable costs. That is, on the basis of this, we can conclude that it is expedient to carry out the optimization of repair in this direction.

To identify the equipment of the GT, which most often fails, in 2018 in the SLD Artyomovsky collected statistical material about defects in the GT1h-002 nodes (Table 4).

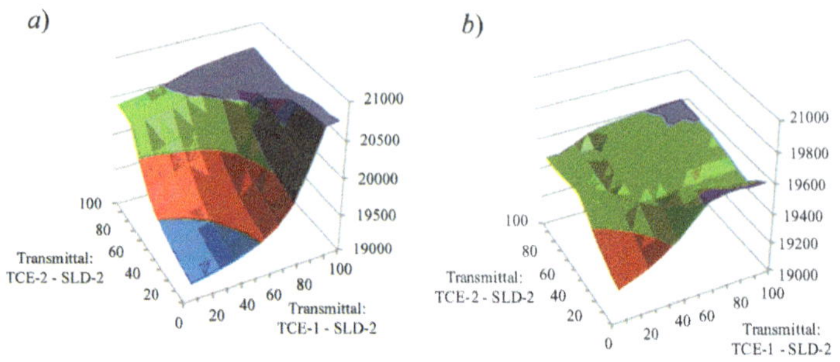

Fig. 3. Charts of specific costs for repair of gas turbine carriers before (*a*) and after (*b*) optimization

Table 4. Classification of defects of gas turbine locomotive GT1h-002

Unit	Number of defects	Accumulated amount of defects	The percentage of defects in the total number of defects	Accumulated interest
Fuel delivery system	19	19	31.2	17.5
Mechanical equipment	14	33	23.0	54.2
Traction motor	11	44	18.1	72.3
Avtotormoznoye oborudovaniye	5	49	8.4	80.7
Auxiliary equipment of GSU	3	51	4.9	85.6
Power electrical circuits, devices	2	53	3.2	88.8
Safety devices	1	54	1.6	90.4
Control system	1	55	1.6	92.0
Traction generator	1	56	1.6	93.6
Signal and lighting devices	1	57	1.6	95.2
Alarm systems	1	58	1.6	96.8
Radio stations	1	59	1.6	98.4
Body	1	60	1.6	100
Gas turbine installation NK-361	0	60	0	100
Total:	61	–	100	–

On the basis of statistical data (Table 4) by the method [13] Pareto diagram (Fig. 4) is constructed.

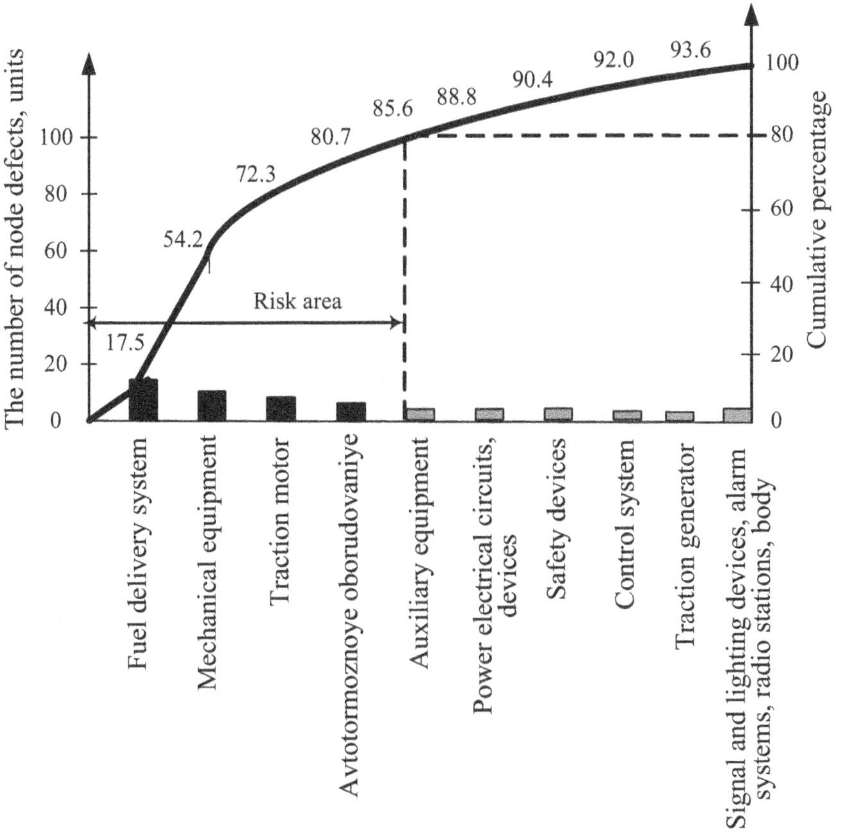

Fig. 4. Pareto diagram for defects in gas turbine locomotive units

From Fig. 4 it can be seen that in the risk area such nodes as the fuel supply system, mechanical equipment, traction engine and auto brake equipment turned out to be only 80.7%. From Table 4 and Fig. 4 it can be seen: the NK-361 gas turbine unit has a significant uptime life (no defects were found during the trips); there is a high risk that the claimed service life of the fuel injection system (cryogenic pump) produced in Switzerland will not be sustained; The bottleneck is the traction motors and auto brake equipment. To solve the problem of failures of traction motors produced in Ukraine (Kharkiv) failed. Questions about the trouble-free operation of the brake system are currently being addressed in USURT and EI JSC "VNIIZhT".

To reduce the number of defects and maintain the locomotive in working condition, it is required to develop methods aimed at improving the reliability of the four nodes of the GT1h-002 gas turbine locomotive. Together with JSC VNIKTI and Lyudinovsky Diesel Locomotive Plant, work is underway to improve the reliability of the gas turbine GT1h-002, as a result of which the technical availability ratio (KTG) was 0.89, and the internal readiness ratio (VHG) 0.95.

Based on the results of processing statistical data according to the developed method [13], after the current repair of TR-1, a diagram of the probability of failure of the equipment of the gas turbine locomotive GT1h-002 (Fig. 5) was constructed.

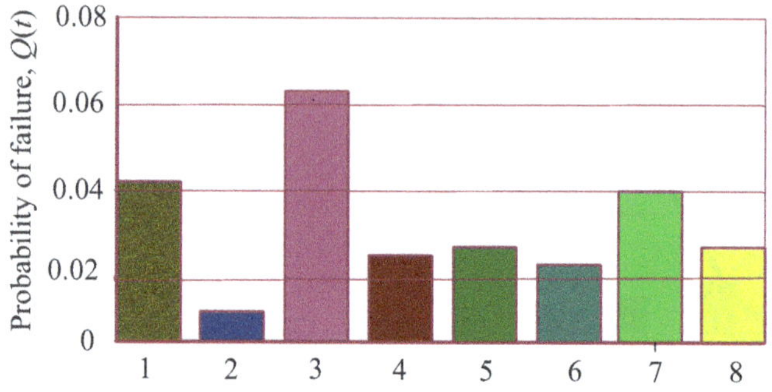

Fig. 5. The probability diagram of equipment failure in 2018: 1 - mechanical equipment; 2 - safety devices; 3 - fuel delivery system; 4 - power circuits, devices; 5 - auto brake equipment; 6 - traction generator; 7 - traction engine; 8 - auxiliary equipment.

When calculating it was taken into account that the complex repair crews work for 12 h. Three variants of work were taken into account: (1) around the clock, without days off and holidays, F = 365 days; (2) around the clock minus holidays, F = 353 days; (3) around the clock with weekends and holidays, F = 249 days.

In addition to the stall sites for inspection, revision and repair of GT during routine repairs, special stalls should be provided for equipment diagnostics. Such stall diagnosing, for example, has SLD Sverdlovsk. The number of calculated stall sites required for the diagnosis and production of the current repair of the TR-1 GT1h gas-turbine locomotive is given in Table 5.

Table 5. Number of stall sites for the maintenance of TR-1

Name	Fund working time stall day		
	365	353	249
Number of stall sites for the maintenance of TR-1	1.36	1.40	2.24
Number of stall sites for diagnosis	0.68	0.70	0.99

In the future, the number of stalls can be reduced if the inspection and revision of equipment on the stall is performed during one shift in 12 h.

With the help of the developed computer program "Optimization of the frequency of repair of locomotives at the railroad test site" [9], the structure of the repair cycle of the GT1h-002 gas turbine locomotive was calculated and optimized. Figure 6a shows the structure of the repair cycle of parts before optimization, and Fig. 6b shows an optimized structure (with a two-fold reduction in the amount of failure of the fuel supply system, mechanical equipment, traction engines and auto brake equipment).

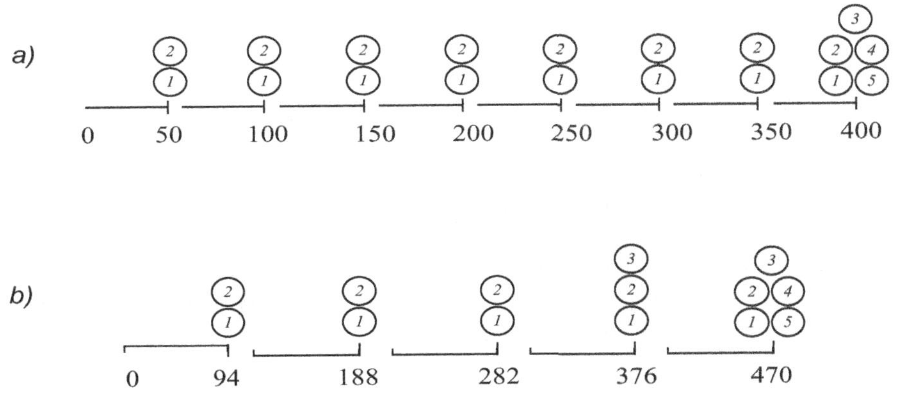

Fig. 6. The structure of the repair cycle before (*a*) and after (*b*) optimization: *1* - fuel supply system; *2* - mechanical equipment; *3* - traction engine; *4* - auto brake equipment; *5* - power circuits, devices

Runs between servicing and repairs of GT after structure optimization are summarized in Table 6.

Table 6. Rates of locomotives between repairs

Locomotive series	Periodicity by repair type, thousand Km				
	TR-1	TR-2	TR-3	SR	KR
GT1h	94	282	470	1410	2820

The obtained optimized structure of the repair cycle allowed us to correct the overhaul run for repairs [14] and to reduce the amount of work on routine repairs of GT1h gas turbine locomotives.

4 Conclusions

1. The method of calculating the number of locomotives based on the volume of the proposed work for the far Eastern landfill of Russian Railways.
2. The questions of the organization and operation of the locomotive running on liquefied natural gas, for the realization of technology export of goods from the far East of the Russian Federation.
3. The placement of equipment objects for locomotives is Determined, the operated fleet of gas turbine locomotives for use on the far Eastern railway is calculated.

Calculated the number of services and repairs of gas turbines, as well as, according to the annual program of repair of gas turbines, taking into account downtime for maintenance and current repairs, the number of required repair items.

4. The Use of gas turbine locomotives with high traction characteristics will solve several problems:

   - to organize the movement of heavy trains in the regions of the far North, on non-electrified sections of the Baikal-Amur mainline, ensuring increasing volumes of cargo transportation while reducing the need for gas turbine locomotives and locomotive crews;
   - to increase the carrying and carrying capacity of Railways on the existing track infrastructure;
   - contribute to the intensive development of the economic regions of the far East;
   - to reduce the negative impact of rail transport on the environment by reducing emissions from locomotives using LNG.

References

1. Chen, X., Zhu, X., Zhou, O., Wong, Y.D.: J. Adv. Transp. **2017**, 3128372 (2017)
2. Yoneda, T., Ryuba, H.: CTL model checking of time petri nets using geometric regions. IEICE Trans. Inf. Syst. **E99-D**(3), 297–396 (1998)
3. Trushnikov, D.N., Belinin, D.S., Shitsyn, Y.D.: Mod. Probl. Sci. Educ. **2**, 95 (2014)
4. Berthomieu, B., Diaz, M.: Modeling and verification of time dependent systems using time Petri nets. IEEE Trans. Softw. Eng. **17**(3), 8–14 (1991)
5. Zhu, G., Chow, T.T., Fong, K.F., Lee, C.K.: Comparative study on humidified gas turbine cycles with different air saturator designs. Appl. Energy **254**, 113592 (2019)
6. Buynosov, A.P., Volkov, D.V., Umylin, I.V.: Definition of a limit difference in the diameters of bandages at one wheel pair of the motor car of an electric train(s). Sovremennye tekhnologii, Sistemnyi analiz, Modelirovanie **56**(2), 8–18 (2017). https://doi.org/10.26731/1813-9108.2018.2(58).90-97
7. Teodorovic, D., Janic, M.: Transportation Engineering. Butterworth-Heinemann, Oxford (2016)
8. Buynosov, A.P., Volkov, D.V., Fedorov, E.V.: Automatic control of train braking system integrity and density. Transp. Urals **4**(59), 77–80 (2018). https://doi.org/10.20291/1815-9400-2018-4-7-80
9. Buynosov, A., Lapshin, V., Smolyaninov, A., Dinislamov, A.: Calculation of hardening process parameters for locomotive parts. In: Polytransport Systems-2018. MATEC Web of Conferences, vol. 216, p. 02020 (2018). https://doi.org/10.1051/matecconf/201821602020
10. Buynosov, A.P., Tsikhalevskiy, I.S., Laptev, S.I.: Organization of gas turbine locomotives "TT1h" exploitation and technical maintenance. Herald Ural State Univ. Railway Transp. **3**(39), 41–55 (2018). https://doi.org/10.20291/2079-0392-2018-3-43-55
11. Girzhon, V.V., Ovchinnikov, A.V.: Metal Sci. Heat Treat. **58**(11–12), 719–723 (2017)

12. Ronkin, M.V., Kalmykov, A.A., Nagovicin, V.S., Buinosov, A.P.: An algorithm for precision railway navigation by FMCW radio frequency identification. In: REIT-Autumn 2018, 4th International Workshop on Radio Electronics and Information Technologies, vol. 2274, pp. 52–61 (2018)
13. Buynosov, A.P.: Doctoral thesis, Ural State University of Railway Transport, Ekaterinburg (2011)
14. Martinez, A.S., Brouwer, J., Samuelsen, G.S.: Comparative analysis of SOFC–GT freight locomotive fueled by natural gas and diesel with onboard reformation. Appl. Energy **148**, 421–438 (2015)

Optimization of Locating of Recycling Facilities for Vehicles in the Region

Elena Kuznetsova[1(✉)], Anastasiya Markina[1], Valentina Parshina[2], and Nikita Amosov[1]

[1] Ural Federal University named after the first President of Russia B.N. Yeltsin, Mira Street 19, Yekaterinburg 620062, Russia
e.y.kuznetsova@urfu.ru

[2] Ural State University of Railway Transport, Kolmogorov Street 66, Yekaterinburg 620034, Russia

Abstract. The situation at the modern market of passenger motor transport needs urgent creation of a recycling system. The purpose of the article is to develop a scientifically based methodology for the locating of infrastructure facilities of the recycling system (return logistics), optimal by the criterion of minimizing the transportation cost of the objects for recycling and ensuring the loading of processing enterprises. The classification of the recycling companies is offered: depending on the implemented scope of functions - from the collection of autowaste for processing to ensure complex utilization - from frame shredding to deep recycling of light fractions (plastic, rubber, elements of upholstery, glass, fiber, minerals). The method of definition the geographical positioning of utilization centers-algorithmically organized search for optimal locations of centers definition for the Sverdlovsk region. To develop the methodology the indicator of suitability of settlements for locating enterprises of the recycling system is calculated, such factors as the population at the settlements, the volume of recyclable vehicles, cost of transport, depending on the category of road are considered (the higher the road category, the lower transportation costs, higher average speed, lower fuel consumption and lower costs for maintenance and repair of transport units). The article considers the corrective factors of qualitative nature that can make adjustments to the calculated results. The method is based on a graph-geographical model of the region and can be applied to a geographical territory of any scale.

Keywords: Utilization of autowaste · Method of locating of objects · Autorecycling · Return logistics

1 Actuality

One of the ways to solve urgent problems of sustainable development is the management of reverse flows. The relevance of the use of reverse logistics for the automotive market is obvious, since the automotive industry is resource-intensive both in terms of natural and cost evaluation indicators - on the one hand; on the other hand, the fleet of passenger cars in operation, everywhere, and especially in Russia, is not new and is operated outside the standard service life (Table 1).

Z. Popovic et al. (Eds.): TransSiberia 2019, AISC 1115, pp. 218–232, 2020.
https://doi.org/10.1007/978-3-030-37916-2_23

Table 1. Summary of the European automotive industry

Country	Car fleet, units	Produced cars, 2018 г., units	Average age of the Car fleet, years	Autocomplex waste processing Shredder facilities, units [1]	Vehicles to be recycled of, units
France	30 850 000	2 191 450	8.3	40	1 583 283
Germany	41 321 171	5 368 278	8.2	47	500 193
Great Britain	30 309 171	1 672 180	6.9	37	1 157 438
Russia	35 057 514	1 770 545	12.5	4	No data

It is necessary to consider both physical and moral obsolescence of cars, which leads to environmental pollution – for this purpose, technical norms of environmental standards for cars with both gasoline and diesel engines have been introduced. Reverse logistics concerns both the sphere of secondary resources (recovery and recycling) and the sphere of customer service (warranty replacements, repairs, recalls, recoveries). The task of reverse logistics in the market of production and operation of cars is so significant that it requires the creation of a special infrastructure for the collection and processing of automotive waste and obtaining secondary resources. The definition of the areas of responsibility of the elements of this infrastructure, and the layout of the territory are equally important. For the successful functioning of the utilization system of vehicles it is necessary to develop a scientifically based method of locating of its infrastructure facilities, effective in terms of minimizing the cost of transportation of the objects for recycling and ensuring the loading of processing enterprises, which is the purpose of this work.

2 Level of the Problem Elaboration

Russian scientists and practitioners, for example [2–4], note that at present there is no waste management system in Russia, there are no specialized production facilities for recycling. In scientific works of domestic and foreign researchers, the question of the car recycling system locating is considered rare, so we'll use the best practices - the models used for facilities management of waste and objects of social and commercial infrastructure.

One of the first scientists who devoted their work to finding the optimal placement of individual industrial enterprises became Launhardt [5] who described the impact of transport costs, taking over the major factors of distance to sources of raw materials and distribution, has created the so-called "locating triangle of Launhardt". This model still has not lost its relevance in solving logistical problems of placement. A. Weber developed the theory of B. Launhardt and suggested expanding the factors influencing the optimal location of the enterprise, divided them into three groups: transportation (cost of transportation of raw materials and finished products), labor (the wage), infrastructure (use of best placement) [6]. Methods of gravitational locating of objects are considered in the works Of W. Isard, W. Reilly, D. Huff and other scientists [7].

These models use population density as the main factor and are often used to select locations for social facilities and commercial enterprises [8]. Further development of placement models led to the need to take into account the influence of market factors and similar enterprises, as a result, it was proposed to consider the locating of objects at the macro - and micro-level, which was reflected in the works of domestic researchers [9, 10].

Korotaev [11] chooses transport costs and capacity utilization as factors influencing the location of waste disposal and consumption facilities. Velikanova and Ladoshkin [12] when choosing the location of solid waste disposal facilities proposed to take into account, along with transport, infrastructure costs (availability of access roads, the cost of energy, the cost of land) and labor costs. Lylin [13] the choice of locations for disposal of equipment (used in agriculture) bases on the number of inhabitants of settlements and the maximum radius of collection of farm machinery.

In solving logistical problems of objects placing the methods and algorithms of mathematical programming, graph theory: Mowder, Galiev and Satarov, Raenko, Maslov [14–17] are used quite often. The placement problem statement are distinguished by the selected optimality criteria and the introduced limitations, which are determined by the specifics of the research area for which the problem is solved.

In [18], the authors Moiseev and Sotnikov propose a classification of location problems and General methods of solving them, based on a generalization of works in different fields. The authors note that there is a wide variety of models often used to solve the same type of problems, and upon this, they conclude that it is necessary to form a common method applicable to solving the problem of optimal object locating. However, the General methods are universal and, as a consequence, are not adapted to the specifics of a particular task. Hence, they can be applied to solving a broad class of similar problems from different areas of economic activity, and this predetermines the lack of specificity and the possibility of flexible transformation of such methods for a specific production and economic situation.

Among the conventional models of objects placement of the market economy it is important to reflect: the gravity model, the model of "Central places" by V. Kristaller, which is considered reasonable for the decision on the locating of agriculture and unacceptable for industrial facilities [19] and model by Lesch on the spatial organization of the economy, spatial networks [20]. Similar tasks are submitted in the works Kondratiev, Ismaylovoy [21, 22]: about the placement of the service objects and social objects using shared methods that are not linked to specific activities hosted objects.

The General methods, as a rule, a number of parameters, on which the solution of the problem depends significantly, are not taken into account. For example, in the standard method of potential [8, 23] the calculations are made on the basis of only two parameters – the population and the distance between settlements, so it is most often used in solving problems of locating of social or commercial objects. The formula of the gravitational model is:

$$P_i = N_i + \sum\nolimits_{j=1}^{k} \frac{N_j}{R_{ij}}, \tag{1}$$

where P_i – potential of the settlement i;
N_i – population at the settlement i;
N_j – population at the settlement j;
R_{ij} – distance between settlements by road.

The solution of the problem of optimization of material flows in the system of waste recycling of the motor transport complex (the closest to our object of study) was considered by Pukhov and Astanin [23], but the issue of locating of recycling enterprises is not considered here.

The common generic methods give approximate solutions, i.e., either the approximate area of location, or multiple locations to choose from. Since conventional methods are not effective for the location of facilities in the field of production and consumption waste management, the researchers suggest the development and refinement of approaches and methods. To solve the problem of recycling centers locating, it is necessary to take into account such parameters as the number of vehicles per capita, the total number of registered vehicles and their age, the quality of roads in the region, the availability of the proposed territory for the location of production facilities, the availability of consumers of secondary materials. The use of numerical indicators will allow to determine the location of objects of the recycling system with greater accuracy, while reducing the subjectivity of the assessment, in comparison with the score indicators.

One of the distinctive features of solving of this task is to offer a two-stage scheme of processing of waste: in the first stage, waste is collected by collection points – in this case, the importance of the availability of collection points for car owners and minimum costs for delivery of waste to processing plants; the second stage, waste is transported to the enterprise waste management – in this case, a significant thing is the equipment loading, the availability of appropriate infrastructure to ensure operating activities and the availability of consumer recycled materials. For solving the issue of optimizing the location of recycling centers, an organizational model is proposed, which is presented briefly in this article, and its detailed presentation is presented in [24, 25].

3 Developing an Organizational Model

Utilization enterprises depending on the realized volume of functions (by analogy with logistics operators) are divided by the authors into four classes (Table 2), which is reflected in [24].

On the basis of this, it is proposed to consider the locating of objects in two stages: the locating of enterprises for the utilization of classes A and B, then the locating of enterprises of classes C and D, based on different sets of the most significant factors. To solve the problem of the small businesses recycling enterprises (C and D) whose primary purpose is the collection of end-of-life vehicles and their wastes, it is necessary to consider their features: the suppliers for these companies will be the last owners of vehicles, public areas and enterprises. As for individuals, enterprises provide a service for the recycling of cars, the solution of the problem of this stage is similar to the solution of the problems of locating of social and commercial facilities [8, 10].

Table 2. Classification of recycling companies [compiled by 24]

Class	Characteristics of recycling companies belonging to the class	Functions performed by companies in the auto-recycling system	The recycling rate	Factors
A	Large recycling companies providing the solution of a full range of tasks of the recycling system and the implementation of high intensity of all interaction flows, the implementation of unique complex projects related to the development of solutions for the recycling properties of the object Locating in areas of cities that are able to provide full load of production capacity and sufficient space. Transport accessibility. Availability of sufficient land and infrastructure to support the operational activities of enterprises. Availability of research organizations to carry out research projects	All functions, interaction with the elements of the overhead system (interests of the recycling system in the transport complex)	till 95%	
B	Large recycling companies that provide the full range of recycling system tasks and most interaction flows, that are local players in the market Location near major cities. Transport accessibility. Availability of manpower and necessary infrastructure for operations	Collecting, transportation, dismantling, recycling of products, recycling of materials	till 80%	
C	Specialized recycling companies that perform a certain segment of recycling work and produce a narrow list of secondary materials for which there is the greatest demand in the market Location on the outskirts of small and medium-sized cities, district centers. Availability of secondary raw materials consumers	Metal collection, transportation, recycling	till 55%	– transportation costs – topographic factors
D	Collection points specialize in execution of acts of utilization and providing transport services. Locating in territories and in suburbs of the large cities for availability to processing enterprises of classes A and B and suppliers of automobile waste. Mobile points of collecting autowaste in areas with the low index of suitability	Collecting, transportation	–	

At the first stage, it is necessary to determine the location of large enterprises (A and B), providing comprehensive processing services, since their location in the region is determined by a large set of factors, and operating activities are quite fund - and resource-intensive.

On the basis of the analysis of research works devoted to the search for the optimal placement of social infrastructure, commercial and industrial facilities, recommendations for the locating of each class of facilities and a group of major factors affecting their location in the region are proposed (also shown at the Table 2).

At the utilization centers of class A and B draining of operational liquids, dismantling of buses and ecologically dangerous components (for processing), dismantling of component parts is carried out (on the spare part, after recovery). Skeletons of cars undergo shredding process, that assumes crushing, cleaning of rust and scale, then electromagnetic, pneumatic, vibration, floatation sorting of components of materials according to groups. Light fractions (plastic, rubber, upholstery elements, glasses, fibers, minerals, etc.) are exposed to deep secondary processing. What is unsuitable to separation and recovery goes to polygons for burial.

The enterprises of utilization system of class D have, as the main objective, collecting the automobile waste and cars which are subject to utilization therefore their locating will be defined by transport expenses: in distance to suppliers and consumers. Such collection points can be combined with the dealer centers, the enterprises of collecting other types of waste, auto repair shops. The enterprises of class C are offered to be placed in the average and small cities, processing of metal parts of autowaste is characteristic of these enterprises, in the cities of Sverdlovsk region with high probability there will be corresponding consumers owing to specifics of its production orientation. Rubber, fabric and plastic details are supposed to be sent for processing in the large multiple-purpose utilization centers.

4 Method of Locating of the Utilization Enterprises of Motor Transport in the Region

The method of definition of geographical positioning of the utilization centers of automobile waste is developed. The similar model - algorithmically organized search of optimum locations of the centers of utilization of autowaste - can be offered for any geographical territory, in our case we were limited to the territory of Sverdlovsk region.

Stages of method of definition of positioning of the utilization centers of autowaste in the region, the following:

1. creation of grafo-geographical model of the region with possible splitting model into shares;
2. formation of the array of factual data – bases for determination of potentials and distances on the graph;
3. definition of indicators of suitability of tops and finding of tops of grafo-geographical model with the maximum indicators of suitability [26].

The technique is based on grafo-geographical model of the region which represents the flat graph with the marked tops and edges of $G = [\{V_i\}, \{E_{ij}\}; \{P_i\}, \{R_{ij}\}]$, where

$\{V_i\}$ – set of tops, $\{E_{ij}\}$ – set of edges graph G; $P_i \in R$ – numerical tag of top; V_i, R_{ij} – numerical tag of edge of E_{ij} connecting tops of V_i and V_j [5]. V_i tops of G graph represent settlements of Sverdlovsk region. Edges of E_{ij} reflect highways between the existing settlements of V_i. The numerical tag of P_i of top of V_i and tag of R_{ij} of edge of E_{ij} graph G is the potential of top of V_i and is defined by distance between tops of V_i and V_j. The substantial sense of these tags is defined below according to positioning of the centers of utilization of automobile waste [26].

Further it is necessary to define tags of tops and edges of grafo-geographical model for the solution of problem of location of the utilization centers of automobile waste. Potential ("weight") of top of V_i decides on the help of linear form [27]:

$$P_i = w_1 * \alpha_1 * N_i + w_2 * \alpha_2 * Z_i + w_3 * \alpha_3 * K_i \tag{2}$$

Indicator for determination of potential is N_i - population in the V_i point. The average salary of the population of Z_i reflects possibility of acquisition by the population of new vehicles and, respectively, potential of utilization of old cars in the V_i point. The number of vehicles (K_i) per 1000 people in the V_i point, reflects automobilization level. This indicator considers the number of vehicles, the need for which utilization arises in the V_i point as the number of recyclable vehicles is the fixed share from total number of the available motor transport in the V_i point.

In the present linear form, in order to determine the vertex potential, the weighting factors w_1, w_2, w_3 give weight to the considered factors N_i, Z_i, K_i. Consider these factors equal to $w_1 = w_2 = w_3 = 1$, although, if necessary, these weights can be given expert values.

In this model, it is required to introduce the normalization coefficients α_1, α_2, α_3 in order to equalize the order of the fold values of the $\alpha_1 N_i$, $\alpha_2 Z_i$, $\alpha_3 K_i$ to prevent the factor with the largest numerical value from dominating and losing the significance of the remaining factors, and equalize the dimension of the fold values so that the resulting potential P_i of the vertex V_i becomes a dimensionless value.

The normalizing coefficients α_1, α_2, α_3 have the following dimensions:

$$[\alpha_1] = \frac{1}{number\ of\ people}; [\alpha_2] = \frac{1}{roubles}; [\alpha_3] = \frac{1}{\text{units}}.$$

Numerical values of coefficients α_1, α_2, α_3 are determined on the basis of maximum existing values (In our case - in the Sverdlovsk region) of indicators N_i, Z_i, K_i:

$$\alpha_1 = \frac{1}{\max\limits_i N_i}; \quad \alpha_2 = \frac{1}{\max\limits_i Z_i}; \quad \alpha_3 = \frac{1}{\max\limits_i K_i}. \tag{3}$$

The possibility and correctness of such determination of regulation coefficients is justified in [28, 29].

Let's define the ways in which we can move the auto waste for our model. In assessing the route between the sending and destination points, the model takes into account both the transportation distance and the quality of the roads in operation.

The distance R_{ij} between vertices V_i and V_j is defined in the model under consideration as follows.

Let

$$L_{ij}(\alpha_1, \alpha_2, \ldots, \alpha_{m-1}) = \{V_i = V_{\alpha_0} \to V_{\alpha_1} \to V_{\alpha_2} \to \ldots \to V_{\alpha_{m-1}} \to V_{\alpha_m} = V_j\} \quad (4)$$

- the way connecting in the graph the top V_i and V_j which contains m of moving through intermediate settlements of $V_{\alpha 1} \to V_{\alpha 2} \to \ldots \to V_{\alpha m-1}$.

Let's designate k_i - the category of the highway (from the first till the fifth in accepted classification) on moving of $V_{\alpha k-1} \to V_{\alpha k}$, t e. coefficient of quality of the highway connecting the $V_{\alpha k-1}$ and $V_{\alpha k}$ points, and D_k - length of this moving of $V_{\alpha k-1} \to V_{\alpha k}$ on the highway, km. The length of transport from vertex V_i to vertex V_j along this path (2) will be called the value

$$S(L_{ij}(\alpha_1, \alpha_2, \ldots, \alpha_{m-1})) = \sum_{r=1}^{m} (k_r \cdot D_r). \quad (5)$$

Thus, the length of the journey from point V_i to point V_j along the path L_{ij} $(\alpha_1, \alpha_2, \ldots, \alpha_{m-1})$ is the sum of the length of the crossings between the intermediate points multiplied by the road quality indicator between these points.

Let us determine the distance R_{ij} between the vertices of the graph (settlements) V_i and V_j - the minimum extent among the extensions of all kinds of paths connecting the vertexes V_i and V_j:

$$R_{ij} = min_{by\ all\ possible\ ways\ L_{ij}(\alpha_1, \alpha_2, \ldots)} * S(L_{ij} * (\alpha_1, \alpha_2, \ldots, \alpha_{m-1})) \quad (6)$$

It can be concluded that the distance R_{ij} is the "price of question," or the "cost" of travelling along the shortest and most convenient path in terms of road quality, between points V_i and V_j. This integrated indicator takes into account both the distance between settlements and the time and convenience of movement between settlements, which is directly dependent on the quality of roads.

The values of the distances between the vertices form a symmetric square matrix of distances of size n × n (n is the number of vertices in the graph G; $R_{ij} = R_{ji}$).

We will return to the declared algorithm of solving the problem of location of waste disposal centers in the region. So, the stated stages:

- formation of the graph, based on the existing geography of the research object (the territory of the oblast, the region, the country, and other);
- determination of usability index of disposal center location for each vertex of the graph

$$W(V)_i = P_i + \sum_{\substack{on\,all\,tops \\ V_j\,of\,the\,graph\,G,\,where\,i \neq j}} \frac{P_j}{R_{ij}} \quad (7)$$

This indicator is of the standard type used in the potential method, but takes into account the specific nature of the problem in question. A settlement with higher own potential Pi in terms of population, purchasing power and volume of motor vehicles is preferred for the location of a recycling center, producing a higher suitability of the location of a motor vehicle recycling center according to the criterion of the volume of cargo to be disposed of and the cost of its delivery

$$\sum_{\substack{onalltops \\ V_j\,of\,the\,graph\,G,\,where\,i \neq j}} \frac{P_j}{R_{ij}} \rightarrow max \tag{8}$$

It is shown the selection of vertex V_i with maximum suitability index $W(V_i) =$ Wmax among all vertixes of graph G as optimal place for location of recycling center. In the case of fixed division of a graph into fractions, such a vertex is for each partition fraction.

5 Application of Model of Location of Infrastructure Facilities of Utilization System of Motor Transport in Sverdlovsk Region

Using the information database about population in settlements, volume and vehicles which are subject to utilization for all settlements, the scheme of the region (Fig. 1) has been constructed. For settlements coordinates on the planes reflecting location of geographical point (the answer to the question "where?"), the third have been set (substantial, or qualitative) the coordinate of point is the population and the number of cars which are subject to utilization (the answer to the question "how many?"). Borders of the territory of Sverdlovsk region were set by "zero heights", or zero volume indicators (Pn points on the scheme).

The scheme demonstrates two points with increased formation of motor vehicles waste and population density, which are intended to become the basis of two centers of recycling of motor vehicles of class A: the first center is intended for servicing the territory of the city district of Yekaterinburg and the Southern and Western administrative-territorial districts of the region, the second center is intended to serve the conglomerate of the city of Nidzny Tagil, Gornoszavodsky, Northern and Eastern districts of the region. This division of the area under study into service areas by waste recycling centers implies the division of the geographical model of the region into so-called "fractions" [27, 30]. The edges of the graph between the split fractions are removed, it implies that there is no transfer of waste from one "fraction" to another. The split of the model of the Sverdlovsk region with the selected fractions is shown in Fig. 2.

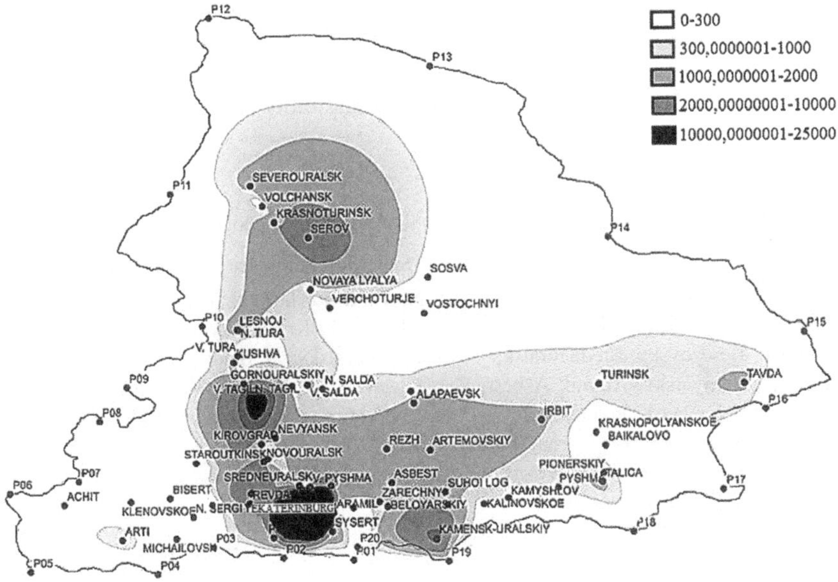

Fig. 1. Waste generation cartoschem according to population density of Sverdlovsk region [26]

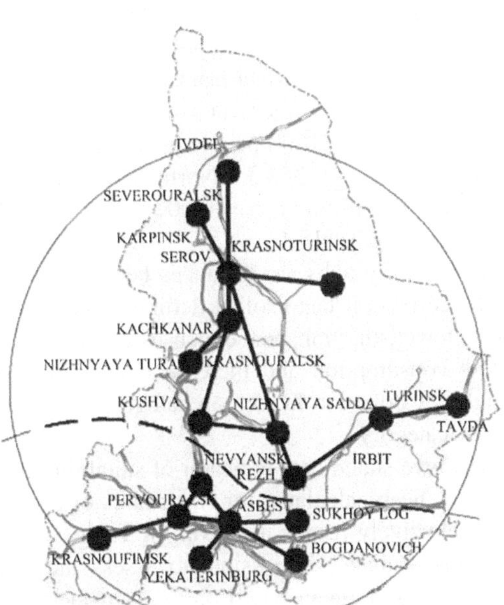

Fig. 2. Division of the graph-geographical model into fractions corresponding to the maintenance areas of recycling centers

The graph-geographical model distributed the existing settlements corresponding to the vertixes of the graphs, as shown in Table 3. The "sphere of influence" of the City of Nidzny Tagil is represented by smaller numbers of settlements, but much more by the territory, which really corresponds to transport accessibility and density of formation of motor vehicles.

Table 3. Division of settlements by shares of graph-geographical model

"Sphere of Influence" Yekaterinburg	"Sphere of Influence" Nidzny Tagil
Yekaterinburg; V. Pishma; Kamensk-Uralsky; Kamyshlov; Verkh-Neyvinsky; V. Dubovo; Pervouralsk; Berezovsky; Sredneuralsk; Novouralsk; Beloyarsky; Aramil; Asbest; V. Tagil; Revda; Nevyansk; Polevskoy; Zarechny; Bogdanovich; Kirovgrad; Talitsa; Rezh; Druzhinino; Artemovsky; Sysert; Pioneersky; N. Sergi; Kalinovskoye; Baykalovo; Pyshma; Bysert; Staroutkinsk; Mikhaylovsk; Arti; Krasnaya Polyana; Klenovskoye; Tavda; Achit; Turinsk	N. Tagil; Lesnoy; N. Tura; V. Salda; Serov; Alapaevsk; Krasnoturyinsk; Irbit; Gornouralsky; N. Salda; Sosva; Svobodny; Severouralsk; Kushva; V. Sinachiha; Verkhoturye; New Lyalya; Volchansk; V. Tura; Vostochny

Further, according to algorithm, we define indicator of suitability of settlements for locating of the enterprises of utilization system as the size of potential of this point and costs of transportation. Using statistical data from official sources [31–33], on formula (1) calculation of capacities of 81 settlements of Sverdlovsk region is carried out. Quantity the avtotransoprt units per 1000 people in the city of Yekaterinburg – 409.5, in other settlements of area – 318.3. Normalizing coefficients respectively: $\alpha_1 = 0.00000068$; $\alpha_2 = 0.00003172$ and $\alpha_3 = 0.002442$. Values of capacities of some settlements of area are given in Table 4.

Distances between tops of graphs are distances between settlements on highways taking into account the correction factor of ki defining category of the highway. The category of the road is lower, the transport expenses connected with drop of average speed, increase in fuel consumption and increase in costs of service and repair of transport units are higher. Correction factors on categories of roads - from 0.8 till 1.8 respectively for the I-V category.

On the basis of the above-stated data indicator of suitability of seating of objects of system of utilization have been calculated for two shares of grafo-geographical model of Sverdlovsk region, selectively results are given in Table 4.

The proposed technique makes it possible to assess the expediency of placing recycling centers in a given settlement based on selected factors.

Table 4. Potentials and indicators of suitability of settlements of Sverdlovsk region (selectively)

Settlement	Population	Average salary, rub.	Potential value	Suitability indicator
Southern part				
Yekaterinburg	1477737	31522	3.0000	3.339
Kamensk-Uralsky	172040	25842	1.7135	1.925
Pervouralsk	148450	25316	1.6809	1.858
Berezovsky	74109	24926	1.6182	1.839
Druzhinino	4511	24783	1.5666	1.668
Polevskoy	70358	24728	1.6094	1.690
Talitsa	44609	22403	1.5182	1.647
Turinsk	7281	22295	1.4895	1.481
Northern part				
N. Tagil	359450	25301	1.8232	1.935
Irbit	37444	2606	1.6296	1.701
Krasnoturyins	63181	22913	1.5469	1.694
Alapaevsk	43756	23960	1.5670	1.687
Severouralsk	41579	23029	1.5360	1.602

6 Findings and Discussion Issues of Development

The model allows to determine the places of disposal centers in the territory, taking into account factors: concentration of settlements, availability of railway network and transhipment stations, class of roads, number and solvency of the population.

However, there are qualitative corrective factors:

1. Strategic priorities for the development of the region - for example, the need to take into account the Program for the Development of Single Towns;
2. The historical situation - the presence at the investigated territory of settlements with city-forming industries or "closed" territories, in the Sverdlovsk region there are four;
3. Ecological and cultural zones - natural and historical parks;
4. Technological conservation zones - water basins of cities and settlements;
5. Established administrative and territorial division.

According to technique, the utilization enterprises of class A have to be located in settlements with the greatest indicators of suitability according to the allocated shares of grafo-cartographic model. Such settlements are Yekaterinburg and Nizhny Tagil. We will not talk about locating of recycling cents of class B for the following reasons yet: the studied experience of foreign countries where there is effective system of vehicle scrappage, says that the centers of class B and C arise as natural stage of development of the utilization centers of class A and D, after exit of their work at full capacity.

In Gornozavodskoy district, northern and eastern part of the area it is expedient to locate the recycling center of class D in number of 3 units (justification of the required production capacities is carried out in work [34]: loading of collection points makes

3–5 thousand units a year), and intensity of leaving of operation of the motor transport in northern part of the area - of 13.5 thousand units a year. In this case it is necessary to consider the adjusting factors, for example, the second for the size of indicator of suitability is Lesnoy, however it is the “closed” city, therefore locating of collection point is impossible in it. Therefore, according to model, we choose the cities of Irbit, Krasnoturyinsk and Alapaevsk with suitability indicators respectively 1.701; 1.694 and 1, 687.

For the southern part of grafo-geographical model with the center of Yekaterinburg which, according to calculations, has the greatest indicator of suitability of locating and the volume of leaving is about 50 thousand units a year that corresponds to the enterprises of class A and B. This indicator is in Yekaterinburg at the level of 27 thousand units a year that speaks about need, except creation of the utilization center of class A, to create at least five collection points for automobile waste in the territory of the city or in the nearby cities located most optimum in terms of availability of car owners (in 50-km zone), presence of floor spaces for collecting and storage of auto-waste, transport availability. These are the cities of Pervouralsk, Verkhnyaya Pyshma, Berezovsky, Nevyansk, Sukhoy Log.

Besides collecting points at Yekaterinburg, 4 more collecting points which location has to be defined by the high level of indicator of suitability and the adjusting factors are necessary for the southern part of the area. Our offers, proceeding from model - Kamensk-Uralsky, Polevskoy, Druzhinino, Talitsa with indicators of suitability 1.925; 1, 690; 1.668; 1.647.

In extreme eastern part of the area population density rather low therefore settlements have coefficient of suitability of locating of the utilization enterprises is less than 1.5. For settlements of this area it is offered to use the mobile collecting office which is based in Kamyshlov; the same is in northern part where also settlements density is low therefore automobile waste is offered to be collected with some frequency and to use also mobile office.

7 Conclusions

Management and optimization of disposal facilities will create favorable conditions for the implementation of sustainable development principles and will create positive experience that can be used in waste management for other industries.

In case the formation of the disposal system of vehicles is decided at the level of the federal district, or the country as a whole, the cartoschem of the population density of the district will differ, but can be created according to the same principle.

References

1. Average Vehicle Age. Official website of the European Association of Motor Manufacturers. http://www.acea.be/statistics/tag/category/average-vehicle-age
2. Bobovich, B.B.: Vehicle and Autocomponents Scrappage. MGIU, Moscow (2010)

3. Trofimenko, Y.V., Vorontsov, Y.M., Trofimenko, K.Y.: Vehicle Scrappage. AKSPRESS, Moscow (2011)
4. Petrov, R.L.: Features and the prospects of utilization of old cars in Russia and comparison with the European practice of technical. Autom. Ind. **1**, 44–49 (2014)
5. Limonov, L.E., Batchayev, A.R., Berezin, M.P., et al.: Regional Economy and Spatial Development. Yurait, Moscow (2015)
6. Granberg, A.G.: Fundamentals of Regional Economy. GU HSE, Moscow (2000)
7. Kuzmenko, Yu.G., Saveleva, I.P., Skatkova E.D.: Logistics of trade services: models and methods of territorial placement of market economy objects - foreign experience. YurSU Vestnic, Ser. Econ. Manag. **9**(3), 159–168 (2015)
8. Yegorov, D.O.: Territorial model of placement of the social infrastructure facility of the region. J. Samara State Econ. Univ. **11**(133), 39–44 (2015)
9. Tsuji, A., Nelson, Y., Kean, A.: Recyclability index for automobiles. In: Proceedings of the 99th Annual Meeting and Exhibition of the Air and Waste Management Association, New Orleans (2006)
10. Kopylova, O.A.: Method of Evaluation of Options for Placement of Regional Logistics Centers. UrGUPS, Yekaterinburg (2014)
11. Korotayev, V.N.: Methodical approaches to improvement of production and logistics system of transportation of production/consumption wastes. J. Ural. State Univ. Railw. Transp. **1**, 83–89 (2009)
12. Velikanova, T.V., Ladoshkin, A.I.: Use of optimization methods in planning the placement of production. J. Samara Munic. Inst. Manag. **2**, 66–73 (2013)
13. Lylin, N.A.: Justification of the Network of Enterprises for the Disposal of Equipment (on the Example of the Moscow Region). WGAU-ISHA, Moscow (2015)
14. Mowder, J., Elmagrabia, S., et al.: Operations Research: Models and Applications, vol. 2. Mir, Moscow (1981)
15. Galiev, S.I., Satarov, A.Z.: Optimization of emergency stations using continuous models. Vestnic Kazan. State Tech. **4**, 39–44 (1996)
16. Raenko, N.V.: Stochastic problem of optimal placement. Methods of solving nonlinear equations and optimization problems. In: 3rd Symposium: Reports and Messages, Tallin, pp. 157–158 (1984)
17. Maslov, O.V.: Combined algorithm of arrangement of elements on printed circuit board using multicriteria optimization. Control Syst. Mach. **3**, 43–48 (1990)
18. Moiseev, V.S., Sotnikov, S.V.: General mathematical model of the problem of optimal placement of objects. J. KGTU named after A. N. Tupolev **3**, 31–36 (2000)
19. Prigozhin, I., Stengers, I.: Order from Chaos. Editorial URSS, Moscow (2001)
20. Lesh, A.: Spatial Organization of the Farm. Ed. by A.G. Granberg. Science, Moscow (2007)
21. Condratiev, V.D.: Methods of Solving the Problem of Location of Service Objects. Management of Large Systems: Collection of Works, pp. 47–56 (2008)
22. Ismagilov, L.A.: Model of territorial placement of objects in the sphere of services. J. UGAT **12**, 3(32), 134–140 (2009)
23. Pukhov, E.V., Astanine, V.K.: Improvement of waste disposal system of enterprises of technical service of transport and technological machines AIC. Technol. Wheel. Tracked Mach. **2**, 41–50 (2014)
24. Kuznetsova, EYu., Akulova, A.A.: Organization of the recycling process in the Russian motor vehicle complex. J. Ural. State Univ. Railw. Transp. **4**(28), 81–90 (2015)
25. Kuznetsova, EYu., Akulova, A.A., Markin, G.A., Yuferova, E.O.: Placement of motor vehicle recycling enterprises in the Ural region. SibADI Vestnic **3**(55), 81–87 (2017)

26. Sizy, S.V., Kuznetsova, EYu., Akulova, A.A.: Grafo-geografical model of the region and a barycentric technique of placement of the utilization enterprises of motor transportation waste in the Ural region. Mess. Ural. State Univ. Railw. Transp. **2**(34), 57–66 (2017)
27. Say, V.M., Sizy, S.V.: Geometric characteristics of organizational networks. World Transp. **4**(32), 10–17 (2010)
28. Sizy, S.V.: Determination of appeal and ranging of regions of Russia in terms of the interests of the JSC "Russian Railways" company by means of technique of linear estimated forms. Econ. Railr. **8**, 20–33 (2010)
29. Sai, V.M., Sizy, S.V.: Education, Functioning and Disintegration of Organizational Networks. Scientific Monograph. USURT, Yekaterinburg (2010)
30. Asanov, M.O., Baransky, V.A., Racine, V.V.: Discrete Mathematics: Graphs, Matroida, Algorithms: Manual for Higher Education Institutions, The 2nd prod. LAN Publishing House, SPb (2010)
31. The average monthly nominal charged wage of workers in 2015 on municipal units of Sverdlovsk region. Statistics. Official site of URA.RU. https://ura.news/news/1052245485
32. Information on roads. Official site of Steering of highways of Sverdlovsk region. http://www.uadso.ru/roads/
33. In 15 years the park of cars in Russia has doubled. Analytical agency "AVTOSTAT". https://www.autostat.ru/news/13246/
34. Volodkin, P.P.: Recycling of the motor transportation. Electron. Sci. Publ. "Sci. Notes TOGA" **4**(4), 1514–1518 (2013)

Solving the Problem of Income Loss in the Networks of the Transport Telecommunications Operator When Providing the VPN Service

Alexandra Radchenko[1], Galina Kolodeznaia[1](✉), and Igor Karpovich[2]

[1] Far Eastern State Transport University, Khabarovsk, Russia
KGV-ATC@yandex.ru
[2] Skyscanner, London, UK

Abstract. Intensive development of the communications industry forces operators to pay special attention to fraud control and cybersecurity issues. New technologies allow fraudsters to improve methods of gaining illegal access to transport network infrastructure and it is becoming increasingly difficult for operators to confront this problem. Significant losses from fraud and cybercrime can cause the company to lose part of its revenue and the company can suffer from the decline in investment attractiveness as well as potential brand damage. To avoid this, it is necessary to use specialized technical means that allow timely detection and prevention of illegal actions of attackers. The authors of the article analysed the points of the income loss in providing the IP VPN service within a network of a transport communications operator. A software solution is proposed that performs the tasks of monitoring the correctness of the provision and accounting of access services to a data network using L2VPN/L3VPN technologies or direct connection to an IP/MPLS network.

Keywords: Income loss · Fraud · Network · VPN

1 Introduction

Intensive development of the communications industry forces operators to pay special attention to fraud control and cybersecurity issues. New technologies allow fraudsters to improve methods of gaining illegal access to transport network infrastructure and it is becoming increasingly difficult for operators to confront this problem.

Significant losses from fraud and cybercrime can cause the company to lose part of its revenue and the company can suffer from the decline in investment attractiveness as well as potential brand damage.

To avoid this, it is necessary to use specialized technical means that allow timely detection and prevention of illegal actions of attackers.

Bernardo Gal-vao Lucas in his article noted: " While most Communication Service Providers (CSPs) focused on launching new services and partnerships with Digital Service Providers (DSPs), my fear is that security and fraud are being sidelined as an

Z. Popovic et al. (Eds.): TransSiberia 2019, AISC 1115, pp. 233–244, 2020.
https://doi.org/10.1007/978-3-030-37916-2_24

afterthought – or that CSPs will believe that their old ways of managing fraud will suffice. Nothing could be further from the truth. Ultimately, new technologies are allowing for new fraud opportunities. Communications service providers must see 2018 as a tipping point for prioritizing their fraud management strategies so that they can enjoy the benefits of the exciting digital era" [1]. A complete analysis of the types of fraud and associated financial losses is given in SoK: Fraud in Telephony Networks Sahin, Francillon, Guptaz, Ahamadx [2].

The Technology Research Institute, in its article, attempts to identify and explain the main control measures that operators introduce in a wide range of areas of anti-fraud activities. The Network Inventory Reconciliation Network Inventory chapter provides two views on this issue. First, Mark Nicholson, CTO of Subex, shows the virtues of network resource scanning and argues that the "configuration of the actual network is far more reliable than any representation of it." Then an opposite strategy is advocated by Suren Nathan of Razorsight who feels carriers, especially small to mid-sized ones, should reconcile their operations by using off-network systems such as ordering and billing [3].

Dan Baker, Yates, Taylor, Anand in their analytical material analyze the types of fraud in communication networks and give recommendations on the control and protection against fraud taking into account both external and internal factors [4].

Frank McDermott and Joe McDermott believe that cloud technology has advantages for inventory and reconciliation processes. Creating one common database helps to prevent loss of business revenue and avoid fraud [5].

Most common network security solutions—intrusion detection systems, firewalls, network behavior analysis systems, etc. are aimed at taming outside threats from hackers, malware, and the like. But as network security improves, criminals will increasingly rely more on inside threats where they gain access to sensitive data by bribing, blackmailing or even tricking employees or trusted contractors [6].

Jans, Lybaert, Vanhoof deem research to reduce internal fraud risk is pivotal. In their paper there present the IFR^2 framework, deduced from both the academic literature and from current business practices, where the core of this framework suggests to use a data mining approach [7].

Kim and Kwon report on the Korean Insurance Fraud Recognition System that employs an unsupervised three-stage statistical and link analysis to identify presumably fraudulent claims. The government draws on this system to make decisions. The authors evaluate the system and offer recommendations for improvement [8].

Tsung et al. apply manufacturing batch techniques to the field of fraud detection [9]. Hoogs et al. present a genetic algorithm approach to detect financial statement fraud. They find that exceptional anomaly scores are valuable metrics for characterizing corporate financial behavior and that analyzing these scores over time represents an effective way of detecting potentially fraudulent behavior [10]. Juszczak et al. apply many different classification techniques in a supervised two-class setting and a semi-supervised one-class setting in order to compare the performances of these techniques and settings [11].

In anti fraud management survey conducted by BAKER TILLY (2008), The Majority of the companies (59%) of the represented by the survey perceive their greatest fraud threat to b from their own employees, the vast majority of the respondents (77%) stated that thy had participated in n investigation, indicting that companies are taking fraud seriously and investigated them. However, only half of the respondents (51%) stated that they have an incident response plane in place, about the company's exposure to fraud, respondents preserved that their companies are most susceptible to internal or employee fraud, and nearly half of the respondents were unable to quantify the impact of fraud in their company, and bout fraud risk management and assessment, less half of the respondents reported having completed some sort of formalized fraud risk assessment. An over helming percentage of respondents feel like they could be doing more regards to fraud risk management [12].

Losses of telecommunication companies from fraud can reach up to 9–12% of total revenue, with up to 10% being lost within the company itself [13]. Infrastructure points monitoring systems allow timely detection of fraud on the operator's networks and prevention of company's revenue loss.

In order to identify the points of income loss when enabling VPN service you need to consider the entire path of receiving and transmitting a request for enabling the service at all stages (Fig. 1).

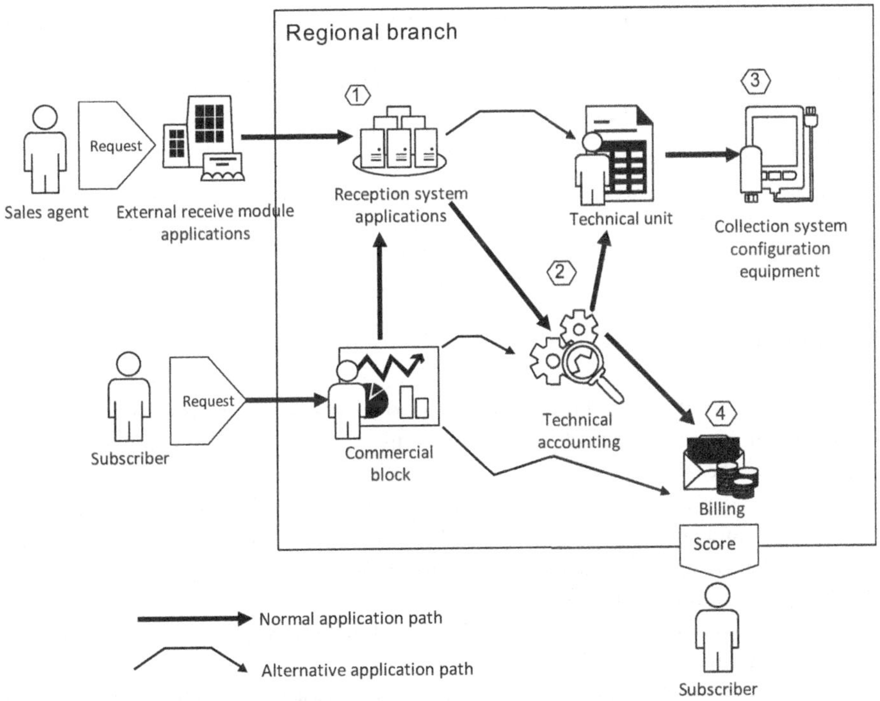

Fig. 1. The path of the application for connecting the service

At the first stage, a set of measures is implemented in order to receive the client's request for enabling the service.

At the second stage, a technical feasibility study is performed.

The goal of the third stage is to carry out a set of measures to assign technical data and to activate the service.

The fourth stage includes entering data on the enabled service into the billing system.

VPN service that the company provides to legal entities, telecom operators and government agencies is highly profitable. The company helps to unite all branches, remote offices and nodes of telecom operators' communication networks into a single secure private network based on IP protocol supporting a full range of telecommunication services with a guaranteed high degree of reliability.

The most vulnerable points of the telecommunications operator infrastructure are considered in the work of N. Abdikeev and his colleagues [14], the points of revenue loss for the VPN service can be:

- missing information about a new subscriber in the billing system;
- incorrect display of services in the billing system;
- connections in the directions marked by the technical accounting system as "technological";
- error in service deactivation while deleting a subscriber record from the billing system;
- activation of the service without entering subscriber's record into the billing system;
- ability to change configuration settings in switching hardware settings (such as port policy).

Methods of dealing with fraud on communication networks and general algorithms for detecting fraud are given in [15]. For the case under consideration, algorithms using the analytical method seem to be the most effective.

2 Methods

To control the points of income loss of a telecom operator, it is necessary to create a reconciliation system to detect discrepancies in systems. The following sources are used to create the algorithm.

Request acceptance module - an automated system for supporting sales and providing communication services, designed to provide automation and information support to the sales management of communication services.

Billing is an automated billing system.

Technical accounting - a system of linear technical accounting.

Configuration - equipment monitoring system.

It is necessary to carry out the following reconciliations to find all the discrepancies of the systems:

- Billing - Technical Accounting;
- Configuration - Technical Accounting;

- Configuration - Billing;
- Request acceptance module - Billing.

To implement all reconciliations, data must be uploaded not only on existing clients with VPN service but also on clients with disabled VPN service and clients that do not have VPN service. Comparing the sources helps to detect a situation when a disabled service is still provided or when the service was initially activated but not connected in billing. Therefore, it is necessary to unload the billing data into three tables:

1. All subscribers of a corporate segment who have services.
2. All subscribers with a valid VPN service.
3. All subscribers with a disabled VPN service.

It is also necessary to upload the configuration files of all equipment used in the telecommunications company from last-mile switches to top-level equipment.

The four systems used in reconciliations do not have one internal identifier that would allow them to be matched by one key. But between certain systems there are keys that allow you to compare data with each other.

All reconciliations are carried out in the forward and reverse direction. In each direction of reconciliation, the first source is always the reference and the second source is the verified one.

The main objective of control is to identify cases of rendering services without payment and/or billing subscribers for services that are not actually provided.

Target reconciliation is performed by comparing billing data associated with technical accounting data and configuration data for the technical parameters of the equipment. The target reconciliation can be extended by matching the virtual number of the subscriber if this number is present in the description of the interface in the configuration.

Non-targeted verification is performed by comparing the configuration data with the billing data at the address.

When reconciling in the forward direction, Billing - Configuration, for each record in the billing system, it searches for the corresponding records in the configuration files. The reconciliation task is to search for entries that are entered into billing but are not in the equipment configuration. In this situation a customer could be billed for services that were not actually provided.

The billing records obtained as a result and enriched with the technical data from technical accounting, are compared with the configuration files data in the IP address fields of the equipment and logical interface. If there is a match, the record receives the "Found in billing" status. Additionally, when reconciling Billing - Configuration, the "Subscriber ID" field is compared directly with the configuration data. If a value of this field is found in the logical interface field in the configuration file, the record receives the "Found by virtual number" status.

The reconciliation algorithm in the form of a block diagram is shown in Fig. 2.

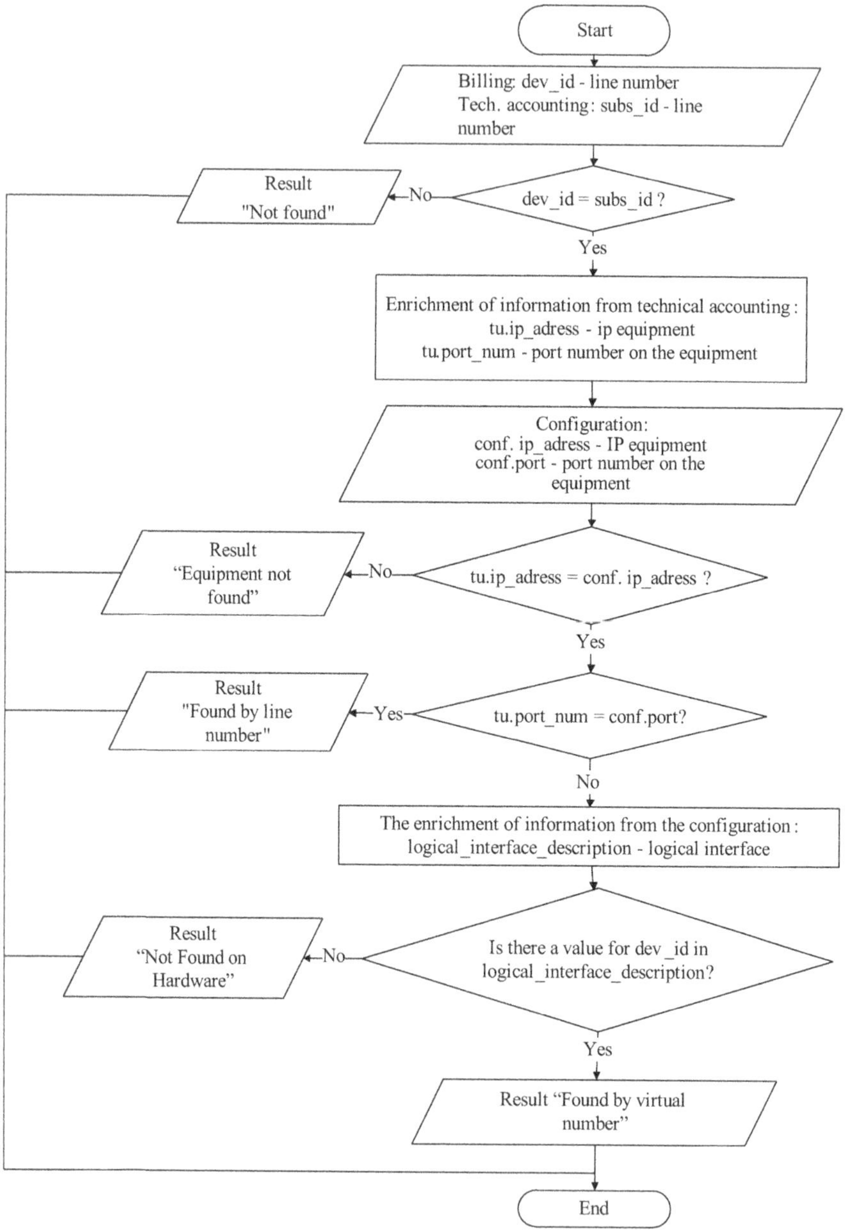

Fig. 2. Algorithm of reconciliation Billing – Configuration

When reconciling in the opposite direction, Configuration - Billing, for each entry in the configuration files, a search is made for the corresponding entries in the billing. The reconciliation task is to search for accounts that are entered in the configuration, but are absent in the billing. This situation means that the service is actually provided, but the customer is not billed, and, accordingly, the service remains unpaid.

For reverse mapping, VPN entries downloaded from equipment configuration files are mapped to VPN entries in billing. They are compared by linking the data from the technical accounting by IP address and port number on the equipment, enriched with the line number information. If the record matches, the result of the co-delivery “Found by the line number in billing” is assigned.

If there is no equipment when searching by the IP address of the equipment, the result of the comparison “No equipment in technical accounting” is assigned.

The address in the description of the logical interface is compared with the address of the subscriber from billing. If the record matches, the result of the matching “Found by address in billing” is assigned.

Additionally the entries from equipment configuration files are compared with the entries by billing subscribers who do not have a VPN attribute.

The address in the description of the logical interface is compared with the address of the ACP subscriber. If the record matches, the result of the comparison “Found in billing among subscribers without a VPN” is assigned.

The reconciliation algorithm is shown in Fig. 3.

The main task of Billing control - Technical Accounting is to increase the reliability of data entered in billing and technical accounting. When reconciling in the forward direction for each billing record, the system searches for the corresponding record in the technical accounting. The task of reconciliation is to find services that are paid by customers, but which are not considered in technical accounting. When reconciling in the opposite direction, Technical Accounting - Billing, for each record in technical accounting, the system searches for compliance in billing. The reconciliation task is to search for services that are registered in the technical accounting, but which are not in the billing.

For matching purposes, the subscriber’s identifier is used from the billing entries that show subscribers with the VPN service. A corresponding subscriber’s record is then searched in the technical accounting. If there is a match, the record receives the “Found in technical accounting” status.

Algorithm for reconciling Billing systems - Technical Accounting, shown in Fig. 4

For backward matching, a search is also made for entries in the billing table by line identifier. If there is a match, the record receives the “Found in billing” status. Additionally, a comparison is made with billing records without the VPN attribute. If there is a match, the record receives the status “Found in billing among subscribers without a VPN”. Algorithm for reconciling systems Technical Accounting - Billing is shown in Fig. 5.

The main task of reconciliation Technical Accounting - Configuration - improving the reliability of data entered in the technical accounting and configuration of equipment.

Target reconciliation is carried out by comparing the technical accounting data and configuration data according to the technical parameters of the equipment. Non-targeted verification is performed by matching configuration data with technical accounting by the address.

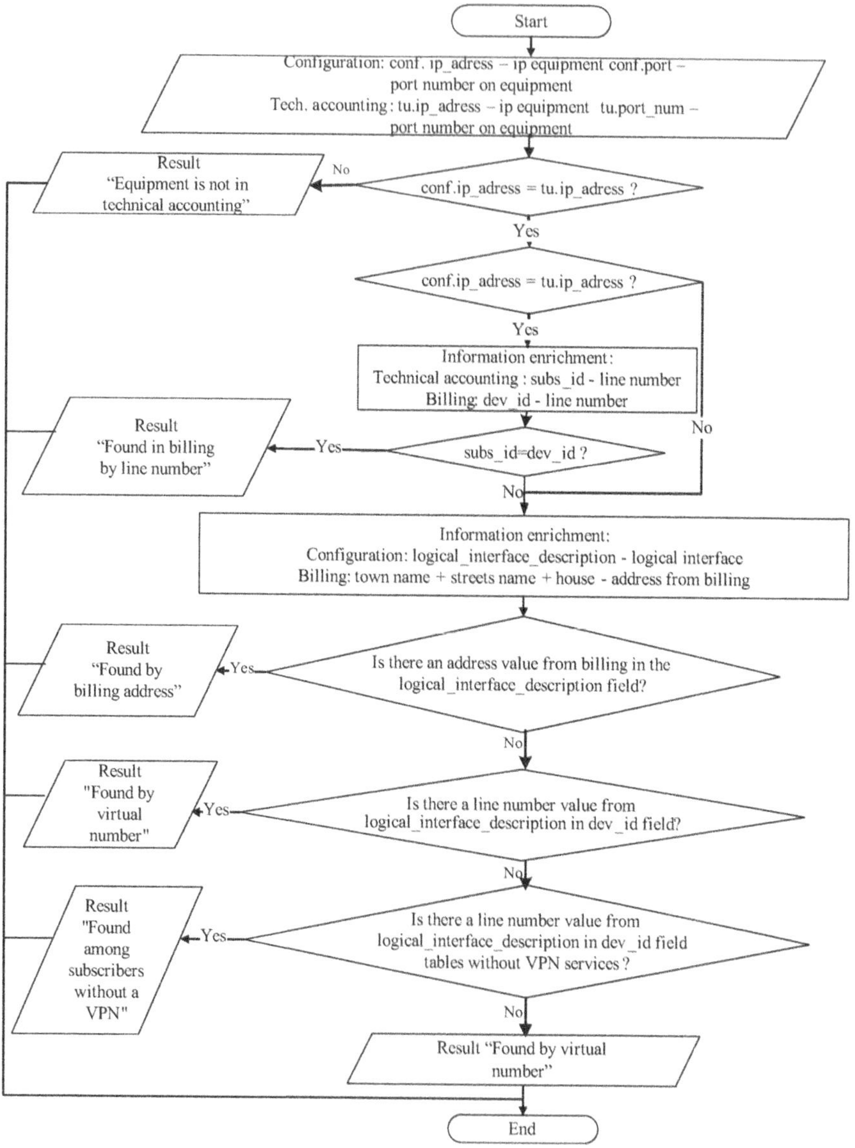

Fig. 3. Algorithm of reconciliation Configuration – Billing

When conducting a reconciliation in the forward direction, for each record in the technical accounting the system searches for the corresponding record in the configuration files, checking that such a record exists. The reconciliation task is to identify situations where although resources are reserved in the system of technical accounting, in fact these resources are not configured on the network and are free to organize other connections.

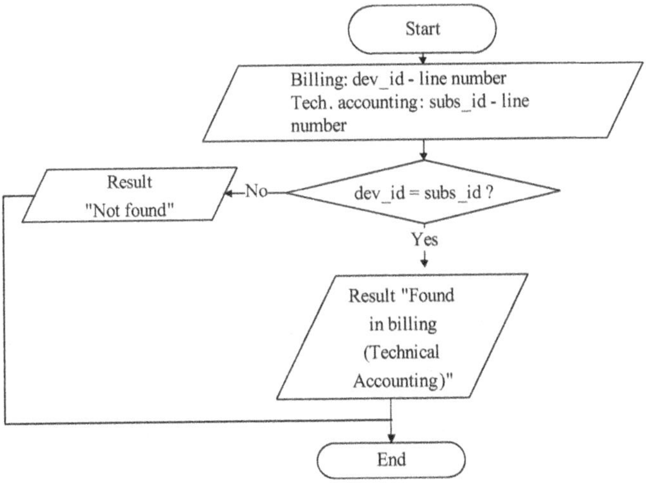

Fig. 4. Algorithm for reconciling Billing systems - Technical Accounting

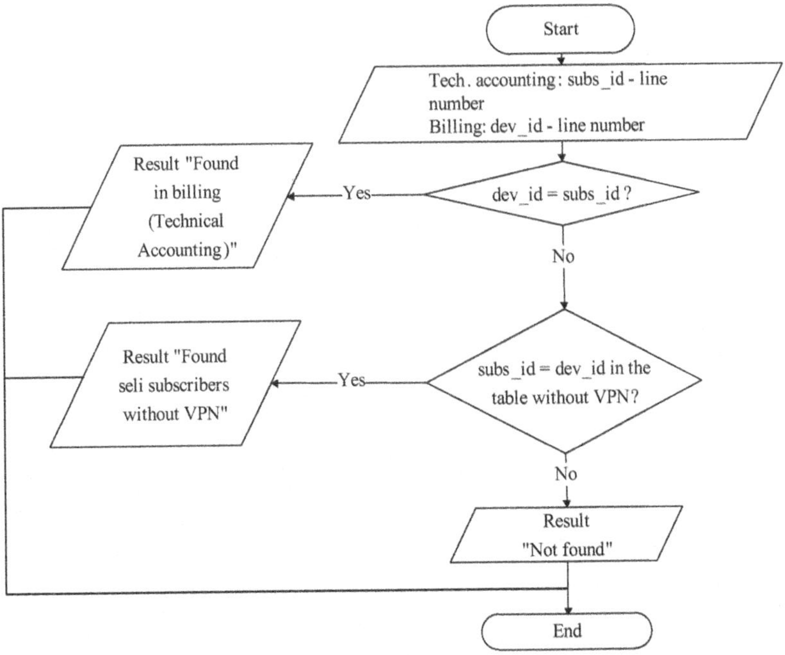

Fig. 5. Algorithm for reconciling systems Technical Accounting – Billing

When reconciling in the opposite direction, Configuration - Technical Accounting, for each entry in the configuration files, the system searches for a similar entry in the technical accounting, checking the existence of such an entry. The reconciliation task is to identify situations where resources reserved on the network are marked as free in the technical accounting system.

When comparing Technical Accounting - Configuration, VPN entries that are in the technical accounting are matched with VPN entries downloaded from the equipment configuration files.

Comparison of technical accounting data with file data is carried out by the value of the part of the equipment data field that contains the name of the equipment. The enriched technical records obtained as a result are compared with the data of the configuration files by fields of the IP address of the equipment and the port number on the equipment. If there is a match, the record receives the status "Found by IP of the manager equipment". If the matching IP address is found in the configuration files, but the required port number is missing, the entry receives the status "Not Found". If the matching IP address was not found in the configuration files, the record receives the status "Configuration file not found", while in fact the record receives the double status "Configuration file not found" and "Not found", since the files the configuration is missing both the required IP and the required port number on the equipment.

3 Results

Based on the results of introducing the program into a telecommunication company, the results of reconciliations are shown in Table 1.

Table 1. Reconciliation results

Direction	Amount VPN		Amount VPN coincided	Amount VPN coincided, %
	Reference	Verifiable		
Configuration - Billing	3535	5323	1243	35.16
Billing - Configuration	5323	3535	1347	25.31
Billing - Technical Accounting	5323	5308	4512	84.76
Technical Accounting - Billing	5308	5323	5308	100
Configuration-Technical Accounting	3535	5308	1424	40.28
Technical Accounting - Configuration	5308	3535	1428	26.9

4 Discussion

According to the results of the reconciliations, it is clear in which systems it is necessary to carry out the work and to identify the undetected connections.

T.O. the studies and proposed solutions allow us to identify the points of revenue loss when connecting to the VPN service.

There is an open question to the business running on why the configuration is the most vulnerable part where most of the changes are not accounted for in other systems.

This shows lack of automation and reliance on manual processes when transferring service data between different components of the system. As an opposite observation, there is an up to a hundred percent match of services between billing and technical accounting which means either strictly follow manual routine or sophisticated service-to-service automated integration.

Configuration quality as such could display low results due to low level of technology tooling used around it. Modern approaches include at least three base principles which should be applied to increase the quality:

1. Using configuration-as-code services. This means storing all configuration for VPN access and switching hardware in version control systems, ideally described in programming languages. This is done by using SDKs provided by hardware vendors or by utilizing REST or other HTTP based APIs. Version control provides auditability of all the changes made together with peer review.
2. Using automated testing which could check that necessary changes applied to the confirmation are safe enough not to cause disruptions and to produce consistent change to other service and accounting related systems.
3. Using of continuous integration and continuous delivery systems which automatically run tests above paired with automated security audit. The tool engineered in this paper could also be included as part of continuous integration process to automatically verify all the changes prior to them being rolled out to hardware.

For large telecom companies with multiple teams responsible for different components of the system some fraud attack vectors could also be reduced by introducing organisational measures. This approach is based on dividing responsibilities for the components so they are owned by different teams. Those teams should not have write access to each other's systems. Customer facing departments should have a single point of data entry so there's no chance for them to (un)intentionally create service accounting mismatch. This approach while being preventive requires significant investment into automation so it might not be available to small or mid-sized operators.

5 Conclusions

The software developed is an automated system that performs the tasks of monitoring the correctness of the provision and accounting of data access transport network services using L2VPN/L3VPN technologies or direct connection to the IP/MPLS network. Using this software as the main monitoring tool allows the operator to solve such problems as providing complete control over the work of departments when creating and maintaining NSIs for connecting customers accessing the data network via L2VPN, L3VPN or direct connection to the IP network/MPLS.

The implementation of the proposed monitoring system improves the customer billing results, excludes the cases of providing services without payment, eliminates incorrectly applied tariffs and minimizes the loss of important billing information.

Such implementation gives immediate results in a form of the economic effect of identifying the points of revenue loss and reducing the level of revenue loss. At the same time, it provides the business unit with a high-quality service for visualization and automation of work and data reporting. This results in improving the quality and safety of providing services to customers of a telecommunications company.

References

1. Bernardo, G.L.: 2018's Top Telecom Fraud Threats. https://www.thefastmode.com/services-and-innovations/11751-2018-s-top-telecom-fraud-threats
2. Sahin, M., Francillon, A., Gupta, P., Ahamad, M.: SoK: fraud in telephony networks. In: 2017 IEEE European Symposium on Security and Privacy, pp. 235–250 (2017). https://doi.org/10.1109/eurosp.2017.40
3. Mark, N., Suren, N.: Network Intventory Reconciliation. Telecom Fraud Management Services Software & Processes, pp. 3–6 (2015). http://www.technology-research.com/products/bizassure/bizassure_toc.php#b14
4. Baker, D.: Telecom Fraud & Business Assurance Solutions, Services, & Strategies (2019). http://www.technology-research.com/products/fraudmgt.php
5. McDermott, F., McDermott, J.: ARM Data Center Software's Cloud-Based Network Inventory Links Network, Operations, Billing, Sales & CRM to One Database (2019). http://bswan.org/single_database_inventory.asp
6. Knight, P.: Protecting Data from the Insider Threat. Veriato Closes the Security Gap with Massive Data Collection & Machine Learning (2019). http://fraudtech.net/veriato_insider_threat.asp
7. Jans, M., Lybaert, N., Vanhoof, K.: A framework for internal fraud risk reduction at IT integrating business processes: the IFR^2 framework. Int. J. Digit. Account. Res. **9**, 1–29 (2009). https://doi.org/10.4192/1577-8517-v9_1
8. Kim, H., Kwon, W.J.: Multi-line insurance fraud recognition system: a government-led approach in Korea. Risk Manag. Insur. Rev. **92**, 131–147 (2006)
9. Tsung, F., Zhou, Z., Jiang, W.: Applying manufacturing batch techniques to fraud detection with incomplete customer information. IIE Trans. **396**, 671–680 (2007)
10. Hoogs, B., Kiehl, T., Lacomb, C., Senturk, D.: A genetic algorithm approach to detecting temporal patterns indicative of financial statement fraud. Intell. Syst. Account. Finance Manag. **15**, 41–56 (2007)
11. Juszczak, P., Adams, N.M., Hand, D.J., Whitrow, C., Weston, D.J.: Off-the-peg and bespoke classifiers for fraud detection. Comput. Stat. Data Anal. **52**(9), 4521–4532 (2008)
12. Baker, T.: Anti – fraud management survey results, magnify your anti fraud management (2008). http://www.bakertilly.com/cms/public/userfiles/40eb699b88a3fbd294f043905ecdb9b1e448003f/00d16e0290b6ce8f3262342142d8ed8f093ec767Antifraud%20management%20survey_2008.04.pdf
13. de Reuver, M., Verschuur, E., Nikayin, F., Cerpa, N., Bouwman, H.: Collective action for mobile payment platforms: a case study on collaboration issues between banks and telecom operators. Electron. Commer. Res. Appl. **14**(5), 331–344 (2015)
14. Glick, W.H.: Business Process Reengineering. Wiley Encyclopedia of Management, pp. 1–2 (2015)
15. Van Vlasselaer, V., Eliassi-Rad, T., Akoglu, L., Snoeck, M., Baesens, B.: Afraid: fraud detection via active inference in time-evolving social networks. In: Proceedings of the 2015 IEEE/ACM International Conference on Advances in Social Networks Analysis and Mining, pp. 659–666 (2015)

Development of an Automatic Locomotive Traction Drive Control System to Reduce the Amount of Wheel Slippage on the Rail

Alexander Buynosov(✉), Boris Sergeev, Anatoly Kalinichenko, and Sergey Antropov

Ural State University of Railway Transport, Kolmogorova Street, 66, Yekaterinburg 620034, Russia
byinosov@mail.ru

Abstract. The subject of the study is the structure of the vector control system. The relevance of this topic lies in the current trends in the development of traction drive control systems that provide stable operation of an electric locomotive. On the basis of the developed and described mathematical model, the basic principles of control of the traction drive with the task of speed and torque are determined. The simulation results reflect the key moments of electric locomotive drive control in the conditions of external disturbing factors. The control system for the traction drive will completely provide the realization of traction properties of a locomotive subject to achieve and automatically maintain specified speed until the construction regardless of the profile path with an acceleration due to the task utilization of time at the wheel of the locomotive.

Keywords: Locomotive · Traction · Drive · Wheel · Rail · Slippage · Control · Target · Reduction · Automatic · System · Development

1 Introduction

Modern traction electric drives of locomotives are complex electromechanical systems, the creation and development of which requires improving the methodology of their research [1].

One of the most promising options for automated traction drives can be called a frequency-controlled electric drive, which includes a bunch of "frequency Converter-asynchronous motor". The use of induction motors with squirrel-cage rotor in the traction drive of an electric locomotive is primarily due to the optimal price/quality ratio, high reliability, low maintenance costs and high efficiency [2].

Currently, the methods of frequency control of the electric drive include [3–6]:

- scalar control;
- vector control.

The most widespread in high-frequency drives, to which the traction drive of electric locomotives belongs, is vector control by flow coupling of the rotor, which can be both with a sensor and without an angular speed sensor. Sensorless control, like any

Z. Popovic et al. (Eds.): TransSiberia 2019, AISC 1115, pp. 245–254, 2020.
https://doi.org/10.1007/978-3-030-37916-2_25

other system, has its advantages and disadvantages, namely: the determination of the position of the rotor flux coupling can be calculated from the output data from the sensors of phase currents and phase voltages. However, at low frequencies of the supply voltage, the accuracy of data from the sensors decreases, which leads to unstable operation of the vector control system (IED) of the Converter.

Therefore, on the rolling stock is often used IED with a sensor of angular frequency of rotation of the rotor on the negative feedback.

It should be noted that vector control, in relation to scalar, has significant advantages, such as a higher level of accuracy in regulating the speed of rotation of the shaft and rapid response to possible changes in the load, provoked by external disturbing factors from the infrastructure and the environment [7].

In this case, vector control is divided into flow-oriented control and direct torque control [8].

2 Mathematical Model of Traction Drive Control System

2.1 The Main Provisions of the Vector Control System

The implementation of flow-oriented control is based on the direct measurement of the flow vector, which can be determined on the basis of data on the angular position of the rotor using a position sensor, or the so-called state observer with Sensorless control, which by means of mathematical transformations based on changes in the flow and stator currents calculates the flow-coupling of the rotor and its angular position.

The main disadvantage of the classical vector control system of the electric drive is its complex structure due to the need to perform operations of converting coordinate systems from stationary to rotating, oriented along the rotor field (x-y), and Vice versa, necessary to control the magnetic field vector by means of a current in the direction of the coordinate x and the quadrature component of the field in the orthogonal direction y. However, modern trends in the development of drive electronics allow us to create relatively cheap and high-performance frequency converters. For traction induction motors with squirrel cage rotor frequency control is the most perfect way of economical speed control over a wide range.

2.2 Construction of the Equivalent Circuit of the Asynchronous Traction Motor

In General, when designing asynchronous traction motors tend to ensure that magnetization losses, compared with active losses in the stator windings, can be neglected. Thus, as a model to illustrate the processes occurring in the engine, often use a simplified t-shaped equivalent circuit (Fig. 1).

In the equivalent circuit (Fig. 1): u_s – generalized vector of input voltage; i_s – generalized vector of stator current.

The stator current vector is divided into two components: i_m – generalized magnetization current vector; i_r – generalized rotor current vector. The values and ratios of motor currents at a given input voltage are determined by the parameters of the equivalent circuit. Stator resistance (R_s) characterizes the active losses in the stator windings and is equal to the resistance of the phase winding of the stator. Stator scattering inductance ($L_{\sigma s}$) characterizes the part of the stator flow that is not coupled to the rotor and does not participate in the creation of the moment. The main inductance (L_m) characterizes the part of the flow that is coupled to the stator and the rotor and participates in the creation of the moment. On the linear portion of the magnetization curve of the motor, the main inductance is a constant. When the saturation of the magnetic motor value of the main inductance decreases. The rotor scattering inductance ($L_{\sigma s}$) characterizes the part of the rotor flow that is not coupled to the stator and does not participate in the creation of the moment. In addition, the R_r/S – parameter characterizes the active losses in the rotor (R_r – rotor resistance in the short-circuit mode (the motor shaft is braked).

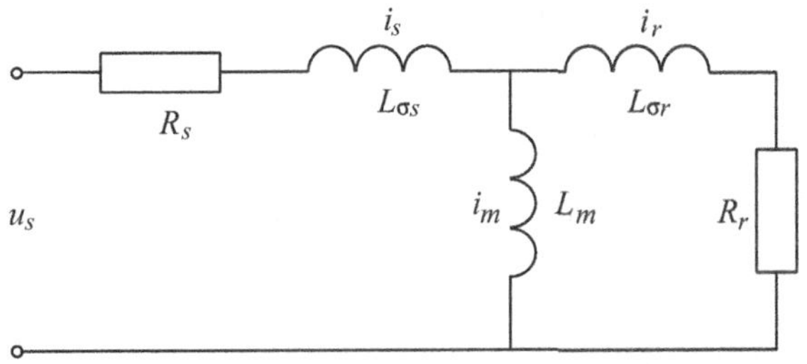

Fig. 1. Induction traction motor equivalent circuit

The slide S is defined as follows

$$S = \frac{n_1 - n_2}{n_1}, \tag{1}$$

where n_1 – the speed of rotation of the field; n_2 – the speed of rotation of the rotor.

Induction motor is characterized by the following state vectors: $\vec{u}_s$ – stator voltage vector; $\vec{i}_s$ – stator current vector; $\vec{i}_r$ – rotor current vector; $\vec{i}_m$ – magnetization current vector; $\vec{\psi}_s$ – stator flow vector; $\vec{\psi}_r$ – rotor flow vector; $\vec{\psi}_m$ – flow vector in the air gap (magnetization flow).

2.3 Vector Diagram of Asynchronous Traction Motor

Figure 2 shows a vector diagram illustrating the relationship of the induction motor state vectors.

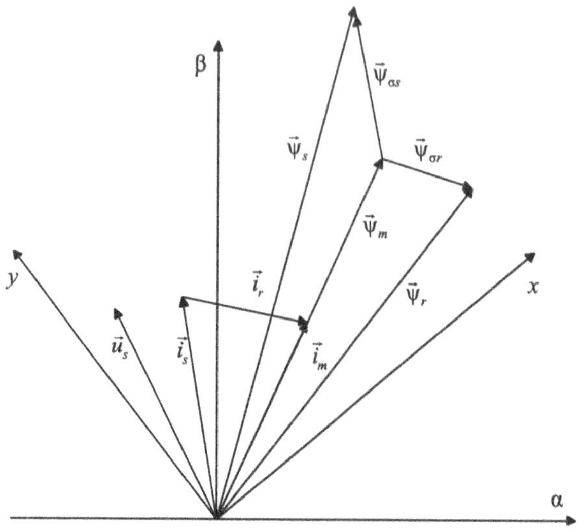

Fig. 2. Relationship of induction motor state vectors

In the presented vector diagram, thc α and β axes are associated with a fixed stator, and the x and y axes are associated with the rotor position and its windings.

During operation of the induction motor, all vectors rotate in the cross-sectional plane of the induction motor around the axis of rotation of the rotor. At constant speed and load moment (in steady-state mode) the amplitudes, phase shifts and velocities of all state vectors remain constant. In dynamics, the amplitudes of state vectors and phase shifts between them change during transients.

Based on the above vector diagram, you can calculate the basic values. So the magnetization current vector is equal to the mousse of the rotor and stator current vectors:

$$\vec{i}_m = \vec{i}_s + \vec{i}_r. \tag{2}$$

The stator flux coupling vector is equal to the sum of the magnetization flux vector and the stator scattering flux:

$$\vec{\psi}_s = \vec{\psi}_m + \vec{\psi}_{\sigma s}. \tag{3}$$

The stator scattering flux is equal to the product of the stator current by the stator scattering inductance:

$$\vec{\psi}_{\sigma s} = L_{\sigma s} \cdot \vec{i}_s. \tag{4}$$

The rotor flux coupling vector is equal to the sum of the magnetization flux vector and the rotor scattering flux:

$$\vec{\psi}_r = \vec{\psi}_m + \vec{\psi}_{\sigma r}. \tag{5}$$

The rotor scattering flux is equal to the product of the rotor current by the rotor scattering inductance:

$$\vec{\psi}_{\sigma r} = L_{\sigma r} \cdot \vec{i}_r. \tag{6}$$

Based on these expressions, you can determine the relationship between the current triangle and the flow triangles. Here the magnetization flux is equal to the product of the magnetization current by the motor magnetization inductance:

$$\vec{\psi}_m = L_m \cdot \vec{i}_m. \tag{7}$$

By means of the above expressions the flows are expressed through the stator and rotor currents as follows:

$$\vec{\psi}_m = L_m \cdot \vec{i}_m = L_m \cdot \vec{i}_s + L_m \cdot \vec{i}_r, \tag{8}$$

$$\vec{\psi}_s = L_s \cdot \vec{i}_s + L_m \cdot \vec{i}_r, \tag{9}$$

$$\vec{\psi}_r = L_r \cdot \vec{i}_r + L_m \cdot \vec{i}_s, \tag{10}$$

where L_s is the inductance of the stator, L_r is the inductance of the rotor.

$$L_s = L_m + L_{\sigma s}, \tag{11}$$

$$L_r = L_m + L_{\sigma r}. \tag{12}$$

The electromagnetic torque of the induction motor arises due to the interaction of the rotor current with the flow coupling in the air gap (magnetization flow). Hence the electromagnetic torque of a three-phase motor is defined as follows:

$$M_{et} = \frac{3}{2} \cdot z_p \cdot \frac{L_m}{L_r} \cdot \psi_x \cdot i_x = \frac{3 \cdot z_p}{2 \cdot R} \cdot \psi_x \cdot \omega, \tag{13}$$

where z_p – the number of pairs of poles; L_m – mutual inductance of the stator and rotor; L_r – inductance of the rotor; ψ_x – flux coupling of the rotor; i_x – stator current; R – resistance of the rotor winding of the motor; ω –angular frequency of the rotor.

2.4 Varieties of Vector Control Systems Traction Drive Locomotive

The traction drive can be controlled in three ways:

- speed control (indirectly by torque);
- torque control;
- adjacent speed and torque control.

Figures 3 and 4 show both traction control structures.

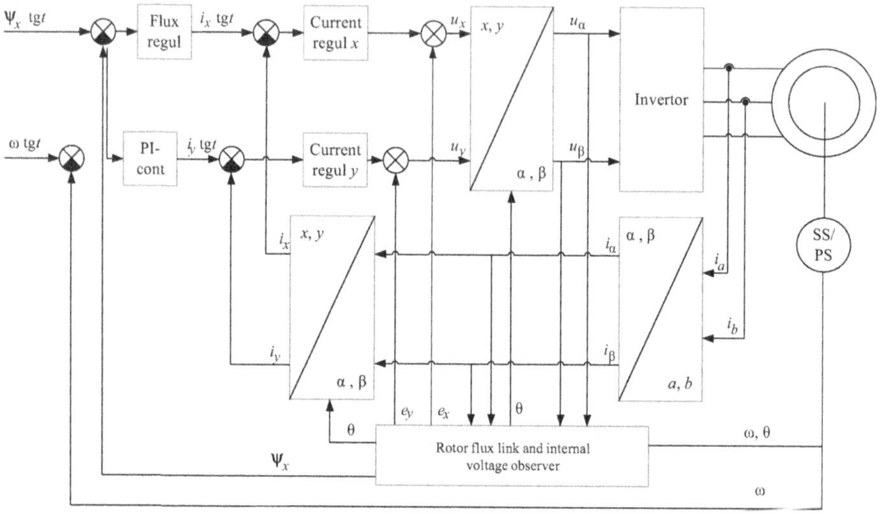

Fig. 3. The structure of the vector control system with the task of the moment

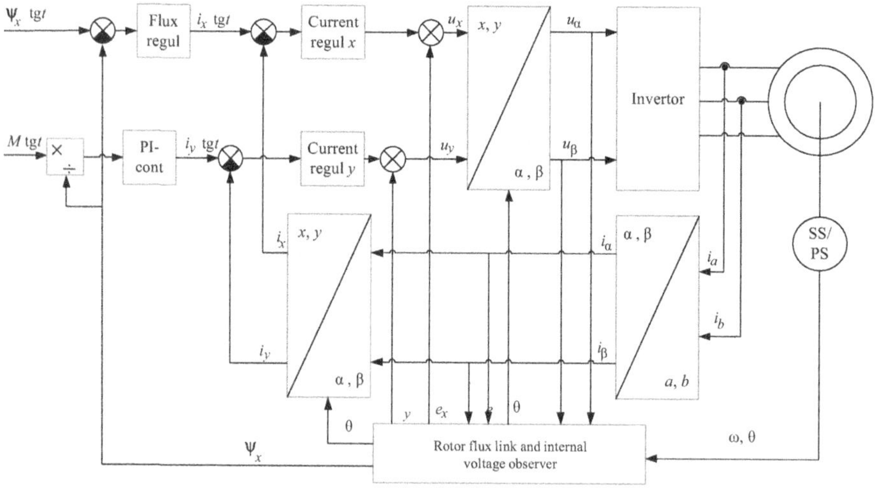

Fig. 4. Structure of vector control system with speed assignment

An important part of the vector control structure is the rotor flux coupling observer and EMF unit, which is used to calculate the amplitude and angular position of the rotor flux coupling vector. In addition, it is optionally able to calculate the components of the stator EMF along the axes in aid of current regulators. Depending on the implementation, the rotor flow coupling observer unit is capable of operating the speed and position of the rotor, the motor phase current and the applied voltages.

Estimation of the angular position used in the coordinate transformation job stresses from the x and y axes in α and β and the inverse transformation of the measured currents of α and β in x and y. Jobs for currents in the x and y axes are supplied from the flow regulator (RP) and from the reference signal point. A flow regulator is required to boost the transients in the magnetization loop. In the end, the current assignment on the x axis will be determined by the formula:

$$\vec{\psi}_r = L_m \cdot \vec{i}_s. \tag{14}$$

However, the transient time is determined by the rotor chain time constant. In relation to the traction motor of electric locomotives, this parameter will reach a few seconds.

To speed up the process, it will be necessary to create a larger current in the stator. Then the rotor current at the first moment of time, when the flux coupling is still equal to zero, will be determined by the expression:

$$i_r = \frac{L_m}{L_r} \cdot i_s \tag{15}$$

Control of the traction drive with an external torque task makes sense when directly regulating the traction force without reference to a given linear speed of the electric locomotive. As mentioned earlier, the traction drive of the electric locomotive before reaching full power is operated in critical operating modes with a limitation on the clutch exceeding the nominal torque [9, 10]. Thus, there are a number of difficulties in the implementation of the maximum traction properties of the drive at speeds below the design [11, 12]. If you control the traction drive of an electric locomotive in this range of speeds with the constancy of the electromagnetic moment, it will lead to underutilization of traction properties, and, as a consequence, reduce the technical speed of the train.

Vector control of the traction drive with the speed task has a greater potential in terms of the implementation of traction properties, since there is no direct control of the motor torque, i.e. it is possible to regulate the electromagnetic torque up to the maximum regardless of the task on the driver's controller. Thus, in case of deterioration of coupling properties of the electric locomotive, or short-term disturbing influences from infrastructure, this system is capable to fulfill with high speed on regulation of the electromagnetic moment of the asynchronous motor.

3 Comparison of Traction Drive Vector Control Systems

To verify these judgments, a simplified model of the electric drive control system in the Matlab software package was developed, which allows to visually assess the features of control with a task by moment and with a task by speed (Fig. 5).

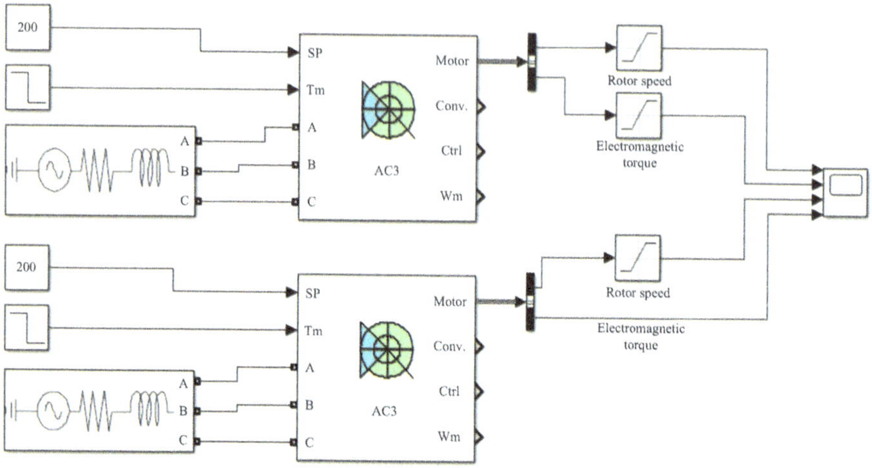

Fig. 5. Mathematical model of traction drive control system with torque and speed assignment

Figure 6 shows the results of the calculation of the mathematical model. The upper graphs show the angular rotational speed of the induction motor rotor with direct torque (left) and speed (right). The lower graphs show the current electromagnetic torque of the motor [13, 14].

At the time of 0–0.4 s, the engine is started with a starting torque of 50 N m. the torque Throw is caused by a jump in the starting current. At 0.4 s in the model with the task time is set torque equal to 200 N m. Thus, the angular frequency of rotation of the rotor of the motor is increasing. The moment at the same time, having an inversely proportional dependence to the angular frequency, is slightly reduced, which in our case can be taken as const [15].

In the model with the speed task, the angular speed of the rotor is controlled by a negative feedback on the speed sensor. At the same time, to achieve a given angular frequency of rotation of the rotor, the control system increases the torque on the engine to a value of 350 N m, while accelerating and then, reaching a steady speed, reduces the torque to a level that ensures the maintenance of a constant angular frequency of rotation of the rotor.

It should be noted that the absolute advantage of the control system with a speed task is its resistance to external disturbing factors. For clarity, in this model at a time of 1.5 s simulated failure of the wheel coupling with the rail ($F_{\mathrm{friction}} < F_{\mathrm{k}}$), in other words, decreased the moment of resistance on the rotor of the engine. As you can see from the left graphs, the task of the moment remained at the same level, the angular speed increased slightly. On the right graphs you can see that when controlling the traction drive with a speed task for 1.5 s with a decrease in the drag torque, the electromagnetic torque of the engine also decreased. The angular frequency of rotation has not increased.

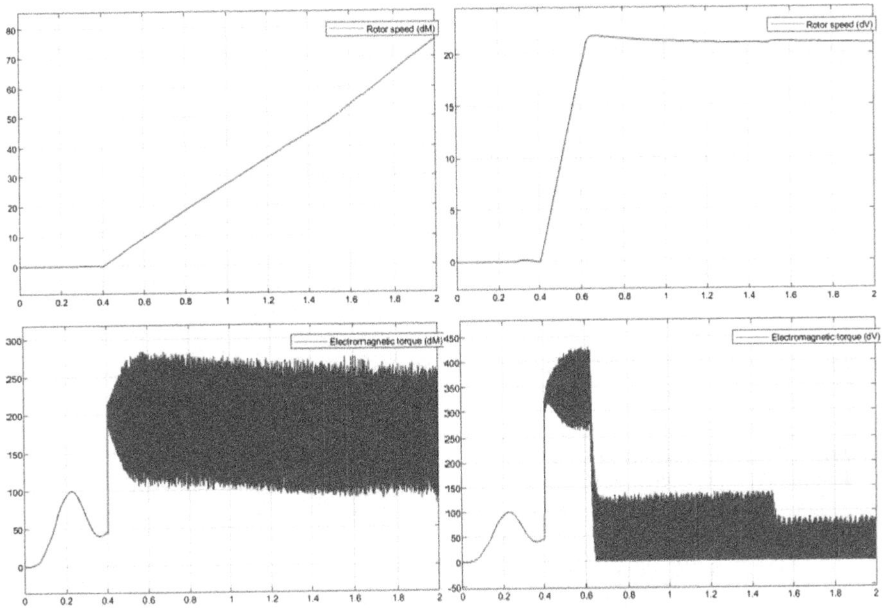

Fig. 6. Simulation results of traction drive control system with torque and speed assignment

4 Conclusions

The analysis of the results showed that the considered methods of controlling the traction drive of electric locomotives allow both adjusting the value of the torque setting on the wheel and automatic acceleration of the electric locomotive to a given speed with its further maintenance without direct torque setting. However, in view of the fact that the greatest operating time of traction rolling stock is in the range from 0 to 40–60 km/h (with a limitation on the clutch), it is advisable to control them adjacent. Thus, the control system for the traction drive will completely provide the realization of traction properties of a locomotive subject to achieve and automatically maintain specified speed until the construction regardless of the profile path with an acceleration due to the task utilization of time on the wheel.

References

1. Takahashi, I., Noguchi, T.: A new quick-response and high efficiency control strategy of an induction motor. IEEE Trans. Ind. Appl. **IA-22**, 820–827 (1986)
2. Takahashi, I., Ohimori, Y.: High-performance direct torque control of an induction motor. IEEE Trans. Ind. Appl. **25**, 257–264 (1989)
3. Krzysztoszek, K., Luft, M., Lukasik, Z.: Digital control model for electric traction vehicles. Procedia Comput. Sci. **149**, 274–277 (2019)

4. Buynosov, A.P., Antropov, S.N.: Study of rail corrugations considering “wheel–ail” friction theory. Transp. Urals **1**(60), 86–89 (2019). https://doi.org/10.20291/1815-9400-2019-1-86-89
5. Drabek, P., Pittermann, M., Los, M., Bednar, B.: Traction drive with MFT - novel control strategy based on zero vectors insertion. IFAC Proc. Vol. **47**(3), 11968–11973 (2014)
6. Buynosov, A.P., Volkov, D.V., Dinislamov, A.R.: Nondestructive control method of depth of surface tempering of details of mechanical part of locomotives as the alternative to the destroying ways. Transp. Urals **1**(56), 26–29 (2018). https://doi.org/10.20291/1815-9400-2018-1-26-29
7. Buynosov, A.P., Volkov, D.V., Fedorov, E.V.: Automatic control of train braking system integrity and density. Transp. Urals **4**(59), 77–80 (2018). https://doi.org/10.20291/1815-9400-2018-4-7-80
8. Liu, Y., Jiang, T., Zhao, X., Wen, Z., Liang, S.: On the wheel rolling contact fatigue of high power AC locomotives running in complicated environments. Wear **436–437**, Article 202956 (2019)
9. Jiqin, Wu: Pantograph and Contact Line System. Elsevier, London (2018)
10. Pyndak, V.I., Novikov, A.E.: Met. Sci. Heat Treat. **3–4**(58), 226–230 (2016)
11. Zhuravlev, G.M., Gvozdev, A.E., Cheglov, A.E., Sergeev, N.N., Gubanov, O.M.: Steel Transl. **6**(47), 399–411 (2017)
12. Kolpakov, A.S., Kardonina, N.I.: Therm. Eng. **2**(63), 150–155 (2016)
13. Trushnikov, D.N., Belinin, D.S., Shitsyn, Y.D.: Mod. Probl. Sci. Educ. **2**(95) (2014)
14. Buynosov, A.P., Volkov, D.V., Umylin, I.V.: Definition of a limit difference in the diameters of bandages at one wheel pair of the motor car of an electric train(s). Sovremennye tekhnologii, Sistemnyi analiz, Modelirovanie **56**(2), 8–18 (2017). https://doi.org/10.26731/1813-9108.2018.2(58).90-97
15. Buynosov, A., Lapshin, V., Smolyaninov, A., Dinislamov, A.: Calculation of hardening process parameters for locomotive parts. In: Polytransport Systems-2018. MATEC Web of Conferences, vol. 216, p. 02020 (2018). https://doi.org/10.1051/matecconf/201821602020

Development of a Model of a Source of Stabilized Battery Charge Current on Electric Trains with Rheostatic Braking

Alexander Buynosov(✉), Andrey Borodin, Alexander Aleksandrov, and Ilya Umylin

Ural State University of Railway Transport, Kolmogorova Street, 66, Yekaterinburg 620034, Russia
byinosov@mail.ru

Abstract. From the analysis of the operation of electric trains of the ET2, ET2M, ED2T, ED4M, ED4MK series, it was revealed that their failures in some cases are associated with battery failure. The aim of the work is to increase the reliability of the technical condition of the fleet of suburban rail vehicles, by developing and improving scientifically based methods aimed at increasing the life of batteries and control circuits in general. The analysis of charging circuits for rechargeable batteries and power supply of control circuits on electric trains is carried out. A scheme has been developed for stabilizing the charging current of batteries on electric trains. The elements of the circuit of a voltage regulator are calculated and its model is compiled. The analysis of the operation of the circuit based on the transients that accompany the operation of the charger in the process of charging the battery. The analysis confirmed the working ability of the circuit and the correct calculation of its elements. Comprehensive all-season operational tests have confirmed the high efficiency of the developed battery charging circuit eliminating overheating, recharging, and in some cases an explosion, as well as eliminating thermomechanical defects in the control circuits. The developed charging scheme was put into continuous operation on an electric train of the ET2M and ED2T series.

Keywords: Electric train · Braking · Rheostat · Accumulator · Battery · Charge · Current · Source · Stabilization · Model · Development

1 Introduction

Rail transport plays an important role in many aspects of the life of the population of Russia. Like passenger transport, in principle, it provides population mobility. Often, it is rail transport that is a priority among passengers, this is due to the massive pendulum movement of the population, passengers also give priority to rail transport due to the absence of traffic jams, as well as its much higher environmental friendliness. A typical example is Moscow, St. Petersburg, Yekaterinburg, Tver. Currently, the suburban fleet is being updated, but along with modern motor-car rolling stock (MVPS), it also continues to regularly operate (data for May 2019) MVPS which require modernization, namely the series [1]:

Z. Popovic et al. (Eds.): TransSiberia 2019, AISC 1115, pp. 255–280, 2020.
https://doi.org/10.1007/978-3-030-37916-2_26

- ET2 (108 sections);
- ET2M (488 sections);
- ED2T (390 sections);
- ED4M, ED4MK (56 sections).

From the analysis of malfunctions of the above electric trains it became clear that the failures of electric trains in some cases are associated with the failure of batteries (AB). The most common causes of failure of the battery is the heating of the electrolyte over 47 °C, leakage of the electrolyte, loss of capacity, overcharging, equalizing currents [2, 3].

In order to minimize failures, follow the train schedule and ensure the proper level of train safety, it is necessary to develop a set of technical measures aimed at improving the reliability of electrical equipment and devices the main cause of failure of which is recharging, burnout of elements, low charge, loss of capacity, and in some cases, the explosion of AB that can have a negative impact on the train crew. The main prerequisites for the failure of the battery at the MVPS are shown in the diagrams (Fig. 1):

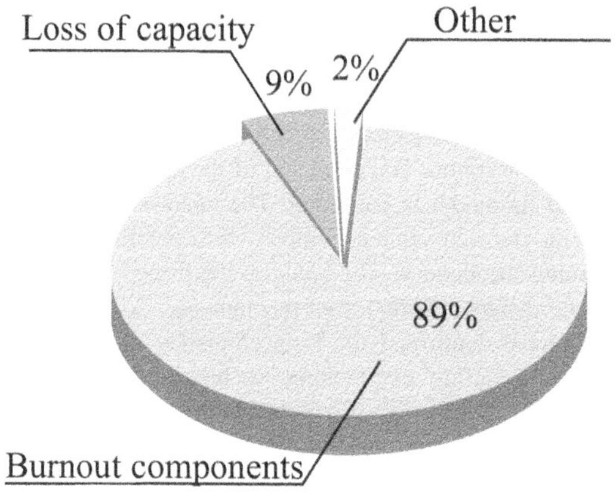

Fig. 1. The distribution of failure of batteries

- burnout of components due to surface electrolysis due to leakage currents (89%);
- loss of capacity (9%);
- other (2%).

The main source of burnout components of AB include:

- accumulation of carbonic salts (carbonates) in the electrolyte in excess of the permissible value;
- AB operation on an electrolyte without the addition of caustic lithium;
- AB operation at an electrolyte temperature exceeding + 47 °C;
- the occurrence of harmful impurities in the electrolyte as a result of chemical reactions;
- AB faults and critical self-discharge.

The loss of battery capacity usually occurs due to the selection of an incorrect mode of operation of the battery [4, 5]:

- long operation of control and lighting circuits without recharging the batteries;
- undercharging of the battery due to an incorrect voltage regulator option.

The accumulation of carbonates in the electrolyte occurs more actively every year from June to August. Another common cause of carbonate is damage or the absence of plugs in the battery sections. Accumulated in the electrolyte, carbonates lower the concentration of alkaline composition, this leads to a decrease in AB capacity.

High electrolyte temperatures in AB operation lead to capacity loss due to passivation of positive electrodes (dissolution of intense mass in the electrolyte). Intensive mass, practically insoluble in the electrolyte under normal operating conditions, dissolves at the highest temperature and settles on the pyus electrode, causing loss of capacity. High electrolyte temperatures, arising as a result of an improperly selected battery charge mode at the MHPS, lead to their periodic recharging [6].

Electrolyte contamination with harmful impurities can occur as a result of accidental ingestion of iron objects into separate sections, for example, when charging batteries with non-distilled water. Short circuits in AB occur in most cases due to the presence of sludge that occurs when the intensive mass is washed out of the electrodes. Excessive leaching of intense mass, especially from the negative (cadmium) electrode, can be caused by periodic recharging of the battery. Overcharging leads to abundant gas evolution during electrolysis of water. Gases carry particles of intense mass along and carry them from the plane of the electrodes into the electrolyte. At the highest electrolyte temperature (55–70 °C), particles of intense cadmium mass partially dissolve in the electrolyte, and then at normal temperature they settle on electrodes and separators in the form of an iron film. The metallization of the separators turns them into conductors of electric current and leads to enhanced self-discharge and short circuit. In addition to short circuits inside the battery, often short circuits appear in the battery itself. This most often goes beyond this time, when the counteraction to isolation of AB is lower than generally recognized standards. The most common mechanical damage to ABs at MVPS is damage to insulating rubber covers and ventilation plugs. During the movement of the electric train, ripple from the carriage frame is transmitted to the battery compartment, in fact, which leads to the loosening of the wedges fastening the batteries. During the operation of the MVPS, grinding of the rubber covers of the batteries occurs. Practice proves that rubber covers have an unnecessary mechanical strength, damage to which leads to leakage of electrolyte from the can and grounding of the battery, current leakage, earth faults, burnout of the battery case, and in some cases explosion [7, 8].

An analysis of the battery charging circuit at the MVPS with rheostatic braking shows that the actual negative operating modes of the batteries are caused by:

- giant surge currents due to defects in the power supply circuit of the control circuits;
- a large value of the charging current.

To eliminate the adverse effects on the operation of the battery on the MVPS of the above grounds, formed due to the lack of a device or an electronic circuit for monitoring and stabilizing the charge current of the batteries, respectively:

- eliminate the likelihood of surge currents;
- limit the charge current of the batteries to the boundaries of the nominal charging current (14 A for AB NK55) with maintaining it within the specified limits.

A test of the tasks that need to be solved by the method of development and implementation of the charger, provides a specific list of criteria for the charger and its circuit [9].

From the introduction of the device-circuit will require:

- making fewer changes;
- the savings of the place to be telecommunications;
- it is likely lower mass-dimensional indicators;
- stabilization of the charging current;
- destruction of surge currents during parallel operation of batteries in the discharge mode;
- automation.

2 Research Methods

Among the variety of electronic devices, a DC/DC converter with a pulse interface and current feedback can be distinguished. Feedback systems (closed systems) due to a number of [6] advantages are extremely widespread in technology. Moreover, the scope of such systems is not limited only to the tasks of automatic regulation. In a closed cycle, many measuring and calculating devices, various amplifiers with feedbacks, etc.

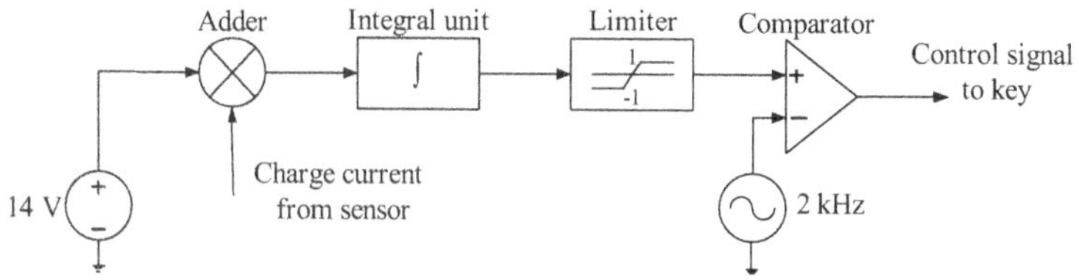

Fig. 2. PSIM modeled control system

A current sensor acts as a sensitive element in the control system of the developed converter, since the current at its output is an object of regulation.

In the adder, the current value is subtracted from the given current value. With the given value, we take the charge current recommended by the manufacturer equal to 14 A. As the setting element, we use a current source of 14 A.

At the next stage, the integral is calculated in the Integral block over a period of 5 s. Due to this, stability is achieved at the initial time when the error is more than 1 A.

Next, in the "Limiter" block, the input signal to the "Comparator" is formed by rounding the error to 1 or –1. Due to intermediate transformations at the positive input of the comparator, we have: 1 V if $y(t) < y_{set}$; 0 V if $y(t) > y_{set}$.

An alternating voltage of ± 1 V is applied to the minus input of the comparator with the required PWM frequency. At the output of the comparator we have either 1 or 0. In the case when the power switch is open at the output of the comparator 1, it is closed otherwise. To limit the battery charge current and maintain it within these limits, you should use a DC/DC converter with a pulse interface and current feedback, which is included in the battery circuit. To study transients, it is necessary to make a model of a charging device, which is based on a circuit of a step-up voltage regulator with stabilization of the charge current. The system was simulated using the PSIM software package.

The model uses the designation of the elements corresponding to the scheme of the electric train. The model includes a control transformer TRu, a rectifier installation consisting of D32–D37 diodes and a converter. The converter is connected to the output terminals of the rectifier installation. The rectifier installation rectifies the sinusoidal voltage of the secondary winding of the control transformer. The model of the circuit using a persistor controller is shown in Fig. 3.

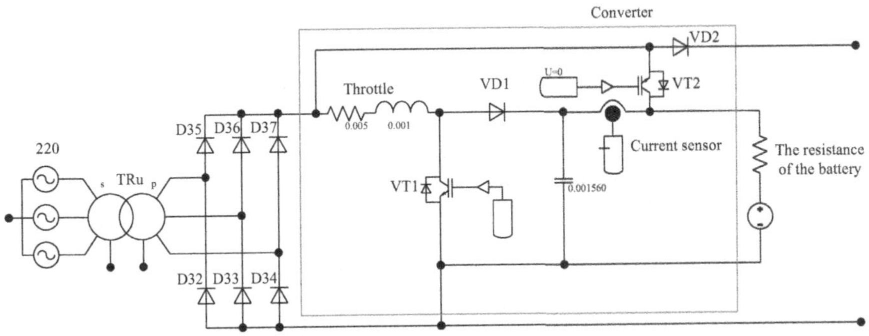

Fig. 3. Model scheme constant current charging

3 Experimental Data and Results

During the battery charge due to the circuit calculated above, the charge current is stabilized at the level set by the battery manufacturer. In this case, the control circuit power is parallel. Simulation of the charger at constant load gives the following results, presented in Fig. 4.

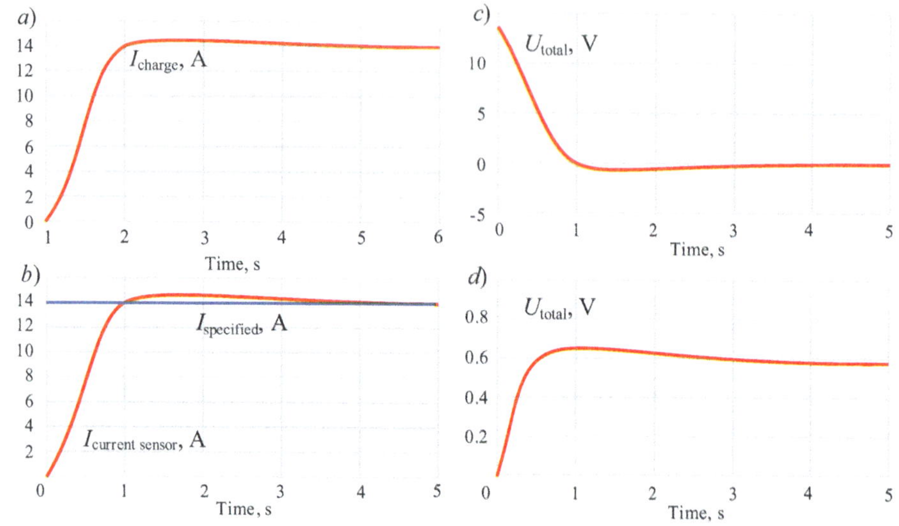

Fig. 4. Constant load charger operation

The diagrams (Fig. 4) show the following values:

(a) is the signal value corresponding to a given battery charge current;
(b) the signal value corresponding to the actual charge current received from the current sensor;
(c) signals at the output of the adder;
(d) is the signal at the output of the integrator.

Let us evaluate the transition process. Figure 5 shows the battery charge current.

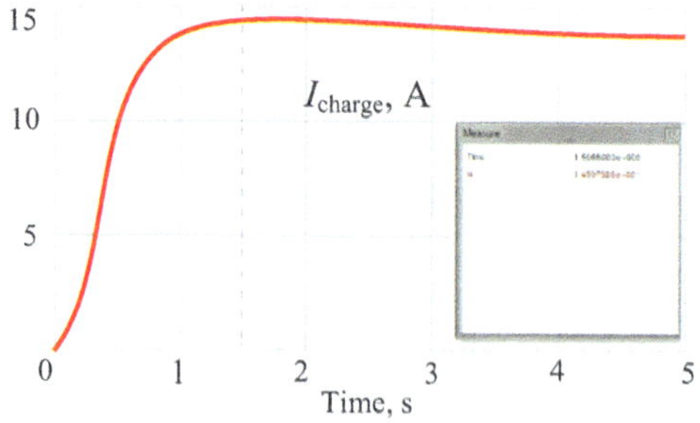

Fig. 5. Battery charge current when the charger is turned on.

By the type of the curve of the transient response, which is the reaction of the ATS to a single step effect, it can be determined that the curve refers to an aperiodic transient with one overshoot.

Overshoot calculation:

$$\sigma_m = \frac{h_{max} - h_{уст}}{h_{уст}} \cdot 100\%, \quad (1)$$

$$\sigma_m = \frac{14.6 - 14}{14} \cdot 100\% = 4.3\%.$$

This value characterizes the ATS as an ATS having a high quality transient.

Simulation of the charger when the input voltage is turned off. In case of power loss at the inlet of the converter, the VT2 transistor opens (Fig. 3) and the control circuits are powered by the battery. There is a discharge mode AB. The diode VD2 eliminates the occurrence of surge currents during parallel operation of the AB several sections.

The battery discharge mode continues until the voltage level on the batteries reaches a certain critical value. At the moment when the battery voltage level reaches a critical value, VT1 will close. Thus, the battery will be protected from excessive discharge.

Figure 6 shows a transient diagram in the event of a loss of input voltage caused by a malfunction of the generator or transformer, as well as an open circuit. The case of the complete absence of voltage at the input of the converter is considered.

The diagram (Fig. 6a) shows the signal emitting disconnection of the input voltage.

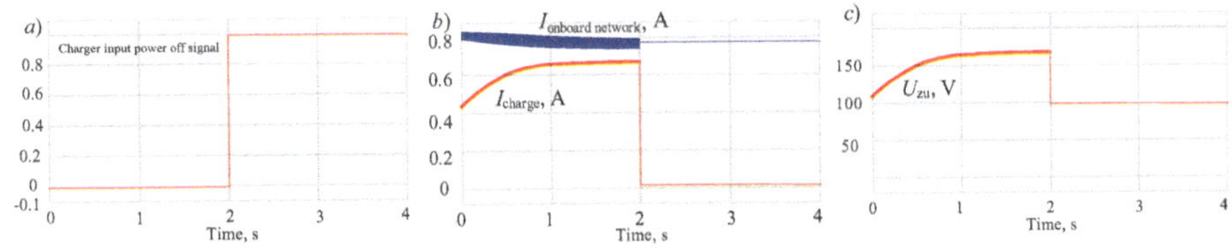

Fig. 6. Transients due to input voltage shutdown

The diagram (Fig. 6b) shows that the battery current has changed sign. The charge current has turned into a discharge current, the magnitude of which is due to the load on-board network consumers. At the same time, the voltage on the batteries decreased, since the circuit of the step-up voltage regulator was turned off. The residual voltage is determined only by the battery level. Also in Fig. 6 it can be seen that the battery discharge current flowing in the converter without taking into account the sign corresponds to the current consumed by the on-board network.

When the converter is turned off, the alternating component of the supply voltage is excluded, which is reflected in the graphs of the voltage at the terminals of the battery and the current consumed by the on-board network.

Simulation of the charger during charging. As you know, during the charge of the battery between the cathode and the anode, redistribution of positively and negatively charged particles occurs, as a result of which the emf of the battery increases. The voltage level of a discharged battery may well reach 90 V. The voltage of a charged battery is 110 V.

Figure 7 shows graphs of currents and voltages during battery charging. Figure 7a shows a graph of the varying voltage across the battery terminals. In the model, the process of increasing the emf of the source is compressed to 10 s. The graph (Fig. 7b) shows the voltage change across the terminals of the charger. The voltage at its terminals rises as the counter-energized battery voltage rises. Due to the increase in voltage at the terminals of the charger, the charge current remains stable (Fig. 7c).

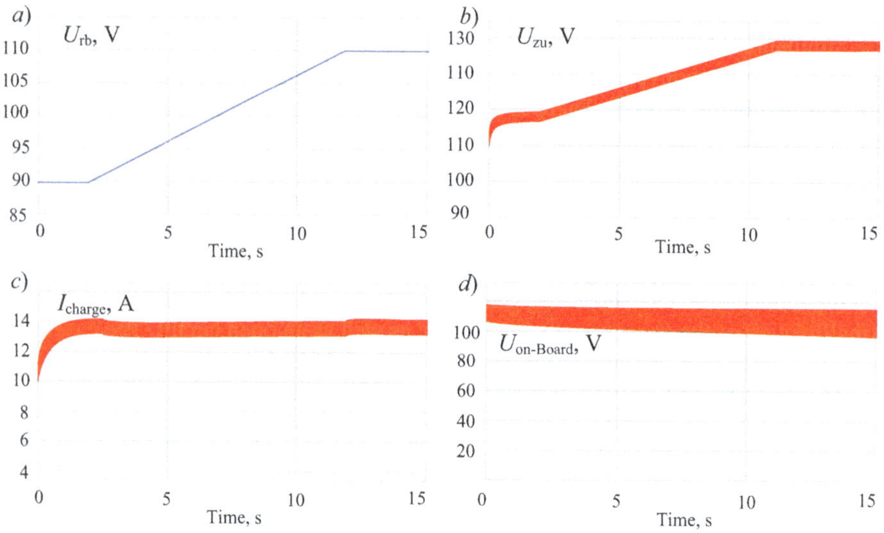

Fig. 7. Battery charge process

At the same time, the voltage of the on-board network remains stable at the level of 108–110 V (Fig. 7d).

Testing the voltage surge at the output of the rectifier installation. Let us compare and analyze the behavior of the existing and developed battery charge schemes using the example of working out a voltage surge of 10 V. The initial circuit parameters with respect to the rectifier installation and the load are the same. At the second second of operation, an overvoltage of 10 V is emitted.

Figure 8 shows the graphs of the charging currents during voltage surge at the output of the rectifier installation. The graph (Fig. 8a) shows the level of the charging current generated by the charger. The surge of voltage led to an increase in the charge current to the level of 19 A and its further stabilization at the level of 14 A. The time for which the system returned to the set parameters was approximately 0.5 s. The graph (Fig. 8b) shows the behavior of the charging current generated by the rectifier installation and the voltdabav. The surge of voltage at the input of the system led to an increase in the average current to the level of 18 A. The charging current remained at that level until the voltage subsided, after which its value returned to the nominal value. In real conditions at MVPS, this would lead to overcharging of batteries and their damage.

Reducing the voltage at the output of the rectifier installation to the voltage level of the battery would lead to the termination of the charging process or to a decrease in the charge current to an ineffective level.

The circuit of a step-up voltage regulator with stabilization of the charge current avoids such situations and normalizes the battery charge process, thereby extending their service life.

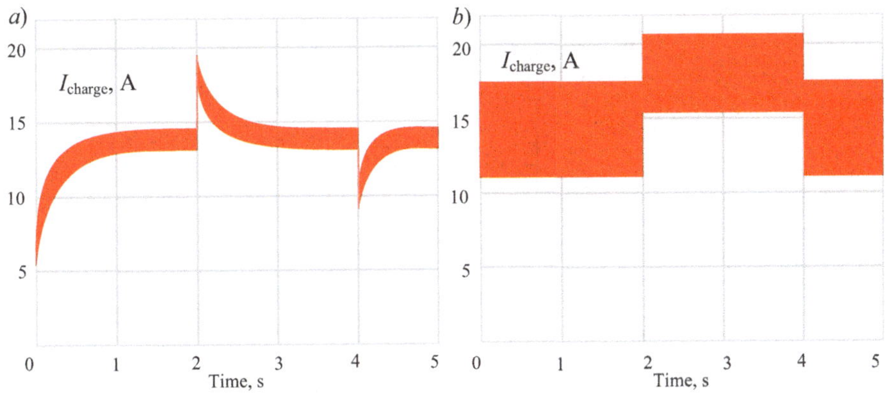

Fig. 8. Surge and voltage drop at the output of the rectifier installation

Based on the study of transients of the battery charging circuit, the following conclusions can be drawn:

- the developed charger circuit meets the initial requirements for eliminating surge currents in the discharge mode of the battery and limiting the charging current to 14 A;
- transients when turning on and off the charger, as well as directly in the process of charging showed the correct calculation of the parameters of the electrical circuit;
- the resulting ATS transient is an aperiodic transient with one overshoot. The overshoot is 5%;
- the developed circuit extends the boundaries of the permissible input voltage range for charging the battery, thereby mitigating the requirements for the stability of the rectified voltage.

4 Calculation of the Sawtooth Generator

Sawtooth generators can operate in auto-generator modes and with external control. In the developed circuit, we accept the widespread design of the generator based on n-p-n transistors (Fig. 9). The main task of the transistor V3 is to discharge the capacitor C1. On transistor V3, a standard pulse switch circuit with a supply voltage of 12 V is assembled. Capacitor C1 charging, provides a linearly changing voltage. However, the nature of the voltage change on C1 is not linear:

$$U_{C1} = E(1 - e^{-t/\tau}), \tag{2}$$

where τ is the charge time constant at $I_k \approx 0$ A. τ = R2C1 s.; t_r is the rise time of the voltage in the pulse, the stroke time, sec., $\tau >> t_r$.

To ensure line voltages, select (Fig. 10).

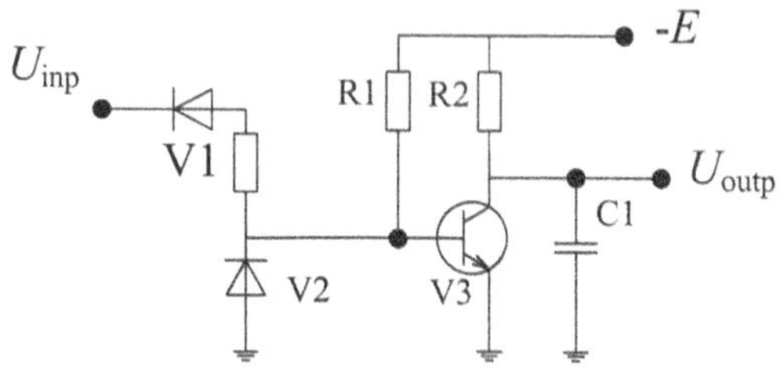

Fig. 9. Sawtooth generator circuit

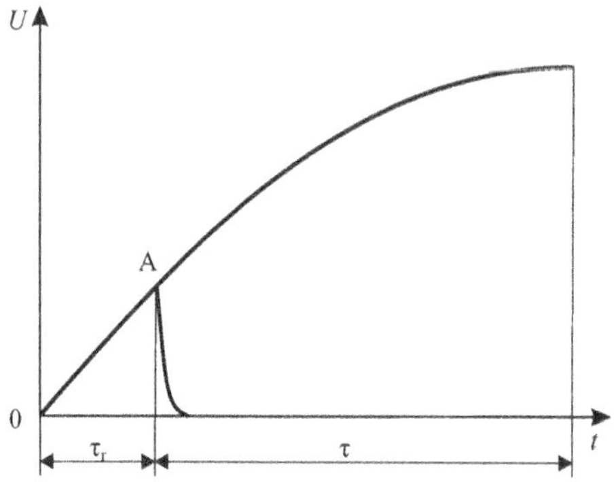

Fig. 10. Sawtooth pulse pattern

Figure 2 shows that in the "0A" section, the voltage increases linearly. For an industrial frequency of 50 Hz, t_r = 0.02 s, But control requires wide pulses with t_r = 0.05 s.

We determine the value of R2. We select the transistor V3 brand KT315B. Parameters: maximum collector current $I_{k\,max}$ = 100 mA. Based on the conditions of reliability and performance at E = 12 V.

$$R2 = \frac{E}{I_{kmax}}. \tag{3}$$

From the expression (3) it follows that

$$R2 \geq 1200\,\text{Ohm}. \tag{4}$$

We take the value C1 = 1 mkF, then to fulfill the condition $\tau > > t_r$, we take $t_r = 0.05$ s. So:

$$R_k = \frac{\tau}{C}, \tag{5}$$

$$R_k = \frac{0.05}{0.000001} = 50000\,\text{Ohm}.$$

We take the value $R_k = 47\,\text{kOhm}$, which simultaneously satisfies conditions (2)–(4), then:

$$I_k = \frac{12}{47000} = 0.00025\text{A}.$$

The state of the transistor is determined by the boundary value of the start of opening and the saturated value of the open transistor:

$$I_{bboun} = \frac{I_k}{\beta_{st}}, \tag{6}$$

$$I_{bboun} = \frac{E_k}{\beta_{st} R_k}. \tag{7}$$

For the transistor to fully unlock, it is necessary that $I_b > I_{boun}$. The saturation coefficient of the transistor is determined by the formula:

$$S = \frac{I_b}{I_{bboun}}. \tag{8}$$

To ensure complete unlocking of the transistor and fast discharge take S = 1,5–3,0. We accept S = 3. In the saturation mode, the base current of the transistor should be:

$$I_b \geq S \cdot \frac{I_k}{\beta_{st}}. \tag{9}$$

The found base current is provided by the input circuit parameters of the key circuit:

$$I_б = \frac{U_{inp} - U_{бэ}}{R_п}. \tag{10}$$

The value of the circuit parameters substituted in expressions (9) and (10) gives the value $I_b \approx 0.0005$–0.0010 A. We accept $I_b = 0.001$ A, hence the value of R_b provided:

$$R_b > > R_{\rm inp}, \quad (11)$$

where $R_{\rm inp}$ is the input resistance of the transistor connected according to the scheme with a common emitter.

$$R_{\rm k} = \frac{E}{I_{\rm b}}, \quad (12)$$

$$R_{\rm k} = \frac{12}{0.001} = 12000\,{\rm O_M} = 12\,{\rm kOhm}.$$

The value of the circuit parameters was verified by practical assembly of the circuit and mathematical modeling was carried out using the Electronics Workbench 5.0 program (Interactive software product). The calculation scheme and the calculation results of the scheme are presented in Fig. 11.

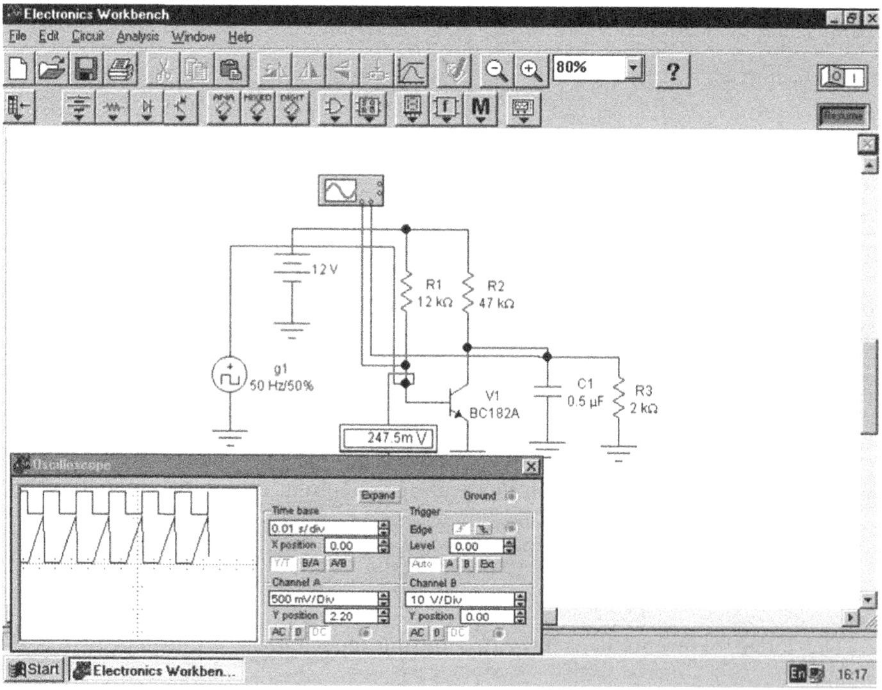

Fig. 11. Design diagram of a sawtooth generator, simulated in the Workbench program

The generator provides sufficient linearity of the output voltage at an amplitude of $U_{\rm outp}$ = 10 V and a value of $t_{\rm p}$ = 0.05 s.

5 Comparator Calculation

Figure 12 shows the comparator circuit.

A comparator is a conventional operational amplifier operating in the Schmit trigger mode, or specialized microcircuits that provide a threshold operation mode. The comparator is made on the operational amplifier K140UD1B. The non-inverting input D1 receives sawtooth pulses from the output of the sawtooth voltage generator through the limiting resistor R3. The amplitude of the pulses at the input is 2 V. The reference voltage from the divider R1 and R2 is applied to the inverse input. The voltage E on the divider is 12 V, stabilized by the power supply. The reference voltage U_{op} is the control voltage and determines the duration of the rectangular pulses at the output of the comparator. The input voltage depends on the difference $U_p - U_{op}$. Figure 13 shows that a rectangular pulse appears at the output of the circuit when U_p reaches the value of U_{op}, therefore, by changing the values of U_{op} one can change the values of t_p.

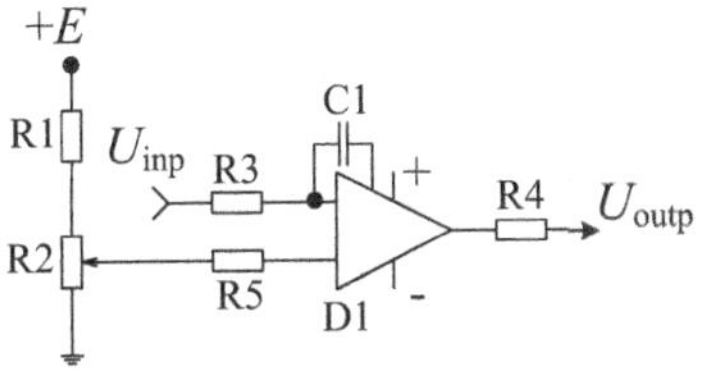

Fig. 12. Comparator circuit

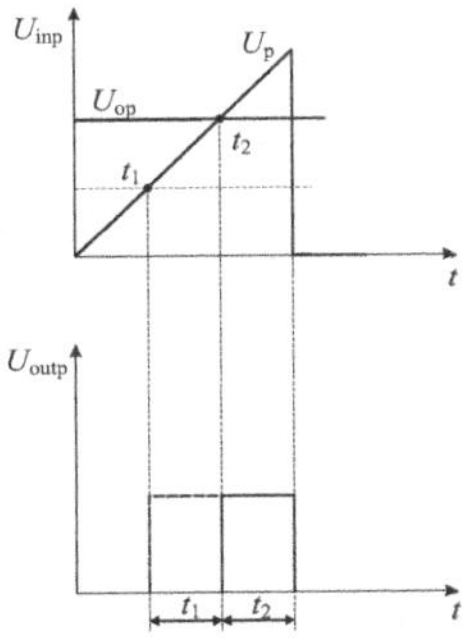

Fig. 13. Rectangular pulse diagram

The calculation of the divider determining the value of U_{op} is reduced to the calculation of the proportion in order to form a constant voltage on R2, based on the condition:

$$U_{R2} \leq U_p. \tag{13}$$

The input impedance of the operational amplifier $R_{inp} \geq 0.5$ mOhm therefore it can be ignored. To get $U_{op} = 0–2$ V, the ratio between R1 and R2 should be 4:1. We take the value of R1 = 820 Ohm, and the value of R2 = 220 Ohm. The operational amplifier is powered by bipolar voltage from the power supply. Checking the operation of the circuit with the above parameters was carried out by mathematical modeling using the Electronics Workbench 5.0 program and practical assembly of the circuit. Figures 14 and 15 show the waveforms of the input and output pulses at different values of t_n.

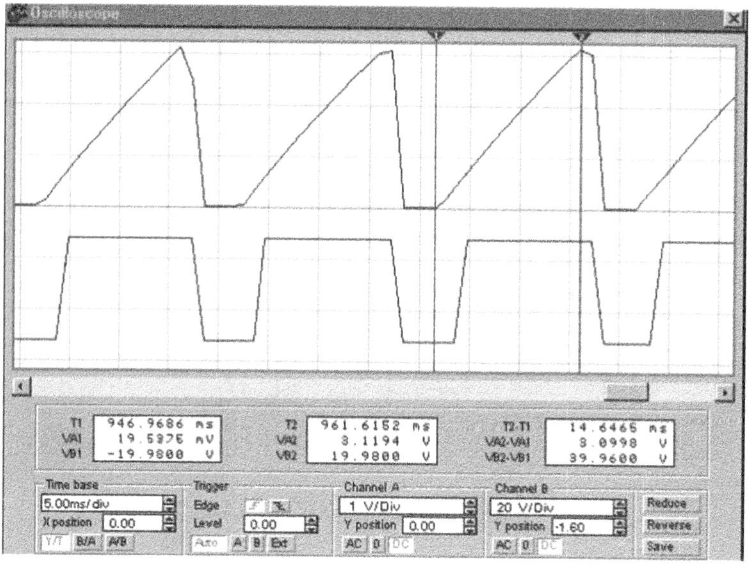

Fig. 14. Oscillogram of input and output pulses with t_n = 10 ms

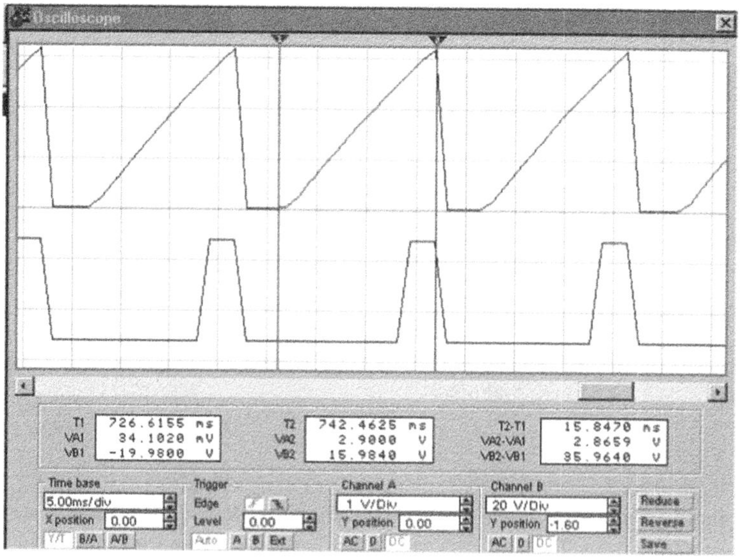

Fig. 15. Oscillogram of input and output pulses with t_n = 2 ms

The value of U_{op} = 23% of $U_{op\ max}$(R2 = 0.23 $R2_{max}$). As a result, rectangular pulses with $t_i \approx$ 10 ms are obtained at the output. Figure 7 shows an oscillogram with the parameters U_{op} = 80% of $U_{op\ max}$(R2 = 0.8 $R2_{max}$). The duration of the output pulses is $t_p \approx$ 2 ms. Practical assembly of the circuit and tests show its full operability at values above the specified parameters.

6 Calculation of the Synchronization Device

Calculation of a synchronizing device is reduced to determining the parameters and selecting industrial transformers of the same type. The output windings of the transformer 1–3 are connected to the control circuits of the transistors of the sawtooth generators and to a three-phase rectifier of the power supply circuit (Fig. 16).

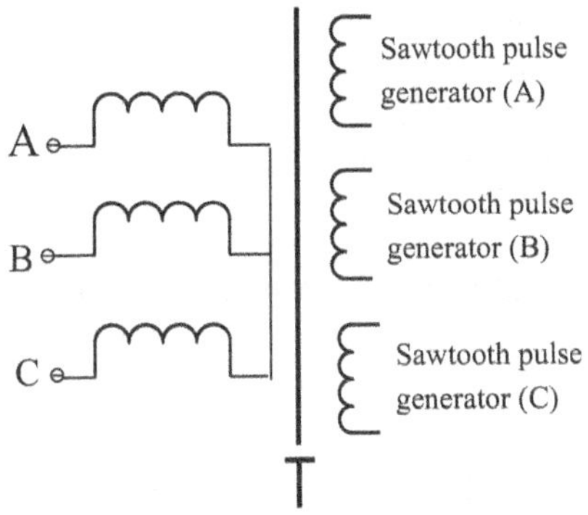

Fig. 16. Sync device diagram

For control circuits, choose $U_{inp} \approx$ 6–8 V. For base currents of transistors in the control circuits $I_b \approx$ 0.001 A. The power is calculated on the load of the input circuit $P_{inp} \leq$ 0.008 W $\leq$ 8 mW. Thus, the secondary windings of control circuits do not limit the power of transformers. Power output windings make up for a three-phase system.

Let's make an approximate calculation of the power consumed by the circuit elements

$$P_n = 3P_{svg} + 3P_{comp} + 3P_{outp}, \tag{14}$$

where P_{svg} is the power consumed by the chains SVG of the sawtooth generator, W; P_{comp} – power consumed by the comparator, W; P_{outp} – power consumed by the output circuit (blocking generator and high-frequency transformer), W.

In the sawtooth generator circuits, the consumed average power at $E \approx$ 12 V is determined by the sum of the collector currents and the base of the transistors, and is no more than $P_{svg} \approx$ 0.12 W.

In the comparator circuit, the main power is consumed by the K140UD1B microcircuit. The nameplate value of the power consumption is $P_{comp} = 108$ mW = 0.1 W. In the input circuit (blocking generator), the power is determined by the collector current $I_k \leq 0.1$ A, thus: $P_{outp} \leq 0.1 \cdot 12 = 1.2$ W.

The total power of the circuit circuits, by expression, is: $P_n = 3 \cdot 0.12 + 3 \cdot 0.1 + 3 \cdot 1.2 = 4.26 \approx 5$ W.

Taking into account the unaccounted current values in the circuit, $P \approx 10$ W. The parametric stabilizers of the circuit have efficiency coefficients η = 50–60%, and hence:

$$P_{outp} = \frac{P}{\eta}, \tag{15}$$

$$P_{outp} = \frac{10}{0.5} = 20 \text{ W}.$$

We accept industrial transformers IT 247-127/220-50 as synchronization device transformers, having the following parameters: $P_{outp\ nom} = 22$ W, $U_1 = 220$ or 127 V, $U_{2.1} = 8$ V, $U_{2.2} = 20$ V, $U_{2.3} = 2.58$ V ($U_{2.1}$, $U_{2.1}$, $U_{2.3}$ – voltage groups of the secondary windings of the transformer).

7 Calculation of the Driver Amplifier Circuit (Blocking Generator)

The power of the signal received from the output of the comparator is small to control the rectifier thyristors. For this reason, the circuit provides an amplifier-shaper (Fig. 17), which provides amplification and formation of pulses before feeding them to the control electrode circuit of the power thyristor.

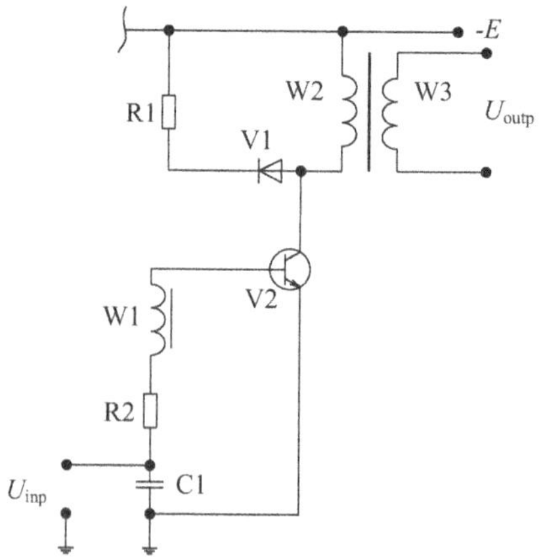

Fig. 17. Shaper amplifier circuit

The unlocking pulses from the output of the shaper are transmitted to the thyristor control electrode circuit using a pulse transformer [10]. The construction of the circuit depends on the requirements for the parameters of the control pulses. The necessary duration of the control pulses depends on the type of converter circuit and the nature of the load. In three-phase controlled thyristor rectifiers, the duration of the control pulse must exceed 60°. This is necessary for starting and creating operability in the intermittent current mode, when the load current decreases to zero by the time the next thyristor is unlocked [11, 12] (Fig. 18). In converters operating on the excitation windings of DC machines and batteries, wide control pulses with a duration of 3.9–6.6 ms are used. Controlling wide pulses creates the specifics of constructing amplifier circuits-shapers.

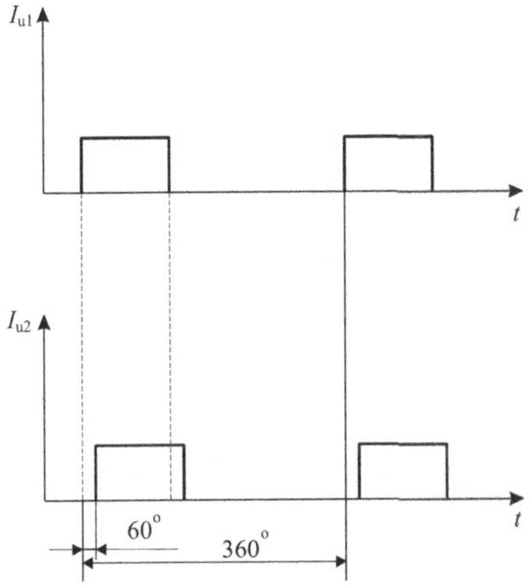

Fig. 18. Thyristor unlocking diagram

The task is complicated by the difficulty of transmitting wide control pulses through a pulse transformer to the thyristor control circuit. This problem is solved by transmitting a packet of bipolar pulses of the required duration through an output transformer with their subsequent rectification, before being fed to the thyristor control circuit. The pulse repetition rate in the packet is chosen 5–10 kHz. This option of generating control pulses is also called "high-frequency" filling (Fig. 19).

For the shaper, power pulsers are usually used, performed according to the blocking-oscillator circuit (Fig. 17). In the device under development, a scheme of a controlled blocking generator is adopted. The generator consists of a transistor V1 and a pulse transformer. The working winding W1 is connected to the feedback winding W2 by a magnetic field. The winding W1 provides the unlocking and locking of the transistor V1. Thanks to the positive feedback, the generator is self-excited. Rectangular current pulses appear in the collector circuit V1, which, passing along the primary winding W2, provide induction of the EMF in the secondary, output winding W3. The output voltage U_{outp} arrives after rectification in the thyristor control circuit.

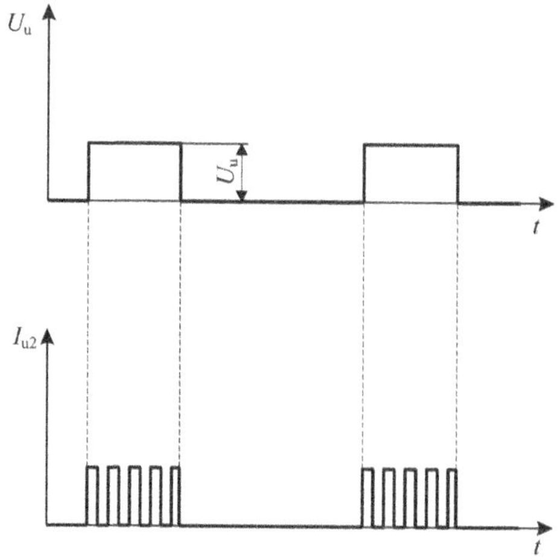

Fig. 19. High frequency fill chart

For normal self-excitation of the blocking generator, it is necessary to ensure the condition of the regenerative process in the circuit. It is accompanied by a mutual increase in the base and collector currents, as a result of which the transistor V1 opens. For this, it is necessary to provide the condition [13] in the feedback circuit:

$$i_b \geq \frac{i_k}{\beta} \tag{16}$$

where i_b is the base current V1; i_k is the collector current; β is the gain of the transistor in a circuit with a common base.

The collector current of the transistor is equal to the sum of the base and load currents reduced to the collector winding [13]:

$$i_k = \frac{i_b}{n_b} + \frac{i_n}{n_n}, \tag{17}$$

where i_b is the base current V1; n_b is the transformation coefficient in the base positive feedback circuit, which is equal to the ratio of the number of turns of the collector winding of the transformer W2 to the number of turns of the positive feedback winding W1 (W2/W1); n_n is the transformation coefficient in the load circuit, which is equal to the ratio of the number of turns of the collector winding of the transformer W2 to the number of turns of the output load winding W3 (W2/W3).

If we take at the stage of the regenerative process the voltage across the collector winding ΔU, then the current in the base circuit:

$$i_b = \frac{\Delta U}{n_b(R2 + r_{inp})}, \tag{18}$$

where r_{inp} is the input resistance of the transistor; R2 is the resistance in the base circuit.

We substitute expression (18) into expression (16) and find the condition necessary for the development of a direct blocking process in the circuit:

$$\beta \geq \frac{1}{n_b}\left[1 + \frac{n_b^2(R2 + r_{inp})}{n_n^2 R_n}\right] \tag{19}$$

where R_n is the resistance of the load circuit.

Since the pulse frequency does not exceed 10 kHz, we select ferrite 2000 NM. For a high-frequency pulse transformer, we select an annular core made of manganese-zinc ferrite of the 2000 NM-A0878 grade of K31 size – 18.5 × 7.0 m with a window area of 268.8 mm^2 and a cross-sectional area of 42.79 mm^2 (S_c).

The number of turns of the annular core is determined by the formula [12]:

$$W2 = 280 \cdot \sqrt{\frac{L \cdot l_{eff}}{\mu_g \cdot S_c}}, \tag{20}$$

where L is the inductance, mH; l_{eff} is the effective length of the magnetic line, l_{eff} = 7.441 sm; μ_g is the dynamic magnetic permeability of the core material, we take μ_g = 2000; S_c is the cross-sectional area, S_c = 0.4279 sm^2.

The inductance of the coil W2 should provide a sufficient value of the inductive resistance that limits the collector current V1, and calculate by the formula:

$$X_L = 2 \cdot \pi \cdot f \cdot L, \tag{21}$$

where f is the frequency, Hz; L – inductance, Gn. We take f = 10000 Hz, L = 0.055 Gn.

$$X_L = 2 \cdot 3,14 \cdot 10000 \cdot 0,055 = 3454\,\text{ohm} = 3,45\,\text{kohm}.$$

As transistor V1 we take KT315B with the maximum allowable collector current $I_{kmax} \leq$ 100 mA = 0.1 A. Resistance in the collector circuit should limit the collector current in a pulsed, dynamic mode:

$$R_k = \frac{U}{I_{max}}. \tag{22}$$

For dynamic mode, we take U = 12 V.

$$R_k = \frac{12}{3454} = 0,0035\,\text{ohm}.$$

The condition for the maximum permissible current is satisfied. We put the obtained values into expression (20):

$$W2 = 280 \cdot \sqrt{\frac{55 \cdot 7.4}{2000 \cdot 0.428}} = 190\,\text{coils'}.$$

Given the values taken by the inductance of the winding and the fulfillment of the condition Ikmax > Ik of the transistor V1, we reduce the number of turns by 20% and take W2 = 160 turns. We take the values $n_b = n_n = 2$ in order to increase the values of the base currents and the load. Then W1 = 80 turns, and W2 = 80 turns. The values of the reduced load currents and base are determined by the formulas:

$$i'_n = \frac{i_n}{n}, \tag{23}$$

where i'_n is the current reduced to the collector winding; i_n – load current.

For the KT315B transistor, the maximum allowable power is P_{max} = 0.1 W. Based on this condition, we accept:

$$P_n \approx P_b << \frac{P_{max}}{2}, \tag{24}$$

$$P_n \approx P_b << \frac{0.1}{2} = 0.05\,\text{W}.$$

With n_n = 2, the voltage in the winding is W3 ≈ 6 V, and the load current:

$$I_H = \frac{P_n}{U} \tag{25}$$

$$I_n = \frac{0.05}{6} = 0.0083\,\text{A}.$$

then the reduced collector load current is found by the formula (23):

$$i_n = \frac{0.0083}{2} = 0.0042\,\text{A}.$$

Base current reduced:

$$i'_b = \frac{i_b}{n}, \tag{26}$$

where i'_b is the base current reduced to the collector winding; i_b – base current.

Magnetic current magnitude of the magnetic circuit:

$$I_{\mu} = (0.05\text{–}0.1) \cdot i'_{\mathrm{n}}. \tag{27}$$

We substitute the values in expression (27) and get I_{μ} = 0.00465 A. The base current is selected from the condition [12]:

$$i_{\mathrm{b}} > i_{\mathrm{k}}\beta = \frac{E_{\mathrm{k}}}{\beta R'_{n}}. \tag{28}$$

The load resistance $R\prime_n$ is determined by the formula:

$$R'_{n} = n_{\mathrm{n}}^{2} \cdot R_{\mathrm{n}}, \tag{29}$$

where R_{n} is the resistance in the load circuit is 20 Ohm, $R\prime_n$ = 2^2·20 Ohm.

Substitute the obtained values in the formula (27):

$$i_{\mathrm{b}} = \frac{12}{100 \cdot 80} = 0.001\ \mathrm{A}.$$

Base Chain Time Constant [7]:

$$\tau = c \cdot \left(R2 + r_{\mathrm{inp}}\right), \tag{30}$$

where τ is the resistance of the base resistor; c is the capacitance of the capacitor; r_{inp} – input impedance (40 Ohm for KT315B).

For the accepted frequency of 10 kHz constant τ = 0.1, then the capacitance:

$$c = \frac{\tau}{\mathrm{R2} + r_{\mathrm{inp}}}. \tag{31}$$

The recommended value of R2 = 1.5–3.0 kOhm [3]. We take the value of R2 = 2 kOhm. Then c = 0.6 mkF. Given the calculation errors, we take c = 1.0–0.5 mkF with a value of R2 = 2 kOhm.

Thus, as a result of the calculation received: V1 – transistor KT315B; R2 = 2 kOhm; c = 1.0–0.5 mkF; W1 = 160 coils'; W2 = 80 coils'; W3 = 80 coils'.

Let us verify the correctness of the calculations by substituting numerical values in expression (16) $\beta \approx 50$. For the adopted transistor KT315B, the value is β = 50–150 and on average is 100. Condition (16) for the regenerative process is fulfilled. Practical assembly and configuration of the circuit confirms the calculated values. The waveform of the voltage pulses on the collector V1 is shown in Fig. 20.

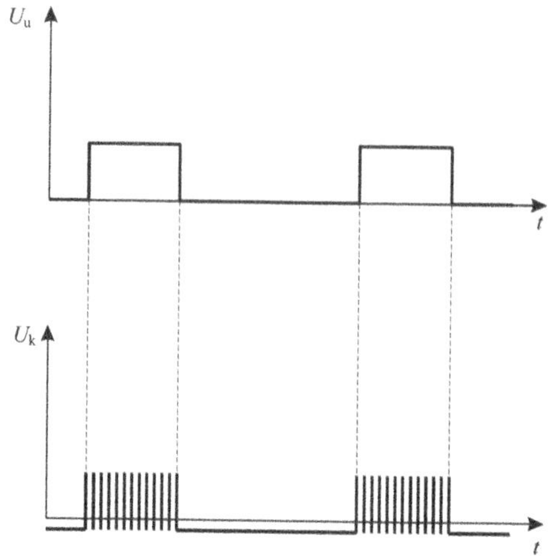

Fig. 20. Waveform of voltage pulses on the collector V1

8 Power Supply Calculation

Figure 21 shows the power supply circuit. The power supply consists of a three-phase two-half-wave rectifier, L-shaped power filters R13, C1 and R14, C2, as well as two parametric voltage stabilizers R13, V17 and R14, V18.

The power supply unit provides the bipolar supply voltage necessary to power the K140UD1B microcircuits and other components of the circuit. Considering the small ripple coefficient after three-phase rectification [1], we take the capacitances C1 = C2 = 200 mkF × 50 V for the L-shaped power filter. The values of R13 and R14 will be determined during the calculation of the parametric voltage stabilizer.

A schematic diagram of a parametric stabilizer is shown in Fig. 22.

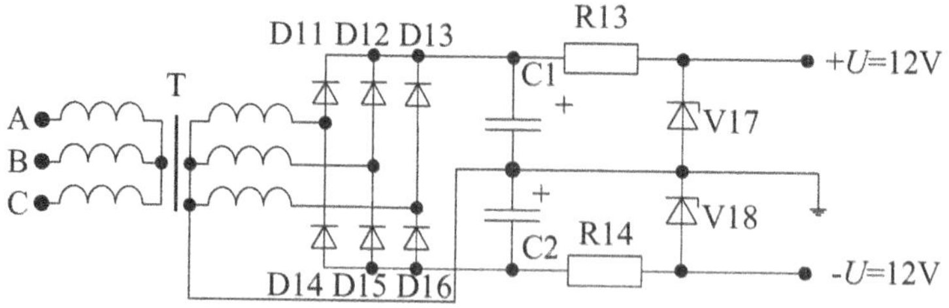

Fig. 21. Power supply circuit

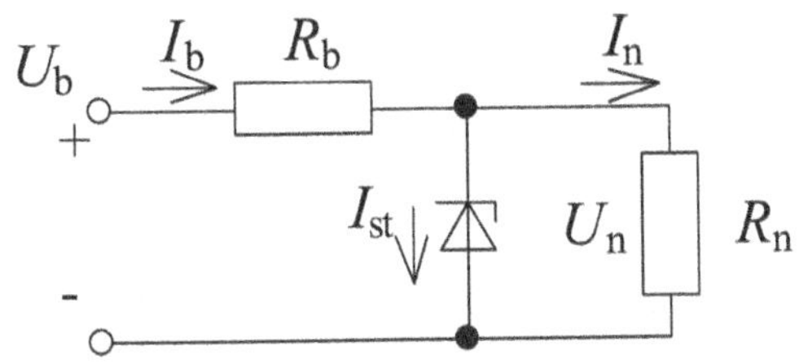

Fig. 22. Schematic diagram of a parametric stabilizer

Given that the voltage in parallel circuits are equal:

$$U_n = U_{st}. \quad (32)$$

According to the law of Kirchhoff

$$I_d = I_{st} + I_n. \quad (33)$$

Voltage drop across ballast R_b

$$U_{Rb} = (I_n + I_{st}) \times R_b. \quad (34)$$

Based on the above expressions (32) and (33), the zener current is:

$$I_{st} = \frac{U_d - U_n}{R_b} - \frac{U_n}{R_n}. \quad (35)$$

The voltage is $U_n = U_{st}$ and varies insignificantly. Then, under conditions of changing load and voltage U_d, the zener diode current I_{st} will change from a certain maximum value to a minimum value (I_{stmin}, I_{stmax}). Figure 23 shows the current-voltage characteristic of the zener diode.

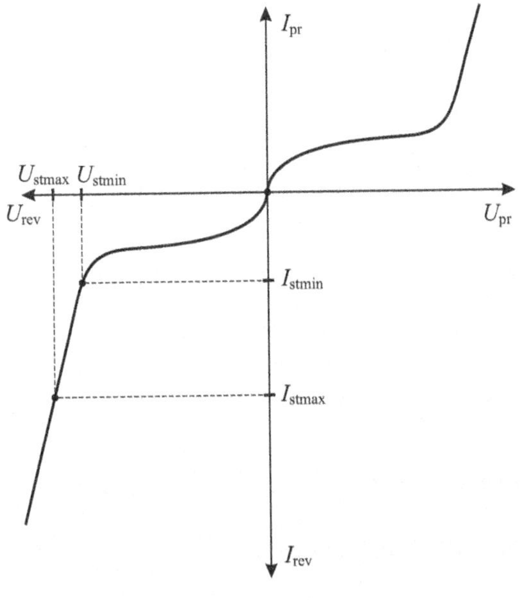

Fig. 23. Current-voltage characteristic of the zener diode

The calculation of the stabilizer circuit is reduced to the choice of the value of R_b. For normal operation of the stabilizer, it is necessary that a current I_{stmin} flowing through R_b corresponds to the beginning of the working section of the current-voltage characteristic. To calculate the ballast resistance, the formula is used:

$$R_b = \frac{U_{dmin} - U_n}{I_{stmin} + \frac{U_n}{R_n}}. \quad (36)$$

Define the parameters for calculation by the formula (36). The value of U_d = 20 V ± 15% = 17–23 V. Thus, U_{dmin} = 17 V. We take U_n equal to 12 V. The value of R_{nmin} corresponds to the maximum power consumption at the load current power P_{noutp} = 10 W, according to Ohm's law:

$$P = U \cdot I. \quad (37)$$

At U = 12 V, the current I_n is found by the formula:

$$I = \frac{P}{U}, \quad (38)$$

$$I_n = \frac{10}{12} = 0.83\,\text{A},$$

$$R = \frac{U}{I}, \quad (39)$$

$$R = \frac{12}{0.83} = 14.45 \approx 15\,\text{Ohm}.$$

As a zener diode according to the reference book, we take D815D for which U_{st} = 12 V and I_{stmin} = 25 mA, I_{stmax} = 650 mA. Substitute the obtained value in the formula (36)

$$R_n = \frac{17 - 12}{0.025 + \frac{12}{15}} = 6\,\text{Ohm}.$$

The pilot assembly and testing of the circuit showed that the values of I_n and P_n were clearly overestimated in the calculations of the transformer and power supply. In order to reduce the idle current of the stabilizer and increase its efficiency, we take the value R_b = 240 Ohm, which is consistent with the values obtained during the test during the test circuit.

9 Calculation of the Thyristor Control Circuit

Figure 24 shows the control circuit of the main thyristors V4–V6.

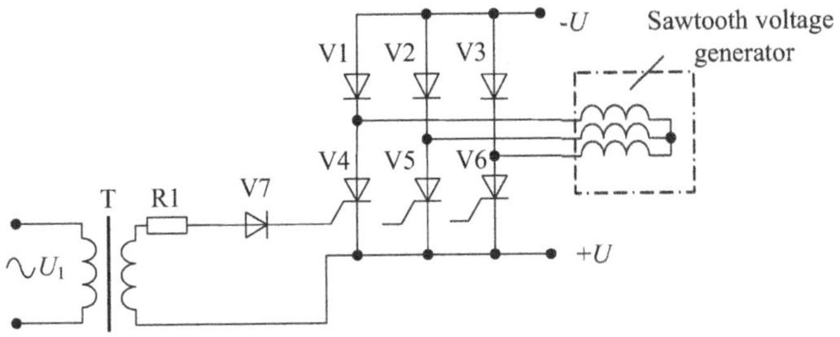

Fig. 24. Thyristor control circuit

From the output of the amplifier-shaper (blocking generator), the control pulses are fed to a high-frequency pulse transformer TR. At the transformer output, the amplitude of the "bundles" of pulses is 6 V. As a power thyristor, thyristors of the TL2-200 brand having the parameters are used:

- the maximum allowable direct pulse control current is 2 A;
- the maximum allowable direct current control – 10 A;
- switching current $\leq$ 1.2 A;
- unlocking control current (at T = 25 °C) $\leq$ 0.028 A;
- unlocking control current at (T = –50 °C) $\leq$ 0.6 A;
- non-unlocking control current – 0.002 A;
- – unlocking constant control voltage (T = 25 °C) $\leq$ 0.6 V.

The presence of a wide variation in control parameters does not allow the application of the results of calculations of one thyristor, for others [12]. To ensure the stable operation of all thyristors, an individual selection of the resistance R1 for each is necessary.

When developing the circuit, we use control diagrams from the reference manual [13]. The calculation according to the diagram shows that the load line passes in the region of admissible values of the power of the control circuit. To limit the current control current, we take the value R1 = 15 Ohms, with the output voltage of the transformer U_{inp} = 6 V. As a result of experimental testing and selection of the resistance value R1 in the thyristor control circuit, it is 15–200 Ohm. During the pilot test, thyristors were selected in order to obtain control parameter discrepancies of not more than 20%. Diode V7 protects the transition "control electrode–cathode" from reverse overvoltages. As a diode V7, a diode D7G, D226B with U_{revmax} = 400 V and forward current I_{prmax} = 0.3 A was used, which corresponds to the parameters of the control circuit [7].

10 Conclusions

The work considers the most significant operating conditions of batteries in order to improve their operation and extend their service life by introducing promising new developments; a way has been found to solve the problem of stabilizing the charging of batteries on a motor vehicle with rheostatic braking. The causes of premature failure of the batteries at the MVPS (burnout of cells and loss of capacity) are identified.

The design and principle of operation of alkaline batteries are considered. The analysis of battery charging circuits and control circuit power supply on electric trains is carried out.

A scheme has been developed for stabilizing the charging current of rechargeable batteries on electric trains in accordance with the requirements for it.

The elements of the circuit of a voltage regulator are calculated and its model is compiled.

The analysis of the operation of the circuit based on the transients that accompany the operation of the charger in the process of charging the battery. The analysis confirmed the efficiency of the circuit and the correct calculation of its elements.

The introduction of a step-up voltage regulator with stabilization at the output of the battery charging current at the MVPS is an effective measure, the implementation of which will extend the life of the batteries and reduce the costs associated with the replacement and purchase of new batteries during depot repairs.

References

1. Chen, X., Zhu, X., Zhou, O., Wong, Y.D.: J. Adv. Transp. **2017**, 3128372 (2017)
2. Yoneda, T., Ryuba, H.: CTL model checking of time petri nets using geometric regions. IEICE Trans. Inf. Syst. **81**, 297–306 (1998)
3. Trushnikov, D.N., Belinin, D.S., Shitsyn, Y.D.: Modern Probl. Sci. Educ. **2**, 95 (2014)
4. Berthomieu, B., Diaz, M.: Modeling and verification of time dependent systems using time Petri nets. IEEE Trans. Softw. Eng. **17**(3), 8–14 (1991)
5. Boşneagu, R., Coca, C.E., Sorescu, F.: EIRP Proc. **10**, 360–368 (2015)
6. Buynosov, A., Lapshin, V., Smolyaninov, A., Dinislamov, A.: Calculation of hardening process parameters for loco-motive parts. In: Polytransport Systems-2018. MATEC Web of Conferences, vol. 216, p. 02020 (2018). https://doi.org/10.1051/matecconf/201821602020
7. Teodorovic, D., Janic, M.: Transportation Engineering. Butterworth-Heinemann, Oxford (2016)
8. Buynosov, A.P., Volkov, D.V., Fedorov, E.V.: Automatic control of train braking system integrity and density. Transp. Urals **4**(59), 77–80 (2018). https://doi.org/10.20291/1815-9400-2018-4-7-80
9. Hernandez-Ramirez, A.G., Martinez-Tavera, E., Rodriguez-Espinosa, P.F., Mendoza-Pérez, J.A., Sujitha, S.B.: Detection, provenance and associated environmental risks of water quality pollutants during anomaly events in River Atoyac, Central Mexico: a real-time monitoring approach. Sci. Total Environ. **669**, 1019–1032 (2019)
10. Buynosov, A.P., Tsikhalevskiy, I.S., Laptev, S.I.: Organization of gas turbine locomotives "ГТ1h" exploitation and technical maintenance. Herald Ural State Univ. Railway Transp. **3** (39), 41–55 (2018). https://doi.org/10.20291/2079-0392-2018-3-43-55
11. Girzhon, V.V., Ovchinnikov, A.V.: Metal Sci. Heat Treat. **58**(11–12), 719–723 (2017)
12. Buynosov, A.P., Volkov, D.V., Umylin, I.V.: Definition of a limit difference in the diameters of bandages at one wheel pair of the motor car of an electric train(s). Sovremennye tekhnologii. Sistemnyi analiz. Modelirovanie **56**(2), 8–18 (2017). https://doi.org/10.26731/1813-9108.2018.2(58).90-97
13. Buynosov, A.P.: Doctoral thesis, Ural State University of Railway Transport, Ekaterinburg (2011)

Thermal Load of a Thermos Car in Transportation of Metallurgical Blanks

Vasily Lapshin(✉), Alexander Smolyaninov, Alexander Buynosov, and Ivan Dobychin

Ural State University of Railway Transport, Kolmogorova street, 66, 620034 Yekaterinburg, Russia
VLapshin@usurt.ru

Abstract. The article describes the task of a mechanical system work while it is thermally loaded with bulky metallurgical blanks. A thermos car is given as an example to illustrate the methods of estimation of the structure bearing component limiting a car safe exploitation. The main heat fluxes are determined which define the condition of the car structure parts, the features of an intra plant rolling-stock exploitation are given. A model analyses of the thermos car is presented in three subsystems "frame – cowl – metallurgical blank". It also takes into account shear stress on a slide surface. On the basis of specifications and technical documentation, three types of loads are described, which have influence on the car in the process of its exploitation. The estimate of the distribution of thermal fields and structure stresses of the thermos car is calculated based on the finite elements method in the program complex ANSYS Workbench. The results of the estimate of the developed model compliance to the real distribution of thermal fields are given. The estimates show the necessity of the modernization of the car frame thermal insulation. The solution is found which allows keeping the tempo of metallurgical blanks cooling within a wide range of temperature of the environment, securing the elements of the running gear and braking equipment, guaranteeing the mechanical safety of the car bearing structure, securing the personnel servicing the hot car elements of the rolling stock.

Keywords: Thermos car · Metallurgical blank · Thermal process · Thermal safety · Finite element method · Numerical experiment · Thermal field · Stress equivalent · Thermal load

1 Introduction

Heat treatment is made to prevent flakes [1] appearance in metallurgy. One of the ways to reduce the time of anti flake processing is to conduct isothermal annealing "with much reduced duration and very long-time forging cooling" [2].

During anti flake processing of bulky blanks [1], some variants of long-time forging cooling are used:

– in wells under sand layer (gravel, slag wool, etc.) [1, 3];
– in thermoses or cowls [4].

Z. Popovic et al. (Eds.): TransSiberia 2019, AISC 1115, pp. 281–290, 2020.
https://doi.org/10.1007/978-3-030-37916-2_27

To reduce transportation expanses, special cars are used, which are engaged in the technological process of products creation as well as cargo transportation. In this case, the cars are used as a vehicle and store room at the same time which cools the blanks to a pre-set temperature. The thermos car [5] consists of a platform on which the blanks are loaded and a cowl with thermal insulation made of industrial felt [6]. There is a fireclay brick lining on the frame. The cowl is set on the platform after hot blanks had been loaded. The cars are rolled out of the production area where they are kept about 20–24 h. When the blanks temperature is lowed to 300°, further processing is conducted in accordance to the technical rout.

The search of the process of the metallurgical blanks cooling is made to solve two tasks:

- analyses of heat flux in the thermos car and technological effect test of the blanks heat treatment;
- valuation of thermal stresses influence on the thermos car structure.

The analyses of the heat fluxes in the thermos car and the testing of the technological effect of the blanks cooling were examined in works [6, 7]. The solution of the task was received due to the heat balance equation [8] in the supposition that the air temperature in the thermos rises quickly equally to the air temperature inside the thermos when the blanks were being loaded, to some temperature after its structure had been warmed. The results of the modeling of the blanks cooling process showed the possibility of their use for the anti flake processing of thermos cars. The dependences of the forging cooling at different parameters of the car thermal insulation were obtained.

The related lining temperature and its parameters in the structure of such cars was determined according to the demands of the blanks production, and the influence of temperature stresses on the car structure was not taken into account. Therefore, on the projecting stage, there is an actual task of quantitative estimation of the range of the extreme stresses on the structure limiting the safe car exploitation.

The thermos car is an intra factory transport with a narrow sphere of use, followed by:

- high temperature stresses on the elements of the car structure (wheel sets, axle box, brakes equipment);
- heavy stresses in the time of the blanks in time of loading;
- the absence of hump yards while being used;
- increased axial load;
- low moving speed;
- the absence of emergency stops.

The main rated forces are: the tensile and squeezing forces of behavior between cars and a locomotive, gravity of the car and cargo. Such combination of stresses is typical only for the thermos car frame. Thus, due to the thermos car structure, the durability of the frame and cowl was estimated separately. The thermal stress is an additional one. Unequal temperature distribution in the car structure elements or the limitation of their heat increase can lead to plastic deformations and, as a result, the destruction of elements [9].

The uniqueness of the situation is that such conditions as the outside temperature and its change while the blanks are being cooled to a pre-set temperature, the lining layer condition, the shape and rigidity of the structure and cowl are not known beforehand. Therefore, the definition of "loading" was defined as "the condition of machinery and mechanisms, caused by external influences and the conditions of functioning" [10]. In our case – the combination of heating and mechanical processes and phenomena. According to [11], the main characteristics of thermal stress of the car parts are the temperature of assembly units and the car parts (axle box, brakes equipment, etc.), which satisfy the demands of the car safety and the stresses in the car structure parts caused by the main and additional stresses.

2 Method of Calculation

We will consider the time of blanks load on the supporting bars and the set of thermally insulated cowl as the starting point. The heat fluxes scheme is represented in Fig. 1. The blanks are cooled due to free convection from the blanks surface to the air Q_1 in the thermos car, and due to heat conductivity through supporting bars Q_2.

To ease mathematical description, a traditional for the tasks of heat engineering method was used: the interaction of a solid body with gas comes down to the process of heat conductivity, and their interaction is considered through boundary values. Lining on the frame, thermal insulation on the cowl, external surfaces of the frame and the cowl parts were considered as coupled surfaces [11]. In this case, convection heat transfer through the elements is the surface load and was modeled with the relevant coefficient of heat transfer on these surfaces. Heat flux contacting with two bodies was determined on condition that the temperature of the contact and respond surfaces is equal, for example, on a blank and the supporting bar of the frame. Heat transfer of the car multilayer elements (for example, thermally insulated walls of the cowl, the car floor with lining) was determined through the equivalent coefficient of heat conductivity which is equal to the scope of layers [12]:

$$\lambda_{\text{equival}} = \sum_{k=1}^{n} h_k \frac{\sum_{k=1}^{n} h_k}{\sum_{k=1}^{n} \frac{h_k}{\lambda_k}}, \qquad (1)$$

where h_k – thickness of the layer; λ_k – coefficient of the layer heat conductivity.

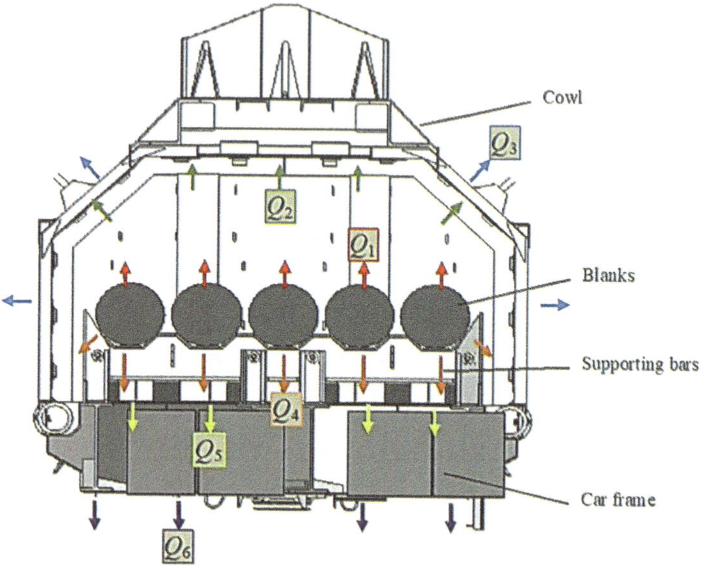

Fig. 1. Scheme of the heat fluxes in the thermos car

Figure 1: Q_1 – the amount of heat transferred from the blanks to the air inside the thermos; Q_2 – the amount of heat transferred from the air through the thermal insulation of the cowl having thickness hk to the external surfaces of the cowl; Q_3 – the amount of heat removed from the external surfaces of the thermos car cowl to the atmospheric air; Q_4 – the amount of heat transferred from the blanks through the supporting bars; Q_5 – the amount of heat transferred through the fire-clay lining having thickness hf to the metallic platform box; Q_6 – the amount of heat transferred from the metallic structure of the car frame to the atmospheric air.

The assessment of the thermos car stress condition was made on the bases of the finite element method [13] in the program complex ANSYS Workbench [14]. The thermos car model analyses is a spatial volume-lamellar scheme (Fig. 2).

The model analyses of the thermos car is represented as three subsystems "frame – cowl – metallurgical blank" and is built on the following assumptions:

- the use of the contact "line – plane" type. Such assumption is used in the places of junctions of the thermos car frame and cowl without relevant shift (rigidly knot to knot);
- in positions 1, 2, 3 (Fig. 2), being conjugate surfaces of the supporting tubes of the cowl and frame arms, supporting bars and frame boxes, metallurgical blanks and supporting bars a friction contact is used "plane – plane";
- in positions 4, 5, 6 (Fig. 2), there were hinged fixing with the possibility of a relevant deformation along pre-set directions. Such solution is the result of the material thermal widening and lets the structure "breath". The accepted type of fixings is used in bolt fixations between the cowl tubes, fixations between the skin panels and cowl bars, fixations between the supporting bars and frame limiters.

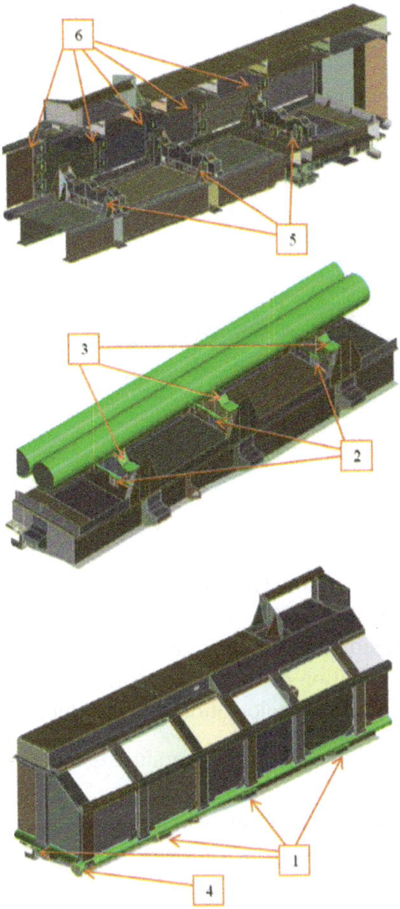

Fig. 2. The Thermos Car Lamellar Model and the Conditions of the Structure Parts Interaction: 1, 2, 3 – frictional contact "plane–plane" type; 4, 5, 6 – hinged fixing with the possibility of a relevant thermal deformation

The modeling of the sheet and roll-formed shape of the cowl and the frame was made with elements SHELL 181 [15], which let set the necessary thickness of a lamellar structure element. The volume final elements SOLID 187 were used for the body center plate approximation, the on-body center plate boxes, the front and back thrust blocks. The contact surfaces were described by the final elements CONTA 174 and TARGE 170 for the contact and responding surfaces relatively. While choosing the target and contact surfaces, we followed the condition that the target surface is less convex or flat.

Table 1 gives the calculating model, formed after the final-element net was placed.

Table 1. The final-element model

The name of the final-element model	The amount of knots	The amount of elements
Frame	352211	197734
Cowl	336122	176940
Metallurgical blank	380980	214581

Solving the contacting task, we considered that the contact surfaces relevantly to each other remove a part of shear stress along the section surfaces. In this case, the equivalent shear stress τ was defined as:

$$\tau_i = \mu N_i + \beta, \tag{2}$$

where μ – traction coefficient; β – sliding resistance (coupling); N_i – pressure on the contact surface, i = 1, 2, 3 in Fig. 2.

If there is an excess of the equivalent shifting stress of some value τ_{max}, the surfaces start sliding relevantly to each other. In the calculating model we considered only sliding in contact and the absence of shear stresses, the amount of which is not large while tensely deformed condition of the entire structure was being assessed.

The procedure of the tensions calculation was conducted within two stages [15]:

- the first stage – the calculation of the temperature fields and the relevant temperature deformations;
- the second stage – vertical loads of the weight of the car parts (frame and cowl) and metallurgical blanks were put.

For the assessment of the frame and the cowl durability, a traditional approach was used, based on comparison to safe tension. Equivalent stresses of the theory of maximum power of shape changes (Mises-Hencky theory) was accepted as calculating stresses. Possible destructions of the lining thermal insulation of the car frame were taken into account by changing the coefficient of the heat transfer in the local areas of the final element model.

3 Results

The calculation of the temperature fields distribution in the parts of the thermos car structure was made with the metallurgical blanks heated up to 550 °C and the outside air 22 °C. Figure 3 shows the distribution of the temperature fields in the mostly thermally stressed elements of the thermos car metallurgical structure after its structure had been warmed up.

The level of the compliance of the developed model to the distribution of the temperature fields was estimated comparing theoretical and experimental data. The data were received in Nizhniy Tagil Iron and Steel Works while examining thermos cars in 24–26 h after the metallurgical blanks had been loaded (Table 2).

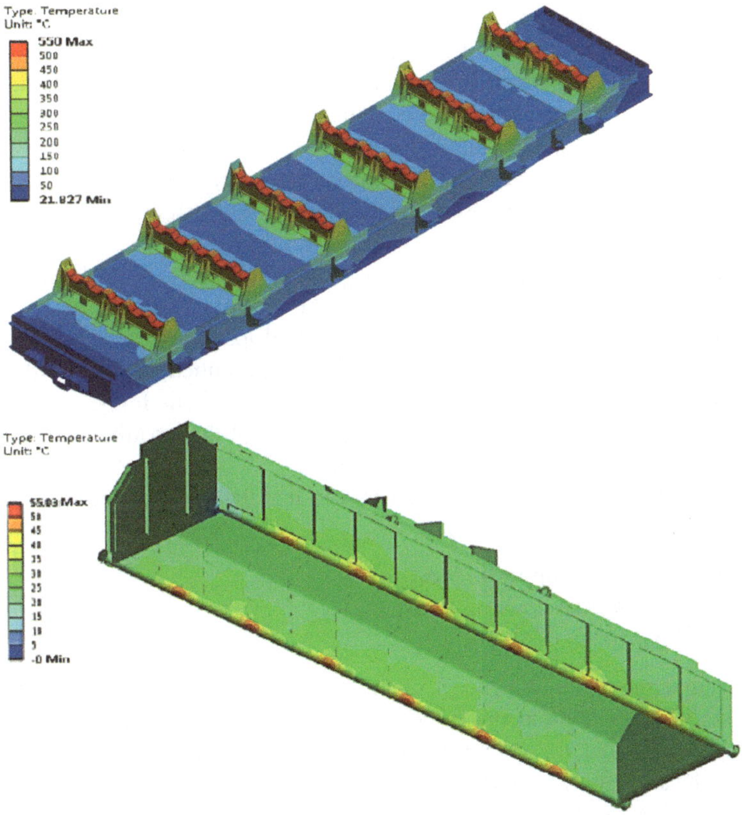

Fig. 3. The distribution of the temperature fields in the thermos car frame

Table 2. The comparison of theoretical and experimental temperature data in the elements of the car frame and the cowl

Car frame element	Experiment, °C*	Theoretical calculation, °C
Spinal beam	73/61	56
The flooring (bottom)	147/131	129
Side binder	77/48	37
Cowl	40	37
Vertical walls	73	54
Supporting bars		

* Numerator – the upper temperature rate, denominator – the average temperature rate of five cars testing.

Relatively large difference of the temperature rates of some parts are explained by a large amount of parameters and factors which influence on the final result. They are the parameters of the thermos car structure, and the parameters of the environment. The surfaces of the supporting bars of the cowl of the real car had initial deflections and bearing elements deformations, local destructions of the lining layer and others. The data in Table 2 are presented under ideal conditions of carrying the numerical experiment out. When the metallurgical blanks were loaded and the cowl was set the thermos car was kept in the preset temperature conditions during 24 h. In real life, most of the time is spent for waiting and shunting operations in a service department where the temperature may differ much from the experimental one. The received temperature fields were accepted as surface loads in the stage of estimating the main element of mechanical safety – strength analyses.

Mechanical safety [12] being influenced by the combination of loads was estimated for steel structure 09G2S with the limit of the material fluidity $\sigma_{flu} = 295$ MPa. Equivalent stresses, by the first two combinations of loads [18] appearing in the elements of the thermos car frame, do not exceed the permitted stress according to the strength condition. The maximum stresses rates are seen in the frame elements (pivot box of the car). The rates are 0.6 σ_{flu}. In the elements of the metallurgical structure of the cowl, the stress rates do not exceed 0.4 σ_{flu}.

As a result of the temperature load, it is seen that the summary of the loads in the local parts of the structure grow a lot:

- spine beam elements up to 0.75 σ_{flu};
- supporting arm of the cowl up to 0.5 σ_{flu};
- supporting bar (0.85–0.90) σ_{flu}.

The main temperature influence is on the vertical sheet of the supporting bar. There are hard stress concentrations 0.9 σ_{flu} on the welded parts of the vertical sheet and the supporting plate under the combination of influence of mechanical and temperature stresses on the structure. Possible deviation from the data of the heat characteristics may cause further growth of the combination of stresses; as a result they reach the critical point of fluidity, that is the transfer to the plastic characteristics of the work of metals. However, this conclusion refers only to the supporting bar (replaceable "sacrifice" part), being in direct contact with hot metallurgical blanks, not contacting with bearing elements of the thermos car, the parts the under-car equipment and running gear.

The situation gets different when the parameters of the thermal insulation of the car frame are changed, caused by the destruction of the lining layer fire-clay brick. Mechanical influence in the process of loading-unloading leads to the destruction of the thermal insulation layer. This finally leads to the necessity of changing the thermal insulation layer. The calculations show, if the thickness of the thermal insulation layer is shortened at 1.5 times in the area of the pivot beam the temperature of the flooring rises up to 180 °C. In this case the parts of the trolley equipped with polymeric elastic parts may be extremely heated. The working temperature of most of the polymeric parts in cars structure is limited by 100–120 °C. The change of the efficient condition of the running gear, caused by increase of the temperature limit, goes along with softening and melting of slips to the frictional wedge, elastic slider, adapter insertion of the axle box.

The temperature distribution analyses along the thermos car showed that the car protecting parts need the development of additional demands concerning thermal safety. The structure of the supporting bars of the car cowl, spine and crossbeams of the frame are located in the area of direct contact of the staff with the hot surfaces. Therefore, there is a necessity of additional demands to exclude the contact of the personnel servicing a rolling-stock with the hot car parts, on thermal safety of the technological process of exploitation and technical service of the described rolling-stock [16].

As a variant, in the thermos car structure modernization, we see the change of the thermal insulation layer from the fire-clay brick to the industrial felt thermal insulation overlapped with a metallic net to protect from mechanic damages. The calculations show that the change of the fire-clay lining reduces heat losses, and warm up time is reduced by 16%. The change of the frame thermal insulation decreases the car container more than 2.5 tons, reduces the temperature of the protecting surfaces of the car frame up to 35–40%. It also provides a stable blanks cooling not depending upon the environmental conditions.

4 Conclusion

The calculations of the thermos car gave a motivated conclusion about the necessity of modernizing the lining thermal insulation of the frame structure. The developed volume-lamellar and final element models of the thermos car can be adapted for the structural changes without heavy expenses. The calculation method of the thermos car structure will not be changed under the use of other materials of the frame lining and thermal insulation of the cowl, including standard variants. The increased complexity of the thermal load calculation is possible if the model of the lining material is changed. As a whole, such approach lets estimate the possibility of using several types of the thermal insulation of the car structure. In difference from the standard approaches for the estimation of non-self-propelled rolling-stock safety [11, 16], the suggested calculation takes into account thermal stress of all car parts.

The considered thermos car structure responds the demands of mechanical safety in accordance to the outside temperature and mechanical effects. The received data prove the correspondence of the rolling-stock to the demands of the knots and car parts safety under the influence of high temperature and the possibility of such cars exploitation for providing technological operations. The guaranteed safe range of the structure exploitation is determined by assumptions limitations and criteria of the limit conditions accepted by the methods. Under the influence of extreme temperature (caused by the destruction of the thermal insulation layer of the car frame), it is necessary to prevent the personnel from contacting with hot parts while servicing and operating a rolling-stock.

References

1. Buynosov, A.P., Volkov, D.V., Fedorov, E.V.: Automatic control of train braking system integrity and density. Transp. Urals **4**(59), 77–80 (2018). https://doi.org/10.20291/1815-9400-2018-4-7-80
2. Shaburov, A.D.: Theoretical and technological aspects of energy-saving anti-flock waste treatment using out-of-furnace delayed cooling in thermoses taking into account the effects of hydrogen capture by traps. Extended Abstract of Cand. Sci. South Ural State University, Chelyabinsk (2013)
3. Kedzierzawski, P., Oriani, R.A., Hirth, J.P., Smialowski, M. (eds.): Hydrogen Degradation of Ferrous Alloys. Noyes Publications (1985)
4. Shaburov, A.D., Mirzaev, D.A., Smolko, D.F.: Bull. MSTU **2**, 65–67 (2011)
5. Buynosov, A.P., Volkov, D.V., Umylin, I.V.: Definition of a limit difference in the diameters of bandages at one wheel pair of the motor car of an electric train(s). Modern Technol. Syst. Anal. Modeling **56**(2), 8–18 (2017). https://doi.org/10.26731/1813-9108.2018.2(58).90-97
6. Lapshin, V.F., Namjatov, A.B., Bulanov, N.V.: Defining consumer properties of heat insulated cars for steel billets. Transp. Urals **4**(47), 48–53 (2015). https://doi.org/10.20291/1815-9400-2015-4-48-53
7. Shaburov, A.D., Mirzaev, D.A., Maltsev, I.V.: Bull. SUSU **13**, 77–80 (2010)
8. Buynosov, A.P., Volkov, D.V., Dinislamov, A.R.: Nondestructive control method of depth of surface tempering of details of mechanical part of locomotives as the alternative to the destroying ways. Transp. Urals **1**(56), 26–29 (2018). https://doi.org/10.20291/1815-9400-2018-1-26-29
9. Lagerev, I.A.: Sci. Tech. Bull. Bryansk State Univ. **1**, 49–53 (2015)
10. Ghobadpour, A., Boulon, L., Mousazadeh, H., Malvajerdi, A.S., Rafiee, S.: State of the art of autonomous agricultural off-road vehicles driven by renewable energy systems. Energy Procedia **162**, 4–13 (2019)
11. Fontanari, V., Benedetti, M., Monelli, B.D., Degasperi, F.: Fire behavior of steel wire ropes: experimental investigation and numerical analysis. Eng. Struct. **84**, 340–349 (2015)
12. Copaci, A., Tanasoiu B.: Proceedings of invited papers. In: 3th WSEAS International Conference on Finite Differences - Finite Elements - Finite Volumes - Boundary Elements (F-and-B 2011), Bucharest, Romania (2010)
13. Buynosov, A., Lapshin, V., Smolyaninov, A., Dinislamov, A.: Calculation of hardening process parameters for locomotive parts: polytransport systems-2018. In: MATEC Web of Conferences, vol. 216, p. 02020 (2018). https://doi.org/10.1051/matecconf/201821602020
14. Anfinsen, M., Lagesen, A.V., Ryghaug, M.: Green and gendered. Cultural perspectives on the road towards electric vehicles in Norway. Transp. Res. Part D: Transp. Environ. **71**, 37–46 (2019)
15. Zhao, M., Wei, M., Tian, G., Song, P.: Simulation of effects of ORC system installation on heavy-duty truck. Appl. Thermal Eng. **128**, 1322–1330 (2018)
16. Kacem, M., Zaghdoudi, K., Morales-Rubio, A., Guardia, M.: Preliminary results on the influence of car characteristics on their gases emissions using gas sensors. Microchem. J. **139**, 69–73 (2018)

Use of Aluminum Matrix Material for Electrospark Alloying of Carbon Steels

Sergey Nikolenko[1(✉)], Leonid Konevtsov[1], Victor Makienko[2], and Evgenii Kim[3]

[1] Institute of Material Studies, Tikhookeanskaya Street 153, Khabarovsk 680042, Russia
nikola1960@mail.ru
[2] Far Eastern State Transport University, Seryshev Street, 47, Khabarovsk 680021, Russia
[3] Pacific State University, Tikhookeanskaya Street 136, Khabarovsk 680035, Russia

Abstract. Shows the results of a study in materialogy the surface of the formation of coatings by the method of electrospark alloying (ESA) using anodic aluminum-matrix composite materials (AMCM). The kinetic experimental and mathematical dependences of the formation of the alloyed layer (AL) are shown, the results of studies of the wear resistance of AL with recommendations for the use of anodic materials and treatment modes are presented. The obtained series of cathode weight gain, erosion resistance of anodic materials, mass transfer coefficient, wear resistance of the surface after ESA, efficiency of formation of the AL and efficiency indicators of ESA taking into account the treatment modes, recommendations for achieving the required properties of AL at ESA AMCM are shown.

Keywords: Materialogy of surface · Electrospark alloying · Alloy layer · Anodic materials · Aluminum matrix materials · Wear

1 Introduction

Surface is an integral property of any means of activity created by nature or humans and studies of surface increasingly raise scientific and practical interest among the research community. Along with differentiated studies of surfaces by different branches of science, a new stage in the development of material science has been undergoing formation lately - materialogy [1] and its important branch – the materialogy of the surface that involves research of surfaces in all aspects. One of the urgent issues of the materialogy of surface is studying the process of formation of coatings for working surfaces of machine parts. Proceeding from the functional purposes of machine parts, their working surfaces are the subject of increasingly rigid standards for hardness, wear resistance, heatproof properties, etc. The desired properties of working surfaces are attained to a greater extent by the method of electrospark alloying (ESA), invented by a Russian scientist B.R. Lazarenko [2]. This method is distinguished by a high level of science linkage [3] and allows the application of coatings in the form of an alloy layer

Z. Popovic et al. (Eds.): TransSiberia 2019, AISC 1115, pp. 291–299, 2020.
https://doi.org/10.1007/978-3-030-37916-2_28

(AL) using anodic materials. The latter are selected according to the functional purposes of the surfaces under study.

Among the anodic materials, the use of aluminium matrix materials, especially anodic aluminium matrix composite materials (AMCM) [4] is under special research and practical interest as they are distinguished by improved wear resistance properties and some other reinforced properties [5]. Our study was aimed at the development and research of new anodic AMCM that improve the efficiency of the ESA process due to decrease in erosion resistance and improvement of the physical, chemical and operational characteristics of coatings owing to the use of composite mixtures (based on Al and dispersion reinforced materials), intended for deposition of anti-friction ESA-coatings.

2 Methodology, Materials and Equipment

We used high quality steel 45 as the material for the base (cathode). This steel is most widely used for production of mechanical engineering articles and is distinguished by sufficient reasonable strength and wear resistance properties. It is used for production of distribution shafts, crank shafts, pinion gear shafts, as well as cog-wheels, rack-wheels, clutches, cylinders, spindles and other machine parts subjected to heat treatment.

We chose aluminium matrix materials as anodic materials in consideration for the properties of aluminium [6], which does not form continuous solid solutions with other elements. Nevertheless, Al forms limited solid solutions with Zr, Cr, Fe, and Ni. It also forms limited solid solutions on the basis of other elements α-Ti, β-Ti, Cr, Fe, Ni, Zr. In addition, Al forms compounds with Ti, Zr, Cr, Fe, Ni, H, C, N, and O. Al also forms eutectic mixtures with Fe and Ni. Taking into account the above mentioned properties of Al, in the laboratories of the Institute for Material Studies of the Khabarovsk Scientific Center of the Far Eastern Branch of the Russian Academy of Sciences, we obtained the following aluminium matrix materials: (1) "A" (№36) NiO-TiO_2-Al; (2) "B" (№37) NiO-Zr-TiO_2-Al; (3) "C" (№65) Ni-Cr-Zr-Al; 4) "D" (№66) Ni-Cr-Zr-Ti-Al. We took into account that Al does not form alloys with the above mentioned elements. Besides that, anodic materials "A" and "B" contain oxide TiO, which reacts with air at temperature of 800 °C forming TiO_2, and oxide NiO, which reacts with carbon recovering to metallic nickel and with oxides of Al and Fe when heated, reducing them to pure metal; and materials "C" and "D" contain Cr.

As a pulse generator, we used the plant "IMEIL" of the Institute for Material Studies with the following modes: pulse rate t_p/r = 50–25 μs, charge pulse duration t_p = 40–80 μs, and frequency f = 500 Hz. Two treatment modes developed experimentally were chosen for comparison: $t_{p/r}$ = 50 μs, t_p = 40 μs (hereinafter the first mode $t_{p/r}/t_p$ = 50/40 is indicated as ′) and $t_{p/r}$ = 25 μs, t_p = 80 μs (the second mode 25/80 is indicated as ″). The values for the cathode weight increase and anode erosion were determined by a gravimetric procedure with electronic balance Shinko Denshi HTR-220 CE accurate within $\pm 2 \cdot 10^{-4}$ g. Microabrasive wear was performed using the plant "Calotest CSM", with ZrO_2 as a sphere counterbody.

3 Results and Discussions

As is seen from Fig. 1, all the curves of mass transfer $\Sigma\Delta_C$, obtained through ESA of steel 45 with anodic AMCM under the modes A′, B′, C′, D′ and A″, B″, C″, D″, refer to the class of "Lazarenko curves" [7].

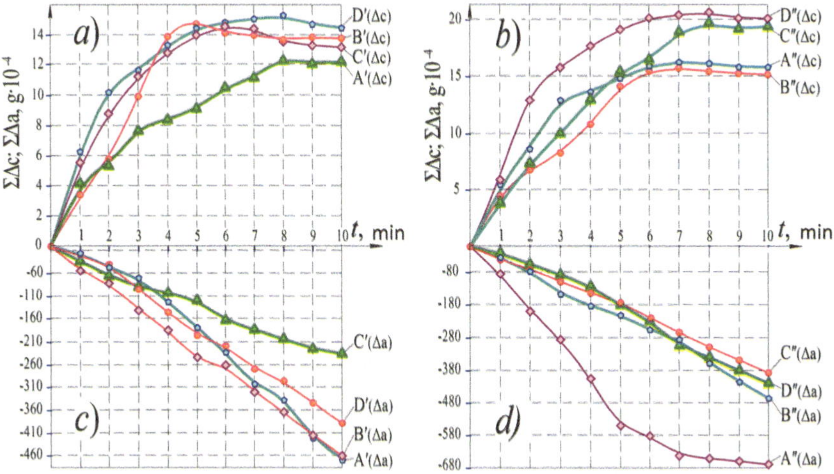

Fig. 1. NURBS curves of the process kinetics for ESA of steel 45 with anodic AMCM A, B, C, D under the ESA modes: (a) $t_{p/r}/t_p$ = 50/40; (b) $t_{p/r}/t_p$ = 25/80.

The maximum value of mass transfer $\Sigma\Delta_C$ was observed at t = 5–8 min when performing ESA of surface specific area of 1 cm^2 and reached the values of $\Sigma\Delta_C$ = 12–20 $g\cdot10^{-4}$. After that up to t = 10 min the ESA process became stabilized and no gain $\Sigma\Delta_C$ was observed. In all cases, the value $\Sigma\Delta_C$ under the modes $t_{p/r}/t_p$ = 25/80 exceeded (1.11-1.54 times) the value under the modes $t_{p/r}/t_p$ = 50/40 and achieved maximum value $\Sigma\Delta_C = 20.6\cdot10^{-4}$g at t = 8 min under ESA with anodic material D″. At t = 10 min, the maximum value $\Sigma\Delta_C = 20.2\cdot10^{-4}$ g was observed under ESA with anodic material D″, the minimum value $\Sigma\Delta_C = 12.1\cdot10^{-4}$ g was observed under ESA with anodic material A′. We obtained the weight gain series for anodic materials with consideration to the changes in treatment modes: D″ → C″ → A″ → B″ → D′ → B′ → C′ → A′ (series 1*), which shows that anode materials D″ and C″ had the maximum weight gain at ESA t = 10 min ($\Sigma\Delta_C$ = 20.2 and 19.6 $g\cdot10^{-4}$ at tp/r/t_p = 25/80 correspondingly) and that materials D″ and C″ had minimum weight gain ($\Sigma\Delta_C$ = 13.1 and 12.1 $g\cdot10^{-4}$ at $t_{p/r}/t_p$ = 40/50 correspondingly).

The values for erosion at t = 10 min of the ESA anodic materials under research reached the values of $\Sigma\Delta_{a(10min)} = 231.7–667\cdot10^{-4}$ g. We obtained the erosion resistance series for anodic materials under the changes in treatment modes: D′ → C′ → C″ → D″ → B′ → A′ → B″ → A″ (series 2*). This series show that anodic materials D′ and C′ are distinguished by the maximum erosion resistance ($\Sigma\Delta_{a(10min)}$ = 231.7 and $385.9\cdot10^{-4}$ g correspondingly) under the ESA modes $t_{p/r}/t_p$ = 50/40; and the materials

B″ and A″ have the minimum erosion resistance ($\Sigma\Delta_{a(10min)}$ = 465.5 and $667 \cdot 10^{-4}$ g correspondingly) at $t_{p/r}/t_p$ = 25/80.

Table 1 presents some of the values resulted from the research of ESA of steel 45 with anodic AMCM.

Table 1. Values for cathode weight gain $\Sigma\Delta_{C(10min)}$; anode erosion $\Sigma\Delta_{a(10min)}$; mass transfer coefficient $\Sigma C_{MTr(av)}$, %; wear ΣW_{AL}, 10^{-4} g and the forming efficiency indicators for AL ($\gamma_{ALF} \cdot 10^{-3}$) and ESA ($\gamma_{eff} \cdot 10^{-3}$)

Mode	A′	B′	C′	D′	A″	B″	C″	D″
$\Sigma\Delta_{C(10min)}$	12.1	13.7	13.1	14.3	15.9	15.2	19.6	20.2
$\Sigma\Delta_{a(10min)}$	457	456.7	385.9	231.7	667	465.5	392.3	408.7
$\Sigma C_{MTr(av)}$, %	6.7	5.5	10.0	11.4	3.2	6.2	10.4	8.5
ΣW_{AL}, 10^{-4} г	42	29	33	25	39	27	31	22
$\gamma_{ALF} \cdot 10^{-3}$	0.37	0.34	0.51	0.38	0.34	0.44	0.80	0.70
$\gamma_{eff} \cdot 10^{-3}$	15.6	10.0	16.7	9.4	13.2	11.8	24.8	15.4

We obtained the following mathematic expressions of polynominal equations for the kinetic dependencies trendlines (Fig. 1) under the ESA of the surface specific area of 1 cm^2 of steel 45 with anodic AMCM depending on the time (t, min):

- changes in the resulting cathode weight gain ($\Sigma\Delta_c$) under the treatment modes A′, B′, C′, D′ (40/50):

$$\Sigma\Delta c(A') = 0.1318t^3 - 4.6966t^2 + 46.974;\ R^2 = 0.9710; \quad (1)$$

$$\Sigma\Delta c(B') = 0.1169t^3 - 3.0233t^2 + 30.354 + 5.6154t;\ R^2 = 0.9856; \quad (2)$$

$$\Sigma\Delta c(C') = 0.2681t^3 - 6.8986t^2 + 54,973t + 2.1399;\ R^2 = 0.9975; \quad (3)$$

$$\Sigma\Delta c(D') = 0.3825t^3 - 8.5836t^2 + 62.215t;\ R^2 = 0.9893; \quad (4)$$

- anode erosion ($\Sigma\Delta_a$) under the treatment modes A′, B′, C′, D′ (40/50) :

$$\Sigma\Delta a(A') = -23.099t^2 - 248.78t + 65.287;\ R^2 = 0.9949; \quad (5)$$

$$\Sigma\Delta a(B') = -0.5723t^2 - 450,3t;\ R^2 = 0.9977; \quad (6)$$

$$\Sigma\Delta a(C') = -2.824t^2 - 371.97t + 136.96;\ R^2 = 0.9946; \quad (7)$$

$$\Sigma\Delta a(D') = -235.55t - 82.364;\ R^2 = 0.9919; \quad (8)$$

– changes in the anode erosion ($\Sigma\Delta_C$) under the modes A″, B″, C″, D″ (25/80) :

$$\Sigma\Delta_C(A'') = 0.2795t^3 - 7.0635t^2 + 58.413t + 2.014;\ R^2 = 0.9941; \quad (9)$$

$$\Sigma\Delta_C(B'') = -2.3287t^2 + 38.378t + 0.7972;\ R^2 = 0.9847; \quad (10)$$

$$\Sigma\Delta_C(C'') = -2.2308t^2 + 42.126t - 1.0979;\ R^2 = 0.9971; \quad (11)$$

$$\Sigma\Delta_C(D'') = 0.3409t^3 - 8.8456t^2 + 74.638t - 0,979;\ R^2 = 0.9983; \quad (12)$$

– and the anode erosion ($\Sigma\Delta_a$) under the modes A″, B″, C″, D″ (25/80) :

$$\Sigma\Delta a(A'') = 6.1051t^3 - 21.703t^2 - 1041.9t - 45.371;\ R^2 = 0.9922; \quad (13)$$

$$\Sigma\Delta a(B'') = -7.5408t^2 - 379.32t - 41.566;\ R^2 = 0.9951; \quad (14)$$

$$\Sigma\Delta a(C'') = -6.2826t^2 - 331.86;\ R^2 = 0.9991; \quad (15)$$

$$\Sigma\Delta a(D'') = -11.788t^2 - 315.76t + 57.455;\ R^2 = 0.9908. \quad (16)$$

The validation criterion of the obtained polynominal trendlines $R^\wedge$ for the resulting weight gain $\Sigma\Delta_C$ lies within the limits 0.9983 > R^2 > 0.9710, and for the anode erosion $\Sigma\Delta_a$ the validation criterion lies within the limits 99.91 > R^2 > 9908.

Under all ESA modes we observed a decrease in the mass transfer coefficient ΣC_{MTr} during the first t = 3–4 min followed by gradual stabilization and slow decrease in its value to t = 10 min (Fig. 2). We have obtained a series of dependencies for the mass transfer coefficient $\Sigma C_{MTr\ (av)}$ with consideration to the changing treatment modes: D′ → C″→ C′ → D″ → A′→ B″ → B′ → A″ (series 3*) which displays the best values of $\Sigma C_{MTr\ (av)}$ for the anodic materials D′ and C″ ($\Sigma K_{п(cp)}$ = 11.4 and10.4 correspondingly) and the worst values of $\Sigma C_{MTr\ (av)}$ for the materials B′ and A″ ($\Sigma\ C_{MTr\ (av)}$ = 5.5 and 3.2 correspondingly).

We have obtained the following polynominal equations for the dependencies of the mass transfer coefficient $\Sigma C_{MTr\ (av)}$ (Fig. 2) under the ESA of the surface specific area of 1 cm^2 of steel 45 depending on the treatment time (t, min):

– using anodic AMCM A′, B′, C′, D′ under the mode 40/50:

$$\Sigma C_{MTr\,(av)}(A') = 0.1939t^2 - 3.4556 + 18.193;\ R^2 = 0.9755; \quad (17)$$

$$\Sigma C_{MTr\,(av)}(B') = -0.0408t^2 - 0.1141t + 7.7048;\ R^2 = 0.9198; \quad (18)$$

$$\Sigma C_{MTr\,(av)}(C') = -0.0759t^3 + 1.7051t^2 - 13.058t + 39.094;\ R^2 = 0.9877; \quad (19)$$

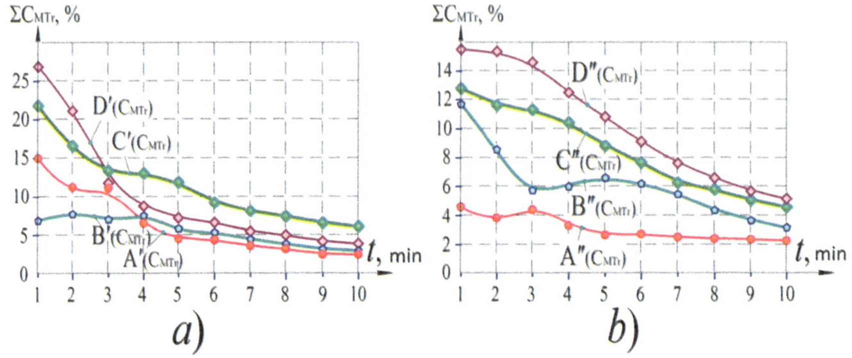

Fig. 2. NURBS – mass transfer coefficient $\Sigma C_{MTr\ (av)}$ curves, % under the ESA of the steel 45 with AMCM A, B, C, D with the modes: (a) A′, B′, C′, D′ (50/40); (b) A″, B″, C″, D″ (25/80)

$$\Sigma\, C_{MTr\,(av)}(D') = 0.1699t^2 - 3.4243t + 23.709;\ R^2 = 0.9644; \qquad (20)$$

– using materials A″, B″, C″, D″ under the mode 25/80:

$$\Sigma\, C_{MTr\,(av)}(A'') = 0.0047t^3 - 0.0475t^2 - 0.2354t + 4.8803;\ R^2 = 0.9044; \qquad (21)$$

$$\Sigma\, C_{MTr\,(av)}(B'') = -0.044t^3 + 0.7993t^2 - 4.8849t + 15.614;\ R^2 = 0.9258; \qquad (22)$$

$$\Sigma\, C_{MTr\,(av)}(C'') = 0.0295t^2 - 0.4753t + 0.795 + 15.357;\ R^2 = 0.9965; \qquad (23)$$

$$\Sigma\, C_{MTr\,(av)}(D'') = 0.021t^2 - 1.195t + 14.256;\ R^2 = 0.9811; \qquad (24)$$

The validation criterion for the obtained polynominal trendlines R^ for the changes in the dependencies of mass transfer coefficient $\Sigma C_{MTr\ (av)}$ lies within the limits of $0.9965 > R^2 > 0.9044$.

4 Investigation of Properties of AL and the Analysis of ESA Efficiency

Research in the wear resistance of the steel 45 surface before and after ESA with anodic AMCM (Fig. 3) showed that wear resistance properties increase 2.4 times on the average. In all cases we observed rapid increase in the wear during the first 2–3 min (burn-in period), then the wear process proceeds uniformly. The best wear resistance values were obtained after ES alloying with material "D", which displayed the best values for all treatment modes under the study: ΣW_{AL}(D″ and D′) 3.27 and 2.88 times correspondingly. We obtained the lowest values for wear resistance for material "A": ΣW_{AL} (A″ and A′) 1.85 and 1.71 times correspondingly. This obtained wear resistance series for materials: D″ → D′ → B″ → B′ → C″ → C′ → A″ → A′ (series 4*) displays the highest wear resistance under ESA with anodic materials D″ and D′ for treatment modes (25/80 and 50/40, $\Sigma W_{AL} = 22 \cdot 10^{-4}$ g and $25 \cdot 10^{-4}$ g correspondingly), and the lowest with materials A′ ($\Sigma W_{AL} = 42 \cdot 10^{-4}$ g). Under similar conditions of wear resistance testing, the wear of steel 45 without coatings was $\Sigma W_{(st.45)} = 72$ 10^{-4} g.

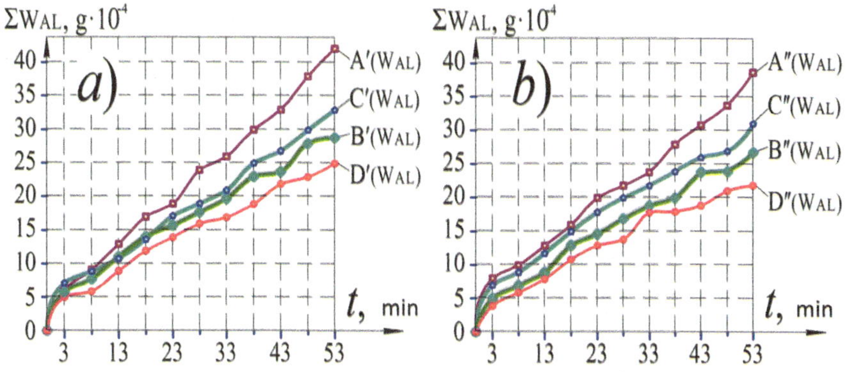

Fig. 3. NURBS curves of AL wear ($\Sigma W_{лс}$) after ESA of steel 45 with AMCM A, B, C, D under treatment modes: (a) A′, B′, C′, D′ (50/40); (b) A″, B″, C″, D″ (25/80)

We have obtained the following polynominal equations for the AL wear dependencies (ΣW_{AL}) after ESA of steel 45 with AMCM A′, B′, C′, D′

– under the ESA treatment modes 50/40:

$$\Sigma W(A') = -0.0077t^2 + 3.7195 - 2.3409;\ R^2 = 0.9957; \tag{25}$$

$$\Sigma W(B') = -0.0692t^2 + 3.3784t - 1.7955;\ R^2 = 0.9914; \tag{26}$$

$$\Sigma W(C') = -0.0277t^2 + 3.1191 - 1.0227;\ R^2 = 0.9901; \tag{27}$$

$$\Sigma W(D') = -0.0659t^2 + 3.011t - 2;\ R^2 = 0.9929; \tag{28}$$

– under the ESA treatment modes A″, B″, C″, D″:

$$\Sigma W(A'') = -0.0227t^2 + 3.4808 - 0.9773;\ R^2 = 0.9882; \tag{29}$$

$$\Sigma W(B'') = -0.0749t^2 + 3.2887 - 2.3182;\ R^2 = 0.9919; \tag{30}$$

$$\Sigma W(C'') = -0.1061t^2 + 3.9078t - 2.0682;\ R^2 = 0.9893; \tag{31}$$

$$\Sigma W(D'') = -0.0919t^2 + 3.1459t - 2.6364; \; R^2 = 0.9928. \tag{32}$$

The validation criterion for the obtained polynominal trendlines R^ of the changes in the aggregate wear lies within the limits $0.9957 > R^2 > 0.9882$.

When forming AL, an important constituent of the ESA assessment is the efficiency of the ESA method (γ_{eff}). The latter depends on the properties of the anodic material and the material of the base. The efficiency is also connected with the diverse character of the following parameters: mass transfer, formation of the AL "secondary structures", destruction of AL under the impact from impulse thermo-mechanical loads, duration of the ESA (t_{ESA}), mean mass transfer coefficient ($C_{MTr\ (av).}$), and the AL material's acquired properties ($\gamma_{pr.AL}$), including wear resistance (W_{AL}), hardness and other parameters [7].

$$\gamma_{ф.лс} = \Sigma\Delta_{к} \cdot t_{ESA} \cdot \Sigma K_{п.ср.}, \tag{33}$$

where: $\Sigma\ C_{MTr\ (av)} = \Sigma\Delta_{к.ср} / \Sigma\Delta_{a.ср.}$;

$$\gamma_{eff} = \gamma_{ф.AL} \cdot \gamma_{pr.AL}; \tag{34}$$

where: $\gamma_{pr.AL} = \gamma_{WAL(st45)} / \gamma_{WAL}$ is the ratio of the properties under consideration (including wear resistance, heat resistance, hardness, etc.) of the base material to those of AL. Some of the efficiency parameters for ESA of steel 45 with anodic AMCM are presented in the Table. In the case under consideration, the parameter γ_{eff} is presumptive for particular parameters $\Sigma\Delta_C$, $\Sigma\Delta_a$, $C_{MTr\ (av)}$, ΣW_{AL}, γ_{ALF} that constitute the efficiency. One of the most important among them is the parameter of AL formation ($\gamma_{ф.AL}$), which significantly depends on the anodic material and the modes of treatment, as well as on the values of pulse rate ($t_{p/r}$) and the charge impulse duration (t_p).

On the basis of comparison of the numerical values in Table 1 we have obtained a series of priority ratios for ESA that are characteristic of the achievement of the most important constituents and parameters of efficiency (series 1* – $\Sigma\Delta_C$, mg; series 2* – $\Sigma\Delta_a$, mg; series 3* – $C_{MTr\ (av)}$, %; series 4* – ΣW_{AL}, $g \cdot 10^{-4}$). These series of the efficiency parameters (1*-4*) do not coincide with the series of the AL formation efficiency γ_{ALF}: $C'' \rightarrow D'' \rightarrow C' \rightarrow B'' \rightarrow D' \rightarrow A' \rightarrow B' \rightarrow A''$ (series 5*) and the series of the ESA efficiency (γ_{eff}): $C'' \rightarrow C' \rightarrow A' \rightarrow D'' \rightarrow A'' \rightarrow B'' \rightarrow B' \rightarrow D'$ (series 6*). Nevertheless, they can be used for forecasting reachable parameters of AL when ES alloying steel 45 with the explored anodic materials with consideration to the applied treatment modes. For instance, the best values of treatment modes for the maximum value of the cathode weight gain $\Sigma\Delta_к$ are as follows: pulse rate $t_{p/r} = 25$ µs, charge impulse duration $t_p = 80$ µs at the frequency of $f = 500$ Hz using anodic material "D" (the second is material "C" with the same ratio of pulse rate to impulse duration: $t_{p/r}/t_p = 25/80$). On the parameter of maximum AL wear resistance ΣW_{AL}, the best variant is material "D" under all treatment modes under research. The best anodic material for AL formation γ_{ALF} is material C″, (material D″ is the second under the same treatment modes 25/80). Yet, in order to achieve the parameter of overall efficiency of ESA (γ_{eff}), that involves aggregate achievement of a series of constituents (individual parameters $\Sigma\Delta_C$, $\Sigma\Delta_a$, $C_{MTr(av)}$, γ_{ALF}), anodic material "C" should be chosen under all treatment modes under research; and anodic material "A" is only the third under the treatment modes $t_{p/r}/t_p = 40/50$.

5 Conclusions

1. Materialogy of surfaces is a vast field of urgent and unexplored research problems, including the issue of AL formation by ESA using aluminium matrix materials.
2. We have obtained experimental dependencies for anode weight gain, erosion resistance of anodic materials, mass transfer coefficient, AL wear resistance and their mathematical expressions with the validity criterion no less than $R^2 > 0.9044$.
3. After the ESA with anodic AMCM, the wear resistance of steel 45 increases 2.4 times on the average; the highest values for wear resistance for all treatment modes under study were obtained for the anodic material Ni-Cr-Zr-Ti-Al (3.3-2.9 times).
4. We have obtained a series [of results] for anode weight gain $\Sigma\Delta_C$, anodic materials erosion resistance $\Sigma\Delta_a$, mass transfer coefficient $C_{MTr\ (av)}$, surface wear resistance after ESA ΣW_{AL}, AL formation efficiency γ_{ALF}, and the ESA efficiency γ_{eff}, which can be used as recommendations for achieving the desired parameters after ESA with AMCM.

References

1. Mansurov, Y.N., Reva, V.P., Mansurov, S.Y., Beloborodov, M.V.: Economic and social basis of material science development in the far east. Tsvet. Metally **11**, 88–93 (2016)
2. De Gennes, P.G.: Superconductivity of Metals and Alloys. CRC Press, Boca Raton (2018)
3. Slobodyan, B.S., Lyashenko, B.A., Malanchuk, N.I., Marchuk, V.E., Martynyak, R.M.: Modeling of contact interaction of periodically textured bodies with regard for frictional slip. J. Math. Sci. **215**(1), 110–120 (2016)
4. Peters, S.T. (ed.): Handbook of Composites. Springer, Heidelberg (2013)
5. Prusov, E.S., Panfilov, A.A.: Properties of cast aluminum-based composite alloys reinforced by endogenous and exogenous phases. Russian Metall. (Metally) **2011**(7), 670–674 (2011)
6. Shakhpazov, E.K., Zaitsev, A.I., Rodionova, I.G.: Current problems in the steel metallurgy and materiology. Metallurgist **53**(3–4), 187–195 (2009)
7. Ivanov, V.I., Burumkulov, F.K., Verkhoturov, A.D., Gordiyenko, P.S., Panin, Y.S., Konevtsov, L.A.: Formation of the surface layer on a low-carbon steel in electrospark treatment. Welding Int. **27**(11), 903–906 (2013)

Flexible Pricing System Implementation in International Passenger Traffic

Aleksandr Butyrkin(✉), Ekaterina Kulikova, Konstantin Yatskevich, and Elizaveta Dmitrieva

Russian University of Transport, Obraztsova Street 9, Moscow 127994, Russia
a_butyrkin@mail.ru

Abstract. Tariff policy in the field of railway transport in international traffic deserves special attention. Passenger carriers operate in conditions of fierce competition in the transport market and the ticket price is often the most important factor affecting the consumer's choice of mode of transport. University experts have developed an integrated model for assessing the elasticity of demand under the influence of various factors. The model was tested using the results of marketing research on specific passenger train routes in international traffic. Demand elasticity curves are constructed with decreasing and increasing fare in a passenger train on the Moscow-Chisinau route.

Keywords: Passenger transportation · Marketing research · Mathematical model · Pricing

1 Introduction

Pricing is an important element of strategy and tactics of running any business, in particular, the organization of passenger rail transportation. Tariff policy in international transportation deserves special attention. Due to the high degree of competition on the part of, primarily, air and road (buses) transport, railway authorities, that are actually not free in determining the price of a ticket, should be guided by market realities.

Currently, railway authorities do not have a unified approach to the formation of the final fare for a passenger. For example, the current documents of the Directorate of the Council on Rail Transport of the Member States of the Commonwealth do not provide for a procedure for implementing a pricing policy, which is advisable to be formed by the involved railway authorities for each traffic, taking into account the competitive environment, train route, purchasing power of the population, travel motivation, etc. [1].

The current documents of the Directorate of the Council for Rail Transport of the Member States of the Commonwealth do not provide for a pricing policy that should be formulated by the involved railway administrations for each service, taking into account the competitive environment, train route, paying capacity of the population, travel motivation, etc. In general, railway administrations do not currently have a unified approach to the formation of the final fare for a passenger [2].

Z. Popovic et al. (Eds.): TransSiberia 2019, AISC 1115, pp. 300–311, 2020.
https://doi.org/10.1007/978-3-030-37916-2_29

The Russian University of Transport (RUT (MIIT) conducted unique marketing research of the transport market in the field of passenger transportation in the Commonwealth of Independent States in 2018. The features and performance indicators of more than 50 passenger trains in international traffic in the CIS and Baltic countries were analyzed in detail, more than 20,000 questionnaires obtained as a result of a survey of passengers and processed. The obtained results lead to the following conclusion: the pricing factor influences individually on the choice of passenger transport mode for each route and direction of passenger trains, requires from the company-carrier point approach in the study of supply and demand balanced decisions flexibility in the formation of tariffs for travel in cars of different classes.

So, for example, in the process of conducting and processing the results of marketing research, the initial hypothesis put forward at the preparatory stage about the dominant influence of the price factor on the process of choosing a mode of transport by passengers was not fully confirmed. The integrated complex analysis of the factors influencing the change in passenger demand depending on the travel time, the railway administration and the route showed the following: for passengers with a travel time of up to 14 h, the key factor is a convenient schedule (16.9%), the second and third places are shared by the optimal travel time (14.3%) and comfortable travel conditions (12.5%), the fare only takes the fourth place (11.1% of passengers).

With an increase in the duration of the trip, the picture changes, and the time factor takes the first place: the share of passengers in trains with travel time of 24–26 h is 16.8% according to this indicator, in trains with travel time of more than 27 h - 15.4%. The influence of factors of schedule convenience and optimal travel time is predictably weakening [3].

In addition, it was noted that a set of factors and the degree of their influence on the elasticity of demand seriously depend on the railway administration for the formation of a passenger train. For example, only for trains formed in the Russian Federation and the Republic of Kazakhstan, the cost factor comes first for passengers, and this is despite the fact that the standard of living in the Russian Federation and the Republic of Kazakhstan (in terms of average wages) is ahead of the corresponding indicators of the Republic of Belarus and the Republic of Uzbekistan.

Obviously, the elasticity of demand under the influence of various factors should be evaluated for each passenger train route taking into account local conditions. Railway administrations (carrier companies) need an appropriate tool to make such calculations.

2 Methods

As a tool that enables railway administrations and carrier companies to put into practice a flexible system of tariff formation, university experts have developed an integrated model for assessing demand elasticity, taking into account various factors. The methodology was tested using the results of marketing research on specific passenger train routes in international traffic [4–6].

The information base for the theoretical construction and further practical use of the economic and mathematical model of the price elasticity of the reaction of consumers (passengers) to changes in ticket prices should be the results of a survey of passengers in the considered direction.

The following symbols will be applied:

X – the number of a passenger's trips on the route per year;
Y – the ticked price in car of certain class.

By the base value of the ticket price (or, more simply, the base price), we mean the actual price of the travel document at the time of the survey. Then, respectively, the base revenue is the product of the base ticket price for the number of trips on this route per year.

Thus, the base revenue B(C) on the route per year is:

$$B(C) = X(C) * Y(C), \;\; \text{rub} \tag{1}$$

with:

X(C) – being the number of trips with base price;
Y(C) – base ticket price.

The number of trips with base price can be described as:

$$X(C) = x1 * n1 + x2 * n2 + x3 * n3, \tag{2}$$

with:

x1 – percent of passengers, travelling more than 6 times per year;
x2 – percent of passengers, travelling 3–5 times per year;
x3 – percent of passengers, travelling with other regularity;

$$(x1 + x2 + x3 = 100\%), \tag{3}$$

n1 – average number of trips in among passengers travelling more than 6 times per year;
n2 – average number of trips in among passengers travelling 3–5 times per year;
n3 – average number of trips in among passengers travelling with other regularity.

The source of information for the values of x1, x2 and x3 is the corresponding indicators of passenger surveys for each route (for example, the answers of passengers to the question "Frequency of travel on this route").

When calculating the values, it is preferable (but not necessary) to carry out specific calculations per 100 or 1000 people.

The values n1, n2, n3 are accepted expertly. For example, the value 7 can reasonably be taken as n1, and the value 4 as n2.

2.1 Making a Curve of Elasticity with a Decrease in Price in Relation to the Base

Let us consider the relative change in revenue with decreasing fares. In this case, it is advisable to ask the passenger the following questions during the survey: "How will the passenger travel on the route change when the fare on this train is reduced by…%?" And "The amount of the discount, which could affect the increase in passenger travel on the railway?".

The number of trips X on a given route increases due to groups of passengers who have answered "I will travel more often" (f1, %) and "I will refuse to travel by other means of transport" (f2, %) with the corresponding option of reducing the price.

Then, in the general case, the number of trips with discount is received (with the specific indicator for 1000 people):

$$X(1,2,3) = 1000 * (1 + \Omega) * (x1 * (f1 + f2) * m1 + x1(1 - f1 - f2) * n1 + x2 * (f1 + f2) * m2 + x2 * (1 - f1 - f2) * n2 + x3 * (f1 + f2) * m3 + x3 * (1 - f1 - f2) * n3) \quad (4)$$

with:

f1, f2 (as a share) – values of changes in number of passenger trip in the corresponding price ranges (less than 10%; 10–20; over 30%);
m1, m2, m3 – average increased number of trips in a group;
Ω – share of passengers, who switched to rail transport from other transport modes as the result of potential fare reduction, share.

Relative revenue change in this case can be calculated as:

$$D(1,2,3) = B(1,2,3)/B(C), \quad (5)$$

or in a final way:

$$D(1,2,3) = X(1,2,3) * (1 - k)/X(C), \quad (6)$$

with k – discount percent/100.

As a result of the calculations, 3 points are obtained (in addition to the base) for the elasticity curve when the price decreases relatively to the base.

As can be seen from the obtained formula, the relative change in revenue does not depend on the initial base value of the price.

2.2 Making a Curve of Elasticity with an Increase in Price in Relation to the Base

Let us consider the relative change in revenue with increasing fares. An important difference from the previous case, namely the case of a decrease in the price relative to the base one, is a change (decrease) in the number of passengers who continue to use railway transport. The specific indicator (per 100 or 1000 people) must be multiplied by the sum (q1 + q2), where these indicators mean, respectively, the percentage of those who answered "I will use another type of transport" and "I will refuse to travel at all" in the required price range.

Then, similarly to the case of a decrease in price relatively to the base, in general case we get the value of the number of trips with an increase in price:

$$\begin{aligned} X(1,2,3) = 1000 * (1 - (q1 + q2))\,((x1 * f3 * m4 + x1 * (1 - f3) * n1 \\ + x2 * f3 * m5 + x2\,(1 - f3) * n2 + x3 * f3 * m6 + x3\,(1 - f3) * n3) \end{aligned} \quad (7)$$

with:

f3 – value «I will travel less often» in the relevant price range, share;
m4, m5, m6 – average reduced number of trips per group.

Relative revenue change can be calculated:

$$D(1,2,3) = B(1,2,3)/B(C), \quad (8)$$

or in a final way:

$$D(1,2,3) = X(1,2,3) * (1 + k)/X(C) \quad (9)$$

with k – percent of price increase/100.

As a result of the calculations, 3 points are obtained (in addition to the base) for the elasticity curve with an increase in price relative to the base.

As can be seen from the obtained formula, as in the previous case, the relative change in revenue does not depend on the initial base value of the price.

3 Results

3.1 Testing the Model for Calculating Price Elasticity. Determining the Amount of Change in Revenue D(C) on the Route Moscow-Chisinau Per Year

The average number of passengers traveling by train at a base price per year can be calculated as:

$$X(C) = x1 * n1 + x2 * n2 + x3 * n3 \quad (10)$$

The initial data for determining the amount of basic revenue was obtained from the results of marketing research, namely, a survey of passengers on the route (Table 1). The frequency of travel by passengers along the route is shown in Fig. 1.

Table 1. Initial data for determining the average number of passenger trips on the train No. 47/48 Moscow-Chisinau per year

Legend	Value
x_1	10%
x_2	36%
x_3	54%
n_1	7 times per year
n_2	4 times per year
n_3	1 times per year

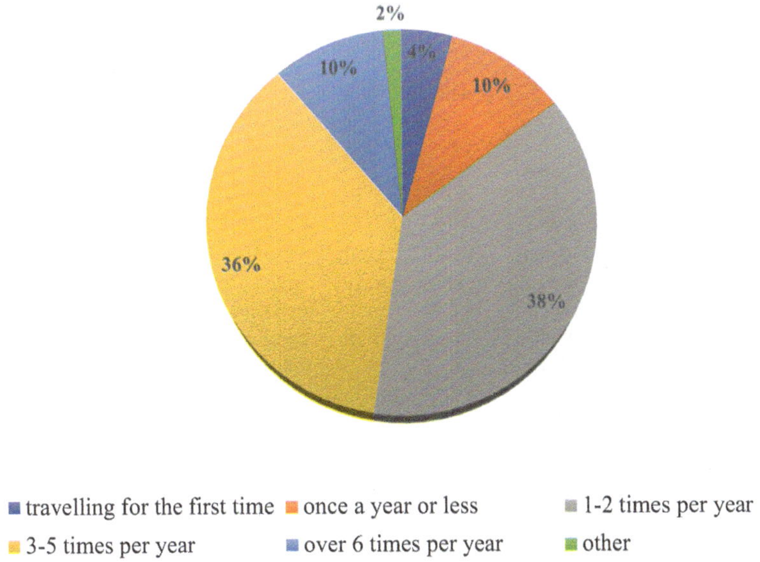

Fig. 1. Frequency of travel of passengers along the route

Then the average number of trips that passengers make by train on the route per year (with the base price option) is:

$$X(C) = 100 * 7 + 360 * 4 + 540 * 1 = 2680$$

3.2 Making a Curve of Elasticity with a Decrease in Price Relatively to the Base

The question "How will the frequency of travel by rail change when the fare is reduced?", which was asked to the train passengers during the survey, offered three answer options: "I will continue to travel at the same frequency", "I will travel more often" and "I will refuse to travel by other means of transport" for various discounts (up to 10%, 10–30%, over 30%). When making the elasticity curve of passenger demand for trips when changing the fare, only "price-sensitive" segments were taken into account - the number of trips along the train route can increase as a whole only due to groups of passengers who answered "I will travel more often" and "I will refuse to travel by others by means of transport" (Fig. 2) with the appropriate option to reduce prices.

The initial data for making the elasticity curve (the proportion of passengers "sensitive" to the corresponding change in fare and the average value of the number of passengers traveling over the base in each group) are listed in Table 2.

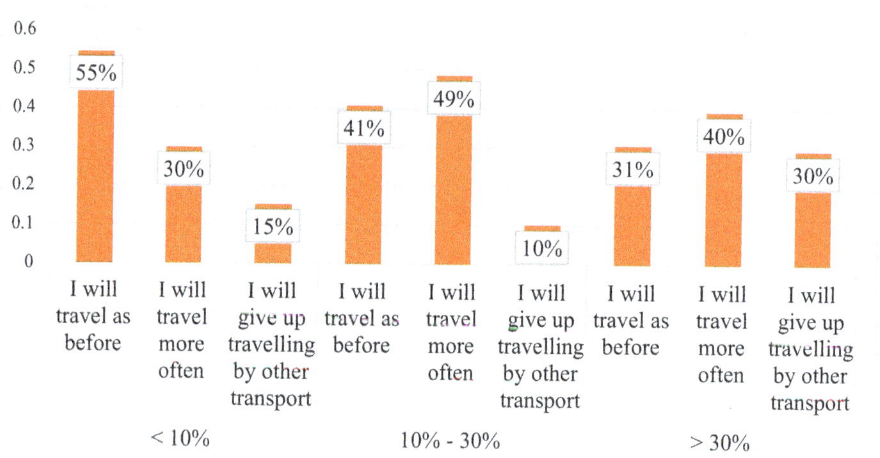

Fig. 2. Change in the frequency of travel of passengers along the route by train while reducing the cost of travel in this train

In this case the total number of trips per year that passengers can potentially take by train along the route in question, with a decrease in fare by 10%, will amount to:

$$X(1) = 1000 * (10\% * (0.3 + 0.15) * 8 + 10\% * (1 - 0.3 - 0.15) * 7 + 36\% * (0.3 + 0.15) * 5 + 36\% * (1 - 0.3 - 0.15) * 4 + 54\% * (0.3 + 0.15) * 1.3 + 54\% * (1 - 0.3 - 0.15) * 1) = 2959.9$$

The relative change in revenue to the base level:

$$D(1) = (2959.9 * (1 - 0.07))/2680 = 2.71\%.$$

The total number of trips per year that passengers can potentially take by train along the route in question, with a decrease in fare by 20%, will amount to:

$$X(2) = 1000 * (10\% * (0.49 + 0.1) * 8 + 10\% * (1 - 0.49 - 0.1) * 7 + 36\% * (0.49 + 0.1) * 5 + 36\% * (1 - 0.49 - 0.1) * 4 + 54\% * (0.49 + 0.1) * 1.3 + 54\% * (1 - 0.49 - 0.1) * 1) = 3078.84.$$

The relative change in revenue to the base level:

$$D(2) = (3078.84 * (1 - 0.2))/2680 = -8.09\%.$$

Table 2. Initial data for making an elasticity curve while reducing train fare

Legend	Value
Price reduction by 10%	
f_1	0.3
f_2	0.15
Price reduction by 20%	
f_1	0.49
f_2	0.1
Price reduction by 30%	
f_1	0.4
f_2	0.3
Average increase of trips per group	
m_1	8
m_2	5
m_3	1.3

The total number of trips per year that passengers can potentially take by train along the route in question, with a decrease in fare by 30%, will amount to:

$$X(3) = 1000 * (10\% * (0.4 + 0.3) * 8 + 10\% * (1 - 0.4 - 0.3) * 7 + 36\% * (0.4 + 0.3) * 5 + 36\% * (1 - 0.4 - 0.3) * 4 + 54\% * (0.4 + 0.3) * 1.3 + 54\% * (1 - 0.4 - 0.3) * 1) = 3153.2$$

The relative change in revenue to the base level:

$$D(3) = (3153.2\,(1 - 0.3))/2680 = -17.64\%.$$

3.3 Making a Curve of Elasticity with an Increase in Price Relatively to the Base

The question "How will the frequency of travel by rail change when the fare increases?", which was asked to the train passengers during the survey, offered six answer options: "I will continue to travel with the same frequency", "I will switch to another type of car", "I will switch to another train", "I will use another type of transport", "I will make trips less often" and "I will refuse to travel at all" for different amounts of tariff extra charges (up to 10%, 20–30%, over 40%). When making the elasticity curve of passenger demand for trips when changing the fare, only "price-sensitive" segments were taken into account - the number of trips along the train route can increase as a whole only due to groups of passengers who answered "I will use another mode of transport" and "I will refuse trips altogether" with the exception of the segment "I will make trips less often" (Fig. 3) with the corresponding option to reduce prices.

The initial data for making the elasticity curve (the share of passengers "sensitive" to the corresponding change in fare and the average value of the number of passengers traveling above/below the base in each group) are listed in Table 3.

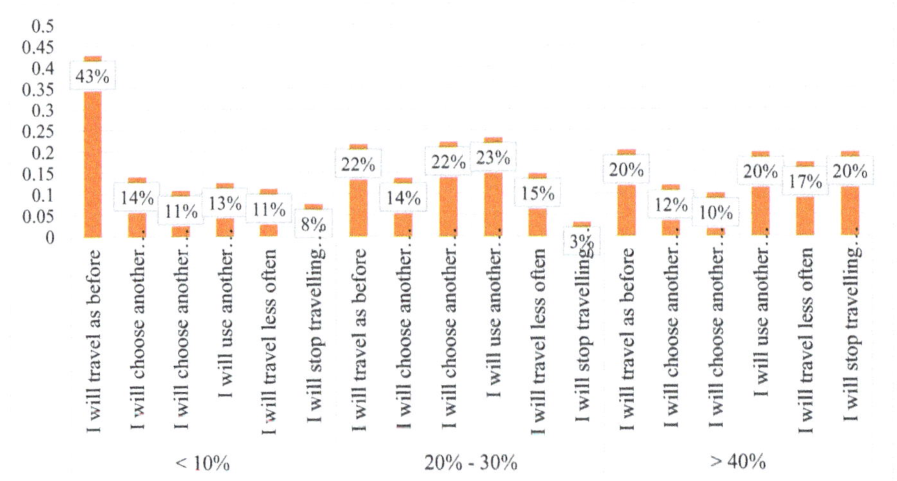

Fig. 3. Change in the frequency of travel of passengers along the route by train while increasing the cost of travel in this train

Table 3. Initial data for making an elasticity curve while increasing train fare

Legend	Value
Price increase by 10%	
q_1	0.13
q_2	0.08
f_3	0.11
Price increase by 20%	
q_1	0.23
q_2	0.03
f_3	0.15
Price increase by 40%	
q_1	0.2
q_2	0.2
f_3	0.17
Average decrease of trips per group	
m_4	6
m_5	3
m_6	0.7

Then the total number of trips per year that passengers can potentially take by train on the route in question, with an increase in fare by 10%, will be:

$$X(1) = (1000 * (1 - (0.13 + 0.08))) * ((10\% * (0.11) * 6 + 10\% * (1 - 0.11) * 7 + 36\% * (0.11) * 3 + 36\% * (1 - 0.11) * 4 + 54\% * (0.11) * 0.7 + 54\% * (1 - 0.11) * 1)) = 2063.15.$$

The relative change in revenue to the base level:

$$D(1) = (2\,063.15 * (1 + 0.1))/2680 = -15.32\%.$$

The total number of trips per year that passengers can potentially take by train on the route in question, with an increase in fare by 20%, will be:

$$X(2) = (1000 * (1 - (0.23 + 0.03))) * ((10\% * (0.15) * 6 + 10\% * (1 - 0.15) * 7 + 36\% * (0.15) * 3 + 36\% * (1 - 0.15) * 4 + 54\% * (0.15) * 0.7 + 54\% * (1 - 0.15) * 1)) = 1914.16.$$

The relative change in revenue to the base level:

$$D(2) = (1\,914.16\,(1 + 0.2))/2680 = -14.29\%.$$

The total number of trips per year that passengers can potentially take by train on the route in question, with an increase in fare by 40%, will be:

$$X(3) = (1000 * (1 - (0.2 + 0.2))) * ((10\% * (0.17) * 6 + 10\% * (1 - 0.17) * 7 + 36\% * (0.17) * 3 + 36\% * (1 - 0.17) * 4 + 54\% * (0.17) * 0.7 + 54\% * (1 - 0.17) * 1)) = 1\,544.56.$$

The relative change in revenue to the base level:

$$D(3) = (1\,544.56\,(1 + 0.4))/2680 = -19.31\%.$$

Based on the data obtained, a demand elasticity curve was made with increasing and decreasing prices for a train on a particular route (Fig. 4).

In addition to the graph of the elasticity curve which was made without taking into account the flow of passengers from other modes of transport ($\Omega = 0$), an additional graph was constructed, based on expert estimates, taking into account the flow of passengers under the condition $\Omega \neq 0$. Ω is expertly assumed to be $0.3 * (q1 + q2)$.

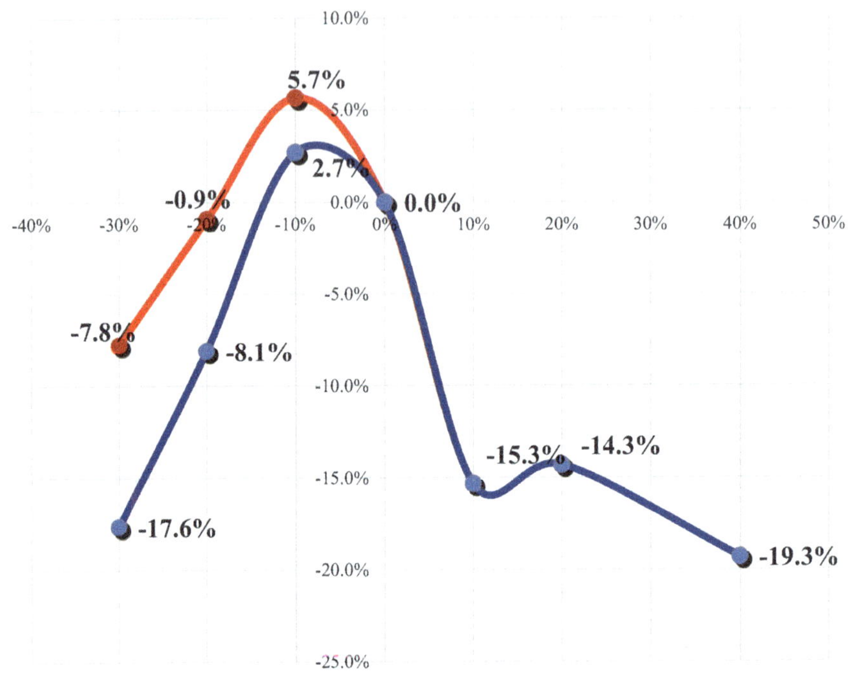

Legend:

- elasticity curve without Ω;

- - elasticity curve with expertly assumed Ω,

Fig. 4. Curve of elasticity of demand with a decrease and increase in the fare on the train on the route

4 Discussion

As noted in a number of original foreign research on pricing in the transport industry, when setting the optimal ticket price, it is advisable to use a modified indicator of demand elasticity of price (in addition to or separately from the traditional indicator of the dependence of the sales volume of a product on the price used in basic industries). There are 3 basic values of this indicator:

1. Demand is elastic - when a ticket price is reduced by 1%, sales revenue increases by more than 1%.
2. Inelastic - when a ticket price is reduced by 1%, demand increases by less than 1%.
3. Single elasticity - with a decrease in the price of goods by 1%, demand increases by 1%.

The generally accepted recommendation is to reduce prices on routes with elastic demand and (subject to additional analysis) on routes with demand of single elasticity. In the general case, without taking into account additional factors, price reduction on routes with inelastic demand is impractical.

5 Conclusions

Based on the conducted marketing research, the results of which, as noted above, do not take into account the flow of passengers from other modes of transport and the constructed financial and economic model of elasticity, we can conclude that demand on the considered route is inelastic. The current price is optimal and its reduction will not lead to an adequate increase in revenue. Moreover, this conclusion for the route in question remains valid even if a corrective assumption is introduced into the financial model that there is an overflow of passengers from air and bus services on the corresponding routes.

References

1. Cross, R.G.: Revenue Management: Hard-Core Tactics for Market Domination. Crown Business, New York (2011)
2. Cox, T., Houdmont, J., Griffiths, A.: Rail passenger crowding, stress, health and safety in Britain. Transp. Res. Part A: Policy Pract. **40**(3), 244–258 (2006)
3. Morfoulaki, M., Tyrinopoulos, Y., Aifadopoulou, G.: Estimation of satisfied customers in public transport systems: a new methodological approach. J. Transp. Res. Forum **46**(1) (2010)
4. Zitrický, V., Gašparík, J., Pečený, L.: The methodology of rating quality standards in the regional passenger transport. Transp. Probl. **10** (2015)
5. Chistik, O.F., Nosov, V.V., Tsypin, A.P., Ivanov, O.B., Permjakova, T.V.: Research indicators of railway transport activity in time series. J. Econ. Manag. Perspect. **10**(3), 57–65 (2016)
6. Chiambaretto, P., Fernandez, A.S.: The evolution of coopetitive and collaborative alliances in an alliance portfolio: the air France case. Ind. Market. Manag. **57**, 75–85 (2016)

Improving the Accuracy of Determining the In-cylinder Pressure of a Diesel Engine When Measured Through an Indicator Channel

Aleksei Konkov[1] and Anton Trunov[2(✉)]

[1] Far Eastern State Transport University, Serysheva Street 47, Khabarovsk 680021, Russia

[2] Pacific National University, Tihookeanskaya Street 136, Khabarovsk 680042, Russia

a_n_t_r_u_n@mail.ru

Abstract. The pressure in the cylinder of a piston engine reflects information about the engine's operating processes to the fullest, therefore, the in-cylinder pressure diagram is widely used in engine diagnostic systems. Under operating conditions the pressure in the engine cylinder is measured through the indicator channel, which introduces additional inaccuracies in the measurement results, distorting the signal waveform. These distortions do not allow to accurately determine such diagnostic parameters as the peak firing pressure, phase positions of representative points on the in-cylinder pressure diagram, and fuel burnup parameters. The paper proposes a method for obtaining a reliable indicator diagram based on the results of measuring the in-cylinder pressure through the indicator channel using the method of identifying the parameters of the heat release model. The report presents the results of an experimental testing of proposed methods, covering the results of comparing the reconstructed diagrams of internal cylinder pressure measured through the indicator channel with the results of direct pressure measurement in the engine cylinder.

Keywords: Piston internal combustion engine · In-cylinder pressure · Indicator diagram · Indicator channel · Heat release · Identification

1 Introduction

The indicator diagram represents the correlation between the pressure in the cylinder of the piston internal combustion engine (ICE) and the crankshaft rotation angle (CRA) or other parameter that is uniquely associated with the phase of the cycle. It reflects information about the processes occurring in the engine cylinder to the fullest extent possible and, therefore, is widely used in engine diagnostic and setting systems [1–7]. The reliability of the results obtained using such systems directly depends on the accuracy of the measurement of the indicator diagram [6–10], which, in turn, depends on many factors [11, 12]. Under operating conditions pressure measurement in order to assess the technical condition of the engine is carried out through the indicator channel

Z. Popovic et al. (Eds.): TransSiberia 2019, AISC 1115, pp. 312–320, 2020.
https://doi.org/10.1007/978-3-030-37916-2_30

- the channel connecting the working propulsion chamber with a pressure sensor. It is well known that the indicator channel has a significant impact on the indicator diagram [11]. The study of these processes and ways to eliminate their influence on the indicator diagram has been the subject of many papers [8–10, 13, 14]. Additional errors in measuring pressure through the indicator channel can be figuratively divided into phase ones and amplitude ones. Phase errors are explained by the disturbance finite velocity along the indicator channel, and the amplitude errors are due to interference of pressure waves and gas throttling in the indicator channel.

The analysis of literary sources showed that there are several approaches to solving the problem of distortions in the indicator channel.

The first and most obvious approach is trying to get rid of the indicator channel completely, or at least reduce its influence. In connection with the above, we should note the work of Nagao [13], which offers recommendations on choosing the geometry of the indicator channel depending on the rotational speed of the crankshaft and the pressure rise rate that are based on the analysis of the system of equations of gas dynamics describing the gas flow in the indicator channel. Obviously, this approach is relevant at the stage of ICE development and cannot always be applied during operation.

The second approach is based upon the correction not of the indicator diagram itself, but of its individual parameters (integral or instantaneous) obtained during its processing. Such an approach, for instance, was implemented by Brown [11], who proposed introducing a correction to the average pressure, determined by the indicator diagram being measured through the indicator channel. Several semi-empirical dependences similar in appearance that allow to estimate errors in phase and amplitude of the maximum combustion pressure and some other parameters of the indicator diagram are presented in following papers [9]. The definition of these corrections is based on the use of simple ratios of acoustics and gas dynamics that do not take into account the geometric characteristics of the indicator channel and the effect of the gas temperature on its physical properties.

The third approach involves the implementation of corrective input on the results of measurement of the indicator diagram through the indicator channel. Oezatay et al. [14] propose to restore the indicator diagram using methods of frequency processing of the pressure signal, determining the transfer-function of the indicator channel in the laboratory in advance. This method does not take into account engine wear and changes in the geometry of the indicator channel. It is also not clear how it will work when changing measuring equipment.

In papers [9, 10], it was proposed to determine the errors of pressure measurement through the indicator channel by mathematical modeling of dynamic pneumatic processes in the channel. This is a fairly time-consuming method requiring highly skilled implementers and careful consideration of many factors. In addition, the authors do not provide experimental data to evaluate the practical effectiveness of their solutions.

Acoustic resonance arising in the indicator channel leads to the appearance of peculiar oscillations in the combustion and expansion area, that can be eliminated by digital filtering [12] of the pressure signal. However, the use of filtering requires caution, as the gas oscillation frequencies in the indicator channel are closely adjacent to the carrier frequencies of the indicator diagram, and their removal from the spectrum leads to distortions of the useful signal [11, 14].

In general, the performed analysis showed that a study aimed at improving the accuracy of determining the indicator diagram when measuring pressure through the indicator channel under operating conditions remains a relevant objective.

2 Methods and Experimental Setup

This paper proposes a method for making the diesel engine indicator diagram by parametric identification of the working process based on the results of measuring the internal cylinder pressure through the indicator channel. Block diagram of the identification process is displayed in Fig. 1.

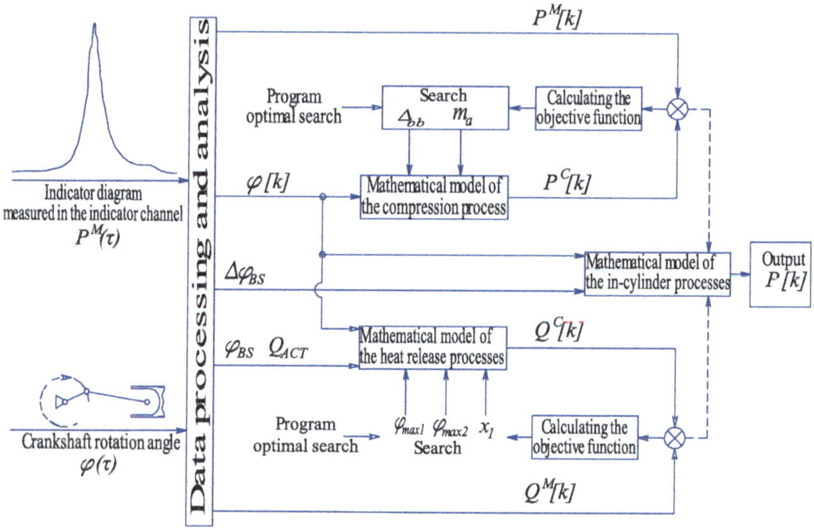

Fig. 1. Block diagram of the identification of the diesel engine internal cylinder process. φ – crankshaft rotation angle, τ – time, P – pressure, Q – integral characteristic of active heat release, index M – parameters obtained in course of measuring process through the indicator channel, index C – parameters obtained in course of mathematical modeling, φ_{BS} – visible combustion onset angle, φ_B – duration of heat release, $\Delta\varphi_{BS}$ – phase error of the measurement of the indicator diagram, Q_{ACT} – amount of heat used in the cycle, Δ_{bb} – estimate area of cylinder leaks, m_a - estimate air mass in the cylinder, x_1, φ_{max1}, φ_{max2} – estimates of the coefficients in the heat release model

The implementation of the proposed method requires: an indicator diagram of the engine cycle synchronized with the position of the crankshaft; knowing the design parameters of the engine needed to calculate the internal cylinder processes. For the proposed method to work correctly the signals when creating the indicator diagram should be quantized in such a way that the step along the crankshaft rotation angle $\Delta\varphi$ would not exceed 0.1°. The minimum required sampling frequency ν is related to the crankshaft rotational speed n by ratio $\nu = 60$ n.

The indicator diagram, measured through the indicator channel, was subjected to processing and analysis. During data processing, the indicator diagram was cleared from high-frequency noise [12] and the experimental characteristic of active heat release was calculated. The characteristic of active heat release was determined on the basis of the first law of thermodynamics by the ratio

$$Q(\varphi) = \int\limits_{\varphi_{BS}}^{\varphi_{BS}+\varphi_B} \frac{k}{k-1} P \frac{dV}{d\varphi} + \frac{1}{k-1} V \frac{dP}{d\varphi} \tag{1}$$

where $k = C_p/C_v$ – is the heat capacity ration, V – cylinder volume, φ_B – combustion period.

The analysis of the indicator diagram determined the parameters of the combustion process and the magnitude of the phase displacement due to the finite velocity of pressure wave emission. Visible combustion onset angle φ_{BS} and the amount of heat that went towards performing mechanical work and increasing the internal energy of the working medium Q_{ACT} were determined by the integrated heat release characteristic of Fig. 2a.

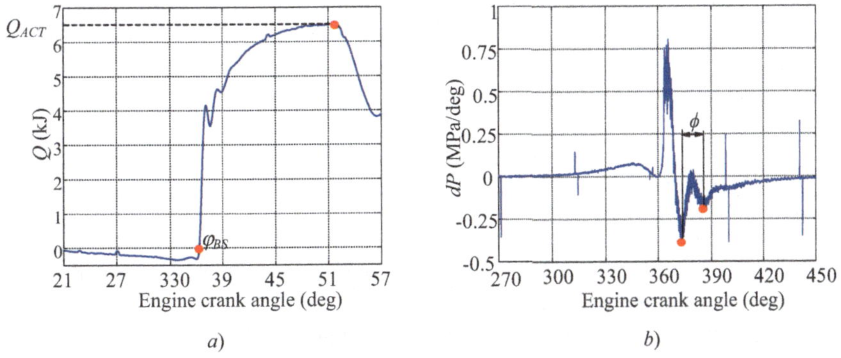

Fig. 2. For the analysis of the measured indicator diagram. a - integral characteristic of active heat release; b - rate of change of internal cylinder pressure

If the heat release characteristic has deformations (oscillations) throughout the entire combustion process, then Q_{ACT} is calculated as the average value between the extrema in the final combustion area. The signal delay in the indicator channel $\Delta\varphi_{BS}$ is determined as follows

$$\Delta\varphi_{BS} = \frac{\phi}{4} \tag{2}$$

where ϕ – period of oscillation (Fig. 2b).

Full-scale and calculation studies of the processes in the indicator channel of a medium-speed diesel engine of 18/22 dimensions showed that the delay of the pressure signal in the compression area of the indicator channel can be neglected. Therefore, this paper proposes to introduce a phase allowance only at the combustion area.

The main idea of the proposed method for identifying the indicator diagram of a diesel engine involves searching for such values of variable parameters for which the results of the data measured and obtained during mathematical modeling are closest to each other (ideally, they coincide). The values of the five parameters were evaluated during the identification process. It is known that the solution of such problems with a large number of variable parameters is fraught with serious difficulties; therefore, the identification of the work process is carried out in two stages.

First stage was dedicated to addressing the compression process. An optimal search for a solution was performed for two parameters: leakage rate through cylinder leaks Δ_{bb} and the mass of charge in the cylinder m_a at the beginning of the compression process. The proximity of the measured and calculated indicator diagrams was estimated based on the maximum pressure deviations in the compression area. The largest absolute pressure difference between the measured $P^M [k]$ and the calculated $P^C [k]$ of the indicator diagram in the compression area was used as the objective function ε.

$$\varepsilon\{P^M, P^C; m_a, \Delta_{bb}\} = \max\left|P^M[k] - P^C[k; m_a, \Delta_{bb}]\right| \tag{3}$$

Second stage was dedicated to addressing the combustion process. An optimal search for the coefficients in a semi-empirical diesel heat release model developed by Pugachev was carried out. The proximity of the experimental $Q^M[k]$ and the calculated $Q^C[k]$ heat release characteristics was estimated using the least-squares method. The objective function ε at this stage was determined as follows

$$\varepsilon\{Q^M, Q^C; x_1, \varphi_{\max 1}, \varphi_{\max 2}\} = \sum_{k=\Delta\varphi}^{N} \left(Q^M[k] - Q^C[k; x_1, \varphi_{\max 1}, \varphi_{\max 2}]\right)^2 \tag{4}$$

The search for a solution at the identification stages shall be performed until an improvement in the quality of the solution is observed.

Indicator diagram modeling was performed using a mathematical model described by Benson and Whitehouse [15].

The semi-empirical Pugachev heat release model used describes a single combustion process with two phases identified: kinetic and diffusion

$$x(\varphi) = \int_0^{\varphi_B} \left[\frac{x_1}{\varphi^2_{\max 1}}\varphi \cdot \exp\left(-\frac{\varphi_B^2}{2\varphi^2_{\max 1}}\right) + \frac{x_2}{2\varphi^2_{\max 2}}\varphi \cdot \exp\left(-\frac{\varphi_B^2}{2\varphi^2_{\max 2}}\right)\right] d\varphi \tag{5}$$

where $x(\varphi)$ – is the relative heat release characteristic, x_1 – portion of heat released in the kinetic phase of combustion; x_2 – portion of heat released in the diffusion phase of combustion; φ_B – combustion period; φ_{max1}, φ_{max2} – crankshaft rotation angle value, reached from the moment the combustion started to achieving maximum heat release $(dx_1/dt)_{max}$, $(dx_2/dt)_{max}$.

The calculated characteristic of active heat release is determined by the following ratio

$$Q^C(\varphi) = x(\varphi) \cdot Q_{ACT} \tag{6}$$

In order to simplify the search for a solution at the second stage, it is necessary to reduce the number of determined model coefficients in the process of identification of combustion. The combustion period φ_B does not affect the course of the heat release process and is assumed to be equal to the crankshaft rotation angle value, reached from the onset of the visible combustion to the opening of the exhaust valve. When calculating the relative heat release characteristics, we can assume that $x_1 + x_2 = 1$. Then, setting the portion of heat released in the kinetic phase x_1, we can calculate the portion of heat released in the phase of diffusion combustion. Thus, it is necessary to find three coefficients of the heat release model x_1, φ_{max1}, φ_{max2} during the identification of the combustion process.

In order to account for the phase error introduced by the indicator channel, it is proposed that the heat input in the cycle when simulating combustion processes be started earlier by $\Delta\varphi_{BS}$, determined in the process of analysing the measured indicator diagram according to (1).

The described technique was experimentally tested on a single-cylinder compartment of a medium-speed diesel engine OChN18/22; the main engine parameters are reflected in Table 1.

Table 1. Main parameters of the engine

Crankshaft speed at maximum power, min^{-1}	750
Stroke	Four-stroke
Cylinder diameter, mm	180
Piston stroke, mm	220
Compression ratio	12
Ratio of the crank radius to the length of the connecting rod	11/60
Fuel supply equipment	split type, direct fuel injection, closed-type fuel injector with hydraulic locking
Type of combustion chamber	Hesselman
Indicator channel	length 180 mm, diameter 8 mm

The measurement of the internal cylinder pressure was carried out by cooled piezoelectric sensors of RFT design. Pressure sensors were installed in the combustion chamber of the engine and on the indicator channel. The Honeywell GT1 Hall Sensor was used as a top dead center marker. Data from the primary transducers was recorded synchronously with a sampling frequency of 12 μs.

The error in recovery of the indicator diagram by the proposed method was estimated by the following ratio

$$\Delta = \max\left|\frac{P_{cyl}(\varphi) - P(\varphi)}{P_{cyl}(\varphi)}\right| \cdot 100\% \tag{7}$$

where P_{cyl} – pressure measured in the combustion chamber.

3 Results and Discussion

A comparison of the data obtained from the measurement results and the identification results by the proposed method is shown in Fig. 3. Figure 3a shows indicator diagrams made using various methods. Two of them were made by measuring the pressure in the cylinder and the indicator channel. The third was made in course of identification by the proposed method. It can be seen that significant discrepancies between measurements appear at the onset of combustion, as we expected. When measuring through the indicator channel, the value of the maximum combustion pressure was underestimated, and the combustion-expansion area shows oscillations that impede the analysis of the indicator diagram, which is consistent with other studies.

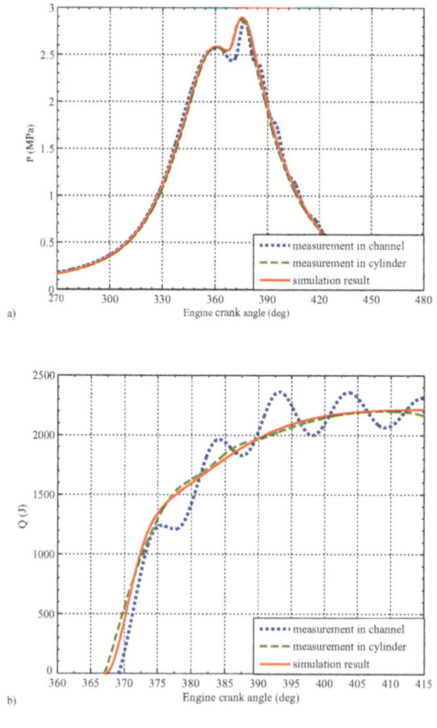

Fig. 3. Results of the identification of the diesel engine internal cylinder process. *a* – pressure; *b* – heat release characteristic

The error in measuring pressure through the indicator channel amounted to 10.8%. Figure 3b shows heat release characteristics that came out during the processing of indicator diagrams according to (2). There is a phase delay between the heat release characteristic in the cylinder and the indicator channel. It can also seen that the oscillations of the curve occur relative to the heat release characteristic that only describes the process in the cylinder with a phase delay. Therefore, the processes in the indicator channel do not affect the heat release processes in the cylinder. The heat release curve made during the identification of the combustion process coincides with the curve made during the processing of the indicator diagram measured in the engine cylinder. This means that it can be used in modeling a real indicator diagram. The maximum relative error in determining the internal cylinder pressure in the compression-combustion-expansion area by the proposed method was 2.53%.

Thus, the proposed method can be used to restore the indicator diagrams of low-speed and medium-speed internal combustion engines, measured through the indicator channel. Further research may be aimed at testing the proposed methodology for its application to high-speed internal combustion engines. It is necessary to further study the pressure delays in the compression area in the channel with a different geometry and at high rates of pressure change.

4 Conclusions

Based on the results of the work, the following conclusions can be made.

1. A technique has been proposed for increasing the accuracy of measuring the internal cylinder pressure by compensating for distortions introduced by the indicator channel based on the parametric identification of compression and heat release processes.
2. It has been established that the phase delay in the propagation of a pressure wave in the indicator channel can be quantified by the results of the analysis of the oscillations of the first-order derivative of pressure excited in the indicator channel during combustion.
3. An experimental test performed on the OChN18/22 diesel engine compartment showed that when using the developed method, the error in determining the pressure decreases from 10.8% to 2.5%.

References

1. Mohamed, E.S.: Performance analysis and condition monitoring of ICE piston-ring based on combustion and thermal characteristics. Appl. Therm. Eng. **132**, 824–840 (2018). https://doi.org/10.1016/j.applthermaleng.2017.12.111
2. Hountalas, D., Antonopoulos, A., Zovanos, G., Papagiannakis, R.: Evaluation of a new diagnostic technique to detect and account for load variation during cylinder pressure measurement of large-scale four-stroke diesel engines. SAE Technical Paper 2012-01-1342 (2012). https://doi.org/10.4271/2012-01-1342

3. Lamaris, V., Antonopoulos, A., Hountalas, D.: Evaluation of an advanced diagnostic technique for the determination of diesel engine condition and tuning based on laboratory measurements. SAE Technical Paper 2010-01-0154 (2010). doi.org/https://doi.org/10.4271/2010-01-0154
4. Corti, E., Moro, D., Solieri, L.: Real-Time evaluation of IMEP and ROHR-related parameters. SAE Technical Paper 2007-24-0068 (2007). https://doi.org/10.4271/2007-24-0068
5. Neumann, S., Varbanets, R., Kyrylash, O., Maulevych, V., Yeryganov, O.: Marine diesels working cycle monitoring on the base of IMES GmbH pressure sensors data. Diagnostyka **20**(2), 21–26 (2019). https://doi.org/10.29354/diag/104516
6. Hountalasa, D.T., Papagiannakisb, R.G., Zovanosa, G., Antonopoulosa, A.: Comparative evaluation of various methodologies to account for the effect of load variation during cylinder pressure measurement of large scale two-stroke diesel engines. Appl. Energy **113**, 1027–1042 (2014). https://doi.org/10.1016/j.apenergy.2013.08.036
7. Abart, M., Bfbeluk, M.: Method for assessing the state of an internal combustion engine WO2018027251 (2018). https://patentscope.wipo.int/search/ru/detail.jsf?docId=WO2018027251&_cid=P12-K1Q814-28683-1
8. Pawletko, R., Polanowski, S.: Influence of gas channels of medium speed marine engines on the accuracy of determination of diagnostic parameters based on the indicator diagrams. J. Polish CIMAC: Diagnosis Reliab. Saf. **7**(2) (2012). http://www.polishcimeeac.pl/Papers2/2012/016.pdf
9. Polanowski, S., Pawletko, R., Witkowski, K.: Influence of pressure sensor location on the quality of thermodynamic parameters calculated from the marine engine indicator diagram. Combust. Engines **154**(3), 319–323 (2013). http://www.combustion-engines.eu/en/numbers/11/313
10. Hountalas, D.T., Anestis A.: Effect of pressure transducer position on measured cylinder pressure diagram of high speed diesel engines. Energy Conv. Manag. **39**, 589–607 (1998). https://doi.org/10.1016/s0196-8904(97)10009-7
11. Brown, W.: Methods for evaluating requirements and errors in cylinder pressure measurement. SAE Technical Paper 670008 (1967). https://doi.org/10.4271/670008
12. Payri, F., Olmeda, P., Guardiola, C., Martín, J.: Adaptive determination of cut-off frequencies for filtering the in-cylinder pressure in diesel engines combustion analysis. Appl. Thermal Eng. **31**, 2869–2876 (2011). https://doi.org/10.1016/j.applthermaleng.2011.05.012
13. Nagao, F., Ikegami, M.: Errors of an indicator due to a connecting passage. Bull. JSME **8** (29), 98–108 (1965). https://doi.org/10.1299/jsme1958.8.98
14. Oezatay, E., Voegelin, P., Obrecht, P., Onder, C.H., Fuchs, P., Fuchs, W.: Malfunction diagnosis at marine diesel engines based on indicator cock pressure data. In: 26th Congres International des Machines a Combustion, Bergen, pp. 1–22 (2010)
15. Benson, R.S., Whitehouse, N.D.: Internal Combustion Engines. Pergamon, Oxford (1979)

Increasing the Reliability of Reversible Converters of AC Electric Locomotives

Stanislav Vlasevskii(✉), Liliya Buzmakova, and Aleksei Blazhnov

Far Eastern State Transport University, Seryshev Street, 47, Khabarovsk 680021, Russia
vlas@festu.khv.ru

Abstract. The aim of the work is to increase the operability of AC electric locomotives during emergency operation of reversible converters. The article talks about the benefits of electrifying railways with alternating current compared to direct current. The main historical stages of the development of electric locomotive converters are considered. The role of the reversible converter in the operation of AC electric locomotives is indicated. In this connection, the question of diagnosing and improving the efficiency of the reversible converter is raised. For analysis, a VIP-4000M reversible converter installed on an electric locomotive VL80R was taken. A typical method for controlling the branches of the converter in various control zones is described. As a result, the principle of controlling the branches of the converter in the event of an emergency mode is proposed. An example of the application of the principle in the construction of the converter control algorithm in emergency mode is given. Conclusions are drawn about the possibility of applying the principle to maintain the process of regulating voltage and rectifying current in emergency mode in order to increase the efficiency of an electric locomotive.

Keywords: AC electric locomotive · Reversible converter · Emergency mode · Branch failure · Branch breakdown

1 Introduction

The use of power electronic converters on electric locomotives is due to the choice of an alternating current system instead of direct current for electrification of railways [1–3].

DC traction has its advantages: the voltage drop is determined only by the active resistance of the network, a slight effect on the communication line, the symmetrical load of the three-phase primary energy supply system. But there are also disadvantages: a large number of traction substations per unit length, high cost of building a contact network, high costs of non-ferrous metal, converters are required at traction substations.

Success in electrification by alternating current was determined by the use of converters on the electric rolling stock of railways. Despite the disadvantages of electrification with alternating current: distortion of the shape of the consumed current

Z. Popovic et al. (Eds.): TransSiberia 2019, AISC 1115, pp. 321–328, 2020.
https://doi.org/10.1007/978-3-030-37916-2_31

when using static converters, the asymmetry of the phases in the power supply system. The idea of installing converters on electrically rolling stock had its advantages: the simplicity of traction substations, the possibility of increasing the voltage of traction substations to 20–25 kV and increasing the distance between them, reducing the cost of building substations, the ability to supply non-traction power consumers from contact network.

Thus, electrification of railways with alternating current turned out to be more profitable than direct current: capital costs are lower by 15–18%, total losses on train traction are lower by 5–6%, and the number of traction substations is 2–3 times less. As a result, the cost of transporting with alternating current is 20% lower than with direct current.

The first sections electrified with alternating current appeared in some European countries in the first half of the 20th century; alternating current of a reduced frequency of 16 2/3 Hz, 15 kV was used to power the traction drive based on a single-phase collector electric drive. In other countries, alternating current with a frequency of 50 Hz, voltage of 25 kV was used to power the collector DC traction electric motors, machine converters were used as converters.

In the USSR in 1932, the Technical Commission analyzed several hundred variants of combinations of current and voltage systems for specific sections of railways. It was concluded that the most effective is a system consisting of the following objects: a single-phase alternating current source with a voltage of 20 kV and a frequency of 50 Hz, a rectifier, a direct current traction motor.

In 1938, the OR22-01, the first AC electric locomotive in the USSR, was built on which twelve-anode mercury rectifiers with mesh control and water cooling were installed. Tests of the electric locomotive OP22-01 ended with the outbreak of war in 1941, the results were positive. The famous scientist B.N. Tikhmenev made a great contribution to the creation and testing of this electric locomotive.

They returned to AC electric locomotives in the early 1950s in connection with the beginning of the fifth five-year plan, according to whose plans mass electrification of railways was provided.

In 1952, the design of a new AC electric locomotive with ignitrons began at the Novocherkassk Electric Locomotive Plant. In 1954, the Novocherkassk Electric Locomotive Plant produced the first 2 electric locomotives (Novocherkassk single-phase, later - VL61). Only eight ignitrons were installed on the electric locomotive, four ignitrons in each truck. A pair of ignitrons formed a full-wave rectifier with a midpoint.

In 1955–1956 The first section of the Moscow-Kursk-Donbass railway Ozherelye - Pavelets was electrified (25 kV, 50 Hz). Eight electric locomotives of the VL61 series were operated on it. As converters, metal anode ignitrons were used, which were switched on in pairs in parallel. It is worth noting that the reliability of ignitrons was low.

The operation of the VL61 electric locomotives was quite successful. Therefore, the People's Commissariat of Railways of the USSR decided to electrify on alternating current immediately the whole direction of the Trans-Siberian railway, namely: Mariinsk - Zima, 1222 km long.

In 1956, serial production of VL60 electric locomotives began at the Novocherkassk Electric Locomotive Plant. Two rectifier units were installed on electric locomotives, each of which had four ignitrons. Electric locomotives VL60 were produced by the plant for more than 5 years and became the main type of freight locomotive on electrified AC lines.

However, ignitrons had certain significant disadvantages:

- when the ignitron body is destroyed, there is a high risk of environmental contamination and poisoning of people and animals;
- devices were needed that monitored the ignitron and turned it off in emergency operation;
- ignitrons do not tolerate strong shocks, which are often in rail;
- ignitrons are sensitive to temperature conditions;
- losses in ignitrons were rather high in comparison with semiconductor devices.

In 1953, it became possible to obtain high-purity silicon and the formation of large silicon disks, which served as the impetus for the creation of power semiconductor electronic devices designed to work with high currents [4]. Based on this technology, a power uncontrolled semiconductor device - a diode - was created. In 1955, a controlled semiconductor device was created - a thyristor.

Silicon rectifier allow rectifier installations to operate in a wide temperature range, have an increased efficiency and require less heat dissipation. In addition, such converters are less sensitive to shaking.

In 1962, the VL62 electric locomotive was launched, on which semiconductor silicon devices - diodes were first used. A bridge circuit with four arms was used to rectify the current. Each arm consisted of three parallel branches, eight diodes connected in series in each branch.

Both for lines electrified with direct current and for lines electrified with alternating current, in addition to six-axle freight electric locomotives, more powerful eight-axle freight electric locomotives were needed. The development of the project of an eight-axle freight AC electric locomotive has begun.

In 1962 - 1963 Novocherkassk Electric Locomotive Plant produces eight-axis AC electric locomotives VL80 with ignitrons and an electric locomotive VL80k with silicon converters.

The advantages of using semiconductors were obvious: thyristors were significantly more compact compared to ignitrons; did not require liquid cooling; did not require heating after parking; were safer for humans.

In addition, before the advent of semiconductor converters on electric locomotives, less effective rheostatic braking was used. During rheostatic braking, the kinetic energy of the moving composition passes into the thermal energy released by special rheostats, due to this, braking occurs.

There was another method of braking - regenerative braking. During regenerative braking, the kinetic energy of a moving train passes into electrical energy and returns back to the contact network. Regenerative braking is based on the property of reversibility of electric machines - the engine can operate in generator mode (and vice versa).

In 1968, the Novocherkassk Electric Locomotive Plant manufactures an experimental electric locomotive with regenerative braking VL80-r, which will later become the prototype of modern domestic AC electric locomotives. On the electric locomotive, thyristor-based converters were used. Due to this, the voltage on the traction electric motors was regulated smoothly, which improved the traction properties of the electric locomotive. During regenerative braking, the current generated by traction motors was converted to alternating current and returned back to the network.

When introducing electric locomotives with semiconductor reversible converters, there were problems in their maintenance, which are based on the use of electronic equipment. For this reason, a new approach was needed in the diagnosis and maintenance of converters.

AC locomotives occupy a large place in the process of rail transportation on Russian roads, especially freight. Their reliable operation and efficiency determine the quality of the transportation process. As mentioned earlier, the reversible converter is an important component of an alternating current electric locomotive, which passes almost all the power consumed from the mains supply through itself. Any change in the technical condition of the reversible converter can lead to a deterioration or improvement in the operational reliability and performance of the entire electric locomotive.

For this reason, over the years, much attention has been paid to ensuring the reliable operation of reversible converters of AC electric locomotives. In particular - the technical diagnosis of reversible converters of AC electric locomotives.

Despite the fact that semiconductor converters of traction rolling stock are quite reliable, according to statistics, the share of failures of reversible converters accounts for up to 15% of failures of all electric equipment of an electric locomotive [5, 6].

A review of literature showed that a large number of works are currently devoted to the development of technical diagnostic methods and the development of technical solutions to increase the efficiency of reversible converters of electric rolling stock. This is most likely due to the following reasons:

- the severe operating conditions of the converters: alternating current and voltage loads, a wide range of temperature changes, mechanical loads, high humidity, dustiness;
- significant economic costs for the restoration of the inverter with an inactivity of electric locomotive;
- particular importance is the reliability of the converter in ensuring traffic safety.

The main malfunctions that occur in reversible converters can be divided into the following groups:

- malfunctions caused by thyristor failures;
- malfunctions caused by a failure of the pulse generation system;
- malfunctions caused by a failure of the control unit of the reversible converter.

All this can lead to a breakdown of voltage rectification, stopping the electric locomotive and economic losses. Therefore, to reduce the accident rate and the time of repair, it is necessary to timely detect the occurring malfunctions.

The authors of the article suggest using the system of working diagnostics of an electric locomotive in real time to track the occurring malfunctions. In the event of a malfunction, the on-board system changes the algorithm for controlling the branches of the reversible converter, reserving the faulty arm with another free serviceable branch for a given half-period of the zone.

2 Materials and Methods

The article discusses the operation of the VIP-4000M reversible converter installed on the VL80R electric locomotive in the current rectification mode in the IV control zone (Fig. 1).

In such a converter, a three-section bridge circuit with section voltages ΔU, ΔU, 2ΔU with four-zone regulation of the rectified voltage is adopted [7]. In traction mode, the converter operates as a controlled rectifier. In turn, the control system generates control pulses supplied to the branches VS1…VS8:

- α_0 – unregulated, supplied at the beginning of each half-cycle, with a phase α_0 corresponding to the minimum opening angle of the thyristors;
- α_{0d} – delayed relative to unregulated pulses by 300–400 μs, with phase α_{0d};
- α_c – phase controllable α_c.

Figure 1 shows: CR - current receiver; TT - traction transformer; a1, 1, 2, x1 - conclusions of the secondary winding of the TT; RC - reversible converter; VS1….VS8 - thyristor branches of reversible converter; SR - smoothing reactor; TE1, TE2 - traction engines; EW1, EW2- excitation windings TD.

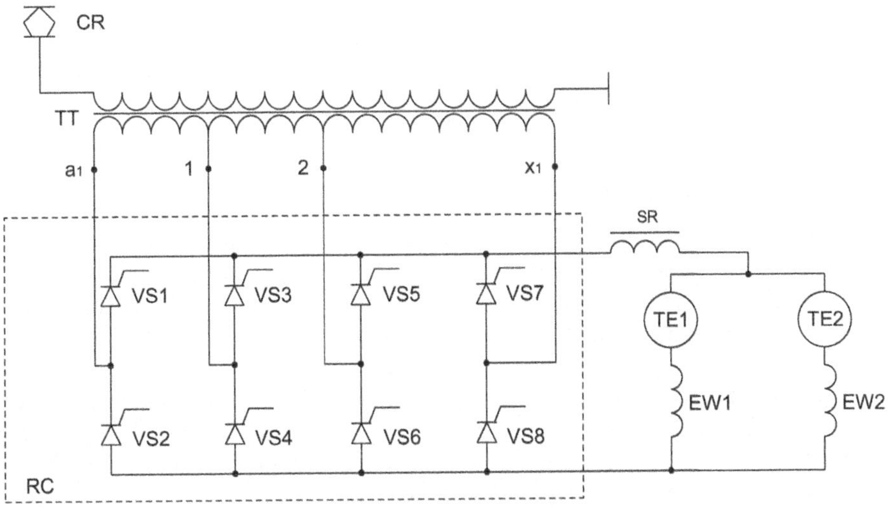

Fig. 1. A simplified power circuit of an AC electric locomotive VL80r as applied to one truck with two traction motors

In zone IV, the bridge formed by the branches VS1, VS2, VS3, VS4, VS7, VS8 is controlled. According to the control algorithm (see Table 1), pulses α_0 are applied to the thyristors of the branches VS7, VS8, α_{0d} to the thyristors of the branches VS3, and α_c to the thyristors of the branches VS1, VS4. Due to the change in the opening phase of the thyristors of the branches VS1 and VS2, the rectified voltage is regulated. Toward the end of the zone, the duration of the small bridge formed by the branches VS3, VS4, VS7, VS8 decreases, and the operating time of the large bridge formed by the branches VS1, VS2, VS7, VS8 increases. As a result, the rectified voltage varies from 3/4 to the rated voltage [8, 9].

In emergency operation due to non-acceptance of the load by the branch or breakdown of the branch, the authors of the article propose to increase the operability of the reversible converter by applying the new principles of controlling the converter branches in emergency mode by reserving the faulty converter branch with another branch that is free and not used during the half-period of the zone.

Table 1. The control algorithm of the reversible converter in traction mode

Zone	Half-cycle	Impulses of control on the branches of the reversible converter in traction mode							
		VS1	VS2	VS3	VS4	VS5	VS6	VS7	VS8
1	←			α_0			α_c		
	→				α_c	α_0, α_c			
2	←	α_c		α_{0d}			α_0		
	→		α_c		α_{0d}	α_0			
3	←			α_c		α_{0d}			α_0
	→				α_c		α_{0d}	α_0	
4	←	α_c		α_{0d}					α_0
	→		α_c		α_{0d}			α_0	

3 Results

The general principle of constructing a new control algorithm is as follows:

1. In the event of failure of the controllable branch of the converter (branch, the pulses of which are applied to the thyristors α_c) at a certain half-period of the zone, it is proposed that the control pulses not be applied to the thyristors of this branch, thus excluding the branch from the control algorithm for this half-period of the zone.
2. In case of failure of the uncontrollable branch of the converter (branch, to the thyristors of which pulses α_{0d} or pulses α_0 are supplied), it is proposed to use another branch (possibly a different combination of branch) instead of a faulted branch in this half-period of the zone, so that another bridge with a voltage lower than the voltage of the full zones, therefore, make a reservation.

Let us consider the application of these principles for constructing the algorithm for controlling the branches of a reversible converter in emergency modes using the example of zone IV.

According to the proposed principles, we will draw up a control algorithm for the reversible converter in the IV zone (see Table 2) in case of failure on one of the branches VS1 - VS6. In the event of a failure on the VS7 or VS8 branch, the algorithm is not applied. This is due to the fact that it is not possible to select such a combination of serviceable and idle thyristor branches so that the voltage is greater than ¾ of the rated voltage - the total voltage of zone III.

Table 2. A new control algorithm for the reversible converter in zone IV

Branch failure	Half-cycle	Impulses of control on the branches of the reversible converter in traction mode							
		VS1	VS2	VS3	VS4	VS5	VS6	VS7	VS8
VS1	←	**X**		α_{0d}					α_0
	→		α_c		α_{0d}			α_0	
VS2	←	α_c	**X**	α_{0d}					α_0
	→				α_{0d}			α_0	
VS3	←	α_c		**X**		α_{0d}			α_0
	→		α_c		α_{0d}			α_0	
VS4	←	α_c		α_{0d}	**X**				α_0
	→		α_c				α_{0d}	α_0	
VS5	←	α_c		α_{0d}		**X**			α_0
	→		α_c		α_{0d}			α_0	
VS6	←	α_c		α_{0d}			**X**		α_0
	→		α_c		α_{0d}			α_0	

4 Discussion

In the existing system, the control of the reversible converter of an AC electric locomotive on each half-cycle of the voltage regulation zone there are unused branches.

The principle of increasing operability proposed by the authors makes it possible in emergency mode to restore the operability of a converter by reserving a failed branch with another free branch in a given half-period of the zone.

The article gives an example of building a new algorithm according to the proposed principle on the example of zone IV. This allows in emergency mode in case of failure in the branch:

- VS1 or VS2 – to restore the efficiency of the reversible converter, with a decrease in the range of smooth regulation of the rectified voltage from 3/4 to 7/8 of the rated voltage;
- VS3 or VS4 – restore the efficiency of the reversible converter, without changing the range of smooth regulation of the rectified voltage.

It is worth noting that there is a circuit for a reversible converter with the inclusion of a diode discharge branch, which increases its performance. In this case, the faulty branch is backed up by the diode discharge branch, which increases the efficiency in emergency mode. Among the disadvantages of this solution is the need to upgrade the structural part of the converter [10, 11].

The proposed solution, changing the control algorithm of the reversible converter of the electric locomotive in emergency mode, can be implemented by changing the software of the control system of the electric locomotive converter. Application of a new control algorithm in emergency operation of the converter will allow to support the rectification process and maintain traction of the electric locomotive during movement.

5 Conclusions

Thus, the study showed that in emergency mode on each zone of the reversible converter there are unused branches. Such free branches can be used to reserve faulty branches of a reversible converter. The results obtained will help to further improve the reliability of an alternating current electric locomotive in general and a reversible converter in particular. In practice, the principles obtained can be implemented in software on-board automated system for working diagnostics of an AC electric locomotive.

References

1. Ushakov, V.Y.: From the history of electrical power engineering. In: Electrical Power Engineering, pp. 13–21 (2018)
2. Rakov, V.: Locomotives of domestic railways (1845–1955). Transport, Moscow (1995)
3. Rakov, V.: Locomotives of domestic railways (1956–1975)
4. Matsunami, H.: Development of power semiconductor silicon carbide technology-a breakthrough and epoch-making steps. In: 2015 15th International Workshop on Junction Technology (IWJT), pp. 11–14 (2015)
5. Buzmakova, L.V., Vlas'evskii, S.V., Ovseichik, S.Z.: A method of diagnosing reversible converters of alternating current electric locomotives. Russ. Electr. Eng. **87**(2), 80–83 (2016)
6. Domanov, K., Shatohin, A., Nezevak, V., Cheremisin, V.: Improving the technology of operating electric locomotives using electric power storage device. In: E3S Web of Conferences, vol. 110, p. 01033 (2019)
7. Florentsev, S.N., Polyukhovich, V.S.: Traction electric equipment set for industrial shunting locomotives. In: 2018 17th International Ural Conference on AC Electric Drives, pp. 1–8 (2018)
8. Bunyaeva, E.V., Skorik, V.G., Vlas'evskii, S.V., Fokin, D.S.: A method for improving the energy efficiency of an alternating current electric locomotive in the regenerative braking mode. Russ. Electr. Eng. **87**(2), 73–76 (2016)
9. Vlas'evskii, S.V., Mel'nichenko, O.V., Malysheva, O.A.: Emergency operation processes in the thyristor rectifier of an alternating current electric locomotive. Russ. Electr. Eng. **87**(2), 68–72 (2016)
10. Vlasevskii, S., Melnichenko, O., Babichuk, A.: Method for control of multizone rectifier of single-phase alternating current. Patent 2322749, Russian Federation (2006)
11. Barinov, I.A., Melnichenko, O.V.: Power IGBTs application in AC-Wire DC-Motor locomotive thyristor-based power circuit for regenerative brake energy efficiency increase. In: 2019 International Conference on Industrial Engineering, Applications and Manufacturing, pp. 1–5 (2019)

Diagnostic Criteria for the Signal of the First-Order Derivative of Diesel Engine in-Cylinder Pressure

Iurii Davydov, Irina Konkova, and Aleksei Konkov(✉)

Far Eastern State Transport University,
Serysheva Street 47, 680021 Khabarovsk, Russia
akonkov1964@gmail.com

Abstract. Sensors using the direct piezoelectric effect are most widely used in internal combustion engine (ICE) electronic indicating measurement systems. Usually, the output signal of the sensor (charge on the electrodes of the piezoelectric element) is converted by the charge amplifiers into voltage proportional to pressure. This paper examines an alternative method, which consists in analysing a signal obtained using a current-voltage converter, the output of which is proportional to the first-order derivative of the measured pressure. The paper proposes new dimensionless criteria defined in the analysis of such a signal. Criteria can be identified without linking the signal of pressure change rate to the phase of the cycle (crankshaft rotation angle, piston position, etc.). Due to the dimensionless properties of the criteria, their use in diagnostic systems does not require calibration of the primary transducers and the entire measuring channel. It has been theoretically and experimentally shown that, using the proposed criteria, it is possible to simply and reliably diagnose the wear of a cylinder-piston group and deviations in the adjustment of the fuel injection advance angle. The main results were obtained during mathematical modeling of the working process of a medium-speed diesel engine and verified experimentally for two types of engines.

Keywords: Diesel · Diagnostics · Diagnostic criteria · Piezoelectric sensor · In-cylinder pressure · Leakage · Fuel timing

1 Introduction

The pressure signal in the cylinder of a piston engine contains valuable information about the course of the working process, which is influenced by many factors, including the technical condition of the engine. Therefore, the dependence of the pressure in the cylinder on the crankshaft rotation angle or on the volume of the cylinder, referred to as the indicator diagram, is widespread in diagnostic systems and engine settings. First of all, this applies to low-speed and medium-speed diesel engines of high power: main and auxiliary marine engines [1–3], diesel generators [4–6]. As a rule, such engines are equipped with an indicator cork, through which a gas pressure wave in the cylinder is transmitted to the sensor membrane. Historically, even since the time of the inventor of the first indicator, James Watt, who patented at the beginning of

Z. Popovic et al. (Eds.): TransSiberia 2019, AISC 1115, pp. 329–339, 2020.
https://doi.org/10.1007/978-3-030-37916-2_32

the 9th century a method for assessing the efficiency of the steam engine work process, diagnostic information is extracted from the pressure signal. In addition to the above, a parameter obtained directly from the pressure signal (maximum combustion pressure, compression pressure, etc.), or during the analysis of the indicator diagram (mean indicated pressure, indicated work, etc.) can be considered the criterion for assessing the technical condition. In some cases, this analysis can take considerable time and is difficult to fully automate. For example, in method [6], the duration of the parametric identification step takes several minutes of machine time for each cylinder. At the same time, successful works, aimed at creating diesel control systems in which the signal from the pressure sensor is processed in order to extract diagnostic information during one cycle of a high-speed engine and later to introduce a corrective action in the next cycle [7], are also known. Of course, in order to achieve such rapid rates of response, its preferable to use methods where it is possible to implement fast, simple and reliable algorithms for obtaining diagnostic information.

The signal of the first-order derivative of pressure is a variant of the signal that meets the stated requirements. This signal can be obtained using piezoelectric pressure sensors, which are widely used in monitoring and diagnostics systems of internal combustion engines due to their ability to operate at temperatures of 800 °C and higher [8]. The output signal of such a sensor is a charge on the electrodes of the piezoelectric element, which is proportional to the force applied to the element and is usually converted in the electric circuits of the measuring device into a voltage proportional to the measured pressure. Another method for measuring alternating pressure is also known, which consists of using a transducer of a current flowing from the electrodes of the piezoelectric sensor into a voltage that will be proportional not to pressure, but to its first-order derivative, i.e. rate of change of pressure. Operating amplifiers (Fig. 1) can be used in both methods.

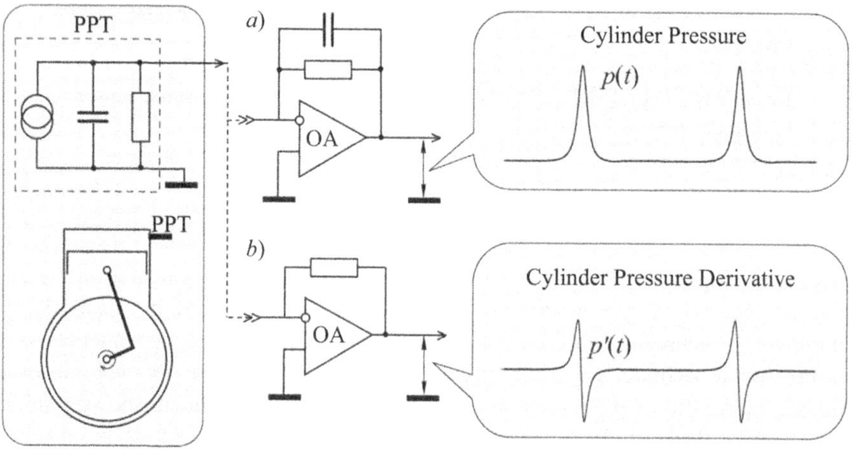

Fig. 1. Conversion of the output signal: (a) charge amplifier; (b) current-voltage converter; PPT – piezoelectric pressure transducer; OA - operational amplifier

Bueno et al. [9, 10] in their works show several advantages of using a current-voltage converter instead of a charge amplifier, in particular, to obtain the characteristics of heat release. A known disadvantage of such a converter is the loss of a direct component of the pressure signal. However, it can be easily compensated by applying the methods of thermodynamic correction of the indicator diagram, the main provisions of which can be found in [11–13].

This paper considers a new approach to the analysis of the curve of the rate of pressure change, the purpose of which was to find diagnostic features that would not require a signal to be attached to the position of the piston or crankshaft, on the one hand, and would be tolerant to a change in signal level, on the other hand The latter requirement is due to the fact that the output signal of the piezoelectric sensor can vary significantly due to changes in the temperature of the sensitive element and some other factors. Achieving these requirements will reduce the level of complexity of the measuring part of the diagnostic system.

2 Cylinder Sealing Criterion

As the cylinder-piston diesel engine group wears out, the leakage of the working fluid increases, which leads to a decrease in pressure and temperature at the final compression and becomes the reason for the difficulties in starting-up of the diesel engine and its unstable operation at idle and low power modes. The decrease in cylinder density can be determined, for example, by the value of the maximum pressure on the indicator diagram in the engine cylinder with the fuel supply in one cylinder or in the starter mode turned off. Obviously, a change in the signal of the piezoelectric sensor due to temperature drift can lead to erroneous conclusions. The solution to the problem, in our opinion, is to use the relative parameter kseal, which we call the cylinder sealing criterion.

As in the case of pressure signal analysis, when testing the engine, there is no fuel combustion in the tested cylinder. The pressure change rate curve $p'(\varphi)$ in this case takes the form shown by the red line in Fig. 2 with distinct maximum $p'_{\max}$ and minimum $p'_{\min}$.

Theoretically, in the absence of leaks and heat transfer, the absolute values of these extrema of the p' curve would be equal. Such a result will be achieved if the process of compression and expansion in the cylinder of a piston engine is considered isentropic. Gas leaks lead to the fact that a smaller amount of the working fluid is involved in the expansion process than it was in the compression process, meaning that the rate of pressure drop in the expansion section is greater (in absolute value) than the rate of increase in pressure in the compression section. The indicated regularity can be taken into account as follows.

$$k_{seal} = \frac{|dp_{\min}| - dp_{\max}}{dp_{\max}} = \frac{\Delta p'_{ex}}{dp_{\max}} \tag{1}$$

where $\Delta p'_{ex}$ – amplitude difference of extremum points.

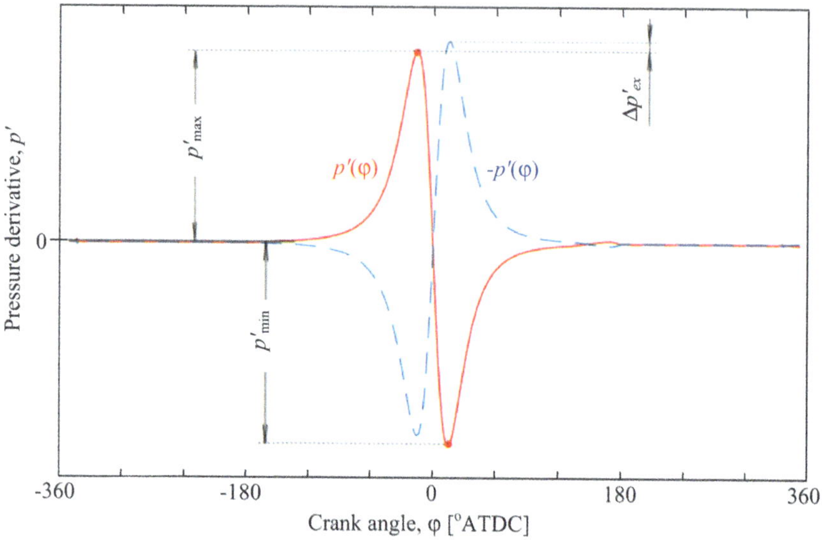

Fig. 2. Scheme for determining the cylinder sealing criterion

Calculated analysis of the effect of leaks on the value of the criterion was performed for a 6ChN18/22 medium-speed marine engine using a mathematical model of the compression-expansion process, the main provisions of which are presented in [14].

During the calculation experiment, the influence of boost parameters and the average temperature of the walls of the working chamber on the value of the cylinder sealing criterion was studied along with the net area of conventional leakage holes. Each of these parameters has its own effect on the value of the k_{seal} criterion. For instance, an increase in air pressure at the p_{inp} inlet leads to an almost twofold increase in the mass charge. For example, at an engine shaft speed of n = 350 min^{-1} the mass charge increases from 6.44 g to 13.47 g. As a result, the maximum pressure rise rate increases from 965 bar/s to 1977 bar/s and the absolute values of the minimum pressure rate from −1096 bar/s to −2238 bar/s, i.e. by more than 100% each. In addition, the value of the k_{seal} criterion decreases from 0.136 to 0.132, which is less than 3%. An increase in the air temperature at the T_{inp} inlet from −10 °C to + 80 °C under otherwise invariant parameters leads to a certain decrease in the mass charge and an increase in the heat transferred to the cylinder walls. Together, these two factors set the actual compression process apart from the isentropic one and lead to an increase in the k_{seal} criterion by 16,7%. With an increase in the wall temperature t_w (from 70 to 140 °C), portion of heat lost in the walls decreases, and the k_{seal} criterion decreases by 1.8%. The leakage area f_{leak} has the most significant effect. The criterion value for the selected area range (from 0.05 to 0.35 cm^2) changed by a factor of almost 30 (increased by 2790%). In order to allow a comparative analysis of the influence of each of the considered parameters (Fig. 3a), the latter were reduced to a dimensionless form.

An experimental check of the revealed patterns was performed on a single-cylinder compartment of the OChN18/22 diesel engine. An RFT type piezoelectric sensor with two series-connected quartz crystals was installed in the cylinder cover of the research unit with the smallest possible indicator channel. The current-voltage converter was made on the basis of the AD623ANZ instrumentation amplifier with a front-end on the BF245A field emission transistor. In order to reduce the influence of the electric capacitance of the measuring channel, the frame of the transmitter was designed to be installed directly on the sensor connector without using a cable.

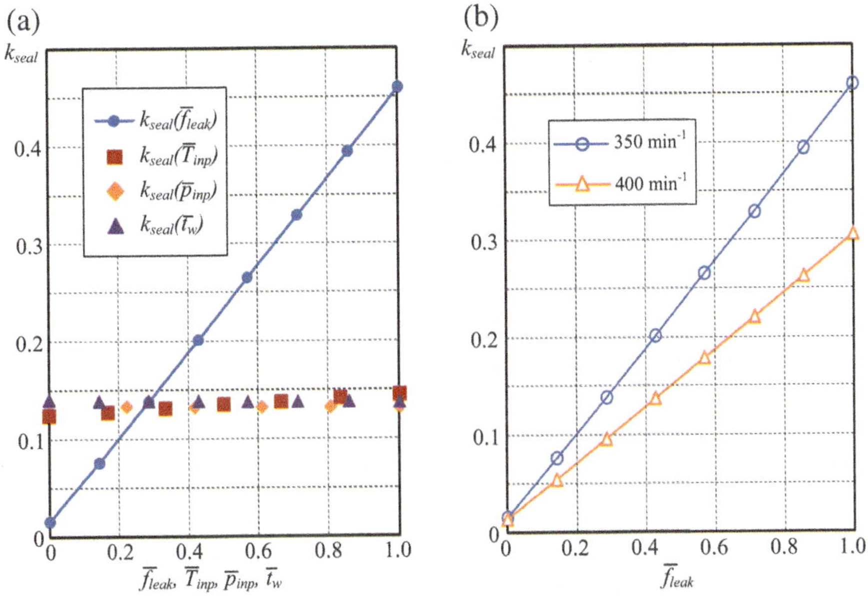

Fig. 3. Correlations between changes in the k_{seal} criterion and the studied parameters at n = 350 min^{-1} (*a*) and the effect of the crankshaft revolution (*b*)

The cylinder cover was designed with a channel to allow the working chamber to communicate with the environment through washers with various internal diameters, through which additional leaks of the working medium were created. Washers with diameters from 1.5 to 4 mm were used in the results presented below.

The signal was recorded using the "DieselLab-001" software and hardware complex in asynchronous recording mode, i.e. without association with the crankshaft rotation angle. The length of the data block was chosen on the basis of recording of at least five consecutive engine cycles with a sampling frequency of 125 kHz, which provided time intervals between adjacent ADC readouts of 8 μs. Data blocks were recorded twice for each experiment (washer), so that in the future it would be possible to process 10 - 12 one-cycle diagrams on average for each experiment. Assuming that the signal output is linearly related to the rate of pressure change, the k_{seal} criterion was calculated on the basis of the signal amplitudes expressed directly in the ADC numerical codes. Figure 4 shows the results of processing the experimental data at a crankshaft speed of 400 min^{-1}.

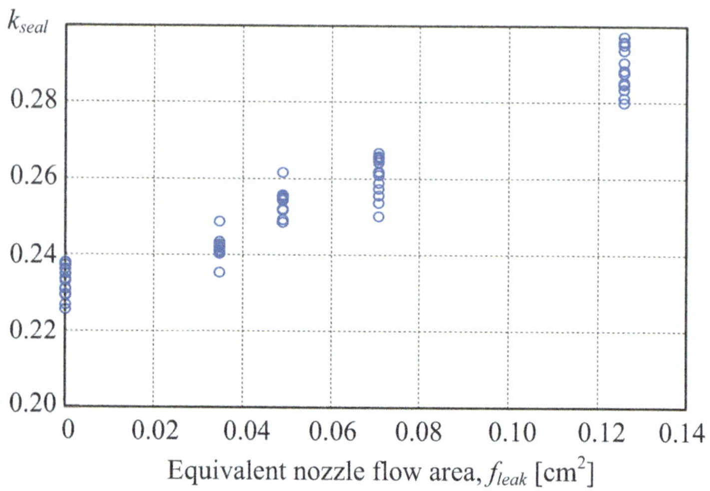

Fig. 4. The results of the experimental testing at n = 400 min^{-1}

3 Fuel Timing Advance Criterion

Determining the phase of the start of combustion (point *i* in Fig. 5) is considered to be more important in the field of technical diagnostics than determining the phase of the fuel delivery commencement. And although the deviation of the ignition phase from the optimal value can be caused by more than just incorrect adjustment of the start of the fuel injection, nevertheless, eliminating this reason is a necessary condition for the economical and safe operation of the diesel engine. It is known that the duration of the ignition delay period depends on the properties of the fuel, temperature and pressure in the cylinder at the time of the start of the fuel delivery. If the assessment of a diesel engine is performed under equal conditions, then it can be fair to expect the existence of an unmistakable relationship between the start of fuel delivery and the start of its ignition. In practice, equality of conditions can be ensured by maintaining a certain diesel mode during its testing. In addition, the relationship between the start of injection and the start of fuel combustion can be established experimentally for various modes of operation of the power plant.

Two strategies can be proposed to identify deviations of the start of combustion from the optimum value. The first involves determining the duration of the process in *c*-*i* section (Fig. 5). Since the offset of point c relative to the TDC is small, the duration of this section can be used to determine the ignition angle relative to the TDC without using data on the position of the crankshaft. A visible drawback of this strategy is the difficulty of accurately determining the phase of point *c*, especially for a noise-contaminated signal.

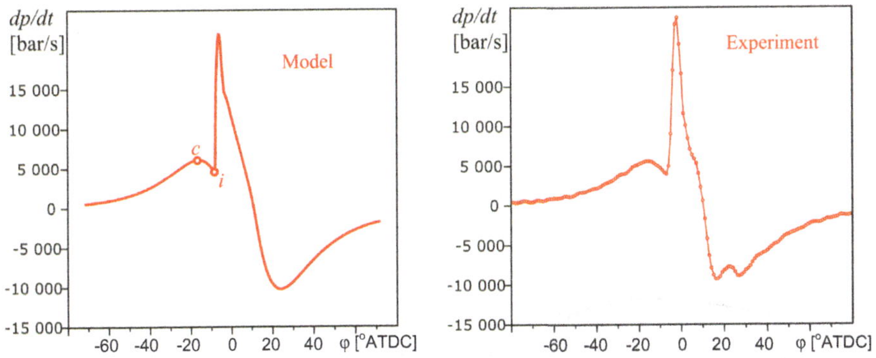

Fig. 5. Signal of the rate of change of pressure in the cylinder of the diesel engine 1A-5D49 (1050 kW mode at 610 min^{-1})

The second strategy is based on the fact that when the visible ignition angle, the position of point *i* also changes, approaching (early combustion) or moving away (late combustion) to point *c*. This allows us to propose a dimensionless criterion for fuel timing advance in the following form

$$k_{inj} = \frac{(dp/dt)_c - (dp/dt)_i}{2(dp/dt)_c} \tag{2}$$

where *(dp/dt)i* – are the values of the pressure velocity signal at point *i*.

With fuel is delivered in advance, the value of criterion (2) decreases. Theoretically, its minimum value is zero and could be achieved if points *c* and *i* coincide, but in practice it can be eliminated due to the impossibility of autoignition when the fuel is delivered too early. For similar reasons, the maximum value of the criterion cannot exceed unity and in practice it will always be less. Since the criterion is dimensionless, it is not necessary to determine the physical value of the pressure velocity to calculate it, the calculation according to (2) can be performed based on the data from the signal, provided the converter is linearly characterized. In this case, calibration of the measuring channel is not required.

To identify the nature of the influence of the fuel injection timing angle (FITA) on the value of criterion (2), a computational study was performed using mathematical models [14, 15]. The calculations were performed for a diesel engine 1A-5D49 operating at 50% diesel engine power at a crankshaft speed of n = 610 min^{-1}. The fuel injection timing angle value varied from 8 to 34° in increments of 2° of crankshaft rotation. The boost parameters and injection characteristics were kept constant for all calculations. As an example, Fig. 6 shows graphs of changes in the pressure rate in the cylinder depending on the crankshaft rotation angle for several fuel injection advance angles. The graphs additionally display the points involved in the calculation of the proposed criterion: point c (common for all graphs) and points i_j, where the number *j* corresponds to the fuel injection advance angle (in degrees). The results of the analysis of the obtained curves are presented in Fig. 6.

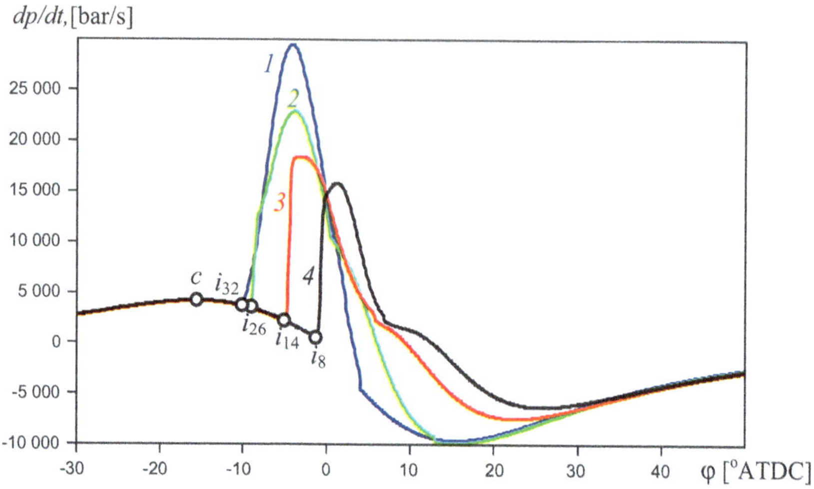

Fig. 6. Simulation results for various fuel injection timing angles (1050 kW mode at 610 min^{-1}): 1 – φ_{inj}= 32°; 2 – φ_{inj} = 26°; 3 – φ_{inj} = 14°; 4 – φ_{inj} = 8°; *c* – maximum pressure rate in the compression area; i_j – points corresponding to the start of combustion at $\varphi_{inj} = j$

The angle of onset of visible combustion shifts in the intended direction, but non-linearly. This is explained by the fact that when the fuel injection advance angle is wide, the ignition delay period increases (Fig. 3) due to the relatively low temperature and air pressure in the cylinder, as a result of which the start of combustion phase does not experience practically any changes when the advance is large enough. Calculations showed that combustion does not start earlier than point c for the selected mode at arbitrarily large advance times, which confirms the thesis stated above.

When ignition starts in the range from 0 to 7.5° to TDC, which corresponds to the interval of the start of injection from 8 to 18°, the dependence of the proposed criterion on fuel injection timing angle becomes almost linear (line 4 in Fig. 7), which makes it possible to implement simple and fast algorithms for determining this parameter for specified conditions.

When ignition starts in the range from 0 to 7.5° to TDC, which corresponds to the interval of the start of injection from 8 to 18°, the dependence of the proposed criterion on fuel injection timing angle becomes almost linear (line 4 in Fig. 7), which makes it possible to implement simple and fast algorithms for determining this parameter for specified conditions.

We compared the data of automatic diagnostics performed by the "Magistral" complex with the results obtained by the described methodology as an experimental justification for the efficiency of the proposed criterion. The experimental data—indicator diagrams—was extracted from the database (DB) of the program, after which it was numerically differentiated. It should be noted that the database of the "Magistral" program stores diagrams averaged over several tens of cycles with a resolution of 1° of the crankshaft rotation angle. Therefore, the correct processing of the pressure velocity curve was not always possible due to distortions in the area of the maximum pressure

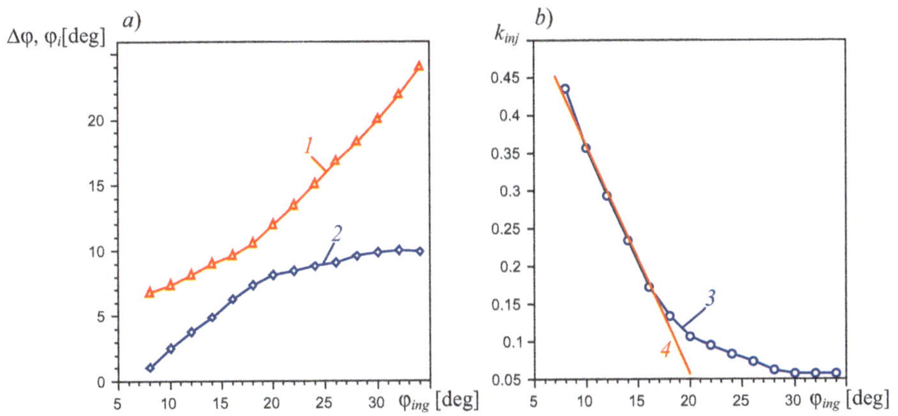

Fig. 7. The results of the experiment: *a* – dependencies of the phase parameters of combustion on the fuel injection advance angle; *1* – ignition delay period $\Delta\varphi(\varphi_{inj})$; *2* – visible combustion onset angle (relative to TDC) $\varphi_i(\varphi_{inj})$; *b* – dependence of the criterion for fuel timing advance on the fuel injection advance angle $k_{inj}(\varphi_{\text{inj}})$: *3* – calculation; *4* – linear approximation for the area $8^o < \varphi_{\text{опт}} < 18^o$

rise rate (in the vicinity of point c). In "Magistral" complex, indicator pressure diagrams are compared with a conditional reference diagram, which is considered to be a diagram of one of the engine cylinders. The purpose of adjustment in this setting is to fine-tune the processes of all engine cylinders to a selected standard. A special algorithm that takes into account signals from vibration and pressure sensors in the cylinder is used to determine the fuel injection timing angle. Figure 8 shows all possible cases: normal, early and late fuel injection in the test cylinder relative to the process in the reference cylinder.

The fuel timing advance criterion (2) was calculated for each pair of signals of pressure change rate, and the fuel injection timing angle was calculated based on the approximation of the linear sector of the curve $f\,(k_{inj})$.

$$\varphi_{inj} = 21,9 - 32,8\,k_{inj}. \tag{3}$$

It should be noted that coefficients in the equation of the straight line (3) were obtained from the results of mathematical modelling and, in all probability, should be refined experimentally. Also, to this date the influence of the speed and load conditions of the engine on the $\varphi_{inj}(k_{inj})$ curve has not been adequately studied. Despite this, the agreement of results of the indirect determination of the fuel injection timing angle from the signal of the first-order pressure derivative with the data of the "Magistral" hardware-software complex can be considered satisfactory. At the same time, at too large fuel injection timing angles the quantitative values of the diagnostic symptom (deviations of the controlled parameter from the reference value) determined by the "Magistral" complex and by the proposed method significantly diverge. This is a completely logical result, since the formula (3) in this dependence domain $\varphi_{inj}(k_{inj})$ is not applicable. A "too early injection of fuel" diagnosis can be proposed for a signal with k_{inj} lower than 0.02 as an option to solve the problem.

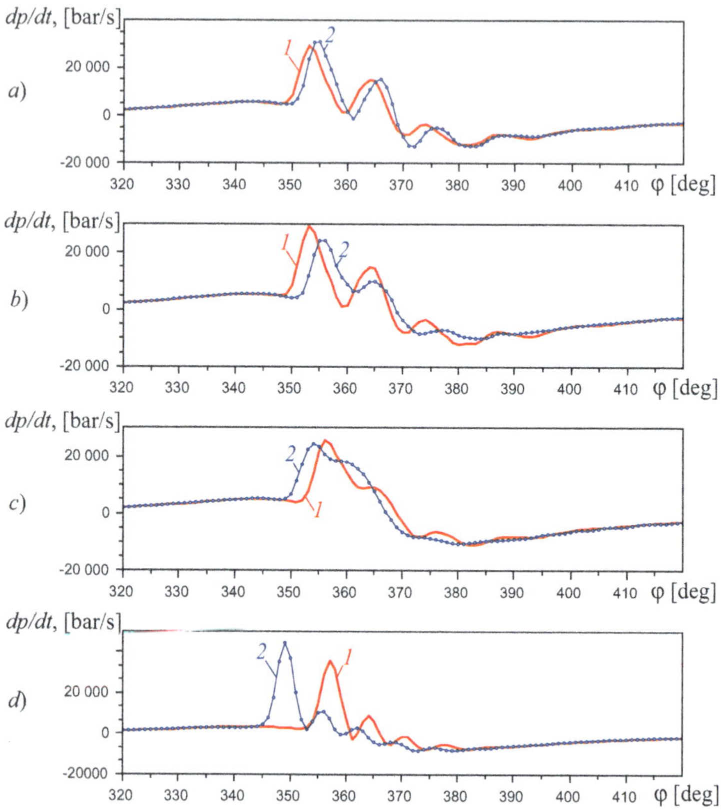

Fig. 8. The results of the experimental verification of the method under various conditions: *(a)* fault-free; *(b)* late injection; *(c)* early injection; *(d)* too early injection; 1 – "reference" cylinder, 2 – test cylinder

4 Conclusion

Criteria for cylinder sealing and advance fuel injection were proposed based on calculation and experimental studies. Both criteria are dimensionless coefficients determined solely by the magnitude of the signal of the first-order pressure derivative at characteristic points of the curve, which are local extrema of the signal. Thus, neither measuring the crankshaft rotation angle, nor calibrating the sensor is required in order to implement diagnostic methods using the proposed criteria. The impact of the engine shaft speed can be taken into account either during a diagnostic test at a selected constant speed mode of engine operation, or when using analytical expressions previously obtained for a given type of engine.

References

1. Lamaris, V.T., Hountalas, D.T.: A general purpose diagnostic technique for marine diesel engines – application on the main propulsion and auxiliary diesel units of a marine vessel. Energy Convers. Manag. **51**(4), 740–753 (2010). https://doi.org/10.1016/j.enconman.2009.10.031
2. Neumann, S., Varbanets, R., Kyrylash, O., Yeryganov, O.V., Maulevych, V.O.: Marine diesels working cycle monitoring on the base of IMES GmbH pressure sensors data. Diagnostyka **20**(2), 19–26 (2019). https://doi.org/10.29354/diag/104516
3. Varbanets, R., Karianskyi, S., Rudenko, S., Gritsuk, I., et al.: Improvement of diagnosing methods of the diesel engine functioning under operating conditions. SAE Technical Paper 2017-01-2218 (2017). https://doi.org/10.4271/2017-01-2218
4. Sakellaridis, N.F., Raptotasios, S.I., Antonopoulos, A.K., Mavropoulos, G.C., Hountalas, D.T.: Development and validation of a new turbocharger simulation methodology for marine two stroke diesel engine modelling and diagnostic applications. Energy **91**, 952–966 (2015). https://doi.org/10.1016/j.energy.2015.08.049
5. Hountalas, D.T., Antonopoulos, A.A., Zovanos, G.N., Papagiannakis, R.G.: Evaluation of a new diagnostic technique to detect and account for load variation during cylinder pressure measurement of large-scale four-stroke diesel engines. SAE Technical Paper 2012-01-1342 (2012). https://doi.org/10.4271/2012-01-1342
6. Hountalas, D.T., Kouremenos, D.A., Sideris, M.: A diagnostic method for heavy-duty diesel engines used in stationary applications. J. Eng. Gas Turbines Power **126**(4), 886–898 (2004). https://doi.org/10.1115/1.1787500
7. Husted, H., Kruger, D., Fattic, G., Ripley, G., Kelly, E.: Cylinder pressure-based control of pre-mixed diesel combustion. SAE Technical Paper 2007-01-0773 (2007). https://doi.org/10.4271/2007-01-0773
8. Akiyama, M., Tabaru, T., Kishi, K.: Development of a pressure sensor using a piezoelectric material thin film. Synthesiol. English Ed. **5**(3), 171–180 (2012). https://doi.org/10.5571/syntheng.5.171
9. Bueno, A.V., José, A.V., Velásquez, J.A., Milanez, L.F.: A new engine indicating measurement procedure for combustion heat release analysis. Appl. Therm. Eng. **29**(8–9), 1657–1675 (2009). https://doi.org/10.1016/j.applthermaleng.2008.07.023
10. Hag, M.Z.: Applied Measurement Systems. InTech, Rijeka (2012). https://doi.org/10.5772/2360
11. Randolph, A.L.: Methods of processing cylinder-pressure transducer signals to maximize data accuracy. SAE Technical Paper 900170 (1991). https://doi.org/10.4271/900170
12. Lee, K., Kwon, M., Sunwoo, M., Yoon, M.: An in-cylinder pressure referencing method based on a variable polytropic coefficient. SAE Technical Paper 2007-01-3535 (2007). https://doi.org/10.4271/2007-01-3535
13. Sun, W., Du, W., Dai, X., Bai, X., et al.: A cylinder pressure correction method based on calculated polytropic exponent. SAE Technical Paper 2017-01-2252 (2017). https://doi.org/10.4271/2017-01-2252
14. Pipitone, E., Beccari, A.: Determination of TDC in internal combustion engines by a newly developed thermodynamic approach. Appl. Therm. Eng. **30**(14–15), 1914–1926 (2010). https://doi.org/10.1016/j.applthermaleng.2010.04.012
15. Kuleshov, A.S.: Model for predicting air-fuel mixing, combustion and emissions in DI diesel engines over whole operating range. SAE Technical Paper 2005-01-2119 (2005). https://doi.org/10.4271/2005-01-2119

Calculation of Tank Car Under Quasirandom Cyclic Loading

Igor Emelyanov[1,2(✉)], Vladimir Mironov[2,3], and Dmitriy Ogorelkov[2]

[1] Ural State University of Railway Transport, Ekaterinburg, Russia
emelyanov@imach.uran.ru
[2] Institute of Engineering Science UB RAS, Ekaterinburg, Russia
[3] Ural Federal University, Ekaterinburg, Russia

Abstract. Alternative method of calculation of the projected design of the tank car for strength and service life is considered. To improve the reliability of calculations, the results of the strength determination solution are carried out for different levels of discretization of the studied object and compared with the available analytical solutions. The resource of the structure is defined for one of the most loaded units of the welded joint. The estimation of resource indicators is carried out on the basis of experimental data on the fatigue limit of welded joints, taking into account the geometry of the node and the method of loading. In this case it is possible to avoid unreasonably high forecasts of durability of products. The prediction is improved by taking into account the rate of cyclic degradation of the material on the basis of original experiments on the degeneration of the static deformation diagram with increasing operating time. The paper presents the results of a numerical experiment to determine the durability of the tank structure at different rates of degradation of the material properties. Simulation of a random process of tensioning changes at a dangerous point of the structure was performed using randomization operation.

Keywords: Tank · Shell · Durability

1 Introduction

The service life of any structure is determined by the degradation processes occurring in the material of the most loaded elements. The direct functional connection between the intensity of degeneration of material properties and the stress level is proved experimentally and presented in many scientific papers. Consequently, the accuracy of determining the operating loads and stress state of the structure largely determines the accuracy of the forecast of the load-carrying capacity and residual life of the entire product.

Currently, in the design of rolling stock structures, computer programs based on the finite element method are widely used, which allow to choose a rational design scheme and simulate possible cases of loading of the structure [1, 2].

In this paper, on the example of the boiler design of a tank car with an improved scheme of support on the frame, the specified fatigue life of one of the structural

Z. Popovic et al. (Eds.): TransSiberia 2019, AISC 1115, pp. 340–348, 2020.
https://doi.org/10.1007/978-3-030-37916-2_33

elements is determined. Methods of numerical analysis were used to determine the stress state of the studied structure. For demonstration of the offered variant of a technique the resource of one of welded joints of a design is defined.

The design of the boiler of the tank car is a shell of rotation lying on the supports of the lodgment type. The stress state of the structure is largely determined by the system of fastening the boiler to the frame (legs and supports). Thus, it is necessary to adequately, mathematically accurately describe the zone of load transfer, i.e. the zone of local loading of the boiler in the area of the legs and contact interaction with the supports, which make a significant heterogeneity in the stress-strain state of the structure.

The general method of calculating the resource of the investigated welded metal structure includes three interrelated tasks. First, the development of a mathematical model describing the stress state of the entire structure, and the definition of the most loaded points of the structure. Secondly, determination of the spectrum of quasi-stationary loads and determination of the service life for the most loaded points of the structure by linear summation of damages. In the third problem, the process of degradation of static properties of the material under unsteady external load is simulated and the specified durability at the same points is determined by an alternative method.

2 Methods of Solution and Resolving Equations

Two numerical methods, the finite element method and the discrete orthogonalization method, were chosen to determine the stress state of the tank car boiler structure. A problem-oriented computer complex is used to implement the finite element method COSMOS/M [3]. The basic equations and methods of solving thin-walled shell using the discrete orthogonalization method are presented in the monograph [4]. The results of solving the problem for different levels of discretization by finite elements of the studied design for different loading modes and comparison with the available analytical solutions are presented in [5, 6]. Determination of the spectrum of quasi-stationary loads and calculation of the service life for the most loaded points of the structure by linear summation of damages are considered in [6, 7]. Figure 1 shows the design scheme of the half (due to the symmetry of the structure) of the tank clothing with support on the frame and the distribution of the Mises stress. The stresses are given for the case of loading of the structure according to one of the design modes. From the analysis of the results obtained, the most stressed zones of the structure for different loading regimes are found.

Since the design of the tank car consists of many elements connected by welding, it is necessary to determine their fatigue life. It is known that more than 80% of all cases of operational destruction of elements of welded metal structures occurs as a result of cyclic loading [8]. Variable magnitude mechanical effects cause irreversible changes in the material of structural elements, which lead to fatigue failure. Therefore, it is necessary to take into account the degradation of the static properties of the material under non-stationary external load in determining the durability of the tank car.

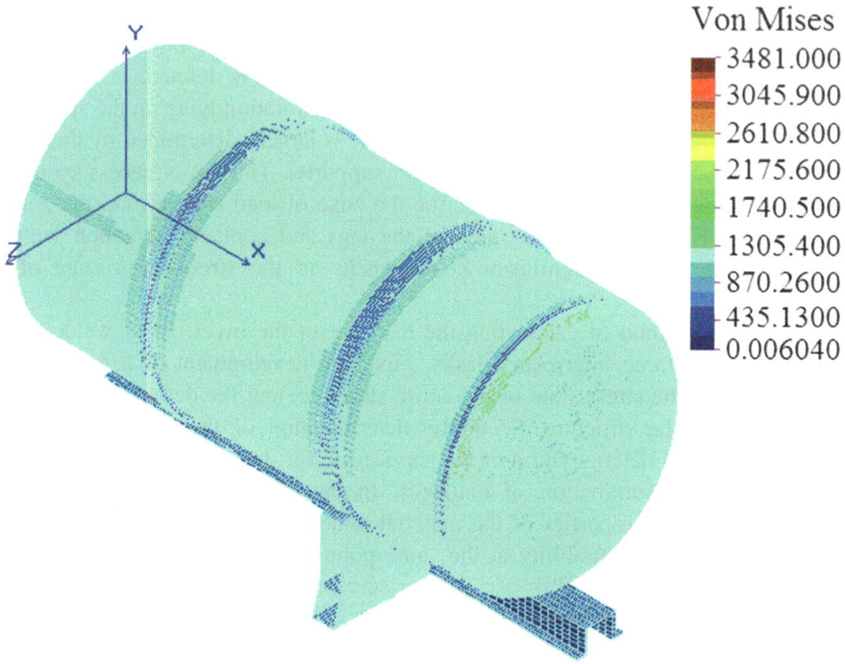

Fig. 1. The stress state of the boiler shell (the dimension of the Mises stress on the pictogram - kg/sm2)

The paper uses both the traditional approach to the fatigue problem based on basis of the linear hypothesis of damage summation and an alternative approach to calculate the service life of welded elements of a tank car [9, 10]. In the latter case, the model includes both new rheological equations of the fatigue process and traditional relations of the classical theory of fatigue [11].
fatigue curve:

$$\sigma^{\alpha} N = \sigma_{RK}^{\alpha} N_0, \tag{1}$$

fatigue limit in asymmetric cycle

$$\sigma_{RK} = \frac{2\sigma_{-1}}{(1-R)K + (1+R)\psi} \le \sigma_T, \tag{2}$$

transition to a complex stress state

$$\sigma_{\Pi}(N) = \sigma_{-1}(N) = \sigma_{-1}, \tag{3}$$

example of reduced stress

$$\sigma_{\Pi} = (\sigma_1^2 + \sigma_2^2 - \sigma_1\sigma_2)^{0,5}, \tag{4}$$

experimental kinetic curve

$$S_B(\sigma_M, n) = S_{BO} - k_{\sigma} n^m, \tag{5}$$

kinetic coefficient

$$k_{\sigma} = \frac{S_{BO} - \sigma_M}{N^m}, \tag{6}$$

fatigue and static failure criterion

$$S_B(\sigma_N, N) = \sigma_M, \tag{7}$$

failure criterion under irregular loading

$$\sigma_{MK} \geq S_B\left[\sigma_M(n), \sum_i^k n_i\right], \tag{8}$$

cyclically equivalent conditions of material

$$S_B(\sigma_{M1}, n_1) = S_B(\sigma_{M2}, n_2), \tag{9}$$

where N_0 – base durability, S_{B0} – initial tensile strength, σ_{Π} – led stress, σ_{-1} – limit of endurance at a symmetric cycle, R – cycle asymmetry coefficient, ψ – coefficient of sensitivity of the material to the asymmetry of the cycle, σ_M – the maximum stress stationary cycle, m- experimental material constant, n_i – number of load cycles per i – stage, α – kinematic coefficient determining the angle of inclination of the fatigue line in double logarithmic coordinates, K – effective stress concentration factor, N – calculated number of fatigue curve cycles.

The experimental kinetic curve given for the case of the force approach is constructed in fatigue tests of thin-walled samples with different degrees of operating time [12]. The criterion of failure is the reduction of the ultimate strength to the level of maximum cycle stress. The transition from one kinetic curve to another when the cycle stress level changes is determined by the condition of cyclically equivalent states of the material.

In the monograph [13] cyclic damage ω_C the model sample was considered from the standpoint of the structural-phenomenological approach and was estimated by the relative number of elements of the structure, shattered from fatigue. A formula linking structural damage to the macroparameter, more precisely, the ultimate strength, is obtained

$$S_B(\sigma_M, \mathrm{n}) = (1 - \omega_C) S_{B0}. \tag{10}$$

Considering the last formula together with formulas (1)–(9), in the case of a stationary loading process, we come to an expression for cyclic damage in the model of Corten-Dollan [12]

$$\omega_C = \frac{S_{B0} - \sigma_M}{S_{B0}} \left(\frac{n}{N(\sigma_M)} \right)^m. \tag{11}$$

Damage at the moment of fatigue failure, i.e. at $n = N$, is not equal to one and is not a constant of the material. It is determined by the level of destructive stress σ_M, and the intensity of damage accumulation depends on the individual properties of the material, i.e. the index m.

3 The Algorithm for Calculating the Durability of a Random Stress Spectrum

To demonstrate the method of calculation of the specified resource, consider the scheme of calculation of a particular structural element for multi-cycle fatigue. The first stage is the formation of the initial stress spectrum in the design zone of the structure. This can be done in the presence of recording devices accurately enough, given the geometric and physical linearity of the system, working in the field of multi-cycle fatigue. Continuous recording and calculation of stresses allow to avoid the proof of ergodicity of the random process of stress changes in the calculated zone. In the illustrative example the randomization operation on the set interval of change of normal stresses in the area of a leg of fastening of a cover of the boiler to a centre girder will be used.

The second stage is associated with the use of known data on the material of the structure. For a welded structure a group of welding assembly in the design zone is determined and for each stress cycle with a given value σ_M is the coefficient of asymmetry R. Then the table value of the base limit of endurance σ_{-1K} and the second formula model (2) calculates the endurance limit σ_{RK} taking into account the stress concentration [14]. Further from the expression of the fatigue curve is the durability $N(\sigma_M)$, which is required to calculate the current value of the tensile strength of the fifth formula of the model (5). Exponent m it is found experimentally [12], for example, for cast steels its value is close to two.

The third stage is associated with the transition to a new level of stress, at which the tensile strength is calculated by the transformed formula of the kinetic curve

$$S_B(\sigma_M; n) = S_{B0} - \frac{S_{B0} - \sigma_M}{N^m} (n_e + 1)^m \tag{12}$$

where n_e- the equivalent number of cycles in which the new stress level achieves the same damage that was obtained before the transition. The calculation stops when the condition is met $\sigma_i \geq S_B(\sigma_{i-1})$, and number of cycles i determines the desired durability at variable stress levels.

The fourth stage consists in the construction of graphs of damage changes found by this method (11) and the linear hypothesis of fatigue damage summation

$$\omega_C = \sum \frac{n}{N} = 1 \tag{13}$$

4 Calculation of Durability of a Lath of Fastening of a Copper to a Backbone Beam

In the above finite element calculation of the tank it is established that the maximum stress in the place of fastening the bottom of the boiler to the centre girder reaches 90 MPa. According to the classification of welded assemblies by stress concentration [15], the fastener assembly to the pipe belongs to the ninth group with the limit of endurance in a symmetrical cycle $\sigma_{-1,K} = 30\,\text{MPa}$. Performance trials of cars show that the stresses in the metal structure change randomly. Having no experimental spectrum of stress changes in the area of the fastening strip, we use the randomization operation on the interval of maximum stress changes $\sigma_M \in [60; 90]$ MPa (Fig. 2). We assume that the average cycle voltage $\sigma_M = 60\,\text{MPa} = const$ and the stress change cycle is symmetrical.

Presented in Fig. 2 a sample of 100 cycles was repeated until the fatigue failure criterion was met, which is equivalent to the assumption of ergodicity of a random process.

The material data is supplemented with the value of the initial strength limit $S_{B0} = 470\,\text{MPa}$. The exponent of dependence (4), which determines the rate of degradation of the strength of the material, is adopted as a parameter in the range of values [0.5; 6]. The indicator of the fatigue curve for steel A 516-55 (09G2S) $\alpha = 3.5$ for welded assemblies adopted by Russian State Standard «General rules for the design of steel structures» SP 53 -102-2004, DEAN, Saint-Petersburg (2007). The choice of design parameters not according to the certificate of steel, but according to the results of testing of welded assemblies allows to take into account the behavior of this material in a particular welded structure.

The kinetic curve formula above (1) was used to calculate the current strength limit, and the damage formulas (11) and (12) were used for the proposed model and the hypothesis of linear summation of fatigue damage. According to the results of calculation of the loading process with a random maximum stress (Fig. 2), using the formulas (1)–(9), (3)–(4) and the data of the numerical example, the graph in Fig. 3.

Depending on the material constant m changing the estimated durability N_*, but the destruction occurs at the maximum or close to the cycle voltage. The maximum accumulated damage ranges around $\omega_{CR} = 0.81$. The previously formulated rule of intersection of the kinetic curve with the maximum cycle stress in the stress spectrum preserves the meaning of the fatigue strength criterion of the structure [9].

Taking into account the cyclic degradation gives a conservative estimate of the durability at random nature of the stress spectrum. Theoretically, only when $m \to \infty$, when the properties of the material do not change up to failure, the predicted number of

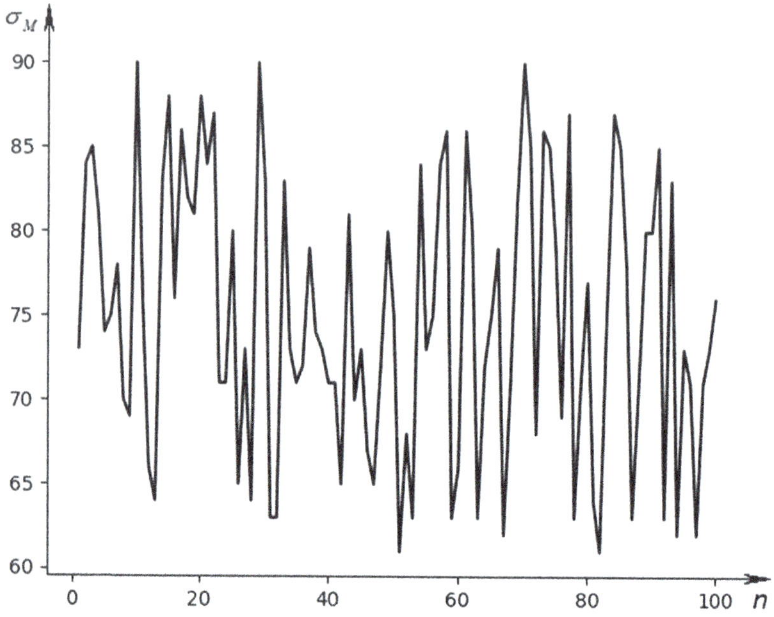

Fig. 2. Distribution pattern of maximum stress at the interval [60; 90] MPa

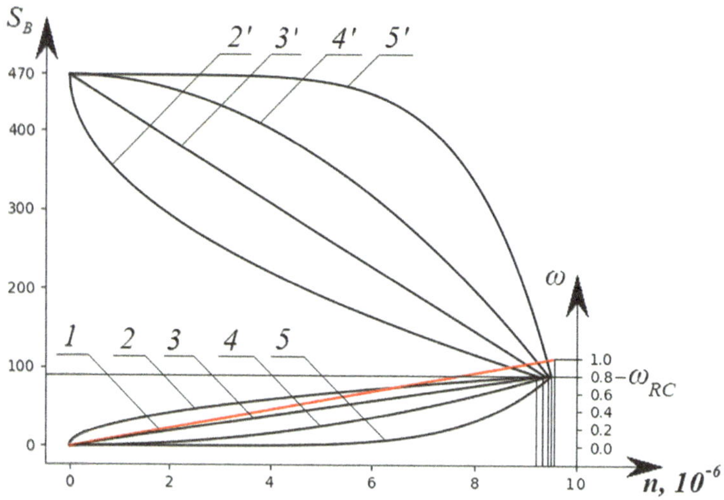

Fig. 3. Kinetic curves of strength decrease (figures with prime mark) and damage increase with operating time increase (1-by the linear hypothesis and by the refined technique 2, 3, 4 and 5 at m = 0.5, m = 1, m = 2 and m = 6)

cycles to failure according to the discussed technique and the linear hypothesis are the same:$N_* = 9.54 \cdot 10^6$. Specific values of design life N_* are shown in Table 1.

Table 1. Setting Word's margins

m	0.5	1	2	6
N_*, 10^{-6}	9.24	9.39	9.46	9.51

Note that with the growth of the numerical value of m, the danger of emergency overloads at the final stage of cycling decreases.

The prospects of this approach to the problem of fatigue are related to the fact that there is a fundamental possibility of developing a technique that combines the stages of origin and development of a fatigue crack. In [16], the above methodology was used to describe the kinetics of fatigue crack propagation in a thin cast-iron plate with a central hole. The numerical calculation was carried out within the framework of the described approach, without involving any hypotheses of the crack theory. Only anomaly of surface layer properties was taken into account. The qualitative and quantitative correspondence of the calculation to the experiment was established.

5 Conclusions

1. With the use of modern computational methods the dangerous points of the tank structure were determined.
2. Selection of cyclic tensile strength $S_{B0}(n, \sigma_M)$ as representative parameters of the fatigue process allowed to reflect the kinetics of fatigue damage accumulation in the calculation of the durability of one of the welded units of the tank.
3. It is of practical importance to assess the residual strength of the fastening unit of the shell to the spine beam at the final stage of cycling.
4. Intersection rule does not impose restrictions on the type of random process and does not oppose the criteria of static and cyclic destruction.

References

1. Copaci, I., Tanasoiu, A., Tanasoiu, B.: On the resistance of tank wagons to the shock caused by collision. In: Proceedings of 3rd WSEAS International Conference on Finite Differences, Finite Elements, Finite Volumes, Boundary Elements, Bucharest, pp. 23–27 (2010)
2. Manescu, T., Zaharia, N., Avram, D., Tiganasu, N., Stroia, M.: Nonlinear finite element analysis used at tank wagons. In: Proceedings of the 4th WSEAS International Conference on Finite Differences, Finite Elements, Finite Volumes, Boundary Elements, Paris, France, pp. 110–113 (2011)
3. Alyamovskij, A.A.: SolidWorks/COSMOSWorks. Finite element engineering analysis. DMK Press, Moscow (2004)

4. Grigorenko, Y.M., Vasilenko, A.T.: Methods of shell analysis. Theory of shells of variable stiffness. Naukova dumka, Kiev (1981)
5. Emelyanov, I.G., Efimov, V.P., Kuznetsov, A.V.: Model of the stress-strain state of the boiler of a tank car with an improved scheme of support on the frame. Heavy Eng. **8**, 44–49 (2005)
6. Emelyanov, I.G., Mironov, V.I., Kuznetsov, A.V.: Determining the stress state and the lifetime of a shell structure load. J. Mach. Manuf. Reliab. **5**(36), 443–449 (2007)
7. Emelyanov, I.G., Mironov, V.I., Kuznetsov, A.V.: Evaluation of the life of a shell construction lying on supports. J. Mach. Manuf. Reliab. **1**(39), 83–88 (2010)
8. Bolotin, V.V.: Forecasting of resource of machines and structures. Mashinostroenie, Moscow (1984)
9. Volkov, S.D., Mironov, V.I.: The use of resistance functions in assessing the durability of structures. Dep. VINITI 1883, 78 (1978)
10. Mironov, V.I., Lukashuk, O.A., Vichuzhanin, D.I.: Method for describing fatigue processes in structural materials. Solid State Phenom. **265**, 815–820 (2017). https://doi.org/10.4028/www.scientific.net/SSP.265.815
11. Gusev, A.S.: Fatigue Resistance and survivability of structures under random loads. Mashinostroenie, Moscow (1989)
12. Emelyanov, I.G., Mironov, V.I.: Durability of shell structures. EPH UB RAS, Ekaterinburg (2012)
13. Struzhanov, V.V., Mironov, V.I.: Deformation softening of material in structural elements. EPH UB RAS, Ekaterinburg (1995)
14. Sokolov, S.A.: Structural mechanics and metal structures of machines. Polytechnic, Saint-Petersburg (2011)
15. Trufyakova, V.I. (ed.): Strength of welded joints under variable loads. Naukova dumka, Kiev (1990)
16. Mironov, V.I., Kuznetsov, A.V., Emelyanov, I.G.: Consideration of cyclic degradation of the material and abnormality of the surface layer mechanical properties in calculating the life of a plate with an opening. Strength Mater. **5**(46), 638–643 (2014). https://doi.org/10.1007/s11223-014-9594-y

Efficiency of Reconstruction of Stations for High-Speed Traffic

Natalya Kukleva(✉) and Denis Kuklev

Far Eastern State Transport University, Seryshev Street, 47, Khabarovsk 680021, Russia
kukleva.natalya@bk.ru

Abstract. Nowadays, more and more attention is paid to the development of high-speed traffic in Russia and in the whole world. Russia has yet to build a high-speed specialized highway but high-speed traffic is already being practised on a number of lines reconstructed at higher speeds. When choosing lines for the introduction of passenger trains high-speed movement, the problem arises of determining the feasibility of reconstructing each separate point with track development for one or another level of maximum speed for passing passenger trains. Due to the large number of small separate points with track development, which, as a rule, are practically all "barrier" places, solving this problem requires a lot of time and money. That is why, at the pre-design stage, it seems appropriate to use the methodology for determining indicative capital investments into the reconstruction of small separate points with track development. Moreover, depending on the reorganization length, it is possible to determine the maximum capital investment into the reorganization of such separate points, which will pay off due to the resulting savings on travel time winning. The authors investigate the structure of capital investments, which are usually determined during the reconstruction for high-speed traffic, coming to the conclusion that some part of them can be "put out of brackets", meaning that the payback periods are reduced. The research can be interesting to specialists and scientific workers, who are involved in justification for the efficiency of increasing the speed of passenger express trains in railway transport.

Keywords: High-speed passenger traffic · High-speed lines · Reconstructing separate point · Capital investments · Railway stations

1 Introduction

Improvement of the quality of the services provided, is the most important tool in the competition of the railway transport for a passenger, in particular by reducing the time spent on the trip, which can be evidenced by the experience of different countries [1–8]. It can be achieved by using the innovative rolling stock, special high-speed lines (HSL) as well as the reconstruction of existing railways at higher speeds. These events can be implemented separately from each other, but the greatest effect on increasing the speeds of passenger trains can be achieved with their complex application [9].

Nowadays in Russia there are no high-speed specialized highways built especially for passenger railway traffic with the speeds of more than 250 km/h. Although plans for

Z. Popovic et al. (Eds.): TransSiberia 2019, AISC 1115, pp. 349–356, 2020.
https://doi.org/10.1007/978-3-030-37916-2_34

its implementation existed even in the times of the USSR, unfortunately they did not get proper support from the state.

In the recent history of Russia, it was supposed that the first high-speed line will be the line St. Petersburg – Moscow, but now on the foreground is the Moscow – Kazan line for the movement of passenger trains at speeds of up to 400 km/h. According to the data of the International Union of Railways, the length of high-speed railways in our country is planned at the level of 762 km [10], which does not allow to go side by side with the world leaders in this sphere. That is why, researches connected with the introduction of high-speed passenger traffic are relevant, both from practical and theoretical points of view.

2 Materials and Methods

The initial data are the studies which were previously carried out by the authors in the field of increasing passenger train speeds justification by reconstructing stations at a number of existing lines, which were supposed to be rebuilt to introduce high-speed traffic. Small stations on such lines were taken as objects, about which statistical information on the number of tracks, switches, platforms and etc., was collected and processed, which allowed applying the developed methodology, based on the typification of stations, to determine approximately the required capital investments for their reconstruction. It should be mentioned that the research was carried out in accordance with the Methodological recommendations for the economic feasibility of infrastructure projects on Russian railways.

It is a well-known fact that the construction of the HSL requires huge investments and needs a larger increase. That is why, one should not forget about the introduction of high-speed traffic on existing railways by rebuilding their speeds (141–200 km/h). During the reconstruction of such railways for the introduction of high-speed traffic the implementation of a large amount of work is required, both on stage and on separate points. Moreover, a significant part of these works, and, respectively, capital investments, will be necessary for the reconstruction of small intermediate separate points with track development (passing points and intermediate stations), since with their large numbers they usually do not satisfy the requirements of high-speed traffic.

Of high importance is the choice of the rational speed level for passing passenger trains through each separate point, at the same time it is necessary to measure the saving in time after the reconstruction of an intermediate station or a passing point, and the investments needed for this reorganization. Only small separate points that do not provide the passenger trains parking, fall under the scope of consideration.

Thus, the feasibility of reconstructing a separate point, and the line in whole, for high-speed movement can be determined by the formula:

$$\sum_{t=0}^{T} R_t \cdot \alpha_t \geq \sum_{t=0}^{T} K_{\text{р.п.}t} \cdot \alpha_t + \sum_{t=0}^{T} C_t \cdot \alpha_t \tag{1}$$

where R_t is the results got after the reconstruction of a separate point (line), expressed in the reduction of operating costs, connected with the decrease in the time of passenger trains movement and the duration of the passengers' trip; C_t – operating costs for maintaining a separate item (line) per year; $K_{p.n.t}$ – capital costs in the reconstruction of a separate point (line) per year; α_t – discount coefficient; T – horizon of calculation.

To choose a rational level of the maximum speed of passenger trains passing through separate points at the pre-design stage, one can use the developed methodology, which will quickly allow to determine the estimated capital investments into the reconstruction of small separate points with track development, depending on the number of pick-up routes and the type of conversion.

According to the methodology, all separate points, depending on the length of the conversion and the magnitude of the main paths displacement relatively to the leads, are divided into five principal types: I – a separate point is moved to a new place, II – the conversion affects both necks of the station, but the existing station platform is used to a large extent, III – one of the necks of the station is under reconstruction, IV – the plan of the main routes does not need any change, and work is performed, not associated with the restructuring of the path plan; V – passenger devices are situated between the main tracks, which must be carried out of the space between the main track sat speeds greater than 140 km/h.

The amount of approximate capital investments for the reconstruction of intermediate separate points with track development depends on a number of factors, the most significant of which are: the number and useful length of pick-up ways, the number of turnouts on the main and pick-up ways, the existing and planned speed of passenger trains.

3 Results

The purpose of the research is determination of the maximum capital investment at which the reconstruction of separate points will be appropriate. As a starting point we take the following:

- type of conversion – III;
- the initial speed of passenger trains – 100 km/h;
- the maximum speed of a separate point with track development – 200 km/h;
- the number of pick-up routes at separate points – up to 10;
- the number of high-speed passenger trains pairs – from 4 to 10;
- the cost of passenger hours – from 50 to 250 rubles.

The calculation results (in base prices) are presented in diagrams (Fig. 1).

From the diagrams (see Fig. 1) it can be seen that the higher the price of a passenger-hour is, the more economical it is to reconfigure a separate point – on the diagrams it is shown by a change in the angle of inclination of the capital investment curve to the abscissa axis, and it is obvious that with the increase in the number of high-speed trains this curve becomes more sloping.

The diagrams presented allow us to determine whether it the reconstruction of a separate point with track development for a specific speed level will be appropriate. For

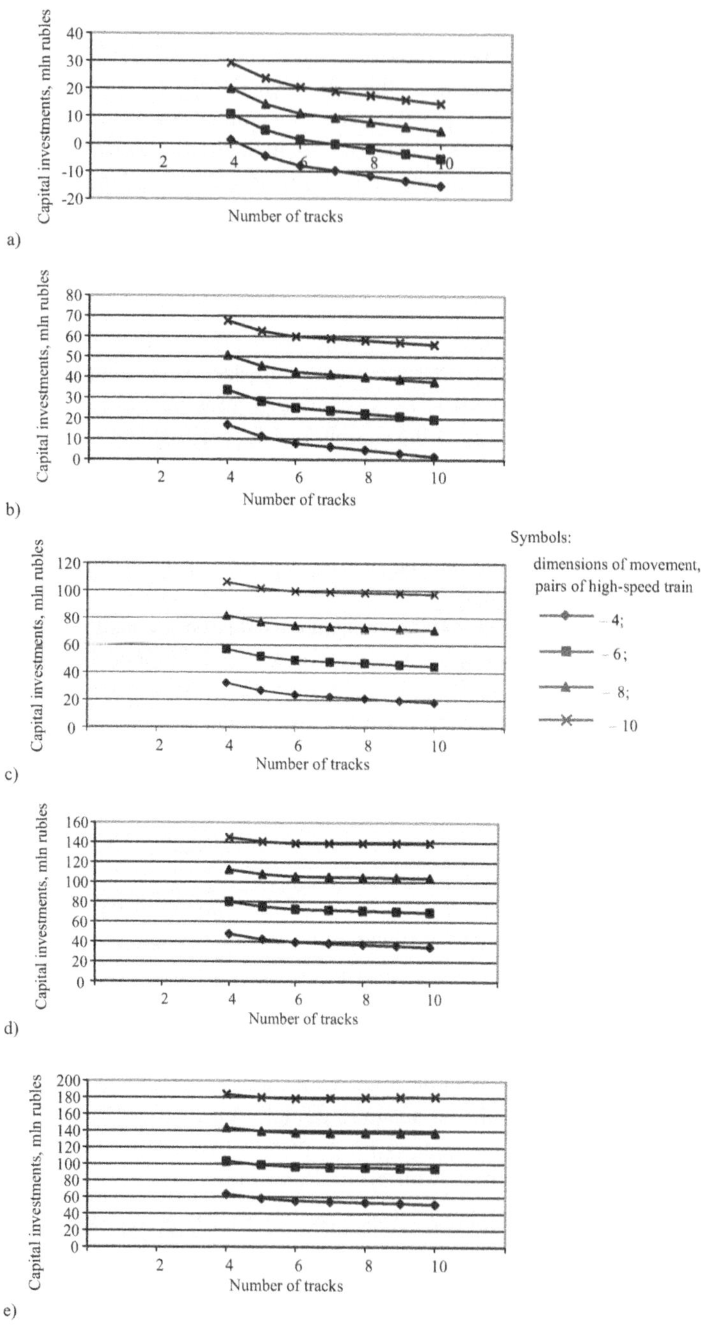

Fig. 1. Diagrams of the maximum capital investment from the number of tracks and the number of high-speed trains with the price of a passenger-hour 50 rubles. (a), 100 rubles. (b), 150 rubles. (c), 200 rubles. (d), 250 rubles. (e)

this, the points corresponding to the approximate capital investments obtained by the methodology must be plotted on the area of the corresponding diagrams: if these points are situated between the curve of boundary capital costs and the abscissa axis, the reconstruction will be advisable. The capital investment values that have negative values are not taken into regard, since they are outside the zone of efficiency of the conversion of separate points with track development for the conditions under consideration.

Also, the feasibility of reconstruction is affected by the section of speed limit – the larger this section is, the more profitable is the conversion of both individual separate points and the line as a whole. Then follows the schedule of boundary capital investments at a passenger-hour cost of 100 rubles and with an increase in the length of the reconstruction site (Fig. 2).

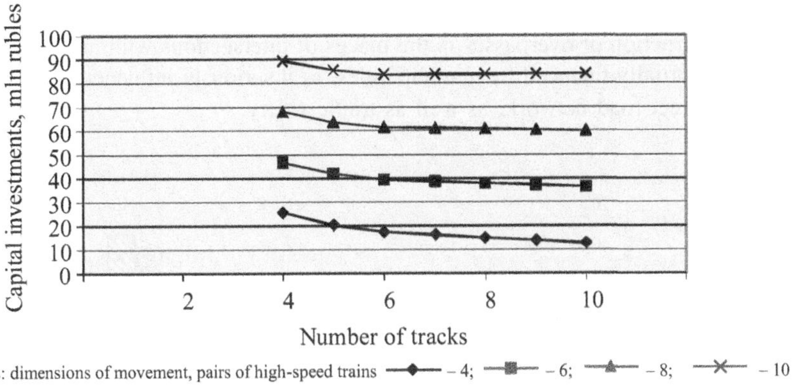

Fig. 2. The schedule of boundary capital investments at a passenger-hour cost of 100 rubles and with an increase in the length of the reconstruction site

Comparing the results presented in Fig. 1b and Fig. 2, it is possible to make a conclusion that for the case under consideration, the increase of the speed limit area by 2 times increases the size of the effective maximum capital investment up to 40% – with minimal track development and more than a few times – with maximum track development for a different number of high-speed passenger trains. Moreover, the smaller the expected size of the speed movement is, the greater influence has the length of the speed limit area on the amount of boundary capital investments.

Studying the question of determining the required capital investments in the reconstruction of separate points with track development, from our point of view it is necessary, noting, undoubtedly, large required investments, to highlight the following. Works with increasing the speeds of passenger train at separate points with track development can be divided into two groups: dependent on speed and independent from it.

The first should include those works, without which the introduction of high-speed traffic is impossible as a matter of principle, for example: changing the plan of the route of main tracks if there are horizontal curves of insufficient radius; strengthening the track's upper structure by laying high-speed turnouts, increasing the number of sleepers per 1 km of track, using reinforced rails, etc.

The second type of work includes those that are not specific for high-speed movement as their implementation is necessary at the existing level of speed. For example, the construction of overpasses; fencing the area of the railway station, passages for passengers at different levels, etc.

At the same time, the following should be meant:

- when replacing the turnouts on the main tracks with special high-speed ones, they can be put in another place, which allows to take into account their return value;
- on conventional lines, especially with the operation of heavy freight trains, laying reinforced turnouts, which are necessary for high-speed movement, will reduce the wear of both turnouts and pairs of wheels of rolling stock;
- the construction of passages at different levels for passengers, fencing the area of a separate point is also a requirement for ordinary railway lines;
- the construction of overpasses in the places of intersections with highways in cities can be partially financed by municipalities, as it seriously influences the congestion of the street road network, as well as traffic safety.

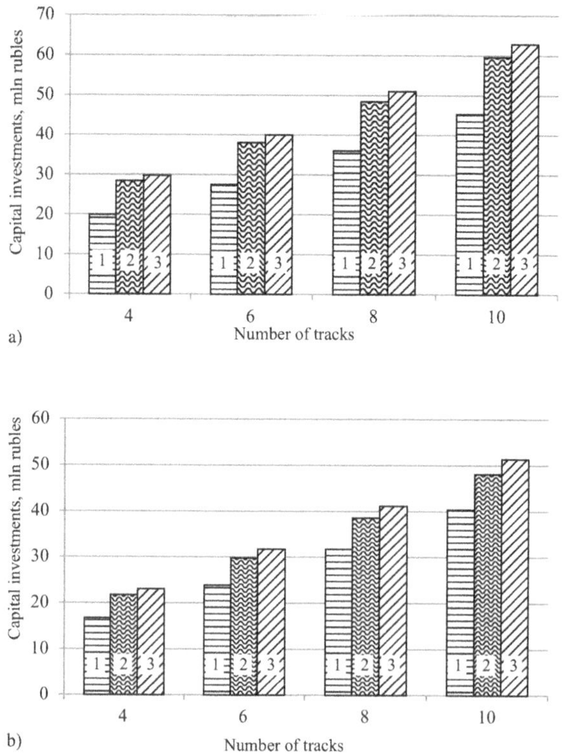

Fig. 3. Diagrams of dependence of estimated capital investments with various accounting for expense items: 1 – 140 km/h; 2 – 160 km/h; 3 – 200 km/h

To illustrate the resulting effect when following the indicated positions, we will consider one of the types of conversion of an intermediate separate point with track development, approximately determining capital investments for different gradations of speed and the number of station tracks (Fig. 3).

4 Discussion

The achieved results are of high importance in the issue of substantiating the effectiveness of the required investments to achieve a certain passenger train speed level. This is especially valuable in conditions where it is necessary relatively quickly, at the pre-project stage, having an estimated amount of investment, to determine the feasibility of reconstructing small separate points for a particular level of maximum passenger train speed. Moreover, the proposal to exclude the assignment of part of the reconstruction costs to works related not only to high-speed traffic, is interesting from a practical point of view.

5 Conclusions

Thus, the efficiency of capital investments, and, consequently, the feasibility of reconstruction of separate points to the maximum speed level of passenger trains are influenced by a number of factors. The article highlights the main of them, such as the price of a passenger-hour, the discount rate, the length of the speed limit section.

It is also offered to take into account in the total volume of capital investments only the costs directly related to high-speed traffic, which will reduce the payback period of projects for the conversion of intermediate separate points with track development for the introduction of high-speed passenger trains movement, and therefore make them more attractive to investors. The given private example shows that the difference in the required investments can increase to several times depending on the initial data.

The further authors' research will be aimed at deeper studies of this question with the generalization of the results depending on the conversion type, the sizes of the track development, and also the estimated maximum passenger train speeds levels.

References

1. Román, C., Martín, J.C.: Integration of HSR and air transport: understanding passengers' preferences. Transp. Res. Part E Logistics Transp. Rev. **71**, 129–141 (2014). https://doi.org/10.1016/j.tre.2014.09.001
2. Yang, H., Burghouwt, G., Wang, J., Boonekamp, T., Dijst, M.: The implications of high-speed railways on air passenger flows in China. Appl. Geogr. **97**, 1–9 (2018). https://doi.org/10.1016/j.apgeog.2018.05.006
3. D'Alfonso, T., Jiang, C., Bracaglia, V.: Air transport and high-speed rail competition: environmental implications and mitigation strategies. Transp. Res. Part A Policy Pract. **92**, 261–276 (2016). https://doi.org/10.1016/j.tra.2016.06.009

4. Liang, X.H., Tan, K.H., Whiteing, A., Nash, C., Johnson, D.: Parcels and mail by high speed rail–a comparative analysis of Germany, France and China. J. Rail Transp. Plan. Manag. **2**, 77–88 (2016). https://doi.org/10.1016/j.jrtpm.2016.04.003
5. Kim, H., Sultana, S.: The impacts of high-speed rail extensions on accessibility and spatial equity changes in South Korea from 2004 to 2018. J. Transp. Geogr. **45**, 46–61 (2015). https://doi.org/10.1016/j.jtrangeo.2015.04.007
6. Wetwitoo, J., Kato, H.: High-speed rail and regional economic productivity through agglomeration and network externality: a case study of inter-regional transportation in Japan. Case Stud. Transp. Policy **4**, 549–559 (2017). https://doi.org/10.1016/j.cstp.2017.10.008
7. Moyano, A.: High speed rail commuting: efficiency analysis of the Spanish HSR links. In: XII Conference on Transport Engineering (2016). https://doi.org/10.1016/j.trpro.2016.12.029
8. Cascettaa, E., Coppola, P.: EWGT2013 – 16th Meeting of the EURO Working Group on Transportation, vol. 111, pp. 176–185 (2014). https://doi.org/10.1016/j.sbspro.2014.01.050
9. Kukleva, N.V., Kuklev, D.N.: On the joint use of high-speed and conventional railway lines. In: Line IOP Conference Series: Earth and Environmental Science, vol. 272 (2019). https://doi.org/10.1088/1755-1315/272/2/022110
10. Railway Statistics Synopsis. 2019th edn. https://uic.org/IMG/pdf/uic-statistics-synopsis-2019.pdf

Usage of Dynamic Programming Method in Transport and Logistics Centers Construction and Development Projects

Vladimir Lamekhov(✉) and Elena Chervotenko

Far Eastern State Transport University, Seryshev Street, 47, Khabarovsk 680021, Russia
lamekhovva@gmail.com

Abstract. This article provides a comparative analysis of the concepts of a transport and logistics center and a multimodal transport and logistics center. An assessment is given of the development of the railway transport infrastructure in connection with trade and economic cooperation with the countries of the Asia-Pacific region. An analysis of development stages methodology based on the dynamic programming method of transport infrastructure objects is carried out and the possibility of applying this technique to projects of multimodal transport and logistics centers is revealed. Introduced the term "transport microsystem", which is used to consider the objects complex interacting with the transport and logistics center. The existing development stages methodologies, studied by Russian scientists, are analyzed. Based on the compared methods, the authors of the article proposed two approaches to the application of the dynamic programming method in the construction and development projects of multimodal transport and logistics centers. In the first approach, these objects are considered separately from each other, but taking into account their mutual influence. In the second approach, the elements of this system are studied in a complex, as part of one enlarged transport object. For clarity of the analysis of approaches, "time-state" diagrams were constructed, their functional features are described. The advantages and disadvantages of these approaches are identified, the conclusion is drawn on the possibility of their application in complex projects of construction and development of multimodal transport and logistics centers.

Keywords: Logistics · Dynamic programming · Stage development methodology · Transport and logistics center · Multimodal

1 Introduction

The transport system of Russia is one of the most developed and extensive transport networks in the world, the basis of which is the transport infrastructure of various types of transport. At the moment, the urgent task is the creation of transport and logistics centers, designed to provide various options for the transport of goods in the interaction of modes of transport. The Transport and Logistics Center (TLC) is a complex of interconnected and interdependent structures located in transport hubs that provides comprehensive freight forwarding services to customers, minimizing the cost of

Z. Popovic et al. (Eds.): TransSiberia 2019, AISC 1115, pp. 357–366, 2020.
https://doi.org/10.1007/978-3-030-37916-2_35

transporting goods for participants in the transport and logistics process by integrating their material, information and financial streams. The Multimodal Transport and Logistics Center (MTLC) is a complex of transport terminals of various types of transport, united by a common technology of interaction. The main functions implemented by MTLC are measures to manage, control, ensure the safety of material streams, their information support, as well as search for the most economic favorable conditions for the interaction of all modes of transport and participants of the transport and logistics process.

Rail transport, which accounts for about 50% of the total volume of transported goods, occupies a leading position in the Russian transport system. At the international level, Russian railways are part of transcontinental transport corridors, providing international production and trade links. Currently, there is a stable trend in trade and economic cooperation and an increase in freight traffic to the countries of the Asia-Pacific region. Based on the analysis of transport infrastructure elements optimal development studies by foreign and Russian scientists, it was revealed that for the efficient management of cargo streams, it is necessary to create a network of multimodal transport and logistics centers that will provide a comprehensive organization and interaction of all types of transport infrastructure. Rail transport infrastructure is the basis for organizing such a network of MTLC.

2 Materials and Methods

The authors of the study propose to make calculations using the development stages methodology based on the dynamic programming method for complex projects to optimize the interaction of transport infrastructure elements of different modes of transport.

The choice of options for the technical condition and technological support of multimodal transport logistics centers in time can be carried out on the dynamic programming method basis.

The choice of options for the technical condition and technological support of multimodal transport logistics centers in time can be carried out on the basis of the dynamic programming method. The term "dynamic programming" first appeared in the mid-50s in the works of R. Bellman. In the early 60s of the twentieth century, works were published proving the effectiveness of the staged development of railway infrastructure and formed the fundamental basis of this methodology. Since the mid-sixties, the ideas of staged development have been applied to optimize construction projects and modernization of precinct and sorting railway stations, their subsystems, transport terminals, railway lines, roads and trunk pipelines. These topics are covered in the works: V. M. Akulinicheva, V. A. Ardashin, A. P. Baturin, A. V. Bykadorov, I. D. Bulavchenko, S. M. Goncharuk, A. V. Gorinov, P. S. Gruntov, V. G. Dzyuba, Yu. I. Efimenko, I. T. Kozlova, D. N. Kuklev, A. V., Mineeva, Yu. P. Nayashkov, N. A. Tuzhilkina. However, at the moment, theoretical studies on the rational stages of the construction and modernization of infrastructure cannot be considered complete. There is no single methodology comprehensively applicable to most objects of the transport network, in particular, the dynamic programming method has not been adapted to MTLC projects.

In essence, the development stages methodology implies a sequence of events to change the technical state and timing of such events that, together with a rational technology of work, ensures a minimum of total costs. The objective function of this methodology takes into account all given total costs, including capital investment and operating costs of infrastructure, throughout the entire modernization period of the facility.

The objective function of this technique in general is written:

$$E = \sum_{t=0}^{n} K_t \alpha_t + \sum_{t=0}^{n} C_t \alpha_t \rightarrow min, \quad (1)$$

in which, E – total costs; K_t, C_t – respectively, capital investments and operating costs at the t-th step; α_t – discounting costs coefficient.

A visual representation of this technique can be presented in the form of a “time-state” scheme, which marks the path of development of the facility, taking into account the possibility of developing all freight traffic and the total minimum of all the costs involved (Fig. 1). In the nodes of this graph, the values of the efficiency criterion, the sum of the total costs, are indicated, and on the lines connecting the nodes - the increment of the efficiency criterion at the corresponding optimization steps.

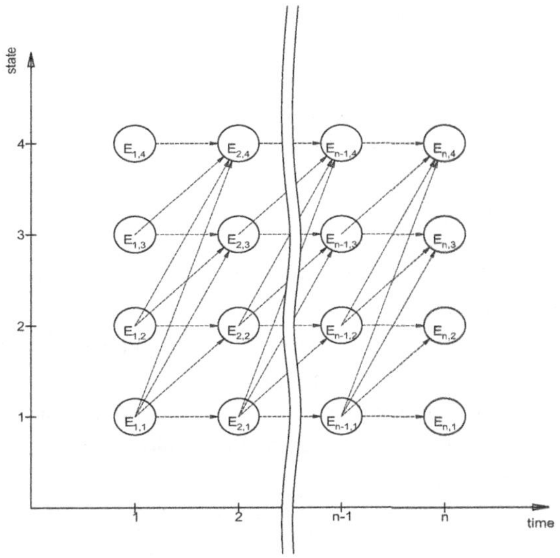

Fig. 1. The “time-state” diagram

To determine the possibility of applying the dynamic programming method to the construction and development projects of MTLC, an analysis of existing methods for railway transport facilities was carried out.

The authors of the article suggest dividing these methods into two conditional groups: the first group considers the objects of transport infrastructure in isolation from each other, in the second group the elements of infrastructure are studied in close relationship taking into account mutual influence and impact. Comparison of these approaches is based on two selected methods: the stages of development of railway stations and junctions by Yu. I. Efimenko and the choice of optimal development of technical equipment of the railway network by A.P. Baturin.

The application of these methods to MTLCs has limitations related to the specifics of their functioning. The first group of methods is designed to use the method of dynamic programming to one object of transport infrastructure and does not take into account the mutual influence between objects. The second group of methods takes this mutual influence into account only through the dynamics of changes in cargo and passenger streams, but does not consider variants of objects work technology at preservation of the sizes of the given streams and changes of technical equipment of one of those objects due to their significant distance from each other.

The first group of methods considers railway stations and junctions opposite from the other infrastructure objects and the optimal criterion for this group takes form:

$$\mathrm{E}_{t_c}^{opt} = \min(\sum\nolimits_{t=0}^{t_c} K_t \rho_t^{(\kappa)} + \sum\nolimits_{t=1}^{t_c} C_t \rho_t^{(C)} - \kappa_{t_c}^{res} \rho_{t_c}^{(\kappa)} \tag{2}$$

in which, K_t, C_t – investments and operating costs in the year t;

ρ_t – reduction coefficient, taking into account the decrease in the significance of costs incurred after t years;

t_c – cost accumulation period;

κ_{tc}^{res} – the sum of all residual values of objects.

It should be mentioned that this method does not consider mutual influence between different objects.

Thus, it is advisable to use these criteria to select management decisions at individual optimization steps of a particular object.

The second group of methods consider railway network as a group of interacting objects and the optimal criterion for these methods takes form:

$$\mathrm{E}(S,T,X(t),Y(t)) = \sum\nolimits_{i=1}^{n} \sum\nolimits_{j=1}^{|s_i|} (s_i) \left[\int_{t_{j-1}}^{t_j} \frac{E_j(X(t),Y(t))dt}{e^{E_{\mathrm{H}}t}} + \frac{K_j}{e^{E_{\mathrm{H}}t}} \right] - \frac{F_{res}(S,T)}{e^{E_{\mathrm{H}}T_p}} \tag{3}$$

in which, $F_{res}(S,T)$ – reduced residual value of fixed assets for the period T_p;

$E_j(X(t),Y(t)) - E_j(X(t),Y(t))$ – differential costs (the intensity of changes in operating costs) at the j-th stage of development of the transport object s_i;

K_j – capital investments given during the construction period at the jth stage of development of the facility s_i;

S – network development scheme, number of development schemes of network objects;

T – estimated period of network reconstruction;

(S, T) – network development plan;

X(t) – network passenger traffic plan;

Y(t) – network freight passenger traffic plan.

Correlation between network objects is carried out by taking into account the impact of the development of network facilities on the organization of cargo and passenger streams.

Thus, it is advisable to use these criteria to select management decisions at individual optimization steps of the entire railway network or a large number of remote objects.

3 Results

The specifics of multimodal transport and logistics centers provides for their continuous interaction with other objects of transport infrastructure. It includes coordination of cargo handling amounts, coordination of vehicle maintenance technologies, reduction of the time for cargoes to be stored in transportation hubs.

Thus, it can be concluded that the close interaction of some elements of transport infrastructure requires their separation into a particular *transport microsystem*, in order to further consider these elements as an indivisible unit. Changes in the technical equipment or technology of one element of the microsystem can cause variations in operating costs and technology of the other elements.

In this paper two approaches to the process of applying the method of dynamic programming for the construction and development of multimodal transport logistics centers are considered.

They are based on the groups of methods of application of development stages methodology.

In *the first approach*, transport infrastructure facilities are considered *separately*. For complex projects of development of MTLC, the application of the first approach has the following limitations of use, namely:

- the need to take into account the development of technical equipment of interacting objects of transport infrastructure and interaction technology when calculating operating costs at each stage of development of the MTLC;
- the presence of a "forced" change in the technology of interaction between objects or the technology of one of the objects during technical or technological changes on another element of the system;
- the complexity of the calculations when using this technique for the simultaneous staged development of all elements of a multimodal transport node is caused by a significant increase in the number of development options for the microsystem.

Thus, when using the first approach, it is necessary to analyze the work and apply the existing methods of staged development to all infrastructure facilities of multimodal transport nodes. Number of options for the staged development, in a simplified form, can be represented in the form of a scheme for the interaction of a railway station and MTLC (Fig. 2).

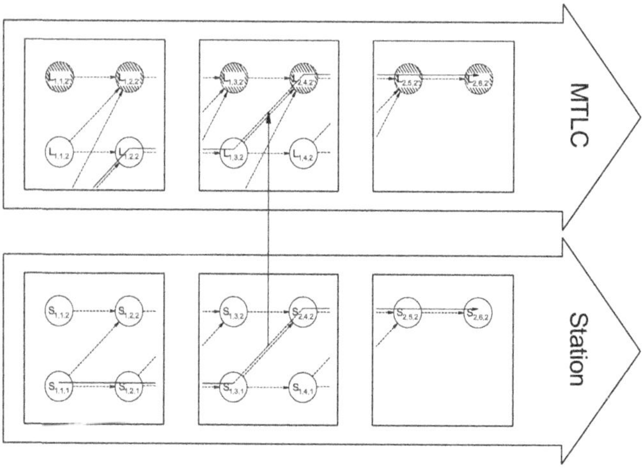

Fig. 2. Fragment of time-state diagrams of the station-logistics center system

For this scheme, the following notation: station values of the efficiency criterion – $S_{z,t,i}$, multimodal transport and logistics center – $L_{z,t,i}$. In which, z – state of the second object in the system; t – system development step, i – stage of development of the object; i′ – stage of development of the object with a change in the total costs. The thick line between the development vectors of the individual elements shows the "forced" transition of the object to the next stage.

The objective function in the MTLC-railway station system should take into account the minimum of total costs at each of the facilities, and the system of restrictions should combine the work technology and the volume of cargo processing. The objective function for this system takes the form:

$$\begin{cases} E^S = \sum_{t=0}^{n} K_t^S \alpha_t^S + \sum_{t=0}^{n} C_t^S \alpha_t^S \to min \\ E^L = \sum_{t=0}^{n} K_t^L \alpha_t^L + \sum_{t=0}^{n} C_t^L \alpha_t^L \to min \end{cases}, \tag{4}$$

in which, E^S, E^L – total costs for the station and the transport and logistics center, respectively.

Constraint system:

$$\begin{cases} X_{(t)}^{SL} = X_{(t)}^{L} = X_{(t)}^{LS} + X_{(t)}^{tr} \\ t_{pr}^{L} = T_n \end{cases}, \tag{5}$$

in which, $X_{(t)}^{SL}$ – freight streams developed by the station, arriving at the logistics centre within the established time period t; $X_{(t)}^{L}$ – freight streams, developed by the logistics centre within the set time period t; $X_{(t)}^{LS}$ – freight streams developed by the station, arriving from the logistics center within the set time period t; $X_{(t)}^{tr}$ – freight streams developed by modes of transport interacting in MTLC; t_{pr}^{L} – duration of freight traffic processing by the logistics center; T_n – established standard value of the duration of freight traffic processing by the logistics center.

This constraint system leads to the need for “forced” transitions of the development stage. The difficulty lies in the fact that such “forced” transitions affect operating costs not only for this stage, but also for all subsequent stages of the development of the system object. Visually for this and for the next stages of development of the facility, this dependence is shown on the graph of the dependence of operating costs on the development step of the facility (Fig. 3).

A second disadvantage of this approach is the need to iterate over a substantial amount of development options microsystems, if the system interacts with a large number of transport infrastructure, and the construction and development of these facilities is carried out in many stages.

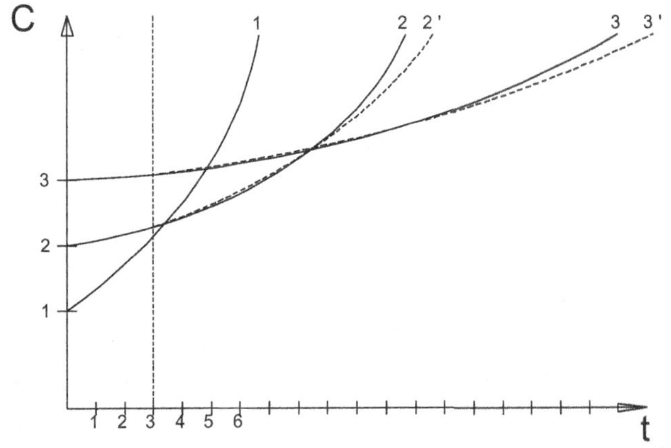

Fig. 3. Dependence of operating costs on the development step of the MTLC - railway station system

When analyzing the possibility of using the dynamic programming method in the construction and development of transport infrastructure *facilities as separate elements*, the following results were obtained:

- analysis of the objects of the transport network in cooperation within the framework of this method creates the need to consider “forced” state transition of objects;
- “forced” transitions complicate the use of the dynamic programming method, as they contribute to the appearance of changes in the total costs at several stages of development of an object of a given system;

- the use of the dynamic programming method leads to the need to recount a large number of options, which complicates the use of the methodology for the development of transport infrastructure facilities.

The *second approach* considers objects and applying the dynamic programming method to them inextricably from each other, as part of one enlarged transport infrastructure object. This approach has the following features of consideration requiring detailed study:

- the possibility of developing freight flows with the given technical development of an enlarged facility in a timely manner;
- the ability to service the required number of rolling stocks of all modes of transport, taking into account the technology of their interaction;
- detailing the stages of development of each object and highlighting those that are more likely to affect the operation of the entire system of objects.

This approach involves the use of the objective function for the entire microsystem, without dividing it into separate objects. In this case, the objective function will take the form:

$$E^{SL} = \sum\nolimits_{t=0}^{n} \left(K_t^S \alpha_t^S + K_t^L \alpha_t^L\right) + \sum\nolimits_{t=0}^{n} \left(C_t^S \alpha_t^S + C_t^L \alpha_t^L\right) \rightarrow min, \tag{6}$$

The system of restrictions for the second approach will take the form:

$$\begin{cases} X_{(t)}^{enter} = X_{(t)}^{exit} \\ t_{pr}^{L} = T_n \end{cases}, \tag{7}$$

in which, $X_{(t)}^{enter}$ – the incoming freight traffic developed by the enlarged transport object in a specified period of time t; $X_{(t)}^{exit}$ – outgoing freight traffic, developed by all modes of transport in a specified period of time t;

In this case, the time-state diagram will be compiled in the singular for the entire system, but it will contain a larger number of development stages. This diagram will take the form shown in Fig. 1, but will have more stages of development of this enlarged object.

The disadvantages of this approach are:

- difficulty in applying this methodology when the objects of the system belong to different owners;
- significant increase in the number of stages of the system or the complexity of each stage, if it is necessary to consider large number of interacting objects;
- the need for a detailed study of each stage of the development of the system, taking into account the technology of interaction of several modes of transport.

4 Discussion

Each of the two approaches considered has its own advantages and disadvantages.

The application of the dynamic programming method with the construction of "time-state" diagrams to the objects of the transport microsystem separately from each other makes it possible to more accurately analyze the moments of transition of each object to the next stages of development. At the same time, it is difficult to take into account the effect of changes in one element of the microsystem on the technical operation and technology of other elements, and also, there is a variability in the construction of time-state diagrams.

If we consider all the elements of this transport microsystem inextricably from each other, then uncertainty is created when accounting for investments in each object of the transport infrastructure, while the continuity of the interaction technology of all elements of this system, the integration of their information streams are necessary conditions for applying this approach.

5 Conclusions

The purpose of this study was to identify the possibility of applying the dynamic programming method to projects of multimodal transport and logistics centers. In the course of the study, it was found that the features of the operation of the MTLC, the tasks assigned to them, require the development of interacting transport infrastructure objects inextricably from each other. In the work, a model of using a transport microsystem to simplify the analysis of the development of objects interacting with MTLC is proposed. The approaches to using the methodology for the staged development of objects within a given system have key features in their practical application, which requires preliminary preparation for choosing one of them.

The existing plans for the development of transport infrastructure provide for the formation of a transport network without gaps and bottlenecks, elimination of unevenness in the level of development of transport infrastructure in certain regions of the country, and the need for the formation of a support network of multimodal transport and logistics centers significantly increases the burden on federal and regional transport budgets. Optimization of the development of this network, as well as its individual elements, should include both the implementation of the required volumes of cargo processing and the lowest required costs for the construction and operation of the infrastructure of the MTLC network. To solve this problem, in particular, one should use the methodology for the development of transport infrastructure facilities. Its application will allow to fulfill the goals set for the MTLC network with the lowest economic costs, to combine the technology of interaction between logistics facilities and transport infrastructure facilities.

References

1. Giray, H., Turkay, R.: Comput. Aided Chem. Eng. **36**, 401–420 (2015). https://doi.org/10.1016/B978-0-444-63472-6.00016-1
2. Minken, H., Johansen, B.: Econ. Transp. **19** (2019). https://doi.org/10.1016/j.ecotra.2019.04.001
3. Ranaiefar, F., Regan, A.: Logistics Operations and Management Concepts and Models, pp. 333–358 (2011). https://doi.org/10.1016/b978-0-12-385202-1.00016-5
4. Sergeyev, V.: The analysis of logistics and supply chain management organizational structures on Russian market. Springer (2019). https://doi.org/10.1007/978-3-030-12450-2_37

Operation of Infrastructure and Rolling Stock at Railway Polygon

Nina Sirina(✉) and Svetlana Yushkova

Ural State University of Railway Transport, Kolmogorova Street, 66, Yekaterinburg 620034, Russian Federation
nsirina@usurt.ru

Abstract. The article analyzes the state of transport infrastructure ensuring the required volume of transportation. The work employs the methods of marketing and strategic analysis and also the provisions on synthesis of adequate adaptive mechanisms with identification allowing to make the rational choice on the use of the infrastructure from a possible set of options, available resources and shippers needs. The carried-out SWOT analysis defined the risks and possible effectiveness of use of the single infrastructure of the railroads polygon regarding further possible additional costs. The transition to the polygon technology of the train traffic control allows JSC Russian Railways to organize rhythmic train traffic providing the «gaps» for the whole polygon in a single section. The adaptation mechanism for the costs variance of the consolidated liability of the transportation process participants allowing to monitor primary variances synchronously with the infrastructure works performance and delivery is offered. The general algorithm of the model of integrative management of the infrastructure and traction resources is formalized and illustrated on the example of the railroads polygon. Their interaction is based on the results of integral estimation which unites a diverse quantity of indicators and estimating factors of the transportation process. The mathematical support of the cost value formation is implemented in «Automated system of loading supervision of a production department». The received results are integrated into «Automated system of operational work supervision of the integrated centre of the infrastructure control» and adapted to transformation and effective use of the industry enterprises capacity.

Keywords: Infrastructure · Rolling stock · Railway polygon · Cargo turnover · Transportation process · Adaptive mechanism

1 Introduction

The key trends in the geo-economic state, which are directly related to transport and logistics activities in Russia, determine the need for effective management of traction resources by organizing a unified technology of transport activities of JSC Russian Railways, which results in rational use of the transport infrastructure, a commensurate increase of loading volumes and attraction of investments in the railways. In the current conditions of dynamism and structural transformations of the Russian transport industry there are significant changes in the technology of the transport process due to

Z. Popovic et al. (Eds.): TransSiberia 2019, AISC 1115, pp. 367–383, 2020.
https://doi.org/10.1007/978-3-030-37916-2_36

the level of infrastructure, traction resources and needs for goods and passengers delivery.

In the international operational length, the share of domestic railways is about 7% (more than 85 thousand km), by the cost of production capital funds - 14%, GDP - 4.9%, the world passenger and cargo turnover - 15 and 24%, respectively [1].

At that, the wear of the track infrastructure reaches 55%, of the electric centralization systems - 76%, of the automatic blocking system – 55%, of the traction substations and the catenary system – 54%, of the locomotive fleet – more than 20% [2]. Until 2025, maintenance of a low rate of modernization of the railway infrastructure having more than 13 thousand km of «bottlenecks» is expected. Elimination of the «bottlenecks» requires an investment of 1.5 trillion rubles. For this reason, in 2020, 196.2 million tons of cargo will not be transported with entailing the budget contribution losses of 1.3 trillion rubles [2].

Railway transport is pursuing the projected growth of transportations - by 2025 the increase of cargo turnover is expected by 21% from the level of 2018 [1], according to its results, by the values of 2017, loading increased by 2.2% to 1289.6 million tons. (3533.2 thousand tons per day, averagely) (Fig. 1a), the average transportation range increased by 35 km (1.9%), up to 1835 km, and loaded cargo turnover reached a historic high of 2596.9 billion tkm (4.2%) (Fig. 1b).

At the same time, the share of railways in the cargo turnover of the country transport system reached more than 87% with an absolute record in the traffic volumes to the peak value of the Soviet period in 1988 - cargo turnover increased by 2.5%. More than 1.157 billion people were involved in the passenger traffic.

The infrastructure of public railway transport (further – infrastructure) represents «the technological complex including public railway tracks and other constructions, railway stations, power supply facilities, communication networks, signaling, centralizing and blocking systems, information complexes and traffic control system and other buildings constructions, devices and equipment providing functioning of this complex» and excludes the rolling stock, using of which means providing a service to a carrier by the operator of the rolling stock – «the legal entity or the individual entrepreneur, possessing the rolling stock… on the right of ownership or another right and providing the rolling stock to legal entities or individuals…» or the owner of the infrastructure – «a legal entity or an individual entrepreneur, possessing the infrastructure on the right of ownership or another right and providing services of its use on the basis of a relevant contract» [The Railway Transport Charter of the Russian Federation: Federal Law of the Russian Federation No. 18-FZ of 10 January 2003]. The major owner of the railway infrastructure of the Russian Federation is JSC Russian Railways (85513 km) which exercises centralized supervisory control of the transportation process. The general principles of providing non-discriminatory access of carriers to the infrastructure are defined in [On approval of regulations of non-discriminatory access to public railway transport infrastructure (Decree of the Government of the Russian Federation No. 710 of 25 November 2003)] on the terms of the public contract of use signed between the carrier and the owner of the infrastructure [On approval of regulations of providing services in use of public transport infrastructure: Decree of the Government of the Russian Federation No. 703 of 20 November 2003].

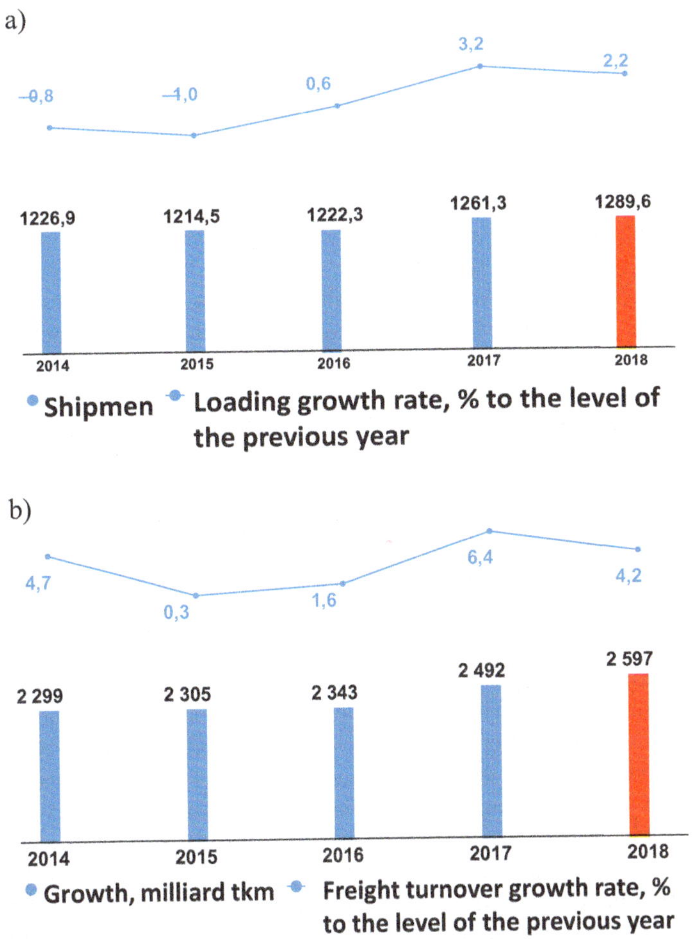

Fig. 1. (a) Loading dynamics, million tons; (b) Loaded cargo turnover dynamics [Annual report of JSC «RZD» 2019 approved by the Government of the Russian Federation (Decree of the Government of the Russian Federation No. 1393-p of 29 June 2019)]

Modeling of transport infrastructure operation and organization of the transport process are represented in the works of B.M. Lapidus, S.M. Rezer, S.N. Kornilov, V.N. Morozov, A.N. Rahmangulov, etc.; the problems of transport systems development and infrastructure formation - in the works of G.V. Veselov, A.P. Abramov, V.M. Buneev, V.G. Galaburda, etc.; analysis of the number of car fleets and harmonization of the track infrastructure capacity - in the studies of P.A. Kozlov, A.S. Misharin, N.E. Aksyonenko, O.N. Dunaev, etc.; the economic problems of locomotive fleet management - in the scientific activity of V.I. Apattsev, A.P. Baturin, K.A. Berngard, I.N. Shapkin.

The above-mentioned works are of considerable theoretical and practical importance both for the improvement of infrastructure and traction tools and for the organization of railway transportation, and at the same time need to be further developed with regard to the modern requirements of the infrastructure and rolling stock.

The uneven distribution of the rolling stock (a half of the total fleet of cargo cars) in the eastern part of the railway network, which has 39% of the capacity of the stations track development, has resulted in the practical exhaustion of the track infrastructure capacity (the average running speed of cargo trains is 50 km/h). Its development requires the implementation of long-term capital-intensive projects, investments in which often pay off slowly [3].

Ensuring economic growth rates higher than the global ones, maintaining macroeconomic stability and introducing modern technologies are difficult to imagine without solving the problems of rapid modernization of the transport complex, renewal of the infrastructure and improvement of transport technologies. The priority area of the development of the production unit of JSC Russian Railways is to replace the regional principles of transportation control with the planning and organization of the train traffic at railway polygons (hereinafter referred to as the polygon), which allows to increase the operation efficiency and clearly distinguish functionalities and responsibilities of functional units.

In 2017 the Eastern Polygon of railroads (further – the Eastern Polygon or the polygon) (Fig. 2), which united the operation processes of the Krasnoyarsk, Trans-baikal, East Siberian and Far East railroads, was created. It is the basis for the formation of the international transport corridors of the Asia-North American Railway. The general coordination of the train traffic flows at the polygon is carried out by the centralized traffic control center (further – CTCC), and regulation of the traction resources is concentrated in the traction resources control centre (further – TRCC).

The cost of the second stage of development of the Eastern Polygon is 464.7 billion rubles (433.2 billion rubles - the state, 31.5 billion rubles - own funds of JSC Russian Railways). At the polygon, with the growth of traffic volumes to the level of last year by 5% in 2018, the repair volume increased by 13.8%, and reconstruction of the track - by 23.6%. More than a quarter of the repair works (28%) were carried out on the improvement of the track, the catenary system with a total length of 111.7 km was reconstructed, 334 facilities were capitally repaired. Automated complexes of the tasks of dispatching planning and train traffic monitoring are being introduced, as well as works are being carried out within the framework of planning and operational regulation of the locomotive fleet systems and modeling railway station operation.

The efficiency of each transportation variation at the polygon is estimated by the method of «accurate analysis of a particular situation» [1], which fully takes into account the environmental conditions, the criteria of the given task, the existing demand and the technology of transportation combined with the capacity of the locomotive traction and infrastructure. The main tools used for selection of an effective strategy are strategic and marketing analyses, which allow to identify the risks of transportation, the state of the infrastructure and the possibilities of railway transport in general. The SWOT analysis is the most common and attractive method.

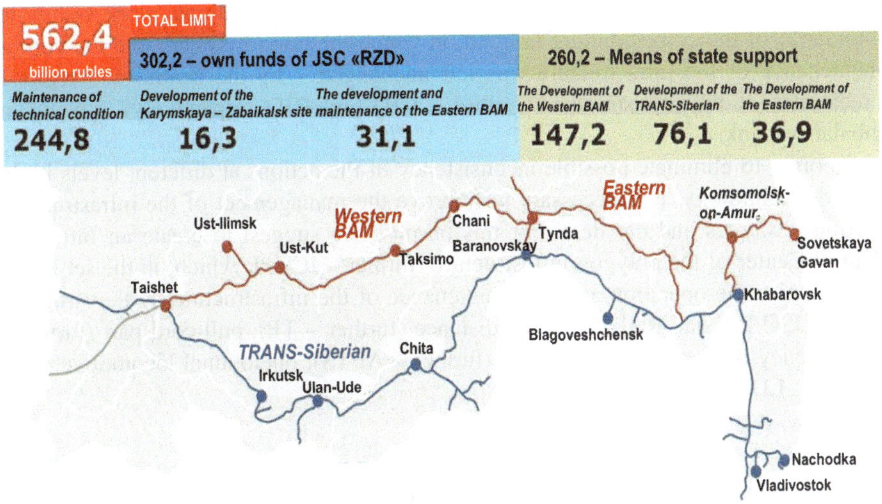

Fig. 2. Development of the Eastern Polygon railway infrastructure at the first stage

2 Methods

The matrix method of marketing analysis – SWOT-analysis – is popular due to its simplicity, versatility and availability, which allow to form a comprehensive view of the business process. It implies an analysis and construction of multidimensional matrices allowing to model situations and the market participants behaviour. The purpose of this analysis is to identify the strengths and weaknesses of a process, the opportunities and threats originating from its immediate environment (competitive environment).

1. Weaknesses - disadvantages;
2. Strengths - advantages;
3. Opportunities – the features of the environment which create advantages in the market;
4. Threats – factors that potentially worsen the market situation.

While analyzing the market, resources and competitive opportunities, four additional parameters are investigated:

(A) strategic problems;
(B) effectiveness of the current strategy;
(C) cost competitiveness;
(D) stability of the position compared to the main competitors.

The polygon technology is based on three-level control of the transportation process, in which decisions on regulation of the locomotive fleet, allocation of loading resources and provision of «gaps» on the infrastructure are made in a rigid hierarchy from the central to regional levels. At that, the regional centers of corporate management (further – RCCM) carry out coordination of operational work within the boundaries of the corresponding railroad on which the control system and the current maintenance of the infrastructure was left unchanged – by the central and regional directorates of the infrastructure (further – CID and RID respectively) – being an unbalanced link.

In order to eliminate possible inconsistency of the actions at different levels in the polygon technology, it is necessary to improve the management of the infrastructure, traction resources and car fleet. For this purpose we suggest to create an Integrated Control Center of the polygon infrastructure (further – ICCPI) which, in the set limits, will organize the operation and the maintenance of the infrastructure by the structural units of JSC Russian Railways (path distance (further – TP), pull-cord part (further – CP), Agency of firm transport service (further – AFTS), operational locomotive depot (further – LD) and other) (Fig. 3).

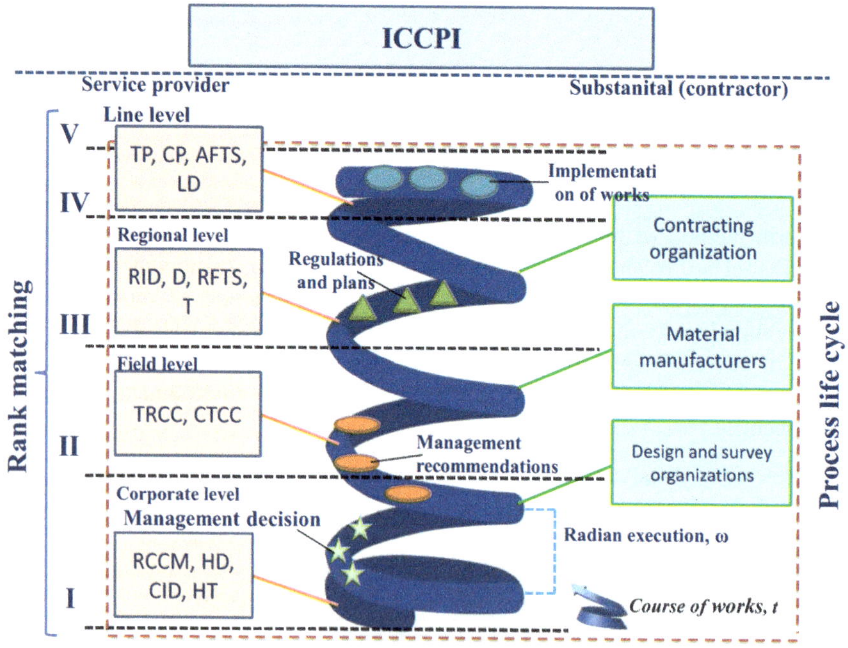

Fig. 3. Model of integrative management of the infrastructure and track resources at the polygon

In the operation of the polygon, ICCPI interacts with TRCC, CTCC, the directorates: of the traffic control, of the traction, of the power supply; with the territorial centers of corporate transport services (further – RFTS), their structural units (AFTS TP, LD and other) and analyzes the efficiency of the infrastructure and locomotives operation on the basis of the active systems theory by means of integral estimation.

The integral assessment consists of evaluation and ranking units, in which the estimation results of heterogeneous productive-economic indicators are combined through convolution matrices - the rank of activity is formed at the intersection of the line and the column (Fig. 4). This mechanism provides for the assessment reliability or the indicator rank: at first, the permissible estimation intervals are ranked by experts - the specialists of the production and economic unit, and then – by experts-competent railway transport specialists of railway.

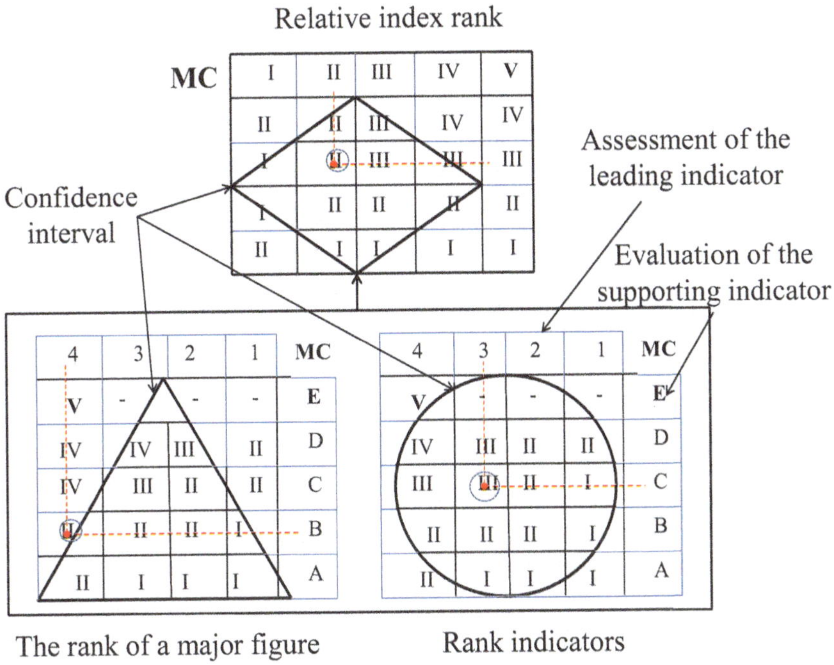

Fig. 4. Formation of the indicator rank

The following indicators are distinguished: the main ones (from 1 to 4), which characterize the execution of planned targets, and the supporting ones (from A to D), which determine the efficiency of production. The rank of the main and additional indicator (from I to V) (Table 1) is characterized by the confidence interval and reflects the quality of the infrastructure and traction operation: the first corresponds to non-fulfillment of the established plan and a low corporate level of interaction at the polygon, and the fifth corresponds to implementation of the targets at the linear level of control:

$$\frac{X_{cp} - tS}{\sqrt{n}} \leq \alpha \leq \frac{X_{cp} + tS}{\sqrt{n}} \tag{1}$$

Where α – the productive-economic indicator;

T – Student's t-distribution parameter;

X_{avg} – an average sample value of the required indicator;

S – the square root of dispersion;
$\sqrt{n}$ – the square root of the total sample.

An expert at each level reviews his own responses only when they fall out of the confidence interval (1), which contains 90% of all the estimates and allows to identify the true value of the statistical characteristic.

Table 1. Formation of the indicator rank

Rank	Integral estimation of the indicator					
I	3*A*	2*A*	1*B*	1*A*		
II	4*A*	3*B*	2*C*	2*B*	1*D*	1*C*
III	4*B*	3*C*	2*D*			
IV	4*C*	4*C*	3*D*	3*D*		
V	4*D*					

In calculating of pricing, cost, evaluation of competitiveness and justification for various transport modifications, this study employs a strategic analysis tool - the Delphi method of expert assessments. It is applied in logistics to monitor shipment quantities and is based on experience, intuition and characteristics measurement. Collecting and summarizing individual assessments of the market situation belonging to various experts, it becomes possible to obtain a collective opinion that will be sufficiently reliable. This method is employed in the programming of the «Automated System of Operation Control of the Integrated Center of the Infrastructure Control».

The polygon control system is analyzed with the help of the above-mentioned methods and the results obtained through studying the existing condition of transport infrastructure and efficiency of its use by the carrier are presented.

3 Findings

The main tool used for selection of an effective strategy of the transport technology is the SWOT analysis, which allows to determine the real assessment of transportation, condition and capabilities of the infrastructure with the traction resources (Table 2).

The level of threats identified by the SWOT-analysis is not critical, and the existing risks can be reduced due to effective technologically built interaction among all the participants of the transport process - JSC Russian Railways, shippers, the state and other market players (Table 3).

Thus, there is a high potential to use the polygon system of transport process control on the railway network, allowing to establish international cooperation and mutually advantageous cooperation of the Russian Federation in the military, economic and cultural spheres.

Table 2. The SWOT-analysis of the polygon traffic control system

Strength 1. Transportation costs reduction 2. Locomotives productivity gains 3. Cargo traffic speedup 4. A single network of terminals and railway tracks 5. The JSC Russian Railways financial position with stable staffing 6. The optimum logistic technology of cargo operations and cargo traffic control 7. Decrease of locomotives single run length 8. Relatively low investments in the polygon development 9. Availability of the automated systems (the software) 10. Favorable directions of the traffic flows 11. Absence of competitors in the market	***Weaknesses*** 1. Updating of technical documentation 2. Changing of the train traffic technology 3. Lack of the required practical experience 4. State tariff regulation of the rail transportation 5. Low level of awareness of JSC Russian Railways employees, counterparties and clients 6. Existence of constantly hampered train traffic sections 7. Severe climatic conditions 8. Additional training of the production staff
Opportunities *1.* Provision of state support 2. Significant market potential 3. Large traffic volumes 4. Stabilization of the delivery time 5. Considerable length of cargo transportations 6. Increase of the section speed 7. Increase of the guaranteed haul distances 8. Time factor for the consumers 9. Rapid implementation of the polygon technology on the railway network	***Threats*** 1. Change of the staff training procedure 2. Inadequate regulatory framework 3. Risks of the train traffic schedule violations 4. Trains and locomotives time spent at technical stations 5. Significant investments 6. Lack of state support throughout the project 7. Changes of the legislation

Table 3. The SWOT-analysis results interpretation

	Strengths	Weaknesses
Opportunities	Development sector 1. Studying and introduction of the modern world transportation technologies 2. Development of mutually advantageous relations with clients 3. Reduction of the transportation cost 4. Reduced number of "dismissed" trains 5. Improvement of the "gaps" provision technology	What to change sector 1. Implementation of a polygon system on the railway network 2. Formation of a management system with clients and staff 3. Solution of the employment problem of JSC Russian Railways released employees 4. Audit of the existing regulatory documents 5. Development of a polygon management program
Threats	Threat compensation sector 1. Competent trained staff 2. Continuous monitoring of the legislation 3. Observance of the train traffic schedules 4. Modernization of the infrastructure and rolling stock	Problems sector 1. Development of measures enhancing the level of service 2. Impromptness at solving urgent tasks in non-standard situations 3. Development of the industry manufacturing products for heavy train traffic

One of the main indicators of the transport process quality is the consumption of electric energy by train traction taking into account the current state of the infrastructure (absence of warnings limiting the predetermined speeds, the traffic capacity, track weight loads, etc.). Taking into account the practical experience of the Eastern Polygon in the volumes and distances of cargo transportation in the territory of the Russian Federation, the application of the polygon transportation control system and traction rolling stock allows to meet the requirements of the transportation process with a smaller amount of traction resource due to the rational use of the infrastructure and increase of the locomotive fleet efficiency.

In order to obtain the target volumes specified by the train traffic standard schedule, daily, weekly, decadal or monthly planning of locomotives and locomotive crews operation is based on calculation of the need and establishment of the standards for maintenance of operated train locomotive fleets taking into account the available number of locomotive crews [4]. Earlier studies [5, 6] have shown that the method of optimal network flow [7, 8] is effective, allowing for real-time operational planning of the transportation process at the polygon. The network flow model of the traction operation technology in cargo traffic is based on the existing schemes of locomotive circulation sections L (further – LCS) and maintenance ranges of locomotive crews N (further – MRLC) (Fig. 5).

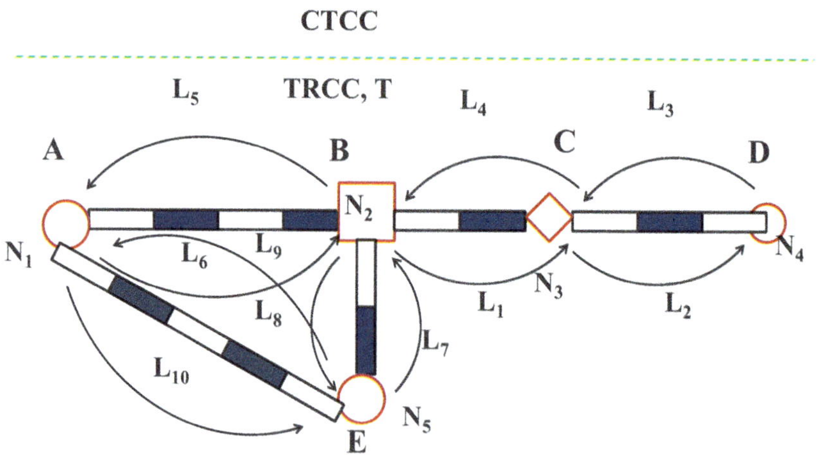

Fig. 5. The scheme of locomotive circulation section $L_1 - L_3$ in directions $N_1 - N_5$

Each LCS where locomotives maintenance (T, by service companies) is carried out is considered as a set of several MRLS where locomotive crews operate from home station B to turnaround stations A, C, D, E and vice versa according to the train paths (with a train, a reserve locomotive) or passengers. The Polygon technology allows TRCC to cover CTCC needs for the traffic flow due to extension of the traction resources circulation sections. In case of train traffic schedule deviations, a considerable gain in resource consumption is achieved due to correction of the trajectory in alternative directions immediately during the trains running. In such cases, ICCPI offers the most optimal (reserve) options for the use of the infrastructure within an alternate scheme of the railway facilities current maintenance. In order to avoid possible conflicts of interests between economic entities at the «joints» of railways and complete the track repair campaign, the regional centre of corporate management carries out technological coordination of transportations in the volume required by the polygon (Fig. 6).

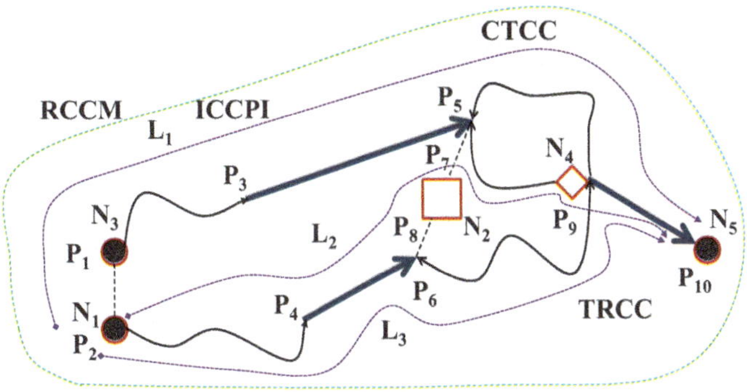

Fig. 6. Assignment of locomotives $L_1 - L_3$ to trains $P_1 - P_{10}$ in direction $N_1 - N_5$

Applying the Methodology guidelines on requirements generation of cargo traffic locomotives and indicators of their scheduled operation (TDL-60), approved by JSC «RZD» 25.06.2014 No. 266, to solve the problem of assignment of locomotives L characterized with different types of traction to trains P in the reserve directions of infrastructure N of different specialization, it is necessary to modify the classic version due to approximation of heterogeneous resources: to have an opportunity to replace an existing resource with a similar one open back on any of the edges of the previous resource (<N_1, N_5 > , > filled L_3, open to L2 through <N1, N2, N5> without the train weight break and changes of the traction type). The application of quasi-classical rollbacks results in a significant increase in the locomotives performance (fewer locomotives are assigned to remove a similar number of trains or cargoes) and the uniformity of loading or repair of the infrastructure.

At the same time, further improvement of the polygon principles of transport process control, operation and infrastructure of JSC Russian Railways is required. In turn, the business unit «Railway Transport and Infrastructure» (hereinafter – RTI) integrates the main production business units engaged in the organization of railway transport, maintenance and development of the infrastructure and locomotive complex. The production and financial result of the transportation process depends on the efficiency of their activities and the level of their technological interaction. The problem of advance prediction and uncertainty of possible trend reversals in the changes of key performance indicators, which makes long-term economic forecasting difficult, especially in such a capital-intensive industry as the railway transport, is extremely serious [9]. Its elimination requires the development of mechanisms ensuring the continuous production and economic operation of railway polygons. In order to increase the profitability of transport activities, it is important, along with efficient operational management, to improve the efficiency of the infrastructure and traction [10] through continuous monitoring of the direct costs (Fig. 7).

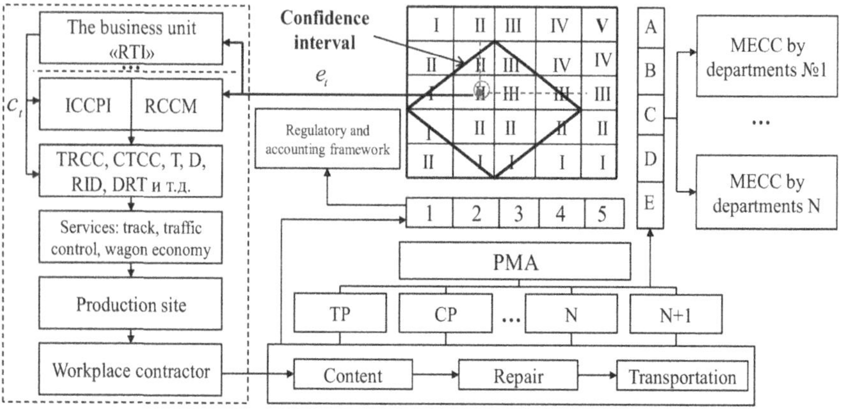

Fig. 7. The mechanism of adaptation of the transport process participants' consolidated liability cost variances

The mechanism of adaptation of the cost variances to the cargo turnover volumes (further – MA) functions on the basis of fundamental approaches to solution of problems [11] on adequate adaptive mechanisms synthesis with identification which allow to exercise control of primary variances synchronously with the infrastructure operation and transportation.

Primary registration information on the variances amount and cost is fixed by the primary adaptation mechanism (further – PMA) which collects the data from functional units, technological sections and workplaces. The production and financial activity of RTI structural units is controlled by the mechanism of the enterprise costs control (further – MECC). In each the input data represents the costs, and in PMA of normatives – the received normatives. Comparing them, the assessment of the normatives changes is conducted. The main objective of the assessment and ranking is formation of the level of influence of each cost item devoted to adaptation of norms and standards of production. On the basis of the obtained information on the order and variances, the priority factors, the key reasons and units, the most influencing costs are defined. Further, administrative recommendations on adjustment of the normative are developed and changes are introduced in the regulatory base on the types of activity and executives.

In PMA of i-resource costs, the assessments and ranks are used for stimulation of the executive. If the executive uses several resources, then complex assessment is formed on the basis of all the values of local ranks in MS.

$$e_t = k_t y_t\,,\ \ k_t = \begin{cases} 1+\nu(x_t/y_t - 1), \text{ если } x_t \le y_t \\ 1+\mu(1 - x_t/y_t), \text{ если } x_t > y_t, \end{cases} \qquad (2)$$

where y_t – the indicator value;
x_t – the normative of assessment;
$\chi(x_t, y_t)$ – the function of stimulation for the deviations of the indicator values from the normative;
ν – the incentive multiplier;
μ – the penalty multiplier, ν, $\mu > 0$.

The cost of the infrastructure and traction operation for t-period is identified:

$$c_t = \sum_{i=t}^{n} k_{it} P_{it} \qquad (3)$$

where k_{it} – the quantity of i-resource actually consumed by the infrastructure for t-period;
P_{it} – i-resource cost;
i – the resource number, $i = 1, 2, \ldots, n$

Decrease in the cost of transportation process is possible due to the correct solution of a technological task on the effective organization of traffic volumes, rational loading of the railway network, rhythmical operation of the polygon, intensive use of technical means and track development of stations and sections. Presence of the person in each point assumes inevitable inertness in decision-making, possible errors, and subjectivity when choosing options of solution for the arisen situations and also complexity and duration of coordination in the chain of command.

The obtained results of the deviations ranking are used by the production and administrative units of JSC Russian Railways for operational elimination of over expenditure (adjustment of the budget, development of additional measures, etc.). The frequency of control satisfies the requirements of technological process and can be continuous. Regular deviations of III, IV ranks are transferred through communication channels to the directorates and centers of JSC Russian Railways, and the primary ones are controlled synchronously with transportation process and arrive to ICCPI from the places of their occurrence – structural units – and are fixed by PMA. Deviations of the loading rhythm of railway stations are presented in «Automated system of loading supervision of a production department» (further – ASLS) [Certificate of state registration of the computer program № 209619025 of 09 July 2019] (Fig. 8).

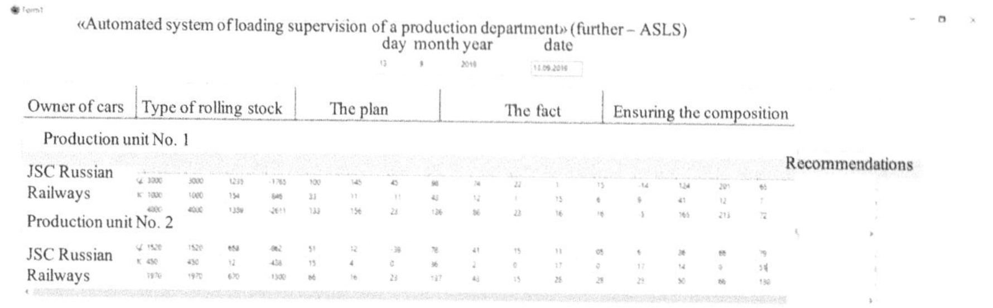

Fig. 8. The interface of ASLS

The program is designed to automatically determine the standard of loading of a production unit for the reporting day and the progressive total for a month and to monitor the implementation of the established plan and «gaps» schedule. ASLS allows to give management recommendations on the loading rhythm taking into account the close and distant approach of the rolling stock to reach the performance targets of railway stations in cases of deviation from the reporting loading standard.

Railway transportation has become an area where efficiency is increasingly dependent on the ability to extract information from complex data sets as well as to make optimal real-time decisions. Effective information and data management is therefore vital for the railway, which is a closely interconnected ontological system of systems where changes in any of its parts can have significant consequences elsewhere. The results obtained in ASLS are used as the input data to select recommendations on making management decisions from ACS of ICCPI [Certificate of state registration of the computer program № 2019612748 of 26 February 2019].

It is developed in the Lazarus programming language and is used for automated coordination and evaluation of actions on management of the production infrastructure and traffic of the traction resources by comparing the parameters of the railway track facilities maintenance to adjacent values (Fig. 9).

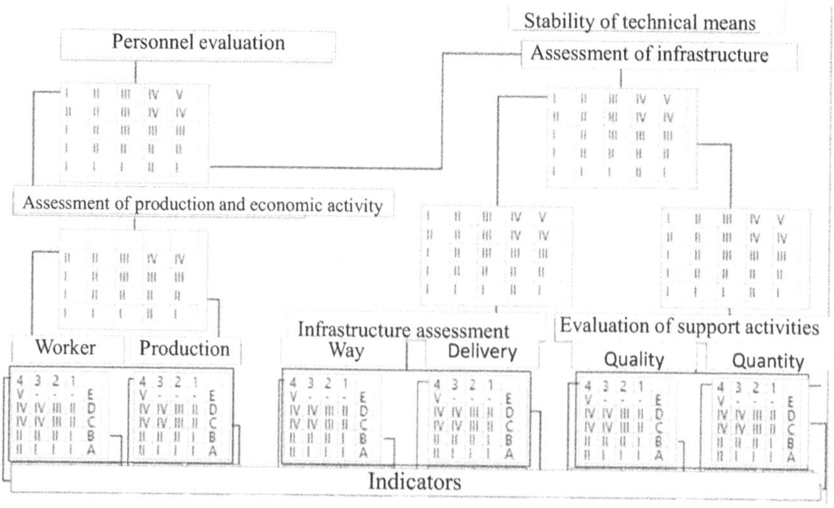

Fig. 9. The operation interface of ICCPI ACS

ICCPI AS with the help of MS ranks of production, economic and management activities of the carrier determines the integral assessment of the transport infrastructure and the process of providing the cargo delivery service as a whole. This program allows to consolidate the transportation process KPIs (including retrospective results) at the polygon by integrating the output data (tabular forms) from the corporate information environment - industry automated systems used in JSC Russian Railways (ASOUP- 2, EC ACS FR, ACS MS, ACS MR, SIRIUS, etc.), and visual representing of the operation of the enterprises servicing the railway infrastructure.

The system operation process is structured as follows:

1. Obtaining the initial data on the railway network, train routes, operational parameters and the infrastructure condition.
2. Setting of program operation mode: the order of the infrastructure objects repair regarding the train traffic schedule.
3. Designing the infrastructure maintenance process at the polygon.
4. Generation of algorithms for solving the set problem.
5. Construction of optimal variants of the polygon operation.
6. Identification of «barriers» and implementation of the established train formation plan in accordance with the route prioritization.

Most existing automated railway management systems are not equipped with the necessary dynamic models of rapid response to market fluctuations and are focused primarily on information services for the operational staff of regional directorates structural units and management. ICCPI ACS is able to provide almost all enterprises with operational information and management recommendations through the possibility of interaction with the upper-level management systems. The indicators analyzed are presented in the form of separate «digital layers» for the main vertically integrated directorates of the Railway Transport and Infrastructure Unit. According to the results of the estimated period, each «digital layer» is ranked in the specified range and with the help of the convolution matrix is located with the adjacent «digital layer» to determine the integral assessment of the polygon operation.

For full-scale implementation, adaptive development of the ICCPI ACS functioning model is needed, which will require sufficient financial investments not only in the system architecture, but also in the main constituents and computer equipment components. At the same time, ICCPI ACS is interlinked in technological interaction and delegation of responsibility - with the train traffic control centers and the infrastructure content management centers; in planning and loading - with the system of company transport services; in organization of cars traffic - with clients.

4 Discussion of Results

In the context of increasing costs of energy, materials and production resources, the transportation process through the infrastructure of the railway network needs to improve the energy efficiency of technological processes. For this purpose, the industry has created conditions for uninterrupted operation of the transport corridors, including the introduction of a new polygon model capable of eliminating or reducing losses from the irrational interaction of the structural units of JSC Russian Railways in the use of the infrastructure and allocation of traction resources.

In the initial stage of the polygon technology application, the key problem in the development of the transport infrastructure is the search for a balance in the use of traction, the state of infrastructure and business needs.

The SWOT-analysis revealed the following directions of transport process development at the polygon:

1. Recharacterization of discharged staff;
2. Audit of transport regulatory documents;
3. Development of new polygon network performance indicators;
4. Introduction of modern technologies to reduce the cost of transport process;
5. Concentration of dispatch personnel in one building;
6. Elimination of artificial barriers in train running (by avoiding «joints»);
7. Concentration of single series of locomotives and spare parts for service depots;
8. Vulnerability of the dispatching control system, in case of military actions;
9. Development of measures to improve the quality of services provided;
10. Integration of logistics development production processes.

However, it is important to understand that each of these areas requires separate solutions and continuous monitoring.

Under the continuous increase of the purchase price of new locomotives, the high wear degree of existing traction resources and the limited investments of JSC Russian Railways, the issue of improving the efficiency of rolling stock operation is particularly relevant. Despite the fact that the network infrastructure restrictions of track development introduce significant limitations on the running speed of cargo traffic, and the financial congruence of the rolling stock restricts satisfaction of the client's needs, innovative systems of transport control using artificial intelligence can expand the «bottlenecks» . ICCPI will strive to provide such an infrastructure capacity together with qualitative repair works, which will ensure the maximum income of JSC Russian Railways and the target passage of trains by existing traction resources.

The productivity of investment measures providing for repair, modernization of capacities, the infrastructure facilities and trains traction to a larger extent depends on the efficiency of management, than on the network load. The model of integrative infrastructure and traction resources management on the polygon meets the requirements to reliability of results and has applicational significance when developing automated systems on railway transport. ICCPI ACS allows to exert automated management of the production infrastructure of the railways and rolling stock.

Based on the results of the research conducted, it can be concluded that the system of traction resources energy efficiency management implemented by JSC Russian Railways, combined with technical and technological readiness of the unified infrastructure, taking into account the improvements proposed in the article, will ensure favorable implementation of polygon communication on the railway network.

References

1. Sirina, N.F., Yushkova, S.S.: Improvement of the railway polygon transport infrastructure control. Mod. Technol. Syst. Anal. Model. **62**(2), 98–108 (2019). https://doi.org/10.26731/1813-9108.2019.2(62).98-108
2. Sirina, N.F., Yushkova, S.S.: Integrative management of infrastructure in railway transportation activities. Transp. Urals **61**(1), 7–17 (2019). https://doi.org/10.20291/1815-9400-2019-1-7-17
3. Macheret, P.D., Macheret, D.A., Savchuk, R.R.: Economic importance of automation of transport infrastructure design. In: Proceedings of the 2019 IEEE Conference of Russian Young Researchers in Electrical and Electronic Engineering, pp. 1430–1431 (2019). https://doi.org/10.1109/eiconrus.2019.8657126
4. Miroshnichenko, O.F., Oginskaya, A.E.: Methodology for calculating financial result for the owner of the infrastructure from increasing carrying capacity of the direction during operation in cars with increased capacity. Vestnik Railway Res. Inst. **78**, 33–40 (2019). https://doi.org/10.21780/2223-9731-2019-78-1-33-40
5. Goldberg, A.V., Tarjan, R.E.: A new approach to the maximum flow problem. In: Proceedings of the Eighteenth Annual ACM Symposium on Theory of computing (1986)
6. Bertsekas, D.P.: Distributed asynchronous relaxation methods for linear network flow problems (1986). Report LIDS-P-1606
7. Goldberg, A.V., Tarjan, R.E.: Solving minimum cost flow problems by successive approximations. In: Proceedings of 19th ACM STOC (1987)
8. Bertsekas, D.P., Castanon, D.A.: The auction algorithm for the minimum cost network flow problem (1989). Report LIDS-P-1925
9. Macheret, D.A., Vallev, N.A.: Scientific toolkit for predictive efficiency management of railway transportation. Vestnik VNIIZHT **77**, 84–91 (2018). https://doi.org/10.21780/2223-9731-2018-77-2-84-91
10. Silyuta, A.G., Kossov, E.E.: Modeling the train operation of the mainline locomotive. Vestnik Railway Res. Inst. **77**, 218–221 (2018). https://doi.org/10.21780/2223-9731-2018-77-4-218-221
11. Tsyganov, V.V.: Emotional expectations and social stability. In: Proceedings of 18th IFAC Conference on Technology, Culture and International Stability. IFAC PaperOnline, vol. 51, no. 30, pp. 112–117 (2018). https://doi.org/10.1016/j.ifacol.2018.11.258

Feedforward Tilt Control on Curves for Gyroscopically Stabilised Monorail Vehicles Using Machine Vision

Yaroslav Zisser(✉) and Irina Zisser

Far Eastern State Transport University, Seryshev st., 47, Khabarovsk 680021, Russia
zisser@yandex.ru

Abstract. Dynamic stabilisation of gyroscopically stabilised single track (monorail) vehicles necessitates accurate tilt angle control on transitions between track sections with different curvature to eliminate or to effectively damp unwanted sideways swaying of the vehicle caused by flywheel nutation and by the control lag arising from feedback control loop. In this paper, a feedforward tilt control method is proposed, which utilises on-the-go running rail image analysis to predict an oncoming rail curvature change and to generate an input for the flywheel gimbal servo drive to get the vehicle cant adjusted to the rail curvature beforehand. This approach will allow such vehicles to be used in high-speed transportation, where their advantages could be manifested in full.

Keywords: Gyroscopic stabilisation · Single-rail vehicle · Gyro-monorail · Feedforward control · Tilt control · Machine vision

1 Introduction

Normally, two-wheeled vehicles (bicycles, scooters, etc.) running on flat surfaces cannot move along a straight line because of precession causing their wheels to follow a wave-like trajectory. The precession, being a key factor of dynamic equilibrium, is maintained by two controls: handle bar (or steering wheel) and sideways adjustment of the centre of mass [1, 2]. Similar vehicles guided by the rail cannot divert from their track, their wheels lack the degree of freedom to precess, thus, an additional actively controlled gyroscope (flywheel in a gimbal) is used to maintain precession. The most famous gyroscopic monorail vehicle was designed by Louis Brennan in early 20th century [3], who succeeded in attracting enough funds to develop his ideas [4] to an operating full-scale rail car. However, at that time, such vehicles were discarded as impractical and unsafe because they were supposed for usage at regular speeds, whilst their drawbacks will likely to be overridden by their advantages at high speeds (Table 1).

Z. Popovic et al. (Eds.): TransSiberia 2019, AISC 1115, pp. 384–391, 2020.
https://doi.org/10.1007/978-3-030-37916-2_37

Table 1. Features of gyroscopically stabilised single-rail vehicles

Advantages	Disadvantages
• Very low rolling friction • Inherent absence of detrimental hunting oscillations [5] • Inherent ability to bank (tilt) on turns so that lateral forces acting on passengers and wheel flanges are eliminated • Reduced track complexity, thus, lower construction and maintenance costs • Increased crashworthiness: derailment does not lead to abrupt toppling or turnover	• Extra energy is consumed to accelerate and maintain rotation of massive flywheel(s) • Impossibility to make a consist of carriages due to mutual interference • Special props are needed to support the vehicle when gyro-stabilisation is off • Difficulties in handling disturbances other than natural toppling moment

Among the issues mentioned in the right column of Table 1, we can mark out the disturbances being crucial for the safety and riding comfort. The most weighty ones affecting vehicle stability are:

1. Gravity force tending to topple the vehicle down;
2. Sideways displacement of the centre of mass caused by load imbalance;
3. Crosswind or other extrinsic side impacts;
4. Change in horizontal track curvature.

For a balanced monorail vehicle, which undergoes no toppling forces except gravity, control action of gimbal servo drive is simple and aims to replenish vehicle's potential energy loss by assisting precession at proper moments depending on vehicle's tilting angle and angular acceleration [2, 6, 7].

Factors 2 and 3 from the list above both have the same nature – accidental side force causing corresponding toppling moment. Although monorail vehicle still may be balanced only with the help of servo drive by adjusting gimbal angle, it appears impractical because in this case the vehicle would reach its equilibrium tilted, which is inconvenient (Fig. 1).

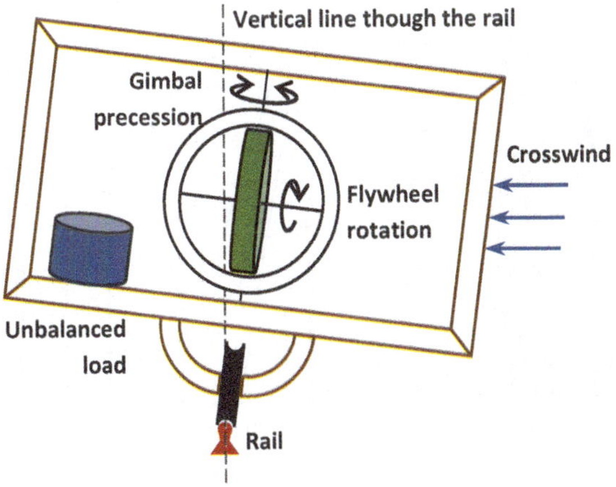

Fig. 1. Unopposed external force leads to undesirable cant

Measures to tackle these disturbances incorporate any means to restore initial balance of the vehicle, namely to shift some counterweight (possibly the gimbal itself) transversely to set vehicle's centre of gravity above the rail and render the car body upright.

Similar tilting effect caused by centripetal force also takes place on curved stretches of the track. However, unlike previous case, this tilt is favourable because it naturally eliminates uncomfortable side force on turns and leaves the only component normal to the vehicle bottom (see Table 1, left column).

An idea arising from these considerations is to timely distinguish causes of tilt in order to suppress undesirable (from accidental side force) and to maintain desirable (from curved track) tilt components, which may occur imposed one on another on curvilinear track sections. In this article, we outline the method to generate inputs for gimbal and counterweight servo drives based on continuous analysis of the rail image in front of the vehicle, and on the signals from speed and tilt sensors. This allows for prediction of forthcoming change in centripetal force directly, which is classified as feedforward control [8], unlike feedback loops only used in early designs to control gimbal angle by the signal obtained either from car tilt angle sensor or from gimbal trunnion angle sensor or both. Such feedbacks are retained in the proposed system as a proven technology but complemented with machine vision, modified control algorithm and hardware (to be employed at the embodiment stage).

2 Monorail Vehicle Tilt Predictive Control System Design

In order to be solved, the problem posed in the introduction should be subdivided into two parts specified below.

2.1 Extracting a Value of Rail Curvature Out of the Image Filmed in Front of the Vehicle

The processing of this image is illustrated by Fig. 2 and implies:

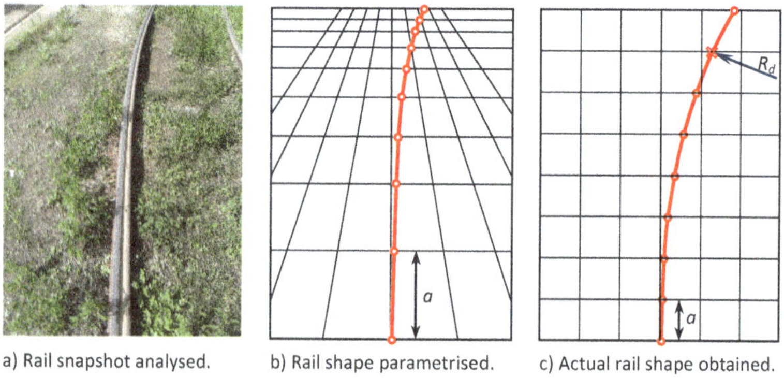

Fig. 2. Tramway rail image processing example

- Choosing suitable neural network [9] and training it to recognise the rail (Fig. 2a);
- Image vectorisation and interpolation to get the parameterised line representing the rail shape (Fig. 2b);
- Affine transformation of the line with necessary corrections accounting for view point, camera aberrations, and current tilt angle to get non-distorted rail shape (Fig. 2c);
- Calculation of the line curvature at the point of interest (cross marker in Fig. 2c).

2.2 Modified Control System with the Value of Curvature Introduced into It

It should be noticed that more detailed description of part 2.1 extends beyond the scope of this article and will be expounded in another publication. Henceforth we shall assume that the first part of the problem is solved, and approximate value of rail curvature κ_d at certain distance d ahead of the vehicle obtained.

Further considerations are related to the control process itself:

- **Control objective is to maintain vehicle's floor horizontally** on a rectilinear track section (zero curvature) and to tilt the vehicle on a curvilinear section by the angle eliminating sideward force component;
- **The distance d to be constantly adjusted** with respect to vehicle speed v.
- **Sudden disturbances** (crosswind, load imbalance, etc.) **are detected by their tilting effect**, which is quantitatively expressed through instantaneous tilt angle θ_t obtained from the tilt (horizontality) sensor(s);
- Both **gimbal and counterweight servo drives must be inverter-fed synchronous motors** to provide high-precision speed and torque control;
- **The servo drives must act concurrently in both feedback and feedforward control loops** where gimbal angle θ_g and counterweight cross travel (transverse displacement) d_c affect the output (target) variable θ_t;
- **Predetermined rail curvature κ_d is used to anticipate disturbance from centripetal force** and to generate corresponding intentional cant θ_a in advance;

Control process diagram based on the above listed principles is shown in Fig. 3. Physically all blue (numbered) blocks may be integrated in one microcontroller or programmable logic controller (PLC) as relatively simple computational subroutines. Rail image processing unit requires either specialised image/vision processor with embedded neural network architecture [10] or general-purpose microcontroller with sufficient performance and specialised software libraries. Block-wise description of the diagram is given below.

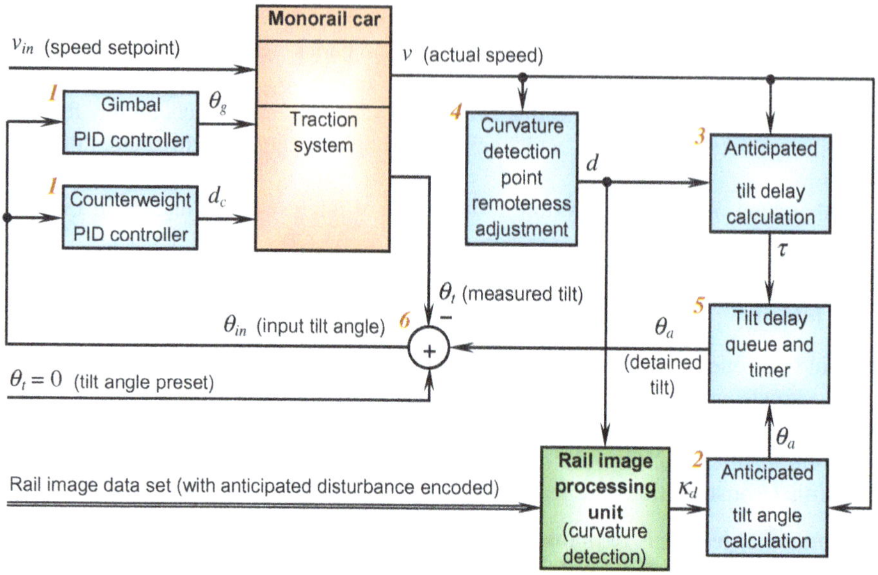

Fig. 3. Generalised block diagram of single-rail vehicle stabilisation system

2.2.1 Blocks of Gimbal and Counterweight PID Controllers

PID controllers of both balancing servo drives have the standard form of transfer function:

$$F(s) = K_p + K_i \frac{1}{s} + K_d s \tag{1}$$

Parameters of either PID controller in Fig. 3 depend on vehicle's inertial properties (dimensions, centre of gravity, mass distribution) and must be tuned experimentally provided that:

- Counterweight PID controller must compensate for imbalance by properly positioned load with respect to vehicle's central axis. To achieve this, its proportional gain K_p must be prevailing over the remaining two (integral and derivative, K_i and K_d) and non-linear, proportional to $\tan(\theta_{in})$ within operational range, to get the angle θ_t converted into linear displacement of counterweight.
- Gimbal PID controller must react on changes of cant rather than on cant itself, therefore its derivative gain K_d must prevail over K_p and K_i, and must be linear (angle-to-angle conversion).

Neither of PID controllers should actively utilise its integral gain K_i except for small values introduced for static error elimination.

2.2.2 Block of Anticipated Tilt Angle Calculation

Anticipated tilt θ_a may be determined from the analysis of forces acting on the monorail car body on a curve (Fig. 4) in assumption that all side forces are continuously and timely neutralised by the counterweight feedback loop, i.e. resultant force is composed only of gravitational and centripetal components:

$$\mathbf{F_r} = \mathbf{F_g} + \mathbf{F_c}. \tag{2}$$

Hence

$$\theta_a = \arctan\left(\frac{F_c}{F_g}\right) = \arctan\left(\frac{\kappa_d v^2}{g}\right). \tag{3}$$

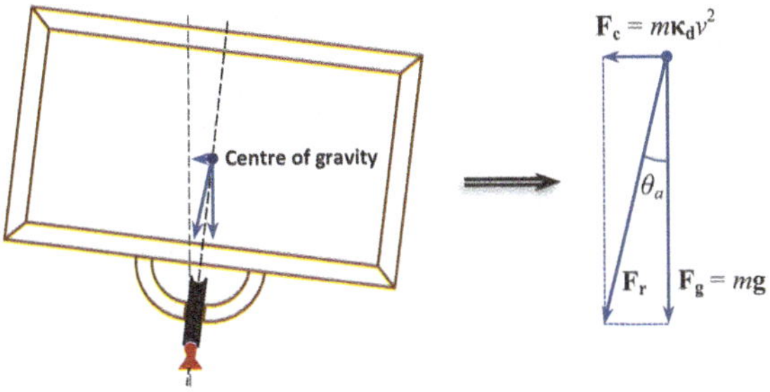

Fig. 4. Diagram of forces acting on the car body

2.2.3 Block of Anticipated Tilt Delay Calculation

The time τ to elapse before the point, for which κ_d was determined, may be found approximately in assumption of v = const, i.e. the speed would not significantly change while the vehicle passing the distance d:

$$\tau = \frac{d}{v} - T_{proc} - T_{cw}, \tag{4}$$

where T_{proc} – processing time needed to get anticipated tilt value ready since the instance of rail image capture, T_{cw} – time needed by the counterweight to react on the change of θ_a.

2.2.4 Block of Curvature Detection Point Remoteness Adjustment

The last two terms in (4) may be assumed constant, but the first ratio may occur troublesome at low or zero speed. To mitigate this circumstance, image processing unit includes adjustment of d in dependence of v.

It would be preferable to get d/v constant that would yield almost constant τ and simpler control, but the value of d is bounded by camera's field of view (Fig. 5).

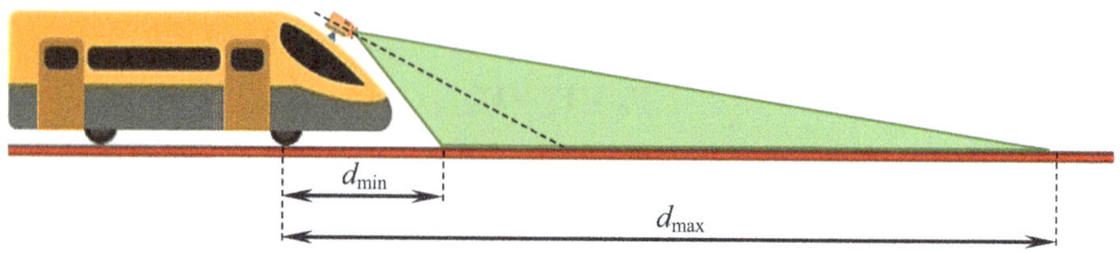

Fig. 5. Distance d ranging within the scope of the camera

For a certain low speed threshold v_{min}, when d/v becomes very large, feedforward loop ceases making sense and must be halted with a fixed output $\theta_a = 0$ at $d = d_{min}$ (Fig. 6).

Likewise, if vehicle speed v overruns certain upper bound, point d has to be confined within rail image margins to retain certainty of the input needed for curvature detection unit. In case of fixed $d = d_{max}$ but increased v, the ratio d/v diminishess and may fall to very small or even negative τ in (4) resulting in control failure. Thus, vehicle speed v must be limited to maximal value v_{max} allowing for minimal time τ sufficient for timely reaction.

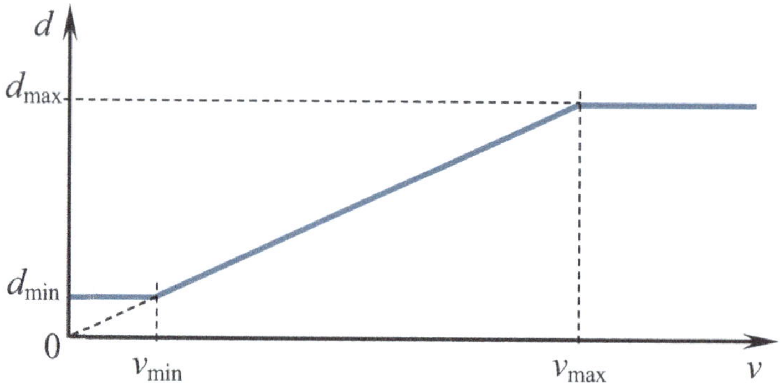

Fig. 6. Non-linear output generated by point d remoteness adjustment unit

The signals for θ_a reset and speed limitation of traction drive are not shown on the diagram in Fig. 3 as related to control robustness rather than to control method. Subsequently, rail image processing software sets the point of interest ahead by d along the detected curve.

Note that reverse motion of the monorail vehicle requires switching to opposite camera facing backwards and working identically.

2.2.5 Block of Tilt Delay Queue and Timer

Every instance of tilt angle θ_a coming from block **2** is associated with the delay τ, by which it is detained before being fed to summator ***6***. The block calculates timestamps for each release of θ_a by adding current time t and the delay τ: $t_{\text{stamp}} = t + \tau$. These pairs (timestamp and θ_a) are stored in memory forming a queue (FIFO – first in, first out).

The earliest timestamp is continuously compared with current time (e.g. every few milliseconds or rarer). The event of $t_{\text{stamp}} \geq t$ triggers retrieval of value θ_a from memory and sending it to the summator. Then next timestamp is extracted from memory and compared with t until coincidence, after which a new corresponding value θ_a is also read from memory and supplied to the summator. So the process is continuously repeated. Every value θ_a is held at summator's input until replaced with next value.

Evidently, the "current time" need not to be absolute "human" time. Any local onboard timer would suffice for reference and sampling.

3 Conclusion

Implementation and improvement of the above described control method for single-rail vehicles is designed to solve one of the major problems preventing such vehicles from broader usage – unwanted swaying and improper lean caused by side forces or track curvature. Microchips specially designed for industrial applications of machine vision and neural networks are the best candidates to implement the system with the least labour and cost.

Deeper analysis, rail curvature detection process, and refined calculations accounting for more factors will be discussed in other papers along with designing a prototype.

References

1. Lot, R., Fleming, J.: Gyroscopic stabilisers for powered two-wheeled vehicles. Veh. Syst. Dyn. **57**(9), 1381–1406 (2019). https://doi.org/10.1080/00423114.2018.1506588
2. Åström, K.J., Klein, R.E., Lennartsson, A.: Bicycle dynamics and control. Control. Syst. Mag. **25**(4), 26–47 (2005). https://doi.org/10.1109/MCS.2005.1499389
3. Eddy, H.T.: The mechanical principles of Brennan's mono-rail car. J. Franklin Inst. **169**(6), 467–485 (1910). https://doi.org/10.1016/S0016-0032(10)90004-5
4. Brennan, L.: Means for imparting stability to unstable bodies. US Patent No. 796893 (1905)
5. Iwnicki, S.: Handbook of Railway Vehicle Dynamics. CRC Press, Boca Raton (2006)
6. Kim, K.Y., Bretney, K., Tsang, L.: Gyroscopic stabilized vehicle. US Patent No. OO8706390B2 (2014)
7. Arnold, R.N., Maunder, L.: Gyrodynamics and its Engineering Applications. Academic Press, London (1961)
8. Altmann, W., Macdonald, D., Mackay, S. (eds.): Practical Process Control for Engineers and Technicians. Newnes, Oxford (2005)
9. Bishop, C.M.: Neural Networks for Pattern Recognition. Clarendon Press, Oxford (1995)
10. Intel Corporation website. https://www.intel.ai/

Development Simulation Model of Traction Network Alternating Current 25 kV

Alexander Onischenko[1(✉)] and Ilya Chernykh[2]

[1] Far Eastern State Transport University, Seryshev str. 47, Khabarovsk, Russia
a.a.onischenko@mail.ru
[2] Ural Federal University, Mira str. 19, Ekaterinburg, Russia

Abstract. Railway transport is the most important part of the country's transport system. Railway transport accounts for about 70% of domestic freight turnover and almost 60% of passenger traffic. More than 50% of cargo transportation is carried out by electric traction. Train safety devices connected to rail circuits are subject to constant interference from traction network power supply. The sources of influence are noise generators and the processes of summing them from several sources (including resonance in contact network). Interference arising in the contact network may appear at the input of the receivers of SCB devices galvanically or inductively. The article covers technical characteristics of electrified Russian Railways JSC "RZD", with particular emphasis given traction AC network of 25 kV, proposed and developed in MATLAB & Simulink a generalized mathematical model of the contact network, electric rolling stock and rail network designed to conduct a comprehensive study of processes resulting from operation of the traction network and interaction between different physical processes.

Keywords: Rail transport · Electrification · Traction network · Contact network · Electric rolling stock · Rail network · Simulation · MATLAB & simulink

1 Introduction

The railway complex is special strategic importance for Russia. It is link of the unified economic system, ensures stable activity of industrial enterprises, timely delivery of vital goods to most remote corners of the country and most affordable transport for citizens.

According 2018, the operational length Railways of Russia for more than 85.5 thousand km, length of electrified lines is approximately 43.7 thousand km. Over 19 thousand km of Railways were electrified by the 3 kV DC system. In recent years, preference has been given to electrification of lines to more advanced alternating current systems of 25 kV or 2 × 25 kV of 50 Hz power frequency.

Figure 1 shows a section of the railway length of 40–50 km electrified AC system 25 kV with two traction power substations (TPSS1, TPSS 2), located near stations A and B. To the power line (transmission line) three-phase alternating current 110/220 kV connected three-phase transformer (TphTr), which lowers the primary

Z. Popovic et al. (Eds.): TransSiberia 2019, AISC 1115, pp. 392–400, 2020.
https://doi.org/10.1007/978-3-030-37916-2_38

voltage 110 kV to 10, 25 or 35 kV. Voltage 25 kV is supplied to the bus (phase A, B, C) supply traction network (STN), voltage 35 or 10 kV - on the power bus adjacent to the substation area [1].

Uniform loading of phases power supply system is provided by delivery in TN at station A and section S1 voltage differing in phase from the voltage of TN S2, which is implemented by connection contact network (CN) of above section and stations to the different phases of tire 27.5 kV: CN of a section S2 through feeder (F3) and appropriate switch feeder contact network (SFCN) connected to the bus phase b; CN station A and section S1 to the bus phase a; rails through rail feeder (RF) to the bus phase c. This connection makes possible connection of the CN of section S1, station A and current collector (CC) of moving electric rolling stock (ERS) and impossible connection to CN station A and section S2 due to joining different phases a and b, which will lead to short circuit (SC) phases transformer (TphTr), so these areas of CN divided by air gaps (AG) and neutral insert (NI), which eliminates possibility of even short circuit phases a and b when passing site ERS. TN at section S2 receives the voltage from TPSS1 and TPSS2 through TphTr1, TphTr2 connected to the power lines according to special rules, enabling two-way power ERS and uniform loading phases of power lines.

After supplying voltage to the TN train driver raises CC and switch on switch (SA), providing AC voltage to the primary winding of step-down traction transformer (TV). On secondary winding of TV, voltage is converted by rectifier (VD) and through smoothing reactor (SR) is fed to the traction motors (TM), through which current begins to flow, driving ERS in motion. On auxiliary power lines from tires of traction voltage 27.5 kV power get and non-traction consumers, which is implemented by connection buses of phases a and b through power switch (PSNTC) two wires mounted on poles, fixed on CN supports from field side. Step-down transformers of consumers (SDTrC) and rail lines are connected to them on system of TWR (two wires – rail). To ensure uninterrupted power supply to consumers, a disconnector (QF) is provided in middle of TWR line, which provides a transition to power from one TN when other is disconnected.

From transformer of own needs (TrON) through buss (BON) get power to the load own needs TN (power control circuits, alarms, lighting, heating, motor loads and other), through a transformer of signalization, centralisation and blocking (TrSCB) voltage is applied to a high-voltage line (HVL SCB) to supply equipment signalling and communication, through using low-power step-down transformers (LPSDTr) and the device is in relay cabinets (RC) are powered traffic lights. Disconnecting switch (QS) enables to power supply the HVL SCB from any of TN (QS closed) or each half of his (open QS). Ensuring safety of train traffic and continuous power supply of SCB devices is achieved by presence of a backup power supply (BUPS), which receives power through step-down single-phase transformers (SDTr) from TWR line. As a reverse wire for passage of reverse traction currents (RTC) electro-technical complex of railway power supply used electrically combined elements: rail line (RL), choke-transformers (DT), microcline and throttle jumper, butt-traction connectors, and between line and between rail jumper, suction feeders (SF) traction substations.

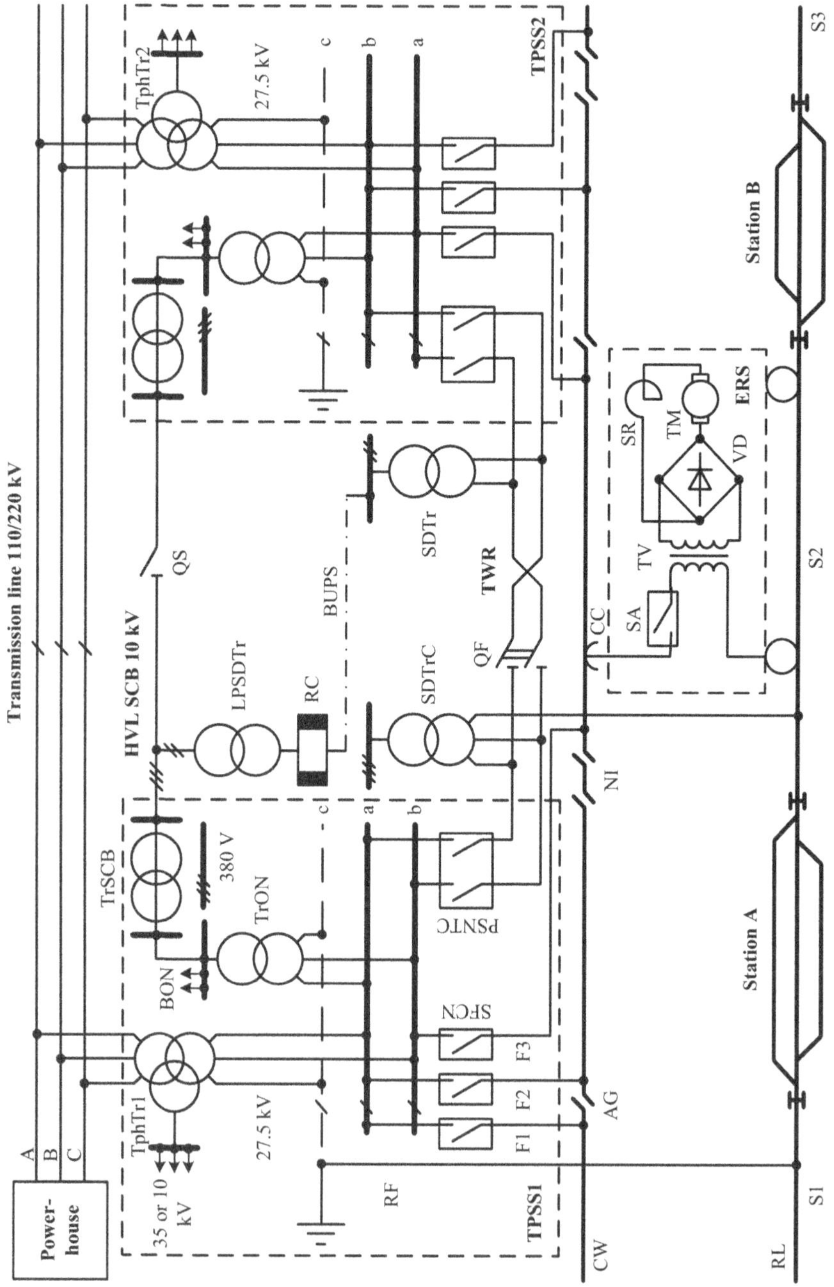

Fig. 1. Scheme of railway section, electrified by AC 25 kV

The advantages AC systems with voltage of 25 kV include higher voltage in the contact network (CN) and simple reduction possibility it by electric locomotive transformer. Electric locomotive with capacity of 6000 kW at direct current (DC) consumes from CN current of 2000 A, at an alternating current (AC) – 300 A, so the DC is more complex and has large number of wires (usually two copper contact wires with a cross section of 100 mm^2 each, a copper carrier cable with a cross section of 120 mm^2 and one or two reinforcing aluminium wires with a cross section of 185 mm^2 each), and at alternating current of CN is less complex and consists of one copper contact wire with a cross section of 100 mm^2 and a bimetallic wire carrying cable Sect. 95 mm^2. The design of AC substation compared to DC substation is simpler due to the absence of rectifying units. The number of substations on lines in the AC system is less, because they are located at long distances. The disadvantages of AC system are increased influence on the communication line (CL), since the alternating current creates an alternating electromagnetic field around the wires. CL passing along the railway must be executed cabling and not overhead cable, as a direct current, which leads to the increase in cost of railway electrification. In addition, there are problems asymmetry of currents and voltages due to the fact that electric locomotives consume single-phase current, transmission line are three-phase. There is a need for installation of neutral inserts at each substation, the presence of which increases the probability of burnout contact wire.

2 Materials and Methods

Figure 2 shows a simplified diagram flow of traction and reverse traction currents by elements of traction network AC 25 kV.

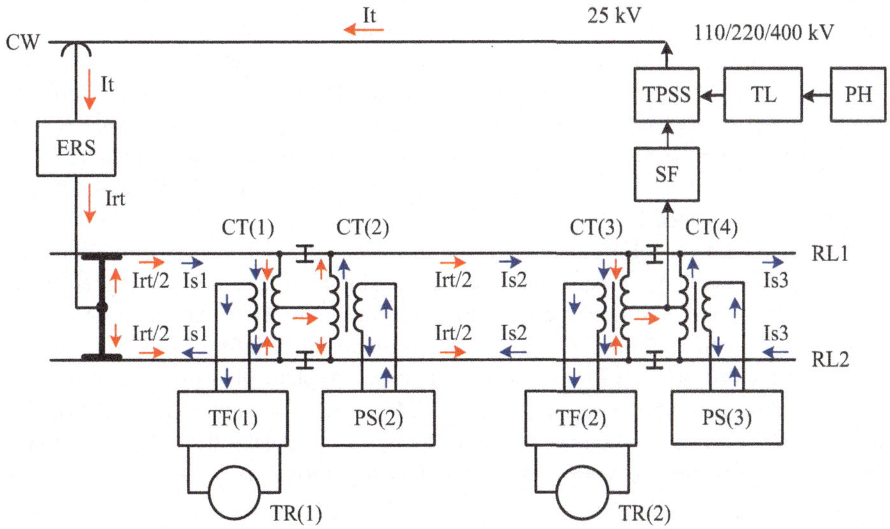

Fig. 2. Scheme of flow currents by elements traction network

The harmonic components of traction current affect rail circuit (RC) only if there is sufficiently large difference between RTC in different RL (asymmetry). Passing through two half-windings CT reverse traction currents (Irt/2) can induce in the secondary winding CT interfering EMF due to the fact that the primary half-windings are included counter for them and consonant to the signal currents (Is). In the presence of resistance asymmetry in RL components of traction current passing through semi-windings may not be equal, and with a large difference can lead to induction of interfering EMF on the secondary winding, on track relay (TR) and locomotive coils in presence ERS on the track section.

The presence of asymmetry of traction current changes resistance of main winding of the CT, which generally changes the input resistance at ends of RC and adversely affects main modes of its operation. Asymmetry can be longitudinal (due to breakage of butt connectors, rail breakage, run of insulating joints, inhomogeneity metal in rails, inequality of resistance throttle jumpers, etc.) or transverse (due to grounding of contact supports on right rail, shorting to ground of one RL, the difference of resistance of the ballast for RL, etc.). A large number of scientific papers have been devoted to study effect of reverse traction network (RTN) asymmetry, for example [2–7].

Figure 3 shows oscillograms, where pause of code signals automatic locomotive signalization continuous type action (ALSC) filled AC voltage with frequency of 50 Hz induced on receiving end of RC current asymmetry, while the pulse code signals are distorted.

3 Modeling

The problem of research in operation devices of a contact network (CN), electric rolling stock (ERS) and rail network (RN) difficult the experiment in context of multicriteria nature interference influencing the useful signals and the functioning of equipment of railway automation and telemechanic not always unambiguous.

A large number of factors affecting operation of traction network devices make it difficult to identify causes of breakdown or failures, their elimination in real time. To conduct experiments and establish interdependencies in practice, it is necessary to simultaneously use special measuring and recording equipment on the way and ERS to attract a large number of specialists, to create conditions under which breakdown or failure will be repeated [8–11].

The development of computer technology has created the preconditions for improving methodology of research work, allowed part of tasks to be assigned to the software, therefore, to simplify experiments, reduce number of staffs, to ensure the accumulation and possibility of mathematical processing of information [12, 13].

Therefore, this article proposes a new approach implemented in the form of data collection and centralized processing: the development of a simulation model of CN, ERS and RC for a comprehensive study of the processes arising from the operation of these systems and the interaction of various physical processes.

The process of modeling AC traction network 25 kV consisted implementation of model contact network of AC 25 kV; simplified circuit EPS AC type 2ES5K «Ermak»; model rail network (coded rail circuit 25 Hz, which is powered by a frequency converter FC-50/25 with track relay IMVSH-110, response voltage of which is 3–4.5 V. Figure 4 shows developed model of AC traction network 25 kV [14–16].

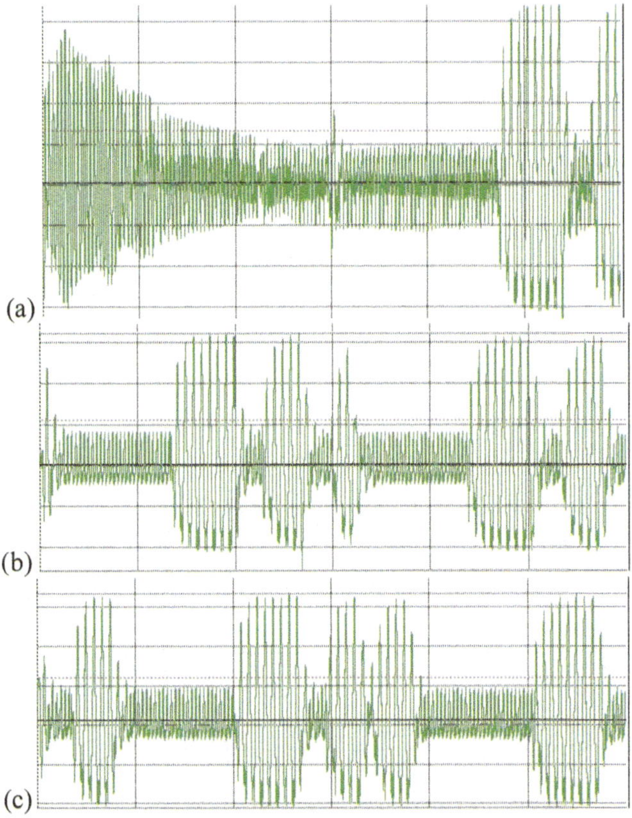

Fig. 3. Oscillograms of ALSN signals under the influence of OTT asymmetry

4 Findings from the Research

The mathematical model of traction AC 25 kV allows to conduct experimental researches on personal computer in laboratory to determine behaviour of various circuit elements during normal operation and during influence of destabilizing factors (disturbances) to develop and implement in simulation environment of MATLAB & Simulink of different designs to increase noise immunity and protection of this complex device, to simulate appearance of possible situations and make a digital processing of measurement results. Further improvement and development of simulation models' complex devices of railway automation and telemechanic (RAT), use of modern simulation environments will create a powerful apparatus for studying influence of interference on normal operation of individual nodes, development of algorithms for protection of distortion transmitted information, evaluation of noise immunity devices under influence of various destabilizing factors. In the future, improvement of simulation model, it is advisable to supplement it with blocks that simulate the influence of destabilizing factors and various disturbances [17].

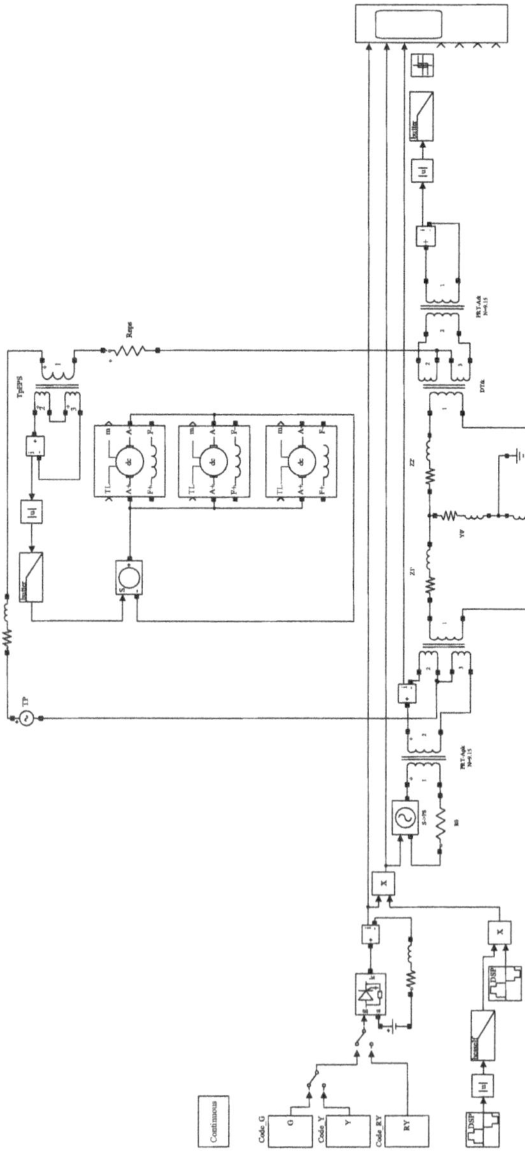

Fig. 4. Model of CNS, ERS, coded RC

Acknowledgment. We Express our deep gratitude to all our colleagues, authors of the works, familiarization with which greatly helped us in conducting research and obtaining relevant results, although they may not agree with all our interpretations and conclusions.

References

1. Grigor'ev, N., Klykov, M., Parfianovich, A., Shestukhina, V.: Increasing the power efficiency of a 25-kV alternating current traction power supply system. Russ. Electr. Eng. **87** (2), 113–115 (2016). https://doi.org/10.3103/S106837121602005X
2. Plakhova, M., Bassam, M., Arboleya P.: Static model of a 2 × 25 kV AC traction system. In: 2015 6th International Conference on Power Electronics Systems and Applications (PESA), pp. 1–5 (2015). https://doi.org/10.1109/pesa.2015.7398917
3. Godyaev, A., Onischenko, A.: Automated monitoring of level distortion coded signals transmitted on rail line. In: International Multi-Conference on Industrial Engineering and Modern Technologies (FarEastCon), 18393174 (2018). https://doi.org/10.1109/fareastcon.2018.8602696
4. German, L., Serebryakov, A., Osokin, V., Subkhanverdiev, K.: Equivalent scheme for calculating short-circuit current in an alternating-current electric traction network. Russ. Electr. Engin. **90**, 516 (2019). https://doi.org/10.3103/S1068371219070058A
5. Mariscotti, A.: Induced voltage calculation in electric traction systems: simplified methods, screening factors, and accuracy. IEEE Trans. Intell. Transp. Syst. **12**, 201–210 (2011). https://doi.org/10.1109/TITS.2010.2076327
6. Tryapkin, E., Ignatenko, I., Keino, M.: Development of "digital substation" technology for power supply of railways. In: Siberian Transport Forum – TransSiberia. MATEC Web Conference, vol. 239, p. 01051 (2018). https://doi.org/10.1051/matecconf/201823901051
7. Serdiuk, T.; Kuznetsov, V., Kuznetsova, Y.: About electromagnetic compatibility of rail circuits with the traction supply system of railway. In: IEEE 3rd International Conference on Intelligent Energy and Power Systems (IEPS), 18323445 (2018). https://doi.org/10.1109/ieps.2018.8559541
8. Buyakova, N., Zakaryukin, V., Kryukov, A., Nguyen, T.: Electromagnetic safety enhancing in railway electric supply systems. In: E3S Web of Conferences, vol. 58, p. 03003 (2018). https://doi.org/10.1051/e3sconf/20185803003
9. Apollonskiy, S.: Problems of electromagnetic safety on electrified railway. Electromagn. Saf. Railw. Altern. Curr. Tract. Power Netw. **58**(03003), 414 (2018). https://doi.org/10.1051/e3sconf/20185803003. In: E3S Web of Conferences, In 2 volumes. T. II
10. Zharkov, Y., Popova, N., Figurnov E.: Accounting power supply schemes for traction substations in the calculation of short circuits in the AC traction network. Vestnik of the Railway Research Institute, vol. 78, JSC Vniizht (2019). https://doi.org/10.21780/2223-9731-2019-78-1-10-18
11. Bellan, D.: Reconstruction of noisy electromagnetic fields by means of compressive sensing theory. Appl. Mech. Mater. **263–266**, 99–102 (2013). https://doi.org/10.4028/www.scientific.net/AMM.263-266.99
12. Figurnov, E., Zharkov, Y., Kharchevnikov, V.: Determination of long-term permissible currents in wires of power supply systems of railways. Vestnik of the Railway Research Institute, vol. 78, JSC Vniizht (2019). https://doi.org/10.21780/2223-9731-2019-78-2-90-95
13. Shamanov, V.: The process of traction-current asymmetry generation in rail lines. Russ. Electr. Eng. **85**(8), 34–37 (2014). https://doi.org/10.3103/S1068371214080124

14. Charalambous, C., Cotton, I., Aylott, P.: Modeling for preliminary stray current design assessments: the effect of crosstrack regeneration supply. IEEE Trans. Power Deliv. **28**, 13596015 (2013). https://doi.org/10.1109/TPWRD.2013.2259849
15. Baranovskii, A.: Computer simulation and experimental investigations of noise levels and spectra for railway transport. Autom. Control. Comput. Sci. **45**(5), 69–78 (2011). https://doi.org/10.3103/S0146411611050026
16. Belyaev, N., Korovkin, N., Frolov, O., Chudnyi, V.: Methods for optimization of power system operation modes. Russ. Electr. Eng. **85**(8), 541 (2014). https://doi.org/10.3103/S1068371214080136
17. Badyor, M., In'kov, Y.: Electromagnetic compatibility of a traction power-supply system and infrastructure elements in areas with high traffic. Russ. Electr. Eng. **85**(8), 488–492 (2014). https://doi.org/10.3103/s1068371214080021

Development of Innovative Railway Rolling Stock Technologies

Dmitry Bannikov(✉) and Nina Sirina

Ural State University of Railway Transport, Kolmogorova 66, Ekaterinburg 620034, Russia
bannikov_d@mail.ru

Abstract. Approaches to the management of maintenance and repair of rolling stock are now increasingly being developed into management decision support systems. The given article is directed on studying of innovative technologies on an example of service maintenance service and repair of a rolling stock. The model of service maintenance and repair of rolling stock is able to dynamically optimize management decisions on management of multi-component system during the whole life cycle. This model is based on monitoring and dynamic grouping of the operations under consideration. Dynamic grouping includes short-term information about the use of rolling stock, as well as unexpected failures of rolling stock. The method of dynamic optimization of service maintenance and repair uses a priori reliability characteristics to update the planning of maintenance or repair at the time of making management decisions. The simulation model will allow to display the system and obtain technical and economic indicators of efficiency of the service maintenance strategy functioning. The modeling results will allow transport companies to evaluate the service maintenance and repair strategy, which will influence the management decision making.

Keywords: Multi-component system · Service maintenance and repair · Maintenance strategy · Agent-based modeling

1 Introduction

Organization of service maintenance and repair of passenger rolling stock is one of the most difficult stages in the planning process, and also plays a key role in the economic costs of transport companies. The organization of maintenance and repair services is becoming a strategic task in management decision making.

Conventional modelling tools are generally limited in their ability to cover and analyze multiple levels of decision-making and system configurations. The creation of a simulation model allows for the creation of an integrated model capable of simulating information flow scenarios and generating multi-level metrics to provide guidance to users in the management decision-making process. These solutions will increase flexibility and productivity by reducing product implementation time.

The scientific support of the maintenance and repair system does not provide a sufficiently accurate assessment to date. There is no possibility to predict the level of

Z. Popovic et al. (Eds.): TransSiberia 2019, AISC 1115, pp. 401–407, 2020.
https://doi.org/10.1007/978-3-030-37916-2_39

readiness of rolling stock at the established limits of financing, the level of passenger traffic, the intensity of failures [1].

Currently, the tools for planning maintenance and repair of passenger rolling stock are static [2, 3]. They do not take into account the dynamic information during the life cycle of the rolling stock. Planning tools are mainly a copy of the maintenance and repair manual [4, 5]. These activities are planned at fixed intervals or actual rolling stock mileage.

Service maintenance and repair of passenger rolling stock leads to the evolution and transformation of strategies from diagnostic to predictive and prescriptive types of maintenance and R. At the conceptual level, fundamental aspects of the service model have been developed to predict future events and to draw up optimal plans taking into account the multimodality and structural heterogeneity of data [1].

The idea of service maintenance and repair is that the system would check and take into account the information about the wear and tear of rolling stock in the model of management decision-making. The model determines which maintenance or repair operations can be extended or postponed.

Implementation of the structure under consideration faces fundamental technological and non-technological problems in the development and implementation [6–8]. Taking into account the researches under consideration, three main aspects are defined:

1 – correct identification of the functional capabilities of the system;
2 – optimal choice of characteristics in accordance with the technological level of security of the enterprise;
3 – enterprise's ability to overcome non-technological problems.

To make effective management decisions in the maintenance and repair of rolling stock, reliable and timely information is required. For this purpose, the management decision-making process is based on the use of digital cloud services and web applications. Digital information management network unites and synchronizes various service and reliability applications for information collection and processing.

2 Methodology

The study of complex systems requires a virtual, abstract representation of the system using a simulation model. The model is not created as a copy of a real system, but is formed with reference to some aspects. This refers to the concept of abstraction and determines the fact that the model is always adapted to provide answers to specific tasks, to make managerial decisions. For this reason, the model being developed is a simplification of reality in order to provide a reasoned basis for understanding the system in question.

The simulation model is essentially a model that builds a trajectory of changes in the state of the system. It can be said that the simulation model is a set of rules according to which the system passes from one state to another. The rules are set with the help of differential equations and state diagrams. The model output data allow to analyze the system behavior in the given parameters [9–13].

Creation of the model is carried out with the use of the apparatus of system dynamics of simulation modeling. This apparatus includes both discrete-event and agent-based simulation in AnyLogic environment. System dynamics is a simulation approach, the methods and tools of which allow you to understand the structure and dynamics of the projected model. System dynamics assumes a high level of abstraction and is possible to use for strategic modeling of the system.

Figure 1 shows the general architecture and components of the service and maintenance simulation model, which consists of:

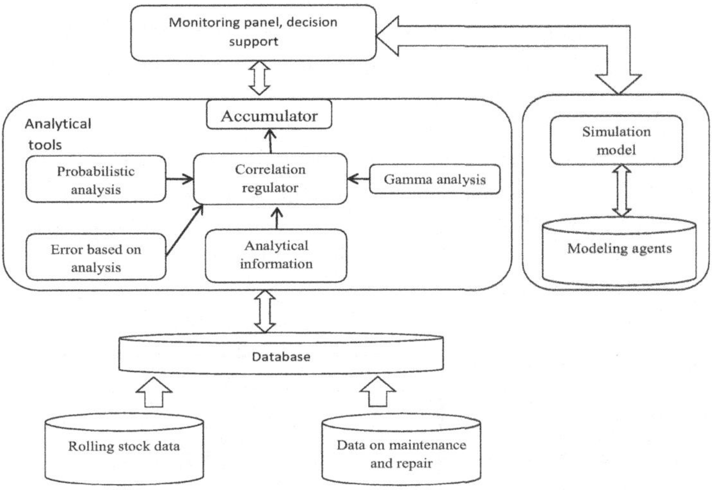

Fig. 1. General architecture of the simulation model

- database management parameters;
- analytical modeling tools;
- components of the simulation model;
- toolbar for monitoring and supporting management decision making.

The simulation model is an object-oriented model developed on the basis of hybrid approach (discrete event modeling) in AnyLogic. The model is able to describe the characteristics of passenger rolling stock.

The discrete event modeling assumes the representation of the simulated system in the form of a process, i.e. the sequence of operations performed with the simulation agents. The model is set graphically in the form of a process diagram, the blocks of which are separate operations.

Agent-based modeling represents an alternative view on the behavior of the simulated system. The structure of the agent model can be described not only graphically, but also with the help of a given scenario. Agent behavior can be defined by different actions when certain events occur.

In order to select a suitable strategy for maintenance and repair of passenger rolling stock, it is necessary to take into account non-metric factors (traffic safety, quality of passenger service) and metric factors (cost of maintenance and repair, cost of spare parts, failure rate). Optimal maintenance and repair strategy will determine the balance between the cost and quality of work performed.

The basis of discrete event modeling is the concept of incoming applications, under which the rolling stock for maintenance or repair is accepted. The discrete event modeling assumes representation of the simulated system in the form of a process, i.e. sequence of operations performed with the modeling agents. The model is set graphically in the form of a process diagram, the blocks of which are separate operations.

Agent-based modeling allows to describe a discrete process system in the form of several subsystems defined by the agents. Each agent of the model under design interacts with other agents, thus forming the external environment of the simulation model. The structure of the agent model can be described not only graphically, but also with the help of a given scenario.

The modelling agent has autonomous behavior, making decisions in accordance with a set of restrictions and rules. The behavior of the agent is related to other interacting agents in the simulation model. The algorithm of the model agents behavior is described by the graph of passenger rolling stock states [14] in the simulated system of service maintenance and repair. Thus, the formalization of a discrete system by simulation agents is presented in Fig. 2.

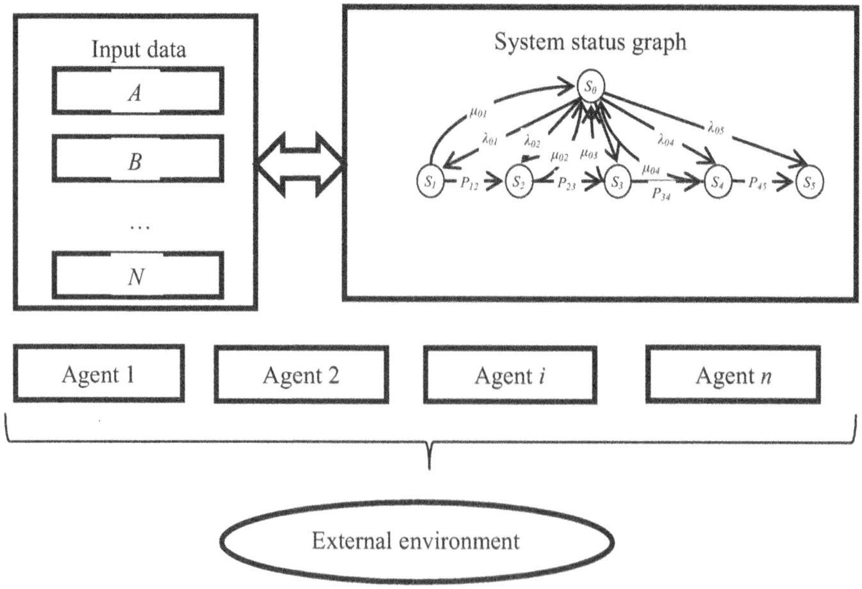

Fig. 2. Formalization of the discrete system by modeling agents

The system states graph is an oriented graph G = (S, P), in which: nodes S are states of the agent, arcs P is the probability of events that transfer the agent from one state to another.

In the developed model the agent is considered as a module of system modeling, controlling the process (Fig. 2). The agent analyses the model over time and receives feedback from the simulation results. The agent then controls the input of the input data to set the modeling task when creating the experiment. The structure of the agent interacts with the modeling process, using the iteration of «agent-boundary-experiment behaviour» to set the appropriate uncertainties and assumptions accepted in the process of modeling the system.

In order to change the uncertainty space, the agent chooses a number of experiments generated from the initial model of agent behavior, groups similar model behavior, showing the results of the behavior of the design model as a whole.

Agent-based modeling provides an alternative view of the behavior of the simulated system. Agent behavior can be defined by different actions when certain events occur.

The next step of the offered strategy of service maintenance service and repair consists in the cumulative account in the process of acceptance of administrative decisions of predictive working capacity of objects and system, and also economic dependences between components. For this purpose, a two-level process of making managerial decisions is proposed, both at the system and component levels (Fig. 3).

At the system level of management decision making, the model considers passenger rolling stock before the previous *R(Tk + 1; xTk)* inspection period. Firstly, all types of failures that have occurred to the rolling stock during the period of T are taken into account in the forecast. Then, the serviceability coefficient is compared with the threshold value R_0 $(0 < R_0 \leq 1)$ to start calculations for decision making, where R_0 is the decision variable. If it is necessary to perform maintenance and repair, then further determination of the algorithm is to find the optimal type of maintenance or repair to maintain the passenger rolling stock in good working order.

At the component level of management decision making the calculation of the algorithm is based on the factor of improvement of the group on the basis of the costs of maintenance and repair. This algorithm allows you to select a specific group. If at the moment of *Tk* modeling maintenance or repair will allow to increase the predicted coefficient of serviceability of the system higher than R_0, the system carries out unscheduled maintenance and repair. Determination of the optimal type consists in the cost of work for a particular type of rolling stock. Based on a variety of possible options, a certain type of maintenance and repair.

In order to achieve the optimal value of the target function of the service maintenance and repair model, it is planned to reduce the costs of maintenance and repair during the entire life cycle of the rolling stock. In order to provide the target function, the mathematical expression of long-term maintenance and repair costs per unit of time is implemented in the model.

Actually the basic forecast of cost of life cycle cost of system of service maintenance service and repair is caused by consequences of decision-making accepted at early stages of designing. Great influence on the efficiency of the system and the cost of the life cycle has an impact on the failure rate. The developed technique of making managerial decisions can influence necessary changes in the technical system.

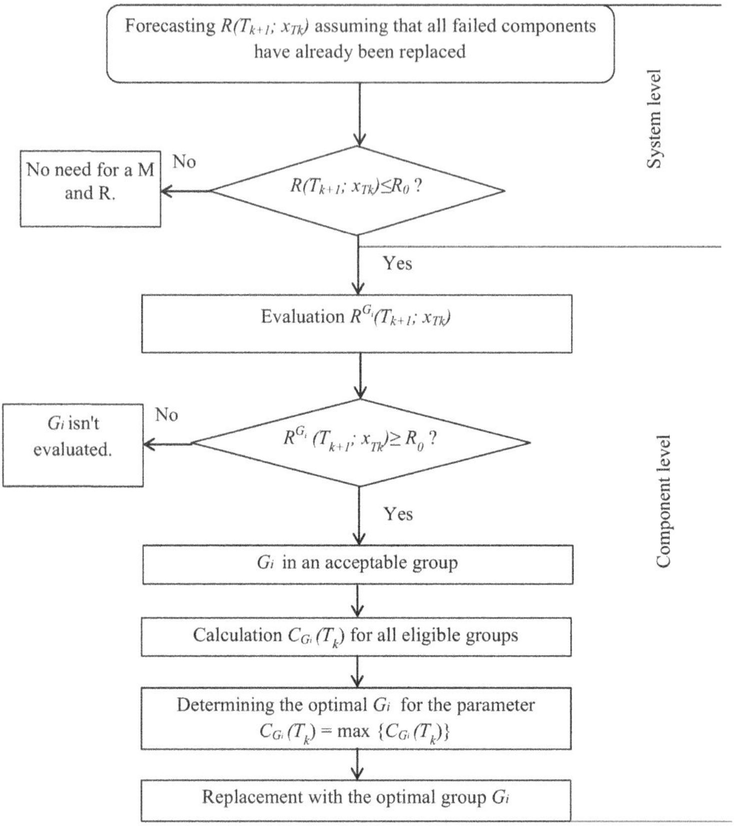

Fig. 3. Procedure for management decision making

3 Conclusions

Discrete event simulation by agents is an excellent method for modeling various production processes of maintenance and repair of passenger rolling stock. Using all the features of AnyLogic, the simulated situation is displayed in full, reflecting the behavior of the model and its characteristics. This model is designed to determine the optimal location of service centers, providing a high availability of rolling stock and minimal time and material costs for maintenance and repair.

Service centers will provide both maintenance and repair of passenger rolling stock. The developed model will be used as a basis for the creation of a simulation model for the assessment of production capacities of passenger car maintenance and repair enterprises. The model will allow predicting a uniform distribution of rolling stock across the enterprises for making managerial decisions. For further development of the model it is planned to work out the optimization model, which will include statistical data of rolling stock failures. This will allow to study the probability of rolling stock idle time in anticipation of repair, as well as the impact on the availability factor and material costs.

The simulation model will make it possible to assess the material component of the efficiency of their work. In a short period of time it is possible to assess the changes introduced, their impact on the system of maintenance and repair, as well as to determine a rational balance of costs for both repair and maintenance of passenger rolling stock.

References

1. Bannikov, D., Sirina, N.: Model of passenger rolling stock maintenance. In: MATEC Web of Conferences Polytransport Systems, vol. 216, p. 022018 (2018). https://doi.org/10.1051/matecconf/201821602018
2. Evseev, D.G.: Evolution of organization of repair of rolling stock of science and transport technique. World Transp. Transp. J. **3**, 134–139 (2014)
3. Sirina, N.F.: Modeling of the car-repair complex control. Vestnik of the Ural State University of Railway Transport **2**, 16–24 (2009)
4. On introduction of a new system of maintenance and repair of passenger cars. Order of the Ministry of Railways of the Russian Federation of 04.04.1997 No. 9C, 13 January 2011
5. Normative service life of machinery and equipment (developed by the Ministry of Heavy and Transport Engineering)
6. Lee, E.A.: What is real time computing? A personal view. IEEE Des. Test **35**(2), 64–72 (2017). https://doi.org/10.1007/978-3-030-00151-3_2
7. Lee, E.A., Seshia, S.A.: Introduction to Embedded Systems. A Cyber-Physical Systems Approach, vol. 54, no. 4, pp. 517–526 (2017) https://doi.org/10.1007/s10559-018-0053-9
8. Maier, R.: Knowledge Management Systems: Information and Communication Technologies for Knowledge Management, Berlin (2009). https://doi.org/10.1007/978-3-540-24779-1
9. Pierskalla, W.P., Voelker, J.A.: A survey of maintenance models: the control and surveillance of deteriorating systems. Naval Res. Logistics Q. **23**, 353–388 (1976). https://doi.org/10.1002/nav.3800230302
10. Osaki, S., Nakagawa, T.: Bibliography for reliability and availability of stochastic systems. IEEE Trans. Reliab. **25**, 284–287 (1976)
11. Sherif, Y.S., Smith, M.L.: Optimal maintenance models for systems subject to failure–a review. Naval Res. Logistics Q. **28**, 47–74 (1981). https://doi.org/10.1002/nav.3800280104
12. Valdez-Flores, C., Feldman, R.M.: A survey of preventive maintenance models for stochastically deteriorating single-unit systems. Naval Res. Logistics **36**, 419–446 (1989)
13. Cho, D.I., Parlar, M.: A survey of maintenance models for multiunit systems. Eur. J. Oper. Res. **51**, 1–23 (1991)
14. Bannikov, D., Galkin, A., Sirina, N.: Calculation algorithm of organization of service technical maintenance and repair of passenger rolling stock. Transp. Urals **4**(55), 31–35 (2017). https://doi.org/10.20291/1815-9400-2017-4-31-35

Microprocessor Control System for a Three-Phase Voltage Inverter with a Modified Algorithm

Alexander Gulyaev(✉), Evgenii Ten, Dmitry Fokin, and Andrei Hohlin

Far Eastern State Transport University, Seryshev st., 47, Khabarovsk 680021, Russia
89143127984@ya.ru

Abstract. In this paper investigates the influence of high frequency switching of power transistors of an autonomous voltage inverter on the distortion of the output quasi-sinusoidal signal. To solve the discovered problem, a modified autonomous voltage control algorithm is proposed, based on the use of different pulse-width modulation frequencies in one half cycle. The standard two-level sinusoidal pulse-width modulation is formed as a result of the intersection of the modulating sinusoid with the reference sawtooth signal. In classical pulse-width modulation, the ratio of the frequency of the reference signal to the modulated frequency is always constant and equal to an integer, i.e., this pulse-width modulation is synchronous. An algorithmic solution is proposed for the formation of a modified sinusoidal pulse-width modulation, when the frequency of the reference sawtooth signal during one period of the modulated sinusoid changes twice from a smaller value to a larger one and vice versa.

Keywords: Induction motor · Three-phase inverter · Pulse-width modulation · Power losses · Microcontroller · Frequency converter

1 Introduction

Currently, a number of established methods are used to control squirrel-cage induction motors, which significantly differ in the quality of regulation, cost and complexity of technical implementation. In modern frequency converters, control systems are built on controllers that programmatically implement various algorithms [1, 2].

The paper discusses the control modes of an induction motor with short-circuited rotor frequency methods. As the simplest and most effective, a frequency converter with a DC link is considered. The paper analyzes the control of an autonomous voltage inverter by means of pulse-width modulation (PWM). Through such control of the inverter, a three-phase voltage of a quasi-sinusoidal shape is formed at its output. Today, two methods for controlling such an inverter are common: a sinusoidal PWM and a spatially vector PWM. The essence of a sinusoidal PWM is to simultaneously control at a given carrier frequency all the inverter switches at once so that at the midpoints of each inverter rack, waves of quasi-sinusoidal output voltage uA, uB, uC

Z. Popovic et al. (Eds.): TransSiberia 2019, AISC 1115, pp. 408–419, 2020.
https://doi.org/10.1007/978-3-030-37916-2_40

are formed, shifted by 120 electrical degrees relative to each other [3]. The main advantage of this type of PWM is the best, in comparison with other control methods of the autonomous voltage inverter, the harmonic composition of the modulated voltage. The main disadvantage of a sinusoidal PWM is the underutilization of the DC link voltage by 14%. The problem of underutilization of the DC link is solved in several ways [4]. A method based on overmodulation, the essence of which is to increase the amplitude of the vector for setting the relative voltage above unity. However, in this case, 5, 7, 9, 13 and other odd harmonics appear in the output voltage of the autonomous voltage inverter, distorting the sinusoidality of the smooth component. The method of pre-modulation of the third harmonic, which consists in adding to the signal of the task of each phase of the signal zero sequence containing the third harmonic in the optimal proportion. Harmonics that are multiples of three are zero-sequence harmonics and are not contained in line voltages.

Based on research in the field of PWM control of the inverter [5–7], a modified inverter control algorithm was proposed to form a three-phase supply voltage of a quasi-sinusoidal motor. The algorithm allows reducing losses in the inverter at those moments when the motor consumes the smallest instantaneous power. Based on the proposed method for controlling the inverter, a frequency converter with microprocessor control is developed.

2 Materials and Methods

According to the definition of IEC 551-16-30, pulse-width modulation is called pulse control, in which the width or frequency of the pulses, or both, are modulated within the period of the fundamental frequency in order to create a certain shape of the output voltage curve. In previous years, PWM for controlling voltage inverters of AC electric drives was implemented using analog equipment. Typically, PWM signals are obtained by comparing the triangular carrier signal and the signal to be modulated. Three-phase systems require three independent PWM channels: one for each phase. The inputs of such a device are the specified phase voltage. These voltages are formed on the basis of the adopted law of frequency control.

Modern control devices for AC electric drives are implemented solely on the basis of digital control technology. Generating PWM signals is a discrete process, and in this regard, the use of the analog method of generating PWM signals is not justified. The integrators and comparators used in analog technology are replaced by digital timers of microcontrollers. Most of the frequency-controlled electric drives with autonomous voltage inverter produced today have a PWM implementation in the direction of a sinusoidal PWM or spatially vector PWM.

The driver is an intermediate link between the control device and power switch and performs the following functions:

- the formation of unlocking and locking pulses in accordance with the MK command on the IGBT shutter;
- compliance with the delay time of switching the power transistor in one shoulder;
- determination of current overload;

- timely shutdown of the power transistor when a short circuit current is detected;
- power voltage control driver.

To open and close an insulated gate bipolar transistor, the driver either performs the gate charge of the transistor to the release voltage or discharges it to the isolation voltage.

When switching from one state of the transistor to another in the circuit between the driver, the resistors in the control circuit and the power switch, a certain power is allocated, W:

$$P_{DRV} = Q_g \cdot f \cdot \Delta V_g, \quad (1)$$

Where Q_g is the gate charge of the transistor, C;
f - switching frequency, Hz;
ΔV_g - voltage swing at the driver output, V.

Another important parameter is the maximum gate control current $I_{out.max}$. The magnitude of this current should be sufficient to control the gate resistance at a selected voltage swing. The current flowing in the gate circuit does not exceed 70% of the maximum first-order current $I_{out.g}$, A:

$$I_{out.max} = 0,7 \cdot I_{out.g} \quad (2)$$

Simplified, the maximum current of the first order is found as, A:

$$I_{out.g} = \frac{\Delta V_g}{R_{g.mi\Pi}} \quad (3)$$

Where $R_{g.mi\Pi}$ is the minimum resistance in the gate circuit of the transistor, Ohm.

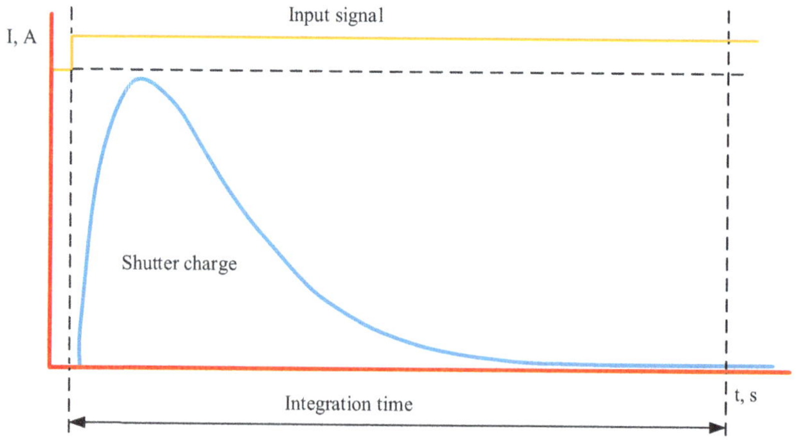

Fig. 1. Curve of the shutter charge

The value of the gate charge should be given in the reference data on the power semiconductor element. If there is no such value, it can be calculated from the charge-charge curve (Fig. 1).

As you can see, the charge curve is very nonlinear, so the charge is obtained by integrating the driver output current, C:

$$Q_g = \int I_{out} dt \tag{4}$$

The integration interval is the time from the moment of receipt of the opening pulse before the output current takes a constant value.

When assessing the quality of electricity, the main indicators are the sinusine-range form and the symmetry of the currents of a three-phase network. When powered by a frequency converter, there are no special problems with the symmetry of currents, in contrast to distortion of their shape.

When operating an electric drive from AIN, to reduce the coefficient of nonlinear distortion, the modulation frequency is increased, thereby reducing losses in an induction motor. However, not any increase in the frequency of a discrete signal leads to a positive effect. Firstly, the disadvantages of this method of suppressing current ripples include an increase in losses in the inverter for switching transistors. Secondly, due to these losses, an additional distortion of the current shape [5–7], and voltage also occurs (Figs. 2 and 3).

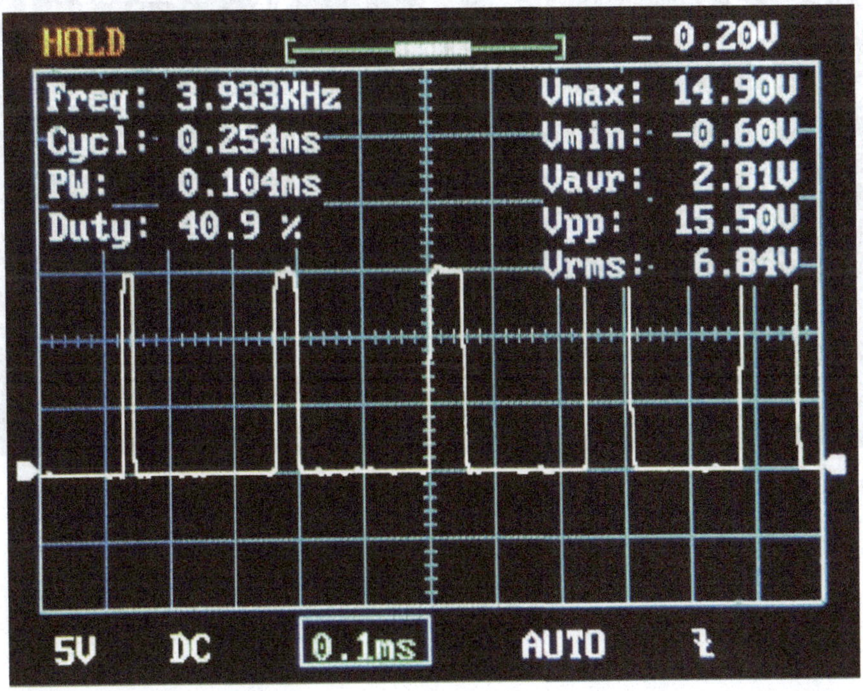

Fig. 2. Oscillogram of the beginning of the half-period of modulation of a 50 Hz sine wave with a PWM frequency of 4 kHz

We will make a calculation to determine the dependence of the ratio of the switching time of the inverter switch to the useful pulse duration during the half-cycle. The calculation was made at a modulation frequency of f_{pwm} = 15000 Hz and at a switching time of 0.3 μs (the time interval between the supply of a control signal from the MC to the complete opening of the power switch) and when the voltage of the nominal frequency is modulated. The number of pulses per half period is determined by the formula $N = f_{pwm}/f_1 = 15000/50 = 150$. Let us determine the dependence of the ratio of the switching time of the inverter switch to the useful pulse duration during the half-cycle for each pulse in the half-cycle according to the formula, %: t_{on}/t_i, where t_i is the pulse duration, determined by the sinusoidal law, with $t_i = T_i \sin(i\pi/N)$. Where i - is the serial number of the pulse; T_i – period of the impulse. Below is a graph of the relationship between the switching time of the inverter switch and the pulse duration during the half-cycle (Fig. 4).

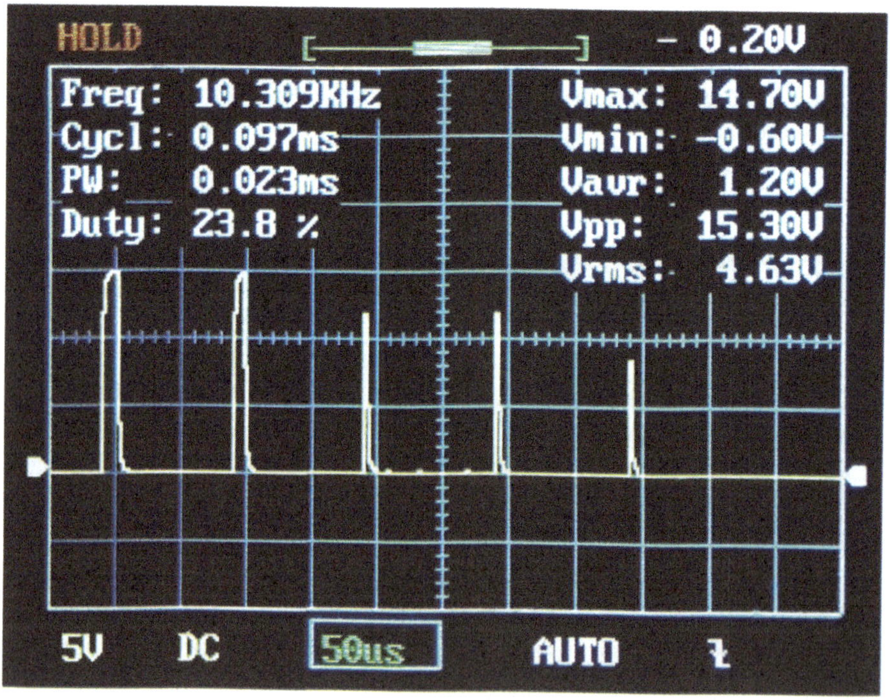

Fig. 3. Oscillogram of the end of the half-cycle of modulation of a 50 Hz sine wave with a PWM frequency of 10 kHz

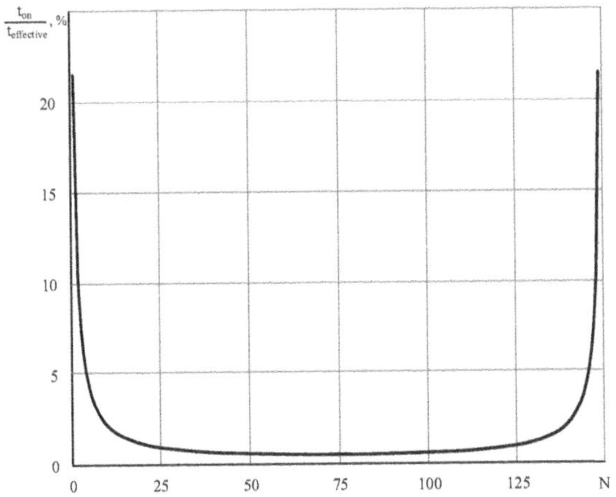

Fig. 4. Graph of the relationship between the switching time of the inverter switch and the useful pulse duration during the half-cycle

3 Results

A solution to the identified problem is proposed by dividing the half-period of the modulated sinusoid into sections in which modulation will be carried out with different frequencies. Namely, sections of a half-period from 0 to 60 and from 120 to 180 electrical degrees will be modulated with a frequency half that of a section of 60–120 electrical degrees.

In addition to reducing the sinusoid distortion by reducing the ratio of switching time to the useful pulse time, this approach is due to a decrease in power losses for switching power switches during the smallest instantaneous power consumed by the engine, i.e., at the beginning and at the end of the half-cycle.

Some modern Ifs have the ability to adjust the PWM frequency, but the modulation frequency in any case remains constant throughout the work. Therefore, the work also proposes the implementation of a frequency converter with the proposed algorithm for controlling the autonomous voltage inverter using modern digital methods.

To coordinate the microcontrollers with power circuits, we use a half-bridge switch driver – IR2104. It is worth noting that it is classically accepted in three-phase inverters to use the drivers of the upper and their switch (for example, IR2101). Half-bridge drivers, as a rule, are used in single-phase power supply networks. In operation, the half-bridge driver controls the three-phase inverter as follows. The IN input is used to select the opening of the upper or lower bipolar transistor with an insulated gate, thereby determining the sign of the applied voltage in the half-period section. Sine-wave PWM signals come to the SD input. The driver implements hardware "dead time" – a short delay between opening one switch and closing another [8] in order to exclude the possibility of simultaneously opening the upper and lower switches and passing a short circuit current through them.

Thanks to this approach, the control system becomes more reliable at the hardware level and at the same time the program code is significantly reduced by reducing the number of channels generating a pulse-width modulated signal. This solution allows you to reduce the amount of microprocessor memory used and safely manage IGBTs.

The software for controlling the inverter in the frequency converter was carried out in a highly integrated software environment designed to develop the MK code of the ARM architecture – CooCox CoIDE.

The program, as expected, begins with the connection of the necessary libraries with the command "#include", after which the file of the connected library is written in quotation marks. Next, we define global macros for simplified perception of calculations in the code structure and quick replacement if necessary. "PERIOD" is the period of one pulse of a sinusoidal PWM signal. "DOUBLE_PERIOD" double period of one impulse. "F_PWM" is the frequency of the sinusoidal PWM. "F_MIN" – the minimum modulated frequency of the sinusoidal voltage supplied to the motor. "PI" is the number of pi rounded to 3.1416. Then we explain the global variables. "I", "j" "k" are counters showing the current serial number of PWM pulses. "Adc_value" – value with ADC, "sine_a [2]", "sine_b [2]", "sine_c [2]" – arrays containing two successive values of pulse duration; "N" is the number of PWM pulses per half-period of modulated frequency; "N_a", "n_c" is the number of the pulse from which the half-cycle of phase B and C of the relative phase A begins; "Flag_a", "flag_b", "flag_c" – flags, the truth of which determines the presence of a signal on ports PB13, PB14 and PB15, respectively. "Flag_count" – flag for the end of the pulse duration calculation. "S" and "pulse_number" are counters for the correct operation of the program. The only constant of the real type "d" contains the resolution of the ADC and is necessary for calculations, its value is 0.011. Next, select the port pin (or several) with the "GPIO_Pin" command and determine the slope of the front/fall – the voltage rise/fall rate with the "GPIO_Speed" command with operating options at frequencies 2, 10 and 50 MHz ("GPIO_Speed_2 MHz", "GPIO_Speed_10 MHz", "GPIO_Speed_50 MHz" Respectively). Having set the maximum steepness, we will be able to switch the foot with a frequency of 50 MHz – but this will increase electromagnetic interference from sharp fronts. To significantly reduce radio interference, we set the minimum slope with a frequency of 2 MHz on all legs. To supply rectangular pulses to the IR2104 driver's IN input to control the polarity of the voltage supplied to the motor, we configure pins PB13-PB15 as an output with two states "GPIO_Mode_Out_PP".

In order for the timer to use its channels as PWM generators supplied to the inverse input of the IR2104 driver SD, we configure its first three channels (PB6-PB8) as an open collector output for working with peripherals with two states "GPIO_Mode_AF_PP". Interrupt management and processing is performed by the controller of priority vector interrupts, which is part of the Cortex-M core [9].

In addition to simply setting the priority of interrupts, the control panel implements the ability to group priorities. Interrupts in a higher priority group can interrupt interrupt handlers in a lower priority group. Interrupts from the same group, but with different priorities within the group, cannot interrupt each other. The priority within the group determines only the order in which the handler is called when both events have been activated.

Timers 2, 3, and 4 (TIM1, TIM2, TIM3) MK are general purpose timers, each of which has four independent channels that can operate in the following modes: signal capture, comparison, PWM generation, single pulse generation. The project uses the first three channels of the fourth timer PB6, PB7, PB8 in the PWM generation mode, each channel is responsible for one of the phases of the modulated sinusoidal PWM. The PWM generator on the timer operates as follows. The timer is clocked from the APB1 bus through the frequency divider and starts counting from 0 to "TIM_Period". When the timer counts to the "TIM_Pulse" value, the value of the timer channel leg is inverted. Counting to "TIM_Period", the timer inverts the channel leg back and starts counting again. After the timer is initialized, a direct reference to the CCRx comparison register will be used to change the duty cycle of the timer channel, where x is the channel number [10]. This is necessary to increase the speed of the program.

Algorithm Description

After clocking the MK core and initializing its periphery, the bits are set into the registers corresponding to the signs of the half-waves of the phases: for phases A and C – positive, for phase B – negative. Then a command is issued to start the timer. In the main cycle, a continuous calculation of the duty cycle of the next pulse is performed according to a sinusoidal law. In the first program cycles, before the first timer overflow interrupt, the value of the "adc_value" variable will be zero, because the ADC is polled during the timer overflow interrupt. In addition, the counter variables of the next PWM pulse are also incremented during the same interrupt. It follows that in an infinite cycle until the first interruption on timer overflow, the first pulse of the sinusoidal PWM will be calculated over and over again for the minimum frequency "F_MIN" – 5 Hz. Let us consider in more detail the main cycle of the program.

In the first line of the cycle, the number of pulses of the PWM signal is calculated for a half-wave of a sinusoid according to the following formula:

$$n = \frac{F\_PWM}{2 \cdot (adc\_value + F\_MIN)} \tag{5}$$

If we take the zero point of phase A as the initial position of the three-phase sinusoid, then at this point phase B will be shifted back by a third of the half-period, and phase C – by two-thirds of the half-period. It follows that it is worth starting the modulation of the sinusoid of phase B with a pulse whose serial number is equal to one third of the pulses in a half-period, and two-thirds for phase C.

Thus, in the following two lines, the values of the additional values "n_b" and "n_c" are calculated, which will be added to the counters of the corresponding phases: "j" and "k".

Then follows the condition for the end of a half-period by comparing the counter and the total number of pulses per half-period. If the counter value is exceeded, and in the case of phases B and C, the values "n_b + j" and "n_c + k", respectively, of the number of pulses, the half-cycle is updated by zeroing the counter of phase A "I = 0" and assigning the following values to phase counters B and C: "j = –n_b" and "k = –n _c", respectively, for the transition from "n_b + j > n" "n_c + k > n" to "n_b + j = 0" and "n_c + k = 0". Also, after updating the half-cycle, a flag is set ("flag_a", "flag_b" or "flag_c"), which is a condition for switching the port C leg of the corresponding phase (PB13, PB14, PB15).

We divide the section of the half-wave of the phase A sinusoid into 3 equal sections of 60 electrical degrees. From the proposed method for modifying the sinusoidal PWM, it follows that in Sect. 1 it is necessary to provide modulation of the output voltage with half the frequency for phases A and C, and with a normal frequency for phase B. In Sect. 2, modulation with half the frequency is necessary for phases B and C, and with a normal frequency for phase B. In Sect. 3, modulation with half the frequency is necessary for phases A and B, and with a normal frequency for phase C.

As can be seen from Fig. 5, the sequence of phases that require PWM generation with a normal frequency is a cycle whose period is the half-period of a modulated sine wave. First B, then A, then C.

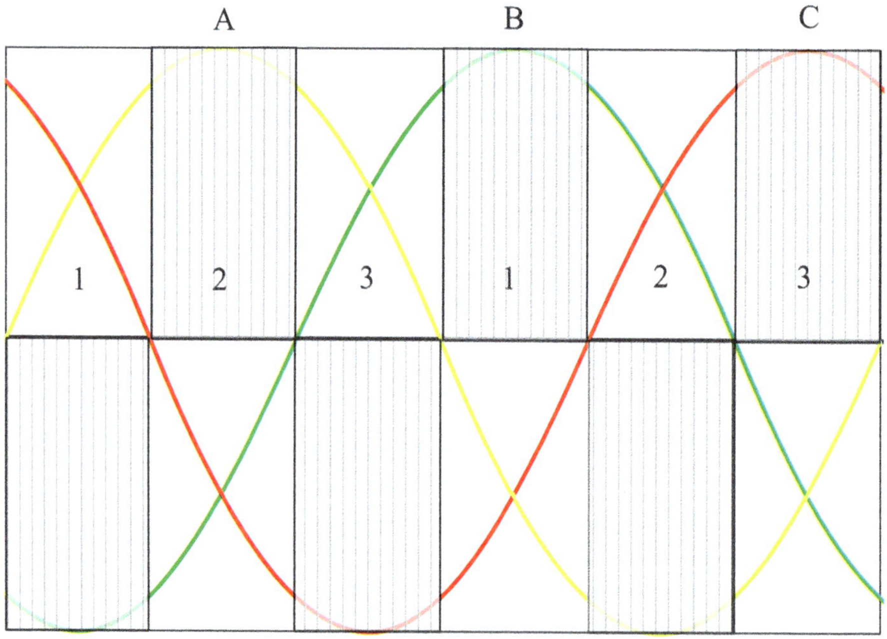

Fig. 5. Division of a three-phase sinusoid into sections

The timer clocks all its channels at a single frequency, i.e. you cannot get the simultaneous operation of two channels of the same timer with different frequencies. The solution to this problem – halving the frequency – will be done by combining the periods of two adjacent pulses.

The next step in the cycle is to calculate the pulse width of a sinusoidal PWM if the value of the counter "pulse_number" is less than one.

The calculation of the pulse duration of half the frequency is carried out as follows.

(1) The pulse duration is calculated with a double period (for example, phase A):

$$\text{sine\_a}[0] = \text{DOUBLE\_PERIOD} \cdot \sin\left(\frac{\text{i} \cdot \text{PI}}{\text{n}}\right), \tag{6}$$

where the sine argument is the product of the serial number of the current pulse "i" and the ratio of the half-cycle to the number of pulses "PI/n", which determines how much of the half-cycle is occupied by one pulse.

(2) If the pulse duration is longer than the pulse period, then the duration of the first pulse is equal to its period, and the duration of the second is the remainder of the division of the calculated duration by the period.
(3) If the pulse duration is less than the pulse period, then the duration of the first is equal to the calculated value, and the duration of the second pulse is zero.

Further, the calculation of pulses of normal frequency applies the formula (for example, phase A):

$$\text{sine\_a}[0] = \text{PERIOD} \cdot \sin\left(\frac{\text{i} \cdot \text{PI}}{\text{n}}\right) \tag{7}$$

Such calculations are presented in the functions "Third_1", "Third_2" and "Third_3" for the first, second and third sections, respectively. After the calculation, the flag of completion of the calculation "flag_count" is set, and pulse counters are also incremented. Next, the counter "pulse_number" is assigned a value of one to enable updating the pulse duration in the interrupt handler.

The function of the interrupt handler for overflowing the timer "void TIM4_IRQHandler ()" begins with the reset of the interrupt bit to be able to interrupt again at the next overflow, and the ADC is polled to the variable "adc_value".

Then, the condition for setting the flag of each phase is checked; if the condition is met, the ODR register bit of the corresponding leg is inverted and the flag is reset to zero.

Then, in turn, in the register of each timer channel: CCR1-CCR3, the value of the pulse width of the sinusoidal PWM calculated from the infinite loop is contained in the first element of the arrays "sine_a (b, c) [2]". The next timer overflow is expected and the next pulse duration from the second elements of the arrays is entered.

Thus, in the main cycle, the duration of two consecutive pulses is calculated, and during the interruption, the calculated duration is entered in the CCRx comparison register.

4 Discussion

The possibilities of using modern electric drives continue to expand constantly due to advances in related fields of science and technology: electrical engineering and electrical engineering, electronics and computer engineering, automation and mechanics. It is expected that the transition from an unregulated electric drive to a controlled one in technologies where this is required can save up to 25–30% of electricity. In one of the technologies - in water-air supply - the transition to a controlled electric drive, as experience has shown, saves about 50% of electricity, up to 25% of water and up to 10% of heat.

Three-phase asynchronous motor with a squirrel-cage rotor is the most common type of electric machine, it is equipped with all modern pumps, compressors and fans, and many other electric drives. Such a widespread use of the electric motor is due to its low cost, high reliability and low operating costs. The main disadvantage of asynchronous electric motors is the difficulty of regulating their speed by traditional methods (by changing the supply voltage, introducing additional resistances into the winding circuit). Controlling an asynchronous electric motor in frequency mode until recently was a big problem, although the theory of frequency regulation was developed back in the thirties. The development of a frequency-controlled electric drive was constrained by the high cost of frequency converters. Now the situation has changed radically: perfect and affordable electronic frequency converters have appeared on the wide market. The system "frequency converter – asynchronous motor with squirrel-cage rotor" will become the main technical solution for the mass controlled electric drive in the coming years.

First of all, a structural diagram of the frequency converter was built. Further, basing on theoretical information and experience, power elements and control elements were calculated and selected. The next step was to create a circuit diagram of the frequency converter. The final stage of the work was devoted to the development and debugging of microcontroller software that provides control of an autonomous voltage inverter. The result of the work is a functioning frequency converter with a modified inverter PWM control algorithm.

5 Conclusions

Frequency converters are an integral part of a modern adjustable electric drive. The control algorithm of the autonomous inverter in the frequency converter directly affects the quality of the output voltage, and, accordingly, the loss in the motor due to voltage non-sinusoidality. The importance of proper control of an autonomous inverter determines the loss in the inverter itself.

The work revealed the problem of distortion of the sinusoidality of the output voltage due to high-speed switching of the inverter power switches and proposed a way to solve this problem. The authors propose an algorithm for controlling an autonomous voltage inverter, which will reduce the distortion of the shape of the output voltage, and will also help to reduce the loss of switching power switches at times when the motor consumes the lowest instantaneous power, i.e. at the beginning and end of a half-period of sinusoidal voltage. In the future, a detailed comparison of the inverter control algorithms is required by comparing the root-mean-square losses in the inverter and the motor.

References

1. Chitra, A., Sultana, R., Vanishree, W.J., Sreejith, S., Jose, S., Pulickan, A.J.: Performance comparison of multilevel inverter topologies for closed loop v/f controlled induction motor drive. Energy Procedia **117**, 958–965 (2017). https://doi.org/10.1016/j.egypro.2017.05.216
2. Gaballah, M., El-Bardini, M.: Low cost digital signal generation for driving space vector PWM inverter. Ain Shams Eng. J. **4**(4), 763–774 (2013). https://doi.org/10.1016/j.asej.2013.02.004
3. El-Barbary, Z.M.S.: Single-to-three phase induction motor sensorless drive system. Alexandria Eng. J. **51**(2), 77–83 (2012). https://doi.org/10.1016/j.aej.2012.05.003
4. Yegane, M.S.O., Sarvi, M.: An improved harmonic injection PWM-frequency modulated triangular carrier method with multiobjective optimizations for inverters. Electric Power Syst. Res. **160**, 372–380 (2018). https://doi.org/10.1016/j.epsr.2018.03.011
5. Gulyaev, A., Fokin, D., Ten, E., Vlasyevsky, V.: PWM algorithms synthesis. Procedia Eng. **165**, 1529–1535 (2016). https://doi.org/10.1016/j.proeng.2016.11.889
6. Gulyaev, A.V., Fokin, D.S., Ten, E.E., Skorik, V.G., Shukharev, S.A.: Research and development of conversion of direct current voltage to the quasi-sinusoidal voltage with pulse-width modulation. Russ. Electr. Eng. **87**(2), 90–92 (2016). https://doi.org/10.3103/S1068371216020061
7. Gulyaev, A., Fokin, D., Shuharev, S., Ten, E.: Effect of different methods of pulse width modulation on power losses in an induction motor. In: IOP Conference Series: Earth and Environmental Science 19. Cep. "Energy Management of Municipal Transportation Facilities and Transport, EMMFT 2017", p. 012042. (2017). https://doi.org/10.1088/1755-1315/90/1/012042
8. Akar, M., Gercekcioglu, H.S.: Instantaneous power factor signature analysis for efficient fault diagnosis in inverter fed three phased induction motors. Int. J. Hydrogen Energy **42** (12), 8338–8345 (2017). https://doi.org/10.1016/j.ijhydene.2017.02.151
9. Data Sheet STM32F103x8, STM32F103x. https://www.st.com/resource/en/datasheet/cd00161566.pdf
10. RM0008 Reference Manual STM32F101xx, STM32F102xx, STM32F103xx, STM32F105xx and STM32F107xx advanced Arm®-based 32-bit MCUs. https://www.st.com/content/ccc/resource/technical/document/reference_manual/59/b9/ba/7f/11/af/43/d5/CD00171190.pdf/files/CD00171190.pdf/jcr:content/translations/en.CD00171190.pdf

Increasing Energy Efficiency and Reliability of Electric Multiple Unit Re-generative Braking

Vladislav Ivanov, Roman Ustinov(✉), and Oleg Melnichenko

Irkutsk State Transport University,
Chernyshevskogo st., 15, Irkutsk 664074, Russia
romust93@mail.ru

Abstract. The article provides a brief overview of a number of promising proposals aimed at improving the technology of regenerative braking of modern domestic AC motor vehicle rolling stock in terms of reliability and energy efficiency. Regenerative braking has become widespread especially in railway transport, it provides the maximum speed of trains on the site due to adjusting braking in automatic mode, significantly reduces the longitudinal-dynamic forces in the train, reduces the wear of wheel sets and brake pads, increases the safety of train movements. However, the application of regenerative braking on electromotive requires solving a number of problems, two main ones can be distinguished - low power factor of rectifier-inverter converters and low operability in case of emergency operation. The article discusses existing technologies aimed at eliminating these problems, and also presents a set of technical solutions aimed at improving the energy efficiency and operational reliability of modern domestic AC car rolling stock. The proposed technical solutions can be implemented both during factory repairs as part of modernization and in the manufacture of new electric trains.

Keywords: Regenerative braking · Converting equipment · Rectifier-inverter converter · Rectifier installation of excitation · Control algorithm · Switching · Power factor · Reliability · Emergency process

One of the tendencies of development of the Russian economy is a growing need for energy resources. To fulfill that requirement, along with increasing electric energy production output, higher energy efficiency is also being pursued, especially in energy-demanding fields such as heavy industry, construction, transport and communication [1–3].

The necessity for introducing energy saving technologies into the railway transport industry is proclaimed by all base development programs adopted by the Russian Railways, including the company's Energy strategy as well as the Strategy of scientific and technical development, set to be relevant until 2030 and 2025 respectively.

In this context, special credit can be given to the widespread application of regenerative braking technology. First introduced on domestic railways in late 1970s, energy regeneration still maintains its progressive potential by saving up to 20% of the energy spent on traction on certain railway track sections with steep profile. However, using the full potential within the regenerative braking can be achieved only after certain downsides inhibiting the technology are overpassed.

Z. Popovic et al. (Eds.): TransSiberia 2019, AISC 1115, pp. 420–426, 2020.
https://doi.org/10.1007/978-3-030-37916-2_41

One of the main disadvantages of regenerative braking is its low power factor in thyristor-based rolling stock, which includes domestic electric multiple units (EMUs). This factor does not exceed 0.71 for energy regeneration, which suggests low energy efficiency in general. High consumption of reactive power by EMUs, which results from low power factor, consequently distorts the voltage sinewave in the overhead line and lowers energy quality of the entire grid. As a result, modern domestic EMUs do not meet the requirements for energy parameters determined by the relevant national standards.

Additionally, the unpractical energy consumption by EMUs leads to increased fares, which in turn hinders population mobility and decreases competitiveness of the railway transport.

Low power factor of EMUs is almost entirely attributable to their power circuit, namely the obsolescence of their main, thyristor-based reversible power converters (RPCs), as well as thyristor-based rectifying excitation devices (REDs) used for regenerative braking mode. Thus, further energy efficiency increase in this field is strongly dependent on modernizing these RPCs and REDs.

The invention and widespread introduction of reliable power IGBTs has opened an opportunity to upgrade the semiconductor base of RPCs and REDs without changing overall mass and size parameters. Other advantages of IGBTs, including their minimization of stray inductances, high stability towards *du/dt* changes, low and temperature-nondependent residual currents etc., also greatly facilitate power equipment working conditions.

A complete modernization of thyristor-based RPCs and REDs with modern, fully controllable transistor semiconductor devices is suggested by the authors. The main circuit of the suggested RPC consists of the same 8-arm bridge arrangement with an inclusion of an additional 9th discharge arm, while prospected RED is based on a full-wave diode rectifier with a series transistor arm, Fig. 1. Such a schematic solution allows to introduce a controllable diode bridge for maintaining and regulating excitation currents of traction motors working in regenerative braking mode. For RPC, protective RC snubbers are also included in parallel with each section of the power transformer's secondary winding [4–7].

In order to organize transistor-based RPCs and REDs, special control algorithms were also developed. To-day, this technical solution has been successfully tested in mathematical simulation environment and on an experimental stand.

As demonstrated by conducted research, total power factor of an EMU with proposed converters during regenerative braking increased up to 0.99. The amount of electric energy returned into the grid by a single driving carriage of an EMU increased by 28.95%, while relative pulsation coefficient diminished by 25.24% as opposed to an EMU with type equipment. Moreover, average values of the voltage supplied to the RPC have risen by 23.52% due to lower losses on converter arm commutation.

Analysis of the obtained results allowed to confirm that the introduction of power transistors into RPCs and REDs in place of thyristors, supplemented by new control algorithms, compensates the phase shift between grid current and voltage sinewaves created by inductances. As a result, the reactive power previously generated within the circuit is practically neutralized, which means higher power factor for the transistor-based EMU.

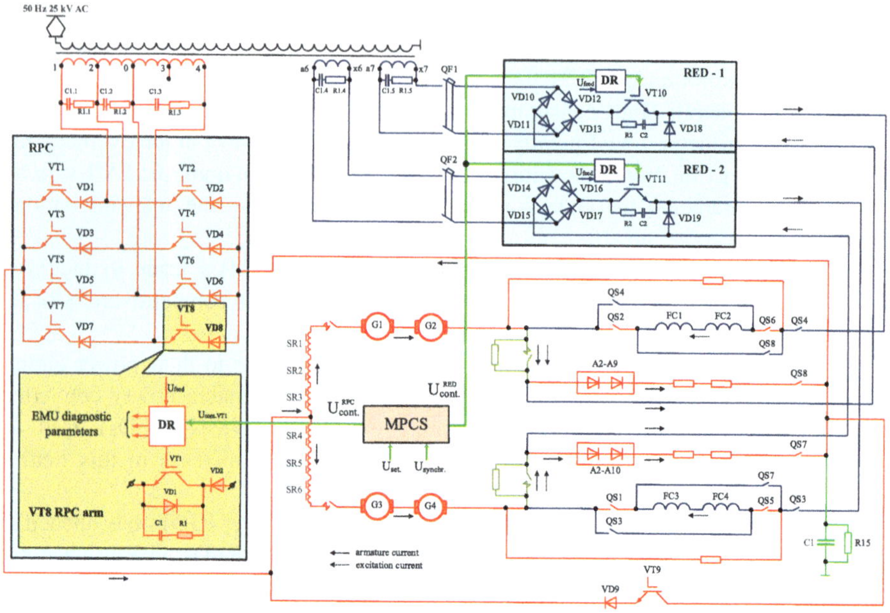

Fig. 1. Basic electric scheme of an EMU power circuit with proposed transistor-based reversible power converter (RPC) and rectifying excitation devices (REDs)

Another advantage of a transistor converter concerns reliability of regenerative braking mode itself. Discarding thyristors and their specific control algorithm allows to eliminate the risk of a catastrophic failure event known as the "invertor turnover", which ensures reliability and safety of energy regeneration.

While a full-scale modernization of converters may be more expedient for new EMU series (such as EP3D), an alternative technical solution, due to its economical feasibility and simplicity, may be more suitable for implementation on currently operational EMUs (like the ED9E series) that apply type, thyristor-based RPCs and REDs. The solution in question implies implementing a discharge arm that shunts the rectified sub-circuit of the EMU, Fig. 2.

The discharge arm is composed of a power diode VD1 and a fully controllable IGBT VT1 in series, which the arm as a whole connected between the cathode and anode busbars of the converter. Switching the arm on and off through control impulses transmitted to the IGBT at certain moments within each half-wave of gird voltage allows to achieve a discharge circuit that accelerates thyristor arm commutation [8–10].

Application of this shunting discharge arm allows to optimize and accelerate commutation processes in RPC thyristor arms, which leads to a decrease in the phase shift between current and voltage sinewaves. Consequently, reactive power decreases and power factor reaches the value of 0.9 for the 4th voltage regulation zone.

Along with increasing energy efficiency of regenerative braking, the problem of ensuring reliable and fail-safe operation of power RPCs remains equally urgent [12–14]. As the analysis of EMU electric equipment failures suggests, the majority of power

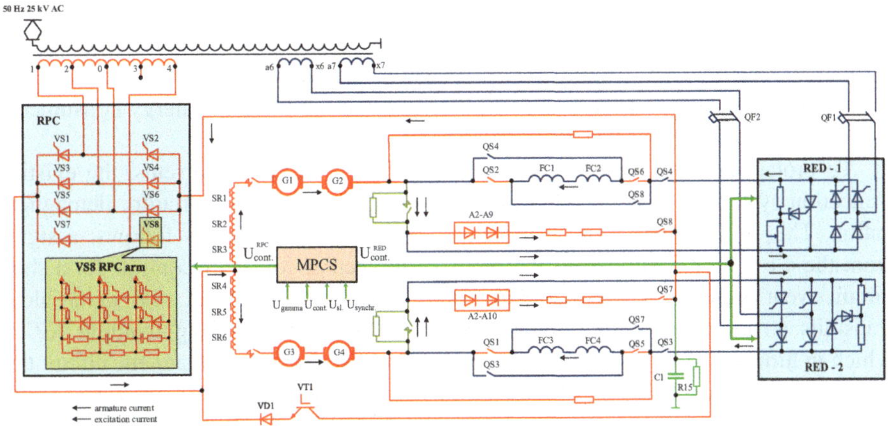

Fig. 2. Basic electric scheme of an EMU power circuit with type RPC and REDs but an additional discharge arm composed of a series diode and IGBT

circuit breakages occurs due to damage of sensible electronic parts and control system failures. Most of these breakages lead to an omission of control impulses sent to converters' thyristor arms, which in turn triggers circuit protection and causes regenerative braking to shut down. Losing braking effect of even a single driving carriage within an EMU creates additional longitudinal dynamic forces that can lead to derailment. Such unreliability of regenerative braking cannot be tolerated, as it is supposed to serve as a reserve braking system in case of the pneumatic brake system failure.

The following description presents the mechanism of a short circuit current event inside an EMU power circuit during regenerative braking and on the 4th zone of voltage regulation. Figure 3 provides an example, where a control impulse meant for the VS2 thyristor arm is omitted.

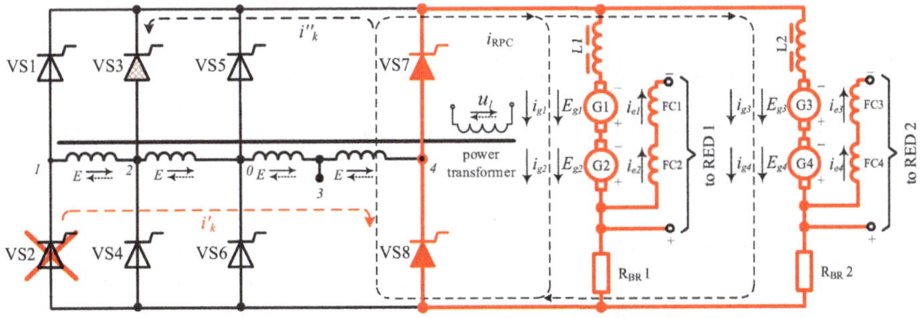

Fig. 3. Simplified electric scheme of an EMU power circuit in regenerative braking mode

Omission of a control impulse meant for the VS2 thyristor arm leads to an absence of the larger commutation sub-circuit $i'k$ and locks another thyristor arm, VS8, in its conducting state. After the smaller sub-circuit ends its commutation $i''k$, a short circuit event occurs as the current now flows through the positive outlets of generators G1-G4, blocks of ballast resistors (BRs), then open thyristor arms VS8-VS7 in series, smoothing reactors L1-L2 and finally the negative outlets of G1-G4. In such a circuitry the secondary winding of the power transformer that serves as an energy consumer is absent, thus leading to a dramatic increase in current.

To protect power circuits from short circuit events and overcurrents, the existing protection system used on ED9E and EP3D EMU series applies circuitry demagnetization. This protection system is presented in Fig. 4 and works as follows: a signal from the T1 sensor is applied to the block of protective contactor control. As a short circuit event occurs, it leads to armature current rising above 650 A, and so the block transmits an impulse for brief power supply of the demagnetizing winding KZT1, which in turn leads to a decrease in flux of holding coils inside the magnetic circuit that holds the contactors' anchor. Consequently, the contactor switches off. Because the main contacts of KZT1 are shunted with a resistor R12, during their breaking the resistor experiences a voltage drop due to the armature current of traction motors. Due to polarity of this current, a protective thyristor VS5 inside the RED thyristor block fires. This creates a circuit for the demagnetizing current to flow through the filed coils (FC) of traction motors FC1-FC2, directed inversely to the excitation current. The described operation results in rapid decrease in the armature current.

At the same time, contacts of the KZT1 contactor remove the control signal from the control system supply block, leading to an end in control impulse formation for RPC and REDs thyristors.

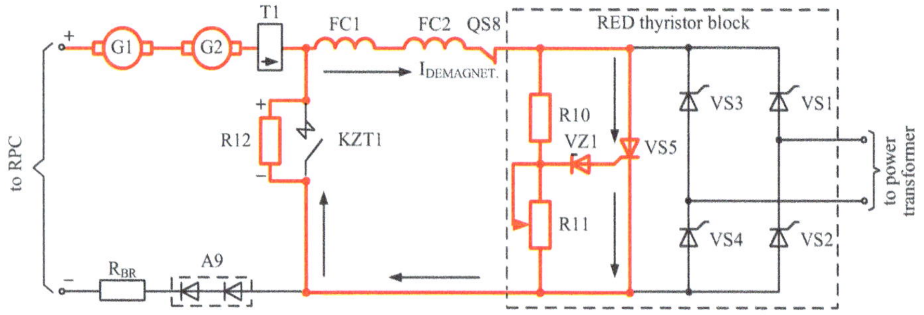

Fig. 4. Simplified electric scheme of a traction motor demagnetizing circuit during a regenerative braking short circuit event

The main advantage of the presented protection system is the presence of a demagnetizing circuit and its ability for intensive decrease of overcurrents, leading to a sufficient rise in power circuit reliability. The crucial disadvantage, however, is that the firing of the protection system shuts down regenerative braking mode of an EMU, effectively eliminating its braking effect.

Most of the earlier inventions aimed to increase durability and reliability of RPCs in regenerative braking mode are based on introducing additional elements and equipment into EMU designs, which entails vehicles further complication and price rise. Thanks to the universal application of microprocessor-based control systems on modern domestic EMUs, many of the reliability issues can now be solved by simple revisions of software.

In order to protect RPCs and the entire power circuit from catastrophic failures caused by control impulse omissions, the authors suggest supplementing the type working algorithm of an RPC control block with following subsystems: a control impulse omission block, that determines the omission event by checking commutation duration Uγ by the $\gamma = \gamma' + \gamma''$ equation, where the total commutation γ consists of commutation in the larger sub-circuit γ' and commutation in the smaller sub-circuit γ''; and a current rising rate control block, detecting an absence of one of the sub-circuits by the amount of current increase [11–15].

The obtained data allows to determine the failed thyristor arm of the converter, so the system can then substitute it with another arm in accordance with the control algorithm, i.e. in Fig. 5 the failed thyristor arm VS2 is substituted by the arm VS4.

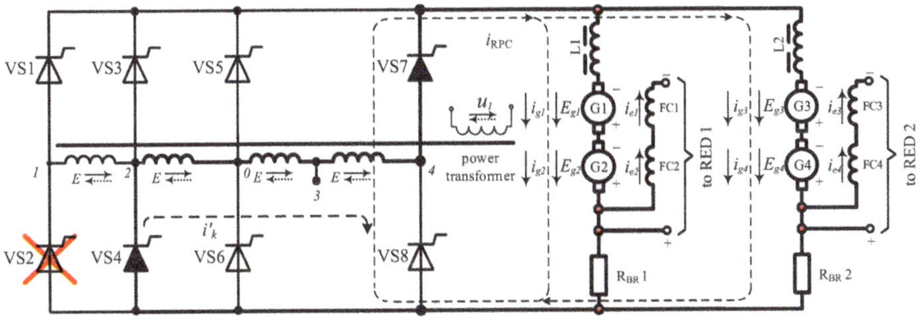

Fig. 5. Simplified electric scheme of an EMU power circuit in regenerative braking mode

1 Conclusions

The paper has described multiple means for elimination of the regenerative braking many disadvantages, achieved by modernizing electric multiple unit power circuits with transistor-based converters and new control algorithms. The suggested technical solutions are verified by mathematical modelling and experimental research.

A power circuit packed with fully controllable semiconductor devices allows to increase power factor to 0.99 and the amount of electric energy returned into the grid by 28.95%, while decreasing relative pulsation coefficient by 25.24% and increasing the average values of voltage supplied to the RPC by 23.52%.

Another suggested method of power factor increase implies only a single diode-transistor arm to be introduced into the power circuit, which shunts the rectified sub-circuit. Application of this technology is most expedient for currently operational EMUs (such as ED9E series) with thyristor RPCs due to its economical feasibility and simplicity. As the results of mathematical modeling demonstrate, the power factor reaches as high as 0.9 with this simple solution.

Along with increasing energy efficiency of regenerative braking, the problem of ensuring reliable and fail-safe operation of power RPCs remains equally urgent. The presented method for RPC reliability increase is based on revisioning type software and its signals and does not require additional equipment, allowing to substitute failed thyristor arms with other arms and maintain uninterrupted regenerative braking, thus ensuring safety for the entire vehicle.

References

1. González-Gil, A., Palacin, R., Batty, P., Powell, J.P.: A systems approach to reduce urban rail energy consumption. Energy Convers. Manage. **80**, 509–524 (2014)
2. Douglas, H., Roberts, C., Hillmansen, S., Schmid, F.: An assessment of available measures to reduce traction energy use in railway networks. Energy Convers. Manage. **106**, 1149–1165 (2015)
3. Brown, M., Rawtani, J., Patil, D.: Troubleshooting variable speed drives. In: Practical Troubleshooting of Electrical Equipment and Control Circuits, pp. 121–172 (2004)
4. Ivanov, V.S., Melnichenko, O.V.: Mathematical modeling of traction transformer of electric train of alternating current of series ED9E simulation matlab Simulink. Mod. Technol. Syst. Anal. Model. **59**(3), 57–58 (2018). https://doi.org/10.26731/1813-9108.2018.3. (in Russian)
5. Single-phase alternating current power rectifier-inverter converter for electric transport and method of its control. Patent RU2700594 (2018). (in Russian)
6. Ivanov, V.S., Melnichenko, O.V.: Analysis of the use of converters in power circuits of domestic AC motor rolling stock. Herald of Irkutsk National Research Technical University, vol. 4, pp. 231–242 (2018). https://doi.org/10.21285/1814-3520-2018-4-231-242. (in Russian)
7. Traction converter of electric rolling stock of alternating current and method of its control. Patent RU2659799 (2018). (in Russian)
8. Tomilov, V., Barinov, I., Melnichenko, O., Shramko, S., Linkov, A.: Improvement of AC locomotive power circuit in regenerative braking mode. In: Proceedings of Irkutsk State Technical University, vol. 23(2), pp. 346–355 (2019). https://doi.org/10.21285/1814-3520-2019-2-346-355. (in Russian)
9. Barinov, I.A., Melnichenko, O.V.: Power IGBTs application in AC-wire DC-motor locomotive thyristor-based power circuit for regenerative brake energy efficiency increase. In: 2019 International Conference on Industrial Engineering, Applications and Manufacturing (ICIEAM). IEEE (2019). https://doi.org/10.1109/icieam.2019.8742933
10. Method for control of a multi-zone rectifier-inverter converter of single-phase AC. Patent RU2689786 (2019). (in Russian)
11. Method for alternating current electric locomotive work capacity increase during regenerative braking mode and device for its realization. Patent RU2659756 (2018). (in Russian)
12. Device for thyristor converter protection. Patent RU2152679 (2000). (in Russian)
13. Device for functional diagnostic and protection of a thyristor converter. Patent SU1690072 (1988). (in Russian)
14. Device for reversible thyristor converter diagnostic and protection. Patent RU2183896 (2002). (in Russian)
15. Ustinov, R.I., Melnichenko, O.V.: Modeling emergency processes of electric locomotive reversible converters under control pulse skip. In: Proceedings of Irkutsk State Technical University, vol. 22(3), pp. 244–254 (2018). https://doi.org/10.21285/1814-3520-2018-3-244-254. (in Russian)

Actual Artificial-Intelligence Based System for Assessment of the Technical State of the Rolling Stock Fleet

Artyom Plyaskin and Alexey Kushniruk(✉)

Far Eastern State Transport University, Seryshev st., 47, Khabarovsk 680021, Russia
alexey.kushniruk@mail.ru

Abstract. The goal of the work is to development of a synchronous-replicated model for the assessment of the technical state of a locomotive as a technological system. When performing the work, method of systems analysis, computer and mathematical modelling, artificial intelligence, and mathematical analysis were used. As a result of the research we have obtained a mathematical synchronous-replicated model for the assessment of the technical state of a locomotive based on multilayer neuron forecasting network. The model developed can be used in systems for monitoring, controlling and diagnosing the technical state of the locomotive fleet. This model possesses such novel specific features as low sampling period between quizzing of monitoring facilities, versatility, adaptability and operability. This suggested model resolves a range of tasks set forth in the Concept for the development of the OJSC Russian Railways connected with the implementation of an actual system for repairs and maintenance according to the current condition of locomotives as well as the digitalization of the advanced fields of the company.

Keywords: Levenberg–Marquardt algorithm · Computer modelling · Diagnostic system · Locomotive fleet · Monitoring data · Neural network

1 Introduction

The technical state of the locomotive fleet is one of the most important constituents of the transportation process. Monitoring, control, diagnosing and genesis of the technical state of locomotives is accomplished by a complex of automated systems for technical diagnosis (ASTD) which currently functions within the frameworks of the automated system for the locomotive fleet reliability management (ASLFRM). However, despite the application of a system approach to maintaining reliability parameters of the locomotive equipment components, the number of sudden failures increases with every year.

Z. Popovic et al. (Eds.): TransSiberia 2019, AISC 1115, pp. 427–442, 2020.
https://doi.org/10.1007/978-3-030-37916-2_42

2 System Analysis of the Existing Complex for Assessment of the Technical State of the Locomotive Fleet

The problem of increase in the number of sudden failures of rail transport is not only directly linked with the technical state of the rolling stock but also with the technology of detecting pre-failure conditions. At present, the process of analysis is based upon the application of an asynchronous-replicated data-base, such as the integrated system for technical monitoring of the locomotive fleet (ISTMLF). As stationary systems for monitoring, controlling, diagnosing and forecasting prevail in the Russian rail transport, the process of detecting incidents is governed by a considerable sampling period. When the latter increases, the risk of emergency repairs rises.

The predominant use of stationary systems is caused by the complexity and identity of diagnosing approaches and methodologies. Thus, with existing mathematical models, the application of vibro-acoustic diagnosis to bearing assemblies and speed transformers as a component of vehicle-mounted (continuous) diagnosis systems is impossible. The reason for this is that the output parameter in this kind of diagnosis is vibro-acoustic acceleration that is subject to significant noise impact during the locomotive operation.

The prevalence of stationary diagnostic systems over vehicle-mounted systems is also determined by the working conditions of certain equipment components. The wear of engine piston rings and cylinder bushing is diagnosed during the engine rig test as installation of pressure gauges in the combustion chamber is irrational due to the aggressive medium of monitoring.

The existing vehicle-mounted diagnostic systems have imperfect mathematical models that in general perform control functions and automated analytical post-processing of data when the locomotive is disposed at depot. Thus, information about excess temperature in the diesel exhaust manifold is transferred by means of a removable storage device (microprocessor system for control, regulation and diagnosis) and is processed upon the arrival of the diesel locomotive at the depot. Or? The data on the overheating of axle equipment and engine-anchor bearings (EAB) in the onboard temperature checkout subsystem (OBTCSS) is also transferred upon the arrival of locomotive at the depot. The presented principles of data transfer show the high level of discontinuity in accomplishing diagnostic works, which causes unscheduled repairs.

Based on the description above, the key flaw of the existing planned preventive system for maintenance, repairs and diagnosis is the high period of discontinuity connected with the employment of stationary diagnosis systems as a result of case-to-case approaches of the diagnosis models, and therefore the assessment of the technical state of the locomotives.

3 Formation of a Continuous System for the Assessment of the Technical Condition of the Locomotive Fleet Based upon the Application of Artificial Intelligence Theory

The development of a universal mathematical model for continuous assessment of the technical state of the locomotives requires employment of an adaptable theory that would be able to interpret and formalize the diverse processes connected with the specific features of the work of locomotive units. These processes are different by nature and possess different properties. Neuron networks of the theory of artificial intelligence have gained wide popularity among the scientific community due to their adaptability to different branches of knowledge [1]. They possess properties for resolution of various practical problems: they realize a universal apparatus of approximation corresponding to multidimensional arrays of information; they have a learning algorithm and adapt to changing conditions; and they can synthesize gained knowledge. On this basis they are referred to as? the systems of artificial intelligence.

"Neuron networks", a branch of the theory of artificial intelligence, is an interdisciplinary area of knowledge that covers bio-cybernetics, electronics, applied mathematics, statistics, automatics and medicine. Artificial neuron networks are interpreted by the process that takes place in the neuron system of living organisms aimed at the development of new technological solutions [2–4].

In general terms the neuron, as the most important network element, has the formalized structure of a biological nerve cell. The generalized structure that includes the basis for the majority of the neuron mathematical models belongs to the McCulloch-Pitts model [5] presented in Fig. 1.

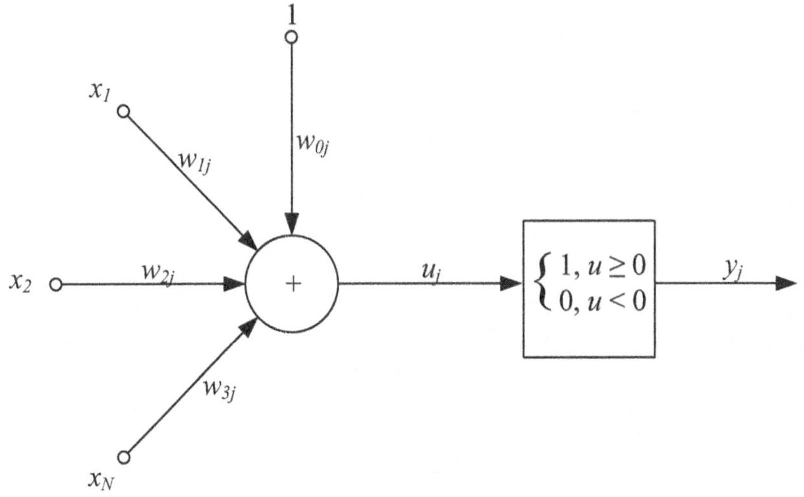

Fig. 1. McCulloch-Pitts neuron (perceptron) model

According to Fig. 1 x_{Nj} are the input signals of the N-th number and the j-th teaching stage – within the systems for the assessment of the technical state registered values of external impacts or the environmental conditions; w_{0j} is the input polarization unit impulse (threshold value); and u_j is the output signal of summarizer:

$$u_j = \sum_{1}^{N} w_{Nj} \cdot x_{Nj} + w_{0j}, \tag{1}$$

w_{Nj} are the weight coefficients of the summarizer inputs; they steer the activation function:

$$y_j(u_j) = \begin{cases} 1, u \geq 0 \\ 0, u < 0 \end{cases}. \tag{2}$$

During the teaching process of the neuron network, this function registers the correspondence of input x_{Nj} and output (within the technical state assessment system – diagnostic) y_j parameters against the teaching sampling. The purpose of the activation function is to determine the grade of membership between the received value of an input network parameter and the output parameter of the teaching sampling using a concrete membership function (linear, sigmoid and others) [6, 7] which is aimed at obtaining a conclusion on the further changes in the output parameter obtained by means of the neuron network.

The mathematical model of a neuron is a function that calculates its value based on several input parameters x_{Nj} conveyed. Each value of input parameter x_{Nj} is associated with a weight coefficient w_{Nj}, which forms the output parameter y_j.

The neuron network teaching process resolves itself into selecting weight coefficients w_{Nj} for corresponding input parameters x_{Nj} to form equations that will result in the correspondence of the teaching sampling on each N-th and j-th output parameter y_j; the output signal of y_i is determined by the function of the weighted sum of the product of the weights and the values of the input parameters x_{Nj}:

$$y_j = f\left(\sum_{1}^{Nj} w_{Nj} \cdot x_{Nj} + w_{0j}\right) \tag{3}$$

Practically, the process of teaching the neural network consists in supplying an input signal on the hidden layer of perceptron. After the output signal is received, training error is calculated which is minimized by means of changing weight coefficients. Then the network teaching is directed according to the formula

$$\Delta y_j = y_j - a, \tag{4}$$

where a is the output parameter, obtained from the teaching sampling (experimental or monitoring data); and changing of weight coefficient is reduced to the formula

$$\Delta w_{Nj} = \lambda_j \cdot \Delta y_j + \alpha \cdot (w_{Nj} - w_{Nj-1}), \tag{5}$$

where λ_j is the minimization pace;

α is the torque coefficient that takes the value within the interval [0, 1].

According to (3) input signals of parameter x_{Nj} are summed up with consideration of the weights w_{Nj} in the summarizer, while further on the resultant value is compared to the sampling specified in the training. If the resultant value complies with the value specified by the training sample, then the equation is recorded. Otherwise, the equation is perceived as insufficiently precise or erroneous within the frameworks of approximation and the correction of weights is required. The weights are changed and the process recurs. After the training is completed, it results in a neural network – a system of equations built on the basis of the input values x_{Nj} and the output values y_i on the whole training sampling with a certain regression.

Based upon the presented mathematical description of the perceptron, the task of the neural network training is reduced to the minimization of error function (formula (4)).

One of the major algorithms for the minimization of the training error function in the theory of neural networks is the steepest descents algorithm. The steepest descents algorithm is based upon determining the minimum value of the function under consideration [8]. The algorithm is accomplished by finding derivatives of the training error function for each perceptron of its weights:

$$\nabla f(w_{Nj}) = \Delta y_j + a \to a = \frac{\partial(w_{0j} + x_1 \cdot w_{1j} + x_2 \cdot w_{2j} + x_N \cdot w_{Nj})}{\partial w_{Nj}} \tag{6}$$

and substitution of arbitrary values of the point of search direction w_{NO} into differentiated function to calculate the values of the gradient vector for this point.

Based upon the values obtained, the breakpoint of the function minimization is checked:

$$\left|\nabla f(w_{Nj})\right| < \varepsilon. \tag{7}$$

where ε is the limit of the function minimization which is accepted based upon the target error.

After the value of the assumed function $f(w_{NO})$ in the accepted setting out point w_{NO} is calculated, a step along the negative gradient direction aimed at the determining the value of function in the new point is performed:

$$w_{N1} = w_{NO} - \lambda_j \nabla f(w_{NO}) \tag{8}$$

Further on, the cycle recurs, beginning with the substitution of values into the differentiated function in a new point until the condition (7) is fulfiled.

Along with the steepest descents algorithm, the Gauss–Newton method [9] is another famous strategy of function optimization. This method consists of the search for minimum error of the network target value by differentiating functions analogously to (6) and determining the value of the latter in the starting point. In this case the optimization is calculated as the difference between the current value of the weight in the point under consideration and the ratio of the value of the initial function to the differentiated one:

$$w_{Nj1} = w_{N0} - \frac{f(w_{N0})}{\nabla f(w_{N0})}, \tag{9}$$

Calculation is accomplished by means of sequential subtraction from the initial starting point to find the subsequent point; the set of weights in the differentiated functions under consideration take the form of the Jacobian matrix:

$$J(w) = \begin{bmatrix} \frac{\partial f_1}{\partial w_{11}} & \frac{\partial f_1}{\partial w_{21}} & \cdots & \frac{\partial f_1}{\partial w_n} \\ \frac{\partial f_2}{\partial w_{12}} & \frac{\partial f_2}{\partial w_{22}} & \cdots & \frac{\partial f_2}{\partial w_n} \\ \cdots & \cdots & \cdots & \cdots \\ \frac{\partial f_j}{\partial w_{1j}} & \frac{\partial f_j}{\partial w_{2j}} & \cdots & \frac{\partial f_j}{\partial w_{3j}} \end{bmatrix}, \tag{10}$$

where f_j is the error function of the j-stage of the network training.

After these operations, the cycle recurs until the sought minimization of the network error function is achieved.

A combined method for function optimization is the Levenberg–Marquardt strategy [10]. This strategy implies the alternate use of the above described algorithms depending on the efficiency of network training. Gradient and approximate Hessian matrix *G(w)* (second-order matrix of differentiated functions of the network errors, i.e. transposed product of the Jacobian matrix employed in the Newton iteration method) is determined as:

$$\nabla f(w_{Nj}) = [J(w)]^T \cdot f_j(w), \tag{11}$$

where ^{T} is the sign of the matrix transposition.

$$G(w) = [J(w)]^T \cdot J(w) + \mu, \tag{12}$$

where μ is the regularization factor, that determines the method for optimization of the network error function (the steepest descents method or the Gauss–Newton method).

The Levenberg–Marquardt optimization strategy consists of the following: at the initial stage of the network training, when the actual values of weights w_{Nj} differ from the optimized values (the values of errors Δy_j are considerable), the value of the regularization factor μ exceeding the value of the Hessian matrix is accepted. In this case the Hessian is actually substituted with the regularization factor μ, and the direction of minimization is selected with the steepest descents method according to the formulas (6, 7, 8). As the error values Δy_j decrease, the parameter μ decreases until the fidelity coefficient q reaches the value of unity ($q \approx 1$):

$$q = \frac{\Delta y_j - \Delta y_{j-1}}{[\Delta w_{Nj}]^T \cdot \nabla f(w_{Nj}) + 0,5 \cdot [\Delta w_{Nj}]^T \cdot G(w_{Nj}) \cdot \Delta w_{Nj}}. \quad (13)$$

At that the value of parameter μ is determined in accordance with the classical Levenberg–Marquardt algorithm:

- if $\Delta y_{j\text{-}1}(\mu_{j\text{-}1}/r) \leq \Delta y_j$, then $\mu_j = \mu_{j\text{-}1}/r$;
- if $\Delta y_{j\text{-}1}(\mu_{j\text{-}1}/r) > \Delta y_j$ and $\Delta y_{j\text{-}1}(\mu_{j\text{-}1}) < \Delta y_j$ then $\mu_j = \mu_{j\text{-}1}$;
- if $\Delta y_{j\text{-}1}(\mu_{j\text{-}1}/r) > \Delta y_j$ and $\Delta y_{j\text{-}1}(\mu_{j\text{-}1}) > \Delta y_j$, then value μ is increased m times simultaneously accepting $\mu_j = \mu_{j\text{-}1}.r^m$, where r is the value reduction factor of the regularization factor μ.

If value $q \approx 1$, then approximation of the target function has a high degree of coincidence with true values which proves the realization of the optimal solution. In this case, the regularization factor μ can be omitted (μ = 0), the process of determining Hessian is reduced to direct first order approximation, and the Levenberg–Marquardt algorithm reduces itself to the Gauss–Newton method (9, 10).

The described process of function minimization is a formalized model of the neuron network teaching and can be used in systems for controlling, diagnosing, forecasting, machine operating and other spheres. The advantages of this model are universality and the adaptability of the approximation functions with diverse data of the teaching samplings.

4 Model of a Forecasting Neural Network Based System for the Assessment of the Technical State

Let us consider the process of the development of a system for assessment of the technical state by the example of axle caps of the locomotive wheel-motor block (WMB). This assessment system is based on the neural network for forecasting output parameters. Temperature $y_{Taxle\ cap}$ is the output parameter of the axle cap that allows for the maximum adequate assessment of the technical state. The most significant influencing variables are the ambient temperature (air) x_{Ta} and the velocity of motion x_{Vd}.

Figure 2 presents monitoring data that shows the dependency of changes in the temperature of the axle cap bearing brass upon the locomotive velocity (and its derivative on time–acceleration, correspondingly) and the ambient air temperature. Network training is accomplished in accordance with the data presented.

Free-forward neural networks with one-way links from input to output layers are used for resolving the tasks of forecasting and approximating non-linear functions.

Teaching of the neural network was performed in the package *Neural Network Training* of the program *Matlab*. A unidirectional multilayered neural network with one hidden layer having activation functions – sigmoid and linear correspondingly – is used as a learning network. The Levenberg–Marquardt algorithm [9] is used as a learning function. For the further assessment of the quality of performance of the extrapolation task by the neural network, the network training is accomplished with regard to the monitoring data presented in Fig. 2, namely on the time interval t: 0–780 min. Assessment of the quality of performance of the extrapolation task is accomplished within the time interval t: 780–1020 min correspondingly.

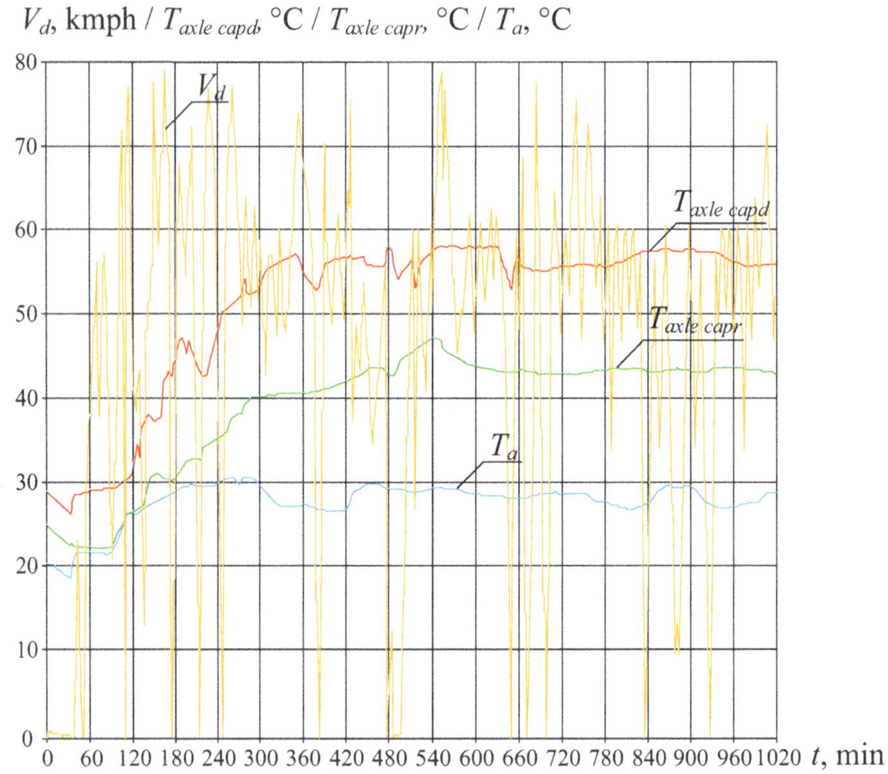

Fig. 2. Monitoring data in complex form, where: V_d is the velocity of the locomotive motion, kmph; $T_{axle\ capd}$ is the defective axle cap temperature, °C; $T_{axle\ capr}$ is the temperature of an axle cap with no defects, °C; T_a – the ambient temperature (air), °C; and t – the locomotive travel time, min

Based on the experimental samplings of network training we have determined the optimal number of perceptrons is 56 and the number of iterations – 215000. The correlation coefficient under the given parameters is R = 0.98484, and the coefficient of determination is R^2 = 0.97 correspondingly. The program code of sequential operations for training a network in the *Matlab* environment looks as follows:

>>P = {Input layer} «Input data of training sampling – $V_d, V'_d = \frac{\partial V_\partial}{\partial t}, T_a$»

>>C = {Target (output) layer} «Output data of training sampling – $T_{axle\ capr}$, $T_{axle\ capd}$»

>>*net* = *newff*([*minmax*(*P*)], [56 2], {'*tansig*' '*purelin*'}, '*trainlm*'); «Creation of unidirectional multilayered network with two hidden layers with sigmoid and linear activation functions»
>>*net.trainParam.epochs* = 215000; «Number of training iterations»
>>*net.trainParam.goal* = 0; «Condition of network training interruption on the deviation from target values (output layer) of training sampling»
>>*net* = *train*(*net*, *P*, *C*); «Network training on input and output layer»

The model of neural network for forecasting and assessment of the technical state of the locomotive axle cap is presented in Fig. 3:

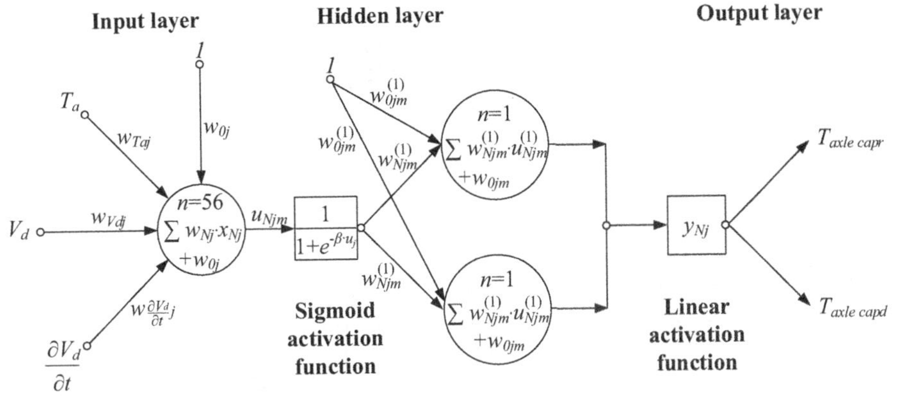

Fig. 3. Model of the forecasting neural network and the assessment of the technical state of the locomotive axle cap, where n is the number of the neurons in a layer, $^{(1)}$ is the ordinal number of a hidden layer; and w_{Taj}, w_{Vdj}, $w\,\frac{\partial V_d}{\partial t} j$ are the weight coefficients of the output parameters of air temperature, motion velocity and acceleration correspondingly

Figures 4, 5 and the Table 1 present the results of the neural network training.

Let us model the output parameter of the network based upon the training sampling under consideration (Fig. 6):

The obtained approximation function for change in diagnosis parameter $T_{axle\ capd}$ ($T_{axle\ capr}$) = $Net(V_d, V_d', T_a)$ describes real $T_{axle\ capd}$ ($T_{axle\ capr}$) = $f(V_d, V_d', T_a)$. There are also noise effects caused by the changing input parameters (specifically, velocity). In order to suppress these noises a filter is used. It excludes those values of the diagnostic parameter that exceed the maximum fixed in the sampling under consideration and at a current point in time and equate them to the previous value.

$$\text{If } \frac{\partial y_j}{\partial t} > \frac{\partial a_{\max(\min)}}{\partial t}, \text{then } y_j = y_{j-1}, \tag{14}$$

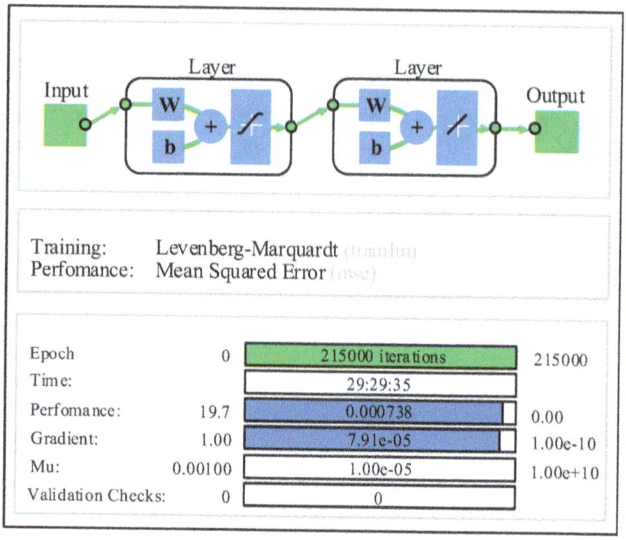

Fig. 4. Results of training of the neural network for forecasting output parameters

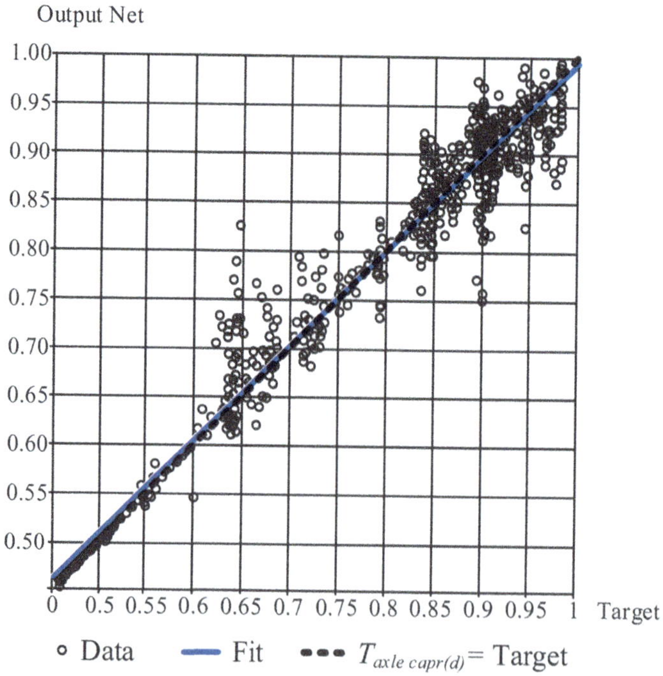

Fig. 5. Distribution of the output parameter of the trained network from the output parameter of training sampling characterized by correlation coefficient

Table 1. Results of training neural network *Net*

Characteristic	Value
Number of network training iterations (epochs) (*Epoch*)	215000
Minimal mean square deviation between the output parameter of the trained network and the output parameter of the training sampling σ_m (*Performance*)	0.000738
Minimal gradient value $\nabla f(w_{Nj}) = a - \Delta y_j$ (*Gradient*)	0.0000791
Correlation coefficient R	0.98484
Coefficient of determination R^2	0.96991

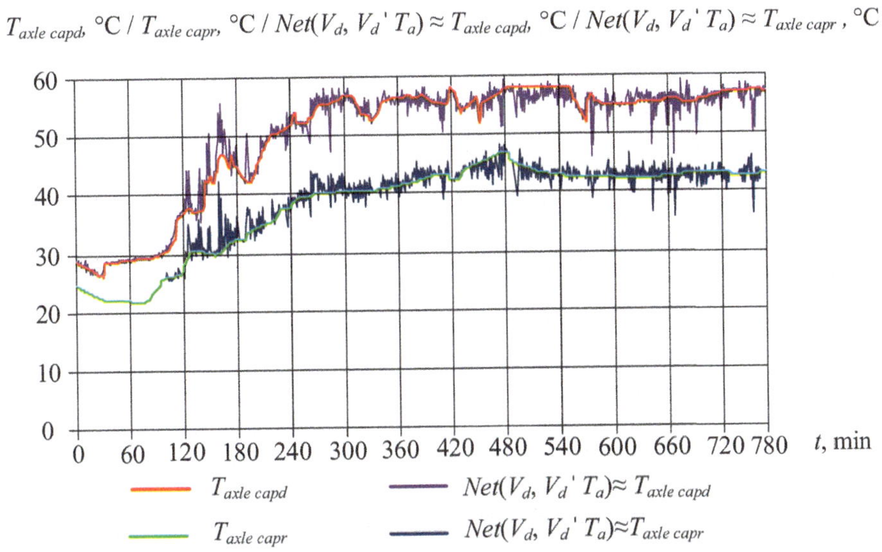

Fig. 6. Results of the approximation of function $T_{axle\ capd}$ ($T_{axle\ capr}$) = $f(V_d, V_d', T_a)$ by the neural network *Net*

Figure 7 presents the results of the function approximation $T_{axle\ capd}$ ($T_{axle\ capr}$) = f (V_d, V_d', T_a) by the neural network *Net* applying a filter smoother for intensity of changing the diagnostic parameter (14):

In the model under consideration, deviations characterized by the coefficient of determination do not exceed the threshold value: for the axle cap bearing brass, the defective temperature is 80 °C, and the temperature of the axle cap must not exceed ambient air temperature by more than 45 °C. The trained neural network allows us to forecast the values of the diagnosis parameters depending on the input data within the frameworks of the approximate sampling. The technical state is assessed by means of building the function of refusal membership based on the comparison of the forecasted and the actual values of the diagnosis parameter.

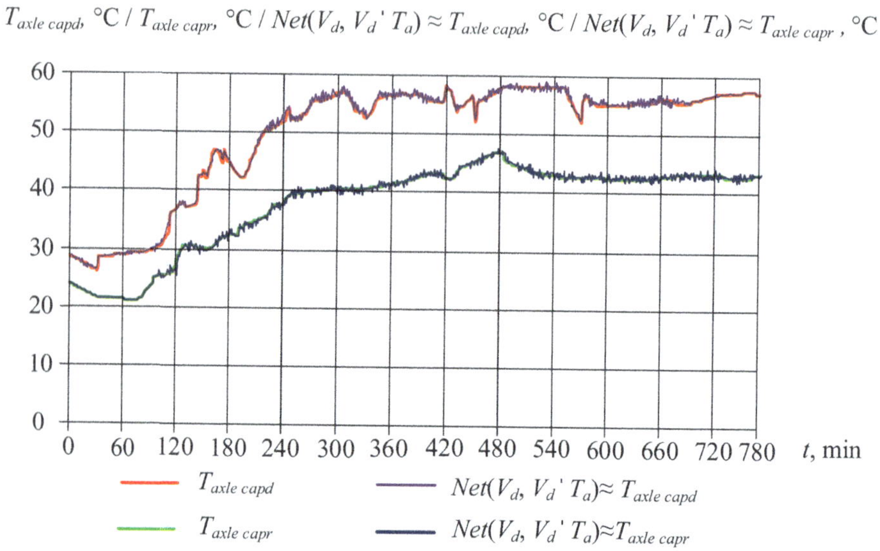

Fig. 7. Results of the function approximation $T_{axle\ capd}$ ($T_{axle\ capr}$) = $f(V_d, V_d', T_a)$ by the neural network *Net* applying a filter for intensity of changing the diagnostic parameter

In addition to the performed approximation task, the extrapolation accuracy is the major indicator of the quality of the network training. Within the frameworks of the task under consideration, extrapolation implies forecasting of the diagnosis parameter based upon the input data beyond the limits of training sampling, i.e. on the basis of unknown input data that were not used during the network training.

Let us model a neural network on the input layer without preliminary training in accordance with the monitoring data (Fig. 2) and obtain an extrapolated function of the neural network:

According to Figs. 7 and 8, a trained neural network is capable of accomplishing both approximation and extrapolation tasks as the functions obtained have a high determination coefficient $R^2 \rightarrow 1$. These properties allow us to conduct integrated control with consideration to changes in related parameters, to forecast changes in the diagnostic parameter and to accomplish indicative assessment of the technical state of the unit under consideration.

Based on the research conducted we can conclude that employment of neural networks in the systems for control, diagnosis and technical state assessment envisages further development. Thus, as compared with the existing control systems, neural networks are capable of controlling not only against the output parameter but also taking into account the impacts from various factors (input parameters). Unlike the existing forecasting systems, neural networks are based upon a sufficiently universal mathematical model. In addition, an outstanding feature of neural networks is their adaptability, confirmed by the accomplishment of the extrapolation tasks.

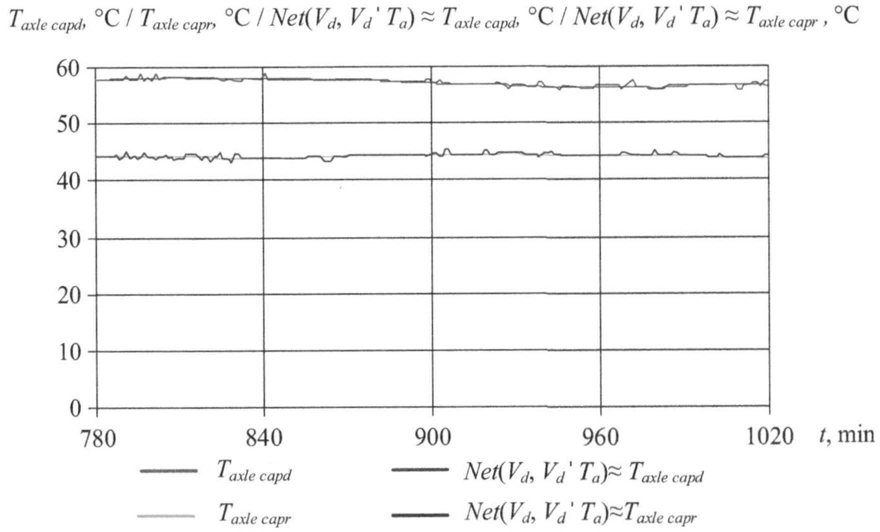

Fig. 8. Results of extrapolation of function $T_{axle\ capd}$ $(T_{axle\ capr}) = f(V_d, V_d', T_a)$ by the neural network *Net* Net applying a filter for intensity of changing the diagnostic parameter

5 Working Algorithm of the Forecasting Neural Network Based Systems for the Locomotive Fleet Technical State Assessment

The properties of the neural networks described above allow us to develop the following working algorithm for the locomotive fleet technical state assessment system (Fig. 9):

The work of the algorithm consists in the employment of sniffers for continuous monitoring of input and output parameters of a technical system. The trained neural network accepts input signals from the monitoring facilities and calculates the output parameters. Comparing the calculated values of the output parameters with the actual values, the model assessment of the technical state is accomplished. In addition, failures are diagnosed on the previously received (trained) defective neural networks of specific failures and control is accomplished on the limiting value of the output parameter a_{cr}, that is determined by the requirements to a technical system, component or element. If the actual value of the output parameter is higher or lower than the limiting value a_{cr}, the failure is apparent and requires immediate elimination, i.e. unscheduled repair. On the contrary, if the value of the actual output parameter lies on the boundary between the limiting and the reference value, then stationary diagnosis works aimed at confirmation of the neural network reply are required; in the case of error the network is retrained depending on the condition. If diagnostic works detect defects, the defected network y_{jd} is trained on the corresponding fact, in case of erroneous detection of failure by the neural network, the latter is retrained to gain the intermediate state. In case of finding a failure confirmed by the diagnosis works, the

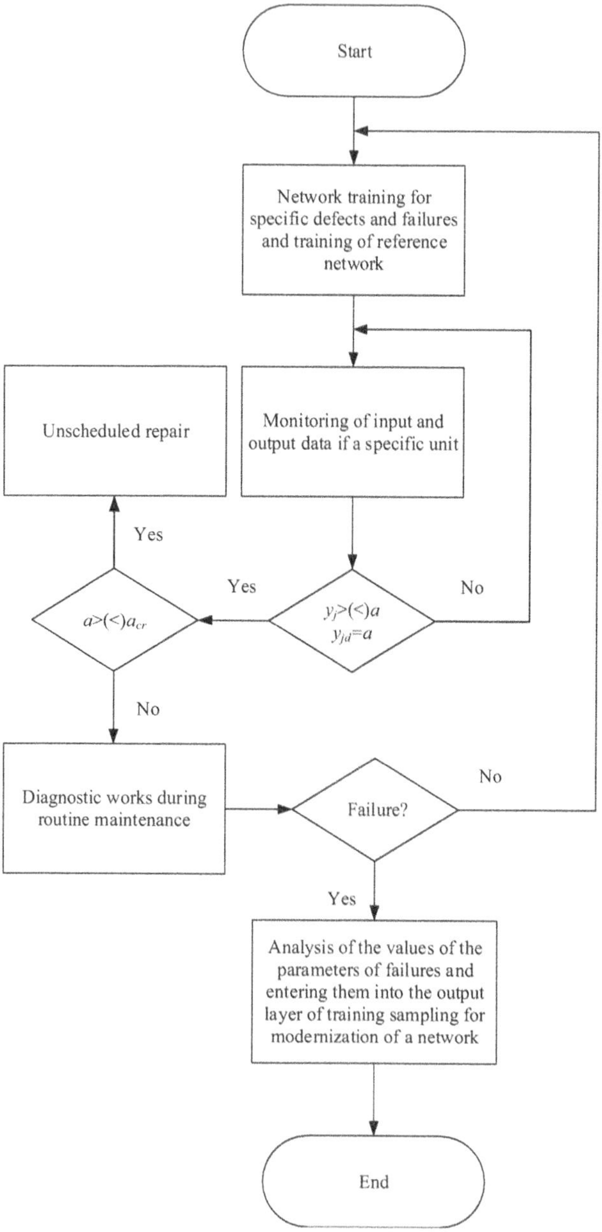

Fig. 9. Working algorithm of the neural network based system for the technical state assessment

network is trained on the defect parameters if there are any for the purposes of complete digitization of the diagnostic nomenclature of components, parts, elements and systems of locomotive fleet.

6 Conclusion

We have presented a set of problems realized within the frameworks of the system for controlling the reliability of the locomotive fleet, as currently this system includes complexes that record failures and their reasons; and record the parameters of defects. Such complexes are:

- A System for controlling locomotive complexes ASUT;
- A complex automated system for recording and handling failures of technical means and their reliability analysis CASART.

In addition to that, a considerable number of continuous on-board means for monitoring both input and output parameters of locomotive work, that are also required to form neural system for the technical state assessment function of the locomotive fleet within the frameworks of the United system for monitoring the technical state of traction rolling stock - Automated systems for technical diagnosis.

The model presented for the technical state assessment of the locomotive fleet is a link between the terminal advanced monitoring, controlling and diagnosing systems and the information systems that form the resulting data for decision-making. The advantages of the model presented are as follows:

- universality regarding output parameters;
- adaptability;
- low period of discretization;
- efficient organization of routine maintenance.

References

1. Golubev, Y.F.: Neural networks in mechactronics. J. Math. Sci. **147**(2), 6607–6622 (2007). https://doi.org/10.1007/s10958-007-0497-3
2. Andreichenko, D.K., Zhadaev, F.M.: Training or neural network regulators for stabilization of combined dynamic systems. News Saratov Univ. New Ser. Math. Mech. Inform. **18**(3), 354–360 (2018). https://doi.org/10.18500/1816-9791-2018-18-3-354-360
3. Kominkova Oplatkova, Z., Senkerik, R., Zelinka, I., Pluhacek, M.: Analytic programming in the task of evolutionary synthesis of a controller for high order oscillations stabilization of discrete chaotic systems. Comput. Math Appl. **66**(2), 177–189 (2013). https://doi.org/10.1016/j.camwa.2013.02.008
4. Daoud, M., Mayo, M.: A survey of neural network-based cancer prediction models from microarray data. Artif. Intell. Med. **97**(5), 204–214 (2019). https://doi.org/10.1016/j.artmed.2019.01.006
5. Hara, K., Okada, M.: On-line learning through simple perceptron learning with a margin. Neural Netwo. **17**(2), 215–223 (2004). https://doi.org/10.1016/j.neunet.2003.08.005
6. Uhobotov, V.I., Mihailova E.S.: On the comparison of fuzzy numbers in the decision-making problems. Bull. Udmurt Univ. Math. Mech. Comput. Sci. **26**(1), 87–94 (2016). https://doi.org/10.20537/vm160108

7. Mohammed, M.A., Naji, T.A.H., Abdulyabbar, H.M.: The effect of the activation functions on the classification accuracy of satellite image by artificial neural network. Energy Procedia **157**(2), 164–170 (2019). https://doi.org/10.1016/j.egypro.2018.11.177
8. Diene, O., Bhaya, A.: Conjugate gradient and steepest descent constant modulus algorithms applied to a blind adaptive array. Signal Process. **90**(10), 2835–2841 (2010). https://doi.org/10.1016/j.sigpro.2010.04.002
9. Chen, J., Weiguo, L.: Convergence of Gauss-Newton's method and uniqueness of the solution. Appl. Math. Comput. **170**(1), 686–705 (2005). https://doi.org/10.1016/j.amc.2004.12.055
10. Maric, I.: Optimization of self-organizing polynomial neural networks. Expert Syst. Appl. **40**(11), 4528–4538 (2013). https://doi.org/10.1016/j.eswa.2013.01.060

Mathematical Model of Traction Rolling Stock Oscillations for the Assessment of Dynamic Loading of Its Components

I. I. Galiev, M. Kh. Minzhasarov, and D. V. Lipunov(✉)

Omsk State Transport University, Karl Marx Ave., 35, 644046 Omsk, Russia
dima-lipunov@mail.ru

Abstract. The paper presents an analysis of the failures of components of the mechanical part of 2ES6 Sinara main electric locomotives in operation at the West Siberian Railway operating domain. The causes and consequences of failures of the most vulnerable components are determined. The task is to determine the level of dynamic loading in the "bogie – rod - traction electric motor" subsystem to reduce the dynamic effects in the "locomotive - track" system. The urgency of the problem of increased dynamic loading of components of the mechanical part is proved, and a mathematical model of vertical oscillations of traction rolling stock is formed on the basis of the second-order Lagrange equation in the form of a system of equations that allows assessing the load of components in operation. The spectral density of random perturbations was chosen—an approximation of random perturbations using the spectral density of the track roughness of Professor A. I. Belyaev. Using a computer, values were calculated, and graphs of the amplitude-frequency characteristics of vertical displacements, maximum accelerations of the body, bogie, traction electric motor and wheelset of the considered conditional uniaxial electric locomotive were built. A comparative analysis of the calculation results and empirical data is carried out. The task is to change the existing design of the suspension system of the traction electric motor of the electric locomotive under consideration and to analyze the oscillations of its components in further studies.

Keywords: Traction rolling stock · Wheel-motor block · Mathematical model · Design diagram · Dynamic loading · Vertical oscillations · Galloping of the traction electric motor

1 Introduction

The total number of unscheduled repairs of the mechanical equipment of 2ES6 Sinara electric locomotives assigned to the locomotive fleet of the TChE-2 Omsk amounted to 1139 cases for four years from 2015 to 2018. The distribution of failures among the components of the mechanical part is shown in Fig. 1.

The analysis of failures of the components of the mechanical part (see Fig. 1) showed that a significant proportion of them belongs to the components of the locomotive wheel-motor block. The axial support suspension of the traction electric motor causes an increased force interaction of the non-sprung mass with the upper track

Z. Popovic et al. (Eds.): TransSiberia 2019, AISC 1115, pp. 443–454, 2020.
https://doi.org/10.1007/978-3-030-37916-2_43

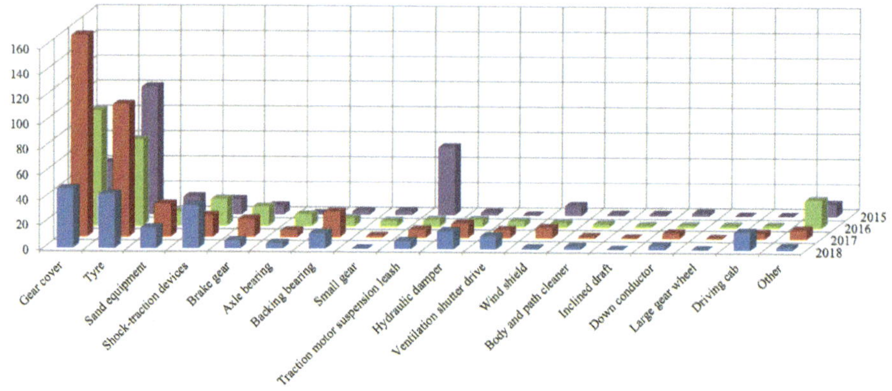

Fig. 1. Failures of the components of the mechanical part of 2ES6 Sinara electric locomotives for 2015–2018

structure, which leads to an increase in the dynamic loading of the locomotive [1]. The "locomotive - track" system is a complex oscillatory circuit, on the dynamic interaction of the subsystems of which, in particular, "unresolved mass is the upper structure of the track", the wear process of the entire system depends [2, 3]. Determination of the level of dynamic loading in the "bogie-rod-traction electric motor" subsystem is an urgent task due to the need to reduce dynamic forces in the system [4–6].

The main reasons for the failure of the mechanical equipment of 2ES6 Sinara electric locomotives for 2015–2018 are related to the failure of the gear casing (339 cases, 29.8%), wheelset tyres (317 cases, 27.8%), and rods for suspension of traction electric motors to the bogie frame (71 cases, 7%) [7]. For comparison, we note that in 2013-2015, failure of the traction engine suspension rod amounted to 127 cases, which is 31.1% of the total number of failures of the mechanical parts. It is worth noting that the main reasons for the failure of the rods of traction electric motors of 2ES6 Sinara electric locomotives remain the same:

- extruding the rubber seal over the end of the rod in the lower head;
- damage to the hinge blocks due to chipping of the polyurethane element of the suspension rod;
- presence of atypical fastening of rods (washer-grover);
- damage to the hinge block due to the expansion of the rubber element of the clutch of the suspension rod.

The decrease in rod failures (71 cases for 2015–2018 versus 127 in 2013–2015) is due to the fact that changes have been made in the next version of the operational documentation for the electric locomotive. The manufacturer stated the need for a complete replacement of rods every 300 thousand km of mileage and with each change of the wheel-motor block (regardless of mileage).

From this it follows that the failure-free operation of rods is not guaranteed even with mileage of less than 300 thousand km. A solution to the problem was found only by a more frequent replacement of rods, which leads to an increase in material costs for the supply of new components, and the complexity of component maintenance increases. Since 2014, the manufacturer has replaced the rubber hinge blocks with polyurethane ones [8]. However, a decrease in failures was not observed in this regard, since the rubber hinged blocks that were used initially had greater operational reliability due to the better elasticity of the material. It should be noted that all these measures are only a struggle with the consequences, while the root cause of failure of the rods is an increased level of dynamic interaction between the bogie and the traction electric motor.

2 Materials and Methods

To study the dynamic loading of suspension of a traction electric motor, it is necessary to formulate a mathematical model of oscillations of the 2ES6 Sinara locomotive, the cars of which are symmetrical in the longitudinal vertical and horizontal planes, which allows considering these oscillations independently of each other, considering them to be unbound [9]. To study the vertical oscillations of the wheel-motor block, it is enough to use the conditional uniaxial model of the locomotive.

In the theory of oscillations of railway cars, it is customary to represent the locomotive and the track as a single mechanical system in which the track model can be selected by the researcher. The simplest of all known track models is an absolutely rigid foundation, the main advantage of which is a significant simplification of the study, since such a statement of the problem reduces the number of degrees of freedom and, as a result, reduces the number of differential equations necessary for modeling the system. But the presence of high-performance computers, as well as special mathematical programs, makes it possible to solve higher-order differential equations within the framework of an engineering error, and allows the use of a more detailed track model, which takes into account the inertial and elastic-dissipative properties inherent in the real track.

In the formation of a mathematical model of the interaction of the locomotive and the track, an important component is the choice of the spectral density of random disturbances. The spectral density that best meets these conditions is the approximation of random perturbations proposed by Professor A. I. Belyaev, which will be used in studies of vertical oscillations of locomotives [9].

In order to simplify the calculations, let's accept a number of assumptions that allow us not to take into account certain parameters and displacements of bodies (Fig. 2):

1. the body, bogie, wheel-motor block and wheelsets are absolutely solid bodies, since their rigidity is many times greater than the rigidity of elastic coupling;
2. the continuous motion of wheelsets on rails is considered;
3. the track model is discrete [10, 11];
4. the locomotive moves at a constant speed [12].

The following notations are made in the design scheme:

z_b - generalized coordinate of the body bouncing; $z_{bog.}$ - generalized coordinate of the bogie bouncing; z_{ws} - generalized coordinate of the wheelset bouncing; z_m -

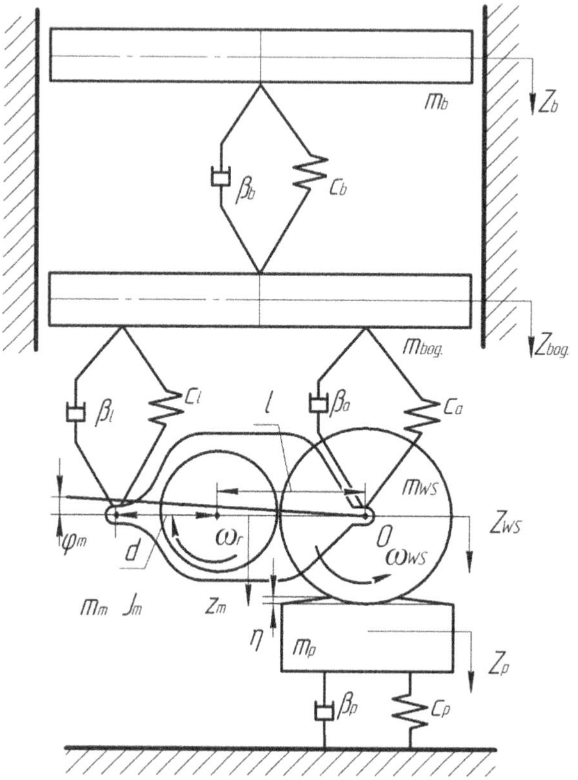

Fig. 2. The design scheme of the conditional uniaxial electric locomotive 2ES6

generalized coordinate of the traction motor bouncing; z_p - moving the reduced mass of the path (track); φ_m - angles of rotation of the traction motor relative to the axis of the wheelset; J_m - moment of inertia of the traction motor; m_b - body mass per one wheelset; $m_{bog.}$ - mass of the compressed parts of the bogie per one wheelset; m_{ws} - mass of the wheelset; m_m - mass of the traction motor; m_p - reduced mass of the path (track); c_b - rigidity of the central suspension stage; c_a - rigidity of the axle box suspension; c_l - rigidity of the leash (rod); c_p - reduced rigidity of the path (track); β_b - coefficient of viscous friction of the central suspension stage; β_a - coefficient of viscous friction of the axle box suspension; β_l - coefficient of viscous friction in the rubber-metal elements of the leash (rod); β_p - reduced coefficient of viscous friction of the path (track); η - random function of the unevenness of the rail under the wheelset; ω_r - angular rotation speed of the reed.

The hypothesis of continuous motion of the wheel along the rail allows applying the second-order Lagrange equations to form a mathematical model of the system under consideration. Conditions (1) allow establishing a relationship between the movement of the reduced mass of the track and bouncing of the wheelset. Conditions (2) establish the relationship between bouncing a wheelset and a traction motor. Condition (3) follows from the fact that the wheelset rotates around its own axis and makes bouncing, this leads to the fact that the rotation speed of the reed of the traction motor will depend on the speed of the locomotive, gear ratio, diameter of the wheelset, and galloping of the traction motor:

$$\begin{cases} z_p = z_{ws} - \eta\,; \\ \dot{z}_p = \dot{z}_{ws} - \dot{\eta}\,. \end{cases} \tag{1}$$

$$\begin{cases} z_m = z_{ws} - l\phi_m\,; \\ \dot{z}_m = \dot{z}_{ws} - l\dot{\phi}_m. \end{cases} \tag{2}$$

$$\omega_r = \frac{2V}{D_t i_g} - \frac{i_g + 1}{i_g}\dot{\phi}_m. \tag{3}$$

The mathematical model of vertical oscillations of the considered system is based on the second-order Lagrange equations:

$$\frac{d}{dt}\left(\frac{\partial T}{\partial \dot{q}_i}\right) - \frac{\partial T}{\partial q_i} + \frac{\partial P}{\partial q_i} + \frac{\partial F}{\partial \dot{q}_i} = Q_i,\ (i = 1\ldots4), \tag{4}$$

where T - kinetic energy of a mechanical system; P - potential energy of a mechanical system; F - scattering function of a mechanical system; Qi - generalized force; q_i-generalized coordinate; $\dot{q}_i$ - generalized speed.

The kinetic energy of the system is the sum of the kinetic energies of the movement of the body, bogie, wheelset, frame of the traction electric motor, and the reduced mass of the path, as well as the rotation of the reed of the traction motor around its own axis. Taking into account the bonds superimposed on the mechanical system, we will have the following expression:

$$\begin{aligned} T = {} & \frac{1}{2}m_b\dot{z}_b^2 + \frac{1}{2}m_{\text{bog}}\dot{z}_{\text{bog}}^2 + \frac{1}{2}m_{ws}\cdot\dot{z}_{ws}^2 + \frac{1}{2}m_m\cdot\dot{z}_{ws}^2 - m_m\cdot\dot{z}_{ws}\cdot l\cdot\dot{\phi}_m + \frac{1}{2}m_m\cdot l^2\cdot\dot{\phi}_m^2 + \frac{1}{2}J_m\dot{\phi}_m^2 \\ & + \frac{1}{2}m_p\cdot\dot{z}_{ws}^2 - m_p\cdot\dot{z}_{ws}\cdot\dot{\eta} + \frac{1}{2}m_p\cdot\dot{\eta}^2 + \frac{1}{2}J_r\left(\frac{2V}{D_t i_g}\right)^2 - J_r\frac{2V}{D_t i_g}\cdot\frac{i_g+1}{i_g}\dot{\phi}_m + \frac{1}{2}J_r\left(\frac{i_g+1}{i_g}\dot{\phi}_m\right)^2, \end{aligned} \tag{5}$$

where J_r – moment of inertia of the reed; V – longitudinal speed of the train; D_t - wheelset tread diameter; i_g - gear ratio of a large gear.

To determine the potential energy of the system in elastic couplings and the dissipative function, we denote and define the deflections and deflection rates:

- body suspension stage: $\Delta_b = z_b - z_{\text{bog}}; \dot{\Delta}_b = \dot{z}_b - \dot{z}_{\text{bog}};$
- axle box springs: $\Delta_a = z_{\text{bog}} - z_{ws};\ \dot{\Delta}_a = \dot{z}_{\text{bog}} - \dot{z}_{ws};$
- leash (rod): $\Delta_l = z_{\text{bog}} - z_{ws} + (l+d)\phi_m;\ \dot{\Delta}_l = \dot{z}_{\text{bog}} - \dot{z}_{ws} + (l+d)\cdot\dot{\phi}_m;$
- upper structure of the path (track): $\Delta_p = z_{ws} - \eta;\ \dot{\Delta}_p = \dot{z}_{ws} - \dot{\eta}.$

Potential energy of the system:

$$
\begin{aligned}
P = & \frac{1}{2}c_b \cdot z_b^2 - c_b \cdot z_b \cdot z_{\text{bog}} + \frac{1}{2}c_b \cdot z_{\text{bog}}^2 + \frac{1}{2}c_a \cdot z_{\text{bog}}^2 - c_a \cdot z_{ws} \cdot z_{\text{bog}} + \frac{1}{2}c_a \cdot z_{ws}^2 \\
& + \frac{1}{2}c_l \cdot z_{\text{bog}}^2 + \frac{1}{2}c_l \cdot z_{\text{ws}}^2 + \frac{1}{2}c_l \cdot (l+d)^2 \cdot \phi_m^2 + c_l \cdot (l+d) \cdot \phi_m \cdot z_{\text{bog}} \\
& - c_l \cdot z_{\text{bog}} \cdot z_{ws} - c_l \cdot z_{ws}(l+d) \cdot \phi_m + \frac{1}{2}c_p \cdot z_{ws}^2 - c_p \cdot z_{ws} \cdot \eta + \frac{1}{2}c_p \cdot \eta^2.
\end{aligned} \tag{6}
$$

Dissipative function of the system:

$$
\begin{aligned}
F = & \frac{1}{2}\beta_b \cdot \dot{z}_b^2 - \beta_b \cdot \dot{z}_b \cdot \dot{z}_{\text{bog}} + \frac{1}{2}\beta_b \cdot \dot{z}_{\text{bog}}^2 + \frac{1}{2}\beta_a \cdot \dot{z}_{\text{bog}}^2 - \beta_a \cdot \dot{z}_{ws} \cdot \dot{z}_{\text{bog}} + \frac{1}{2}\beta_a \cdot \dot{z}_{ws}^2 \\
& + \frac{1}{2}\beta_l \cdot \dot{z}_{\text{bog}}^2 + \frac{1}{2}\beta_l \cdot \dot{z}_{\text{ws}}^2 + \frac{1}{2}\beta_l \cdot (l+d)^2 \cdot \dot{\phi}_m^2 + \beta_l \cdot (l+d) \cdot \dot{\phi}_m \cdot \dot{z}_{\text{bog}} \\
& - \beta_l \cdot \dot{z}_{\text{bog}} \cdot \dot{z}_{ws} - \beta_l \cdot \dot{z}_{ws}(l+d) \cdot \dot{\phi}_m + \frac{1}{2}\beta_p \cdot \dot{z}_{ws}^2 - \beta_p \cdot \dot{z}_{ws} \cdot \dot{\eta} + \frac{1}{2}\beta_p \cdot \dot{\eta}^2.
\end{aligned} \tag{7}
$$

Let's substitute (5), (6) and (7) into the second-order Lagrange Eq. (4), taking the corresponding derivatives with respect to the selected four generalized coordinates. We obtain the system of equations, which is presented below:

$$
\begin{cases}
m_b \cdot \ddot{z}_b + \beta_b(\dot{z}_b - \dot{z}_{\text{bog}}) + c_b(z_b - z_{\text{bog}}) = 0; \\
m_{\text{bog}} \cdot \ddot{z}_{\text{bog}} + \beta_b(\dot{z}_{\text{bog}} - \dot{z}_b) + \beta_a(\dot{z}_{\text{bog}} - \dot{z}_{ws}) + \beta_l(\dot{z}_{\text{bog}} + \dot{\phi}_m(l+d) - \dot{z}_{ws}) \\
\quad + c_b(z_{\text{bog}} - z_b) + c_a(z_{\text{bog}} - z_{ws}) + c_l(z_{\text{bog}} + \phi_m(l+d) - z_{ws}) = 0; \\
\left[m_m l^2 + J_m + J_r((i_g+1)/i_g)^2\right] \cdot \ddot{\phi}_m - m_m l \cdot \ddot{z}_{ws} + \beta_l\left[(l+d)^2\dot{\phi}_m + (l+d)\dot{z}_{\text{bog}} - (l+d)\dot{z}_{ws}\right] \\
\quad + c_l\left[(l+d)^2\phi_m + (l+d)z_{\text{bog}} - (l+d)z_{ws}\right] = 0 \\
(m_{ws} + m_m + m_p)\ddot{z}_{ws} - m_m \cdot l \cdot \ddot{\phi}_m + \beta_a(\dot{z}_{ws} - \dot{z}_{\text{bog}}) + \beta_l(\dot{z}_{ws} - \dot{z}_{\text{bog}} - (l+d)\dot{\phi}_m) + \beta_p \dot{z}_{ws} \\
\quad + c_a(z_{ws} - z_{\text{bog}}) + c_l(z_{ws} - z_{\text{bpg}} - (l+d)\phi_m) + c_p z_{ws} = m_p \cdot \ddot{\eta} + \beta_p \dot{\eta} + c_p \eta.
\end{cases} \tag{8}
$$

In the resulting system, the first equation describes the oscillations of the body bouncing, the second describes the bouncing of the bogie, the next two describe the galloping of the traction motor and the bouncing of the wheelset, respectively. To determine the transfer functions, it is necessary to convert the system according to Laplace:

$$
\begin{cases}
(m_b \cdot s^2 + \beta_b \cdot s + c_b) \cdot Z_b(s) - (\beta_b s + c_b) \cdot Z_{\text{bog}}(s) = 0; \\
-(\beta_b s + c_b) \cdot Z_b(s) + (m_{\text{bog}} s^2 + \beta_b s + \beta_a s + \beta_l \cdot s + c_b + c_a + c_l) \cdot Z_{\text{bog}}(s) \\
\quad + (\beta_l(l+d) \cdot s + c_l(l+d)) \cdot \phi_m(s) - (\beta_a s + \beta_l \cdot s + c_a + c_l) \cdot Z_{ws}(s) = 0; \\
(\beta_l(l+d) \cdot s + c_l(l+d)) \cdot Z_{\text{bog}}(s) + (\left[m_m l^2 + J_m + J_r((i_g+1)/i_g)^2\right] \cdot s^2 + \beta_l(l+d)^2 s \\
\quad + c_l(l+d)^2) \cdot \phi_m(s) - (m_m \cdot l \cdot s^2 + \beta_l(l+d) \cdot s + c_l(l+d)) \cdot Z_{ws}(s) = 0; \\
-(\beta_a s + \beta_l s + c_a + c_l) \cdot Z_{\text{bog}}(s) - (m_m \cdot l \cdot s^2 + \beta_l(l+d) \cdot s + c_l(l+d)) \cdot \phi_m(s) \\
\quad + ((m_{ws} + m_m + m_p) \cdot s^2 + \beta_a s + \beta_l \cdot s + \beta_p s + c_a + c_l + c_p) \cdot Z_{ws}(s) = (m_p s^2 + \beta_p s + c_p) \cdot H(s).
\end{cases} \tag{9}
$$

We transform system (9) into a matrix form, which is more descriptive, and obtain the following matrices:

1. the main matrix of the system, consisting of coefficients with generalized coordinates:

$$\vec{A}(s)=\begin{pmatrix} \begin{matrix} m_b\cdot s^2 \\ +\ \beta_b\cdot s+c_b \end{matrix} & -(\beta_b s+c_b) & 0 & 0 \\ -\ (\beta_b s+c_b) & \begin{matrix} m_{\text{bog}}s^2+\beta_b s \\ +\ \beta_b s+\beta_l\cdot s \\ +\ c_b+c_a+c_l \end{matrix} & \begin{matrix} \beta_l(l+d)\cdot s \\ +\ c_l(l+d) \end{matrix} & -\begin{pmatrix} \beta_a s+\beta_l\cdot s \\ +\ c_a+c_l \end{pmatrix} \\ 0 & \begin{matrix} \beta_l(l+d)\cdot s \\ +\ c_l(l+d) \end{matrix} & \begin{matrix} \begin{bmatrix} m_m l^2+J_m \\ +\ J_r((i_g+1)/i_g)^2 \end{bmatrix}\cdot s^2+ \\ \beta_l(l+d)^2 s \\ +\ c_l(l+d)^2 \end{matrix} & -\begin{pmatrix} m_m\cdot l\cdot s^2 \\ +\ \beta_l(l+d)\cdot s \\ +\ c_l(l+d) \end{pmatrix} \\ 0 & -\begin{pmatrix} \beta_a s+\beta_l s \\ +\ c_a+c_l \end{pmatrix} & -\begin{pmatrix} m_m\cdot l\cdot s^2 \\ +\ \beta_l(l+d)\cdot s \\ +\ c_l(l+d) \end{pmatrix} & \begin{matrix} ((m_{ws}+m_m+m_p)\cdot s^2 \\ +\ \beta_a s+\beta_l\cdot s+\beta_p s \\ +\ c_a+c_l+c_p) \end{matrix} \end{pmatrix} \tag{10}$$

2. matrix vector of unknown generalized coordinates:

$$\vec{Z}(s)=\begin{pmatrix} Z_{\text{b}}(s) \\ Z_{\text{bog}}(s) \\ \phi_m(s) \\ Z_{\text{ws}}(s) \end{pmatrix}. \tag{11}$$

Let's multiply the matrix $\vec{A}(s)$ by the vector $\vec{Z}(s)$. The system can be represented as follows:

$$\vec{A}(s)\cdot\vec{Z}(s)=\vec{B}(s), \tag{12}$$

where $\vec{B}(s)=\begin{pmatrix} 0 \\ 0 \\ 0 \\ (m_p\cdot s^2+\beta_p\cdot s+c_p)H(s) \end{pmatrix}$ – vector matrix consisting of elements located on the right side of the system.

Let's introduce the vector of transfer functions of generalized coordinates:

$$\vec{W}(s) = \frac{\vec{Z}(s)}{\vec{H}(s)} = \begin{pmatrix} \vec{W}_{Z_b}(s) \\ \vec{W}_{Z_{bog}}(s) \\ \vec{W}_{\phi_m}(s) \\ \vec{W}_{Z_{ws}}(s) \end{pmatrix} \tag{13}$$

3 Results

The transfer function is determined by the Cramer's formulas. Below are graphs of the frequency response of the body, bogie, TEM galloping, and wheelset bouncing (Figs. 3, 4, 5 and 6).

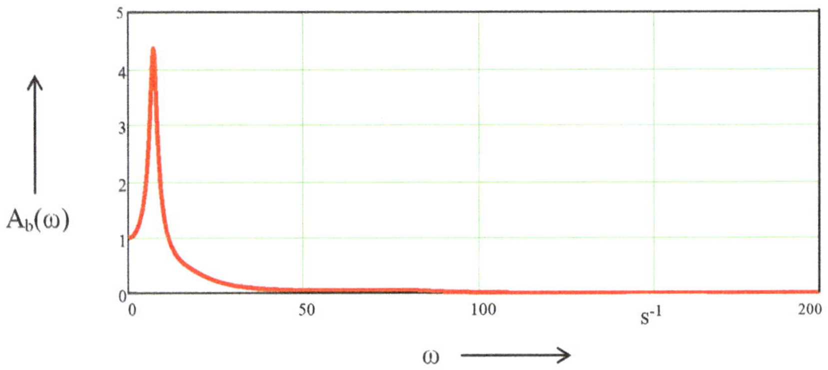

Fig. 3. Frequency response of body oscillations

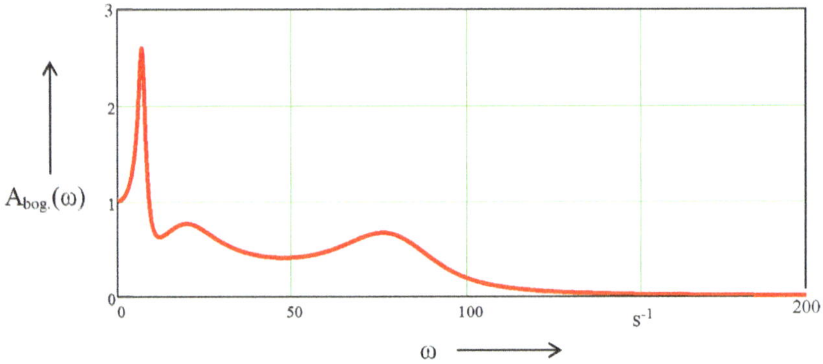

Fig. 4. Frequency response of bogie oscillations

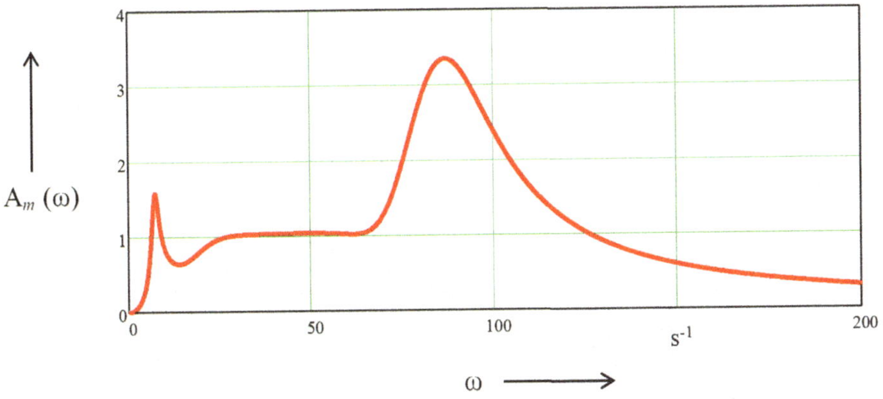

Fig. 5. Frequency response of traction electric motor galloping

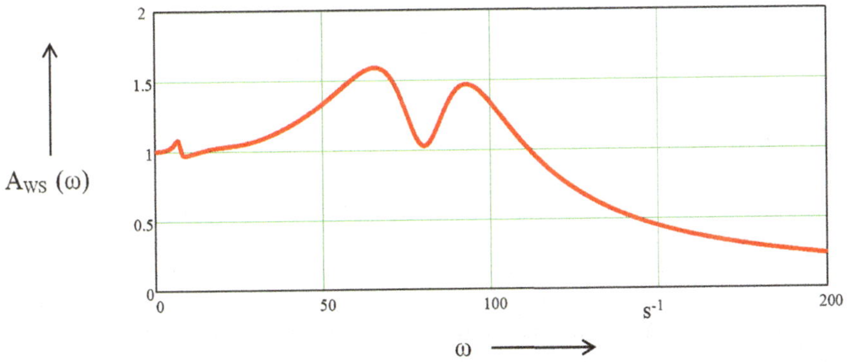

Fig. 6. Frequency response of wheelset bouncing

For statistical studies of random oscillations of a locomotive, that frequency response is important, which is a module of the frequency transfer function [12]. Frequency response depends on the inertial and elastic-dissipative parameters of the railway train. The frequency response graphs show resonance maxima: for vertical movements of the body - ω = 10 s-1; bogie - 20 s-1; TEM galloping - 85 s-1, and wheelset bouncing - 65 s-1. It should be noted that there is a mutual influence of the oscillations of the locomotive components, for example, the resonant peak of the body at a frequency of 10 s-1 can be clearly seen on all other frequency responses, similarly, on the frequency response of bogie movements, there are peaks from the resonant frequency of the wheelset 65 s-1.

After determining the frequency response of the locomotive components, using the spectral density of the track roughness proposed by Professor A. I. Belyaev, we calculate the spectral density of the vertical accelerations of the components, their standard and maximum deviations, the graphs of which are presented in Fig. 7.

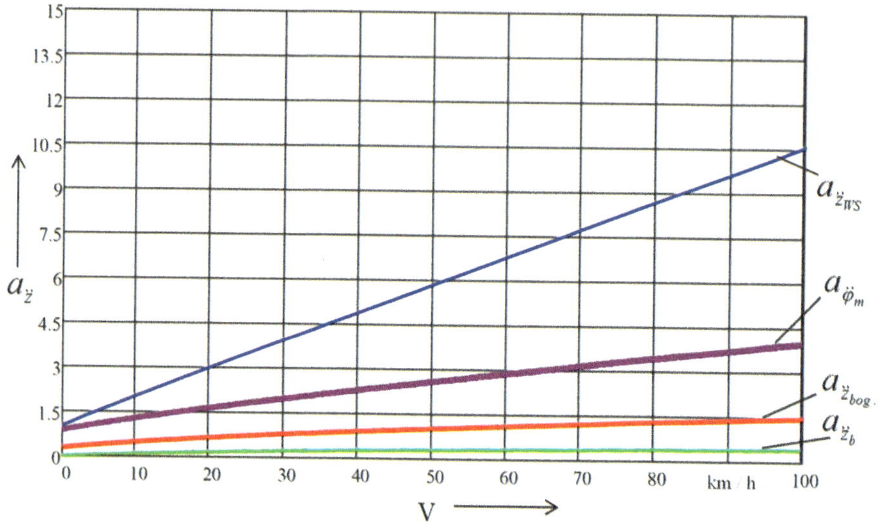

Fig. 7. Maximum acceleration of locomotive components

4 Discussion

To verify the reliability of theoretical studies and the adequacy of the obtained mathematical model, it is necessary to conduct a comparative analysis of the calculation results and empirical data. The experimental data are the results of vibration tests obtained by NIKTI LLC on 2ES10 Granit electric locomotives in the Belovo locomotive depot of the West Siberian Railway [13]. Vibration tests were carried out on electric locomotives with serial numbers No. 095 and No. 090. The need for testing was caused by increased vibrations in the body of these locomotives. In order to identify the causes of vibration, four experimental trips were conducted for various types of operation (in single unit operation and with freight trains), various modes of the train driving (traction, slowing and braking), and various types of track structure (wooden and reinforced concrete sleepers). The obtained test results have a wide range for these locomotives. The cars of 2ES10 Granit and 2ES6 Sinara electric locomotives are unified, with the exception of the type of traction electric motors and a large gear [14, 15]. The results of theoretical calculations and experimental data are summarized in Table 1.

Based on a comparative analysis, it can be said that the considered mathematical model of oscillations of the 2ES6 Sinara electric locomotive is adequate and will allow determining the dynamic loading of the locomotive for the entire range of operating speeds. In further studies, it is proposed to make constructive changes to the system “truck - rod - traction electric motor” in order to reduce the dynamic loading of this subsystem, and, as a result, reduce the number of failures of the mechanical parts.

Table 1. Values of maximum vibration acceleration of electric locomotives, m/s^2

Name	Experimental				Theoretical	
Electric locomotive	2ES10-095A		2ES10-090B		2ES6 model	
Speed km/h	50	60	50	60	50	60
Maximum wheelset acceleration	36.7	31	63	55	55.1	66
Maximum traction motor acceleration	21.9	23.7	27.3	32.1	23.5	29.4
Maximum bogie acceleration	8.1	17.7	19.6	16.9	9.5	12
Maximum body acceleration	1.42	2.44	0.85	0.83	1.5	2.1

References

1. Galiev, I.I.: Methods and means of vibration protection of railway trains. Educational and methodological center for education in railway transport, Moscow, 340 p. (2010)
2. Eliseev, S.V.: Transport suspensions. mathematical models. the choice of coordinate systems, modern technologies. system analysis. Modeling **2**, 8–17 (2011)
3. Galiev, I.I.: Improvement of spring suspension of rolling stock. Railway Transp. **11**, 59–61 (2015)
4. Alpatov, V.: Numerical studies of the nodal connections of metal spatial frames. In: IOP Conference Series: Materials Science and Engineering, vol. 463(3), p. 032103 (2018). https://doi.org/10.1088/1757-899x/463/3/032103
5. Smirnova, O.M., Belentsov, Y.A., Kharitonov, A.M.: Influence of polyolefin fibers on the strength and deformability properties of road pavement concrete. J. Traffic Transp. Eng. (Eng. Ed.) **6**(4), 407–417 (2019). https://doi.org/10.1016/j.jtte.2017.12.004
6. Alpatov, V.: Features of application of spatial metal farms in the production and public buildings. In: IOP Conference Series: Materials Science and Engineering, vol. 463(3), p. 032102 (2018). https://doi.org/10.1088/1757-899x/463/3/032102
7. Bozhuk, S., Maslova, T., Kozlova, N., Krasnostavskaia, N.: Transformation of mechanism of sales and services promotion in digital environment. In: IOP Conference Series: Materials Science and Engineering, vol. 497(1), p. 012114 (2019)
8. Report on the proposed measures to increase the reliability level of locomotive components of the 2ES6 series. STM-Service LLC, 19 p. (2016)
9. Galiev, I.I.: Scientific directions of the school of P. P. Pakhomov for 50 years (1961–2011). Omsk State Transport University, Omsk, 175 p. (2012)
10. Selivanov, K.M.: Computer simulation of the movement of the train along an uneven path. Young Sci. **6**(41), 18–23 (2012)
11. Minzhasarov, M.Kh.: A brief analysis of the features of railway trains with nonlinear elas-tic elements. Izvestiya Transsiba, Omsk State Transport University, Omsk, vol. 1(17), pp. 99–109 (2014)
12. Minzhasarov, M.Kh.: Improving the dynamic qualities of a locomotive by improving the axle box spring suspension: thesis of Ph.D. in Engineering, 05 December 2007, Omsk, 166 p. (2016)
13. Preliminary conclusion on the results of vibration tests of 2ES10 electric locomotives in the locomotive depot TChE-14 Belovo of the West Siberian Railway No. 14-15-05. - VNIKTI OJSC", 12 p. (2015)

14. Cargo electric locomotive of direct current 2ES10 (Granit) with asynchronous traction electric motors. Operation manual: in 8 parts. Part 3. Mechanical equipment. Publishing house of the Design and Research Center of STM OJSC, 94 p. (2009)
15. Cargo electric locomotive of direct current 2ES6 (Sinara) with traction collector electric motors. Operation manual: in 9 parts. Part 6. Mechanical equipment and ventilation systems. - UZZhM OJSC Publishing house, 97 p. (2008)

Spectral Analysis of the Results of Mathematical Modeling of a Nonlinear Mechanical System with a Rigid Cubic Force Characteristic with Kinematic External Disturbance

Viktor Nekhaev, Viktor Nikolaev(✉), and Marina Safronova

Omsk State Transport University, Marx av., 35, 644046 Omsk, Russia
NikolaevVA@omgups.ru

Abstract. The dynamics of a nonlinear mechanical system with a rigid cubic force characteristic for vibration protection of a human operator is studied. A numerical modeling of a system similar to the Duffing equation for kinematic excitation is performed. To analyze the results, an improved method of spectral analysis based on the representation of the correlation function on a small time interval by a square polynomial was used. It has been established that in the pre-resonance and resonance regions, the general solution should consist of three components of the 1/3-order subharmonic, the fundamental harmonic and the third harmonic. In the resonance zone, only the 1/3-order subharmonic and the fundamental harmonic are important. The numerically constructed module of the transfer function of the system in absolute motion indicates the possibility of an amplitude jump, which is demonstrated in laboratory experiments. The most sensitive parameter is the acceleration of the object protected from vibration. Therefore, at the spectral power of the displacement acceleration, in addition to the main harmonic, the third harmonic is also distinguishable. When studying even simple nonlinear mechanical systems, it is necessary to use both approximate analytical and numerical methods, but in combination with spectral analysis.

Keywords: Mechanical system · Stiff cubic force characteristic · Duffing equation · Approximate analytical methods · Mathematical modeling · Spectral density (power) · Subharmonics

1 Introduction

Increasing the productivity of various transport and technological vehicles entails an increase in the level of vibrational effects on a human operator (driver of a heavy truck, locomotive driver, machine operator, etc.). It is known that one of the most effective ways of vibration protection for a human operator is the creation of systems based on the application of the principle of compensation of external disturbances [1]. The design scheme of such a system is shown in Fig. 1.

Z. Popovic et al. (Eds.): TransSiberia 2019, AISC 1115, pp. 455–462, 2020.
https://doi.org/10.1007/978-3-030-37916-2_44

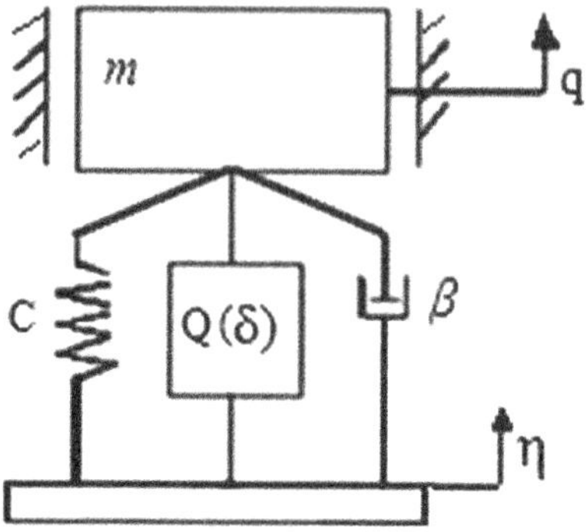

Fig. 1. The design scheme of the system for vibration protection of a human operator

The following notation is introduced in this figure: m – the mass of the protected object, C – the stiffness of the main elastic element, β – the viscous friction coefficient of the system, q – the movement of the load, η – the kinematic external excitation, $Q(\delta) = a_1\delta + a_3\delta^3$ – power characteristic of the compensating device, $\delta = z - \eta$ – suspension deflection. Thus, the dynamics of this system is described by the Duffing equation.

It is well known that, under certain conditions, subharmonics appear in the solution of a nonlinear equation of Duffing type. These are components whose frequency is an integer number of times less than the frequency of an external disturbance. In the system described by the Duffing equation, it is easier to experimentally obtain a subharmonic whose frequency is equal to one third of the frequency of the action. Subharmonics of other orders can also be obtained experimentally, and their appearance is predicted as a result of a theoretical study [2–4]. So, the experiment shows that the starting conditions are of great importance in order to obtain a subharmonic resonance in the system. At the same time, the amplitude and frequency of the external excitation must be within certain specified limits, and the system itself must satisfy certain initial conditions. Due to such a strong dependence on the initial conditions, it is obvious that subharmonic oscillations are related to what is usually understood as transient processes in a linear oscillatory system.

The oscillatory system, whether it is linear or nonlinear, can be made to make oscillations with the help of some initial impulse. Since energy losses exist in any physical system, any oscillations caused by the pulse will decay sooner or later. In a nonlinear system, the oscillations are non-sinusoidal and contain higher points, i.e. components whose frequency is an integer number of times greater than the fundamental frequency. It seems possible, under certain conditions, to maintain steady oscillations of the system by supplying energy at the frequency of one of these harmonics. Moreover, since the frequency of external influence is an integer number of times greater than the fundamental frequency of oscillations of the system, this frequency is subharmonic with respect to the frequency of external excitation. These are the conditions for the appearance of subharmonic oscillations. We will carry out numerical modeling of a nonlinear mechanical system in combination with spectral analysis. So, let's write the differential equation of a single-degree mechanical system with a rigid cubic force characteristic under kinematic excitation in the Cauchy form, which is most suitable for numerical integration of systems of differential equations:

$$\begin{cases} \dot{X}_1 = X_2; \\ \dot{X}_2 = \omega^2 \eta_0 \sin \omega t - 2nX_2 - k_0^2\left(1 + hX_1^2\right)X_1. \end{cases} \tag{1}$$

To avoid a differential equation with variable coefficients, it is written in the deflections for the main elastic suspension element. It is quite simple to switch to absolute coordinates by adding to the result obtained by a numerical method the value of the external disturbance at a given time. The numerical integration of system (1) was carried out using the Runge–Kutta method written in the Mathcad system with the addition of the right sides of differential equations in the resulting list, because the standard function of this software package does not have this property.

2 Results

Figure 2 shows the absolute value of the system transfer function module (to get this picture even more ideal, it is necessary to use a smaller frequency step; in this case, 37 points were used to construct the transfer function module).

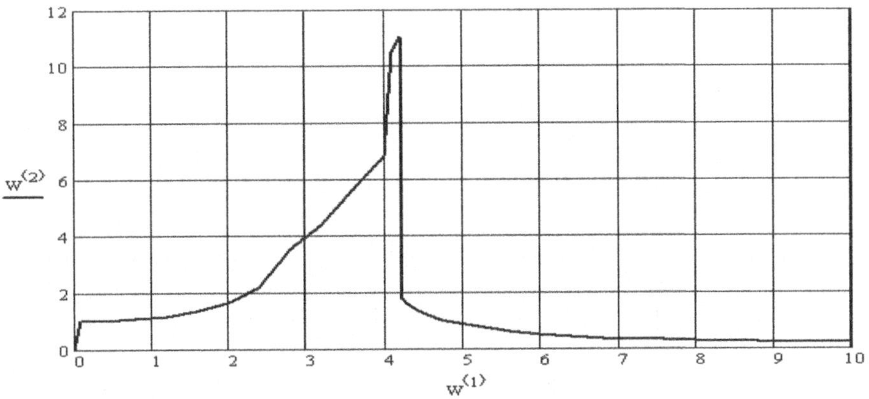

Fig. 2. The transfer function module of the system for vibration protection of a human operator in absolute measurement (frequency response)

In the presented figure, the magnitude of the transfer function module was plotted along the ordinate axis, and the frequency of external excitation along the abscissa axis. It clearly shows a jump in the amplitude of oscillations of the system from the upper branch of the frequency response to the lower one. The middle branch of the frequency response, which is almost not implemented by numerical methods due to instability, is also exist. The presence of viscous friction, as is known, rounds the left and middle branches and creates a point at which the derivative of the oscillation amplitude with respect to the disturbance frequency is equal to infinity. Indeed, the transfer function module instantly changed from 11.015 to 1.8118, i.e. six times. Further, the system moves along the lower branch of the transfer function, asymptotically approaching the zero value. Figures 3 and 4 show the relative and absolute displacements of the protected object - the human operator's seat (locomotive driver).

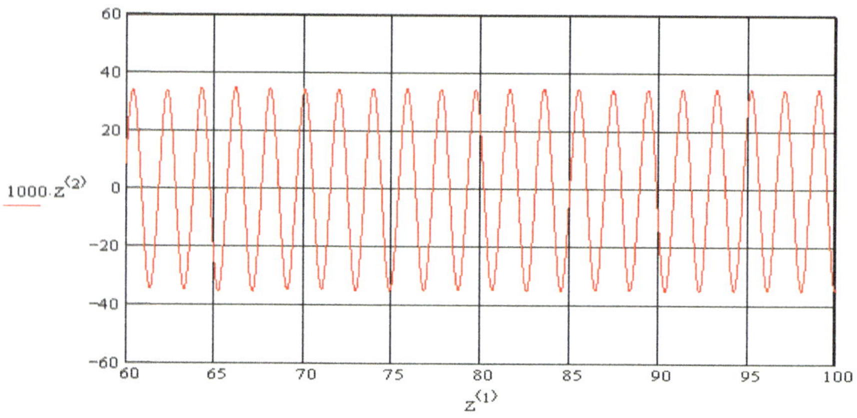

Fig. 3. Relative displacement of the object of vibration protection, mm, at resonance, $\omega \approx 3.2519$ rad/s

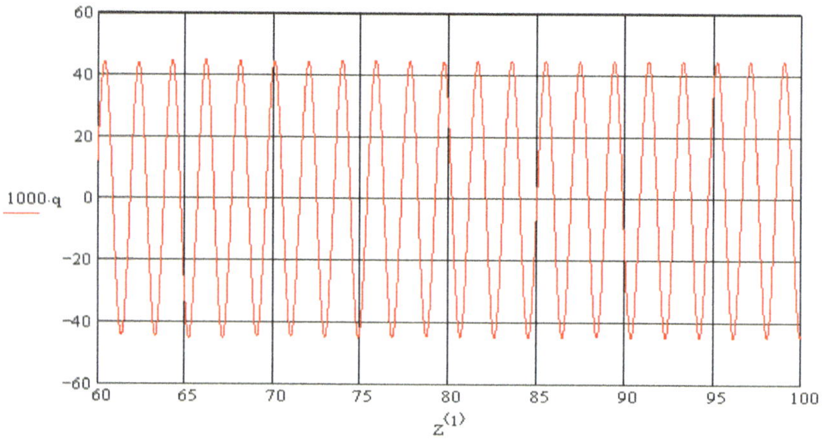

Fig. 4. Absolute displacement of the human operator's seat, mm, at resonance, $\omega \approx 3.2519$ rad/s

Let's make a remark on the used concept of resonance, since for nonlinear systems, it does not mean that the frequency of the external disturbance coincides with the natural frequency of the system. For a nonlinear system, the concept of resonance indicates the equality of the work of friction forces and disturbing forces. But in this unprincipled case, the authors had in mind the classic concept.

At first, it seems that they occur at the same frequency, but the spectral analysis performed below will show that the solution consists of three components: the main, 1/3-order subharmonics and the third harmonic. In Figs. 5 and 6, the results on the relative and absolute accelerations of the human operator's seat at resonance are presented.

Figures 4 and 5 are slightly different from each other. But it is almost impossible to see the difference.

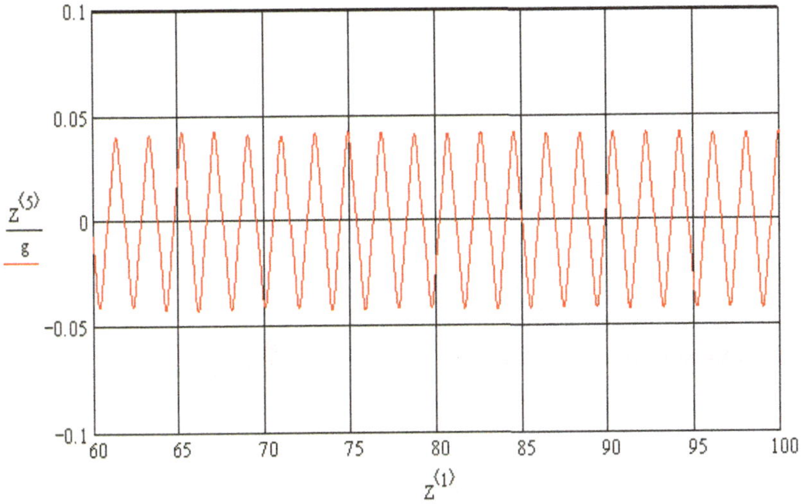

Fig. 5. Relative acceleration of the human operator's seat, in fractions of g, at a resonance $\omega \approx 3.2519$ rad/s

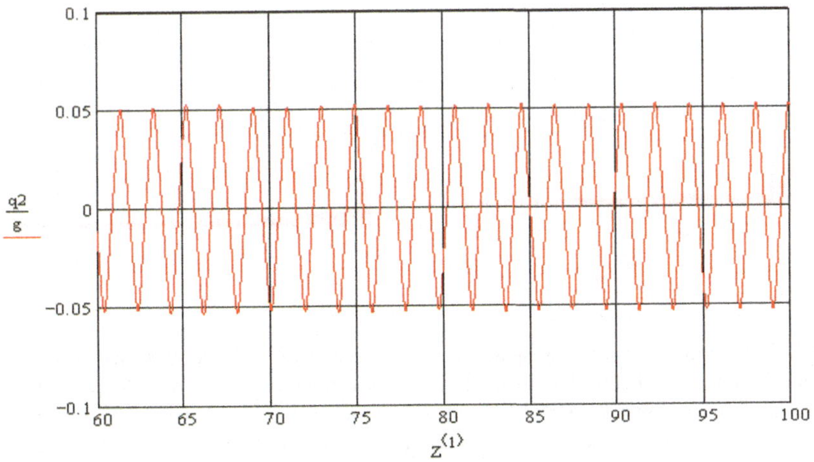

Fig. 6. Absolute acceleration of the human operator's seat, in fractions of g, at a resonance $\omega \approx 3.2519$ rad/s

Therefore, the authors developed a spectral analysis algorithm using the Filon method, which consists in the fact that the correlation function was approximated over small time intervals by a polynomial of the second degree, although the degree of the polynomial can be taken higher. Therefore, the accuracy of calculating the spectral density is now determined by the accuracy of the representation of the correlation function, because the corresponding integrals are now calculated exactly, since they are tabular. Figure 7 shows the spectral power of the displacement and acceleration of a human operator's seat in absolute motion only. The relative motion of the studied mechanical system was not subjected to spectral analysis. At the same time, displacements and accelerations were plotted along the axes of ordinates, and frequency - on the abscissa axis.

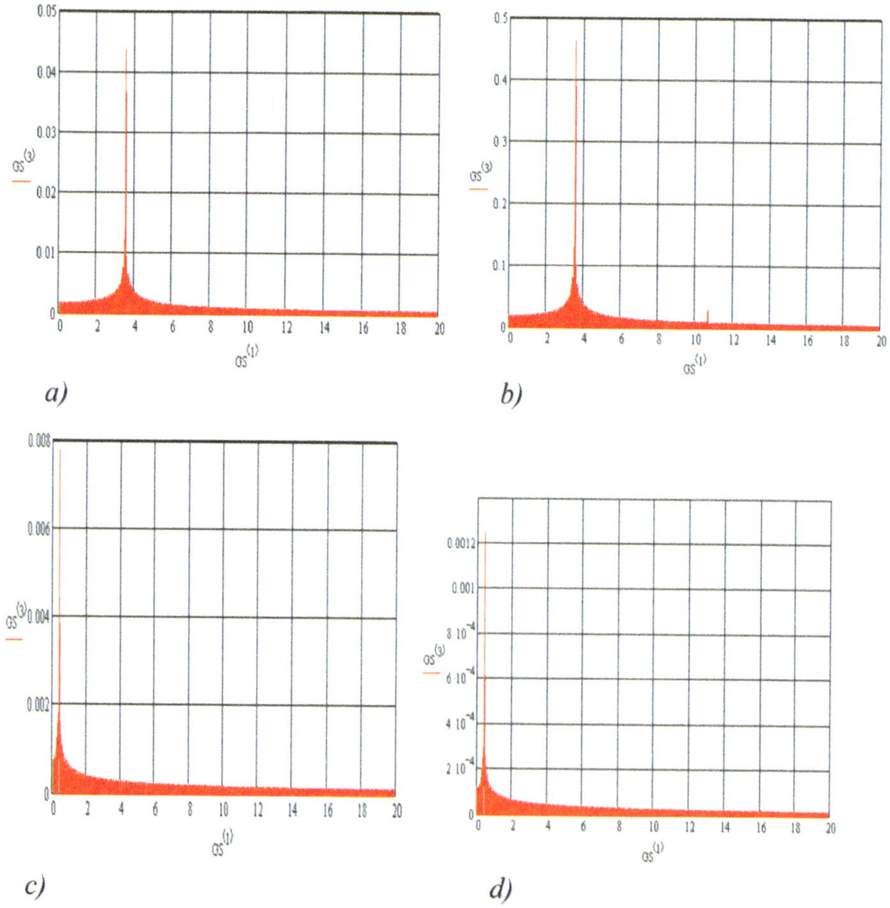

Fig. 7. The spectral power of displacement and acceleration of the protected object — the human operator's seat in absolute motion: *(a)* - displacement of the seat and *(b)* - acceleration of the seat at resonance (ω = 3.251914 rad/s), *(c)* - displacement of the seat and *(d)* - acceleration of the seat in the pre-resonance zone (ω = 0.4 rad/s), *(e)* - displacement of the seat and *(f)* - acceleration of the seat in the resonance zone (ω = 4 rad/s), *(g)* - displacement of the seat and *(h)* - acceleration of the seat in the above-resonance zone

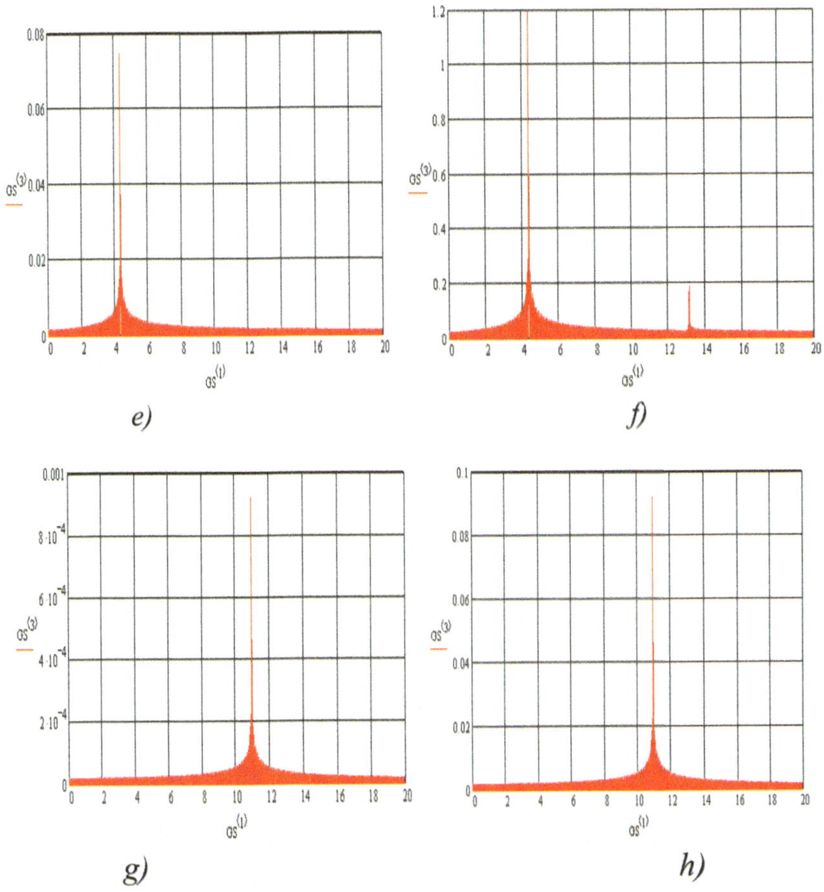

Fig. 7. (*continued*)

3 Conclusions

The analysis of the results of mathematical modeling of the behavior of a nonlinear mechanical system with a rigid cubic force characteristic presented in this figure leads to the following conclusions:

- in the pre-resonance and resonance zones, the solution consists of a subharmonic, fundamental oscillation and a third harmonic: at the disturbance frequency $\omega = 3.2519$ rad/s, there are still components with frequencies of 3.58 and 10.73 and amplitudes of 0.4638 and 0, 03075 m/s^2; at the disturbance frequency $\omega = 0.4$ rad/s, we obtain components with frequencies of 0.44 and 1.32 and quite close amplitudes - 0.0078098 and 0.00124957 m/s^2; at a disturbance frequency $\omega = 4$ rad/s, frequencies of 4.4 and 13.2 with amplitudes of 1.2 and 0.18944 m/s^2 are detected;

- in the resonance zone at ω = 10 rad/s, there is still a component with a frequency of 11 and an amplitude of 0.092385 m/s^2;
- it is impossible to find the subharmonic component and the third harmonic on the spectral power of displacements of the object protected from vibration against the noise generated due to the fact that the analyzed implementations, although they had 172718 points, are nevertheless finite, but these components are clearly detected in the analysis of accelerations of the human-operator's seat.

It should also be noted that the amplitude of the third harmonic compared with the fundamental one is 6.25 times smaller in the resonance zone, 15.084 times smaller in the resonance zone at ω = 3.251914 rad/s, and 6.334 times less in the resonance zone at ω = 4 rad/s. In the above-resonance zone, the third harmonic is not detected.

Thus, in a detailed study of nonlinear mechanical systems, even simple ones, it is necessary to use both approximate analytical and numerical methods in combination with spectral analysis.

References

1. Nekhaev, V., Nikolaev, V., Rauba, A.: On the issue of effective protection of objects from vibration and shock, 27 November 2018. https://doi.org/10.1051/matecconf/201823901053
2. Lukin, A.O., Alpatov, V.Y.: Influence of a metal beam of a corrugated wall geometrical parameters on size and nature of internal tension distribution. Mater. Sci. Forum **931**, 247–251 (2018). https://doi.org/10.4028/www.scientific.net/MSF.931.247
3. Zubkov, V., Lukin, A., Alpatov, V.: Experimental research of beams with corrugated web. In: MATEC Web of Conferences, vol. 196, p. 01005 (2018). https://doi.org/10.1051/matecconf/201819601005
4. Alpatov, V., Soloviev, A.: Thin-walled profiles and their joint assembly units built with screws: numerical studies of load bearing capacity. In: MATEC Web of Conferences, vol. 196, p. 01008 (2018). https://doi.org/10.1051/matecconf/201819601008

Modeling of Heat Exchange Processes in the Locomotive Cooling System

Sergei Ovcharenko, Oleg Balagin(✉), and Dmitrii Balagin

Omsk State Transport University, Karl Marx Avenue, 35, 644046 Omsk, Russia
balaginov@mail.ru

Abstract. A theoretical model for calculating the locomotive cooling system is presented, which allows determining the parameters of heat exchangers taking into account their technical condition. The interrelation of hydraulic, thermal and aerodynamic characteristics of the locomotive cooling system is studied.

Keywords: Diesel locomotive · Heat exchangers · Radiator sections · Temperature · Pressure · Level of contamination

1 Introduction

The effective operation of diesel locomotives largely depends on the quality of the cooling system, which must provide the necessary heat removal, stability of the coolant temperature, prevent deterioration of the quality of the working process in the cylinders of a diesel engine, decrease or increase in the coolant temperature to critical values, overheating of parts of the cylinder-piston group of a diesel engine, increase in power consumption for the refrigerator fan drive, etc. [1–3].

The solution to the problem of assessing the technical condition of heat exchangers of diesel locomotives can be obtained by building a mathematical model of the cooling system, which is a tool for analyzing the influence of operational factors considered during experimental studies on its performance and the selection of informative control parameters. The problems of mathematical modeling of heat transfer processes in a cooling system can be solved using methods of numerical modeling, the basic equations of heat balance, heat transfer, aerodynamics and hydraulics. The equations of aerodynamics and hydraulics are related to the equations of heat transfer, so they must be solved together. The work of [4–6] and several other authors are devoted to the studies of the parameters of the cooling system of diesel locomotives in operation.

This paper presents a theoretical model for calculating the cooling system of a diesel locomotive, which allows determining the parameters of heat exchangers taking into account their technical condition. The interrelation of hydraulic, thermal and aerodynamic characteristics of the locomotive cooling system is investigated. In the future, it is planned to study all modes of operation of a diesel locomotive with various technical conditions of heat exchangers and diagnostic parameters to assess the efficiency of the cooling system.

Z. Popovic et al. (Eds.): TransSiberia 2019, AISC 1115, pp. 463–472, 2020.
https://doi.org/10.1007/978-3-030-37916-2_45

2 Materials and Methods

The modeling of the operation of heat exchangers of a diesel locomotive (radiator sections, a water-oil heat exchanger, a charge air cooler) should be started with a hydraulic calculation of the cooling system, since the network resistance in the circulation circuits, which is overcome by the pump, determines the flow of coolant, which affects the heat transfer rate and heat transfer coefficients of heat exchangers.

The method developed at the Department "Locomotives" of the Omsk State Transport University [7] allows determining the actual coolant flow rate required when performing heat calculation of a locomotive cooling system. In the calculations, it is assumed that there is no water flow through the fuel heater and inter-circuit bypass. The head and rate specification of a water pump with a sufficient degree of reliability is presented in the form of a polynomial of the third degree [1, 10, 11]:

$$H_i = a \cdot V_i^3 + b \cdot V_i^2 + c \cdot V_i + d \tag{1}$$

where i – locomotive cooling circuit number;

a, b, c, d – regression coefficients;
V – water flow rate, m^3/s.

The regression coefficients used for a variable water flow rate V in Eq. (1) are shown in Table 1.

Table 1. The results of approximating the characteristics of water pumps

Characteristics of a water pump	Regression coefficients							
	First circuit				Second circuit			
	a	b	c	d	a	b	c	d
According to technical specifications	764.57	−279.79	7.433	0.422	87.54	−270.01	5.863	0.380
According to experimental data	−912.20	−143.73	4.740	0.453	2484.80	−424.29	8.087	0.357

In general, the characteristics of the network of cooling circuits can be represented as follows [4, 5, 8]:

$$H_{1,2} = (\Sigma\xi_k + Z_{1,2}^{-n} \cdot \xi_{c1} \cdot V_{1,2}^{n-2}) \cdot V_{1,2}^2 \tag{2}$$

where ξ_k – reduced coefficient of hydraulic resistance of the k-th network element;

$Z_{1,2}$ – the number of parallel connected sections of the i-th circuit;
ξ_{r1} – reduced coefficient of hydraulic resistance of a section, $MPa{\cdot}s^2/m^6$;
n – an exponent.

As a result, we obtain a system of equations defining the condition for the joint work "pump - hydraulic network" [1, 5, 12]:

$$H_i = (\Sigma\xi_{kconst} + Z_i^{-n} \cdot \xi_{r1} \cdot V_i^{n-2}) \cdot V_i^2 \quad (3)$$

$$H_i = a \cdot V_i^3 + b \cdot V_i^2 + c \cdot V_i + d \quad (4)$$

The relative coefficient of hydraulic resistance of a section is determined by the expression [4, 5]:

$$\bar{\zeta}_{r1} = \bar{\zeta}_{1l} \cdot \omega^{n-1,78} \quad (5)$$

where $\bar{\zeta}_{1l}$ – relative coefficient of hydraulic resistance at a water speed of 1 m/s.

The values of the relative resistance coefficients for sections with varying degrees of contamination of its water cavity are given in Table 2.

Table 2. The values of the relative coefficients of hydraulic resistance ($\bar{\zeta}_{1l}$) for sections with different technical conditions of the water cavity, depending on the degree of contamination of the section (ε).

ε	20	30	40	50	60	70	75
$\bar{\zeta}_{1l}$	0.858	0.933	1.024	1.14	1.302	1.573	1.941

The relationship between the exponent and the relative coefficient of resistance is approximated by the expression [9]:

$$n = 1,812 + 0,1649 \cdot \ln\bar{\zeta}_{1l} - 6,119 \cdot 10^{-2} \cdot (\ln\bar{\zeta}_{1l})^2 \quad (6)$$

The calculation procedure is as follows. Initial values of the water flow along the circuits and the relative coefficient of hydraulic resistance from Table 2 for a given degree of contamination are set. By the expression (6), the value of the exponent n is found, and according to the given $\bar{\zeta}_{1l}$ and calculated n, the relative coefficient of hydraulic resistance $\bar{\zeta}_{r1}$ is determined by (5). Next, the system of Eqs. (3), (4) is solved with respect to a given flow rate. The obtained value of V1,2 is compared with the set at the initial stage. When the discrepancy between the values is greater than the specified accuracy, the calculation is repeated at new flow rates.

An objective assessment of the degree of contamination of the heat transfer surfaces of the section, both on the air and water sides, is the heat transfer intensity, i.e. the value of the heat transfer coefficient under given operating conditions.

As is known, during operation under the influence of operational factors, a significant decrease in the heat transfer coefficient is observed. This decrease occurs for two reasons. The first one is the increase in thermal resistance to heat transfer from the walls of the tubes and cooling plates to the air due to contaminating deposits and a violation of the contact between tubes and plates. The second one is a decrease in the velocity of water in the tubes (a decrease in the heat transfer coefficient from water to the tube wall) and air in the intertubular space (a decrease in the heat transfer coefficient from the tube walls and cooling plates to the air). The total thermal resistance of contaminating deposits is determined by the ratio of Kern [10]:

$$R_c = R_{c\,max} \cdot [1 - \exp(-b_c \cdot \tau)] \tag{7}$$

where $R_{c\,max}$ – asymptotic (at maximum thickness of deposits) thermal resistance of contamination, m^2·K/W;
b_c – experimental coefficient depending on the speed, temperature of the coolant, material and geometry of the heat transfer surface;
τ – time, hours.

The next stage of modeling is to solve a system of linearized algebraic equations [9, 10]. When compiling systems of equations, it is necessary to observe the basic rule: none of the equations contained in the system should be obtained from a combination of any others.

Based on the basic equations of heat balance, heat transfer and a series of transformations, the following systems of equations were obtained (using 2TE10M diesel locomotive as an example):

for the first circuit:

$$t_{11} - t_{12} = \frac{Q_w}{W_{w1}} \tag{8}$$

$$\frac{e^{\beta_1'}}{e^{\beta_1'} - 1} \cdot t_{12}' - \frac{1}{e^{\beta_1'} - 1} \cdot t_{11} = \tau_0 \tag{9}$$

$$\frac{e^{\beta_1''}}{e^{\beta_1''} - 1} \cdot t_{12} - \frac{1}{e^{\beta_1''} - 1} \cdot t_{12}' = \tau_0 \tag{10}$$

$$\tau_{01}' - \frac{W_{w1}'}{W_{01}'} \cdot t_{11} + \frac{W_{w1}'}{W_{01}'} \cdot t_{12}' = \tau_0 \tag{11}$$

$$\tau_{01}'' - \frac{W_{w1}''}{W_{01}''} \cdot t_{12}' + \frac{W_{w1}''}{W_{01}''} \cdot t_{12} = \tau_0 \tag{12}$$

for the second circuit:

$$t_{21} - t_{22} = \frac{Q_{CA} + Q_O}{W_{w2}} \tag{13}$$

$$\frac{e^{\beta_{21}'}}{e^{\beta_{21}'} - 1} \cdot t_{23}' - \frac{1}{e^{\beta_{21}'} - 1} \cdot t_{21} = \tau_0 \tag{14}$$

$$\frac{e^{\beta_{21}''}}{e^{\beta_{21}''} - 1} \cdot t_{23} - \frac{1}{e^{\beta_{21}''} - 1} \cdot t_{23}' = \tau_0 \tag{15}$$

$$\frac{e^{\beta'_{2r}}}{e^{\beta'_{2r}}-1}\cdot t'_{22}-\frac{1}{e^{\beta'_{2r}}-1}\cdot t_{21}=\tau_0 \tag{16}$$

$$\frac{e^{\beta''_{2r}}}{e^{\beta''_{2r}}-1}\cdot t_{22}-\frac{1}{e^{\beta''_{2r}}-1}\cdot t'_{22}=\tau_0 \tag{17}$$

$$\tau'_{021}-\frac{W'_{w21}}{W'_{021}}\cdot t_{21}+\frac{W'_{w21}}{W'_{021}}\cdot t'_{23}=\tau_0 \tag{18}$$

$$\tau''_{021}-\frac{W''_{w21}}{W''_{021}}\cdot t'_{23}+\frac{W''_{w21}}{W''_{021}}\cdot t_{23}=\tau_0 \tag{19}$$

$$\tau'_{02r}-\frac{W'_{w2r}}{W'_{02r}}\cdot t_{21}+\frac{W'_{w2r}}{W'_{02r}}\cdot t'_{22}=\tau_0 \tag{20}$$

$$\tau''_{02r}-\frac{W''_{w2r}}{W''_{02r}}\cdot t'_{22}+\frac{W''_{w2r}}{W''_{02r}}\cdot t_{22}=\tau_0 \tag{21}$$

$$t_{2w}-t_{22}=\frac{Q_{CA}}{W_{w2}} \tag{22}$$

where $\beta^j_{is}=\frac{W^j_{0is}}{W^j_{wis}}\cdot(1-h)$, $h=\frac{1}{e^m}$, $m=\frac{Z_i\cdot k^j_{is}\cdot F^j}{W^j_{0is}}$;

W^j_{0is}, W^j_{wis} – water equivalents by air and water;
Q_W, Q_O, Q_{CA} – diesel heat dissipation into water, oil, heat removal from charge air, kW;
i – the cooling circuit number;
j – a group of sections (′ - upper group of sections,″ - lower group of sections);
s – a group of sections for the second circuit (l - on the left side of the chamber, r - on the right side of the chamber);
τ_0 – ambient temperature, °C;
τ^j_{0is} – air temperature after passing through a group of sections, °C;
t_i – water temperature, °C, (second digit of the index: 1 - at the entrance to the group of sections; 2 - at the exit from the group of sections of the first circuit; 3 - at the exit from the group of sections of the second circuit);
k^j_{is} – heat transfer coefficient, W/(m²·K).

Equations (8), (13), (22) represent the heat balance in water; (11), (12), (18)–(21) - the heat balance in the air. Dependencies (9), (10), (14)–(17) are based on the balance of heat transfer from water to air through the cooling surface in groups of sections.

The obtained systems of equations are nonlinear due to the dependence on unknown temperature values: the thermophysical characteristics of the coolants and the heat transfer coefficient of the sections. During the calculation, nonlinearities in the indicated parameters are corrected according to the coolants' temperature actually obtained in the previous iterative cycle. The purpose of the next stage of the calculation is to verify the reliability of the choice of the mass velocity of the air flow Uair for a particular condition of the air side of the section and, therefore, the correspondence of the results of hydraulic and thermal calculations to the joint operation of the diesel engine and the cooling system [5]. When performing the aerodynamic calculation of a cooling device, first of all, it is necessary to take into account external contamination of the heat exchange surface, which is subject to all the radiator sections of the diesel locomotive without exception. The relative coefficient of aerodynamic resistance of the section $\overline{\zeta}_{2l}$ at a mass air speed of 1 kg/(m^2·s), the values of which for water-air sections with various degrees of contamination of its air side are given in Table 3, are taken as an objective indicator of the degree of contamination of the external heat transfer surface of the section, characterizing the degree of decrease in air velocity in the intratubular space.

Table 3. The values of the relative coefficient of resistance of the section and the formulas for determining the aerodynamic resistance of sections with different technical conditions of the air side

Conditional section status number	Coefficient $\overline{\zeta}_{2l}$	Formula for determining ΔP_{r2}	Characteristic of the condition and method of contamination of the air side of the section
1	0.92	$11.833 \cdot U_2^{1.6}$	The initial (clean) state of the air side of the new section
2	1.00	$12.852 \cdot U_2^{1.6}$	Section limiting resistance according to Russian State Standard GOST 20556-75
3	1.001	$12.865 \cdot U_2^{1.63}$	The first degree of contamination of the new section (oiling of the air side)
4	1.010	$12.981 \cdot U_2^{1.66}$	The second degree of contamination of the new section (spraying 0.5 kg of sand)
5	1.021	$13.120 \cdot U_2^{1.70}$	The third degree of contamination of the new section (additional spraying of 0.5 kg of sand)
6	1.029	$13.231 \cdot U_2^{1.85}$	The third degree of contamination of the new section and crushing of 30% of the plates
7	1.558	$20.024 \cdot U_2^{1.90}$	Section in operational condition after flushing the air side

Note: U_2 – average air speed along the height of the front of the locomotive refrigerator chamber.

The first stage of aerodynamic calculation is the determination of the peripheral speed of the ends of the fan blades by the known frequency of rotation of the fan wheel and the mass air flow rate at its predetermined mass speed at the refrigerator front and the geometric characteristics of the sections.

The aerodynamic resistance of the sections is calculated from the accepted value of the relative coefficient of aerodynamic resistance of the studied operational state of the section and mass air velocity.

$$\Delta P_{r2} = 12,852\overline{\zeta}_{2l}U_2^{0,15\,\mathrm{th}(115\cdot\overline{\zeta}_{2l}-117,64)+1,75} \tag{23}$$

Next, we calculate the static pressure of the fan, the density of the air in the chamber, the volumetric flow rate and the axial velocity of the air at the exit of the fan, the dynamic and total pressure of the fan. Using the calculated values of the flow and pressure coefficients, we calculate the values of the dimensionless flow and pressure values.

$$\overline{H}_f = \frac{H_t}{\rho_{ch} \cdot \upsilon^2} \tag{24}$$

where $\upsilon = \frac{\pi D_f n_f}{60}$ – peripheral speed of the outer edges of the blades of the fan wheel, m/s.

$$\overline{V} = \frac{V_{air}}{F_f \cdot \upsilon} \tag{25}$$

where $F_f = \frac{\pi \cdot D_f^2}{4}$ – fan wheel area by outer diameter, m^2.

If the discrepancy between the values is greater than the specified accuracy, the calculation is repeated with the new value of the mass air velocity.

The result of the above aerodynamic calculation is to determine the power spent on the fan drive at different levels of contamination of the water-air sections.

The power consumed by the refrigerator fan is determined by the formula:

$$N_f = \frac{V_{air} \cdot H_t}{\eta_f} \tag{26}$$

Thus, modeling the operational conditions of the radiator sections, it is possible to determine the increase in power to the fan drive.

3 Results

The modeling algorithm contains the main stages of the calculation: hydraulic calculation, heat transfer conduction calculation, aerodynamic calculation [5].

The calculation results of the locomotive cooling system with increasing hydraulic resistance of the radiators are presented in Fig. 1.

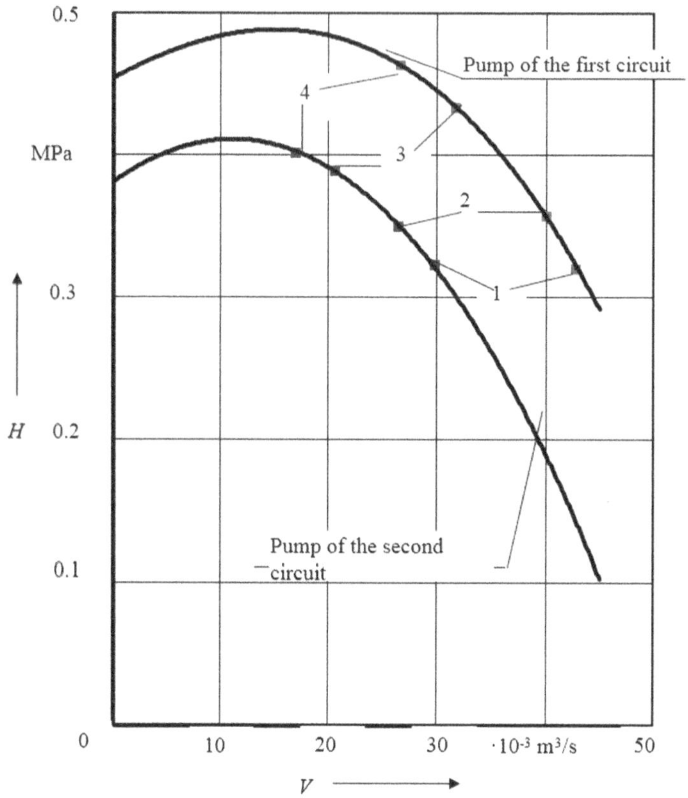

1 - initial state of sections; 2 – contamination degree of sections 20%; 3 - contamination degree of sections 40%; 4 - contamination degree of sections 60%

Fig. 1. The results of the hydraulic calculation of the cooling system of a 2TE10M diesel locomotive

The results of modeling the temperature parameters of the cooling system at nominal conditions with different technical conditions of the radiator sections are presented in Fig. 2.

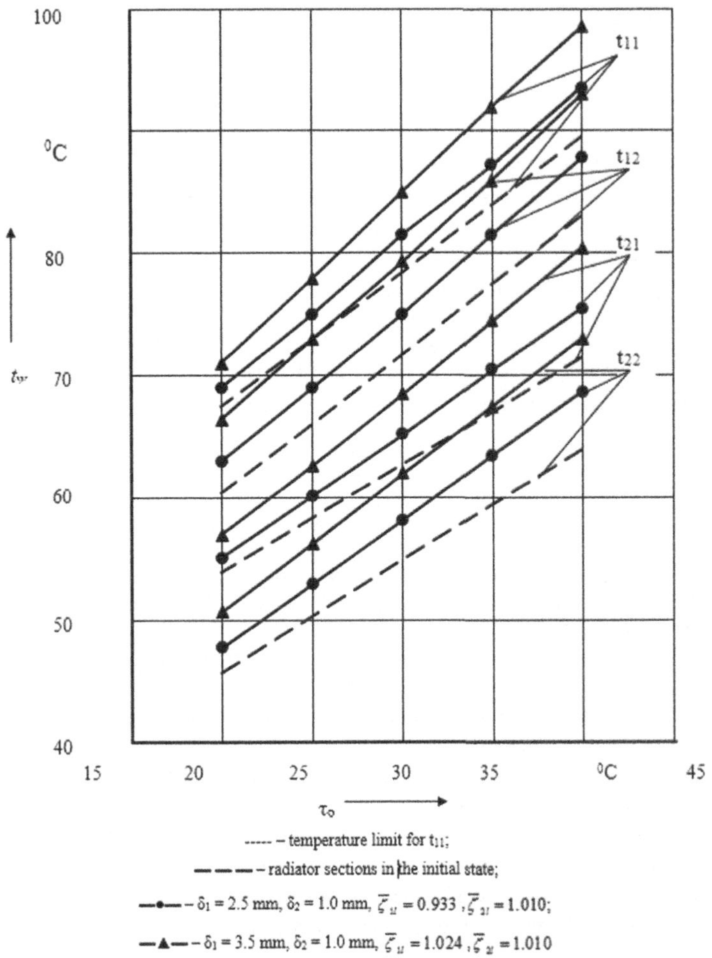

Fig. 2. The effect of increasing the degree of contamination of refrigerator sections on the performance indicators of the cooling system

4 Discussion

From Fig. 1 it follows that an increase in the hydraulic resistance of the sections shifts the point of joint work "pump-network" to a zone of lower costs, a decrease in which causes a decrease in the speed of water in the tubes of the sections and, as a result, a decrease in the coefficient of heat transfer from water to the tube wall.

The analysis of the presented dependencies in Fig. 2 shows that at a certain moment of operation, there may be a limit on the realized capacity of the diesel generator unit. In addition, from Fig. 2 it can be seen that the degree of water cooling is reduced by 2–4 °C.

5 Conclusions

1. As a result of theoretical studies:
   - a mathematical model has been developed that allows the calculation of the thermal characteristics of the radiator sections and the cooling system of the diesel locomotive as a whole for various values of contamination indicators;
   - data were obtained that allowed proceeding to the calculation of temperature fields on the surface of heat exchangers and the development of new models for the diagnosis of their technical condition.
2. The reliability of the modeling results is confirmed by experimental data obtained in the Karasuk operational locomotive depot of Russian Railways JSC when testing five 2TE10M diesel locomotives under various initial conditions.

References

1. Gorin, A.V.: Methods of monitoring the heat conduction condition of cooling devices of diesel locomotives in operation. Dissertation of the Ph.D. in Engineering, Moscow, VNIIZHT, 238 p. (2016)
2. Gorin, V.I.: Cooling device for modern diesel locomotives: what should it be? Locomotive **7**, 27–29 (2014)
3. Sklifus, Ya.K.: Reduction of power consumption of the cooling system of a diesel engine of a locomotive by changing a functional diagram and method of heat transfer. Dissertation of the Ph.D. in Engineering. Rostov-on-Don, RSTU, 159 p. (2015)
4. Ovcharenko, S.: Evaluation of fuel equipment operability of diesel locomotive engine with use of infrared receivers. In: IOP Conference Series: Materials Science and Engineering, vol. 327, p. 042073 (2018)
5. Ovcharenko, S.: Cooling system operation efficiency of locomotive diesel engine. In: IOP Conference Series: Earth and Environmental Science, vol. 90, conference 1, p. 012011 (2017)
6. Ovcharenko, S.M.: Improving the efficiency of the cooling system of diesel locomotives in operation, Izvestia Transsiba. Scientific Journal, Omsk State Transport University, Omsk, vol. 1, no. 29, pp. 27–35 (2017)
7. Volodin, A.I.: Thermal imaging control of the technical condition of the units of rolling stock. Science and Technology of Transport, vol. 1, pp. 27–31. ROAT MIIT, Moscow (2009)
8. Mohyla, V.: The cooling device of locomotive with vaporizing coolant. Commission of Motorization and Power Industry in Agriculture "TEKA", LUBLIN. vol. XIA, pp. 169–176 (2011)
9. Mogila, V.: The prospects of increasing the effectiveness of the cooling device of a diesel locomotive. Commission of Motorization and Power Industry in Agriculture "TEKA", LUBLIN, vol. XC. pp. 198–203 (2010)
10. Sklifus, Y.: The results of the experimental research of the heat transfer coefficient during steam condensation in the tubes of the diesel radiator sections. An International Journal on Motorization, Vehicle Operation, Energy Efficiency and Mechanical Engineering. TEKA, Lublin–Lugansk, vol. 12, no. 4, pp. 264–267 (2012)

Identification of Operation Modes of Locomotive Diesel Engines

Vitaly Chetvergov, Alexander Anisimov(✉), and Igor Chernyshkov

Omsk State Transport University, Marx Avenue, 35, Omsk 644046, Russia
anisimovas1971@mail.ru

Abstract. The paper describes the method that allows indirect determination of the main characteristics of the process of fuel combustion in the cylinder of a locomotive diesel engine. The method is based on the use of a mathematical model of the working cycle and a model of the equilibrium composition of diesel exhaust gases. Using the method makes it possible to obtain data on the actual condition of locomotive power plants on the basis of an array of data quickly recorded by on-board systems during operation.

Keywords: Diesel locomotive · Diesel engine · Mathematical modeling · Control of the exhaust gases composition · In-place assessment of the technical condition

1 Introduction

For railway transport, issues related to the development of methods for assessing the technical condition of traction rolling stock are becoming increasingly important. Among the methods developed by many authors, the most interesting are those that do not require large time expenditures and the decommissioning of locomotives.

Nowadays, an increasing number of locomotives on domestic railways are equipped with microprocessor based on-board systems designed primarily to control the fuel level in the fuel tank and normalize fuel consumption for train and shunting operations. However, in addition to monitoring the fuel level in the diesel engine's fuel tank, locomotive on-board systems are equipped with many sensors that record in real time the values of physical quantities characterizing the status of various components and assemblies of the diesel locomotive, as well as devices that make it possible to record readings. This opens up the possibility of accumulating large amounts of data on the condition of the locomotive fleet and makes it possible to widely apply methods for the rapid determination of the technical condition of locomotive units using appropriate software.

In this regard, the development of algorithms to assess the technical condition of locomotives on the basis of the analysis of data arrays recorded by locomotive on-board systems can be called an important area of scientific research.

The locomotive diesel engine is a complex unit that requires constant monitoring. However, the complexity of the processes occurring in the cylinders of diesel engines

Z. Popovic et al. (Eds.): TransSiberia 2019, AISC 1115, pp. 473–483, 2020.
https://doi.org/10.1007/978-3-030-37916-2_46

significantly complicates the process of timely detection of malfunctions. However, the development and widespread use of mathematical models opens up new possibilities in monitoring the condition of the locomotive fleet.

2 Experimental Section

The main indicator determining the efficiency of the locomotive is the quality of the processes of compression, combustion and expansion of the working fluid in the diesel cylinder. It is possible to determine the nature of the course of these processes both on the basis of experimental data and indirectly from the calculation of thermodynamic parameters, such as pressure and temperature. By comparing the current values of pressure and temperature in the engine cylinder with standard values, it is possible to determine the technical condition without decommissioning the locomotive.

Nevertheless, for modeling the operation of a diesel engine at the current operating mode, it is necessary to have a sufficient set of parameters characterizing the course of thermodynamic processes. Moreover, not all parameters can be measured in operation without interruption of the locomotive operation.

For the use in the identification of diesel engine operating modes, a mathematical model of the working process of a locomotive diesel engine was developed, which is based on the equations:

volume balance [1]:

$$dp = \frac{kp_g}{V_\varphi}(\partial_S V + \partial_B V + \partial_M V + \partial_Q V - \partial V) \tag{1}$$

where dp – change in gas pressure in the cylinder, MPa;
k – adiabatic coefficient;
p_g – current pressure in the cylinder, MPa;
V_φ – current cylinder volume, m^3;
$\partial_S V$ – volume change as a result of gas entering the cylinder, m^3;
$\partial_B V$ – volume change as a result of gas discharge from the cylinder, m^3;
$\partial_M V$ – volume change as a result of a change in the number of molecules, m^3;
$\partial_Q V$ – volume change as a result of external heat supply to gases, m^3;
∂V – volume change as a result of the piston movement, m^3;

equations of two-phase heat release of Pugachev [2, 3]:

$$\frac{dx}{d\varphi} = \frac{x_1}{\varphi_1^2}\varphi e^{-\frac{\varphi^2}{2\varphi_1^2}} + \frac{x_2}{\varphi_2^2}\varphi e^{-\frac{\varphi^2}{2\varphi_2^2}}, \tag{2}$$

where x_1 – proportion of heat released in the first phase;
x_2 – proportion of heat released in the second phase;

φ_1 – duration from the start of combustion to the moment of maximum heat release rate in the first phase, s;
φ_2 – duration from the start of combustion to the moment of maximum heat release rate in the second phase, s;

equation of the heat transfer in a diesel cylinder [4]:

$$dQ_o^\varphi = \alpha_\Sigma \left(T_g - T_{st}\right) F_{st} \frac{d\varphi}{6n_d}, \tag{3}$$

where α_Σ – total averaged over the surface coefficient of heat transfer from the working fluid to the cylinder walls, W/(m^2·K);
T_g – temperature of the working fluid, K;
T_{st} – surface temperature of the combustion chamber, K;
F_{st} – heat exchange surface area, m^2;
$d\varphi/6n_d$ – heat transfer time, s;

the formula for the heat transfer coefficient obtained by G. Woschni, refined and transformed by Hohenberg [5]:

$$\alpha_\Sigma = 130 V_\varphi^{-0.06} T_g^{-0.4} p_g^{0.8} [c_m + 1.4]^{0.8}, \tag{4}$$

where V_φ – current volume of a cylinder, m^3;
p_g – current pressure in a cylinder, bar;
T_g – temperature of the working fluid, K;
c_m – average speed of the piston, m/s.

A mathematical model of the engine's working cycle makes it possible to obtain data arrays characterizing the change in the pressure and temperature of the gases in the diesel cylinder for each moment of time during the compression, combustion, and expansion processes, as well as the coefficient of performance, fuel consumption, and average effective pressure. The algorithm implemented in the presented model is shown in Fig. 1 [6].

However, for the successful use of modeling in assessing the condition of a locomotive in operation, it is necessary to have a large array of initial data, without which an accurate assessment of the course of the working cycles in the cylinder is impossible (Fig. 2). Besides, the registration of all necessary operation parameters of the engine in operation is often difficult or requires the installation of new measuring instruments.

To obtain additional information about the course of the fuel combustion process, an attempt was made to use data on the equilibrium composition of exhaust gases, for which a mathematical model was developed.

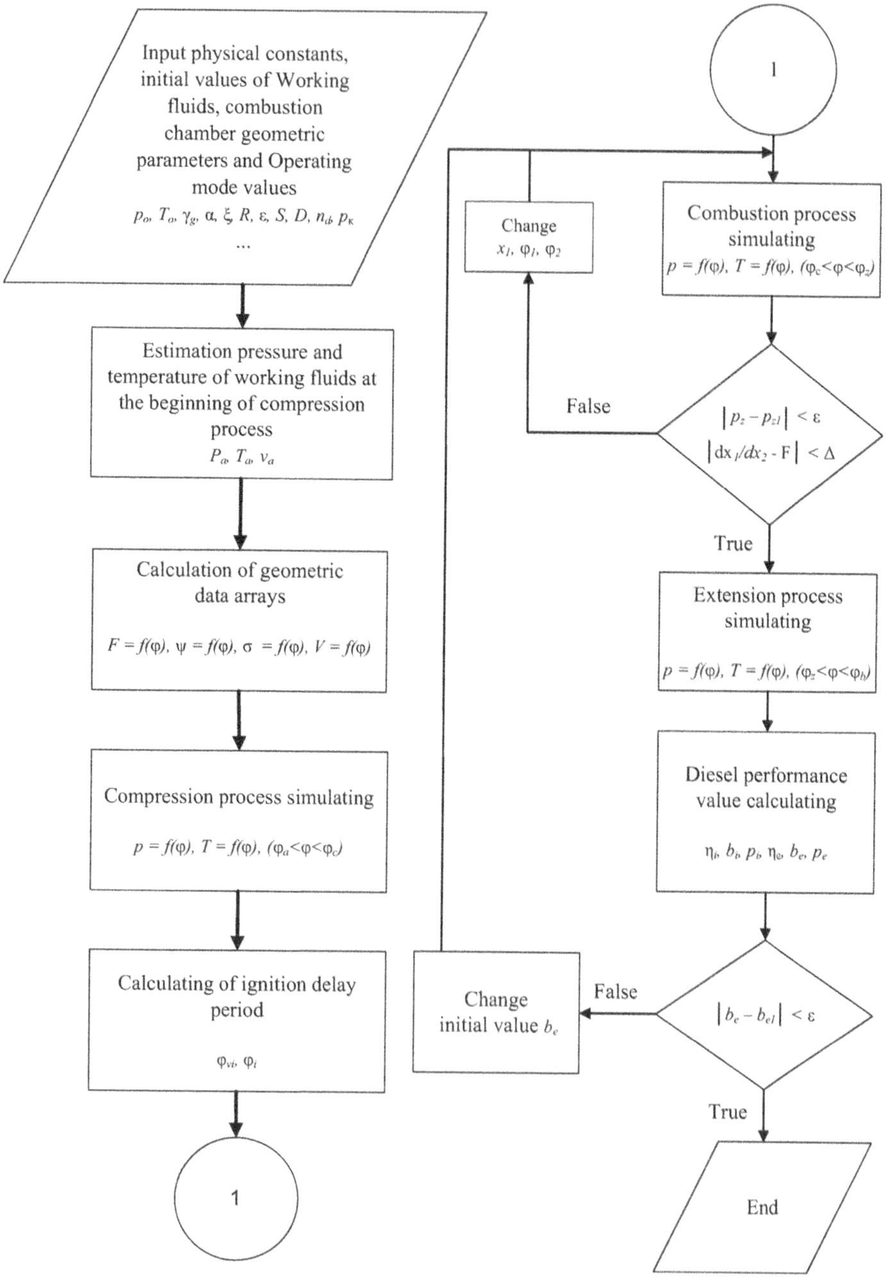

Fig. 1. Algorithm for calculating the working process of a diesel engine

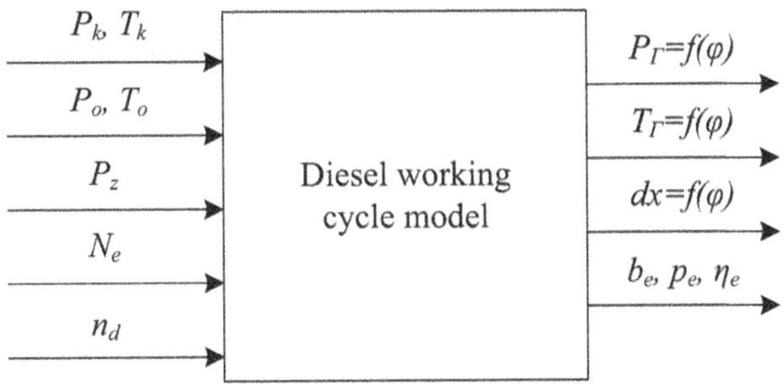

Fig. 2. Scheme of work of a mathematical model of the locomotive diesel working cycle

The purpose of the simulation was to analyze the effect of maximum combustion pressure and cyclic fuel supply on the composition of exhaust gases. The model is based on the equations of material balance and takes into account eleven elements in the composition of the exhaust gases [7, 8]:

$$\begin{cases} \alpha_P(P_O + P_{O2} + P_{OH} + P_{H2O} + P_{NO} + P_{CO} + P_{CO2}) = \alpha(P_{CO} + P_{CO2}), \\ \beta_P(P_N + P_{N2} + P_{NO}) = P_O + P_{O2} + P_{OH} + P_{H2O} + P_{NO} + P_{CO} + P_{CO2}, \\ \gamma_P(P_H + P_{H2} + P_{OH} + P_{H2O}) = P_{CO} + P_{CO2}, \\ P = P_O + P_{O2} + P_H + P_{H2} + P_{OH} + P_{H2O} + P_N + P_{N2} + P_{NO} + P_{CO} + P_{CO2}, \end{cases} \quad (5)$$

where α_p, β_p, γ_p – equilibrium coefficients depending on the elemental composition of the fuel;
α – total coefficient of excess air;
P – total pressure of the gas mixture;
Pi – partial pressure of the i-th element of the combustion products.

The presented system (5) is solved for the entire combustion line using the pressure and temperature values known from the calculation of the working cycle, taking into account the heat release characteristic [9]:

$$m_i(\varphi) = \sum_{j=\varphi_0}^{\varphi_Z} m_{ij}(p_j, T_j) dx_j, \quad (6)$$

where m_i – change in mass of the i-th element of the gas mixture at the j-th moment of time;
m_{ij} – mass of the i-th element of the gas mixture at the time of combustion, corresponding to the j-th angle of rotation of the crankshaft;
p_j – pressure of gases in the diesel cylinder at the j-th moment of the combustion process;
T_j – temperature of gases in the diesel cylinder at the j-th moment of the combustion process;
dx_j – proportion of burnt fuel in the diesel cylinder at the j-th moment of the combustion process.

The mathematical model of the equilibrium composition of the combustion products, structurally presented as a black box, is shown in Fig. 3. The calculation algorithm used in the model is shown in Fig. 4.

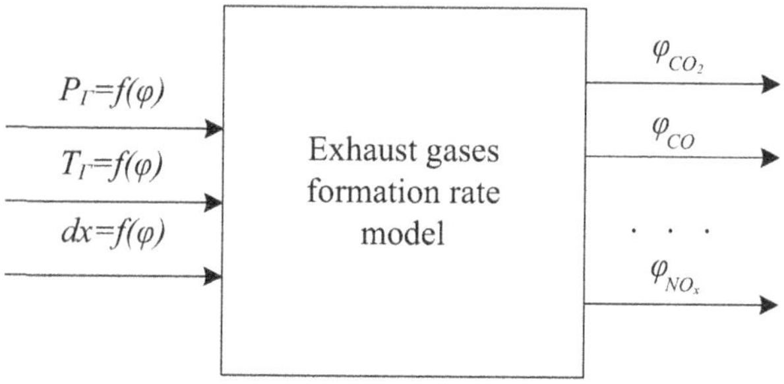

Fig. 3. Scheme of work of a mathematical model for determining the composition of exhaust gases

The main part of the input parameters required for the use in the presented models can be obtained using locomotive on-board systems, which are actively developing and are increasingly being used on diesel locomotives.

An exception and the greatest difficulty here are such parameters as the effective diesel power and maximum combustion pressure, the cyclic fuel supply and the lead angle of fuel supply. It is proposed that the first two parameters should be determined using a mathematical model based on data on the composition of exhaust gases of the diesel engine. Obviously, using indirect methods, it is difficult to obtain absolutely exact values of these parameters, which depend on many factors caused by the course of internal cylinder processes. Therefore, the algorithm for determining the effective diesel power and the maximum combustion pressure should allow obtaining approximate values with sufficient accuracy to determine the technical condition of the diesel locomotive.

3 Results and Discussion Section

Figure 5 shows the content of carbon dioxide (Fig. 5a) and nitrogen oxides (Fig. 5b) in the exhaust gases of a 1-PD4D diesel engine at nominal operating conditions for various values of effective power N_e and maximum combustion pressure p_z.

The algorithm for identifying the parameters of the working process is implemented by comparing the data recorded on the diesel locomotive and the data obtained during modeling data on the content of combustion products in the composition of diesel gases at several most likely operation modes [10].

The data characterizing the composition of the gases is a function that depends on the operating mode of the locomotive diesel engine:

$$\varphi = f(N_{ei}, P_{zj}), \tag{7}$$

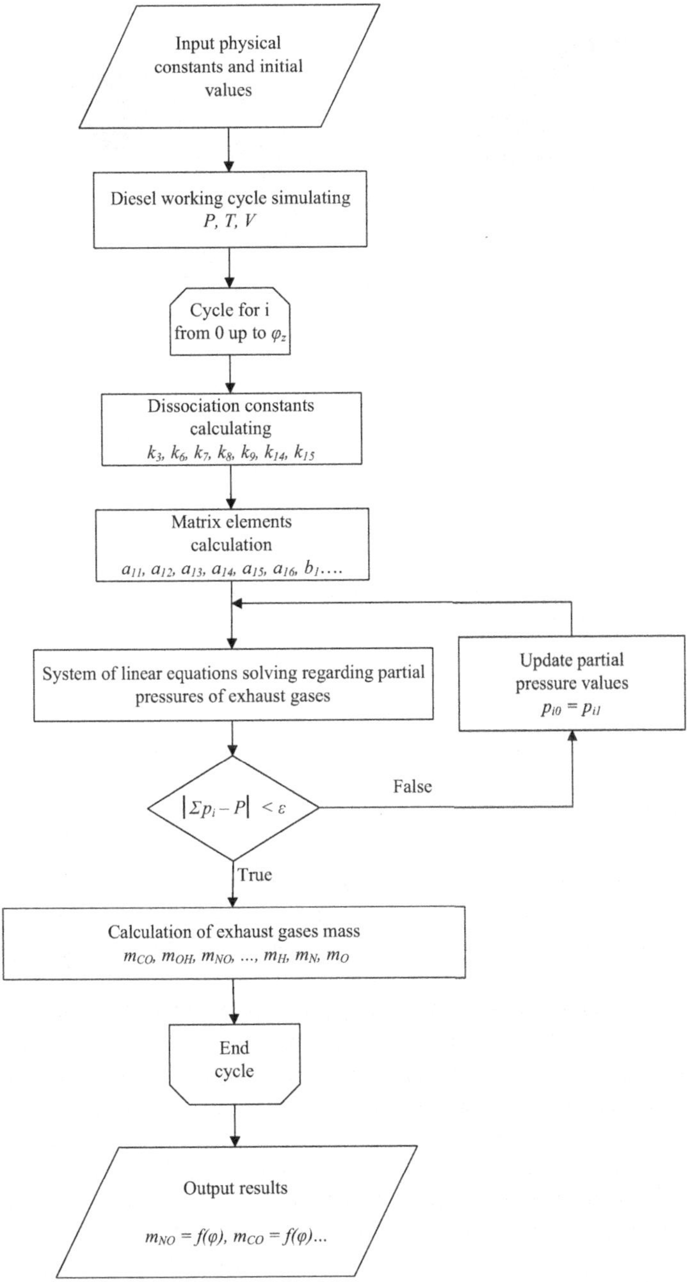

Fig. 4. Algorithm for determining the composition of exhaust gases of the diesel engine

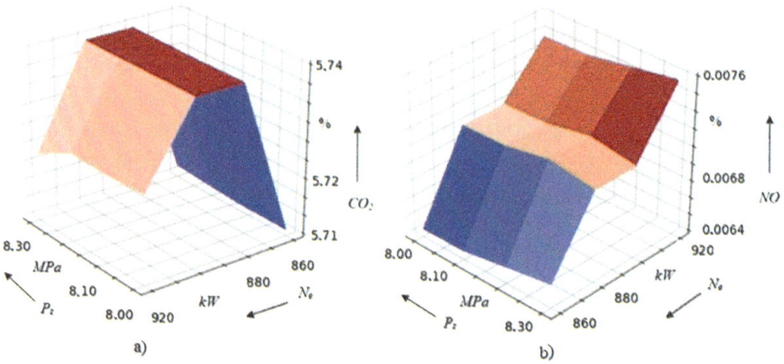

Fig. 5. Volume proportion of carbon dioxide (a) and nitrogen oxides (b) in the exhaust gases of a diesel engine at different values of the effective power and maximum combustion pressure

where i, j – position number of the parameter in the array of typical values for a given position of the driver's controller:

$$\begin{pmatrix} [N_{e\min}, P_{z\min}] & [N_{esr}, P_{z\min}] & [N_{e\max}, P_{z\min}] \\ [N_{e\min}, P_{zsr}] & [N_{esr}, P_{zsr}] & [N_{e\max}, P_{zsr}] \\ [N_{e\min}, P_{z\max}] & [N_{esr}, P_{z\max}] & [N_{e\max}, P_{z\max}] \end{pmatrix} \quad (8)$$

where N_{ei} – the minimum, average and maximum value of the effective diesel power encountered in operation at a certain position of the locomotive driver's controller;

P_{zj} – the minimum, average and maximum value of the maximum combustion pressure encountered in operation at a certain position of the locomotive driver's controller.

In the composition of the exhaust gases, it seems most appropriate to consider the volume proportion of nitrogen oxides NO_x, since its dependence on the temperature of the gases in the diesel cylinder and environmental parameters is easy to record in operation.

To solve the identification problem in accordance with the algorithm, for each pair of N_e and P_z values presented in matrix (8), the pressure and temperature are calculated during fuel combustion, obtaining the corresponding data for the entire range of possible changes in the operation parameters:

$$\begin{pmatrix} [P_{11}(\varphi), T_{11}(\varphi), dx_{11}(\varphi)] & [P_{12}(\varphi), T_{12}(\varphi), dx_{12}(\varphi)] & [P_{13}(\varphi), T_{13}(\varphi), dx_{13}(\varphi)] \\ [P_{21}(\varphi), T_{21}(\varphi), dx_{21}(\varphi)] & [P_{22}(\varphi), T_{22}(\varphi), dx_{22}(\varphi)] & [P_{23}(\varphi), T_{23}(\varphi), dx_{23}(\varphi)] \\ [P_{31}(\varphi), T_{31}(\varphi), dx_{31}(\varphi)] & [P_{32}(\varphi), T_{32}(\varphi), dx_{32}(\varphi)] & [P_{33}(\varphi), T_{33}(\varphi), dx_{33}(\varphi)] \end{pmatrix} \quad (9)$$

where P_{ij} – an array of values of the pressure in the cylinder during the combustion process in the corresponding operating mode;

T_{ij} – an array of values of the temperature in the cylinder during the combustion process in the corresponding operating mode;

dx_{ij} – an array of values of the rate of fuel combustion in the cylinder in the corresponding operating mode.

For each data set from matrix (9), the value of the volume proportion of NO_x is calculated:

$$\begin{pmatrix} [\varphi_{NO_x}]_{11} & [\varphi_{NO_x}]_{12} & [\varphi_{NO_x}]_{13} \\ [\varphi_{NO_x}]_{21} & [\varphi_{NO_x}]_{22} & [\varphi_{NO_x}]_{23} \\ [\varphi_{NO_x}]_{31} & [\varphi_{NO_x}]_{32} & [\varphi_{NO_x}]_{33} \end{pmatrix}, \tag{10}$$

where $[\varphi_{NOx}]_{ij}$ – the volume proportion of NO_x in the composition of the combustion products corresponds to the i-th effective diesel power and the j-th combustion pressure.

The choice of the desired pair of values of N_e and P_z is made on the condition that the values of the volume proportion of nitrogen oxides are close:

$$\left| [\varphi_{NO_x}]_{ij} - [\varphi_{NO_x}]_{el} \right| \to \min. \tag{11}$$

Below are the results of identification of the operating modes of TEM18DM diesel locomotives and an assessment of their efficiency. As the initial data, the results of measuring the composition of the exhaust gases at the environmental control station and the data of the on-board system were used (Table 1). The tests were carried out under various weather conditions at the rated operating mode of the power plant.

Table 1. Recorded parameters of diesel locomotives

Diesel locomotive no.	P_a, kPa	T_a, K	n_d, min^{-1}	N_g, kW	P_k, MPa	φ_{NOx}, *ppm*
1	100	298	745	718	0.146	1125
2	100	286	746	735	0.151	1000
3	101	265	748	761	0.162	970
4	102	258	751	749	0.152	822
5	99	262	749	747	0.158	1063

Table 2. Arrays of values obtained during the implementation of the identification algorithm

Modes for modeling [P_z, MPa; N_e, kW]				Calculated values of nitrogen oxides NO_x, *ppm*			
	$N_{e\ min}$	$N_{e\ sr}$	$N_{e\ max}$		$N_{e\ min}$	$N_{e\ sr}$	$N_{e\ max}$
$P_{z\ min}$	[7.8; 855]	[7.8; 874]	[7.8; 892]	$P_{z\ min}$	999	1003	1006
$P_{z\ sr}$	[8.0; 855]	[8.0; 874]	[8.0; 892]	$P_{z\ sr}$	1046	1051	1054
$P_{z\ max}$	[8.2; 855]	[8.2; 874]	[8.2; 892]	$P_{z\ max}$	1113	1118	1120

Table 2 presents the arrays of values obtained using the described algorithm, which are calculated for the diesel locomotive No. 5 from Table 1.

The proportion of NO_x in the composition of the combustion products is 1063 *ppm* (Table 1) and is closer to the value of 1054 *ppm* (Table 2) corresponding to the mode N_e = 892 kW and P_z = 8.0 MPa. This mode is used to determine the technical condition of the diesel engine.

The developed algorithm allows the identification of data on the course of in-cylinder processes, the determination of the values of N_e and P_z according to the known equilibrium composition of the fuel combustion products. This will expand the capabilities of monitoring the condition of diesel locomotives and complement the functionality of on-board systems of locomotives.

4 Conclusions

Based on the use of mathematical modeling, the algorithm is designed to facilitate the identification of the operating processes of locomotive diesel engines in operation, which expands the possibilities for an operational assessment of their technical condition.

Further research in this area is aimed at developing an algorithm for diagnosing locomotive diesel engines using the presented identification method.

References

1. Konkov, A.Yu.: Diagnostics of a diesel engine based on identification of work processes: monograph, p. 365. Dalnauka, Vladivostok (2014)
2. Dyachenko, N.Kh.: Theory of internal combustion engines, Leningrad, p. 552 (1974)
3. Kavtaradze, R.Z.: Theory of Piston Engines: Special Chapters: a Textbook for High Schools. Bauman MSTU, Moscow, 720 p. (2008)
4. Volodin, A.I., Skovorodnikov, E.I., Anisimov, A.S.: A comprehensive analysis of the thermodynamic, economic and environmental characteristics of locomotive diesel engines in operation: monograph. Omsk State Transport University, Omsk, p. 166 (2011)
5. Kavtaradze, R.Z.: Local heat transfer in piston engines: a training manual, p. 472. Bauman MSTU, Moscow (2007)
6. Anisimov, A.S., Chernyshkov, I.V.: Modeling the process of fuel combustion in a locomotive diesel engine using the equation of two-phase heat release. Izvestiya Transsiba **1** (25), 2–10 (2016)
7. Volodin, A.I.: Methods for assessing the technical condition, operational efficiency and environmental safety of diesel locomotives: monograph, p. 264. Zheldorizdat LLC, Moscow (2007)
8. Skovorodnikov, E.I., Anisimov, A.S., Minakov, V.A., Chernyshkov, I.V.: Analytical assessment of the environmental safety of locomotive diesel engines. Izvestiya PGUPS **1**, 118–130 (2019)
9. Alpatov, V., Soloviev, A.: Thin-walled profiles and their joint assembly units built with screws: numerical studies of load bearing capacity. In: MATEC Web of Conferences, vol. 196, p. 01008 (2018). https://doi.org/10.1051/matecconf/201819601008

10. Anisimov, A.S., Chernyshkov, I.V., Ivanova, I.N., Dvorko, N.Yu.: Method for determining indicators of the energy efficiency of diesel locomotives in operation based on data from on-board systems. In: Innovative Projects and Technologies in Education, Industry and Transport: materials of a Scientific Conference, pp. 137–142. Omsk State Transport University, Omsk (2018)

Model of Formation of Wear Debris Concentration in Diesel Engine Oil During Operation

Sergei Ovcharenko and Vitalyi Minakov(✉)

Omsk State Transport University, Karl Marx Ave., 35, 644046 Omsk, Russia
vitalya_13@mail.ru

Abstract. During the operation of a locomotive diesel engine, gradual wear of parts of the cylinder-piston group and the slider and crank mechanism occurs. Periodic in-place assessment of the degree of wear of parts allows determining their operability. The difficulty lies in assessing, based on the analysis of current values of wear debris concentration, the amount of metal worn from parts, since external factors influence the process of concentration formation. The paper presents a mathematical model of the formation of the metal volume worn from rubbing parts of a diesel in engine oil.

Keywords: Diesel · Engine oil · Wear of details · Spectral analysis · Diagnostics · Reliability · In-place assessment of a technical condition

1 Introduction

In all types of autonomous rolling stock, diesel engine is one of the least reliable components of a diesel locomotive, which accounts for more than 40% of failures and unplanned repairs. About half of the diesel engine malfunctions are associated with the consequences of the wear of parts of the cylinder-piston group and the slider and crank mechanism. Improving the reliability of diesel engine operation by preventing failures during operation is an urgent task at the present stage. One of the ways to increase reliability is to use diagnostic methods and tools that can significantly increase reliability, eliminate unplanned repairs, protect the diesel engine from negative consequences, and, accordingly, reduce operating costs [1, 2].

A diagnostic method based on the use of the results of spectral analysis of engine oil is considered one of the most effective methods of in-place assessment of the current technical condition of diesel parts during operation. Spectral analysis of engine oil allows assessing the amount of metal entered the engine oil as a result of mechanical wear of the rubbing parts. During the operation of a diesel engine in the lubricant, the concentration of wear debris increases, the value of which characterizes the degree of wear of parts [3, 4]. For an in-place assessment of the degree of wear of parts according to the concentration of wear debris, it is necessary to take into account a number of features:

Z. Popovic et al. (Eds.): TransSiberia 2019, AISC 1115, pp. 484–492, 2020.
https://doi.org/10.1007/978-3-030-37916-2_47

- during the operation of a diesel engine, redistribution of wear debris occurs in engine oil, therefore only a part of them forms the current (measured) value of concentration;
- such factors as burn-out loss, leakages, oil change affect the volume of engine oil in the system;
- materials of different groups of parts partially contain the same chemical elements, for example, iron is contained in cylinder liners, compression rings, piston pins, etc.

2 Materials and Methods

In order to carry out a qualitative assessment of the technical condition of diesel engine parts on the basis of the results of spectral analysis of engine oil, it is necessary to carry out theoretical studies and develop a mathematical model of the process of formation of wear debris concentration in engine oil.

It is necessary to take into account that during the operation of the diesel engine, the mass transfer of wear debris occurs in the oil system (Fig. 1) [5].

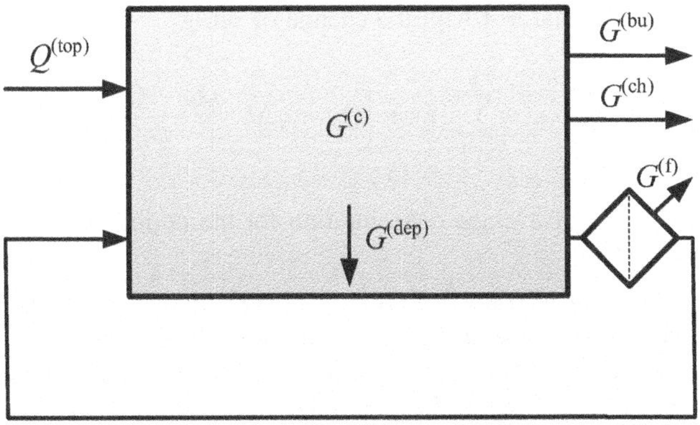

$G^{(c)}$ - intake of wear debris in engine oil; $G^{(bu)}$ - removal of wear debris with burn-out loss or oil leak; $G^{(ch)}$ - loss with oil change; $G^{(f)}$ - deposition in the filtration system; $G^{(dep)}$ - deposition of metal on the walls of the crankcase and pipeline.

Fig. 1. Mass transfer scheme in a diesel crankcase

Particles of metal detached from the surface of the parts during friction get into the engine oil, circulating through the system. The concentration of chemical elements in engine oil is defined as the ratio of the number of elements currently contained in engine oil in grams to the volume of engine oil in tons.

The total amount of the chemical element entered the oil system as a result of wear of parts for the considered interval is calculated by the formula, g:

$$G = G^{(c)} + G^{(bu)} + G^{(ch)} + G^{(f)} + G^{(dep)} \tag{1}$$

The concentration of metal in the engine oil at the time of control is determined as, g:

$$G^{(c)} = K_N(Q_N - q(L_N - L_{N-1})) \tag{2}$$

where K - metal concentration at the time of control is determined, g/t:
N - periodic concentration control number;
Q - oil volume in the diesel system, t;
q - the intensity of oil removal from the system, t/thousand km;
$L_N - L_{N-1}$ - locomotive mileage between control points, thousand km

Calculation of the amount of metal lost during burn-out loss and oil leaks, at intervals of topping up oil, g:

$$G^{(bu)} = K_N \cdot \frac{q(L_N - L_{N-1})}{Q_N - Q^{(top)}} \cdot Q_N \tag{3}$$

where $Q^{(top)}$ - the amount of oil topping in the system, t.
The amount of metal lost with the change of oil, g,

$$G_{ch} = \sum_{j=1}^{N} (K_N) \cdot \left[Q_N - q\Delta L \cdot \frac{Q_N - Q^{(top)}}{Q_N} \right] \tag{4}$$

where $K_{av}^{(i)}$ - values of average concentration for the periods between control points, g/t.

The amount of metal deposited on the walls of the crankcase and the pipeline can only be established empirically. Based on the results of the study, we take the amount of deposition equal to 5% of the total volume of metal worn from parts.

The amount of metal retained by the filters, g,

$$G_f = k_f \cdot \sum_{i=1}^{N} (K_N) \cdot \frac{(Q_0 - Q^{(top)})}{Q_0} \tag{5}$$

where G_f' - the amount of metal retained by the filters during the period between controls, g;
k_f - engine oil cleaning ratios.

The studies of the fractional composition of particles of wear debris in waste oil showed that the bulk of the particles (up to 80%) has a size of up to 10 microns, while the ability to clean with filters: 20–25 microns - for digital filter; 30–35 microns - for fine filter; from 100 μm - for deep bed filter [6, 7].

Using formulas 1–5 allows determining the amount of wear debris received from diesel parts for the period between control points. To improve the accuracy of the assessment, it is necessary to take into account the dynamics of the concentration of impurities between the control points, which depends on such factors as the period and the number of topping up of engine oil between the concentration control points [8].

The rate of change in engine oil contamination, g/t thousand km:

$$a = \frac{K_{n+1} - K_n}{L_{i+1} - L_i} \tag{6}$$

where K_{n+1} – measured value of oil concentration with impurities at the moment of control n+1, g/t;

K_n – measured value of oil concentration with impurities at the moment of control n, g/t;

L_i – operating time for the previous control step, thousand km;

L_{i+1} – operating time for the next control step, thousand km

The concentration after topping up oil:

$$k_i = k'_{i+1} + a(L_{i+1} - L_i) \tag{7}$$

where k'_{i+1} – concentration value before engine oil is topped up.

The value of the concentration of impurities in the oil to before topping up:

$$k'_{\mathrm{i}} = k_{\mathrm{i}} \cdot \frac{Q_0 - Q_i^{(t)}}{Q_0} \tag{8}$$

where Q_0 – the maximum amount of oil in the system, t;

$Q_{\mathrm{i}}^{(t)}$ – amount of oil topped up, t.

To solve the genesis problem, to calculate the intermediate values of the concentration of wear debris in engine oil at the moments of topping up, a system of equations that is solved by an iterative method is proposed:

$$\left|\begin{array}{c} k_0 = 0 \\ k'_1 = (k_0 + a\Delta L_1) \cdot \frac{Q - Q_1^{(t)}}{Q} \\ k'_2 = (k_0 + a\Delta L_1) \cdot \frac{Q - Q_1^{(t)}}{Q} + (k'_1 + a\Delta L_1) \cdot \frac{Q - Q_2^{(t)}}{Q} \\ \cdots\cdots \\ k'_i = (k_0 + a\Delta L_1) \cdot \frac{Q - Q_1^{(t)}}{Q} + (k'_1 + a\Delta L_1) \cdot \frac{Q - Q_2^{(t)}}{Q} \; \ldots\ldots (k'_{i-1} + a\Delta L_{i-1}) \cdot \frac{Q - Q_{i-1}^{(t)}}{Q} \\ \cdots\cdots \\ k'_j = (k_0 + a\Delta L_1) \cdot \frac{Q - Q_1^{(t)}}{Q} + (k'_1 + a\Delta L_1) \cdot \frac{Q - Q_2^{(t)}}{Q} \; \ldots\ldots (k'_{i-1} + a\Delta L_{i-1}) \cdot \frac{Q - Q_{i-1}^{(t)}}{Q} \ldots\ldots \left(k'_{i+j-1} + a\Delta L_{i+j-1}\right) \cdot \frac{Q - Q_{i+j-1}^{(t)}}{Q} \end{array}\right| \tag{9}$$

where $k_{1,2,\ldots,i,\ldots j}$ – concentration of wear debris in engine oil at the time of control, g/t;
Q – amount of oil un the system, kg;
$Q^{t}_{1,2,\ldots,i,\ldots j}$ – amount of topped up oil, kg;
$\wedge L = L_i - L_{i-1}$ – operating time of the diesel between topping up oil.

Thus, knowing the amount of toppings and the period of engine oil change, it is possible to restore the process of changing the concentration of wear debris in diesel engine oil by control points according to formula 9.

The algorithm for calculating the restoration of oil concentration values by wear debris at all points of topping is presented in Fig. 2.

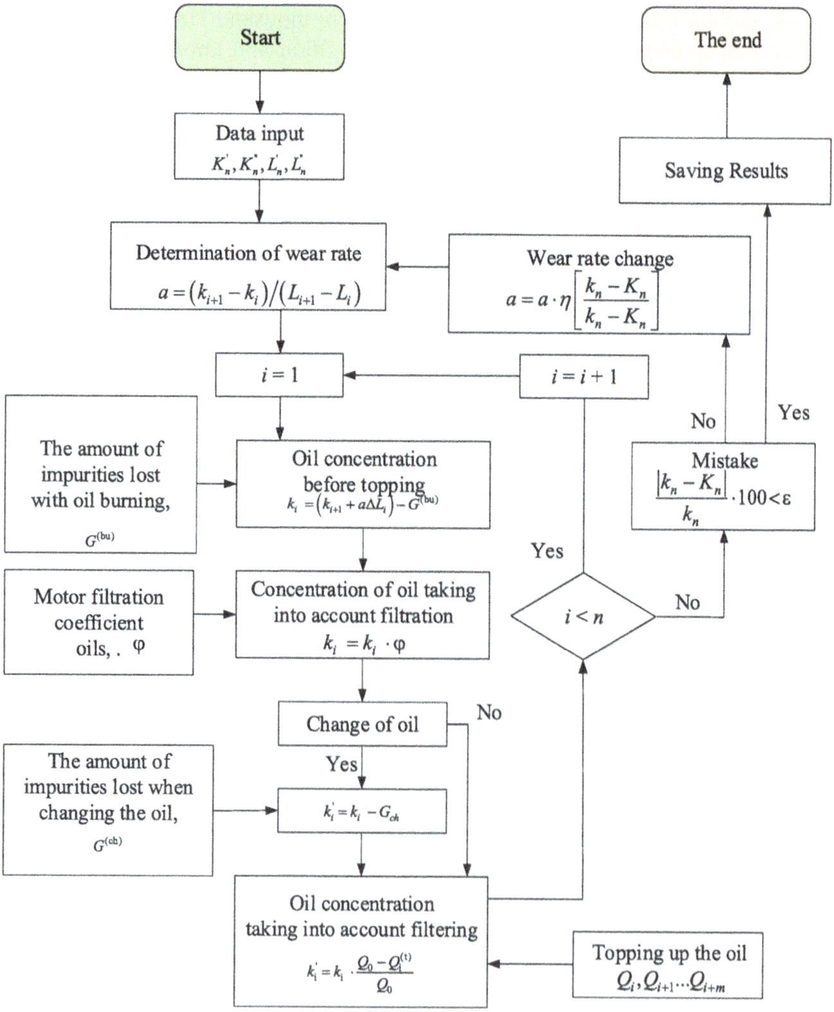

Fig. 2. The algorithm for calculating the restoration of the values of wear debris concentration, according to the points of topping up oil

As an example, the dynamics of wear debris concentration in diesel engine oil is presented (Fig. 3). It should be noted that when solving the system of equations, the assumption was made that the intensity of the intake of wear debris into the engine oil remains constant over the design interval.

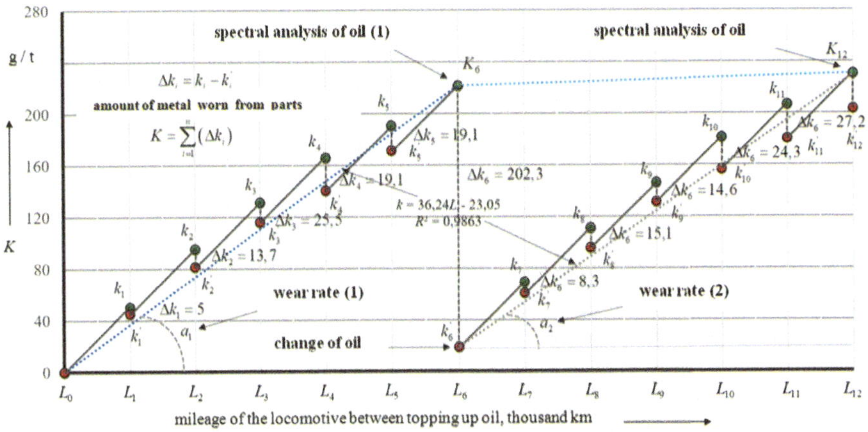

Fig. 3. Dynamics of wear debris concentration in diesel engine oil

To separate the volumes of metal in oil by groups of chemical elements, a model of the process of formation was performed in the MatLab Simulink software [9, 10].

The volume of metal worn from the *i*-th part at L_o operating time is expressed as, mm^3:

$$V_i = f(I_i) \cdot L_o \tag{10}$$

When parts wear out, all particles of wear get into the oil. Then, the total volume of elements worn from parts is expressed as, mm^3:

$$V_{L_p} = \sum_{i=1}^{n} V_i \tag{11}$$

where n - the number of parts in the system.

The amount of *j*-th element is determined as, g:

$$M_j = \rho_j V_{L_p} s_j \tag{12}$$

where s_j - the percentage of the j-th type of chemical element in the i-th type of parts.

Figure 4 shows a fragment of the system for calculating the amount of metal worn from a cylinder liner, using an example of a diesel engine of D49 type, by the groups of chemical elements, which is implemented in the *MatLab* Simulink software and hardware environment.

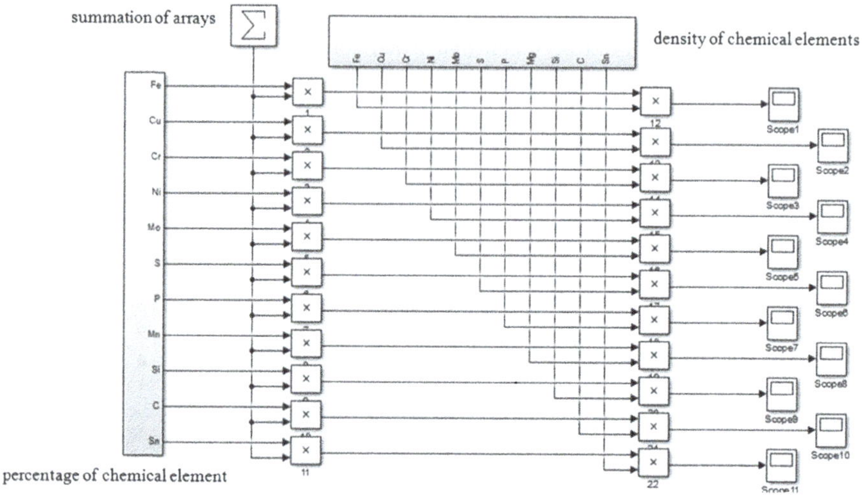

Fig. 4. Calculation of the amount of metal worn from cylinder liners by the groups of chemical elements

The described model allows calculating the process of formation of metal volumes in diesel engine oil, depending on the wear rate, one or more groups of parts.

3 Results

Figures 5, 6, 7 and 8 show the surfaces obtained by modeling the formation of metal volumes in engine oil in the absence of intense wear, i.e. during normal stable operation of the diesel engine. The graphs show the intake of chemical elements into the oil, depending on the operating time (running hours) of the controlled parts.

Figures 5, 6, 7 and 8 show the names of the

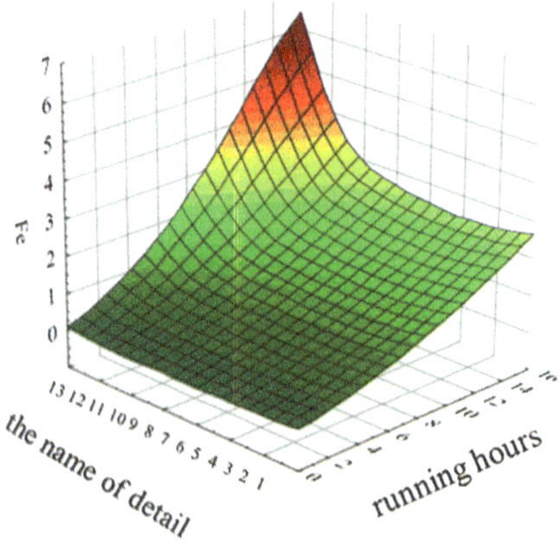

Fig. 5. Distribution of (Fe) iron

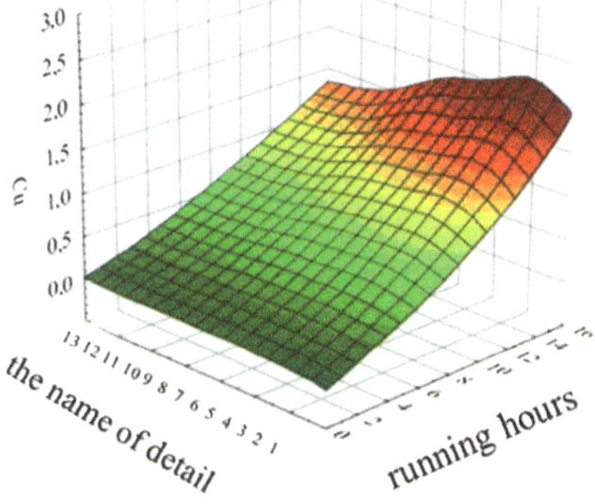

Fig. 6. Distribution of (Cu) copper

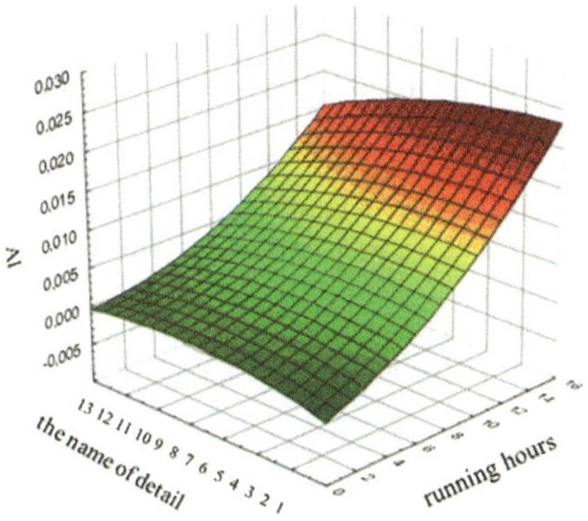

Fig. 7. Distribution of (Al) aluminum

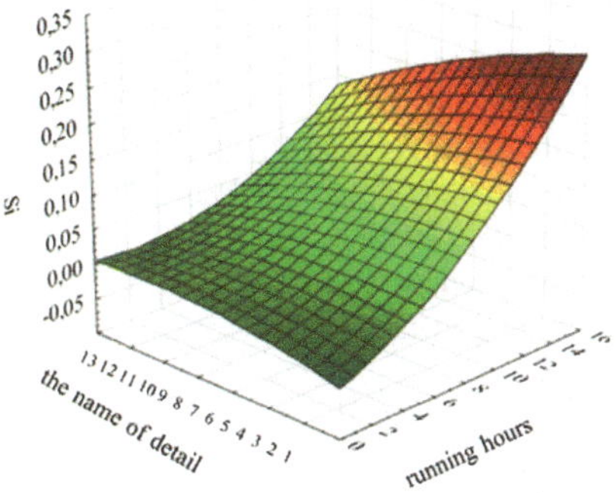

Fig. 8. Distribution of (Si) silicon -

details: 1 oil scraper rings; 2 - main bearing; 3 - bronze liner of a knuckle pin; 4 - piston skirt; 5 - bronze liner of the piston insert; 6 - knuckle pin; 7 - crank bearing; 8 - crank pin; 9 - main pin; 10 - compression rings; 11 - piston head; 12 - piston pin; 13 - cylinder liner.

4 Discussion

Diesel engine oil is a valuable source of information about its current technical condition. The values of wear debris concentration in engine oil make it possible to assess the degree of wear of its parts. The reliability of the diagnostic results is significantly affected by the accuracy of the assessment of the amount of metal wear from the controlled parts. The paper proposes a refined model for calculating the volume of metal worn from parts by groups of chemical elements. The continuation of the study is aimed at improving the in-place method for assessing the technical condition of diesel parts according to the results of monitoring the concentration of wear debris in engine oil.

5 Conclusions

The proposed model for calculating the amount of metal worn from diesel parts allows, based on intermediate concentration values, to assess the rate of metal intake into the engine oil during operation and to assess the amount of wear at the time of control.

References

1. Nadeev, A.I., Bui Khan, G., Svirepov, F.V.: Diagnostics of the technical condition of marine diesel engines based on data mining. Vestn. Astrakhan State Tech. Univ. Ser. Mar. Eng. Technol. **4**, 105–110 (2011)
2. Bodnar, B.E., Ochkasov, O.B.: Methods of in-place diagnosis of diesel engines in operation of a rolling stock. Sci. Transp. Prog. Vestn. Dnipro Natl. Univ. Rail Transp. **2**(68), 56–60 (2017)
3. Ovcharenko, S.M., Minakov, V.A.: Theoretical and experimental study of the patterns of wearing details of a diesel engine of D49 type. Izv. Transsiba: Sci. Tech. J. **3**(15), 55–61 (2013)
4. Liu, Y., Xie, Y.B., Zhao, F., Yao, Z.G.: Revision to the concept of equilibrium concentration of particles in lubrication system of machines. Wear **215**, 205–210 (1998)
5. Ovcharenko, S.M.: Modeling and assessing the effectiveness of engine oil cleaning systems of various series of diesel. Vestn. RSTU **1**, 21–27 (2008)
6. Chankin, V.V., Pakhomov, E.A.: The dynamics of changes in the concentration of impurities in diesel oils. Vestnik VNIIZhT **6**, 31–34 (1964)
7. Beloglazova, N.A., Minakov, V.A., Noskov, V.O.: External stationary system for cleaning engine oil of diesel locomotives. Vestn. Sci. Tech. J./Rostov State Transp. Univ. Rostov Don **1**(65), 42–46 (2017)
8. Ovcharenko, S.M., Minakov, V.A.: Modeling the process of accumulation of wear debris in the engine oil of a locomotive diesel D49. Izv. Transsiba: Sci. Tech. J. **3**(19), 31–36 (2014)
9. Jiang, Q.Y., Wang, S.N.: Abrasive wear of locomotive diesel engines and contaminant control. J. Tribol. Trans. **41**, 605–609 (2008)
10. Ovcharenko, S.M., Minakov, V.A.: Assessment of the technical condition of diesel parts using neural network data processing. Control Diagn.: Sci. Tech. J. Russ. Soc. Non-Destr. Test. Tech. Diagn. Publ. House "Spectrum" LLC **2**(212), 46–50 (2016)

Study of Electromechanical Oscillations in the System of Traction Electric Motor

Sergey Shantarenko(✉), Evgeniy Ponomarev, and Alexander Vaganov

Omsk State Transport University, Karl Marx Ave., 35, 644046 Omsk, Russia
nich.omgups@mail.ru

Abstract. The paper studies electromechanical oscillations in the traction electric motor system. The relative movement of the anchor and the frame of the traction electric motor in the coordinate system, which is rigidly connected to the frame, has been modeled. The level of anchor vibration acceleration has been assessed as well as dynamic additives of eccentricity and the degree of radial load on the anchor bearing. The calculated kinematic schemes of the anchor's spatial oscillations take into account technological parameters of the locomotive drive system and the internal design parameters of the traction electric motor. Influence of centrifugal moment and gyroscopic forces on the dynamic behavior of the anchor of the traction electric motor was assessed.

Keywords: Traction electric motor · Locomotive drive dynamics · Electromechanical oscillations · Anchor vibration acceleration · Frame · Anchor bearing · Wheel-motor block · Gearbox · Centrifugal moments · Gyroscopic forces

1 Introduction

The operation of traction electric motors (TEM) takes place under the conditions of continuously changing non-stationary modes. Dynamic behavior of TEM elements is formed by its own electromechanical parameters, as well as the parameters that determine the dynamics of the locomotive in general.

The electric locomotive's traction engine is the centerpiece of a complex electromechanical system that ensures the normal functioning of rolling stock in various operating conditions [1–3]. Mathematical modeling of such a system as a single whole (based on the generalized equations of Lagrange-Maxwell) leads to systems of non-linear differential equations of large dimensions relative to heterogeneous generalized coordinates. Detailed analysis of such systems, even with a small number of parameters, becomes almost impossible.

Theoretical study of dynamics of rolling stock [4] shows that it is possible to present a single mathematical system as a set of computational models connected by kinetic and elastic-dissipative relations. That allows performing mathematical description in the most beneficial coordinate system for each separate occurrence with the help of the best-suited modeling principles.

Z. Popovic et al. (Eds.): TransSiberia 2019, AISC 1115, pp. 493–503, 2020.
https://doi.org/10.1007/978-3-030-37916-2_48

Performance of TEM is largely determined by the operating conditions of motor-anchor bearings and changes in the air gap under the main poles. Therefore, it is necessary to consider the relative oscillations of the anchor and the traction engine stator in the wheel-motor block system.

2 Materials and Methods

The anchor of the traction engine is connected with the rest of the wheel-motor block mainly through motor- anchor bearings, through the gearbox gears and through the forces of the magnetic field in the air gap under the main poles.

Radial elastic forces Fb and the forces of rolling friction Mb occur in the supports of the bearing. We can accept the following with the sufficient accuracy [5]

$$F_b = C_b|\delta_b|^{3/2}; M_b = fF_b, \quad (1)$$

where δ_b – elastic deformation of bearing material; C_b – coefficient of material rigidity; f – coefficient of roiling friction.

The presence of the air gap (ε_b) can be taken into account with the following function:

$$F_b = \frac{C_b}{2}(1 + sign(|\delta_b| - \varepsilon_b))(\delta_b - \varepsilon_b sign\delta_b)^{3/2} \quad (2)$$

The occurrence of shock effects is possible in this situation. However, the anchor's axis neck is mostly pressed against the inner ring of the bearing. Therefore, while calculating most modes of operation of TEM (except for the passage of rail connection joints), it is possible to accept elastic characteristic F_b from the expression (1). The air gap (ε_b) value should be used for calculations of sufficient number of load receiving rollers and determination of the highest loaded roller.

The drive gearbox experiences elastic forces of F_R and moments of M_R under the influence of forces from the wheel pair and the rotating moment of the TEM anchor. They can be described as:

$$F_R = C_R\delta_R^{3/2}; M_{R1} = r_{g1}F_R; M_{R2} = r_{g2}F_R, \quad (3)$$

where δ_R – elastic deformation of the gear teeth; r_{g1}, r_{g2} – the radii of the leading and driven gears, respectively; C_R – contact rigidity coefficient.

Free movement of gear teeth within the gap (ε_R) may occur during the operation mode switch of the motor. As a result, force characteristics during operation mode switches can be defined as:

$$F_R = \frac{C_R}{2}(1 + sign(|\delta_R| - \varepsilon_R))(\delta_R - \varepsilon_R sign\delta_R)^{3/2} \quad (4)$$

Formulas (3) and (4) allow estimating the maximum force generated by the wheel-motor block gearbox. Since elastic deformations of δp depend on the dynamic components of the speeds of the anchor and wheelset, the dynamics of the anchor must be considered in conjunction with the progressive movement of the wheelset.

The operation of the traction motor anchor is significantly influenced by the heterogeneity of the radial forces of the magnetic field in the air gap [6]. The air gap transforms in the plane of the perpendicular axis of the anchor and its length along the main poles and the anchor in the process of dynamic oscillations. These transformations are dependent of the values of technological eccentricity (ε_0) caused by assembly imperfections and dynamic eccentricity (ε_d) that is caused by the shift of center of gravity of anchor in relation to the frame symmetry. The intensity of magnetic attraction P_{at} correlated to anchor length value can be presented with empirical equation [7]:

$$P_{at} = K_{tm} \cdot 2R_a \frac{\varepsilon}{\delta_{gir}} \tag{5}$$

where ε – total eccentricity; K_{tm} – a constant coefficient for the chosen type of motor.

The radial forces of the magnetic field are distributed along the line on the surface of the anchor, the points of which lie in the smallest distance to the surface of the stator in sections perpendicular to the axis of the motor.

In addition, the change in the air gap under the main poles leads to heterogeneity of the magnetic flow, which in turn has a significant impact on the conditions of the commutation of TEM [8].

Since the anchor has significant mass that leads to high angular velocity of rotation, it is necessary to take into account the gyroscopic forces acting on the supporting bearings of the anchor [9]. The effect of these forces may manifest in the presence of anchor eccentricity and gaps in bearings, as well as in curved sections of the path.

The gyroscopic moment is defined by the following equation:

$$\bar{M}_0^{gir} = J_a\, \bar{\omega}_a \times \bar{\omega}_1 \tag{6}$$

where $\bar{\omega}_1$ - angular velocity of anchor precession (the speed with which the anchor rotates around its center of gravity); J_a - anchor inertia moment; $\bar{\omega}_a$ - angular anchor velocity.

In order to assess the factors that determine the work of the TEM anchor, it is necessary to know the law of its position change in relation to other nodes. We need to define the equations of the spatial oscillations of the anchor [10].

To calculate mathematical models, we will use the calculated kinetic schemes shown in the picture as well as Lagrange's equations of the second kind

$$\frac{d}{dt}\left(\frac{\partial L}{\partial \dot{q}_i}\right) - \frac{\partial L}{\partial q_i} + \frac{\partial PS}{\partial \dot{q}_i} = Q_i$$

where $L = T - E_p$; T, E_p, PS, Q_i - respectively are: kinetic energy, potential energy, scattering function and generalized force and moment corresponding to the generalized coordinate q_i (Fig. 1).

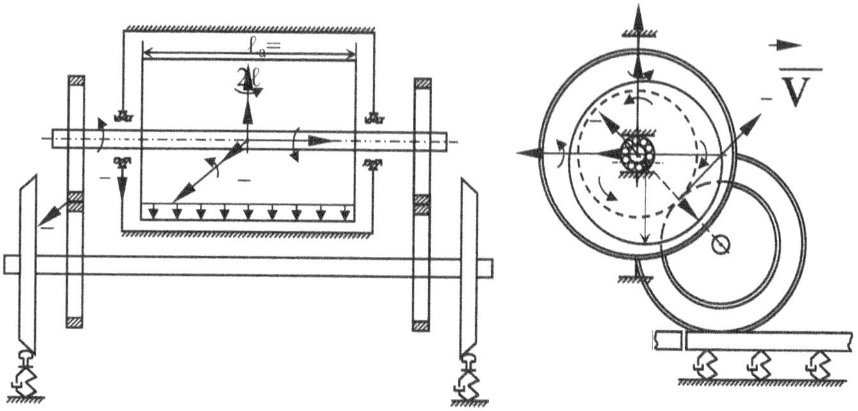

a) b)

Fig. 1. Analytic model of spatial oscillations of TEM anchor

The axes of coordinates were chosen as follows:

- OY axis is compatible with the stator axis (axis, that goes through the centers of anchor bearings);
- OX axis and OZ axis were placed in a perpendicular plane that goes through one of the centers of gravity of the anchor.

Simultaneously we choose the right system of coordinates, and the angles of the turn, speed of rotation and moments will be considered positive if they are directed counterclockwise.

We choose the following to be the generalized coordinates:

Z_a, X_a, Y_a - vertical- and longitudinal movements of the anchor axis; θ_a, φ_a, ψ_a - dynamic components of the anchor axis rotation around the OX, OY, OZ, respectively.

Then we find expressions of kinetic (T), potential (Ep) energies and scattering function PS through generalized coordinates:

$$T = \tfrac{1}{2} m_a \bar{V}_0^2 + \tfrac{1}{2} \iiint\limits_{(m)} \left[\omega^2 r^2 - (\bar{\omega} \cdot \bar{r})^2 \right] dm$$
$$= \tfrac{1}{2} m_a \left(\dot{X}_a^2 + \dot{Z}_a^2 \right) + \tfrac{1}{2} \left[J_{aX} \dot{\theta}_a^2 + J_{aY} \dot{\varphi}_a^2 + J_{aZ} \dot{\psi}_a^2 - 2 J_{XZ} \dot{\theta}_a \dot{\psi}_a \right]. \quad (7)$$

Moments of inertia relative to the selected axes are:

$$J_{aY} = m_a\left(\frac{R_a^2}{2} + \varepsilon_0^2\right)$$

$$J_{aX} = m_a\left[\left(\frac{R_a^2}{4} + \frac{\ell^2}{12}\right) + \varepsilon_0^2\cos^2\varphi\right]$$

$$J_{aZ} = m_a\left[\left(\frac{R_a^2}{4} + \frac{\ell^2}{12}\right) + \varepsilon_0^2\sin^2\varphi\right]$$

where φ - angle of anchor rotation around its axis.

Centrifugal moments J_{YZ} and J_{YX} equal 0, and J_{XZ} is calculated with a triple integral

$$\begin{aligned} J_{XZ} &= \iiint\limits_{(V)} \gamma\, zx\, dX\, dy\, dz \\ &= \gamma \iiint\limits_{V} (\rho\cos\alpha + \varepsilon_0\sin\varphi)(\rho\sin\alpha + \varepsilon_0\cos\varphi)\,\rho\, d\alpha\, d\rho\, dy \\ &= \gamma\tfrac{\pi R_a^2\ell}{2}\varepsilon_0^2\sin 2\varphi = \tfrac{m_a}{2}\varepsilon_0^2\sin 2\varphi. \end{aligned} \tag{8}$$

Kinetic energy

$$\begin{aligned} T = \tfrac{1}{2}m_a\Big[&\dot{x}_a^2 + \dot{z}_a^2 + \Big\{\Big(\tfrac{R_a^2}{4} + \tfrac{\ell^2}{12}\Big) + \varepsilon_0^2\cos^2\varphi\Big\}\dot{\theta}_a^2 \\ &+ \Big\{\Big(\tfrac{R_a^2}{4} + \tfrac{\ell^2}{12}\Big) + \varepsilon_0^2\sin^2\varphi\Big\}\dot{\psi}_a^2 - 2\varepsilon_0^2\sin 2\varphi\,\dot{\psi}\,\dot{\theta}\Big]. \end{aligned} \tag{9}$$

This equation includes φ - TEM anchor rotation. Its oscillations $\varphi_a = \omega_a t$ can be neglected. Thus, this angle is a variable parameter and will change according to the operation modes of TEM.

The potential energy of the system is determined by elastic forces in motor-anchor bearings [11, 12] and gearbox, and radial forces of the magnetic field

$$E_p = \int\limits_0^{\delta_{bleft}} F_{bleft}\, d\delta_{bleft} + \int\limits_0^{\delta_{bright}} F_{bright}\, d\delta_{bright} + \int\limits_0^{\delta_{Rleft}} F_{Rleft}\, d\delta_{Rleft} + \int\limits_0^{\delta_{Rright}} F_{Rright}\, d\delta_{Rright} + E_{pB}.$$

We find dependencies from the movement δ_b, δ_R and ε from generalized coordinates.

Elastic deformation of bearings of the right and left supports:

$$\begin{aligned} \delta_{bleft} &= \sqrt{(z_a - l_R\sin\theta_a)^2 + (x_a + l_R\sin\psi_a)^2} \\ \delta_{bright} &= \sqrt{(z_a + l_R\sin\theta_a)^2 + (x_a - l_R\sin\psi_a)^2} \end{aligned} \tag{10}$$

where l_R – distance between the centers of motor-anchor bearings.

Elastic pressing of the gearbox gears:

$$\delta_{Rright(left)} = r_{g1}\varphi + r_{g2}\varphi_{pw}$$

Eccentricity of the center of gravity of the anchor section, perpendicular to the axis of the engine and detached from the beginning of coordinates at a distance of y:

$$\varepsilon_{left} = \sqrt{(\varepsilon_0 \cos\varphi + z_a - y\sin\theta_a)^2 + (\varepsilon_0 \sin\varphi + x_a + y\sin\psi_a)^2}; \; (-\ell < y < 0)$$
$$\varepsilon_{right} = \sqrt{(\varepsilon_0 \cos\varphi + z_a + y\sin\theta_a)^2 + (\varepsilon_0 \sin\varphi + x_a - y\sin\psi_a)^2}. \; (0 < y < \ell)$$

Angles ψ_a and θ_a are small values, so we can assume that $\sin\psi_a \approx \psi_a$, $\sin\theta_a \approx \theta_a$. Thus, neglecting the small values of the second order, we get the following expressions for deformations:

$$\delta_{bleft} = \sqrt{x_a^2 + z_a^2 - 2l_R(z_a\theta_a - \theta_a\psi_a)}$$
$$\delta_{bright} = \sqrt{x_a^2 + z_a^2 + 2l_R(z_a\theta_a - \theta_a\psi_a)}$$
$$\varepsilon_{left} = \sqrt{z_a^2 + x_a^2 + \varepsilon_0^2 + 2\varepsilon_0(z_a\cos\varphi + x_a\sin\varphi) - 2y[\theta_a(z_a + \varepsilon_0\cos\varphi) - \psi_a(x_a + \varepsilon_0\sin\varphi)]}$$
$$\varepsilon_{right} = \sqrt{z_a^2 + x_a^2 + \varepsilon_0^2 + 2\varepsilon_0(z_a\cos\varphi + x_a\sin\varphi) + 2y[\theta_a(z_a + \varepsilon_0\cos\varphi) - \psi_a(x_a + \varepsilon_0\sin\varphi)]} \quad (11)$$

The radial force of the F_B magnetic field will be represented by an integral to the length of the anchor

$$F_B = \int_{-\ell}^{0} \frac{K_{tm} \cdot 2R_a}{\delta_{gir}} \varepsilon_{left} dy + \int_{0}^{\ell} \frac{K_{tm} \cdot 2R_a}{\delta_{gir}} \varepsilon_{right} dy$$

Potential energy of the radial forces of the magnetic field

$$E_{pB} = -\frac{K_{tm} \cdot 2R_a}{\delta_{gir}} \left[\int_0^{\varepsilon_{left}} \left(\int_{-\ell}^{0} \varepsilon_{left} dy \right) d\varepsilon_{left} + \int_0^{\varepsilon_{right}} \left(\int_0^{\ell} \varepsilon_{right} dy \right) d\varepsilon_{right} \right] \quad (12)$$

If we change the order of integration, we get:

$$\begin{aligned} E_{pB} &= -\frac{K_{tm} 2R_a}{\delta_{gir}} \left[\int_{-\ell}^{0} \frac{\varepsilon_{left}^2}{2} dy + \int_0^{\ell} \frac{\varepsilon_{right}^2}{2} dy \right] \\ &= -\frac{2\ell K_{tm} 2R_a}{2 \cdot \delta_{gir}} \{ x_a^2 + z_a^2 + \varepsilon_0^2 + 2\varepsilon_0(z_a\cos\varphi + x_a\sin\varphi) \\ &\quad + [\theta_a(z_a + \varepsilon_0\cos\varphi) - \psi_a(x_a + \varepsilon_0\sin\varphi)]\ell \}. \end{aligned} \quad (13)$$

The loss of the energy of vibration of the traction engine will occur mainly due to internal friction forces in the materials of the anchor bearings and gear teeth, as well as due to the friction forces in the bearings [13, 14].

The function of energy scattering of elastic deformations is:

$$PS = \int_0^{\delta_{bleft}} \beta_b \dot{\delta}_{bleft}\, d\dot{\delta}_{bleft} + \int_0^{\delta_{bright}} \beta_b \dot{\delta}_{bright}\, d\dot{\delta}_{bright} + 2\int_0^{\delta_R} \beta_g \dot{\delta}_R\, d\dot{\delta}_R \tag{14}$$

where β_b and β_g – damping coefficients of bearing and gears materials.

Deformation velocities $\dot{\delta}_{пл}$, $\dot{\delta}_{пп}$ and $\dot{\delta}_p$ can be presented through the velocities of generalized coordinates:

$$\begin{gathered}\dot{\delta}_{пл} = \sqrt{(\dot{z}_a - l_p\dot{\theta}_a)^2 + (\dot{\theta}_a + l_p\dot{\psi}_a)^2} \\ \dot{\delta}_{пп} = \sqrt{(\dot{z}_a + l_p\dot{\theta}_a)^2 + (\dot{\theta}_a - l_p\dot{\psi}_a)^2} \\ \dot{\delta}_p = r_{ш1}(\dot{\varphi}_a + \dot{\omega}_a) + r_{ш2}\dot{\varphi}_{кп}\end{gathered} \tag{15}$$

The forces of friction and slip are included in the equation as generalized forces and moments [15].

Let's find the expressions of derivatives in the Lagrange equation:

(a) Coordinate Xa

$$\frac{\partial L}{\partial \dot{x}_a} = m_a\dot{x}_a;\quad \frac{d}{dt}\left(\frac{\partial L}{\partial \dot{x}_a}\right) = m_a\ddot{x}_a$$

$$-\frac{\partial L}{\partial x_a} = \frac{F_{пл}}{\delta_{пл}}(x_a + l_p\psi_a) + \frac{F_{пп}}{\delta_{пп}}(x_a - l_p\psi_a) - \frac{4K_д R_a \ell}{\delta_г}\left(x_a + \varepsilon_0\sin\varphi - \frac{\ell\psi_a}{2}\right) \tag{16}$$

$$\frac{\partial \Phi}{\partial \dot{x}_a} = 2\beta_{оп}\dot{x}_a$$

(b) Coordinate Za

$$\frac{\partial L}{\partial \dot{z}_a} = m_a\dot{z}_a;\quad \frac{d}{dt}\left(\frac{\partial L}{\partial \dot{z}_a}\right) = m_a\dot{z}_a;$$

$$-\frac{\partial L}{\partial x_a} = \frac{F_{пл}}{\delta_{пл}}(z_a - l_p\theta_a) + \frac{F_{пп}}{\delta_{пп}}(z_a + l_p\theta_a) - \frac{4K_l R_a \ell}{\delta_г}\left(z_a + \varepsilon_0\sin\varphi - \frac{\ell\theta_a}{2}\right); \tag{17}$$

$$\frac{\partial \Phi}{\partial \dot{z}_a} = 2\beta_{оп}\dot{z}_a$$

(c) Coordinate θa

$$\frac{\partial L}{\partial \dot{\theta}_a} = m_a\left\{\left(\frac{R_a^2}{4}+\frac{\ell^2}{3}+\frac{\varepsilon_0^2}{2}\right)\dot{\theta}_a + \frac{\varepsilon_0^2}{2}\cos 2\varphi\ \dot{\theta}_a - \varepsilon_0^2 \sin 2\varphi\ \dot{\psi}_a\right\}$$

$$\begin{aligned}
&\frac{d}{dt}\left(\frac{\partial L}{\partial \dot{\theta}_a}\right) = m_a\left\{\left(\frac{R_a^2}{4}+\frac{\ell^2}{3}+\frac{\varepsilon_0^2}{2}\right)\ddot{\theta}_a + \frac{\varepsilon_0^2}{2}\cos 2\varphi\ \ddot{\theta}_a - \varepsilon_0^2 \sin 2\varphi\ \ddot{\psi}_a - \right.\\
&\left. - \varepsilon_0^2\left(\dot{\varphi}_a + \omega_a\right)\left(\dot{\theta}_a \sin 2\varphi + \dot{\psi}_a 2\cos 2\varphi\right)\right\};\\
&-\frac{\partial L}{\partial \theta_a} = \frac{\partial \Pi}{\partial \theta_a} = F_{пл}\frac{\partial \delta_{пл}}{\partial \theta_a} + F_{пп}\frac{\partial \delta_{пп}}{\partial \theta_a} + \frac{\partial \Pi_т}{\partial \theta_a} = z_a l_p\left[\frac{F_{пп}}{\delta_{пп}} - \frac{F_{пл}}{\delta_{пл}}\right] - \\
&-\frac{2K_д R_a \ell}{\delta_{гп}}\ell\left(z_a + \varepsilon_0 \cos\varphi\right);
\end{aligned} \tag{18}$$

$$\frac{\partial \Phi}{\partial \dot{\theta}_a} = 2\beta_{оп} l_p^2 \dot{\theta}_a$$

(d) Coordinate Ψa

$$\frac{\partial L}{\partial \dot{\psi}_a} = m_a\left\{\left(\frac{R_a^2}{4}+\frac{\ell^2}{3}+\frac{\varepsilon_0^2}{2}\right)\dot{\psi}_a + \frac{\varepsilon_0^2}{2}\cos 2\varphi\ \dot{\psi}_a - \varepsilon_0^2 \sin 2\varphi\ \dot{\theta}_a\right\}$$

$$\begin{aligned}
&\frac{d}{dt}\left(\frac{\partial L}{\partial \dot{\psi}_a}\right) = m_a\left\{\left(\frac{R_a^2}{4}+\frac{\ell^2}{3}+\frac{\varepsilon_0^2}{2}\right)\ddot{\psi}_a + \frac{\varepsilon_0^2}{2}\cos 2\varphi\ \ddot{\psi}_a - \varepsilon_0^2 \sin 2\varphi\ \ddot{\theta}_a - \right.\\
&\left. - \varepsilon_0^2\left(\dot{\varphi}_a + \omega_a\right)\left(\dot{\theta}_a 2\cos 2\varphi - \dot{\psi}_a \sin 2\varphi\right)\right\};\\
&-\frac{\partial L}{\partial \psi_a} = \frac{\partial \Pi}{\partial \psi_a} = F_{пл}\frac{\partial \delta_{пл}}{\partial \psi_a} + F_{пп}\frac{\partial \delta_{пп}}{\partial \psi_a} + \frac{\partial \Pi_т}{\partial \theta_a} = x_a l_p\left[\frac{F_{пп}}{\delta_{пп}} - \frac{F_{пл}}{\delta_{пл}}\right] + \\
&+\frac{2K_д R_a \ell}{\delta_г}\ell\left(x_a + \varepsilon_0 \sin\varphi\right);
\end{aligned} \tag{19}$$

$$\frac{\partial \Phi}{\partial \dot{\psi}_a} = 2\beta_{оп} l_p^2 \dot{\psi}_a$$

(e) Coordinate φ

$$\frac{\partial L}{\partial \dot{\varphi}} = m_a\left(\frac{R_a^2}{2} + \varepsilon_0^2\right)\dot{\varphi} = m_a\left(\frac{R_a^2}{2} + \varepsilon_0^2\right)\left(\dot{\varphi}_a + \omega_a\right)$$

$$\frac{d}{dt}\left(\frac{\partial L}{\partial \dot{\varphi}}\right) = m_a\left(\frac{R_a^2}{2} + \varepsilon_0^2\right)\ddot{\varphi} = m_a\left(\frac{R_a^2}{2} + \varepsilon_0^2\right)\ddot{\varphi}_a \tag{20}$$

$$-\frac{\partial L}{\partial \varphi} = -\frac{\partial T}{\partial \varphi} + \frac{\partial \Pi}{\partial \varphi}$$

Let's find derivatives $\frac{\partial T}{\partial \varphi}$ and $\frac{\partial \Pi}{\partial \varphi}$:

$$-\frac{\partial T}{\partial \varphi} = m_a\left\{\frac{\varepsilon_0^2}{2}\left(\dot{\theta}_a^2 - \dot{\psi}_a^2\right)\sin 2\varphi + 2\varepsilon_0^2\dot{\psi}_a\dot{\theta}_a\cos 2\varphi\right\} \tag{21}$$

$$\frac{\partial \Pi}{\partial \varphi} = -\frac{4K_д R_a \ell}{\delta_г}\left\{2\varepsilon_0\left(x_a\cos\varphi - z_a\sin\varphi\right) - \ell\varepsilon_0\left(\psi_a\cos\varphi + \theta_a\sin\varphi\right)\right\} + 2F_p r_{ш1}$$

$$\frac{\partial \Phi}{\partial \dot{\varphi}_p} = 2\beta_ш r_{ш1}\left(\dot{\varphi}_a + \omega_a + \frac{r_{ш2}}{r_{ш1}}\dot{\varphi}_{кп}\right) \tag{21}$$

These generalized moments are: gyroscopic moments of rotation around OX and OZ axes, moment of forces in the anchor bearing while rotating around OY axis, and the rotating moment of the linkage:

$$M_{гир}(\theta_a) = J_{ay}\dot{\varphi}\,\dot{\theta}_a$$

$$M_{гир}(\psi_a) = J_{ay}\dot{\varphi}\,\dot{\psi}_a \tag{22}$$

$$M_{оп}\left(\dot{\varphi}\right) = f_к\left(F_{пл} + F_{пп}\right)sign\dot{\varphi}\ ;\ M_{эм} = c_м\Phi\left(i_в\right)i_a\cdot$$

If we include found equations into Lagrange equation, we get:

$$
\left\{
\begin{aligned}
&m_a \ddot{x}_a + x_a\left(\frac{F_{\text{пл}}}{\delta_{\text{пл}}} + \frac{F_{\text{пп}}}{\delta_{\text{пп}}} - \frac{4K_{\text{д}} R_a \ell}{\delta_{\text{г}}}\right) + 2\beta_{\text{оп}} \dot{x}_a + \ell \psi_a \left(\frac{F_{\text{пл}}}{\delta_{\text{пл}}} - \frac{F_{\text{пп}}}{\delta_{\text{пп}}} + \frac{2K_{\text{д}} R_a \ell}{\delta_{\text{г}}}\right) - \\
&- \frac{2K_{\text{д}} R_a \ell}{\delta_{\text{г}}} \varepsilon_0 \sin\varphi = 0; \\
&m_a \ddot{z}_a + z_a\left(\frac{F_{\text{пл}}}{\delta_{\text{пл}}} + \frac{F_{\text{пп}}}{\delta_{\text{пп}}} - \frac{4K_{\text{д}} R_a \ell}{\delta_{\text{г}}}\right) + 2\beta_{\text{оп}} \dot{z}_a + \ell \theta_a \left(\frac{F_{\text{пл}}}{\delta_{\text{пл}}} - \frac{F_{\text{пп}}}{\delta_{\text{пп}}} + \frac{2K_{\text{д}} R_a \ell}{\delta_{\text{г}}}\right) - \\
&- \frac{2K_{\text{д}} R_a \ell}{\delta_{\text{г}}} \varepsilon_0 \cos\varphi = 0; \\
&m_a\left[\left(\frac{R_a^2}{4} + \frac{\ell^2}{3} + \frac{\varepsilon_0^2}{2}\right)\ddot{\theta}_a + \frac{\varepsilon_0^2}{2}\cos 2\varphi\, \ddot{\theta}_a - \varepsilon_0^2 \sin 2\varphi\, \ddot{\psi}_a\right] - m_a \varepsilon_0^2\left(\dot{\varphi}_a + \omega_a\right) \times \\
&\times\left(\sin 2\varphi\, \dot{\theta}_a + 2\cos 2\varphi\, \dot{\psi}_a\right) + z_a l_p \left[\frac{F_{\text{пп}}}{\delta_{\text{пп}}} - \frac{F_{\text{пл}}}{\delta_{\text{пл}}}\right] - \frac{2K_{\text{д}} R_a \ell}{\delta_{\text{г}}} \ell\left(z_a + \varepsilon_0 \cos\varphi\right) + \\
&+ 2\beta_{\text{оп}} l_p^2 \dot{\theta}_a - m_a\left(\frac{R_a^2}{2} + \varepsilon_0^2\right)\dot{\varphi}\dot{\theta}_a = 0; \\
&m_a\left[\left(\frac{R_a^2}{4} + \frac{\ell^2}{3} + \frac{\varepsilon_0^2}{2}\right)\ddot{\psi}_a + \frac{\varepsilon_0^2}{2}\cos 2\varphi\, \ddot{\psi}_a - \varepsilon_0^2 \sin 2\varphi\, \ddot{\theta}_a\right] - m_a \varepsilon_0^2\left(\dot{\varphi}_a + \omega_a\right) \times \\
&\times\left(\dot{\theta}_a \cos 2\varphi + \dot{\psi}_a \sin 2\varphi\right) + x_a l_p \left[\frac{F_{\text{пп}}}{\delta_{\text{пп}}} - \frac{F_{\text{пл}}}{\delta_{\text{пл}}}\right] - \frac{2K_{\text{д}} R_a \ell}{\delta_{\text{г}}} \ell\left(x_a + \varepsilon_0 \sin\varphi\right) + \\
&+ 2\beta_{\text{оп}} l_p^2 \dot{\psi}_a - m_a\left(\frac{R_a^2}{2} + \varepsilon_0^2\right)\dot{\varphi}\dot{\psi}_a = 0; \\
&m_a\left(\frac{R_a^2}{2} + \varepsilon_0^2\right)\ddot{\varphi} + 2F_p r_{\text{ш}1} - \frac{4K_{\text{д}} R_a \ell}{\delta_{\text{г}}}\left[2\varepsilon_0\left(x_a \cos\varphi - z_a \sin\varphi\right) - \varepsilon_0 \ell\left(\psi_a \cos\varphi + \theta_a \sin\varphi\right)\right] + \\
&+ m_a\left[\frac{\varepsilon_0^2}{2}\left(\dot{\theta}_a^2 - \dot{\psi}_a^2\right)\sin 2\varphi + 2\varepsilon_0^2 \dot{\psi}_a^2 \dot{\theta}_a^2 \cos 2\varphi\right] + 2\beta_{\text{ш}} r_{\text{ш}1}\left(\dot{\varphi}_a + \omega_a + \frac{r_{\text{ш}2}}{r_{\text{ш}1}}\dot{\varphi}_{\text{кп}}\right) + \\
&+ f_{\text{к}}\left(F_{\text{пл}} + F_{\text{пп}}\right)\operatorname{sign}\dot{\varphi} = M_{\text{эм}}.
\end{aligned}
\right.
$$

Thus, we have a system of five nonlinear differential equations, which is a mathematical model of the dynamics of the traction motor anchor.

3 Results

The Eq. (23) model relative movement of the anchor and the frame of the traction electric motor in the coordinate system, which is rigidly connected to the frame. That model allows assessing the level of anchor vibration acceleration, dynamic additives of eccentricity, and the degree of radial load on the anchor bearing.

The calculated kinematic schemes of the anchor's spatial oscillations also take into account technological parameters of the locomotive drive system and the internal design parameters of the traction electric motor. Influence of centrifugal moment and gyroscopic forces on the dynamic behavior of the anchor of the traction electric motor was also assessed.

References

1. Lavrinovich, V.A.: Research of traction electric motors of electric locomotive 2ES6 based on a comprehensive approach. Fundam. Res. **6**(2), 278–283 (2013)
2. Andrienko, P.D.: Research of transitional modes with the sequentially connection of serial electric motors DC. Electrotech. Electro Energ. **1**, 10–16 (2009)
3. Andrienko, P.D.: Research of braking modes in the system of pulse control of the series electric motor. Electrotech. Electro Energ. **2**, 11–14 (2007)
4. Pavlenko, A.P.: Dynamics of traction motors of line locomotives, p. 192. Mechanical Engineering, Moscow (1991)
5. Spitsyn, N.A.: Supports of axis and shafts of machines and devices, p. 520. Mechanical Engineering, Moscow (1970)
6. Neiman, L.A.: Application of conductivity method to account for the force of the one-sided magnetic attraction of the asymmetric electromagnet. Herald of the IRGTU, no. 2 (97), pp. 214–218 (2015)
7. Luzin, V.M.: Development of methods of improvement of switching reliability of high-used DC machines, p. 253. Dissertation of PhD in Engineering, Moscow (1987)
8. Scobelev, V.E.: Pulsating current motors, Leningrad, p. 208 (1985)
9. Andrienko, P.D.: Research of the dynamics of a serial electric motor with various impulse circuits of regulation. Electrotech. Electro Energ. **1**, 4–8 (2007)
10. Filkin, N.M.: Mathematical modeling of the dynamics of mechanical systems with non-holonomic connections using the Lagrange equation of the second kind. Successes Mod. Nat. Hist. **2**, 24–25 (2005)
11. Ivanov, V.A., Yerkaev, N.V.: Analysis of elastic deformations in the sliding bearing. Fundam. Res. 241–245 (2015)
12. Steam, E.F.: Method of calculating the characteristics of elastic-damper suspension from MR material for sliding bearing. Her. Samara State Aerosp. Univ. Part 1 **5**(47), 68–74 (2014)
13. Afanasov, A.M.: Choosing methods for determining mechanical losses in traction electric motors of constant and pulsating current, Science and Progress of Transport, Herald of the Dnipropetrovsk National University Railway. Transport Academician V, Lazaryan, pp. 151–154 (2010)
14. Bondarenko, L.N.: Moment of friction in sliding bearings. Her. Transnistrian State Acad. Constr. Arch. **9**(222), 36–41 (2016)
15. Drozdov, Y.N., Yudin, E.G., Belov, A.I.: Applied Tribology (Friction, Wear, Lubrication) ed. by Y.N. Drozdova, 604 p. Eco-Press, Moscow (2010)

The Use of Electrophysical Methods for Diagnosing Pre-defect States of Highly Loaded Units and Parts of Rolling Stock

Yury Matyash, Yury Sosnovsky(✉), and Evgeniy Kondrikov

Omsk State Transport University, Karl Marx Avenue, 35, Omsk 644046, Russia
sosnovskyym@mail.ru

Abstract. The paper presents the results of numerous laboratory experiments aimed at establishing a relationship between mechanical bending and thermo-EMF of steel samples used for the manufacture of units and parts of rolling stock. The mechanism of the occurrence of permanent deformation is studied. The effect of the loading form of the samples on the value of thermo-EMF recorded experimentally is shown. It has been suggested that it is possible to control the elastic properties of rolling stock elements through the value of thermo-EMF.

Keywords: Electrical and physical methods · Pre-defect state · Rolling stock · Residual life · Specific coefficient of thermo-EMF

1 Introduction

The problem of determining the residual life of complex technical systems, which include railway transport, due to the lack of understanding of the physical processes associated with the characteristics of the occurrence of fatigue and corrosion phenomena that affect aging and destruction of metals, is quite relevant. The methods currently used (ultrasonic, magnetic particle, etc.) are not aimed at establishing the causes of defects, but are focused on identifying defects of various origins that have already arisen. These methods do not take into account structural changes in metals that occur both during their natural aging and when exposed to mechanical and thermal loads that occur during operation.

Recently, computer methods have been developed for the needs of the railway, both for monitoring the maximum permissible (critical) geometric dimensions during the train ride, and for detecting cracks of various origins.

In Russia, as part of the Digital Railway, an Integrated Post for Automated Receiving and Diagnostics of Rolling Stock was developed, which includes the following subsystems: set of technical monitoring tools; complex of technical measurements; WILD; Technovisor and others. Their implementation required significant financial costs, but did not significantly affect the improvement of railway safety.

It can be said that the diagnostic system at maintenance points and during repair in the railway depot does not take into account structural changes in the steels from which

Z. Popovic et al. (Eds.): TransSiberia 2019, AISC 1115, pp. 504–518, 2020.
https://doi.org/10.1007/978-3-030-37916-2_49

the most critical components of the rolling stock are made (automatic coupler, rail, side frame, etc.), caused by operational factors [1].

A similar situation is observed abroad, in Western Europe and the USA. For example, in Finland, the system for monitoring the vertical forces of the interaction of wheels with rails is designed to check the condition of the wheels and prevent defects that could affect the safety of movement: sliders, holes, asymmetric wheels, etc., which can cause damage to the rail and/or cars. In the United States and other countries, computer vision technology is used, which is based on monitoring the design of cars. This system allows determining and controlling the position of the friction wedges, the presence or absence of brake pads, the height difference of the axial beams of the automatic coupler and much more [2, 3]. In the Netherlands, the Quo Vadis system for weighing rolling stock in motion is widespread [4].

At the same time, the theory of kinetic strength developed by S.N. Zhurkov explained the reasons for the natural aging of metals based on thermodynamic ideas about the structure of the crystal lattice and its defects. So, for example, it was shown that the destruction process consists of elementary fluctuations in the breaking of interatomic bonds, caused primarily by the thermal motion of atoms located in the nodes and interstitial sites of the lattice. According to this theory, the destruction process should be considered as a process in which atoms, as a result of thermal fluctuations, overcome the energy barrier U_0 and transfer to another most favorable energy state with lower energy. It should be noted that external mechanical stress can lower the energy barrier and thereby create a condition for the most frequent transitions:

$$\tau = \tau_0 \exp\left(\frac{U_0 - \gamma\sigma}{kT}\right) \tag{1}$$

where τ - time between two successive fluctuations, τ_0 - period of natural thermal vibrations of atoms, usually $\sim 10^{-13}$ s, γ - structural or volume factor, σ - external mechanical stress, k - Boltzmann constant, T - absolute temperature.

It can be argued that such transitions lead to a loss of the strength properties of metals. Since the application of rigorous theoretical calculations to the description of this situation is very difficult for objective reasons, in recent years, semi-empirical relationships between the physical parameters of substances and structural friability, which qualitatively characterizes the energy of interatomic interaction, have begun to appear in the scientific literature [5]. The physical meaning of friability is similar to entropy: the stronger the interatomic bonds, the stronger the effect of pulling or retention of atoms in the crystal structure, which corresponds to a decrease in both friability and entropy [6].

Based on the foregoing, the aim of the study was formulated: to establish a relationship between structural changes in metals under the action of mechanical stresses with any physical structurally sensitive characteristic and create a theoretical and experimental justification for diagnosing pre-defect states of real metal products in order to increase the reliability of estimating the residual life.

Indirect confirmation of structural changes caused by operational factors are the results presented in [7], in which the author analyzed the destruction of more than 300 parts of automatic couplers. As a result, it was found that the "process of degradation" or embrittlement is associated with a change in its hardness: during operation, the hardness of the parts under study increased.

An additional confirmation of this assumption is the results presented in [8]. The object of study in which was the reaction column of one of the petrochemical plants. In particular, it was noted that under the influence of operational factors, a change in the structure and mechanical properties of the material occurs (embrittlement). It was also noted that these changes occur not only under the influence of alternating but also static loads. A change (loss) in the ductility of the materials from which the most nodes and details of rolling stock are made leads to the appearance of stress concentration zones. It is in these zones that the probability of product destruction is high, therefore, the task of determining the degree of loss of ductility and the detection of these zones is very important.

In the scientific literature, the metal magnetic memory method [9] based on the Villari effect discovered in 1865 is widely represented. In Russia, this method is included in the list of basic non-destructive testing methods that are recommended for use in hazardous industrial facilities. At the same time, studies carried out using this method relate, as a rule, to boiler rooms and steam pipes, and do not yet have practical results for rail transport. As a structurally sensitive physical characteristic, the specific coefficient of thermo-EMF α of materials, of which the most critical components and parts of the rolling stock were made, was selected.

As is well known, the physical mechanism of the occurrence of thermo-EMF in metals is based on three main sources: the bulk component of the EMF, the contact component of the EMF, and the phonon drag of electrons.

The volume component of the EMF depends on the presence of a temperature gradient in the conductor. The presence of a temperature gradient causes a pressure drop of the electron gas, to balance which an appropriate electric field must arise. The intensity of this field is determined by the EMF, which is called the volume component. The contact component of the EMF depends on a change in the position of the Fermi level caused by the heating of the metal. If we compose an electric circuit of two different metals, then a contact potential difference arises on the two junctions, which is caused by the difference in the Fermi levels. It is identical in magnitude and opposite in sign if the junctions are at the same temperatures. If the junctions will be at different temperatures, the Fermi levels are shifted, and the contact component of the EMF arises.

The effect of phonon drag of electrons becomes noticeable at sufficiently low cryogenic temperatures. The theoretical concepts of thermoelectricity from the day of discovery to the present can be lined up: Seebeck theory (thermomagnetic effect); Oersted's theory (thermoelectricity); Kelvin thermodynamic theory; theory of electron gas; quantum theory of thermoelectricity; theory of quasi-thermodynamic irreversible processes.

In the scientific literature, the application of this phenomenon is represented quite widely and varied.

For example, thermo-EMF is used for the study steels with a large margin of ductility [10] and for non-destructive assessment of the aging effects [11].

Thermocouples can be used at very high temperatures to study the chemical composition, etc. [12–15].

The control over the change in the structure of welded joints and the microstructure of the Al – Mn – Cu alloy using thermo-EMF was described in [16, 17]. The effect of elastic deformation on thermo-EMF was considered in [18] using wire stretching as an example.

In [19, 20], the thermoelectric method is used in a wider range of tensile deformations without failure. In particular, it was noted that stretching to plastic deformation leads to a decrease in thermo-EMF.

In our previous work [21], the relationship between the specific coefficient of thermo-EMF (including) of steel samples α, which were subjected to bending with a frequency of 50 Hz, was studied. The number of vibrations was 2–3 million, and the amplitude of mechanical loading did not exceed 1% from the limit of elasticity.

In this work, the authors continue to study pre-defect states, aimed at establishing a relationship between structural changes in steels caused by operational factors and structurally sensitive physical characteristics.

The magnitude of loading was significantly increased until the onset of noticeable plastic deformation. In addition, two types of loading were used: a smooth and intermittent (variable) change in the applied force.

2 Materials and Methods

The general principle of all measurements presented in our work was as follows: we used two thermocouples, one of which, with a known specific coefficient of thermo-EMF, controlled the temperature difference at certain points of the sample, and in the other thermocouple, the studied sample was one of its conductors, subject to mechanical deformation.

Figure 1 shows the shape of the samples used. The test samples were made in the form of a rectangular parallelepiped of steel used for the manufacture of automatic couplings (sample No. 1), rails (sample No. 2) and steel 45 (sample No. 3).

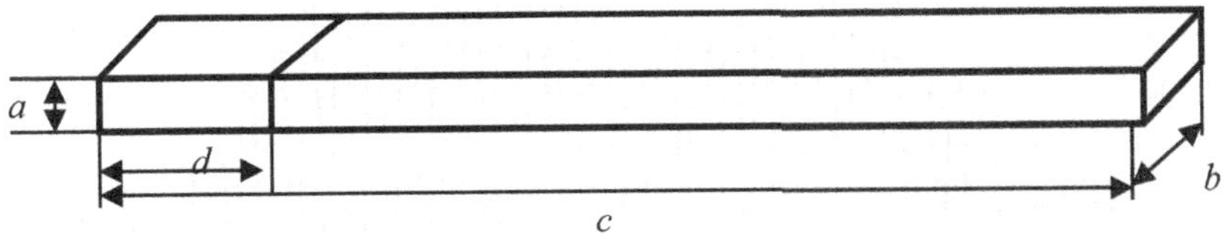

Fig. 1. Shape and designation of the sample: c - length of the sample, b - width of the sample, a - height of the sample, d - clamping area of the sample in the fixing device

The test samples at one end are rigidly fixed in the fixing device (part d), and loads of different weights were suspended from the free end. Under the influence of the weight of the load F, the free end deflected by Δ (Fig. 2).

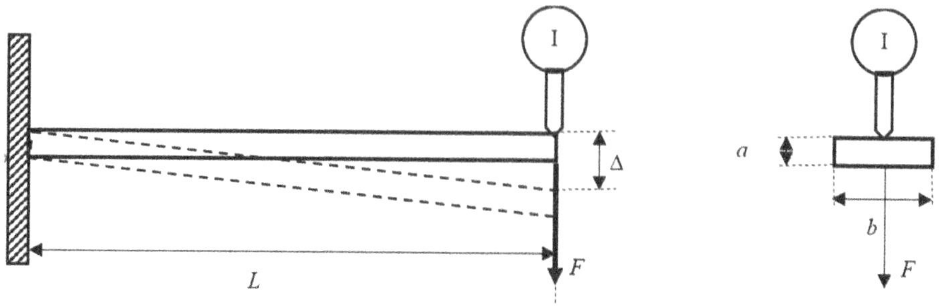

Fig. 2. Scheme of the experiment: L - the arm of the applied force F, Δ - the deflection, I - the indicating gage.

Within the limits of elastic deformation, the mechanical stress σ in the sample was calculated by the formula:

$$\sigma = \frac{6FL}{a^2 b} \tag{2}$$

where σ - the mechanical stress (Pa), F - the magnitude of the applied force (N), L - the arm of the force (m), a - the height of the sample, b - the width of the sample. The calculated value of the maximum mechanical stress was 450 MPa.

Figures 3 and 4 show a typical time dependence of the applied force to the free end of the sample in two cases.

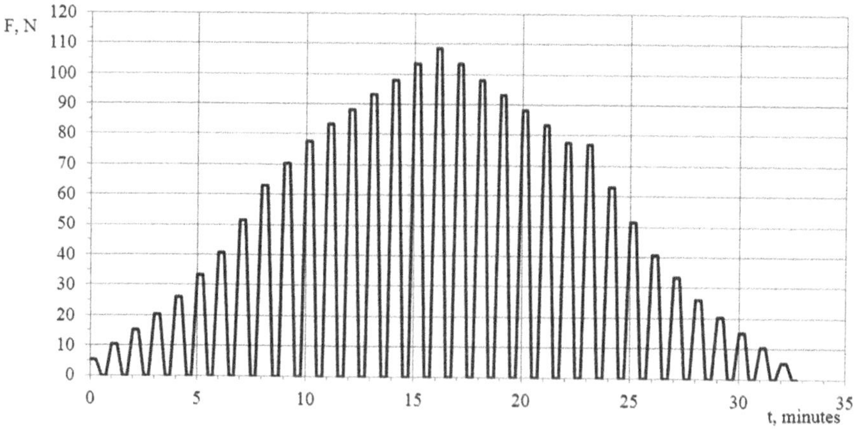

Fig. 3. Time dependence of the applied force in the first form of loading.

The choice of the type of loading was determined by the peculiarities of the operating conditions of particularly critical parts of freight cars. During operation, molded parts such as a side frame, bolsters, axle box, wheelset, coupler, etc. perceive significant static and dynamic loads (from the weight of the car and load, from shock

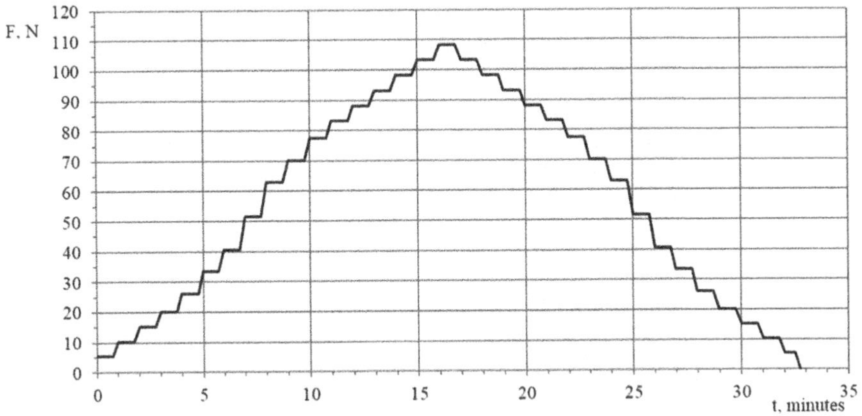

Fig. 4. Time dependence of the applied force in the second form of loading.

when passing rail joints, change in traction when moving or collisions of cars, the impact of torque when fitting into curves). It should be noted that the main part of the dynamic vertical loads is cyclical. This was the reason for the choice of the loading mode of the samples.

Figure 5 shows the diagram of stresses in the samples under elastic bending.

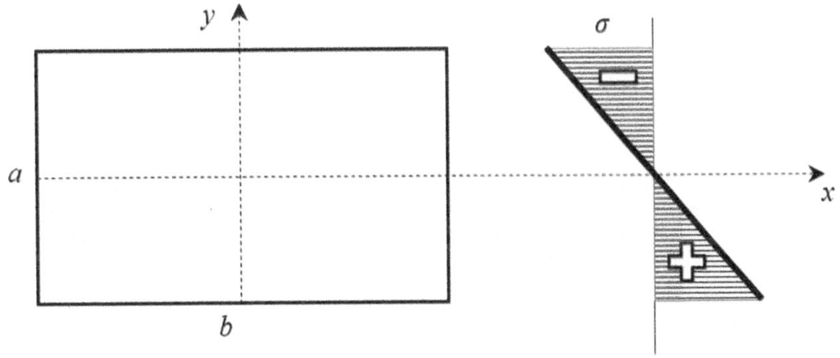

Fig. 5. Diagram of stresses in samples under elastic bending.

It can be seen from Fig. 5 that the upper part of the test sample is subject to stretching, and the lower - to compression. Moreover, in the sample, the mechanical stress is zero along the X axis.

Carrying out the experiment, the readings on the indicator were recorded twice: in the first case, according to the first form of loading, with and without load. In the second case, according to the second form of loading, the readings on the indicator were recorded once when a new load was added. Before each cycle of the experiment, zero value was set on the displacement indicator. The maximum value of the applied force was determined by the onset of plastic deformation of the fixed sample, i.e. in such a situation, when it was impossible to record a stable value of the displacement Δ. All measurements were carried out at room temperature.

Using the ARGON-5SF emission spectrometer, a spectral analysis of the samples was previously carried out in the research laboratory of OmSTU. The results of spectral analysis for one of the chemical elements are shown in Table 1.

Table 1. Results of preliminary chemical analysis of used samples.

Sample no.	Sample	Mass fraction of carbon, C (%)
1	Auto coupler	0.215
2	Rail	0.761
3	Steel 45	0.271

Figure 6 shows an electrical diagram of a laboratory setup for measuring the specific coefficient of thermo-EMF. A heating ring was attached to the free end of the test sample mounted in a massive frame acting as a cooler, to which heat from the heater was transferred through a flexible copper bus.

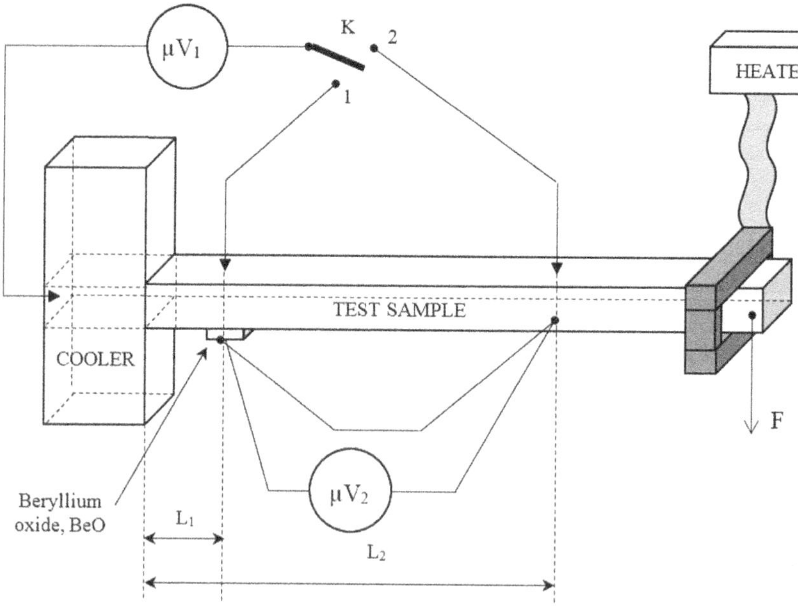

Fig. 6. Circuit diagram of the laboratory setup for measuring the specific coefficient of thermo-EMF.

Using the clamping contacts, a copper-constantan thermocouple was attached to the lower face of the test sample, the specific thermo-EMF coefficient of which was known:

$\alpha_T = 41.2$ μV/deg. To prevent a short circuit between the ends of the thermocouple through a metal sample, a Beryllium oxide (BeO) plate was placed between one of the ends and the sample. Beryllium oxide is a good dielectric at room temperature and has a fairly high thermal conductivity. The readings of the lower thermocouple were recorded using a microvoltmeter - μV_2.

One of the ends of the upper thermocouple, one of the conductors of which was the studied sample, was connected to the upper face of the test sample through silver point contacts made of the same material and through a key K at room temperature. The other end of the thermocouple consisted of a fixed end of the test sample and a conductor having reliable electrical contact with it.

The presence of two contact pads on the upper face of the sample suggested the study of the effect of voltage in the sample on the specific thermal-EMF coefficient.

Table 2. List of measuring instruments and testing equipment

n/n	Name	Type	Limit of measurement	Scale interval	Accuracy class
1	Microvolt Nano-Ammeter	F 136	5 mV	0.1 mV	1.5
2	Microvolt Nano-Ammeter	F 136	500 μV	10 μV	1.5
3	Microvolt Nano-Ammeter	F 136	250 μV	10 μV	1.5
4	Microvolt Nano-Ammeter	F 136	100 μV	2 μV	1.5
5	Microvolt Nano-Ammeter	F 136	50 μV	1 μV	1.5
6	Microvolt Nano-Ammeter	F 136	10 μV	0.2 μV	1.5
7	Ammeter	M 109	10 A	0.1 A	0.5
8	Ammeter	M 109	1 A	0.01 A	0.5
9	Micrometer	–	5 cm	0.01 mm	–
10	Dial gauge	–	1 cm	0.01 mm	–

The readings of the upper thermocouple were recorded using another microvoltmeter—μV_1. Switch K made it possible to connect the upper thermocouple to the test sample at two points: in the first case, the hot contact of the upper thermocouple was located on the upper face of the test sample at a distance L_1, and in the second case, at a distance L_2 from the cooler (see Fig. 6).

The sample was heated until the temperature difference between the contacts of the lower thermocouple reached 12–14 °C, and the readings of the μV_2 microvoltmeter became stable. Thus, the temperature gradient in the studied sample was a priori considered linear. Knowing the readings of a μV_2 microvoltmeter, the specific coefficient of thermo-EMF of a copper-constantan thermocouple α_T taking into account the linearity of the temperature gradient, the temperature difference was determined between the cooler and the upper contact located at a distance of L_1 or L_2. Based on this temperature difference and the readings of the microvoltmeter μV_1, the specific coefficient of thermo-EMF of the test sample α_1 or α_2 was determined:

$$\alpha_1 = \frac{\alpha_T(L_2 - L_1)}{L_1} \cdot \frac{U_1}{\varepsilon_T} \tag{3}$$

$$\alpha_2 = \frac{\alpha_T(L_2 - L_1)}{L_2} \cdot \frac{U_2}{\varepsilon_T} \tag{4}$$

where α_T - the specific coefficient of thermo-EMF of the copper-constantan thermocouple, U_1 and U_2 - the readings of the microvoltmeter μV_1 at two positions of the switch K, ε_T - the EMF of the lower thermocouple, determined by the readings of the microvoltmeter μV_2, L_1 and L_2 - the parameters of the laboratory setup.

Thus, the laboratory setup made it possible to determine the specific coefficient of thermo-EMF of the material from which the test sample was made in two cases: α_1-according to the readings of the microvoltmeter μV_1 connected to the sample through pin 1 in switch K, and α_2 - according to the readings of the microvoltmeter μV_1 connected to the sample through pin 2 in switch K.

The list of measuring instruments and certified test equipment is given in Table 2.

The relative measurement error of α_1 and α_2 did not exceed 17% and 5%, respectively.

3 Results

In total, more than one hundred and fifty cycles of measurements were carried out and, in the aggregate of all experiments, the following conclusions were made.

1. For all samples No. 1, No. 2, and No. 3 under loading in the first form (see Fig. 3), the mismatch of the dependence of the deflection magnitude of the sample on the applied force (stress) was characteristic during an increase and decrease in load. Figure 7 shows the dependence of the magnitude of the deflection in the forward and reverse directions for sample No. 1. Under small loads (see Fig. 7a), the dependences of the deflection in the forward and reverse directions almost coincide, which corresponds to a linear dependence or feasibility of Hooke's law. As the load increases and the transition to the region of plastic deformation occurs, the mismatch of the dependence of the deflection of the sample on the applied force in the forward and reverse directions increases (see Fig. 7b, c, and d).

Of particular interest is the appearance of residual deformation $\Delta_{residue}$ in the same sample, shown in Fig. 8. The sample was not removed from the setup, the indicator readings were recorded relative to the initial position. It can be seen from Fig. 8a that with linear deformation of the sample, in the zone of Hooke's law fulfillment, residual deformation was practically not observed. As the applied force increases (stress increases), the value of the residual deformation becomes significant (see Fig. 8b, c, and d).

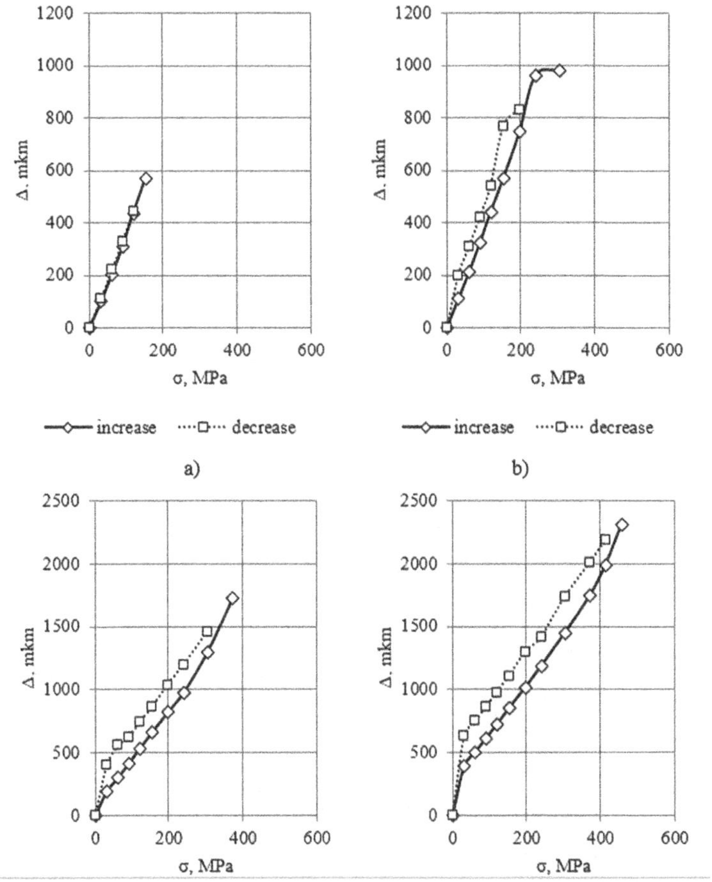

Fig. 7. The dependence of the magnitude of the deflection Δ in the forward and reverse directions for sample No. 1.

Similar values of the residual deformation were characteristic of all samples except sample No. 2.

Analyzing the graphs shown in Fig. 8, it is clearly seen that the change in the residual deformation $\Delta_{residue}$ increases with increasing mechanical stress. So, for example, from Fig. 8b it can be seen that the change in the permanent deformation begins at about 160 MPa. From Fig. 8c and d it can be seen that the same values are already approximately 240 and 320 MPa, respectively.

Figure 9 shows the dependence of the deflection Δ and the residual deformation $\Delta_{residue}$ for sample No. 2.

If for sample No. 1 the maximum value of residual deformation was approximately 500 μm (see Fig. 8d), then for sample No. 2 this value was approximately 9 μm (see Fig. 9b). This difference is explained by the higher hardness of rail steel due to the high carbon fraction in this sample: 0.761% versus 0.215% (see Table 1). The mass fraction of carbon in steel is nonlinearly related to the deflection Δ. For example, an increase in

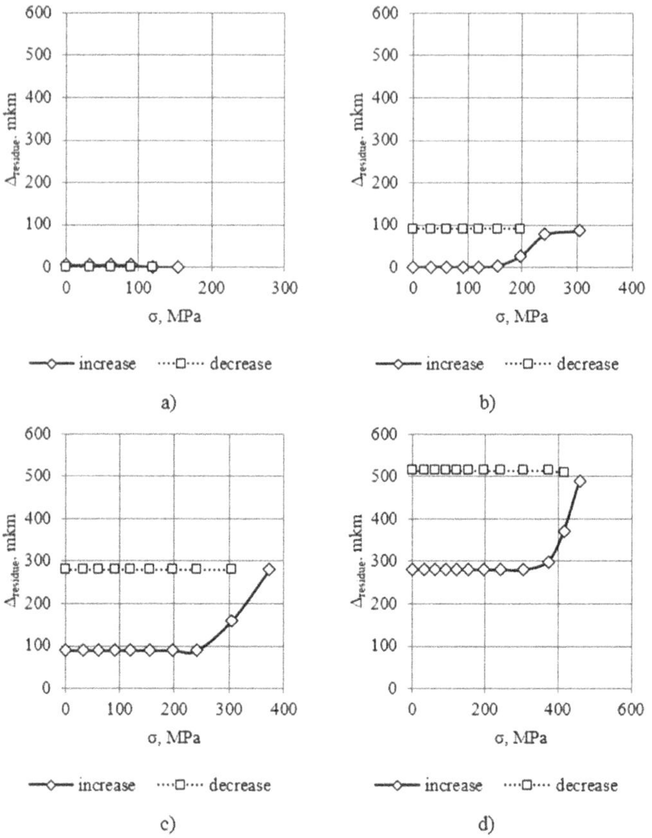

Fig. 8. The value and dynamics of changes in residual deformation $\Delta_{residue}$ for sample No. 1.

the percentage of carbon by about three times leads to the fact that the deflection Δ decreases by about ten times: at a stress of 400 MPa, the deflection for sample No. 1 is approximately 2000 μm (see Fig. 7d), and for sample No. 2, at the same stress, 200 μm (see Fig. 9a).

2. Figure 10 shows the average values of the specific coefficient of thermo-EMF α of the samples used in the experiment. It is noteworthy that the highest value α = 17.4 μV/K corresponds to sample No. 1, which was previously subjected to substantial elastoplastic deformation. It should be noted that according to the data given in [21], the value of the specific coefficient of thermo-EMF for this sample was approximately 5 μV/K. A significant change in the specific coefficient of thermo-EMF (more than three times) occurred after the sample was subjected to significant elastoplastic deformation.

In sample No. 2, in which there is almost no residual deformation, the smallest value of the specific coefficient of thermo-EMF is α = 10.6 μV/K. The average value of α was calculated as the arithmetic mean between α_1 and α_2 calculated by formulas (3) and (4).

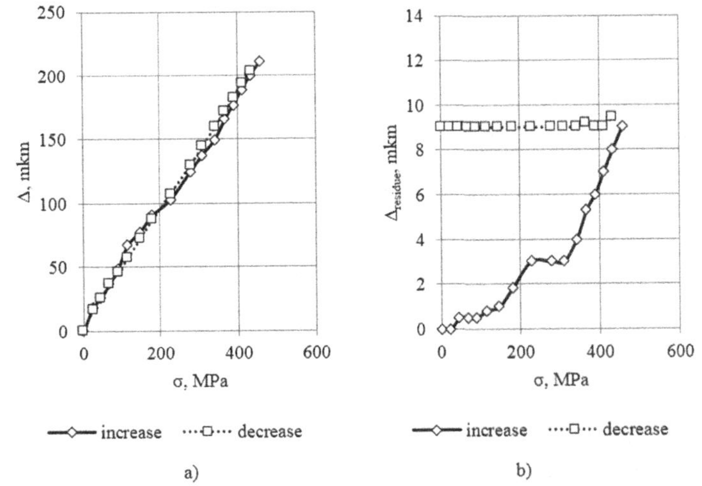

Fig. 9. The dependence of the magnitude of the deflection Δ and residual deformation $\Delta_{residue}$ for sample No. 2.

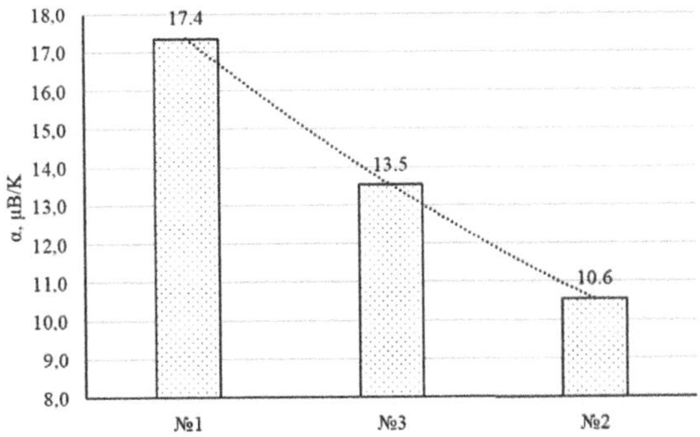

Fig. 10. Average values and dynamics of changes in the specific coefficient of thermo-EMF of various samples.

3. A change in the nature of the load (see Figs. 3 and 4) affected the scatter of the values of α1 and α2 with increasing and decreasing loading for almost all samples. Under loading in the first form, the largest experimental scatter of the values of α1 and α2 is characteristic. Figure 11 shows a typical example of a change in α1 and α2 for sample No. 3. Trends in changes of α1 and α2 are the same. The insignificant difference between α1 and α2 is connected, from our point of view, with the fact that the temperature boundary in the sample at the clamping point (relative to which the distance to the contacts was calculated) is not an ideal plane.

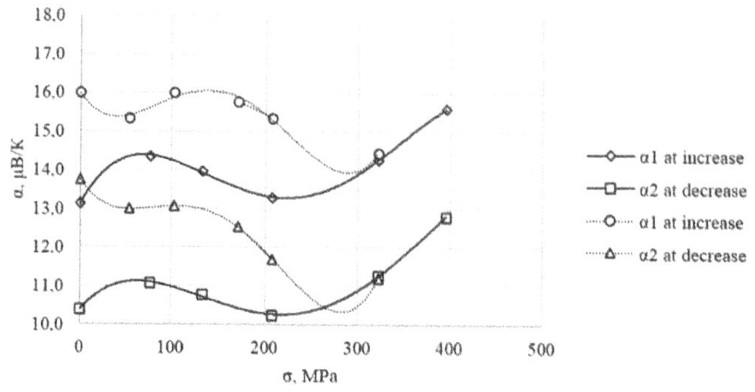

Fig. 11. Sample No. 3. Loading in the first form. The nature of the changes in α_1 and α_2.

Under loading in the second form (see Fig. 4), the scatter became much smaller for all samples. Figure 12 shows a typical example of a change in α_1 and α_2 for sample No. 2. This sample is characterized by the most significant alignment of α_1 and α_2 with a change in the form of loading.

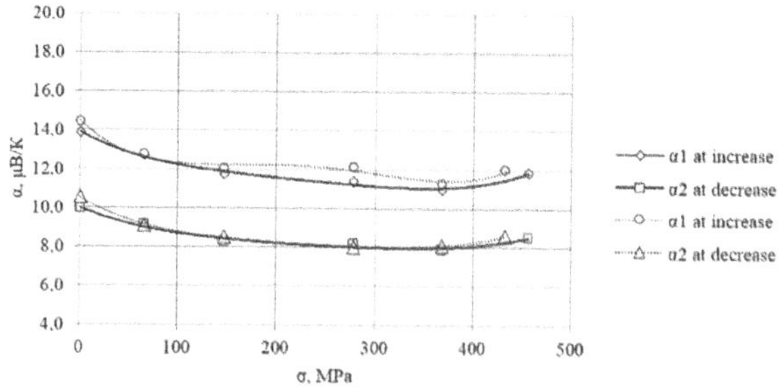

Fig. 12. Sample No. 2. Loading in the second form. The nature of the changes in α_1 and α_2.

4 Discussion

Based on the foregoing, we can draw the following conclusions:

1. The residual deformation and residual stresses increase the specific coefficient of thermo-EMF of the operated parts.
2. The dynamics of changes in the specific coefficient of thermo-EMF of parts subject to varying loads and parts subject to static loads is different. Under the conditions of the experiment, during deformation of the samples, tension appears above and compression below, which can lead to two different mechanisms of influence on the thermo-EMF of the sample as a whole.
3. The loss of elastic properties of nodes and parts of the rolling stock is associated with a violation of the operating mode. Therefore, it is possible to control the change in thermo-EMF to determine the moment of loss of the elastic properties of

the part without involving additional mechanical loads, which are currently used to monitor the state of the side frames of cars.

4. As a result of the studies, it was found that structural changes are associated with a change in the elastic properties of the studied materials. Consequently, there is a fundamental possibility of establishing a connection between the operation mode of parts and their residual life.
5. For the practical use of the results, there is a wide possibility of using certified precision instruments, which can significantly increase the reliability of determining the residual life of the studied objects.

5 Conclusions

The authors express their sincere gratitude to the associate professor of the Department of Technology of Transport Engineering and Rolling Stock Repair, OMSU, PhD in Engineering, Dmitry Valerievich Muravyov, for the help in conducting a chemical analysis of the materials under study.

References

1. Petrakova, A.G.: Increasing the service life of solid-rolled wheels of freight cars by choosing a rational interval of their hardness. Thesis of PhD in Engineering [Text], Omsk State Transport University, Omsk, p. 18 (2008)
2. Hart, J.M., Ahuja, N., Barkan, C.P.L., Davis, D.D.: A Machine Vision System for Monitoring Railcar Health: Preliminary Results. Technology Digest: TD-04-008. Association of American Railroads, Pueblo, Colorado (2004)
3. Edwards, J.R., Hart, J.M., Todorovic, S., Barkan, C.P.L., Ahuja, N., Chua, Z., Kocher, N., Zeman, J.: Development of Machine Vision Technology for Railcar Safety Appliance Inspection. In: Proceedings of the International Heavy Haul Conference Specialist Technical Session - High Tech in Heavy Haul, Kiruna, Sweden, pp. 745–752 (2007)
4. den Buurman, G.: Measuring Quo vadis system. Eur. Railway Rev. **3**, 80–86 (2005)
5. Zuev, V.V., Potselueva, L.N., Goncharov, Yu. D.: Crystalloenergetics as a basis for evaluating the properties of solid materials (including magnesia cements). Saint-Petersburg, Publishing house of the "Alfapol" plant, p. 139 (2006)
6. Matyash, Y.I., Sosnovsky, Yu.M., Razhkovsky, A.A., Kondrikov, E.M.: Change of physical and chemical properties of steel units and details of railway transport as a method for determining the residual resource. Omsk Sci. Herald Omsk **4**(154), 9–13 (2017)
7. Levkovich, T.I.: Prediction of the probability of dangerous brittle fractures of coupling casings of cars. Thesis of PhD in Engineering. Bryansk, p. 22 (2010)
8. Zakirnichnaya, M.M., Kuzeev, I.R., Berdin, V.K., Kirillova, N.Yu.: Embrittlement of steel 20 in the process of long-term operation. Oil Gas Bus. Mach. Devices **4**(1), 207–214 (2006)
9. Dubov, A.A.: The method of magnetic memory of metals. TechNadzor **9**(130), 44–46 (2017)
10. Massardier, V., Merlin, J., Le Patezour, E., Soler, M.: Mn-C interaction in Fe-C-Mn steels: study by thermoelectric power and internal friction. Metall. Mater. Trans. A **36**(7), 1745–1755 (2005)

11. Noemi, O.L., Ruiz, R., Rubio, C., Ambriz, R.R.: Nondestructive assessing of the aging effects in 2205 duplex stainless steel using thermoelectric power. NDT&E Int. **44**(5), 463–468 (2011)
12. Ito, T., Yotsuji, J., Nagamune, A.: Development of pig iron and molten slag level measurement technique for blast furnace. ISU Int. **54**(11), 2618–2622 (2014)
13. Lavaire, N., Massardier, V., Merlin, J.: Quantitative evaluation of the interstitial content (C and/or N) in solid solution in extra-mild steels by thermoelectric power measurements. Scripta Mater. **50**, 131–135 (2004)
14. Abdala, M.R.W.S., Garcia de Blas, J.C., Barbosa, C., Acselrad, O.: Thermoelectrical power analysis of precipitation in 6013 aluminum alloy. Mater. Charact. **59**, 271–277 (2008)
15. Rana, R., Massardier, V., Singh, S.B., Mohandy, O.N.: Effect of the pre-treatment on copper-alloyed interstitial free steel studied by thermoelectric power measurement. Metall. Mater. Trans. A **44**(6), 186–200 (2013)
16. Steklova, E.O., Soloviev, S.D., Kim, S.L.: Application of the thermoelectric power method in the study of the chemical composition and structural state of welded joints. Welding Diagn. **2**, 10–13 (2011)
17. Udalaya, K.R., Belenky, A.M., Alabin, A.N.: Monitoring the change in the microstructure and copper content in Al-Mn-Cu system alloys by thermopower. Factory Lab. **81**(3), 26–31 (2015)
18. Kochkin, Y.P., Solntsev, A.Y.: The nature of the changes in the thermopower with a small elastic deformation of carbon steel. Metal Form. **1**(18), 54–56 (2016)
19. Soldatov, A.A.: Control of the plastic deformation by thermo-electric method. J. Int. Sci. Publ. Mater. Methods Tehnol. **5**(3), 148–155 (2011)
20. Soldatov, A.A., Seleznev, A.I., Fiks, I.I., Soldatov, A.I., Kröning, Kh.M.: Nondestructive proximate testing of plastic deformations by differential thermal EMF measurements. Russ. J. Nondestr. Testing **48**(3), 184–186 (2012)
21. Yury, M., Yury, S., Evgeniy, K., Dmitriy, M.: Study of the structurally sensitive physical characteristics of materials used to fabricate highly loaded elements of rolling stock. In: Issue MATEC Web Conference, 2018 Siberian Transport Forum - TransSiberia 2018, Section Mechanical and Energy Transport Systems, vol. 239, p. 10, article number 01027 (2018) https://doi.org/10.1051/matecconf/201823901027

Assessment of Energy Efficiency of Race Cars of Shell Eco-Marathon Competitions

Sotskov Andrey Vladimirovich(✉), Khaziev Anvar Askhatovich, and Izmaylova Diana Ansarovna

Moscow Automobile and Road Construction State Technical University, 125319 Leningradsky Prospect 64, Moscow, Russia
sotskov-andrey@mail.ru

Abstract. The paper provides an assessment of the results of research conducted by the team of MADI on the results of international competitions for fuel efficiency - Shell Eco-Marathon. They take place on sections of city streets in Rotterdam, the Netherlands, and in London, England. These competitions involve vehicles with power plants operating on both traditional energy sources, such as gasoline and diesel fuel, and alternative ones - hydrogen, solar energy, biofuels, and electricity. The race car developed by the students of MADI uses petrol with an octane rating of 95 as fuel. In their studies, they assessed the energy efficiency of the car depending on its mass, aerodynamic properties, rolling resistance, tire pressure, road conditions, driving conditions, etc. The dependences of the change in the fuel consumption of the car on the mass, overall dimensions, tire pressure, as well as the effect of tire pressure on the vehicle's overrun were obtained. Data for determination of the optimal driving modes on the track in London, England was obtained. For further study of influence degree of the power plant systems, chassis, and also maintenance supplies, additional studies are needed.

Keywords: Shell Eco-Marathon · Fuel consumption · Mass · Tire pressure · Competition · Aerodynamics · Overrun · Vehicle · Race car · Driving modes

1 Introduction

Nowadays, considerable attention is paid to improving energy efficiency and reducing the harmful effects of vehicles on the environment [1]. To this end, Shell, one of the leaders in the oil and gas industry, organized an international student competition - Shell Eco-Marathon, participants of which compete to achieve maximum fuel efficiency in the vehicles developed by them (Fig. 1) [2]. More than 200 teams from 26 countries take part in it. Shell Eco-Marathon competitions have been held for more than 30 years [3].

As energy sources, both traditional gasoline and diesel fuel are used as well as alternative ones - hydrogen, solar energy, biofuel, electric power [3].

The students who created their race cars compete on the tracks, which are sections of city roads blocked for other vehicles (Fig. 2) [4].

Z. Popovic et al. (Eds.): TransSiberia 2019, AISC 1115, pp. 519–526, 2020.
https://doi.org/10.1007/978-3-030-37916-2_50

Fig. 1. Race car for Shell Eco-Marathon.

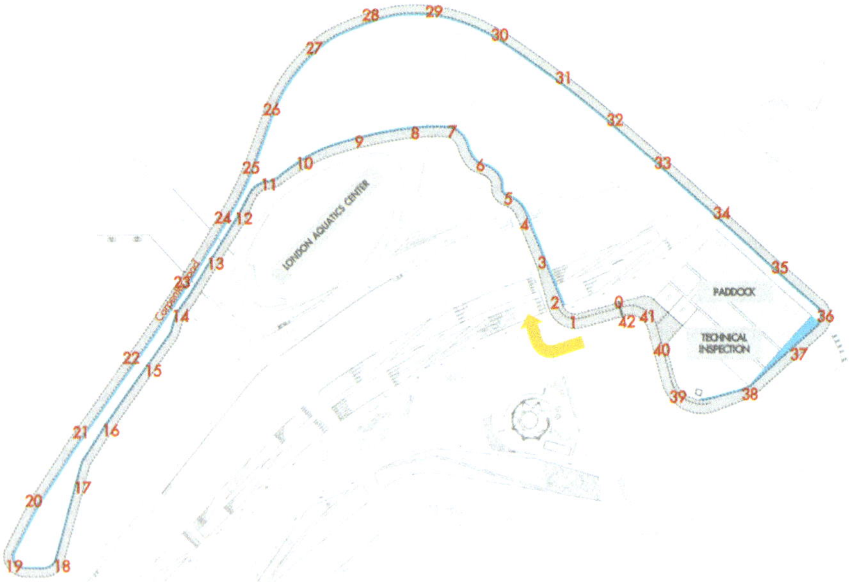

Fig. 2. Racing track in London, England.

To improve fuel efficiency, students optimize the performance of all systems of their cars.

2 Materials and Methods

During the development of the vehicle, the engineers of MADI team had a task to create a car with a low aerodynamic drag coefficient, which, during the calculation, turned out to be equal to 0.12 (Fig. 3). The team made studies of changes in fuel consumption depending on the geometric dimensions of the car and its shape. For example, an increase in the vehicle's width by 10 cm increases fuel consumption by 4.2%.

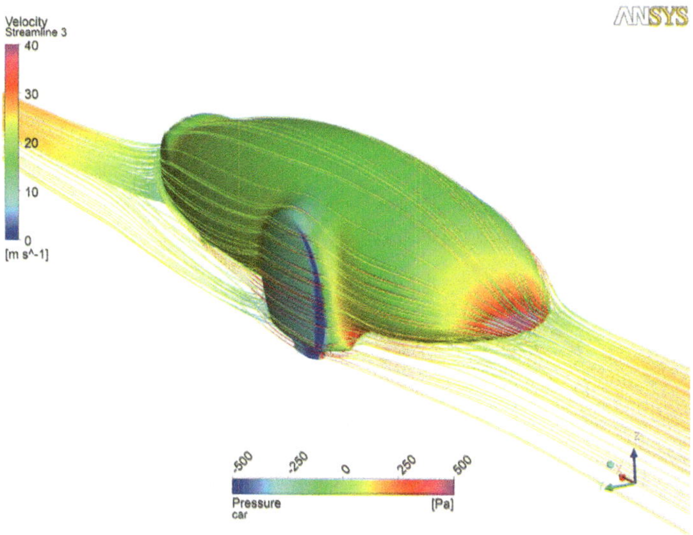

Fig. 3. Graphic display of the aerodynamic drag of the MADI car.

The graphical dependence of an increase in fuel consumption depending on the aerodynamic drag coefficient Cx, obtained for the race car of the MADI team during the study, is presented in Fig. 4.

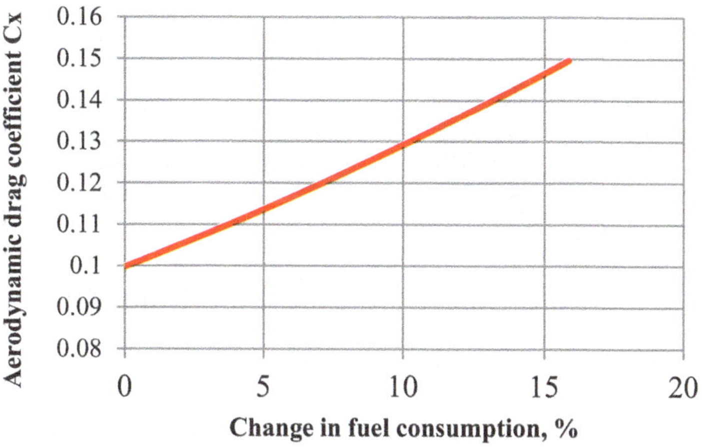

Fig. 4. The dependence of fuel consumption on the aerodynamic drag coefficient for the car of the MADI team.

The designed car has a mass of 40 kg, while an increase in its mass leads to a nearly linear increase in fuel consumption (Fig. 5).

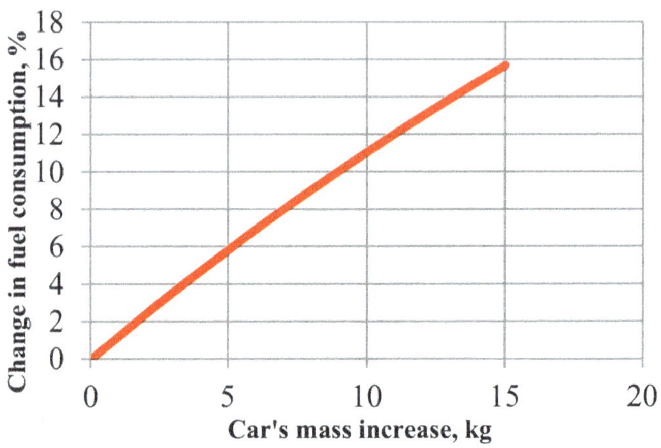

Fig. 5. The dependence of fuel consumption on the car's mass.

The lightest car in competition weighs 23.5 kg, while the team takes 7th place in its category, and the winning team has a vehicle weighting 27 kg (Fig. 6) [5].

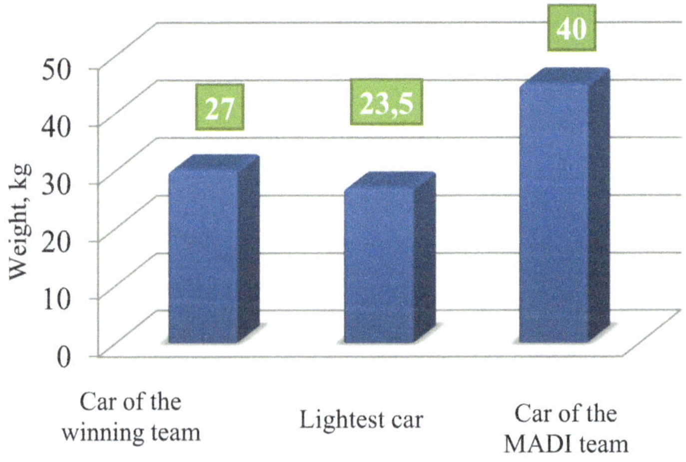

Fig. 6. The weights of cars participating in Shell Eco-Marathon.

Specially for Shell Eco-Marathon, Michelin developed 45/75R16 tires with a low rolling resistance coefficient (Fig. 7) [6]. These tires have a maximum speed limit of 70 km/h, a maximum load limit of 100 kg, and a maximum pressure limit of 0.7 MPa.

During the test runs on the track of Shell Eco-Marathon, the team of MADI conducted studies on the degree of influence of tire pressure on fuel consumption. For this, races with tire pressure of 0.35 MPa and 0.75 MPa were carried out, while other parameters did not change. The result is shown in Fig. 8. Similarly, the dependence of the car overrun from a speed of 40 km/h was obtained depending on the tire pressure (Fig. 9).

Fig. 7. 45/75R16 Michelin tire for Shell Eco-Marathon.

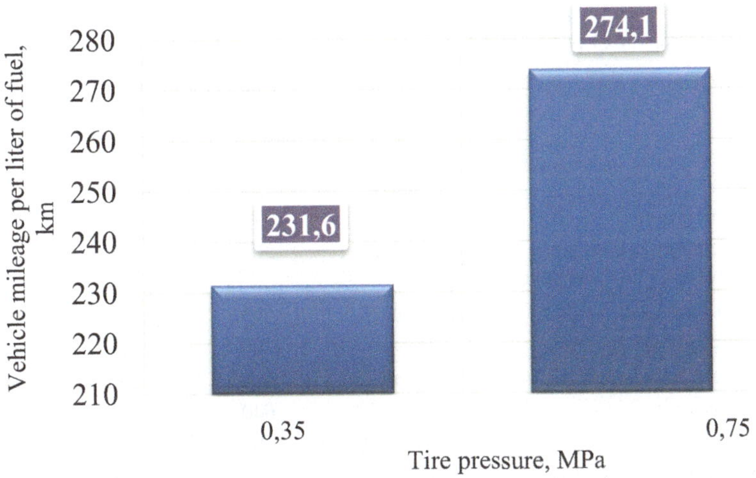

Fig. 8. The mileage of the MADI team car on 1 l of fuel on the track of Shell Eco-Marathon, depending on tire pressure.

After changing the track in 2016, the driving conditions changed: the number of turns increased by 44%, the elevation difference increased from 3 to 8 m. This caused a significant deterioration in fuel economy for most of the participants (Table 1).

To obtain data on driving modes, a device was installed on the car for recording data with the possibility of recording coordinates using the GPS Starlane Athon XS signal.

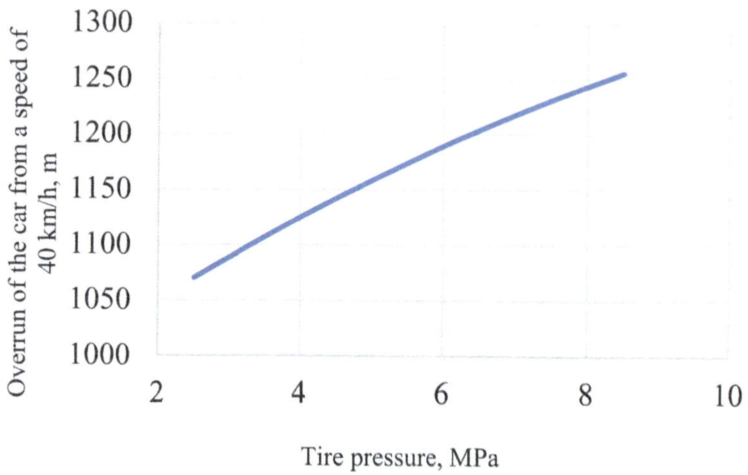

Fig. 9. Influence of tire pressure on the car overrun from a speed of 40 km/h.

Table 1. Comparison of the results of Shell Eco-Marathon on the tracks in Rotterdam and London in the category of petrol prototypes

No	Team name	Country	Results on the tracks, km		Change, %
			Rotterdam	London	
1	TED	France	2308	2300	−0.36
2	Remmi-Team	Finland	1972	1685	−16.99
3	Association Marathon Shell IUT Aix en Provence	France	1509	1709	11.68
4	Proto Insa Club	France	1165	1020	−14.26
5	Team ENSTA Bretagne	France	760	858	11.50
6	Hanseatic Racing Organisation	Germany	664	750	11.44
7	Iron Warriors	Poland	640	471	−36.05
8	LTAM	Luxemburg	593	479	−23.64
9	High-Efficiency-Karlsruhe	Germany	564	466	−17.4
10	Zenith	France	412	282	−46.31
11	SjF TUKE	Slovakia	354	406	12.68
12	ESAF-Mec	Portugal	353	394	10.30
13	ENIM LEADERS	Morocco	240	92	−161.3
14	SCB-MADI	Russia	231	193	−19.57

3 Results

As a result, the characteristics of the car's motion on the track were obtained: speed and time of driving along the track, longitudinal and transverse accelerations, engine crankshaft speed, etc. It was established that the car moved 42% of the time with the engine running (Fig. 10). Since the most economical is the mode of acceleration to a given speed and subsequent free running with the engine turned off, the question arises about the optimal moment of starting and shutting down the engine. It is significantly affected by gradients of slope on descents and ascents, as well as the number of turns and their characteristics. At competitions in Rotterdam and London, the moment of turning on the engine was chosen by the pilot subjectively.

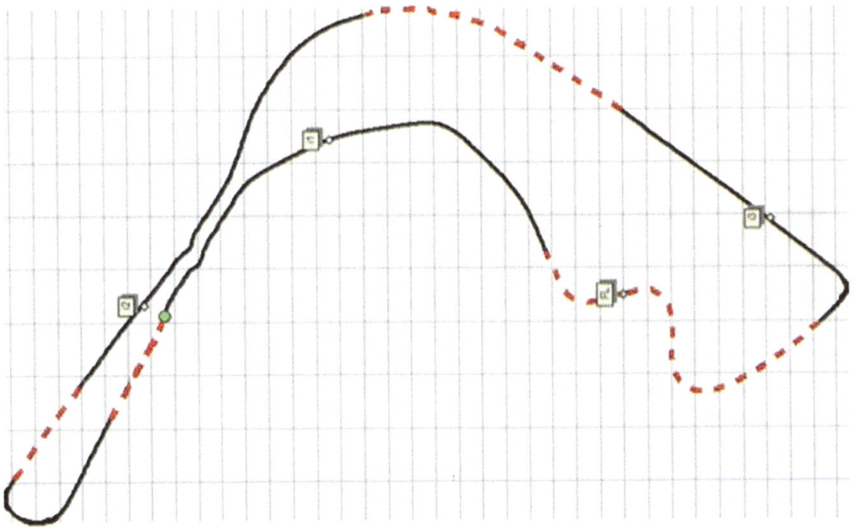

Fig. 10. Graphic display of data on engine starting, superimposed on the coordinates of the track in London (dashed line - engine running, solid – free running).

The studies conducted by MADI during the test races of the car allowed establishing the dependence of energy efficiency of the vehicle on:

- aerodynamic characteristics;
- mass characteristics;
- total rolling resistance of wheels;
- driving modes;
- driving conditions.

4 Conclusions

To analyze the degree of influence of various parameters and systems of the power plant, chassis, and maintenance supplies on energy efficiency of Shell Eco-Marathon race cars, additional studies are needed, which is the subject of further research.

References

1. Lackner, M.: Energy Efficiency: Comparison of Different Systems and Technologies. Handbook of climate change mitigation and adaptation 2nd (edn.) (2016)
2. Sotskov, A.V., Chekanov, A.Yu, Guly, V.V.: International student projects with participation of student teams. Gruzovik (5), 20–22 (2013)
3. The official web site of Shell Eco-Marathon international competition. https://www.shell.com/energy-and-innovation/shell-ecomarathon/about.html
4. Sotskov, A.V.: On the driving modes on the tracks of Shell Eco-Marathon international competitions
5. Izmaylova, D.A.: Problems of technical operation and car service of the car rolling stock. In: Collection of scientific works of the Department of Road transport operation and car service, dedicated to the 90th anniversary of the Honored Worker of Science and Technology, D. Science in Engineering, Professor Evgeny Semenovich Kuznetsov, based on the proceedings of the 77th Scientific, Methodological and Research Conference of MADI (2019)
6. Lyndon, B., McLening, Ch., Burgess, J.: ECO-CAR: a perfect vehicle for technical design teaching. In: DS 78: Proceedings of the 16th International conference on Engineering and Product Design Education (E&PDE14), Design Education and Human Technology Relations, pp. 030–035. University of Twente, The Netherlands, 04–05 September 2014
7. Official Michelin informational internet resource. https://thetiredigest.michelin.com/michelin-ultimate-energy-tire

Selection of the Best Location for RFID Wagon Monitoring Device on Serbian Railways Based on FUCOM-TOPSIS Method and Fuzzy Set Theory

Milan Milosavljević(✉), Dušan Jeremić, and Sanjin Milinković

University of Belgrade, Vojvode Stepe, 305, 11000 Belgrade, Serbia
milan.milosavljevic@vzs.edu.rs

Abstract. The railway is one of the key factors in the sustainable development of modern society. To improve its competitiveness against other modes of transport, it is necessary to constantly improve it and use innovative technical solutions and technologies. We consider the use of specific information technology, namely Radio Frequency Identification (RFID) technology. The RFID technology is used in various systems, but in this paper, we use the RFID device for monitoring the freight wagons. The RFID systems provide automated tracking of rail vehicles via RFID tags, readers and integrated middleware, and thus enable wagons and locomotive identification and location information available for further analysis and tracking. We present the two-phase model for optimizing the RFID locations on Serbian railways for monitoring and collecting the data from freight trains. In the first phase we use the FUCOM-TOPSIS model for determining the macro location on the railway network, and in the second phase we use the fuzzy logic model for determining the micro-location of the RFID device on the rail line.

Keywords: RFID · Railway · Location problem · FUCOM-TOPSIS · Fuzzy logic

1 Introduction

Bad state of infrastructure and outdated work technologies have led to trains having incredibly long delays, which can sometimes be described in days. In addition, due to the lack of planned activities for the use of wagon resources, a large number of wagons remain in intermediate stations for a very long time or have large movement of empty wagons, which is a great loss. This loss is reflected in a decrease in the efficiency and reliability of the system, which further leads to a reduction in the number of users. If we also take into account the fact that the majority of the wagons that operate on the Serbian railways are owned by other railway administrations (leased and private wagons), the loss is also reflected in the cost of unnecessary wagon layover.

Therefore, the use of technologies for collecting data about wagon location is needed to reduce the above losses and increase the level of competitiveness through

Z. Popovic et al. (Eds.): TransSiberia 2019, AISC 1115, pp. 527–539, 2020.
https://doi.org/10.1007/978-3-030-37916-2_51

greater automation of work in marshaling yards, at border crossings, and in intermediate stations. The advantage of RFID technology, as data collection technology, is its ability to capture data in a very simple way, without requiring traffic interruption or the construction of special infrastructure facilities. There is a lot of papers based on the implementation of RFID technology on the railway system [1]. Unfortunately, there isn't a practical implementation of this technology on Serbian railways. Only one case that shows the practical usage of RFID in the railway system is the pilot project presented in [2].

Many authors deal with location problems [3] and railway yard layout designing using different approaches and techniques [4, 5]. There is a large number of location problems are solved using multi-criteria decision-making methods. Unlike conventional methods and techniques of operational research, these methods do not provide "objectively the best" solution. These methods are based on mathematical algorithms that are developed to help decision-makers in choosing the most suitable variant. Methods for multi-criteria decision-making models are basically divided into methods for determining the weights of coefficients such as presented in [6], and methods for determining the best possible solution in a complex multi-criteria environment [7, 8]. More recently, multi-criteria decision-making techniques are combined with other techniques like fuzzy set theory, grey numbers or rough set theory [9, 10]. An example of usage of this hybrid method like fuzzy-PROMETHEE method for wayside train monitoring at Serbian railways is presented in [11], or selecting the location of the hospital is presented in [12].

On the other hand, fuzzy logic is not a very common technique for determining location problems. A usage of fuzzy set theory in solving the location problems is presented in [13].

The paper is organized as follows. After the introductions in Sect. 1, a brief overview of the FUCOM and TOPSIS method and the theory of fuzzy sets is given in Sect. 2. Section 3 presents the problem formulation and finding the best location of the RFID device on the macro and micro levels. The final Sect. 4 presents conclusions.

2 Methods

The problem is observed from two aspects, so it is necessary to create two models. This chapter provides a brief overview of used methods, two multi-criteria decision-making methods, the FUCOM method for determining weight coefficients of criteria and the TOPSIS method which gives us the final result of the first model. Also in this chapter, the theory of fuzzy sets is presented, which shows where to locate an RFID device.

2.1 FUCOM Method

Full Consistency Method is a method for determining weight coefficients of criteria [14]. FUCOM provides the ability to validate the model by calculating the error value for the obtained weight vectors by determining the DFC. Mathematical computation of this method is presented through steps as following:

Step 1. Criteria ranking based on a predefined set of evaluation criteria.

The ranking is performed according to the significance of the criteria, it starts from the criterion that is expected to have the greatest impact, i.e. the highest value of the weight coefficient up to the criteria with the least significance. If there is two or more criterion with the same significance than the "equal" sign is placed instead ">" between these criteria.

Step 2. In the second step, a comparison of the ranking criteria is carried out and the comparative priority of evaluation criteria is determined $\varphi_{k/(k+1)}$ $k = 1, 2, \ldots n$, where k represents the rank of the criteria). The comparative priority of the evaluation criteria $\varphi_{k/(k+1)}$ is an advantage of the criterion of the $C_{j(k)}$ rank compared to the criterion of the $C_{j(k+1)}$ rank. Thus, the vectors of the comparative priorities of the evaluation criteria are obtained, as in the expression (1):

$$\Phi = \left(\varphi_{1/2}, \varphi_{2/3}, \ldots, \varphi_{k/(k+1)}\right) \tag{1}$$

where $\varphi_{k/(k+1)}$ represents the significance (priority) that the criterion of the $C_{j(k)}$ rank has compared to the criterion of the $C_{j(k+1)}$ rank.

Step 3. In the third step, the final values of the weight coefficients of the evaluation criteria $(w_1, w_2, \ldots, w_n)^T$ are calculated. The final values of the weight coefficients should satisfy the two conditions:

that the ratio of the weight coefficients is equal to the comparative priority among the observed criteria $(\varphi_{k/(k+1)})$ defined in Step 2, i.e. that the following condition is met:

$$\frac{\omega_k}{\omega_{k+1}} = \varphi_{\frac{k}{k+1}} \tag{2}$$

In addition to the condition (2), the final values of the weight coefficients should satisfy the condition of mathematical transitivity. Thus, yet another condition that the final values of the weight coefficients of the evaluation criteria need to meet is obtained, namely:

$$\frac{\omega_k}{\omega_{k+2}} = \varphi_{k/(k+1)} \otimes \varphi_{(k+1)/(k+2)} \tag{3}$$

Based on the defined settings, the final model for determining the final values of the weight coefficients of the evaluation criteria can be defined.

$$\begin{gathered} \mathrm{Min}\chi \\ \left|\frac{\omega_k}{\omega_{k+1}} - \varphi_{k/(k+1)}\right| \leq \chi \\ \left|\frac{\omega_k}{\omega_{k+2}} - \varphi_{k/(k+1)} \otimes \varphi_{(k+1)/(k+2)}\right| \leq \chi \qquad \sum_{j=1}^{n} \omega_j = 1; \omega_j \geq 0, \forall j \end{gathered} \tag{4}$$

By solving model (4), the final values of the evaluation criteria $(w_1, w_2, \ldots, w_n)^T$ and degree of DFC (χ) are generated.

2.2 TOPSIS Method

The TOPSIS method is a method which compares alternatives based on their distance from positive and negative ideal solution. Characteristic for this method is calculation the weighted normalized decision matrix and formulation the positive and negative ideal solution. Also, this method is based on the concept that the chosen alternative should have the shortest distance from the positive ideal solution and the longest distance from the negative ideal solution. The theoretical background of this method was presented in [15].

Mathematical computation of this method are well known and it is presented through several steps as following: Step 1. Creating the initial decision matrix X. Step 2. Normalization of the elements of the initial decision matrix X. Step 3. Calculation of the weighted matrix elements V. Step 4. Creation a positive ideal and negative ideal solution. Step 5. Calculate the distance (Euclid distance) of each alternative from a positive ideal and negative ideal solution.

$$S_i^+ = \sqrt{\sum_{j=1}^{n} \left(v_{ij} - v_j^+\right)^2} \tag{5}$$

$$S_i^- = \sqrt{\sum_{j=1}^{n} \left(v_{ij} - v_j^-\right)^2} \tag{6}$$

where: v_{ij} is weighted performance of ith alternative in relation to jth criteria, v_j^+ is a positive ideal solution, v_j^- is negative ideal solution, S_i^+ is alternative distance from positive ideal solution, and S_i^- is an alternative distance from a negative ideal solution.

Step 6. Calculation the relative closeness of alternatives to an ideal solution. The relative closeness is calculated as the quotient of alternative distance from negative ideal solution and sum of the alternative distance from positive and negative solution. This parameter is in the range 0–1.

$$C_i = \frac{S_i^-}{S_i^{\pm} + S_i^+} \qquad 0 \le C_i \le 1 \tag{7}$$

Step 7. Rank the alternatives. Alternatives are ranked in descending order of Ci. The greater the value of Ci, the higher is the priority of the alternative. If Ci is 0, then the alternative is a negative ideal solution, and if Ci is 1, the alternative is a positive ideal solution.

2.3 Fuzzy Sets Theory

Standard Fuzzy Inference System (FIS) consists of four elements: fuzzification, fuzzy rules, conclusion, and defuzzification. [16]. There are two basic FIS types: Mamdani and Takagi-Sugeno. Mamdani's approach, which is used in this paper, is characterized by the fact that the final exit from all fuzzy rules is fuzzy set selected with a minimum strength of rules. The output from the Mamdani type is a fuzzy set, which requires the aggregation process in the defuzzification process. Takagi-Sugeno is very similar to Mamdani, there is the same approach in the fuzzification of the input variable. The main difference is observed in the type of output function, which in the case of the Sugeno model, only appears as a linear function or constant [17].

This rule can be represented as follow:

$$\textit{If } x_1 \textit{ is } A \textit{ and } x_2 \textit{ is } B \textit{ then } y = f(x) \tag{8}$$

where A and B are fuzzy sets in premise, $y = f(x)$ is a faded function in the sequel. Usually, $f(x)$ is a polynomial of constant, but there can be any other function that adequately describes the system output in the fuzzy domain defined by the premise of the rule. The output of each rule yi, is hampered by the strength of the rule wi, which represents the degree of satisfaction of the premising part of the rule, also known as the degree of fulfillment.

For example, for one "and rule" the strength of the rule is:

$$w_i = \textit{AndMethod } (\mu A(x_1), \mu B(x_2)) \tag{9}$$

where $\mu A(x_1)$ and $\mu B(x_2)$ are membership functions for x_1 and x_2, respectively. The final output of the system can be calculated as follow:

$$KI = \frac{\sum_{i=1}^{N} w_i \cdot y_i}{\sum_{i=1}^{N} w_i} \tag{10}$$

3 Problem Formulation

The observed problem consist in selecting the most suitable location for RFID device for monitoring railway wagons on Serbian Railways. To determine the exact location for the RFID device, we solve two models, one for macro and one for micro-location using methods described in Sect. 2.

In order for the whole system to function, first, it is necessary to equip all railway wagons with appropriate RFID tags and ensure that the whole system has all the necessary elements [18]. The biggest problem that arises is the equipping of rail wagons with tags. This is very important because all wagons must be equipped according to current standards and protocols, so that there is no difference, and that the same system can be used everywhere, i.e. in all countries through which wagons pass.

Each tag should contain information about the wagon identification number, which collects data about wagon series, subseries, technical data (maximum length over buffers, load area, capacity, speed and etc.) as well as information about wagon ownership, whether they can traffic in international traffic or not. Using the electronic consignment note, it is possible to determine the exact information about the sender and the consignee, freight type and mass, as well as transport routing. Therefore, with the information collected in this way, the train, wagon or goods contained in the wagon can be easily identified throughout the entire transport process.

3.1 FUCOM-TOPSIS Model

As a potential location for this RFID device, railway organizational units for transport operations were used. Total numbers of alternatives are 11, although Serbian railway network is divided into twelve organizational units: Požarevac, Lapovo, Niš, Zaječar, Kraljevo, Užice, Pančevo, Zrenjanin, Novi Sad, Subotica, and Ruma. Belgrade railway section was not taken into consideration, because we assume that this section already have RFID monitoring device.

Through analysis of the volume of traffic on the Serbian railway network, it is clear that the majority of wagons cross the border crossings. Basing on this analysis and through discussion with railway experts, the first assumption is that RFID devices should be installed at each border crossing (Dimitrovgrad, Subotica, Šid, Ristovac, Vršac, and Prijepolje teretna).

In addition, it implies the installation of the RFID monitoring device in the area of the Belgrade railway junction, more precisely in the marshalling yard "Belgrade marshalling yard - Makiš", so, for this reason, Belgrade was not taken into consideration.

Considering all the input variables and the assumptions given, it is necessary to create two models. For the first model, the macro location model, we use the FUCOM-TOPSIS method for determining which organizational unit we can locate the RFID device. The second model represents a fuzzy logic model for determining the exact part of the railway line where to locate the RFID device.

We defined four main criteria:

- C_1 - the volume of transport (wagons/year) - this criterion is expressed by the number of wagons passing the train lines in the organizational unit. The total number of wagons includes all freight wagons, in domestic and international traffic, i.e. loading, unloading, and transit.
- C_2 - participation of main train lines (number of lines) - this criterion is defined through a number of main and regional lines and train line categories in each organizational unit.
- C_3 - number of international freight trains (number of trains/day) - this criterion is defined as a number of regular and optional international freight trains that can operate each day.
- C_4 - number of domestic freight trains (number of trains/day) - this criterion is defined as a number of regular and optional domestic freight trains that can operate each day.

Each of the criteria needs to be maximized, which is logical, because by increasing the volume of transport, the number of trains and the number of stripes served by the section, it has a greater need for this system as more wagons will move in the area of that section.

Criteria for comparison and selecting the best alternative are described in the previous section and their values are shown in Table 1.

Table 1. Initial decision making matrix.

Alternative	C_1	C_2	C_3	C_4
Požarevac	63891	1	55	35
Lapovo	23562	3	71	60
Niš	34799	4	74	44
Zaječar	58602	1	3	28
Kraljevo	25845	1	10	30
Užice	27952	1	29	22
Subotica	12381	1	82	46
Novi Sad	23675	1	76	64
Zrenjanin	11312	1	16	26
Pančevo	43849	2	38	27
Ruma	30398	2	62	33

Determining the weight coefficients of the criteria is one of the most delicate tasks that can significantly affect on decision-making process and the final rank of the alternatives, so special attention is paid to the determination of weight coefficients. In this paper, weight coefficients are determined by FUCOM method, and final results are presented in Table 2.

Table 2. Values of weight coefficients.

	C_1	C_2	C_3	C_4
w_i	0,332	0,167	0,255	0,246

Based on previously defined input parameters and criteria weighting, the results of TOPSIS method shows where to locate RFID device. The complete ranking of alternatives using the TOPSIS method is shown in Table 3.

Table 3. Ranking lists.

Alternative	C	Rank	Alternative	C	Rank
Požarevac	0,6170	2	Subotica	0,4153	8
Lapovo	0,5341	3	Novi Sad	0,4921	4
Niš	0,6218	1	Zrenjanin	0,0880	11
Zaječar	0,4660	6	Pančevo	0,4705	5
Kraljevo	0,1997	10	Ruma	0,4527	7
Užice	0,2607	9			

Solving the model as the most suitable alternative, the alternative A3 (Niš) was obtained. Alternative A1 is second best, and alternatives A5 and A9 are worst. The model was also tested in the case of 12 alternatives when in addition to the existing eleven organizational units, the Belgrade organizational unit was included. In that case, the alternative Niš was the second choice immediately after Belgrade in both methods.

Obtaining a final solution using a multi-criteria analysis method is not a sufficient indicator that this alternative is the most appropriate. In addition, it is necessary to examine how the obtained solution changes depending on the change of the weight coefficients of the criteria. Changes in the final ranking of alternatives in the MABAC method is shown in Table 4.

Table 4. Alternatives ranking for different weight criteria scenarios.

	A_1	A_2	A_3	A_4	A_5	A_6	A_7	A_8	A_9	A_{10}	A_{11}
S_1	1	7	4	2	9	6	10	8	11	3	5
S_2	5	2	1	8	10	9	7	6	11	4	3
S_3	6	2	1	10	11	8	3	4	9	7	5
S_4	5	3	1	7	8	11	4	2	10	9	6
S_5	3	2	1	8	10	9	7	4	11	6	5
S_0	2	3	1	6	10	9	8	4	11	5	7

Total number of different scenarios in the sensitivity analysis is 5 (S1–S5). Scenarios S1 to S5 shows the changes in ranking of alternatives when only one criterion is prioritized and all other weights of criteria are the same (for example S1 - criterion C1 has a value 0,7 and all the others have equal values 0, 1). Scenario S5 means that all criteria have uniform weights, and scenario S0 is one with real weight coefficients obtained by FUCOM method.

As we can see from Table 4 there is relative stability of obtained results on changes in weighting criteria. Alternative A_3 is the best solution in 4 of 5 scenarios (83%) and in one scenario this solution is in second place. Alternatives A_5, A_6 and A_9 are mostly in the last three places so that alternatives are not be considered as a potential solution. Ranking of the other alternatives is more dependent on changes in values of the weight coefficients.

The stability of the obtained results is also tested through other MCDM methods: EDAS, ARAS, COPRAS, WASPAS, and MABAC. All the input values and weighted coefficients are the same for all used methods, and comparison is shown in Fig. 1.

Based on Fig. 1 we can see that there is a very small difference between results for all methods. Alternative A_3 is the best solution for all six methods, which is an adequate verification of the proposed model, an RFID device should be located in operational unit Niš.

3.2 Fuzzy Set Theory Model

Each organizational unit on Serbian railways is divided into sections and it covers more than one railway line. The borders stations of railway lines in the organizational unit

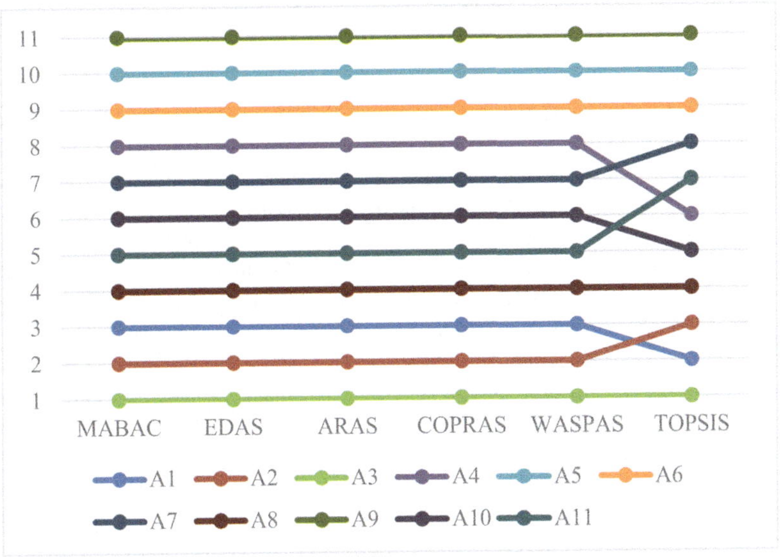

Fig. 1. Alternatives ranking by tested MCDM methods

Niš, as the best solution obtained through the macro location model, are Dimitrovgrad, Preševo and Kosanicka Rača. The model did not take into account the part of the railway line to Lapovo, because a very small part of this line is under the jurisdiction of the organizational unit of Niš.

The fuzzy set theory model represents micro-location model, which has four input variables and one output variable. All input and output variables are defined by triangular and trapezoidal fuzzy numbers. The input variables in the model are:

- x_1 - condition of railway line (points) - input variable which is described as "bad", "good" and "excellent" section. This variable depends on many parameters like condition of superstructure, condition of substructure, number, and condition of railway objects, tunnels, number of restricted speed, maximum speed on the line, minimum radius of curvature, incline, and track maintenance.
- x_2 - number of loaded and unloaded wagons (wagons/day) - input variable described as: "very low", "low", "medium", "high" and "very high".
- x_3 - average number of wagons (wagons/day) - input variable described as "small" and "large".
- x_4 - distance from the border station (km) - this input variable is described with three fuzzy numbers: "low", "medium" and "high". Because of the assumption that there are already installed RFID devices on border stations, sections more distant from the border are the better solution.

The input variables of the fuzzy model for determining micro-location are shown in Table 5.

After defining the input variables, the output variable called railway section (y) is defined. The output variable is represented through three fuzzy numbers "bad", "good"

Table 5. Input variables.

Railway line	Railway section	x_1	x_2	x_3	x_4
NIŠ - DIMGRAD	NIŠ - Niška Banja	Bad	Very low	Large	Medium
	Niška Banja - Dolac	Bad	Medium	Large	Low
	Dolac - Bela Palanka	Bad	Very low	Large	Low
	Bela Palanka - Pirot	Bad	Very high	Large	Low
	Pirot - DIMITROVGRAD	Good	Very low	Large	Low
NIŠ - PREŠEVO	NIŠ - Doljevac	Good	High	Large	High
	Doljevac - Pečenjevce	Excellent	Very low	Large	Medium
	Pečenjevce - Leskovac	Good	High	Large	Medium
	Leskovac - Vladičin Han	Bad	Very low	Large	Medium
	Vladičin Han - Vranje	Bad	Medium	Large	Low
	Vranje - PREŠEVO	Bad	Low	Large	Low
NIŠ - KOSANIČKA RAČA	NIŠ - Doljevac	Good	Very low	Small	Medium
	Doljevac - Žitorađa	Bad	Very low	Small	Medium
	Žitorađa - Prokuplje	Bad	Very low	Small	Medium
	Prokuplje - Kuršumlija	Bad	Very low	Small	High
	Kuršumlija - KOSANIČKA RAČA	Good	Very low	Small	High

and "excellent", and takes a value from the range of 0 to 10. The lowest rating indicates that it is the worst, and the highest rating indicates that it is the best section for the location of an RFID device. The output variable provides the location for the RFID device to be installed, based on all values of the input variables. The division domain for all input variables and the corresponding membership functions are shown in Fig. 2.

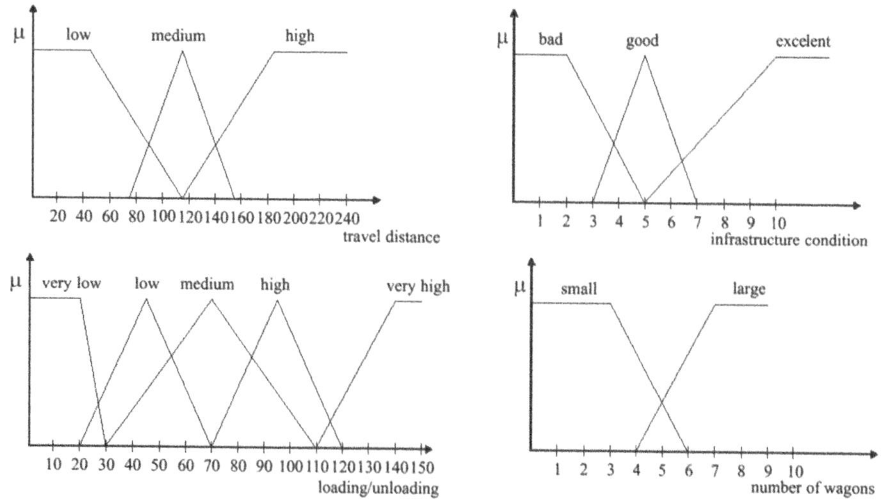

Fig. 2. Membership functions for input and output variables.

After determining all the values and the corresponding membership functions, fuzzy rules database are defined. Since RFID still does not have its application on Serbian Railways, the rules are defined based on the author's experience and knowledge, as well as in consultation with employees of Serbian railways. The total number of rules in the model is 90. All rules are the same importance, with a value of 1. Defining the rules is as follows:

R1: IF (condition of railway line is "bad") and (number of loaded and unloaded wagons is "very small") and (average number of wagons is "small") and (distance from the border station is "short") THEN (railway section is "bad").

The model was tested for known values, obtained by the organizational unit Niš. The most suitable location is the railway section Niš - Doljevac, which takes a value of 7.62 and belongs to the membership function "excellent". This value was obtained for the following input values: the condition of railway line is "satisfactory" (value 1.5), the average number of wagons per section during 2018 was 8.21 trains per day, the section distance from the border station was 157 km, and the number of loaded and unloaded wagons per section is 116 wagons annually.

After determining all the values and the corresponding membership functions, fuzzy rules database are defined. Since RFID still does not have its application on Serbian Railways, the rules are defined based on the author's experience and knowledge, as well as in consultation with employees of Serbian railways. The total number of rules in the model is 90. All rules are the same importance, with a value of 1. Defining the rules is as follows:

R1: IF (condition of railway line is "bad") and (number of loaded and unloaded wagons is "very small") and (average number of wagons is "small") and (distance from the border station is "short") THEN (railway section is "bad").

The model was tested for known values, obtained by the organizational unit Niš. The most suitable location is the railway section Niš - Doljevac, which takes a value of 7.62 and belongs to the membership function "excellent". This value was obtained for the following input values: the condition of railway line is "satisfactory" (value 1.5), the average number of wagons per section during 2018 was 8.21 trains per day, the section distance from the border station was 157 km, and the number of loaded and unloaded wagons per section is 116 wagons annually.

4 Conclusions

Consideration of the entire transport chain is extremely important in order to improve the existing supply and to make the rail system more competitive and attractive. The technologies of automatic marking and identification of products and other segments in the work process have become an integral part of modern business information systems of a large number of companies.

The results obtained from both models answer the question of where to locate the RFID device. However, this is only a small contribution to the implementation of such a rail system. First of all, it is necessary to include other services in order to make full use of the system, to facilitate operation and to achieve greater automation of the system. The introduction of such a system would lead to a better allocation of vehicular resources, and with appropriate interaction with other systems, it would be very easy to automate the whole system to such an extent that the necessary paperwork accompanying the train and records is automatically filled out, above all the ranking lists, and bill of lading.

Statistics obtained through the use of this technology can show the highest values for idling a wagon, that is, it could have the effect of increasing the bandwidth of the line and identifying the problematic points where the wagon is most likely to stay on its transport route. This can have the effect of reducing the waiting time of the wagon, reducing the cost and affecting the timetable and improving and adapting it to users.

Further directions of development are reflected first and foremost in a more detailed analysis of all the data that was needed to build the model. Specifically, it is necessary to analyze the condition of the line in more detail, expand the analysis of the parameter of the number of trains, in order to fully determine the number of trains operating on the line (optional and regular) and to examine the cost-effectiveness of such a system.

The introduction of the RFID system will result in simpler procedures at border crossings, will facilitate the work in marshalling yards as well as in the intermediate stations, but will also facilitate the tracking of shipments as loading and unloading of all wagons and types of trains throughout the railway network.

References

1. Sucharitha, G.: Applications in railway industry. Int. Mag. Eng. Technol. Manag. Res. **2**, 544–548 (2015)
2. Đordjevic, Ž.: Batajnica checkpoint for wayside train monitoring. In: Proceedings of the XIV Scientific-Expert Conference on Railways, pp. 189–192 (2014)
3. Márton, P., Milinković, S., et al.: Solving a container terminal location problem using decision support systems. Transp. Res. Procedia **40**, 1459–1464 (2019)
4. Belošević, I., Ivić, M., et al.: Railroad transshipment yards: layouts and rail operation. Horizons Series, Aćimović **3**, 559–569 (2016)
5. Belošević, I., Ivić, M., et al.: Challenges in the railway yards layout designing regarding the implementation of intermodal technologies. In: 2nd Logistics International Conference. Aćimović, pp. 62–67 (2015)
6. Fazlollahtabar, H., Smailbašić, A.: FUCOM method in group decision-making: selection of forklift in a warehouse. Decis. Making: Appl. Manag. Eng. **2**, 49–65 (2019)
7. Milosavljević, M., Bursać, M., et al.: Selection of the railroad container terminal in Serbia based on multi-criteria decision-making methods. Decis. Making: Appl. Manag. Eng. **1**, 1–15 (2018)
8. Hamrucu, M., Eren, T.: An application of multicriteria decision-making for the evaluation of alternative monorail routes. Mathematics **7**, 1–17 (2019)
9. Belošević, I., Kosijer, M., et al.: Group decision making process for early stage evaluations of infrastructure projects using extended. VIKOR method under fuzzy environment. In: European Transport Research Review, vol. 10, p. 43 (2018)
10. Belošević, I., Milinković, S., et al.: A fuzzy group decision making for a rail-road transshipment yard micro location problem. In: MATEC Web of Conferences, vol. 235, p. 00019 (2018)
11. Milinković, S., Vesković, S., et al.: A site selection model for wayside train monitoring systems at serbian railways. Communications **17**(2), 49–54 (2015)

12. Senvar, O., Otay, I., et al.: Hospital site selection via hesitant fuzzy TOPSIS. IFAC-Papers OnLine **49**, 1140–1145 (2016)
13. Batanović, V., Petrović, D., et al.: Fuzzy logic based algorithms for maximum covering location problems. Inf. Sci. **179**, 120–129 (2009)
14. Pamučar, D., Stević, Ž.: A new model for determining weight coefficients of criteria in MCDM models: Full consistency method (FUCOM). Symmetry **10**(9), 1–22 (2018)
15. Hwang, C.L., Yoon, K.: Multiple Attribute Decision Making: Methods and Application. Springer, Berlin (1981)
16. Zadeh, L.A.: Fuzzy sets. Inf. Control **8**, 338–353 (1965)
17. Milosavljević, M., Jeremić, D., et al.: Application of fuzzy logic for train braking distance determination. Horiz. Series **5**, 79–91 (2018)
18. Mašek, J., Kolarovszki, P., et al.: Application of RFID Technology in railway transport services and logistic chains. Procedia Eng. **134**, 231–236 (2015)

Methods of Increasing of Onboard Tracking System with Optical Device Dynamic Accuracy

Valery Sheval(✉) and Nikolay Rozhnin

Moscow Aviation Institute (National Research University), Volokolamskoe sh., 4, Moscow 125993, Russia
sheval@list.ru

Abstract. In this article, some hardware and system conditions for formation of the automatic tracking system in two-channel onboard structures class are considered. For that, common spatial target tracking task is separated into two subtasks (by means of reasonable decomposition): autonomous flights in horizontal and vertical axis from which last case is selected in this work. Using features of quadcopter flight in vertical axis, procedure of differential equations during its formulation is realized. Also structure of feedback actuator functioning in one plane is developed in its work representation. Methods of functioning matching of onboard feedback actuator and aircraft vehicle control structures be means of combined control two-channel structures using are suggested. For studying of methods of increasing of onboard automatic tracking system dynamic accuracy of class under consideration, mathematical model of such complex system is created and approved. In experimental imitation mode, possibilities of automatic tracking system dynamic accuracy improvement are verified.

Keywords: Automatic tracking · Feedback actuator · Optical direction finder · Quadcopter · Complex technical system · Decomposition · Correction of structure · Dynamic accuracy

1 Introduction

At the moment, field of technical aircraft tasks expands; its hardware solution is based on unmanned aerial vehicle (UAV) using. Among them, tasks required of tracking for movable targets with high indexes of dynamic accuracy play important role in terms of functioning. Solution of such tasks is based on modern control methods of UAV and using of light helicopter type aircrafts – quadcopters (QC), particularly. While solution of technical tasks of rapid maneuvers targets tracking, complex problems of defining of QC flight control corresponded with the same target tracking modes of onboard equipment are appeared. These problems are repeatedly investigated in technical sources [1–4]. In this work, well-known methods of QC flight control synthesis are considered [5–8], attending problems of structures in developing automatic tracking systems (ATS) under consideration.

As one of the particular cases, type of ATS made on base of feedback actuator with optical device (tracking error measurer) placed on QC is considered in this paper. It is

Z. Popovic et al. (Eds.): TransSiberia 2019, AISC 1115, pp. 540–549, 2020.
https://doi.org/10.1007/978-3-030-37916-2_52

demonstrated that among positive characteristic of such ATSs possibility of transition to potentially perspective two-channel structures for dynamic accuracy increasing in structural aspect is present in case of fulfillment of certain technical conditions.

In practice task of ATS dynamics research is appeared. Such research implements via onboard optical device (OOD) which placed on QC.

2 Research Methods

It is rational to consider QC movements only in vertical axis for research of movable targets tracking dynamic errors formation. This condition does not decrease value of received results and allows to carry out such experiments with more simple equipment at the same time.

Flight control system (FCS) implements quadcopter position control, and optical load feedback actuator (OL FA) controls OOD observing line. Figure 1a shows kinematics of formation of input signals and disturbances, functional block diagram of ATS control is depicted in Fig. 1b.

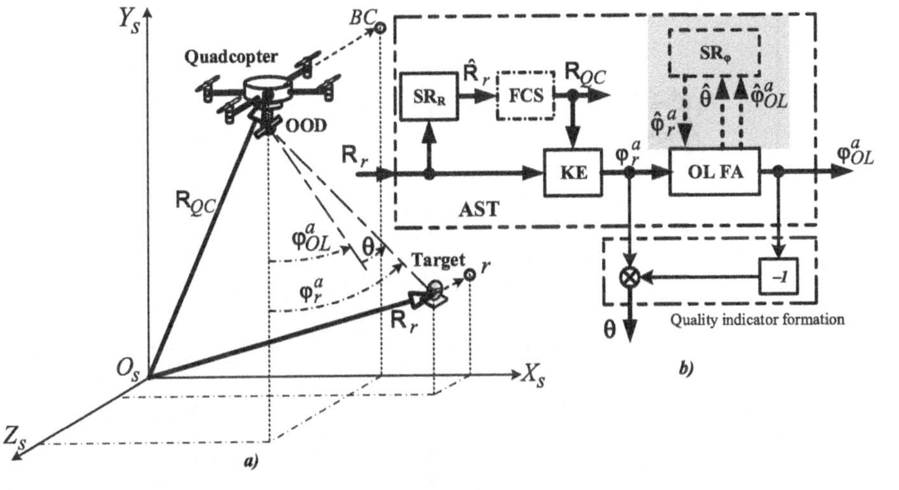

Fig. 1. Schematic representation of tracking mission: (a) kinematics of bodies mutual arrangement in tracking mission, (b) generalized functional block diagram of ATS

Figure 1 has following designations: R_r and R_{QC} are radius vectors of target and quadcopter attitudes, respectively; φ_r^a is angle of target sight, φ_{OL}^a is angle of OL rotation, θ is error between these two angles; SR_R and $SR\varphi$ are signal regenerators of $\hat{\mathbf{R}}_r$ and $\hat{\varphi}_r$, respectively; KE is kinematic element of formation of φ_r^a.

ATS is complex technical spatial control system which consists of two subsystems – FCS and OL FA. Thus, during systems design it is advisable to consider the task of lower hierarchy level, namely task of one of two tracking plane dynamics research. For this, kinematics of angle φ_r^a formation is represented while motion in $O_SX_SY_SZ_S$ axes wherefore mass centers of quadcopter and target are displaced to BC and r points (Fig. 1). In this case, flat version of quadcopter with half mass will be named «bicopter».

According to the mission of bicopter plane flight [10], forces applied to bicopter are depicted in Fig. 2. Figure 2 has following designations of plane axes: $O_SX_SY_SZ_S$, $O_1X_gY_g$, $O_1X_VY_V$, $O_1X_1Y_1$ are start, ground, aerodynamic and body axes, respectively. Besides, O_1 point is bicopter mass center. Also F_a and F_b are aerodynamic and drag forces, respectively; ϑ_{BC} and θ_{BC} are tilt angles of bicopter thrust force vectors F_T and its airspeed V_{BC} in ground axes which is parallel to start axes, respectively.

According to D'Alembert's principle one can write differential equation in aerodynamic axes:

$$\begin{cases} m_{BC}\dot{V}_{BC} = F_T \cos\alpha - m_{BC}g\cos\theta_{BC} - F_b, \\ m_{BC}V_{BC}\dot{\theta}_{BC} = F_T \sin\alpha - m_{BC}g\sin\theta_{BC} - F_a. \end{cases} \tag{1}$$

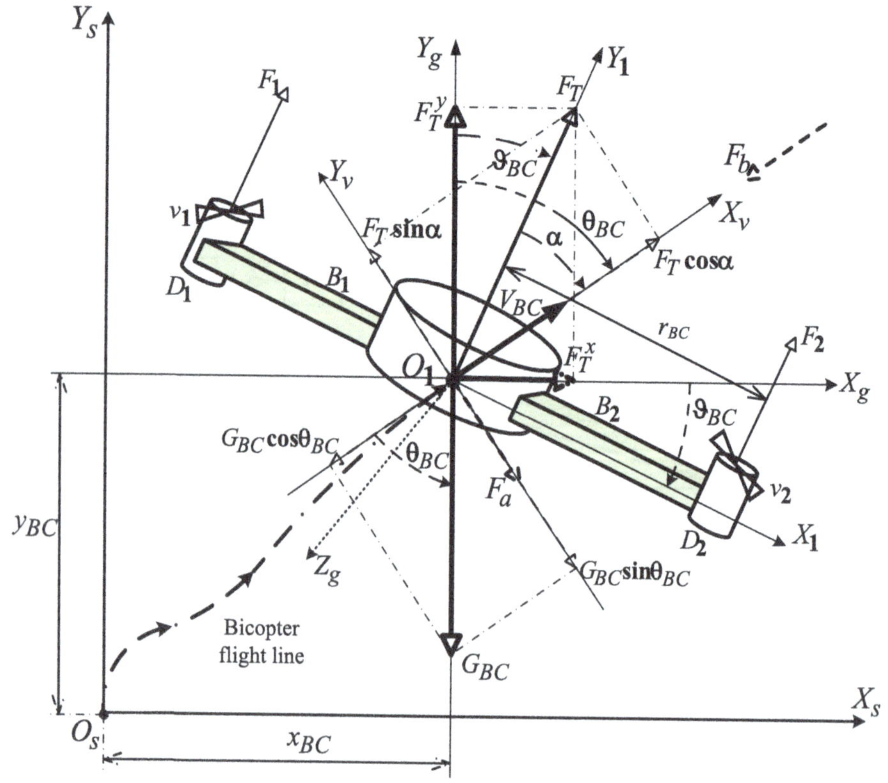

Fig. 2. Scheme of forces applied to bicopter

Equation of bicopter rotational motion around fictitious axis OZ0 is formulated according to D'Alembert's principle as well:

$$I_{BC}\frac{d\dot{\vartheta}_{BC}}{dt} = (F_1 - F_2)r_{BC} + M_a + M_b, \quad (2)$$

where r_{BC} is bicopter radius of gyration; M_a and M_b are aerodynamic and air bicopter torques which depend on angle of attack $\alpha = \vartheta_{BC} - \theta_{BC}$, its velocity $\dot{\alpha}$, bicopter airspeed V_{BC} and air density nonlinearly.

Equations (1) and (2) are nonlinear and its direct solution is complicated problem.

It is advisable to use linearization methods of differential equations or simplified practical method of linearization on the first stages of systems design of control systems where plane representation of quadcopter spatial flight is employed.

For this, all aerodynamics effects applied to bicopter set to zero $F_a = F_b = M_a = M_b = 0$, and also assume that bicopter flight corresponds to level flight condition $F_T \cos\vartheta_{BC} \approx m_{BC}g$. If bicopter flight along O_SX_S and O_SY_S axis is considered autonomous, then while moving along O_SX_S axis $\theta_{BC} = 90°$ and this angle one can consider constant, thus, $\dot{\theta}_{BC} = 0$; while moving along O_SY_S axis θ_{BC} and this angle one can consider constant as well ($\dot{\theta}_{BC} = 0$).

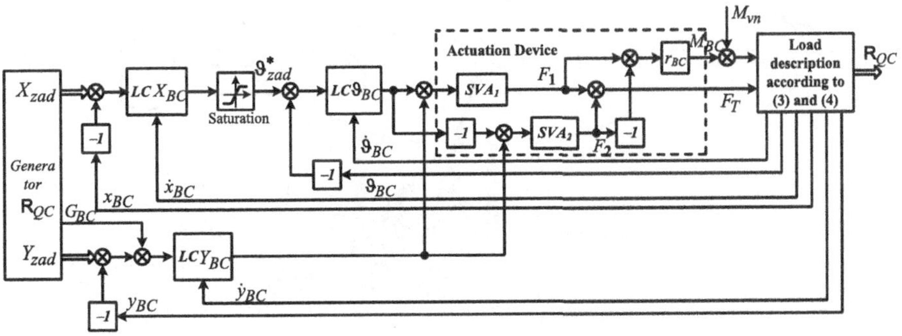

Fig. 3. Structural scheme of bicopter FCS

Therefore, Eqs. (1) and (2) are formulated as:

$$\begin{cases} m_{BC}\dot{V}_{BC} = F_T\cos(90° - \vartheta_{BC}) = F_T\sin\vartheta_{BC} = F_T^x, \\ m_{BC}\dot{V}_{BC} = F_T\cos\vartheta_{BC} - m_{BC}g, \\ I_{BC}\frac{d\dot{\vartheta}_{BC}}{dt} = (F_1 - F_2)r_{BC}. \end{cases} \quad (3)$$

We will add kinematics conditions to system of differential equations (3) with aforementioned assumptions:

$$\begin{cases} x_{BC} = x_{BC}(0) + \int V_{BC}^x dt, \\ y_{BC} = y_{BC}(0) + \int V_{BC}^y dt. \end{cases} \quad (4)$$

Based on [11], structural scheme of bicopter FCS one can represent as shown in Fig. 3 where SVA is stable velocity actuator; LC is load controller; M_{load} is external moment applied to the bicopter; $\mathbf{R}_{BC} = x_{BC}\mathbf{x}_S^0 + y_{BC}\mathbf{y}_S^0$.

Figure 4 shows scheme of design relationships of OL rotation center, OL mass center and bicopter mass center, relative position of which determines the structure and parameters of equations of ATS load. We will introduce following designations: Ox_1y_1 is basis of OL initial position (axes are parallel to body axes) and Ox_2y_2 is basis after OL rotation to the φ_{OL}^{BC} angle, respectively; r_{vr} is distance between quadcopter mass center and OL rotation center; r_m is distance between OL rotation and mass centers.

Absolute angular motion of bicopter will be defined by instantaneous angular velocity $\boldsymbol{\omega}_{BC} = \dot{\boldsymbol{\vartheta}}_{BC}$ and infinitesimal rotation angle $d\boldsymbol{\vartheta}_{BC}$. Relative angular motion of OL will be defined by instantaneous relative velocity $\dot{\boldsymbol{\varphi}}_{OL}^{BC}$ and infinitesimal relative rotation angle $d\boldsymbol{\varphi}_{OL}^{BC}$.

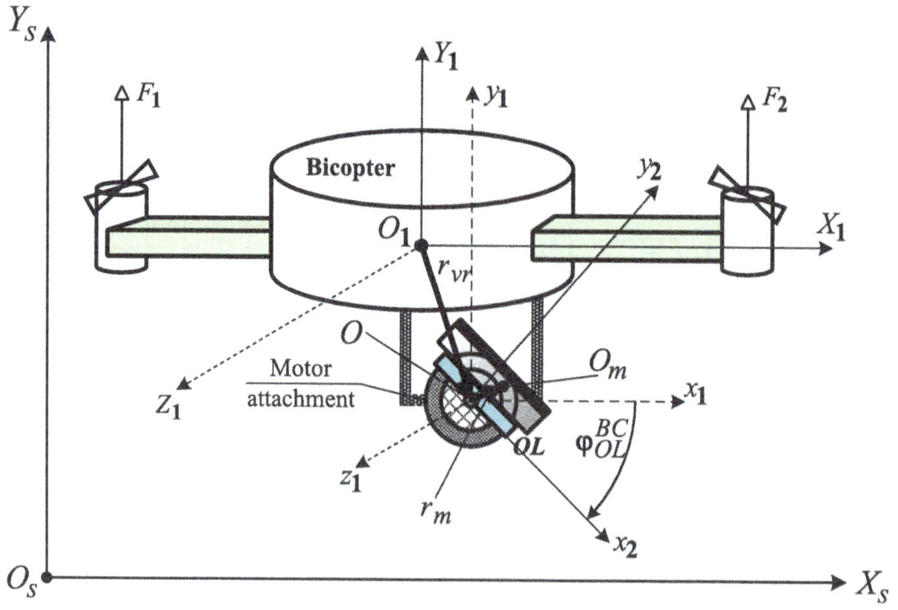

Fig. 4. Scheme of design relationships between solid bodies rotation and mass centers

One can write relationships for absolute and relative angular velocities of bicopter and OL and also for absolute and relative infinitesimal rotation angles:

$$\boldsymbol{\omega}_{OL}^{a} = \boldsymbol{\omega}_{BC} + \dot{\boldsymbol{\varphi}}_{OL}^{BC} = \boldsymbol{\omega}_{BC} + \boldsymbol{\omega}_{OL}^{BC};$$
$$d\boldsymbol{\varphi}_{OL}^{a} = d\boldsymbol{\vartheta}_{BC} + d\boldsymbol{\varphi}_{OL}^{BC}.$$

We will use Lagrange equation of the second kind relatively independent velocity $\omega_{OL}^{BC} = \dot{\varphi}_{OL}^{BC}$ and independent angle $d\varphi_{OL}^{BC}$ for formulation differential equations of motion of mechanical subsystem (load for OL FA):

$$\frac{d}{dt}\left(\frac{\partial L}{\partial \omega_{OL}^{BC}}\right) = \frac{\partial L}{\partial \varphi_{OL}^{BC}}, \tag{5}$$

where $L = T_{OL} + U_{OL}$ is Lagrange function; T_{OL} and U_{OL} are kinetic energy and potential function of the system, respectively.

Kinetic energy of this system can be written as

$$T_{OL} = \frac{1}{2} m_{OL} V_O^2 + \frac{1}{2} J\left(\omega_{OL}^a\right)^2 + m_{OL} V_O V_{O_m}, \tag{6}$$

where $V_O = \omega_{BC} \cdot r_{vr}$; $V_{O_m} = \left(\omega_{BC} + \omega_{OL}^{BC}\right) \cdot r_m$; O_m is OL mass center.

Potential function of this system can be written as

$$U_{OL} = \int dA_{OL}, \tag{7}$$

where $dA_{OL} = M_{dv} \cdot d\varphi_{OL}^{BC}$ is infinitesimal work of active forces (torques) of the system.

It is worth to notice that kinetic energy (6) do not depends on infinitesimal rotation angles and potential function (7) do not depends on instantaneous angular velocities. Therefore, formula (5) can be written as:

$$\frac{d}{dt}\left(\frac{\partial T_{OL}}{\partial \omega_{OL}^{BC}}\right) = \frac{\partial U_{OL}}{\partial \varphi_{OL}^{BC}}. \tag{8}$$

Considering relations for kinetic energy (6) and potential function (7) Eq. (8) can be formulated as:

$$J\frac{d\omega_{OL}^{BC}}{dt} + J_{OL}^{*}\frac{d\omega_{BC}}{dt} = M_{dv}, \tag{9}$$

where M_{dv} is torque of ATS actuator motor; $J_{OL}^{*} = m_{OL} \cdot r_{vr} \cdot r_m + J_{OL}$; $J_{OL} = J_{dv} q^2 + J_{zer}$; J_{dv} and J_{zer} are moments of inertia of motor and deflected optical element, respectively; q is gear ratio.

Structural scheme of OL FA incorporating to ATS subsystem is shown in Fig. 5 where W_{pel} is transfer function of optical bearing finder; W_{reg} is transfer function of OL FA controller; W_{um} is transfer function of actuator amplifier; W_{dv}^{em} is transfer function of electromagnetic subsystem of motor; W_{cor} is transfer function of feedback compensator; W_{gd} is transfer function of gyroscopic sensor; W_{du} is transfer function of angular position transducer.

Accept hypothesis that tracking of ground target moving horizontally with constant velocity V_T is implemented. In that case, OL FA input signal (equation for KE block in Fig. 1b) can be described by

$$\varphi_r^a = \operatorname{arctg}\left(\frac{x_r(\mathbf{0}) + V_r t - x_{BC}}{y_{BC}}\right).$$

Further, we assume that operation of onboard ATS devices implements on y'_{BC} height on the assumption of tactical task condition. Coordinates of $\hat{\mathbf{R}}_T$ vector in $\mathrm{SR_R}$ block correspond to relation (considering introduced condition $y_T = 0$):

$$\hat{x}_r = \hat{x}_{BC} + \hat{y}_{BC}\operatorname{tg}\left(\hat{\varphi}_{OL}^{BC} + \hat{\boldsymbol{\theta}}\right) \tag{10}$$

Signals for formula (10) are measured by angular position transducer $\left(\hat{\varphi}_{OL}^{BC}\right)$, optical bearing finder $\left(\hat{\boldsymbol{\theta}}\right)$ and GLONASS sensor ($\hat{x}_{BC}$ and $\hat{y}_{BC}$).

Therefore, ATS can be described by presence of cross couplings: from OL FA to FCS along motor load response torque $M_{load} = -M_{dv}$ signal, from FCS to OL FA along input signal $\varphi_r^a(x_{BC}, y_{BC})$, and disturbance ϑ_{BC}.

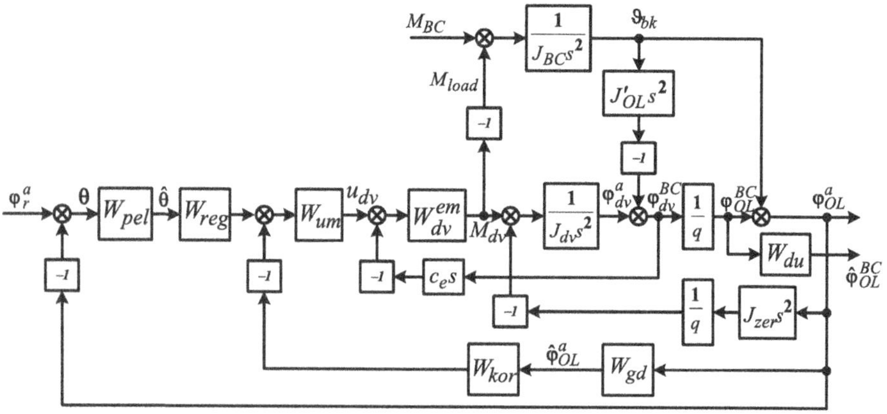

Fig. 5. Structural scheme of OL FA

3 Results

Main performance criterion of ATS (quality indicator) is tracking error $\boldsymbol{\theta}$ value of which is criterion of accomplishment efficiency of functional mission of specific aircraft complex. Presence in ATS structure of above-mentioned cross couplings leads to $\boldsymbol{\theta}$ increasing, thus, one of the useful methods of dynamic accuracy rising is different structural methods to parry such couplings.

We will employ structure, which corresponds to combined control principles. It is necessary to regenerate signal for implementation of this structure:

$$\hat{\varphi}_r^a = \hat{\varphi}_{OL}^a + \hat{\theta}. \qquad (11)$$

In this case, the structure including blocks and connections which are depicted by dashed lines in Fig. 1b is appeared (signal $\hat{\varphi}_r^a$ regenerator SR_φ makes Eq. (11)). Signals values for Eq. (9) are measured by gyroscopic sensor ($\hat{\varphi}_{OL}^a$) and optical bearing finder ($\hat{\boldsymbol{\theta}}$).

Figures 6 and 7 show plot of error $\boldsymbol{\theta}$ changing for the structure showing in Fig. 1b not including blocks and connections which are depicted by dashed lines; besides, FCS load is described by Eqs. (3) and (4), and OL FA load is described by Eq. (9). Also, Figs. 6 and 7 show plot of error $\boldsymbol{\theta}$ for above-suggested structural compensation ATS structure.

ATS initial structure settling time is 0.4 s, error $\boldsymbol{\theta}$ oscillation amplitude in steady state is 1.72 arcmin. ATS compensated structure settling time is 0.37 s, error $\boldsymbol{\theta}$ oscillation amplitude in steady state is 0.86 arcmin.

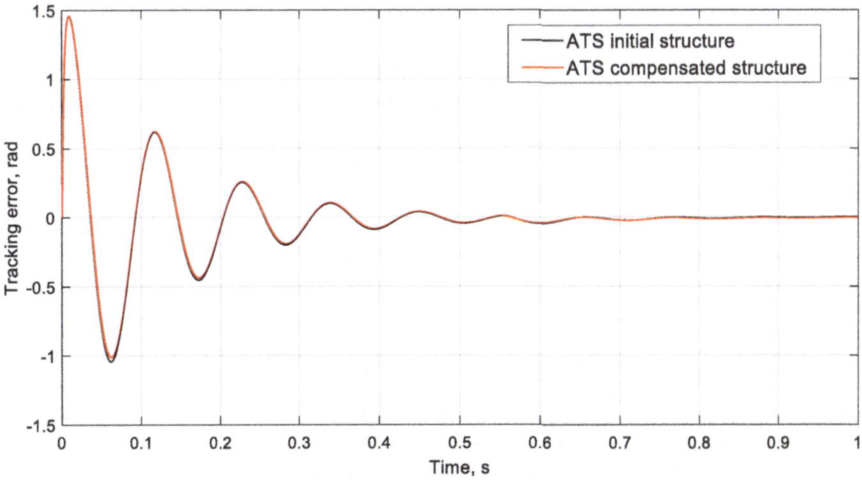

Fig. 6. Error signal plot of ATS (in the interval 0 ÷ 1 s)

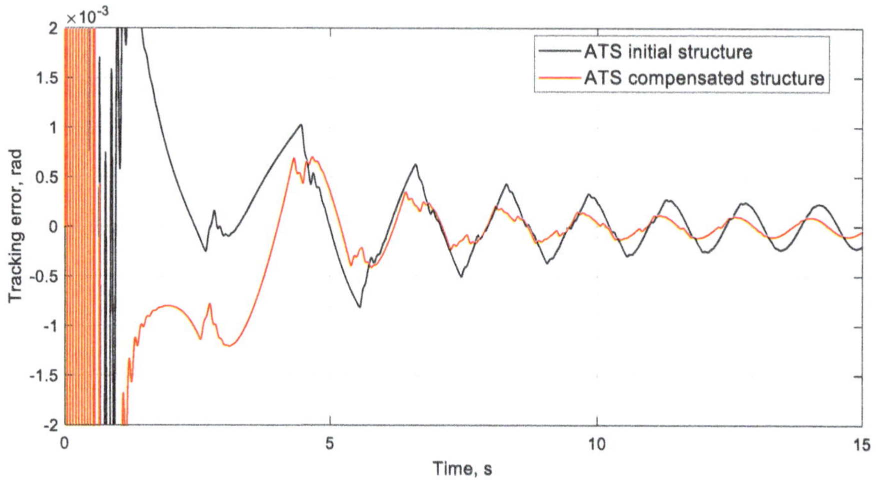

Fig. 7. Error signal plot of ATS (in the interval 0 ÷ 15 s)

4 Discussion

Structure of plane ATS suggested in this paper allows to carry out comparative researches of this system dynamics in accordance with error $\boldsymbol{\theta}$ value criterion on this level.

Suggested ATS structure creation method in one plane allows to introduce various real factors of ATS devices production and consideration of disturbances.

Mathematical modelling verifies improvement possibility of dynamic accuracy of ATS considered type.

References

1. Castillo, P., Dzul, A., Lozano, R.: Real-time stabilization and tracking of a four-rotor mini rotorcraft. IEEE Trans. Control Syst. Technol. **12**(4), 510–516 (2004)
2. Mellinger, D., Michael, N., Kumar, V.: Trajectory generation and control for precise aggressive maneuvers with quadrotors. Int. J. Robot. Res. **31**(5), 664–674 (2012)
3. Gong, X., Bai, Y., Peng, C., Zhao, C., Tian, Y.: Trajectory tracking control of a quad-rotor UAV based on command filtered backstepping. In: Third International Conference on Intelligent Control and Information Processing (ICICIP), pp. 179–184. IEEE Publications (2012). https://doi.org/10.1109/icicip.2012.6391413
4. Logachev, V.G., Minin, I.V.: Fundamental researches. Tech. Sci. **11**(1), 85–91 (2015)
5. Bemporad, A., Pascucci, C.A., Rocchi, C.: Hierarchical and hybrid model predictive control of quadcopter air vehicles. In: 3rd IFAC Conference on Analysis and Design of Hybrid Systems, Zaragoza, Spain, pp. 14–19 (2009)
6. Bresciani, T.: Modeling, identification and control of a quadrotor helicopter. Master's thesis, Department of Automatic control, Lund University, p. 170 (2008)

7. Raffo, G.V., Ortega, M.G., Rubio, F.R.: An integral predictive/nonlinear H∞ control structure for a quadrotor helicopter. Automatica **46**(1), 29–39 (2010). https://doi.org/10.1016/j.automatica.2009.10.018
8. Dierks, T., Jagannathan, S.: Output feedback control of a quadrotor UAV using neural networks. IEEE Trans. Neural Netw. **21**(1), 50–66 (2010). https://doi.org/10.1109/TNN.2009.2034145
9. Yatsun, S.F., Emel'yanova, O.V., Popov, N.I.: Actual problems of technical sciences. In: II International Science Conference, Perm, pp. 66–69 (2013)
10. Ogoltsov, I.I., Rozhnin, N.B., Sheval, V.V.: Izvestiya of Tula State University. Tech. Sci. **1**, 47–55 (2012)
11. Ogoltsov, I.I., Rozhnin, N.B., Sheval, V.V.: Trudy MAI, p. 83. MAI Publishing House, Moscow (2015)

Research of the Kinematic Error of a Wave Gear with Rolling Bodies

Mikhail Vasiliev[1,2] and Vilen Stepanov[2(✉)]

[1] Irkut Corporation, Leningradsky Avenue, 68, Moscow 125315, Russia
[2] Moscow Aviation Institute (National Research University), Volokolamskoe shosse, 4, Moscow 125993, Russia
stevilen@mail.ru

Abstract. Aim of this research is to assess the influence of the wave gear with rolling bodies (further WGRB) design parameters on the kinematic error. In the research, method of estimating kinematic error using MSC.ADAMS, convergence of which is shown by the results of the experiment, was used. The article suggests possible actions to reduce the kinematic error without significant increase in the cost of production; affection of design parameters of WGRB parts to the kinematic error is analyzed; influence of WGRB parts dimension deviations on the kinematic error is defined. According to the results of the research, the following conclusions can be made: methodology for determining the kinematic error of WGRB parts depending on deviations of the size and shape of the parts working surfaces is obtained and verified; static component of the kinematic error is inversely proportional to the wave former diameter; smoothness of the gear increases with increase in the number of rolling bodies; tightening tolerance on the diameter of the rolling body leads to significant decrease in the magnitude of the static component kinematic error; 3 angular minutes value of the kinematic error of the WGRB is obtained in the manufacture of working surfaces of parts with 6 tolerance grade.

Keywords: Wave gear with rolling bodies · MSC.ADAMS · Kinematic error

1 Introduction

Research of the design parameters influence of the WGRB on the kinematic error is carried out in the framework of assessing the practical significance of methods for calculating the kinematic error of the WGRB (similar tasks were studied in [1–9] for other mechanisms with clearances).

Previously, the authors considered the following methods for calculating the kinematic error of WGRB, as well as their advantages and disadvantages are considered:

(1) method for simulating the kinematic error of WGRB in a Siemens NX CAD system;
(2) method for simulating the kinematic error of WGRB in the MSC.ADAMS CAE system (for details on working in MSC.ADAMS with reference to this task, see [10, 11]);
(3) analytical method for determining the kinematic error of the WGRB.

Z. Popovic et al. (Eds.): TransSiberia 2019, AISC 1115, pp. 550–558, 2020.
https://doi.org/10.1007/978-3-030-37916-2_53

At the moment, there are no researches that would allow to estimate the value of kinematic error at the stage of design documentation development for WGRB depending on the selected parameters of WGRB (number of rolling bodies, main dimensions of gear, etc.).

2 Research Methods

The research of the WGRB design parameters influence on the kinematic error is carried out with the method of simulating the functioning of WGRB in MSC.ADAMS.

To confirm that the method of simulating of WGRB functioning in MSC.ADAMS is reliable, computational calculations and experimental work are carried out.

For the experimental determination of the WGRB kinematic error, the "Diakin-3" kinematomer, which consists of two high-precision angle sensors (encoders) with information acquisition and processing system, is used. Encoders are connected using clutches to the drive and driven shafts of the WGRB and uniform rotation of the drive shaft is implemented, first in one direction and then in the opposite direction, so that the driven shaft made at least one full turn in both directions. Based on discrepancy between the angles of rotation of the drive and driven shafts, kinematic error is plotted. Figure 1 shows the graphs of kinematic errors that are obtained from the results of experimental work and via MSC.ADAMS.

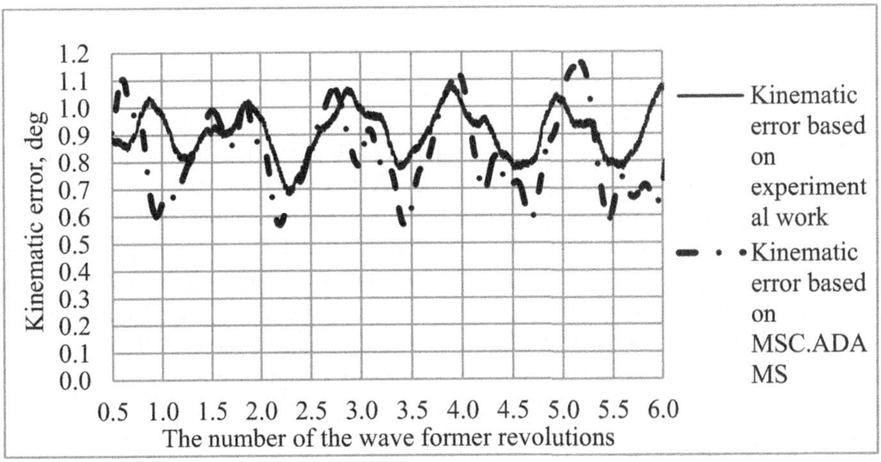

Fig. 1. Graphs of kinematic errors obtained from the results of the experimental work and with via MSC.ADAMS

Analyzing the kinematic error graphs obtained during experimental work and via modeling in MSC.ADAMS allows to determine that the static component differs by no more than 8%, the harmonic component (associated with the speed of wave former rotation) differs by no more than 20%. From this, it follows that for the problem of tolerances selection in order to ensure the required accuracy of the gear, the simulation method of the WGRB in the MSC.ADAMS provides a sufficient degree of convergence.

3 Results

In order to assess how the design parameters of the WGRB affect on the kinematic error, electronic models of the product (hereafter, the EMP) of the WGRB have been developed with six different sets of design parameters (hereafter referred to as types). Table 1 shows the nominal WGRB parameters for these types.

Table 1. Design parameters of WGRB

Design parameters	Type 1	Type 2	Type 3	Type 4	Type 5	Type 6
Diameter of rolling body, mm	10	20	10	10	20	20
Type of rolling body	Roller	Roller	Roller	Roller	Roller	Roller
Diameter of wave former, mm	52	104	110	84	220	168
Size of hole in separator, mm	10	20	10	10	20	20
Eccentricity, mm	2	4	2	2	4	4
Number of rolling bodies	9	9	18	14	18	14
The number of the rigid wheel periods	10	10	19	15	19	15

For each of the design types in accordance with Table 1 MSC.ADAMS model is prepared. To investigate the influence of the WGRB design parameters on the kinematic error of gear, three series of computational experiments are carried out.

The first series of computational experiments is aimed to determining the minimum possible value of the static component of the WGRB kinematic error when selecting tolerances of with 6 tolerance grade for the following deviations: deviation by the diameter of the rolling body, deviation by the diameter of the wave former, deviation by the size of the separator hole diameter, inaccuracy of rigid wheel manufacturing. According to this computational experiment, it was determined that fifth-type design has a minimum static component of the WGRB kinematic error with deviations of 6 tolerance grade (about 0.048°, which corresponds to 3 angular minutes, the kinematic error graph is shown in Fig. 2). For comparison, Fig. 3 also shows the graph of the kinematic error of the WGRB for the first-type design according to 6 tolerance grades. The magnitude of the static component for this type of design is 0.108° (6.48 angular minutes).

Modeling showed that without significant increase in the cost of part production for the WGRB (the machine park existing at the machine-building enterprises allows to produce parts of tolerance grade 6 with acceptable cost) it is possible to achieve kinematic error of the WGRB not more than 3 angular minutes by the static component. Simulations also showed that with increase in the number of rolling bodies, the

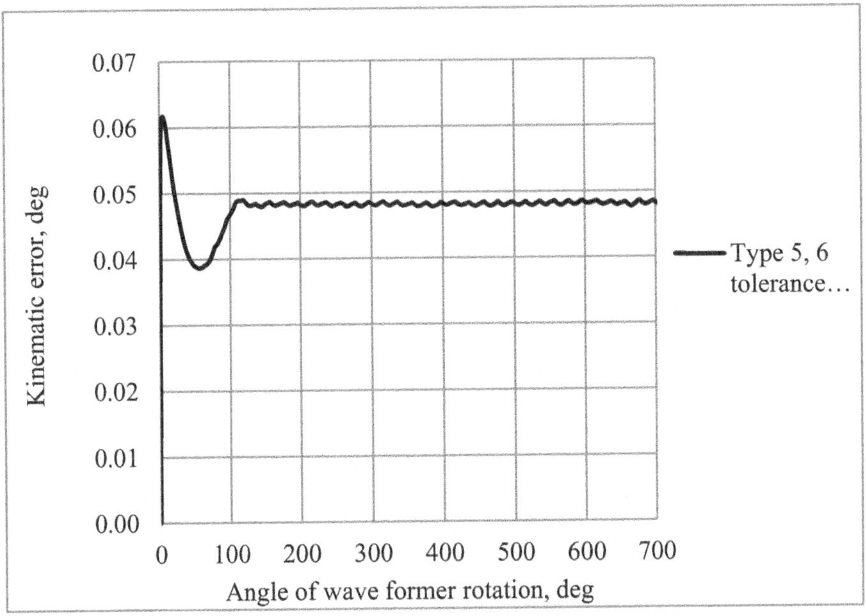

Fig. 2. The graph of the kinematic error for the fifth-type design according to 6 tolerance grade

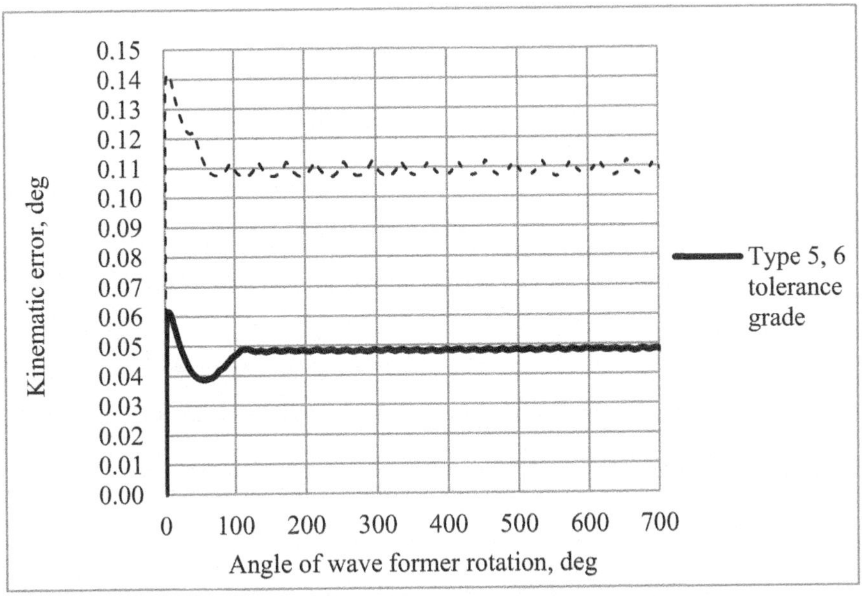

Fig. 3. Graphs of the kinematic error for the first- and fifth-type design according to 6 tolerance grade

high-frequency component decreases (it is shown for example in Fig. 3 for types 1 and 5: for type 1, the number of rolling bodies is 9, for type 5 the number of rolling bodies is 18).

In the second series of computational experiments for type 1, deviations of 6 tolerance grade are introduced for the following elements: separator hole size (deviation equals to 0.009 mm), wave former diameter (deviation equals to 0.019 mm), rolling body diameter (deviation equals to 0.009 mm).

Further, the above-mentioned deviations are introduced into all remaining types of WGRB designs in order to determine the influence of the gear design parameters on the value of WGRB kinematic error.

Figure 4 shows that the value of the wave former diameter influences on the static component significantly: with increase in the diameter of the wave former the magnitude of the kinematic error decreases. Also Fig. 4 shows that the diameter of the rolling bodies does not significantly affect on the magnitude of WGRB kinematic error: with increase in the diameter of the rolling body by 2 times (type 2 and type 3) with practically unchanged diameter of the wave former, kinematic error on the static component has a slight change. This means that the larger the diameter of the wave former, the smaller the kinematic error on the static component can be achieved. The allowable size of gear limits the diameter of the wave former. It is also obvious that with increase of rolling bodies number the high-frequency component of the kinematic error decreases (the smoothness of the transmission increases).

Figure 4 shows graphs of the kinematic error for various types of WGRB designs.

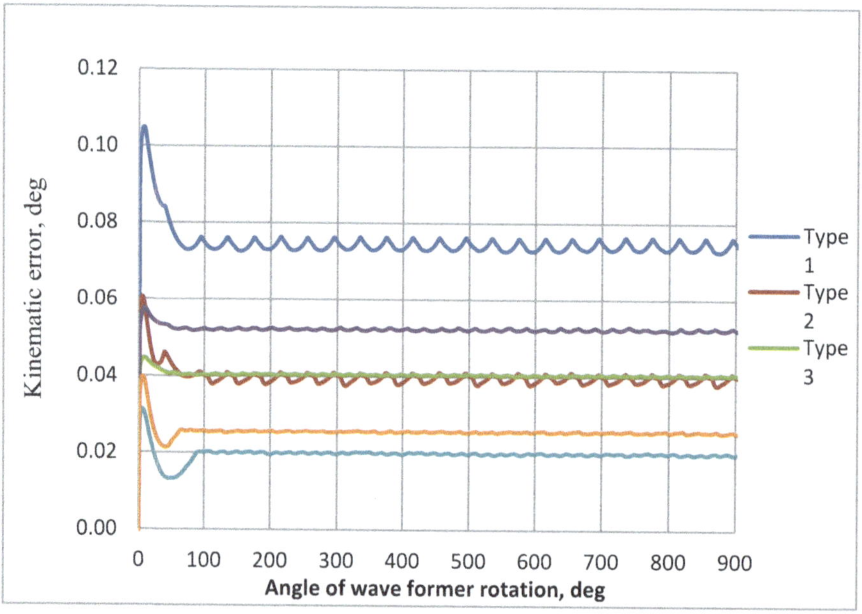

Fig. 4. Graphs of kinematic error for various types of designs

In the third series of computational experiments, second-type design is simulated with the following deviations: the inaccuracy of rigid wheel manufacturing (in the direction of increasing kinematic error) is 0.024 mm (which corresponds to 6 tolerance grade), the deviation for the size of hole in separator is 0.021 mm (which corresponds to 7 tolerance grade), the deviation by the value of the diameter of the rolling bodies is equal to 0.021 mm (which corresponds to 7 tolerance grade), the deviation by the value of the diameter of the wave former is equal to 0.035 mm (which corresponds to 7 tolerance grade). Call the tolerance set No. 0. Figure 5 shows a graph of the kinematic error for given set of tolerances.

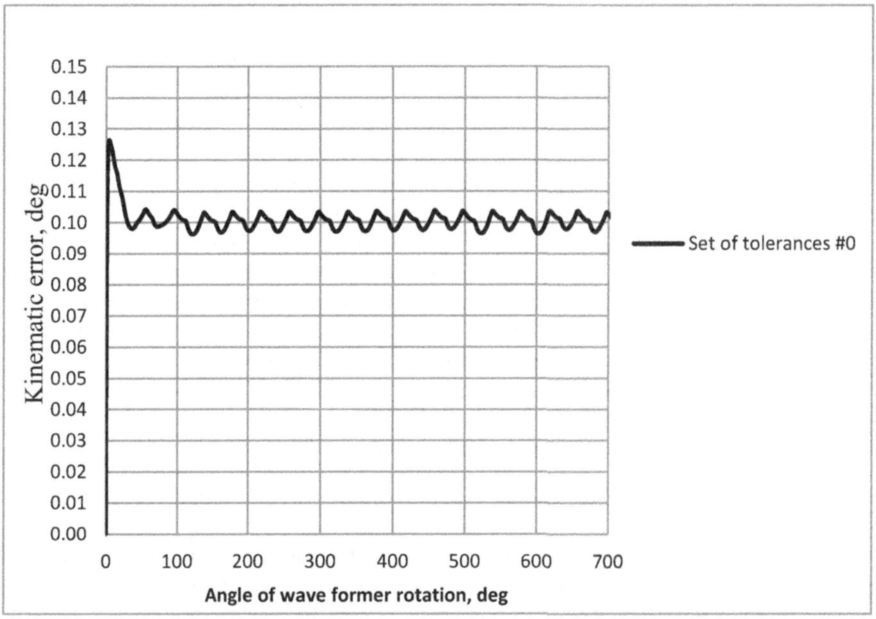

Fig. 5. Graphs of kinematic error for tolerance set No. 0

The next goal is pursued in the simulation: tightening of which tolerance will maximally reduce the value of the static component of kinematic error. The reduction of inaccuracies in the production of a rigid wheel by 0.012 mm is taken as a base, thus the inaccuracy of rigid wheel manufacturing is 0.012 mm (which corresponds to 4 tolerance grade). Such tightening of the tolerance will lead to significant complication of the manufacturing process of a rigid wheel, as well as increase in the cost of manufacturing a rigid wheel and controlling the made dimensions. The remaining sets of tolerances also have tightening on one of the types of tolerances on the value of 0.012 mm.

Table 2 shows the following sets of tolerances (for each set, tightening the tolerance is carried out only for one size): set of tolerances No. 1 (tightening the tolerance on the size of the separator hole), set of tolerances No. 2 (tightening the tolerance for inaccurate manufacturing of the rigid wheel), set of tolerances No. 3 (tightening the tolerance on the diameter of the wave former), set of tolerances No. 4 (tightening the tolerance on the diameter of the rolling bodies).

Table 2. Tolerance set description

	Tolerance set				
	No. 0	No. 1	No. 2	No. 3	No. 4
Tolerance value for inaccuracies in the manufacture of a rigid wheel	0,024	0,024	0,012[b]	0,024	0,024
Tolerance value on the size of the separator hole	0,021	0,009[a]	0,021	0,021	0,021
Tolerance value for wave diameter	0,035	0,035	0,035	0,023[c]	0,035
Tolerance value for rolling bodies diameter	0,021	0,021	0,021	0,021	0,009[d]

Comments:
[a]for a separator, tightening the tolerance to 0.009 mm corresponds to 5 tolerance grade, this will lead to significant increase in the cost of manufacturing the separator;
[b]for a rigid wheel, tightening the tolerance to 0.012 mm corresponds to 4 tolerance grade, making a rigid wheel with such parameters is either not possible with the existing technological base, or will increase the cost of manufacturing the rigid wheel by orders of magnitude;
[c]for a wave former, tightening of the tolerance to 0.023 mm corresponds to 6 tolerance grade, this will lead to insignificant (possibly not changing the cost) increase in the cost of manufacturing of the wave former;
[d]for rolling bodies, tightening the tolerance to 0.009 mm corresponds to the choice of I class of accuracy rolling bodies, the use of standard rolling bodies will not significantly increase the cost of manufacturing of the gear.

Figure 6 shows the graphs for different sets of tolerances. Figure 6 shows that the least preferred solution is to select the tolerance set No. 1 (tightening the tolerance on the size of the separator hole): with insignificant reduction in the kinematic error, the cost of manufacturing of the separator will increase by several times.

Using set of tolerances No. 2 or No. 3 is also impractical due to slight decrease in kinematic error with increase in the cost of production of the gear (in the case of tightening the tolerance for the manufacture of rigid wheel, the production of rigid wheel with the specified tolerance is almost impossible). The most preferable solution is set of tolerances No. 4: using standard rollers of class I accuracy significantly reduces the kinematic error (compared to other sets of tolerances) without significantly increasing the cost of the production process.

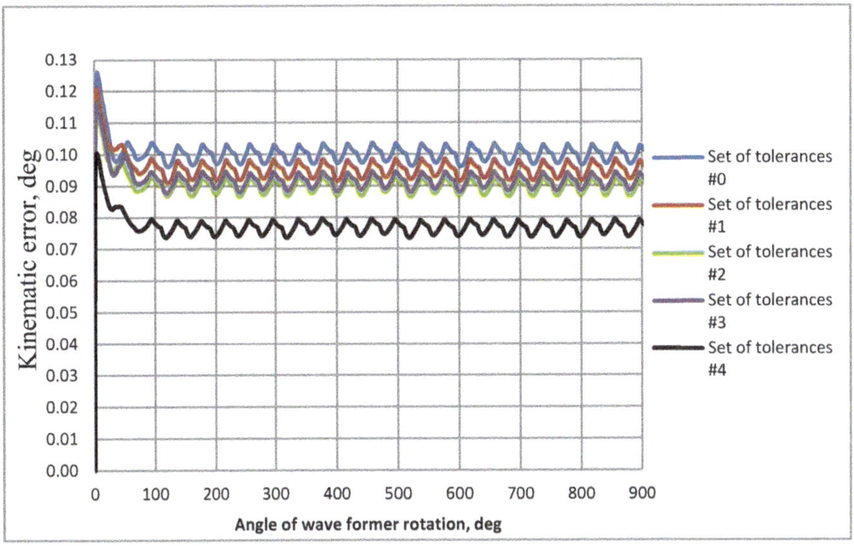

Fig. 6. Graphs of the kinematic error of the WGRB for different sets of tolerances.

4 Discussion

The given method of determining the kinematic error of WGRB depending on deviations of the dimensions and shape of the working surfaces of its parts is verified by experiment with acceptable accuracy. According to the results of the research, the following conclusions can be made:

- there is an opportunity to manufacture a WGRB with deviations in details of 6 tolerance grade, the kinematic error of which in the static component will not exceed 3 angular minutes;
- with an increase wave former diameter, the kinematic error in the static component decreases significantly: for example, with an increase in the wave former diameter by 4 times, the static component of the kinematic error also decreases by 4 times;
- with increase in the number of rolling bodies increases the smoothness of the gear;
- tightening of the tolerance for the diameter of the rolling bodies leads to significant decrease in the kinematic error of the WGRB in terms of the static component.

These conclusions can be used by designers as recommendations when designing gears based on WGRB in order to obtain the required kinematic error. This article considers the limited set of deviations, and therefore in future researches it is planned to carry out a research of other types of deviations by the magnitude of the kinematic error.

References

1. Flores, P., Leine, R., Glocker, C.: Modeling and analysis of planar rigid multibody systems with translational clearance joints based on the non-smooth dynamics approach. Multibody Syst. Dyn. **23**, 165–190 (2010)
2. Crowthera, A.R., Singha, R., Zhangb, N., Chapman, C.: Impulsive response of an automatic transmission system with multiple clearances: formulation, simulation and experiment. J. Sound Vib. **306**, 444–466 (2007)
3. Flores, P., Ambrosio, J., Claro, J.C.P., Lankarani, H.M.: Dynamic behaviour of planar rigid multi-body systems including revolute joints with clearance. Proc. Inst. Mech. Eng. K J. Multi-Body Dyn. **221**(2), 161–174 (2007)
4. Tsai, M.J., Lai, T.H.: Kinematic sensitivity analysis of linkage with joint clearance based on transmission quality. Mech. Mach. Theory **39**, 1189–1206 (2004)
5. Lin, K.-S., Chan, K.-Y., Lee, J.-J.: Kinematic error analysis and tolerance allocation of cycloidal gear reducers. Mech. Mach. Theory **124**, 73–91 (2018)
6. Ma, J., Qian, L.N.: Modeling and simulation of planar multibody systems considering multiple revolute clearance joints. Nonlinear Dyn. **90**(3), 1907–1940 (2017)
7. Lankarani, H.M., Nikravesh, P.E.: A contact force model with hysteresis damping for impact analysis of multibody systems. J. Mech. Des. **112**(3), 369–376 (1990)
8. Sommer, A., Meagher, J., Wu, X.: An advanced numerical model of gear tooth loading from backlash and profile errors. In: Rotary Machinery, Structural Health Monitoring, Shock and Vibration, Volume 5, pp. 191–201 (2011)
9. Wu, X., Meagher, J., Sommer, A.: A differential planetary gear model with backlash and teeth damage. In: Rotary Machinery, Structural Health Monitoring, Shock and Vibration, Volume 5, pp. 203–215 (2011)
10. Ben Abdallah, M.A., Khemili, I., Aifaoui, N.: Numerical investigation of a flexible slider-crank mechanism with multijoints with clearance. Multibody Syst. Dyn. **38**, 173–199 (2016)
11. Blundell, M., Harty, D.: Multibody systems simulation software. In: The Multibody Approach to Vehicle Dynamics, pp. 75–130 (2004)

Changes in the Structure of Rail Steel Under High-Frequency Loading

Ekaterina Gridasova[1(✉)], Pavel Nikiforov[1], Alexey Loktev[2], Vadim Korolev[2], and Irina Shishkina[2]

[1] Far Eastern Federal University, Sukhanova St., 8, 690091 Vladivostok, Russia
olvin@list.ru
[2] Russian University of Transport (MIIT), Chasovaya Str., 22/2, 125190 Moscow, Russia

Abstract. The article deals questions of the behavior of various materials of construction of the permanent way and engineering structures under high-frequency vibration exposure, which occurs during high-speed movement of the carriages, possible changes in the microstructure rail steel in the field gigacycles loadings. Some traditional materials when exposed to a large number of cycles of load application (10^8–10^{10}) are destroyed by voltages that were previously considered acceptable. In the modern normative documents is considered basic, if the material is able to withstand 10^7 cycles of loading, the level of the relevant stress is the fatigue limit of the material, the material itself is not destroyed by the increasing number of cycles. Nonetheless, the results of current research demonstrate the values of fatigue limits gigacycles loadings, as well as the emergence of new failure mechanisms under high-frequency effects. High frequency dynamic load can occur not only under the action of external loads corresponding frequency, but the imposition of harmonics of different frequencies, for example, under the action of vibration from the moving crew, temperature or vibration of the elements of the railway track, and artificial structures that are traversed by this path. The impact gigacycle load and study the obtained results allow modeling the life cycle of rails by skipping the required train load and to build a forecast of the behaviour of whips in different areas when skipping to 1000 million gross tons.

Keywords: High-frequency vibration exposure · Changes in the microstructure · Rail steel · Gigacycles loadings · Fatigue limit · Modeling the life cycle of rails

1 Introduction

The development of modern transport systems is characterized by the increased speeds of vehicles, this trend is evident for the railroads. Optimization of railway operations connected with the increase in the load on the wheelset axles, increase in the number of cars, which in General leads to intensification of use of transport infrastructure, including such item as a rail-sleeper grid. A pressing issue is the condition of occurrence of the limit state design and its individual elements. An important role in this

Z. Popovic et al. (Eds.): TransSiberia 2019, AISC 1115, pp. 559–569, 2020.
https://doi.org/10.1007/978-3-030-37916-2_54

formulation is played by mechanical properties of materials, even well-studied, but used in the new loading conditions. Insufficiently studied is the question of the behavior of various materials of construction of the permanent way and engineering structures under high-frequency vibration exposure, which occurs during high-speed movement of the carriages. Little-studied possible changes in the microstructure rail steel in the field gigacycles loadings.

Gigacycle phenomenon of fatigue of structural materials, including steel aroused scientific interest a few decades ago [1, 2], but these earlier studies were more theoretical. Currently, in connection with increasing the speed of repetition of crews and, in General, more intensive use of linear transportation facilities, such studies are becoming more practical. Many researchers have noted that some traditional materials when exposed to a large number of cycles of load application (108–1010) are destroyed by voltages that were previously considered acceptable [3]. In the modern normative documents is considered basic, if the material is able to withstand 107 cycles of loading, the level of the relevant stress is the fatigue limit of the material, the material itself is not destroyed by the increasing number of cycles [4–6]. Nonetheless, the results of current research [1–4] demonstrate the values of fatigue limits gigacycles loadings, as well as the emergence of new failure mechanisms under high-frequency effects. High frequency dynamic load can occur not only under the action of external loads corresponding frequency [7–10], but the imposition of harmonics of different frequencies, for example, under the action of vibration from the moving crew, temperature or vibration of the elements of the railway track, and artificial structures that are traversed by this path (Fig. 1(a)).

The impact gigacycle load and study the obtained results allow modeling the life cycle of rails by skipping the required train load and to build a forecast of the behaviour of whips in different areas when skipping to 1000 million gross tons. In terms of laboratory facilities Moscow State University of Transport and Far Eastern Federal University, this might take a few weeks instead of several years in landfill conditions.

2 Research Technique and Results

Statistics of detected rail defects shows that the most significant are the operational defects originating from the impact of rolling stock on the rail and manifested in the form of peeling, chipping metal head rail and fractures, due to a lack of contact fatigue strength of the metal (Fig. 2). In pic. 2. and provided the code defect 30G.1 in the new NTD is it a defect 30.1 – longitudinal cracks in the head, and pic. 2.b – defect code 21.2 – transverse fatigue crack in the head in the form of a light or dark spot, arising from internal or external longitudinal cracks formed due to lack of contact fatigue strength of metals, and led to the failure of the rail after missing the warranty of the tonnage out of the joint. However, the causes of such defects and, especially, methods of their detection at early stages, do not seem to be fully studied, as previously were not considered high-frequency components of the loading and behavior of materials elements of paths when gigacycles influences.

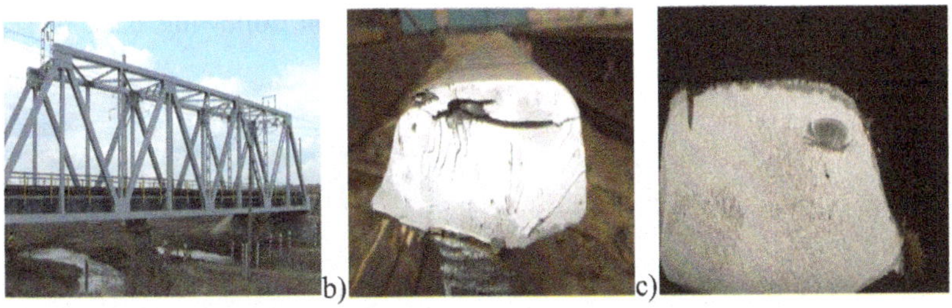

a) b) c)

Fig. 1. (a) Metal railway bridge across the river at the site of the Privolzhskaya railway; (b) Defects of rails: defect code 30G.1; (c) Defects of rails: defect code 21.2.

To study the effect of high frequency impacts was taken a fragment through-hardened rail with a welded seam. Samples from different zones of welded rail were investigated by the method of atomic-emission analysis to determine elemental composition. The results of the analysis are shown in Table 1.

Table 1. Results of the elemental analysis of the welded rail, mass %.

Item	Base metal	The weld metal
C	0.7334…0.7628	0.7698…1.0643 (average value 0.8570)
Mn	0.8237…0.8387	0.7107…0.8105
Si	0.2964…0.3330	0.6843…0.9038
Cr	0.0697…0.0856	0.1128…0.1203
Mo	0.0031…0.0074	0.0246…0.0277
W	0.0010…0.0022	0.0020…0.0069
Ti	0.0035…0.0052	0.0010…0.0037
V	0.0354…0.0377	0.1433…0.1609
Ni	0.0685…0.0900	0.0580…0.0879
Cu	0.1175…0.1343	0.0558…0.0750
S	0.0100…0.0126	0.0059…0.0151
P	0.0123…0.0138	0.0143…0.0177
	About the brand −70G or 75G	About the brand 80GSL, increased content of vanadium

Thus, as in the case of domestic rail steels, roughly correspond to compositions of spring steels and steels for wear-resistant castings. Typical of such steels structure contains about 100% pearlite. Consider two sample RSM-ZS and the RSM-OM cut from different zones of the initial segment of the rail. In the course of sample RSM-3C microscopes to get a Fig. 3.

This sample has a pearlite structure. The typical size of pearlite colonies – 50… 100 mkm. In addition, the sample has considerable porosity, characteristic pore size of about 10 mkm. The microhardness of the sample (Table 2) is quite uniform, corresponds to troostite.

Examining a sample of the RSM-OM, we get the following data, presented in Fig. 3 and in Table 3.

a) 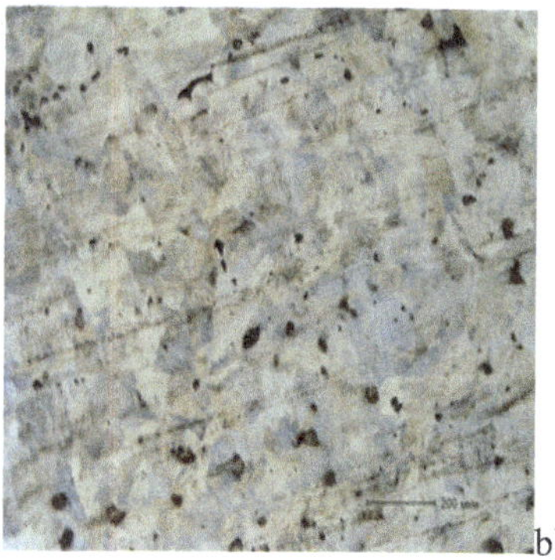 b) 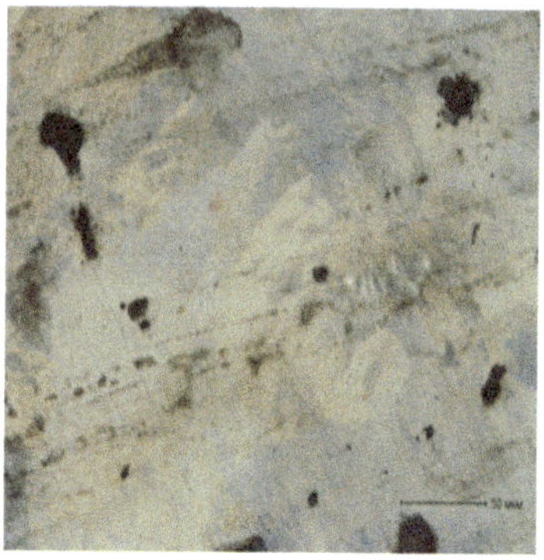

Fig. 2. Sample RSM-3C: (a) an increase of 100^x, (b) an increase 500^x.

Table 2. Microhardness of the sample RSM-3C.

Center	100 mkm	200 mkm	300 mkm	400 mkm	500 mkm	600 mkm	700 mkm	800 mkm	900 mkm	1000 mkm	Shank
427	460	435	407	407	378	458	412	471	491	457	497
385	397	430	402	460	427	509	448	429	389	427	440
478	370	429	471	395	445	464	432	424	380	415	478
480	415	406	421	437	440	424	419	422	395	400	389
482	362	446	406	433	396	413	376	405	380	425	440

a) 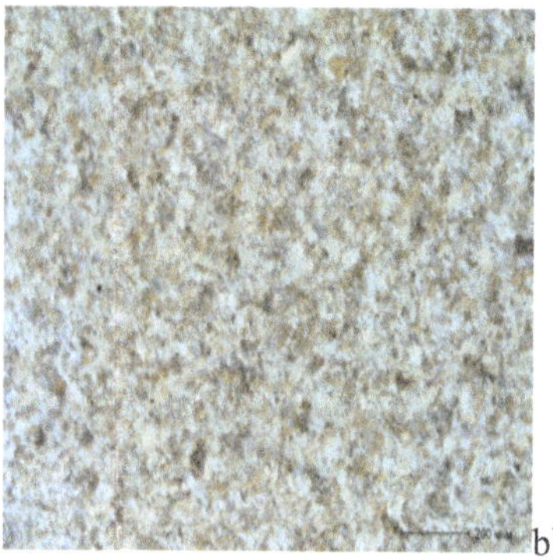 b) 

Fig. 3. Sample RSM-OM: (a) an increase of 100^x, (b) an increase 500^x.

Table 3. The microhardness of the sample RSM-OM.

Center	100 mkm	200 mkm	300 mkm	400 mkm	500 mkm	600 mkm	700 mkm	800 mkm	900 mkm	1000 mkm	Shank
346	343	355	380	368	359	354	342	332	355	341	384
366	362	348	371	385	365	355	361	362	372	356	372
372	350	375	375	360	365	359	357	350	360	350	373
331	348	353	370	368	361	359	340	347	341	370	365
376	371	359	357	351	346	339	338	358	357	351	361

The sample has a pearlite structure with a characteristic size of colonies 30… 50 mkm. Slightly porous, characteristic pore size less than 5 mkm. The microhardness of the sample is uniform, corresponds to the sorbitol or troostite.

Primary structure analysis can be done and the determination of microhardness can draw the following conclusions: loading the welding zone of the rail with the amplitude of the voltage σ_a = 370 MPa on the basis of $1.16 \cdot 10^9$ cycles didnt lead to occurrence of fatigue cracks or other structural changes. However, to ensure that this mode of loading for welded joints of rails is safe, it is necessary to study a larger number of samples.

If we consider the work of the rail under the influence of mounted wheels, we can say that the main areas of concentration alternating stresses are concentrated along the rail head at a depth of from 3 to 14 mm from the tread surface. It is this part of the track design often experiences a high frequency of exposure, which may be significantly intensified in the presence of defects in the wheels (weld-on deposit, slid flat, etc.). In the presence of defects of mounted wheels changes the nature of the interaction between wheel and rail, and the Hertz contact model (1), generally fair to small speeds of movement, must be replaced with other dependencies, linking the strength of the interaction, the local deformation of the materials of the contacting bodies, move with connection do not see into account of wave and oscillatory processes, e.g., viscoelastic model with an exponential kernel of relaxation (2) or a viscoelastic model with fractional derivatives of Riemann-Liouville (3):

$$P = k\alpha^{3/2}, \tag{1}$$

$$P(t) = E_1(\alpha - w) - \frac{E_1}{\tau_1}\int_0^t (\dot{\alpha} - \dot{w})e^{-\frac{t-t^1}{\tau_1}}dt', \tag{2}$$

$$P + \tau^y D^y P = E_1\tau^y D^y(a - w),\ D^y P = \frac{d}{dt}\int_0^t \frac{p(t-t')}{\tilde{A}(1-y)t'^y}dt', \tag{3}$$

here α - local wrinkling of the material of rail and wheel, k is a coefficient determined by the geometry of the contacting bodies and the elastic characteristics of the material, E1 is the modulus of elasticity of the region of interaction between mounted wheels and rail, w – move the bottom edge of the rail, $\sigma 1$ – Poisson's ratio for mounted wheels [2, 8], $\tau_1 = n_1/E_1$, $\tau 1$ is the relaxation time in the case of the viscoelastic model, $\eta 1$ – coefficient of viscous resistance, γ $(0 < \gamma \leq 1)$ – parameter fragmentation. In the presence of defects on the tread surface of mounted wheels is not only a change in the solution of the contact problem and stress distribution near the region of interaction of wheel and rail, but changes the whole picture of the wave processes in the assembled rails and sleepers, which can lead to an increase in the intensity fluctuations and gigacycle loads on the individual elements of the permanent way. One of the objectives of the present work is to study the influence of high frequency vibration (f = 20000 Hz) on the structure of low-carbon steel in low-cycle and gigacycle fields with amplitudes in excess of and equal to the tensile strength. As a model of low-carbon steel was chosen as the steel St3sp (GOST 535-88). After high-frequency loading of the samples was investigated and their microstructure and microhardness in the axial section.

For fabrication of specimens (Fig. 4) used steel St3sp (GOST 535-88) in the form of a rod with a diameter of 12 mm. Fatigue tests samples were performed on the ultrasonic testing machine USF-2000 (Shimadzu, Japan) with a frequency of 20 kHz.

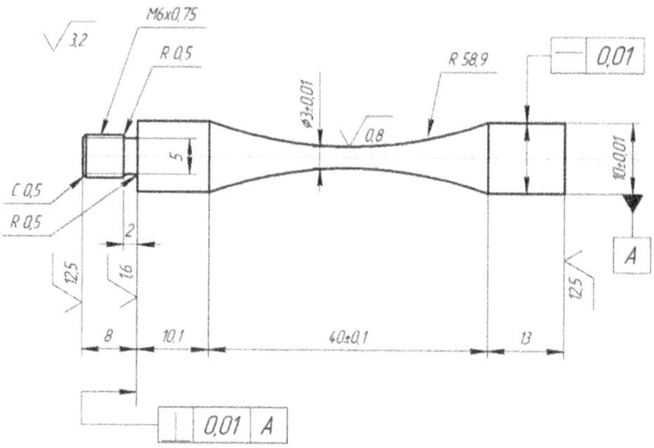

Fig. 4. Sample for high frequency testing.

In Fig. 4 the center of the spigot shaft corresponds approximately to the left side of the picture, the area of the spigot shaft – 0–300 mkm from the center; the transition zone of 300–600 mkm from the center; the "unmodified" area – more than 600 mkm from the center.

For conducting metallographic studies of the samples were cut the middle part with a length of 27 ± 1 mm (subject to maximum loads) and bush in acrylic resin. The pressing was performed on press machine MECAPRESS 3 (Presi, France) at a temperature of 150 °C. Then successively polished on abrasive paper, grit from P180 to P1200, and polished on diamond suspension with a particle size of 9 mkm to 1 mkm. Processing was performed in such a way that the result obtained in the axial section closest to the axis of the specimen. Grinding and polishing were performed on grinding and polishing machines MECATECH 234 (Presi, France) using modes and consumables recommended by the manufacturer.

After polishing the samples were washed under running water, then distilled water, then removal of fat with 50% isopropyl alcohol. Then the samples were pickling 4% solution of nitric acid in isoamyl alcohol, washed with distilled water and dried with a cotton swab. Optical metallographic examination was carried out on EclipseMA200 inverted-stage microscope (Nikon, Japan) at magnifications 50×, 100×, 200×, 500× and 1000×.

Afterwards the samples were investigated by scanning electron microscope Carl-ZeissUltra+. After the study of the microstructure was measured by microhardness testing of the samples on the microhardness tester HMV-G-FA-D (Shimadzu, Japan). In this work were investigated two samples selected from the test series with different stress amplitudes in a symmetrical cycle of loading.

Sample No. 1 was tested at an amplitude loading of 690 MPa, number of cycles to the exit of the resonance $N = 1{,}5 \times 10^{4\cdot}$ After the test, the sample was observed with a spigot shaft with a diameter 0.1 mm smaller than the original, and on the spigot shaft and the contiguous zone (out of a total width of 8 mm) observed color of a tint, mostly purple and blue. This indicates that, despite the continuous cooling of the compressed air in the process of loading, this area was heated up to 280–300 °C [5].

Sample No. 2 was tested at an amplitude of 390 MPa, and after 2.0159×10^9 loading cycles it imploded and didn't even have visible signs of deformation.

For the study the microstructures of the samples after the destruction of the high frequency gigacycle load applied optical and electron microscopy. At the beginning we present the results of photographing by means of an optical microscope image thus obtained can be processed using modern software systems control and monitoring for the presence of certain structures, defects and inclusions.

On Fig. 5a shows the microstructure of sample No. 1 after the test (increase of 100x), obtained using optical microscopy. The left side of picture corresponds to the neck, right side – section near the end of the sample. This picture illustrates very well the processes of change in the steel structure that occurs when an increase in peak stresses under high-frequency load. Conditionally it can be divided into three zones:

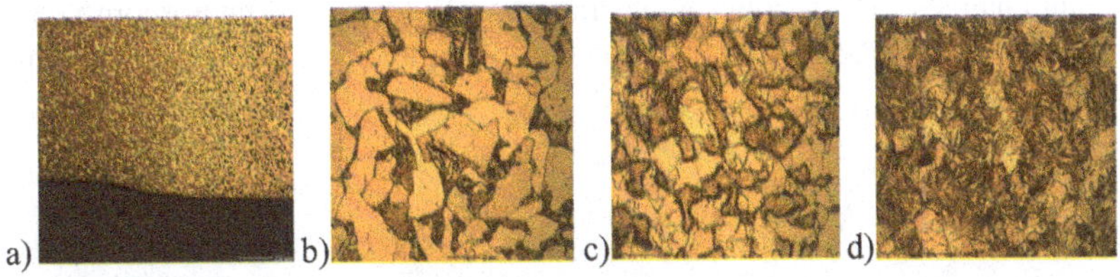

Fig. 5. Microstructure of sample №1 (optical microscopy) after the test: (a) 100^x; (b) «unchanged structure» 1000^x; (c) «the intermediate zone» 1000^x; (d) the «neck» 1000^x.

1. the area on the right edge of the picture corresponds to the section with the largest area, that is having a smaller load. In this zone, the structure meets St3sp (GOST 535-88) (conditionally – «unchanged structure»);
2. in the middle of the ferrite grain is mostly still retain their shape (although it starts crushing), pearlite noticeable changes occur. In particular, changes the color of the pearlite colonies after etching (conventionally, «the intermediate zone»);

3. the left part of picture corresponds to the thinnest part of the specimen experiencing the maximum load and deformed with a necking. In this zone there is an intensive crushing of the ferrite grains, as well as further changes in the structure of pearlite and expansion of pearlite colonies at the expense of neighboring ferrite grains (suspended – «neck»).

The structure of these three zones with increasing 1000^x (optical microscopy) is shown in Fig. 5b, c, d.

In Fig. 6a shows the microstructure of sample No. 2 after the test (increase of 100^x), obtained using optical microscopy. Shooting location corresponds exactly to the geometric center of the working part of the specimen, i.e. the place where was supposed to form the neck.

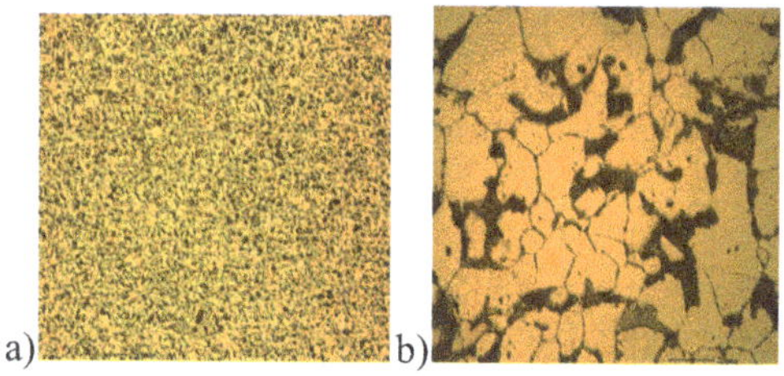

Fig. 6. The microstructure of sample No. 2 (optical microscopy) after the test: (a) 100^x; (b) 1000^x.

As you can see from this picture, corresponds to the structure of the steel St3sp (GOST 535-88), the impact loading is not observed. It should be noted that peak stress when testing this sample approximately matches the tensile strength of St3sp when tested according to standard procedures, was therefore carried out a targeted search for structural changes across the longitudinal section of the specimen. Despite this, any significant changes (which could be unambiguously fix optical microscopy) could not be found. In Fig. 6b shows the microstructure of sample No. 2 at the same point when you zoom 1000x.

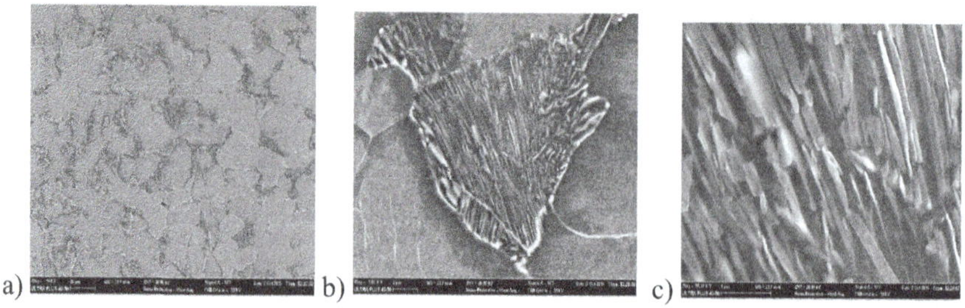

Fig. 7. Microstructure of specimen No. 1 (electron microscopy) in the area of «original structure» by increasing: (a) 500^x; (b) 5000^x; (C) 20000^x.

The capabilities of modern electron microscopy give much greater opportunity to present the microstructure of the material, as in Fig. 7a, b, c shows photographs of the microstructure of sample №1, is made approximately 2 mm from the cervical (in the area of «original structure») at magnifications of 500^x, 5000^x and 20000^x. The photograph with magnification of 500^x (Fig. 8a) allows to assess the structure of this point in general: the characteristic grain size of ferrite is 15-30 microns, pearlite is about 25% of the total area of the image. Image at higher magnifications (Fig. 7b, c) allow to estimate the structure of the pearlite at this point: crystals of cementite constituent of pearlite, have a form of plates with a typical thickness of 0.2–0.3 mkm and a length of 1–5 mkm.

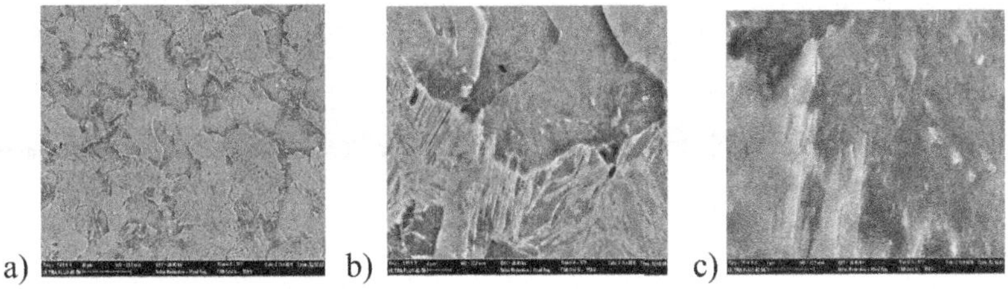

Fig. 8. The microstructure of sample No. 1 in the «twilight zone» by increasing: (a) 1000^x; (b) 5000^x; (c) 20000^x.

In Fig. 8a, b, c shows photographs of the microstructure of sample No. 1 in the intermediate zone at magnifications of 1000^x, 5000^x and 20000^x.

As can be seen from Fig. 8a, b, c in the «intermediate zone», the following changes occur: (1) there is fragmentation of the plates of cementite included in the pearlite; (2) the field of pearlite increases due to the field of ferrite; (3) decreasing the characteristic size of grains of ferrite.

In the cervical area (Fig. 9a, b) observed the same changes as in the «intermediate zone», but to the maximum extent. Thus, the characteristic grain size of ferrite in the cervix 5–15 microns (2–2.5 times less than in the unmodified zone) and the pearlite is about 50–60% of the area of the image.

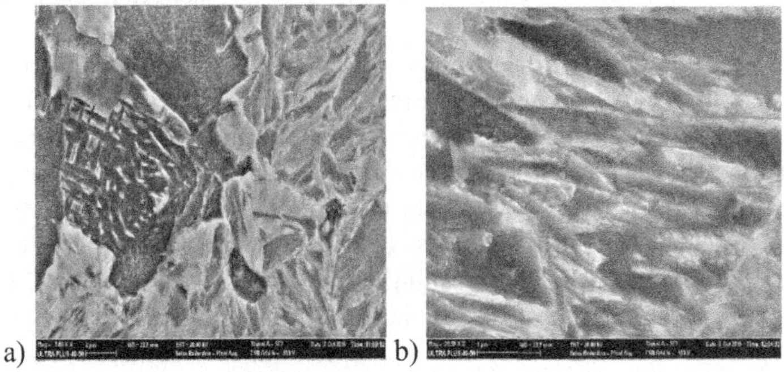

Fig. 9. The microstructure of sample No. 1 in the area of the cervix by increasing: (a) 5000^x; (b) 20000^x.

In Fig. 10a, b shows the microstructure of sample No. 2 at the thinnest part (where presumably had formed the neck) at magnifications of 1000^{x}, 5000^{x} and 20000^{x}.

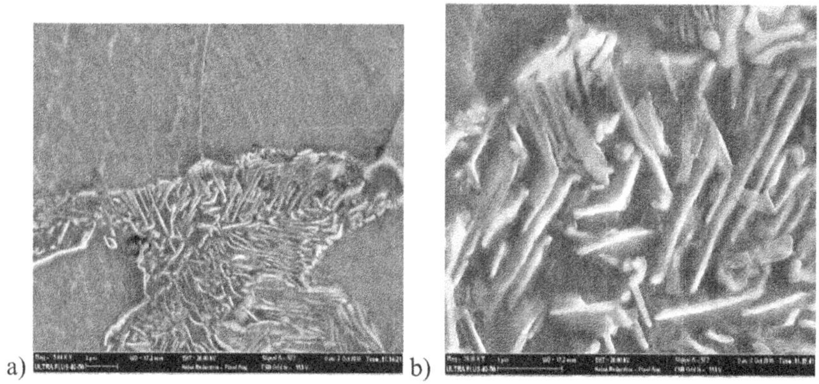

Fig. 10. The microstructure of sample No. 2 in place of the alleged spigot shaft with increasing: (a) 5000^{x}; (b) 20000^{x}.

As can be seen from Fig. 10a, b microstructure of sample No. 2 is subjected to the most significant stresses that are almost similar to the microstructure of sample No. 1 in the «unmodified» area and does not have any changes, similar to those observed in the spigot shaft and the intermediate zone of sample No. 1.

3 Discussion of the Results

On the basis of the conducted researches and obtained results we can draw the following conclusions.

1. With increasing amplitude of stress changes in the structure of rail steel begins with perlite. Likely, the first is the chattering of cementite as the most fragile phase. Then the shattered crystals of cementite shift to the area initially occupied by free ferrite, also the crushing of grains of ferrite. The nature of the changes in the perlite visible in particular on Figs. 5b, 6b, 7b;
2. Compared to the microhardness of pearlite at 1100 mkm from the center of the spigot shaft, the microhardness of pearlite in the zone of the spigot shaft is higher in 2.5–3 times, in the transition zone – about 2 times. To emphasise the high microhardness of pearlite in the zone of the spigot shaft, can be compared with hardness values of steel U8, having in the annealed condition is purely a pearlite structure. Standard steel has a hardness of HRC 58–59 (which roughly corresponds to the HV 710–740 [6]) after quenching and tempering at 200–220 °C [7] or after isothermal annealing at 250 °C [8], i.e. such values roughly correspond to the hardness of the martensite tempering or upper limit for bainite.

3. At the distance of 600–900 mkm from the center of the spigot shaft microhardness commensurate with the transitional zone, increasing the image of a section of the sample to $1000^{\times}$ shows that the structure is similar to that observed at 1100 mkm from the center of the spigot shaft. This phenomenon needs additional study.
4. Microhardness of ferrite in all areas about the same and is in the range of 200–230 HV.
5. Loading St3sp with an amplitude of 390 MPa and a frequency of 20 kHz on the basis of 2×10^9 cycles does not cause fatigue damage and even changes in the microstructure, that shows the dependence of fatigue characteristics of materials on the frequency of loading.
6. In general, the method employed allows to investigate changes in metal elements subject to action of high-frequency gigacycle load and can be used when selecting materials for different designs of high-speed transport systems.
7. The study gigacycle dynamic effects on the rail will allow you to build a prediction model of its state for the entire life cycle up to pass 1 billion tons gross train load.

References

1. Gridasov, A.V., Gridasova, E.A.: Gigacycle fatigue of welded joints of structural materials (a review). In: Advanced Engineering Forum, vol. 17, pp. 14–30 (2016)
2. Siddique, S.V., Imran, M., et al.: Fatigue assessment of laser additive manufactured AlSi12 eutectic alloy in the very high cycle fatigue (VHCF) range up to 10^9 cycles. In: APMAS 2015, 5th International Advances in Applied Physics and Materials Science Congress & Exhibition, Oludeniz, Turkey, 16–19 April 2015, pp. 26–32 (2015)
3. Loktev, A.A., Shishkina, I.V., et al.: Modeling the dynamic behavior of the upper structure of the railway track. In: Transportation Geotechnics and Geoecology, TGG 2017, Saint Petersburg, Russia, May 2017. https://doi.org/10.1016/j.proeng.2017.05.022, Proc. Eng. **189,** 133–137
4. Glusberg, B., Shishkina, I., et al.: Calculation of track component failure caused by the most dangerous defects on change of their design and operational conditions. In: MATEC Web of Conferences, vol. 239, p. 01054 (2018). https://doi.org/10.1051/matecconf/201823901054
5. Loktev, A.A., Shishkina, I.V., et al.: High frequency vibrations in the elements of the rolling stock on the railway bridges. In: IOP Conference Series: Materials Science and Engineering, vol. 463, p. 032019 (2018). https://doi.org/10.1088/1757-899x/463/3/032019
6. Loktev, A.A., Korolev, V.V., et al.: Mathematical modeling of antenna-mast structures with aerodynamic effects. In: IOP Conference Series: Materials Science and Engineering, vol. 463, p. 032018 (2018). https://doi.org/10.1088/1757-899x/463/3/032018
7. Savin, A., Kogan, A., et al.: Evaluation of the service life of non-ballast track based on calculation and test. Int. J. Innov. Technol. Explor. Eng. (IJITEE) **8**(7) (2019). ISSN 2278-3075, G5991058719/19©BEIESP
8. Glusberg, B., Shishkina, I., et al.: New lining with cushion for energy efficient railway turnouts. In: Advances in Intelligent Systems and Computing, vol. 982, pp. 556–570 (2020). https://doi.org/10.1007/978-3-030-19756-8_53
9. Glusberg, B., Shishkina, I.: Calculation of heat distribution of electric heating systems for turnouts. In: Advances in Intelligent Systems and Computing, vol. 982, pp. 337–345 (2020). https://doi.org/10.1007/978-3-030-19756-8_31
10. Loktev, A., Shishkina, I., et al.: Modeling of railway track sections on approaches to constructive works and selection of track parameters for its normal functioning

Assessment of Transport and Storage Systems

Oksana Pokrovskaya[1] and Roman Fedorenko[2(✉)]

[1] Emperor Alexander I St. Petersburg State Transport University (PGUPS), Moskovsky pr. 9, St. Petersburg 190031, Russia
[2] Samara State University of Economics, Sovetskoy Armii st. 141, Samara 443090, Russia
dorenko083@yandex.ru

Abstract. The aim of the study is to develop a system of indicators for a comprehensive assessment of transport and storage systems in general and individual logistics facilities. The main object of evaluation is terminal and warehouse infrastructure of railways. The complexity of the assessment consists in determining the number and location of the nodes of the terminal network, their composition, development stage and functionality. The assessment considers the state of the flow processes of technological and economic origin, as well as the impact of the multiplicative effect. The Logistic rationing system uses the methods of logistics, economics, planning the operational work of railway transport. In addition, it integrates the transport, logistics and economic indicators of transport and storage systems, it reflects the logistic work in the terminal warehouse cargo service, which was not previously considered in the statistics of railway transport. Within the framework of the new applied methodology - terminalistics - a system of indicators has been proposed for rationing the activity of the terminal and warehouse infrastructure facilities. This system was automated in the software developed by the authors, which made it possible to formulate several functional dependencies. The results of this study can be applied by railway companies in assessing the current operational status of the terminal and warehouse infrastructure; in the rational design of the terminal network and in planning the work of rail transport in general.

Keywords: Logistics · Transport and storage system · Rationing system

1 Introduction

The logistics infrastructure of the region is the basis of its successful development. The ability of regional enterprises to integrate into market relations, both nationally and internationally, depends on the state of the terminal and warehouse network. The importance of the development of logistics infrastructure and its ability to influence the integration of regions into the system of world economic relations is noted in the authors' early works [1, 2]. The size of logistical costs directly depends on the state of the logistics infrastructure. The reduction in logistics costs has a direct impact on the expansion of interregional cooperation [3] and increases the competitiveness of the state in the international arena [4].

Z. Popovic et al. (Eds.): TransSiberia 2019, AISC 1115, pp. 570–577, 2020.
https://doi.org/10.1007/978-3-030-37916-2_55

Today, not less than 5 thousand logistic entities (LE) of various types work on the Russian railways network. Their share in the "portfolio" of Russian Railways' business is more than 20%. Despite the wide range of terminal services provided, the activities of these objects are not systematically regulated and normalized, which complicates the control of the logistics chain. Obviously, it is necessary to solve a number of diverse tasks on integrated design, audit and regulation of their activities.

The complexity of this approach is related to the multidimensional nature of the LE as a complex transport and storage system (TSS), their multifunctional role in the logistics chain and the variety of types and formats of doing business [5].

Analysis of the literature showed the following. Basically, the objects of the terminal and warehouse infrastructure are not considered as transport and storage systems, but only in certain aspects/perspectives - as transport hubs, warehouse distribution systems [6] or complex logistics management systems [7].

At the same time, the tasks of design, integration with global transport systems and corridors [8], economic and spatial development, etc. are solved separately. It remains an opened question to develop an integrated assessment methodology for the TSS. This problem can be solved within the framework of an integrated applied methodology for the comprehensive assessment of logistics facilities of all formats. A comprehensive assessment of the logistics infrastructure at the national level was carried out by researchers from various countries. An assessment of the logistics infrastructure of Vietnam is presented. Markovitz et al. in their works they assessed the logistics infrastructure of Hungary and other countries of the European Union. The aim of the study is to develop a system of indicators for the comprehensive assessment of transport and storage systems in general and individual logistics facilities in particular, which are part of the terminal and warehouse infrastructure of railways, in the framework of the new applied methodology - terminalistics. The scientific novelty of our work consists in isolating an independent interdisciplinary section from logistics and in developing a system of indicators for the logistic rationing of the parameters of the transport and storage systems.

2 Research Methods

Before proceeding to the description of the methodology for comprehensive assessment of railway TSS, let us introduce several terms.

A new interdisciplinary methodology for the integrated study and evaluation of transport and storage systems will be called terminalistics. One of its applied tools is the system of logistic rationing of the terminal and warehouse infrastructure. This methodology is designed to link the organization of transportation with the work of its facilities (warehouses, cargo terminals and other LEs) by applying the methods and tools of transport, economic and logistics sciences. The complexity of the assessment consists in determining the number and location of the nodes of the terminal network, their composition, development stage and functionality, considering the flow processes of technological and economic origin, as well as the impact of multiplicative effect.

Logistics considers as a subject of study all types of flows in the delivery chain, regardless of their origin. Terminalistics, being its independent direction, limits the subject of study only to flows that are generated and transformed in the logistics infrastructure. Terminalistics as a methodology is close to the foreign teachings of transport geography and supply chain management. However, it is not limited only to the configuration of transport and logistics systems from the standpoint of logistics and geography and is aimed at solving a complex of tasks in terms of designing and evaluating different TSS using transport, economic and logistical theories.

Logistics entities as a type of TSS physically provide transport and customer service and are geographically concentrated in one area. The authors introduce a new concept – the Logistic rationing system (LRS). This is a new system of indicators for assessing the key parameters of the functioning and development of the terminal warehouse infrastructure of railway transport. The well-known parameters of the TSS characterize only the quality of design solutions and technology of cargo processing and are often only economic (accounting for operating or investment costs, etc.) or technical (interior zoning, calculation of cargo fronts, etc.) indicators. Therefore, they do not reflect the interests of all participants in the delivery system (customers, warehouse tenants) and do not assess the size of the added value of goods, the location of the LE, the complexity of the service, the efficiency of logistic work, etc. LRS differs from the well-known rationing system of operational indicators by its complexity and customer-oriented approach to the assessment of the TSS logistics activity. Using the proposed indicators, it is possible to plan, consider, monitor, analyze and evaluate the activities of the TSS for the purposes of both operational management (for the owner or operator of LE) and entrepreneurship (for the customer of the customer or the cargo owner). In Russia, the issue of conducting a comprehensive assessment of transport and storage systems as the basis of the logistics infrastructure has not been fully studied. This is largely due to the underdevelopment of the domestic logistics infrastructure and the shortage of high-quality Class A warehouse space (according to the international classification of the Knightfrank company). The Logistic rationing system uses the methods of logistics, economics, planning the operational work of railway transport. In addition, it integrates the transport, logistics and economic indicators of the functioning of the TSS, it reflects the logistics work in the terminal and warehouse service, which was not previously considered in the statistics of railway transport. For example, LRS allows to evaluate the work of LE and TSS from the standpoint of their economic efficiency, structure, complexity of logistics operations with a unit of cargo (and, accordingly, the increase in the cost of added value for the end customer).

3 Research Results

The calculating device LRS as the mathematical software of terminology includes 85 indicators, 41 of them are copyrighted. Figure 1 shows the composition of the LRS as a system of indicators, which can be grouped by 8 key features.

Let us consider some of them as a tool for recording the peculiarities of logistic activity in the operational work of railway terminal and warehouse infrastructure facilities.

Logistics work, QLW, shows the number of single logistic operations performed by LE per ton of cargo per unit of time (ton operations/day). Unlike the well-known

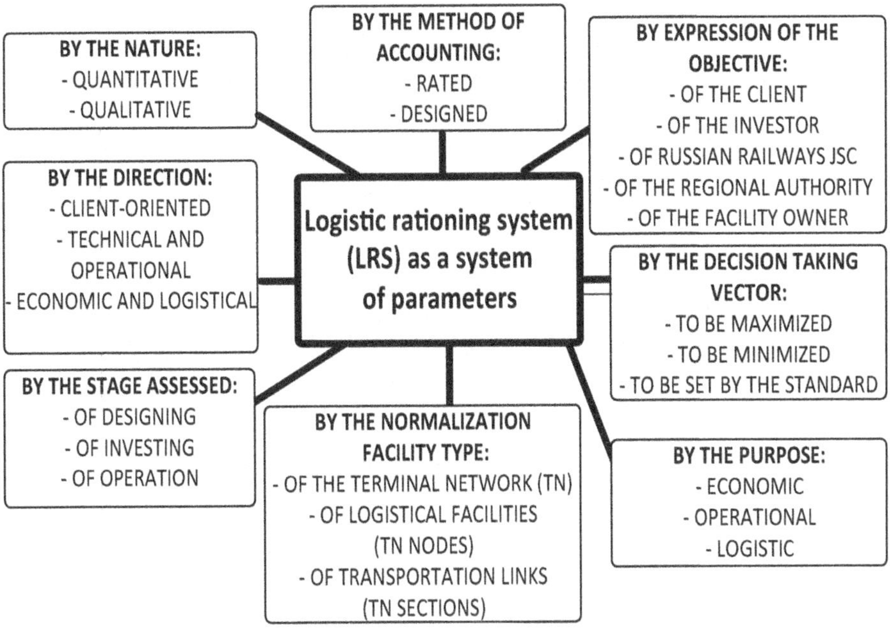

Fig. 1. The composition of LRS indicators

overload factor, it covers cargo operations not only during the overload process, but also during its entire stay in the TSS, including value- added operations.

This allows the client to conclude whether the given LE corresponds to the complexity of the service for the specifics of its cargo. At the same time, both for the client and for the owner of the LE, this indicator is useful in order to calculate the full logistic cycle, which affects the delivery time of the cargo and its price for the final buyer. Logistics performance speed, VLP, is the ratio of the magnitude of the logistic work to the total duration of the logistic cargo handling at the LE. According to this indicator, the client-customer can estimate how productively LE performs the complex of services with a certain load. For the owner of the LE indicator is useful because it reflects the current operational performance of a typical set of services.

It can be increased by improving the technology of cargo handling and warehousing of cargo at the LE. The higher its value, the more services can be implemented per unit of time, and the more profitable is the logistics work of the LE.

The key customer-oriented indicator - the coefficient of logistic work, RLWi - is an integrated indicator by which the client can determine whether his LE is satisfied with the time and money costs, as well as the ability to provide the necessary amount of logistic service. As part of this indicator there is a dislocation factor, reflecting the convenience of its geographical location relative to transport corridors and large transport hubs. In addition, the coefficient of logistic work considers changes in the value added of the cargo during its terminal processing.

The higher is the RLWi value, the more successful is its deployment, and the higher is the speed of the logistics work and the complexity of LE's service.

Logistic utility coefficient, LOG, characterizes the ability of LE independently and fully implement complex logistic services for cargo. It shows the ratio of the total value of goods processed at LE to the change in their added value, considering the number of logistic operations performed per unit of cargo.

Figure 2 shows the graphical dependence of the value of the logistic utility factor (LOG) on the useful storage area of LE (the number of pallets, Qp).

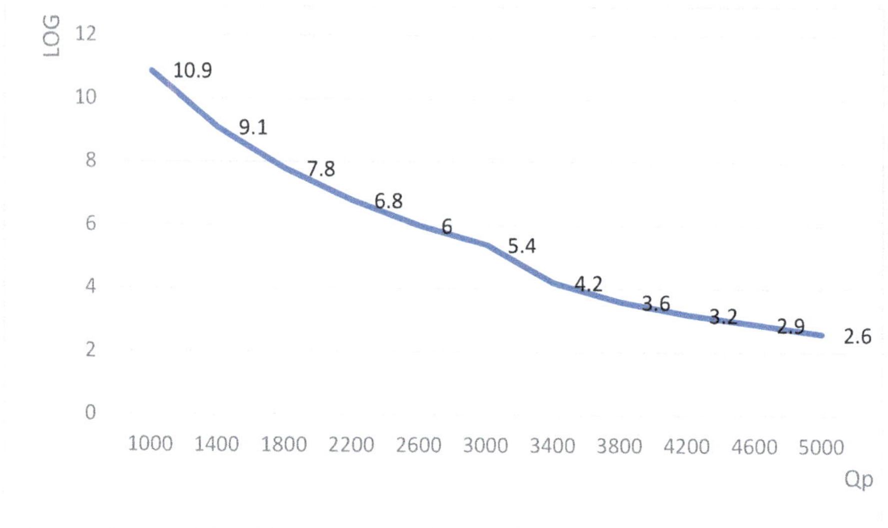

Fig. 2. Graphical dependence of the value of the logistic utility factor (LOG) on the useful storage area (Qp)

The indicator is proposed to be normalized according to planned indicators for loading and unloading. In this case, the higher is its value, the more efficient is the logistics work of the LE. It reflects the increase in the added value of the cargo and the number of operations with it, which LE can perform on its own.

The indicator of the complexity of transport technology services for an individual client at LE, Clx, characterizes the logistic independence of this object. It represents the volume of services performed by in-house LE, as well as the assortment of elementary logistic operations included in the standard package of its services.

Unlike the well-known logistic chain signaling ratio, it estimates not the number of retail chains, but the number of co-contractors of the logistics service, considering the autonomy of each. Clx also allows the owner of the LE to assess the controllability of the logistics chain, which is built through this LE. Obviously, the higher is the value of the indicator, the more complex transport and logistics services for the cargo will be received by each client at the LE. The index of provision with the region's logistics infrastructure, INF, reflects the availability of the terminal and warehouse infrastructure in this economic and geographical region. The higher is its value, the closer the LEs are to the key customers and the greater is the number of available to customers in a particular section of the terminal network. The logistic rating of a section of the railway network, R, is measured in points. In its qualitative form, this indicator reflects the density and congestion of the supporting terminal network throughout the entire range of railways (thousand rubles per unit of cargo per year). In quantitative form, it characterizes the equipment of the railway landfill with cargo handling capacities considering the availability of logistics infrastructure (the number of high-quality logistics facilities per kilometer of railways). High-quality LEs are considered as possessing class A, A+ storage facilities according to the international classification.

Figure 3 shows the graphical dependence of the value of the potential number of LEs (NLE, in units) on the logistic rating of a section of the railway network, R.

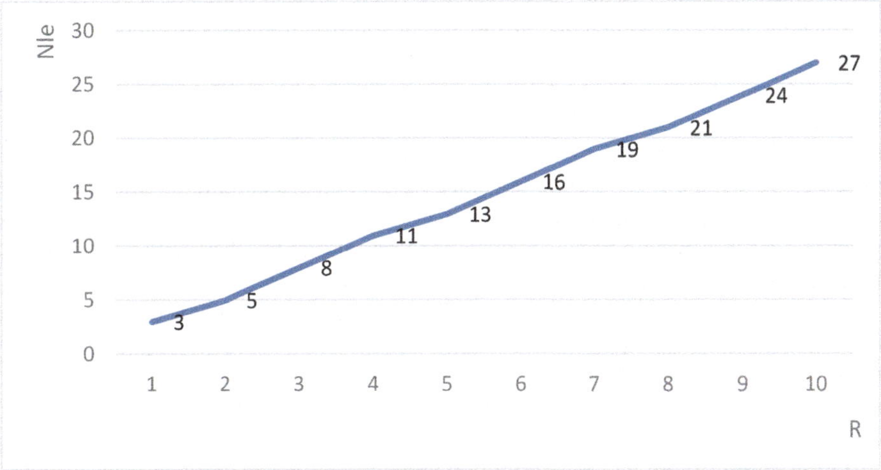

Fig. 3. Graphical dependence of the value of the potential number of LEs (N_{LE}) on the logistic rating of a section of the railway network (R)

The higher is the value of the indicator, the higher is the availability of high-quality logistics services to customers of a given economic and geographic area and, accordingly, the more efficiently the supporting terminal network operates in this area or landfill.

4 Results and Discussion

The peculiarity of the author's indicators is their integration: each of them, to a certain extent, considers a whole complex of economic, technical, and operational parameters, reflecting both the current operational state of the LE and its functionality for the customer.

Thus, LRS synthesizes tools for assessing customer focus and complexity and relates them to the technical standards of operational work of railway transport (from system-wide parameters to specific features of the railway terminal network).

The authors have developed a computer program that implements the proposed system for estimating TSS according to LRS indicators in the C# programming environment for practical testing of the calculation procedures.

Let's consider the composition of the program. The working interface of this software product contains tabs. These are the "Author's Indicators", "Visualization", "Map" and "Methodological Support". In the tab "Author's Indicators" the user enters into the empty fields the values of known source data. At the same time there is an opportunity to make a sample of the necessary indicators (for individual indicators and for their groups) in accordance with the objectives of a comprehensive assessment. After clicking the "Calculate" button, the program provides the calculation results in a pop-up window, and allows you to upload the results to an MS Excel file and build graphical dependencies of the indicators. The "Help" button brings up a context menu on which the user can get detailed methodological instructions on LRS and instructions for working with the program.

In addition, it is possible to detail the calculation to individual transport sections or nodes that are part of the terminal network. By filling in the empty fields with relevant data, the user receives an express assessment of the work of the LE on integrated groups of indicators. It is possible to autocomplete fields according to the data entered earlier in the "Author's indicators" tab. According to the results of calculations, the program can visually present the results in the form of graphs, charts and dependencies, the type and ratio of which is determined by the user in the "Visualization" tab.

In the "Map" tab there is the possibility of forming an interactive map of the location and state of the terminal network with a specified degree of detail. The map is necessary for visualizing the control and accounting of the activities of the terminal and warehouse infrastructure. With the help of LRS indicators, the map displays the current state of each LE for individual sections of the vehicle.

In the future, it will be necessary to integrate the proposed software with network-wide technologies, as well as to clarify and supplement the system with new indicators. The theoretical results of the study are the characterization and composition of the LRS as a tool in the theory and methodology of terminalistics. The main theoretical results also include approbation of the proposed system for assessing and rationing TSS activities in the author's software product, which automates decision-making on all indicators. It can be assumed that the practical results of this study can be used by railway companies in solving the following applied problems:

- identification of the type and condition of LE with regard to the design, type of storage, size, technical equipment, etc.;
- customer focus of LE activities;
- integrated assessment of terminal warehouse infrastructure of railways;
- LE design;
- rail transport planning.

The result of a comprehensive assessment will be the identification of bottlenecks in the logistics infrastructure and the identification of areas for its further development. Modern world requires a progress in transport technologies and qualitative logistics are a necessary condition for investors to operate efficiently. The development of logistics creates conditions for increasing trade volumes, reducing the total costs of local producers, and increasing the region's export potential. Transport and logistics services facilitate international trade and play an important role in the growth and development of the local economy. Foreign researchers acknowledged that infrastructures are important in the development of national logistics systems. The assessment methodology proposed by the authors can be the basis for identifying opportunities for improving the logistics of goods flows at the level of interregional and international trade.

Acknowledgments. The reported study was funded by RFBR and FRLC according to the research project № 19-510-23001.

References

1. Fedorenko, R.: Modern issues of development of the customs and logistics infrastructure of the international north-south transport corridor. In: Contributions to Economics 1, pp. 63–77 (2019). https://doi.org/10.1007/978-3-030-11754-2_5
2. Pokrovskaya, O., Fedorenko, R.: Evolutionary-functional approach to transport hubs classification. In: Advances in Intelligent Systems and Computing 1, pp. 356–366 (2019). https://doi.org/10.1007/978-3-030-19756-8
3. Popova, N., Kvint, M., Dementyev, A.: Interaction in transportation as the basis for implementation of intergovernmental cooperation. In: MATEC Web Conferences, vol. 216, pp. 1–14 (2018). https://doi.org/10.1051/matecconf/201821601014
4. Dang, V.L., Yeo, G.T.: Weighing the key factors to improve Vietnam's logistics system. Asian J. Shipp. Logist. **34**(4), 308–316 (2018). https://doi.org/10.1016/j.ajsl.2018.12.004
5. Pokrovskaya, O.: Terminalistics as the methodology of integrated assessment of transportation and warehousing systems. In: MATEC Web of Conferences, vol. 216 (2018). https://doi.org/10.1051/matecconf/201821602014
6. Dybskaya, V., Vinogradov, A.: Promising directions for the logistics service providers development on the Russian market in times of recession. Transp. Telecommun. **19**(2), 151–163 (2018). https://doi.org/10.2478/ttj-2018-0013
7. Likhacheva, A., Makarov, I., Pestich, A.: Building a common Eurasian infrastructure: agenda for the Eurasian Economic Union. Int. Organ. Res. J. **13**(3), 97–112 (2018). https://doi.org/10.17323/1996-7845-2018-03-06
8. Nguyen, L., Notteboom, T.: Public-private partnership model selection for dry port development: an application to Vietnam. World Rev. Intermodal Transp. Res. **6**(3), 229–250 (2017). https://doi.org/10.1504/WRITR.2017.086208

Analysis of Trends and Processes of Auto Service Promotion

Valeriy Zenchenko and Mikhail Grigoriev(✉)

Moscow Automobile and Road Construction State Technical University, Leningradsky Pr., 64, Moscow 125319, Russia
{V.Zenchenko,M.Grigoriev}@madi.ru

Abstract. The paper presents the main approaches and principles of planning and managing service promotion, as well as organizing effective production processes for maintaining cars in a safe and sound technical condition. The article presents a mathematical model for forecasting the supply and demand of services for servicing automobiles with justification of the required production capacities and another mathematical model for estimating the incoming flows to service enterprises, taking into account the intensity of operation of automobiles and their reliability, which varies over time, allowing optimization of the necessary production resources for the functioning of existing as well as emerging service enterprises. The conducted studies allow us to formulate and solve the problems of ensuring the possibility of assessing the optimal levels of profitability of the promoted projects for the development of service enterprises at specified time lags, considering market conditions and possible risks.

Keywords: Technical operation of ground vehicles · Promotion of auto service · Auto service life cycle · Forecasting supply and demand

1 Introduction

In recent years, there has been a growth in the vehicle fleet in the Russian Federation, including passenger vehicles. It was determined by the positive dynamics of population welfare, an increase in consumer properties of the vehicles, convenient conditions of supply, purchase, use of the vehicles, and a number of other factors. At the same time, attention should be paid to the growth in the promotion of premium-class vehicles equipped with high-tech systems to ensure traffic safety, comfort, autonomous vehicle (AV) control systems, as well as hybrids and electric cars, cars using alternative energy sources, etc. Forecast estimates of innovations (Fig. 1) in the field of road transport show a potential growth in the volume of the market for such vehicles.

In these conditions, it is important to solve problems aimed at promoting auto services in the framework of improving the functioning of production and technological systems of enterprises of automotive industry in the areas of their business activity.

Z. Popovic et al. (Eds.): TransSiberia 2019, AISC 1115, pp. 578–590, 2020.
https://doi.org/10.1007/978-3-030-37916-2_56

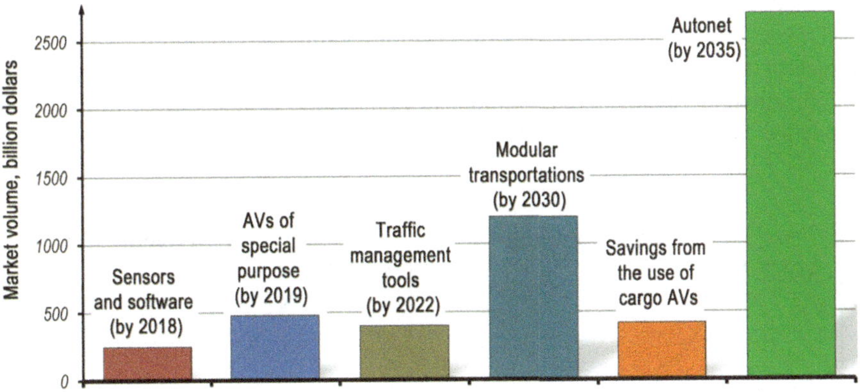

Fig. 1. Forecast of the market dynamics of innovations in the field of road transport

2 Research Methods

Analysis of research in the field of economic development theory [1] shows the convergence of prospects for the creation and promotion of high-tech vehicles (see Fig. 1) and the cyclical nature of technological structures (5th and 6th cycles) according to the Schumpeter – Kondratieff wave theory (Fig. 2), reflecting the growth of economic activity in the short and long term due to innovative approaches in the field of automotive engineering.

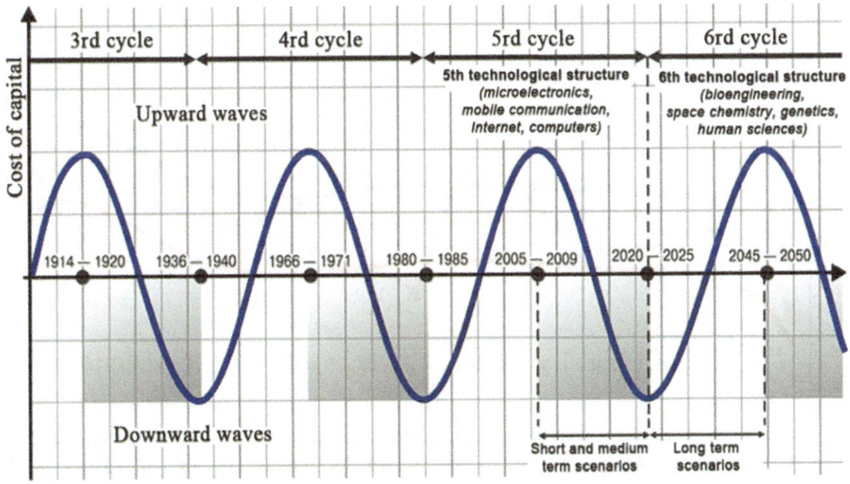

Fig. 2. Schumpeter-Kondratieff's innovative-cyclical theory of economic development

Nowadays, there is already a development of promising areas and new technologies focused on the creation and promotion of vehicles of a new generation, equipped with automated autonomous control systems (AACS) [2] (according to the Action Plan - the "road map" of "Autonet" National Technology Initiative, Appendix No. 1 to the minutes of the meeting of the Presidium of the Presidential Council for Economic Modernization and Innovative Development of the Russian Federation), which implies a reduction in the cost of production and the life cycle of such cars (Fig. 3).

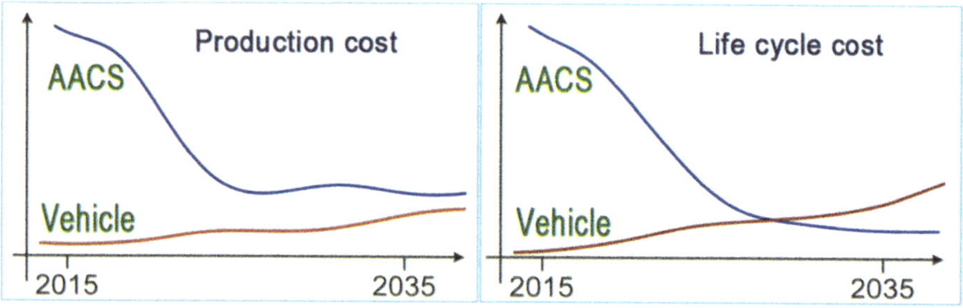

Fig. 3. Predictive assessment of changes in production costs and the life cycle of the promotion of the vehicle with AACS

The studies carried out at the department of "Operation of road transport and car service" of Moscow Automobile and Road Construction State Technical University (MADI) allowed predicting the growth of sales of passenger cars after 2016 and confirm the resulting model for predicting sales in the retrospective period [3–5] (see Fig. 4). In this figure, Q_t is the trend in sales, and $Q^u_{t_\gamma}$ and $Q^l_{t_\gamma}$ are its upper and lower confidence limits for a given probability γ.

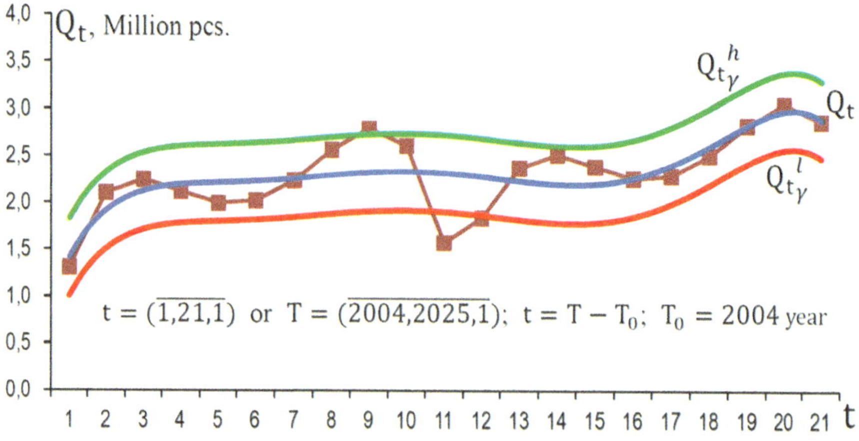

Fig. 4. Modeling of sales volumes Q_t, $Q^h_{t_\gamma}$, $Q^l_{t_\gamma}$ using the time lag t

The functioning of an auto service company (ASC) in a market economy predetermines the need to obtain objective and reliable information that, in the process of solving problems of development and improvement of production and commercial activities, allows:

- gaining competitive advantages in the auto repair service market; reducing financial risks and hazards for the image of ASC;
- determining consumer relations; following the external environment;
- assessing production activities in the field of ensuring the performance of vehicles;
- increasing the credibility of advertising;
- receiving support in decisions;
- strengthening intuition;
- improving efficiency, etc.

At the same time, it is necessary to develop and adopt a system of measures to influence the market, consumer demand, taking into account the available material resources and prospects for profit.

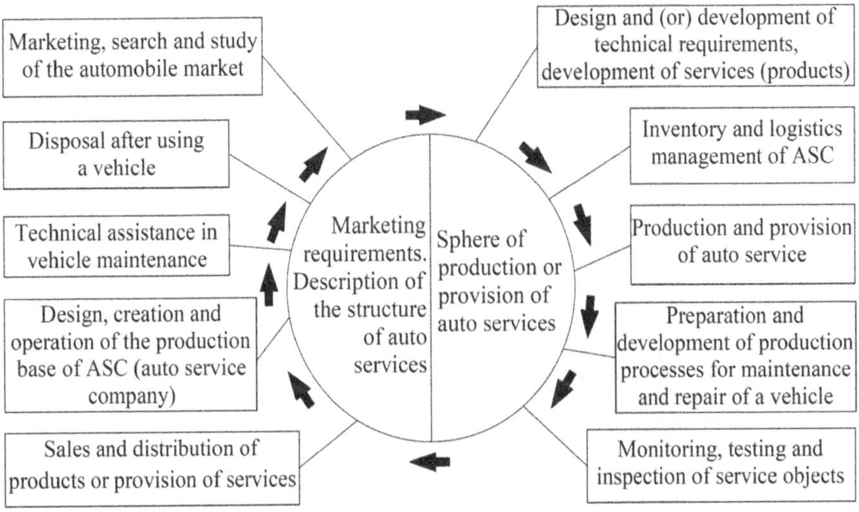

Fig. 5. Place of marketing in the system of production – circulation

In accordance with the requirements of ensuring the quality of services (products), which are based on the Russian State Standards ISO of 9000 and 9004 series, the place of marketing in the system of production or provision of services (including auto services) can be generally presented in the form of the scheme shown in Fig. 5.

Sources of marketing information are: market for services; internal production environment; external macro environment associated with the production and sales of services.

Marketing analysis, encompassing the entire production and economic activity of the enterprise, should ultimately lead to the advancement of new ideas and goals, the development and assessment of ways to achieve them, the relevant strategic directions of development, and the management's decision on their implementation. The scheme of planning and management of marketing research is presented in Fig. 6. The basis of marketing planning should be information about the resources of ASC, actions and intentions of competitors, the development of the situation in the auto repair service market [3, 4].

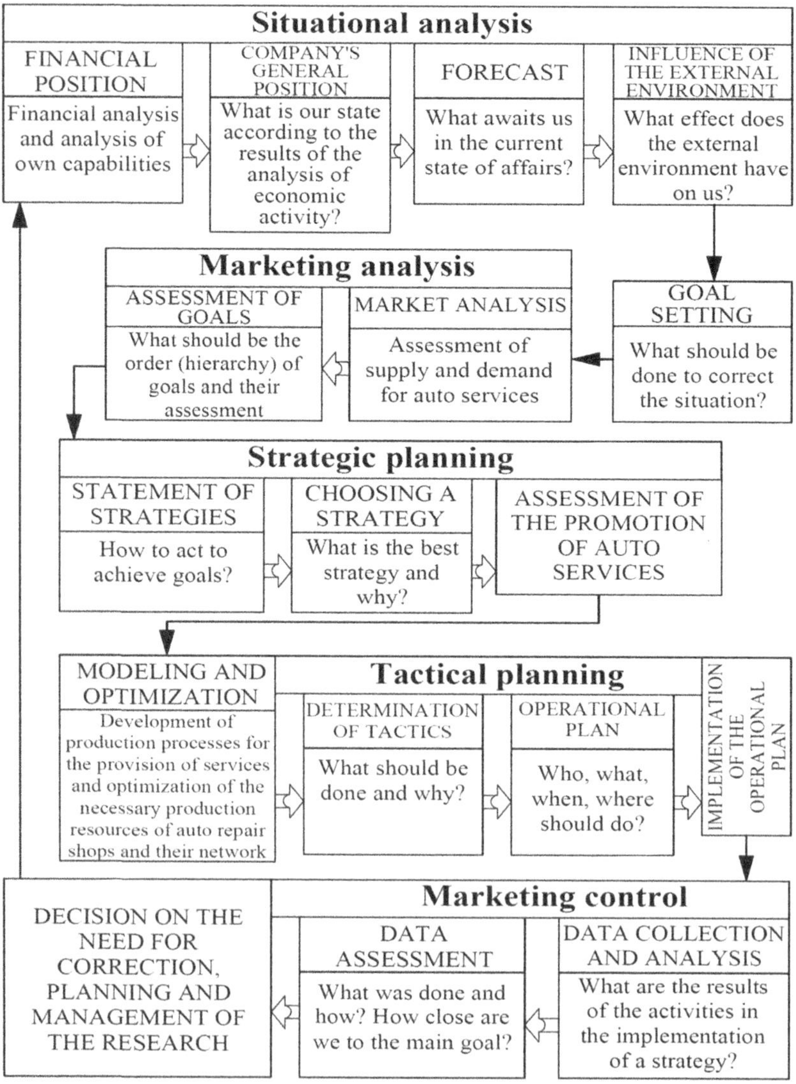

Fig. 6. The general scheme of marketing research aimed at the management and promotion of services

As objects of situational analysis, the following should be indicated: market for services; types of services of their consumers and competitors; pricing policy; the system of formation of demand and sales promotion of services; advertising. At the same time, it is necessary to take into account the availability of financial and other resources of the enterprise, to clearly define the stages of the "life cycle" [5], where each type of service is located in the considered segments of the auto repair service market (Fig. 7). This figure shows the market periods of promotion of services using the time lag T, for which there is a change over time of the scope of promoted services Q and profit PR.

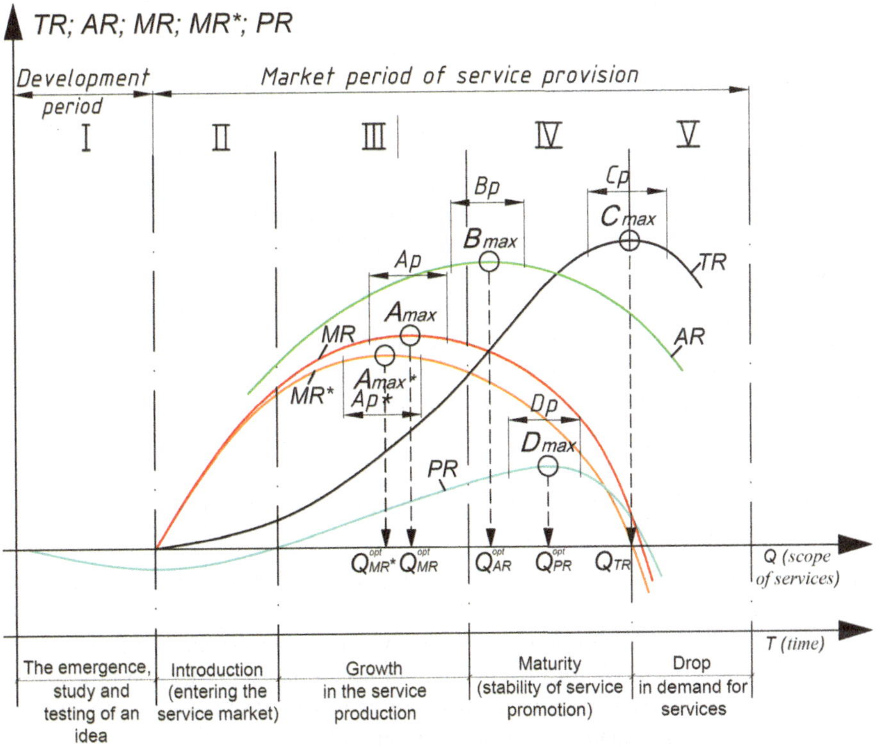

Fig. 7. Schematic representation of the life cycle of services

At varying prices of auto services P(Q) = var, depending on their scope Q, total TR and average AR revenue, as well as the increase in income MR* from changes in the scope of services and price, and the marginal income MR from promoting services are determined from the following equations:

$$TR(Q) = P(Q) \cdot Q \tag{1}$$

$$AR = \frac{TR(Q)}{Q} = P(Q) \tag{2}$$

$$MR^* = \frac{dP(Q)}{dQ} \cdot Q \tag{3}$$

$$MR = \frac{d[P(Q) \cdot Q]}{dQ} = \frac{dP(Q)}{dQ} \cdot Q + AR(Q) \tag{4}$$

In Fig. 7, A_{max}, A^*_{max}, B_{max}, C_{max}, D_{max} are the maximum values of MR, MR*, AR, TR, and PR, respectively, for the considered stages of the market period (life cycle) of the provision of auto repair services, and the ranges A_p, B_p, C_p, D_p - rational period intervals ensuring the achievement of the largest MR, MR*, AR, TR, and PR.

Promotion of auto repair services should include market segmentation by identifying homogeneous groups of potential consumers, followed by the formation of appropriate types of service offers for them. At the same time, when assessing the performance of ASE, it is necessary to analyze its competitiveness. The study of the market and competitors is of particular importance in conditions of rapid expansion of the range of services, high profitability, and should be directed to those areas that are the subject of an analysis of ASE own potential. An analysis of competition in the market is carried out within the framework of the general system of collecting and processing information that operates at the enterprise. Identifying the strengths and weaknesses of competitors' activities and formulating directions and ways to counter them is the final result of a marketing research of market competition, which concentrates the result of an analysis of all aspects of production, finance, and strategies of competing auto service enterprises.

In the process of conducting research, principles and approaches were developed [6] that are aimed at:

- an assessment of the supply and demand of auto repair services and their forecasting;
- working out the principles and approaches of promoting the marked services;
- development of methods for organizing high-performance production processes of auto repair service and optimization of the necessary resources for the effective functioning of auto repair shops and their network;
- development and practical implementation of methods for optimizing the need for spare parts for the dealer network of auto repair shops in the areas of its business activity, which is reflected in general form in Fig. 8.

As part of the marketing research, issues addressed to solving the problems of forecasting the promotion of auto repair services are considered, taking into account the structure and interrelation of external (**Y**) and internal (**X**) factors and their subfactors $\{\mathbf{Y}_{i\alpha}\}$, $\{\mathbf{X}_{j\beta}\}$, shown in general form in Fig. 9 [2].

In general form, the interrelation of external (Y) and internal (X) factors of the environment (with the corresponding sets of their subfactors $\{y_{1t}\}$,...,$\{y_{6r}\}$, $\{y_{1v}\}$,...,$\{y_{6w}\}$), affecting the promotion of auto repair services (Z) when forecasting supply and demand, is shown in Fig. 10.

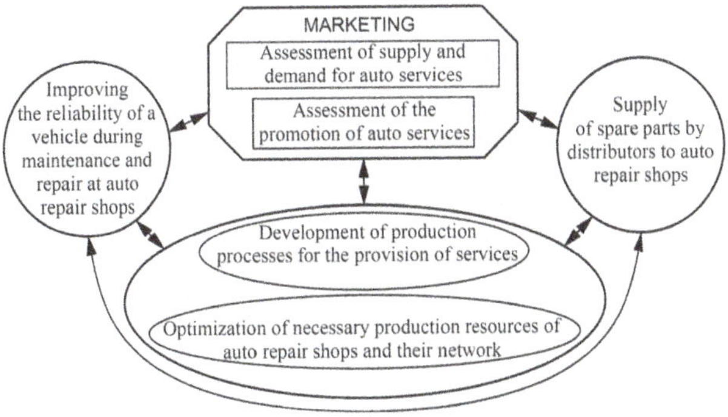

Fig. 8. The relationship of marketing with the requirement of unification of vehicle elements, working out the system of supply of spare parts, and optimization of production processes and resources of auto repair shops

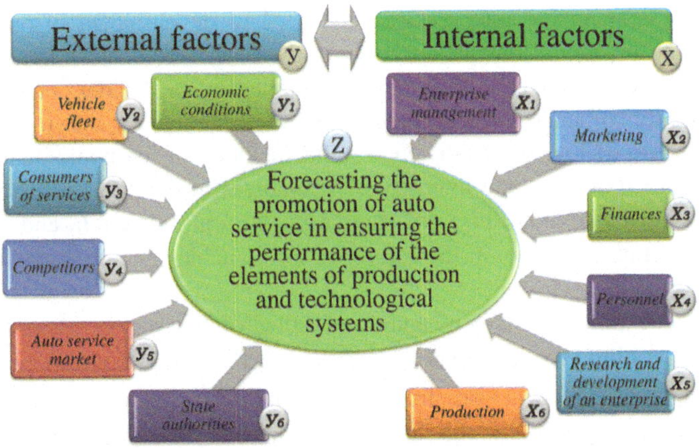

Fig. 9. The influence of factors of external and internal environment on forecasting the promotion of auto repair services

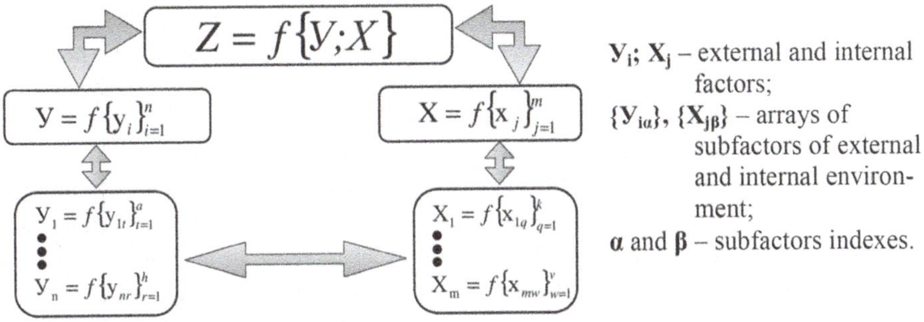

Fig. 10. Diagram of the relationship of factors of external and internal environment, affecting the promotion of auto repair services

For example, factor X_6 (production) acting as an internal factor includes many subfactors: X_{61} (modern technological equipment), X_{62} (flexibility of production processes), X_{63} (quality of production planning and management), X_{64} (production technology), etc. It has a direct impact with a certain degree of significance (weight) on the promotion of auto repair services. Dealing with the interrelationships between internal and external factors allows a targeted approach to solving the problems of forecasting the demand and supply of auto repair services (both for existing and newly created auto repair shops with an assessment of their potential production capacities and necessary production resources) in the area of business activity of auto service systems.

When solving this task, it is necessary to take into account the parameters characterizing: the prospects for population change and the dynamics of saturation of the area with passenger cars at the current moment and for the forecast time period; the dynamics of changes in the share of owners using auto repair services; probabilistic distribution of cars serviced at auto repair shops (by model); operational reliability of vehicles and modes of their maintenance; the intensity of car operation, the level of supplying the demand for auto services; the pace of development of operating auto repair shops in the area of business activity in the short and medium term, taking into account fluctuations in the economic situation on the time lags under consideration, etc.

The procedure of analyzing and forecasting the demand y_t for auto repair services is based on the use of logistic dependencies and methods of harmonic analysis, taking into account the dynamics and intensity of y_t changes in the past (in the range of the retrospective period from $\boldsymbol{t}_0$ to $\boldsymbol{t}_p$), the state at the current moment $\boldsymbol{t}_i$, and indicators of population saturation by cars in the future up to the achievement of maximum future demand M_p (see Fig. 11).

At the same time, assessment and forecasting of the supply of services by the network of operating auto repair shops (trend M_B as a function of variable sets $M_B = f\left[\{M_{yk}\}; \{\alpha_{ck}\}; \{k\}_1^K\right]$ and its possible fluctuations within the upper and lower confidence limits). The array M_{yk} is the satisfied demand for the k-th auto repair shop from the considered network, and α_{ck} is a forecast for an increase in the satisfied demand also for the k-th auto repair shop.

In order to determine whether it is necessary to carry out development or construction and commissioning of a service enterprise (or a group of auto repair shops for their network) in a given business area, it is necessary to assess the expected gap Δ by the time $\boldsymbol{t}_i$ between the required scope of services y_{t_i} and the one that can be provided by the existing network of service enterprises after its modernization $M_B(\boldsymbol{t}_i)$. At the same time, the additional demand for services M_{ad} in the area at the time of commissioning the designed auto repair shop is defined as the difference $(y - M_B)$ at the time $\boldsymbol{t}_i$. At the current time $\boldsymbol{t}_p$ (end of the retrospective period), identifying the array of scope of offers N_k^B for auto repair services in the existing auto repair shops in the area under consideration allows, given the forecast of an increase in meeting the demand α_{ck}, predicting the growth of N_k^B at the time of launching a new auto repair shop (or their network) and assessing the patterns of distribution of $f(N_k^B)$ with the identification

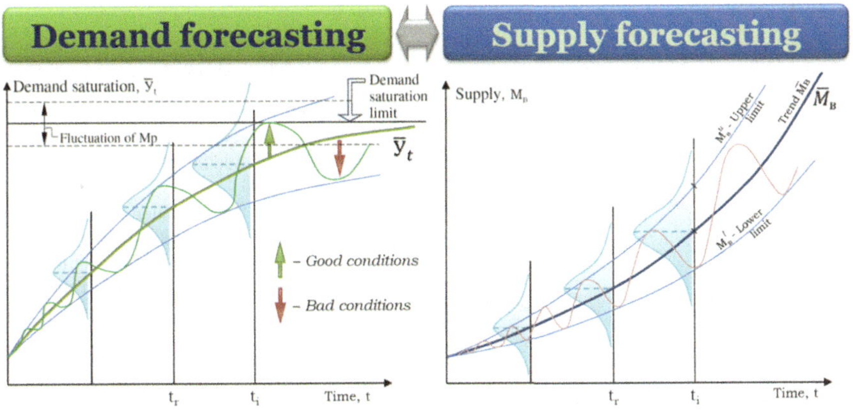

Average demand

$$\bar{y}_t = \frac{M_p \cdot M}{M+(M_p-M)\cdot exp[-\varphi M_p(t-t_r)]},$$

where: t_r – final year of the retrospective period;
φ – coefficient of proportionality;
M_p – maximum prospective demand;
M – total demand for the end of the retrospective period.

$M_в$ – offer of services by a network of existing auto repair shops

$$M_в = f[\{M_{yk}\}; \{\alpha_{ck}\}; \{k\}_1^K],$$

where: M_{yk} – satisfied demand for the ***k**-th* auto repair shop;
α_{ck} – forecast of an increase in satisfied demand for ***k**-th* auto repair shop.

Fig. 11. The mechanism of demand forecasting and service promotion

of their upper $\bar{N}_u^B$ and lower $\bar{N}_l^B$ tolerant boundaries for a given probability γ. The received information, taking into account a preliminary assessment of the production capacity limit N_{max}^B (excluding possible risks in the competition between the network of operating and newly established auto repair shops), creates conditions for an informed decision on the accepted value of the scope and the supply of services $\tilde{N}^B$ for the newly created auto repair shop and determining the conditional number of fixed cars N_c for it (taking into account the arrays of annual mileages L_A and usage L_j in cases of handling when using the j-th level of production organization, as well as the share of the car β using the services of auto repair shops) (see. Figure 12).

3 Research Results

The implementation of the aforementioned stages will allow a target approach to the optimization of the necessary production resources in organizing the maintenance and repair of vehicles at auto repair shops, the procedure of which is generally reflected in Fig. 13.

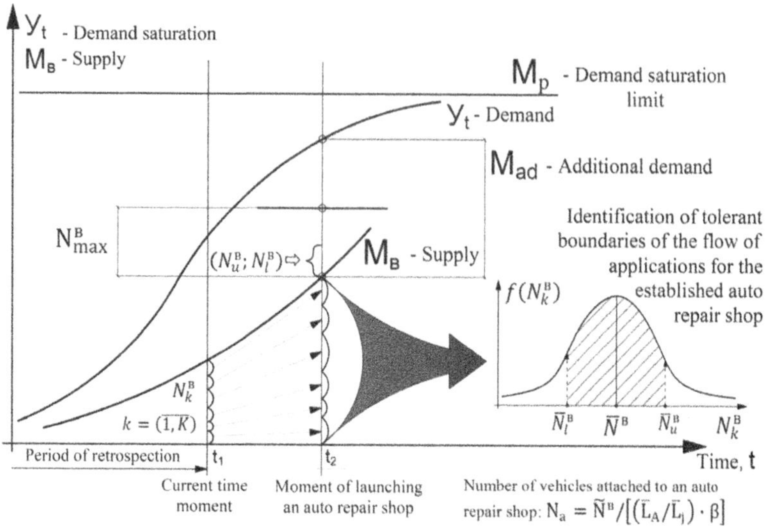

Fig. 12. The procedure for assessing the allowable limit of the flow of applications N^{B}_{max} at the created service station

Depending on the levels of production organization X_j (used technologies, equipment, qualification of personnel, management systems for maintenance and repair of cars, etc.. [6]), modeling and formation of arrays of average complexity of servicing the applications $\bar{t}_j$ for combinations of factors X_j and identification of average (realized) complexity of servicing the applications $\bar{t}$ can be done. This will make it possible to assess, through the correction parameter K_j, the change in production capacity M_j, which in turn takes into account the duration of production during the day T_D, the number of maintenance personnel R at the maintenance and repair workplaces, and the number of these workplaces Nwp. It should be borne in mind that the production capacity M_j is influenced by incoming flows of requirements to the auto repair shop N_{oj} (ΔL) in the mileage range of cars ΔL, depending on the average daily ℓ_{ad} vehicle mileage (characterizing the intensity of operation), the conditional number of cars N_c assigned to the created auto repair shop, and the values of the leading functions of the flow of failures (circulations) of the car $\Omega_j(\Delta L)$ on the mileage ΔL when using a given level of production organization X_j, etc. The obtained results allow approaching the assessment of the reliability of service workplaces P_j and the production units (system) as a whole Pc_j (i.e. the probabilities of fulfilling the specified scope of work t_{d_j} for the assigned directive time T_D with the availability of production resources M_j). In this case, the indicator Pc_j is modeled for the range of workplaces Nwp from their minimum N^{min}_{wp} to the maximum N^{max}_{wp} values (taking into account possible combinations $C^{N^{min}_{wp}}_{N_{wp}}$ of the performance level of service workplaces, determined by the reliability of the process equipment and the intensity of its recovery, the turnover of workers, and other factors).

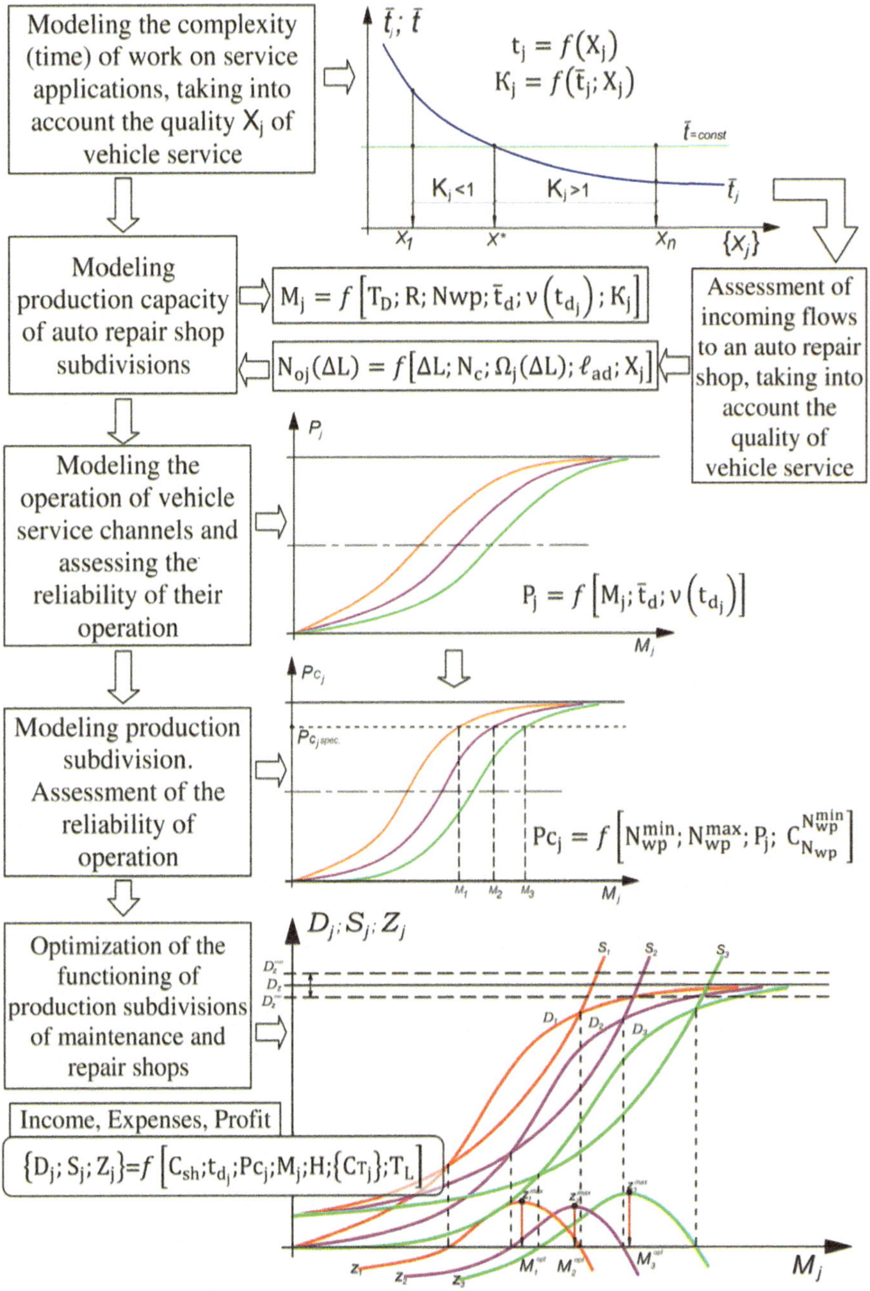

Fig. 13. The procedure for optimizing the necessary production resources in the organization of maintenance and repair of vehicles at auto repair shops

The modeling results and the obtaining of the indicators Y_t, $M_B(t_i)$, which characterize the demand for services and their supply by the network of auto repair shops, make it possible to identify the conditional number of cars N_c and the car flow parameters $\tilde{N}^B$ at the created auto repair shop, to assess the indicators $\bar{t}_j$, K_j, M_j, $\Omega_j(\Delta L)$, $N_{oj}(\Delta L)$, and Pc_j, and implement on their basis the modeling and optimization of maintenance and repair of vehicles (taking into account the specified level of its organization X_j). As the final result, this optimization will provide an opportunity to obtain estimates of optimal levels of income D_j, expenses S_j, and profitability Z_j of projects being promoted at the time lag T_L, taking into account cost of standard hour C_{sh}, tax deductions H and an array of refinancing rates, inflation expectations and risks C_{T_j} (see Fig. 13).

References

1. Glazyev, S.: Discovery of regularities of changes of technological orders in the central economic and mathematics Institute of the Soviet Academy of Sciences. Ekonomika i matematicheskie metody **54**(3), 17–30 (2018). https://doi.org/10.31857/S042473880000655-9
2. Saikin, A.M., Bakhmutov, S.V., Terenchenko, A.S., et al.: Tendency of creation of “driverless” vehicles abroad. Biosci., Biotechnol. Res. Asia **11**(Spl. Edn.), 241–246 (2014). https://doi.org/10.13005/bbra/1469
3. Ageev, E.V., Alekhin, Yu.G., Grashcow, S.A., Sarigo, N.V.: Reconstruction of infrastructure LLC “Avtomir”. Proc. Southwest State Univ. **21**(6(75)), 32–43 (2017). https://doi.org/10.21869/2223-1560-2017-21-6-32-43, https://science.swsu.ru/jour/article/view/263/261, (in Russia)
4. Lysanov, D., Nazmutdinov, A.: The demand and operation of car service centers. Int. J. Appl. Eng. Res. **11**(6), 4314–4317 (2016)
5. Wiesner, S., Freitag, M., Westphal, I., Thoben, K.-D.: Interactions between service and product lifecycle management. Proc. CIRP **30**, 36–41 (2015). https://doi.org/10.1016/j.procir.2015.02.018
6. Ageeva, E.V., Sabelnikov, B.N., Shcherbakov, A.V., Pykhtin, A.I.: Improving the efficiency of the process of technical operation of vehicles by using the distance diagnostic method. Proc. Southwest State Univ. **22**(6(81)), 6–13 (2018). https://doi.org/10.21869/2223-1560-2018-22-6-6-13, https://science.swsu.ru/jour/article/view/407/392#, (in Russia)

Distributing Forces and Capacities of the Vibrocleaver Operating Process for Compacted Snow

Gennadiy Voskresenskiy(✉) and Evgeniy Kligunov

Pacific National University,
Tihookeanskaya st., 136, Khabarovsk 680035, Russia
004179@pnu.edu.ru, kes70@bk.ru

Abstract. The mathematical model for the movement process of compact vibrocleaver for compacted snow, which provides mechanized cleaning of sidewalks and roads from snow-ice formations, is presented. The device uses high-frequency impact of the cutting tool, which is called vibrocutting. The interaction of the cutting tool is considered within the time interval: first the introduction into the destructible medium, then the cleavage of the snow, a sharp decrease in load, the return of the blade to its original position. The model that takes into account the inertial forces, the interaction resistance of the cutting tool in the form of a triangular impulse function. It is adopted that the maximum value of the driving force is limited by the adhesion of the driving wheels to the movement surface. It is assumed that the component forces of operating process are determined by the values of the inertial forces, the driving forces and the forces of impulse interaction between the cutting tool and compacted snow. The variability depends on the interaction force of the cutting tool with compacted snow. The capacity distribution of the vibrocleaver operating process according to the maximum values of the interaction forces between the cutting tool and snow creates correlation between the engine power and the power to overcome inertia forces. Also, it makes it possible to determine the engine power taking into account restrictions on the adhesion condition of the driving wheels and the movement surface.

Keywords: Compacted snow · Vibrocleaver · Mechanized cleaning of sidewalks and roads · Vibrocutting

1 Introduction

The problem of compacted snow on roads, sidewalks and other transport facilities is of great importance for the climatic conditions of the Russian Federation. Finding ways to improve the efficiency and performance of snow removers remains an urgent task. In order to solve this problem, experts of many countries on the winter road maintenance and transport facilities have conducted significant studies, which are presented in the materials of the Second International Symposium on fighting snow and ice slicks in Hanover, New Hampshire, U.S., in 1978, as well as described in detail [1–10].

Z. Popovic et al. (Eds.): TransSiberia 2019, AISC 1115, pp. 591–599, 2020.
https://doi.org/10.1007/978-3-030-37916-2_57

2 Methods

In most works, the greatest attention is paid to the implementation experience, the search for new chemical reagents and the norms of their distribution [11, 12, 13].

The difficulties of winter cleaning are primarily related to the metamorphism of snow, i.e. ability to change its properties under the influence of climatic conditions (temperature, humidity) as well as the movement of paving facilities and pedestrians. Work on cleaning roads and sidewalks should be carried out in the shortest possible time, under the influence of wheels of transport, movement of pedestrians and temperature fluctuations. The snow on the roadway is quickly compacted, and then it turns into a snow-ice run-up or ice. The strength of compacted snow is 17–33 times higher than the strength of sand snow.

The most rational snow cleaning technology in heavy traffic conditions is the complex use of mechanization tools and chemical materials, solutions of which are characterized by low eutectic temperatures. At the same time, the presence of reagents in the snow retains its loose state and prevents compaction, therefore it is easily raked and swept away from the road by plow-brush snow removers [2]. At the same time, the recommendations on the use of sand-salt mixtures with the addition of sodium and calcium chlorides are given. The effectiveness of such mixtures is limited by −19 °C, −20 °C air temperature. At lower temperatures, a layer of compacted snow so far can be removed only mechanically with the help of knife tools of graders, bulldozers, front loaders, single-bucket hydraulic excavators with significant energy costs. For the first time, the reduction of energy costs during the destruction of high-strength rocks under vibration exposure was established in 1934 by G.I. Pokrovsky and his staff [4]. A.N. Zelenin found that giving the direction to additional vibrations of the cutting tool, coinciding with the direction of cutting the soil at low cutting speeds, allows decreasing the pulling force by 3–4 times compared to the cutting force of this soil with a non-vibrating cutting tool [4]. D. Kumabe in his researches also confirms the effect of reducing cutting forces during vibration cutting of metals.

Vibrocleavers for compacted snow and ice using high-frequency pulsed knife oscillations were developed and successfully tested at Pacific National University.

Further studies have shown that the selection of rational vibrocleaver parameters depends on specifying the ratio of the operating forces and capacities of the working process. The aim of the research is to develop a mathematical movement process model of the vibrocleaver for compacted snow and to determine the ratios of the operating forces and capacities of the working process on the basis of a numerical solution.

3 Results

The workflow of the compacted snow vibrocleaver, intended to destroy the snow-ice formations on the sidewalks in a mechanized way, can be characterized by a pulsed loading mode. The cutting force depends on the following parameters of the snow layer being cut: density, depth and width of the cut, cutting angle, and air temperature. In addition, the cutting force increases at the beginning of the interaction process between the knife blade and snow, and then drastically decreases when the snow particles are chipped, which is common for the destruction of brittle solid media. New design of small-size vibrocleaver for compacted snow compactor has been developed for experimental research. It can become a commercial prototype based on pneumatic-wheel machines. The power cultivator engine capacity is 2.95 kW, the rotational speed is 3000 rpm. It is adopted that the drive power of the hydro-pulse mechanism is 2 kW, 0.95 kW for the travel mechanism.

The motion equation of the vibrocleaver is as follows

$$m\ddot{x} = P_{dv} - P(t) - P_f \tag{1}$$

where m is the mass of the vibrocleaver; P_{dv} is driving force; $P(t)$ - force of resistance to cutting, P_f - force of resistance to rolling of wheels.

Driving force is determined by the characteristic of the internal combustion engine

$$P_{dv} = \frac{M_e \cdot i_{tr} \cdot \eta_{tr}}{r_c} \tag{2}$$

where M_e is the engine torque, Nm; i_{tr} - transmission ratio; η_{tr} - transmission efficiency; r_c - power radius of the drive wheels, m

A mathematical model of the movement of compact compacting vibratory snow with the assumption of limited driving force.

The driving force is determined for linearized torque characteristic of the internal combustion engine, taking into account the power outlet for the hydraulic drive.

The maximum value of the driving force is determined from the condition of the realized engine power on the forward movement:

$$P_{dv}^{\max} = \frac{100 \cdot N_e}{V_{\min}} \tag{3}$$

where N_e is the engine power realized on the forward movement of the vibrocleaver.

Wheel rolling resistance

$$P_f = m \cdot g \cdot f \tag{4}$$

where f is rolling resistance coefficient, f = 0.03–0.05.

Driving force is limited by surface traction

$$P_{dv} \leq m \cdot g \cdot \varphi \tag{5}$$

where φ is the adhesion coefficient, φ = 0.75–0.85.

Engine power will be as follows

$$N_{dv} = \frac{m \cdot g \cdot \varphi \cdot V}{1000 \cdot \eta_{tr}} \tag{6}$$

where V is the movement speed of the vibrocleaver, V is assumed as 0.5 $V_{\max}$; $V_{\max}$ - maximum movement speed, $V_{\max}$ = 1.0 m/s; η_{tr} - transmission efficiency, η_{tr} = 0.85–0.9.

The pulsed cutting force P(t) is decomposed into a Fourier series with seven decomposition members in order to study the motion process of the vibrocleaver.

$$P(t) = P_0 \left[\alpha_0 + \sum_{i=1}^{7} (a_i \cos(i \cdot \omega \cdot t) + b_i \sin(i \cdot \omega \cdot t)) \right] \quad (7)$$

where P_0 is the maximum cutting force; α_0, a_i, b_i are the decomposition coefficients of a Fourier series:

$a_0 = 0.25$, $a_1 = -0.203$, $a_2 = 1.49 \cdot 10^{-4}$, $a_3 = -0.023$, $a_4 = 1.49 \cdot 10^{-4}$, $a_5 = -8.25 \cdot 10^{-3}$, $a_6 = 1.49 \cdot 10^{-4}$, $a_7 = -4.28 \cdot 10^{-3}$, $b_1 = 0.318$, $b_2 = -0.159$, $b_3 = 0.106$, $b_4 = -0.08$, $b_5 = 0.064$, $b_6 = -0.053$, $b_7 = 0.045$.

The cutting forces with the values P_o = 435 N, 567 N, 700 N are presented in Fig. 1.

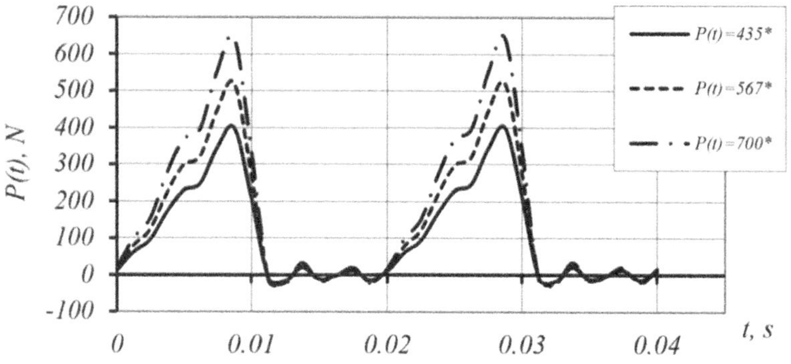

Fig. 1. Realization of cutting forces considering expansion in a Fourier series

The driving force can be determined using the static characteristic:

$$P_{dv} = (V_n - V) \cdot tg\beta \quad (8)$$

where V is the current movement speed value of the vibrocleaver;

β - the inclination angle of the driving force characteristic.

$$tg\beta = \frac{P_{dv}^{\max}}{V_n - V_{\min}} \quad (9)$$

where V_n is the nominal speed movement value, V_n takes the value of 1.0 m/s; $V_{\min}$ is the minimal steady movement speed, $V_{\min}$ = 0,5 m/s.

The nominal movement speed is determined using the following equation:

$$V_n = \frac{\pi \cdot D \cdot n_e}{i_{tr} \cdot 30} \quad (10)$$

where D is the outer diameter of the driving wheels; n_e – the nominal rotational speed value of the internal combustion engine, $n_e = 3000$ rpm; i_{tr} – gear ratio of transmission drive of leading wheels, $i_{tr} = 69$.

The Eq. (1) can be exposed in the following form:

$$m \cdot \frac{dV}{dt} = \frac{(V_n - V)}{V_n - V_{\min}} \cdot P_{dv}^{\max} - P(t) - P_f. \tag{11}$$

After the transformations, the Eq. (1) takes the following form:

$$\frac{dV}{dt} = A - B \cdot V - \frac{1}{m} P(t) \tag{12}$$

where $A = \frac{V_n \cdot P_{dv}^{\max}}{(V_n - V_{\min}) \cdot m} - \frac{P_f}{m}$; $B = \frac{P_{dv}^{\max}}{(V_n - V_{\min}) \cdot m}$.

The movement process power of the vibrocleaver is determined using the Eq. (11):

$$N_j = N_{dv} - N_p - N_f \tag{13}$$

where $N_j = m\frac{dV}{dt} \cdot \frac{V}{1000}$ is the power to overcome the forces of inertia; $N_p = \frac{P(t) \cdot V}{1000}$ is the power to overcome the forces of resistance to cutting compacted snow; $N_f = \frac{P_f \cdot V}{1000}$ - movement power of the vibrocleaver for compacted snow.

The numerical solution of Eq. (13) was obtained for ω frequencies in the range from 100 to 314 rad/s and efforts P_0 ranging from 435 N to 1500 N in the MATLAB software suite (Fig. 2).

The results of calculating the movement speed are presented in Fig. 3

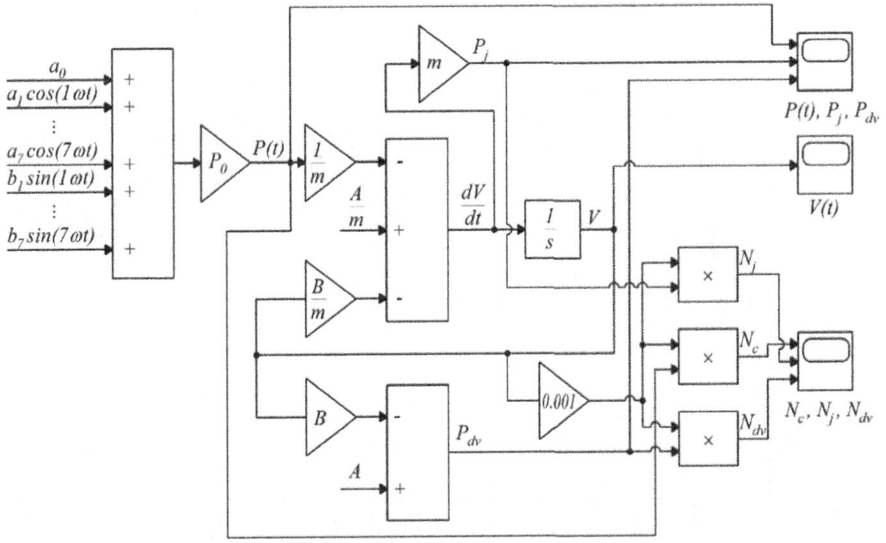

Fig. 2. Equation set diagram of a mathematical model of the vibrocleaver motion process

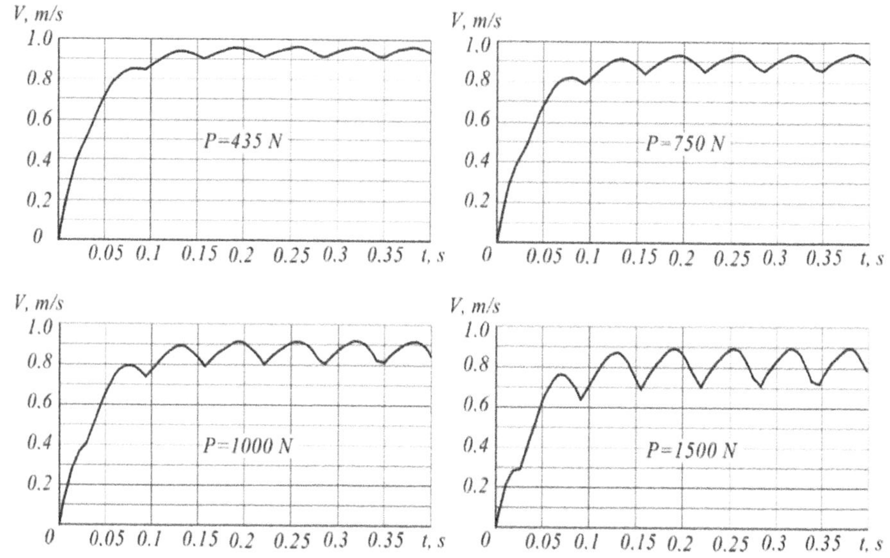

Fig. 3. The results of calculating the speed with frequency $\omega = 100$ rad/s

The implementation of the obtained scheme was carried out in following stages: the movement beginning of the vibrocleaver for compacted snow (Fig. 3) and the steady operation mode.

In order to determine the parameters more accurately, the steady state motion is limited by time from 0.37 s to 0.48 s.

The engine power of the initial and steady state operation is shown in Fig. 4.

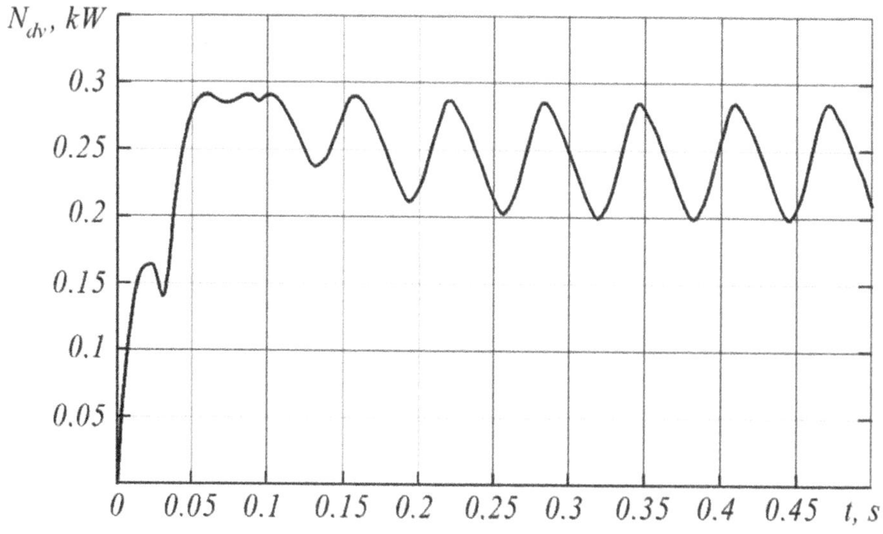

Fig. 4. The engine power diagram of the vibrocleaver for compacted snow $\omega = 100$ rad/s, $P(t) = 1500$ N

According to the graphs (Fig. 3), the vibrocleaver speed is characterized by acceleration at the initial stage and steady acceleration with fluctuations determined by the frequency and force value of the resistance to cutting of compacted snow. The engine power in the acceleration mode reaches a value of 0.29 kW, and in steady state the average power is 0.241 kW with a swing of 0.041 kW (Fig. 4, 6).

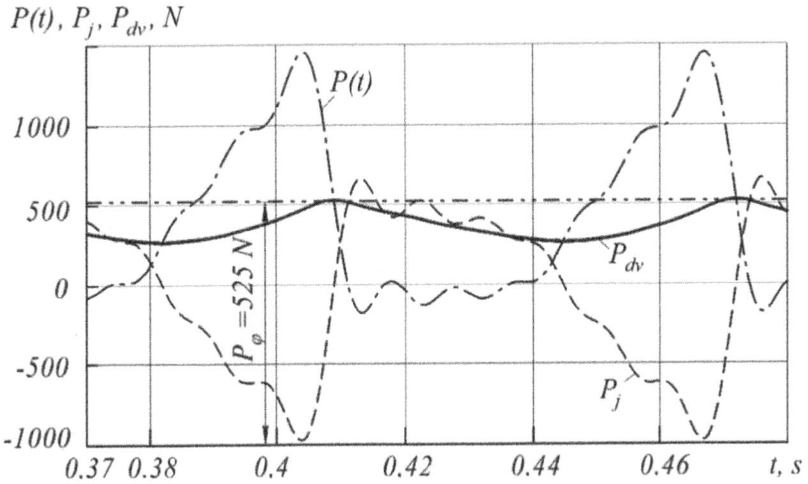

Fig. 5. Dependency network of the driving force P_{dv}, inertial force P_j and cutting resistance force $P(t)$ in steady state for $\omega = 100$ rad/s, $P(t)_{\max} = 1500$ N

In the steady state mode of the vibrocleaver, an inertia force P_j arises when the speed decreases. It is determined by the mass and acceleration (deceleration) of motion, which adds up to the driving force and helps to overcome the resistance force $P(t)$ according to the force balance condition

$$P_{dv} + P_j = P(t) \tag{14}$$

After the cleavage of the compacted snow particle is completed, the force $P(t)$ sharply decreases and it is necessary to restore the nominal movement speed; as a result, the driving force is expended to overcome the inertia forces.

The driving force is limited by the adhesion conditions of the drive wheels with the road coating and for the coefficient of adhesion between the metal wheel and metal lugs $\varphi = 0.75$ is $P_{\varphi} = 525$ N, which ensures movement in the marginal conditions (Fig. 5) without slipping.

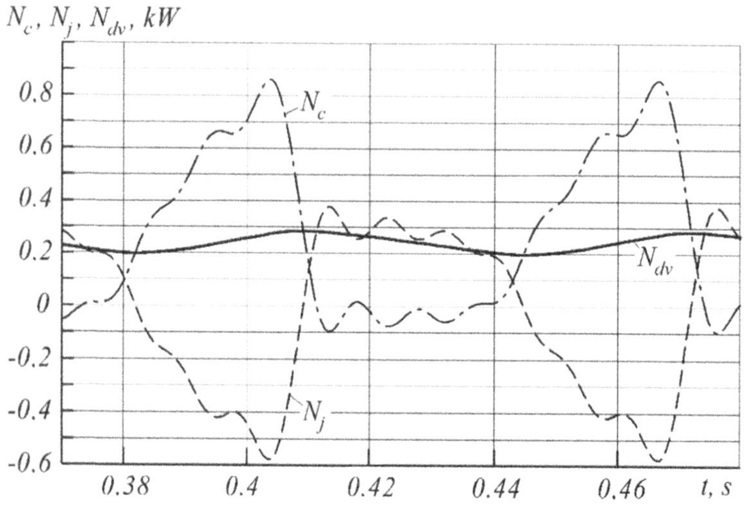

Fig. 6. Ratio diagrams of the power of driving forces N_{dv}, inertia forces N_j and resistance to the cutting of compacted snow N_c

After investigating the power ratio N_{dv}, N_j, N_c it can be stated that in this case the condition of power balance is maintained (Fig. 6).

$$N_{dv} + N_j = N_c \tag{15}$$

It can be observed while investigating the power ratio (Fig. 6), that the power to overcome the inertia forces complements the engine power in the deceleration mode. When the compacted snow breaks, $P(t)$ tends to zero, i.e. the load on the cutting tool is absent and then the engine power is realized for acceleration of the vibrocleaver for compacted snow.

The calculations were made for the frequency range ω from 100 rad/s to 314 rad/s and from 435 N to 1500 N, which preserved the obtained parameter ratios with their stabilization with an increase in the oscillation frequency.

The received power and characteristics of the engine provide a steady motion mode for the vibrocleaver of compacted snow.

References

1. Sadati, S., Kristen, C., Ceylan, H., Sassani, A., Kim, S.: Energy and thermal performance evaluation of an automated snow and ice removal system at airports using numerical modeling and field measurements. Sustain. Cities Soc. **43**, 238–250 (2018). https://doi.org/10.1016/S221067071830996X
2. Perrier, N., Langevin, A., Campbell, J.F.: A survey of models and algorithms for winter road maintenance. Part IV: vehicle routing and fleet sizing for plowing and snow disposal. Comput. Oper. Res. **34**(1), 258–294 (2007). https://doi.org/10.1016/S0305054805001565

3. Fu, L., Trudel, M., Kim, V.: Optimizing winter road maintenance operations under real-time information. Eur. J. Oper. Res. **196**(1), 332–341 (2009). https://doi.org/10.1016/S0377221708002427
4. Perrier, N., Langevin, A., Campbell, J.F.: A survey of models and algorithms for winter road maintenance. Part II: system design for snow disposal. Comput. Oper. Res. **33**(1), 239–262 (2006). https://doi.org/10.1016/S0305054804001637
5. Gudac, I., Marović, I., Hanak, T.: Sustainable optimization of winter road maintenance services under real-time information. Proc. Eng. **85**, 183–192 (2014). https://doi.org/10.1016//S1877705814019092
6. Perrier, N., Langevin, A., Campbell, J.F.: The sector design and assignment problem for snow disposal operations. Eur. J. Oper. Res. **189**(2), 508–525 (2008). https://doi.org/10.1016/S0377221707005164
7. Labelle, A., Langevin, A., Campbell, J.F.: Sector design for snow removal and disposal in urban areas. Socio-Econ. Plan. Sci. **36**(3), 183–202 (2002). https://doi.org/10.1016/S0038-0121(01)00024-6
8. Kramberger, T., Žerovnik, J.: A contribution to environmentally friendly winter road maintenance: optimizing road de-icing. Transp. Res. Part D: Transp. Environ. **13**(5), 340–346 (2008). https://doi.org/10.1016/S1361920908000345
9. Hinkka, V., Pilli-Sihvola, E., Mantsinen, H., Leviäkangas, P., Aapaoja, A., Hautala, R.: Integrated winter road maintenance management—new directions for cold regions research. Cold Reg. Sci. Technol. **121**, 108–117 (2016). https://doi.org/10.1016/S0165232X15002542
10. Min, J., Fu, L., Usman, T., Tan, Z.: Does winter road maintenance help reduce air emissions and fuel consumption? Transp. Res. Part D: Transp. Environ. **48**, 85–95 (2016). https://doi.org/10.1016/S1361920915302170
11. Vignisdottir, H.R., Ebrahimi, B., Booto, G.K., O'Born, R., Brattebø, H., Wallbaum, H., Bohne, R.A.: A review of environmental impacts of winter road maintenance. Cold Reg. Sci. Technol. **158**, 143–153 (2019). https://doi.org/10.1016/S0165232X18303938

Joint Work of Dissimilar Electrohydraulic Actuators

Artem Alekseenkov, Sergey Ermakov, and Aleksey Naidenov(✉)

Moscow Aviation Institute (National Research University), Volokolamskoe sh., 4, Moscow 125993, Russia
naydenov.alexey@gmail.com

Abstract. This article presents the results of the study of joint work of electro-hydraulic actuator with combined speed control of the output link and electro-hydraulic servovalve actuator, obtained using the method of mathematical modeling. It is shown that at joint work of the above mentioned heterogeneous electro-hydraulic actuators in a mode of summation of efforts on a uniform control surface, on the output links of actuators there is a force fighting that negatively affects a resource of a design and the quality of regulation of actuators at the area of a low input control signals. The article presents the results of mathematical modeling of the system of force equalization (FES) that was developed by the authors and shows the possibility of significant reduction of mutual power load in the actuators when including the above-mentioned force equalization system in the control loop. As the base of data to mathematical model were used a real characteristics of actuators.

Keywords: Actuator · Electro-Hydraulic · Force fight

1 Introduction

At the moment, the term «more electric aircraft» is increasingly often used in modern technical publications [1–4]. This term is understood as an aircraft in which one or more centralized hydraulic systems are replaced by power electrical systems. From the point of view of the executive part of the control system, such trend leads to the need to develop and implement new types of actuators capable of carrying out their own power supply from the power supply system of the aircraft, while possessing high static, dynamic and energy characteristics, as well as the required level of dependability and safety.

Among the existing electrohydraulic actuators used or planned for use on board of «more electric aircraft», the following main types can be identified:

1. Electro-Hydrostatic Actuator – EHA [1].
2. Electro-Hydraulic actuators with combined speed regulation – EHA-CSR [4, 5].
3. Electrical-Backup Hydraulic Actuator – EBHA [6, 7].

As the review of modern civil aircrafts complex control system design shows [1, 8, 9], the world's leading aircraft manufacturers consider the variants of installation of several

Z. Popovic et al. (Eds.): TransSiberia 2019, AISC 1115, pp. 600–606, 2020.
https://doi.org/10.1007/978-3-030-37916-2_58

heterogeneous electro-hydraulic actuators on one control surface (including the main control surfaces). In addition, on the existing passenger aircraft Airbus A-380 actuators of the main control surfaces are dissimilar both in power supply and in the method of regulating the speed of the output link of the actuators [2, 6, 10]. So, for example, rudder sections are deflected by EBHA actuators [7, 11], EHA and EBHA are installed on the elevators, ailerons or MFSpoilers.

The above-mentioned trend shows that the implementation of modern aircraft complex control systems designs already considers the potential for the organization of joint work of dissimilar electro-hydraulic actuators on a single control surface. It means that the question of research of joint work and determination of ways to decrease actuator force fighting is relevant and demanded.

2 Research Methods

As objects of research the authors of the work have chosen electrohydraulic actuator with combined speed regulation (hereinafter referred to as EHSA-CSR) as the most promising, according to the authors [5, 12, 13], type of electrohydraulic actuators with electric power supply, and throttle speed regulation actuator as the most frequently used type of steering actuators on board of the existing manned aircraft. The above-mentioned actuators are dissimilar by type of power supply.

Research of joint work was carried out by method of mathematical modeling: the mathematical model of an electrohydraulic actuator with the combined regulation of speed [4, 12] which general structure is resulted in Fig. 1, and mathematical model of a typical electrohydraulic actuator with throttle regulation of speed which structure is resulted in Fig. 2 have been developed in MATLAB Simulink.

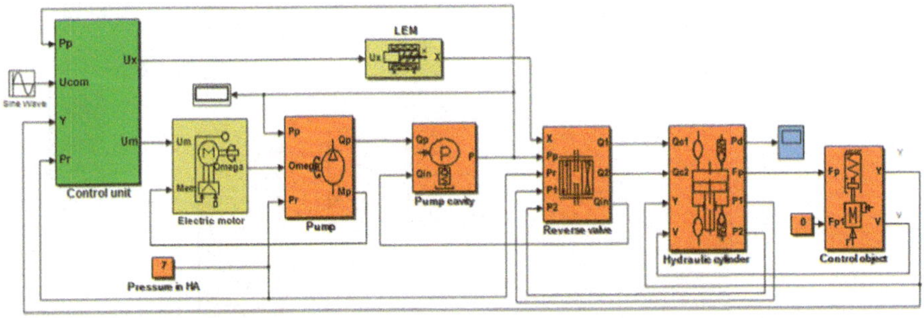

Fig. 1. Structure of mathematical model of electrohydraulic actuator with combined speed regulation. Designations: HA – hydraulic accumulator, LEM – linear electric motor

In the research of joint work, it was assumed that the actuators would be installed on the same control surface and would have the same or similar (not more than 10% of the difference) nominal idling speed and braking forces, as well as the stroke of the piston rod. In addition, to achieve the goals of the research, the authors consider the worst case of the operation of steering actuators, in which their output links are rigidly

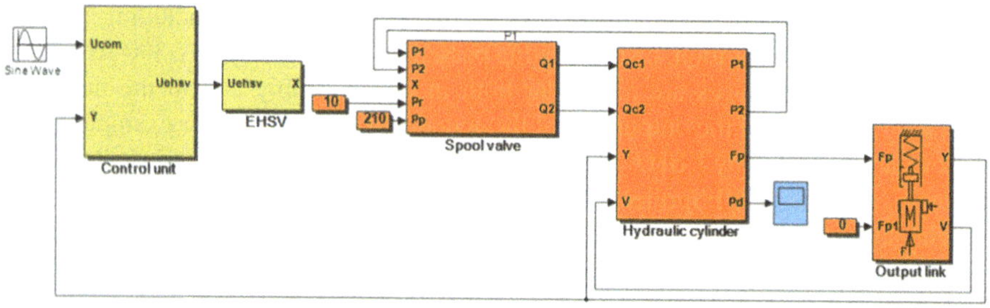

Fig. 2. Structure of mathematical model of electrohydraulic actuator with throttle speed control. Designation: EHSV – electrohydraulic servo valve

connected to each other (rod to rod). With such connection force fighting does not decrease at the expense of final rigidness of knots of fastening of EHSA to control surface and twisting of a power beam that will allow to define the maximum size of force fighting and to develop the most effective force-fight equalization system (FES) [2, 14].

Figure 3 shows the mathematical model of joint work of the considered actuators at rigid connection of their output links with each other.

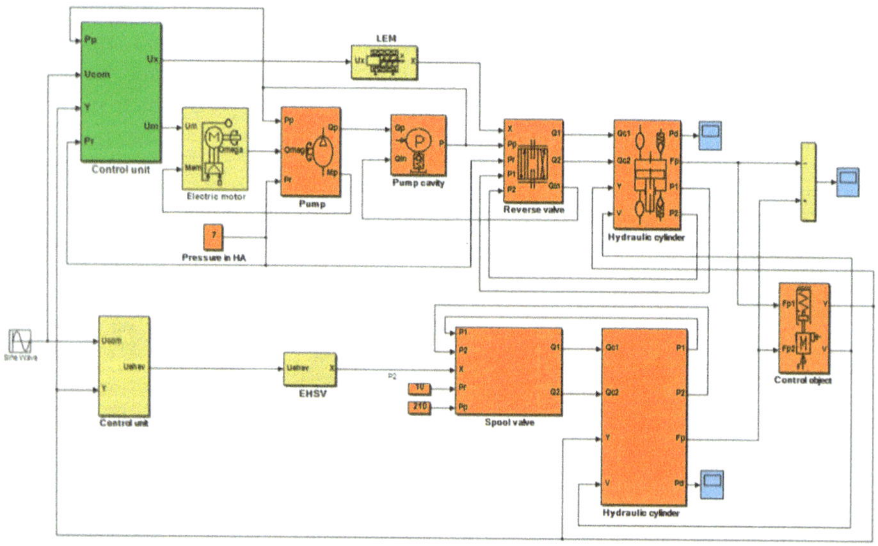

Fig. 3. Structure of mathematical model of joint work of electro-hydraulic steering actuator with throttle control of speed and with combined speed regulation.

3 Results

In the process of movement of EHSA rods it is impossible to provide identity of speeds of movement, because of what at working off of actuators of the harmonic or other control signal without introduction of FES to a control loop there is a considerable force fighting on output links of actuators, registered by pressure difference sensors in chambers of each hydraulic cylinder. Force fighting can also occur in a static state, e.g. in the case of a zero-position offset of the feedback sensors of each actuator.

To equalize actuator force fighting authors have designed the FES in each of actuators and research of performance with various schemes of its inclusion in a control loop of actuators has been carried out:

- "active - passive", where the passive is the EHSA and the active is EHSA-CSR. Correction in this case is realized only in the EHSA;
- "active – passive", where the passive is the EHSA-CSR and the active is EHA. Correction in this case is realized only in the EHSA-CSR;
- FES is realized in both actuators.

An example of FES implementation in the actuator control loop with throttle speed control is shown in Fig. 4. Figure 5 shows the value of force fighting of the actuators during the development of various variants of FES, including the limitation of the maximum value of correction for the stroke of the rod. It is also worth noting that in the course of the study it was assumed that the object of application of the drives under consideration is the steering surface of the non-agile aircraft, and therefore, the actuators do not require high dynamic capabilities, and the control frequencies for the harmonic control signal were frequencies up to 3 … 5 Hz.

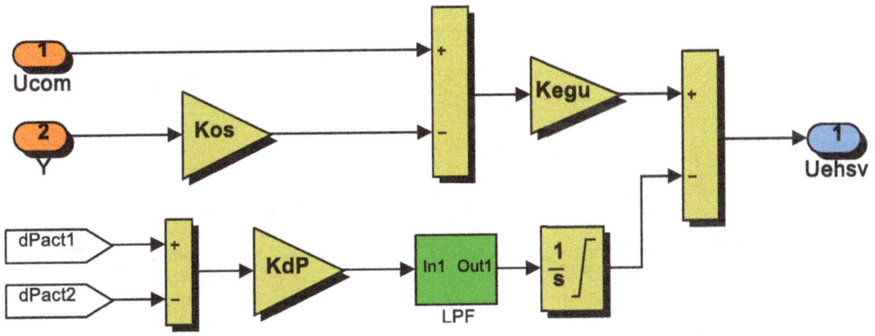

Fig. 4. Implementation of the force equalization system on the example of the EHSA control algorithm. In figure following designations are entered: Ucom – control signal, Y – position of actuator rod, dPact1 – pressure drop in a hydraulic cylinder of EHSA, dPact2 – pressure drop in a hydraulic cylinder of EHSA-CSR, Kos – feedback coefficient of EHSA, Kegu – error signal gain, KdP – gain of disagreement by pressure, SwdP – a sign of inclusion of correction by pressure, Uehsv – actuator's electrohydraulic amplifier control signal, LPF – Low-pass filter

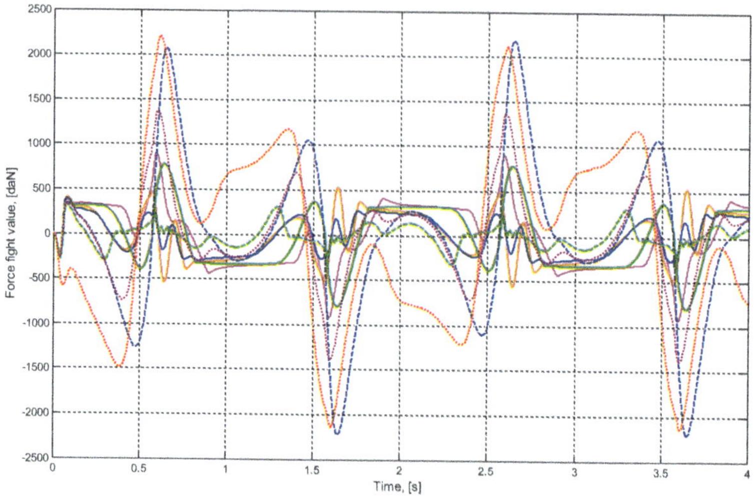

Fig. 5. Force fighting of the actuators under consideration in various implementation types of the forces equalization system and development of the harmonic control signal by actuators

Figure 6 shows the force fighting in the researched actuators with the selected FES settings for the three selected cases: FES enables only for EHSA-CSR, FES enables only for EHSA and FES enables in two actuators at once. As it can be seen from the figure, the force equalization system decreases the force fighting in the actuators from 4600 daN to 400 … 900 daN, which shows its effectiveness.

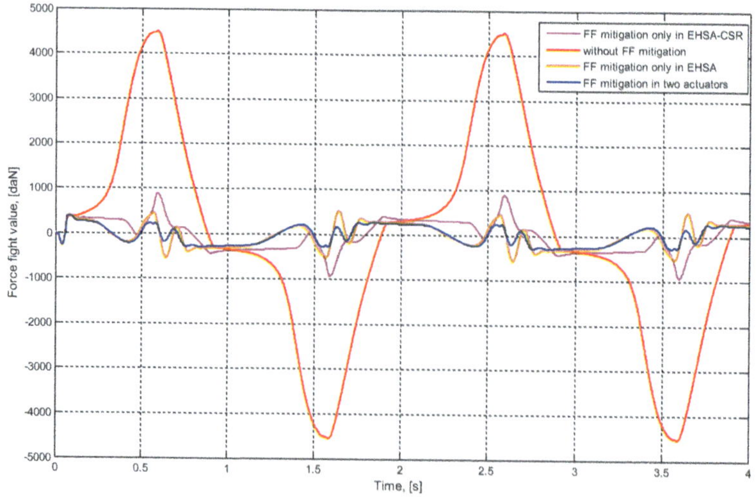

Fig. 6. Force fighting (FF) of the actuators under consideration at the selected schemes of FF mitigation during the control of harmonic signal by actuators

It is worth mentioning that the authors do not give the development of step input signals by the actuators in the article due to the limitation on the volume of publication. Nevertheless, the above-mentioned cases were considered, and the maximum value of force fighting of actuators for a short time could make up to 2700 … 3500 daN (at the moments of EHSA interconnections), and in the steady-state mode was not more than 600 daN at once in two EHSA.

4 Discussion

As a result of the conducted research it has been established that force fighting appears while work of dissimilar electro-hydraulic steering actuators on a uniform control surface.

With the purpose of reduction of force fighting and correction of output characteristics of actuators FES being a part of EHSA control algorithm has been designed, and following cases are considered:

- "active - passive", where the passive is the EHSA and the active is EHA-CSR. Correction in this case is realized only in the EHSA;
- "active - passive", where the passive is the EHSA-CSR and the active is EHA. Correction in this case is realized only in the EHSA-CSR;
- FES is realized in both actuators.

The results show:

- The reduction of the actuators force fighting is possible by means of controls without any modification of the actuator design (it was assumed that the design already includes pressure sensors in the hydraulic cylinder);
- In an "active - passive", where the passive is the EHSA and the active is EHA-CSR force fighting dropped to 16% of maximum power of each actuator;
- In an "active - passive", where the passive is the EHSA-CSR and the active is EHA force fighting dropped to 21% of maximum power of each actuator;
- When correcting two actuators at once, the force fighting was reduced to 10% of the maximum force developed by each of the actuators.

The above suggests that it is more rational to make a two-wheel actuator levelling system or, if this is not technically possible, to build an "active - passive" structure where EHSA is passive (because of the better dynamic characteristics) and EHSA – CSR is active.

It is worth mentioning that in the research conducted there are no cases of actuators performing under loading, however, as the experience of authors shows, force fighting of actuators decreases with growth of the external loading influencing on EHSA through object of control.

References

1. Mare, J.C.: Aerospace Actuators 2: Signal–by–Wire and Power–by–Wire. ISTE Ltd, Hoboken (2017)
2. Bossche, D.: The A380 Flight control electrohydrostatic actuators, achievements and lessons learnt. In: ICAS (2006). https://icas.org/ICAS_ARCHIVE/ICAS2006/PAPERS/048.PDF
3. Kowalski, R.: Force fight in parallel-redundant electro-mechanical actuation systems. In: Proceedings of More Electric Aircraft, Bordeaux (2017)
4. Alekseenkov, A.S., Ermakov, S.A., Konstantinov, S.V., Kuznetsov, V.E., Obolensky, Yu. G., Redko, P.G.: Systems of electro-hydraulic actuators of the aircraft control complex. In: S. V. Konstantinova (ed.). Publishing house of SPbGU, SPb (2019)
5. Ogoltsov, I., Samsonovich, S., Selivanov, A., Alekseenkov, A.: New developments of electrically powered electrohydraulic and electromechanical actuators for the more electric aircraft. In: ICAS (2014). https://icas.org/ICAS_ARCHIVE/ICAS2014/data/papers/2014_0747_paper.pdf
6. Evolution of Powered Flight Controls, MOOG's official presentation 10 February (2012). https://ru.scribd.com/document/202144348/Evolution-of-Powered-Flight-Controls-Seminar
7. Le Trone, X.: A380 flight controls overview. Hamburg University of Applied Sciences (2007). https://www.fzt.haw-hamburg.de/pers/Scholz/dglr/hh/text_2007_09_27_A380_Flight_Controls.pdf
8. Liebherr: NextGen Integrated Flight Controls System, AST Workshop (2019)
9. Aircraft Group Overview Executive Summary. MOOG presentation paper, SIAE (2019)
10. Bildstein, M.: A-380 spoiler electrical back-up hydraulic actuator (A380 spoiler EBHA), SAE A-6 Symposium, Toulouse, France 17 October (2002)
11. Recent Advances in Aerospace Actuation Systems and Components. INSA, de Toulouse (2018)
12. Alekseenkov, A.S.: Improvement of dynamic properties and research into operation processes of aviation elevator hydraulic combination rate control actuator under an increased outer load. Candidate's dissertation in technical sciences (2014)
13. Alekseenkov, A.S., Ermakov, S.A., Kiselev, M.A., Konstantinov, S.V., Obolensky, YuG: Temperature measurements at critical points of dual-mode electrohydraulic actuator with combined rate control. Periodico Tche Quimica **15**(Special Issue 1), 396–404 (2018)
14. Wang, L.: Force equalization for active/active redundant actuation system involving servo-hydraulic and electromechanical technologies. Doctoral thesis, Toulouse (2012)

Influence of Parameters of Rail Wheel Flanges on the Railway Operation Safety on Pointworks

Boris Glyuzberg(✉)

JSC Railway Research Institute, 3rd Mytischinskaya Street, 10, Moscow 129626, Russia
Glusberg@mail.ru

Abstract. The probabilistic approach for assessing the railway operation safety was proposed in order to evaluate the possibility and frequency of adverse events occurring when trains move on turnouts. Norms and tolerances in the "rail track - wheel track" system are interconnected. Based on this relationship, an analysis for the adverse events likelihood is made for cases when wheelsets having flanges with various wear values are passing through pointworks.

Keywords: Wheel track · Rail track · Wheel flange wear · Safety criteria · Probabilities of adverse events · Design norms of pointwork geometry

1 Introduction

Norms and tolerances in the system of geometric dimensions "rail track - wheel track" are interconnected. Proposals for changing the regulatory framework regarding geometry of the rail track and the rolling stock truck, in particular the rejection dimensions of the minimum thickness of the rail wheel flanges, should be based on analyzing the rolling stock safety along the track and pointworks. Arbitrary interference in this dimension system can lead to a threat to the safety of the rolling stock passage through pointworks and to large material losses.

Specific events may occur during the passage of wheelsets with different wheel track parameters through the pointworks, including adverse traffic safety conditions.

The thickness of the rail wheel flanges has a great influence on the safe passage through the switch blades by the rolling stock. When the wheels climb the bent out switch blade, disorders of the rail track occur, the switch blade do not adhere to the point rail. In the case when such occur frequently, bends and even kinks in the connecting rods, which directly threaten the safety of the pointwork can appear.

Safe railway operation along the crosspiece junction of ordinary pointworks is provided only if the following events are excluded:

- hits of the wheels and the "idle" guard rail to the throat;
- hits of the wheels and the catching parts (mouths) of guard rails and counter rails;
- wheel bumps at the tip of the core;
- thrust of the wheelset with a guard rail and a counter rail.

Z. Popovic et al. (Eds.): TransSiberia 2019, AISC 1115, pp. 607–620, 2020.
https://doi.org/10.1007/978-3-030-37916-2_59

In order to fulfill these requirements, the required gauge and groove widths in the crosspiece are established based on analysis of the most unfavorable combinations of wheel pairs and gauge sizes in the crosspiece, taking into account their possible repeatability and accepted tolerances for these sizes.

2 Research Methods

It is impossible to consider all the movement features of the rolling stock on the pointworks within the framework of one article. As an example, the conditions for the passage of a pair of wheels through a cross-piece assembly are considered.

Figure 1 shows a diagram of the cross-head assembly of a single ordinary pointwork.

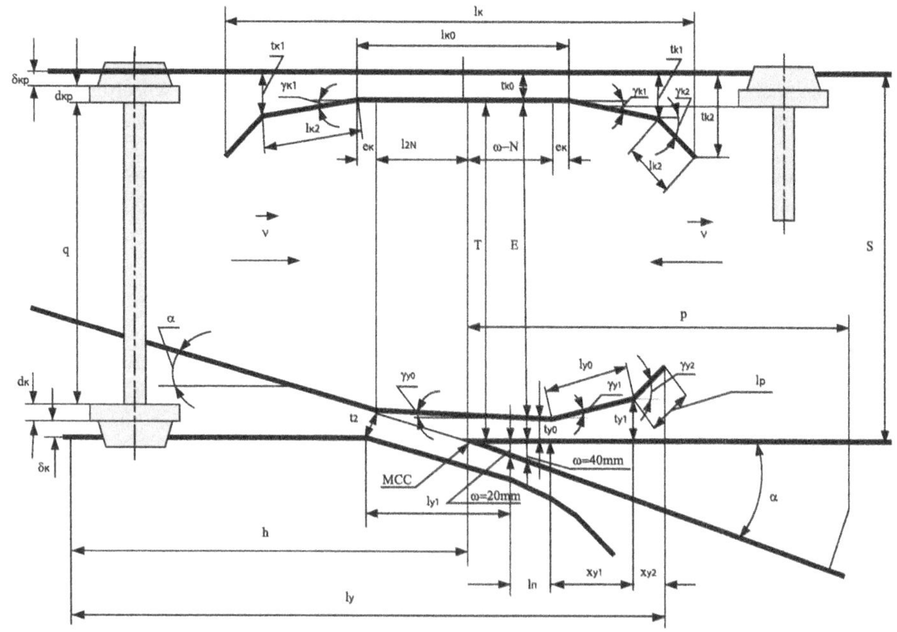

Fig. 1. A geometric diagram of the cross-head assembly of a single ordinary pointwork

Using the notation of Fig. 1, let us compose a system of relations ensuring the fulfillment of the safety conditions listed above.

$$\begin{cases} (\delta_{cr} + d_{cr} + q + \Delta) < S - t_t \\ (\delta_{cr} + d_{cr} + q + \Delta) < S - t_{g1} \\ (\delta_c + d_c + q + \Delta) < S - t_{c1} \\ (q + 2\Delta + d_c) < (S - t_{k0}) = T \\ q + 2\Delta < \left(S - t_{c0} - t_{g0}\right) = E \end{cases} \tag{1}$$

where S is the gauge width in the cross-head;

T is the distance between the counter rail and the core of the cross-head;
E is the distance between the counter rail and the guard rail of the cross-head;
d_{cr} and d_c – the flange thickness of the wheelset from the side of the counter rail and cross-head, respectively;
q – wheelset mounting;
Δ is the gap between the vertical and the back side of the carriage wheel flange at the measurement level (Δ = 1 mm);
t_t, t_{g0}, t_{g1} – gutters for cross-heads in the throat, middle part of the guardrail and at the entrance to its bent part, respectively;
t_{c0} and t_{c1} – gutters in the counter rail in its middle part and at the entrance to the bent part, respectively;
δ_{cr} and δ_c are the gaps between the wheel flange and the rail of the cross-head, and the wheel flange and the guardrail of the cross-head, respectively.

Fulfilling the system (1) requirements guarantees the safety of movement through the crosspiece switch unit. Given the size of the wheelset, the dimensions of the track and gutters on the cross can be calculated, or vice versa, having the dimensions of the cross-head assembly, the possible proposals for changing the sizes of the wheelsets can be determined.

The simplest method of analysis is the limit combination method. In order to perform calculations using this method, it is necessary to add the following relations to the system of relations (1)

$$\left\{\begin{array}{c} d_{cr}^{max} = d_{\kappa}^{max} = [d]^{max} \\ d_{cr}^{min} = d_{\kappa}^{min} = [d]^{min} \\ S^{max} = [S]^{max} + S^{g} \\ S^{min} = [S]^{min} \\ q^{max} = [q]^{max} \\ q^{min} = [q\ ]^{min} - q^{g} \\ \delta^{max} = S^{max} - \left(q^{min} + 2[d\ ]^{min}\right) \\ \delta^{min} = 0 \end{array}\right. \tag{2}$$

The system of relations (1) + (2) is complete and allows solving the problem of determining the nominal values and tolerances in the “wheel – rail track” system.

In particular, after simple transformations it is possible to obtain relations for the normalized sizes of the gutters in the cross-head assembly:

$$\left\{\begin{array}{c} t_t^{min} = t_{c0}^{min} = t_{g0}^{min} > ([S\ ]^{max} + S^{g}) - \left([q\ ]^{min} - q^{g}\right) - [d\ ]^{min} \\ E^{\ max} < [q\ ]^{min} - q^{g} \\ T^{min} > [q\ ]^{max} + [d\ ]^{max} \\ \left(t_{c1} + t_{g1}\right)^{max} > [S\ ]^{max} - \left([q\ ]^{min} + q^{g}\right) \end{array}\right. \tag{3}$$

This method is used in several countries even today. It was used on Russian railways until the mid-60s of the last century. Considering all the clarity and simplicity of this method, it guarantees only "theoretical" safety of moving through the cross-head assembly. Taking into account the operating practice specifics of Russian railways showed that the ratios (1 + 2) contain contradictions that make this system incompatible, and make the calculations using the limit combinations method conditional.

For example, mass measurements of wheelsets, repeatedly performed by the "Strelochnoe xozyajstvo" laboratory of VNIIZhT in 1964–2002, have convincingly proved that there are practically no cases when both flanges are of the smallest size on one wheel pair.

Thus, according to the results of measurements taken in 2002, for wheelsets where the flange thickness of one wheel was 25 mm, the average flange thickness of the second wheel was 31.0 mm, and for wheelsets where the flange thickness of one wheel was 23 mm, it was 30.6 mm, respectively. The high correlation level between the presence of "thin" flanges and uneven wear of the ridges of one wheelset can be noted. Depending on $(d_1 - d_2) = f(d_1 + d_2)$, the correlation coefficient is 0.76 and higher.

These data allow concluding that the uneven wear of the flanges of one wheel pair is associated either with the design flaws of the carriage trolleys or with the poor condition of the carriage of the carriages in operation.

Understanding this led to a number of indications made by the Ministry of Railways of Russia in the 90s of the last century. According to the indications, it was proposed to turn around the wheelset relative to the truck in order to increase the resource of wheelsets, when one of the wheel flanges reached the maximum wear.

The most important factor requiring the search for other approaches to standardizing dimensions in the "wheel pair – rail track" system is that in practice of operation on the roads of our country, the actual dimensions of both wheelsets and rail track go beyond the limits established by regulatory documents. So, according to measurements in 2002, six wheels from each thousand measured wheelsets had a flange thickness less than the minimal (25 mm), and the minimum flange thickness detected was 22.8 mm. The situation is similar with the rail gauge on pointworks. In these conditions, focusing on assessments of "theoretical" safety does not allow us to assess the actual possibilities of adverse events in operational practice.

In order to assess the practical (actual) traffic safety, a probabilistic approach was proposed to assess the possibility and frequency of adverse events that occur when trains move on pointworks. The main idea of this approach is that the norms of dimensions and tolerances in the "wheel pair – rail track" system should be so the probability of dangerous cases is less than one time during the pointwork period of operation (as a rule, it is a lifespan) with actual combinations of sizes taking place on a road network.

The authorship in the development of this method belongs to the "Strelochnoe xozyajstvo" VNIIZhT laboratory. In 1960, Ph.D. G.I. Ivashchenko published the first research on this issue. In the future, the development of methods based on a probabilistic approach was carried out by Ph.D. L.G. Krysanov, he developed the method of probabilistic compositions and the author of this article (developed the method of conditional probabilities) [1]. Based on these methods, the norms of the pointwork geometric parameters of 1520 mm gauge and the current standards, enshrined in the regulatory documents of the Ministry of Railways of the Russian Federation, and then

JSC Russian Railways, were formed. Several decades of operational work have confirmed the correctness of the developed methods.

The main method that is practically used at present is the "conditional probability method" developed in the "Strelochnoe xozyajstvo" laboratory in the 90s of the last century [2]. One of the distinguishing features of the method is the consideration of the correlation between the parameters included in the relations (1.a + 1.b) when analyzing the probability of the considered event. Thus, elastic changes in the gauge and gutter widths are considered as a function of their initial values. And the flange thickness of the second wheel in the wheelset is considered as a function of the distance between the heel rims at the level of measuring the gauge and flange thickness of the first wheel.

The method also takes into account the actual forms of the working surfaces of the wheels and the pointwork elements, as well as a number of other features.

The technique of using the method is as follows. Initially, according to the actual distributions of the gauge and wheel pair parameters, the distribution of the positions of the latter in the rail track is determined, taking into account the distribution of its possible positions when entering the pointwork. Then, according to the initial parameters of the track and gutters, the probability and places of the wheel running on the pointwork elements are determined.

Such calculations are carried out for each of the possible (occurring in operation) combinations of rail gauge parameters. Based on the calculation results, the probability density distributions of the studied events are built and the probabilities of the phenomena under consideration are calculated by integrating these dependencies within the given limits.

Since the present article deals with the permissible values of the minimum thickness of the wheel flanges, the calculations for wheelsets having a minimum flange thickness of 25 mm (current standard) and 23 mm (one of the proposed options) are carried out.

The complete set of data regarding distributions of rail gauge parameters on pointworks and wheel pair parameters should be based on large-scale measurements (not less than 5,000 wheelsets and at least 150 pointworks). Due to the fact that studies have not been carried out in the required volumes for a quarter of a century and there are no data for a full review, the problem can be solved for a more individual case:

- the calculation is performed for a cross-head unit, in which the gauge and gutter dimensions are at the limit of current standards, that is, $[S]^{max} = 1523$ mm; $[t_{к1}]^{min} = 62$ mm; $[t_г]^{min} = 61$ mm;
- the second flange size of the wheelset is adopted according to the regression dependences obtained in the 2000s after the limited measurement results (500 units) of the wheelsets.

With values close to 23–25 mm, the regression equation for determining the flange thickness of the second wheel pair takes the following form according to the measurements made in 2002:

$$d_2^{cp} = 0.204\, d_1^{cp} + 25.910 \quad (4)$$

- the minimum distance between the flanges of the wheels at the level of measuring the gauge width, based on previously obtained $q^{min} = 1435.5$ mm;
- the distribution of the elastic deformations of the track and wheelsets is considered normal (which practically coincides with the distributions obtained from the tests).

The largest total clearance of the wheelset in the rail track

$$\delta^{max} = (S^{max} + \Delta S_{din}) - (Q - \Delta q_{din}), \quad (5)$$

After substituting the necessary data in (5) and analyzing the obtained values for all possible combinations, the distribution of the wheel pair clearances in the rail track of the cross-head is obtained. The thickness of one of the flanges is 25 and 23 mm (the flange thickness of the second wheel is determined according to the following correlation)

$$\delta^{max}_{25\,mm} => M(\delta_{25\,mm}) = 30.12\,mm,\ \sigma(\delta_{25\,mm}) = 1.36\,mm, \quad (6)$$

$$\delta^{max}_{23\,mm} => M(\delta_{23\,mm}) = 32.51\,mm,\ \sigma(\delta_{23\,mm}) = 1.37\,mm, \quad (7)$$

where M and σ are the average value and standard deviation of the distributions, respectively.

If the wheel pair is not pressed against one of the track rails, then δ^{max} is the total gap between the wheel set and the rail track. According to the data (5), the probabilities of the total clearance for wheelsets with one 25 and 23 mm wheel flange are determined.

The calculation results are presented in Table 1.

Table 1. The probability of the total clearance values on the cross-head assembly

Clearance δ_{max}, mm	Flange thickness 25 mm		Flange thickness 23 mm	
	n_σ	$P(\delta^{max})$	n_σ	$P(\delta^{max})$
23	−4.5	0.000005	−6.26	–
23–25	−3.03	0.001223	−4.79	0.0000006
25–27	−1.84	0.031661	−3.32	0.000045
27–29	−0.09	0.43106	−1.85	0.031707
29–31	+1.38	0.453092	−0.38	0.319816
31–33	+2.85	0.081007	+1.09	0.51017
33–35	+4.32	0.002183	+2.56	0.132623
35–37	+5.79	0.000005	+4.03	0.005232
37–39	–	–	+5.5	0.000003
>39	–	0	–	0

According to the data in the table, it can be seen that the greatest probability of a total clearance for wheels with a flange thickness of 25 mm reaches 0.45 with a total clearance of 29–31 mm, and reaches 0.51 for wheels with a flange thickness of 23 mm, with a total clearance of 31–33 mm.

It is possible to obtain the probability distribution of the flange position of one wheel in the rut (one-sided clearance) [2] by using the values of the total clearance in the rut. Then, the composition of the distributions of the one-side clearance and flange thickness can be calculated, as well as the probability of the position of the back of the flange running onto the cross piece (for example, counter rail).

The calculation data of the probability distribution for the position of the wheels with a ridge thickness of 23 and 25 mm are shown in Table 2.

Table 2. The probability distribution of the wheel position in the rut

Clearance, mm	Wheel flange thickness			
	d = 25 mm		d = 23 mm	
	(δ + d_2), mm	Probability	(δ + d_2), mm	Probability P (δ + d_2)
0–2,5	31–33.5	0.16965779	30.6–33.1	0.16057094
2.51–5.0	33.5–36.0	0.09002034	33.1–35.6	0.083515024
5.01–7.5	36.1–38.5	0.06485213	35.61–38.1	0.061795832
7.51–10.0	38.51–41.0	0.05920676	38.11–40.6	0.053495528
10.01–12.5	41.01–43.5	0.06263181	40.61–43.1	0.058314712
12.51–15.0	43.51–46.0	0.05951140	43.11–45.6	0.053495972
15.01–17.5	46.01–48.5	0.06035550	45.61–48.1	0.059871588
17.51–20.0	48.51–51.0	0.06235826	48.11–50.6	0.057601172
20.1–22.5	51.1–53.5	0.06037101	50.61–53.1	0.058597696
22.51–25.0	53.51–56.0	0.06950821	53.11–55.6	0.054131966
25.01–27.5	56.01–58.5	0.11452278	55.61–58.1	0.071048410
27.51–30.0	58.51–61.0	0.11356360	58.11–60.6	0.114929960
30.0–32.5	61.01–63.5	0.01313576	60.61–63.1	0.092603430
32.51–35.0	63.51–66.0	0.00305620	63.11–65.6	0.019090630
35.01–37.5	66.01–68.5	0.0000001	65.61–68.1	0.00052359
37.51–40.0	68.61–71.0	0	68.11–70.6	0
Σ		1.0		1.0

The climbing of the wheel on the catching part (mouth) of the counter-rail and on the idle guard rail up to the throat will occur, if the total flange thickness and one-sided clearance in the rut exceeds the groove size in the throat and the mouth of the counter-rail.

Given that the minimum size of the groove at the entrance to the bent part of the counter rail at the present time is 62 mm, the position probability of the rear rim part of the running wheel at which P (d_{cr} + δ_{cr}) > 62 mm is unacceptable.

According to the calculation results, it is obtained that for wheels with a flange thickness of 25.0 mm, the probability of violating this ratio is P_{25} = 0.00819. For wheels with a flange thickness of 23.0 mm, this probability is P_{23} = 0.04184.

Thus, when wheels with a flange thickness of 23.0 mm are allowed for implementation, the probability of wheel hits with the catching part of the counter-rail increases by more than 5 times.

When designing the pointworks both in Russia and in world practice, one of the factors determining the structural speed of a pointwork is the transfer of kinetic energy by the rolling stock wheels to the reverse gear elements when the wheels hit and climb on them. The parameter associated with the loss of kinetic energy, W, is determined by the following formula

$$W = V \cdot \sin \beta, \tag{8}$$

where V is the design speed;

β is the angle of deviation (climbing) on the switch blade, counter-rail, guardrail, etc., under which the wheel climbs onto the pointwork element.

When the wheels run onto the catching part of the counter rail, the parameter W, which characterizes the transfer of energy from the wheels to the counter rail, increases. It increases in proportion to the sine of the angle, at which the corresponding counter rail part deviates from the line parallel to the active face of the cross rail.

The correlation for the transmission energy parameter of the wheels impact to the counter rail can be obtained by substituting the corresponding angles (see Fig. 1) in (8) and by taking the relation for the case of a collision into a mouth and climbing onto designated counter rail part

$$\frac{W_1}{W_2} = \frac{\sin \gamma_{c1}}{\sin \gamma_{c2}} \tag{4}$$

and movement speeds through the pointwork

$$V_2 = V_1 \cdot (\sin \gamma_{\kappa 1} / \sin \gamma_{\kappa 2}). \tag{10}$$

For the P65 pointwork type, grade 1/11, the increase in exposure will be about 6 times. Therefore, in order to maintain the level of impact on the pointwork incorporated in the design, the transferring speed for the underframe, which wheelsets climb onto the catching parts of the cross, must be reduced by the same amount.

An increase in the likelihood of such effects by more than 5 times will have an extremely negative impact on the strength of the pointwork elements and the stability of the rail gauge parameters on it.

The proper use of relation (6) in the design of railroad pointworks has been proven by many decades of the operation of railways in different countries and on various pointwork designs. Nevertheless, the question of the permissible rail and wheel track dimensions arises regularly.

In 1994, a direct experiment was conducted to determine the effect of differences in wheel and rail gauge dimensions on the passage of underframes through turnouts and operational possibility testing of reducing the rejection size of worn wheel flanges for Russian railways.

An experiment was carried out on the VNIIZhT experimental ring in order to determine the effect of differences in wheel and rail track sizes on the passage of underframes through pointworks.

For this purpose, a test train was formed using the VL-80 locomotive and 12 cars (Fig. 2). The test train included carriages with an unworn undercarriage, loaded to 25–27 t/axle. The thicknesses of the flanges were the same (new wheel profile). The wheel track parameters of the carriage wheelsets varied due to the different mountings of the wheels over a wide range, from 1422 mm to 1450 mm. On 10 carriages, the mounting of the wheelsets of each carriage was the same, and on two it was mixed. Thus, the total static clearance of the wheel pair in the rail track (excluding elastic deformations) varied from 4 mm to 32 mm. The test train traveled through a pointwork at speeds of up to 70 km/h.

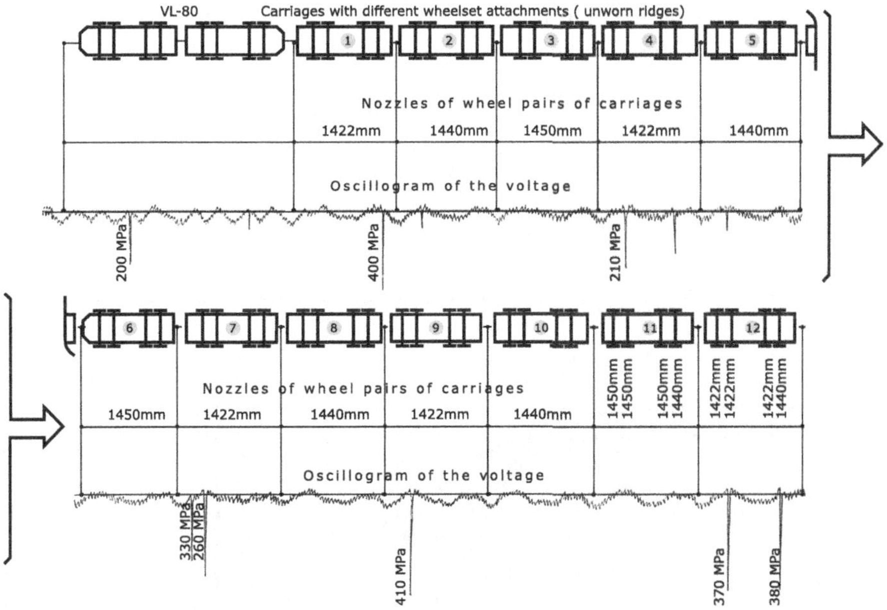

Fig. 2. Oscillogram of stresses in the most loaded section of the counter rail during the passage of wheelsets with different clearance values in the rail track

The pointwork included counter-rail units with counter rails made of the special RK65 profile. On the pointwork, the stresses in the counter rail and in the counter-rail bolts in the place where the wheels climbed onto the counter rail were recorded.

The test results are represented by a copy of the stress oscillogram at the most loaded point of the counter rail (see Fig. 2). As can be seen from the oscillograms, wheel pairs with a gap in the rail track of 22–32 mm had minor effects on the counter rail (up to 70 MPa), with an allowable 330 MPa. This is due to the fact that the wheels of such wheelsets climb onto the counter rail within its bent part, where it is provided by the pointwork design. Wheel pairs having a small clearance in the rail track pass the counter-rail assembly, climbing onto (hitting) the catching part of it. The impact of these wheelsets on the counter rail is quite significant [3, 4]. Stresses in the counter rail from their influence reach 410 MPa, which significantly exceeds the permissible value. The data on the effect on counter-rail bolts have similar patterns, where the stresses from the wheels in the wheelsets with a large gap in the rail track are up to 1.5 times higher than the permissible ones. (It should be noted that for counter-rail assemblies made of the RK65 special profile, a break of two counter-rail bolts in a row threatens safety and requires movement stop.)

3 Results of the Research

In general, according to the test results, it was found that wheels of wheelsets with a minimum clearance in the rail track cause over-stresses in the counter rail by an average of 22%, and in counter-rail bolts by an average of 15%. The number of such wheelsets having an increased effect on the counter-rail assembly is 18% more than of wheelsets having a gap of 22–32 mm.

The Ministry of Railways of Russia gave temporary permission to operate such wheelsets for the purpose of operational testing the possibility of using wheelsets with a 23 mm thickness of one of the flanges in 1994. The operating results were as follows.

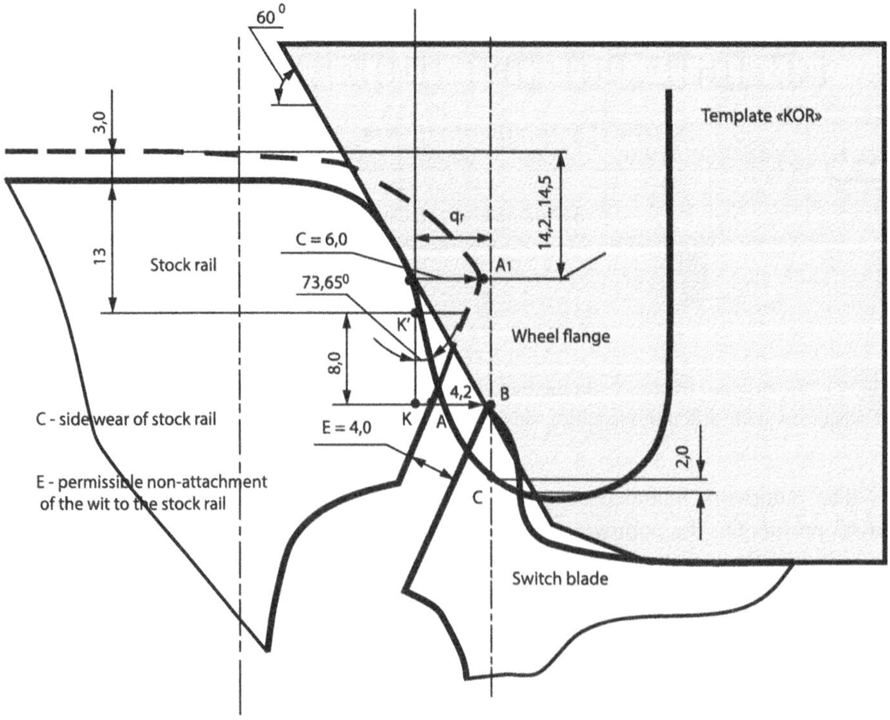

Fig. 3. Wheel passing the switch blade edge with a worn point rail

According to official data provided by 10 roads, the number of breaks of counter-rail bolts increased by 10–50% during the validity period of the 23 mm minimum thickness standard for wheel flanges. The costs of replacing the elements of the counter-rail units and adjusting the parameters of the rail gauge at the pointworks increased by 10–20%. Deviations of the rail track and reverse gears on the pointworks also increased.

The change in the standard for the minimum wheel flange thickness on the pointworks operated in curved sections of the track had an extremely negative effect. According to the Sverdlovsk Railway, the cost of their maintenance increased by an average of 39%, and a speed limit had to be introduced for 76 pointworks.

Thus, the test results and trial operation fully confirm the above-mentioned theoretical analysis.

Studies show that the greater the wear of the wheel flange, the greater is the angle of its inclination. At large flange inclination angles, it becomes possible to roll the wheel onto the switch blade in the area of its edge [5]. Figure 3 shows a diagram of the wheel passing through the switch blade edge.

The condition for safe passage of the switch blade edge by the wheel is determined by its position relative to the point rail. As the flange and point rail wear out, the conditions for safe passage of the switch blade edge by the wheel deteriorate. With the greatest allowable wear of the point rail and the largest allowable gap between the point rail and the switch blade, the wheel will not climb onto the switch blade, if the inclination of the flange generatrix corresponds to $q_r \geq 6.5$ mm (see Fig. 3).

$$q_r = AB + AK = 4.2 + \left(8 / \mathrm{tg}\, 73.65^0\right) = 4.2 + 2.3 = 6.5\,\mathrm{mm} \qquad (11)$$

In the operation practice of European railways, the safety of wheels passing the switch blade edge is provided by monitoring the relative position of the switch blade - point rail and monitoring the parameter q_r on the wheels of the rolling stock.

Monitoring the relative position of the switch blade and point rail using a special template (KOR template) was introduced on Russian railways in the mid-90s and is now considered mandatory. If the requirements of the KOR template are not fulfilled, track workers should grind the switch blade, which ensures the necessary position of its operating surface.

The control of the inclination of flange generatrix according to the parameter q_r is considered mandatory on European railways, however, on Russian railways it is carried out only for locomotive wheels. One of the reasons for this is that the control measurements of the wheelsets, carried out in 2002, showed that even with the 25 mm thickness of the flanges, some of the wheels have the shape of flanges that violate the q_r requirement. These wheelsets should be sent to the turning regardless of the minimum permissible thickness of the flanges, which will require additional costs.

Thus, after the introduction of requirements according to the **KOR** template and the refusal to put the requirements on q_r into effect, the security requirements for the passage of carriages through the pointwork are completely transferred to the workers of the track facilities. However, such a one-sided approach, does not fully ensure compliance with security requirements as can be seen from the analysis.

Particularly unfavorable is the passage of the switch blade edge by the wheels with sharp flanges (Fig. 4). Therefore, controlling the development of the sharp flanges should be mandatory.

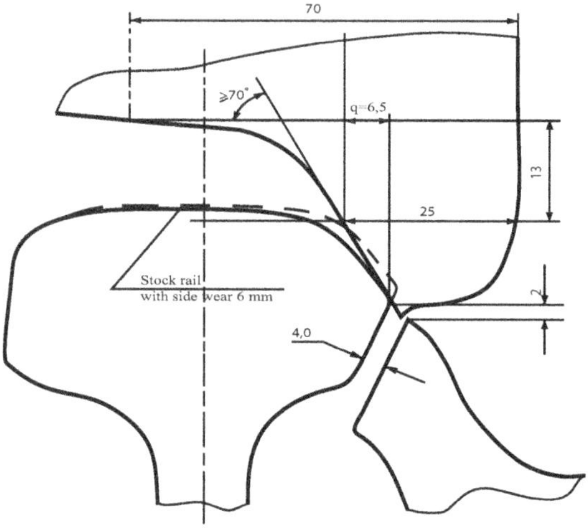

Fig. 4. Wheel with a sharp flange climbing onto a switch blade

According to the results of studies carried out in 80–90s of the last century, significant work was done to improve the structural strength of the main pointwork assemblies in order to improve traffic safety. In particular, a transition was made in mass constructions of turnouts, from counter-rail units based on special profiles of the Republic of Kazakhstan to counter-rail units based on the "angle of the counter rail SP-850". Unfortunately, this allowed changing the situation for the better only for a while. Since the beginning of the 2000s, the outages rate of elements of counter-rail units began to increase again. The number of defects that cannot be eliminated along the way by the linear workers has increased. The railways are forced to order the supply of spare parts, "repair kits of counter-rail assemblies". So in 2017, according to the track management of JSC Russian Railways, about 5,000 of such repair kits were purchased for operation needs.

Elements of counter-rail units fail due to the increased impact of rolling stock on them. One of the reasons for this, possibly, is a change in the state of the running gear of carriages.

As previously noted, with an increase in the flange wear, the inclination of its generatrix increases. Therefore, the issue of reducing the rejection size of the thickness of the wheel flanges requires serious research. The actual distribution of the sizes of the wheel pairs and the rail track in operation should be taken into account, as well as the patterns of their change during wear process.

4 Conclusions

1. The introduction of changes to the "wheel pair – rail track" dimension system should be based on an analysis of the actual condition of the wheel pairs and rail track parameters on the pointworks. In particular, the following steps are necessary in order to determine the possibility of changing the rejection size of the minimum thickness:

- to carry out mass measurements of wheelsets and pointworks in operation;
- to process the obtained data and identify the dependencies between different combinations of sizes, including determining the correlation between the flange thickness and the inclination angle of its generatrix;
- to revise the methodology for calculating the probability of occurrence of dangerous events when passing the pointworks by wheelsets, taking into account modern modeling tools and accumulated knowledge;
- to perform dynamic strength tests in order to determine the dependence of stresses and strains of pointwork elements when passing by wheelsets with different flange thicknesses and inclination angles;
- to identify the possibilities for changing in the "wheelset – rail track" dimension system, taking into account the strength and reliability of pointworks.

2. Simultaneously with conducting these works, a new methodology for calculating adverse events occurring when passing the pointworks by wheelsets should be developed, taking into account the accumulated knowledge and capabilities of modern computing facilities.

References

1. Guo, M., Lily, H., Ye, L.: Cognition and driving safety: how does the high-speed railway drivers' cognitive ability affect safety performance? Transp. Res. Part F Traffic Psychol. Behav. **65**, 10–22 (2019). https://doi.org/10.1016/j.trf.2019.07.006
2. Glusberg, B., Korolev, V., Shishkina, I., Loktev, A., Shukurov, J., Geluh, P., Loktev, D.: Calculation of track component failure caused by the most dangerous defects on change of their design and operational conditions. In: MATEC Web of Conferences, vol. 239, p. 01054 (2018). https://doi.org/10.1051/matecconf/201823901054
3. Loktev, A.A., Korolev, V.V., Shishkina, I.V.: High frequency vibrations in the elements of the rolling stock on the railway bridges. In: IOP Conference Series: Materials Science and Engineering, vol. 463, no. 3, p. 032019 (2018). https://doi.org/10.1088/1757-899X/463/3/032019
4. Loktev, A.A., Korolev, V.V., Poddaeva, O.I., Chernikov, I.Y.: Mathematical modeling of antenna-mast structures with aerodynamic effects. In: IOP Conference Series: Materials Science and Engineering, vol. 463, no. 3, p. 032018 (2018). https://doi.org/10.1088/1757-899X/463/3/032018

5. Qian, W., Qi, T., Zhao, Y., Le, Y., Yi, H.: Deformation characteristics and safety assessment of a high-speed railway induced by undercutting metro tunnel excavation. J. Rock Mech. Geotech. Eng. **11**(1), 88–98 (2019). https://doi.org/10.1016/j.jrmge.2018.04.014
6. Glusberg, B., Savin, A., Loktev, A., Korolev, V., Shishkina, I., Alexandrova, D., Loktev, D.: New lining with cushion for energy efficient railway turnouts. In: Advances in Intelligent Systems and Computing, vol. 982, pp. 556–570 (2020). https://doi.org/10.1007/978-3-030-19756-8_53

Guard Rail Operation of Lateral Path of Railroad Switch

Vadim Korolev(✉)

Russian University of Transport, Chasovaya St., 22/2, Moscow 125190, Russia
korolevadim@mail.ru

Abstract. The article discusses four different options for entering the guide wheels to the guard rail bends: when driving along a straight path of railroad switches with a straight or curved frog, the rolling stock has a free arrangement in a track; the flanges of its wheelsets either fall into the guard rail gutter without hitting the bends, or run into the bend at any point in it; when moving against the direction and along the direction when leaving the frog curve of the same direction with the railroad switch curve, the front axle of each bogie runs along the outer wheel flange onto the outer rail; the rear axle, depending on the insertion mode, is either pressed against the inner rail, or occupies an intermediate position, or pressed against the outer rail; in case of a directly movement from a straight path to a curved frog, the rolling stock is in a track in a free position and a impact is possible either in the working face of the core, or in the guard track bend; when directly entering the railroad switch with a curve, the reverse railroad switch impact with the front axles of the rolling stocks or bogies will occur in the check rails, the wheels of the rear axles can hit the bends of the guard rails. In all variants of the impact on the bends of guard rails when moving on the side track, they are determined by the number of strokes, the mass participating in the impact, the angle of impact, as well as the amount of transverse displacement of the rolling stock when hitting a guard rail, characterizing the work extinguished by the guard rail, and the smoothness of movement in the frog zone, methods calculating the determination of the length of the straight bends of the guard rails with straight frogs, determining the length of the rectilinear bends of the guard rails with curved frogs, determining the length of the curved bends to guard rails having a constant angle of inclination to the roadway with curvilinear frogs. Guard rails of the sideways of railroad switches take significantly greater impacts than guard rails of the forward direction. The only way to reduce the impact on these bends is to reduce the impact angles and collisions. For railroad switches with curved frogs, as well as for railroad switches with straight grogs with a railroad switch curve extending to the guard rails zone, it is best to use guard rails with curved bends that have the smallest length and provide the most secure entry to the bend.

Keywords: Railroad switch · Guard rail · Bend · Frog · Lateral direction · Angle of impact · Rolling stock lateral displacement · Variants of entrance schemes

Z. Popovic et al. (Eds.): TransSiberia 2019, AISC 1115, pp. 621–638, 2020.
https://doi.org/10.1007/978-3-030-37916-2_60

1 Introduction

As it is known, the interaction of the rolling stock and the guard rails of the lateral direction of the railroad switch in the horizontal plane determines their work and size.

An analysis of the entrance schemes of rolling stocks when moving along the lateral direction of railroad switches (with rectilinear and curvilinear frogs) suggests that there are four different options for entrance the steering wheels to the guard rail bends.

In all cases, the impact on the bends of guard rails when moving on a side track is determined by the number of impacts, the mass participating in the impact, the angle of impact, as well as the lateral displacement of the rolling stock when hitting a guard rail, characterizing the work extinguished by the guard rail and the smoothness of movement in the frog zone [1].

2 Materials and Methods

V a r i a n t 1. When moving along a direct path of railroad switches with a straight or curved frog (Fig. 1, position 1), the rolling stock has a free arrangement in a track; the flanges of its wheelsets either fall into the guard rail gutter without hitting the bends, or run into the bend at any point [2]. In the first case, there will not be any changes in the rolling stock's movement, in the second case, the wheelset will begin to be shifted by the guard rail bend across the path until the flange enters the guard rail gutter. The magnitude of this lateral displacement is relatively small and, depending on the location of the collision with the bend, can range from 0 to 20–23 mm.

When determining the impact forces arising in this case, it is necessary to take into account the gaps in the running gear of the rolling stock, due to which the masses of the wheelset, bogie, and then the body participate in the impact in series. If the displacement of the wheelset caused by the collision with the guard rail bend is less than the transverse run-ups in the axle box, then the body mass will not take part in the impact at all [3].

The angles of the bend offsets of the counter rails in this case are assumed to be equal to the permissible angle of impact into the wit, so that the bumps in the counter rail will not be stronger than the impacts in the rail point.

V a r i a n t 2. When moving against the direction and along the direction when exiting the frog curve in the same direction with the railroad switch curve (Fig. 1, positions III and IV), the front axle of each bogie runs along the outer wheel flange onto the outer rail; depending on the entrance mode, the rear axle is either pressed against the inner rail (static forced entrance), or occupies an intermediate position (static free entrance), or pressed against the outer rail (dynamic entrance). In all these cases, under the action of centrifugal forces, all the run-ups and clearances in the incidence running gear of the rolling stock turn out to be selected (used), and when hitting a guard rail bend, the entire mass of the rolling stock participates in the impact [4].

For approximate comparisons, it can be assumed that when moving along a straight path, only the mass of the bogie with wheelsets is involved in the impact; when moving on the side path to the mass of each bogie is added half the mass of the body. For

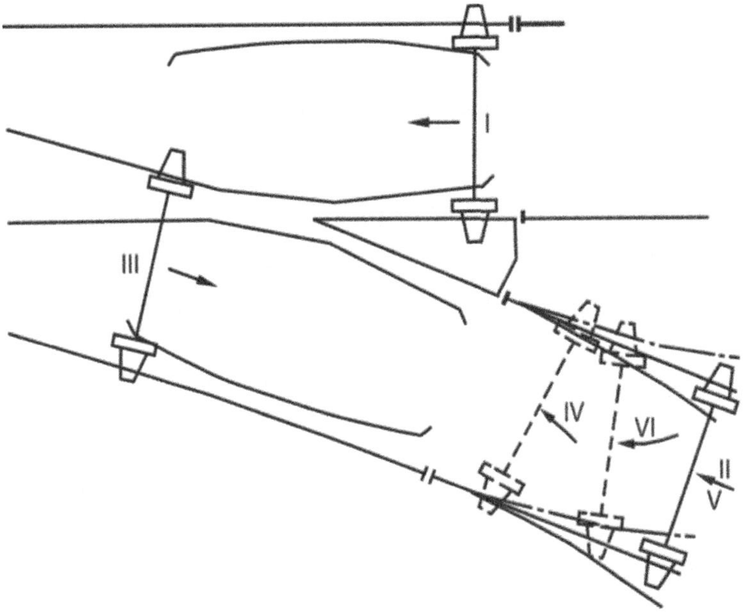

Fig. 1. Possible schemes of the approach of the wheel to the bends of the guard rails.

comparison, let's take a sixty-ton four-axle car with a weight of 4.65 tons of bogies, a body weight of 15.0 tons and an average load of 30 tons. Then, when driving along a straight path, a mass of 4.65 tons will participate in the impact, and 27 tons along the lateral track; that is, six times as large [5].

When moving along a lateral path, the incidence angle on the guard rail bend also changes. This happens because the straight section of the path between the end of the railroad switch curve and the beginning of the guard rail is usually assigned less than the rigid base of the rolling stock. At the same time, the rolling stock meeting with the guard rail bend occurs at an angle greater than the calculated one by the value of the angle of incidence on the guide rail $\Delta\gamma$ (Fig. 2).

When considering the entry conditions for the guard rail of lateral track, one should distinguish between the angle of impact in the guard rail and the angle of incidence on it. When approaching the guard rail along a curve, the rolling stock is in a state of steady curvilinear movement with a certain constant angle of incidence on the outer rail $\Delta\gamma$. Hitting a bend causes an additional dynamic effect - an impact, characterized by the difference in the angles of the collision of the bend and the incidence on the outer rail, equal to the angle between the track rail and the bend of the guard rail at the point of approach ψ.

After the impact, the rolling stock's direction is made by guard rail, the incidence angle on the bend of which will be equal to the sum of the angles of impact in the guard rail and the incidence on the track rail before the bend meeting.

Thus, the angle $\Delta\gamma$ will not affect the dynamics when meeting with a bend, but the impact on the guard rails at the same time increases due to an increase in the directing

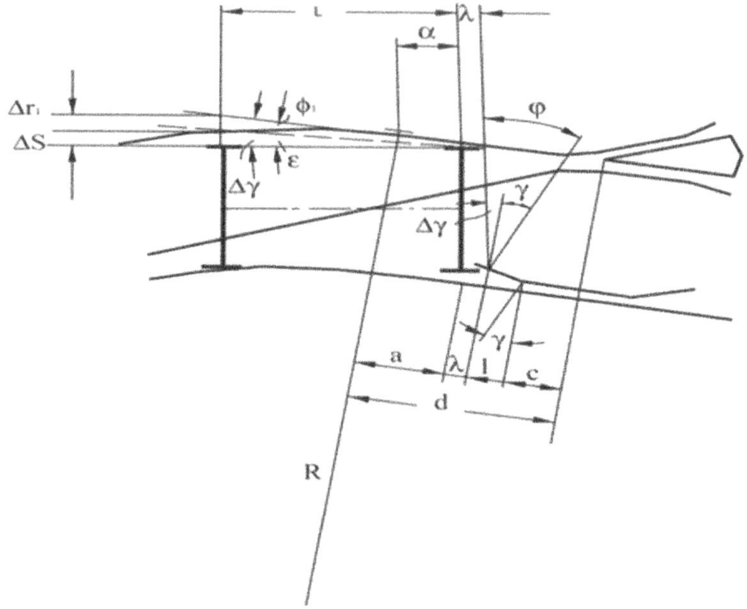

Fig. 2. The design scheme for determining additional impact angles and lengths of bends with straight frogs.

forces directly proportional to the angle of incidence. In other words, the magnitude of the additional dynamic forces depends on the angle of the bend offset γ, and the magnitude of the guiding forces depends on the sum of the angles of the bend offset γ and the angle of incidence on the guide rail $\Delta\gamma$.

The angle $\Delta\gamma$ depends only on the radius of the curve, track gauge, speed and rigid base of the rolling stock and does not depend on the form of the guard rail bend.

The ratios of these values determine how the rolling stock entrance into the curve: statically or dynamically, which affects the method for determining the angle $\Delta\gamma$ [6].

Regardless of the type of entrance in determining the angle $\Delta\gamma$, two cases are possible: with curved frogs (as well as with straight frogs in the case when a impact to the guard rail bend occurs before the beginning of the straight part of the frog), the additional collision angle $\Delta\gamma$ will be equal to the angle of the bogie incidence in the curve φ; with rectilinear frogs, when the impact to the guard rail bend occurs after the part of the bogie enters the frog straight line, the angle $\Delta\gamma$ can be defined as the angle of incidence of the bogie in the curve φ, reduced by a certain amount $\Delta\varphi$, depending on the length a of the straight part paths in front of the point of impact in the bend (see Fig. 2). Due to the small value of the angle $\Delta\gamma$, the size a is assumed equal to the outer and inner paths.

In the first case, the angle of incidence in the curve φ with a static entrance is determined by the known radius of the railroad switch curve R, the distance to the instantaneous center of rotation x and the ridge incidence λ:

$$\phi = \arcsin\frac{x+\lambda}{R}$$

The value of x is determined by conventional methods.

With dynamic entrance:

$$\phi = \arcsin\frac{L+2\lambda}{2R}$$

In the second case, with a static entrance, the initial angle φ is determined in the same way as in the first. To determine the angle $\Delta\varphi$, we will consider the incident angle as a composite formed by the rotation of the rolling stock with respect to the direct path due to the gap in the track (angle ε) and the curvature of the path in front of the frog (angle ψ) (see Fig. 2).

The value of the angle ε, depending on the ratios of the radius of the curve, track width and speed, can vary from ε = 0 (with dynamic entrance) to $\varepsilon = \arcsin\frac{\Delta S}{L}$ (with static forced entrance). If, when exiting the curve, the track width and speed are constant, then it can be considered unchanged.

Therefore, when determining the angle $\Delta\varphi$, we will be interested in the angle ψ, which varies depending on the magnitude of the straight part of the path in front of the point of impact in the bend a:

when the rolling stock moves in a curve

$$\psi = \arcsin\frac{\Delta r}{L+\lambda}$$

where

$$\Delta r \cong \frac{(L+\lambda)^2}{2R}$$

when entering the straight line at the moment of impact in the bend (see Fig. 2)

$$\psi_1 = \arcsin\frac{\Delta r_1}{L+\lambda}$$

$$\Delta r_1 \cong \frac{(L+\lambda-a)^2}{2R}$$

Here a is the distance from the front axle at the moment of impact in the bend to the end of the curve; in calculations, it must be used as much as possible, that is, equal to the distance from the end of the curve to the offset of the guard rail (section Cof), since the direct insert is obtained in this case minimal.

Since it was established above that the magnitude of the angle $\Delta\varphi$ depends only on a change in the angle ψ, it can be found:

$$\Delta\phi = \psi - \psi_1 = \arcsin\frac{(L+\lambda)^2}{2R(L+\lambda)} - \arcsin\frac{(L+\lambda-a)^2}{2R(L+\lambda)}$$

Given the small magnitude of the considered angles, it can be written:

$$\Delta\phi = \arcsin\frac{(L+\lambda)^2 - (L+\lambda)^2 + 2(L+\lambda)a - a^2}{2R(L+\lambda)} = \arcsin\left[\frac{a}{R} - \frac{a^2}{2R(L+\lambda)}\right]$$

Additional incidence angle

$$\Delta\gamma = \phi - \Delta\phi = \arcsin\frac{x}{R} - \arcsin\left[\frac{a}{R} - \frac{a^2}{2R(L+\lambda)}\right] \cong \arcsin\frac{x - a + \frac{a^2}{2(L+\lambda)}}{R}$$

With dynamic entrance ΔS = 0, then:

$$\Delta\gamma = \psi_1 = \arcsin\frac{\Delta r_1}{L+\lambda} = \arcsin\left[\frac{L+\lambda}{2R} - \frac{a}{R} + \frac{a^2}{2R(L+\lambda)}\right]$$

In this case, the value of the additional entrance angles will be minimal.

The additional impact angles will take maximum values at the extreme skew of the rolling stock in the rail track when the value of ΔS will be maximum.

In this case, the angle Δγ is defined as the sum of the rolling stock rotation angles formed due to the gap in the track (angle ε) and the curvature of the path in front of the frog (angle ψ). According to the scheme (Fig. 2) it can be found:

$$\varepsilon = arctg\frac{\Delta S}{L}$$

$$\psi = arctg\frac{\Delta r}{L}$$

Knowing the value of the direct insertion in front of the mathematical edge of the frog core d, the length of the straight part of the guard rail in front of it c, the length of the bend l and the flange incidence λ, the value Δr can be found by the expression:

$$\Delta r = \frac{(L-a)^2}{2R} = \frac{(L-(d-c-l-\lambda))^2}{2R}$$

Finally maximum additional angle of impact

$$\Delta\gamma = \varepsilon + \psi = arctg\frac{\Delta S}{L} + arcth\frac{\Delta r}{L} \cong arctg\frac{\Delta S + \Delta r}{L}$$

According to the formulas obtained, additional impact angles are calculated that take place in existing railroad switches of different brands. The minimum values of Δγ obtained during dynamic entrance are shown in Table 1. The maximum values (for the maximum skew of the rolling stock) are shown in Table 3. Table 2 shows the values of Δγ for the most probable static speed at existing speeds on the lateral entrance path.

Table 1. Angles of impact in the guard rail during dynamic entrance of rolling stock.

Railroad switch brand	Offset bend angle	Bend length l, mm	Radius of the railroad switch curve R, mm	a, mm	x, mm	Additional incidence angle Δγ	Actual collision angle on the bend φ	The increase in the actual collision angle compared to the offset angle, %
1/22	0°27′30″	3000	1440000	–	–	0° 4′10″	0°31′40″	15
1/18	0 34 33	2400	960000	–	–	0 6 16	0 40 39	18
1/11	0 53 32	1400	297000	1387	–	0 7 22	1 00 54	14
1/9	1 17 09	980	180000	907	–	0 18 20	1 35 29	21
1/7	1 50 03	687	115432	425	–	0 40 16	2 30 19	36
1/5	2 00 34	627	54802	539	–	1 42 08	3 42 42	85

Table 2. Angles of impact in the guard rail with a static entrance of rolling stock.

Railroad switch brand	Offset bend angle	Bend length l, mm	Radius of the railroad switch curve R, mm	a, mm	x, mm	Additional incidence angle Δγ	Actual collision angle on the bend φ	The increase in the actual collision angle compared to the offset angle, %
1/22	0°27′30″	3000	1440000	–	3500	0° 8′22″	0°35′52″	30
1/18	0 34 33	2400	960000	–	3500	0 12 32	0 46 55	37
1/11	0 53 32	1400	297000	1387	3500	0 27 38	1 21 10	51
1/9	1 17 09	980	180000	907	3084	0 43 56	2 1 05	57
1/7	1 50 03	687	115432	425	2607	0 5 46	1 55 49	60
1/5	2 00 34	627	54802	539	2157	1 44 6	3 44 40	87

Table 3. Angles of impact in the guard rail at the extreme skew of the rolling stock.

Railroad switch brand	Offset bend angle	Bend length l, mm	Radius of the railroad switch curve R, mm	a, mm	x, mm	Additional incidence angle Δγ	Actual collision angle on the bend φ	The increase in the actual collision angle compared to the offset angle, %
1/22	0°27′30″	3000	1440000	–	–	0° 29′44″	0°57′14″	108
1/18	0 34 33	2400	960000	–	–	0 31 50	0 6 13	93
1/11	0 53 32	1400	297000	1387	–	0 32 55	1 26 27	62
1/9	1 17 09	980	180000	907	–	0 43 55	2 1 4	57
1/7	1 50 03	687	115432	425	–	0 5 48	2 55 51	60
1/5	2 00 34	627	54802	539	–	1 44 4	3 44 38	87

When calculating in all cases, $\Delta S = 26$ mm was taken as the average value observed in practice, L = 3500 for the bogie of a promising mass rolling stock (six-axle gondola car) and $\lambda = 0$.

Comparison of the average and maximum actual angles of collision to the bend (data from Tables 2 and 3) reveals the equality of their values for railroad switches of grades 1/5–1/9, calculated by different methods. This happens because the calculated rolling stock with these types of railroad switches entrances into the railroad switch curve forcibly (x < L) and always occupies an extremely distorted position in their railroad switch curves.

For high-speed railroad switches of grades 1/18 and 1/22, the obtained values of the incidence angles are somewhat overestimated. Bends designed rectilinear become curvilinear, since they are located in the zone of the railroad switch curve.

However, it is difficult to practically bend the track rail together with the guard rail when laying the railroad switch, and therefore in the zone of the guard rail the radius of the railroad switch curve will be larger than the design, and the incidence angles will be larger than the calculated ones.

When moving to a lateral path, not only the angles increase, but also the number of impacts to the bend and the work paid off by it with each impact. According to G.I. Ivashchenko, when driving along a direct path, a guard track bends impacts 46% of the axles, and the average shift of these axes across the path ΔC_{av}^{ax} is 3.4 mm.

When driving on a lateral track under conditions of static entrance ($\upsilon < 50$ km/h), each front axle of all the bogies impacts into the bend, i.e. 50% of the axles; their average shift across the path can also be determined on the basis of G.I. Ivashchenko's data on the measurement of wheelsets and track width. According to these data, the average size of the blind nozzle can be taken 1440 mm, the average track width in the area of the frog - 1524 mm. The average flange thickness based on full-scale measurement of 1617 wheelsets at the Chelyabinsk Metallurgical Plant was set at 29.88 mm.

With these data, the average value of the transverse shift will be:

$$\Delta C_{av}^{ax} = 1524 - 29,88 - 1440 - 44 = 10.1 \text{ mm}$$

(44 - the size of the guard rail gutter).

With a dynamic entrance, all axes will impact (100%) into the bend, since when approaching the guard rail, all wheelsets will be pressed against the outer rail. The average value of the lateral shift will remain the same as with the static entrance.

The obtained ratios of the rolling stock masses participating in the impacts, the number of impacts, the values of the incidence angles and the lateral displacements allow us to tentatively compare the impacts on the guard rail bends when driving along straight and lateral tracks with equal speeds. It was determined above that when driving on a lateral track under conditions of static incidence, the impact is carried out by six times the mass, the incidence angle on the bend (Tables 2 and 3) is 30–87% more, the number of impacts on the bend is 50–46% more and the magnitude lateral shift is 2.97 times greater [7].

With dynamic entrance, the increase in the angle of incidence on the bend (Table 1) will be from 15 to 85% of the estimated angle of impact, and the number of axes impacting to the bend will double (100%).

These comparisons suggest that lateral path guard rails experience significantly greater effects than direct-path guard rails.

V a r i a n t 3. When the direct movement from a straight path to a curved frog (in position V, Fig. 1), the rolling stock is made a free position is in a track in and a impact is possible either in the working face of the core or in the guard track bend [8]. The largest shock-dynamic impacts will be realized upon impacts in the guard rail bend. In this case, the incidence and impact angles are equal to each other and exceed the angle of the bend offset by the angle ξ, the value of which depends on the radius of the frog curve, the distance from the tail of the frog to the inflection point of the guard rail and the shape of the bend (Fig. 3).

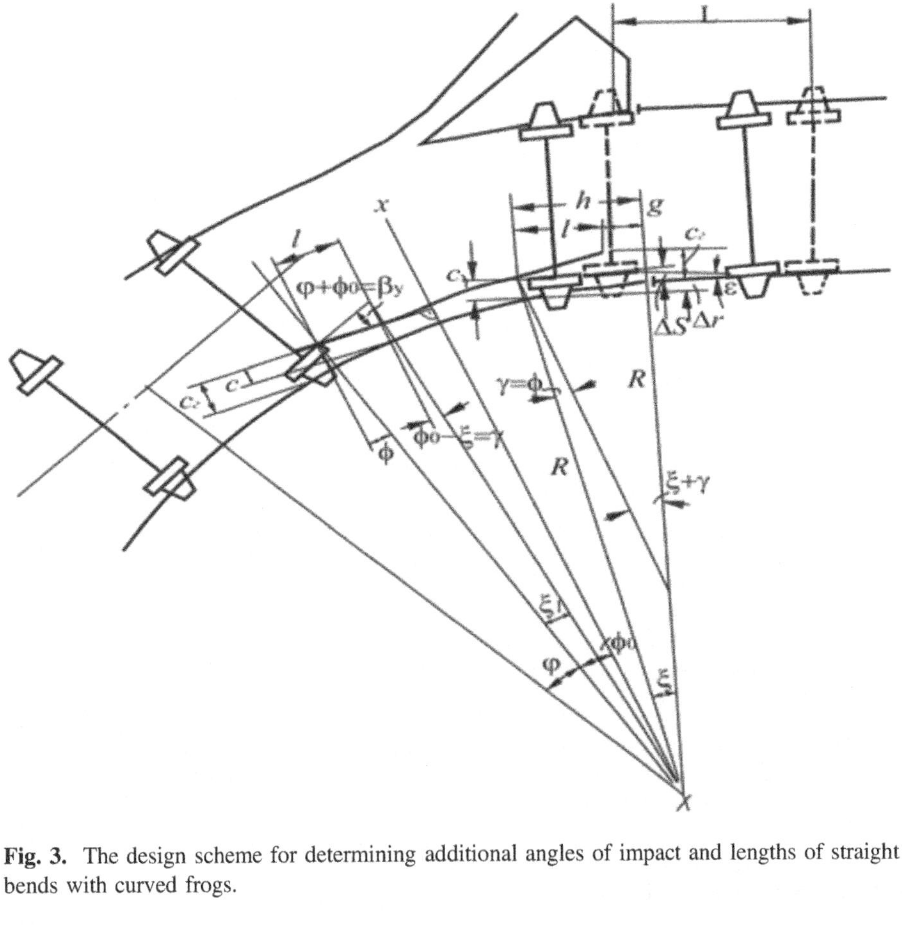

Fig. 3. The design scheme for determining additional angles of impact and lengths of straight bends with curved frogs.

With a direct bend, the angle of impact (collision) does not depend on the place of impact in the bend and is equal to the sum of the angles of the bend offset and the central angle ξ.

$$\xi = arctg\frac{h}{R}$$

where h is the distance from the inflection point of the guard rail to the end of the frog;

R is the radius of the track rail in the zone of the frog.

Calculations show that for the current railroad switches, the angle ξ is about 20% of the angle γ.

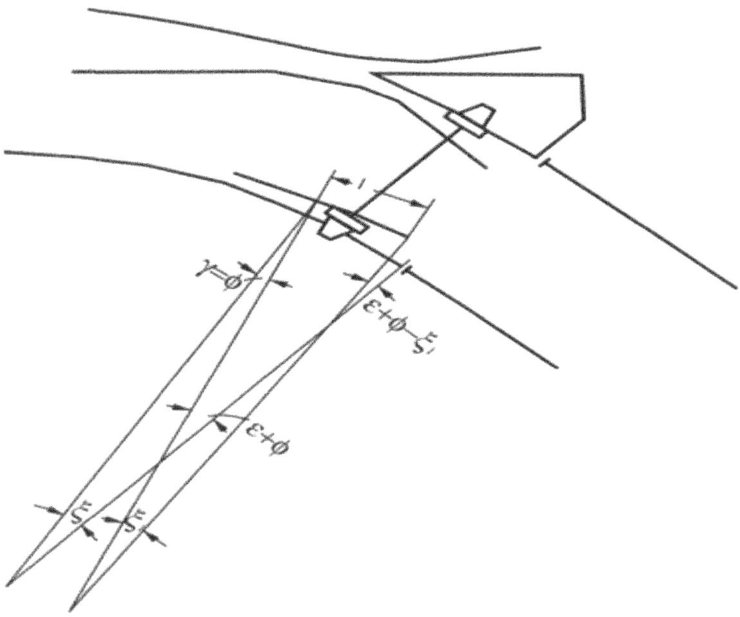

Fig. 4. The design scheme for determining the length of curved bends with curved frogs.

With a curvilinear bend, the angle of impact (collision), depending on the location of collision, can be in the range from $\xi + \gamma - \xi 1$ to $\xi + \gamma$ (Fig. 4), here $\xi = arctg\frac{l}{R}$. Thus, when approaching the location of the impact to the inflection point, the angle of impact increases, the change is equal to the value of the central angle contracting bend $\xi 1$.

However, when calculating the bends, one can proceed from the condition:

$$\xi + \gamma - \xi_1 \leq \gamma_{\max},$$

where γmax is the maximum permissible angle of impact.

This is explained by the fact that when collision a bends, the rolling stock or the bogie can be skewed due to the angle between the rail and wheel track ΔS by an angle (Fig. 3).

$$\varepsilon = arctg\frac{\Delta S}{L},$$

where L is the rigid base of the bogie or rolling stock.

Table 4. The angle of rotation of the rolling stock bogies due to the gap in the track.

Four-axle freight car	Passenger railway car	Six-axle freight car	
L = 1800 mm	L = 2700 mm	L = 3050 mm	L = 3500 mm
0°49′39″	0°33′05″	0°29′18″	0°25′32″

The values of ε calculated for some mass types of rolling stock are shown in Table 4. in the calculations is usually larger than the change in the angle ξ1 (ξ1 $\leq$ 0°20′); at In the calculation, ΔS is taken equal to 26 mm.

Despite the significant value of the angle ε, it was never taken into account in the calculations of railroad switches. In the case under consideration, the angle ε will also vary depending on the location of the impact in the bend, but inversely with the law of change in the collision angle, decreasing from the value of ε at the socket to zero as the point of impact moves from the beginning of the guard rail (in the bend area, the impact can only occur when the crest is pressed to the inner edge (see Fig. 3), that is, in the case when ε = 0. The magnitude of the change in the angle ε not taken into account increasing the angle ξ1, the angle ε decreases and the total value of the angle of impact remains less than angle ε, therefore, when calculating it is possible not to take into account the angle ξ1.

Variant 4. When the direct entrance to the railroad switch from a curve opposite the railroad switch curve (position VI, Fig. 1), the front axles of the rolling stocks or bogies will impact on the check rails; the wheels of the rear axles can impact in the bends of the guard rails [9]. At the same time, the angle of impact in the bends of the guard rails will be less than the offset angle due to the curvature of the path before the bend and the skew of the rolling stock when moving along the curve. Such conditions for the passage of guard rails are the best in comparison with previous cases.

If the radii of the frog and the railroad switch curves are equal, the magnitude of the decrease in the angle of impact in the bend of the guard rail will be equal to the value of the additional angle of impact in the second variant.

From the considered cases of guard rails passing, it can be seen that the second option is the most difficult in all indicators (Fig. 1, positions III and IV), and the third case in terms of the angle of impact for the combination of a curved frog and guard rails with straight bends (Fig. 1 position V). However, in the third case, the impact force decreases due to the sequence of participation in the impact of the masses of the wheelset, bogie and body, so it is impossible to say that the impact on the bends in this case is greater than in the second. The true value of the impacts on the bends can only be determined empirically. Consider methods for calculating bends for both cases.

1. Determination of the lengths of straight bends of guard rails with straight frogs

Reducing the impact of rolling stocks on the guard rails of the lateral track in railroad switches with straight frogs and straight bends can be achieved only by reducing the angle of the bend offset (increasing its length).

For example, for existing 1/7 railroad switches of the type P50 R = 115 432, l = 687 mm, s = 594, d = 1706, l1 = 1°50′03″. If you extend the bend to l1 = 1100 mm, then for L = 3500 and λ = 0 we get: al = d-c-l1, = 1706-594-1100 = 12 mm; x = 2607 (see Table 1) and according to the formula derived above:

$$\Delta\gamma_1 = \arcsin\frac{x - a + \frac{a}{2L}}{R} = arcsin\ 0.02248076 = 1^{\circ}17'18'' = 1.28153^{\circ};$$

$$\gamma_1 = \arcsin\frac{C_{of} - C}{l} = arcsin\ 0.02090909 = 1^{\circ}11'52'' = 1.1978^{\circ};$$

$$\gamma_1 + \Delta\gamma = 1^{\circ}17'18'' + 1^{\circ}11'52'' = 2^{\circ}29'10''.$$

Moreover, despite an increase in Δγ (for comparison, see Table 1), guiding forces decrease by 1.18 times, and dynamic forces - by 2.34 times.

By choosing the length of the bend in this way, it is possible to achieve the desired reduction in the impact on the lateral path counter rail.

2. Determination of the length of the straight bends of the guard rails with curved frogs

For the second option, the largest value of the angle of impact will be with a static entrance. For this condition, the calculation is given (see Fig. 3).

The length of the bend, taking into account the small angle, can be defined as the length of the arc:

$$l = R\xi_1,$$

where R is the radius of the curve in the zone of the frog;

ξ1 is the angle between the radii of the railroad switch curve drawn through the beginning and end of the bend, radians.

To find this angle, we compose the projection on the perpendicular to the bend axis X - X:

$$R\cos(\psi_0 - \xi_1) - R\cos\psi_0 = C_{of} - C,$$

where Cof is the size of the offset of the counter rail;

C is the size of the guard rail gutter;

ψ0 is the angle between the track rail and the guard rail at the offset.

When assigning the angle ψ0, it should be borne in mind that when approaching the guard rail the rolling stock is in a state of steady curvilinear motion. Collision on the bend causes an additional dynamic effect, characterized by the difference between the angles of the collision on the bend and the incidence on the outer rail, equal to the angle between the track rail and the bend of the guard rail ψ [10].

Therefore, the angle $\psi 0$ should be assigned from the condition of limiting the magnitude of the additional dynamic impact, and in all cases it should not exceed the permissible angle of impact in the bend of the guard rail of the direct path.

To solve the resulting transcendental equation, we expand the cosine in a series, and, discarding all terms to a degree higher than the second, we obtain:

$$\cos(\psi_0 - \xi_1) = 1 - \frac{(\psi_0 - \xi_1)^2}{2}, \cos\psi_0 = 1 - \frac{\psi_0^2}{2}$$

After transformations

$$\frac{\xi_1^2 R}{2} - \xi R\psi_0 + (C_{of} - C) = 0$$

where from

$$\xi_1 = \frac{\sqrt{(R\psi_0)^2 - 2R(C_{of} - C)}}{R}$$

Bend length:

$$l = R\xi_1 = R\psi_0 \pm \sqrt{(R\psi_0)^2 - 2R(C_{of} - C)}$$

Using this formula, we plotted the dependence of the length of the direct bend on the radius of the railroad switch curve (Fig. 5, line 1). The calculations were performed at $\psi 0 = 1°20'00''$ and Cof - C = 20 mm. The length of the direct bend with a constant value of $\psi 0$ depends on the radius of the railroad switch curve and the frog.

In a railroad switch with a curved frog and straight bends of guard rails, the maximum angle of incidence to the bend will take place at the end of the bend. When collision a bend at a point closer to the gutter, the collision angle will decrease by the amount of rotation of the rolling stock relative to the bend due to its movement along the railroad switch curve; in total, this angle can change by $\xi 1$. If you set the maximum angle of impact into the bend $\psi 0$ and reduce the radius of the railroad switch curve, the length of the bend will increase and, ultimately, a further decrease in the radius will be impossible, since the track rail will not be able to approach the bend at a distance of the width of the gutter C.

$$I - l_{max} = 1719; II - l_{max} = 2119$$

In Fig. 6 line 3 shows the position of the track rail with a minimum radius of the railroad switch curve. The value of this radius can be determined from the equation

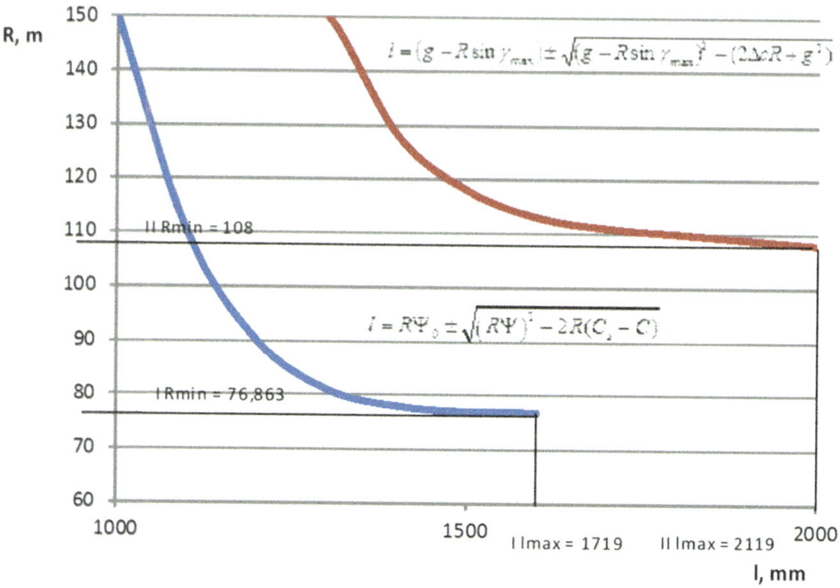

Fig. 5. The graph of the dependence of the length of the direct bend on the radius of the railroad switch curve at $\psi_0 = 1°20'00''$.

obtained by the projection of the elements of the railroad switch curve on the direction of the radius R3

$$R_{min} - R_{min}\cos\psi_0 = C_{of} - C$$

where

$$R_{min} = \frac{C_{of} - C}{1 - \cos\psi_0}$$

Maximum bend length will be:

$$l_{\max} = (R_{\min} + C)\sin\psi_0$$

In this case, the angle ψ0 = ξ1

The values of Rmin and lmax can also be determined from the equation:

$$l = R\psi_0 \pm \sqrt{(R\psi_0)^2 - 2R(C_г - C)}$$

As the radius decreases, the radical expression decreases. Equating it to zero, you can find the value of the minimum radius:

$$(R_{min}\psi_0)^2 - 2R_{min}(C_{of} - C) = 0$$

where

$$R_{min} = \frac{2(C_{of} - C)}{\psi_0^2}$$

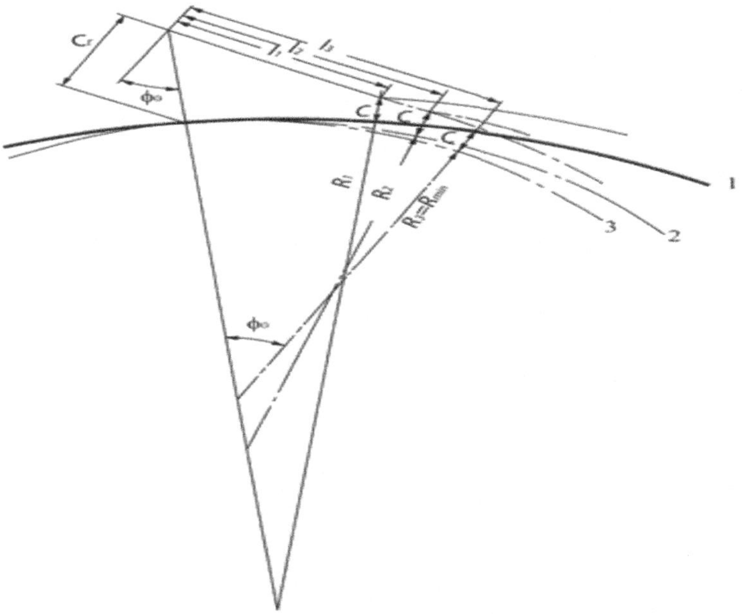

Fig. 6. Scheme for changing the length of the direct bend when changing the radius of the curve (ψ_0 = const).

Maximum length of direct bend:

$$l_{\max} = R_{\min}\psi_0$$

When calculating the bend for the third option, we set the condition that the angle of impact in the bend should not exceed the maximum accepted for the calculated railroad switch, i.e. $\xi + \gamma \leq \gamma max$ (Fig. 3); then it can be written:

$$\gamma_{max} \geq arctg\frac{C_{of} + \Delta r - C}{l}$$

$$\Delta r = \frac{(l+g)^2}{2R}$$

$$C_{of} - C = \Delta C$$

where l is the length of the bend;

g is the distance from the bend end of the guard rail to the end of the frog;

$$\gamma_{\max} = arctg\frac{\Delta C + \frac{(l+g)^2}{2R}}{l} = arctg\frac{\Delta C2R + (l+g)^2}{2Rl}$$

or

$$tg\gamma_{\max}2Rl = 2\Delta CR + (l+g)^2;$$
$$l^2 + 2l(g - Rtg\gamma_{\max}) + 2\Delta CR + g^2 = 0$$

where from

$$l = -(g - Rtg\gamma_{\max}) \pm \sqrt{(g - Rtg\gamma_{\max})^2 - (2\Delta CR + g^2)}.$$

Here, as in the previous case, the length of the bend depends on the radius of the railroad switch curve, therefore, in the same way, the minimum radius can be found at which a direct bend can be installed.

Equate the root expression to zero:

$$[g - Rtg(\gamma_{\min})]^2 - [2\Delta CR + g^2] = 0$$

where

$$R_{\min} = \frac{2(gtg\gamma_{\max} + \Delta C)}{tg^2\gamma_{\max}}$$

The maximum length of the direct bend in this case:

$$l_{\max} = -(g - R_{\min}tg\gamma_{\max}) = R_{\min}tg\gamma_{\max} - g$$

The angle of guard rail inflection:

$$\gamma = \gamma_{\max} - \xi,$$

where

$$\xi = arctg\frac{l+g}{R}$$

The graph of the dependence of the length of the direct bend on the radius of the railroa bends, the length of the latter should be increased in comparison with the length of d switch curve for this case is shown in Fig. 5 line II. It can be seen from the graph that, under the conditions of preserving the given value of the angle of impact during the direct entrance from a straight lateral path to a curved frog with straight the bends obtained for the second variant of the entrance to the guard rail [11].

The given graphs shown also that the use of direct bends of guard rails for curved frogs is limited by the fact that they cannot be used with small radii of the railroad switch curves, as well as with a large length that depends on the radii of the railroad switch curve and the frog.

3. Determination of the length of the curved bends of the guard rails having a constant angle of inclination to the track rails with curved frogs

If you create a bend having the same inclination to the track rail at any point, then its length will not depend on the radius of the curve. Such a bend is a curved bend, curved along the radius of the frog curve. The angle of inclination of this bend to the track rail ψ is the angle of impact in the bend.

To calculate the length of the bend, we take the radii of the railroad switch and frog curves, as well as the radius of the bend equal to infinity, and the angle between the bend and the track rail is equal to the calculated angle of impact in the lateral path bend (see Fig. 4) [12]. Then bend length:

$$l = \frac{C_{of} - C}{\sin \psi}$$

The angle of inflection of the guard rail γ will also be equal to the angle ψ. When the track rail is curved and bent to the radius of the railroad switch curve, the angle ψ characterizing the dynamics of the entrance to the bend does not change, which allows determining the length of the bend regardless of the radius of the curve.

For example, we define the length of the branch at $\psi = 1°20'00''$ and Cof - C = 20 mm:

$$l = \frac{C_{of} - C}{\sin \psi} = \frac{20}{0.02327105} = 859 \text{ mm}$$

The length of such bend will be significantly less than the length of the straight bend defined for any radius of the railroad switch curve at the same angle of impact ψ (see Fig. 5).

Curved bends are already practically used in railroad switch of brands 1/22 and 1/18, where the railroad switch curve enters the grog zone and, when properly laid, the bends are curved along with the track rails.

3 Conclusions

1. Guard rails of the lateral paths of railroad switches take significantly greater impacts than guard rails of the forward direction.

2. The only way to reduce the impact on these bends is to reduce impact and collision angles.
3. For railroad switches with curved frogs, as well as for railroad switches with straight frogs with a railroad switches curve extending to the guard rails zone, it is best to use guard rails with curved bends that have the smallest length and provide the most secure entrance to the bend.

References

1. Glusberg, B., Savin, A., et al.: Counter-rail special profile for new generation railroad switch. In: Advances in Intelligent Systems and Computing, vol. 982, pp. 571–587. https://doi.org/10.1007/978-3-030-19756-8_54
2. Loktev, A., Korolev, V., et al.: Modeling of railway track sections on approaches to constructive works and selection of track parameters for its normal functioning. In: Advances in Intelligent Systems and Computing, vol. 982, pp. 325–336. https://doi.org/10.1007/978-3-030-19756-8_30
3. Loktev, A.A., Korolev, V.V., et al.: High frequency vibrations in the elements of the rolling stock on the railway bridges. In: IOP Conference Series: Materials Science and Engineering, vol. 463, p. 032019 (2018). https://doi.org/10.1088/1757-899x/463/3/032019
4. Glyuzberg, B.E., Korolev, V.V.: Corner rail. Patent for utility model RUS 108760, 06 August 2011
5. Loktev, A.A., Korolev, V.V., et al: Modeling the dynamic behavior of the upper structure of the railway track transportation geotechnics and geoecology. In: TGG 2017, Procedia Engineering Saint Petersburg, Russia, vol. 189 pp. 133–137 (2017). https://doi.org/10.1016/j.proeng.2017.05.022
6. Glusberg, B. Korolev, V. et al: Calculation of track component failure caused by the most dangerous defects on change of their design and operational conditions. In: MATEC Web of Conferences, vol. 239, p. 01054 (2018). https://doi.org/10.1051/matecconf/201823901054
7. Loktev, A.A., Korolev, V.V.: Mathematical modeling of antenna-mast structures with aerodynamic effects. In: IOP Conference Series: Materials Science and Engineering, vol. 463, p. 032018 (2018). https://doi.org/10.1088/1757-899x/463/3/032018
8. Savin, A., Kogan, A., et al.: Evaluation of the service life of non-ballast track based on calculation and test. Int. J. Innov. Technol. Explor. Eng. (IJITEE) **8**(7) (2019). ISSN 2278-3075. G5991058719/19©BEIESP
9. Glusberg, B., Savin, A.: New lining with cushion for energy efficient railway railroad switchs. In: Advances in Intelligent Systems and Computing, vol. 982, pp. 556–570 (2020). https://doi.org/10.1007/978-3-030-19756-8_53
10. Glusberg, B., Loktev, A., et al.: Calculation of heat distribution of electric heating systems for railroad switchs, In: Advances in Intelligent Systems and Computing, vol. 982, pp. 337–345 (2020). https://doi.org/10.1007/978-3-030-19756-8_31
11. Loktev, A.A., Korolev, V.V. et al.: Mathematical modeling of aerodynamic behavior of antenna-mast structures when designing communication on railway transport. Vestnik of the Railway Research Institute, Russia (2018). https://doi.org/10.21780/2223-9731-2018-77-2-77-83
12. Loktev, A.A., Korolev, V.V. et al.: Perspective constructions of bridge overpasses on transport main lines. Vestnik of the Railway Research Institute, Russia (2018). https://doi.org/10.21780/2223-9731-2018-77-6-331-336

Techniques of Armature Magnetic Induction Vectors Forming for Two-Phase Four-Sectional Brushless Direct Current Motor

Alexander Krivilev, Evgeniy Dunich(✉), and Sergey Penkin

Moscow Aviation Institute (National Research University), Volokolamskoe Rd. 4, 125080 Moscow, Russia
evgeniydunich@mail.ru

Abstract. Systematization of techniques to form armature magnetic induction base vector sets for two-phase brushless direct current (DC) motor with four-sectional winding is proposed. Each set is defined by section connection scheme and phase energizing technique; it has an influence on motor stall torque. The armature equivalent circuit is coordinated and considered on a complex plane, the motor's phases are directed along its axis. The techniques are grouped by participation of the phases, by sections within one phase, in the base vector forming process during an intercom mutation interval. A terminology is introduced which allows to consider base vector forming techniques from unified positions for *n*-phase brushless DC machines applied on perspective vehicles. A name for each set is proposed and a mathematical description is obtained. Amplitude values of the vectors at different connection schemes, at different energizing techniques within one scheme, are obtained and collated one another. With the help of the proposed systematization the quantity of possible base vector sets at a certain connection scheme is derived, their graphical representation diagrams are built. Based on the obtained results, vector forming techniques characterized by different power efficiency are discovered. Recommendations for drive developers to use a certain technique are given.

Keywords: Two-phase brushless DC motor · Brushless DC motor control · Phase winding sections · Phase commutation · Phase connection scheme

1 Introduction

Nowadays vehicles using electrical power to feed tractive and auxiliary drives are increasingly more shared. To convert electrical power from electrical into mechanical form electric motors are used. To use an electric motor on vehicles it is needed to provide a high power efficiency with keeping of load capability. Usability of a motor in an electric drive to implement required moving law is determined by its static characteristics, they depend but constructive parameters also on phase connection scheme, commutation and impulse control method [1, 2].

Z. Popovic et al. (Eds.): TransSiberia 2019, AISC 1115, pp. 639–647, 2020.
https://doi.org/10.1007/978-3-030-37916-2_61

In this paper a two-phase brushless DC motor (BLDC) with four-sectional winding is considered. Its two phase windings are allocated on 90 electrical degrees, each of them is divided into two sections. To determine possibility to use a motor of considered type data about its static and energy characteristics at a certain phase commutation method is needed.

At the initial state of researching physical processes in two-phase BLDC and obtaining its static characteristics following problems appear:

(1) Existing of four sections causes variety of connection schemes which in part gives amount of additional commutation techniques.
(2) Considering the motor on the phase level does not enable to determine stall torque beforehand in the case when more than one section exist in one phase.

Attempts to systemize and compare two-phase motor control methods are being made by researchers. In their works physical processes are being investigated and static characteristics are given. However, traditionally, the meanings of phase and winding section are distinguished fuzzily, consideration of one section as a basic calculation unit is absent, that does not enable to determine the quantity of possible motor control techniques and the view of its characteristics at each of them.

In papers [3, 4] physical processes are investigated and an efficient impulse control method is proposed, however a dividing of the phases into sections is absent, that makes difficult the transferring into speed-torque characteristics end their analysis.

In paper [5] only one armature magnetic induction vector set forming technique is given: combined at the radial winding connection scheme with common point (by the proposed systematization), and nothing is said about existing of other base vector set forming techniques. In paper [6] a two-phase BLDC motor is investigated and the dividing of the phase windings into sections is given. However there a special problem is solved: efficiency increasing by passive elements (voltage booster chokes), alternative connection schemes and energizing switching sequences are not considered. Thus, nowadays a system approach where one winding section is used as a basic calculation unit to consider base vector set forming techniques, and also an analysis of their influence on speed-torque characteristics, is absent.

To implement each technique, it is needed to create its mathematical description and, on its base, to determine power amplifier's switching element commutation functions (control functions). Such a work was carried out for three-phase machines in [7].

The proposed systematization represents a progress of the sectional approach proposed by the authors in [8, 9] and covers armature induction base vector set forming techniques on machine's electrical period at absence of the back-EMF. Only basic cases are considered, without the pulse control. The systematization is based on representation of the machine's armature in a rectangle coordinate system with allocation of the phases along its axis and on the calculation of each phase's contribution into creation of the moving torque, and within one phase—contribution of each section.

With help of the proposed systematization the quantity of possible sets at a certain connection scheme is determined. Amplitude values of the vectors are obtained and collated one another at different connection schemes and within one scheme—at different energizing techniques. Based on the obtained results recommendations for drive developers to apply a certain technique are given.

2 Systematization

2.1 Terms and Definitions

By the systematization of the techniques it has been required to extend the term base used at the current time to work with brushless electric machines. The following term definitions are proposed:

- *section*—a part of the armature winding giving a magnetic induction vector with certain direction and having own leads for connection to external circuitries;
- *reference sectional vector*—armature magnetic induction vector formed by one section when it is energized by the full power supply voltage and the back-EMF is absent;
- *sectional vector*—armature magnetic induction vector formed by one section at conditions different from the previous item;
- *base vector*—armature magnetic induction vector acting during an intercommutation interval, it is the result of the summation of section vectors acting on this interval;
- *phase*—circuitry unit consisting of the sections which are forming collinear armature magnetic induction vectors;
- *section connection scheme*—a variant of electrical section connection one with another.

2.2 Systematization Principles

The systematization is based on enumeration of all possible section connection schemes, determination of all possible energizing techniques at each of the scheme and calculation of base vectors in a rectangle coordinate system (as which a complex plane is used). In the Fig. 1 the equivalent circuit diagram of the armature winding composed in [8, 9] is shown. It is allocated on a complex plane, its real axis is directed to the Right along the line C – A, the imaginary—to the Up, along the line D – B. Due to the perpendicularity of the phases a complex plane exists whose axis are collinear with the motor's phases. In this case, absolute values of sectional vectors' projections on corresponding axis are equal to one, on perpendicular axis—to zero. It enables to consider direction cosines of the section vectors as equal to one without necessity to write them explicitly; and the sectional vectors belonging to different phases—as independent one on another.

On the diagram the following designations are used: $A - D$—points for connection to the power supply unit;

$a - h$—leads of the sections; $L_{ab} - L_{gh}$, $R_{ab} - R_{gh}$, $E_{ab} - E_{gh}$—correspondingly the inductance, ohmic resistance and back-EMF of each section.

As the positive counting direction and the forward direction of the motor's rotor the counter-clockwise direction is selected. The leads of the sections and the points for connection to the power supply unit are also directed counter-clockwise.

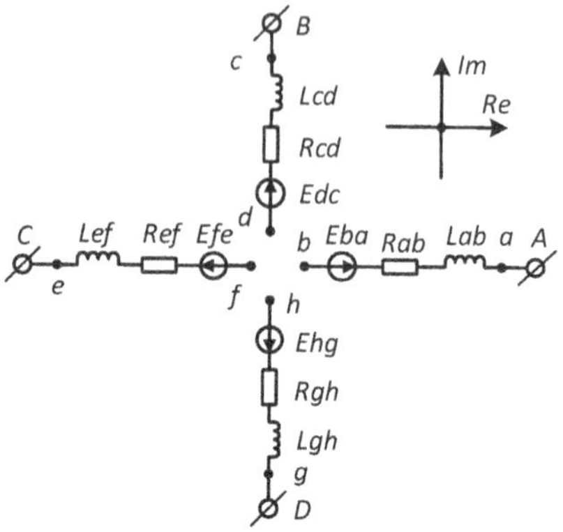

Fig. 1. Equivalent scheme of the armature winding of the two-phase BLDC motor

By the composing of the equivalent circuit diagram the following assumptions are made:

1. Resistances and inductances of all sections are equal.
2. The back-EMF values of the sections in each phase are sign-opposite.
3. The mutual inductance between phases is neglected.

In the Fig. 2 the section connection schematic diagrams are shown: with unconnected sections (a); serial (b); parallel (c); closed: the same scheme is shown in the coordinate view keeping the sections' allocation (d) and looking as a square (e); radial schemes: "star" with neutral point which is not connected (f) and with common point connected either to the supply bus or to the ground (g). The introduced designation and sections' position on the complex plane are kept on the diagrams; thanks to this, it is simple to determine the participation of each section in the vector forming along a certain axis at a certain connection scheme.

The phase energizing technique does not depend on the section connection scheme. At any of them the sections, therefore, the sectional vectors are allocated along the axes of the complex plane, that enables to accept the reference sectional vector as the basic calculation unit and to calculate any base vector as a sum of sectional ones. So, by quantity of the acting sections amplitude and direction of a base vector can be immediately determined.

The phase energizing technique does not depend on the section connection scheme. At any of them the sections, therefore, the sectional vectors are allocated along the axes of the complex plane, that enables to accept the reference sectional vector as the basic calculation unit and to calculate any base vector as a sum of sectional ones. So, by quantity of the acting sections amplitude and direction of a base vector can be immediately determined. At all of the connection schemes there are two possible energizing techniques which are different by phases in which the sections taking part in the magnetic field vector forming during an intercommutation interval are allocated:

"Separate" (the separate energizing)—the sections allocated only in one phase take part (collinear);

"Joint" (the joint energizing)—sections belonging to both phases take part simultaneously (perpendicular).

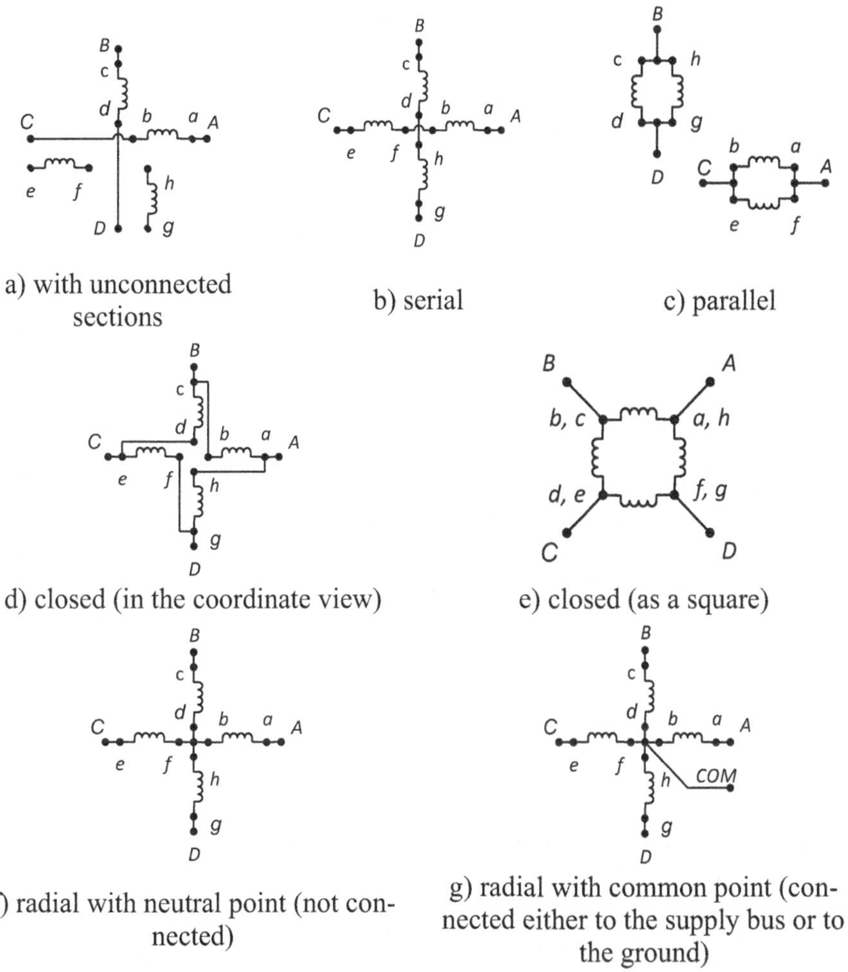

Fig. 2. Section connection schemes

There exists a possibility to apply the separate and joint energizing techniques in adjacent intercommutation intervals alternately during the full electrical period. Such an energizing technique became in this paper the name "Combined" (the combined energizing)—base vector sets formed by combined techniques are alternating. At each section connection scheme there are own combined techniques depending on the quantity of active or passive sections in each phase.

At the closed or the radial scheme with neutral point the techniques take place in which either one section is not connected or it is connected to the same power supply bus through both leads, and also when two sections belonging to one phase are connected oppositely. In such cases additional energizing techniques appear:

"With passive section in one phase"—one lead of a section is not connected directly to any power supply bus, the vector formed by this section is equal to zero, this section does not participate in the moving torque creation.

"With opposite energizing of sections in one phase"—the magnetic induction vectors of one phase's sections are directed opposite one to other, and all the phase does not participate in the resulting base vector creation.

A mathematical expression for each of the sets is obtained. It consists of the description of a vector in the first quarter and its rotating by an angle divisible by $\pi/2$. In the first quarter the vector is calculated as a sum of sectional vectors participating in the base vector forming taking into consideration the sign of the back-EMF according the equivalent circuit (Fig. 1) and the direction of the complex plane axis. For example, the section allocated on the Left has negative back-EMF but the axis direction also is negative, thus the vector directed into the Right formed by this section has the positive sign. Analogically the sign of the vector formed by the section directed into the Down is determined. To implement the rotating, the coefficient k showing the quantity of the turnings and by it determining number of the quarter is introduced.

The names of the base vector forming techniques and the amplitudes of corresponding vectors are shown in the **TABLE** where the following designations are used: V_k—base vector set resulted in by a certain forming, k—number of an intercommutation interval, $k = \overline{1,4}$; $p = \{-1, 1\}$—coefficient by the power index determining whether in the nominator or the denominator of a fraction is its base (Table 1).

Table 1. Base vector forming techniques and their amplitude values.

Number	Name	Mathematical expression of a base vector set
1. Scheme with unconnected sections		
1.1	Separate with passive section in a phase	$\overline{V_k} = e^{i\left[(k-1)\frac{\pi}{2}\right]}$
1.2	Joint with passive section in each phase	$\overline{V_k} = \sqrt{2} \cdot e^{i\left[\frac{\pi}{4} + (k-1)\frac{\pi}{2}\right]}$
1.3	Combined 1.1 and 1.2	
2. Parallel connection of sections in each phase		
2.1	Separate	$\overline{V_k} = 2 \cdot e^{i\left[(k-1)\frac{\pi}{2}\right]}$
2.2	Joint	$\overline{V_k} = \sqrt{2} \cdot e^{i\left[\frac{\pi}{4} + (k-1)\frac{\pi}{2}\right]}$
2.3	Combined 2.1 and 2.2	
3. Serial connection of sections in each phase		
3.1	Separate	$\overline{V_k} = e^{i\left[(k-1)\frac{\pi}{2}\right]}$
3.2	Joint	$\overline{V_k} = \sqrt{2} \cdot e^{i\left[\frac{\pi}{4} + (k-1)\frac{\pi}{2}\right]}$
3.3	Combined 3.1 and 3.2	

(continued)

Table 1. (*continued*)

Number	Name	Mathematical expression of a base vector set
4. Radial connection with common point (connected either to the supply bus or to the ground)		
4.1	Separate with passive section in a phase	$\overline{V_k} = e^{i\left[(k-1)\frac{\pi}{2}\right]}$
4.2	Joint with passive section in each phase	$\overline{V_k} = \sqrt{2} \cdot e^{i\left[\frac{\pi}{4} + (k-1)\frac{\pi}{2}\right]}$
4.3	Combined 4.1 and 4.2	
5. Radial with neutral point (not connected)		
5.1	Separate	$\overline{V_k} = e^{i\left[(k-1)\frac{\pi}{2}\right]}$
5.2	Joint	$\overline{V_k} = \sqrt{2} \cdot e^{i\left[\frac{\pi}{4} + (k-1)\frac{\pi}{2}\right]}$
5.3	Combined 5.1 and 5.2	
5.4	Joint with opposite energizing of sections in each phase	$\overline{V_k} = e^{i\left[(k-1)\frac{\pi}{2}\right]}$
5.5	Joint with passive section in one phase	$\overline{V_k} = \frac{\sqrt{10}}{3} \cdot e^{i\left[(k-1)\frac{\pi}{2} + arctg(3^p)\right]}$
5.6	Joint with passive section in each phase	$\overline{V_k} = \frac{1}{\sqrt{2}} \cdot e^{i\left[\frac{\pi}{4} + (k-1)\frac{\pi}{2}\right]}$
5.7	Joint with opposite energizing of sections in one phase and passive section in the other phase	$\overline{V_k} = \frac{2}{3} e^{i\left[(k-1)\frac{\pi}{2}\right]}$
	Combined 5.6 and 5.7	
6. Closed section connection		
6.1	Separate	
6.2	Joint	$\overline{V_k} = \sqrt{2} \cdot e^{i\left[\frac{\pi}{4} + (k-1)\frac{\pi}{2}\right]}$
6.3	Combined 6.1 and 6.2	
6.4	Joint with opposite energizing of sections in one phase	$\overline{V_k} = \frac{4}{3} \cdot e^{i\left[(k-1)\frac{\pi}{2}\right]}$
6.5	Joint with passive section in one phase	$\overline{V_k} = \frac{\sqrt{10}}{2} \cdot e^{i\left[(k-1)\frac{\pi}{2} + arctg(3^p)\right]}$
6.6	Joint with passive section in each phase	$\overline{V_k} = \sqrt{2} \cdot e^{i\left[\frac{\pi}{4} + (k-1)\frac{\pi}{2}\right]}$
6.7	Combined 6.4 and 6.6	

It is seen from the Table that when there is a passive section or a phase, the description of such set is more complex, the absolute value of the vector becomes a view of in irrational fraction and the argument—values which are different from ones divisible by $\pi/4$. The cause is that the base vector consists in an unequal quantity of sectional vectors in different phases. The full vector set picture at each forming technique is shown in a graphical representation in the Fig. 3. Each picture (a–f) contains all totality of the sets in all the quarters and corresponds to one whole item of the Table without dividing into sub-items.

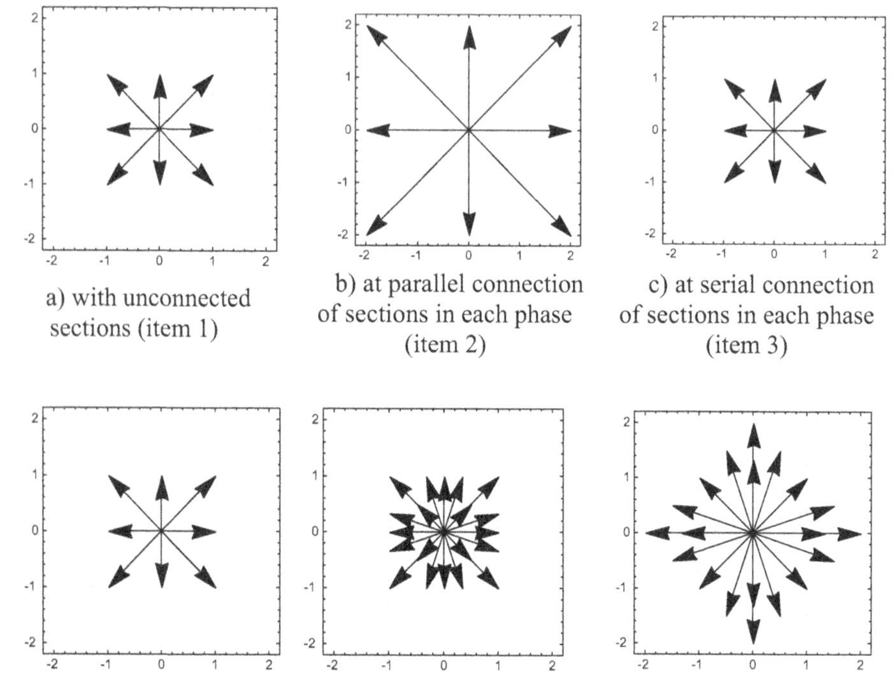

a) with unconnected sections (item 1)

b) at parallel connection of sections in each phase (item 2)

c) at serial connection of sections in each phase (item 3)

d) at radial connection with common point (item 4)

e) at radial with neutral point (item 5)

f) at closed section connection (item 6)

Fig. 3. Base vector sets at different section connection schemes

It is seen from the Fig. 3 that there are equal base vector sets at different connection schemes. However, as the sets were obtained at different schemes, the complexity of their energizing's implementation will be other.

As among of the sets there are ones obtained at the opposite section vector direction in one phase, a current takes place which does not contribute into the moving torque, thus a useless power consumption increases. At the radial and close connection schemes there are more sets than at the other ones. Base vectors allocated at angles different from $\pi/4$ are seen. The result of the presence of such vectors is an unevenness of the moving torque distribution on the circle (increased torque ripple) and therefore an unevenness of the rotating speed at load torque values near to the stall torque.

3 Conclusions

1. A systematization of base vector set forming based on section connection schemes and phase energizing techniques is proposed which enables to bring a sole set of vectors whose amplitude corresponds to the stall torque of the BLDC motor into accordance with each technique.

2. The term base used at the current time to work with brushless electric machines is extended taking in consideration the influence of the sections on physical processes in the BLDC motor.
3. Names to each base vector forming technique are given.
4. Amplitude values of base vectors having an influence on the stall torque are determined.
5. At different connection schemes the obtaining of equal base vector sets is possible, differencing by implementation complexity of a certain energizing technique.
6. At some techniques sections belonging to one phase are energized oppositely that causes an increasing of common consumed current but gives no increasing of torque, thus such techniques are not recommended to be used in drives.

References

1. Krivilev, A.V.: Automation of construction of characteristic curves in impulse control problems for the power amplifier - servomotor system. I. speed-torque characteristics. J. Comput. Syst. Int. **52**(3), 252–265 (2013)
2. Krivilev, A.V.: Automation of construction of characteristic curves in impulse control problems for the power amplifier - servomotor system. II. Energy characteristics. J. Comput. Syst. Int. **52**(3), 458–467 (2013)
3. Lin, H., Kwon, B.: New PWM technique for two-phase brushless DC motor drives. J. Electr. Eng. Technol. **8**(5), 1107–1115 (2013). http://dx.doi.org/10.5370/JEET.2013.8.5.1107
4. Lin, H., You, Y.-M., Cheon, S.-R., Kwon, B.: Digital implementation of PWM techniques for two-phase eight-switch inverter fed brushless DC motor drives. J. Electr. Eng. Technol. **8** (2), 295–303 (2013). https://doi.org/10.5370/JEET.2013.8.2.295
5. Katzman, M.M.: Electric Machines: Textbook for Studies of Electrotech. Spec. Techn., 2nd edn., revised and extended. Vysh. Sch., p. 463 (1990)
6. Kosulin, V.D., Mikhailov, G.B., Omelchenko, V.V., Putnikov, V.V.: Brushless DC Motors for Industrial Robots, p. 184. Energoatomizdat, Leningrad (1988)
7. Krivilev, A.V.: Automated synthesis of boolean control functions of mechanotronic drive module with three-phase valve motor. J. Comput. Syst. Int. **59**(2), 318–328 (2010)
8. Krivilev, A.V., Dunich, E.A., Penkin, S.S.: Sectional approach to obtaining of static characteristics of brushless DC, questions of the innovational development of the airspace complex of Russia. In: Materials of the First All-Russian Science-Practical Conference, pp. 132–137. Publishing house “Dobroe Slovo” (2018)
9. Krivilev, A.V., Dunich, E.A., Penkin, S.S.: Sectional approach to researching of two-phase BLDC motor. In: IOP Conference Series: Materials Science and Engineering, vol. 537, p. 062093 (2019). https://iopscience.iop.org/article/10.1088/1757-899X/537/6/062093/pdf

Stability of the Continuous Welded Rail on Transition Sections

Alexander Savin[1(✉)], Oleg Suslov[1], Vadim Korolev[2], Alexey Loktev[2], and Irina Shishkina[2]

[1] Joint Stock Company "Scientific Research Institute of Railway Transport", 3rd Mytishchinskaya st. 10, 129626 Moscow, Russia
savin.aleksandr@vniizht.ru

[2] Russian University of Transport (RUT - MIIT), Chasovaya St. 22/2, 125190 Moscow, Russia

Abstract. The approaching section of a conventional ballast path to a ballast free path poses a potential hazard for several reasons: the path section on a bridge or in a tunnel represents a kind of support or barrier for the longitudinal movement of tracks; no transverse movements of the tracks; the transition section usually has an increased degradation of the ballast layer. These factors significantly reduce the resistance of the rail track panel to shear across the axis of the path in the transition section. To predict the stability of a continuous welded rail on the transition sections, the following dependencies were calculated using the finite element method: the dependence of the critical temperature of the path on the number of sagging sleepers that have lost their grip to the ballast; the dependence of the critical temperature of the path on the percentage of weakening the pressure of the rail foot to the base; the dependence of the maximum transverse displacements to achieve a temperature close to the critical one, on the length of additional stiffening elements inside the track (additional rails).

Keywords: Transport · Transition sections · Welded rail

1 Introduction

The most problematic places for all experimental structures are transition sections between the ballast free path and the path on the ballast [1–3]. Subsidence in these places reach 40 mm and require periodic straightening work by tamping the sleepers with a rectifying-tamping-straightening machine [4]. With such subsidence, sagging of the sleepers take place, i.e. loss of grip rail foot to the ballast. This phenomenon significantly reduces the stability of the continuous welded rail. The problem of transition sections was also solved by a number of foreign experts [5–9].

A continuous welded rail on a ballast free base seems to be quite stable, since the rail track panel does not have the possibility of transverse movement. The rails are attached to the under-rail base, which is not able to move in the transverse or longitudinal direction. The same situation is observed on the approaches to a ballast free path for high-speed traffic. The approach of the usual path on the ballast to the ballast

Z. Popovic et al. (Eds.): TransSiberia 2019, AISC 1115, pp. 648–654, 2020.
https://doi.org/10.1007/978-3-030-37916-2_62

free path presents a potential hazard for several reasons. First, the section of ballast free path is a stop or barrier for the longitudinal movement of the rails, since there are no longitudinal movements of the sleepers in the ballast and the rails are more reliably fixed in the rail fastening assemblies. Thus, an elevated level of longitudinal stresses is created in front of the section of a ballast free path, which are not able to be redistributed along the rail. This situation is most dangerous when additional compressive stresses create a braking train on the way to the ballast free path. Secondly, there are no transverse movements of the rails. In the case of a usual rail track panel, the path has the ability to shift outward of the curved section of the path, thus producing a "self-discharge" of thermal stresses, and the rails on a ballast free base have no such possibility. Thirdly, the transitional section usually has an increased degradation of the ballast layer due to differences in the rigidity of the concrete and gravel base. There is always a subsidence on the first 3–5 sleepers and, as a consequence, sagging of the sleepers. These factors significantly reduce the resistance of the rail track panel to shear across the axis of the path.

2 Research Methods

One of the ways to increase the stability of the continuous welded rail in transition sections is to place additional rails inside the track [2]. This, firstly, increases the resistance to transverse movement, and, secondly, provides a smooth change in the rigidity of the base, which reduces the frustration of the ballast in this section.

Additional rails inside the track on the experimental transition sections were applied during testing on the Experimental ring of JSC VNIIZhT Art. Shcherbinka on designs Rheda 2000, LVT, NBT (Alstom). It should be noted that the use of typical domestic shuttle sleepers for the transitional sections of LVT and NBT for these purposes was an irrational constructive solution, since the under-rail sites do not coincide in level with a ballast free path and fastening threaded rods do not securely fasten the rail foot to the sleeper. It remains an open question about the length of these additional rails, so that in addition to a smooth change in the rigidity of the path, they ensure an increase in the stability of the continuous welded rail on the transition sections. To determine the required length of additional rails, the following approach is proposed. To ensure the safety of train traffic in transition sections, it is necessary to determine the extent to which the condition of the path at the transition section can be relaxed so that there is no danger of throwing off the continuous welded rail. In addition, it is necessary to calculate how the length of additional elements inside the track in the form of a shuttle from a corner or additional rail affects the stability of the path. Figure 1 shows the scheme for calculating the stability of the continuous welded rail in the transition section.

To predict the stability of the continuous welded rail on the transitional sections by the finite element method, the following dependencies are calculated:

- the dependence of the critical temperature of the path on the number of sagging sleepers that have lost their grip to the ballast;

- the dependence of the critical temperature of the path on the percentage of weakening the pressure of the rail foot to the base;
- the dependence of the maximum transverse displacements to achieve a temperature close to the critical, on the length of the additional stiffening elements inside the track (additional rails).

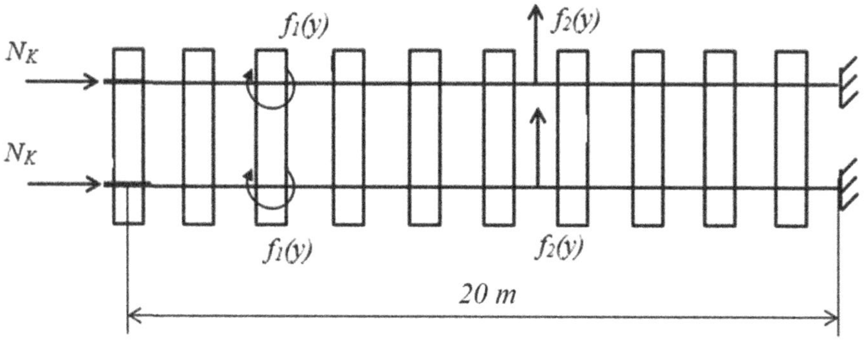

Fig. 1. Scheme for calculating the stability of the continuous welded rail on the transition section

The system of equations interlinking the critical longitudinal force, the magnitude of the irregularity and the length of the irregularity, has been proposed by Doctor of Technical Science Professor Kogan A. Ya and looks like:

$$
\begin{aligned}
& B^{*} = B - \frac{2P_{tk}}{\pi R}; \\
& b^{*} = b\left(1 - \frac{2P_{tk}}{\pi BR}\right)\sec^{2}\frac{P_{tk}}{BR}; \\
& P_{tk} = 2\sqrt{\frac{2EIB^{*}b^{*}(\sqrt{b^{*2}+C^{2}}-b^{*})}{C^{2}\sqrt{b^{*2}+C^{2}}}} + q + \frac{M}{\sqrt{r^{2}+\omega^{2}C^{2}}} \\
& \quad + \frac{M}{\sqrt{r^{2}+\omega^{2}C^{2}}}\left(\frac{r-\sqrt{r^{2}+\omega^{2}C^{2}}}{\omega C}\right)^{2}; \\
& \omega = \sqrt[4]{\frac{2B^{*}b^{*}(\sqrt{b^{*2}+C^{2}}-b^{*})}{EIC^{2}\sqrt{b^{*2}+C^{2}}}}; \\
& C_{6} = \frac{EI\omega^{4}C^{2} + \left(q + \frac{M}{\sqrt{r^{2}+\omega^{2}C^{2}}} + \frac{M}{\sqrt{r^{2}+\omega^{2}C^{2}}}\left(\frac{r-\sqrt{r^{2}+\omega^{2}C^{2}}}{\omega C}\right)^{2}\right)\omega^{2}C^{2} +}{P_{tk}C\omega^{2}} \\
& \quad \frac{+2B^{*}(\sqrt{b^{*2}+C^{2}}-b^{*})}{P_{tk}C\omega^{2}}.
\end{aligned}
\tag{1}
$$

where

B and b – the coefficients in the approximation function of the resistance of the sleeper to transverse displacements $f_1(y) = B \cdot arctg(y/b)$;
M, r and q – the coefficients of the approximating function of the resistance to rotation of the rails relative to the sleeper, depending on the angle of rotation of the rail $f_2(y) = qy' + M \cdot arctg(y'/r)$;
R – the radius of the curve;
E – the elastic modulus of rail steel;
J_z^o – the moment of inertia of the rail relative to the main vertical axis;
$C_б$ – the amplitude of the household irregularity, equal to the sum of the amplitudes of the unstressed C_o and the stressed C irregularities;
N_K – critical longitudinal force in the rail;
ω – the frequency of the irregularity corresponding to the length of the irregularity $L = 2/\omega$;
Knot of a rail fastening on the ballast free path – Vossloh W-30;
Knot of rail fastening on the ballast path – RCS 65-S;
Rail type – R65;
Sleeper pattern – 1840.

The modeling proceeded from the following considerations: on approaches to a ballast free path there will be a subsidence, therefore, there will be sagging sleepers, therefore, resistance to shear on the base of the sleepers will be lost, but resistance to the ends of the sleepers and torsion at the fastening points will remain. At the same time, from the additional vertical oscillations at the site of the subsidence, a weakening of the pressure of the rail foot to the sleeper will be observed.

In other words, for the transition section, it is necessary to determine how three factors influence the critical longitudinal force: the length of the section with sagging sleepers, the tightening of the terminals, the length of the additional rails inside the track.

The model described in [10] was adopted as the model of stability of the continuous welded rail.

Figure 2 shows the layout of the additional rails inside the track.

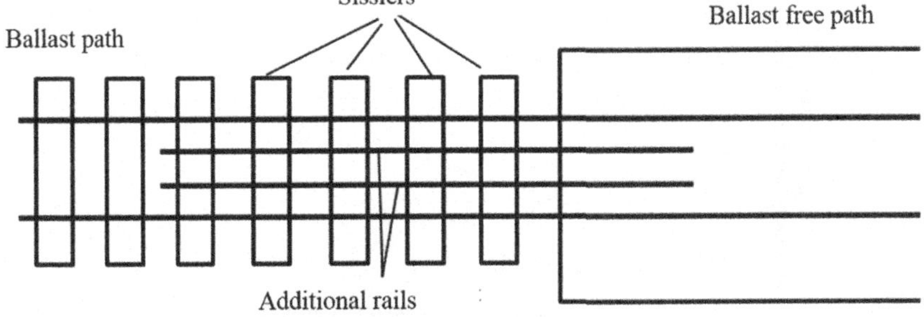

Fig. 2. Layout of additional rails inside the track

The solution of these problems is carried out in the environment of finite element modeling [11]. As a baseline, a 200-meter-long section of the rail track is presented, one half of which is located on the ballast free structure of the path, the other on crushed-stone ballast.

3 The Results of the Study

Figure 3 shows the dependence of the critical temperature on the number of sagging sleepers.

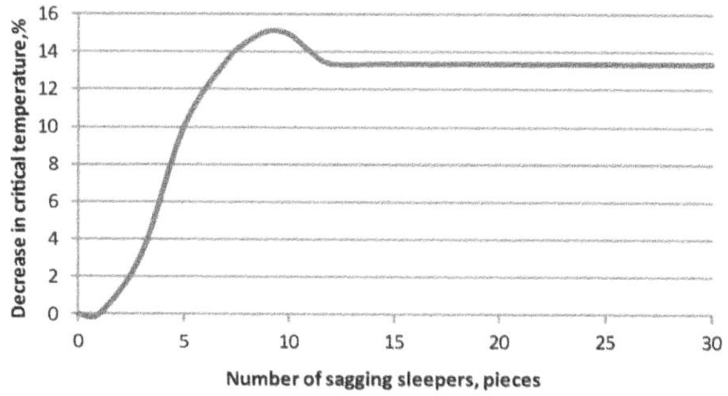

Fig. 3. Dependence of the critical temperature on the number of sagging sleepers

The plot of the critical temperature versus the number of sagging sleepers shows that the critical temperature decreases sharply with an increase in the number of sagging sleepers from 0 to 10. The abrupt decrease in the critical temperature in the region of 8–9 sagging sleepers is explained by the fact that the form of buckling coincides with the initial curvature of the rail track panel due to different transverse and longitudinal resistance of the ballast layer. Figure 4 shows the calculated approximating dependence of the reduction of the longitudinal force, resulting in the release of the reduction of the pressing force of the rail to the sleeper in the fastening unit in the range

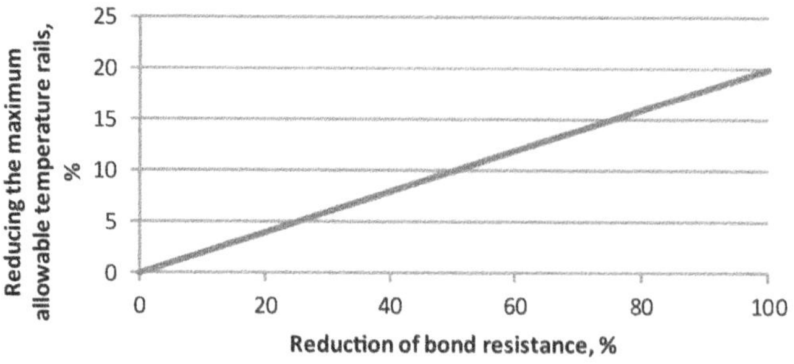

Fig. 4. Dependence of the critical temperature on the percentage of weakening the pressure of the fasteners

from standard to 0 in the length of the transition section equal to 15 m. The decrease in the maximum permissible temperature rise is no more than 20%, which indicates a relatively insignificant effect of this parameter on the stability of the path within the transition section.

Figure 5 shows the dependence of the maximum transverse displacements within the transition section upon reaching a temperature close to the critical one (in this case, this temperature is 230°) versus the length of two additional rails inside the track.

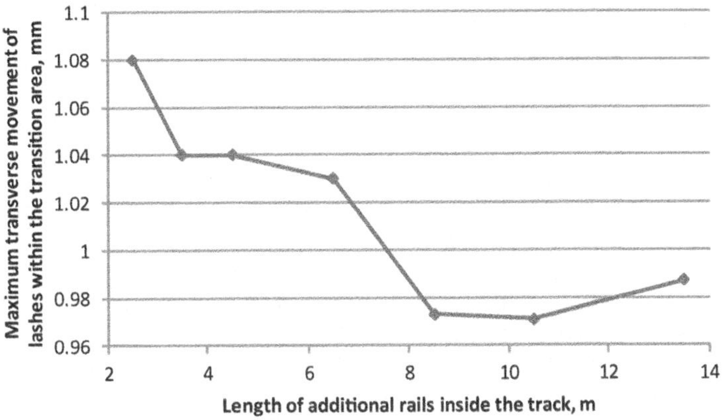

Fig. 5. Dependence of transverse movements within the transition section on the length of two additional rails inside the track

The calculations show a decrease in the maximum transverse displacements with an increase in the length of the additional rails to 10 m. A further increase in the length of the rails inside the track does not lead to an increase in the stability of the continuous welded rail within the transition section from a ballast free to ballast structure.

Thus, the optimal length of additional rails under the conditions for the current content of the ballast layer and intermediate rail fasteners is the length of 9–11 m. The results of the calculations are confirmed by the results of tests on the Experimental ring of JSC VNIIZhT Art. Shcherbinka in the period 2014–2018 [12]. During this period, a tonnage of 1.1 billion tons of gross has been missed.

References

1. Christopher, W.: Research Results Digest 79-Design of Track Transition, Transit Cooperative Research Program, TCRP Project D-7/Task, vol. 15, pp. 1–36 (2006)
2. Coenraad, E.: Modern Railway Track, MRT-Productions, pp. 157–170 (2001)
3. Xiaoyan, L., Lijun, M.: Dynamic response analyses of vehicle and track coupled system on track transition of conventional high speed railway. J. Sound Vibr. **271**, 1133–1146 (2004)
4. Sun, Y.Q., Dhanasekar, M.: A dynamic model for the vertical interaction of the rail track and wagon system. Int. J. Solids Struct. **39**, 1337–1359 (2002)

5. Kang, Y.S., et al.: Study on Dynamic Calculation and Analysis of Track At Bridge/Embankment Joining Section Miscellany of Spring Congress, Korea Railway Society, Seoul, pp. 250–255 (2002)
6. Alexey, L., Vadim, V.K., Irina, V.S., Dmitry, A.B.: Modeling the dynamic behavior of the upper structure of the railway track. Procedia Eng. **189**, 133–137 (2017). https://doi.org/10.1016/j.proeng.2017.05.022. Transportation Geotechnics and Geoecology, TGG 2017, 17–19 May 2017, Saint Petersburg, Russia
7. Glusberg, B., Korolev, V., Shishkina, I., Loktev, A., Shukurov, J., Geluh, P.: Calculation of track component failure caused by the most dangerous defects on change of their design and operational conditions. In: MATEC Web of Conferences, vol. 239, p. 01054 (2018). https://doi.org/10.1051/matecconf/201823901054
8. Loktev, A.A., Korolev, V.V., Shishkina, I.V.: High frequency vibrations in the elements of the rolling stock on the railway bridges. In: IOP Conference Series: Materials Science and Engineering, vol. 463, p. 032019 (2018). https://doi.org/10.1088/1757-899x/463/3/032019
9. Loktev, A.A., Korolev, V.V., Poddaeva, O.I., Chernikov, I.Y.: Mathematical modeling of antenna-mast Structures with aerodynamic effects. In: IOP Conference Series: Materials Science and Engineering, vol. 463, p. 032018 (2018). https://doi.org/10.1088/1757-899x/463/3/032018
10. Alexander, S., Alexander, K., Alexey, L., Vadim, K.: Evaluation of the service life of non-ballast track based on calculation and test. Int. J. Innovative Technol. Exploring Eng. (IJITEE) **8**(7) (2019). G5991058719/19©BEIESP. ISSN: 2278-3075
11. Loktev, A.A., Korolev, V.V., Poddaeva, O.I., Stepanov, K.D., Chernikov, I.Y.: Mathematical modeling of aerodynamic behavior of antenna-mast structures when designing communication on railway transport. Vestnik Railway Res. Inst. **77**(2), 77–83 (2018). https://doi.org/10.21780/2223-9731-2018-77-2-77-83. (in Russian)
12. Loktev, A.A., Korolev, V.V., Loktev, D.A., Shukyurov, D.R., Gelyukh, P.A., Shishkina I. V.: Perspective constructions of bridge overpasses on transport main lines. Vestnik Railway Res. Inst. **77**(6), 331–336 (2018). https://doi.org/10.21780/2223-9731-2018-77-6-331-336. (in Russian)

Using the Coefficient of Concavity in the Analysis of the Quality of Filling the Tracks of the Hump Yard

Konstantin Kornienko(✉), Sergei Bessonenko, and Iuliia Tanaino

Siberian Transport University, Dusi Koval'chuk Street, 191, 630049 Novosibirsk, Russia
Kkonstantini@mail.ru

Abstract. Increasing the processing capacity and lowering the operating costs of the humps are associated with improving the track filling quality of the hump yard. The track filling quality depends on the curvature of the profile. To analyze the profile deviation, it is proposed to use the concavity coefficient, which is the ratio of the cross-sectional area of the actual profile to the normative profile values. For the study, the developed simulation model of track filling of the hump yard was used, based on calculating the movement speed of the cut and taking into account the possibility of unhooking the cut after stopping, the error of the speed control devices operation of the uncoupling cut on the way and the retarder operation, the possibility of starting the train cars when the next cut collides with them. Studies have shown that the concavity coefficient can be used as a numerical criterion in assessing the deviation of the profile from the normative values. There is a dependence between the concavity coefficient and the quality indicators of the track filling. The obtained results can be used to assess the effectiveness of the track profile alignment.

Keywords: Hump yard · Simulating · Profile concavity coefficient · Sorting yard · Hump · Occupancy quality · Reverse movement

1 Introduction

One of the most important elements of transport infrastructure is hump yards and humps [1–6]. The sorting capacity of the hump and its operating costs depend on many factors, one of which is the hump yard. The profile of the hump yard changes during operation [7–10]. This leads to a decrease in the quality of track fillings, an increase in the probability of the window formation and the collision of the cuts with an excess of the collide velocity [10]. This reduces the processing capacity of the hump yard complex and leads to monetary costs due to the train cars damage. Therefore, profile alignment of the hump yard tracks [7–10] is necessary. The alignment of the slope profile is associated with large economic and time costs; therefore, the validity of this event is required.

Z. Popovic et al. (Eds.): TransSiberia 2019, AISC 1115, pp. 655–662, 2020.
https://doi.org/10.1007/978-3-030-37916-2_63

The current methodology for calculating the track profile alignment efficiency of the hump yard [11] does not take into account such factors as the structure of the traffic flow, the cut flow, destination power, current track slopes and others. In this connection, the calculations obtained by this method may be inaccurate. Therefore, one of the priorities is the development of new ways of the profile alignment effectiveness assessment. The purpose of this article is to study the possibility of using the concavity coefficient in analyzing the track filling quality of the hump yard.

2 Research Methodology

In articles [12–15], it is proposed to use the concavity coefficient to analyze the descent part profile of the hump. In general, this is a characteristic of the hill profile, which depends on the values of the slopes and its individual elements, determined by the formula:

$$\mu = 1 - \frac{P}{P_{max}} \quad (1)$$

where P - the area of the longitudinal section of the hump with a certain profile;

P_{max} - the maximum possible area of the longitudinal section of the hump (with a uniform slope from the top of the hill to the calculated point).

The use of this coefficient for analyzing the profile of the descent part is difficult, since the comparison is made not with the normative values, but with the maximum possible area of the longitudinal section of the hill. Due to the fact that the profile of the descent part of the hump is not homogeneous (consists of a large number of slopes of different size and length due to the minimum rolling time from the hump of the hill), the best rolling performance was achieved with a concavity coefficient of 0.5–0.6.

Hump yard according to the normative should consist of a slope of one magnitude, therefore the deviation of the actual profile from the normative affects the change in the concavity coefficient. Figure 1 shows an example of the actual and normative profile of the hump yard track.

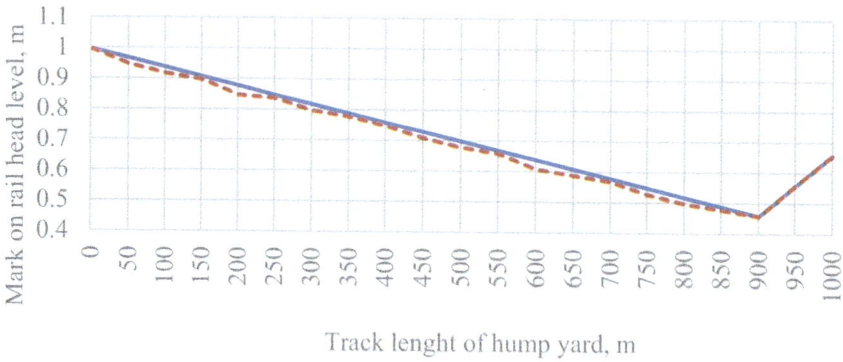

Fig. 1. Comparison of the normative and the actual profile of the hump yard track

According to Fig. 1, it can be concluded that when applying formula (1) in the hump yard, the concavity coefficient characterizes the deviation of the profile from the normative value. Knowing the height of the marks and the distance between them, formula 1 can be written in the form:

$$\mu = 1 - \frac{\sum_{i=1}^{n-1} (h_i + h_{i+1}) * l_i}{(h_1 + h_{norm}) * l_y} \qquad (2)$$

where h_i, h_{i+1} - the height at the beginning and end of the i-th section; l_i - the length of the i-th section; h_1 - the height of the track beginning of the hump yard, m; h_{norm} - the normative height of the slope end of the hump yard, m; n- the number of sections, pc; l_y - the normative length of the track slope, m.

This article discusses the dependence of the quality parameters of the hump yard track filling on the concavity coefficient for the case when the rail head levels changes on the 1000-m-long track at 450 m of the rail, while the heights of the beginning and end of the descent remain unchanged. In this case, formula (2) takes the form:

$$\mu = 1 - \frac{h_1 + 2 * h_2 + h_3}{(h_1 + h_3) * 2} \qquad (3)$$

The concavity coefficient can take both negative and positive values. Positive values mean that the profile subsided during operation. Negative values of this coefficient can take when profile swelling. Profile swelling is common in the spring-autumn period, when groundwater freezes through the ballast section.

For the study, a simulation model of track filling was used [16, 17]. This model takes into account the possibility of starting the cut after a stop, the error in the operation of devices for controlling the speed of cut release on the way and the retarder operation, the possibility of starting the train cars when the next cut collides with them. The structure of the train car flow and cut flow was specified by a random distribution. The outdoor temperature, wind speed and direction were set for the Novosibirsk region, taking into account the correction for the track axis. The track was considered within the park brake position, to the insulating joints at the track end, the profile shoot was taken every 50 m. Normative values correspond to [18, 19].

The object of the study is the track of the hump yard. The subject of the research is the influence of the track profile on the filling quality.

3 The Results of the Study

The track filling quality of the hump yard is characterized by two factors:

- minimizing “windows”;
- collide of the cuts with permissible speed.

Increasing the slope of the track sections leads to an increase in the speed of cut movement in these sections. According to [19], the number of damaged train cars can be calculated using the average speed of cut collisions. Figure 2 shows the dependence of the average speed of cut collision on the concavity coefficient.

Fig. 2. Dependence of the average speed of cut collision on the profile concavity coefficient

From Fig. 2 it can be concluded that the average speed of cut collision depends on the profile concavity coefficient. The greater this coefficient, the higher the average speed of cut collision, and, consequently, the greater the number of damaged train cars. Figure 3 shows the dependence of the damaged train car number on the profile concavity coefficient calculated by [19].

Fig. 3. Dependence of the damaged train car number on the profile concavity coefficient

According to Fig. 3, we can conclude that an increase in the concavity coefficient leads to an increase in damage to the train cars, and, consequently, to a decrease in the safety of the dissolution of the train cars. From this we can conclude that a reduction in the concavity coefficient will lead to a decrease in operating costs in the event of damage to the goods.

Figures 2 and 3 show that traffic safety is inversely related to the profile concavity coefficient.

The second important parameter characterizing the track filling quality is the size of the windows. One way of estimating the size of the windows between cuts is to use the track filling ratio. In general terms, this coefficient shows the application of the useful length of the hump yard track, and the higher it is, the smaller the size of the windows between the cuts. To assess the track filling quality, we used the average coefficient of the track filling quality, obtained after all dissolutions. The dependence of the track filling quality on the profile concavity coefficient is shown in Fig. 4.

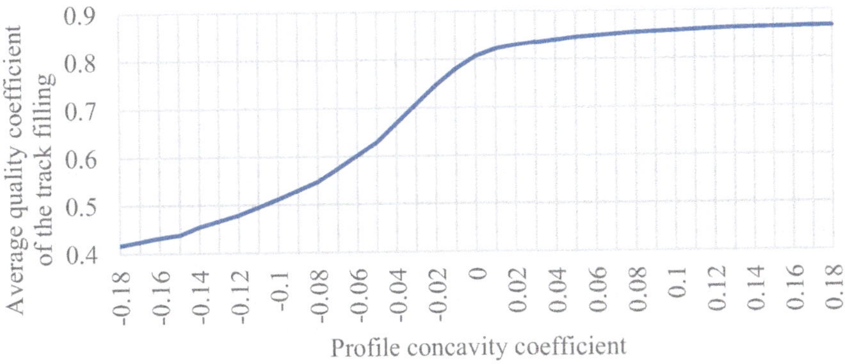

Fig. 4. Dependence of the average quality coefficient of the track filling on the concavity coefficient of the track profile

According to Fig. 4, we can conclude that with a decrease in the concavity coefficient, the quality coefficient of the track filling also decreases. According to the normative, this coefficient should be in the range of 0.8–1 [20, 21]. Therefore, with a zero concavity coefficient, the quality coefficient of the track filling is within the normative values.

Also, in Fig. 4, we can conclude that the subsidence of the profile will not lead to an increase in the quality coefficient of the track filling. After analyzing Figs. 2, 3 and 4, we can conclude that the subsidence of the profile will lead to a deterioration in the track filling quality, since the average collision speed will increase, which will lead to a decrease in traffic safety.

At negative values of the concavity coefficient, the quality coefficient decreases sharply, which means that a decrease in the slope value leads to an increase in the windows between the cuts.

From Fig. 4 it can be assumed that the average number of backups necessary for the accumulation of one train composition also depends on the concavity coefficient. This dependence is shown in Fig. 5.

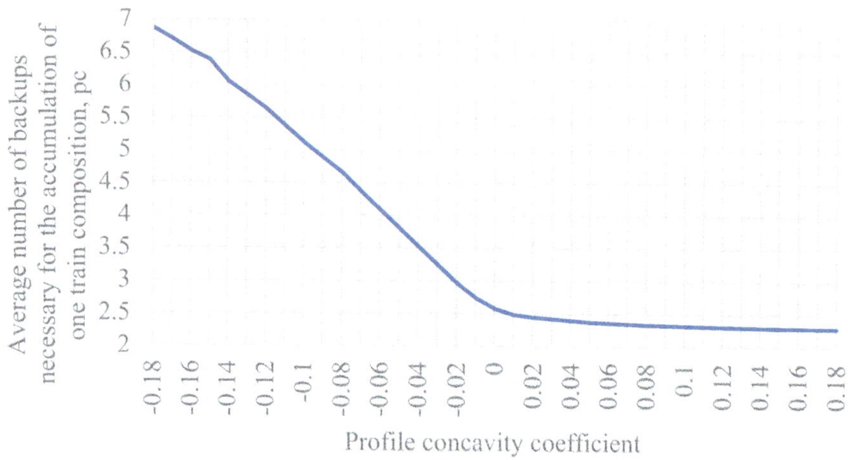

Fig. 5. Dependence of the average number of backups on the concavity coefficient of the track profile

In Fig. 5, we can conclude that there is also an inverse dependence between the concavity coefficient and the average number of backups. With positive values of the concavity coefficient, the decrease in the number of backups occurs at a low speed. The difference between the number of backups necessary for the accumulation of one train composition, with concavity coefficients 0 and 0.18, is only 0.29. At negative values, an increase in the coefficient leads to a sharp decrease in the number of backups.

4 Conclusions

The article presents a new way to assess the profile deviation from the normative value. For this purpose, the concavity coefficient is used, which shows the ratio of the cross-sectional area of the hump yard track profile to the normative sectional area.

As an example, the dependence of the track filling quality on the concavity coefficient for the track is considered. For this, a change in the rail head level in the middle of the slope was investigated. It was revealed that the deviation of the concavity coefficient from zero to a higher side (profile backup) leads to a sharp increase in damaged train cars, and the quality coefficient of track filling and the average amount of backups grow slightly, not exceeding 10%. When the coefficient decreases from zero, there is a sharp decrease in the average speed of cut collides, the average filling quality coefficient, and as a consequence, the increase in the average number of backups necessary for the accumulation of one train composition.

The obtained results allow us to conclude that with an increase in the profile concavity coefficient to zero, the track filling quality increases, a further increase in the coefficient will lead to a decrease in the track filling quality. The concavity coefficient can be used as one of the tools in assessing the alignment effectiveness of the hump yard track profile. The backup of the profile leads to an increase in the concavity coefficient, the alignment of the profile will lead to increased traffic safety.

References

1. Gapanovich, V.A., Shabelnikov, A.N.: On the development of automated sorting systems. Railway Transp. **8**, 23–25 (2010). ISSN 0044-4448
2. Dmitrenko, A., Karaseva, A., Osipov, D.: Regulation of traffic flows by railway stations with a change in the number of main tracks on the hauls. In: MATEC Web of Conferences, vol. 239, p. 02005 (2018) https://doi.org/10.1051/matecconf/201823902005. Influence of opposite elevation on the occupancy level of the tracks of hump yard
3. Tanaino, I., Yugrina, O., Zharikova, L.: Assessment criteria for decisions in the field of rail freight transportation. In: MATEC Web of Conferences, vol. 216, p. 02015 (2018). https://doi.org/10.1051/matecconf/201821602015
4. Medvedev, V., Teslenko, I., Karasev, S.: Method for determining the acceptability of transport operations involving dangerous goods. In: MATEC Web of Conferences, vol. 216, p, 02013 (2018). https://doi.org/10.1051/matecconf/201821602013
5. Zhang, L.: Macrolevel classification yard capacity modeling, transportation research record. J. Transp. Res. Board. **2608**(1), 125–133 (2017)
6. Belošević, I.: Variable neighborhood search for multistage train classification at strategic planning level. Comput. Aided Civ. Infrastruct. Eng. **33**(3), 220–242 (2018)
7. Zhang, C.: Research on multi-objective optimization of vertical section of the rolling down zone on hump. In: 2011 Seventh International Conference on Natural Computation (ICNC): 2011 Seventh International Conference on Natural Computation, pp. 1270–1274. IEEE (2011)
8. Klimov, A., Gunbin, A., Osipov, D., Burdyak, P.: Method for calculating the maximum number of railway cars in a cut. In: MATEC Web of Conferences, vol. 216, p. 02011, (2018). https://doi.org/10.1051/matecconf/201821602011
9. Starostina, T.I.: Analysis of the design features of the longitudinal profile of the hump yard tracks on the operating hump yards, innovative technologies of the new millennium: collection of articles. In: Articles of International Scientific-Practical Conference, Kirov, in 3 Parts. Part 2, 5 May 2016, pp. 81–83. Aeterna, Ufa (2016)
10. Kornienko, K.I.: Investigation of the influence of the hill profile on the movement speed of cuts in the hump yard with a fair wind [Text]. Bull. Siberian State Univ. Ways Commun. **1** (40), 13–18 (2017)
11. Bessonenko, S., Kornienko, K., Tanaino, I.: Influence of opposite elevation on the occupancy level of the tracks of hump yard. In: MATEC Web of Conferences, vol. 239, p. 03002 (2018). https://doi.org/10.1051/matecconf/201823903002
12. Karasev, S.V.: Influence of the hill construction, the structure of the train car flow and the external environment on the track filling quality of the hump yard. Abstract of a technical science candidate dissertation, 22 May 2008, Siberian State Transport University, SGUPS, Novosibirsk, 24 p. (2003)
13. Starostina, T.I.: Determination of the economic efficiency of adjusting the track profile of the hump yard. Transp. Infrastruct. Siberian Reg. **1**, 197–201 (2017)
14. Karasev, S.V.: Ensuring safety in the application of simplified normative and conditions for the dissolution of train cars with dangerous goods for automated humps. Transp. Infrastruct. Siberian Reg. **1**, 64–67 (2018)

15. Alpatov, V., Sakharov, A.: Rapid assessment of the economic efficiency of domes of buildings. In: IOP Conference Series: Materials Science and Engineering, vol. 365, no. 6, p. 062022 (2018). https://doi.org/10.1088/1757-899x/365/6/062022
16. Chernyshev, D., Alpatov, V.: Research of the flow-over of the mechanical core of four pipes by a wind current, using Control Volume Method. In: MATEC Web of Conferences, vol. 117, p. 00028 (2017). https://doi.org/10.1051/matecconf/201711700028
17. Kornienko, K.I.: Improving the simulation methodology for hump yard track filling. Transp. Urals **2**(57), 35–42 (2018). ISSN 1815-9400
18. Kornienko, K., Bessonenko, S.: Effect of the hump yard track profile change on the it's occupancy quality at train humping. In: MATEC Web of Conferences, vol. 216, p. 02012 (2018). https://doi.org/10.1051/matecconf/201821602012
19. Rules and regulations for sorting devices design on 1520 mm gauge railways, p. 168. Techinform, Moscow (2003)
20. Zubkov, V., Lukin, A., Alpatov, V.: Experimental research of beams with corrugated web. In: MATEC Web of Conferences, vol. 196, p. 01005 (2018). https://doi.org/10.1051/matecconf/201819601005
21. Shabelnikov, A.N.: Determination of the target exit speed of a cut from a park braking position. Autom. Commun. Comput. Sci. **3**, 9–11 (2009). ISSN 0005-2329

Calculation of Load Distribution in a Roller Bearing of a Locomotive Traction Engine

Alexander Buynosov(✉), Vasily Lapshin, Boris Argannikov, and Yaroslav Mishin

Ural State University of Railway Transport, Kolmogorova Street, 66, 620034 Yekaterinburg, Russia
byinosov@mail.ru

Abstract. The purpose of this study is to determine the effect of the radial clearance on the load between the rolling elements in the bearing of the locomotive traction motor. A method has been developed for calculating the distribution of the load between the rolling elements of a roller bearing of a traction motor, which makes it possible to take into account the radial clearance in the bearing, as well as the radial movement of the inner ring relative to the outer ring. To determine the load perceived by the most loaded roller, it was proposed to solve the equilibrium equation for the inner ring of the bearing using the method of simple iterations. The implementation of the method was carried out in the Matlab software package. Using the proposed technique, the load distribution between the bearing rollers in the locomotive traction motor was calculated. To determine the load acting on the bearing, the Sarga – Sabic section was selected at the Sverdlovsk railroad test site, which has a complex rail track profile with large slopes. Analysis of the results showed that with an increase in the radial clearance of the bearing, the load on the most loaded roller increases.

Keywords: Load distribution · Roller bearing · Locomotive traction engine

1 Introduction

In bearing units of electric locomotives, roller bearings are used. During operation, the bearings of the traction motors in the radial direction act force, which consists of constant and variable, as well as dynamic loads.

The constant load is determined by the weight of the traction motor, the variable load depends on the tractive effort and the force of magnetic attraction. The dynamic components of the load depend on the vertical and horizontal (along the axis of the path) accelerations of the traction motor, on the imbalance of the armature, on the inaccuracy of manufacture, gear wear and angular accelerations of the locomotive's wheel pair relative to the armature. Motor-anchor bearings of traction motors fail several times faster than the estimated service life [1, 2]. Premature failure of bearings occurs due to fatigue damage to the rolling elements or raceways of the inner or outer ring. At present, in the generally accepted methods of calculating the durability of traction motor bearings, the resource is determined without taking into account the load distribution between the rolling elements and without taking into account the radial

Z. Popovic et al. (Eds.): TransSiberia 2019, AISC 1115, pp. 663–669, 2020.
https://doi.org/10.1007/978-3-030-37916-2_64

clearance. When considering a bearing as a system to which a radial load is applied, the inner ring of the bearing and the finite number of rolling elements on which it rests is a statically indefinable system. This is explained by the fact that even with commonly accepted assumptions, it is not easy to establish a connection between the external load and the elastic displacements. For the first time, the problem of load distribution between rolling elements in a rolling bearing was solved by Shtriebek [3]. The decision made a number of assumptions:

- there is no radial clearance in the bearing;
- only contact deformations are taken into account.

Based on the solution of the problem, he proposed a formula for determining the load on the most loaded rolling body:

$$P_{\max} = \frac{K \cdot Q_r}{z} \tag{1}$$

where P_{max} is the load perceived by the most loaded rolling body, N; z – the number of rollers in the bearing, pcs; Q_r – radial load acting on the bearing, N.

As noted above, formula (1) is valid only for a bearing with zero clearance. In practical calculations, the effect of the radial gap is usually taken into account through the correction factor K by changing it from 4.06 to 5 [4, 5].

Conducted in [6, 7], studies have shown that the adoption of the coefficient K in the interval from 4.06 to 5 is not always permissible. Coefficient values can be taken only for some specific radial clearances. This is due to the change in the gap after its installation and the temperature difference between the elements of the bearing.

The purpose of this study is to determine the effect of the radial clearance on the load between the rolling elements in the bearing of the traction motor. The scientific novelty of the work is to develop a method for calculating the distribution of the load between the rolling elements in a roller bearing, taking into account the radial clearance.

2 Method of Calculation of Load Distribution in Roller Bearing

The effects of the radial clearance on the load distribution between the rolling elements of the bearing can be taken into account through the condition of the balance of forces for the inner ring. For a roller bearing (Fig. 1) under load Qr, we write the condition of the balance of forces [6]:

$$Q_r = \sum_{\psi=0}^{\psi=\pm\psi_l} P_\psi \cdot \cos\psi \tag{2}$$

where P_ψ is the load acting on the roller in position ψ, N; ψ is the angular distance between the rollers, °.

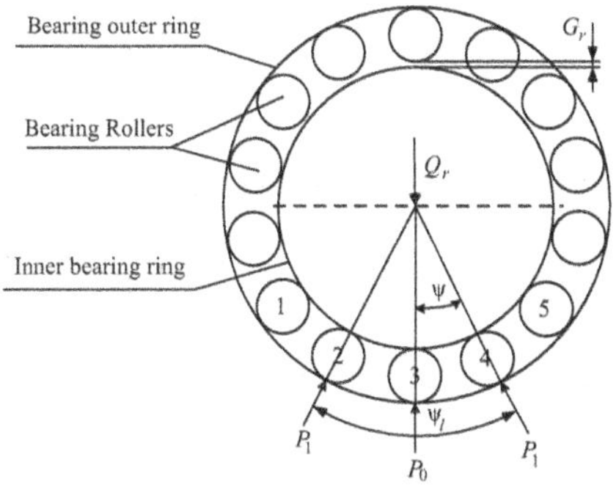

Fig. 1. Load distribution pattern between bearing rollers.

Expression (2) can be written in integral form:

$$Q_r(\varepsilon) = z \cdot P_{\max} \cdot \frac{1}{2\pi} \int_{-\psi_l}^{+\psi_l} \left[1 - \frac{1}{2\varepsilon}(1 + \cos\psi)\right]^{1.11} \cos\psi \, d\psi \tag{3}$$

or

$$Q_r = \mathrm{z} \cdot P_{max} \cdot J_r(\varepsilon), \tag{4}$$

where $P_{\max}$ is the load perceived by the most loaded roller, N; z – the number of rollers in the bearing, pcs; $J_r(\varepsilon)$ – radial load integral; ε – load distribution coefficient.

The load on the most loaded roller is defined as:

$$P_{\max} = K_n \cdot (\delta_r - \frac{1}{2} \cdot G_r)^{1.11}, \tag{5}$$

where δ_r – movement of the inner ring relative to the outer in the radial direction, mm; G_r – radial clearance, mm; K_n is the coefficient characterizing the curvature of the contacting parts of the bearing at the points of contact and the elastic properties of the material, $N/mm^{1.11}$.

For a roller bearing, the K_n coefficient is determined by the following formula:

$$K_n = \left[\frac{1}{\left(\frac{1}{K_l}\right)^{1/1.11} + \left(\frac{1}{K_l}\right)^{1/1.11}}\right]^{1.11} \tag{6}$$

$$K_l = 7.86 \cdot 10^4 l^{8/9} \tag{7}$$

where l is the working length of the roller, mm.

The load distribution coefficient is as follows:

$$\varepsilon = \frac{1}{2}\left(1 - \frac{G_r}{2\delta_r}\right) \tag{8}$$

Analyzing the components of the expression (8) shows that in the load distribution coefficient depends on the radial clearance.

Taking into account all the components of formula (4), we write the equilibrium condition in the following form:

$$Q_r = z \cdot K_n \cdot (\delta_r - \frac{1}{2} \cdot G_r)^{1.11} \cdot J_r(\varepsilon) \tag{9}$$

The main task of the calculation is to determine the distribution of the radial load between the rollers at different values of the radial gap, therefore the constituent formulas (9), such as Q_r, z, K_n and G_r are known, then we represent the formula (9) with respect to the known components:

$$\frac{Q_r}{z \cdot K_n} = (\delta_r - \frac{1}{2} \cdot G_r)^{1.11} \cdot J_r(\varepsilon) \tag{10}$$

From the expression (8) we define the displacement of the inner ring relative to the outer in the radial direction:

$$\delta_r = \frac{G_r}{4 \cdot (0.5 - \varepsilon)} \tag{11}$$

The angle of the loading zone is determined by the expression:

$$\psi_l = \arccos(1 - 2 \cdot \varepsilon) \tag{12}$$

Since Q_r, z, K and G_r are known, in formulas (10) and (11) we introduce the following notation:

$$\frac{1}{2} \cdot G_r = A, \ \frac{G_r}{4} = B \text{ and } \frac{Q_r}{z \cdot K_n} = S$$

Taking into account the above notation, the equilibrium equation and the formula (11) will take the form:

$$\begin{cases} S = (\delta_r - A)^{1.11} \cdot J_r(\varepsilon) \\ \delta_r = \frac{B}{(0.5 - \varepsilon)} \end{cases} \tag{13}$$

The equations can be solved using the two methods "trial and error method" or "iteration method". The use of the "trial and error method" is limited and is used to solve problems of medium complexity (more than 20–30 trial and error) and it is almost impossible to solve complex problems (more than 1000 trial and error). Therefore, to solve the system of Eq. (13), we apply the method of simple iterations.

The essence of the method lies in the sequential search for the root of the equation with a given accuracy [7, 8]. Equation (13) includes the load distribution coefficient ε, which, based on research, is taken from the interval from 0 to 10 [9–12]. The value of S can be both fractional and integer, so the calculation specifies the required accuracy, which determines the number of iterations. In practical calculations of bearings, the value of S is equal to a fractional value and the number of iterations will be immeasurably large, which will require significant computing power from the electronic computer. Therefore, it is necessary to limit the number of iterations by sequentially approximating the root. The calculation algorithm is described as follows:

1. input the source data to find the value of ε, namely A, B, S, the required accuracy, and the known ε_{max} and ε_{min};
2. calculating the value of δ_r at a given accuracy;
3. determining the interval ε and checking the condition $S = (\delta_r - A)^{1.11} \cdot J_r(\varepsilon)$ with one decimal place. If the condition is satisfied, then the new value of ε_{max} and ε_{min} is accepted, in which the desired value of ε is located. Thus, the process will continue as long as $S = (\delta_r - A)^{1.11} \cdot J_r(\varepsilon)$;
4. the conclusion of the result.

To find the load perceived by the most loaded roller, we write the equilibrium equation for the inner ring in the following form:

$$P_{\max} = \frac{Q_r}{z \cdot J_r(\varepsilon)} \tag{14}$$

We find the load on the roller in the position положении (Fig. 1) by the formula:

$$P_{\psi} = P_{\max}\left(1 - \frac{1}{2\varepsilon}(1 - \cos\psi)\right)^{1.11} \tag{15}$$

3 Research Result

We apply the developed method of calculating the load distribution between rolling elements to a real locomotive bearing. As a test object, a 42330ЛМ bearing was selected, which is installed in traction motors of VL11K electric locomotives. The size of the radial clearance was taken in accordance with the instruction from 0.1 mm to 0.29 mm. The magnitude of the load acting on the bearings was determined by the method described in [13]. To determine the mode of operation of the engine, the Sarga – Sabik experimental section was chosen, which has a complex rail track profile with high slope slopes at the Sverdlovsk railway test site.

The given Eq. (13) was solved by an iterative method implemented in the Matlab software package. Based on the results of calculations, for each radial clearance value, the load perceived by the most loaded bearing roller was determined (Fig. 2).

The dependence (Fig. 2) shows that with an increase in the radial clearance, the load on the roller increases. This can be explained by the fact that with increasing radial

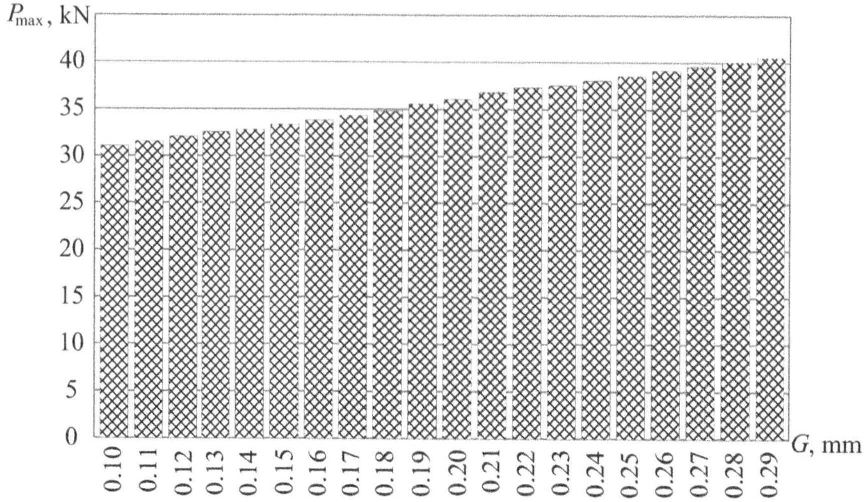

Fig. 2. Dependence of the load perceived by the most loaded bearing roller at different values of the radial clearance.

clearance, the load zone ψl decreases (see Fig. 2). The calculations showed that the radial load acting on the bearing is perceived only with 3 rollers (see Fig. 1).

In Fig. 3 shows the dependences of the load distribution between the bearing rollers and the radial clearance of the bearing.

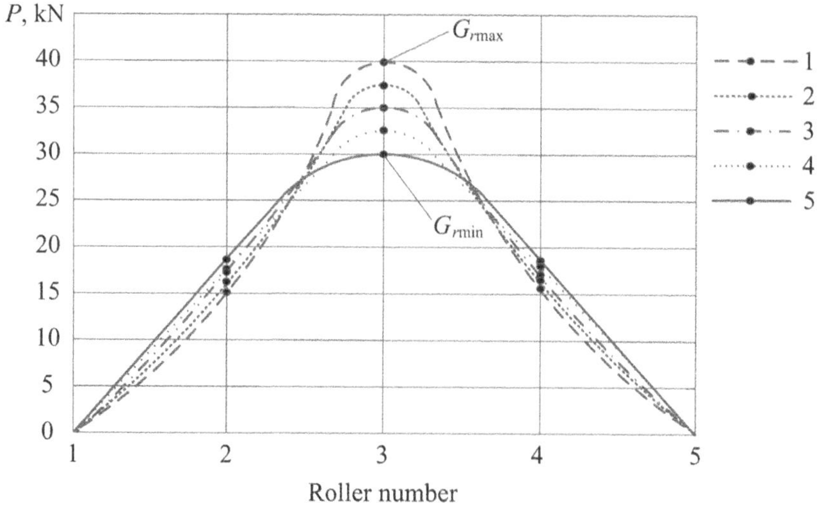

Fig. 3. Load distribution between the roller with a different radial clearance: 1 – G_r = 0.29; 2 – G_r = 0.26; 3 – G_r = 0.2; 4 – G_r = 0.15; 5 – G_r = 0.1.

Analysis of the results showed that with an increase in the radial clearance of the bearing, the load on the most loaded roller increases.

4 Conclusions

Thus, a methodology, an algorithm and a computer program for calculating the load distribution in a roller bearing with regard to the radial clearance have been developed. By the described method, the calculation of the roller bearing of the traction motor, for an electric locomotive operated on the Sarga – Sabic section having a mountain track profile, was made. As a result of the calculations, it was found that with an increase in the radial clearance, the load on the third roller increases. The load acting on the bearing is distributed along an arc limited by an angle equal to $\psi_l = 54°$ with a minimum radial clearance and $\psi_l = 39°$ at the maximum, while the load is perceived only with 3 rollers.

References

1. Buynosov, A.P., Mishin, Y.A.: Scientific problems of transport in Siberia and the Far East **1**, 166–169 (2015)
2. Buynosov, A.P., Mishin, Y.A.: Sci. Tech. Volga Reg. Bullet. **3**, 81–84 (2014)
3. Stribeck, R.: Kugellager fur beiebiege Belastungen Z., Bd. 45, VDI (1901)
4. Meiselman, R.D., Tsypkin, B.V.: Rolling Bearings. Mechanical Engineering, Moscow (1975)
5. Buynosov, A.P., Mishin, Y.A.: Sci. Tech. Volga Reg. Bullet. **3**, 81–84 (2014)
6. Harris, T.A., Kotzalas, M.N.: Essential Concepts of Bearing Technology, New York, p. 376 (2006)
7. Kovalev, M.P., Narodetskiy, M.Z.: Calculation of High-Precision Ball Bearings. Mechanical Engineering, Moscow (1975)
8. Nakhodkin, M.D., Vasilenko, G.V.: Design of Traction Electric Machines. Transport, Moscow (1976)
9. Mishin, Y.A.: Herald of the Ural State University of Railway Transport, vol. 3, no. 35, pp. 118–127 (2017)
10. Cheremisin, V.T., Komyakov, A.A., Ivanchenko, V.I.: Omsk Sci. Bullet. **6**(150), 77–81 (2016)
11. Thursdays, V.A., Ovcharenko, S.M., Greif, K.I.: Izvestiia Transsiba. J. Transsib Railway Stud. **2**(10), 51–58 (2012)
12. Borodin, A.V., Izvestiia, D.V.: Transsiba. J. Transsib Railway Stud. **4**(4), 12–15 (2010)
13. Buynosov, A., Lapshin, V., Smolyaninov, A., Dinislamov, A.: Polytransport systems-2018. In: MATEC Web of Conferences, vol. 216, p. 02020 (2018)

Non-destructive Method of Controlling the Depth of Tempering of Parts of the Mechanical Part of Locomotives

Alexander Buynosov(✉), Alexander Alexandrov, Anatoly Kalinichenko, and Albert Dinislamov

Ural State University of Railway Transport, Kolmogorova Street, 66, 620034 Yekaterinburg, Russia
byinosov@mail.ru

Abstract. The article discusses ways to improve the ultrasonic method developed in USURT together with NPO Introtest JSC to monitor the depth of surface hardening of parts of the mechanical part of locomotives. The non-destructive ultrasound method is developed as an alternative to costly destructive methods. The essence of the method is considered on the example of a spring suspension roller, one of the dynamically loaded surface-hardened parts of the mechanical part of a locomotive. The processes occurring in the system "piezoelectric transducer - object of control", energy losses accompanying these processes, as well as the process of reflection and transmission of the ultrasonic wave at the border "prism - object of control" are considered. The results of experimental studies on the installation of "Etalon", as well as the results of measuring the absorption coefficient for a prism made of various materials. The energy transparency coefficients for the "prism - spring suspension roller" border are calculated. The best material is chosen for the prism of a piezoelectric transducer - crystalline polystyrene.

Keywords: Non-destructive method of controlling · Ultrasonic method · Transport · Locomotive

1 Introduction

The issues of increasing the operational reliability of the parts of the mechanical part of locomotives are highly topical today. High competition in the transport sector dictates the updated standards of speed of transportation and reliability in the operation of rolling stock. At the same time, the diagnostic parameters of one of the most dynamically loaded surface hardened parts of the mechanical part are not controlled or controlled by outdated destructive methods. These methods are expensive, have insufficient efficiency and require considerable time, do not provide control of every detail of the party. To solve this problem, we have developed a non-destructive ultrasonic method for monitoring the depth of the heat-strengthened profile of surface-hardened parts of the mechanical part of locomotives [1]. Consider the essence of the method on the example of the control roller spring suspension TEM2U.35.30.102-01.

Z. Popovic et al. (Eds.): TransSiberia 2019, AISC 1115, pp. 670–676, 2020.
https://doi.org/10.1007/978-3-030-37916-2_65

Roller spring suspension of the locomotive is the object of control (OK). An ultrasonic wave is introduced into the roller, which propagates in a fine-grained, hardened layer without scattering. At that moment, when the wave reaches the transition zone, where the grain size begins to increase, the ultrasonic beam is scattered at the grain boundaries. The depth of the hardened layer is determined by the time of arrival of the scattered ultrasonic wave back to the piezoelectric transducer.

2 Methods of Increasing the Ratio «Signal/Noise»

The amplitude of the scattering wave closely approaches the maximum sensitivity of the detector of the flaw detector, so the useful scattering signal may not be resolved with electrical and structural noise on the A-scan. To increase the signal-to-noise ratio, it is necessary: to apply the methods of signal correction and hardware signal correction [2]; reduce the loss of ultrasonic energy in the system "piezoelectric transducer - object of control" ("probe-OK").

Let us consider the processes occurring in the PEP – OK system and the energy losses accompanying these processes. Ultrasonic energy is radiated by a piezoplate into a prism. A longitudinal ultrasonic wave arises in the prism. The wave propagates up to the border of the prism - a thin layer of contact liquid - metal, partially attenuated. Further, at the boundary of the energy reflected back into the prism, and the rest of the ultrasonic energy goes into the layer of contact fluid and then into the object of control. Therefore, in order for the maximum amount of energy transferred from the converter to "OK", the following conditions must be met:

1. the attenuation of the ultrasonic wave in the prism should be minimal;
2. the energy transparency coefficient at the "prism – contact liquid – metal" boundary should be maximum.

This can be achieved by selecting the optimal material for the prism of the ultrasonic sensor. The weakening of the ultrasonic wave occurs due to the divergence of the beam, as well as due to the attenuation of the wave in the propagation medium.

The divergence of the beam has a strong negative effect on the amplitude of the signals only at a distance x, greater than the Fresnel zone x > N, which is defined as:

$$N = \frac{d^2}{4 \cdot \lambda} \tag{1}$$

where d is the diameter of the piezoplates, λ is the wavelength in the prism.

For the frequency $f = 15$ MHz and $d = 6$ mm [3], we obtain for plexiglas $N = 50$ mm. What more path ultrasonic waves in the prism. That is, $x < N$, the wave propagating within the Fresnel zone, therefore, the wave is flat. Thus, the divergence of the ultrasonic wave can be neglected.

The attenuation of the ultrasonic wave in the medium is described by the attenuation coefficient δ, it shows at what distance x from the emitter the initial acoustic pressure p_0 will weaken e times [4]:

$$p = p_0 \cdot e^{-\delta x} \quad (2)$$

The attenuation coefficient of ultrasound δ consists of two components:

$$\delta = \delta_p + \delta_r, \quad (3)$$

where δ_p is the absorption coefficient, δ_r is the scattering coefficient.

As a material for a prism, various plastics are mainly used; they have a micro-heterophase structure, that is, formed by particles and aggregations of the microdispersed phase, distributed in a homogeneous dispersion medium. Even for sufficiently high frequencies, the condition is satisfied:

$$\lambda << D_n \quad (4)$$

where D_n is the average particle size of the microdispersed phase.

For plexiglas, D_n = 0.1–2.5 μm [4], which is three orders of magnitude less than the wavelength of 15 MHz. Thus, the effect of dispersion in a prism can also be neglected. Absorption is the direct conversion of ultrasonic energy to heat. In our case, the main contribution to attenuation is the absorption of ultrasonic energy by the medium. Determine how much of the ultrasonic energy gets from the prism to the test object. In Fig. 1 schematically shows the process of reflection and transmission of an ultrasonic wave at the "prism - object of control" border.

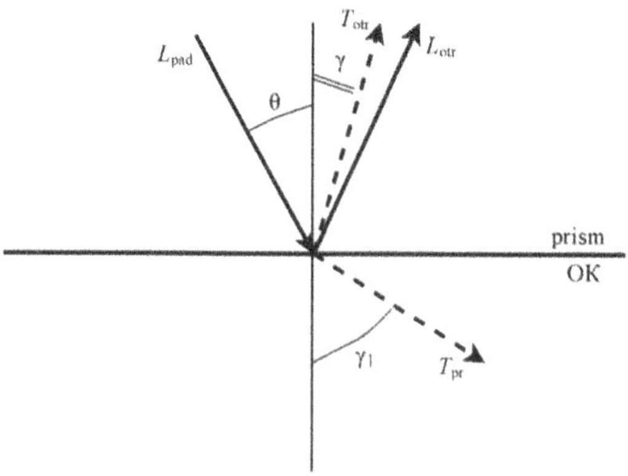

Fig. 1. Diagram of a linearly polarized wave incident on the boundary of two media.

A linearly polarized wave (L_{pad}) falls on the interface between the media. Further, part of the wave is reflected from the surface, transforming in this case into a reflected linearly polarized wave (L_{otr}) and a transversely polarized wave (T_{otr}). The remaining energy passes through the boundary, transforming into a transversely polarized transmitted wave (T_{pr}). A linearly polarized wave in the control object (L_{pr}) is not excited, since the L_{pad} wave falls between the first and second critical angles.

In our case, the angle of refraction γ_1 is defined [5] and is 45°. The angles θ and γ are determined from the Sneliuss ratio (11). Characteristics of elastic media, prism and object of control: ρ and ρ_1 - density of media; k, $\varkappa$, k_1, $\varkappa_1$ are wavenumbers of longitudinal and transverse waves; c, b, c_1, b_1 are the velocities of the longitudinal and transverse waves, respectively. Then the coefficient of transparency by energy E_{lt} (or the ratio of transformation by energy) for a transversely polarized wave in the roller of spring suspension T_{pr} is defined as [6, 7]:

$$E_{lt} = \frac{\rho_1 \cdot \mathrm{tg}\theta}{\rho \cdot \mathrm{tg}\gamma_1} \cdot |W_{lt}|^2 \tag{5}$$

where W_{lt} is the amplitude transparency coefficient.

The constituents of formula (5) are defined as:

$$W_{lt} = \frac{x_1^2}{\Delta\xi^2} \cdot \left(\mathrm{A}_2 + \frac{a_1 \mathrm{B}_1}{\beta_1}\right) \tag{6}$$

$$\Delta = \mathrm{A}_1^2 + \beta \cdot \frac{\mathrm{A}_2^2}{a} + \frac{a_1}{\beta_1} \cdot \left(\mathrm{B}_1^2 + \beta \cdot \frac{\mathrm{B}_2^2}{a}\right) + m \cdot \left(\frac{a_1}{a} + \frac{\beta_1}{\beta}\right) \cdot \left(\frac{x_1^4}{4 \cdot \xi}\right) \tag{7}$$

$$\mathrm{A}_1 = n^2 - m \cdot \frac{\rho_1}{\xi},\ \mathrm{A}_2 = \left(n^2 \cdot \rho - m \cdot \rho_1\right) \cdot \frac{1}{\beta} \tag{8}$$

$$\mathrm{B}_1 = \left(n^2 - m\right) \cdot \frac{\beta_1}{\xi},\ \mathrm{B}_2 = \frac{n^2 \cdot \rho}{\xi} - m \tag{9}$$

$$m = \frac{\rho_1}{\rho},\ \left(n = \frac{b}{b_1}\right) \tag{10}$$

$$\xi = k \sin\theta = k_1 \sin\theta_1 = x \sin\gamma = x_1 \sin\gamma_1 \tag{11}$$

$$a = k \cos\theta,\ a_1 = k_1 \cos\theta_1 \tag{12}$$

$$\beta = x \cos\gamma,\ \beta_1 = x_1 \cos\gamma_1 \tag{13}$$

$$\rho = \frac{x \cos 2\gamma}{2 \sin\gamma} \tag{14}$$

$$\rho_1 = \frac{x_1 \cos 2\gamma_1}{2 \sin\gamma_1} \tag{15}$$

3 The Results of Experimental Studies

Experimental studies were conducted on the installation "Etalon" (Fig. 2).

Fig. 2. Installation for determining the attenuation coefficient "Etalon".

The sample was placed in an immersion bath and the relative amplitude drop between the first and second bottom signals was measured. A schematic diagram of the measurements is shown in Fig. 3.

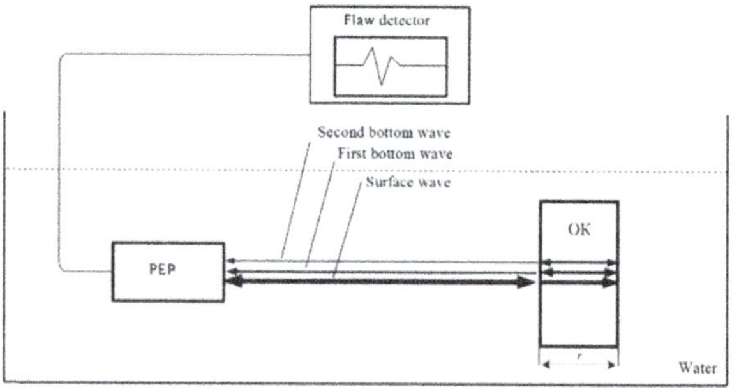

Fig. 3. Schematic diagram of measuring the absorption coefficient.

The probe and specimen of thickness r are placed in an immersion water bath. The piezoelectric transducer emits an ultrasonic wave into the immersion medium [8, 9]. Having reached the control object, the wave is partially reflected from the front surface of the OC and gets back to the probe (surface signal), partially penetrates the OC. Further, the wave propagates in the OK and reaching the bottom surface is reflected from it. Having reached the front surface, the ultrasonic wave partially passes through it into the immersion medium and then onto the probe (first bottom signal, AI), is partially reflected from it and moves to the bottom surface [10]. Then again there is a reflection from the bottom surface and the second bottom signal AII falls on the probe. The absorption coefficient (attenuation) is calculated by the formula:

$$\delta = \frac{A_I - A_{II}}{2 \cdot r} \tag{16}$$

The a-scan of the bottom signals is presented in Fig. 4.

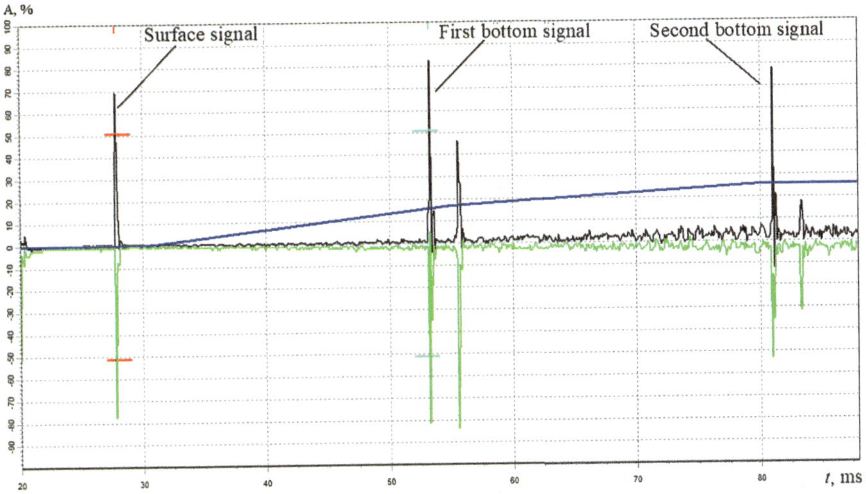

Fig. 4. A-scan bottom signal in the sample.

The following plastics were used as samples: crystalline polystyrene, plexiglas, plexiglass plexiglas, polyacetal, ED-20 [11]. The results are presented in Table 1.

Table 1. Measurement results of absorption and transparency coefficients for various materials.

Material	Absorption coefficient, δ, dB/mm	Energy transparency coefficient, E_{lt}, %
polystyrene	0.324	20.6
plexiglass	1.411	22.6
plexiglass	1.253	21.6
polyacetal	–	26.7
ED-20	–	24.5

For samples made of polyacetal and ED-20, the absorption coefficient was not determined, since it was not possible to measure the second bottom signal, due to the large attenuation in these materials. The minimum value of the absorption coefficient of 0.324 dB/mm corresponds to crystalline polystyrene [12, 13]. As for the energy transparency coefficient, the maximum value of 26.7% corresponds to polyacetal [14, 15]. The minimum value of 20.6% characterizes polystyrene [16, 17]. In this case, due to the small difference in the transparency coefficient and a significant difference in the absorption coefficient, it is preferable to use polystyrene as the probe prism.

4 Conclusions

As a result of the study, the best material for the prism of the piezoelectric transducer was determined. The results of measuring the attenuation coefficients of ultrasound and the transparency coefficients in energy for various materials of the probe are presented. The minimum attenuation $\delta = 0.324$ dB/mm at a frequency of 15 MHz is observed in crystalline polystyrene, the same sample corresponds to the minimum coefficient of transparency $E_{lt} = 20.6\%$. Due to the insignificant difference between the coefficients of transparency (maximum 26.7%) and a significant difference in attenuation, crystalline polystyrene was chosen as the best material of the probe.

In the future, it is necessary to carry out a computer simulation of the scattering process, to estimate the scattering intensity of the ultrasonic wave from the hardness gradient in the transition zone.

References

1. Buynosov, A.P., Volkov, D.V., Dinislamov, A.R.: Transp. Urals **1**(56), 26–29 (2018)
2. Buynosov, A.P., Dinislamov, A.R.: Transp. Urals **2**(57), 69–72 (2018)
3. Buynosov, A.P., Volkov, D.V., Dinislamov, A.R.: Scientific problems of transport in Siberia and the Far East 1, 104–108 (2018)
4. Radchenko, S., Dorokhov, D., Gryadunov, I.: J. Chem. Technol. Metall. **50**, 104112 (2015)
5. Buynosov, A.P., Dinislamov, A.R.: Vestnik transporta Povolzhya **1**, 7–10 (2018)
6. Korotkov, V.A., Zlokazov, M.V.: J. Friction Wear, 133136 (2014)
7. Zhang, M., Chen, C., Wang, X., Hou, J., Zhang, S., Zhu, G., Jing, H.: Acta Metallurgica Sinica (English Letters) **27**, 521529 (2014)
8. Ivantsivsky, V., Parts, K., Popov, V.: Appl. Mech. Mater. 788 (2015)
9. Pyndak, V.I., Novikov, A.E.: Met. Sci. Heat Treat. **58**, 226230 (2016)
10. Zhuravlev, G.M., Gvozdev, A.E., Cheglov, A.E., Sergeev, N.N., Gubanov, O.M.: Steel Translation **47**, 399411 (2017)
11. Kolpakov, A.S., Kardonina, N.I.: Thermal Eng. **63**, 150155 (2016)
12. Trushnikov, D.N., Belinin, D.S., Shitsyn, Y.D.: Mod. Probl. Sci. Educ. **2**, 95 (2014)
13. Buynosov, A., Lapshin, V., Smolyaninov, A., Dinislamov, A.: Polytransport systems-2018. In: MATEC Web of Conferences, vol. 216, p. 02020 (2018)
14. Petrov, S.Y., Kokorin, R.V., Baban, S.M.: Tekhnologiya Mashinostroeniya. Technol. Mech. Eng. **4**, 37–44 (2017)
15. Muravyev, V.V., Petrov, S.Y., Volkova, L.V.: Defektoskopija. Russ. J. Nondestr. Test. **10**, 64–74 (2015)
16. Pilipenko, O.V., Radchenko, S.J., Dorokhov, D.O., Gryadunov, I.M.: Int. J. Appl. Eng. Res. (2017)
17. Girzhon, V.V., Ovchinnikov, A.V.: Met. Sci. Heat Treat. **58**, 719723 (2017)

Optimization of Vehicle Diagnostic Algorithms at Equipment Design Stage

Valery Vasilyev[1(✉)], Viktor Ovsyannikov[1], and Rudolf Kovalev[2]

[1] Kurgan State University, Sovetskaya Str., 63, Building 4, 640020 Kurgan, Russia
vik9800@mail.ru

[2] Ural State Forest Engineering University, Sibirskiy Trakt Str., 37, 620100 Yekaterinburg, Russia

Abstract. The paper shows that today a worker can be considered as an integral part of the "human-machine" system in case of using modern diagnostic equipment. Therefore, in order to ensure the effectiveness of diagnosing vehicle systems, it is necessary to use engineering psychology methods at the stage of equipment design. Traditionally, the description of "human-machine" systems functioning is implemented in the form of an algorithm. Thereat, coefficients of stereotyped and logical complexity act as the criteria of the algorithm rationality. However, at the design stage, it is necessary to estimate the parameters at the preliminary stages, when the algorithm is not finally formulated; therefore, the situation of lack of initial data occurs. The paper considers the possibility of using the fuzzy logic apparatus to overcome this problem. The data determined with implementation of real diagnostic algorithms are used as initial data. The authors elaborated models for estimating the parameters of logical complexity and stereotype that give an error of no more than 10%. Based on fuzzy logic, the model was created that enables formulating recommendations for ensuring a given level of compatibility of the elements of the "human-machine" system when designing diagnostic equipment of transport enterprises.

Keywords: Design · Equipment of transportation enterprises · Fuzzy logic

1 Introduction

The widespread introduction of mechanization and automation facilities substantially changes the structure of personnel activity. The transport industry is no exception. In particular, it is worth mentioning that the informational and psychological stress on operators of diagnostic equipment is continuously increasing. The reason is that functions associated with decision-making and information processing lie with a human. Thus, he or she can be considered as an operator within the framework of a "human-machine" system. This circumstance requires consideration of not only physiological, but also psychological aspects of labor. Moreover, the last ones are even more significant due to the reduction of physical component of labor.

Z. Popovic et al. (Eds.): TransSiberia 2019, AISC 1115, pp. 677–684, 2020.
https://doi.org/10.1007/978-3-030-37916-2_66

Traditionally, engineering psychology uses a number of indicators that characterize the efficiency of the "human-machine" systems [2, 3]. The most significant ones are speed, reliability, timeliness of decision-making and safety [2, 3]. The reliability and safety of a system depend on the number of errors made by an operator, while the speed performance and timeliness depend on the speed and quality of information processing [2, 3]. Thus, a human is the most problematic element in terms of ensuring the effectiveness of systems regarding the class under consideration. Consequently, when designing diagnostic equipment, it is necessary to create such psychological conditions for a human operator that would ensure the minimum number of errors and the most efficient processing of information. The algorithmic description of operator's activity is typically used in engineering psychology [2, 3]. In addition, a numerical assessment of the intensity of labor is possible using the following criteria [2–4]:

1. Stereotype coefficient:

$$Z = \frac{1}{N}\sum_{i=1}^{n}\frac{m_{0i}^2}{m_i} \tag{1}$$

where N is the total number of operators of the algorithm; n is the number of groups of members of the algorithm; m is the number of operators in a group; m_{0i} is the number of elementary operators in a group (actions that do not imply a choice).

2. The coefficient of logical complexity:

$$L = \frac{1}{N}\sum_{j=1}^{n}\frac{m_{lj}^2}{m_j} \tag{2}$$

where m_{lj} is the number of logical conditions in a group.

However, the use of expressions 1 and 2 when evaluating the parameters of algorithms during the process of designing equipment is impossible because the algorithm has not yet been formulated in its final form.

Objective: to develop solutions that will enable optimizing the algorithms for diagnosing transport systems at the equipment design stage.

The tasks include the development of a model for evaluating the indicators of logical complexity and stereotype under conditions when the algorithm is not formulated in its final form, as well as elaboration of solutions enabling determination of directions for optimizing algorithms for ensuring the required level of compatibility of the elements of the "human-machine" system.

2 Research Methods

As mentioned above, the main problem in completing the task is the uncertainty of data that is necessary for estimating the parameters of algorithms characterizing the compatibility of elements of "human-machine" system. Analysis of sources [1, 5] shows that the use of fuzzy logic is the effective way out of this situation, since on the one hand, it enables operating with such categories as "not quite", "approximately", etc., and on the other hand, there is an opportunity of conversion of quality indicators into quantitative parameters through defuzzification [1, 5].

Fuzzy logic operates with sets. Most often, the set is defined as the characteristic function [1, 5], which indicates that the element x belongs to the set A:

$$\mu_A = \begin{cases} 0, x \in A \\ 1, x \notin A \end{cases} \quad (3)$$

Fuzzy sets are a generalization of ordinary sets, whereby the characteristic function can take any value within the interval [0, 1]. In terms of fuzzy logic, the characteristic function is called the membership function, while its μA values refer to the degree of element's belonging to a fuzzy set. There are various membership functions. The most widely used ones are triangular, trapezoidal and Gaussian membership functions. The parameters of the algorithms for the process of diagnosing car systems [4] were used as initial data for tuning the model. The example of the fragment of a car is given in Table 1.

Table 1. Functional diagnostics of power supply system of carburetor engine (fragment)

Algorithm actions	Type of action
Visual inspection	
1. Check the presence of fuel in the tank, the absence of fuel leakage, etc.	Stereotyped
2. Pump up gasoline with the fuel pump	Stereotyped
3. Check fuel level	Stereotyped
4. Warm the engine	Stereotyped
5. Determine the performance of a system by external signs	Logical
Functional diagnostics of the power supply system by cyclic fuel consumption	
1. Place the vehicle on test set K-453	Stereotyped
2. Wire up devices to a vehicle	Stereotyped
3. Adjust the exhaust gas analyzer	Logical
4. Set a stopwatch of flow monitor to 0	Stereotyped
5. Set the tachometer measurement range of switching device to 1500	Stereotyped
6. Start the engine	Stereotyped
7. Set the desired crankshaft speed	Stereotyped
8. Measure fuel consumption	Stereotyped
9. Determine cyclic fuel consumption	Logical
Functional diagnostics of power system on the test set with roller dynamometer	
1. Switch the measurement limit switching device of the tachometer to 7500	Stereotyped
2. Turn on the test set	Stereotyped
3. Get car up on roller dynamometers	Stereotyped
4. Load the car by pressing the "below" button	Logical
5. Read the tachometer	Stereotyped
6. Turn off the test set	Stereotyped
7. Calculate the average effective engine pressure	Logical

Indicators of logical complexity and stereotype for this algorithm equal: $L = 0.07$; $Z = 0.59$.

3 Result of Research

Analysis of expressions (1) and (2) leads to the conclusion that it is advisable to choose number of algorithm operators and the relative density of stereotypic actions (logical conditions) as input variables for elaborating a model of estimation of algorithm parameters based on fuzzy logic. Therefore, the structure of the model for assessing the stereotype coefficient is (Fig. 1):

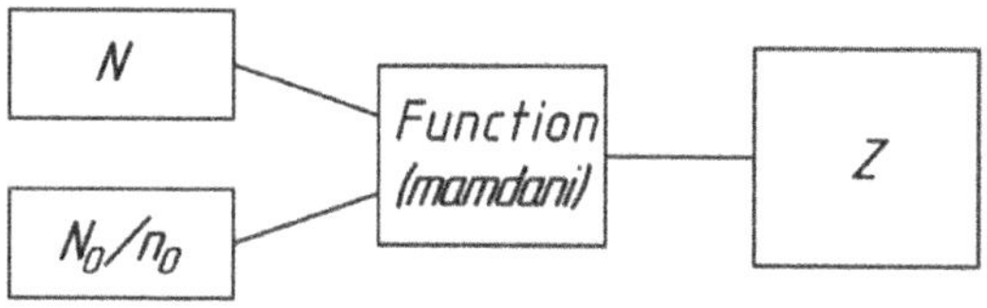

Fig. 1. The structure of assessment model of the stereotype coefficient

Triangular terms are used as membership functions [1, 5]. The Mamdani function, which binds input and output parameters, is set by a system of rules like this:

1. When N = "small" and m/m_0 = "small", then Z = 0.5;
2. When N = "average" and m/m_0 = "small", then Z = 0.05;
3. When N = "large" and m/m_0 = "small", then Z = 0.05;
4. When N = "average" and m/m_0 = "average", then Z = 0.5;
5. When N = "large" and m/m_0 = "average", then Z = 0.05;
6. When N = "small" and m/m_0 = "large", then Z = 0.95;
7. When N = "average" and m/m_0 = "large", then Z = 0.5;
8. When N = "large" and m/m_0 = "large", then Z = 0.05.

In this case, the criterion values of indicator Z are determined based on the analysis of typical diagnostic algorithms. The model for estimating the coefficient of logical complexity has the similar structure and elaboration principles. The use of models enabled obtaining functions that connect the values of the coefficients Z and L with input parameters.

The average inaccuracy in determining the coefficients:

$$\Delta Z = \frac{Z_{\text{mod}} - Z_{theor}}{Z_{\text{mod}}} \times 100\% = \frac{0.861 - 0.83}{0.861} \times 100\% = 4\%$$

$$\Delta L = \frac{L_{\text{mod}} - L_{theor}}{L_{\text{mod}}} \times 100\% = \frac{0.06 - 0.04}{0.06} \times 100\% = 3\%$$

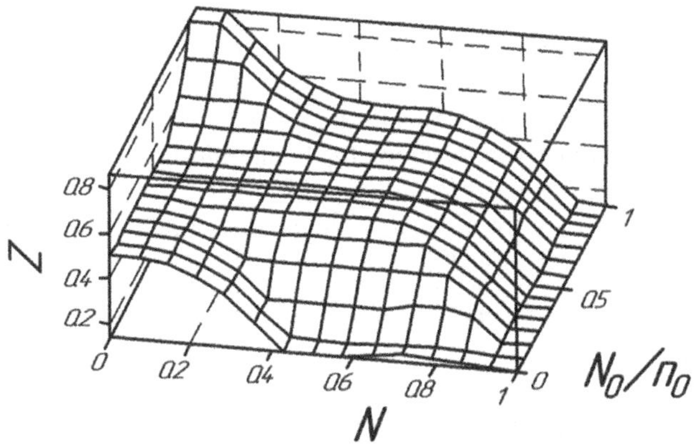

Fig. 2. The function to estimate coefficient Z

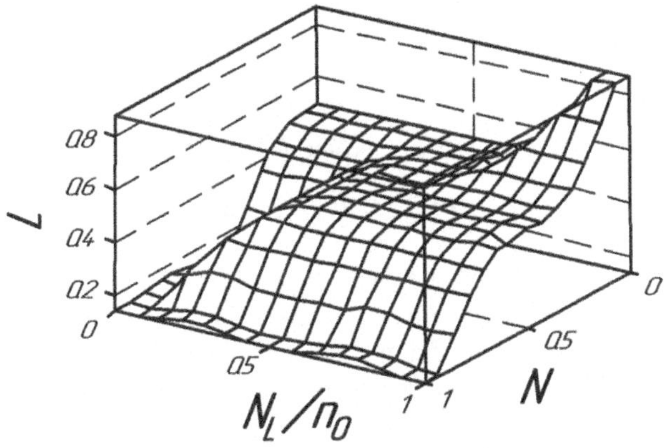

Fig. 3. The function to estimate coefficient L

4 Discussion

As was shown above, the use of fuzzy logic effectively solves the problem of estimating the parameters of the algorithm at the equipment design stage, but by themselves these values do not allow achieving the goal of optimization of the algorithms set by the condition of providing the required level of compatibility of the elements of the human-machine system. It is necessary to formulate specific recommendations on how to change the algorithm, and to do this also in the conditions of insufficient amount of initial data. This means that the optimization of the algorithms should also be carried out using functions of fuzzy logic. The function is defined by the following rule system (Figs. 2 and 3):

1. If the variable Z and the variable L lie within acceptable limits, then the function f (Z, L) takes the value "all parameters are normal";

2. If the variable Z is outside the bounds of the permissible values, and the variable L lies within the permissible limits, then the function f (Z, L) takes the value “excessive stereotype”;
3. If the variable Z does not go beyond the bounds of permissible values, and the variable L goes beyond the tolerance limits, the function f (Z, L) takes the value “excessive logical complexity”;
4. If both variables are out of acceptable limits, then the function f (Z, L) takes the value “both parameters exceed the norm”.

The appearance of the function is shown in Fig. 4.

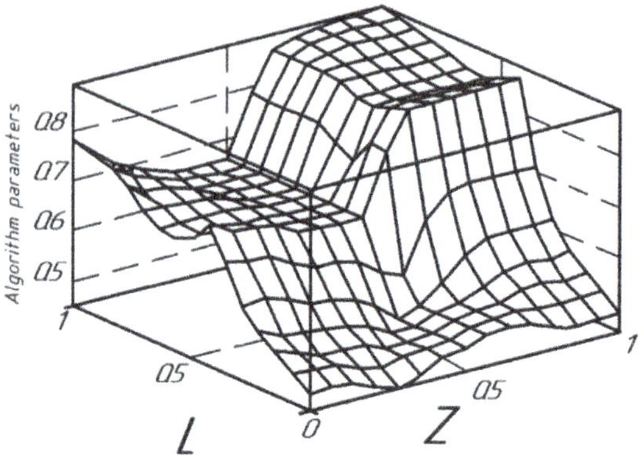

Fig. 4. The function to evaluate the parameters of the algorithm

Based on the analysis performed, the data on the parameters of logical complexity and stereotypedness can be divided into 6 zones, depending on the excess of the optimal values (see Table 2 and Fig. 5).

Table 2. Data on values of coefficients of logical complexity and stereotype

L Z	Z = 0...0.25	Z = 0.25...0.85	Z = 0.85...1
L = 0...0.2	Area 1 0.51 $\leq$ f(Z, L) $\leq$ 0.53	Area 2 0.48 $\leq$ f(Z,L) $\leq$ 0.51	Area 3 0.53 $\leq$ f(Z, L) $\leq$ 0.56
L = 0.2...1	Area 4 f(Z, L) $\geq$ 0.62	Area 5 0.56 $\leq$ f(Z, L) $\leq$ 0.62	Area 6 f(Z, L) $\leq$ 0.48

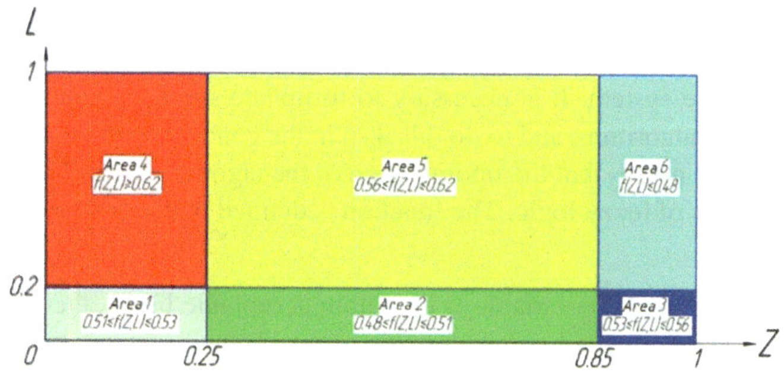

Fig. 5. Coefficient value assorting

5 Conclusions

The following recommendations on the development of algorithms for activities of operators of technological equipment for road transportation enterprises can be formulated depending on finding coefficient values in following particular areas:

1. Area 1 - in this case, an increase in values of the stereotype coefficient is required. This can be achieved by automating the intermediate stages of information processing, thus the chains of stereotypical actions will be longer;
2. Area 2 - in this case, the parameters correspond to optimal ones, the algorithm does not need to be processed;
3. Area 3 - in this case, excessively long chains of stereotypical actions take place, therefore it is necessary to automate their implementation;
4. In area 4, there is a simultaneous excess of the permissible values of the logical complexity coefficient and an insufficient stereotype coefficient value, therefore it is necessary to automate the processing of information and change the structure of the algorithm so that the chains of logical actions are shortened and the stereotypical ones are lengthened;
5. In area 5, there is an excess of the values of the logical complexity coefficient in case of the normal stereotype coefficient, so it is necessary to automate the processing of information in order to shorten the sequence of logical actions;
6. In area 6, values of the coefficients of logical complexity and stereotype are exceeded; therefore, it is necessary to automate both information processing and stereotyped actions.

Thus, the models obtained enable estimating the parameters of the algorithms under conditions of a high degree of uncertainty in the source data and determining the directions for optimizing the algorithm in order to ensure the required level of compatibility of the elements of the human-machine system. For effective practical implementation, it is advisable to combine these models and integrate them into a common expert system, which allows evaluating the parameters of the algorithms at the design stage of diagnostic equipment of transport complex enterprises. Thereat these models will be used as a logical unit of this system [6].

References

1. Zak, Y.A.: Prinyatiye resheniy v usloviyakh nechetkikh i razmytykh dannykh, Moscow, "LIBROKOM", p. 352 (2013). (in Russian)
2. Ovsyannikov, V.Y.: Algoritmicheskiy analiz protsessov diagnostirovaniya sistem avtomobiley, Vestnik KuzGTU, vol. 2, no. 108, pp. 109–111 (2015). (in Russian)
3. Shtovba, S.D.: Proyektirovaniye nechetkikh sistem sredstvami MATLAB. Goryachaya liniya Telekom, Moscow, p. 288 (2007). (in Russian)
4. Kalinina, O.V., Lopatin, M.V.: Strategic and innovative aspects of transition to progressive scale in income taxation. Actual Prob. Econ. **182**(8), 392–405 (2016)

5. Sergeev, S., Kirillova, T., Krasyuk, I.: Modelling of sustainable development of megacities under limited resources. In: E3S Web of Conferences, vol. 91, p. 05007 (2019)
6. Kosnikov, S.N., Khaibullina, I.V., Ignatskaya, M.A., Bakharev, V.V., Pinchuk, V.N.: Characteristic of economic indicators of reproduction of fixed capital. Int. J. Appl. Bus. Econ. Res. **15**(13), 243–253 (2017)

Evaluation of the Parameters of the Unsteady Process of Deceleration of Railway Rolling Stock

Alexander Buynosov(✉), Alexander Smolyaninov, Ivan Dobychin, and Evgeny Fedorov

Ural State University of Railway Transport, Kolmogorova Street, 66, 620034 Yekaterinburg, Russia
byinosov@mail.ru

Abstract. Currently widely used methods for determining the stopping distance of railway rolling stock have a number of drawbacks that affect the accuracy of the calculation results. The aim of the study is to develop a calculation method that allows to take into account the main parameters of the process of unsteady braking (the speed of propagation of the brake wave; feature of the brake cylinder filling diagram; brake mode; the initial speed at which the calculation of the actual stopping distance begins). The methods used in the work allowed us to obtain equations for calculating the specified parameters of the nonstationary mode of the brakes. The results of computer simulation of the stopping distance of the train are presented and the analysis of the parameters obtained using various methods for calculating the stopping distance is carried out. The values of the stopping distance of the train, preparation of brakes and the actual stopping distance are determined. Comparison of the calculation results for the proposed method was performed using computer simulation and the method of integrating the equation of motion over the speed intervals. The proposed method allows more accurate calculations of the stopping distance of railway rolling stock.

Keywords: Unsteady process of deceleration · Railway rolling stock · Nonstationary mode

1 Introduction

Safety of the transportation process is one of the main tasks of the railway industry. The solution of this task is based on the improvement of the whole complex of technical devices and measures aimed at preventing and preventing dangerous situations [1]. The main means of ensuring the safety of trains is the brakes of railway rolling stock. Brakes are a system of technical devices aimed at creating controlled forces of resistance to the movement of a train in order to stop it, reduce speed or maintain it at a constant level [2, 3].

One of the main indicators of traffic safety and braking efficiency is the path traveled in the process of braking rolling stock. Currently, the most common method of calculating the stopping distance for speed intervals. The basis of this method is the

Z. Popovic et al. (Eds.): TransSiberia 2019, AISC 1115, pp. 685–692, 2020.
https://doi.org/10.1007/978-3-030-37916-2_67

conditional splitting of the stopping distance into two parts: preparatory (characterizes unsteady braking) and real (steady braking), calculations are performed at speed intervals, while the specific braking force and resistance to movement are conditionally considered constant and equal for the average speed in a given interval [4–6]. This method has a number of significant drawbacks, for example, the method gives low accuracy at speeds of up to 40 km/h and on descents of more than 20. If the composition stops before the brake cylinders are completely filled, the calculation using this method will lead to an erroneous (overestimated) value of the stopping distance [7–10]. In addition, the method used does not take into account the change in the speed of movement when preparing the automatic brakes for action, and when calculating the way the brakes act, the initial speed is substituted before braking. Also, the presence of an initial pressure jump in the brake cylinder, the time of propagation of the brake wave, and the length of the train [11, 12] are not taken into account.

The aim of the study is to assess the parameters in the unsteady process of deceleration of railway rolling stock. The scientific novelty of the work is to develop a method for calculating the stopping distance, which allows to take into account the main parameters of the braking process, the speed of propagation of the braking wave; feature of the brake cylinder filling chart; brake mode; the initial speed with which the calculation of the steady process of braking begins.

2 Choice of Calculation Methodology

The current rules for braking calculations conditionally divide the process of gradually increasing braking force in a train into two conditional sections: the complete absence of braking force at the beginning of the braking process and the instantaneous action of the braking force of the train. But the purpose of the calculations is to find the stopping distance during unsteady braking process [13, 14].

In Fig. 1 shows the growth pattern of the specific braking force of the train but during the preparation of the automatic brakes for action, which, according to the classifier prof. B.L. Karvatsky, characterized by three preparatory and one steady-state phase of inhibition:

I is the time tv from the beginning of the automatic braking actuation to the appearance of pressure in the brake cylinder of the tail car;

II – the time from the beginning of the operation of the brake of the last car to the complete braking of the head car;

III – the time from the complete braking of the head car to the full braking of the tail car;

IV – this phase corresponds to the movement of a completely inhibited train.

In general, the movement of rolling stock when braking, according to the second law of Newton, is determined by the expression:

$$M \cdot a = B_{\mathrm{T}} + W_{\mathrm{O}}, \qquad (1)$$

where M is the mass of the composition, kg; a – acceleration, m/s^2; B_{T} – the brake force of the train, N; W_{o} –resistance to movement, N.

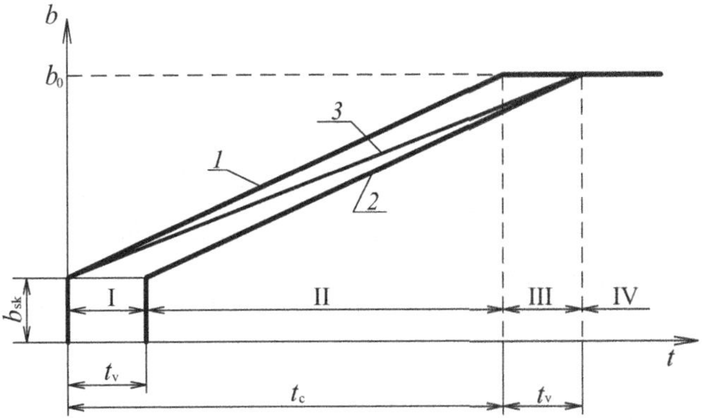

Fig. 1. Chart of increase of the specific braking force in the train: *1* – specific braking force of the head car; *2* – specific braking force of the tail car; *3* – average specific braking force.

After mathematical transformations, the movement of the train under the action of the specific braking force at the site can be represented by the formula:

$$a = (b_T + w_O)g, \tag{2}$$

where b_T, w_o – respectively, specific braking force and specific resistance to train movement, N/kN; g – ravitational acceleration, m/s^2.

When the train is moving along the slope, the expression (2) takes the form:

$$a = (b_T + w_O \pm i_C)g, \tag{3}$$

where i_c is the rectified slope of the path, ‰;

Then the initial speed V_n, at which the train goes into phase IV of steady braking (see Fig. 1), will be determined by the formula:

$$V_n = V_0 - 3.6a(t_c + t_v), \tag{4}$$

where t_c, t_v – respectively, the time of filling the brake cylinder and the time of propagation of the brake wave in the train, s.; a is the average acceleration proportional to the average specific increasing brake force of the train, m/s^2.

The average effective acceleration value is calculated by the formula:

$$a = \lfloor b_p + w_o \pm i_c \rfloor g, \tag{5}$$

where b_p is the average specific braking force during the preparation of the automatic brakes, N/kN.

The value of the braking force is using the proposed method based on the well-known expression, for example, for composite pads:

$$B_t = 0,44\,mK\frac{0,1K+20}{0,4K+20}\left(\frac{V+150}{2V+150}\right), \tag{6}$$

where m – the number of brake pads on the car with pressing K.

The first of the factors of formula (6) determines the braking force of the car in the static state $B_0 = 0,44\,mK\frac{0.1K+20}{0.4K+20}$ and the second factor determines the speed function $f(V) = \frac{V+100}{5V+100}$. Then the braking force of the composition, with the same brake pads, can be expressed by the formula:

$$B = \sum B_0 \cdot f(V). \tag{7}$$

It is convenient to calculate the brake force of a train for various types of braking - shoe, disc, electric brake, etc. Then the average specific braking force during the preparation of the automatic brakes for action is determined (according to Fig. 1) by the formula:

$$b_p = 0.5(b_0 + b_{sk}) \cdot f(V). \tag{8}$$

The value of the specific braking force is presented for one car, and this value, if auto mode is present, can apply to the whole train, because the purpose of auto mode is to ensure the constancy of the calculated braking coefficient.

In this case, the preparation of the brakes will be equal to:

$$S_p = 0.28\,V_0(t_c + t_v) - 0.5a(t_c + t_v)^2. \tag{9}$$

Or after transformation

$$S_p = 0.28\,V_0(t_c + t_v) - 0.005\left[\frac{B_0 + B_{sk}}{Q} \cdot f(V) + w_0 \pm i_c\right] \cdot (t_c + t_v)^2. \tag{10}$$

The speed at which the braking process in the train goes into the steady-state braking phase and from which the calculation of the actual stopping distance begins by the method of summation over the speed intervals is determined by the expression:

$$V_n = V_0 - 0.018\left(\frac{B_0 + B_{sk}}{Q} \cdot f(V) + w_0 \pm i_c\right) \cdot (t_c + t_v). \tag{11}$$

Calculation of stopping distance S_d is performed by the method of numerical integration from the initial speed V_n to a full stop according to the recommended formula:

$$S_d = \sum \frac{16.7V_n}{(b_0 \cdot f(V) + w_0 \pm i_c)}. \tag{12}$$

The calculations start from the initial speed, which is determined when calculating the preparatory path, and continue at a specified interval $\Delta V = 2$ km/h until the train stops ($V = 0$). For this, an algorithm should be compiled and, possibly, an auxiliary table of values of the velocity function $f(V_n)$ with an interval of velocity values $\Delta V =$ 2 km/h.

3 Calculation of the Parameters of the Braking Process

To verify the expressions obtained in the program "Universal Mechanism" (UM), a train model was compiled and computer modeling of the braking process was performed. Characteristics of the model are presented in Fig. 2.

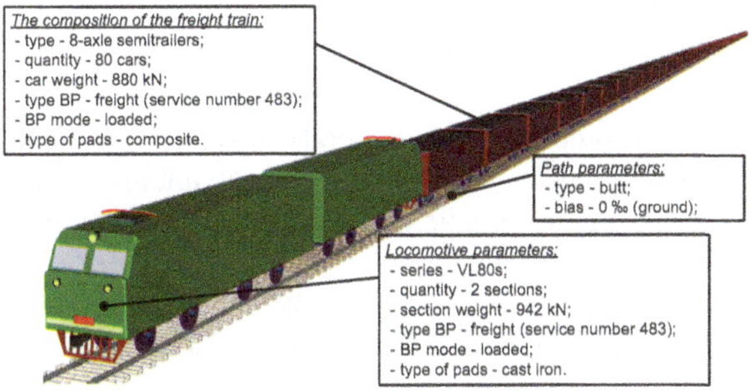

Fig. 2. The parameters of the computer model of the train.

When simulating the train braking process with an initial speed of 80 km/h and to a full stop, dependences of the force of K pads on the wheel [15], the braking force of B_{T} on the deceleration time are recorded (Figs. 3 and 4).

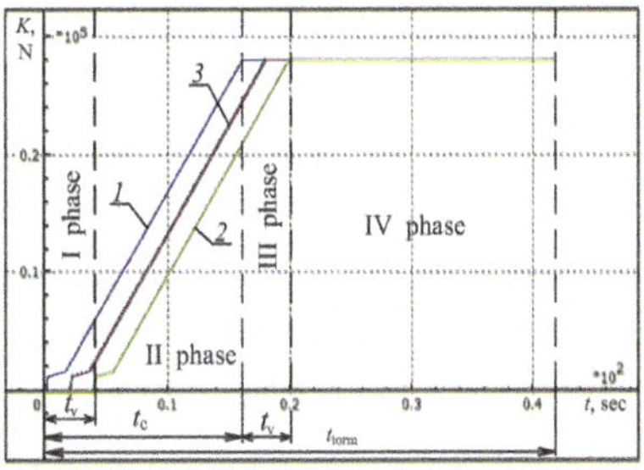

Fig. 3. The dependence of the force pressing the pads on the wheel from the braking time: *1* – for the head car; *2* – for the tail car; *3* – for the average car in the composition.

The following parameters of the braking process were obtained: the time of propagation of the brake wave tv = 4 s (modern air distributors No 483 of the cargo

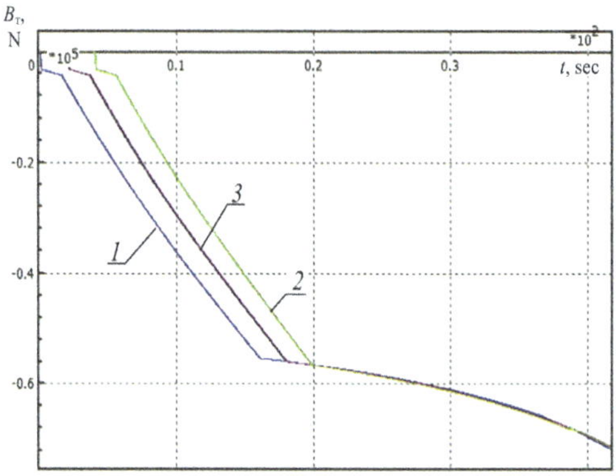

Fig. 4. Dependence of the braking force on the braking time: *1* – for the head car; *2* – for the tail car; *3* – for the average car in the composition.

type realize the speed of propagation of the brake wave v_v = 300 m/s and above), the time of filling the brake cylinder from zero to the maximum pressure t_c = 16 s, time from the moment of turning the handle of the driver's crane to a full stop of the train t_{torm} = 42 s.

Also in Figs. 5 and 6 shows the graphs of the change in speed, and the stopping distance from the braking time, obtained by computer simulation.

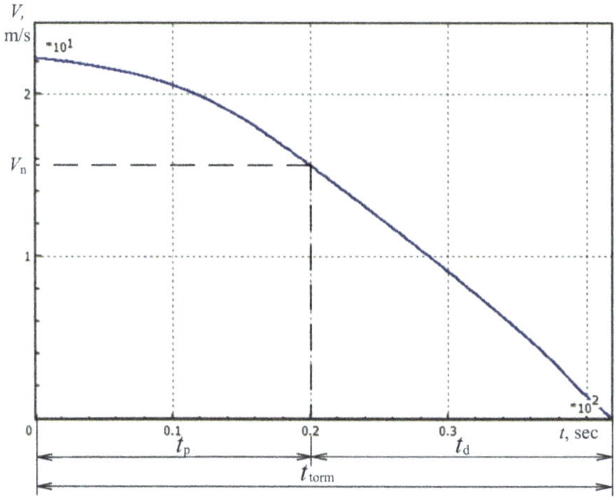

Fig. 5. Schedule of changes in speed during braking.

From Fig. 5 that under the given braking conditions, the speed with which the calculation of the actual stopping distance begins is V_n = 54 km/h.

In the end, with the initial conditions used and the characteristics of the braking system, the stopping distance of the train was determined by computer simulation,

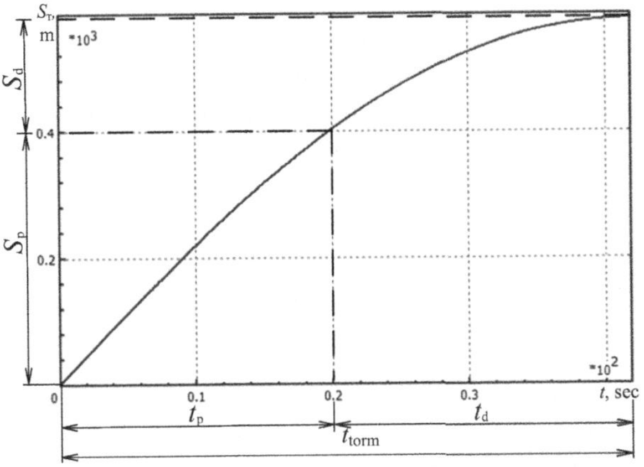

Fig. 6. The graph of braking distance from the braking time.

which amounted to S_T = 576 m. In this case, the preparation path of the brakes S_p = 400 m, and the actual stopping distance is S_d = 176 m.

Under similar initial conditions and parameters of the brake system, the calculation of the braking distance of a train was carried out using the method of integrating the equation of motion over the speed intervals (the method of calculation used), as well as the proposed method of calculation. For convenience of analysis, the calculation results are summarized in Table 1.

Table 1. The results of the calculation of the length of the braking path.

Options	Calculation methods		
	Computer modelling	Proposed method	Calculation of stopping distance for speed intervals
The speed at which the braking processes in the train go into the fourth phase of braking V_n, km/h	54	58	–
Preparatory braking distance S_p, m	400	384	266
Actual braking path S_d, m	176	205	258
General stopping distance S_T, m	576	589	524

4 Conclusions

As a result of the research, a calculation method was proposed that allows one to establish parameters for an unsteady inhibition process. In general, this method will allow more accurate calculations of the stopping distance of railway rolling stock, taking into account the following train characteristics:

- the length of the rolling stock;
- speed of propagation of the brake wave;
- feature of the indicator diagram of the brake cylinder and brake mode;
- the initial speed with which the calculation of the valid braking put begins.

References

1. Buynosov, A.P., Volkov, D.V., Dinislamov, A.R.: Transport Urals **1**(56), 26–29 (2018)
2. Buynosov, A.P., Fedorov, E.V.: Izvestiia Transsiba. J. Transsib Railway Stud. **1**(33), 13–22 (2018)
3. Buynosov, A.P., Fedorov, E.V.: Transport Sci. Equipment Manage. **8**, 7–11 (2018)
4. Buynosov, A.P., Fedorov, E.V.: Sci. Tech. Volga Reg. Bull. **9**, 39–41 (2018)
5. Buynosov, A.P., Volkov, D.V., Fedorov, E.V.: Transport Urals **4**(59), 77–80 (2018)
6. Paukert, H.: Eur. Railway Rev. J. E **2**, 3841 (2005)
7. Buynosov, A., Lapshin, V., Smolyaninov, A., Dinislamov, A.: Polytransport systems-2018. In: MATEC Web of Conferences, vol. 216, p. 02020 (2018)
8. Petrov, S.Y., Kokorin, R.V., Baban, S.M.: Tekhnologiya Mashinostroeniya. Technol. Mech. Eng. **4**, 37–44 (2017)
9. Pyndak, V.I., Novikov, A.E.: Metal Sci. Heat Treat. **58**(226230), 3–4 (2016)
10. Korotkov, V.A., Zlokazov, M.V.: J. Friction Wear **2**, 133136 (2014)
11. Muller, Ch.: Der Eisenbahningenier **6**, 50–53 (2017)
12. Osyaev, A.T., Isaev, I.P., Gorsky, A.V.: Bull. All-Union Sci. Res. Inst. Railway. Transport **6**, 31 (1991)
13. Trushnikov, D.N., Belinin, D.S., Shitsyn, Y.D.: Mod. Probl. Sci. Educ. **2**, 95 (2014)
14. Girzhon, V.V., Ovchinnikov, A.V.: Metal Sci. Heat Treat. **58**, 719723 (2017)
15. Ronkin, M.V., Kalmykov, A.A., Nagovicin, V.S., Buinosov, A.P.: REIT-autumn 2018. In: 4th International Workshop on Radio Electronics and Information Technologies, vol. 2274, pp. 52–61 (2018)

Assessing Remained Service Time of Contact-Line Support Under the Constant Load

Alexander Galkin, Alexey Kovalev, and Alexander Okunev

Ural State University of Railway Transport (USURT), Kolmogorov Str., 66, Ekaterinburg 620034, Russia
{AGalkin, AKovalev, AVOkunev}@usurt.ru

Abstract. Implementing various methods in order to forecast the fault-free state of reinforced concrete contact-line supports is currently a necessary condition for ensuring the safety of train traffic. Such a need arose due to the presence of supports with extended service life in operation, along with the absence of load reduction and without any kind of redundancy. The method for predicting the fault-free state of the contact-line supports is exposed in the article. The exploitation problem, according to which it is rather difficult to reveal the failure system of the considered object under the influence of many factors, needs to be solved. The probability of finding the supports of the contact-line in a certain state is expressed through their statistical properties, based on the change in the carrying capacity. The way to predict the operational status of contact-line supports is developed. Predicted values of carrying capacity reduction below the permissible value allow determining the targeted measures in order to exclude traffic accidents. The developed method based on statistical data allows thoroughly describing the reduction rate of the carrying capacity throughout the life cycle, while excluding distortion of indicators caused by environmental factors.

Keywords: Support · Contact-line · Life cycle · Service life · Railway transport

1 Introduction

The problem of operating contact-line supports with extended service life is not a random correlation, but only the consequence of mass electrification. The complexity of conducting diagnostic tests leads to various mistakes in the diagnosis when describing the operational status. Lack of redundancy complicates ensuring the safety of rolling stock. It is quite difficult to identify the failure system of the contact-line supports. Moreover, in spite of the fact that the influence of various factors, such as changes in atmospheric conditions, technogenic effects, and temperature effects on reinforced concrete contact-line supports is well studied, it is still random. This leads to an increase in instrumental errors of parameters determining the technical state. Using probabilistic approaches in conjunction with statistically accumulated information, it is possible to describe the processes of changing the technical characteristics of reinforced

Z. Popovic et al. (Eds.): TransSiberia 2019, AISC 1115, pp. 693–702, 2020.
https://doi.org/10.1007/978-3-030-37916-2_68

concrete contact-line supports. Therefore, it is necessary to consider the destructive processes leading to the destruction of concrete in more detail.

It is difficult to select parameters that will fully describe the destruction processes of the supports when predicting the health of the reinforced concrete contact-line supports. For example, such parameter as the resistance of the contact-line support, which allows determining the magnitude of current leakage and the electrical corrosion rate of metal reinforcement makes it impossible to predict the health of reinforced concrete support after several years. The reason is that the actual current leakage flow becomes more complicated due to the saturation of concrete with moisture [1].

It was decided to consider the load-bearing capacity in more detail after analyzing the parameters of the contact-line support, which are monitored by diagnostic devices. The load bearing capacity of the contact-line support is defined as strength with respect to the applied load. It can be concluded that such a parameter does not depend on environmental factors, but describes its technical condition. During operation, the load bearing capacity of the supports does not increase. Thus, it is possible to adopt that there is a gradual failure during the process of changing the bearing capacity of the contact-line supports throughout the service life, like one of a non-recoverable object. The tests of reinforced concrete contact-line supports at the site of the Sverdlovsk railway were conducted. Electrocorrosionaly hazardous supports are considered in more detail, due to the concern that the load bearing capacity reduction of such supports will proceed more intensively. It has been obtained that the minimum transfer interval of the load bearing capacity from a higher value to a lower one in the selected operation area occurs no more than once every six months [2, 3].

The table for determining the load bearing capacity according to the indications of the UK-1401 device (Table 1) is used based on Appendix 3 [3]. It can be seen that the indicator P2 increases, which characterizes the decrease in the bearing capacity of the supports in relation to the normative moment. Considering the fact that reinforced concrete supports, despite all their shortcomings, are reliable component of the contact network, and the decrease in performance is slow (can reach 70 years), then the results of measurements of electrocorrosive supports are accepted for the experiment. This choice is caused by the fact that, with a long declared life cycle of the contact network supports, the selected observation interval will not allow obtaining significant dynamics of the process of reducing the bearing capacity under consideration. The influence of additional destructive effects on the supports leads to an acceleration of the process of accumulation of internal damage due to the influence of electrical corrosion.

Table 1. Load bearing capacity of supports depending on the P2 indicator

	P2 indicator							
	1.1	1.15	1.2	1.25	1.3	1.35	1.4	1.45
The multiplicity with respect to the specified torque of supports	1.6	1.5	1.4	1.3	1.2	1.1	1	0.9
Load bearing capacity ratio according to state	7	6	5	4	3	2	1	0

The value of multiplicity with respect to the specified torque determines the intervals between the contact-line supports. Let us take the determined boundaries of the intervals as the boundaries of the measured load bearing capacity, according to which it is possible to quantify the distribution of contact-line supports in the selected area in the first and second half of the year.

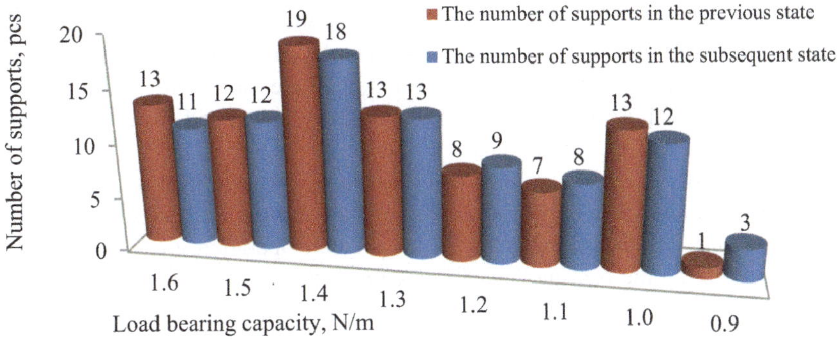

Fig. 1. The distribution histogram of the measurements obtained for the bearing capacity boundaries for the first and second halves of the year.

It can be seen from the histogram that the load bearing capacity of the considered supports changed in the second half of the year in comparison with the first half. Accordingly, the number of hits in the intervals was redistributed (Fig. 1).

2 Materials and Methods

Accepting the assumption that the supports under study in the selected section of the railway are in the same operating conditions, it becomes possible to establish common boundaries of the bearing capacity corresponding to each specific condition. A decrease in the bearing capacity with respect to the normative moment below 1.0 will correspond to the state of failure of the support or significantly increase the likelihood of such an event, and therefore, is unacceptable. In this regard, a column is added to Table 3 that reflects the failure state of the support of the contact network or the "0" state. The indicator P2, having a value of 1.1 or less, characterizes the greatest ability of the support to perceive the applied loads. As a result, the supports, having the indicator P2 equal to 1.1 or less, have the least probability of rejection of the applied loads, and it indicates a good "7" state of the support. States "1–6" will characterize the transition of the support from serviceable to the state of failure; accordingly, they can be assessed as pre-failure [2].

Unrecoverable objects are characterized by a gradual transition from a healthy state to a failure state, overcoming the boundaries of all pre-failure states. The intensity of the transition to each subsequent state ε_0–ε_6 reflects a gradual failure caused by the accumulation of damage in the structure of the support. In turn, sharp transitions from a fault-free and any pre-failure state to a failure state characterize a sudden failure, and the speed of such transitions is expressed in the failure rate λ_0–λ_6.

Accumulated damage in concrete allows concluding that the reinforced concrete support can be considered an unrecoverable object.

The determination of the observation interval plays an important role in solving the problem of forecasting the technical condition. Taking the intervals too long, frequent sudden transitions appear between the fault-free state and the states of pre-failure. In this case, an additional variable will participate in solving the system of equations of the described process, which will also complicate the calculation process.

If the observation interval is taken too small, then the failure will manifest itself in the form of a gradual decrease in the bearing capacity index, and the compiled model will not take into account sudden failures.

During the tests, it was determined that measurements performed once a quarter are optimal for the studied area of operation. Taking into account the accepted assumptions and designations, for the supports in the considered operation area, a graph of the process of changing the bearing capacity is compiled (Fig. 2):

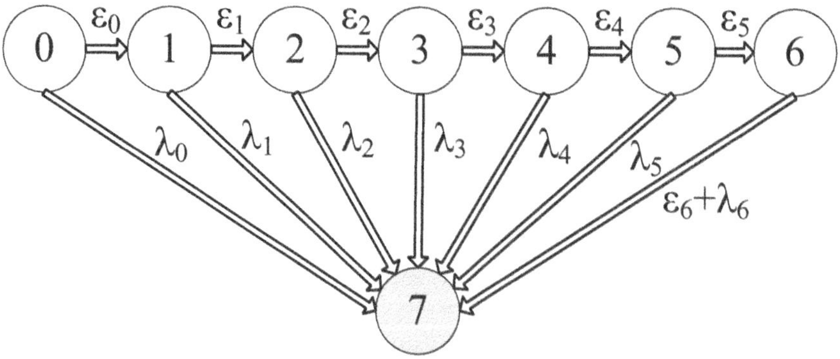

Fig. 2. State and transition graph for reinforced concrete contact-line support

Values of ε_0–ε_6 characterizing gradual failure reflect a gradual decrease in the load bearing capacity with the transition from the operating condition to each subsequent pre-failure condition until the values fall below the normative value (let us conditionally call this state a failure). Values of λ_0–λ_6 characterize a sharp transition from a healthy and pre-failure conditions to failure.

After composing a system of equations in order to define the support in each of the conditions:

$$\begin{cases} \frac{dP_0(t)}{dt} = -(\varepsilon_0 + \lambda_0)P_0(t) \\ \frac{dP_1(t)}{dt} = \varepsilon_0 P_0(t) - (\varepsilon_1 + \lambda_1)P_1(t) \\ \frac{dP_2(t)}{dt} = \varepsilon_1 P_1(t) - (\varepsilon_2 + \lambda_2)P_2(t) \\ \frac{dP_3(t)}{dt} = \varepsilon_2 P_2(t) - (\varepsilon_3 + \lambda_3)P_3(t) \\ \frac{dP_4(t)}{dt} = \varepsilon_3 P_3(t) - (\varepsilon_4 + \lambda_4)P_4(t) \\ \frac{dP_5(t)}{dt} = \varepsilon_4 P_4(t) - (\varepsilon_5 + \lambda_5)P_5(t) \\ \frac{dP_6(t)}{dt} = \varepsilon_5 P_5(t) - (\varepsilon_6 + \lambda_6)P_6(t) \\ \frac{dP_7(t)}{dt} = \lambda_0 P_0(t) + \lambda_1 P_1(t) + \lambda_2 P_2(t) + \lambda_3 P_3(t) + \lambda_4 P_4(t) + \lambda_5 P_5(t) + (\varepsilon_6 + \lambda_6)P_6(t) \end{cases} \quad (1)$$

Given the initial conditions: for $t_0 = 0$, $P_0 = 1$, we obtain the probability of the system being in the state P_0(t):

$$P_0(0) = e^{-K_0 t} \quad (2)$$

Substituting solution (20) into the second equation of system (19), we determine the probability of finding the support in the following state P_1(t):

$$P_1(t) = \frac{\varepsilon_0}{Z_1 - Z_0} e^{-Z_0 t} + \frac{\varepsilon_0}{Z_0 - Z_1} e^{-Z_1 t} \quad (3)$$

Then, in general terms, the probability of finding a contact network support in each of the states has the form:

$$P_n(t) = \frac{\varepsilon_0 \cdot \ldots \cdot \varepsilon_{n-1}}{(Z_1 - Z_0) \cdot \ldots \cdot (Z_{n-1} - Z_0)} \cdot e^{(-Z_0 t)} + \frac{\varepsilon_0 \cdot \ldots \cdot \varepsilon_{n-1}}{(Z_0 - Z_n) \cdot \ldots \cdot (Z_{n-1} - Z_n)} \cdot e^{(-Z_n t)} \quad (4)$$

where coefficients Z_i is determined by the formula:

$$Z_i = \varepsilon_i + \lambda_i \quad (5)$$

where i = 1, 2, 3…n - states of reinforced concrete support.

Over a three-year time interval, the values of gradual and sudden failures for each state are determined, after which the probabilities of finding the contact network supports in each of the pre-failure states are calculated. Excluding the probability of the support being in a state of failure, i.e. in the seventh state, the values of the reliability function are determined:

$$F(t) = P_0(t) + P_1(t) + P_2(t) + P_3(t) + P_4(t) + P_5(t) + P_6(t) \quad (6)$$

The obtained mathematical model allows describing the process of reducing the bearing capacity of the contact network support and determine the remaining service life.

Figure 3 visually depicts the process of reducing the reliability *F(t)* of the reinforced concrete supports of the contact network during operation in the studied area.

Since the supports in the selected area were already in operation and were highly susceptible to electrical corrosion, the dynamics of reducing the reliability of the supports in the area is most pronounced. Thus, Fig. 3 reflects the time scale of the service life of the contact network support. Based on local operating conditions, knowing the elapsed service life of the support and using the presented model, it is possible to determine the remaining service life [4].

According to the results of the obtained values, the dependence of the redistribution of the bearing capacity of the contact network supports from the region of the fault-free state to the region of the fault state was established.

It is possible to compose matrices according to the measurement results, where the values of bearing capacities obtained in the first half of the year characterize the previous state and ones obtained in the second half of the year characterize the next state.

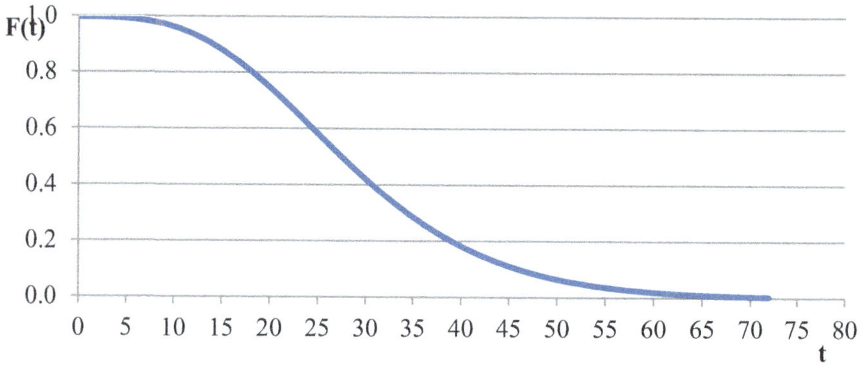

Fig. 3. The process of reducing the reliability of concrete contact supports during operation

Transition probability is the conditional probability of such an event in which the system moves from the previous state *(b)* to the subsequent state *(c)*. The probability of transition from state *(b)* to state *(c)* is determined by the formula (7):

$$p_{bc}(\tau) = \frac{n_{bc}(\tau)}{n_b(\tau - 1)}, \tag{7}$$

where $n_{bc}(\tau)$ is the number of contact-line supports, which are in state *(b)* at the time $(\tau - 1)$ and in state *(c)* at times τ, $n_b(\tau - 1)$ is the number of contact-line supports observed in state *(b)* in moment $(\tau - 1)$.

3 Results

Let us compose the matrix of transition probabilities and a vivid transition of the load bearing capacity of the contact-line supports in the first and second halves of the year (Table 2). The transition probability matrix of change in the load bearing capacity in decimal fractions is composed (Table 3).

The transition probability matrix P = {pbc} characterizes the probability of the transition from the current state to the next state, while the sum of the values of the probabilities of transitions from one state to another should be equal to one, and the process at time moment n is in state (c). The value of P_n can be obtained using formula (8):

$$P_n = P_1^n, \tag{8}$$

where P_n – transition probability matrix in n steps;

P_1 – transition probability matrix in one step.

To obtain the predicted value of the bearing capacity of the support for the third quarter of 2013, we raise the resulting matrix to the second degree.

In Table 3, the transition probabilities are rounded to two decimal places, so that their sum in the line equals one. The building of transition probability matrices confirms the stationarity of the process of reducing the bearing capacity of the supports of the contact network and the determination of the timing for reaching the limit values. When comparing the predicted values with the experimental values of the bearing capacity, the error did not exceed 3%.

According to the analyze of the resulting transition probability matrix, it can be concluded that the bearing capacity of the support does not transfer to the previous state;

Table 2. Matrix of quantitative redistribution of contact-line supports in the first and second halves of the year

Number of interval	Subsequent condition							
	1	2	3	4	5	6	7	8
Previous condition	1.6	1.5	1.4	1.3	1.2	1.1	1	0.9
1.6	11	2	0	0	0	0	0	0
1.5	0	10	2	0	0	0	0	0
1.4	0	0	16	3	0	0	0	0
1.3	0	0	0	10	3	0	0	0
1.2	0	0	0	0	6	2	0	0
1.1	0	0	0	0	0	6	1	0
1	0	0	0	0	0	0	11	2
0.9	0	0	0	0	0	0	0	1

Table 3. The transition probability matrix of the load bearing capacity of the contact-line supports from the first half of the year to the second

Previous condition	Subsequent condition							
	1.6	1.5	1.4	1.3	1.2	1.1	1	0.9
1.6	0.85	0.15	0	0	0	0	0	0
1.5	0	0.83	0.17	0	0	0	0	0
1.4	0	0	0.84	0.16	0	0	0	0
1.3	0	0	0	0.77	0.23	0	0	0
1.2	0	0	0	0	0.75	0.25	0	0
1.1	0	0	0	0	0	0.86	0.14	0
1	0	0	0	0	0	0	0.85	0.15
0.9	0	0	0	0	0	0	0	1.00

4 Discussion

The results of calculating the transition probability matrices confirm the possibility of predicting the further technical condition of the reinforced concrete supports of the contact network, thereby identifying a failure system that depends only on the technical condition of the supports without additional mechanical impact on them. In other words, the process of failure of the supports of the contact network can be visually represented in the form of two functions that change in time (Fig. 4).

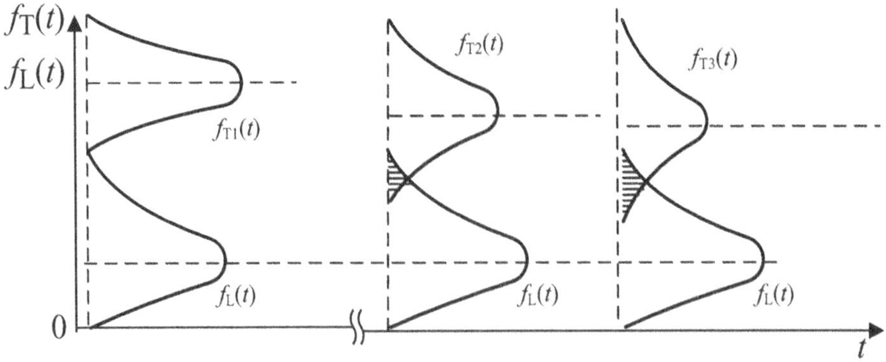

Fig. 4. Failure model of the contact-line support at constant load.

Figure 4 visually represents the process of reducing the strength under constant load, at the time point when the strength value is greater than the load, the graphs do not overlap. If at the initial moment of time the distribution of the values of the strength function of the contact network supports in the section is higher than the load function, then this characterizes the normal operation in the operation area and the absence of disturbance in the movement of trains [5, 6]. The intersection of the function graphs indicates the appearance of supports whose bearing capacity is lower than the permissible value, which increases the likelihood of failure in the operation area due to the inability to sustain the load applied to it. Over time, the bearing capacity of the support decreases due to the accumulation of microdamage in the protective layer of concrete and a decrease in the strength of metal reinforcement. This is evidenced by the increase in the area under intersecting graphs of functions.

5 Conclusions

When checking the possibility of forecasting on the basis of changes in the bearing capacity of reinforced concrete supports of the contact network, on the basis of experimental data obtained during field measurements, it was found that a decrease in the carrying capacity characterizes the wear of the contact network support as an object of operation.

Using Markov chains to simulate the process of changing the bearing capacity, it became possible to describe the process of failure of the contact network supports and depict the process of changing reliability, determining the probabilities of finding the contact network supports in each of the states at a certain point in time.

Since a large number of supports in the study area has been in operation for more than 35 years, it can be said from the calculation results that to eliminate their failure, it is necessary to carry out work on replacing the supports or reconstructing the site with replacing the supports within five years, since the probability of catching the supports in working condition is greatly reduced.

Applying the developed methodology and mathematical model in practice, it becomes possible to identify the failure system of reinforced concrete supports as an object that directly affects the safety of the transportation process. The process of assessing the remaining service life of the contact network supports to which a constant load is applied during operation is also simplified. Due to this, in the future, specialists will be able to assess the need to extend the service life or replace the contact network support.

References

1. Skorobogatov, S.M.: The principle of information entropy for the description of fracture at multiple dimensions in real materials (concrete, rock) of buildings and rock seams. Int. J. Fract. **1**, 335–343 (2014)
2. Kovalev, A.A., Kudriashov, P.I., Kiselnikov, A.I., Startsev, I.M.: Version control system of CAD documents and PLC projects. J. Phys. **1015**, 042020 (2018)
3. Galkin, A.G., Kovalev, A.A., Shaiukhov, T.T.: Integrated approach to the planning of energy consumption by non-traction railway consumers. EDP Sciences
4. Komyakov, A.A., Ponomarev, A.V., Komyakova, O.A.: Izvestiia Transsiba. J. Transsib Railway Stud. 1(5) (2011)
5. Kovalev, A.A., Nesmelov, F.S., Mikawa, A.V., Kardapolov, A.A., Isakov, N.: Transport Case Russia **4**, 24–26 (2013)

6. Skorobogatov, S.M.: Determination of durability of bending structures with the help of serviceability. In: Dhir, R.K., Newlands, M.D., Whyte, A. (eds.) Proceedings of the International Conference on Application of Codes, Design and Regulations 2005 International Congress - Global Construction: Ultimate Concrete Opportunities. Cep. “Application of Codes, Design and Regulations - Proceedings of the International Conference” sponsors: Institution of Civil Engineers, American Concrete Institute, Japan Society of Civil Engineers, University of Dundee, UK, University of Dundee, Concrete Technology Unit, Dundee, Scotland, pp. 529–540 (2005)

Mathematical Description of the Car's Movement on the Descent Part of the Hump

Khabibulla Turanov(✉), Elena Timukhina, and Andrey Gordienko

Ural State University of Railway Transport, Kolmogorova Street, 66, 620034 Yekaterinburg, Russia
turanov@yandex.ru

Abstract. Based on the application of a single impulse, a mathematical model has been developed for the movement of a car along the entire length of the descent part of the hump. The description of the dynamics of the car rolling from the hump, including sections of brake positions, in a generalized form was made for the first time. The change in the instantaneous speed of movement along the entire length of the descent part of the hump is presented in the form of a step function graph for the first time. The proposed model of a generalized mathematical notation of an instantaneous change in the speed of a car rolling down the descent part of the hump has a practical significance. This model allows calculating instantaneous values of the speeds of movement of the car from the top of the hump to the design point in a continuous mode, which makes it possible to accelerate the process of building graphs of changes in accelerations, speeds, and time of car's movement. The resulting model allows quickly analyzing the mode of shunting cars from the humps, the combination of power of brake positions and improve the accuracy of determining the permissible velocity of impact of cars in the sorting yards. This paper is the most important step for solving a promising task of designing an automated system for calculating the dynamic characteristics of a car in a hump yard.

Keywords: Railway · Station · Marshalling hump · Wagon · Heaviside discontinuous functions

1 Introduction

A series of publications are devoted to the problem of calculating and designing the profile of humps, for example [1–11, 12, 13]. The existing method of calculating the humps [4] is based on the use of the concept of "power of brake positions - h_{br} or brake tools h_{brt}". The speed of movement of the car on all sections of the hump, including brake positions, is determined in [1–6, 9] by the formula $v = \sqrt{2g'h}$ (where g'- the acceleration of a freely falling body taking into account the inertia of rotating parts, $h = h_{\text{h}}$ - the height of the fall, and for sections of brake positions: $h = h_{\text{br}}$ - the power of braking positions) (see p. 186 in [1]), applicable only for *ideal constraint* [14]. Thus, the fallacy of determining the energy height of the hump h_{h} is the use of the concept of an *ideal* constraint that is incompatible by the physical meaning to solving the problems of the humps, on which the constraints are in fact *non-ideal*.

Z. Popovic et al. (Eds.): TransSiberia 2019, AISC 1115, pp. 703–716, 2020.
https://doi.org/10.1007/978-3-030-37916-2_69

In [13], it was noted that in [2, 3], the same method was used to determine the energy height of the hump h_h. They made the assumption that at any point on the inclined plane, the energy of a rolling body of mass M is equal to the sum of the potential E_p and the kinetic energy E_c. It is assumed that this energy is spent on the work of the forces of resistance to the motion A_r, i.e. $E_c + E_p = A_r$ (see p. 8 in [2] and formula (6) in [3]). From here determine the energy height of the hump h_h. However, this approach is contrary to the law of energy conservation [14]. As can be seen from studies [1–5, 8], nowadays, the design of the energy height of the hump h_h is performed using the concept of "power of brake positions h_{br} or brake tools h_{brt}". The power of brake positions h_{br} is chosen according to the method [4], in which the free-fall velocity of bodies is used $v = \sqrt{2g'h}$, although $h = h_{br}$ is the height of the braking zone of the brake section of the hump, which is an unknown value and should be determined.

In [10], in order to take real account of the operational conditions of the humps, it is recommended to use the parameters of specific resistances to the movement w, which reflect the generalized characteristics of the modern car fleet and the hump yards. Taking into account this factor, formula (2) is given in [10], which, according to the authors of the paper [10], has a developed universal form:

$$v_f^2 = v_{in}^2 + 2g'(i - w)10^{-3} \cdot l - 2g'h_b. \tag{1}$$

Where

$v_f = v_e$ – final or estimated speed of the car in the corresponding section of the hump, depending on the normalized value $[v_{av}]$ (see Table 4.7 in [4]), m/s;

$v_{in} = v_{or}$ – initial speed or speed of rolling the car from the top of the hump, depending on the power of the hump (see Table 4.6 in [4]), m/s;

g' – acceleration of a freely falling body, taking into account the inertia of rotating parts, m/s^2;

i – slope of the studied section of the track, ‰;

w – specific resistance to the movement of the car in the studied sections of the track, kgf/t;

l – length of the studied section of the track, m;

h_b – height of a zone of braking of a brake section of a hump, m

However, as the results of studies showed in [9, 12], formula (1) contains a number of inaccuracies and gross errors in its components. Although in [10], it is noted without substantiated evidence supported by calculations that formulas (1) and (2) in [10] can be used on any sections of the humps with a slope i, taking into account the presence of certain values of specific resistance to movement w and power of braking positions hbr (i.e. heights of sections of brake positions) (see the first paragraph of the last column on page 36 in [10]). In the opinion of the authors of the paper [10], the calculations of the humps, which simulate the conditions of movement of the designed runners with different running properties, are performed from this expression (see the first paragraph of the last column on page 36 in [10]). Also, in [10], it is indicated that "... any new proposed designed models for the movement of cars" should be compared with formulas (1) and (2) in [10] (see the second paragraph of the last column on page 36 in [10]).

Let us reveal some of the disadvantages of the formula (1). So, for example, it, as noted in [11], contains two completely disparate mathematical expressions describing the movement of a car in various parts of the hump, where the minuend values are valid for non-ideal constraints, and the subtrahend - for an ideal plane (constraint). Such an approach, as noted in [11], contradicts the elementary principles for solving engineering problems of theoretical mechanics: first, in order to simplify, it is necessary to solve the problem either for an ideal constraint (of course, not taking into account the inertia of rotating masses in the sections of brake positions, since when hump retarders are triggered, there is no rolling, but pure sliding of the wheelset), which is of no scientific and practical value; secondly, or for a non-ideal constraint, which is of scientific and practical interest.

Otherwise, the fallacy of the mathematical notation of formula (1) (see formula (2) in [10]) is that it cannot be given a universal form, "mechanically" combining the minuend and the subtrahend.

Thus, it was revealed that the question of exact mathematical modeling of the movement of a car on the descent part of the humps is still relevant.

Objective:

describe the speed of movement of the car at each section of the descent part of the hump, using the principles of classical mechanics;

get a generalized model of the speed of movement of the car at the estimated point of the hump, which allows speeding up the process of building graphs of the change of the kinematic characteristics of the car's movement;

present in the form of a graph of the step function the change in the instantaneous speed of movement of the car along the entire length of the descent part of the hump.

2 Research Methods

Research methods are based on the basic law of the dynamics of a point with a non-ideal constraint (d'Almbert principle) [15] and include the following stages:

- at the first stage, the acceleration of the car's movement in all sections of the profile of the hump as the most important kinematic parameter of the car's movement, the magnitude of which directly determines the other movement parameters (time, speed and path of the studied sections), determined on the basis of the d'Alembert principle [6, 7, 11, 13–15], assuming that all active (gravitational force of a car with a load) and reactive (friction, resistance to the wind, resistance to movement from curves and switches, etc.) forces as forces in fractions of the gravitational force of the car with a load are computable values;
- at the second stage, the speed of movement of the car in each section of the hump profile [4] was found using well-known physics formulas [7, 11, 13] based on the fact that the entering speeds of the car for each section and the acceleration of movement in these sections are known;

- at the third stage, at the edges of the sections of the hump, the formulas of the instantaneous speed of movement of the car are interconnected by the method of matching ("stitching"), which is known in mechanics [13, 15–17];
- at the fourth stage, the change in the instantaneous speed of movement on the descent part of the hump is presented in the form of a step function graph [16, 18, 19];
- at the fifth stage, a generalized mathematical notation of the change in the instantaneous speed of a car rolling down the discharge section of the hump is presented in a compact form [16, 18, 19].

3 Research Results

The formulas of the instantaneous speeds of the car on each section of the hump, according to the simplified method adopted in [6, 7, 11, 13, 17], are written in the form convenient for their calculation. At the same time, these formulas of the instantaneous speed of movement of the car at the edges of the sections can be interconnected by the matching method known in mechanics.

So, for example, let us present the formulas of the speed of the car for each i section of the hump (where i = 1,…,9 are the numbers of the sections of the hump) in the form:

- for the first high-speed section of the hump

$$v_{\mathrm{f1}}^2 = v_{\mathrm{in}}^2 + a_1 t_1 \le [v_{\mathrm{av2f}}]; \tag{2}$$

- for the second high-speed section of the hump to the switch

$$v_{\mathrm{f2}}^2 = v_{\mathrm{f1}}^2 + a_2 t_2 \le [v_{\mathrm{av2}S}]; \tag{3}$$

- for the second high-speed section of the hump after the switch

$$v_{\mathrm{f2S}}^2 = v_{\mathrm{f2}}^2 + a_{2S} t_{2S} \le [v_{\mathrm{av1br}}]; \tag{4}$$

- for the section of the first brake position

$$v_{\mathrm{1br}}^2 = v_{\mathrm{2f}}^2 - a_{\mathrm{1br}} t_{\mathrm{1br}} \le [v_{\mathrm{av4}}]; \tag{5}$$

- for the intermediate section of the hump to the switch

$$v_{\mathrm{f4}}^2 = v_{\mathrm{1br}}^2 + a_4 t_4 \le [v_{\mathrm{av4}S}]; \tag{6}$$

- for the intermediate section of the hump after the switch

$$v_{\mathrm{f4S}}^2 = v_{\mathrm{f4}}^2 + a_{4S} t_{4S} \le [v_{\mathrm{av2br}}]; \tag{7}$$

– for the section of the second brake position

$$v_{2br}^2 = v_{f4S}^2 - a_{2br}t_{2br} \leq [v_{av6}]; \tag{8}$$

– for the switching area of the hump to the first switch

$$v_{f6}^2 = v_{2br}^2 + a_6t_6 \leq [v_{av6S1}]; \tag{9}$$

– for the switching area of the hump after the first switch

$$v_{f6S1}^2 = v_{f6}^2 + a_{av6S1}t_{6S1} \leq [v_{av6S2}]; \tag{10}$$

– for the switching area of the hump to the second switch

$$v_{f6S2}^2 = v_{f6S1}^2 + a_{6S2}t_{6S2} \leq [v_{av6S3}]; \tag{11}$$

– for the switching area of the hump to the third switch

$$v_{f6S3}^2 = v_{f6S2}^2 + a_{6S3}t_{6S3} \leq [v_{av3br}]; \tag{12}$$

– for the first section of the marshalling yard track

$$v_{f7}^2 = v_{f6S3}^2 + a_7t_7 \leq [v_{av3br}]; \tag{13}$$

– for the section of the third brake position

$$v_{f3br}^2 = v_{f7}^2 - a_{3br}t_{3br} \leq [v_{av9}]; \tag{14}$$

– for the second section of the marshalling yard track

$$v_{f9}^2 = v_{f3br}^2 + a_9t_9 \leq [v_{avEP}]. \tag{15}$$

where

$v_{in} = v_{or}$ – as in the formula (1), the initial speed or speed of rolling the car from the top of the hump, depending on the power of the hump (see Table 4.6 in [4]), m/s;

v_{ei} – the estimated (or instantaneous) speed of the car in the corresponding i section of the hump, m/s;

a_i – the acceleration of the car's movement in the corresponding i section of the hump, calculated value [7, 11, 13, 17] (see, for example, formula (10) in [6] for the high-speed sections of the hump, for sections of brake positions - (6) in [11]), m/s;

t_i– the time of movement of the car in the corresponding i section of the hump, calculated in the high-speed sections of the hump according to the formula of elementary physics (see formula (11) in [7], and in sections of brake positions - according to formula (11) in [11]), s;

$[v_i]$ – permissible entering speed of the car to the studied section of the hump (see Table 4.7 in [4]), m/s.

Note that in (2)–(4), (6), (7), (9)–(13), (15) the acceleration of the car a_i is calculated by the formula (10) in [7], and in (5), (6), and (14) - according to the formula (6) in [11], while according to the methodology of works [1–5, 8, 10], there is even no mention of this.

The existing method for calculating humps (see formula (2) in [10]) and method proposed by the authors [7, 11, 13, 17] have distinctive features. Each of them, naturally, has some assumptions, as it is usually in mathematics and mechanics.

Instead of formulas (2)–(15), let us present a generalized view of the mathematical model of the simplified method of the authors' hump calculations [7, 11, 13], considering that the first, second, and third sections of brake positions (1BP, 2BP, and 3BP), for example, can be divided into three conditional sections (the zone of the car's wheelbase entrance (WB), the braking zone of the car (BZ) up to the stop, and the remaining section (AB) after braking, corresponding to the brake position of the car retarder), and representing the movement of the car along the descent of the hump in the form of a Heaviside unit impulse (and/or jump) $f(t)$ or $\sigma_0(t)$ [16, 18, 19].

It should be borne in mind that the task of determining the time t_{br} and the path l_{br} of braking in the braking zone of the car (BZ) until it stops ($v_{br} = 0$) was solved in [11] for the first time.

Considering that the car can be braked, *firstly*, directly when, for example, the front wheelset enters brake positions and/or, *secondly*, when the wheelset of the front carriage enters brake positions, the braking zone can also consist of two sections - the car's braking zone (BZ) (v_{br} = const) up to the stop ($v_{br} = 0$) and the remaining section (AB) after braking (v_{AB} = const).

In addition, in the first and second brake positions (1BP and 2BP), where, as a rule, two car retarders are installed, the braking zone can consist of five sections (car wheelbase entry area (WB) (v_{WB} = const), zone of the first braking of the car (1BZ) (v_{1br} = const) up to the stop ($v_{br} = 0$), and the remaining Section (1AB) after braking (v_{1AB} = const), the zone of the second braking of the car (2BZ) (v_{2br} = const.) up to the stop ($v_{2br} = 0$), and the remaining Section (2AB) after braking (v_{2AB} = const). Also, the braking zone may consist of four sections - zones of the first braking (1BZ) (v_{1br} = const) up to the stop ($v_{1br} = 0$) and the remaining Section (1AB) after braking (v_{1AB} = const), the area of the second braking of the car (2BZ) (v_{2br} = const.) up to the stop ($v_{2br} = 0$), and the remaining area (2AB) after braking (v_{2AB} = const).

In sections of the third braking position (3BP), in which one car retarder is usually installed, the braking zone can consist of two sections: the car braking zone (BZ) (v_{3BP} = const) up to the stop ($v_{3BP} = 0$) and the remaining section (AB) after braking (v_{AB} = const).

The change in the instantaneous speed of movement along the entire length of the descent part of the hump can be represented as a graph of the step function. The graph of the unit impulse $f(t)$ in the form of a step function can be represented as an example in Fig. 1.

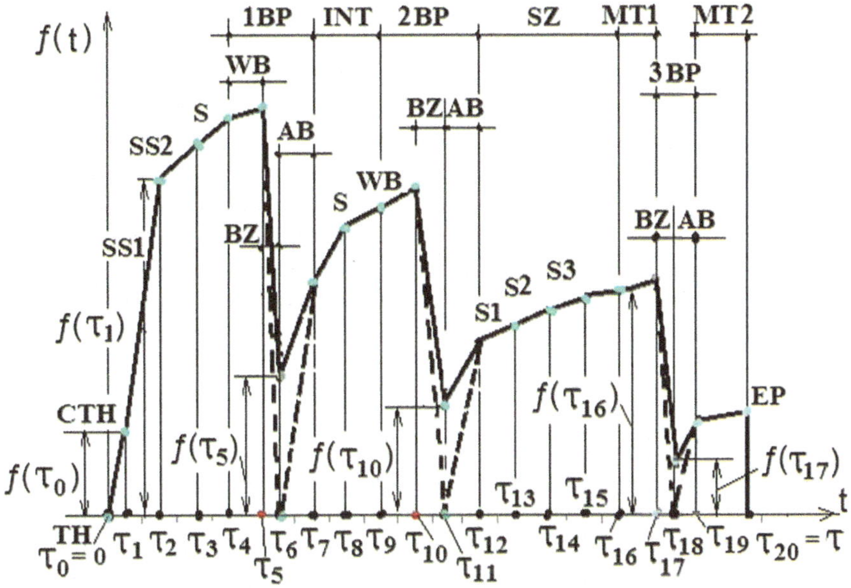

Fig. 1. Graph of the step function representing the change in acceleration and speed of movement along the entire length of the descent part of the hump

In Fig. 1 as well as in Table 1, Fig. 1–3 in [13], the following is denoted:

TH – top of the hump;
CTH – top of the hump, represented by the unit impulse $f(\tau_0)$;
SS1 and SS2 – the first and second speed sections of the hump, represented by unit impulses $f(\tau_1)$ and $f(\tau_2)$;
1BP, 2BP, and 3BP – the first, second, and third brake positions of the hump, represented by unit impulses $f(\tau_4) - f(\tau_6)$, $f(\tau_9) - f(\tau_{11})$, $f(\tau_{16})$, and $f(\tau_{17})$;
INT – intermediate section of the hump, represented by unit impulses $f(\tau_8)$ and $f(\tau_9)$;
SZ – switching zone of the hump, represented by unit impulses $f(\tau_{13}) - f(\tau_{15})$;
MT1 and MT2 – the first and second sections of the marshalling yard track, represented by unit impulses $f(\tau_{16})$ and $f(\tau_{19})$;
S – separation switches represented by unit impulses $f(\tau_3)$ and $f(\tau_7)$;
S1, S2, and S3 – the first, second, and third switches represented by unit impulses $f(\tau_{12})$, $f(\tau_{13})$, and $f(\tau_{14})$;
WB – sections of accounting for the length of the car's wheelbase, represented by unit impulses $f(\tau_4)$ and $f(\tau_9)$;
BZ – car's braking zones represented by unit impulses $f(\tau_5)$, $f(\tau_{10})$, and $f(\tau_{17})$;
AB – remaining parts of the braking positions, represented by unit impulses $f(\tau_6)$, $f(\tau_{11})$ and $f(\tau_{18})$, and corresponding to the braking position of the car retarder;
EP – estimated point represented by the unit impulse $f(\tau_{19})$, with the exception of t - time of the car's movement; τ_j - fixed numbers (j = 1,…, 19 - numbers of sections of the track profile of the hump)

Besides, the dash-dotted lines in the zones of direct braking (BZ) of the car in the brake positions (1BP, 2BP, and 3BP) correspond to the uniformly decelerated motion of the car (with acceleration $|a_{1br}| = - a_{1br} = - a_5 < 0$, $|a_{10br}| = - a_{2br} = - a_{10} < 0$, and $|a_{17br}| = - a_{3br} = - a_{17} < 0$) and with full use of the power of brake positions, when it is necessary to achieve a complete stop of the car, i.e. in cases when $v_{1br} = v_{2br} = v_{3br} = 0$, and solid lines - to a partial braking of the car when the car moves with acceleration $|a_{1br}| = a_5 < 0$, $|a_{10br}| = a_{10} < 0$, and $|a_{17br}| = a_{17} < 0$ and with the speed $v_{1br} = v_{1BZ} = \text{const} > 0$, $v_{2br} = v_{2BZ} = \text{const} > 0$, and $v_{3br} = v_{5BZ} = \text{const} > 0$.

In this case, we use the fact that the unit impulse (and/or jump) $f(t)$ [18] can be considered as the difference between two unit and/or discontinuous Heaviside functions [13, 18, 19]:

$$\boldsymbol{\sigma}_0(t) \text{ and } \boldsymbol{\sigma}_0(t-\tau), \text{ i.e.} f(t) = \boldsymbol{\sigma}_0(t) - \boldsymbol{\sigma}_0(t-\tau),$$

and $\tau = \tau_1$, $\tau = \tau_2 > \tau_1$; $\tau = \tau_3 > \tau_2, \ldots\ldots, \tau = \tau_{19} > \tau_{18}$.

Also, we will keep in mind that when $t < 0$: $f(t) = 0$ - the origin of coordinates;

in the time interval $0 \leq t \leq \tau_1$: $f(t) = f(\tau_0)$ – top of the hump (TH);

in the time interval $\tau_1 \leq t \leq \tau_2$: $f(t) = f(\tau_1)$ – first speed section (SS1);

in the time interval $\tau_2 \leq t \leq \tau_3$: $f(t) = f(\tau_2)$ – second speed section (SS2);

in the time interval $\tau_2 \leq t \leq \tau_3$: $f(t) = f(\tau_3)$ – separation switch (S), …;

in the time interval $\tau_5 \leq t \leq \tau_6$: $f(t) = f(\tau_5)$ – braking zone of the first brake position (1BP), ….;

in the time interval $\tau_8 \leq t \leq \tau_9$: $f(t) = f(\tau_8)$ – intermediate section (INT), ….;

in the time interval $\tau_{10} \leq t \leq \tau_{11}$: $f(t) = f(\tau_{10})$ – braking zone of the second brake position (2BP), ….;

in the time interval $\tau_{17} \leq t \leq \tau_{18}$: $f(t) = f(\tau_{17})$ – braking zone of the third brake position (3BP);

in the time interval $\tau_{18} \leq t \leq \tau_{19}$: $f(t) = f(\tau_{18})$ – remaining section (AB) of the third brake position (3BP);

in the time interval $\tau_{19} \leq t \leq \tau_{20}$: $f(t) = f(\tau_{19})$ – second section of the marshalling yard track

and when $t > \tau_{20}$: $f(t) = 0$ – estimated point (EP).

In addition, we note that unit impulses импульсы $f(\tau_0)$, $f(\tau_1)$, …., $f(\tau_4)$, $f(\tau_6)$, …., $f(\tau_9)$, $f(\tau_{11})$, $f(\tau_{16})$, $f(\tau_{18})$, and $f(\tau_{19})$ characterize the uniformly accelerated motion of the car with acceleration $a_k > 0$ (where $k = 0, 1, \ldots, 4, 6, \ldots, 9, 11, \ldots, 16, 18$, and 19), at which it is accelerated in time intervals $\tau_0 = 0 \leq t \leq \tau_1$, $\tau_1 \leq t \leq \tau_2$, …, $\tau_3 \leq t \leq \tau_4$, $\tau_6 \leq t \leq \tau_7$, …, $\tau_9 \leq t \leq \tau_{10}$, $\tau_{11} \leq t \leq \tau_{12}$, …, $\tau_{16} \leq t \leq \tau_{17}$, $\tau_{18} \leq t \leq \tau_{19}$, and $\tau_{19} \leq t \leq \tau_{20}$.

Unit impulses $f(\tau_5)$, $f(\tau_{10})$, and $f(\tau_{17})$ characterize the uniformly decelerated motion of the car with acceleration $|a_j| < 0$ (where $j = 5, 10$, and 17), at which it is braked in the time interval $\tau_5 \leq t \leq \tau_6$, $\tau_{10} \leq t \leq \tau_{11}$, and $\tau_{17} \leq t \leq \tau_{18}$.

Below we show the mathematical notations of the instantaneous speed of the car in the most characteristic sections of the descent part of the humps.

Let's assume that, for example, the speed of car rolling $v_{in}(t) = f(t)$ before the time moment $t = 0$ was equal to zero ($v_{in}(t) = f(t) = 0$), and then, in the time interval $\tau_0 = 0 \leq t \leq \tau_1$, it took the value $v_{in}(t) = f(t) = \text{const}$, and, starting from the moment

$t = \tau_1$, the car starts its motion uniformly accelerated with acceleration $a_1 > 0$, picking up speed $v_{SS1} = \text{const} > 0$ (see the initial part of Fig. 1).

Mathematically, this can be written in the following form [13, 16, 18, 19]:

$$v_{in}(t) = f(t) = \begin{cases} 0 & \text{when } t = \tau_0 = 0, \\ f(t) = f(\tau_0) = v_{in} = \text{const} & \text{when } \tau_0 = 0 \le t \le \tau_1, \\ f(\tau_1) = v_{SS1} & \text{when } t > \tau_1. \end{cases} \quad (16)$$

Let us give a mathematical notation of the change in the instantaneous speed of the car v_{SS1} in the first speed section (SS1) of the hump (see Fig. 1). The car enters this section with the initial speed $v_{in} = \text{const}$, and then, in the time interval (τ_1, τ_2), it takes the value $f(t) = f(\tau_1) = v_{SS1} = \text{const} > 0$, moving *uniformly accelerated* with acceleration $a_2 > 0$, and, starting from the moment $t = \tau_2$, the car continues its movement equally accelerated with acceleration $a_3 > 0$, picking up speed $v_S = \text{const} > 0$. In this case, the mathematical notation has the following form [13, 16, 18, 19]:

$$v_{fSS1}(t) = f(t) = \begin{cases} f(\tau_1) = v_{in} & \text{when } t < \tau_1, \\ f(t) = f(\tau_1) = v_{SS1} = \text{const} & \text{when } \tau_1 \le t \le \tau_2, \\ f(\tau_2) = v_S & \text{when } t > \tau_2. \end{cases} \quad (17)$$

Let us describe the change in the instantaneous speed of the car using the example of the first brake position (1BP), bearing in mind that this section consists of three zones: the wheelbase zone of the car (WB), the braking zone (BZ), and the remaining section after braking (AB) (see 1BP section in Fig. 1).

The speed of the car entrance to WB v_{en4WB} is equal to the speed of exit from the section of the separation switch (S) $v_{exS} = \text{const} > 0$, and in the time interval $\tau_4 \le t \le \tau_5$, the car moves uniformly accelerated with acceleration $a_4 > 0$ and speed $f(t) = f(\tau_4) = v_{4WB} = \text{const} > 0$.

Further, a car retarder is switched on for braking a car that moves with an initial speed $f(t) = f(\tau_5) = v_{in1br} = v_{in5WB} = \text{const} > 0$, after which it moves *uniformly decelerated* with acceleration $a_{1br} = a_5 = \text{const} < 0$ (where $|a_5| = -a_5$) and with the speed $f(t) = f(\tau_5) = v_{1br} = v_{5BZ} = \text{const} > 0$ if the full power of the car retarder was not used, otherwise, the equality $v_5 = v_{5BZ} = 0$ should be observed.

In Fig. 1, the case when the condition $v_{5br} = v_{5BZ} = \text{const} > 0$ is met is shown by a linearly decreasing solid line, and the case when the equality $v_{5br} = v_{5BZ} = 0$ is observed is represented by a linearly increasing dash-dotted line.

Further, the car moves along the remaining length of the section after braking (AB) (see Sect. 1BP in Fig. 1).

The speed of the car entrance to the AB v_{en6AB} is equal to the speed of its exit from the BZ, i.e. $v_{en6AB} = v_{exBZ} = \text{const} > 0$, and in the time interval $\tau_6 \le t \le \tau_7$, the car moves uniformly accelerated with acceleration $a_6 = \text{const} > 0$ and speed $f(t) = f(\tau_6) = v_{6AB} = \text{const} > 0$.

In these cases, the mathematical notation of the instantaneous speeds of the car is as follows:

– when the car moves in the section equal to the wheelbase of the car (WB)

$$v_{\mathrm{fWB}}(t) = f(t) = \begin{cases} f(\tau_4) = v_S & \text{when} \quad t < \tau_4, \\ f(t) = f(\tau_4) = \text{const} & \text{when} \quad \tau_4 = 0 \le t \le \tau_5, \\ f(\tau_5) = v_{1\mathrm{br}} & \text{when} \quad t > \tau_5; \end{cases} \tag{18}$$

– when the car is braked (BZ) in the case of incomplete use of the power of the brake positions (is represented by a solid line in Fig. 1)

$$v_{\mathrm{f1br}}(t) = f(t) = \begin{cases} f(\tau_5) = v_{\mathrm{WB}} & \text{when} \quad t < \tau_5, \\ f(t) = f(\tau_5) = v_{1\mathrm{br}} = \text{const} & \text{when} \quad \tau_5 \le t \le \tau_6, \\ f(\tau_6) = v_6 & \text{when} \quad t > \tau_6; \end{cases} \tag{19}$$

– when braking the car (BZ) in case of full use of the power of the brake positions (it is represented by the dash-dotted line in Fig. 1)

$$v_{\mathrm{f1br}}(t) = f(t) = \begin{cases} f(\tau_5) = v_{\mathrm{WB}} & \text{when} \quad t < \tau_5, \\ f(t) = f(\tau_5) = 0 & \text{when} \quad \tau_5 \le t \le \tau_6, \\ f(\tau_6) = v_6 & \text{when} \quad t > \tau_6; \end{cases} \tag{20}$$

– when the car moves along the remaining length of the brake positions (AB)

$$v_{\mathrm{fAB}}(t) = f(t) = \begin{cases} f(\tau_6) = v_6 & \text{when} \quad t < \tau_6, \\ f(t) = f(\tau_6) = \text{const} & \text{when} \quad \tau_6 \le t \le \tau_7, \\ f(\tau_7) = v_S & \text{when} \quad t > \tau_7. \end{cases} \tag{21}$$

Similarly, it is possible to write down the instantaneous speed of movement of the car in other sections of the hump.

Let us also describe the change in the instantaneous speed of the car along the second section of the marshalling yard track (MT2) (see the last section of Fig. 1). The speed of the car entrance to this section of the hump is equal to the exit speed from the AB section of the third brake position (3BP) $v_{3\mathrm{AB}} = v_{\mathrm{en}19} = \text{const} > 0$, in case if the full power of the car retarder was not used, otherwise, the entrance speed of the car would be $v_{\mathrm{en}19} = 0$.

In Fig. 1, cases in which the conditions $v_{\mathrm{en}19} = v_{3\mathrm{AB}} = \text{const} > 0$ and $v_{\mathrm{en}19} = v_{3\mathrm{AB}} = 0$ are met are shown by linearly increasing solid lines and dash-dotted lines, respectively.

In the time interval $\tau_{19} \le t \le \tau_{20}$, the car moves uniformly accelerated with acceleration $a_{20} > 0$ and with speed скоростью $f(t) = f(\tau_{19}) = v_{\mathrm{MT2}} = \text{const} > 0$, and, starting from the moment $t = \tau_{20}$, it stops its movement after estimated point (EP).

A generalized mathematical notation of the simplified method of the authors' hump calculations [8, 12, 14], corresponding, in a particular case, to the graph of impulse functions in Fig. 1, which characterizes the change in the instantaneous speed of rolling (and in the deceleration sections - sliding speed) of the car along the descent part of the hump $v(t)$, unlike the unsuccessfully and incorrectly presented expanded universal form of formula (2) in [10], we present in the form of a graph of a step function [16, 18, 19]:

$$
\begin{aligned}
v_{\mathrm{f}}^2 &= v_{\mathrm{in}}^2(\sigma_0(t) - \sigma_0(t-\tau_0)) + a_1 t_{1\mathrm{SS}1}(\sigma_0(t) - \sigma_0(t-\tau_0-\tau_1)) \\
&+ a_{2\mathrm{SS}2} t_{2\mathrm{SS}2}(\sigma_0(t) - \sigma_0(t-\tau_0-\tau_1-\tau_2)) \\
&+ a_{3\mathrm{S}} t_{3\mathrm{S}}(\sigma_0(t) - \sigma_0(t-\tau_0-\tau_1-\tau_2-\tau_3)) \\
&+ a_4 t_4(\sigma_0(t) - \sigma_0(t-\tau_0-\tau_1-\tau_2-\tau_3-\tau_4)) \\
&- a_{1\mathrm{br}} t_{1\mathrm{br}}(\sigma_0(t) - \sigma_0(t-\tau_0-\tau_1-\tau_2-\tau_3-\tau_4-\tau_5)) \\
&+ a_6 t_6(\sigma_0(t) - \sigma_0(t-\tau_0-\tau_1-\tau_2-\tau_3-\tau_4-\tau_5-\tau_6)) \\
&+ a_{7\mathrm{S}} t_{7\mathrm{S}}(\sigma_0(t) - \sigma_0(t-\tau_0-\tau_1-\tau_2-\tau_3-\tau_4-\tau_5-\tau_6-\tau_7)) \\
&+ a_8 t_8(\sigma_0(t) - \sigma_0(t-\tau_0-\tau_1-\tau_2-\tau_3-\tau_4-\tau_5-\tau_6 \\
&- \tau_7-\tau_8)) + a_9 t_9(\sigma_0(t) - \sigma_0(t-\tau_0-\tau_1-\tau_2-\tau_3-\tau_4 \\
&- \tau_5-\tau_6-\tau_7-\tau_8-\tau_9)) - a_{2\mathrm{br}} t_{2\mathrm{br}}(\sigma_0(t) - \sigma_0(t-\tau_1-\tau_2-\tau_3 \\
&- \tau_4-\tau_5-\tau_6-\tau_7-\tau_8-\tau_9-\tau_{10})) + a_{11} t_{11}(\sigma_0(t) - \sigma_0(t \\
&- \tau_1-\tau_2- -\tau_3-\tau_4-\tau_5-\tau_6-\tau_7-\tau_8-\tau_9-\tau_{10}-\tau_{11})) \\
&+ a_{12\mathrm{S}1} t_{12\mathrm{S}1}(\sigma_0(t) - \sigma_0(t-\tau_1-\tau_2-\tau_3-\tau_4-\tau_5-\tau_6 \\
&- \tau_7-\tau_8-\tau_9-\tau_{10}-\tau_{11}-\tau_{12})) + a_{13\mathrm{S}2} t_{13\mathrm{S}2}(\sigma_0(t) - \sigma_0(t-\tau_1 \\
&- \tau_2-\tau_3-\tau_4-\tau_5-\tau_6-\tau_7-\tau_8-\tau_9-\tau_{10}-\tau_{11}-\tau_{12}-\tau_{13})) \\
&+ a_{14\mathrm{S}3} t_{14\mathrm{S}3}(\sigma_0(t) - \sigma_0(t-\tau_1-\tau_2-\tau_3-\tau_4-\tau_5-\tau_6-\tau_7-\tau_8 \\
&- \tau_9-\tau_{10}-\tau_{11}-\tau_{12}-\tau_{13}-\tau_{14})) + a_{15} t_{15}(\sigma_0(t) - \sigma_0(t-\tau_1 \\
&- \tau_2-\tau_3-\tau_4-\tau_5-\tau_6-\tau_7-\tau_8-\tau_9-\tau_{10}-\tau_{11}-\tau_{12}-\tau_{13} \\
&- \tau_{14}-\tau_{15})) + a_{16} t_{16}(\sigma_0(t) - \sigma_0(t-\tau_1-\tau_2-\tau_3-\tau_4 \\
&- \tau_5-\tau_6-\tau_7-\tau_8-\tau_9-\tau_{10}-\tau_{11}-\tau_{12}-\tau_{13}-\tau_{14}-\tau_{15}-\tau_{16})) \\
&- a_{3\mathrm{br}} t_{3\mathrm{br}}(\sigma_0(t) - \sigma_0(t-\tau_1-\tau_2-\tau_3-\tau_4-\tau_5-\tau_6-\tau_7-\tau_8 \\
&- \tau_9-\tau_{10}-\tau_{11}-\tau_{12}-\tau_{13}-\tau_{14}-\tau_{15}-\tau_{16}-\tau_{17}) \\
&+ a_{18} t_{18}(\sigma_0(t) - \sigma_0(t-\tau_1-\tau_2-\tau_3-\tau_4-\tau_5-\tau_6-\tau_7-\tau_8 \\
&- \tau_9-\tau_{10}-\tau_{11}-\tau_{12}-\tau_{13}-\tau_{14}-\tau_{15}-\tau_{16}-\tau_{17}-\tau_{18})) \\
&+ a_{19} t_{19}(\sigma_0(t) - \sigma_0(t-\tau_1-\tau_2-\tau_3-\tau_4-\tau_5-\tau_6-\tau_7-\tau_8 \\
&- \tau_9-\tau_{10}-\tau_{11}-\tau_{12}-\tau_{13}-\tau_{14}-\tau_{15}-\tau_{16}-\tau_{17}-\tau_{18}-\tau_{19})).
\end{aligned}
\tag{22}
$$

and/or let us present a generalized mathematical notation of the change in the instantaneous speed of a car rolling down the descent part of the hump in a more compact form [18, 19]:

$$\begin{aligned} v_{\mathrm{f}}^2 = {} & v_{\mathrm{in}}^2 f(\tau_0) + a_1 t_{1\mathrm{SS1}} f(\tau_1) + a_{2\mathrm{SS2}} t_{2\mathrm{SS2}} f(\tau_2) + a_{3\mathrm{S}} t_{3\mathrm{S}} f(\tau_3) \\ & + a_4 t_4 f(\tau_4) - a_{1\mathrm{br}} t_{1\mathrm{br}} f(\tau_5) + a_6 t_6 f(\tau_6) + a_{7\mathrm{S}} t_{7\mathrm{S}} f(\tau_7) \\ & + a_8 t_8 f(\tau_8) + a_9 t_9 f(\tau_9) - a_{2\mathrm{br}} t_{2\mathrm{br}} f(\tau_{10}) + a_{11} t_{11} f(\tau_{11}) \\ & + a_{12\mathrm{S1}} t_{12\mathrm{S1}} f(\tau_{12}) + a_{13\mathrm{S2}} t_{13\mathrm{S2}} f(\tau_{13}) + a_{14\mathrm{S3}} t_{14\mathrm{S3}} f(\tau_{14}) \\ & + a_{15} t_{15} f(\tau_{15}) + a_{16} t_{16} f(\tau_{16}) - a_{3\mathrm{br}} t_{3\mathrm{br}} f(\tau_{17}) + a_{19} t_{19} f(\tau_{18}). \end{aligned} \tag{23}$$

It should be noted that the mathematical notation of the instantaneous speeds of the car (22) and/or (23) corresponds to the case when the car is moving relative to the top of the hump (TH) uniformly accelerated with the set speed of rolling (v_{in} = const > 0, for example, v_{in} = 0.8, …, 1.7 m/s, depending on the hump capacity (see Table 4.6 in [4]).

In this case, the time interval $\Delta\tau_0 = \tau_1 - \tau_0$ is very small and can be considered almost equal to zero (see Fig. 1). The case is also considered when the car moves uniformly decelerated ($v_{\mathrm{br}k}$ = const $\leq$ 0) in the i sections of brake positions (1BP, 2BP, and 3BP) with the turned on car retarder in the braking zones.

Here, the time intervals $\Delta\tau_8 = \tau_9 - \tau_8$, $\Delta\tau_{10} = \tau_{11} - \tau_{10}$, and $\Delta\tau_{17} = \tau_{18} - \tau_{17}$ are very small (for example, from 1 to 3 s) and can be considered almost equal to zero, since for a negligibly small period of time, the loaded car picks up speed $v_{\mathrm{br}i}$ = const and also quickly stops to $v_{\mathrm{br}i} = 0$, continuing to pick up speed $v_{i\mathrm{AB}}$ = const in the remaining sections (AB) of the brake positions (see Fig. 1).

On the other speed sections of the hump (SS1, SS2, S, WB, AB, INT, BZ, MT1, and MT2), the calculated runner moves uniformly accelerated with average speeds $v_{\mathrm{av}i}$ = const not exceeding the established average speeds of the car $[v_{\mathrm{av}}]$ depending on the hump capacity (see Table 4.7 in [4]) for a sufficiently noticeable period of time, for example, $\Delta\tau_2 = \tau_3 - \tau_2 >> 1$ s.

Also note that in the mathematical notation of the instantaneous speeds of the car (23), as well as in formulas (2)–(4), (6), (7), (9)–(13), (15), the acceleration of the car a_k, according to the d'Alembert principle, is calculated by formula (10) in [7].

The time of movement of the car t_i is found according to the formula of elementary physics (see formula (11) in [7]) from the dependence $t_i = f(v_{0i}, |a_i|, l_i)$ (where l_i is the length of the section under study) in the i-th section of the descent part of the hump, except parts of the brake position. Usually, in the zones of braking, the full power of the brake positions is used, ensuring full stop of the car. Therefore, the time of movement $t_{\mathrm{br}i}$ and the path of braking $l_{\mathrm{br}i}$ of the braked car are found from the condition that the braking speed is zero, i.e. $v_{\mathrm{br}i} = 0$ (see formulas (10) and (11) in [11]).

As can be seen, the mathematical notation (22) and/or (23) of the instantaneous speeds of a car of the simplified method of the authors' hump calculations [7, 11, 13], presented in a generalized form, has a significant difference from the expanded universal form of formula (2) in [10], which has significant inaccuracies.

4 Discussion

Thus, on the basis of the conducted studies, we especially note the following results:

1. The formulas for the instantaneous speeds of the car for each section of the hump yard are presented in a convenient form for practical use.
2. The change in the speed of movement of the car on the entire profile of the descent part of the hump is presented in the form of a step function graph.
3. Using Heaviside unit functions in a compact, simplified form, a generalized mathematical notation of the change in the instantaneous speeds of the car rolling down the descent part of the hump is presented.

The presented paper summarizes the results of previously published papers (see, for example, [7, 11, 13]).

The proposed model of a generalized mathematical notation of an instantaneous change in the speed of a car rolling down the descent part of the hump enriches the theory of rolling a car along the descent part of the hump. This model is of practical importance, since it allows calculating instantaneous values of speeds of the car's movement from the top of the hump to the estimated point in a continuous mode, allowing us to speed up the process of plotting the graph of changes in accelerations, speeds, and time of car's movement (see, for example, [17]). The presented model allows quickly analyzing the mode of rolling cars from the humps, the combination of capacity of brake positions, and improving the accuracy of determining the permissible rates of collision of cars in the hump yards. This paper is the most important step for solving a promising task of designing an automated system for calculating the dynamic characteristics of a car in the hump yard.

References

1. http://ousar.lib.okayama_u.ac.jp/file/15404/Mem_Fac_Eng_OU_27_2_41.pdf. Accessed 10 Dec 2018
2. Apatsev, V.I., Efimenko, Yu.I.: Railway stations and Junctions: manual (FSBEI) «Railway transport educational-methodical center», Moscow (2014)
3. Turanov, Kh.: Analytical investigation of wagon speed and traversed distance during wagon hump rolling under the impact of gravity forces and head wind. Global J. Res. Eng.: Mech. Mech. Eng. **14**(1), 1–9 (2014). Version 1.0, New York
4. Turanov, K.T., Gordienko, A.A.: Calculation of time of movement and speed of a car on the intermediate section of the hump yard under tail wind. World Transp. **14**(4(65)), 86–91 (2016)
5. Rudanovsky, V.M., Starshov, I.P., Kobzev, V.A.: On an attempt to criticize the theoretical positions of the dynamics of rolling the car down the slope of the hump. Transp. Inf. Bull. (6 (252)), 19–28 (2016). ISSN 2072-8115
6. Turanov, Kh.T.: On the attempt to prove a new approach to the study of the movement of the car on the descent part of the hump. Transp. Inf. Bull. **10**(256), 19–24 (2016). ISSN 2072-8115

7. Pozoisky, Yu.O., Kobzev, V.A., Starshov, I.P., Rudanovsky, V.M.: On the question of the movement of the car on the slope of the railway track. Transp. Inf. Bull. **2**(272), 35–38 (2018). ISSN 2072-8115
8. Turanov, Kh.T., Gordienko, A.A.: Mathematical description of the movement of the car in sections of brake positions of the hump. Transp. Urals (2(57)), 3–8 (2018). https://doi.org/10.20291/1815-9400-2018-2-3-8. ISSN 1815-9400
9. Turanov, Kh.T., Gordienko, A.A.: A critical analysis of the theoretical positions of the movement of the car from the hump (Part II). Transp. Inf. Bull. (12(282)), 12–18 (2018). ISSN 2072-8115
10. Khabibulla, T., Andrey, G.: Movement of a railway car rolling down a classification hump with a tailwind. In: MATEC Web of Conferences, vol. 216, p. 02027 (2018)
11. Turanov, Kh.T., Gordienko. A.A.: The certificate of official registration of computer software programs RU № 2017614017, 05 April 2017

Unmanned Aerial Vehicles as a Supporting Tool of Classic Land Surveying in Hard-to-Reach Areas

M. Mrówczyńska[1](✉), B. Grzelak[2], and J. Sztubecki[3]

[1] University of Zielona Góra, Faculty of Civil Engineering, Architecture and Environmental Engineering, ul. Prof. Z. Szafrana 1, 65-516 Zielona Góra, Poland
m.mrowczynska@ib.uz.zgora.pl

[2] ABG Limited Liability Company, ul. Kożuchowska 20C, 65-001 Zielona Góra, Poland

[3] UTP University of Science and Technology, Faculty of Civil and Environmental Engineering and Architecture, ul. Prof. S. Kaliskiego 7, 85-796 Bydgoszcz, Poland

Abstract. The paper presents issues related to the use of unmanned aerial vehicles (UAV) to monitor natural forest resources and to estimate the post-fire surfaces located in forests, as an example of hard-to-reach areas for classic surveying surveys. The use of drones as a tool for geodetic measurements is becoming more and more popular, however, processing data acquired in this way requires proper analysis. The article proposes a procedure for processing images provided by drones to estimate the effects of natural disasters and exemplifies it with experimental data obtained in a forest area located in western Poland (Europe). The presented methodology concerned the automatic identification of post-fire areas in natural forest resources and verification of the results by comparison with the results of classical geodetic tachymetric measurements. The results prove the usefulness of images acquired by UAVs for identifying and determining the size of areas struck by natural disasters. At the same time, data from UAVs can be a reference base for geodetic measurements and a database for institutions managing endangered areas.

Keywords: Engineering photogrammetry · Unmanned aerial vehicle · Natural resource monitoring

1 Introduction

The use of unmanned aerial vehicles (drones, gyroplanes) for geodetic surveys began in the first years of the 21st century. At that time, they were primarily a source of spatial images. At present, unmanned aerial vehicles are used to provide both images and spatial data with adequate accuracy. The use of drones for measuring and monitoring spatial phenomena reduces the time of measurements and increases the possibility of visualizing phenomena in comparison to traditional measurement methods. Moreover, unmanned aerial vehicles can be used to perform measurements in difficult conditions and to measure highly detailed objects.

Z. Popovic et al. (Eds.): TransSiberia 2019, AISC 1115, pp. 717–729, 2020.
https://doi.org/10.1007/978-3-030-37916-2_70

At present, drones are more and more frequently used in geodesy and spatial planning for work carried out by contractors and design companies as well as for statutory tasks of state and local government institutions. The latter use images provided by drones for: creating orthophotomaps with an accuracy of 2–3 cm for implementation of statutory tasks and construction of numerical terrain models (an accuracy of about 5 cm) [1], monitoring coastal zones (including, for example, tracking construction activities, checking the condition of erosion protection and determining the condition of the sea ecosystem) [2], checking the technical condition of hydrotechnical equipment and structures (e.g. flood embankments), updating information on agricultural and forest areas [3–6], checking property tax declarations and determining road traffic intensity [7].

In the safety assessment of various objects, it is necessary to properly combine measuring techniques, calculations and experiments. Technical assessment and modeling of the object's behavior allow for a more comprehensive analysis [8–10]. In the land surveying drones are also used to perform measurements for updating databases of large-scale maps, to provide information for other thematic databases [11, 12], as a source of auxiliary information for property evaluation, to create maps for design purposes and to construct numerical terrain models and numerical models of water reservoir and watercourse beds.

In construction and spatial planning low-altitude photogrammetric images provided by drones are used for monitoring during the construction and operation of engineering structures, making records of hard-to-reach surfaces and cubic objects, e.g. for making records of roofs (including asbestos roofs), cooling towers, tower buildings, dumps, storage sites and landfills [13] as well as for records of natural objects, caves and protected areas. They provide information used in the process of sustainable planning and spatial management in such areas [14, 15]. The images are also used for comprehensive monitoring and precise assessment of the general character and shape of landslides and areas affected by such phenomena, making records of areas and objects of cultural and architectural heritage, recording the dendrological status of parks or forest areas, creating orthomaps and models for open cast mines, which use them to obtain useful information for monitoring the status of the deposits, storage piles of bulk materials as well as for volume measurements.

The purpose of the paper is to present the possibility of using unmanned aerial vehicles as a tool aiding classical geodetic measurements in areas where measurements are difficult to perform.. An example of this problem are identified post-fire and fire areas. It is possible to use UAV, remote sensing and image processing in fire monitoring and detection systems as well as in tactical firefighting operations [16]. Yuan, Liu, Zhang [17] proposed to use remote sensing images recorded by the camera on board an unmanned aircraft and fuzzy logic to detect and classify smoke in forest areas. Wardihani et al. [18] has developed a real-time forest fire monitoring system using UAV equipped with five sensors: temperature, infrared, barometer, GPS and inertia measurement sensor. The platform they use can detect surface temperatures from a height of 20 m, and send data to the server to be displayed on the fire information system website. The results obtained by them also show that the coordinates of the reference points are characterized by an average error of about 1 m from the actual coordinates of the fire. Ma et al. [19], who used an unmanned aircraft to receive images, also had interesting results, but complemented the procedure for a new image filtering algorithm, based on principal component analysis (PCA) and the low-rank matrix recovery (LRMR) method, allowing early and effective smoke detection. This approach is particularly important due to the fact that smoke appears earlier than the flames. This allows earlier identification of fireplaces. The authors [20, 21] point out that the presented solution using UAV has enormous potential and it is worth continuing work on its development. At the same time, they emphasize that by comparing the results with classic methods, the proposed systems using UAV introduce a more accurate fire detection indicator and a smaller false detection rate.

In our work, we present issues related to the identification of hard to reach areas, where the fire was extinguished, in correlation with land surveying solution, in order to determine its extent and estimate losses. As a case study, an area located in western Poland (Europe) was analysed, where the work to be done included geodetic and photogrammetric measurements with the use of drones. It was also emphasized that experience in piloting a drone and knowledge of the process of generating and processing the acquired information were necessary to create high resolution spatial images. The paper concludes with a short discussion of the use of UAVs as a tool for collecting data in hard-to-reach areas and suggests directions for further research.

2 Methods

Photogrammetric methods are used to recognize and determine the location of objects that are directly registered in aerial photographs, and in the case of UAVs, which are used in this study, low-altitude aerial photographs. In most cases, the final result is a numerical map describing the spatial location of objects in a particular area in 2D or 3D [22]. The basic factors that determine the accuracy of the final product are [23, 24]: errors in the identification of points in the area, the scale and quality of photogrammetric images, the precision of the instrument, the method of data processing and correction of systematic errors.

The popularity of unmanned aerial vehicles as a tool used in land surveying has increased significantly over the last decade. Software and hardware also provide an increasing number of applications and technologies. In land surveying, we often encounter problems related to situational and altitude measurements in difficult and inaccessible terrain, and measurements in natural disaster zones (e.g. in post-fire areas) are also dangerous to perform and time-consuming. A good solution is to use low-altitude photogrammetric images. Certainly, the use of modern technologies requires proper planning and data processing so that the results are accurate enough to deal with the complexity of the problem. Considering the need to obtain a $\pm$ 10 cm accuracy of measurements performed with the use of unmanned aerial vehicles while estimating the surface of post-fire areas, a diagram of work to be done has been prepared and shown in Fig. 1.

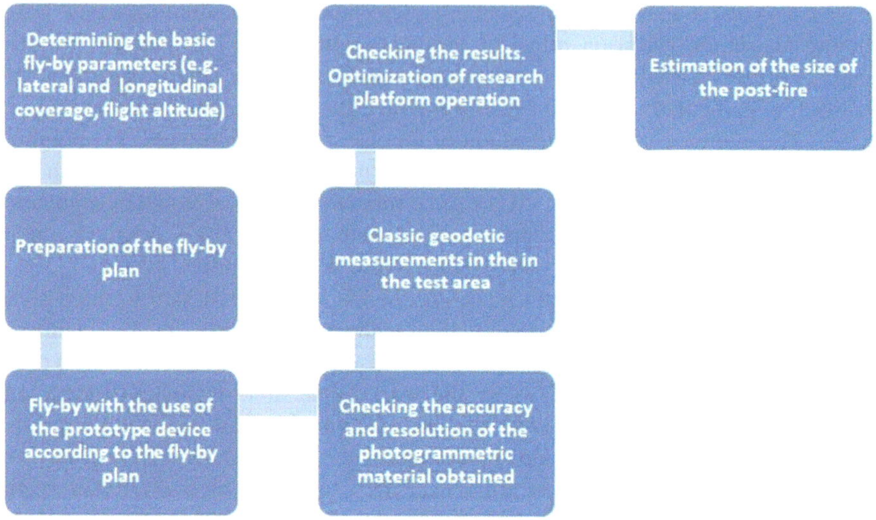

Fig. 1. A block diagram of the measurement procedure for the post-fire areas with the use of a prototype device

Based on available technology and a constructed research platform (Fig. 2, Table 1) consisting of a Matrice 200 DJI drone and an X4S video camera, a case study was prepared and researched to show that the technology based on low-altitude photogrammetry meets the accuracy requirements and can be used as a tool to identify post-fire areas. The study was conducted in a test (pilot) area, located in the Forest Administration District of Przytok (Europe, Poland), which was struck by fire in 2018. This Forest Administration District is interested in using low-altitude photogrammetry for estimating areas struck by fire and post-fire damage in forests.

Fig. 2. The Matrice 200 DJI drone and the X4S camera [picture provided by the producers]

Table 1. Abridged technical characteristics of the drone and the camera [data provided by the producers]

UAV		Camera	
Matrice 200 DJI		X4S	
Dimensions	887 × 880 × 378 mm	Lens	F/2.8-11, 8.8 mm (equivalent of 35 mm: 24 mm format)
Flight duration	38 min.	Photo resolution	3:2, 5472 × 3648 4:3, 4864 × 3648 16:9, 5472 × 3078
Max. signal range	7 km (3.5 km in EU)	Matrix	CMOS, 1″
Max. climb rate	5 m/s	Weight	253 g
Max. take-off weight	6.14 kg	Gimbal control range	Tilt +30° do −90°; Pan: ±320°
Battery	TB50 - LiPo 4280 mAh		

The data obtained from the research platform were processed using a modern, commercial Pix4D programme for processing photogrammetric data. A wide range of internal modules makes it possible to fully automatically combine both terrestrial and aerial photos and to fully position images with the use of georeference modules. Pix4D software makes it possible to obtain: clouds of points, numerical models of terrain cover, filtered numerical terrain models and aerial orthophotomaps. Pix4D software can use data provided by the drone, computer visualizations, image analysis, and machine training to analyse existing situations on objects and to make decisions.

The approach to measurements based on the use of unmanned aerial vehicles was verified by performing classic geodetic measurements (tachometric measurements and GNSS measurements). The tachometric measurements were performed with a Trimble S7 total station, which combines scanning, imaging and measuring in one device, and its use in the research made it possible to verify the results obtained with the use of drones. The measurement data obtained with the use of the Trimble S7 total station were processed with the use of CAD Trimble Business Center (TBC) geodetic software. As part of the research work, GNSS measurements were performed using a Trimble R2 receiver to determine the horizontal coordinates and the height of the reference points. A Trimble R2 receiver tracks a number of GNSS satellite constellations and auxiliary systems. Being inherently equipped with a Trimble Maxwell ™ 6 chip with 220 channels, it provides the highest precision and positioning accuracy. The receiver ensures the highest accuracy in real time thanks to the flexibility of the selection of correction sources - from traditional RTK and VRS networks to Trimble RTX ™ services, which provide access to updates via satellite and through the internet.

3 Case Study

The chapter presents a detailed case study based on photogrammetric images obtained to confirm and demonstrate the practicality of methods for recording, processing and identifying post-fire areas in forests with the use of UAVs. The following subchapters present a description of the whole experiment and analytical process.

3.1 Estimating Fire Damage in Forest Areas

Estimating the amount of fire damage in forests is very difficult but necessary to assess, for example, the economic effectiveness of expenses on fire prevention and fire-fighting. Information about the amount of damage should be passed on to the public and this applies in particular to the lost non-economic functions of the forest. The formal basis for estimating losses resulting from premature felling (loss) of trees is the act on the protection of agricultural and forest land [25]. The methodology for estimating fire damage in forests in Poland is specified in the Ordinance by the Minister of the Environment on single compensation for premature felling of trees [26]. The method presented in this ordinance does not take into account losses resulting from the reduction of non-economic forest functions, losses occurring on the rims of fire zone or losses caused by greenhouse gas emissions and the reduction of the carbon dioxide absorption area.

Poland is regarded as an area of medium fire risk, for example during 12 years at the turn of the 20th and 21st centuries 44.5 million tons of CO_2 were emitted into the atmosphere as a result of fire. Moreover, approximately 492 thousand tons of carbon monoxide was emitted as well as 188 thousand tons of solid and liquid particles (fumes), 47 thousand tons of hydrocarbons and about 9 thousand tons of nitrogen oxides. The calculated average value of 1 ha of forest in Poland is approximately €8300. The value of forests in the areas struck by fire only in 2005 and 2006 amounted to €60 million and €61 million, respectively. The amount of direct losses was estimated at €13.4 million and €9.3 million, which was 23% and 15% respectively of the forest value. Moreover, the amount of losses caused by fire in undeveloped areas in those years was estimated at €13.2 million and €5.4 million [27]. All these factors indicate that there is a need to create a method for identifying post-fire areas, which will make it possible to estimate losses with minimum human involvement. The approach based on the use of UAVs described in this paper may be such a method.

3.2 Research in the Pilot Area

The test area was located in western Poland (Fig. 3). In 2018 the pilot area was struck by fire, which spread over an area of 0.86 ha. The location of the test area in a forest may cause difficulties in performing geodetic measurements (trees, covered horizon, no geodetic network and firebreaks). Foresters performing measurements in a classical way may be in danger, and in some cases there may be a threat to human health and life. This type of location and the existing threat determines the accuracy of identification of the area, its size and the location of control points, the use of drones is also preferable. In order to verify the data provided by UAVs, classical geodetic measurements were performed using a Trimble S7 total station and a GNNS Trimble R2 receiver.

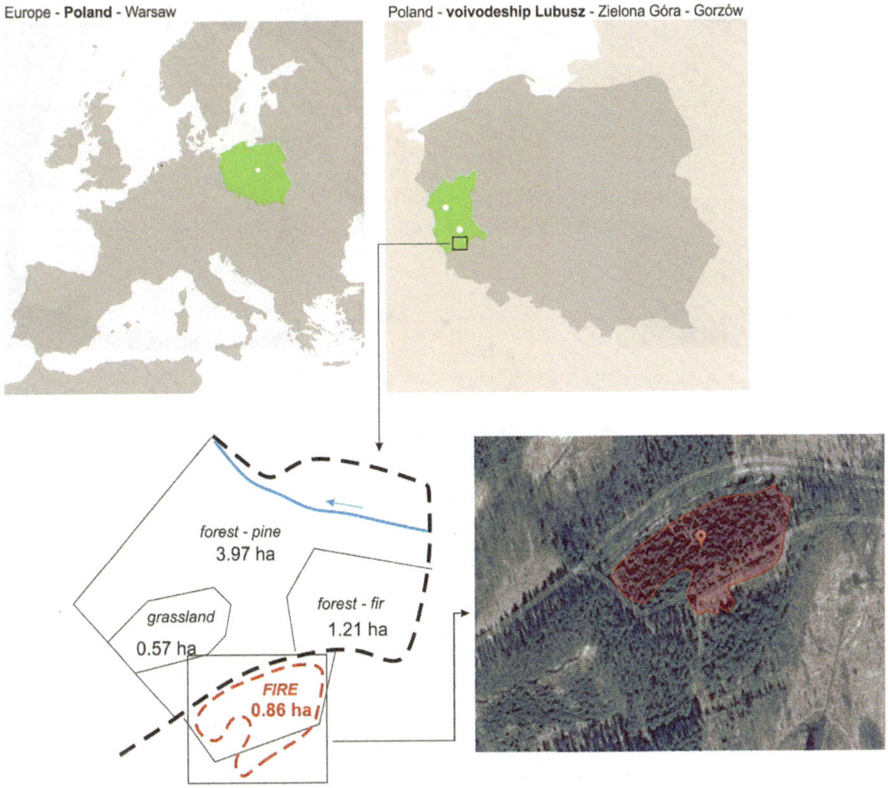

Fig. 3. The location of the test area struck by fire

In order to correctly acquire photogrammetric data and process it properly, the following actions were taken:

- photogrammetric missions were planned with regard to the UAV range, data related to the size of the matrix and lens parameters, the desired size of the field pixel (5 cm) and longitudinal coverage (75%) and transverse coverage (65%);
- a photogrammetric network was established (photo-points) - 12 points within the pilot area; the network has been signaled in a way allowing its unambiguous identification in the area and in the photographs and measured with GPS technology;
- a photogrammetric flight was made using a research platform;
- in the next step, using Pix4D software, the photogrammetric data obtained with the UAV application was analyzed and computer visualizations were made.

Based on the photogrammetric data, a numerical land cover model was created, the location of the post-fire area was identified and the size of the post-fire area was calculated (Fig. 4), which can be used to estimate the loss of trees in the pilot area.

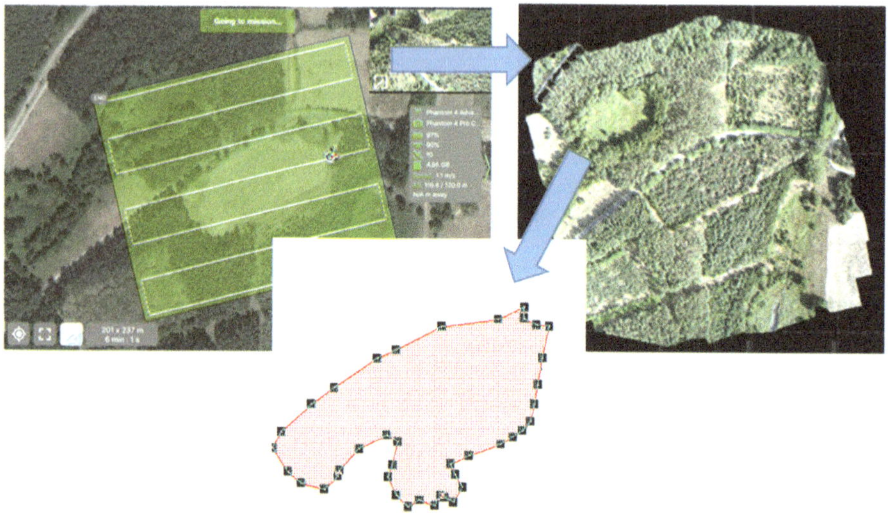

Fig. 4. A fly-by plan, a numerical model of the land cover and calculation of the size of the post-fire area based on photogrammetric images

Based on the research, the size of the pilot area was estimated, which, depending on the estimation method, was: measurements with the use of UAVs −0.6915 ha, tachometric measurements −0.6988 ha and measurements carried out by foresters −0.6500 ha. It can be seen that the results obtained with classical geodetic methods and the ones obtained with the use of drones are similar, which means that the measurements were performed correctly. At the same time, it should be noted that these results are much more accurate than the results of the measurement performed by the foresters. Moreover, the values of plane coordinates and the height of points on the surface of the post-fire area obtained with classical methods and with the use of UAVs were analysed. Maximum differences i.e. the point location error was ±0.23 m, and ±0.46 m for height. These are completely satisfactory accuracy in the identification of forest emergency areas and several times greater than those obtained by foresters using traditional methods used in the economy and forest management [27]. The mean square error was also determined according to the formula [28]

$$RMSE = \sqrt{\frac{\sum_{i=1}^{n} v_i^2}{n}} \tag{1}$$

where v – the difference between the coordinates obtained from geodetic measurements and the coordinates obtained from aerial images, n – the number of control points located in the pilot area. For the horizontal location of points this error was: $RMSE_P$ = ±0.09 m, and for the vertical location it was: $RMSE_H$ = ±0,14 m.

A detailed distribution of the values of the point location errors, height errors and errors at selected control points is shown in the diagrams in Fig. 5. Figure 6 shows a numerical model of the land surface obtained from the research and the height errors in the area under analysis.

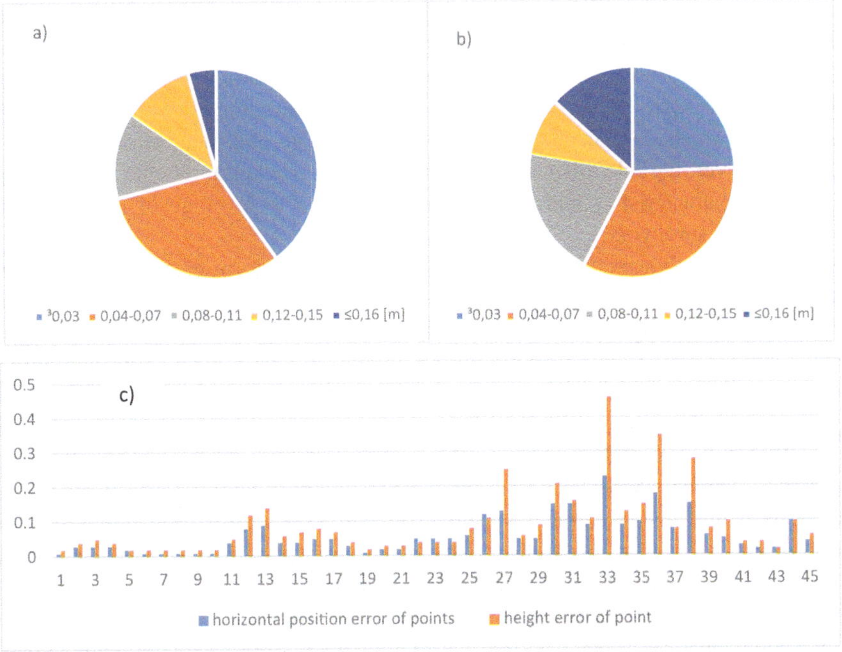

Fig. 5. The values of the point location errors (a), the height errors (b) and the errors at selected points (c)

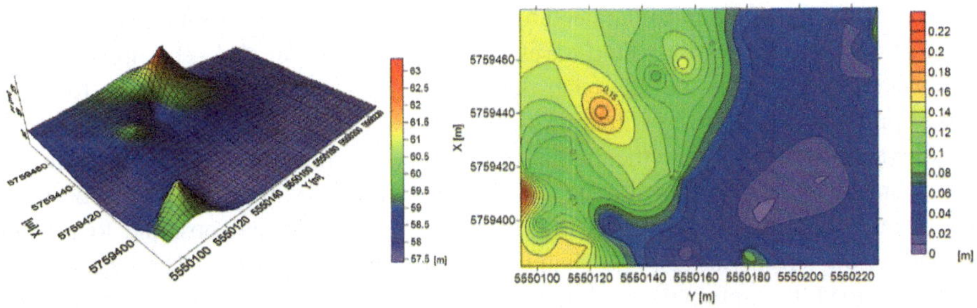

Fig. 6. A numerical terrain model and the height error in the area under analysis

4 Discussion

Currently available photogrammetric and remote sensing techniques provide very high resolution data, which are a source of valuable information about the development of areas, changes occurring in them and ongoing investments. They are also a source of data that would be difficult to obtain due to the specificity of the area (e.g. landslide areas, open-cast mines, coastal areas) or factors directly affecting human life and health (e.g. post-fire areas, floodplains) [29, 30]. The emergence of new photogrammetric technologies for acquiring low-altitude aerial images has made photogrammetric methods useful in more and more studies, and the proportion between the scale of studies and the scale of aerial images is gradually changing. This means that photogrammetric methods using drones are becoming more and more attractive economically. Moreover, digital cameras, which are used, have better and better geometric and radiometric characteristics and they are available to a wide range of potential users, including land surveyors.

Low-altitude photogrammetry makes it possible to perform measurements in small areas with local problems. The photogrammetric technology has been used in geodetic applications for a long time but the invention of drones provides a lot of new possibilities. Drones, while taking photos in the visible spectrum, have simplified both the way of recording and processing images [23]: flights can be planned anywhere and at any time, they can be repeated a number of times, and image resolution depends only on the flight altitude and distance to the target. Certainly, this approach also has its drawbacks, which include: limited flight duration and thus a limited size of the area that can be measured, bad weather conditions (in particular rain, strong winds, fog) and dense vegetation reduce the ability to identify terrain details. It is also worth paying attention to the complexity of legal provisions related to the use of drones, which are different in different countries in Europe and in the world.

It should also be noted that the use of unmanned aerial vehicles equipped with appropriate sensors is also applicable to the detection of fires, especially in areas where traditional fire monitoring and detection techniques are not very effective [31]. Yuan et al. [31] indicate the high potential of using UAV to identify inflammatory points in forests, which results in limiting the number of false alarms. However, there are researchers [32] who point to the need to analyse the real economic efficiency of drone application, which in various field cases is variable. Laszlo et al. [32] emphasize that fundamentally wrong is the approach that uses new products or procedures for marketing purposes, while overestimating their capabilities.

The approach to measurements that makes use of unmanned aerial vehicles in fire struck areas, as described in this paper, has been verified with classic geodetic measurements. The results prove that this technology is useful and possible to use for modelling the terrain surface and changes occurring in the test area. The research has also confirmed the usefulness of photogrammetric methods for automatic identification of fire areas and for determining their size without the need for participation of measuring teams and foresters. As a result of the research a training programme for drone operators was prepared, which included the use of drones for land surveying. The research also resulted in an application for a patent on a method for identifying post-fire areas in forests with the use of photogrammetric data and a proposal of a new and useful idea for construction and arrangement of the components of a platform for acquiring low-altitude photogrammetric images.

5 Conclusions and Future Works

The research has shown that there is a high potential for the use of drones in geodetic work and in sustainable forest management, especially for determining the size of forest areas struck by fire. With extended flight duration, the use of additional sensors and such advantages as manoeuvrability and the possibility of planning a flight route adapted to local problems, drones are becoming an economical and reliable solution for identifying areas struck by fire in forests. For this reason, drones can seriously compete with other classical systems. Taking into account the continuous development of both measurement techniques and computational methods, further research is planned on the use of drones for identifying floodplains [33] and areas threatened by landslides [34, 35]. The research will focus on determining the size of affected areas and attempts to estimate material losses and the risk of occurrence and extent of unfavourable and dangerous phenomena.

Because the presented research confirmed the competitiveness of classical measurement techniques in areas that are difficult to measure, future work will also focus on improving the results of site identification by using artificial intelligence methods. One such method that can be potentially effective will be the use of the support vector machines (SVM) and the use of Kalman filters for image analysis and methods for optimizing tasks using the Pareto principles [36]. Particularly, the SVM technique seems to be very much used to identify fire areas or sources of fire. A certain problem, which we will have to pay attention to, will be the use of classifiers in real time, due to the need for high computing power and the cost of the investment.

Considering the rapid development of technologies and measurement methods using low-level photogrammetry, in the next few years we will be able to obtain even higher accuracy in identifying points and areas in areas that are difficult to access [37]. In fact, due to the rapid technological progress and miniaturized components, a wider spectrum of sensors, including multispectral, hyperspectral measurement systems based on UAV will become more and more available to both the geodesy and construction industry, local government organizations and organizations dealing with environmental protection and monitoring.

References

1. Anderson, K., Gaston, K.J.: Lightweight unmanned aerial vehicles will revolutionize spatial ecology. Front. Ecol. Environ. **11**, 138–146 (2013)
2. Burdziakowski, P.: Evaluation of open drone map toolkit for geodetic grade aerial drone mapping – case study. In: 17th International Multidisciplinary Scientific GeoConference SGEM 2017, Photogrammetry and Remote Sensing (2017)
3. Cunliffe, A.M., Brazier, R.E., Anderson, K.: Ultra-fine grain landscape-scale quantification of dryland vegetation structure with drone-acquired structure-from-motion photogrammetry. Remote Sens. Environ. **183**, 129–143 (2016)
4. Duffy, J.P., et al.: Spatial assessment of intertidal seagrass meadows using optical imaging systems and a lightweight drone. Estuar. Coast. Shelf Sci. **200**, 169–180 (2018)

5. Elarab, M., et al.: Estimating chlorophyll with thermal and broadband multispectral high resolution imagery from an unmanned aerial system using relevance vector machines for precision agriculture. Int. J. Appl. Earth Obs. Geoinf. **43**, 32–42 (2015)
6. Entrop, A.G., Vasenev, A.: Infrared drones in the construction industry: designing a protocol for building thermography procedures. In: 11th Nordic Symposium on Building Physics. Energy Procedia, vol. 132, pp. 63–68 (2017)
7. Helman, D., et al.: Using time series of high-resolution planet satellite images to monitor grapevine stem water potential in commercial vineyards. Remote Sens. **10**(10), 1615 (2018)
8. Sztubecki, J., Mrówczyńska, M., Sztubecka, M.: Deformation analysis of the Czersko Polskie roller dam steel drum using conventional geodetic methods and neural networks. In: Conference Proceedings, vol. 17, no. 22, pp. 211–218 (2017)
9. Zaczek-Peplinska, J., Osińska-Skotak, K.: Concrete surface evaluation based on the reflected TLS laser beam's intensity image classification. Studia Geotechnica et Mechanica **40**(1), 56–64 (2018)
10. Szewranski, S., et al.: Socio-environmental vulnerability mapping for environmental and flood resilience assessment: the case of ageing and poverty in the City of Wrocaw. Poland. Integr. Environ. Assess. Manage. **14**(5), 592–597 (2018)
11. Jayathunga, S., Owari, T., Tsuyuki, S.: The use of fixed–wing UAV photogrammetry with LiDAR DTM to estimate merchantable volume and carbon stock in living biomass over a mixed conifer–broadleaf forest. Int. J. Appl. Earth Obs. Geoinf. **73**, 767–777 (2018)
12. Królewicz, S., et al.: Geodetic evaluation of the orthophotomap accuracy and digital model of the surface area of the WNGIG campus made on the basis of photographic images from the UAV. In: Works in the Field of Geography and Geology, vol. 2, pp. 17–30 (2015). (in Polish)
13. Khan, M.A., et al.: Unmanned aerial vehicle-based traffic analysis: a case study for shockwave identification and flow parameters estimation at signalized intersections. Remote Sensing **10**(3), 458 (2018)
14. Kurczyński, Z., Preuss, R.: Basics of Photogrammetry. Warsaw University of Technology Publishing House, Warsaw (2000). (in Polish)
15. Mrówczyńska, M.: Construction of a numerical terrain model for planning construction projects. Build. Mater. **8**, 116–117 (2017). (In Polish)
16. Sherstjuk, V., Zharikova, M., Sokol, I.: Forest fire monitoring system based on UAV team, remote sensing, and image processing. In: IEEE Second International Conference on Data Stream Mining & Processing (DSMP), pp 590–594 (2018)
17. Yuan, C., Liu, Z., Zhang, Y.: Learning-based smoke detection for unmanned aerial vehicles applied to forest fire surveillance. J. Intell. Rob. Syst. Theory Appl. **93**(1–2), 337–349 (2017)
18. Wardihani, E.D., et al.: Real-time forest fire monitoring system using unmanned aerial vehicle. J. Eng. Sci. Technol. **13**(6), 1587–1594 (2018)
19. Ma, S., et al.: An early forest fire detection method based on unmanned aerial vehicle vision. In: Chinese Control and Decision Conference (CCDC), pp. 6344–6349 (2018)
20. Hristov, G., et al.: Emerging methods for early detection of forest fires using unmanned aerial vehicles and lorawan sensor networks. In: 28th EAEEIE Annual Conference, EAEEIE (2018)
21. Dandge, N., Sanghavi, J.: Forest fire detection in forest video using image processing. J. Eng. Appl. Sci. **12**(8), 8478–8484 (2017)
22. Näsi, R., et al.: Estimating biomass and nitrogen amount of barley and grass using UAV and aircraft based spectral and photogrammetric 3D features. Remote Sens. **10**(7), 1082 (2018)
23. Nex, F., Remondino, F.: UAV: Platforms, regulations, data acquisition and processing. In: Remondino, F., Campana, S. (eds.) 3D Re-Cording and Modelling in Archaeology and Cultural Heritage. BAR, Oxford (2014)

24. Rokhmana, C.A.: The potential of UAV-based remote sensing for supporting precision agriculture in Indonesia. In: The 1st International Symposium on LAPAN-IPB Satellite for Food Security and Environmental. Procedia Environmental Sciences, vol. 24, pp 245–253 (2015)
25. The Act of February 3, 1995 on the protection of agricultural and forest land. Dz.U. 1995 No. 16 (78). http://prawo.sejm.gov.pl/isap.nsf/DocDetails.xsp?id=WDU19950160078. Accessed 28 Mar 2019. (in Polish)
26. Regulation of the Minister of the Environment of 20 June 2002 regarding a one-off compensation for premature felling of trees. Dz.U. 2002 No 99 (905). http://prawo.sejm.gov.pl/isap.nsf/DocDetails.xsp?id=WDU20020990905. Accessed 28 Mar 2019. (in Polish)
27. Wiler, K., Wcisło, P.: Protection of forests against fires. Development and Implementation Center of State Forests, Bedoń, Poland (2013). (in Polish)
28. Juszczyk, M., Leśniak, A.: Modelling construction site cost index based on neural network ensembles. Symmetry **11**(3), 411 (2019)
29. Szulwic, J., et al.: Maritime laser scanning as the source for spatial data. Polish Maritime Res. **22**(4) (2015)
30. Twidwell, D., et al.: Smokey comes of age: unmanned aerial systems for fire management. Front. Ecol. Environ. **14**(6), 333–339 (2016)
31. Yuan, C., et al.: Unmanned aerial vehicle based forest fire monitoring and detection using image processing technique. In: IEEE Chinese Guidance, Navigation and Control Conference, pp. 1870–1875 (2017)
32. Laszlo, B., Agoston, R., Xu, Q.: Conceptual approach of measuring the professional and economic effectiveness of drone applications supporting forest fire management. Procedia Eng. **211**, 8–17 (2018)
33. Jakovljevic, G., et al.: Accuracy assessment of deep learning based classification OF LiDAR and UAV points clouds for DTM creation and flood risk mapping. Geosciences **9**(7), 323 (2019)
34. Pellicani, R., et al.: UAV and airborne LiDAR data for interpreting kinematic evolution of landslide movements: the case study of the montescaglioso landslide (Southern Italy). Geosciences **9**(6), 248 (2019)
35. Skrzypczak, I., et al.: Methods of measuring and mapping of landslide areas. In: Proceedings of the Conference: 3rd World Multidisciplinary Earth Sciences Symposium (WMESS), Prague, Czech Republic (2017)
36. Nowogońska, B.: Preventive Services of Residential Buildings According to the Pareto Principle. In: IOP Conference Series: Materials Science and Engineering, vol. 471, no. 10 (2019)
37. Ventura, D., et al.: Mapping and classification of ecologically sensitive marine habitats using unmanned aerial vehicle (UAV) imagery and object-based image analysis (OBIA). Remote Sens. **10**(9), 1331 (2018)

Electrical Supply of Railway Transport Infrastructure Objects at High Frequency

Dmitriy Ermolenko[1(✉)], Leonid Uferev[2], and Oleg Roschin[2]

[1] Joint Stock Company “Scientific Research Institute of Railway Transport”, Moscow 129626, Russia
conf-transport@mail.ru

[2] Federal Government Research Facility, «Federal Research Agricultural-Engineering Center VIM» (FGBNU FNAZ VIM), Moscow 109428, Russia

Abstract. Higher frequency single cable power supply equipment was developed. Resonance based power supply equipment may be utilized for streetlights with HPS and LED lighting systems. Resonant light systems may variate the power delivery thus controlling luminance power. Usage of resonant single cable power supply systems may reduce the consumption of nonferrous materials for production, decrease capital investment costs, rule out the possibility of line short-circuit and increase the reliability of lighting systems for residential and production buildings, as well as lighting of stations, transfer lines, pedestrian crossings, towns, roads and pathways. Test results of using a resonant single cable power supply for induction heating of rail turnout switch in difficult winter climate conditions are provided in the article. A resonant single cable system was used in an experiment to power the induction rail heaters at the turnover switch. Measurements of electrical and temperature parameters of the induction heating system were carried out. The benefits of resonant induction heating system over a tubular heating element (THE) system were also compared.

Keywords: Resonant single cable system · Frequency converter · Resonant circuit · Induction heating · Fuko currents · Heating · Tubular heating element (THE)

1 Introduction

Revolutionary advances in electrical engineering of power supply electronics as well as in digital and analogue control systems lead to development of a resonant single cable higher frequency power supply system and development of induction heating that is witnessed in conducting mediums due to excitement of electrical currents of higher frequency with an electromagnetic field [1–3]. Starting from 1992, FGBNU VIESH (FGBNU FNAZ VIM since 2016) has been developing higher frequency resonant power supply system over single cable or air [1–4] based on the concepts proposed by inventor Tesla [5–8].

Scientist of VIESH have a long history of collaboration with railroad professionals. VIESH specialists participated in such works as: “Development of methodic of

Z. Popovic et al. (Eds.): TransSiberia 2019, AISC 1115, pp. 730–736, 2020.
https://doi.org/10.1007/978-3-030-37916-2_71

detection of asymmetry of currents and voltages and its influence on primary electric power supply devices and local users" (1956–1959), "Theoretical basis for choosing of efficient primary and traction energy electric power supply" (1957–1958), "Research of power and voltage and their influence on the primary power supply in locomotives with ionic converter. Development of methods for harmonics decrease and their negative influence on power supply and non-traction users" (1957–1959). In 2012 FBNU VIESH in collaboration with JSC "VNIIJT" and with involvement of JSC "RZD" Technical Policy Department, JSC "RZD" Central Infrastructure Department of electrification and power supply and "Promavtomatika" LTD carried out research and testing of 3 kVt resonant single cable power supply system.

Implementation of such system in the development of railway infrastructure may have great significance. As this system is an alternative to conventional systems it may be applied as a reserve power supply system; as part of a lighting system; as an electrical distribution system for low power users (CCTV cameras, traffic lights, automatic systems); as a rail turnout induction heating system; etc.

Resonant electrical higher frequency single cable or airline power supply systems consist of frequency converter device, electrical power line, and a backwards converter for voltage conversion to nominal values for end user requirements.

The operating principle of a resonant power supply system is based on the use of two power converters working in higher frequency of 5…15 kHz and a single cable with a voltage of 1…10 kV in resonant mode. The transferring converter is resonant, and it sets the operating frequency while the receiving converter is a broadband lowering one. At the same time a second cable is used as Ground, otherwise two conducting containers C3, C4 (large enough to transfer electrical power) are attached to converter coils and may be used as condenser armatures if there is dielectric ground between transferring and receiving parts.

Resonant transferring converter is comprised of a resonant contour C2L1 (Fig. 1) and increasing/lowering armature L2 [1]. Since the receiving converter doesn't affect the resonant frequency of power supply system they may be included in the system in any quantity, unless its overall power is lower than that of a transferring converter.

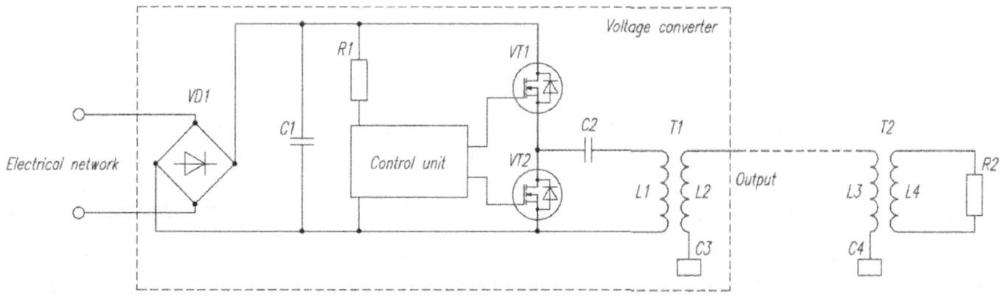

Fig. 1. Structural schematics of resonant power transfer system

The main blocks of resonant converter are power keys K1, K2, control unit CU and a resonant transformer comprised of container C2 and armatures L1 and L2 (Fig. 1). The resonant frequency is generally defined by parameters of the elements C2L1 as well as cable capacity that is added through transformer Tp1 to the lowering condenser C2 and thus lowers C2L1 contour's own resonant frequency.

Power transfer in a resonant system occurs with a higher voltage and higher frequency in resonant mode. Higher voltage allows lower losses through power transfer and higher frequency allows electrical equipment to be smaller in size [9].

A transferring converter with a resonant contour is installed in the start of a power line. Using reactive power, it can "pull" the electricity to the power supply cable. Automated systems work with controlled resonance so voltage in the system gets stabilized and is not influenced by input voltage or frequency.

A receiving converter with a broadband transformer is installed at the end of the line so it transforms all the electricity, as well as harmonics, to the electric power of a secondary armature. An invertor with a standard output voltage of 230 V or 400 V is installed at the output.

In such a system electricity is converted once at the input and then converted second time in the output with a total power loss of 10–12%.

The most efficient single cable power supply system is a higher frequency resonant system (HFRS) among all the single cable power supply systems.

2 Results of the Study

Lightning System. Resonant electrical lighting system consists of: power generator (1), invertor (2), resonant transformer (3, 4), electrical cable (6) and LED lights (8). The invertor is controlled frequency converter connected either to resonant transformer or a series resonant contour which output is connected to the resonant power supply cable.

Stationary power supply lines or autonomous power generators working on local or renewable fuels may be used as power generators.

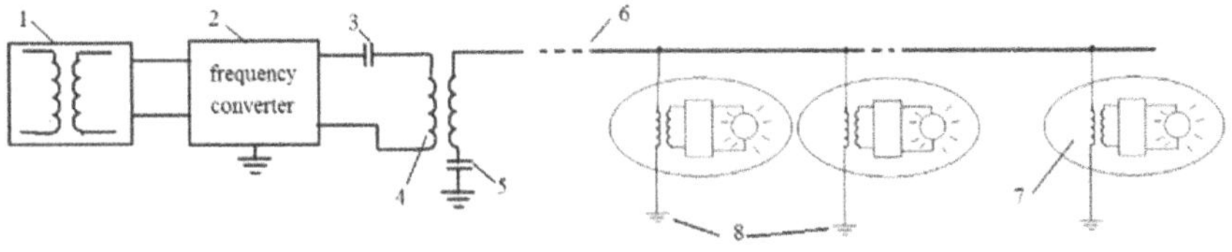

Fig. 2. Scheme of a luminescent lamp lighting system

A resonant lighting system operates as follows: a voltage from the line or any other power generator (2–600 V) (1) is fed to the invertor (2), than it is converted to high frequency voltage and fed to the resonant Tesla transformer (4), the transformer consists of input (3) and output (5) containers and it outputs high voltage with high frequency to the resonant cable (6). The second output of the transformer is connected to the Ground through a condenser (7). Lights are connected in parallel with one output to the high-voltage cable and the other to the ground (8) [10].

Electromagnetic power in the form of electrical currents and voltage moves from the output with high potential though the lights to the natural container with a lower potential. The difference in potentials makes the current go through the bulbs emitting light.

A lighting system with resonant single cable mode is used for lighting of spacious rooms, interiors of various buildings, train stations, railway stops, exhibition areas, train cars as well as civil, industrial, sport, railway agricultural buildings and areas, remote streets and railway stations.

The use of a single cable as a wave transfer medium for transferring of electromagnetic energy with high frequency is based on the ability of an open-circuit line to induct electrical charges on the surface of the conducting material thus allowing for power transfer.

Traditional street lighting system (Fig. 3.) is comprised of: 3-phase high-voltage electrical line (1) with 6…10 kV voltage; a number of transforming substations (2) 6… 10/0,4 kV; 2..3 km low-voltage electrical power lines; gas-discharge lamps with throttle powering. Throttle powering usually has cosφ of 0,4…0,55 which leads to higher consumption of power from transformers and power lines which in turn requires cables with bigger area coils and installation of individual compensators of reactive power.

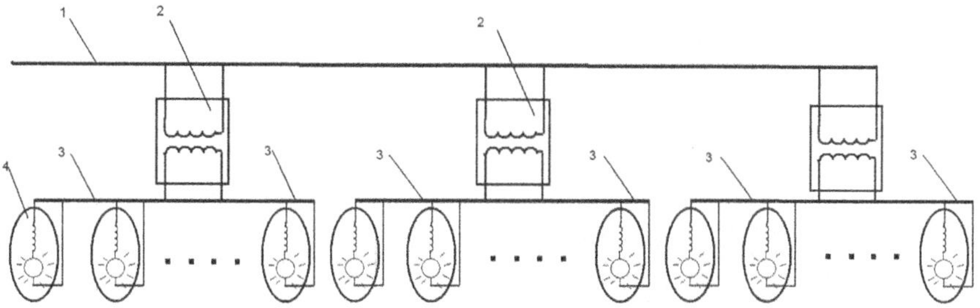

Fig. 3. Conventional street lighting system

In that case 6 power cables (3 high-voltage, 3 low-voltage) should be installed in parallel to the lighting line, not counting the Neutral and Ground lines.

A resonant street lighting system (Fig. 2.) consists of one transforming substation 6…10/0,4…0,6 (1), frequency converter (2), resonance transformer (4), single cable line (6) and lights (7) with reverse transformers.

The voltage of the power source fed to the voltage transformer is transformed into voltage of higher frequency and then fed to the single cable line. Lights are connected to the line. Lights may be luminescent or LED.

As an example, a standard power lighting line with 20 kV power and length of 6 km requires:

- 3-phase power line with 6…10 kV voltage and 3 km length, with minimal coil area of 70 $mm^2 \times 3$ (min 35 mm^2 for branching 6…10 kV lines);
- transforming substation 6–10/0,4, 25 kV of power (cosφ = 0,9);
- two 3-phase low-voltage power lines with 0,4 kV, length of 3000 m, coil area of 150 $mm^2 \times 3$, with voltage losses up to 8%;

- Lights with throttle power (0,9 kg of copper each) and compensators of reactive power;
- HPS or Mercury-discharge lamps.
- A resonant lighting power line with 20 kV of power and 6 km length requires:
- 6–10/0.4–0.66 transformer, 25 kV power (cosφ = 0,9 and transformation efficiency of 0,9);
- 20 kV frequency convertor;
- resonant transformer (10 kg of copper);
- single cable 6 km power line with coil area of 6 mm^2;
- Lights with reverse convertors and electronic ballast (0,12 kg of copper),
- LED or compact luminescent lamps.

Rail turnout heating system. Snowfall and changes in temperatures may provoke ice forming on rails thus affecting the rail switching [11]. To mitigate this effect, snow may be cleaned of by hand, or using cold or hot air blow-machines or using Tubular Heating Elements to melt it off. It is well known that induction heating is more efficient than heating using tubular heating elements.

Figure 5 illustrates the schematics of turnout heating devices connected to a single cable power line [12].

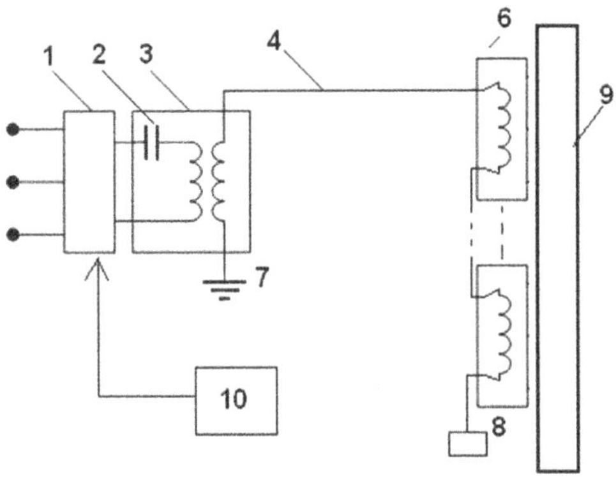

Fig. 4. 1 – Higher frequency generator, 2 – resonant increasing transformer, 3 - container, 4 – single cable power line, 6 - inductors, 7 - ground, 8 - ground, 9 – heated rail, 10 – external environment sensor.

The device (Fig. 4) works as follows: voltage from the power line is increased by generator 1 up to the level of 1–100 kHz; than the voltage is fed to the resonant increasing transformer 2, where it is increased through container 3. Resonant increasing transformer outputs the voltage to the single cable power supply line 4 with inductors 6 connected to it in parallel with resulting operational voltage that equals to the output voltage of the increasing transformer 2. Inductors 9 are installed on the heated rail with or without a gap. Automatic control is maintained by an external envrionment sensor that signals the frequency convertor to turn on.

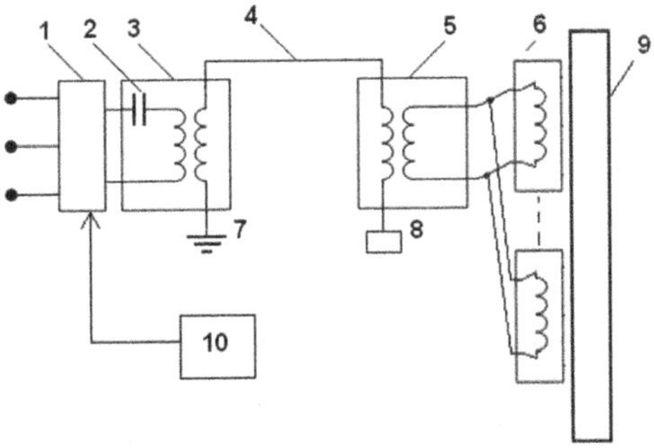

Fig. 5. 1 – Higher frequency generator, 2 – resonant increasing transformer, 3 - container, 4 – single cable power supply line, 5 – lowering transformer, 6 - inductors, 7 - ground, 8 - ground, 9 – heated rail, 10 – external environment sensor.

A principal scheme of an turnout induction heating device in which inductors are installed in parallel to the single cable power supply line through a lowering transformer is given in Fig. 5.

The device (Fig. 5) operates as follows: voltage from the power line is increased by generator 1 up to the level of 1–100 kHz; than the voltage is fed to the resonant increasing transformer 2, where it is icreased through container 3. Resonant increasing transfomer outputs the voltage to the single cable power supply line 4 with a lowering transformer 5 with inductors 6 installed at the output armature of the transformer. The output voltage of the lowering transfomer equals to the operating voltage of the inductors installed with or without a gap at the heated rail. Automatic control is maintained by an external envrioment sensor that signals the frequency convertor to turn on. Inductors are made with ferrite or metall cores encased in insulating coils.

Power efficiency is maintained by induction heating and automated control, that turns the heating only when needed.

Power required for operation may is about 100–500 V/m, with the rail warming up to 30–100 °C.

Laboratory testing supported the possibility and effectives of induction heating using higher frequency single cable resonant power supply systems. Induction heating is possible in metall due to the excitation of electrical currents with alternating magnetic fields. During induction heating only the conducting material heats and not the heating element. 86% of all of the power is generated at the outer layer which is called "penetration depth" which helps save electricty. Active and reactive energy is used in resonace.

Main technical features of the innovative railway turnout induction heating system based on a higher frequency single cable resonant power supply system:

- Power voltage of a transfering converter – 380 V, 50 Hz;
- Output power of a transfering converter – 6-20 kV;
- Volatge in the power supply cable – less than 1 kV;
- Size of transfering converter – 500 × 400 × 350 mm;
- Cable length between transfering converter and point of heating - less than 1500 m;

- Relative power of rail heating – 500 V/m;
- Rail heating temperature – at 50 °C;

3 Conclusions

1. Single cable power supply systems operating in higher frequency are more effective than lines with one ground cable operating at a constant current and alternating current of industrial frequency.
2. Power transfer with a resonant power supply was experimentally proven.
3. Use of a cable with a single thin core allows for a decreased in use of nonferrous metals and overall capital operating costs of an electrical distribution system.
4. Resonant lighting system may be well used in economical lighting of buildings as well as stations, parts of the roads and pathways.
5. Resonant systems will allow significant decrease in the cost of heating of rail turnout systems as well as increase in the reliability of heating systems due to direct induction heating.

References

1. Uferev, L.Y., Roschin, O.A., Strebkov, D.S.: Experimental models of resonant systems of electrical energy, Moscow, VIESH, 180 p. (2010)
2. Strebkov, D.S., Avramenko, S.V., Nekrasov, A.I.: Investigation of electric energy transmission process in non-metallic conducting channels. New Energy Technol. **4**(7), 24–27 (2002)
3. Uferev, L.Y., Mikhalev, A.A., Sokolov, A.V., Proshkin, Y.A.: Use of resonant energy efficient lighting system in poultry breeding facilities. Youth and Innovation, Gorky, part 2, Belarus, 25–27 May 2011
4. Uferev, L.Y., Roschin, O.A., Alexandrov, D.A.: Main problems and their solutions in designing GLS. Innovations Agric. **1**(3), 18–24 (2013)
5. Tesla, N.: World System of Wireless Transmission of Energy. Telegraph Telephone Age, 283 (1927). Selected works. Articles Beograd, p. 283 (1999)
6. Tesla, N.: My Inventions. electrical experimenter, February, March, June and October 1919. Selected works Articles, Beograd, p. 56 (1999)
7. Tesla, N.: Lectures. Patents. Articles. Museum Beograd, 715 p. (1956)
8. Tesla, N.: Colorado Springs Notes 1899–1900. Published by Nolit Beograd, 437 p. (1978)
9. Yuferev, L.: The resonant power transmission system. Handbook of research on renewable energy and electric resources for sustainable rural development (Federal scientific agroengineering center VIM, Russia), pp. 534–560 (2018). https://doi.org/10.4018/978-1-5225-3867-7.ch022
10. Strebkov, D.S., Uferev, L.Y., Roschin, O.A., Mikhalev, A.A.: Resonant LED lighting systems. Achievements of Science and Technology APK, № 10, pp. 20–21 (2009)
11. Kostenko, M.P.: Electrification of transport with AC/DC current of industrial frequency of higher voltage/comp. with supervision of Academy. Publishment house of AS USSR, Moscow, 35 p., Questions of Soviet Science (1957)
12. Ermolenko, D.V., Uferev, L.Y., Roschin, O.A., Kirin, O.Y.: Device for induction heating of rails and turnouts (variants). Patent RF № 135655, 20 December 2013

Optimization of Insensitivity Rate of Speed Control Systems

Victor Kochergin(✉) and Sergey Glushkov

Siberian Transport University (STU), 191 Dusi Kovalchuk Street, 630049 Novosibirsk, Russia
utanayno@gmail.com

Abstract. The present work is aimed at developing a method for estimating the optimal insensitivity parameters of the automatic speed control systems. When operating vehicles, unsteady loads applied on the engine are perceptible, which leads to the decrease in power and efficiency of the engine also reasoned by the existence of dead zone of the speed control system. Reducing the degree of insensitivity of the regulator positively influences the characteristics of regulatory processes, yet the significant reduction causes a decrease in the stability of the system. There is a problem of assessing the need to respond not only to external influences, but also to intra-cycle changes in internal combustion engines, leading to a change in rotational speed. In case of using modern microprocessor systems, the discreteness of receiving and processing initial information greatly influences the regulatory process. In order to optimize the insensitivity rate of regulators, it is proposed to estimate the allowable reduction of the dead zone until self-induced vibrations and resonance phenomena occur. An expression for determining the frequency of self-induced vibrations in control systems causing resonance is defined. The implementation of the proposed method will improve the efficiency of operation and increase the service life of transport-and-technological machines.

Keywords: Internal combustion engine · Speed controller · Insensitivity rate

1 Introduction

Rolling stock of railway transport is operated under conditions of continuously changing load on engine due to constant changes in the track grading and speed. The constant change in load characterizes not only land transport and technological machines, but also ship power plants. The fluctuations of the angular velocity caused by change in the moment of resistance of the external load are summed with the oscillations arising due to the cyclical nature of the internal combustion engine's working process. As a result, the operation of an engine under unsteady modes leads to a decrease in its power and efficiency. Firstly, the filling ratio of engine cylinders, the indicator efficiency and the excess air ratio are reduced. Secondly, in case of using diesel engines equipped with all-frequency speed controllers as a power source, phase shifts occur between the output and input coordinates of the control system [1].

Z. Popovic et al. (Eds.): TransSiberia 2019, AISC 1115, pp. 737–746, 2020.
https://doi.org/10.1007/978-3-030-37916-2_72

These phase shifts consist in the lag of the rotational speed in phase from the change in the moment of resistance and, accordingly, in the lag of the engine torque from the movement of the executive body of the change in fuel supply. The movement of the fuel control body begins with a certain shift in time when the resistance moment changes due to the presence of the regulator's dead zone. Consequently, the beginning of the change in the effective torque of the engine in relation to the beginning of the change in the moment of resistance occurs with a delay that is in some cases a significant value. These factors lead to an incomplete use of power and increase in the specific fuel consumption of a power plant.

The size of the dead zone of an automatic control system is influenced by frictional and backlash forces in kinematic pairs of mechanical elements of regulators, gaps in contacts of electrical control systems, the viscosity of working fluid in hydraulic control systems, the settings of electronic control systems and a plenty of other reasons. The engine itself is also an element of dead time between the moment of fuel flash in the cylinder and the production of net torque. The duration of dead time is determined, among other things, by the angular velocity of a crankshaft and a number of cylinders [2]. It is obvious that reducing the insensitivity rate of the controller positively affects the dynamic and static characteristics of regulatory processes due to the decrease in the amplitude and period of self-oscillations in the system. At the same time, the significant insensitivity rate reduction leads to the probability of arising of high-frequency oscillations and self-oscillations that cause the decrease in stability of control system. For each particular regulation object under given boundary conditions, there is the only duration of the transition process determined by speed of the regulation system that implies the minimal fuel consumption of a power plant [3].

Therefore, in order to improve the efficiency of vehicle operation, it is necessary to optimize the insensitivity parameters of speed controllers and develop measures to implement an additional effect on the controlled value depending on the value of the derivative of the input signal [4]. In modern conditions, the management of fuel supply and speed control is carried out by microprocessor devices, the main advantage of which is the possibility of implementing complex control laws. However, in such control systems, there is also a delay in the receiving and processing of control signal, which significantly affects the quality of control. At the same time, the noises of electronic devices of various origins that occur during operation introduce some signs of randomness into the control process and inconsistency in the operation of the system links. Moreover, in this case, one of the major problems is the problem of assessing the need for the automatic speed control system's responding not only to external influences, but also to cyclic changes in the angular velocity of a crankshaft, caused, for example, by ignition of fuel in the engine cylinders. The accuracy of regulation depends on the frequency and way of processing the initial information on the current value of rotation speed.

One of the modern types of control of fuel supply processes in automatic speed control systems is the process of controlling the position of the fuel pump rail via pulse-width modulation (PWM). The signal generated by the PWM controller is a discrete sequence of rectangular pulses characterized by amplitude, frequency, and duration of the signal. Figures 1 and 2 show the processes of regulating the rotational speed of the crankshaft n for a time t during a load shedding transition process and when the engine is running in stationary mode.

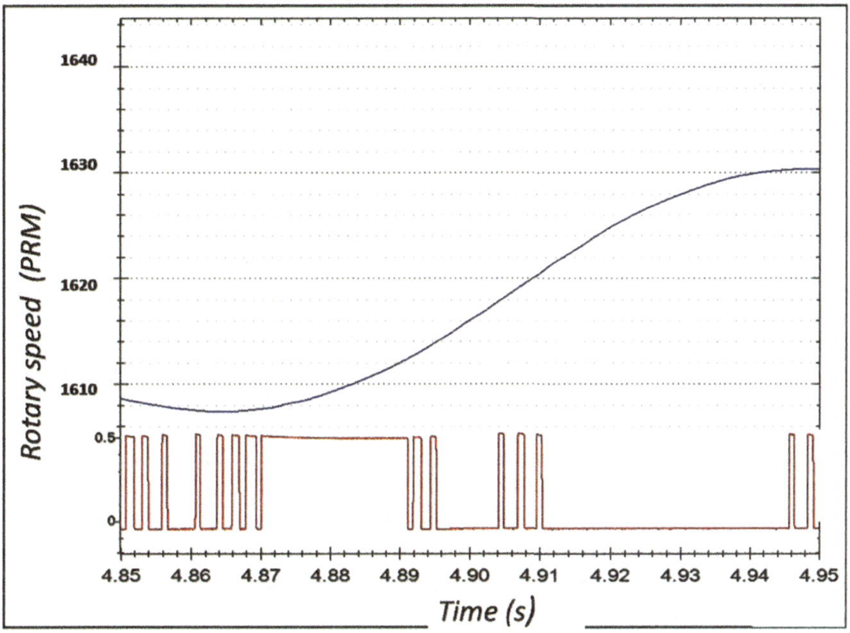

Fig. 1. The process of regulating the rotational speed of a diesel engine equipped with a PWM speed controller in the transition process of load shedding

In order to implement all capabilities of such systems, it is necessary to ensure optimal discretization in signal processing, which determines the accuracy of restoring the original analog signal on the rotational speed value at a specific point in time. The shorter the sampling interval, the more accurate the measurement process is, however, the complexity and cost of the recording equipment increase significantly. Digital signal processing methods fully allow evaluating the technical condition and interaction of individual elements of complex mechanical systems [6]. However, since they are based on restoration of a continuous function from its discrete values, the problem of objectivity and timeliness of obtaining information arises in case of using electronic control systems.

According to one of the interpretations of the Kotelnikov theorem, random signal $u(t)$ with an oscillation spectrum limited by a certain frequency value F_B can be fully restored only by following sample values with a time interval determined by the following expression:

$$\Delta t = \frac{1}{2F_B} \tag{1}$$

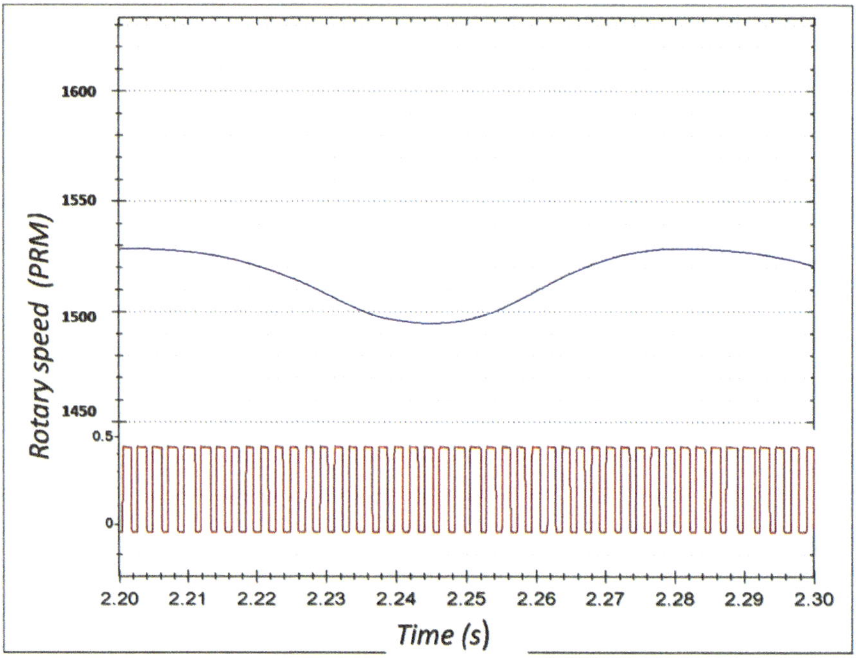

Fig. 2. The process of regulating the speed of diesel engine equipped with PWM speed controller, when working in steady state mode

Proceeding from the above, the wrong conclusion can be made that any harmonic signal with frequency F_B can be represented as a sequence of indications with frequency corresponding to obtaining two measurements of the signal for a period. However, it is impossible even theoretically to restore the original harmonic signals by two indications over a period. Therefore, in practice the restoration of signals with a wide range of oscillations even with the use of Fourier transform algorithms is a difficult task, becoming almost unmanageable in case when the maximum frequency in the signal exceeds half the sampling frequency.

Solving the problem of actually objective signal processing is possible when using the Hilbert transform, the physical meaning of which is to rotate the initial phases of all components of the spectrum of the initial signal $s(t)$ by the same angle $-0,5\pi$. This turn equals to the arising of a common factor $\exp(-i\pi/2)$. Instrumental processing of discrete signals using a phase shifter could help solve the problem of optimizing the degree of insensitivity of the regulator, but the use of this method is hard to implement in mass production and operation of vehicles. Consequently, there is a need for methods to optimize parameters of control systems, that would not depend on the discretization of control signals, if possible.

2 Research Methods

To solve the problem of optimizing the value of the insensitivity rate of rotational speed controllers, it is necessary to determine the criteria for ensuring the highest efficiency of the regulatory system, which means adaptation to achieving the set goal. In this case, it seems preferable to use multiplicative criteria for which the objective function takes the following form, where the plus sign implies restrictions that require the maximum increase of the function, and the minus sign means restrictions that require minimization of the function:

$$f(x) = \frac{\prod_{j=1}^{g} y_j^+(x)}{\prod_{i=j+1}^{m} y_i^-(x)} \tag{2}$$

In this case, as a positive limitation, one can take a response that provides the required gain of the system, that is, the ratio of the amplitude of oscillations of the controlled parameter φ and the amplitude of the disturbing action $f(t) = A \sin \omega t$.

As proved by studies in the field of theory of oscillations and vibration insulation of heat engines [8, 9], friction forces do not significantly affect the amplitudes of nonlinear oscillations, therefore, for example, when analyzing the characteristics of control systems with mechanical speed controllers, friction forces in the controller mechanism and in the rack drive The fuel pump can be neglected. This is all the more likely in the case of microprocessor-based automatic control systems where there are no mechanical moving parts and friction units.

However, it should be noted that external influences on the regulator and the fuel supply control unit of a diesel internal combustion engine contain several harmonic and possibly subharmonic components of a complex spectral composition. With such polyharmonic excitation, the zones of instability increase substantially as the individual elements of the system move. Therefore, as negative constraints in expression (2), which describes the essence of the multiplicative criterion, one can use friction forces that are still objectively present in the regulator. If the rotational speed is controlled by electronic systems, which, like mechanical devices, have their own oscillation frequency, then the control signal lag can be represented as an analogue of the friction force. Thus, as a criterion for optimizing the magnitude of the degree of insensitivity of the regulatory system, it is proposed to use the permissible degree of reduction of the dead zone until the moment of possible self-oscillation and resonance phenomena occurs. The effect of minimizing value ε_P on the behavior of speed control systems is currently not well understood. Based on the well-known theoretical principles of the theory of mechanical oscillations, we present the mechanism of the fuel supply control body of a power plant in the form of an elastic oscillatory system with a natural frequency of oscillations consisting of a regulator with its own conventional stiffness Cr and a drive of the executive control body of the fuel supply with mass m and stiffness cd. Hardness in this case should be considered an analogue of the level of resistance to movement of the controls of the fuel supply due to the presence of internal friction, that is, an analogue of the degree of insensitivity. If, in order to simplify the calculations, we assume the law of the steady-state motion of the mechanism of the fuel supply control organ linear, then we can assume that it is sinusoidal in nature:

$$f(t) = Z_0 \cdot \sin \omega t \tag{3}$$

Where z –oscillation amplitude of the fuel supply control drive; ω –circular oscillation speed.

3 Results of the Research

Differentiating the law of change of the disturbing force on the coordinate of displacement z, we obtain the following expression to describe the oscillatory process of the rail of the high-pressure fuel pump:

$$m\ddot{z} + (C_d - C_r) \cdot z = Z_0 \sin \omega t \tag{4}$$

Introduce the designation of natural oscillations of the vibrating object:

$$v = \sqrt{\frac{C_d - C_r}{m}} \tag{5}$$

Then the expression (4) takes the following form:

$$\ddot{z} + v^2 z = \frac{Z_0}{m} \sin \omega t \tag{6}$$

Using known methods for solving such equations, the following expression is obtained for the amplitude of the steady state of forced oscillations:

$$z = \frac{Z_0}{C_d - C_r} \cdot \frac{1}{1 - \omega^2/v^2} \tag{7}$$

The natural oscillations of a vibrating object can be represented as:

$$v^2 = \frac{C_d - C_r}{m} = v_d^2 (1 - \frac{C_r}{C_d}) \tag{8}$$

where v_d^2 –natural oscillations of the fuel control drive.

Taking into account the expression (8), the amplitude of the forced oscillations of the drive of the fuel control body without friction forces will be determined as follows:

$$z = \frac{Z_0}{C_d} \cdot \frac{1}{1 - \frac{C_r}{C_d}} \cdot \frac{1}{1 - \left(\frac{\omega^2}{v_d^2}\right) \cdot \left(1 - \left[\frac{C_r}{C_d}\right]\right)} \tag{9}$$

In this expression, the multiplier $\frac{Z_0}{C_d}$ is the movement caused by the maximum perturbing force, when it is applied statically, while the other two factors determine the dynamic effect of this force. Their absolute values can be considered the dynamic coefficient of the system:

$$K_g = \frac{z}{Z_0} = \frac{1}{\left(1 - \frac{C_r}{C_d}\right) - \frac{\omega^2}{\nu_n^2}} \tag{10}$$

Figure 3 shows the change in the dynamic coefficient K_g depending on the circular oscillation frequency ω for different values of the conditional stiffness coefficient of the $C_r$controller. When $C_r = 0$, (when conditional elasticity is absent (resistance to movement of the regulator)), the forced oscillations in the system are similar to the oscillations of the load on the spring. If the ratio C_r/C_d reaches1, the natural oscillation frequency decreases, leading to a shift of the amplitude-frequency characteristic of the oscillations to the left along the axis of the abscissas and, consequently to a decrease in the range of those frequency values ω at which $K_g \rangle 1$. When the frequency of the perturbing force approaches the frequency of free oscillations of the system, then the value of the dynamic coefficient rapidly increases.

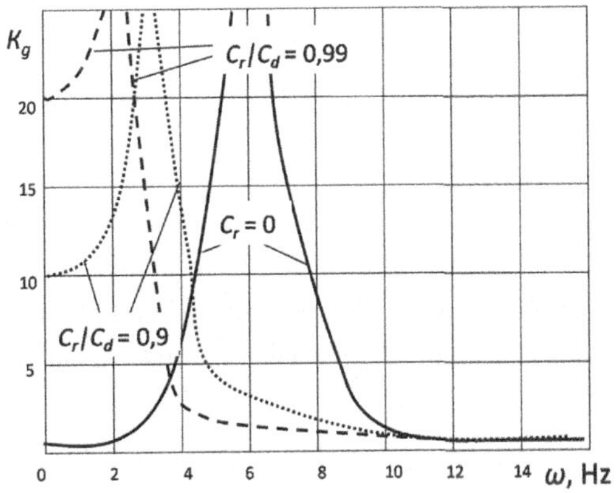

Fig. 3. Influence of conditional stiffness of the regulator on the value of the dynamic coefficient

The value of the dynamic coefficient K_g reaches its maximum under the following conditions:

$$\frac{\omega}{\nu_d} = \sqrt{1 - \frac{C_r}{C_d}} \tag{11}$$

The minimum values of ω, at which K_g values are less than 1, always lie in the super-resonance area, or in other words in the area where resonance is no longer possible. These values are smaller, when the total system stiffness $C_\Sigma = C_d - C_r$ gets smaller values. When the system has zero stiffness, $|C_d| = |C_r|$, then Eq. (10) takes the following form:

$$K_g = \frac{v_d^2}{\omega^2} \tag{12}$$

The obtained expressions show that when the condition $|C_d| = |C_r|$ is met, the system does not have a natural frequency of oscillation, and the resonant peak is located on the ordinate axis. Therefore, in this case oscillations always occur in the super-resonance area.

Now suppose that in the oscillatory system considered above there are forces of dry friction. In this case, the expression (4) looks as follows:

$$m\ddot{z} + 2fsCsign\ddot{z} + (C_d - C_r)z = Z_0 \sin \omega t \tag{13}$$

Where f – sliding friction coefficient; s – contact surface area.

Equation (15) is the usual equation of forced vibrations with dry friction and allows determining the values of amplitude Z_0 and phase shiftφ:

$$z = \frac{Z_0}{C_d - C_r} \cdot \frac{\sqrt{1 - \frac{4 \cdot 2frC_r}{\pi \cdot Z_0}}}{1 - \frac{m\omega^2}{C_d - C_r}} \tag{14}$$

$$tg\varphi = \frac{1}{\sqrt{\left(\frac{\pi \cdot Z_0}{4 \cdot 2frC_r}\right)^2 - 2}} sign\left(1 - \frac{m\omega^2}{C_d - C_r}\right) \tag{15}$$

Using expression (14) and introducing the concept of relative friction coefficient $\gamma = \frac{2fs}{Z_0}$, the expression is obtained for determining the amplitude of oscillations in the control system, taking into account the presence of friction forces:

$$z = Z_0 \cdot \frac{\sqrt{1 - \frac{16}{\pi^2}\gamma^2 \cdot \left(\frac{C_r}{C_d}\right)^2 \cdot \left(\frac{v_d^2}{\omega^2}\right)^2}}{\frac{v_d^2}{\omega^2}\left(1 - \frac{C_r}{C_d}\right) - 1} \tag{16}$$

The dynamic coefficient in this case equals to:

$$K_g = \frac{z}{Z_0} = \frac{\sqrt{1 - \frac{16}{\pi^2}\gamma^2 \cdot \left(\frac{C_r}{C_d}\right)^2 \cdot \left(\frac{v_d^2}{\omega^2}\right)^2}}{\frac{v_d^2}{\omega^2}\left(1 - \frac{C_r}{C_d}\right) - 1} \tag{17}$$

When $f = 0$, then the expression for determination of K_g will take the form:

$$K_g = \frac{1}{\frac{v_d^2}{\omega^2}\left(1 - \frac{C_r}{C_d}\right) - 1} \tag{18}$$

Comparing expressions (18) and (10), it can be concluded that they are identical, but having the difference consisting in the referring of Eq. (18) to a different base value. This expression takes into account the change in the amplitude of the perturbing force with a change in frequency. As the ratio C_r/C_d approaches 1, the amplitude-frequency characteristic of the control system shifts towards smaller values of ω, thus the resonance zone expands. In addition, at the initial values of frequency of the perturbing force, a transition through resonance occurs, but only minor amplitudes of oscillations develop, since there is not enough energy available to significantly shake the system. If the conventional values of the controller stiffness and drive rigidity are equal in values, then the dynamic coefficient equals 1 and there are no resonance phenomena in the system. At the same time, the oscillations remain constant throughout the entire frequency range and will not exceed the amplitude values of Z_0.

Analysis of the curves presented in Fig. 4 and characterizing the expression (18) for different values of ω, C_r/C_d and f, shows that friction forces corresponding to the presence of a dead zone in a controller reduces the resonant peaks of oscillations and allows the system to remain in a "locked" state when a certain frequency of resonant oscillations is reached.

Non-stop self-oscillations begin with a frequency, called the breakaway frequency (the beginning of the resonance), and is determined by the formula:

$$\omega = \sqrt{\frac{4}{\pi}\gamma\frac{C_r}{C_d}v_d^2} \tag{19}$$

Therefore, the limiting value of the relative coefficient of friction is the value determined by the expression:

$$\gamma = \frac{\omega^2 \pi C_d}{4 C_r v_d^2} \tag{20}$$

4 Discussion

The results of calculations and theoretical studies show that for the stability of a control system and the absence of resonant oscillations, the resistance to movement of the regulator's mechanisms or, in other words, the conventional stiffness of the control action should be more than or equal to the conditional movement stiffness of the fuel delivery control (regulatory action). In particular, it is shown that self-oscillations in such systems leading to resonance phenomena, begin with a frequency determined by the formula $\omega = \sqrt{(4/\pi)\cdot\gamma\cdot(C_p/C_n)\cdot v_n^2}$. The analysis of frequency characteristics allows optimizing or changing the sensitivity parameters of automatic control systems, which, in case of using digital control systems, is preferable compared to control of the sampling processes of monitored values.

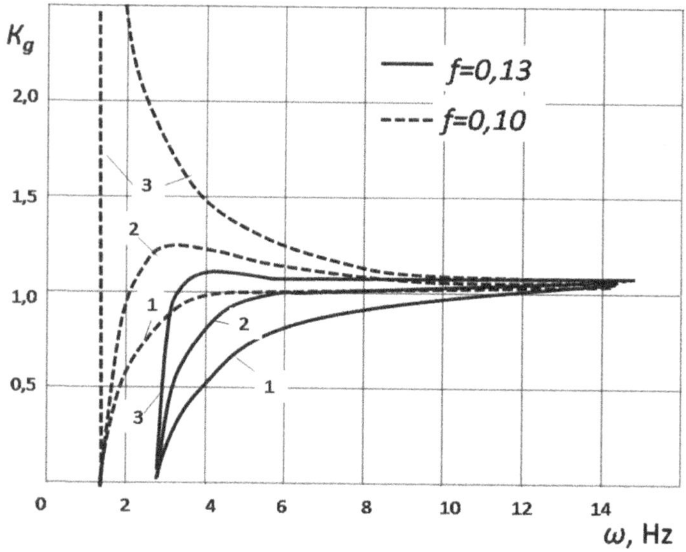

Fig. 4. The influence of friction on the dynamic coefficient for different values of ratio C_r/C_d: 1,0 - curves 1; 0.95 - curves 2; 0.9 - curves 3

In case of applying the proposed method of estimating the appropriate parameters of insensitivity during the design and adjustment of the controllers, the opportunity arises to apply of the most reasonable response to a change in rotational speed during the movement of vehicles and the performance of technological operations. As a result, operational efficiency can be significantly increased and the service life of transport and technological machines can be extended.

References

1. Zhilenkov, A., Efremov, A.: Evaluation of rotation frequency gas-diesel engines when using automatic control system. In: IOP Conference Series: Earth and Environmental Science, vol. 50, pp. 1–7 (2017)
2. Naplatarov, K.K.H.: Optimizatsiya upravleniya protsessami goreniya v promyshlennykh ustanovkakh. Avtomatika i telemekhanika, № 1, pp. 177–186 (2000). (in Russian)
3. Alpatov, V., Lukin, A., Laguta, I.: The creation of a computational model of corrugated beams using the author's program "GOPRO". In: MATEC Web of Conferences, vol. 251, p. 04007 (2018). https://doi.org/10.1051/matecconf/201825104007
4. Lukin, A.O., Alpatov, V.Y.: Influence of a metal beam of a corrugated wall geometrical parameters on size and nature of internal tension distribution. Mater. Sci. Forum **931**, 247–251 (2018). https://doi.org/10.4028/www.scientific.net/MSF.931.247

Mathematical Models of the Process of Air Removal from the Airtight Transport Pipeline During Vehicle Movement

Oleg Larin[1(✉)], Alexander Bokov[2], and Nikolay Goryaev[3]

[1] Moscow State University of Railway Engineering, 9b9 Obrazcova Street, 127994 Moscow, Russia
larin_on@mail.ru

[2] Plekhanov Russian University of Economics, Stremyanny lane 36, 117997 Moscow, Russia

[3] South Ural State University, 76 Lenin Prospect, 454080 Chelyabinsk, Russia

Abstract. The article presents the results of mathematical modeling of the process of air discharge from the hermetic transport pipeline (tunnel) during vehicle movement. In order to organize the effective operation of high-speed vacuum transport system of pipe type, it is required to pump air from the internal cavity of the conveying structure, taking into account the dynamics of the air environment when rolling stock moves. The intensity of the exhaust air should be adjusted taking into account the actual speed of the vehicle and its actual location on a particular stretch of track. Air is diverted by means of compressor units through the ducts installed in the transport case. The characteristics of the air vent system depend on the number, configuration and location of the air ducts, and the nominal capacity of the compressors. To regulate the process of air discharge, it is necessary to assign the operating parameters of each compressor unit during the entire transport cycle. For the construction of mathematical models of the air pumping process, the continuity equation from the theory of gas dynamics is used. On the basis of the developed models, numerical experiments were conducted in the process of air discharge, which allowed establishing the nature of changes in the load on the elements of the air exhaust system in the process of the vehicle moving through the tunnel.

Keywords: High-speed transport systems · Vacuum pipeline · Mathematical modeling · Gas dynamics · Air discharge · Pressure · Speed

1 Introduction

Currently, the development of high-speed tunnel and pipe-type transport systems for the carriage of goods and passengers at a speed of more than one thousand kilometers per hour is underway. The most famous projects in this area are created on the basis of airless (vacuum) transport systems (VTS) with airtight transport lines (evacuated tube transport), which are able to create competition for existing types of transport, first of all, rail, road and aviation [1].

Z. Popovic et al. (Eds.): TransSiberia 2019, AISC 1115, pp. 747–755, 2020.
https://doi.org/10.1007/978-3-030-37916-2_73

Researchers draw attention to a number of advantages of the VTS, which these systems have in comparison with other types of transport: high vehicle speed at relatively low operating costs, low levels of noise and pollution, a high degree of safety of the transport process and protection of the objects being transported. However, the researchers note that the VTS will not be able to completely replace any type of transport, but will complement them, as currently existing types of public transport complement each other. The greatest effect from the operation of vacuum transport systems can be obtained when they are integrated with urban transit systems, in particular, the metro, suburban trains [2–5].

The basis of the VTS is the transport pipeline (tunnel), which is isolated from the external environment, which consists of the body and the transport route used to move the rolling stock, as a rule, the train. Air is pumped out of the pipeline through separate exhaust ducts using compressor systems. Air ducts are tightly attached to the pipeline along the perimeter of the through holes in its case. Air is pumped out of the transport pipeline both before the start of the movement of the vehicle and in the process of moving the train. The limits of pressure reduction are established by constructive features of the VTS. The first prototypes of the VTS appeared more than a hundred years ago. For example, the Russian scientist Boris Weinberg and the American engineer Robert H. Goddard independently proposed to organize the movement of the vehicle at high speed inside the pipe, from which air should be evacuated [6–8]. During the twentieth century, conceptual approaches to the design of military-technical cooperation developed rapidly. Various modifications of the structures of the transport pipeline were proposed, innovative power plants for the drive of the vehicle were adapted. For example, the American inventor Daryl Oster patented a combined construction of several vacuum pipelines, along which the vehicle should move in opposite directions [4, 5]. Modern models of VTS are divided into two main structural types, depending on the degree of air pressure reduction in the internal cavity of the transport pipeline. First, systems with a deep vacuum (hard vacuum), inside which the pressure drops less than 1 Pa. Secondly, systems with forevacuum (vorvakuum), in which the pressure drops below 100 Pa. According to many experts, the most promising are forevacuum-transport systems, since the costs of their creation and operation are significantly lower compared with the required capital investments in deep-vacuum transport systems, and the differences between the two types of data transport systems are not significant in terms of technical and operational performance.

To create a forevacuum inside the pipelines, stationary compressor installations are used. Given the considerable length of the structures, it is advisable to place such installations both at the places of the beginning and end of the movement of rolling stock, as well as along the way. The number and performance of operating compressor installations should be sufficient to reduce the air pressure in the internal cavity of the pipeline. As the train progresses, the compressor units may turn off. In connection with the change in the number of active pumping devices, it is an urgent task to organize the regulation of the power of operation of compressor units, depending on their location relative to the moving vehicle, in order to ensure the necessary air pressure. The analysis of domestic and foreign scientific works shows that insufficient attention is paid to this issue in scientific research [8].

2 Methods

For the arrangement of the discharge systems and regulation of the process of their work, it is necessary to determine a number of parameters, including the minimum capacity values of the pumping compressor units and their number. To solve this problem, the dependences of the change in the intensity of air evacuation (flow rate w) on the train speed in individual sections of the transport pipeline should be known, taking into account possible changes in speed patterns. Since the establishment of appropriate dependencies experimentally is very problematic, in the framework of this study, an attempt was made to develop mathematical models of the process of pumping air from the transport pipeline when the vehicle is moving.

The movement of a vehicle in an isolated transport pipeline is a rather complicated physical process. When mathematical modeling of the parameters of the process of pumping air out of the pipeline during vehicle movement, mathematical physics, hydrodynamic equations for compressible media, etc. were used. Calculation systems were used to solve the equations for cylindrical coordinate systems [2].

At this stage of the study, taking into account the complexity of the simulated system, the limited working prototypes of the VTS and experimental data on the results of their testing, preliminary estimates of the allowable operating modes of the equipment (compressor units) were made using a simplified mathematical model. Figure 1 shows a schematic depiction of the movement direction of the vehicle along the Ox axis at a speed v (m/s) and pumping out air with an intensity w (cubic m/s).

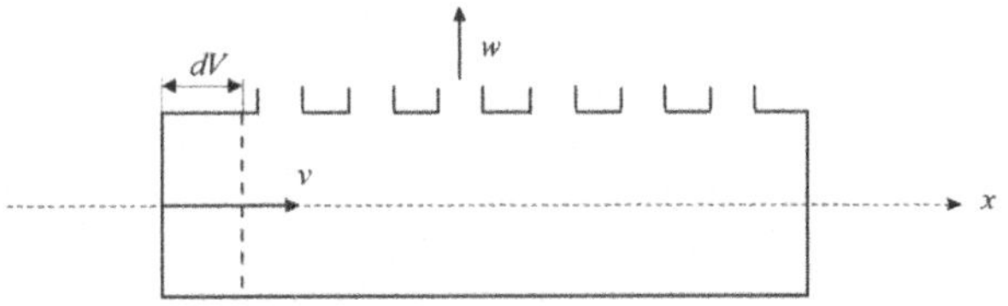

Fig. 1. Driving directions of the vehicle and pumping air

The aggregate (potential) performance of the exhaust system W_e is determined taking into account the nominal capacity of the compressor units included in its composition w_e (cubic m/s) and their number n_e:

$$W_e = w_e \cdot n_e, \text{ cubic m/s.} \tag{1}$$

where n_e – the total number of components of the exhaust system (ducts), units.

The actual capacity of the W_a of the exhaust system is determined by taking into account the number of active (working) air ducts n_a and the actual power of the compressor units w_a:

$$W_a = w_a \cdot n_a, \text{ cubic m/s.} \tag{2}$$

Moreover, the actual capacity of the compressor units w_a can be set below the nominal value of w_e in order to regulate the intensity of the air exhaust depending on the current speed of the rolling stock v [1, 8].

At the moment the train starts moving, the number of active components n_a is equal to their total number in the design of the transport pipeline: $n_a = n_e$. However, as the vehicle moves through the high-speed section, the number of active components may decrease ($n_a \rightarrow 0$), and the exhaust units left "behind" the rolling stock become "inactive", i.e. are transferred to the non-operating mode. In this case, the balance of the components is maintained: $n_e = n_a + n_u$. The total number of components of the exhaust system is not permanent and is determined by the design features of the VTS. According to the locations of the components of the exhaust system, the entire track of the pipeline with a length L is divided into the i-th number of segments Li ($i = 1, \ldots n_e$).

To simulate the process of pumping air used the continuity equation:

$$\frac{\partial \rho}{\partial t} + \nabla \cdot (\rho \vec{v}) = 0$$

where ρ – density, $\vec{v}$ – speed (vector value), t – time, ∇ – the operator "nabla" here denotes the divergence, $\operatorname{div}(\rho \vec{v})$,

or in Euler variables (see, for example, [7]):

$$\frac{d\rho}{dt} + \rho \operatorname{div} \bar{v} = 0 \tag{3}$$

In the first approximation, for conditions of balanced volumes of air pumped out of the transport pipeline and moved by a moving vehicle, we consider the medium incompressible, ρ = const. With this assumption, we get:

$$\operatorname{div} \vec{v} = 0 \tag{4}$$

Let us determine the actual pumping rate (flow rate) of air w_a by one duct as a function of train speed v (scalar value) along the transport pipeline. This dependence is obtained from equality (4) on the basis of the balance between the air injected by the train and the flow rate through the exhaust ducts. On the one hand, for the air transported by the vehicle, the condition is met:

$$|dV| = -V'dt = v \cdot S\,dt \tag{5}$$

$$\text{at } dV < 0,$$

where dV – volume change before the train for an infinitely small period of time dt; S – is the cross-sectional area of the vehicle (taking into account the accepted assumption that the tunnel configuration does not affect the dynamics of air movement), square m.

On the other hand, for the air pumped out by the compressor unit the following condition is met:

$$|dV| = w_a \cdot n_a \, dt \tag{6}$$

Equating (5) and (6) we get:

$$w_a = -\frac{V'}{n_a} = \frac{v \cdot S}{n_a} \tag{7}$$

Expression (7) characterizes the w_a dependence on the train speed, its geometrical parameters, as well as the number of air ducts. Accordingly, for different traffic modes and design features of the tunnel, the characteristics of the flow rate w_a will differ significantly. We illustrate possible variants of the dependence w_a for v = const. The flow function (7) takes the following form:

$$w_a = w(n_a) = [v \cdot S] \frac{1}{n_a} \tag{8}$$

We translate the calculations in dimensionless values. We introduce the variable conditional speed:

$$\upsilon = v/v_{\max} \tag{9}$$

and denote:

$$D = \frac{v_{\max} \cdot S}{n_e} \tag{10}$$

where v_{ma} – maximum vehicle speed in the tunnel, m/s.

Then the expression (7) is rewritten as follows:

$$w_a = w(n_a) = \frac{v \cdot S}{n_e - n_u} = \frac{\upsilon \cdot v_{\max} \cdot S}{n_e(1 - z)} = D \cdot \upsilon \cdot \frac{1}{1 - z} = D_1 \cdot \frac{1}{1 - z} \tag{11}$$

$$\text{at } z = n_u/n_e, 0 < z < 1, 0 \le \upsilon \le 1.$$

Variable $D_1 = D \cdot \upsilon$ introduced to simplify the notation, since $D \cdot \upsilon$ = const.

Expression (11) shows that the consumption of w_a should increase as the train moves through the tunnel to the limit values ($w_a \to w_{\max}$), since the number of active exhaust ducts before the train will decrease ($n_a \to 0$). Under these conditions, it is important to take into account the possible increase in the load on the operation of the compressors (especially if their performance is limited). It is possible to reduce the load by increasing the total number of ducts n_e and (or) by limiting the speed of the vehicle $v \ll v_{\max}$ in the respective sections of the track.

To characterize the change in the intensity of the diversion w_a depending on the length of the section L_i along which the train moves, we will rewrite the function (7) as follows:

$$w_a = w(\xi) = D \cdot \upsilon \frac{1}{1-\xi} = D_1 \frac{1}{1-\xi} \tag{12}$$

$$\text{at } 0 \leq L_i \leq L, \quad \bar{v} = \text{const}, \xi = L_i/L, 0 \leq \xi \leq 1, \ i = 1, \ldots n,$$

where Li – the current coordinate of the location of the rolling stock, if we assume that the train moves in the tunnel from the point with the coordinate L0 = 0 to the point with the coordinate Ln = L; ξ – conditional "dimensionless" coordinate of the location of the train on the i-th section of the tunnel.

Coordinate ξ specifies the movement of the train in the tunnel in relative coordinates from the starting point ξ = 0 to the final one with coordinate ξ = 1.

According to (12), with ξ → 1, the flow rate is wa → ∞. Figure 2 shows the dependence of the intensity of the exhaust air wa on the distance traveled by the train ξ at a constant speed v = const.

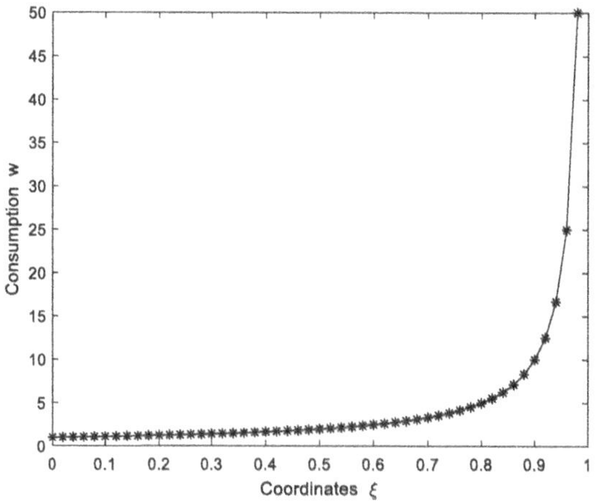

Fig. 2. The dynamics of the flow w_a at v = const.

At the same time, in practice, as a rule, the train slows down the speed as it gets closer to the end point. Therefore, the actual intensity of the exhaust air can also be reduced. Therefore, it is proposed to break the path of the transport pipeline into characteristic traffic sections with different speed modes, for example, initial (acceleration) – [0, ξ1), intermediate (speed) – [ξ1, ξ2), final (deceleration) – [ξ2, 1].

Therefore, we can assume that in the initial and final segments $v \neq$ const, in the intermediate segment v = const, and $v \leq v_{\max}$. We also assume that the velocity v in the initial and final segments is a function of the coordinate ξ. Let us consider how the intensity of the exhaust air w_{av} will change on characteristic sections of the path L. For example, take the speed $v(\xi)$ as a linear function: $v = Ax + B$:

$$v(\xi) = \begin{cases} A\xi, & 0 \le \xi < \xi_1, \\ C = A\xi_1 = \text{const}, & \xi_1 \le \xi < \xi_2, \\ -A_1\xi + B_1, & \xi_2 \le \xi \le 1. \end{cases} \tag{13}$$

Then the consumption w_a will be determined as follows:

$$w_a = \begin{cases} D\frac{A \cdot \xi}{1-\xi} = DA\left(\frac{1}{1-\xi} - 1\right), & 0 \le \xi < \xi_1, \\ D\frac{AL_1}{1-\xi} = D\frac{C}{1-\xi}, & \xi_1 \le \xi < \xi_2, \\ D\frac{-A_1 \cdot \xi + B_1}{1-\xi} = DA_1\left(1 + \frac{B_1/A_1 - 1}{1-\xi}\right), & \xi_2 \le \xi \le 1. \end{cases} \tag{14}$$

According to expression (14), the flow rate w_a approaches a constant value $DA_1(w \rightarrow DA_1)$ at $B_1/A_1 \rightarrow 1$. This dependence is explained by the dynamics of the speed mode $v(\xi)$ in the area of slow motion. The flow diagram w_a in various sections ξ for the linear function of velocity $v(\xi)$ is shown in Fig. 3.

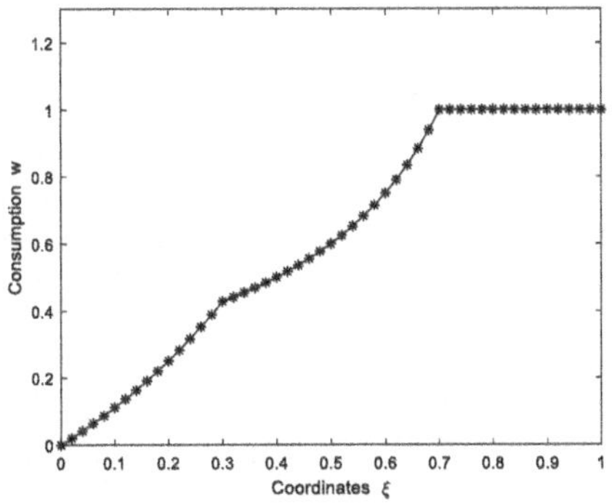

Fig. 3. The dynamics of the flow rate w_a taking into account the change in $v(\xi)$ on characteristic sections

According to expression (14), the flow rate w_a approaches a constant value DA_1 ($w \rightarrow DA_1$) at $B_1/A_1 \rightarrow 1$. This dependence is explained by the dynamics of the speed mode $v(\xi)$ in the area of slow motion. The flow diagram w_a in various sections ξ for the linear function of speed $v(\xi)$ is shown in Fig. 3.

Similar models of the process of air discharge can be constructed for nonlinear speed functions $v(\xi)$, for example.

$$v(\xi) = Ae^{B(\xi - C)^k} \tag{15}$$

$$\text{at } B > 0,$$

or:

$$v(\xi) = A\xi^k + B\xi + C \tag{16}$$

$$\text{at } A > 0,$$

where A, B, C, k – dimensionless parameters of the model.

3 Results

On the basis of such dependencies, it is possible to simulate the operating conditions of the exhaust systems, which makes it possible to determine the technical requirements for equipment used in the VTS, in particular, with regard to the nominal capacity of the compressor units, the number and location of the air ducts, their cross-section configuration, etc. For example, a study was made of the dynamics of the flow ratew_a with an increase in the number of ducts n_e. According to (10), an increase in n_e. leads to a decrease in D_1. Moreover, $D_1 = 1$ is taken for the initial number of ducts n_e. In Fig. 4 shows the variation in the flow rate wa for different values of D_1. The simulation results allowed to establish the nature of the change in air flow for a wide range of values of n_e. The obtained data is recommended to use when determining the requirements for the nominal performance of compressor installations, at which the target value of the speed of the rolling stock is achieved.

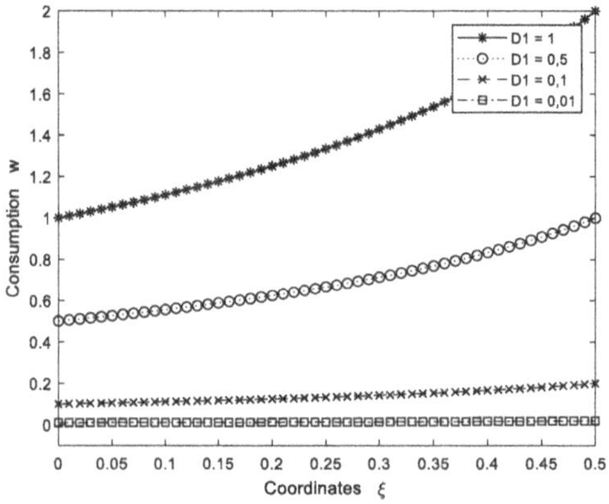

Fig. 4. The dependence of the flow rate w_a on D_1 with increasing n_e.

4 Conclusion

The mathematical models developed by the authors, using the example of basic model functions of speed, characterize in a generalized form the dynamics of various modes of the "track-vehicle-duct" system operation and allow one to obtain simplified (preliminary) estimates for organizing control of this system. According to the simulation results, the nature of the dependence of the intensity of air flow on the speeds of rolling stock movement was revealed. On the basis of the established dependencies, recommendations can be made for limiting the speed limit for certain sections of the route, taking into account the limiting pumping capacities of the compressor units.

The results of the study presented in this article are not exhaustive and do not cover the whole range of problems associated with ensuring the effective operation of high-speed vacuum transport systems of the pipe type that have significant prospects for use for the carriage of goods and passengers. However, the authors believe that the research results presented in the article will make a definite contribution to the development of the theory and practice of organizing such transport systems and mathematical modeling of the parameters of their work. In the future, it is planned to expand the scope and nature of researches in this area, it will be necessary to improve the accuracy of mathematical models of the air pumping process, to develop a software package for carrying out numerical experiments to calculate pressure fields and the velocity of air masses under conditions of forced convection.

References

1. Larin, O.N., Bokov, A.V.: Decreasing of profile air drag to the train movement inside the tube transport. Transp. Syst. Technol. **5**(2), 47–59 (2019). https://doi.org/10.17816/transsyst20195247-59
2. Larin, O.N.: Methodology of Organization and Functioning of Transport Systems of the Regions: Monograph. Publishing House of SUSU, Chelyabinsk (2007)
3. Janzen, R.: TransPod ultra-high-speed tube transportation: dynamics of vehicles and infrastructure. Procedia Eng. **199**, 8–17 (2017)
4. Vakulenko, S.P., Larin, O.N., Levin, S.B.: Theoretical aspects of interaction mechanisms in transport systems. World Transp. **6**(55), 14–27 (2014)
5. http://veinberg.o7.ru/pdf/no_friction_motion.pdf. Accessed 10 Dec 2018
6. Zhang, Y., Daryl, O.: A New Industrial Era Coming: Initial Dialogue on Evacuated Tube Transportation (ETT). Publishing House of Tsinghua University, China (2004)
7. Zhang, Y., Li, Y.: Role and position of ETT in the future comprehensive transportation system, pp. 2796–2803 (2007). https://doi.org/10.1061/40932(246)459
8. Zhang, Y., Yu, X.: On the safety of evacuated tube transportation. J. Transp. Eng. Inf. **4**, 42–50 (2006)

Handle Oscillations of a Pneumatic Hammer with Zero-Hardness of a Basic Elastic Element

Sergey Glushkov(✉) and Yuriy Pudovkin

Siberian Transport University (STU), 191 Dusi Kovalchukst, 630049 Novosibirsk, Russia
direct@midel.me

Abstract. Reducing the risk of occupational diseases of workers is one of the main industrial sanitation tasks. The priority task is to reduce the vibration of a hand-held pneumatic tool. Hardness compensation of the basic elastic element of a pneumatic hammer is one of the effective ways to reduce the handling vibrations. The article dwells upon the results of theoretical studies. The loading diagrams presented for the handle and the case of the air-distribution percussive mechanism of a pneumatic riveter. The calculation of elastic forces and dissipative forces acting on the handle is exposed. A kinematic diagram of the interaction between the compensating device spring coil and the inner surface of the hammer handle and the convex ellipsoid surface of the intermediate link is shown. The differential equation of hammer handle oscillations was obtained on the basis of the conducted research.

Keywords: Hardness compensation · Vibrating insulation · Oscillations · Riveter

1 Introduction

The negative effect of vibration on the human body is a well-known fact, which has been repeatedly proved by doctors and confirmed by the presence of occupational diseases among workers who come into contact with a vibrating tool every day [1]. In order to limit the harmful vibration effects from hand tools, a variety of vibrating insulation designs for the pneumatic hammer handle have been developed. Work on improving the vibration-proof tool is actively being carried out to this day [2]. Earlier, a team of scientists from the Siberian State University of Water Transport developed a vibrating insulation mechanism with ball hardness correction of the basic elastic element, which showed good vibrating insulation properties. This design had one major drawback, the clamping of the balls between the convex ellipsoid surface of the vibrating insulation device and the hammer handle resulted in large surface stresses with deformation in the contact zone of the ball and the convex ellipsoid surface with subsequent cracks in the intermediate link. In order to avoid this drawback, it was proposed to replace hard balls with an elastic spring. This measure allowed the surface stresses to be reduced due to the greater number of contacts of the compensating device (coils of the coil spring) with the inner surface of the handle and the intermediate link convex ellipsoid surface. The scope of the research is to reduce the oscillations of the

Z. Popovic et al. (Eds.): TransSiberia 2019, AISC 1115, pp. 756–767, 2020.
https://doi.org/10.1007/978-3-030-37916-2_74

pneumatic riveter handle by introducing a device with zero-hardness into the design of the air spring vibrating insulator system. Compensating the hardness of the basic elastic element is necessary in order to limit the vibration impact on the operator through the palmar hand surfaces.

2 Theoretical Information

2.1 Flow Diagram of the Air Spring Vibrating Insulation System of Riveter

The flow diagram (Fig. 1) is a passive vibrating insulation system with one degree of freedom. Here the position of the object (handle 5 with mass m) in relation to the base (intermediate link 1) is determined by the relative coordinate U. The coordinate origin is selected in the position of static equilibrium.

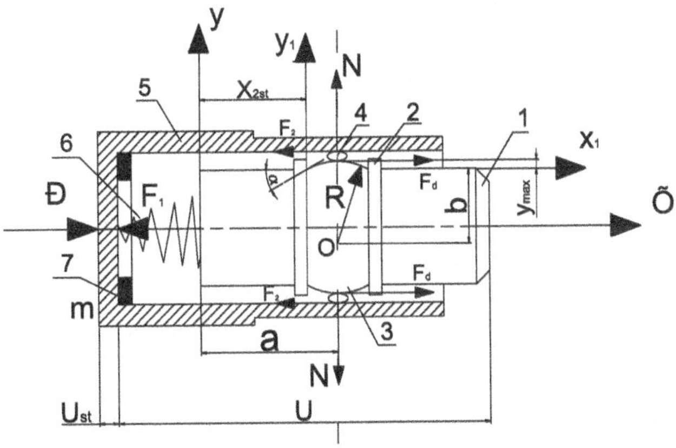

Fig. 1. The flow diagram of the vibrating insulation air spring system of a riveter 1 – intermediate link, 2– limit stop of the compensating element, 3 – convex ellipsoid surface, 4 – compensating element (coil spring), 5 – handle, 6 – basic elastic element, 7 – shock absorber

The following forces acting on a handle of mass m are showed in Fig. 1:

P – pressing force;
F_1 – the elastic force of the basic elastic element;
F_2 – the elastic force of the compensating element;
F_d – dissipative force;
f(t) – motion law of the body;
R – radius of the roller profile;
b – the distance between the coordinate origin of a convex ellipsoidal surface and the Rradius center;
a – the distance between the coordinate origin and the radius center;

x_{2s} – the static position of the spring coil on the ellipsoidal surface relative to the coordinate origin

Let us consider the change in the forces acting on the handle of mass m in relative motion.

The elastic force F_1 of the conical coil spring, taking into account the hardness c_s and the coordinate U_{st} corresponding to the preliminary static tension of the spring, is obtained using the following equation:

$$F_1 = c_s(U + U_{st}). \tag{1}$$

Let us determine the correlation between the elastic force of the compensating device F_2 and a convex ellipsoid surface for a system, in which the ordinate axis passes through the vertex of the ellipsoid surface.

The elastic force of the compensating device F2, depending on its position on the roller (Fig. 1), is determined by the elastic force of the spring coil N and the α inclination angle of the tangent to the ellipsoid surface at the contact point of the spring coil as follows:

$$F_2 = N \tan \alpha, \tag{2}$$

taking into account that

$$\tan \alpha = \frac{dy_1}{dx_1} \tag{3}$$

Since the elastic force N is proportional to the displacement y_1, then

$$N = \frac{y_1}{k}, \tag{4}$$

where k is the coil hardness of the transversely loaded cylindrical spring.

Expression (2) takes the following form:

$$F_2 = \frac{y_1}{k}\frac{dy_1}{dx_1} \tag{5}$$

Taking into account that the profile of a convex ellipsoidal surface is formed by a radius and described by the circle equation, the following can be obtained:

$$(y_1 + b)^2 + x_1^2 = R^2, \tag{6}$$

from (6) the following is obtained:

$$y_1 = \sqrt{R^2 - x_1^2} - b, \tag{7}$$

$$\tan\alpha = \frac{dy_1}{dx_1} = -\frac{x_1}{\sqrt{R^2 - x_1^2}} \tag{8}$$

After substituting expressions (7), (8) into (6), we obtain:

$$F_2 = \frac{bx_1}{kR\sqrt{1 - \frac{x_1^2}{R^2}}} - \frac{x_1}{k}. \tag{9}$$

2.2 Kinematics of Coil Spring and Ellipsoidal Surface

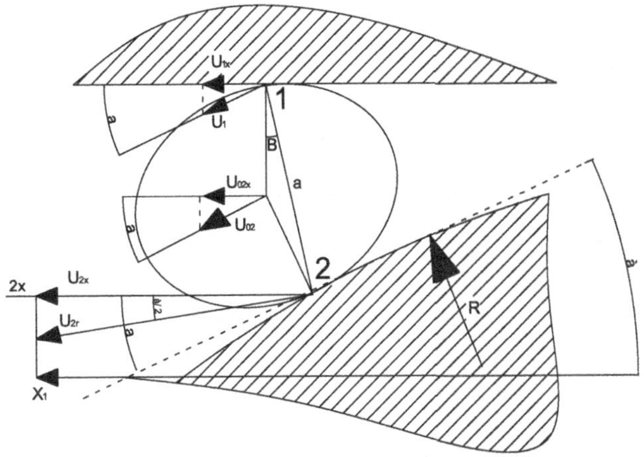

Fig. 2. Kinematic interaction diagram of the coil spring with a roller and the inner surface of the handle.

In order to express the force F_2 through the coordinate of the relative motion U, let us consider the interaction kinematic of the spring coil with the profile roller and the hammer handle (Fig. 2).For this purpose, the projections ratio of relative speeds of points 2 and 0 on the x-axis is determined, i.e.

$$\frac{V_{2x}}{V_{02x}} = \frac{V_{2x} - V_{1x}}{V_{02x}} \tag{10}$$

where V_{2x} is the projection of relative speed of point 2 on the x-axis with respect to the speed of point 1,

V_{02x} – the projection of the relative point 0 (the center of the ball) on the x-axis with respect to point 1,

V_{2x} – velocity projection of point 2 on the x- axis,

V_{1x} – thevelocity projection of point 1 on the x-axis. According to the geometric relations, it can be noted that

$$\beta = \frac{\alpha}{2}. \tag{11}$$

Then

$$V_{2x} = V_2 \cos\frac{\alpha}{2}, \tag{12}$$

V_2 – the velocity of point 2

$$V_2 = a\omega_2, \tag{13}$$

a – the shortest distance from point 2 to point 1.
ω_2 – the angular velocity of spring rotation relative to point 1

Considering that

$$a = 2r\cos\beta = 2r\cos\frac{\alpha}{2} \tag{14}$$

then after substituting the values (13), (14) into expression (12), the following can be obtained:

$$V_{2x} = 2r\omega_2\cos^2\frac{\alpha}{2}. \tag{15}$$

Let us determine V_{1x}

$$V_{1x} = V_1\cos\beta. \tag{16}$$

Since the velocity of point 1

$$V_1 = \omega_1 a = \omega_1 2r\cos\beta = 2r\omega_1\cos\frac{\alpha}{2}, \tag{17}$$

ω_1 – the angular rotation velocity of the coil spring relative to point 2

Substituting the value (17) into (16), one can get

$$V_{1x} = 2r\omega_1\cos^2\frac{\alpha}{2}. \tag{18}$$

Now let us define V_{02x}

$$V_{02x} = r(\omega_1 \cos\alpha - \omega_2) \tag{19}$$

Substituting the values (15), (18), (19) into (10), after the transformations one can obtain

$$\frac{V_{2x}}{V_{02x}} = \frac{2\cos^2\frac{\alpha}{2}(\omega_2 - \omega_1)}{\omega_1 \cos\alpha - \omega_2} \tag{20}$$

$$\frac{V_{2x}}{V_{02x}} = \frac{2}{\tan^2\frac{\alpha}{2} - 1} \tag{21}$$

Taking into account the largest angle $\alpha \leq 8°$, it can be noted that up to the third sign

$$\frac{V_{2x}}{V_{02x}} = 2$$

From here it can be noticed that the coordinate of the point where the spring coil contacts with the ellipsoid surface is two times less than the relative movement coordinate of the handle. Considering that the new coordinate origin is at point 0 (Fig. 2), one can get

$$x_1 = \frac{a}{2} + x_{2st} - a \tag{22}$$

The law of change of the friction force F_d will be derived from the assumption of its proportionality to the coil spring elastic force N (Fig. 1), therefore

$$F_d = k_0 N \tag{23}$$

k_0 – coefficient of rolling friction.

The elastic force of the coil spring N is proportional to displacement y_1, therefore

$$N = \frac{y_1}{k} \tag{24}$$

where $\frac{1}{k}$ is a proportionality coefficient or elasticity coefficient of the coil spring

Defining y_1 from (6), the following can be obtained:

$$y_1 = \sqrt{R^2 - x_1^2} - b. \tag{25}$$

Substituting the value of expressions (24), (25) into the expression for F_d (23), we obtain

$$F_d = \frac{k_0}{k}\left(R\sqrt{1 - \frac{x_1^2}{R^2}} - b\right) \tag{26}$$

It is a reminder that

$$x_1 = \frac{a}{2} + x_{2st} - a.$$

2.3 Derivation of the Motion Equation

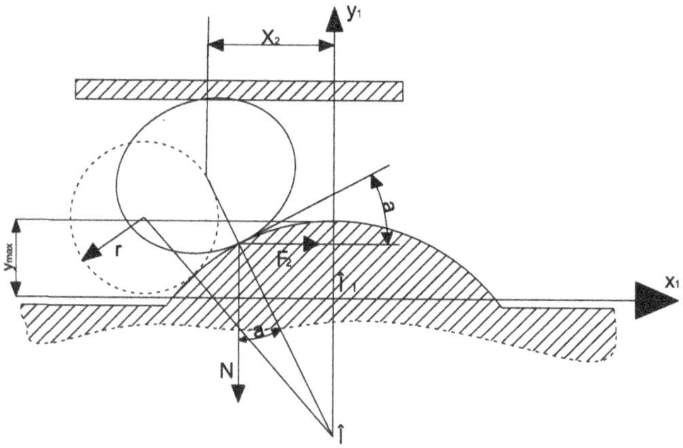

Fig. 3. Interaction diagram of a spring coil and an ellipsoidal surface.

According to the Newton's second law, the equation of motion for an isolated handle with mass m can be composed for the passive system in relative coordinates u. In this case, the schematic diagram (Fig. 1) is used (Fig. 3).

Therefore

$$m(\ddot{u} + f(t)) = P - F_1 - 2F_2 - 2F_d. \tag{27}$$

Transforming (20) to the standard equation form of forced oscillations, it can be noted

$$m\ddot{u} + 2F_d sign\dot{u} + F_1 + 2F_2 - P = -m\ddot{f}(t). \tag{28}$$

Substituting the values F_1, F_2, F_{Tp} in (27), the nonlinear differential equation of forced oscillations is obtained

$$m\ddot{u}+2\frac{k_0}{k}\left(R\sqrt{1-\frac{x_1^2}{R^2}}-b\right)sign\dot{u}+c(U+U_{st})+2\left(\frac{bx_1}{kR\sqrt{1-\frac{x_1^2}{R^2}}}-\frac{x_1}{k}\right)-P=-m\ddot{f}(t). \tag{29}$$

Adopting the motion law of the base as sinusoidal, i.e.

$$f(t)=z_0\sin\omega t. \tag{30}$$

In view of (30), the right-hand side of Eq. (29) takes the following form

$$-m\ddot{f}(t)=mz_0\omega^2\sin\omega t. \tag{31}$$

Here the amplitude of the driving force: $mz_0\omega^2=Q_0$

$$-m\ddot{f}(t)=Q_0\sin\omega t. \tag{32}$$

2.4 Transforming the Motion Equation of the Handle in Order to Solve the Problem

It can be noted that the expression $\frac{x_1^2}{R^2}$ in Eq. (29) is small compared to 1. Considering further calculation, it can be seen that

$$\frac{x_1^2}{R^2}<0.01,$$

Therefore, the Eq. (29) is transformed into the following form

$$\ddot{u}+2k_0R\left(\frac{b}{kR}-\frac{1}{k}\right)sign\dot{u}+(c_k+c_c)(U+U_{st})+2\left(\frac{b}{kR}-\frac{1}{k}\right)\left(\frac{U}{2}+x_{2st}-a\right)-P=Q_0\sin\omega t. \tag{33}$$

It can be noted that the expression $\frac{b}{kR}-\frac{1}{k}$ is the hardness c_k of the compensating device with $x_1=0$. This is proved by differentiating the law of changing the force F_2 (9) by x_1. After transforming Eq. (33), considering that

$$\frac{b}{kR}-\frac{1}{k}=c_k, \tag{34}$$

It can be obtained

$$mu+uck+cc-2k0Rc1signu+ccUSt+2ckx2st-a-P=Q0\sin\omega t. \tag{35}$$

Denoting

$$c_c U_{st} + 2c_k(x_{2st} - a) - P = L, \tag{36}$$

and introducing the new variable

$$U_1 = U + \frac{L}{c_k + c_c}, \tag{37}$$

$$\dot{U}_1 = \dot{U} \tag{38}$$

$$\ddot{U}_1 = \ddot{U}, \tag{39}$$

It can be obtained

$$m\ddot{U}_1 + (c_k + c_c)U_1 + (-2k_0 R c_1) sign\dot{u} = Q_0 \sin \omega t. \tag{40}$$

Equation (40) is a standard equation with dry friction in relative displacement. The solution of such kind of equation by the harmonic balance method for amplitude and phase shift is exposed in [4, 5] by linearizing the equation by replacing dry friction with internal one.

For the amplitude A_1 of forced oscillations in the relative displacement and for the phase shift φ [5, 6], the following formulas are given:

$$A_1 = \pm \frac{P}{k} \frac{\sqrt{1 - (4F/\pi P)^2}}{1 - \frac{\omega^2}{p^2}}; \tag{41}$$

$$\tan \varphi = \pm \frac{4F}{\pi P} \cdot \frac{1}{\sqrt{1 - (4F/\pi P)^2}}. \tag{42}$$

For Eqs. (40), symbols (41), (42) have the following meaning:

$$P = m z_0 \omega^2, \tag{43}$$

$$k = c_k + c_c, \tag{44}$$

$$p^2 = \frac{c_k + c_c}{m}, \tag{45}$$

$$F = -2k_0 R c_k. \tag{46}$$

The solution of the forced vibrations equation with dry friction for amplitude A_1 and phase shift φ is written in the following form:

$$A_1 = \frac{\sqrt{A_0^2 - (4a/\pi)^2}}{\left|1 - \frac{\omega^2}{p^2}\right|} \tag{47}$$

$$\tan\,\varphi = \frac{1}{\sqrt{\left(\frac{\pi P_0}{4R_0}\right)^2 - 1}}, \tag{48}$$

Applied to Eq. (40), the symbols of formulas (47), (48) have the following meaning:

$$A_0 = \frac{Q_0}{c_k + c_c} = \frac{mz_0\omega^2}{c_k + c_c}, \tag{49}$$

$$a = \frac{-2k_0 Rc_1}{c_k + c_c} \tag{50}$$

$$p^2 = \frac{c_k + c_c}{m}, \tag{51}$$

$$R_0 = -2k_0 Rc_1. \tag{52}$$

After transforming Eqs. (41), (42) considering (43), (44), (45), (46) and Eqs. (47), (48) considering (49), (50), (51), (52), the expression for A_1, φ is obtained:

$$A_1 = \frac{z_0\sqrt{1 - \left(\frac{-8k_0 Rc_k}{\pi\, mz_0\omega^2}\right)^2}}{\frac{c_k + c_c}{\omega^2 m} - 1} \tag{53}$$

$$\tan\,\varphi = \frac{1}{\sqrt{\left(\frac{\pi\, mz_0\omega^2}{-8k_0 Rc_k}\right)^2 - 1}}\, sign\left(1 - \frac{\omega^2 m}{c_k + c_c}\right). \tag{54}$$

Passing from the new variable U_1 to the old variable U and using expressions (36), (37), the following can be obtained for the amplitude A in relative displacement

$$A = \frac{z_0\sqrt{1 - \left(\frac{-8k_0 Rc_k}{\pi\, mz_0\omega^2}\right)^2}}{\frac{c_k + c_c}{\omega^2 m} - 1} + \left|\frac{-c_c U_{st} + 2c_k(x_{2st} - a) - P}{c_k + c_c}\right| \tag{55}$$

In order to eliminate the pressing force P, the following can be obtained from the equilibrium conditions, using the diagram (Fig. 1) and expression (9)

$$x_1 = x_{2st} - a$$

$$c_c U_{st} + 2F_2 = c_c U_{st} + 2\left| \frac{b(x_{2st} - a)}{k\sqrt{R^2 - (x_{2st} - a)^2}} - \frac{x_{2st} - a}{\mathrm{k}} \right| \tag{56}$$

Substituting the value of P from (56) into (55), after transformation the following is obtained

$$\mathrm{A} = \frac{z_0\sqrt{1 - \left(\frac{-8k_0 R c_k}{\pi m z_0 \omega^2}\right)^2}}{\frac{c_k + c_c}{\omega^2 m} - 1} + \frac{(x_{2st} - a)\left[\left(1 - \frac{b}{\sqrt{R^2 - (x_{2st} - a)^2}}\right)\frac{2}{\mathrm{k}} + 2c_k\right]}{c_k + c_c} \tag{57}$$

In the expression (57) for the amplitude *A* of the forced oscillations in the handle in relative displacement, the left part of the sum represents the dynamic deflection, and the right part represents the static deflection.

3 Obtained Results

In order to confirm theoretical studies, oscillation measurements of the IP-4009 hammer handle were carried out. Measurement of vibration accelerations was carried out using a VSHV-003-M2 device. The results are presented in Tables 1 and 2.

Table 1. Vibration acceleration value depending on the air pressure in the pneumatic system (before installing the compensating element)

P, atm	6.0	5.5	5.0	4.5	4.0	3.5	3.0
a, m/s^2	2.5	2.3	2.2	1.8	1.6	0.8	0.5
L_a,dB	128	127	127	125	124	118	114

Table 2. Vibration acceleration value depending on the air pressure in the pneumatic system (after installing the compensating element)

P, atm	6.0	5.5	5.0	4.5	4.0	3.5	3.0
a, m/s^2	2.0	1.8	1.7	1.4	1.2	0.6	0.4
L_a,dB	126	125	125	123	122	116	112

After analyzing the data given in Tables 1 and 2, it can be noted that the vibration of the handle was reduced to 2 dB as a result of applying the kinematic model of the above-mentioned vibrating insulation mechanism. The working air pressure in the pneumatic system P = 6.0 atm compared to the vibrating insulation mechanism without a hardness compensator.

4 Conclusions

1. It is possible to achieve arbitrarily small, even zero hardness of the vibrating insulation mechanism by adjusting the parameters of the compensating device.
2. The introduction of a compensating device into the vibrating insulation air spring system design of the pneumatic hammer does not change the external dimensions of the hammer and does not affect the weight change.
3. The frequency of blows and the energy of a single blow of a pneumatic hammer do not change after the introduction of the compensating element into the vibrating insulation air spring system design of the pneumatic hammer.

References

1. Glushkov, S.P.: Vibroizolyaciya teplovy`x dvigatelej (nauchnoeizdanie). Novosibirsk, NGAVT, 217 p. (1999)
2. Edwards, D.J., Holt, G.D.: Hand-arm vibration exposure from construction tools: results of a field study. Constr. Manag. Econ. **24**, 209–217 (2006). ISSN 0144-6193
3. Amitkumar, K., Tayde, M.M.: Design and analysis of jigsaw and damper for effect of vibration reduction in hand held power tool. Int. J. Emerg. Technol. Adv. Eng. **5**(7), 65–73 (2015)
4. Kazanskaya, L.F., Smirnova, O.M.: Supersulphated cements with technogenic raw materials. Int. J. Civil Eng. Technol. **9**(11), 3006–3012 (2018)
5. Belentsov, Yu.A, Smirnova, O.M.: Influence of acceptable defects on decrease of reliability level of reinforced concrete structures. Int. J. Civil Eng. Technol. **9**(11), 2999–3005 (2018)
6. Alpatov, V.Y., Veremeenko, O.Y., Sakharov, A.A., Shirokov, V.S.: Trial design of a dome roof for an church. In: MATEC Web of Conferences, vol. 86, p. 02015 (2016). https://doi.org/10.1051/matecconf/20168602015

Features of Work of a Single-Phase Thyristor Pulse Converter with a General Magnetic Circuit

Anatoly Kalinichenko[1](✉), Petr Kozlov[2], and Alexander Smolyaninov[3]

[1] NPK YUG - INTEGRAL, Novy Arbat, 36, of 30/2, E, 121099 Moscow, Russia
integralcompany@mail.ru
[2] Production Holding NPK STRATEG, ul. Nizhegorodskaya, House 32, Block 15, Moscow, Russia
[3] Ekaterinburg Ural State University of Railway Transport (USURT), ul. Kolmogorova, 66620034 Yekaterinburg, Russia

Abstract. Peculiarities of operation of traction and non-traction, power thyristor converters consist, first of all, in the need of using sufficiently large setting parameters of switching equipment to obtain acceptable performance indicators, for example, permissible levels of current ripples in the load and in the power source, and the absence of interfering effects on communication lines and systems ensure the safety of train traffic. In any case, a reduction in current and voltage ripple is required. In this case, we consider the use of chokes with a common magnetic core, which allows us to change the picture of the appearance of ripples in the output current and to obtain controllability of this process. This work for multiphase systems was performed for the first time and shows the reality of the direction of development of multiphase systems, up to and including obtaining a smooth component with pulsed selection of power, which is very important for transmission over large powers.

Keywords: Thyristor pulse converter · Electric rolling stock · Bridge circuit · Separating choke · Multi-phase converter

1 Introduction

The implementation of high loads on the electric rolling stock of direct current with thyristor-pulse voltage regulation with existing types of power valves is possible in two ways: using a single converter with several branches of thyristors working in parallel or using multiphase converters operating on a common load with one branch of thyristors in each phase.

The parallel operation of the thyristors of the converter places high demands on the control system and requires the use of special measures to ensure a uniform division of the currents between the valves connected in parallel. In addition, the low frequency at the input filter and at the load with such a circuit solution significantly affects the dimensions of the input filter and the smoothing reactor.

Z. Popovic et al. (Eds.): TransSiberia 2019, AISC 1115, pp. 768–777, 2020.
https://doi.org/10.1007/978-3-030-37916-2_75

When using multiphase converters, it is possible to increase the frequency at the input filter and to reduce the current ripple in the common load circuit, as well as eliminate the parallel connection of the converter power thyristors [1]. The high frequency of the converter operation is achieved due to the alternate operation of the phases of the pulse-width converter, which is equal to

$$f_i = nf$$

Where

n – the number of phases of the converter;
f – operating frequency of the inverter thyristors

In Fig. 1, a simplified diagram of a six-phase thyristor-pulse converter using a bridge circuit [2] is shown. To ensure the uniform operation of the phases and to simplify control of the converter thyristors, it is necessary to alternately open the thyristors of the phases in a certain order, for example, first all the even phases of the converter T2, T4, T6, and then the odd phases T1, T3, T5. Such a solution when using a bridge circuit of a thyristor-pulse converter allows recharging the switching capacitor without any difficulty.

To ensure normal operation of the converter phases when operating on a common load, it is imperative to use split reactors L1 – L6 designed to smooth the current ripple in the load circuit, to separate the cathodes of the thyristors of the converter phases, to prevent the converter from disrupting and to perform the voltage divider (Fig. 1).

Given that the consideration of such a structure and the analysis of electromagnetic processes is performed for the first time, it is advisable to dwell on the characteristic features of the course of electromagnetic processes at various stages of the converter operation.

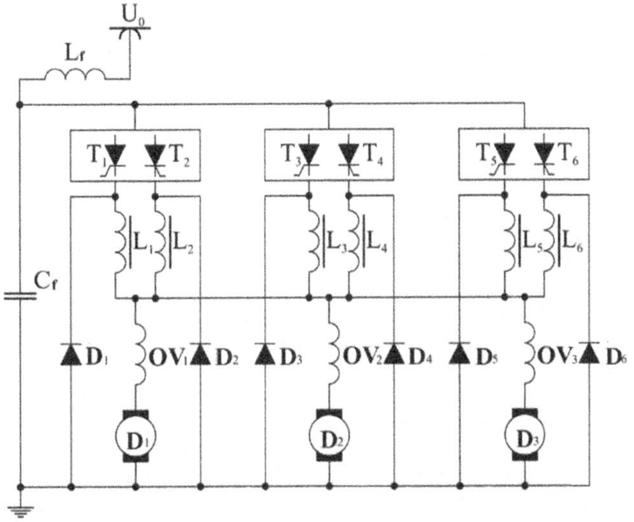

Fig. 1. Simplified six-phase converter circuit.

Depending on the inductance of the separating choke and the inductance of the load circuit, as well as the switching frequency of the thyristors, several operating modes of the converter phases are possible [3]:

1. intermittent current mode;
2. boundary mode;
3. continuous current mode.

The most characteristic are the 1st and 3rd modes. The mode of continuous phase currents occurs when the repetition period of the voltage pulses is smaller than the electromagnetic constant of the phase circuit.

Especially unfavorable is the mode of intermittent currents, since when this mode occurs, there is a danger of overloading the power thyristors into the inverter phase that is being put into operation.

When multi-phase thyristor-pulse converters are in operation, two characteristic points of operation of converters should be distinguished:

$$0 < t_i < T',$$

$$T' < t_i < T_f,$$

where

T_f – period of voltage phase pulses;
t_i – pulse duration;
T' – pulse repetition period at the load

$$T' = \frac{T_f}{n}$$

The most unfavorable condition for the operation of chokes is their operation at $t_i < T'$. [5] In this case, the separating choke of the switched-on phase takes over the maximum voltage in amplitude U_L (Fig. 2), which is equal to

$$U_L = U_0\left(1 - \frac{1}{n}\right) \tag{1}$$

where

U_0 – power supply voltage; while the load is perceived

$$U_i = U_{rd} - U_L$$

where

U_{rd} – voltage on the reverse diode

As the pulse duration of the output phase voltage t_i increases further, and when the last value of $t_i < T'$ is reached, voltage is applied to the adjacent phases of the converter (Fig. 3), i.e. at the same time two phases of the converter will be switched on.

As can be seen from Figs. 2, 3 with pulse duration $t_i < T'$ the impulse voltage is applied to the load with amplitude

$$U_n = \frac{U_0}{n}, \tag{2}$$

and if $t_i < T'$ - pulsating voltage with the amplitude of the variable component is also equal to (2).

When the pulse voltage of an output voltage reaches a value multiple to the period of following pulses on the load, i.e. at $t_i = T', 2T', \ldots nT'$ the amplitude value of the voltage on the load will be equal to the corresponding average value, i.e.

$$U_{av} = U_n, 2U_n, \ldots nU_n$$

in this case, the ripple of voltage and load current will theoretically be equal to zero. The voltage on the load reaches a value equal to the supply voltage at

$$t_i = T_f = nT'$$

As the number of simultaneously operating phases increases, the separating choke of the switched on phase perceives a decreasing voltage in the amplitude U_L (Fig. 3), while the load increases accordingly, due to the fact that the separating chokes simultaneously play the role of voltage dividers.

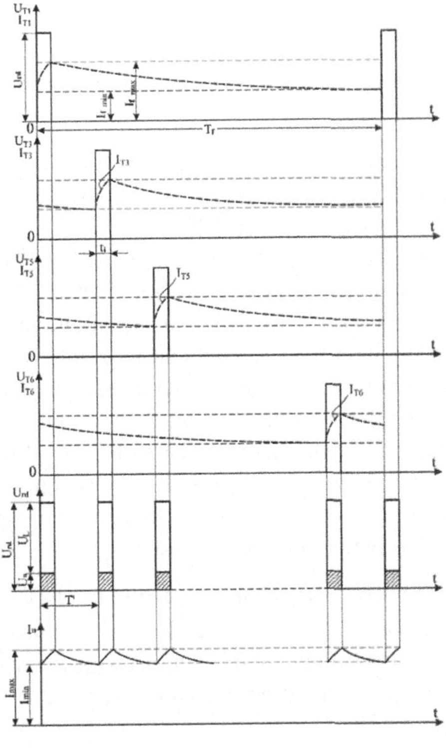

Fig. 2. Diagram of voltages and currents in the converter at $t_i <$ T.

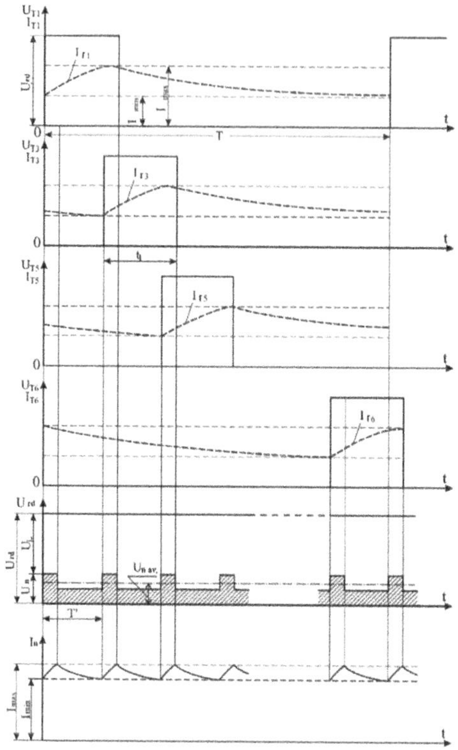

Fig. 3. Diagram of voltages and currents in the converter at $t_{и} > T'$.

Let's analyze the work of a multi-phase converter (Fig. 1) at $t_i < T'$ (Fig. 2). In this case, only one phase of the converter is turned on, in the remaining dividing chokes the current is maintained, decreasing exponentially.

Since the instantaneous values of the currents in all dividing chokes of the non-switched phases are different, the inductance of the chokes $L_2 - L_6$, taking into consideration that they are with steel, have different meanings. In the continuous current mode, these values differ within the accepted current ripple in the choke.

To simplify the reasoning, we assume the inductance of chokes is constant and the same. As can be seen from the simplified scheme of Fig. 4a, the inductors of the non-working phases are interconnected in parallel due to the fact that a current through the reverse diodes is supported by electromagnetic energy accumulated at the moment of impulse. According to the assumption made above, we simplify the replacement scheme. In Fig. 4b the following notation is used:

L_1 – inductance of the separating chokes of working phases;

L_2' – inductance of chokes of non-operating phases, if we take the inductance of one separating choke equal to L, then taking into account the above and conditionally assuming that one phase is turned on, we can write:

$$\frac{1}{L_2'} = \frac{1}{L_2} + \frac{1}{L_3} + \ldots + \frac{1}{L_6}$$

i.e.

$$L_2' = \frac{1}{n-k}L \quad L_1 = \frac{L}{n-1}$$

where

n – total number of converter phases;
k – the number of simultaneously included phases.

We will analyze the electromagnetic processes, for which we compose a system of differential equations for the time intervals corresponding to the state of the converter phases. In doing so, we accept the following assumptions:

1. valves have an ideal characteristic;
2. the voltage at the converter input is perfectly smooth;
3. during one half-period the anti-electromotive force of the engine is constant;
4. the resistance of the separating chokes is zero.

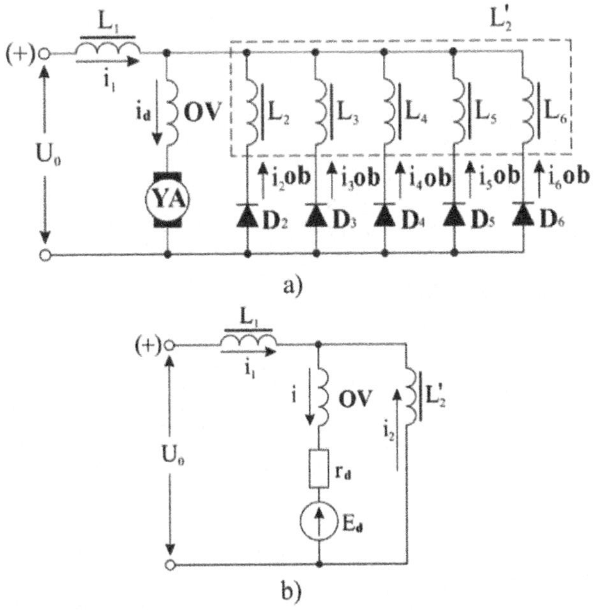

Fig. 4. (a) Instantaneous six-phase converter with single phase operation. (b) The substitution circuit corresponding to (a).

Let's make system of the differential equations according to the equivalent circuit of Fig. 4b.

$$\begin{cases} L_1 \frac{di_1}{dt} + L_d \frac{di_{д}}{dt} + i_d\, r_d + E_d = U_0 \\ L_2 \frac{di_2}{dt} + L_d \frac{di_d}{dt} + i_d\, r_d + E_d = U_0 \\ i_d = i_1 + i_2 \end{cases} \tag{3}$$

where

i_1 – current in the separating choke of the included phase;
i_2 – current in the equivalent choke of non-operating phases;
i_d – load circuit current;
L_1 – inductance of the choke of the switching-on phase;
L_2' – equivalent inductance of the choke of non-working phases;
L_d – load circuit inductance;
E_d – electromotive force of the engine;
U_0 – power supply voltage.

By solving a system of differential Eq. (3) regarding currents i_1 and i_2, we have:

$$\begin{aligned} i_1 = {} & U_0 \frac{\beta_1}{\omega_1 L_1 (1+\gamma_1)} \left(e^{-\frac{\omega_1}{\beta_1}t} + \frac{\omega_1}{\beta_1} t - 1 \right) + I_{10}\, e^{-\frac{\omega_1}{\beta_1}t} \\ & + \left\{ [(1+\gamma_3) U_0 - E_d] \frac{\beta_1 \gamma_1}{L_1 (1+\gamma_1)} + \frac{I_{10} - \gamma_1 I_{20}}{1+\gamma_1} \right\} \left(1 - e^{-\frac{\omega_1}{\beta_1}t} \right) \end{aligned} \tag{4}$$

$$\begin{aligned} i_2 = {} & I_{20} e^{-\frac{\omega_1}{\beta_1}t} - U_0 \frac{\beta_1}{L_1 \omega_1 (1+\gamma_1)} \left(e^{-\frac{\omega_1}{\beta_1}t} + \frac{\omega_1}{\beta_1} t - 1 \right) \\ & + \left[(-\gamma_2 U_0 - E_d) \frac{\beta_1}{L_1 (1+\gamma_1)} + \frac{\gamma_1 I_{20} - I_{10}}{1+\gamma_1} \right] \left(1 - e^{-\frac{\omega_1}{\beta_1}t} \right) \end{aligned} \tag{5}$$

where the designations are taken:

$$\gamma_1 = \frac{L_2'}{L_1}$$

$$\gamma_2 = \frac{L_d}{L_1}$$

$$\gamma_3 = \frac{L_d}{L_2'}$$

$\gamma_1, \gamma_2, \gamma_3$ – proportionality coefficients;

$$\omega_1 = \frac{\left(1 + \frac{1}{\gamma_1}\right)}{1 + \gamma_2 + \gamma_3}$$

$\beta_1 = \frac{L_1}{r_d}$ – time constant

For continuous current mode in phases is correct that:

$$\begin{cases} I_{10}(0) = i_{2(T_f - t_i)} \\ I_{20}(0) = i_{1(t_i)} \end{cases} \tag{6}$$

where

fi – period of the open state of the thyristor phase;
T_f – pulse repetition period of one phase converter

To find expressions for currents I_{10} and I_{20}, we solve together the Eqs. 4–6. As the $I_{10} = I_{min}$, $I_{20} = I_{max}$, then

$$I_{min} = \frac{U_0\left[\varphi - \frac{1}{1+\gamma_1} B\, e^{-\frac{\omega_1}{\beta_1}T_f}\left(e^{\frac{\omega_1}{\beta_1}t_i} - 1\right) - \gamma_1(C + T_f)\right.}{2L_1\left[1 + \gamma_1 - \left(e^{-\frac{\omega_1}{\beta_1}(T_f - t_i)}\right.\right.} \\ \frac{\left.\times\left(e^{-\frac{\omega_1}{\beta_1}t_i} - 1\right) - D\right] - E_d\left[1 - e^{-\frac{\omega_1}{\beta_1}(T_f - t_i)} - \frac{\gamma_1}{1+\gamma_1}(N + Z)\right]}{\left.\left.+ \gamma_1 e^{-\frac{\omega_1}{\beta_1}t_i}\right)\right]} \tag{7}$$

$$I_{max} = \frac{U_0\left[F + \frac{1}{1+\gamma_1} B\left(e^{-\frac{\omega_1}{\beta_1}t_i} - e^{-\frac{\omega_1}{\beta_1}T_f}\right) + C\left(e^{-\frac{\omega_1}{\beta_1}(T_f - t_i)}\right.\right.}{2L_1\left[1 + \gamma_1 - \left(e^{-\frac{\omega_1}{\beta_1}(T_f - t_i)}\right)\right.} \\ \frac{\left.-1) - \varphi\right] + E_d\left[\beta_1\gamma_1\left(e^{-\frac{\omega_1}{\beta_1}t_i} - 1\right) + \frac{1}{1+\gamma_1}(K + Z)\right]}{\left.\left.+ \gamma_1 e^{-\frac{\omega_1}{\beta_1}t_i}\right)\right]}$$

where the following notations are used

$$A = \left(\frac{\omega_1}{\beta_1} - \gamma_2\right)\left(1 - e^{-\frac{\omega_1}{\beta_1}(T_f - t_i)}\right) - (T - t_i)$$

$$B = \beta_1\left(\frac{1}{\omega_1} - \gamma_1 - \gamma_1\gamma_3 - \gamma_2 + \frac{\gamma_1}{\omega_1}\right)$$

$$C = \gamma_2 - \beta_1\omega_1 - \beta_1\gamma_1\gamma_3 - i$$

$$D = t_и\left(\gamma_1 + e^{-\frac{\omega_1}{\beta_1}(T-t_i)}\right)$$

$$N = (\beta_1 - 1)e^{-\frac{\omega_1}{\beta_1}T}\left(1 - e^{\frac{\omega_1}{\beta_1}t_i}\right)$$

$$F = \beta_1\left[\frac{1}{\omega_1} - \gamma_1(1+\gamma_3)\right]\left(e^{-\frac{\omega_1}{\beta_1}t_i} - 1\right) + t_i$$

$$K = \gamma_1(\beta_1 - 1)\left(e^{-\frac{\omega_1}{\beta_1}t_i} - e^{\frac{\omega_1}{\beta_1}T}\right)$$

$$\varphi = (T - t_i)\left(\gamma_1 e^{-\frac{\omega_1}{\beta_1}t_i} + 1\right)$$

$$Z = (1 + \beta_1\gamma_1)\left(e^{-\frac{\omega_1}{\beta_1}t_i} - 1\right)$$

Ripple current in the phase inductors can be determined according to (7) as

$$\Delta I = I_{max} + I_{min}$$

2 Conclusions

To improve the performance of traction converters, first of all, to reduce the size of the input and output filters, to reduce the ripple of currents in the load and in the input filter, it is advisable to use multi-phase traction converters with throttles on the common magnetic core.

When the output voltage of the phase voltage reaches a value that is multiple to the pulse repetition period on the load in multiphase converters with a common magnetic circuit, the voltage ripples and currents in the load circuit are zero, which stimulates an increase in the number of phases of the traction converter to implement the specified multiphase converter [6].

These studies for multiphase (in this case, six-phase) pulsed systems using separating chokes on a common magnetic core are performed for the first time, and the results confirm the feasibility of using such system design principles at high drive powers and achieving output current ripples power selection with the possibility of using them for transmission over long distances with a smooth pulsation component.

The use of the considered features of building multiphase pulse systems makes it possible to ensure the absence of voltage and current ripples on the input filter and on the load in the case of the use of multiphase converter power thyristors, whose turn-on time is equal to the turn-off time of the thyristors, which is currently real and stimulates, in addition, applications in converters of solid semiconductor structures. The latter is implemented using digital control systems that provide a clear division of the time intervals developed by the authors for the implementation of any adaptation control methods. The use of multiphase converters with separator throttles on a common magnetic drive in a traction drive will reduce losses in traction motors, reduce the effect of harmonics on traffic safety systems operating in a wide range of 50 Hz–555 Hz, and improve the operating modes of the contact network.

References

1. Zhao, Y., Crossley, P.: Impact of DC bias on differential protection of converter transformers. Int. J. Electr. Power Energy Syst. **115**, 105426 (2020). https://doi.org/10.1016/j.ijepes.2019.105426
2. Xu, S.-Z., Peng, Y.-F., Li, S.-Y.: Suppression effectiveness research on multi-level EMI filter in thermal electromagnetic interactive filed of explosion-proof three-level NPC converter. Case Stud. Thermal Eng. **15**, 100510 (2019). https://doi.org/10.1016/j.csite.2019.100510
3. Hernández, F., Chiang, L.E., Corbalán, P.: A general architecture for electric power management of small scale NCRE converters: design methodology and validation. Energy. Sustain. Dev. **41**, 128–138 (2017). https://doi.org/10.1016/j.esd.2017.09.001
4. Belentsov, Y., Smirnova, O.M.: Influence of acceptable defects on decrease of reliability level of reinforced concrete structures. Int. J. Civil Eng. Technol. **9**(11), 2999–3005 (2018)
5. Kalinichenko, A.Ya.: On the issue of reducing the weighty indicators of thyristor converters of electric rolling stock. Sci. J. "Transp. Ural", 83–87 (2014). Ekaterinburg
6. Kalinichenko, A.Ya.: The impact of the discrete component in the digital control of the electromotive structure on the train traffic safety system. Collection "Questions of the theory of security and sustainability of systems". M.EC of the Russian Academy of Sciences, pp. 148–158 (2006)

Methods of Monitoring Longitudinal Stresses in Rails Using Acoustoelastic Effect

Ludmila Stepanova(✉), Aleksandr Kurbatov, and Evgeniy Tenitilov

Siberian Transport University (STU),
191 Dusi Kovalchuk Street, 630049 Novosibirsk, Russia
stepanova@stu.ru

Abstract. Two methods of monitoring longitudinal mechanical stresses in rails by ultrasonic scanning using the acoustoelastic effect have been developed. According to the first method, stresses in a rail are monitored through temporal parameters of ultrasonic waves measured in an unloaded rail piece and in a loaded rail of a continuous welded rail track. This is done using two 18° angle beam piezoelectric transducers arranged in a pitch-catch configuration of rail scanning. This way, longitudinal and transverse waves are both excited simultaneously in the rail. Transducer acoustic axes lie in the same plane and face each other over the rail length. Upon reaching the back surface of the rail base, ultrasonic waves reflect from it and transform. The temporal parameters obtained from the reflected longitudinal and transverse waves as well as the waves transformed as a result of reflecting from the rail back surface were used to determine the longitudinal stresses. It was found that the measurement errors for the longitudinal mechanical stresses in the first method of monitoring is 2… 6% on average. The second method for monitoring longitudinal stresses determines mechanical stresses more accurately as its measurement results are less influenced by rail height. The acoustoelastic coefficient of the transformed waves, k_{LT}, has been determined experimentally. The methods of monitoring longitudinal stresses in rails by ultrasonic scanning using the acoustoelastic effect have been verified by measurements with a multichannel microprocessor strain-gauge system MMTS-64.01.

Keywords: Rail · Continuous welded rail track · Ultrasound · Acoustoelasticity · Mechanical stresses

1 Introduction

A railway track is a complex engineering structure that is constantly experiencing various stressing factors. Under the effect of environment, the impact of rolling stock, upon recovery and maintenance works, the rails laid into a track develop longitudinal, bending, and internal stresses. The rails experience vibration, exposure to aggressive media and mechanical loads, and, as a result, develop stresses, structural defects, and fatigue defects that are detrimental to their strength properties and lead to failure [1]. To ensure rail traffic safety, a continuous track monitoring must be organized to uncover

Z. Popovic et al. (Eds.): TransSiberia 2019, AISC 1115, pp. 778–787, 2020.
https://doi.org/10.1007/978-3-030-37916-2_76

regions with high stress levels. Over straight track runs the rails develop mostly longitudinal stresses due to environmental temperatures. In a continuous welded rail track an exposure to temperature produces longitudinal compression stresses (upon rail heating above the stress-free temperature) and tensile stresses, if the temperature goes below the stress-free temperature. An increased temperature generates a compressive force reaching 600–1,000 kN, which is capable of inducing track buckling. Upon temperature decrease, the tensile forces reach 1,200–1,700 kN, causing cracking and failure of rail welds.

Along with temperature-induced stresses that are accounted for in track designs, additional "off-design" stresses also arise due to other reasons: rail creepage, working loads from rolling stock, loosening of ballast section during works without subsequent sufficient punning, deformations caused by repair works, and other factors. All this (together with the existing residual stresses in rail welds and adjacent zones) can lead to formation and development of defects that, if not discovered in due time, can result in rail breakage. The purpose of this work is to develop methods for monitoring longitudinal stresses in rails using the acoustoelastic effect.

2 Research Methods

Rail stresses can be determined using various non-destructive testing (NDT) methods, with the most common being ultrasonic scanning (US). Recent studies examined the use of the acoustoelastic effect for this purpose [1–4]. Acoustoelastic stress monitoring is based on measuring the properties of elastic waves propagating in strain-elastic medium. The main information bearing parameters are the dependences of elastic wave velocity, amplitude, and frequency on mechanical stress.

At the stress-free temperature, there are no compressive or tensile stresses in the rail, i.e. the longitudinal force is zero. The relation between the stress-free temperature and the longitudinal stresses in the rail is calculated by the following expression [1]:

$$\sigma_T = \alpha \cdot E \cdot \Delta T \approx 2.5 \cdot \Delta T \tag{1}$$

where σ_T is the stress in the rail caused by tensile thermal forces arising at rail temperatures below its stress-free level, MPa; α is the linear expansion coefficient of rail steel ($\alpha = 0.0000118\ \text{deg}^{-1}$); E is the elastic modulus of rail steel; ΔT is the difference between the temperature at which the stress is being determined and the stress-free temperature for the rail on its sleepers, °C

$$\Delta \mathrm{T} = \mathrm{T_r} - \mathrm{T_z}, \tag{2}$$

where T_r is the rail temperature at which the longitudinal stress is being determined, °C; T_z is the stress-free temperature for the rail on its sleepers, °C.

Rail stress monitoring by ultrasonic scanning can use different wave types. It is well known that elastic wave propagation velocity is affected by the temperature of a material under study. The elastic wave propagation velocities are determined by elastic moduli, which in turn depend on temperature. The maximum thermal stress $[\sigma]$ for thermostrengthened rails permissible in a continuous welded rail track with traffic loads and the required rail safety factor taken into account can be found from the following expression:

$$[\sigma] \geq \sigma_T + k_n \cdot \sigma_k, \tag{3}$$

where k_n is the safety factor; σ_k are the stresses at the edges of the rail base due to bending and torque from the load exerted by rolling stock wheels, MPa.

The safety factor k_n for first service life rails and for used rails that underwent diagnostics and recovery in shop conditions or profile grinding on the way is k_n = 1.3.

To simplify the procedure of rail stress diagnostics by ultrasonic scanning using the acoustoelastic effect, an additional unloaded piece of an R65 rail placed on a rail cart was employed. The internal mechanical stresses in rails in a functioning fragment of a railway track and in its unloaded counterpart were measured by mounted emitting piezoelectric transducers (PET). Then ultrasonic pulses of longitudinal and transverse waves were introduced into the loaded rail and its unloaded counterpart at an entry angle of 18°. The receiving transducers received the transmitted signals; their transition times were measured and used to calculate the internal mechanical stresses. The transition times of the waves in loaded and unloaded rails were determined from measurements. Additionally, the transverse waves transformed from the longitudinal waves incident at the rail under study and the longitudinal waves transformed from the transverse waves incident at the object under study were also received. The stress was determined from the change in the transition time of the transmitted waves using the following expression [8]:

$$\sigma = \frac{\Delta\tau_L - \Delta\tau_{LT}}{\beta_T \tau_T \cos \alpha_T} \tag{4}$$

where β_T is the acoustic coefficient for a transversal wave in a loaded rail; τ_T is the delay time for a transversal wave in a loaded rail; $\Delta\tau_L$ is the change in delay time for longitudinal waves in loaded and unloaded rails; $\Delta\tau_{LT}$ is the change in delay time for longitudinal waves transformed into transversal waves in loaded and unloaded rails; α_T is the entry angle for transversal waves in loaded rail.

3 Research Results

Figure 1 shows the scheme of sounding in an ultrasonic method for determination of internal mechanical stresses in a rail, where 1 is the rail; 2 is the emitting transducer (GE); 3 is the receiving transducer (GR); H is rail height. Figure 2 shows scope traces of the probing signal 1, the signal from the longitudinal wave 2, the signal from the longitudinal wave transformed into the transverse wave 3, and the transverse wave signal 4. Figure 3 shows the graph of stress distribution obtained with a certified multichannel microprocessor strain-gauge system MMSS-64.01 (type approval certificate RU.C.34.007.A No. 44412, registration number No. 21760-01) of accuracy class 0.2 [5]. When a uniaxial strained state of the rail is created in a hydraulic loading unit, the stress field is oriented along the object [6]. Figure 1 shows the entry angles of ultrasonic oscillations for the longitudinal αL and transverse αT waves. Point A marks the entry point of ultrasonic oscillations for the longitudinal and transverse waves. Point B corresponds to exiting ultrasonic signals for waves transmitted through the rail. At points C, D, ultrasonic waves reflect and transform.

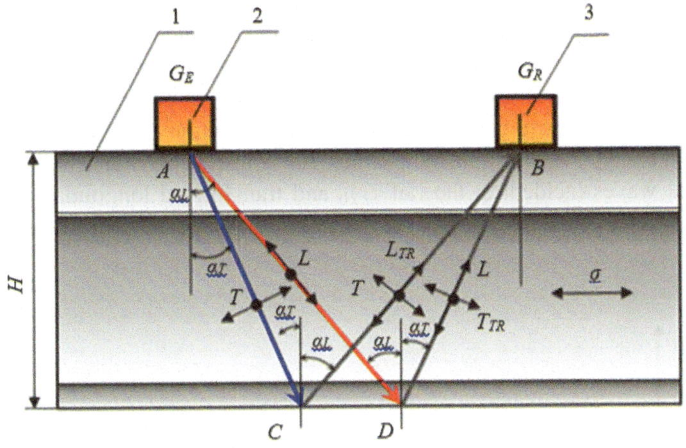

Fig. 1. Scheme of sounding in the ultrasonic method for determination of internal mechanical stresses in the rail.

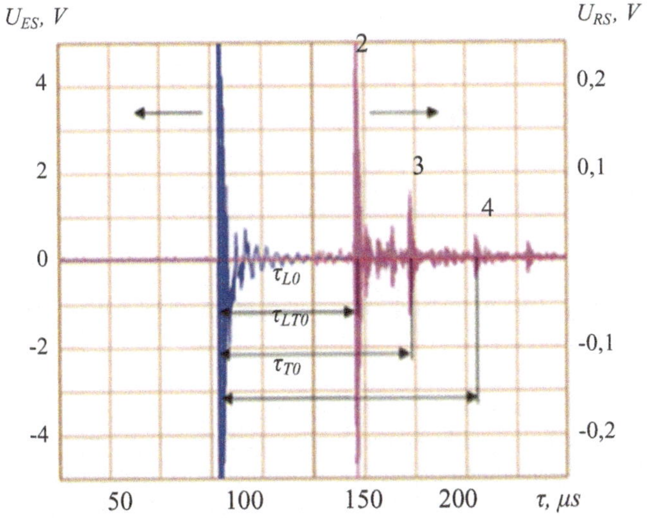

Fig. 2. Scope traces of the probing and received ultrasonic signal.

In Fig. 1, L, T, L_{TR}, T_{TR} mark the directions of the displacement vectors for oscillations of the reflected longitudinal, transverse (L,T), transformed longitudinal L_{TR} and transformed transverse T_{TR} waves. By sending pulses of longitudinal and transverse ultrasonic waves into an unloaded rail at point A at angles α_L and α_T and receiving them at point B, transition times $\tau_{L0}, \tau_{T0}, \tau_{LT0} = \tau_{TL0}$ $\tau_{L_0}, \tau_{T_0}, \tau_{LT_0} = \tau_{TL_0}$ of longitudinal, transverse, longitudinal transformed in the rail into transverse and transverse transformed in the rail into longitudinal waves were measured.

When studying a loaded rail and its unloaded counterpart with the receiving transducer 3 (Fig. 1), transformed transverse waves T_{TR} from longitudinal waves incident at the rail L and transformed longitudinal waves L_{TR} from transverse waves incident at the rail T were additionally received. Transition times for longitudinal τ_{L0} and transverse τ_{T0} waves, transition times for transformed transverse waves from longitudinal waves incident at the rail τ_{LT0} and transformed longitudinal waves from transverse waves incident at the rail τ_{TL0} in an unloaded rail counterpart were measured.

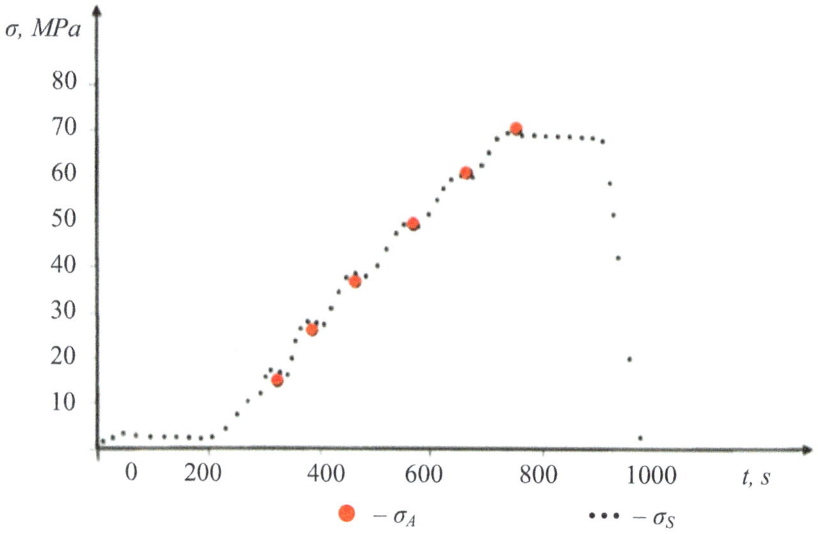

Fig. 3. Graph of time dependence of the distribution of measured and calculated stresses in a rail.

Then the emitting transducer 2 and receiving transducer 3 were mounted on a rail laid into a track, scanning with the receiving transducer 3 and finding the maximum amplitude from the transmitted pulse of longitudinal wave transformed into transverse wave LT. Pulse amplitudes of the longitudinal L (pulse 2 in Fig. 2) and transverse T (pulse 4 in Fig. 2) waves were registered on the scope trace, and the delay times of the longitudinal τ_L, transverse τ_T and transformed into transverse τ_{LT} waves in the loaded rail were determined. Knowing the delay times τ_L, τ_T, τ_{LT} in the loaded rail and its unloaded counterpart, the changes in the delay times $\Delta\tau_L$, $\Delta\tau_T$, $\Delta\tau_{LT}$ were determined, and afterwards the stress σ was calculated from Eq. (3).

The internal mechanical stresses in an R65 rail were measured with type PKS-12 certified resistance strain gauges with resistance $R = 120\ \Omega$, base $l = 12$ mm, strain gauge factor $K = 2.12$ (Certificate RU.C.28.007.A № 30935, state register entry No. 37343-08). The gauges were glued to rail head, web, and base. Prior to monitoring the rail under study, the transition times for ultrasonic signals in an unloaded rail counterpart $\tau_{L0}, \tau_{T0}, \tau_{LT0} = \tau_{TL0}$ were determined (Fig. 1). Scanning the surface of the unloaded rail counterpart with the receiving transducer 3, the maximum amplitude of the transmitted pulse of the longitudinal wave transformed into transverse wave *LT* were registered with a *TDS*-2014 scope. Pulse amplitudes of the longitudinal *L* and transverse *T* waves, delay times of the longitudinal τ_{L0}, transverse τ_{T0} waves, and delay times of longitudinal waves transformed into transverse waves τ_{LT0} were registered on the scope trace (Fig. 2).

After that the emitting transducer 2 and receiving transducer 3 were mounted on a rail laid into a railway track (Fig. 1), and amplitudes of the longitudinal *L* and transverse *T* waves were registered on the scope trace, and the delay times of the longitudinal τ_L, transverse τ_T waves and the delay time of the longitudinal wave transformed into a transverse wave τ_{LT} in the loaded rail were determined. Then the changes in these delay times $\Delta\tau_L$, $\Delta\tau_T$, $\Delta\tau_{LT}$ were determined as $\Delta\tau_L = \tau_L - \tau_{L0}$; $\Delta\tau_T = \tau_T - \tau_{T0}$; $\Delta\tau_{LT} = \tau_{LT} - \tau_{LT0}$. The acoustoelastic coefficient for transverse wave is known and is $\beta_T = 7.2\ \text{TPa}^{-1}$. Entry angle of the transverse wave was determined from the sine law:

$$\alpha_T = \arcsin\left(\frac{C_T \sin\alpha_L}{C_L}\right) = 9.7^\circ$$

Using the known velocity of longitudinal wave C_L in steel, equal to 5,910 m/s, the velocity of transverse wave C_T, equal to 3,230 m/s, and the entry angle of longitudinal wave $\alpha_L = 18^\circ$, the internal stress is calculated from Eq. (3). Thus, calculated values of internal mechanical stresses in loaded rail σ_A and the values measured with a strain-gauge system MMSS-64.01 σ_S are listed in Table 1.

As seen in Table 1, the error of stress measurement and calculation is 2…6% on average, which is acceptable for practical testing. The results of stress measurement with a microprocessor strain-gauge system MMSS-64.01 σ_S are shown in Fig. 3 as a stepwise dependence $\sigma(t)$. The points show the calculated stress values σ_A, which were obtained from the temporal parameters of ultrasonic waves measured with a *TDS*-2014 scope. The suggested method has the advantage of a lower cost of monitoring, as it uses standard production transducers with wave entry angle 18°. This allows measuring stresses in rails laid into a track over their entire length.

Table 1. Measurement errors for mechanical stresses determined with acoustoelastic technique.

Loading number	1	2	3	4	5	6
Error, % $\frac{\lvert\sigma_S-\sigma_A\rvert}{\sigma_S}\cdot 100\%$	13.74	5.92	2.69	0.2	1.22	2.08

A method for improving the accuracy of determination of the mechanical stresses due to lower influence of rail height on the measurement results has also been developed. The pulses transmitted through the rail were received with one normal beam twin transducer probe and three angle beam receiving transducers placed at distances S_1, S_2, and S_3 from the emitting transducer, given by the following expressions:

$$S_1 = 2H \operatorname{tg} \alpha_T \tag{5}$$

$$S_2 = H(\operatorname{tg} \alpha_T + \operatorname{tg} \alpha_L) \tag{6}$$

$$S_3 = 2H \operatorname{tg} \alpha_L \tag{7}$$

where α_L, α_T are the entry angles of the longitudinal and transverse waves of the emitting angle beam transducer.

The transition times for these waves were measured in a loaded rail and in its unloaded counterpart, and the stress σ was determined from equation [9]:

$$\sigma = \frac{\frac{\tau_{T3}-\tau_{T1}}{\tau_{LT2}} - \frac{\tau_{T03}-\tau_{T01}}{\tau_{LT02}}}{k_{LT2}} \tag{8}$$

where τ_{T01}, τ_{T1} are the propagation times of the transverse wave signal from the emitting transducer to the first receiving transducer for the loaded rail and its unloaded counterpart, ns; τ_{T03}, τ_{T3} are the propagation times of the transverse wave signal from the emitting transducer to the third receiving transducer for the unloaded counterpart and the loaded rail, ns; τ_{LT02}, τ_{LT2} are the propagation times of the transformed wave signal from the emitting transducer to the second receiving transducer for the unloaded counterpart and the loaded rail, ns; k_{LT2} is the acoustoelastic coefficient of the transformed waves from the emitting transducer to the second receiving transducer, MPa^{-1}.

Rail height H was determined as:

$$H = \frac{C_{Ldir}\tau_{Ldir}}{2}$$

where C_{Ldir} is the velocity of longitudinal ultrasonic wave in the rail under the normal beam twin transducer probe, m/s; τ_{Ldir} is the propagation time of longitudinal ultrasonic wave in the rail, ns.

Figure 4 shows the scheme of sounding that comprises: G_K is the emitting and receiving the longitudinal (L_{dir}) ultrasonic wave normal beam twin transducer probe (P112-2.5-12); G_E is the emitting the longitudinal (L_{ang}) and transverse (T) ultrasonic waves angle beam transducer with wave entry angle 18° (P121-2.5-18-002); G_{R1}, G_{R2}, G_{R3} are the receiving the ultrasonic waves angle beam transducers with entry angle 18° (P121-2.5-18-002).

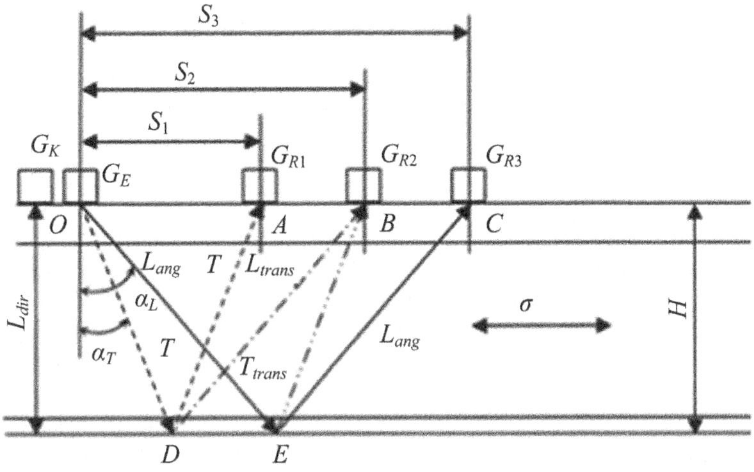

Fig. 4. Scheme of sounding for determining internal mechanical stresses in a rail.

Transducers G_{R1}, G_{R2}, G_{R3} were mounted at points *A*, *B*, *C* at distances S_1, S_2, S_3 from the emitting transducer G_E. The wave entry angle α_L of the longitudinal wave of the transducer G_E is known, while the wave entry angle of the transverse wave in s loaded rail is constant and can be found from the following equation:

$$\alpha_T = \arcsin(0.55 \sin \alpha_L) = const.$$

The distances S_1, S_2, S_3 were found from expressions (4), (5), (6) to determine the locations for the best reception of signals from transverse (*T*), transformed (*LT*) and longitudinal (L_{ang}) waves both in loaded and in unloaded rails kept in the same temperature environment. At point *A*, the arrival times of longitudinal (*L*), transverse (*T*) and transformed (*LT*) waves (τ_{L1}, τ_{T1}, τ_{LT1}) were measured. Similar measurements were also performed at points *B* and *C* to determine times τ_{L2}, τ_{T2}, τ_{LT2}, τ_{L3}, τ_{T3}, τ_{LT3}.

As rail height *H* changed, the twin transducer probe G_K registered a change in the propagation time of longitudinal wave $\tau_{L\text{long}}$, and the distances between the emitting G_E and receiving transducers G_{R1}, G_{R2}, G_{R3} changed by ΔS_1, ΔS_2, ΔS_3. Rail height *H* was determined from the measured propagation time of longitudinal wave τ_{Ldir} of the normal beam twin transducer probe G_k at a given propagation velocity of longitudinal ultrasonic wave. At each reception point *A*, *B*, and *C*, the arrival times of transverse and transformed waves τ_{T1}, τ_{LT2} and τ_{T3} for the loaded rail under study and τ_{T01}, τ_{LT02} and τ_{T03} for its unloaded counterpart were measured. The internal mechanical stresses σ in a rail were calculated from Eq. (7). The acoustoelastic coefficient of transformed waves k_{LT2} had been preliminary determined experimentally.

The internal mechanical stresses σ in a thermostrengthened rail R65 with a height of 180 mm were determined experimentally. Simultaneously, the stresses σ_S were measured with a certified strain-gauge system MMSS-64.01. To measure the stresses in the rail under study, type PKS-12 strain gauges were glued along the applied load in the middle of rail web at both sides.

The distances S_1, S_2 and S_3 were calculated from Eqs. (4), (5), (6). The velocity of longitudinal ultrasonic wave (C_{Ldir}) for the normal beam twin transducer probe G_K is 5.910 m/s; the propagation time of longitudinal wave τ_{Ldir} was measured with a TDS-2014 scope. The distances S_1, S_2 and S_3 from the emitting transducer G_E to the receiving transducers G_{R1}, G_{R2}, G_{R3} were calculated to be 61 mm, 89 mm, 117 mm. At each reception point A, B, and C of the receiving transducers the arrival times for signals of transverse (τ_{T1}, τ_{T3}) and transformed (τ_{LT2}) waves for the loaded rail under study and τ_{T01}, τ_{T03}, τ_{LT02} for its unloaded counterpart were measured with a TDS-2014 scope. The acoustoelastic coefficient k_{LT2} of the transformed waves from the emitting transducer to the second receiving transducer G_{R2} located at point B was determined experimentally and found to be 0.0000114 MPa^{-1}. The measured temporal parameters of ultrasonic waves at different load steps in the hydraulic loading unit were inserted in Eq. (7), and stress values σ_A were calculated.

The relative error between stresses σ_S measured with a strain-gauge system MMSS-64.01 and the calculated stresses σ_A was determined. The results of calculations are shown in Table 2.

Table 2. Measured and calculated mechanical stresses.

Calculated data	Rail load step number					
	0	1	2	3	4	5
τ_{T01}, mcs	113.767					
τ_{T1}, mcs	113.767	113.776	113.784	113.793	113.801	113.809
τ_{T03}, mcs	123.546					
τ_{T3}, mcs	123.546	123.545	123.543	123.541	123.538	123.535
τ_{LT02}, mcs	89.754					
τ_{LT2}, mcs	89.754	89.760	89.764	89.768	89.773	89.778
k_{LT2}, MPa^{-1}	−0.0000114					
σ_A, MPa	0.00	10.41	20.61	31.78	43.06	54.34
σ_S, MPa	0.00	10.22	20.93	32.12	43.60	55.13
Error $\frac{\lvert\sigma_S-\sigma_A\rvert}{\sigma_S}\cdot 100\%$	0	1.9	1.5	1.0	1.2	1.4

The calculated stresses in a R65 rail obtained using the suggested method were experimentally confirmed using a certified strain-gauge system MMSS-64.01. The relative error in the determination of stresses did not exceed 2%, which is acceptable for industrial testing at a railway track.

4 Discussion

1. Using the acoustoelastic effect, longitudinal mechanical stresses developed in the rails of a railway track under the effect of environmental exposure, the impact of rolling stock, upon recovery and maintenance works were measured. The stresses were calculated using the acoustoelastic effect from the measured temporal

parameters of ultrasound. The suggested method of determining longitudinal stresses in the rails of a continuous welded rail track provided monitoring for the entire rail cross-section and length.

2. The developed method of ultrasonic monitoring using longitudinal (*L*) and transformed (*LT*) waves transmitting through an unloaded control rail piece and through a rail under study laid into an operated track, which are kept in identical temperature environments, reduces the measurement error in the determination of longitudinal stresses.
3. A method of improving the accuracy of the mechanical stresses determination by reducing the influence of rail height on the measurement results has been developed. Rail height was determined using an additional normal beam longitudinal wave piezoelectric transducer, which improved the accuracy of the receiving angle beam piezoelectric transducer relative to the emitting one and the accuracy of determination of the temporal parameters of ultrasonic waves. By measuring the height of the rail under study, the errors due to rail wear and its grinding were excluded.
4. The results of ultrasonic measurements of longitudinal stresses in rails using the acoustoelastic effect are confirmed by strain-gauge measurements, the relative error between the measurements is 2…6% on average, which is acceptable for practical testing.

References

1. Stepanova, L.N., Kurbatov, A.N., Tenitilov, E.S.: Investigation of longitudinal stresses in rails using the acoustoelastic effect on an operated railway track. Control Diagn. (2), 14–21 (2019). (in Russian)
2. Gokhale, S., Hurlebaus, S.: Monitoring of the stress free temperature in rails using the acoustoelastic effect. Rev. Quantitative Nondestr. Eval. **27**, 1368–1373 (2008)
3. Vangi, D., Virga, A.: A practical application of ultrasonic thermal stress monitoring in continuous welded rails. Exp. Mech. **47**, 617–623 (2007)
4. Gokhale, S.: Determination of applied stresses in rails using the acoustoelastic effect of ultrasonic waves. M.Sc. thesis, Texas A & M University, 112 p. (2007)
5. Reale, S., Pezzati, A., Vangi, D., Rizzo, L.: Thermic tensity monitoring with ultrasonic technique in the long-welded rails. Ingegneria Ferroviaria **60**(12), 991–998 (2005)
6. Stepanova, L.N., Kurbatov, A.N., Tenitilov, E.S.: Determination of longitudinal stresses in rails using the acoustoelastic effect and strain-gauge measurements. Bull. RGUPS (3), 104–111 (2016). (in Russian)
7. Seryoznov, A.N., Stepanova, L.N., Kabanov, S.I., et al.: Strain-gauge measurements in transport engineering. Nauka, Novosibirsk, 272 p. (2014). (in Russian)
8. Stepanova, L.N., Kurbatov, A.N., Tenitilov, E.S.: Patent RF No. 2 619 842, IPC No. 2619842, IPC G 01 N 29/07. Ultrasonic Method of Determining Mechanical Stresses in Rails, Bulletin no. 14, 18 May 2017. (in Russian)
9. Stepanova, L.N., Kurbatov, A.N., Tenitilov, E.S.: Patent RF No. 2655993, IPC G 01 N 29/04. Ultrasonic method of determining internal mechanical stresses; Bulletin, no. 16, 30 May 2018. (in Russian)

Impact of Calorific Intensity on the Efficiency of an Internal Combustion Engine Operating Cycle

Sergey Glushkov and Veronika Zhidkikh(✉)

Siberian Transport University (STU), 191 Dusi Kovalchuk st., 630049 Novosibirsk, Russia
zhidkihveronika@mail.ru

Abstract. The processes of heat exchange in an internal combustion engine serve as indicators of its good technical condition and efficient operation. The processes of heat exchange and the calorific intensity of engine parts affect the engine reliability and operating life as well as its operating cycle efficiency. The mean effective and mean indicated cycle pressure are the universal criteria of calorific intensity, as a load increase results in an increase in the temperature of the cylinder-piston group. The engine indicator diagram is the most common and efficient way to obtain the required and reliable information on the operating cycle—the process of air-fuel mixture combustion. A thermodynamic analysis of the diagram will allow obtaining more detailed information, which promotes more accurate conclusions with respect to the operating cycle and the technical condition of the engine and its parts. To this end, a simplified method for calculating the amount of heat released during fuel combustion is considered. The method involves presenting the heat release as a sum of heat transferred to gas and heat transferred to the walls of the combustion chamber. This paper provides calculations of heat for each of these processes. Several models for calculating the heat transfer coefficient are presented, and possible errors and inaccuracies of the result obtained using the suggested method are outlined.

Keywords: Internal combustion engine · Cylinder-piston group · Calorific intensity · Heat exchange

1 Introduction

In a working engine, an oil film is formed between the working parts (e.g., the cylinder-piston group) in the combustion chamber. The oil film in the combustion chamber prevents friction between the interacting parts, performing a lubricating function, and contributes to working surfaces cooling. Its thermal resistance is directly proportional to film layer thickness. One of the most efficient ways to reduce oil consumption is maintaining the engine in good technical condition, which decreases oil losses for evaporation and burning. Oil burning in cylinders results in the formation of soot particles, which in turn leads to certain undesirable consequences [1]. Soot formation

Z. Popovic et al. (Eds.): TransSiberia 2019, AISC 1115, pp. 788–796, 2020.
https://doi.org/10.1007/978-3-030-37916-2_77

has a detrimental effect on the heat exchange process in the engine by increasing the proportion of radiative heat exchange, as it is fair to say that heated soot particles are one of the main radiators in the engine cylinder [1]. This problem, i.e. the presence of soot in the cylinder, results in an increased oil surface layer, which affects its cooling properties, and can also promote local overheating in the cylinder due to enhanced radiative heat exchange.

Note that oil hydrodynamics in the gap between the piston and cylinder liner plays an important role in the flow of oil into the engine combustion chamber. Figure 1 shows the oil flow diagram [2]:

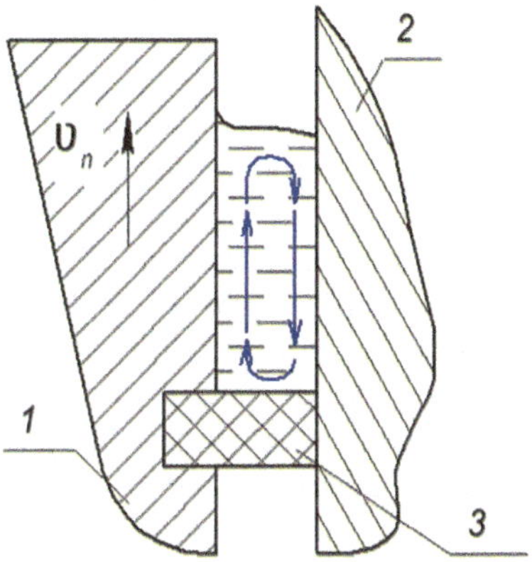

Fig. 1. Oil flow in the gap between the piston and cylinder liner: 1—piston, 2—cylinder wall, 3—piston ring, υ_n—piston velocity.

The flow of oil along the trajectory shown in Fig. 1 can be explained by the limited length of the channel between the piston head and cylinder liner; hence, the end portions of this volume form liquid turnover zones. At the piston wall, the oil flows with velocity equal to piston velocity. At the wall of the cylinder liner, the oil is stationary (due to the sticking effect). In the lower part of the gap, the compression ring reverts the direction of oil flow to the top.

As already mentioned above, the oil performs two important functions: lubricating and cooling. However, the volume of oil entering the combustion chamber should not exceed permissible levels. The flow of oil into the engine combustion chamber is known to occur mostly when the piston is in the top dead center (TDC). By examining Fig. 1, it can be concluded that in a properly functioning engine, it is necessary that, in the middle of a piston stroke, the oil level in the gap does not reach the piston edge. In this case, splashes of oil will be deposited on the walls and then get into the main oil bulk [2].

2 Research Methods

It is important to track the technical condition of the engine in time, in order to avoid the occurrence of failures and breakdowns and to prevent the growth of the flow of failures during periods that are not included in the timing of the scheduled maintenance and repairs [3].

One of the key factors of a proper and efficient engine operation and its good technical condition are heat exchange processes in the cylinder.

In evaluating the technical condition of the cylinder-piston group of an internal combustion engine, the processes of heat exchange in the engine and the calorific intensity of its parts have a huge influence on the engine's reliability and operating life, as well as its operating cycle efficiency. However, due to the technical difficulties in the determination of the thermal state of engine parts, it is necessary to determine either the heat flux to the part or the temperature of the part under study and its oscillations [4].

To study the thermal stresses of the parts of the cylinder-piston group, the authors of the study [4], Sinitsyn and Nechaev, use "indirect criteria" of calorific intensity that reflect certain parameters of the thermal state of the object. Such criteria should naturally be similar for the parts of similar engines. Some examples of such criteria are: mean effective pressure (P_e) or mean indicated pressure (P_i), excess air coefficient α, hourly fuel consumption (G_T) at constant rpm, temperature of exhaust gases for each cylinder. In the opinion of the authors of the studies [4, 5], the mean effective pressure or mean indicated pressure are the most universal criteria, as a load increase results in an increase in the temperature of the parts of the cylinder-piston group. "One general conclusion can be drawn: in a wide range of variations of Pe (Pi), the temperatures of such parts as piston, cylinder head (lid), valves, cylinder liner follow dependences that are close to linear [4]". The dependence of temperature (T) on the mean effective pressure (P_e) can be presented by the following equation:

$$\mathrm{T} = \mathrm{a} + \mathrm{b} \cdot \mathrm{P_e}. \tag{1}$$

The process of combustion, distribution, and transfer of energy and heat in the combustion chamber plays a significant role in the consideration of the issue of calorific intensity of the engine. Furthermore, combustion is a key process that has a decisive influence on the engine characteristics (power, torque, specific fuel consumption), as well as on the engine impact on the environment, i.e. the composition of exhaust gases and noise properties.

Note that one of the most common methods used to obtain the required information about the engine operating cycle is recording cylinder pressure, in other words, the engine indicator diagram. The information obtained in the course of recording cylinder pressure can provide data on the progress of fuel combustion, peak pressure and its position relative to the crankshaft position, pressure in the bottom and top dead centers at different strokes of the cycle. A thermodynamic analysis of the diagram can provide more detailed information on the operating cycle when certain calculations are performed, such as: heat losses and gas exchange losses, combustion rate of fuel-air mixture, etc.

The authors of the study [5] give the following solution for calculating the heat released during combustion:

According to the first law of thermodynamics for an open system,

$$dQ = dU + pdV \tag{2}$$

where dQ is the amount of heat entering or leaving the system, dU is the change in the internal energy of the system, pdV is the mechanical work exerted on the piston.

Heat dQ can be presented as consisting of the following elements: heat released during the combustion of the fuel-air mixture dQ_f, heat transferred to the walls of the combustion chamber dQ_w, and heat taken from the system as a result of flow of mass over the boundaries $Sh_i dm_i$, where h_i and m_i are the heat content (enthalpy) and mass fluxes, respectively. Thus, heat can be expressed as:

$$dQ = dQ_f - dQ_w + \sum_i h_i dm_i \tag{3}$$

If the change in internal energy is presented as

$$dU = d(mu) = mdu + udm \tag{4}$$

and Eqs. 3 and 4 are substituted into Eq. 2, we arrive at the following representation of heat released from the combustion of the fuel-air mixture:

$$dU + pdV = dQ_f - dQ_w + \sum_i h_i dm_i$$

$$dQ_f = mdu + pdV + udm + dQ_w - \sum_i h_i dm_i \tag{5}$$

If the abovementioned conditions are considered specifically in relation to the internal combustion engine, then, taking into account the fact that the fuel in the ICE is injected in the cylinder closer to the end of the compression stroke, we may write the following:

$$dm = dm_f - dm_{cr}; \sum_i h_i dm_i = h_f dm_f - hdm_{cr} \tag{6}$$

where dm_{cr} are the mass fluxes of the slit region.

The heat value of the injected fuel is significantly higher than its enthalpy, h_f=> 0, hence, the latter may be neglected. Substituting the data from Eqs. (6) into (5), we obtain:

$$dQ_f = mdu + pdV + udm_f - udm_{cr} + dQ_w + hdm_{cr}$$

$$dQ_f = mdu + pdV + udm_f + hdm_{cr} - udm_{cr} + dQ_w \tag{7}$$

Let us express Eq. (7) in crank angles:

$$\frac{dQ_f}{d\alpha} = m\frac{du}{d\alpha} + p\frac{dV}{d\alpha} + u\frac{dm_f}{d\alpha} + (h - u)\frac{dm_{cr}}{d\alpha} + \frac{dQ_w}{d\alpha} \tag{8}$$

For further calculation, we need to present the ideal gas law in differential form:

$$pV = mRT$$

$$\frac{1}{p}\frac{dp}{d\alpha} + \frac{1}{V}\frac{dV}{d\alpha} = \frac{1}{m}\frac{dm}{d\alpha} + \frac{1}{R}\frac{dR}{d\alpha} + \frac{1}{T}\frac{dT}{d\alpha} \tag{9}$$

The thermodynamic properties of the gas cylinder are functions of pressure, temperature, and gas composition. Let us express it as an equation and present it in differential form:

$$u = u(p, T, \lambda); R = R(p, T, \lambda)$$

$$\frac{du}{d\alpha} = \frac{\delta u}{\delta p}\frac{dp}{d\alpha} + \frac{\delta u}{\delta T}\frac{dT}{d\alpha} + \frac{\delta u}{\delta \lambda}\frac{d\lambda}{d\alpha}; \frac{dR}{d\alpha} = \frac{\delta R}{\delta p}\frac{dp}{d\alpha} + \frac{\delta R}{\delta T}\frac{dT}{d\alpha} + \frac{\delta R}{\delta \lambda}\frac{d\lambda}{d\alpha} \tag{10}$$

After substituting Eqs. (9) and (10) into Eq. (8), Eq. (8) can be solved numerically, and the rate of heat release during combustion can be estimated.

Note that the method described above, which leads to Eq. (8) and allows us to calculate the rate of heat release during combustion together with Eqs. (9) and (10), which describe gas properties, calls for solving a rather complex equation that will furthermore produce a somewhat approximate result due to certain simplifications and assumptions (e.g., gas temperature and composition are assumed to be uniform, etc.).

3 Solutions

To overcome the technical difficulties in calculations, the authors of the study [6] consider a simplified method for calculating the heat released during combustion.

A method to determine the heat released from combustion using the engine indicator diagram is described in detail in the study [6]. The engine indicator diagram is a series of pressure values recorded as a function of the crank angle. These values can also be transformed into "pressure-volume" data.

As gas state changes, heat from the combustion of mixture dQ_f is released between two consecutive points 1 and 2. Neglecting the effect of slit flux, this heat can be written as the sum of heat transferred to gas dQ_n and heat transferred to the walls of the combustion chamber dQ_w:

$$\Delta Q_f = \Delta Q_n + \Delta Q_w \tag{11}$$

Let us consider both heat transfer processes separately [6].

First, let us consider the case of heat transfer to gas.

Figure 2 shows a p-α diagram that illustrates the change in gas state from point 1 to point 3. The process consists essentially of two phases: during the first phase, an adiabatic expansion (or compression) from point 1 to point 2 takes place, the process involves changes in volume (from V_1 to V_2) and does not involve heat transfer to gas. During the second phase, the heat released into gas during combustion is transferred at a constant volume (V_2 = const) and the gas passes from point 2 to point 3.

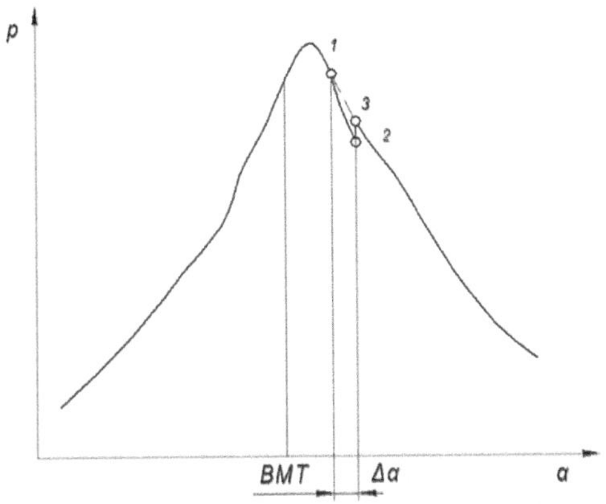

Fig. 2. Diagram showing the change in gas state from point 1 to point 3.

The process parameters required for calculation can be obtained using the generalized equation for a polytropic process [7]:

$$P_2V_3^n = P_1V_1^n \quad (12)$$

where n is the polytropic exponent.

As this process is an adiabatic expansion (or compression), we may write

$$n = k;$$

$$k = \frac{c_p}{c_v}$$

substituting this equality into Eq. (12), we obtain:

$$P_2 = P_1\left(\frac{V_1}{V_3}\right)^k \quad (13)$$

where k is the adiabatic exponent.

The temperature can be determined in the beginning and in the end of the process using the following equations:

$$T_1 = \frac{P_1 V_1}{mR}; T_3 = \frac{P_3 V_3}{mR};$$

$$T_2 = \frac{P_2 V_3}{mR} = \frac{P_1 V_1^k V_3^{1-k}}{mR}$$

$$\Delta Q_n = \Delta Q|_1^3 = mc_v(T_3 - T_2) = \frac{c_v}{R} V_3(P_3 - P_2) \tag{14}$$

Finally, after substituting Eqs. (13) into (14), we obtain the dependence for determining the amount of heat released during combustion and transferred to gas between points 1 and 3:

$$\Delta Q_n = \frac{c_v}{R} V_3 (P_3 - P_1 (\frac{V_1}{V_3})^k) \tag{15}$$

Now let us consider the case of heat transfer to the walls.

In this case, the predominant heat transfer is made via convection. This can be expressed with the following equation:

$$\frac{dQ_w}{dt} = k_w F_w (T - T_w) \tag{16}$$

where k_w is the heat transfer coefficient averaged over the entire surface of the combustion chamber, F_w is the surface area of the combustion chamber, T_w is the average temperature of the wall surface of the combustion chamber, and T is the average gas temperature.

The heat transfer coefficient can be found from the following equation:

$$k_w = \frac{1}{\frac{1}{\alpha_1} + \frac{\delta}{\lambda} + \frac{1}{\alpha_2}} \tag{17}$$

where α_1, α_2 are the heat transfer coefficients of hot and cold heat transfer agents, δ is the wall thickness, λ is the thermal conductivity coefficient ($\frac{\delta}{\lambda}$ is taken as the intrinsic thermal resistance for wall thermal conductivity).

The temperature and heat transfer to the walls surrounding the combustion chamber are considered separately depending on specific parts, as for each surface the temperature and heat transfer may vary significantly. These parts may be, for example, a piston head, a cylinder head or a liner, etc. As a consequence, the heat transfer coefficient is most often taken as the average over the entire surface of the combustion chamber.

$$\frac{dQ_w}{dt} = k_w \sum_i F_{wi}(T - T_{wi}) \tag{18}$$

From Eq. (18) it follows that the amount of heat transferred between points 1 and 2 can be calculated as:

$$\Delta Q_w = (k_w \sum_i F_{wi}(T - T_{wi}))\Delta t = (k_w \sum_i F_{wi}(T - T_{wi}))\frac{\Delta\alpha}{6n} \tag{19}$$

where $\Delta\alpha$ is the angle increment and n is the engine rpm rate.

4 Concluding Remarks

To calculate the heat transfer coefficient of internal combustion engines, several so-called α-equations suggested by different researchers may be applied. Some of them are given below:

G. Hohenberg's model

$$\alpha_w = 0,013V^{-0,06}P^{0,8}T^{-0,4}(c_m + 1,4)^{0,8};$$

R. Z. Kavtaradze's model

$$\alpha_w = \frac{b}{\sqrt{\Delta\tau}}[C_1 + C_2\frac{H_u \cdot \Delta x}{C_p(T - T_w)}];$$

G. Woschni's model

$$\alpha_w = 130\frac{(P \cdot 10^{-5})^{0,8} \cdot \omega^{0,8}}{T^{0,53} \cdot D^{0,2}};$$

W. Annand's model

$$\alpha_{wсм} = 0,26 \cdot \lambda \cdot \frac{\rho_{см}^{0,7} \cdot C_m^{0,7}}{\mu_{дин.см.}^{0,7} \cdot D^{0,3}} + 2,1 \cdot 10_w^{-13} \cdot (T^4 + T^4)$$

In the case studied, it would be appropriate to use G. Hohenberg's formula, which is believed to be the most common and convenient:

$$\alpha_w = 0,013V^{-0,06}P^{0,8}T^{-0,4}(c_m + 1,4)^{0,8} \tag{20}$$

Where V is the cylinder volume, P is pressure, T is temperature, c_m is average piston velocity.

Note that the heat value computed taking into account the heat transfer coefficient calculated using Hohenberg's formula will only be approximate. Its accuracy depends on many factors, the major ones being: insufficient experimental data for the applied model to calculate the heat transfer coefficient and the inability to generalize this model to include all engine types, the inability to accurately determine the temperature of the combustion chamber walls. These and other reasons result in calculation errors, which in turn lead to an inaccurate final result.

Furthermore, the overall result can also be affected by other factors, such as inaccurate pressure readings, erroneous or imprecise synchronization of pressure to engine crank angle, estimate of residual gas mass. As has already been mentioned, the model adopted to calculate the heat transfer coefficient can also be selected incorrectly, without due account of the individual properties of a specific engine, and the employed model should not be taken as a universal one for all engine types and categories. The accuracy of measurement results can be improved by using a PC and specialized software to exclude inconsistencies and errors when taking readings and performing calculations.

References

1. Sinitsyn, V.A.: Physical conditions and mathematic modeling of local heat exchange in ICE. Dissertation of Doctoral Science (Engineering), 04 April 2002. Vladimir Aleksandrovich Sinitsyn, Barnaul, 377 p. (1995). (in Russian)
2. Lebedev, B.O.: The thermophysical basis of oil burning in diesels and the development of operational procedures for its reduction. Dissertation of Doctoral Science (Engineering), 05 April 2002, Boris Olegovich Lebedev. Novosibirsk, 230 p. (In Russian) (2001)
3. Manakov, A., Kolarzh, S., Mashkov, A.: Organization of railway track tamping work using the software and hardware complex SMTC. In: MATEC Web of Conferences, vol. 239, p. 04002 (2018). https://doi.org/10.1051/matecconf/201823904002
4. Sinitsyn, V.A., Nechaev, L.V.: Indirect criteria of the thermal state of parts of the CPG of internal combustion engines. Polzunov Anthol. **4**, 107–112 (2009). (in Russian)
5. Tomic, M.V.: A quick, simplified approach to the evaluation of combustion rate from an internal combustion engine indicator diagram. Therm. Sci. **12**(1), 85–102 (2008)
6. Glushkov, S.P.: Thermal Engineering. Lecture notes Siberian University for Consumer Cooperation. Novosibirsk, p. 123 (2009). (in Russian)
7. Kabanov, A.N.: Selecting a technique for calculating heat transfer in a gas engine with spark ignition. Automob. Transp. **30**, 96–102 (2012). (in Russian)

Vertical Sediment of a Ballastless Track

Aleksander Savin[1,2], Vadim Korolev[1,2], Alexey Loktev[1,2], and Irina Shishkina[1,2(✉)]

[1] Scientific Research Institute of Railway Transport, 3rd Mytishchinskaya St., 10, 129626 Moscow, Russia
savin.aleksandr@vniizht.ru, shishkinaira@inbox.ru
[2] Russian University of Transport, Chasovaya St., 22/2, 125190 Moscow, Russia

Abstract. At the Experimental ring of JSC "VNIIZHT" at Shcherbinka station, the comparative tests of four types of ballastless structures of the LVT track (Russian Railways, Russia), FFB (MaxBögl, Germany), NBT (Alstom. France), EBS (Tines, Poland) had been completed. The tonnage passed through the experimental section of four structures amounted to 600 million tons gross. The tests allowed confirming the operability of the ballastless track for the conditions of the Russian Railways, identifying the features of the current content of each of the structures and giving recommendations for their improvement. The paper analyzes the intensity of ballastless subsidence under the train load. Tests have shown that ballastless track can be used not only for a speed and a high-speed traffic, but also for combined and cargo traffic. Significant risks are that the ballastless track is more demanding for adhering to the construction technology, the composition of concrete mixtures and the quality of the preparation of the foundation.

Keywords: Ballastless track · Strength · Subsidence · Drawdown · Tests · Transition sections · Current content

1 Introduction

The solution to the problem of mastering the growing volume of cargo and passenger traffic in the absence of reserves of carrying and passing capacity in a number of sections and entire directions necessitates the search for new designs of the track and turnouts, providing increased weight, length and speed of trains, as well as minimizing time for all types of railway works.

In modern operating conditions, with increasing the axial load, weight and length of trains, one of the main problems is the deformability of the track. The use of a ballastless track can be one of the options for increasing the stability of the track.

The world has experience of operating the high-speed traffic on ballast and ballastless track superstructures [1, 2], however, the percentage of newly erected lines is significantly shifted towards the ballastless track [3–6]. The question of the appropriateness of the widespread use of track and turnouts on a ballastless basis for the conditions of the Russian Railways, including for cargo traffic, remains open.

Z. Popovic et al. (Eds.): TransSiberia 2019, AISC 1115, pp. 797–808, 2020.
https://doi.org/10.1007/978-3-030-37916-2_78

Tests in the Research Institute of Railway Transport (JSC "VNIIZHT") at the Experimental Ring at Shcherbinka Station of four types of ballastless track became one of the main stages in the study of the possibility of using such a design for the conditions of the Russian railways. The aim of the study is to determine the conditions for the use of ballastless structure to increase the efficiency of the transportation process by reducing the cost of maintaining the track.

Track construction with minimal technical maintenance costs is required. It is needed to determine the conditions under which it is possible and advisable to use a ballastless structure to reduce the deformability and maintenance costs of the track and, as a result, to increase the efficiency of transportation.

2 Research Methods

The aim of research is achieved using the following methods:

- by generalization and analysis of domestic and foreign experience in the development and operation of ballast and ballastless track structures;
- by mathematical modeling of stresses and deflections of ballastless track under a train load;
- by experimental measurements of ballastless draft during prolonged exposure to train load;
- by full-scale experimental studies of the stress-strain state and operational reliability of the ballastless track (structures Rheda, LVT, MaxBögl, Tines, Alstom), including the use of the Doppler shift effect in fiber-optic diagnostic systems;
- by expert assessment of the influence of a large number of parameters on the behavior of complex systems using the ballastless example;
- by development of a model for the transition from full-scale tests on the Experimental Ring to predicting the reliability of a ballastless structure on a current line;
- calculation of the duration and cost of the life cycle of a ballastless track in various application conditions, taking into account the existing risks.

This paper discusses the question of the intensity of vertical setting of experimental sections of ballastless track structures under a train load on the Experimental Ring.

The length of each of the four ballastless structures was 75 m (without transition sections). The length of the entire experimental plot is 500 m.

For all four structures, the subgrade was constructed according to one technology, on the entire front of work 500 m, by one contractor on the existing section of the second ring road. After the excavation of natural soil, two specially prepared layers of 7 m wide were laid with a slope on top in cross section of 0.04.

The first layer, 50 cm thick, is chemically reinforced soil by mixing with polyphilizers (freeze-dried soil). The average value of the deformation modulus along the second branch of loading, taking into account the reprocessing of the soil, is 146 MPa. Design values of the deformation modulus of reinforced soil should be at least 80 MPa. The obtained data of static and dynamic tests of the deformation modulus of freeze-dried soil correlate well with each other. The second layer, 70 cm thick, is crushed stone-sand-gravel mix TU 5711-284-01124323-2012. The average value of the deformation modulus along the second branch of loading is 181.7 MPa with a design value of at least 120 MPa.

Tests of the experimental section with four ballastless structures were carried out in accordance with the Program and methodology approved by the "Russian Railways" OJSC. Tests of four types of ballastless construction tracks (Tines, ALSTOM, Max-Bogl, LVT) were carried out on the straight section of the second ring track of the Experimental Ring of VNIIZhT JSC, Shcherbinka Station in the period from December 3, 2014 to October 20, 2016. The tonnage passed through at the experimental plots amounted to 606.7 million tons gross. The tonnage was provided by a train consisting of an electric locomotive VL80 and 85 gondola cars with a load of 23.5 tons per axle. The speed was 70 km/h. The daily operating time of the tonnage was 1.0 … 1.2 million tons gross.

3 Results of the Research

Consideration of strength as the first limiting state of the railway track and turnouts structures by calculation and experimentally showed that for given operating conditions there is no excess of the maximum permissible values [7, 8, 9, 12, 13]. According to the first limiting state, the values of deflections and stresses experimentally obtained at the Experimental Ring at Shcherbinka station during heavy traffic with a load of 235 kN per axis for all types of ballastless track is approximately two times lower than critical values, which allowed us to conclude that freight traffic is one of the application areas [10, 11].

The second limiting state, deformability, was investigated in the process of resource tests at the Experimental Ring at Shcherbinka station in the amount of 600 million tons gross of passed cargo for four types of LVT ballastless track (RZHDStroi, Russia), FFB (MaxBögl, Germany), NBT (Alstom, France), EBS (Tines, Poland).

Subsidence of the Tines design, depending on the missed tonnage, is characterized by an average value from −0.5 to 2.4 mm, while the standard deviation (MSD) ranges from 0.9 to 5.7. In the transitional sections, the average subsidence is from −13 to 5 mm, the standard deviation is from 0.5 to 12. In general, the slab subsidence is uniform by 2 mm. The greatest value of subsidence up to 5 mm is observed in places of a fixed "splash" from under the concrete bearing slab.

In the Alstom structure, the average subsidence from the passed tonnage is from −2.5 to 11.7 with a standard deviation of 0.7 to 2.9. In transitional sections, the average value is from −19.7 to 3.5 mm, the standard deviation is from 0.7 to 11.5. Compared with the initial position of the slab, it has a fairly uniform subsidence of 18 mm along the entire length of the plot.

MaxBögl ballastless track has an average subsidence value from −2.5 to 5.5 mm, standard deviation from 0.9 to 4.6. Transitional sections have an average subsidence value from −19.7 to 10 mm with a standard deviation from 1.2 to 11.5. Compared with the initial position of the slab, the subsidence is 10 mm generally, with the exception of the latter 20 m along the direction of movement, where the subsidence reaches 16 mm.

The LVT structure has an average subsidence of 0.5 to 1.8 mm with a standard deviation of 0.9 to 3.7. Transitional sections are characterized by an average value of subsidence from −15.8 to 3.8 mm, standard deviation from 1.4 to 9.3. Compared to the initial position of the slab, the subsidence is uneven from 2 to 10 mm, possibly this is due to the different type of blocks in different parts of the structure.

These values indicate a significant scatter in the average subsidence indicators and their standard deviations, indicating significant structural features and their influence on the ballastless track behavior. It should be noted that the Alstom and MaxBögl slab structures of the track have a more uniform subsidence of the entire section by 16–18 mm, and the Tines and LVT block structures have uneven subsidence along the length of the slab, but by a significantly smaller amount. It is quite difficult to assess the subsidence of transitional sections, since they are periodically straightened and knocked out by a VLR machine and electric tampers. The amount of lifting reaches 40 mm. There were no significant differences in the size and number of taps on various structures.

Table 1 presents the average values of subsidence and standard deviation in areas of ballastless structures at the different tonnage.

Figure 1 shows the dependences of subsidences of the middle part of the slab of ballastless track structures in comparison with the ballast track subsidence.

Dependences of the subsidences of ballastless structures in Fig. 1 are approximated by the functions given in Table 2. The subsidence of the track on the ballast is approximated by a logarithmic function with 0.9686 reliability of approximation. The subsidence of four types of ballastless track is approximated by linear functions with approximation reliability from 0.8162 to 0.9723.

From the diagrams in Fig. 1 it can be seen that the subsidence of all types of ballastless track is significantly lower than the subsidence of the ballast track. Permissible draft values of 15 mm exceeded for Alstom and MaxBögl structures.

Table 3 presents the values of subsidence values and the mean square deviation of subsidence at the ballastless sections in comparison with ballast ones.

The lowest subsidence values are demonstrated by the Tines and LVT structures. Subsidence of Tines after passing the 600 million tons gross is 3.3 times less than for the ballast track. Subsidence of LVT is 4.7 times less.

Significantly higher subsidence values are observed for Alstom and MaxBögl structures. Subsidence of Alstom after passing the 600 million tons gross is only 1.3 times less than for the ballast track, and MaxBögl is 2.1 times less.

The greatest non-uniformity of subsidence is observed for the Tines (MSD = 5.79) and Alstom (MSD = 6.12) structures with a maximum value of subsidence of not more than 14 and 35 mm, respectively. The maximum standard deviation is observed for subsidence measurements performed with a pass of 300 million tons gross.

The MSD value for the ballast track is 3.6 with a maximum subsidence value not exceeding 45 mm. The maximum standard deviation is observed for subsidence measurements performed with a pass of 300 million tons gross. This tonnage was skipped across all structures in December 2015.

The subsidence intensity rate shows that for ballastless structures, stabilization occurred after passing the 300 million tons gross.

The maximum subsidence intensity is observed in the Alstom structure, it is 0.1179 mm/mln. tons gross and is observed at a tonnage of 200 million tons gross. For Tines, the highest intensity is 0.0247 mm/mln. tons gross and is observed at a tonnage

Table 1. Average values of subsidence and standard deviations in areas of ballastless structures with different tonnage.

Tonnage passed through, mln. t gross			Tines			Alstom			MaxBogl			LVT		
			Transition	Slab	Transition	Transition	Slab	Transition	Transition	Slab	Transition	Transition	Slab	Transition
0–100	subsidence	r.	1.33	2.13	5.00	−13.0	1.93	−19.75	5.5	−1.23	−2.5	1.2	1.37	-6.0
		l.	4.00	1.91	13.20	−6.25	2.71	−13.0	10.75	−0.85	1.5	5.8	1.89	−1.92
	MSD	r.	1.15	1.50	2.24	2.16	2.37	1.59	4.65	4.22	6.4	9.36	1.21	1.77
		l.	1.00	1.73	2.59	1.26	1.59	3.74	2.87	5.27	9.33	5.5	1.56	2.31
100–200	subsidence	r.	−0.33	2.47	−8.8	−14.5	−2.50	−3.25	−0.25	2.69	−18.75	−15.8	1.87	0.46
		l.	1.33	2.63	−1.6	−9.75	−1.14	−2.00	2.5	4.54	−13.5	−10.8	0.79	0.54
	MSD	r.	1.53	0.97	3.7	1.29	2.93	2.63	2.63	6.12	8.5	6.38	0.91	4.24
		l.	0.58	0.67	3.58	2.75	3.37	4.08	3.7	5.11	9.33	2.77	1.40	4.71
200–300	subsidence	r.	4.33	−0.59	−13.8	3.5	11.79	−6.75	−2.25	10.0	−8.25	−3.80	0.82	−4.38
		l.	6.67	−2.03	−16.8	5.75	12.79	−2.5	0.50	8.31	−8.50	−2.8	1.26	−0.58
	MSD	r.	12.86	5.79	8.11	5.26	2.67	11.53	4.65	5.37	4.92	7.6	3.7	3.06
		l.	7.02	5.82	5.81	9.32	3.31	7.33	5.74	4.01	6.24	5.97	3.36	2.42
300–400	subsidence	r.	−3.67	2.00	2.60	2.0	1.71	−0.5	0.25	0.46	−0.25	0.80	0.53	0.19
		l.	−3.33	1.5	3.6	2.75	1.82	−0.75	0.75	0.15	0.00	−.60	0.79	0.62
	MSD	r.	0.58	2.13	0.55	0.82	0.73	1.91	0.96	1.2	2.5	3.11	1.22	1.44
		l.	0.58	3.33	1.34	0.5	0.67	1.5	1.26	1.41	1.41	0.55	0.92	0.90
400–500	subsidence	r.	−2.33	1.44	2.20	1.50	1.57	−0.25	0.75	0.77	0.50	0.40	0.58	0.77
		l.	−1.0	1.06	1.60	2.0	1.68	0.0	1.00	0.69	0.50	−1.00	0.84	0.92
	MSD	r.	0.58	0.51	0.84	0.58	0.65	1.50	1.26	1.36	1.73	1.82	0.51	0.82
		l.	1.73	1.77	0.55	0.82	0.46	1.15	1.41	1.11	1.73	0.71	0.83	0.80
500-600	subsidence	r.	−1.67	2.0	2.40	2.00	1.36	−0.25	1.5	1.15	0.75	1.40	0.84	0.92
		l.	−1.67	1.81	1.80	1.50	1.61	1.25	1.25	1.46	0.75	−0.60	0.84	0.81
	MSD	r.	1.15	0.52	0.55	0.00	0.5	1.5	0.56	1.28	1.89	0.55	0.37	0.39
		l.	0.58	1.22	0.45	0.58	0.49	0.50	0.50	0.52	1.89	0.89	0.60	0.49
0–600	subsidence	r.	−2.33	9.44	−10.4	−18.5	15.86	−30.75	5.50	13.85	−28.50	−15.80	6.00	-8.04
		l.	6.00	6.88	1.80	−4.00	19.46	−17.0	16.75	14.31	−19.25	−11.0	6.42	0.38
	MSD	r.	12.66	3.52	9.96	9.15	4.47	9.91	4.12	3.76	4.93	13.75	3.68	4.41
		l.	9.64	3.52	10.23	5.48	4.09	7.07	2.63	2.66	4.92	13.60	3.72	4.29

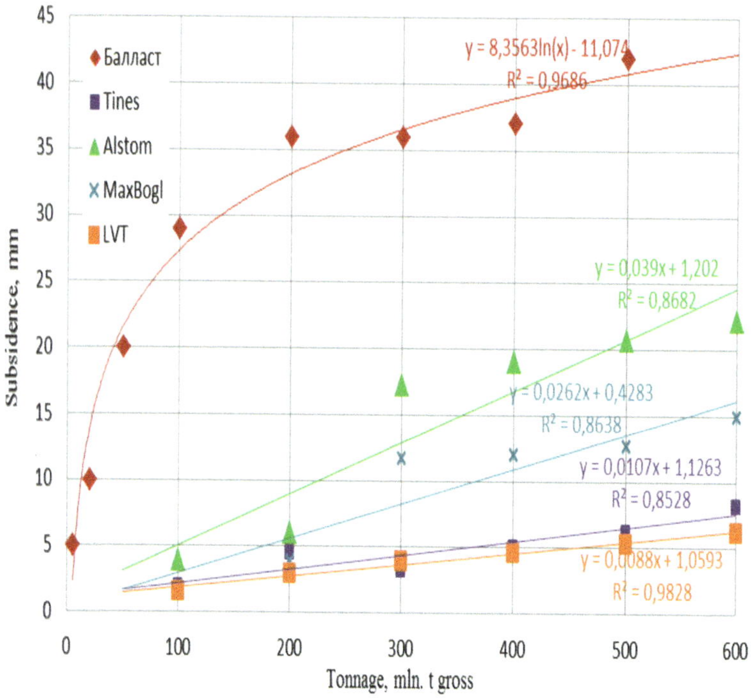

Fig. 1. Dependence of subsidence on the tonnage passed through.

Table 2. Approximating functions of subsidences from the tonnage passed through.

Structure name	Tonnage passed through, mln. t gross	Approximating function	Approximation reliability R^2
Ballast track	750	y = 8,3563ln(x) – 11.074	$R^2 = 0.9686$
Tines	750	y = 0.0107x + 1.1263	$R^2 = 0.8528$
Alstom	750	y = 0.039x + 1.202	$R^2 = 0.8682$
MaxBögl	750	y = 0.0262x + 0.4283	$R^2 = 0.8638$
LVT	750	y = 0.0088x + 1.0593	$R^2 = 0.9828$

of 300 million tons gross. For LVT, the highest intensity is 0.0187 mm/mln. tons gross and is observed at a tonnage of 200 million tons gross. For MaxBögl, the intensity indicator is the smallest and amounts to 0.1 mm/mln. tons gross, is observed at a tonnage of 300mln. tons gross.

Testing of the subgrade during the operation of experimental ballastless track structures was carried out using a fiber-optic signaling system for the state of infrastructure facilities. This system was installed together with the CJSC Laser Solutions company during the construction of pilot plots. The system is designed to track the movement of subgrade layers, which is especially critical for ballastless track structure.

Table 3. Values of subsidence and mean square deviation of subsidence.

Structure name	Tonnage passed through, mln. t gross	Approximating function	Approximation reliability R^2
Ballast track	750	y = 8,3563ln(x) − 11.074	$R^2 = 0.9686$
Tines	750	y = 0.0107x + 1.1263	$R^2 = 0.8528$
Alstom	750	y = 0.039x + 1.202	$R^2 = 0.8682$
MaxBögl	750	y = 0.0262x + 0.4283	$R^2 = 0.8638$
LVT	750	y = 0.0088x + 1.0593	$R^2 = 0.9828$

The fiber-optic system for signaling the state of infrastructure facilities consists of linear and hardware-software parts. The linear part, in turn, consists of two subsystems. The first is deformational, designed to monitor the movement of the subgrade. The second is auxiliary, designed to connect sensors to the analyzer and to compensate the temperature effects. The principles of the system operation are based on the diagnostics of the state of the fiber-optic sensor by measuring the distribution of temperature and deformation along its entire length. The DITEST STA-R analyzer, which is a pulsed optical reflectometer that measures the signal of stimulated Brillouin scattering from each point of the optical fiber of the sensor, is the key instrument of the hardware and software parts of the system. Analysis of the stimulated Brillouin scattering signal, depending on the type of the sensor connected, allows you to measure the distribution of temperature or voltage along the entire length of the sensor.

At the pilot plot of the Experimental Ring, a system with limited functionality was installed (only fiber-optic sensors), the analyzer and the server were not installed at the plot, but will be connected during periodic measurements after passage of every 100 mln. tons gross. The movement of the subgrade in the vertical plane caused by subsidence of the soil is recorded by longitudinal sensors due to the horizontal component, that is, the longitudinal extension of one or another part of the sensor signals the vertical movement of the soil layer. Vertical movements must be controlled at three levels: concrete, protective layers, subgrade. The topmost concrete layer is controlled by the geodesic surveying devices. To control the lower layers, two layers of fiber optic cable were laid (Fig. 2). The first layer was laid under a freeze-dried layer, the second - under a layer of crushed stone-sand-gravel mix.

In the horizontal plane, two layers of optical fiber are laid under the rails. The cross-coupling for connecting the recording equipment is installed at a distance of 2–5 m from the axis of the track.

The configuration of the sensors in the subgrade is a key factor that allows measurements and interpretation of parameters with a given accuracy. Two longitudinal deformation sensors and two temperature sensors were installed at the experimental site. Anchors are installed on the longitudinal deformation sensors with a step of 2 m, which transmit the longitudinal displacement of the soil to the sensor.

We consider the vertical subsidence of ballastless structures at three levels: according to the level of the upper concrete slab, under the crushed stone-sand-gravel mix and under the freeze-dried soil layer.

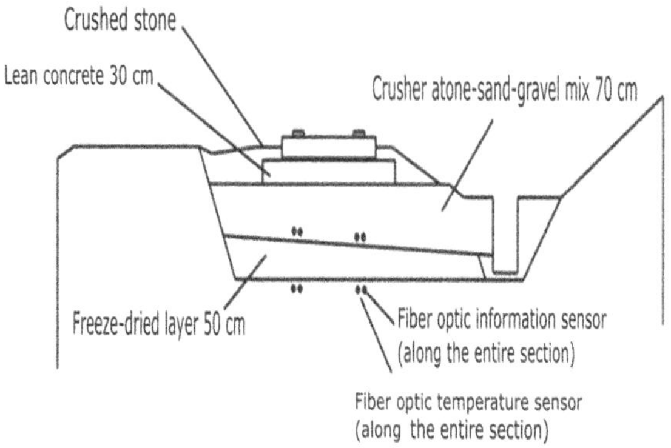

Fig. 2. The arrangement of sensors in the subgrade.

Plots on concrete bearing slabs behave quite stably and have a uniform vertical subsidence of up to 13 mm. The difference on the left and right rail thread does not exceed 2 mm. The sections of variable rigidity and the track on the gravel have more intense subsidence up to 40 mm and require periodic straightening by tamping with the straightening and leveling machine. The tamping is carried out approximately once every three months. Subsidence at the level of the joint between the crushed stone-sand-gravel mix layer and the layer of freeze-dried soil is controlled by the upper layer of the fiber-optic diagnostic system Subsidence is fairly uniform and does not have a pronounced character relative to the location of concrete load-bearing slabs or sections of variable stiffness. In magnitude, they range from 0 to 2 mm.

Subsidence under a layer of freeze-dried soil is monitored using the bottom layer of a fiber-optic diagnostic system. At this level, the subsidence picture is the most interesting. Values range from 0 to 11 mm, and on two experimental structures it is about 2 mm, on the other - from 7 to 11 mm. The maximum values of subsidence are not correlated with the boundaries of concrete load-bearing slabs and are not tied to the location of sections of variable rigidity as at the level of the rail head.

In one of the sections, the location of the peak values of subsidence from measurement to measurement changed their position by tens of meters. In this case, the subsidence value could not only increase, but also decrease, for example, in the fourth section from the penultimate to the last measurement, the subsidence decreased by 3 mm. Between left and right thread the difference is up to 6 mm. The greatest displacements are observed during the thawing period in spring.

An analysis of the possible causes of such a movement of the lower natural layer shows that all the pilot plots are located in the zone of high groundwater levels. This is confirmed by the results of pre-design geological surveys. In addition, in spring, there is a difference in the levels of meltwater on the left and right sides of the mound, which leads to flooding of the base. It should be noted that the areas with the greatest displacements of the lower layer are located lower, and the groundwater level in this place is much higher.

The displacement of the upper layer located above the fortified soil is significantly less. This is explained by the fact that the freeze-dried soil works like a beam or "shell" that keeps the upper layers from moving down. In this case, the displacements in no way correlate with the location of the transition sections and in no way depend on the location of the sections with maximum subsidence from above at the level of the rail head. This indicates a uniform distribution of the load with a layer of lean concrete (hydraulically bonded layer) and a layer of the crushed stone-sand-gravel mix (frost protection layer).

During the tests, displacements (subsidence) of natural soil located under a strengthened (freeze-dried) layer were revealed. The location of these displacements does not depend on the location of the boundaries of the concrete base slab and sections of variable rigidity. Moreover, the greater the moisture content, the greater the subsidence of the lower layer.

The crushed stone-sand-gravel mix layer has the smallest subsidence, which is independent of the sediment of the lower and upper layers. The freeze-dried layer protects it from the movements of the underlying moistened natural soil, and it protects the layer of lean concrete from the tops from movements caused by the movement of the rail. Drawdowns of transitional sections between ballastless and traditional ballast tracks were analyzed separately.

The most intense subsidences up to 35 mm were observed in the first month of operation, which required their weekly dressing with manual electric tampers and straightening-tamping-leveling machine. The maximum drawdowns recorded by the track car during operation amounted to 16 mm. In this case, the subsidence of the input transitional sections is much greater than the output.

The worst condition for rubble of crushed stone is observed at the inlet transitional section of the LVT plot, structurally made with a metal console of variable cross-section in ballast of crushed stone. In this case, the best condition was recorded at the output transitional section of the LVT with rubberized soles of sleepers and of the Alstom in the replacement of rail fasteners for these rails was required.

The smallest subsidence of the transitional sections is observed in the MaxBogl design.

The geometry of the rail track meets the requirements of traffic safety on all four structures. The largest deviations according to the results of the passage of the measuring car are as follows:

- 9 mm broadening at the transition sections of Tines and MaxBogl;
- narrowing is not detected;
- 14 mm skew on the Tines slab and 13 mm on the ALSTOM transition section;
- 14 mm drawdown on transition sections of Tines, ALSTOM, LVT and on ALSTOM slab;
- 9 mm realigning in the transition sections of Tines and ALSTOM;
- 9 mm level on the ALSTOM slab.

Additional measurements of the gauge and level manually with the gauge template show that the gauge is stable on all structures (a change in the range of 500–600 million tons was from −0.7 to 0.46 mm, with standard deviations from −0.33 to 0.04 mm), due to the type of intermediate rail fasteners used.

Vertical rail wear on all structures ranged from 0.011 to 0.024 mm, with MSD values of 0.002 to 0.007 mm. Lateral transitional section with under-sleeper pads.

It should be noted that Alstom transition sections have additional rails inside the track. During operation, a complete wear on all structures ranged from 0.068 to 0.118 mm, with MSD values of 0.006 to 0.02 mm.

The best rail track geometry has the MaxBogl structure. This is due to the factory manufacturing of slabs with high accuracy.

The worst condition in terms of concrete layers is observed in Tines due to the intensive splash from under the load-bearing slab due to the destruction of lean concrete. For this reason, there is a transverse kink of the load-bearing slab at the plot of greatest splash. In addition, there are delamination of blocks from track concrete and numerous cracks of more than 0.5 mm.

All experimental structures are characterized by the allocation of cement components in the form of an aqueous suspension ("splash") at the junction of the first and second concrete layers, as well as from cracks in the concrete layer and at the joints of slabs. This phenomenon has not yet affected traffic safety, but will affect the durability of structures. The most deplorable state in terms of splashes is observed on the ballastless Tines structure.

One of the reasons for the ingress of water between the first and second concrete layers, that is, between the upper concrete load bearing slab and the underlying layer, is the arrangement of ballastless structures on one of the tracks of the double-track section. In the period of heavy rains or snowmelt, water from the inter-blast seeps under the concrete slab towards the drainage system.

Rail Fasteners Condition

The tightening torque of the screws is in the specified ranges and amounts to the following values:

- Tines 315…420 Nm;
- ALSTOM 280…360 Nm;
- MaxBogl 250…400 Nm;
- LVT 250…400 Nm.

The smallest loosening of the fastening screws for rail fasteners on a concrete slab is observed in the Tines structure.

The defectiveness of the fastener elements is as follows:

- ALSTOM - replacement of 16 broken bolts of fastening the additional rails on the transitional section, replacing all the fastening nodes of the additional rails on the concrete slab, kink W300 lining - 1 pc., a gap of 1 … 3 mm between the W300 liner and the concrete slab at all W300 bond points;
- Tines - none;
- MaxBogl - kink of terminals 3 pcs., kink of side stop 1 pc.
- LVT - replacement of 64 blocks with Schwihag fastening (cracks in the dowel hole and in the fillet). The causes of failure are investigated further.

The output of fastening elements to ALSTOM is due to the imperfection of the transition section structure.

The labor costs for the current maintenance are approximately comparable for all designs.

The most problematic places for all experimental designs (Tines, ALSTOM, MaxBogl, LVT) are transitional sections between the ballastless track and the ballast track. Drawdowns in these places reached 40 mm and were corrected in a timely manner by knocking down the railroad ties with a with the straightening and leveling machine. In the worst condition (the greatest subsidence and abrasion of crushed stone) is the first transitional section of the LVT with a metal console of variable cross-section in the ballast. Intensive destruction of the lower part of the sleepers, resting on special platforms of the metal console, required their replacement in the amount of 5 pcs. In addition, crushed stone was replaced at this site.

The smallest subsidence is observed in the second transitional section of the LVT with rubberized soles of sleepers.

During operation, the following repair works were performed:

LVT: Numerous cracks in the concrete layer were sealed. Replaced 64 blocks with Schwihag staple. Replacement of 5 sleepers and gravel in the first transitional section.

MaxBogl: The longitudinal joint of the slabs and the asphalt pavement of the interblock were sealed. Replacing broken terminals 3 pcs. and side stop 1 pc. on concrete slab MaxBogl.

Alstom: Replacing of 16 bolts securing additional rails on the sleepers of transitional sections. All fastenings of additional rails on a concrete slab were replaced.

Tines: Elimination of splashes by sealing the seam of a concrete base slab and lean concrete. Level adjustment on adjustment cards in the W30 bond - 13 pcs.

The labor costs for the current maintenance of ballastless track structures during the period of production of 600 million tons gross amounts to the following values:

- Tines 279.5 man hours; 15.5 machine hours;
- ALSTOM 180.5 man hours, 6.5 machine hours;
- MaxBogl 112.5 man hours, 9.0 machine hours;
- LVT 210.0 man hours, 16.5 machine hours.

The least value of labor costs for the current maintenance has the MaxBogl structure. This is due to the fact that track concrete slabs are prefabricated. In addition, this is the only structure at the pilot plot, in which the interblock is asphalted to reduce the ingress of rain and melt water under the slab.

4 Conclusions

1. An analysis of the intensity of the track subsidence on the gravel base and the intensity of the subsidence of the ballastless track shows that the subsidence of the ballastless track is 4.8 times slower, except for transition sections.
2. To determine the technical feasibility of using the ballastless track in Russia, the empirical dependence of subsidence on the passed tonnage is approximated to find the conditions for achieving the second limit state.
3. According to the first limiting state, the values of deflections and stresses experimentally obtained at the Experimental Ring at Shcherbinka station during the intense traffic with a load of 235 kN per axis for all types of ballastless track, is approximately two times lower than the critical values, which allowed us to conclude that one of the areas of application is cargo traffic.

4. According to the second limiting state, an analysis of the intensity of the track subsidence on the crushed stone base and the intensity of the ballastless track showed that the ballastless track settles 4.8 times slower than the ballast track, with the exception of transitional sections. However, the subsidence occurs unevenly. At a section of 75 m after the passage of 600 million tons gross subsidence varies from 2 to 11 mm.
5. It has been experimentally proved that the failure of the ballastless track occurs according to the second limiting state. Maximum subsidence occurs in the wetting areas of the lower layers of the structure.
6. The condition of the geometric parameters of the rail track and rail fastenings meet the specified requirements and ensure traffic safety in the considered time interval.

References

1. Darr, E., Fiebig, W.: Feste Fahrbahn. Konstruktion und Bauarten für Eisenbahn und Straßenbahn (2006)
2. Nigel, O., Franz, Q.: Innovative track systems criteria for their selection. Project funded by the european community, directorate general energy and transport under the 'competitive and sustainable growth' programme ProMain. TÜV Intertraffic, Cologne (2001)
3. Lichtberger, B.: Track compendium, 1st edn. Eurail Press, Hamburg (2005)
4. Bezin, Y., Farrington, D., et al.: The Dynamic Response of Slab Track Constructions and their Benefit with Respect to Conventional Ballasted Track. Taylor & Francis, Manchester (2010)
5. Giannakos, K.: Requirements of tomorrow's rail transport infrastructure. In: 2nd Traffic & Transportation conference, OSE, Vossloh AG (2004)
6. Schilder, R., Diederich, D.: Installation quality of slab track – a decisive factor for maintenance. RTR Special, Austria (2007)
7. Glusberg, B., Savin, A., et al.: New lining with cushion for energy efficient railway turnouts. In: Advances in Intelligent Systems and Computing, vol. 982, pp. 556–570 (2020). https://doi.org/10.1007/978-3-030-19756-8_53

The Destruction of the Electrification Contact Network Caused by Accidents on Rolling Stock Due to Chloride DIR Exposure

Vladimir Popov, Philipp Sukhov, and Svetlana Churiukina(✉)

Russian University of Transport RUT (MIIT), Moscow, Russia
churukinasv@mail.ru

Abstract. The results of studies of the chloride deicing reagents (DIR) influence on the main materials (concrete, reinforced concrete, copper) that the railway transport facilities consist of are presented. The main railroad emergencies are identified that will be 100% likely to occur when using chloride DIR in the fight against winter slippage. The relationship between emergencies is considered. Measures to ensure the safety of rolling stock and passengers are presented.

Keywords: Chloride deicing reagents (DIR) · Rolling stock safety · Airlines of the railway power contact network · Concrete destruction · Metal corrosion

1 Introduction

Railway transport is considered as the safest mode of transport. It used for both long-distance and short-distance trips, for a variety of goods transportation. For all types of transport, different types of wagons, tanks, locomotives are provided, which can be defined as rolling stock.

Currently, more than a half of all Railways in Russia are electrified. The safety of trains, namely rolling stock – an urgent problem of our time. There are many factors that affect the safety of railway transport in general and its passengers in particular. The safety of rolling stock is closely related to the operation of the railway electrification contact network, as most roads are electrified. In case of failure of overhead power lines there are interruptions in the operation of railway transport and a rolling stock downtime. In turn, airlines of the contact network are very often damaged because of the rolling stock failure. In this article, we will discuss non-usual factors that can cause damage to both rolling stock and airlines of power contact network - chloride de-icing reagents (DIR) common used on objects of passenger railway transport infrastructure.

When the question about the usage of chloride deicing reagents in the Russian railway transport for the safety of passengers arose, the staff of the Department of Chemistry and Engineering ecology of RUT MIIT conducted serious studies and tests of the selected reagents. The studies were aimed at the limits of chloride DIR exposure identifying. It was found out that the reagents affect not only the objects of passenger facilities, but also the airlines of the power contact network [1, 2], and the railway (rails

Z. Popovic et al. (Eds.): TransSiberia 2019, AISC 1115, pp. 809–815, 2020.
https://doi.org/10.1007/978-3-030-37916-2_79

and sleepers). The results of the research of the effect of chloride DIR on the materials of the studied objects were very different from the statements of reagent manufacturers.

2 Brief Literature Review

In carrying out a comprehensive study of the chloride DIR impact on the environment, people, animals, materials that the objects of passenger facilities (passenger platforms high and low, pedestrian bridges, pedestrian railway crossings), rails and sleepers were made from, first of all, the works devoted to this problem were studied. It was found that the direct effect of chloride DIR on various objects, materials and environments has not been studied earlier. There are works and publications related to the influence of chlorides on concrete and reinforced concrete [3], but without taking into account the simultaneous impact of negative temperatures and cyclic temperature fluctuations relative to the freezing point of water (zero degrees), characteristic of the autumn-spring transition periods. The materials from which the studied objects are made include concrete, steel, copper products (wires). It was believed that the copper in normal conditions does not react with chlorides. Scientists from Egypt [4] in their studies have shown that in countries with hot climates located in the sea area, there is corrosion of copper and mixed wires from chlorides transported by air masses from the salt seawater layers.

Corrosion of metal structures, which include such a chemical element as iron (Fe) occurs according to the classical scenario [5].

3 Research Results

The authors of the article, when planning their experiments, took into account two facts. First of all, the cold period of the year with negative temperatures in Russia may last a long time. Secondly, temperature fluctuations (cyclic transitions from plus to minus) are possible not only at the season's junction, but can occur repeatedly in the winter months, even for one day. Therefore, the materials from which the main objects of study are made, namely, concretes, were exposed to chloride DIR models with the simultaneous influence of negative temperatures and temperature cycles. The brine concentration in the experiments corresponded to that in real life when the object contacts with the reagents. The results of the studies, as noted above, were significantly different from the promises made by the DIR manufacturer. The samples of concrete simultaneously applied with the temperature cycles (temperature changes from plus to minus) and the chloride brines were almost completely destroyed (see Fig. 1).

Metal plates also showed the beginning of corrosion (Fig. 2).

The study of a copper samples corrosion in a chloride meDIR was carried out taking into account the fact that the wires of power network in railway transport are heated up to about 200 °C according to the Lenz-Joule law. The copper wire samples appearance after heat treatment in chloride meDIR is shown in Figs. 3 and 4.

Fig. 1. Kind of concrete sample after the test

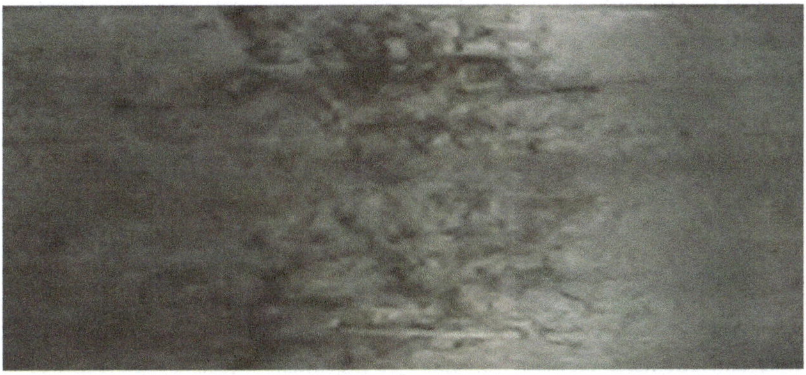

Fig. 2. Appearance of a metal plate tested in a solution of model chloride DIR. Heterogeneity is visible

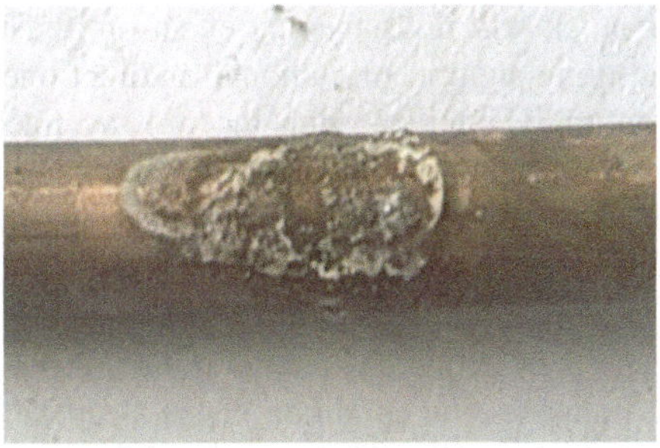

Fig. 3. A fragment of damage to a copper wire sample treated with a model solution of chloride DIR at 200 °C in a laboratory oven

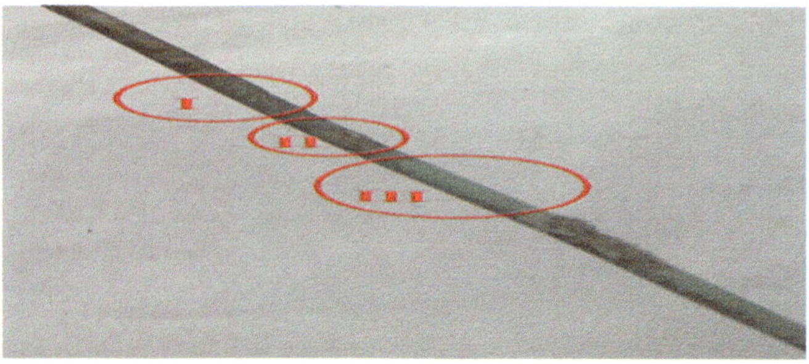

Fig. 4. Appearance of copper wire sample treated with a model solution of chloride DIR after temperature treatment at 200 °C in a laboratory oven

Thus, as a result of laboratory tests, the statements of manufacturers about the safety of chloride DIR for materials (concrete, reinforced concrete, metals) were refuted.

4 Identification of the Relationship Between Accidents Occurring on the Rolling Stock and Airlines of the Railway Power Contact Network, Under the Influence of Chloride DIR

The effect of chloride DIR on the elements of airlines of the power contact network of railway was being considered earlier [2]. It has been proved that chlorides cause corrosion of iron (objects, which include this element), copper and concrete. These materials and chemical elements are part of the main objects of the airlines of the railway power contact network.

It is necessary to find out whether DIR affects the rolling stock, and to identify the relation between accidents on the rolling stock and the breakdown of airlines of the power contact network. It cannot be argued that deicing reagents that do not have direct contact with the rolling stock will not have any effect on it. This statement is wrong. The impact will be, and quite significant, although indirect one.

The rolling stock naturally interacts with the railway tracks. This interaction is called "wheel – rail system". This system allows for contact between the rolling stock and the upper structure of the railway tracks [6]. Thereby in order to ensure the safety of passengers and rolling stock functioning, to reduce accidents and emergencies, as well as the number of failures in the operation of rolling stock, it is necessary that the Railways have no defects, their geometry, including the profile of the rails, would not be violated. For freight transport, the condition of the sleepers and the rubble beneath them is of particular importance. On lines with high load capacity, they use wider sleepers. The safety of operation of the rolling stock, including passengers inside it, will be affected by the condition of rails and sleepers. Their change (the external defects formation), and even more destruction, will lead in the first case to rapid wear and deformation of wheel pairs, in the latter – to accidents and crashes, bringing not only economic damage, but also to human casualties.

The causes of railway accidents are different [7]. Derailment of trains is a type of accidents that occur in case of the sleepers destruction and rails corrosion caused by the usage of chloride DIR in particular. And it will happen sooner or later with a 100% probability, because every our experiment ended with the destruction of a concrete sample tested in a medium of model solutions of deicing reagents with simultaneous application of freezing/thawing cycles. Also there was metal samples corrosion during the tests. Consequently, the use of chloride reagents as a means of preventing injuries to the population will lead to serious problems.

A buoyant force causes the descent of the rolling stock from the rails. Under the influence of this force, the wheel breaks away from the surface of the rail and an accident occurs. Corrosion of the rails strongly contributes to the buoyancy force uprising. Upon the sleepers destruction the geometry of the path will be broken and significantly violated, that will cause such force emergence and development. Such changes in the upper structure of the track will lead to an accident – to the descent of the rolling stock (train derailment). The probability of such accidents increases when using chloride DIR to combat winter slipperiness at passenger facilities, because chlorides cause the destruction of many materials. The beginning of the destruction process is very difficult to notice. In the case of concrete, there is almost spontaneous destruction of the material after a freeze/thaw cycle. Furthermore, there is a Rebinder effect, which is characterized instant destruction by immeDIRte and total destruction of the material under the influence of the adsorption-active medium with simultaneous application of an external force. In this case there is such an effect: concrete products (sleepers) instantaneous destruction, while the impact of the brine of chlorides and disjoining pressure arising in the process of freezing sleepers during the cold period of the year. As mentioned earlier, the occurrence of sleepers destruction at the same conditions occurs with almost 100% probability.

As mentioned above, the rails corrosion and the sleepers destruction causes both types of accidents. In addition, accidents often occur due to a malfunction of the wheel pairs. The malfunction of wheel pairs in turn can occur due to rusting process when using chloride DIR.

It should be noted that very often a destruction of the supports of airlines of the power contact network by rolling stock takes place because of railway transport derailment. It turns out that such accidents are interrelated. There is a rather complex chain of accidents caused only by the use of chloride DIR in the cold period (see Table 1) [8].

Unfortunately, the format of this article does not allow to specify all statistical data on railway accidents occurred, but even with the above given data it is possible to lay down a picture of the existing problems.

Table 1. Statistics of accidents at Russian railways over a 20-year period from 1999 to June 2019

№	Accident type	Date	Place	Cause	Consequences
1	The mail-baggage train No. 940 crash	03.02.2001	Lesnaya—Ingoda railroad of the Chita branch of the Baikal railway	The railway track broadening (due to poor maintenance)	Damaged: 5 train cars (require overhaul and maintenance). Downed: 3 supports of contact network; destroyed: 300 m of railway, 750 m of wires. Without victims
2	Derailment and fire of 21 the tank with gasoline AI-92 of freight train	13.08.2003	Khrustalnaya - Pervouralsk railroad near Novotalitsa	Train wreck	Corrupted: almost 500 m of the railway, 500 m of contact wire
3	Derailment of 18 gondola cars with coal	13.12.2014	Syrostan – Flyusovaya railroad of the South Ural railway near the platform 1988 km	The fault of the way	Damaged: 300 m of the railway track, 3 power line supports. No casualties. All passenger trains running between Ufa and Chelyabinsk were detained for a day. Movement on an even path was restored on December 14, in both directions - on December 15. Loss - 11 million rubles
4	3 cars of the freight train No. 2181 derailment	16.07.2015	Arkhara – Domikan railroad	Unsatisfactory condition of ways	Not specified
5	16 wagons of a freight train № 2714 derailment	15.02.2016	Arteushka – Penykovaya railroad of the TRANS-Baikal railway	Not specified	Broken: the envelope of the adjacent track; damaged: 4 supports of contact network
6	29 cars of freight train No. 2002 derailment	12.01.2019	Delur—Tyret railroad of the double-track electrified section of the Zima—Cheremkhovo of the East-Siberian railway	The wheel disc damage	Damaged: contact network supports, 300 m of railway track in both directions. Detained: 2 fast trains

5 Conclusion

The studies have revealed the destructive effect of chloride deicing reagents (DIR) on various materials that are part of very important objects of railway transport:

- concrete and reinforced concrete from which many elements of the airlines power network are made (in particular, the support or the supports base), sleepers, objects of passenger facilities;
- constructions that include iron. These are the elements of the contact network, rails, elements of passenger facilities and many parts of the rolling stock;
- copper from which wires of a power contact network are generally made in Russia.

Corrosion of the above materials leads to accidents. All accidents are interrelated: one accident can provoke another, with more severe consequences.

The arrangement of warm platforms, covered platforms and covered passenger railway bridges across the tracks can be the main alternative to any anti-icing reagents, including chloride,. When using such engineering solutions, the occurrence of emergency situations in the application of chloride DIR will be reduced to zero.

References

1. Popov, V, Churiukina, S: The effect of chloride deicing reagents on concrete structures. In: International Conference Proceeding, vol. 4, no. 22, pp. 25–28, part 1. Educatio (2016)
2. Popov, V., et al.: Identification of the effect of chloride deicing reagents on the contact network of overhead power transfer lines for land transport. Int. Res. J. **7**(61), 65–68 (2017)
3. Rozental, N.: Corrosion resistance of cement concrete with low and particularly low density. Doctoral dissertation, Moscow (2004)
4. Nady, H.M., El-Rabiei, M.M., Badawy, W.A.: Stability of some copper ternary alloys in chloride solutions polluted by sulfide ions. Chem. Process Eng. Res. **20**, 35–44 (2014)
5. Todt, F.: Corrosion and corrosion protection. Chemistry, Leningrad (1966)
6. https://scienceforum.ru/2018/article/2018005669
7. Classification of railways accidents. http://www.eng.dieselloc.ru/railway-engineering/classification-of-railways-accidents.html
8. List of Russian rail accidents. https://en.wikipedia.org/wiki/List_of_Russian_rail_accidents

Analysis of Dynamic Processes in Maritime Engines of Ships

Aleksandr Bordug[1](✉), Aleksandr Yashin[2], Nadejda Smetuch[1], and Inna Antipenko[1]

[1] Kerch State Maritime Technological University, Kerch 298309, Russian Federation
sergiiblack@mail.ru

[2] Saint Petersburg Electrotechnical University "LETI", ul. Professora Popova 5, 197376 St. Petersburg, Russian Federation

Abstract. Low-speed turbocharged engines (SSE) due to their high efficiency, reaching 54%, resources up to 100,000 h, the possibility of using cheap fuels and the possibility of direct transfer of power to the propeller are the most preferred when choosing the main engines of marine vessels, which confirmed by modern diesel and shipbuilding. The engine is a source of mechanical energy, which determines the area of its application. The technical level of any machine, including the engine, is characterized by the ability to fulfill the functional purpose and conformity of the machine design and its manufacturing technology to the achieved level of science, engineering and technology.

Keywords: Slow speed engine · Propulsion · Turbocharger · Torque · Inertia

1 Introduction

The high level of acceleration of the working process of modern SSE led to the presence of high thermal and mechanical stresses in the details of the cylinder-piston group. Therefore, to ensure reliable operation of SSE when determining the range of operating conditions, it is necessary to exclude the possibility of engine overload.

The processes occurring in a piston internal combustion engine, it is customary to study on models of various levels of complexity.

2 Slow Speed Engine Model Algorithm

The technical level of the engine is characterized by a number of indicators that can be divided into groups [1–3]:

1. Reliability indicators characterizing the engine's ability to perform a set task, while maintaining its operational parameters and characteristics. Quantitatively, reliability is assessed by indicators, the nomenclature of which is chosen taking into account the features of the engine, the modes and conditions of its operation, the consequences of failures.

Z. Popovic et al. (Eds.): TransSiberia 2019, AISC 1115, pp. 816–824, 2020.
https://doi.org/10.1007/978-3-030-37916-2_80

2. Profitability indicators characterizing the rationality of the use of operating materials (fuel, lubricants and other consumables), and also the amount of material and energy used in the production of the engine. Profitability is characterized by efficiency factor, specific fuel and oil consumption, time to oil change, specific power, material consumption, specific weight, costs of production, etc.
3. Processability indicators assessing the adaptability of the engine to the conditions of production, perfection and rationality of technological processes; the main indicator of processability is the specific labor intensity of manufacturing the product.
4. Ergonomic indicators characterizing ease of operation, maintenance, and management of the engine. Ergonomics depends on the compliance of the engine design with the physiological and anthropometric features of a person. For a quantitative assessment, a complex ergonomic indicator is used.
5. Indicators of standardization and unification characterize the degree of application of standardized products in the design, the repeatability of individual parts or elements of the machine. The higher the standardization and unification of the product, the more reliable the product, on the one hand, and on the other hand - simpler and cheaper to operate and maintain.
6. Environmental indicators assessing the degree of harmful effects of engine operation on the environment, including specific emissions of nitrogen oxides, hydrocarbons, carbon oxides, particulate matter.
7. Indicators of patent law purity characterize the degree of security of a machine's design by copyright certificates or patents, as well as the legitimacy of using developments and decisions of third-party authors in the design. Violation of patent law purity impedes the sale of products both domestically and abroad.
8. Safety indicators characterize the degree of adverse effects on the maintenance personnel, labor safety in the management and maintenance of the engine, and compliance of working conditions with sanitary and hygienic standards. Indicators include noise and vibration levels, a comprehensive safety indicator, etc.
9. Transportability indicators characterize the ease of transportation and storage of the engine. These indicators are evaluated by the size and weight of the engine.
10. Aesthetic indicators characterize the individual perception of the engine's appearance, the rationality of the design, its informational expressiveness, the perfection of the production performance of the engine. The quantitative characteristic is a complex aesthetic indicator.

All indicators of the technical level of the engine can be determined in one of three ways: by calculation, by experiment, and by method of expert assessments. Each of the methods is more or less applicable to the determination of individual indicators, but the most reliable is the experimental method.

The operation of the SSE using a thermodynamic model, quasi-stable model, cycle-average description model. Each simulation step continues in one of these three phases, shown in Fig. 1 schematically, in the form of a grouping of the equations.

The mission of the <I> phase is to calculate the intermediate mass flow rate of air and the exhaust gas pressure of the set (row) $INTER = \{\dot{m}_A, p_E\}$. In particular, starting from the set of values of the variables min-1.

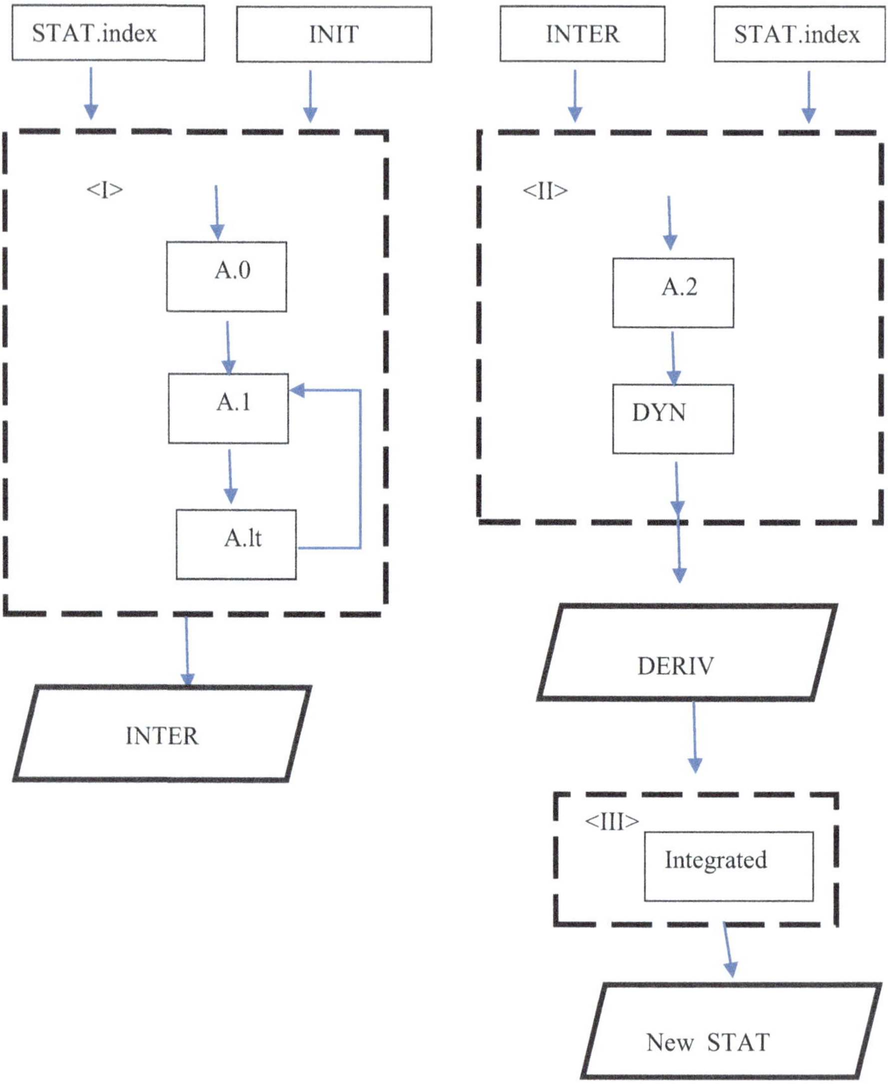

Fig. 1. Slow speed engine model calculating algorithm.

STAT = {nE, nTC} and the position of the fuel rail (index = FR), the values of variables in the INTER set can be calculated using the equations of groups A.0, A.1 and A.

Equations A.0 directly solved regarding to mass fuel consumption ($\dot{m}F$) and blowing pressure (p1). Therefore, STAT and PCC values are distribute through A.0. and provide values for the flow rate of the fuel mass and blowing pressure.

From the other side, the A.I equations make it possible to calculate pressure, air temperature, exhaust gases, etc. based on the values for STST, CCP and

INTER = {ṁA,pE}. Therefore, a numerical iterative procedure should be used to determine the values of variables in the INTER set, since these values depend on those calculated by A.1 in an entangled and nonlinear system of algebraic equations. This carried out in phase <I>. An additional input set of values, namely, INIT, is required for the iterative solution procedure used for groups A.I and A. This set of values includes the initial assumptions for INTER = {ṁA,pE}, the values of INIT, and the numerical solution method used the convergence of influence to the solution [3, 4].

After the values of the variables for the INTER set are determined, the simulation proceeds to phase <II> , which includes the distribution of the values of STAT, PCC and the values of the variable INTER via the equations of group A.2., which leads to the calculation of the equations of engine QE, turbine QT and torque of compressor QC. In fact, the values of the variable torque allow us to calculate the acceleration of the shaft of the turbocharger, n = ṅE и ṅTC. Finally, in phase <III> , new values computed for the variables STAT (new_STAT). This done for using a numerical integration scheme in a discrete time domain. Typical numerical integration schemes used in the practice of control modeling are the inverse difference, direct difference, and Tastin's method (trapezoidal rule). Usually, the inverse difference is used to simulate an engine with a time step equal to 60/(2πnMCR), that is, the minimum duration in seconds of one machine cycle, thus corresponding to the time step specified for the cyclical model. The difference in the opposite direction has a practical interpretation, since the discrete-time controller's interface with the Zero – Order – Hold (ZOH) D/A scheme. That is why the direct difference was chosen as an integration method for implementing an engine model [4, 5].

For the crankshaft rotational speed (nE) or propeller shaft (n), the differential equations are as follows:

$$\dot{n}_E(t) = \frac{Q_E - Q_L}{I_{total}}, \tag{1}$$

where Q_E is the engine average torque value, Q_L is the average cycle moment value of the loaded rowing unit and I_{total} is the moment of inertia of the rowing unit.

It is assume that when connecting the shaft directly to the load (without transmission and clutch):

- there is no gear ratio between engine and propeller speeds, $n \equiv n_E$;
- the load on the engine can be calculated by the formula:

$$Q_L(n_E) = K_Q \cdot n_E^2 = (K_{Q0} + \Delta K_Q) \cdot n_E^2, \tag{2}$$

where K_Q is the coefficient of the moment of resistance of the screw, which is obtained as a result of summing up the indefinite k_Q and the nominal value of K_{Q0}.

The moment of inertia of the propulsion unit can be express by the following equation:

$$I = I_E + 1.15 \cdot I_{\text{ycT}} + \Delta I = I_0 + \Delta I, \tag{3}$$

the sum of the inertia forces of the crankshaft of the engine I_E and inertia the screw I_{prop}, as well as the inertia of the water trapped by the propeller of 15% (nominal value), as analyzed above, is considered the pressure value corresponding to the engine power P N:

$$P = p_e \cdot z_c \cdot V_h \cdot \frac{n}{60}, \tag{4}$$

However, there is the following power expression:

$$P = Q_E \cdot \frac{2\pi}{60} n. \tag{5}$$

From the above equation:

$$Q_E = \frac{Z_c \cdot V_h}{2\pi} \cdot p_e. \tag{6}$$

3 Analysis of Dynamic Processes in Low-Speed Turbocharged Diesel Engines of Ships

After changing the position of the fuel ramp FR, there is a lag τ associated with the inertia of the change in engine torque. The fact is that there is a change in torque after the fuel is inject into the engine cylinders and combustion occurs. Consequently, a new value of torque, depending on the changed position of the fuel rail, will be achieve after a complete revolution of the crankshaft of the two-stroke engine. The change in torque also depends on the sequence of operation of the engine cylinders. The lag in the change in torque of a two-stroke engine is describe by the following expression:

$$\frac{1}{4 \cdot n_E} < \tau < \frac{1}{4 \cdot n_E} + \frac{1}{z_c \cdot n_E}, \tag{7}$$

τ in minutes.

However, this effect (delay) is significant only at low engine speeds, usually below 60 RPM. Moreover, in the worst case:

$$\tau = \frac{1}{4 \cdot n_E} + \frac{1}{z_c \cdot n_E < \frac{1}{n_E}}, \tag{8}$$

for any $z_c \geq 2$.

In conclusion, the effect of engine torque (τ) can be neglect in engine models of average cycle values, because this delay is less than the sampling time interval (stroke) of this model. Travel time is the period of one revolution of the crankshaft (in min.), $1/n_E$.

The dynamics of a turbocharger determined by the following differential equations, similar to the propeller shaft:

$$\dot{n}_{TC}(t) = \frac{Q_T + Q_C}{I_{TC}}, \tag{9}$$

Q_T – turbine torque, Q_C – compressor torque at load. I_{TC} – the moment of inertia of the turbocharger, which includes the inertia of the shaft of the turbocharger, as well as the inertia of the turbine and the compressor. There is no need to include additional parameters in this system, since inertial forces do not change.

Q_T and Q_C is thermodynamic variables from the following algebraic relations:

$$Q_T = \frac{\eta_{iT} \cdot C_{P,exh} \cdot T_E \cdot \dot{m}_E}{2\pi/60 \cdot n_{TC}} \cdot \left[1 - \left(\frac{p_a}{p_e}\right)^{\frac{\gamma_E - 1}{\gamma_E}}\right], \tag{10}$$

$$Q_C = \frac{C_{P,air} \cdot T_a \cdot \dot{m}_A}{\eta_{TC} \cdot 2\pi/60 \cdot n_{TC}} \cdot \left[1 - \left(\frac{p_1}{p_a}\right)^{\frac{\gamma_A - 1}{\gamma_A}}\right]. \tag{11}$$

Permanent γ_A and γ_E – specific heat of air and exhaust gas, respectively:

$$\gamma_A = 1,4,\ \gamma_E = 1,34.$$

It should be note that the above parameters depend on the air and exhaust gas temperatures, respectively. However, the above values are typical for normal air temperature (288 K or 15 ° C) and for exhaust gas temperature (400–1100 K).

$C_{P,air}$ and $C_{P,exh}$ are specific heat capacities at constant exhaust air pressure, respectively. Generally, C_P calculated based on the following mathematical expression:

$$C_p = \frac{\gamma - 1}{\gamma} \cdot \frac{R}{M_{mol}}, \tag{12}$$

where R – ideal gas constant gas ($R = 8.314$ J/(mol K) and M_{mol} – molecular weight of the gas in question. In the case of air and exhaust gases, it is assume that:

$$C_{P,air} = 1005 \frac{I}{kg \cdot K}, \tag{13}$$

$$C_{P,ex} = 1117 \frac{I}{kg \cdot K}. \tag{14}$$

The constant η_{TC} ultimately includes the mechanical efficiency η_{mTC} of the turbocharger and the isentropic efficiency of the turbine η_{iC}. Both of these parameters should be constant and almost equal:

$$\eta_{TC} = \eta_{mTC} \cdot \eta_{iC} \overset{\eta_{mTC} \approx 0.99}{\Rightarrow} \eta_{TC} \approx \eta_{iC}. \tag{15}$$

It is possible to achieve a substantial similarity between the ratio of torque of the turbine and compressor. That is why they are calculated taking into account the adiabatic air compression (in the compressor) and the adiabatic expansion of exhaust gases (in the turbine). Based on these theoretical thermodynamic processes, it is possible to obtain ratios for the torques of a turbine and a compressor: Q_T>0 и Q_C<0. As in the case of the compressor $p_I > p_a$, and in the case of the turbine $p_E > p_a$, while in both cases $(\gamma-1)/\gamma > 0$.

4 Analysis of Dynamic Processes in Low-Speed Turbocharged Diesel Engines of Ships

Modern engines are characterized by such areas of improvement as increasing operating parameters (power, boost pressure, average effective pressure, crankshaft rotation speed), performance indicators (specific effective fuel consumption, net efficiency), obtaining small dimensions and weight (power-to-volume ratio, mass-to-power ratio), increasing the degree of automation, reducing harmful emissions from exhaust gases, combining engines into complex power plants with a single control. The increasing complexity of engines and the tightening of requirements for them lead to the need of increasing their reliability and durability.

Reliability is one of the main indicators of the quality of any product. In accordance with the Russian National Standard 27.002-83 "Reliability in engineering. Terms and definitions", it is considered as a property of an object to preserve in time and within the established limits the values of all parameters characterizing the ability to perform the required functions in given modes and conditions of use, maintenance, repairs, storage, and transportation. In this case, the object has a broad generalizing meaning. In practice, the reliability of any product is determined using the appropriate reliability analysis scheme [4]. The simplest and most widely used structure of reliability analysis is system analysis. Here the product is considered as a system consisting of individual subsystems and elements.

The system includes a set of jointly operating technical devices intended for the independent performance of certain tasks in the process of its application for the intended purpose. An element is a part of the system that does not have independent operational significance and is intended to perform specified functions. In the theory of reliability, the concept of "element" is quite broad. Depending on the type of tasks being performed, the same device can be considered both as a system and as an element.

Systems and elements can be recoverable and non- recoverable, repairable and non-repairable. A recoverable object is an object for which, in the situation under consideration, the restoration of working condition is provided for in the design, regulatory and technical documentation. A non-recoverable object is an object for which the restoration of working capacity is not provided for in the design, regulatory and technical documentation. A repairable object is an object for which repairs are provided for in the design, regulatory and technical documentation. A non-repairable object is an object for which repairs are not provided for in the design, regulatory and technical documentation.

The term "reliability" as applied to the engine is a complex aggregate property consisting generally of failure-free performance, maintainability, durability, and persistence individually or in a certain combination of these properties.

Reliability is the most important characteristic of the technical level of the engine, since an insufficiently reliable engine cannot function efficiently. Each of its stops due to damage to individual components or reduced technical performance below the permissible level usually entails large material losses, and in some cases, this can have disastrous consequences. Only an engine with a high level of reliability can be competitive.

Currently, the problem of reliability is receiving increasing attention, since even the industry of developed countries is suffering huge losses due to the lack of reliability of manufactured products. According to the available statistical data for the entire period of operation, the costs of restoration and repair exceed the costs of manufacturing in aircraft industry by 5 times, in machine tool industry by 8 times, in automotive industry by 6 times.

Reliability features include the fact that it is inextricably connected with all the main stages of the engine life cycle, its design, manufacture and use, starting from the moment when the idea of creating a new engine is formed and substantiated, and ending with the decision of its unserviceability. Each of the stages contributes to the solution of the difficult problem of creating an engine of the required level of reliability with minimal expenditure of material, labor and time resources. The main decisions on reliability made at the stage of designing or manufacturing an engine directly affect its operational and economic indicators.

When designing and calculating the engine, the required level of reliability is taken into account. It depends on the chosen design of the engine, its parts and assemblies, the materials used, methods of protection against harmful effects, reliability of lubrication and cooling, maintainability, and other design features.

Reliability of the engine is ensured in the manufacturing process. It is ensured by strict observance of technological processes, high qualification of workers and labor discipline, and depends on the quality of manufacturing parts and components, control methods in the manufacturing process, build quality, and methods for testing finished products.

5 Conclusion

At the operation stage, reliability is shown most adequately. Indicators of reliability and durability depend on the methods and operating conditions of the engine, the adopted system of its maintenance, the characteristic modes of operation, and other operational factors.

The problem of reliability is of complex nature. It involves the design, production and operation of engines. Various branches of knowledge are involved to solve it. It requires to make both technical and organizational decisions.

Based on the analysis of scientific literature and technical documentation of manufacturers of low-speed marine diesel engines, it is shown that ensuring the reliability and efficiency of their operation as the main engine of ship power plants for marine displacement ships is impossible without improving control systems for operating modes of low-speed diesel engines. Nowadays, there are no means of ensuring stable operation of the low-speed marine diesel engines in a given mode, under conditions of varying operating factors and significant fluctuations in the load on the propeller side. A review of scientific and technical information indicates that improving the reliability of shipboard low-speed diesel engines can be ensured by improving their control systems, aimed at improving the quality of transient processes under destabilizing effects on the engine and preventing the overrun of modes of low-speed marine diesel engines beyond the nominal in case of abrupt changes in modulus of resistance on the propeller side.

Simplified quasistationary models describe the change in crankshaft rotational speed. In more complex engine models, gas exchange processes are also take into account. Consequently, differential ratios for air and exhaust gas flow should also be included, increasing the number of differential equations. However, a medium-cycle method is sufficient for controlling the engine; provide that sufficient simulation accuracy guaranteed, because the controlling effect affects the operation of the engine once per cycle. Below, the algebraic interdependence of the thermodynamic variables presented in the previous sections for calculating engine, turbine and compressor torques will be analyze. A sequential view shows the flow of gases through each major part of the power plant.

References

1. Sokolov, S., Zhilenkov, A., Chernyi, S., Nyrkov, A.: Assessment of the impact of destabilizing factors in the main engine shaft of the adaptive speed controller. Procedia Comput. Sci. **125**, 420–426 (2018)
2. Chernyi, S.: Techniques for selecting topology and implementing the distributed control system network for maritime platforms. AKCE Int. J. Graphs Comb. **15**, 219–223 (2018)
3. Zhukov, V., Sherban, S., Melnik, O., Sokolov, S., Kolesnichenko, S.: Improvement of methods and means of thermal control of ship power plants. In: 2019 IEEE Conference of Russian Young Researchers in Electrical and Electronic Engineering (EIConRus) (2019)
4. Zhukov, V., Sherban, S., Melnik, O., Sokolov, S., Kolesnichenko, S.: Improvement of methods and means of thermal control of ship power plants. In: 2019 IEEE Conference of Russian Young Researchers in Electrical and Electronic Engineering (EIConRus), pp. 389–392. IEEE (2019)
5. Zhukov, V., Pulyaev, A., Melnik, O., Nyrkov, A.: Ensuring the permissible temperature state of parts of the cylinder-piston group of forced diesels. In: 2019 IEEE Conference of Russian Young Researchers in Electrical and Electronic Engineering (EIConRus), pp. 385–388. IEEE (2019)

Temperature Stresses in CWR – Experience of Serbian Railways

Zdenka Popović[1(✉)], Nikola Mirković[1], Ljiljana Brajović[1], Dragan Rakić[2], Luka Lazarević[1], and Slobodan Petričević[3]

[1] Faculty of Civil Engineering, University of Belgrade, 11120 Belgrade, Serbia
{zdenka, llazarevic}@grf.bg.ac.rs
[2] Faculty of Engineering, University of Kragujevac, 34000 Kragujevac, Serbia
[3] School of Electrical Engineering, University of Belgrade, 11120 Belgrade, Serbia

Abstract. The railway track with continuously welded rails is an essential part of the modern railway superstructure around the world. Calculation of the temperature stresses in continuously welded rails has great significance for the planning of maintenance activities in order to maintain traffic safety (risk of derailment due to either rail break, or track buckling). This paper presents the finite element model for calculation of rail stresses and displacements due to the temperature changes, which was developed by the authors. The monitoring of the track sections, which are significant regarding traffic safety, was suggested in this paper.

Keywords: Railway · Continuously welded rail · Finite element model · Temperature stress · Temperature displacement

1 Introduction

Compared to the conventional railway superstructure with rail joints, the railway superstructure with Continuously Welded Rails (CWR) has great economic advantages, as well as advantages for urban environment and ride comfort.

In Serbia, CWR is applied from the fifties of the 20th century. Nowadays, the modern structure of the ballasted track includes continuously welded rails with steel grade at least R260 [1], concrete sleepers [2], and elastic rail fastening system [3, 4].

On the other hand, the temperature changes in CWR could lead to track buckling (Fig. 1) or rail break and could jeopardize the traffic safety. The both track buckling and rail break could result in significant human, environmental and economic loss. For this reason, modern track with CWR implies specific maintenance technology and planning of maintenance activities on the level of railway network and railway sections. This applies in particular to railway sections with:

- uneven temperature changes along CWR [5],
- vehicle/track/bridge interaction [6–8],
- tight curves (radius less than 600 m) [9], and
- poor or inadequate maintenance of railway structure [9, 10].

Z. Popovic et al. (Eds.): TransSiberia 2019, AISC 1115, pp. 825–833, 2020.
https://doi.org/10.1007/978-3-030-37916-2_81

Fig. 1. Track buckling on Belgrade – Vrbnica railway line (left) and in Subotica railway station (right) in Serbia.

The longitudinal (tensile and pressure) stresses in CWR primarily depend on:

- sleeper/ballast interaction described by the resistance to the longitudinal displacement of the sleepers through the ballast, and
- rail/sleeper interaction described by longitudinal rail restraint [4] to the elastic creep of rail in the fastening assembly (relative to the sleeper).

In this paper, the temperature stresses along CWR in terms of possible longitudinal displacements of rail (due to either rail/sleeper or sleeper/ballast interaction) are considered and explained.

Furthermore, paper presents the finite element (FE) model of track with:

- prevented rotation and creep of sleepers in the ballast, and
- prevented longitudinal rail displacement for determination of temperature stresses in the middle section of CWR.

2 Temperature Stresses in CWR

High longitudinal forces are induced by temperature changes due to the prevented longitudinal expansion and contraction in the middle section of CWR, which is sufficiently away from the mechanical and/or expansion rail joint. Longitudinal force P (tensile or pressure) in the middle section of CWR is given by Eq. (1) [11].

$$P = \alpha \cdot E \cdot F \cdot \Delta T \tag{1}$$

where:

α - coefficient of temperature expansion of steel, which varies considerably across the relevant literature, for example: (a) $\alpha = 12.0 \times 10^{-6}$/K [8, 12], (b) $\alpha = 11.8 \times 10^{-6}$/K [15], (c) $\alpha = 11.5 \times 10^{-6}$/K [11], (d) $\alpha = 10.0 \times 10^{-6}$/K [10], etc.,

E - Young's modulus of steel, which depends on temperature [12],
F - cross-sectional area of rail,
ΔT - total temperature change from the stress-free (neutral) temperature Tn.

Serbian Infrastructure Manager prescribed the relevant temperature values as follows:

- minimum value of temperature in rail during wintertime Tmin = −30° C,
- maximum value of temperature in rail during summertime Tmax = +65 °C, and
- neutral (free stress) temperature Tn = 0.5·(Tmin-Tmax) + 5 = 22.5 °C.

Prescribed neutral temperature directly influences the temperature stress in CWR and the danger of track buckling, as well as the gap due to rail break (Fig. 2).

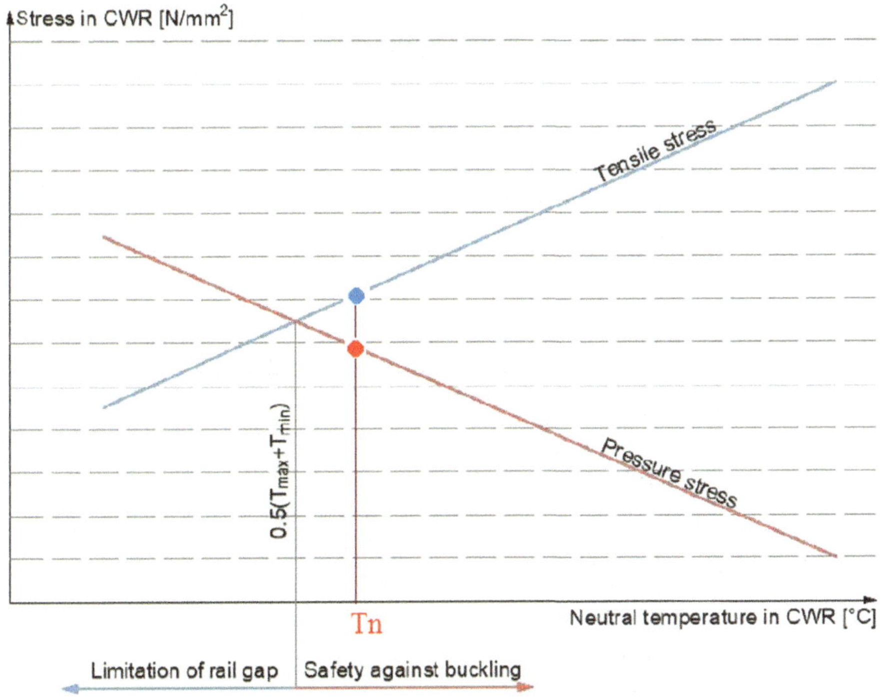

Fig. 2. Neutral temperature in safety zone against buckling

Accordingly, the longitudinal stress σ in each rail in track (in the middle section of CWR with prevented longitudinal expansion and contraction) is given by Eq. (2).

$$\sigma = \alpha \cdot E \cdot \Delta T \tag{2}$$

Therefore, CWR includes the middle section where expansion and contraction are prevented (Fig. 3a) and two "breather" sections with either expansion or contraction (left "breather" section and right "breather" section in Fig. 3b and c).

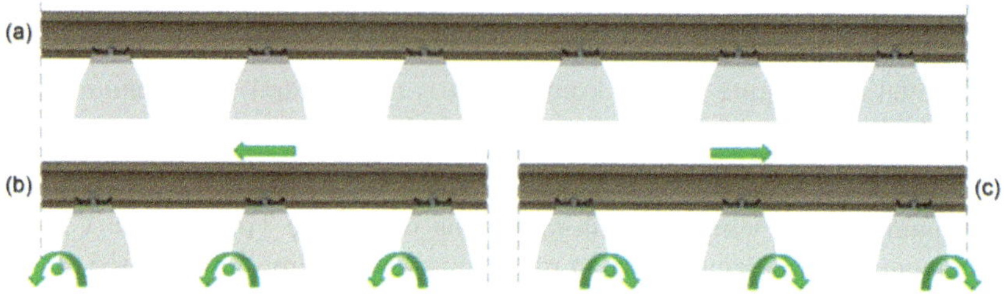

Fig. 3. Sections of CWR in terms of possible longitudinal displacement of rail.

2.1 Longitudinal Resistance to the Displacement of Rails in Track

Real behaviour of the track with CWR in the longitudinal direction is essentially dependent on the resistance to longitudinal displacement of the track structure through ballast, as well as the longitudinal rail restraint in rail fastening system [4]. Track or rail displacements are non-linear, but for practical use they can be approximated with the bilinear function (Fig. 4) [13, 14]. The resistance values in Fig. 3 refer to track consisting of: (a) two 60E1 rail profiles, (b) B70 concrete sleepers at a distance of 0.60 m, (c) minimum thickness 30 cm of a compacted ballast below the sleeper, as well as (d) longitudinal rail restraint 2 × 9 kN (for two track rails: (2 × 9 kN)/0.60 m = 30 N/mm, in Fig. 4).

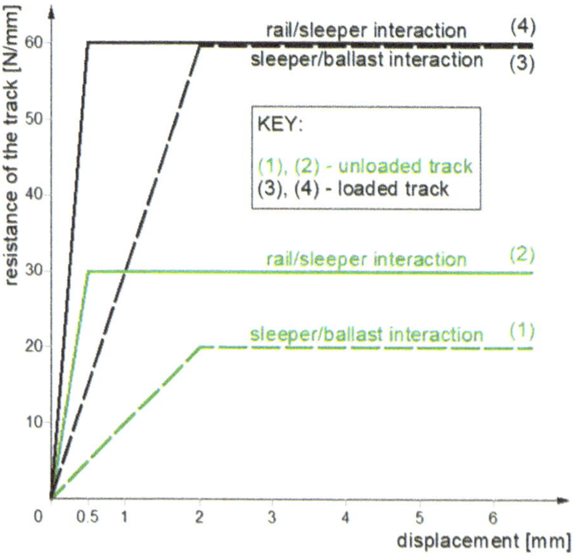

Fig. 4. Track resistance (sleeper/ballast interaction) and rail restraint (rail/sleeper interaction).

The resistance value depends on whether the track is loaded or unloaded (Fig. 4 and Table 1). The resistance to the displacement of the loaded track is higher than the resistance of the unloaded track and strictly depends on the axle load [9]. Table 1 shows the lower longitudinal resistance to CWR displacement based on the comparison of rail/sleeper interaction with sleeper/ballast interaction. In accordance with service conditions (Table 1), the lower value of longitudinal resistance is relevant for determining the stresses and displacements in the CWR. For displacements and stresses in CWR due to temperature changes, the lower resistance of unloaded track is relevant.

Table 1. The relevant (lower) values of longitudinal resistance to CWR displacement

Service and structure conditions		Rail/sleeper interaction	Sleeper/ballast interaction
Relevant conditions for the analysis of temperature stresses			
Unloaded track	Normal conditions (ballast is not frozen)		Relevant interaction
	Frozen ballast	Relevant interaction	
Relevant conditions for the analysis of stresses due to braking and acceleration of the vehicle			
Loaded track	Light axle load		Relevant interaction
	Heavy axle load	Relevant interaction	

2.2 Temperature Displacements in the Unloaded Track

In the unloaded ballasted track, the longitudinal track resistance is considerably lower than the longitudinal rail restraint [4].

Depending on the maintenance of the railway superstructure, the longitudinal resistance of the sleepers in the ballast could vary in range from 20 N/mm (for well maintenance) to 12 N/mm (for poor maintenance).

Under the influence of high temperature changes in CWR, sleepers could move through the ballast. Furthermore, during longitudinal displacement of the track, the following process occurs (Fig. 3b and c):

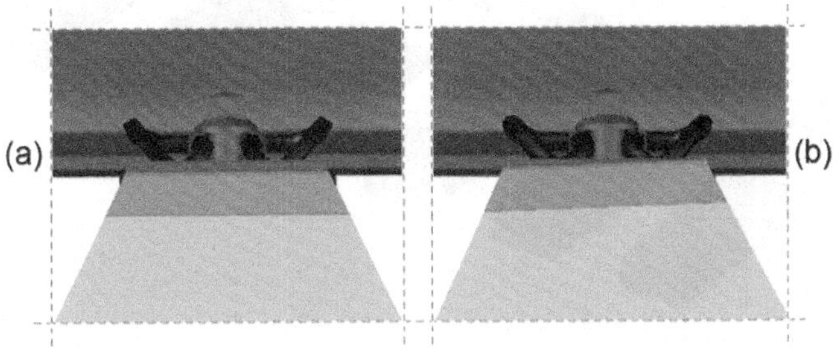

Fig. 5. Rail fastening system with elastic rail pad (a), and limited deformation of elastic rail pad due to the sleeper rotation without rail moving (b).

- Step 1: limited deformation of elastic rail pads (green element in Fig. 5),
- Step 2: displacement due to the rotation of the sleepers around their lower edge in the direction of track displacement (Fig. 3b and c),
- Step 3: displacement due to the creep of the sleepers through ballast in the direction of track displacement.

3 FE Model for Middle Section of CWR

FE model has the prevented rotation and creep of sleepers in the ballast, as well as prevented rail displacements. This model represents the behaviour of middle section of CWR. Due to the longitudinal symmetry of the considered problem, one half of the model was created using the symmetry boundary conditions.

FE model was created using Femap software. Stress analysis and post-processing of the results was performed using the same software. Moreover, numerical simulations were performed using NX Nastran Solver with Advanced Nonlinear Static.

This model was created using hexahedral 8-node finite elements. FE model consists of 87532 nodes and 70740 elements. The displacement of the rail ends in the longitudinal x direction was fixed.

4 Results of FE Model

In the middle section of CWR on FE model, the calculated pressure stress values are in the range of 100.7 MPa to 112 MPa for maximum value of temperature change $\Delta T = 65\ °C - 22.5\ °C = 42.5\ K$ in summertime in the Serbian climate zone (Fig. 6).

Fig. 6. Stresses in rail on FE model (maximum pressure stress $\sigma = -106.35$ MPa for $\Delta T = 42.5$ K).

Fig. 7. Zero displacements on FE model in the middle section of CWR.

Furthermore, Fig. 7 shows calculated corresponding rail displacements on this FE model.

Considering the results on FE model (Figs. 6 and 7), the following effects are observed:

- the results of FE model coincide with the stress values according to the analytical formula (2),
- FE model could be used for making general decisions regarding maintenance planning on the railway network level,
- this FE model cannot be used for making decision regarding maintenance planning for track section with possible CWR instability.

5 Conclusion

Numerical models, such as FE model presented in this paper, could be used for making general decisions regarding maintenance planning at the level of the railway network.

However, the modelling of CWR instability requires accurate values of: (a) "constants" for rail steel (coefficient of temperature expansion, Young's modulus, shear modulus, free stress temperature, etc.), especially during temperature increase, (b) unevenly distributed track properties [7, 15], and (c) unevenly distributed temperature along the rail (Fig. 8).

An open point can be the wide range of prescribed values of rail residual stresses (generated during the manufacturing process of rails) up to 250 MPa [1]. According to European standard [1], destructive saw-cutting measurement method is prescribed as a standard measuring method for residual stresses in the foot of Vignole rail. Unfortunately, this method cannot be used for measurement of residual stresses in CWR in service. Moreover, the residual stress changes during the service life of rail [6, 16]. There are several non-destructive methods for residual stress measurement in rails in service.

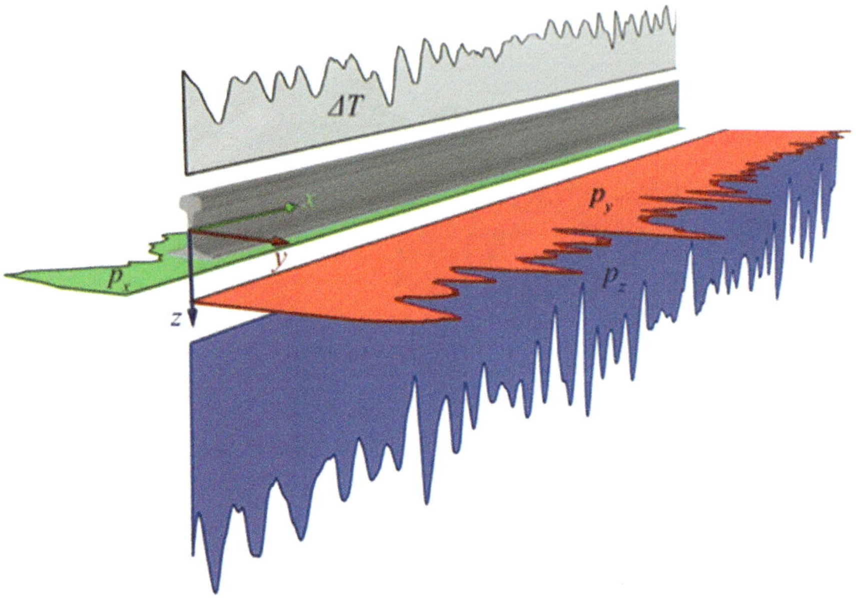

Fig. 8. Uneven distributions of temperature changes ΔT and resistance p against rail displacements in x, y and z direction along the rail.

Correct interpretation and evaluation of the measurement results is quite complex and still not suitable for practical application.

Considering that measured air temperatures were above 42 °C in some places in Serbia during last several years, it is necessary to perform the real-time monitoring of the rail displacements on track sections, which are significant regarding traffic safety (tight curves, transition zones to bridges or tunnels, switches, crossings, expansion joints, tunnels, bridges, sections with poor maintenance, etc). The authors are involved in the development of system for real-time measuring of rail temperatures and displacements in track, which should be a part of monitoring system on the sections significant for track safety [16–18].

Acknowledgement. This research was supported by the Ministry of Education, Science and Technological Development of Republic of Serbia through the research project "Research of technical, technological, staff and organizational capacity of Serbian Railways, from the viewpoint of current and future European Union requirements" (No. 36012).

References

1. CEN: EN 13674-1: 2011 + A1: 2017, Railway applications, Track, Rail, Vignole railway rails 46 kg/m and above, Brussels (2017)
2. CEN/TC 256: EN 13230, Railway applications - Track - Concrete sleepers and bearers, Parts 1, 2, and 4 (2017)

3. CEN/TC 256: EN 13481, Railway applications - Track - Performance requirements for fastening systems, Parts 1-2 (2012)
4. CEN/TC 256: EN 13146, Railway applications - Track - Test methods for fastening systems, Parts 1-9 (2014)
5. Popović, Z., Lazarević, L., et al.: Influence of climate conditions on stresses in CWR on the bridges in Serbia. In: VII International Scientific Conference of the Faculty of Transport Engineering, Pardubice, Czech Republic, pp. 200–207 (2018)
6. Mirković, N.B., Popović, Z.J., et al.: Management of stresses in the rails on railway bridges. FME Trans. **46**(4), 636–643 (2018). https://doi.org/10.5937/fmet1804636M
7. Lazarević, L., Vučković, D., et al.: Application of seismic tomography for assessment of the railway substructure condition. Struct. Health Monit. **18**, 792–805 (2018). 1475921718774778
8. Kupfer, R.: Auswirkungen von Beschleunigungs- und Bremskräften auf die Längsbewegungen des Gleisrostes. Dissertation, Technische Universität München (2004)
9. Popović, Z., Lazarevic, L., et al.: Railway gauge expansion in small radius curvature. Procedia Eng. **117**, 841–848 (2015)
10. Zeng, K., Qiu, T., et al.: Identification of ballast condition using SmartRock and pattern recognition. Constr. Build. Mater. **221**, 50–59 (2019)
11. UIC CODE 720: Laying and Maintenance of CWR Track. UIC, Paris (2005)
12. Bao, Y., Zhang, H., et al.: Measurements of young's and shear moduli of rail steel at elevated temperatures. Ultrasonics **54**(3), 867–873 (2014)
13. International union of railways (UIC), UIC Code 774-3 - Track/bridge Interaction – Recommendations for calculations, Paris (2001)
14. DIN: Fachbericht 101, Einwirkungen auf Brücken, 2. Auflage (2009)
15. Oscarsson, J.: Dynamic track-train-ballast interaction with unevenly distributed track properties. Veh. Syst. Dyn. Suppl. **37**(1), 385–396 (2002)
16. Mirković, N., Brajović, L., et al.: Measurement methods for residual stresses in CWR. In: Energy Management of Municipal Transportation Facilities and Transport, pp. 346–355. Springer, Cham (2018)
17. Popović, Z., Lazarević, L., et al.: Managing rail service life. Metalurgija **53**(4), 721–724 (2014)
18. Lazarević, L., Vučković, D., et al.: Assessment of sleeper support conditions using microtremor analysis. Proc. Inst. Mech. Eng. Part F: J. Rail Rapid Transit **230**(8), 1828–1841 (2016)

Determination of Contact-Fatigue of the Crosspiece Metal

Irina Shishkina(✉)

Russian University of Transport, Chasovaya Street, 22/2, 125190 Moscow, Russia
shishkinaira@inbox.ru

Abstract. The paper discusses the determination of the contact-fatigue characteristics of the crosspiece metal, which is required when assessing the work of the crosspieces in various conditions and finding effective ways to reduce the susceptibility of the crosspieces to contact damage. The methodology provides a universal approach to determining the intensity of damage caused by spalling in the rolling zone of crosspieces with cores made of steel 110G13L and allows quantifying the damage characteristics by spalling of crosspieces in various operating conditions, taking into account the shape and size of the contacting surfaces and casting manufacturing technology. Using this approach to the distribution of tensile strength, it is possible to compare the quality of the manufacture of crosspieces by various plants and substantiate the requirements for its improvement.

Keywords: Crosspiece · Contact damage · Spalling · Assessment · Damage · Manufacture quality · Improvement requirements

1 Introduction

Modern operating conditions impose high requirements for the operability of turnouts and especially their crosspieces. The durability of the crosspieces is determined mainly by their resistance to wear and contact fatigue [1, 2].

The damage to the working surfaces of crosspieces by contact damage on the road network is great and tends to increase due to the increasing complexity of operating conditions. Of the 200 examined crosspieces, 81 had contact damages in the rolling zone. To assess the operation of the crosspieces under various conditions and to find effective ways to reduce the susceptibility of crosspieces to contact damage, it is necessary to know the contact-fatigue characteristics of the metal.

Using the qualitative conclusions of the metallographic analysis, the results of operational observations of the crosspieces and the reliability theory methods, a number of contact-fatigue characteristics of the crosspieces can be obtained [3].

Z. Popovic et al. (Eds.): TransSiberia 2019, AISC 1115, pp. 834–844, 2020.
https://doi.org/10.1007/978-3-030-37916-2_82

2 Methods of Research

Intensive wear of the crosspieces successively reveals layers of metal having different operating times and different loading conditions. Thus, the opportunity is created for a constant analysis of the work of the metal crosspieces in transit during operation. Data on the work of a sufficient number of crosspieces allow such an analysis to be carried out in a probabilistic aspect [4].

The dependence of the probability of defect-free operation of the crosspieces as a function of the crosspiece wear F(h) [5] is shown in Fig. 1. The graph is based on the results of 768 examinations of 316 crosspieces. The points of the graph for the crosspieces with wear less than 2 mm and more than 10 mm do not have sufficient statistical representativeness due to the small amount of data. Therefore, the graph in these zones is drawn with dash line. This graph is a typical dependence of the reliability of an element on operating time, and failure in this case is the appearance of spalling on working surfaces in the rolling zone.

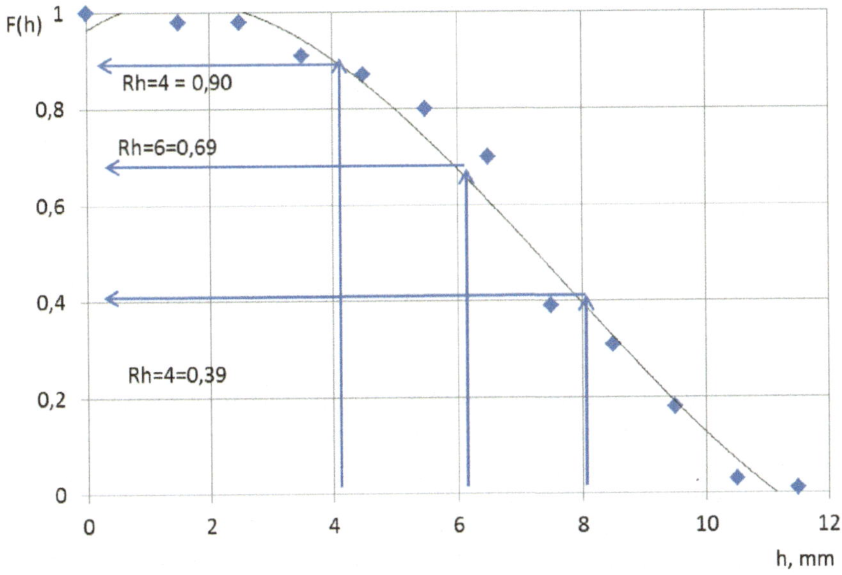

Fig. 1. The probability of operation of the crosspieces without the occurrence of spalling on working surfaces with various wear.

Let's consider the work of a metal layer located at a certain depth from the surface of the crosspiece. The process of the formation of contact defects of the crosspieces is fatigued. Therefore, the scheme of the metal layer can be represented as follows. As the operation of the cross is increased, the metal gradually wears out a certain part of the fatigue strength resource, i.e. strength indicators are reduced in accordance with the fatigue curve. At the same time, there is metal wear bringing this layer closer to the surface of the crosspiece. In this case, the load indicators (components of the stress tensor) in the layer gradually increase. If the metal of the layer manages to fully develop its fatigue life before it appears on the surface of the crosspiece, defects form in it and begin to develop. As the layer approaches the surface, when the size and position of the defects become critical, a micro-fracture of the metal located between the surface and the defect occurs, the defects open and become visible [6].

The probability of the absence of a defect on the surface at a given moment of operation of the crosspieces can be determined if the laws of distribution of strength and load indicators are known at this moment [7].

The choice of strength and load indicators that determine the contact damage of the crosspieces will be made based on the qualitative picture of the metal operation under the action of the contact load. The process of formation of contact damage is characterized by the exhaustion of the metal plasticity. Macro destruction of a metal is possible under the action of tensile or shear deformation. In the case of compression deformation or contact indentation, exhaustion of deformation capacity occurs due to violation of continuity along the direction of shear lines. Therefore, fracture in the contact is associated with the formation of shear cracks. If the compression deformation of the ductile metal does not cause shear fracture, then the ductility of the metal along the corresponding axis is unlimited [8, 9]. Based on the foregoing, it is advisable to take a function of tangential stresses acting on the sites of the most probable shifts as a load indicator, and a function of the tensile strength of the material as a measure of strength. Let's assume that the metal of the crosspiece is isotropic. This will allow, with sufficient accuracy for engineering purposes, to simplify the analysis of the crosspieces' efficiency.

For the transition corresponding to the first limiting state, the condition for the loss of strength was formulated by Mises. It has the form:

$$(\sigma_1 - \sigma_2)^2 + (\sigma_2 - \sigma_3)^2 + (\sigma_3 - \sigma_1)^2 \leq 6K^2 \tag{1}$$

where σ_1, σ_2, σ_3—main stresses;

K—an indicator of the material properties, corresponding to the transition through the first limit state. It is identified with the maximum tangential stress at yield in a state of pure shear.

As an indicator of the stress state, Mises used the octahedral tangential stresses acting on the sites equidistant to the directions of the main stresses. When using octahedral stresses and yield strength, the strength criterion takes the following form:

$$\tau_{oct} \leq \sqrt{\frac{2\sigma_t}{3}} = 0,47l\sigma_t$$

In other words, a linear function of the yield strength of steel is taken as an indicator of the properties of the material that determines the ultimate transition. In this case, Eq. (1) is written as follows:

$$(\sigma_1 - \sigma_2)^2 + (\sigma_2 - \sigma_3)^2 + (\sigma_3 - \sigma_1)^2 \leq 2\sigma_t^2 = (1,4l\sigma_t)^2 \tag{2}$$

To characterize the second limiting state, one should use the criteria and provisions of the fracture mechanism, in particular, the critical values of stress intensity coefficients corresponding to the fracture toughness of steel 110G13L. However, in relation to the efficiency of the crosspieces, these provisions are only just beginning to be developed. Therefore, the limiting state at the crack mouth and its critical dimensions are not considered in this study. With this in mind, the results of metallographic analysis were used for the second limiting state.

The studies performed in the Ural branch of VNIIZHT (Railway Research Institute) showed that local fracture (spalling) is caused by the fact that metal reaches critical hardening, the limiting value of which is linearly related to the tensile strength of steel 110G13L. Therefore, by analogy with the Mises condition, the condition for passing through the second limiting state is formulated as follows:

$$(\sigma_1 - \sigma_2)^2 + (\sigma_2 - \sigma_3)^2 + (\sigma_2 - \sigma_1)^2 \leq (\alpha\sigma_b)^2,$$

where $\alpha\sigma_b$—linear function of ultimate tensile strength;

α—a coefficient characterizing the working conditions of the metal for contact endurance with comprehensive uneven compression and dynamic loads.

We will take this condition as a criterion for the strength of the metal of the cross.

The tensile strength of the metal of new crosspieces is a random variable that varies depending on the chemical composition and technology of making crosspieces. The distribution of the ultimate tensile strength is close to normal with a mathematical expectation of 82.2 kgf/mm^2 and a standard deviation of 7.6 kgf/mm^2. (These figures are given for the products of the Novosibirskiy Strelochnyy Zavod (Novosibirsk Switch Plant).

It is known that fatigue processes have three stages: low-cycle fatigue, limited durability, and long-term endurance. At the stage of low-cycle fatigue, the number of cycles to failure is determined by the magnitude of the plastic deformation of the metal. It is also known that the tensile strength under static tension is equivalent to the destructive cyclic stress corresponding to 1/4 of the loading cycle [10]. This follows from the energy analysis of the cyclic dependence of strain on stress. Due to the fact that the accepted criteria assume the already developed plastic deformation in the metal, and there are no data on tests of 110G13L steel for low-cycle fatigue, we use the generalized fatigue dependence. Let's consider a typical fatigue curve. The equation of this curve in the field of limited durability has the form:

$$(\sigma_i/\sigma_j)^m = N_j/N_i,$$

where N_i и N_j—the number of cycles to failure at stresses of the cycles σ_i and σ_j, respectively;

m—a coefficient characterizing the slope of the fatigue curve.

In the generalized endurance curve for N−1, the ordinate of the curve is σ_b.

Let's compare the 1st and n-th loading cycles. For them, the equation of the generalized fatigue curve has the form:

$$(\sigma_b/\sigma_n)^m = n$$

From here it is easy to obtain:

$$\sigma_{ni} = \sigma_{bi} n^{-1/m}.$$

Consequently, the value of the highest stress that the metal is able to withstand during n cycles decreases $n^{-1/m}$ times.

Thus, the strength indicator is a random value for each crosspiece and a known function of the initial value for various cycles of its loading. Such quantities are called fixed random variables.

The stress state of the metal of the crosspieces is determined by the influence of the wheels of passing trains and the location of the considered point in cross section. The stresses in the layer from the effects of the wheels are determined by the trajectory of their rolling along the crosspiece, and the probability of the occurrence of stresses of this magnitude is determined by the probability of the trajectory implementation. To determine the stresses at specific points of the layer, it is necessary to take into account the location of the contact point on the working surface.

The distribution of dynamic forces in the contact of the wheels and crosspieces was obtained by the results of computer calculations. Histograms of these distributions for crosspieces with different wear are shown in Fig. 2. Table 1 shows the highest values of tangential stresses over the areas of probable shifts depending on the depth of the metal layer hdep. As the design points of the layers, we selected points located (vertically) under the points of location of the most probable contact. For a cross section of 30 mm, these are points lying on a vertical located at a distance of 1/3 from the edge of the core closest to the crest. To determine the stress values by the polarization and

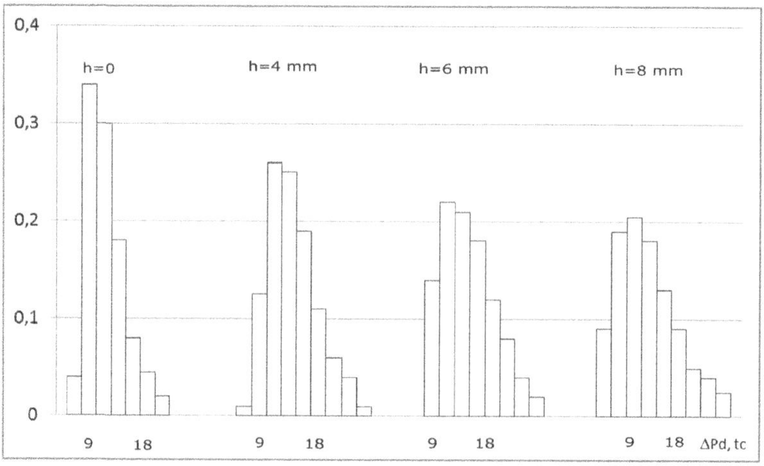

Fig. 2. Dynamic additives of contact forces at the crosspieces (cross section 30 mm) for crosspieces with various wear.

Table 1. The highest values of shear stresses on the areas of probable shifts depending on the depth of the metal layer $h_{dep.}$

Layer depth, mm	The average of the maximum octahedral tangential stresses (numerator) and the root mean square deviations (denominator), kgf/mm², in the layers of metal of the crosspieces in the cross section of 30 mm in case of wear, mm							
	0	1	2	3	4	5	6	7	8
4	$\frac{113.0}{6.4}$	$\frac{119.2}{6.4}$	$\frac{126.2}{6.4}$	$\frac{94.0}{4.8}$	$\frac{58.5}{2.8}$				
6	$\frac{82.5}{4.7}$	$\frac{83.0}{4.6}$	$\frac{83.8}{4.4}$	$\frac{100.5}{5.0}$	$\frac{118.6}{5.8}$	$\frac{90.8}{4.5}$	$\frac{56.5}{2.7}$		
8	$\frac{57.5}{3.2}$	$\frac{57.8}{3.1}$	$\frac{58.2}{3.0}$	$\frac{66.2}{3.2}$	$\frac{79.0}{3.8}$	$\frac{95.6}{4.6}$	$\frac{113.0}{5.6}$	$\frac{87.1}{3.8}$	$\frac{49.4}{2.2}$
10	$\frac{45.0}{2.5}$	$\frac{45.1}{2.4}$	$\frac{45.5}{2.4}$	$\frac{54.2}{2.6}$	$\frac{71.5}{3.4}$	$\frac{82.3}{4.0}$	$\frac{91.0}{4.1}$	$\frac{97.0}{4.4}$	$\frac{102.0}{4.4}$

optical method, stress diagrams were built in cross sections at various depths, and then the lines of the influence of stresses.

The probability of specific stress values is determined by the probability of the location of the contact point, the probability of the wheel position, and the probability of a given value of dynamic force [10].

The probability of the appearance of the dynamic impact of the wheel on the crosspiece can be determined by the formula:

$$F_p(P_{\text{d}}) = F_n(\text{R}_{\text{w}})F_\Delta(\Delta)F_h(h)$$

where $F_n(\text{R}_{\text{w}})$—the probability that the wheel has a rolling Rw; $F_\Delta(\Delta)$—the probability of this gap between the wheel flange and the wing rail of the crosspiece; $F_h(h)$—the probability that the crosspiece has wear h in a given wear range.

The position of the contact point corresponding to the probability $F_p(P_{\text{d}})$ was determined by superimposing the profile of the wheel with rolling P on the diameter of the crosspiece with wear h.

The calculated stresses were determined by the formula:

$$\tau_P = \tau_{0i}P_{\text{d}i},$$

where τ_{0i}—the ordinate of the line of influence of stresses at the design point arising from a unit contact force applied at the i-th point of the surface.

According to the results of calculations of stresses in a cross section of 30 mm for the design points of each layer, statistics of their distributions were determined. Stresses are calculated for points located vertically below the place corresponding to the highest probability of wheel contact with the core. The data obtained are summarized in Table 2.

Table 2. Estimated mean stresses and standard deviations in case of crosspiece wear.

Layer depth, mm	Estimated average stresses (numerator) and standard deviations (denominator), kgf/mm^2, in case of crosspiece wear, mm								
	0	1	2	3	4	5	6	7	8
4	42.0 / 13.3	44.0 / 14.0	46.3 / 15.1						
6	30.4 / 9.6	30.5 / 9.9	36.6 / 10.3	36.7 / 11.6	43.3 / 14.3				
8	30.8 / 6.7	20.9 / 6.5	20.9 / 6.5	23.7 / 7.1	28.5 / 9.0	34.7 / 11.3	41.3 / 13.3		
10	16.1 / 4.8	16.0 / 4.9	16.2 / 5.1	19.7 / 6.1	25.7 / 7.5	29.8 / 9.4	33.0 / 10.4	35.0 / 11.3	36.7 / 11.8

Due to the fact that the probabilities of the parameters that determine the calculated stresses do not depend on strength indicators, the calculated stresses in the considered problem are independent random variables with distributions corresponding to Table 2.

As indicated above, the contact strength indicator of new crosspieces is a fixed random quantity with a known distribution. The strength of each crosspiece after n loading cycles is expressed as a deterministic monotonically decreasing function of the initial random value of strength and value of n. Thus, we have:

$$\sigma_{ni} = \varphi_n[\sigma_{bi}; n],$$

where φ_n—monotonically decreasing function.

If the values of the stress state index τ_P are independent random variables τ_0, τ_1, τ_2, ..., τ_n with known distribution functions depending on n, $f_0(\tau)$, $f_1(\tau)$, ..., $f_n(\tau)$, where f - the stress distribution density τ, then the probability that defects in the form of spalling will not appear on the surface in any of the loading cycles from 1 to n will be:

$$R_n = P\{[\tau_1 < \alpha\sigma_{b1}] \cap [\tau_2 < \alpha\sigma_{b2}] \cap \ldots \cap [\tau_n < \alpha\sigma_{bn}]\},$$

or

$$R_n = P\{[\tau_1 < \varphi_1] \cap [\tau_2 < \varphi_2] \cap \ldots \cap [\tau_n < \varphi_n]\},$$

where P—probability of the event being investigated.

Let the initial value of the strength of some crosspieces be in the interval $d\sigma_b$ from the value σ_{bi}. Then, due to the independence of the stress values, the probability of their spalling to appear will be:

$$R_{ni} = \prod_{k=1}^{n} P_k[\tau_k < \varphi_k\sigma_{bi}]$$

Each of the probabilities P_k is the probability that the calculated stresses in a given loading cycle will not exceed the strength of the metal.

If the density distribution function of the stresses τ in the layer in the k-th loading cycle is $f_k(\tau)$, and the strength index $q = \alpha\sigma_{bi}k^{-1/m}$, then the probability P_k of the non-occurrence of spalling in the layer in this cycle is the probability that τ < q. It can be defined as an integral from the probability density of stresses from the minimum value to the value of strength in the k-th loading cycle, i.e.

$$P_k = \int_0^q f_k(\tau)d\tau.$$

The value of the strength index σ_{bi} is only one of its possible values realized within the limits of the known distribution with probability $\varphi(\sigma_B)$.

In order to get the probability of work of crosspieces without spalling for all possible values of tensile strength, it is necessary to sum all R_{ni} within all possible values of σ_{bi} for new crosspieces [9]:

$$R_n = \int_0^\infty \varphi_0(\sigma_b)\left[\prod_{k=1}^n \int_0^q f_k(\tau)d\tau\right]d\sigma_b \tag{3}$$

where $\varphi_0(\sigma_b)$—distribution density of the metal strength index of new crosspieces; $q = \alpha\sigma_{bi}k^{-1/m}$.

The last equation expresses the probability of defect-free operation of the crosspiece elements for the case when the strength indicator is a fixed random variable with a known initial distribution and a known form of dependence on the number of loading cycles, and the stress is an independent random variable with a distribution division changing in a known manner [11].

In the case under consideration, the dependence of the strength index, its initial distribution and the distribution of the stress index are known. In addition, various values of Rn are known from the operational data (Fig. 1). In the equation, only the coefficients α and m that characterize the fatigue properties of steel under contact loading of the crosspieces are unknown. These coefficients can be found by compiling two equations for different layers and solving them together.

Due to the fact that the values of Rn obtained in operation may have a dispersion, to find the parameters α and m, a system was solved not of two, but of three equations of the form (3) for layers with an initial depth of 4, 6 and 8 mm. The following calculation method was used. For each of the layers, several values of the coefficients α and m were set. Then, the probabilities of defect-free operation Rn were calculated. As a result, the values of m were determined by the iteration method, which for each given value of α give a value of Rn corresponding to that obtained from the operation, i.e. they are solutions of the corresponding equation of the form (3) for a given layer.

Dependences $\alpha(m)$ satisfying the equations of the type (3) for the layers are plotted (Fig. 3). The lines on the graph intersect, since the stress state of the metal layers during the operation of the crosspiece is different. The intersection point of the graphs $\alpha(m)$ for two layers gives values that satisfy both equations, i.e. it uniquely determines α and m. Usually the intersection point of the graphs $\alpha(m)$ for all layers should be one.

However, due to the dispersion of specific operating conditions for each crosspiece that cannot be taken into account in full, the intersection points also have a dispersion. The dispersion of α and m is 7.5 and 4.9%, respectively. Within the triangle of intersections of graphs by the iteration method, the values of α and m were obtained that best satisfy all three equations of the form (3): α = 8.1; m = 13.2.

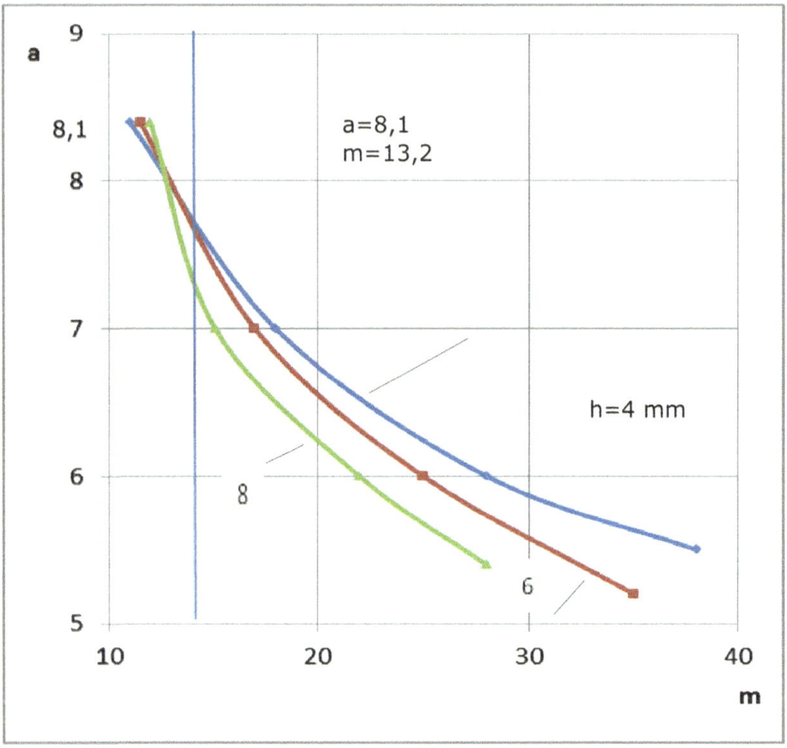

Fig. 3. Determination of the coefficients *a* and *m* for the zone of rolling crosspieces.

The obtained value of the coefficient α does correspond to the physical features of the considered operating conditions of the metal, since the average value of $\alpha\sigma_b$ is 665.8 kgf/mm^2, which is close to the maximum stresses that the samples from manganese steel can withstand during compression tests before cracking (535–600 kgf/mm^2 during static tests) [1]. The larger value of $\alpha\sigma_b$ is explained by the dynamic application of the load on the crosspieces.

Using the coefficients α and m the distribution of the tensile strength of steel 110G13L, it is possible to build contact-fatigue relationships for crosspieces having different initial values of the tensile strength taking into account the probability of these values. As an example, Fig. 4 shows such a dependence for crosspieces with a tensile strength equal to the average, which corresponds to a probability of 50% (standard fatigue curve). Using this dependence, it is possible to determine the number of loading cycles Nb that cause the appearance of spalling at a given amount of wear in various ways, for example, by the method of summing damages, for any wear of the crosspieces. By comparing this number with the number of Ni cycles necessary for the crosspiece to wear out by a predetermined amount, we can conclude whether there will be spalls at the crosspieces for a given amount of wear or not. If Nb < Ni, then spalls will appear, but if Nb > Ni, they will not appear.

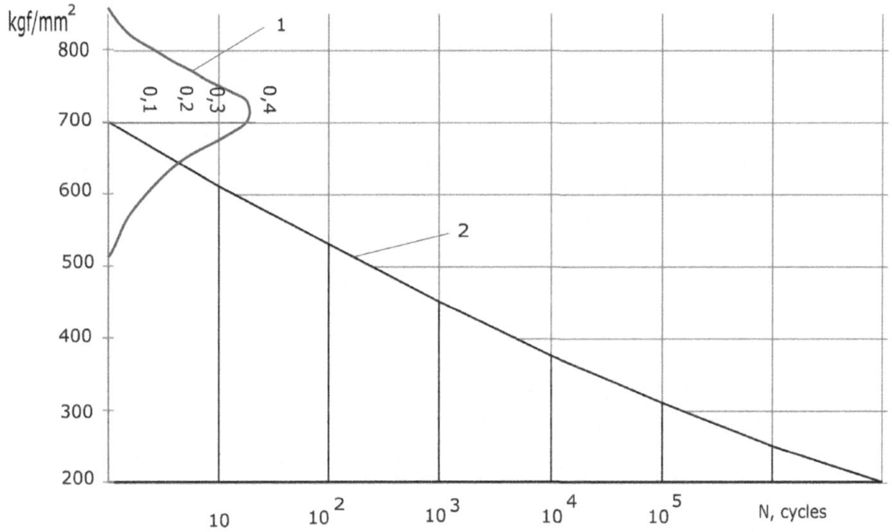

Fig. 4. Contact-fatigue dependence for crosspieces with a tensile strength equal to the average value of 82.2 kgf/mm^2.

Using the distribution of the strength index (Fig. 4), one can build fatigue dependences for crosspieces with different values of σ_b. In addition, each dependence will correspond to a certain proportion of crosspieces corresponding to the probability of the appearance of a given value of σ_b. According to these dependencies, it is possible to judge not only whether defects will appear, but also how much the crosspieces will be affected by them, i.e. have a complete picture of the crosspiece operation.

3 Conclusions

1. The proposed methodology provides a universal approach to determining the intensity of damage caused by spalling in the area of rolling crosspieces with cores made of steel 110G13L and allows us to quantify the damage characteristics of spalling crosspieces under different operating conditions, taking into account the shape and size of the contacting surfaces and casting manufacturing technology.

2. Using this approach to the distribution of tensile strength, it is possible to compare the quality of manufacturing crosspieces by various plants and substantiate the requirements for its improvement.

References

1. Loktev, A.A., Shishkina, I.V., et al.: Modeling the dynamic behavior of the upper structure of the railway track. In: TGG 2017 «Transportation Geotechnics and Geoecology», Saint Petersburg, Russia, 17–19 May 2017, vol. 189, pp. 133–137 (2017). https://doi.org/10.1016/j.proeng.2017.05.022. Procedia Engineering
2. Glusberg, B., Korolev, V., et al.: Calculation of track component failure caused by the most dangerous defects on change of their design and operational conditions. In: MATEC Web of Conferences, vol. 239, p. 01054 (2018). https://doi.org/10.1051/matecconf/201823901054
3. Loktev, A.A., Shishkina, I.V., et al.: High frequency vibrations in the elements of the rolling stock on the railway bridges. IOP Conf. Ser. Mater. Sci. Eng. **463**, 032019 (2018). https://doi.org/10.1088/1757-899x/463/3/032019
4. Loktev, A.A., Korolev, V.V., et al.: Mathematical modeling of antenna-mast structures with aerodynamic effects. IOP Conf. Ser. Mater. Sci. Eng. **463**, 032018 (2018). https://doi.org/10.1088/1757-899x/463/3/032018
5. Savin, A., Kogan, A., et al.: Evaluation of the service life of non-ballast track based on calculation and test. Int. J. Innovative Technology and Exploring Engineering (IJITEE) **8**(7) (2019). ISSN: 2278-3075 G5991058719/19©BEIESP
6. Glusberg, B., Savin, A., et al.: New lining with cushion for energy efficient railway turnouts. In: Advances in Intelligent Systems and Computing, vol. 982, pp. 556–570 (2020). https://doi.org/10.1007/978-3-030-19756-8_53
7. Glusberg, B., Loktev, A., et al.: Calculation of heat distribution of electric heating systems for turnouts. In: Advances in Intelligent Systems and Computing, vol. 982, pp. 337–345 (2020). https://doi.org/10.1007/978-3-030-19756-8_31
8. Loktev, A., Korolev, V., et al.: Modeling of railway track sections on approaches to constructive works and selection of track parameters for its normal functioning. In: Advances in Intelligent Systems and Computing, vol. 982, pp. 325–336. https://doi.org/10.1007/978-3-030-19756-8_30
9. Glusberg, B., Savin, A., et al.: Counter-rail special profile for new generation railroad switch. In: Advances in Intelligent Systems and Computing, vol. 982, pp. 571–587. https://doi.org/10.1007/978-3-030-19756-8_54
10. Loktev, A.A., Korolev, V.V., et al.: Mathematical modeling of aerodynamic behavior of antenna-mast structures when designing communication on railway transport. Vestnik of the Railway Research Institute, Russia (2018). https://doi.org/10.21780/2223-9731-2018-77-2-77-83
11. Loktev, A.A., Korolev, V.V., et al.: Perspective constructions of bridge overpasses on transport main lines. Vestnik of the Railway Research Institute, Russia (2018). https://doi.org/10.21780/2223-9731-2018-77-6-331-336

Digital Transformation of Airline Management as the Basis of Innovative Development

Alexander Sukhorukov[1], Nikolay Koryagin[1], Julia Sulyagina[2], Nataliya Ulitskaya[3], and Sergey Eroshkin[2(✉)]

[1] Plekhanov Russian University of Economics, Moscow, Russia
[2] Russian State Social University (RSSU), Moscow, Russia
EroshkinSIu@rgsu.net
[3] Moscow Automobile and Road Construction State Technical University (MADI), Moscow, Russia

Abstract. The article presents the results of the analysis of the use of management information systems in airlines. The main technologies that will form the basis of digital transformation of management in the coming years are described.

Keywords: Digital economy · Airline information management · Information systems management · Digital transformation of the airline

1 Analysis of Existing Digital Airline Management Information Systems

1.1 General Purpose Digital Airline Management Information Systems

By definition, the innovative development of any transport company can include only those innovations that bring a tangible economic effect. This condition applies to process and organizational innovations in the management of airlines that are developing their digital information systems.

To date, the airline successfully operates digital platform forms from various classes of information systems [1]. In the Table 1 shows an exemplary hierarchical list of the use of widely known general management information systems.

Each of the above classes of information systems is used at a certain level of the airline management hierarchy. Table 1 shows the hierarchical structure of information systems in management, which covers the entire spectrum of airline automation tasks from strategic management to levels of control of technological processes of production and product management of production (services) throughout the life cycle. The listed systems can be built on various methodological approaches [2, 3] and form a single platform. This makes the proposed structure flexible and versatile.

Z. Popovic et al. (Eds.): TransSiberia 2019, AISC 1115, pp. 845–854, 2020.
https://doi.org/10.1007/978-3-030-37916-2_83

At the highest level of the airline's strategic management, the main goal of which is to create competitive advantages, can be used systems such as Business Intelligence (BI), Business Process Management (BPM), Corporate Performance Management (CPM), Governance Risk Compliance (GRC), Project Portfolio Management (PPM), Knowledge Management (KM).

Table 1. The hierarchy of management information systems company

Functions and users	Types of information systems
Strategic management and marketing (top management and marketing)	BI, BPM, CPM, GRC, PPM, CRM, KM
Tactical management and financial management (functional units)	ECM, ERP, CRM, SRM, SCM, BPM, PM
Operational production management (production managers)	MES, CMMS, EAM, WMS
Automated control of technological processes (line managers workshop managers)	SCADA, CAD
Management of the product throughout its life cycle (all production users plus consumers of the product and operator)	PLM

BI systems use tools such as Online Analytical Processing (OLAP) and Data Mining, or Knowledge Discovery in Databases (KDD).

BPM systems for formalizing and modeling business processes (projects) use various notations and languages, such as BPMN, EPC, IDEF0, IDEF3, UML and others.

The popular Balanced Scorecard (BSC) methodology, which uses the Key Performance Indicator (KPI) as a measure of reachability of strategic goals, is fully implemented in a separate class of CPM systems that implement system integration of strategic control processes. In these systems, a link is made between the strategic goal-planning contour and the tactical and operational planning contour, the implementation of the strategic program and business plan is monitored, the entire management cycle is supported.

GRC systems manage risks and support internal control. PPM systems manage project portfolios, and KM systems manage the knowledge needed to successfully implement business strategies.

Online Transaction Processing (OLTP) systems, such as Enterprise Resource Planning (ERP) end Customer Relationships Management (CRM), in which BI, BPM concept elements are implemented, can also be located at the strategic level.

The next level can be attributed to the operational-tactical, the purpose of which is to maximize current financial indicators. At this level, Enterprise Content Management (ECM) systems, full-value ERP systems, isolated CRM, Supplier Relationships Management (SRM) end Supply Chain Management (SCM) are designated. Also noted here are BPM systems that solve the problems of implementing the concept of process (project) management of an airline and use the Business Process Execution Language (BPEL) notation. It should be noted here that BPM tools can be implemented in other systems, for example, in ERP and CRM. PPM systems are strategic, while PMs are operational. CPM's strategic business development planning systems only support the operational management of business processes. Operational management systems are also built on the basis of modern methodological approaches to management.

Further down the hierarchical structure are systems of operational production management, such as Manufacturing Execution System (MES), Computerized Maintenance Management System (CMMS), Enterprise Asset Management (EAM), Warehouse Management System (WMS), etc. And close the hierarchy below Supervisory Control and Data Acquisition (SCADA), Computer-added Design (CAD) and Product Lifecycle Management (PLM) systems.

Given the enormous variety of tasks solved by these systems and the use of various methodological approaches to management in their construction, such a hierarchy can be considered conditional, but convenient for systematizing the concept of information management.

A modern ERP system, for example, can independently solve individual tasks of strategic and operational production management. In practice, as a rule, ECM-systems, ERP-systems are in information connection with professional strategic management software products. ERP-systems allow you to dock with lower-level systems such as MES, CAD. Information flows in the control systems listed above are closely intertwined and depend on the system, process, quantitative and other approaches implemented in them. Depending on this, the general information structure of the airline is being formed, which can integrate entire classes of various systems implemented by software products from different manufacturers.

At present, it is impossible to imagine a successful large airline without system information management at various levels (strategic, tactical and operational) [4–7].

The data on the practical application of general information systems in Russian and foreign airlines are presented in Table 2.

Table 2. Application of general-purpose management information systems in Russian and foreign airlines

Airline company	Information systems management
Aeroflot	SAP (ERP), Oracle Siebel (CRM) + Oracle Airline Data Management (BI), ARIS (BPM), MS Project (PPM end PM)
VIM Airlines	SAP (ERP)
Rossiya Airlines	Microsoft Dynamics Axapta (ERP)
Utair	SAP (ERP)
S7	Oracle E-Business Suite (ERP+CRM)
Air France	SAP (ERP)
Air India	Oracle E-Business Suite (ERP+CRM)
Alitalia	SAP (ERP, CRM), MicroStrategy (BI)

(*continued*)

Table 2. (*continued*)

Airline company	Information systems management
American Airlines	SAP (ERP), Infor CRM Epiphany
Austrian Airlines	mySAP Business Suite (HR, SRM, CRM) Business Information Warehouse BW Cognos (BI)
British Airways	SAP (ERP), Oracle HR
China Eastern Airlines	Oracle E-Business Suite (ERP+CRM)
Finnair	SAP (ERP)
Japan Airlines	SAP (ERP), SAP NetWeaver
Lufthansa	SAP (ERP), SAP NetWeaver, SAP BW (BI)
Saudi Arabian Airlines	SAP (ERP)
Singapore Airlines	SAP (ERP)
Ryanair	SAP (ERP)
Scandinavian Airlines	SAP (ERP), SAS Business Intelligence (BI)

As the analysis of the table shows, at present the majority of airlines use information systems of manufacturers (SAP, Microsoft and Oracle), however, 1C: ERP is gaining more and more popularity in Russia. Since 2015, the number of projects implementing the 1C: ERP platform in Russia has become comparable with the number of SAP ERP implementations. The financial information systems 1C: AVIA and 1C: BUDGET-AVIA are gaining popularity.

1.2 Digital Management Information Systems for Specialized Airlines

In addition to general-purpose management information systems, a number of specialized systems are integrated into the digital platform, which are used only in the airline industry. These include information systems and mobile applications that ensure the automation of airline business processes, as well as the provision of a range of information, including mobile, services to air passengers.

The main directions of increasing the efficiency of airline companies through the use of specialized information systems that implement the concept of "smart airports" are associated with the implementation of:

- geographic information (geolocation) systems for solving problems:
  - monitoring the location of airport employees;
  - aircraft movement monitoring;
  - airport passenger traffic management;
  - increase the intensity of use of vehicles.
- automated control systems for production activities and data exchange designed for the operational management of ground equipment and apron personnel at airports;
- automated control systems for handling air cargo;
- automated flight dispatch control systems;

- automated flight schedule management systems;
- automated systems for the operational categorization of passenger traffic according to the degree of potential danger;
- mobile applications that provide airport employees with access to information about unplanned changes in flights;
- mobile applications that provide automated passport control using a biometric passport;
- specialized mobile services Electronic Flight Bag (EFB), providing aircraft pilots with access to flight cards, maneuvering schemes in areas of airfields around the world in electronic form on their tablets;
- specialized mobile services in the passenger cabin of the aircraft, providing flight attendants the opportunity to provide personalized services to passengers based on the information that they indicated when booking tickets;
- intelligent parking systems.
- systems for providing automatic voice announcements, mobile kiosks and other IT-tools for notifications;
- mobile applications or kiosks for self-registration of air passengers;
- automated baggage reception systems, which provide air passengers with the possibility of self-printing of baggage tags and baggage drop-off with glued tags without the assistance of airport personnel;
- baggage control systems from the moment of their registration to the final delivery point along the entire route, which can go through several airports and include flights by several airlines, and its search in case of loss;
- mobile applications providing passengers with the opportunity to receive information about the status of their baggage;
- mobile applications for navigating air passengers in terminals;
- mobile applications for online informing passengers on the day of departure, for example, about the traffic situation or about the queues at the terminal, as well as about the predicted waiting time in queues at control points or about the duration of the walk to the gate;
- mobile applications that provide sales of additional non-aviation services to airline passengers, such as ordering a taxi, VIP lounge, railway transportation, car rental, government hotels, etc.
- specialized services providing mobile communications on board the aircraft (sms, mms, e-mail);
- mobile applications providing services to air passengers via social networks.

A high level of IT solutions and industry analytics is presented in the systems developed by the international leader in the field of aviation information technologies - SITA. There are also known systems of complex automation of production activities at the COBRA airport, 1C system: airport rates and fees, etc.

The presented brief analysis of the trends in the development of specialized information technologies in the air transport industry shows that the application of various types of application software is one of the most effective ways to increase the competitiveness of airlines in the modern information society [8].

2 Advanced Information Technologies for Digital Transformation of Airline Management

The ongoing processes of digital transformation of management are based on information technology, the development of which is indicated in the list of technologies of The Fourth Industrial Revolution (Industry 4.0). This list is very similar to the listed end-to-end technologies in the national project "Digital Economy of Russia". In Table 3, these technologies are summarized in a common matrix.

Table 3. Relevant information technologies for the development of the digital economy

Artificial intelligence	Big data	Robotics
Distributed ledger technology (Blockchain)	Quantum technology	Additive manufacturing
Industrial Internet of Things	Wireless communication	Virtual reality augmented reality

Below are the technologies that are most effectively used in airlines and form the basis of digital transformation of management.

2.1 Artificial Intelligence End Big Data

Artificial Intelligence (AI) can be considered as digital brain, and big data will represent the memory of the digital world. By modern AI development is meant the development of a number of overlapping technologies, such as:

- Machine Learning (ML), the solved problem of classification, clustering, regression, etc. using methods:
  - Supervised Learning (SL), for exemple, Deep Learning (DL);
  - Unsupervised Learning (UL);
  - Reinforcement Learning (RL).
- Machine Vision (MV) end Optical Character Recognition (OCR) and etc.
- Big data technologies are closely related to artificial intelligence. Big data is usually characterized as "three V": Volume; Velocity, Variety.

The amount of digital information in the world is growing rapidly. According to the data for 2017, about 2.7 zettabytes were accumulated. This is a huge amount of information. For comparison, if all this information is recorded on 4.7 GB DVDs and put into one stack, then the thickness of this stack will take the distance to the Moon and back. The amount of this digital information is growing exponentially. It increases 2.2 times every 2 years, and the amount of business information doubles even faster - in 1.2 years. These are ambitious figures that are yet to be comprehended. Never in the entire history of the development of mankind has there been anything like it. Humanity enters a new digital age in which much will change.

Already, big data in this direction can be called a successful rapidly developing project that brings tangible results, for example, when making decisions in customer loyalty programs.

In 2018, all airlines in the world sold 4.5 billion airline tickets. Over the 6 months of 2019, more than 57 million people used the services of Russian airlines alone. On average, every second inhabitant of the plan flies for a year. Airlines CRM systems store and process information about such a huge number of customers. If previously only structured customer information was stored using multidimensional OLAP cubes and processed in BI systems using Data Mining End Knowledge Discovery in Databases technologies, now innovative big data processing systems that have unstructured information have a big effect. The volumes of such information are orders of magnitude higher and require significantly larger resources for storage and processing. Moreover, the preprocessing of big data itself is already carried out using AI, built, for example, on Artificial Neural Networks (ANN) or connectionist systems implemented in neuroprocessors that differ in their architecture from the usual von Neumann architectures.

One of the leading technologies related to the big data class is the open Hadoop platform, which allows you to process huge amounts of data in a distributed environment. There are special solutions that transform Hadoop's huge distributed information into structured information in OLAP cubes for its business intelligence in traditional BI systems. Big data analysis involving AI will help airline executives plan their new projects taking into account the accumulated information about past results. We especially expect the effect of using AI technologies and big data in risk assessment. Risks in management always exist, and the human mind doesn't do very well with risk-based probability management, especially when you need to consider many different probabilities. Nowadays, the technology of optimizing business processes is becoming popular by replacing the functions that people used to perform in them with software robots. For airline operations, this means that the cost of business processes will decrease.

Robotics Process Automation (RPA) is one of the branches of artificial intelligence. RPA is combined with other AI branches. Most often, RPA is used with machine vision, OCR engines, for recognizing characters and information in pfd documents in order to automate workflow [9].

The accumulation of structured data in project management has gained particular relevance with the development of the PLM approach. This approach also requires huge resources for data storage. In world practice, there are already giant data centers that store information in exabytes and zetta bytes.

The new profession associated with data is rapidly becoming popular. If in 2012 the position of director according to the data existed in only 12% of companies, then over the past seven years the number of such companies has increased to 68%.

2.2 Robotics

The next cluster of information technology related to AI and big data is robotics. These concepts should not be confused with the previously considered concept of RPA, which implied only software robots that were created on top of existing management information systems. Robotics mean physical robots that replicate the functions of human organs, so their use in airline management will be on those life cycles in which the use of such technological machines is possible.

The digital world has several advantages over the physical world, since there are no material limitations in it, there are no problems of distances and time. Any distances can be covered without loss of time. Objects in the digital world can be copied and reproduced without much effort, the boundaries between digital copies of real objects and fictions become blurred. A digital world with unlimited possibilities in the form of the Digital Twin concept will be embodied in the real world with the help of developing robotics technologies.

2.3 Industrial Internet of Things

The Industrial Internet of Things (IIoT) is used to synchronize the digital and real worlds. Cisco analysts believe that IIoT became global between 2008 and 2009. It was at this time that the number of devices connected to the global network exceeded the population of the Earth and "Internet people" turned into "Internet of Things". Ericsson predicts that by 2021, out of approximately 28 billion connected devices in the global global network, about 16 billion will be connected as part of the IIoT concept. The introduction of IIoT has a significant impact on the economy of airlines, contributes to increased labor productivity, and has a positive effect on working conditions and professional growth of employees. Already, many airlines are using IIoT in managing a fleet of aircraft, tracking air transportation of goods and passengers. There are airports with "smart" service for passengers.

2.4 Distributed Ledger Technology

Distributed Ledger Technology (DLT) is a set of methods and tools for using decentralized digital registries that contain blocks that are connected by a network of computer nodes.

A registry - is an immutable database in which transactions are committed forever.

Distributed - this is when the database has many copies. The more copies, the more reliable the data.

The registry is shared. The more organizations use a shared distributed registry, the greater its value.

The registry is safe. It uses cryptography to protect transactions from fraud and helps counterparties avoid disagreements by using a special protocol.

The distributed registry has become very popular due to its use in cryptocurrencies. However, already now these technologies have become widely used in Smart Contracts, special file storages, and Internet protocols. There are many examples of using DLT in the areas of local and cross-border payments, transactions with securities, commodity assets and derivative financial instruments, and the exchange of messages.

There are various types of distributed registries and the most famous of them is Blockhain. The most common Blockchain alternative is the Directed Aciclic Graph (DAG), which solves the same problems, but is a network of blocks instead of a chain of blocks.

The introduction of Blockchain technology in airlines simplifies the exchange of information, allows you to comply with legal requirements and minimize risks. The use of smart contracts gives transparency and "synchronization", and also protects the private and confidential information of all airlines and their employees. For example, the Blockchain solution was implemented at Aeroflot to collect and exchange data on certification of employees.

2.5 Virtual Reality and Augmented Reality

Virtual Reality (VR) is a world created with the help of information technology, which is transmitted to a person through an impact on his sensations: vision, hearing, smell, touch, and others. Virtual reality simulates both exposure and response to exposure. To create a convincing complex of sensations of reality, a computer synthesis of the properties and reactions of virtual reality is performed in real time.

Currently, VR technology is widely used in airlines in flight simulators and simulators, marketing and advertising.

Augmented Reality (AR) - the introduction of any virtual sensory data into the field of perception of real objects in order to supplement information about the environment and improve the perception of information.

An example is the indicator on the windshield of an airplane. With it, pilots see additional information on the front windshield, which they use to improve control. Also, there are already many AR software products for mobile devices, which allow using AR to get the necessary information about the environment. Such technologies, for example, are already used at airports by passengers in various services.

3 Conclusions

The technologies discussed above, as well as others, such as quantum technologies and wireless communications form the basis of the innovative development of airline management. They allow digital transformation of all business processes implemented in information systems at a new qualitative level.

Already, there is a clear tendency to combine existing airline management information systems such as ERP, CRM, RM, BPM, BI, etc. into single digital platforms. In this case, the introduction of innovative technologies can bring tangible effects. A systematic approach to the use of modern digital technologies will make it possible in the foreseeable future to implement the grandiose Digital Twin concept - an important managerial component of the digital economy.

References

1. Eroshkin, S.Y., Kameneva, N.A., Sukhorukov, A.I., Kovkov, D.V.: Conceptual system in the modern information management. Procedia Comput. Sci. 609–612 (2017). https://doi.org/10.1016/j.procs.2017.01.079

2. Koryagin, N.D., Sukhorukov, A.I., Medvedev, A.N.: Implementation of Modern Methodological Approaches to Management in Management Information Systems. MSTUCA, Moscow (2015)
3. Sukhorukov, A.I., Shuhong, G.U.O., Koryagin, N.D., Eroshkin, S.Y.: Tendencies of information management development in the conditions of the origin of a new ecosystem of the digital economy. In: 2018 Eleventh International Conference Management of large-scale system development (MLSD, Moscow, Russia, pp. 1–4 (2018). https://doi.org/10.1109/MLSD.2018.8551859
4. Budiyanto, D., Setyohadi, D.B.: Strategic information system plan for the implementation of information technology at Polytechnic 'API' Yogyakarta. In: 5th International Conference on Cyber and IT Service Management, CITSM 2017, pp. 1–5. Conference: 2017 5th International Conference on Cyber and IT Service Management (CITSM) (2017). https://doi.org/10.1109/citsm.2017.8089267
5. Palominos, F.E., Duran, C.A., Córdova, F.M.: Multidimensional data model for the analysis of information of productive, scientific or service processes. In: 7th International Conference on Computers Communications and Control, ICCCC 2018 - Proceedings, pp. 17–22. Conference: 2018 7th International Conference on Computers Communications and Control (ICCCC) (2018). https://doi.org/10.1109/icccc.2018.8390431
6. Gaitán Bolaños, H., Acosta Ruiz, V.J., Gabalán Coello, J., Vásquez Rizo, F.E.: Relationship between organizational knowledge variables and the result of organizational objectives in a manufacturing company. Espacios (2018)
7. Duran, C.A., Palominos, F.E.: Methodological proposal for decision making soported on ANP and multidimensional database. In: IEEE ICA-ACCA 2018 - IEEE International Conference on Automation/23rd Congress of the Chilean Association of Automatic Control: Towards an Industry 4.0 – Proceedings, pp. 1–10. IEEE, Chile (2019). https://doi.org/10.1109/ica-acca.2018.8609716
8. Koryagin, N.D.: Strategic directions of ensuring competitiveness of the airlines in today's information society. Innovations Civil Aviat. **1** (2016). http://www.mstuca.ru/biblio/magazin.php. Accessed 07 Nov 2016
9. Robotic Process Automation Market, RPA Industry Report, 2018–2025. https://www.grandviewresearch.com/industry-analysis/robotic-process-automation-rpa-market

Control of Idle Losses in Power Transformers of Distribution Electric Networks

E. A. Tretyakov(✉) and V. T. Cheremsin

Omsk State Transport University, Marksa pr., 35, 644046 Omsk, Russia
eugentr@mail.ru, cheremisinvt@gmail.com

Abstract. For lightly loaded electrical networks and for seasonally changing loads, it is proposed to reduce the idle losses in power transformers by changing the connection scheme of the primary and secondary winding coils provided that the transformation coefficient is constant. In this case, the transformer windings must have branches corresponding to the number of coils on the high and low voltage sides. The quantitative estimates of reducing the idle losses in power transformers by the proposed method are based on the methods of a full-scale experiment using the example of a single-phase transformer and simulation modeling of a three-phase power transformer. Circuit designs for sectioning the windings of power transformers are developed. For automatic control of losses in the transformer, an expression is obtained for reducing the total losses in the transformer by changing the connection scheme of the coils of the primary and secondary windings when the load changes. From a practical point of view, a transformer with a change in the connection schemes of its windings using the proposed method can find application as a power transformer of distribution networks with a voltage of 35, 10 (6)/0.4 kV with a load of less than 20% (summer and cottage villages, seasonal loads of industrial enterprises).

Keywords: Power transformer · Reduction of losses · Experiment · Simulation modeling · Sectioning of state windings

1 Introduction

Improving the efficient performance and energy efficiency of electric power transmission in traction and distribution electric networks is achieved primarily by reducing energy losses. A significant share in the structure of technical losses of distribution networks is occupied by the idle losses in transformers at a voltage level of 6–10 kV, reaching 70% of technical losses with a small load on the electric network.

In the distribution networks of railways, oil transformers with a capacity of 100 to 1000 kV•A are mainly installed. Losses of idle electricity within the boundaries of the West Siberian Railway in power transformers with a voltage of 6–10 kV amounted to about 9.59 million kW•h in 2018.

Nowadays, the problem of reducing idle losses in power transformers is solved mainly by turning off one of two parallel transformers working at light loads (if possible), replacing transformers with lower installed capacity, which corresponds to the load, or with an energy efficiency of a higher class [1]. Also, the idle losses in the

Z. Popovic et al. (Eds.): TransSiberia 2019, AISC 1115, pp. 855–863, 2020.
https://doi.org/10.1007/978-3-030-37916-2_84

transformers depend on the voltage level on its primary winding, but the installed control means do not allow for "loss control".

For lightly loaded electrical networks and for seasonally changing loads, it is proposed to reduce the idle losses in power transformers by changing the connection scheme of the primary and secondary winding coils provided that the transformation coefficient is constant. In this case, the transformer windings must have branches corresponding to the number of coils on the high and low voltage sides.

2 Theoretical Part

According to the European standard [2], the following energy classes of transformers are distinguished: A, B, C, D, E. Class A transformers have the best energy efficiency.

Modern domestic transformers mainly belong to class D and E in accordance with the European standard [2]. For example, a transformer with rated voltages of 10/0.4 kV of TMG type (three-phase transformer with oil and air cooling, totally enclosed) with a power of 400 kV•A (Sverdlovsk Works of Current Transformers JSC, Russia) has an idle loss power of 900 W, a short circuit power of 4900 W, and a Shingle transformer of AoAk type (ABS MINEL-TRAFO, Serbia) - idle power loss of 370 W, short circuit power of 3000 W.

A significant increase in the energy efficiency of transformers is ensured by the use of magnetic cores made of amorphous alloys and windings made of foil or high-temperature superconducting materials [3].

Obviously, the economic effect of the proposed measures to reduce losses in power transformers can be achieved on busy sections of the distribution network.

The problem of optimizing the mode parameters to minimize power losses in the electric network is usually solved in relation to load losses of electricity or power [4–6]. Conditionally constant losses are not controlled in electric networks. Under low load conditions, the issue of reducing idle losses in power transformers is particularly acute.

Figure 1 shows a diagram of a single-phase transformer with coils on the low and high voltage sides, and a control system. A three-phase transformer with a different number of winding coils (layers) having their own terminals (branches) is implemented in a similar manner. When the load of a transformer is close to nominal, the primary and secondary coils (layers) of the windings 1 are connected in parallel, providing the nominal parameters of the transformer, including the designed magnetic flux. With a decrease in the load power, it is possible to reduce the power of the transformer by simultaneously switching the connection circuit of the coils on the low and high voltage sides from parallel to serial (in-phase) or serial-parallel (with the number of coils more than four). In this case, the cross section of the windings is inversely proportional to the number of coils, i.e., the mass of active materials of the whole and sectioned winding (excluding terminals) is the same.

As is known [7], the induction in the magnetic core of a transformer is determined by the expression, T:

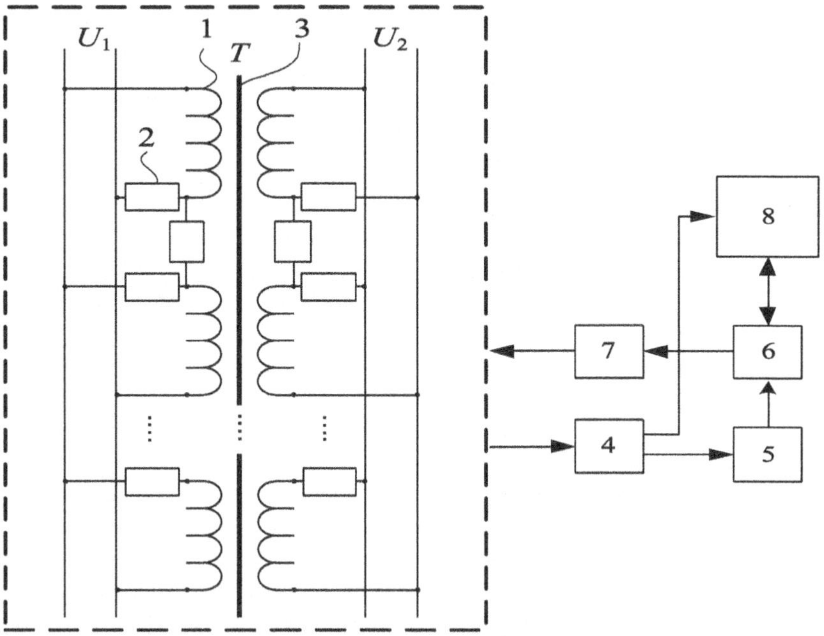

1 – coil (layer) of the transformer winding; 2 - switching devices; 3 - magnetic core; 4 - measuring module; 5 - state assessment module; 6 - coordination module; 7 - control actions generating module; 8 - top-level (network) control system

Fig. 1. Transformer scheme.

$$B = \frac{U_t}{4.44 \cdot f \cdot P_r}, \quad (1)$$

where f – frequency of supply voltage, Hz; Pr - active section of the rod, m2; Ut - turn voltage, V:

$$U_t = \frac{U_f}{w}, \quad (2)$$

Uf – winding voltage, V; w - the number of turns per winding voltage.

The idle power loss in a transformer mainly consists of magnetic losses (for hysteresis and eddy currents) and electric losses in the primary winding from the idle current [7]. In this case, the magnetic losses Pmag are proportional to the induction B and frequency f according to the relation:

$$P_{mag} = B^n f^{1.3}, \quad (3)$$

where n = 1.7–2.8, depending on the steel type of a magnetic core.

Thus, switching the coils of the primary and secondary windings from parallel to a serial circuit at low loads will reduce the voltage per turn of the primary winding, and, consequently, reduce the induction and magnetic flux by half, idle losses in the transformer by about four times. In this case, the nominal (for the new circuit) power will be reduced by half. This will increase electrical losses relative to the parameters of the original circuit at the time of switching by four times. Therefore, the circuit should be switched when the load factor of the transformer is less than 0.5, at the moment when the electrical loss will be equal to the idle loss.

It should be noted that to prevent the occurrence of equalizing currents, the voltage at the branches of the secondary windings should not differ from each other by more than 3% [8].

The expansion of the functionality of the transformer can be ensured by automatically controlling the change of winding sectioning schemes. For this, data from measuring transducers (currents, voltages, powers) and position sensors of switching devices are continuously transmitted to measuring module 4 of the device (Fig. 1). The received data array is continuously fed to the state assessment module 5, in which the analysis of the compliance of the received data with the established requirements of the rated mode of the adjustable transformer is performed. Based on the obtained data and the logic of the implementation of technological functions (for example, the minimum loss in the transformer) of the control, control actions are generated that are coordinated in module 6 with the control system of the upper level (network) 8. In the control actions implementation module 7, a coordinated signal is generated to control switching devices 2.

To solve the set problems, the methods of full-scale experiment and simulation modeling were used.

3 Practical Part

The experimental research program was carried out taking into account the provisions of Russian State Standard GOST 3484.1-88 [9] and included the measurement of idle losses, determination of the parameters of the T-shaped equivalent circuit, and other characteristics of the SOBS-2A single-phase transformer with different winding connections (Fig. 2) [7]. The measurements were carried out for idle mode, short circuit, mode with active-inductive load of the transformer.

The signal single-phase armored dry (SOBS) transformer is designed for powering traffic lights and has the following technical characteristics: power 135 V•A, rated voltage of the primary winding 220 V, rated current of the primary winding 0.7 A, rated current of the secondary winding 3.86 A, rated voltage of the secondary windings 28 V.

The results of experimental studies of the transformer are presented in Tables 1 and 2.

Based on the obtained parameters of the equivalent circuit of the SOBS-2A single-phase transformer, its simulation modeling was performed in Matlab Simulink taking into account hysteresis losses, eddy currents and core saturation (Fig. 3). The main assumptions made during modeling are: stepwise setting the saturation curve of the transformer magnetic core, setting transformer parameters (r0, x0, rk, xk,) based on the

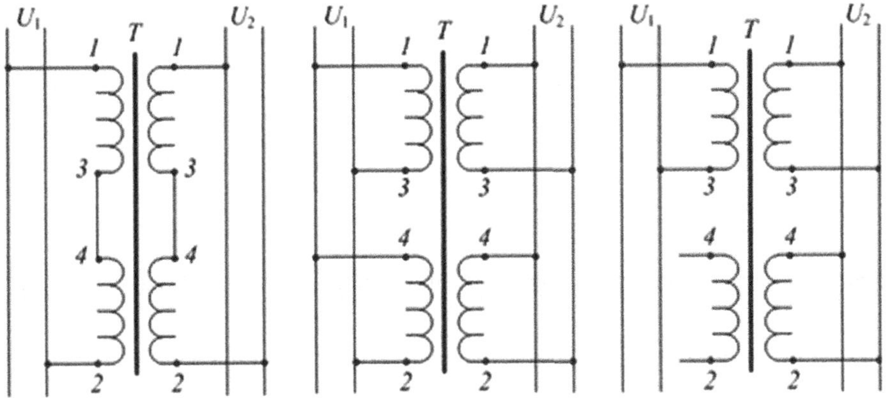

Fig. 2. Connection diagram of transformer windings.

Table 1. The results of measurements and calculations in an idling mode.

№	Diagram in Fig. 2	U_1, V	I_{10}, A	P_{10}, W	Q_{10}, var	U_{20}, V	r_0, O	x_0, O	z_0, O	$\cos\varphi_0$
1	a	220	0.04	3.2	6.7	29.5	2000	5123	5500	0.36
2	b	220	0.28	12	53	29.5	153	771	786	0.19
3	c	220	0.28	13	54	29.5	166	768	786	0.21

Table 2. The results of measurements and calculations in a shirt circuit mode.

№	Diagram in Fig. 2	U_{1k}, V	I_{1k}, A	P_{1k}, W	I_{2k}, A	r_k, O	x_k, O	z_k, O	$\cos\varphi_k$
1	a	7.3	0.19	1.3	0.95	36.01	13.39	38.42	0.95
2	b	13.8	0.71	5.1	3.82	10.12	16.60	19.44	0.52
3	c	14.0	0.73	5.8	3.87	10.88	15.79	19.18	0.57

T-shaped equivalent circuit [8]. The reliability of the results is confirmed by the experiment conducted in accordance with Russian State Standard GOST 3484.1-88 [9] and the coincidence of the obtained values with simulation modeling with an accuracy of about 5–10%.

Analysis of the results allows drawing the following conclusions:

- reduction of idle losses P10 in the diagram in Fig. 2a with respect to the diagram in Fig. 2b is about 4 times, which corresponds to relation (3);
- a slight increase in idle losses in the diagram in Fig. 2c with respect to the diagram in Fig. 2b is caused by an increase in electric losses in the primary winding from the open circuit current; therefore, the use of this diagram is impractical;
- reduction of short circuit losses P_{1k} in the diagram in Fig. 2a with respect to the diagram in Figs. 2b and c is caused by a decrease in the current (available power) of the windings and an increase in their resistance.

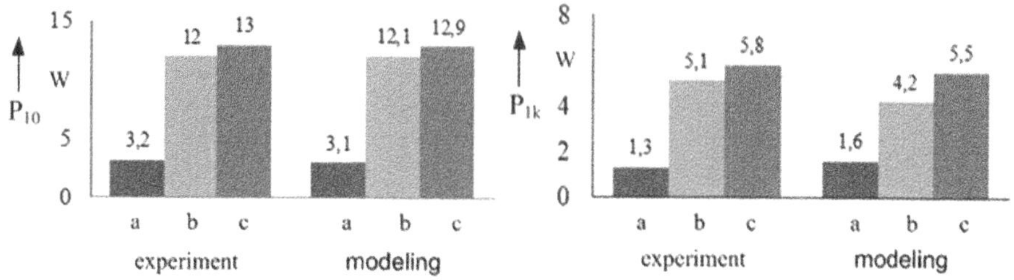

Fig. 3. The results of the experiment and simulation modeling of the SOBS-2A transformer.

Due to the proposed change in the connection scheme of the coils of the primary and secondary windings under the condition of a constant transformation coefficient (Fig. 2a, b), idle losses are reduced, but the equivalent resistance of the windings is increased by four times. Therefore, short circuit losses increase (in copper). Thus, the total losses in the transformer under consideration at a certain load can not only decrease due to idle losses, but also increase due to an increase in electric losses in the windings.

To assess the reduction in losses in a three-phase power transformer TM 400 with a voltage of 10/0.4 kV by changing the connection schemes of its windings in accordance with the proposed method (Fig. 2a, b), its simulation was performed in Matlab Simulink.

Figure 4 shows idle losses (no-load losses), short-circuit losses, and total power losses of TM 400 transformers with a voltage of 10/0.4 kV according to passport data at rated load (Fig. 4a) and with a load factor of 0.5 (Fig. 4b), as well as with the winding connection diagram according to Fig. 2a with a serial connection of windings split into two coils (TM 400, a = 2) with a load factor of 0.5 (Fig. 4c).

In order to assess a load at which the total losses of the transformer with the serial connection of two (a = 2), three (a = 3) and four (a = 4) split coil windings will be lower than with parallel connection (a = 1), a simulation modeling of the change in the total power losses in the TM 400 power transformer with the corresponding winding connection was performed (Fig. 5).

The load factor of the transformer, within which a reduction in losses due to a change in the winding connection diagram is achieved, is determined by the expression:

$$\kappa_{\mathrm{lmax}} = \frac{1}{\mathrm{a}}\sqrt{\frac{P_0}{P_k}}, \tag{4}$$

where a – number of transformer winding coils connected in series.

Expression (4) is obtained based on the equality of the total losses of the transformer with parallel and serial connection of the coils of its winding.

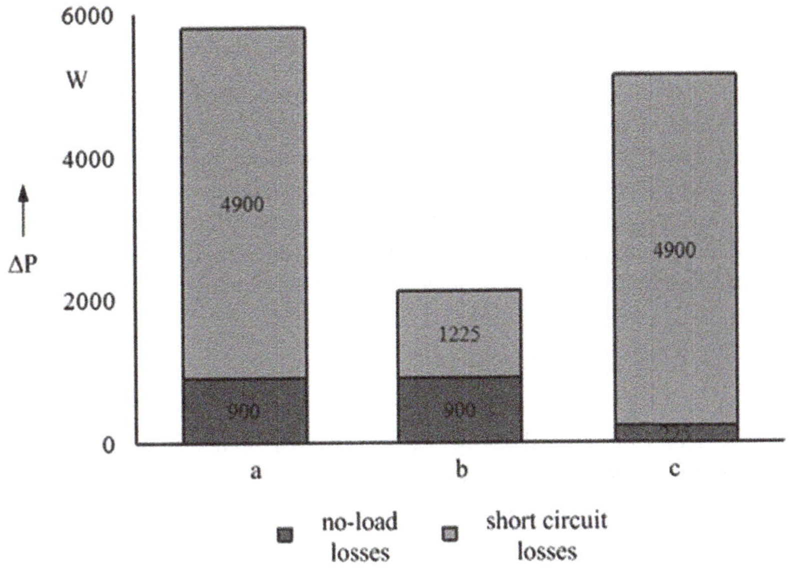

Fig. 4. Power losses in power transformers.

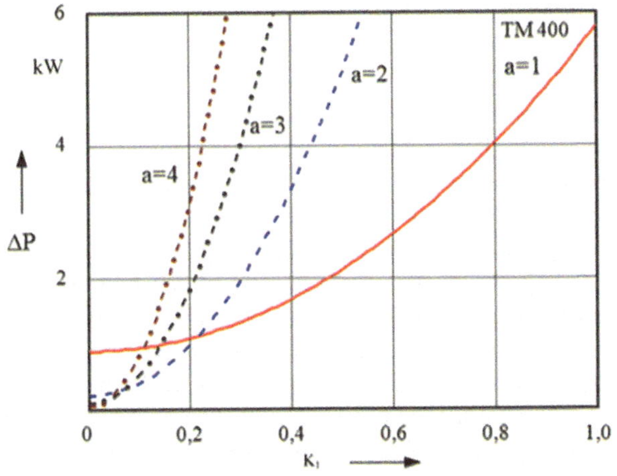

Fig. 5. Change in power loss in power transformers.

An analysis of the ratios of idle and short circuit losses (Figs. 4, 5, formula (4)) indicates that the reduction of losses in the TM 400 transformer (a = 1), and, consequently, the regulation efficiency, is ensured with a load factor not exceeding:

0.214 with serial connection of windings split into two coils (a = 2);
0.143 with a serial connection of windings split into three coils (a = 3);
0.107 with a serial connection of windings split into four coils (a = 4).

Obviously, when a = 4, the number of combinations of connecting the coils of the transformer windings increases, for example, series-parallel connection is possible.

From the expression (4) it also follows that the maximum possible depth of regulation of power losses in the transformer corresponds to the equality of the idle and short-circuit losses at a load factor of 0.5, i.e., with an increase in idle losses of transformers, the depth of regulation of power losses increases.

For automatic control of losses in the transformer (Fig. 1), the expression for reducing the total losses in the transformer due to a change in the connection scheme of the coils of the primary and secondary windings with a change in load was obtained:

$$\Delta P(\kappa_1) = (\mathrm{a}^2 - 1)\left[\frac{P_0}{\mathrm{a}^2} - \kappa_1^2 P_k\right]. \tag{5}$$

It can be seen from formula (5) and Fig. 5 that the saving of power losses occurs only when the load factor is less than 0.2.

4 Conclusion

It is obvious that the transformer under consideration with a change in the winding connection schemes will have a higher cost due to changes in the design of the windings and their insulation, while the overall dimensions of the transformer will increase slightly, mainly due to the longitudinal insulation of the windings. The mass of active materials, i.e., windings and magnetic core, will remain the same. It is possible to change the ratio between the width and height of the transformer magnetic core due to an increase in the longitudinal insulation of the windings [7].

A more detailed assessment of the characteristics of the transformer under consideration can be performed according to the design results.

From a practical point of view, a transformer with a change in the connection schemes of its windings using the proposed method can find application as a power transformer of distribution networks with a voltage of 35, 10 (6)/0.4 kV, in which the load will be less than 20% for a significant part of the time (summer and cottage villages, seasonal loads of industrial enterprises). In addition, the above figure and formula show that the use of four windings will give an effect only when the transformer is loaded less than for 2.5%. Therefore, it is rational to use only two windings. The use of the proposed sectioning of the transformer windings will allow increasing its efficient performance by reducing losses at low loads. Also, the commissioning of the transformer, first with serial, and then with the parallel connection of the windings, will significantly limit the short circuit current (inrush magnetization current).

Automatic control of changes in winding sectioning schemes will expand the functionality of an automated process control system of substations in the task of reducing energy losses. The main components of such a system when changing load curves in real time are discussed in more detail in [5].

The obtained formulas (4) and (5) can be used in the design of the considered transformers with split windings and a feasibility study for controlling power losses in transformers according to real load curves.

References

1. Savintsev, Y.M.: The planned development of the market of power transformers: Utopia or necessity. Electr. Market, (1 – 2), 37–38, 39–42 (2011)
2. EN 50464 – 1:2007. Three-phase oil-immersed distribution 50 Hz, from 50 kVA to 2500 kVA with highest voltage for equipment not exceeding 36 kV. General requirements (2007)
3. Lucija, S.I., Jafarov, E.A.: Superconducting Transformers. Austenitized Publ., Moscow (2002)
4. Venikov, V.A.: The electrical network. Higher school Publ. Power system, vol. 2, Moscow, Russia (1971)
5. Tretyakov, E.A.: Management of Power Quality in Distribution Networks of Railways: Monograph. Ostu Publ, Omsk (2013)
6. Solomin, V.A., Zharkov, Y.I. et al.: Adjustable transformer. Patent RU 2364972. Stated 18.12.2007, published 20.08.2009 (2009)
7. Tikhomirov, P.M.: Calculation of transformers. In: Proceedings of Manual for Schools. Energoatomizdat Publ., Moscow (1986)
8. Voldek, A.I.: Of the Electric machine. In: Textbook for Students of Higher. Tech. Institutions. Energiya Publ., Leningrad, Russia (1978)
9. GOST 3484.1-88. Power transformers. Methods of electromagnetic testing

Mechanical Engineering

Optimization of the Fleet of Multi-turn Fasteners at the Metallurgical Enterprise

Sergey Kornilov(✉), Oleg Fridrihson, and Aleksandr Rakhmangulov

Nosov Magnitogorsk State Technical University, 38, Prospekt Lenina, 455000 Magnitogorsk, Chelyabinsk Oblast, Russia
kornilov_sn@mail.ru, fridrikhsonov@yandex.ru

Abstract. The state of the metal products market is considered, the market position of PSJC Magnitogorsk Iron and Steel Works is defined. The analysis of the fleet of multi-turn fasteners (metal frames) for the delivery of export cargo to seaports was performed. The economic-mathematical model was developed to optimize the fleet of metal frames. A list of measures to shorten the rotation cycle of returnable containers is presented.

Keywords: Metal products · Loading plan · Multi-turn fasteners · Rotation cycle · Optimization of the fleet · Economic and mathematical modeling

1 Introduction

Global steel production in 2018 had increased by 2.3% compared to 2017, to 1.73 billion tons (Fig. 1). China remains the leader in steel production with a total output of 808.4 million tons [1]. As can be seen from the graph, annually there is an imbalance between the volume of production and consumption of metal products (a negative consumption balance of more than 70 million tons). Therefore, the product competition between metallurgical companies is increasing not only in the domestic market, but also in the world one. Russian metallurgical facilities, being one of the most modern in the world, allow to produce high-quality products and almost completely meet the needs of Russian consumers, successfully displacing imports. Russia remains the fifth largest steel producer in the world with a production volume of 71.8 million tons, which is about 5% of the global production (in 2018). The top three metal manufacturers in the domestic market are NLMK (Novolipetsk Iron and Steel Works), MMK (Magnitogorsk Iron and Steel Works) and Severstal (Fig. 2).

Over the past years, PJSC "MMK" demonstrates consistently high rates of production and financial activities. In 2018, the company produced and sold more than 11.41 million tons of finished metal products [2].

About 21% of the plant's products are exported. Generally, these are products of deep processing, and, therefore, the most expensive (Fig. 3). Products manufactured at PJSC "MMK" are becoming more and more demanded abroad every year due to a reasonable quality-price ratio.

Z. Popovic et al. (Eds.): TransSiberia 2019, AISC 1115, pp. 867–876, 2020.
https://doi.org/10.1007/978-3-030-37916-2_85

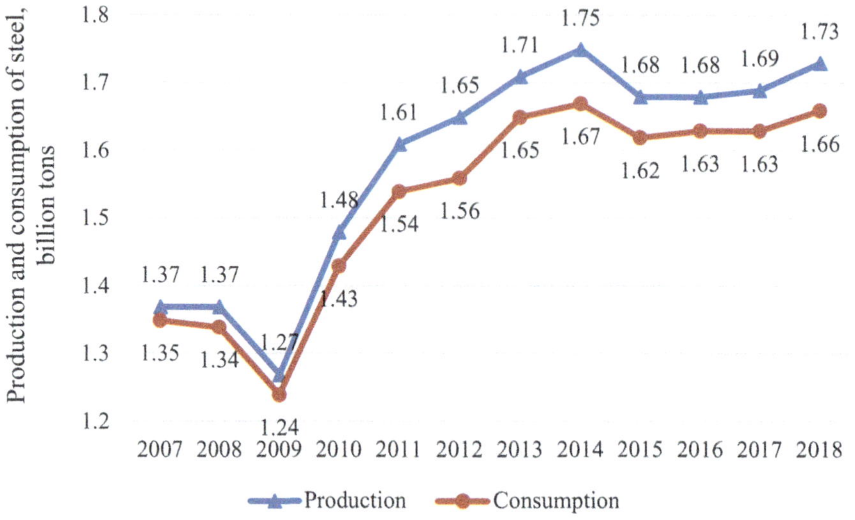

Fig. 1. World production and consumption of steel, billion tons

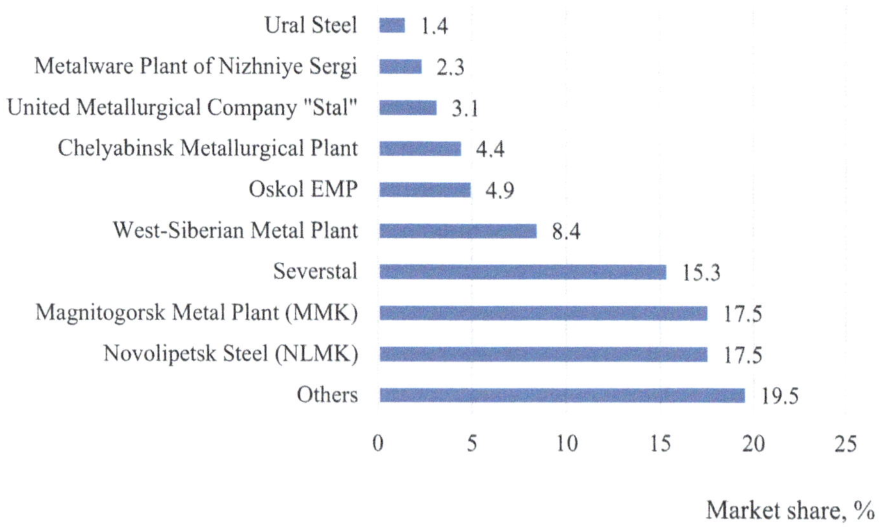

Fig. 2. The structure of metal production in Russia

However, currently, PJSC “MMK”, like other Russian iron and steel producers, faces the issue of cutting the production cost while maintaining sales profitability, since the average price of metal products exported from Russia is comparable with the cost of items from China and India. Metallurgical industry is one of the most material-intensive, the share of raw materials in the structure of production costs is 70–74%, the share of transport and logistics costs is 5–12%. The cost of shipping finished products

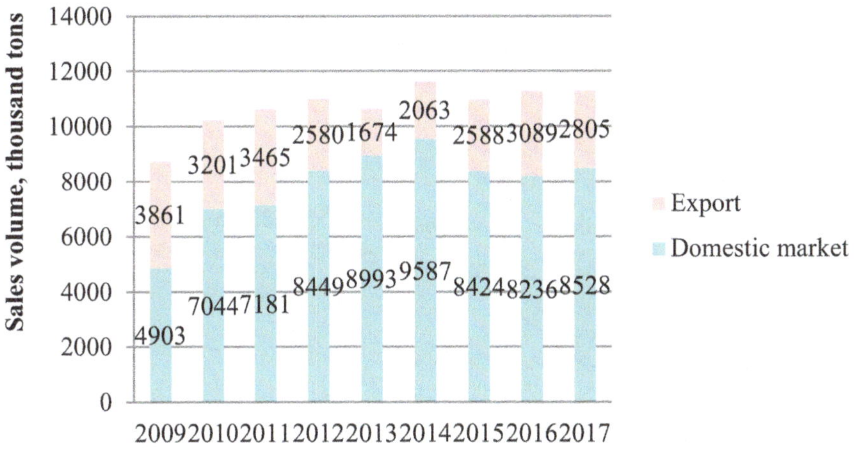

Fig. 3. Manufacture of finished metal products of PJSC “MMK”

to the consumer (including the costs of the industrial and external transport system) is the stage of a potential advantage gaining, and measures, which should be taken to eliminate bottlenecks in transport and logistics processes, are of considerable relevance.

The railway and sea combined scheme is the main export delivery. Metal products are transferred to Russian ports in railway rolling stock, then, on the basic terms of delivery, CIF or FOB (according to Incoterms 2010) the party is loaded onto seagoing vessels (from coasters (deadweight 3–6 thousand tons) to handysize vessels (deadweight 15–35 thousand tons), after which the cargo continues to move along the sea routes of communication. In order to ensure the safety of finished products during transportation in railway gondola cars and on platforms to the seaports, PJSC “MMK” uses multi-turn fasteners (MSK) - metal frames (Fig. 4).

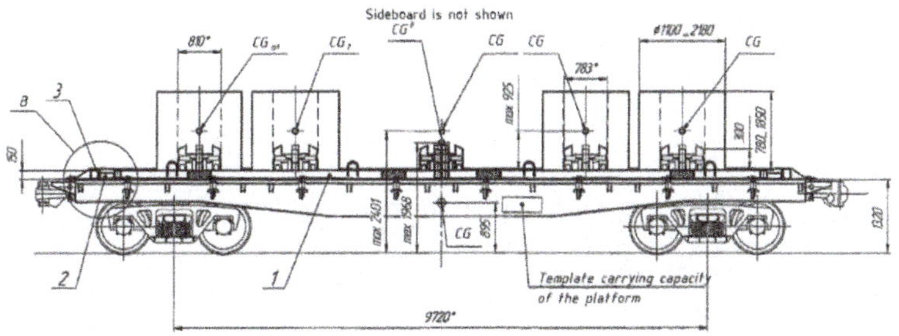

Fig. 4. Front plan view of the scheme of loading coils on a platform equipped with MSK

Loading is carried out daily in accordance with the sales plan of PJSC “MMK”. The coils are transported by railway, in accordance with the local technical standards No. MO-103/587, approved by the South Ural Territorial Center of branded transport services dated February 25, 2016, with the use of MSK (multi-turn fasteners) [3].

Currently, the fleet of metal frames of the enterprise amounts to more than 23 thousand units; more than 2 thousand new frames are produced annually, in order to replace the ones, that have been scrapped and have not been returned by consignees (Fig. 5).

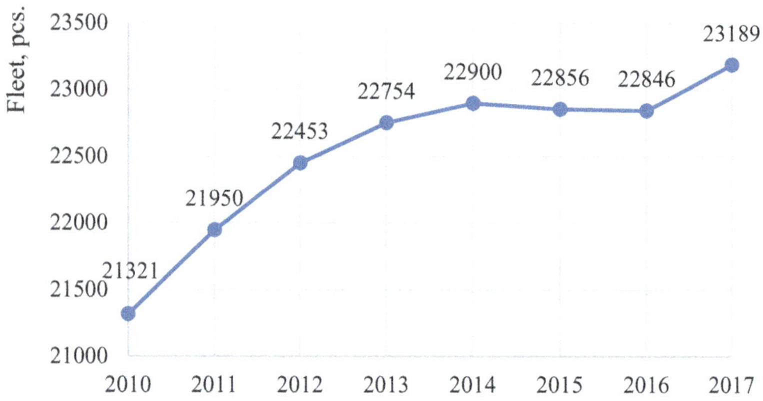

Fig. 5. The fleet dynamics of MSK [4]

At the Russian seaports, after reloading the coils onto ships, metal frames are accumulated for return transportation to PJSC "MMK". In 2017, transportation to ports was carried out in five main areas: Novorossiysk (1972.7 thousand tons), Zhiguli Sea (255.8 thousand tons), Vladivostok (203.1 thousand tons), the Right Bank (24.7 thousand tons), Aktau - port (19.5 thousand tons). So, in 2017, 46,090 multi-turn fasteners were sent to Novorossiysk, and 5976 pcs - to Zhiguli Sea.

Return transportation of frames is carried out in accordance with the local technical standards No. MO-103/387 of 11.11.2012 by railway rolling stock (Fig. 6).

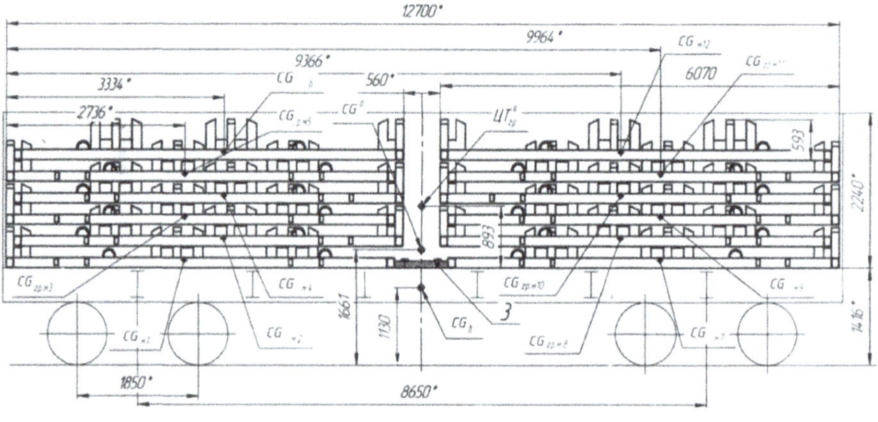

Fig. 6. Front plan view of the scheme of loading frames

2 Problem Statement and Development of Methodological Tools for Its Solution

As it has been noted above, there are about 23 thousand metal frames of different designs today in rotation. The structure of the fleet consists of frames of 23 different sizes.

'The analysis of these accounting records of MSK showed that the average turnover of the frame (export operations in all directions of transportation) is 97 days, in the port of Novorossiysk - 35 days, which exceeds the standard values of the cycle for metal frames by 12–30%., Therefore, the company is forced to increase the number of its fleet in order to compensate the deviations from the normative period for the return of MSKs in the context of increasing exports. It should be noted that today the inventory value of the multi-turned fasteners in use exceeds 1 billion rubles. The expenses of maintenance and depreciation form independent cost items.

In [3] it is noted that railway rolling stock and MSK are among the most cost-effective and limited railway transport assets, which increases the role of tools of its effective use. Information sources [5] contain recommendations for determining the composition of operations that set the length of the MSK cycle (the time from the moment of loading the finished product onto the metal frame until the next loading).

In order to optimize the MSK fleet, an economic-mathematical model has been developed, the objective function of which is presented below.

$$F = C_{\mathrm{f}} + RC_{\mathrm{crnt}} - C_{\mathrm{opt}} - RC_{\mathrm{opt}} - SC \rightarrow max, \text{ when } Q_{\mathrm{crnt}} \leq Q_{\mathrm{lim}} \tag{1}$$

where

C_f – current value of MSK, thousand rubles;

RC_{crnt} – current costs for the maintenance of the MSK fleet, thousand rubles.;

C_{opt} – cost of the MSK fleet of optimal number, thousand rubles;

RC_{opt} – cost for maintenance of the MSK fleet of optimal number, thousand rubles;

SC – total costs for the implementation of measures to cut the time of execution of MSK processing operations, thousand rubles;

Q_{crnt} – current volume of transportation of export products to the seaports, thousand tons;

Q_{lim} – volume of transportation of export products to the seaports after the implementation of measures, thousand tons

In turn, the elements of the objective function are calculated by the expressions

$$\mathrm{C_f}\left(\mathrm{C_{opt}}\right) = \sum_{i=1}^{n} c_i \cdot n_i \tag{2}$$

$$RC_{\mathrm{crnt}}\left(RC_{\mathrm{opt}}\right) = \sum_{i=1}^{n} n_i \cdot l_i \tag{3}$$

$$SC = \sum_{i=1}^{m} r_i SC = \sum_{i=1}^{m} r_i \quad (4)$$

$$n_i = \frac{Q_i}{N_{\text{rot}i} \cdot q_i} n_i = \frac{Q_i}{N_{\text{rot}i} \cdot q_i} \quad (5)$$

$$N_{\text{rot}i} = \frac{365}{\text{T}_{\text{rot}i}} \quad (6)$$

Where c_i – MSK price of the i-th type, thousand rubles;
n_i – number of MSKs of the i-th type, thousand rubles;
l_i – total costs for maintaining the MSK of the i-th type thousand rubles;
r_i – costs for implementing the measures to cut the time of MSK processing operations, thousand rubles;
N_{roti} – number of rotations of the MSK of the i-th type in the j-th direction of transportation, days, rotations per year;
T_{roti} – average rotation cycle of MSK of the i-th type in the j-th direction of transportation, days;
Q_i – volume of transportations of finished products at the MSK of the i-th type in the j-th direction, thousand tons;
q_i – average payload of MSK of the i-th type, tons

The rotation cycle is the sum of the duration of the following technological operations.

$$T_{rot} = t_0 + t_{install} + t_{shunt} + t_{load} + t_{select} + t_{trans1} + t_{unload} + t_{acc} + t_{trans2} + t_{unload}, \quad (7)$$

where t_0 – time to select the necessary frames (production), hours;
$t_{install}$ – time to install the frames in the car, hours;
t_{shunt} – time for shunting operations, hours;
t_{load} – time to load the product, hours;
t_{select} – time to select suitable car for loading, hours;
t_{trans1} – time of transportation to the port, hours;
t_{unload} – time to unload the product, hours;
t_{acc} – time of accumulation of frames at the port, hours;
t_{trans2} – time to return frames to the plant, hours;
t_{unload} – time to unload frames from cars to the warehouse, hours

As can be seen from the objective function, the cost of the MSK fleet, as well as the cost of its maintenance, can be cut by reducing the rotation period. The list of measures to shorten operations with reusable containers is presented in Table 1.

In order to check the economic and mathematical model for adequacy, and to fulfill the usefulness of the developed tools, it is necessary to calculate the optimal number of MSK fleet serving the export directions of transportation of finished products of PJSC "MMK".

Table 1. Measures to reduce the MSK rotation period

Element	Type of work performed	Recommendations for reducing the rotation period
1	2	3
t_0	Selection of the necessary frames in the CPV warehouse, or production of new ones	Preliminary approval of applications for the loading of metal products between CPV and sheet rolling shops
$t_{install}$	Installation of frames in the car	Improving the technology of frames installation
t_{shunt}	Sorting cars in the following directions: sheet rolling shops - 4, 5, 10, 11, coated metal production workshops, removing of empty cars	Coordination of the procedure for submission of cars to the dead ends of the CPV with the railway station
t_{load}	Loading of the finished product on the frames	–
t_{select}	Shunting work on the selection of suitable cars (by owners, by suitability for loading, by directions of the CIS, Far East, Sakha Republic, Kazakhstan)	–
t_{trans1}	Transportation in public routes to ports	Speed change from cargo to high-speed service
t_{unload}	Unloading products at the port	Purchase of own port facilities, agreement and payment of priority handling of PJSC "MMK" cargo
t_{acc}	Accumulation of frames at the port	Determination of the necessary batch of frames for sending, coordination of standards for the time the frames are at the port
t_{trans2}	Transportation of frames in public railway to PJSC "MMK"	Speed change from cargo to high-speed service
t_{unload}	Unloading frames to CPV warehouses	–

3 Performing the Model Exercise

Based on the data on the number of accumulated frames at the port of Novorossiysk, we will calculate the batch size of return frames with an average accumulation time of 7 days. The dynamics of the accumulated frames for return at the port of Novorossiysk in March 2017 is presented in Fig. 7.

As can be seen from Fig. 7, period of accumulation of frames at the port, equal to 7 days, is not optimal. Given the technical ability to load 22 frames in one gondola, it is proposed to reduce the accumulation period to 5 days. The dynamics of the accumulated frames for return with a period of accumulation of 5 days is presented in Fig. 8. When comparing the benchmark - the number of frames at the port of destination, a decrease in the range of 10–47% was established (depending on the intensity of the arrival of rail cars with finished products at the port).

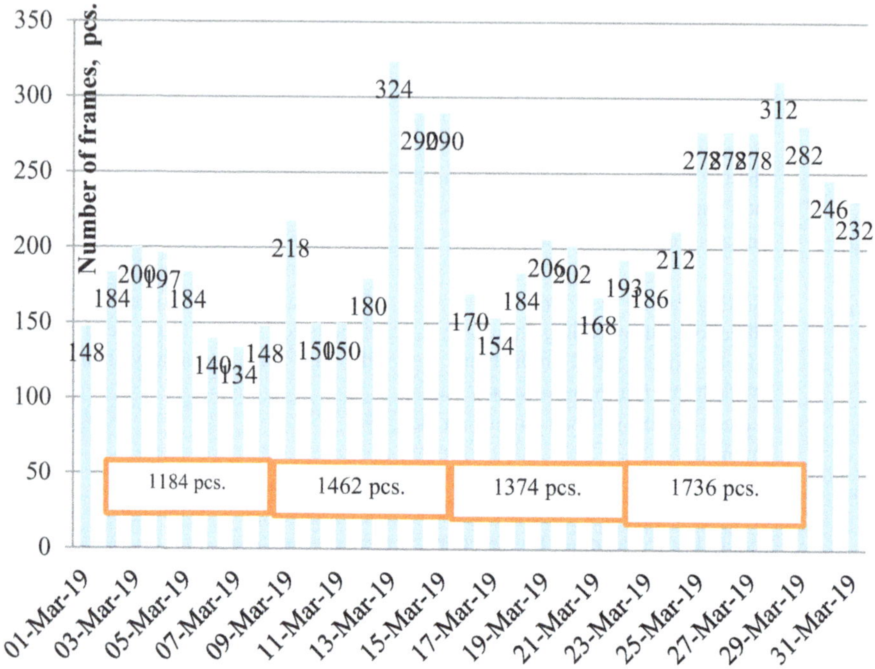

Fig. 7. Dynamics of accumulation of frames at the port of Novorossiysk in March 2017

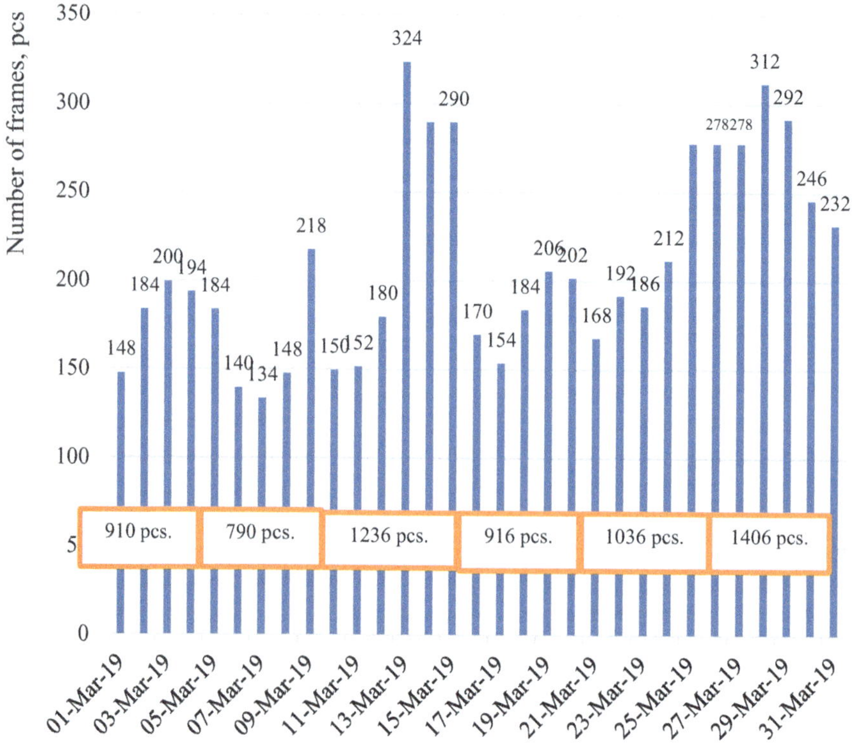

Fig. 8. Dynamics of accumulation of frames at the port of Novorossiysk in March

The calculations for the optimal accumulation period are presented in Table 2.

Table 2. Calculation of the optimal period for accumulation of MSKs at ports

Ports	Current accumulation period, days	Proposed accumulation period, days
1	2	3
Novorossiysk	7	5
Zhiguli Sea	10	9
Vladivostok	12	9

On the basis of calculating the cost of the MSK fleet and the cost of its maintenance for the actual number and estimated optimal conditions, the economic effect arising from the implementation of measures to reduce the processing time of MSKs on the track was estimated (Table 3).

Table 3. Technical and economic indicators

Indicator	Current accumulation period			Optimal accumulation period		
	Novorossiysk	Zhiguli Sea	Vladivostok	Novorossiysk	Zhiguli Sea	Vladivostok
1	2	3	4	5	6	7
Necessary MSK fleet, pcs	4249	459	872	3966	433	872
Cost of the fleet, thousand rubles	204 973.4	24008.5	39267.4	160 212	22650.8	39267.4
Maintenance expenses, thousand rubles	64 584.8	6976.8	13254.4	61 620	6581.6	13254.4
Measure implementation expenses, thousand rubles				2436.3	87.6	0
Economic effect, thousand rubles				47 726.2	1 752.9	0

4 Conclusion

On the basis of the proposed economic and mathematical model, it was proved that changing the accumulation period of MSKs at Russian ports, as well as introducing measures to accelerate the transportation of metal frames, will result in optimizing the frame fleet of 9–10% and cutting the expenses of its operation. Promising areas of research include the development of a dynamic simulation model based on the study of the stochastic characteristics of the process of delivering finished products in railroad cars equipped with MSK, developing software to support managerial decision-making for specialists of transport and logistics departments of enterprises.

References

1. Reuther, M.: Mathematical optimization of rolling stock rotations Ph.D. thesis. Technische Universität, Berlin (2016). https://doi.org/10.14279/depositonce-5865
2. Borndörfer, R.: Recent success stories on integrated optimization of railway systems. Transp. Res. Part C: Emerg. Technol. **74**, 196–211 (2017). ISSN 0968-090X. https://doi.org/10.1016/j.trc.2016.11.015
3. Cacchiani, V., Huisman, D., Kidd, M., Kroon, L., Toth, P., Veelenturf, L., Wagenaar, J.: An overview of recovery models and algorithms for real-time railway rescheduling. Transport. Res. Part B: Methodol. **63**(May), 15–37 (2014). https://doi.org/10.1016/j.trb.2014.01.009
4. Cacchiani, V., Toth, P.: Nominal and robust train timetabling problems. Euro. J. Operat. Res. **219**(3), 727–737 (2012). https://doi.org/10.1016/jejor.2011.11.003
5. Corman, F., Meng, L.: A review of online dynamic models and algorithms for railway traffic management. IEEE Trans. Intell. Transport. Syst. **99**, 1–11 (2014). https://doi.org/10.1109/TITS.2014.2358392

Acoustic Emission Monitoring of the Destruction Process of Carbon Fiber Reinforced Plastic Samples in Different Temperature Ranges

Ludmila Stepanova(✉) and Valentina Chernova

Siberian Transport University (STU), Dusi Kovalchuk St. 191, Novosibirsk 630049, Russia
stepanova@stu.ru

Abstract. The paper provides the results of a study of the destruction process of carbon fiber reinforced plastic samples under simultaneous impact of static loads and positive or negative temperatures using acoustic emission (AE). Samples of *T*800 CFRP with dimensions 100 × 600 × 0.9 mm and a lamination pattern of nine monolayers at [±45/90/0_3/90/±45] were statically loaded. Heating or cooling was applied in the central region, where a stress concentrator as an opening with a diameter of 12 mm was provided. The destruction process was monitored with a certified digital acoustic emission diagnostic system DADS-16.10 with variable selection thresholds. The real time location of the sources of AE signal allowed monitoring sample destruction in the conditions of limited visual access. The main information bearing parameters of the AE signals (amplitude, dominant frequency and structural coefficient) were analyzed from their load dependences. The registered AE signals were evaluated using wavelet transformations, which also provided the structural coefficient. It was found that, at temperatures −20 °C and +20 °C, an increase in loading resulted in the most pronounced variations in AE signal parameters. A pre-destruction state of the material was revealed by characteristic points on the load dependences of the information bearing parameters. The developed monitoring technique may be used for AE diagnostics of composite structures operated at positive and negative temperatures.

Keywords: Carbon fiber reinforced plastic samples · Static loads · Positive or negative temperatures

1 Introduction

The current production quality of composite materials (CM) based on carbon fiber reinforced plastic (CFRP) makes them an attractive option for application in the transport sector [1–8]. Current composite materials are not inferior in terms of their strength properties to their metal counterparts, while having high rigidity, low density, corrosion resistance, durability, and high fatigue strength. This makes them suitable for manufacturing critical elements, including structural ones [1]. However, operation of

Z. Popovic et al. (Eds.): TransSiberia 2019, AISC 1115, pp. 877–884, 2020.
https://doi.org/10.1007/978-3-030-37916-2_86

transport units leads to a gradual deterioration of their mechanical properties due to external factors (shock loads, overloads, humidity, temperature).

Thermal and mechanical stresses experienced by a composite structure result in accumulation of deformations and are associated with an increased risk of fatigue cracks [7]. With time, the fatigue damage builds up in CFRP, compromising their strength properties, service life and rigidity [8, 9]. Such damage may result in a destruction of the composite structure caused by matrix cracking, broken reinforcing fibers, and delamination. Therefore, the problem of an early detection of CM item destruction requires solution and calls for modern methods and means of diagnostics [1].

A major step in developing a technology for manufacturing various CM elements is designing the methods for non-destructive examination (NDE) of their quality. Every CM destruction is accompanied by acoustic signals emission. Therefore, acoustic emission (AE) is widely used for continuous monitoring of CFRP structures in the process of loading. It is capable of localizing defects in real time, does not require scanning the surface of the monitored object, is highly sensitive and efficient for monitoring early stages of defect development in CM.

A drawback of this method is that it picks up noise during loading. The noise affects the accuracy of results and contributes to the scattering of the points of AE signal location. A frequency filtration performed at the level of electronic circuitry cannot remove the noise and interference [9, 10]. This problem has to be solved to improve reliability of information obtained from the AE measurements.

The purpose of this paper is to adopt AE for monitoring the destruction process of CFRP samples under simultaneous impact of static loads and positive or negative temperatures.

2 Research Methods

Static tests of samples made of T800 CFRP were performed at temperatures T1 = −60 °C, T2 = −20 °C, T3 = +20 °C, T4 = +100 °C; the samples had dimensions 100 × 600 × 0.9 mm and a lamination pattern of nine monolayers at [±45/90/03/90/±45]. The positive or negative temperatures affected the central region of the samples, where a stress concentrator as an opening with a diameter of 12 mm was provided. To maintain a positive temperature during loading, a heater was used, while a chamber supplied with liquid nitrogen was employed for cooling the samples. Sample temperatures were measured with a chromel-alumel thermocouple fastened to the sample surface and connected to a TRM-10 controller, which maintained the temperature within (1–1.5) °C of the preset value.

The static tensile load was applied to the samples using an MTS-100 loading machine equipped with hydrogrips. At each temperature, two samples were loaded to destruction. The AE signals were registered with a rectangular piezoprobe located outside the zone of thermal impact and consisting of four acoustic emission transducers (AET) type PK 01-07 with a bandwidth of (100–700) kHz. The AE data were processed with a certified digital acoustic emission diagnostic system DADS-16.10 with floating selection thresholds (Certificate RU.C.27.007.A No. 40707, State Register of Measuring Equipment registration number 45154-10).

Analysis of transient signals emitted upon deformation and destruction of solids requires a method of decomposing the AE signals with respect to both frequency and time. A frequency decomposition reveals the low frequency component, while decomposition with respect to time provides the high frequency components. This problem was solved using wavelet transformations.

Structural coefficients reflecting changes in the ratio of energies in the spectrum of AE signals were calculated. This corresponded to the process of position shifting of the time-frequency energy maximum, which was a sign of changes in the type of sample failure [8].

The structural coefficient of the AE signal was calculated as:

$$P_{Dij}(f) = \frac{\max D_i}{\max D_j}, \qquad (1)$$

where D_i, D_j are sets of coefficients of wavelet decomposition of the i-th and j-th level of refinement, obtained for incoming signal sampling rate $f = 2$ MHz.

3 Research Results

The AE data were registered in the course of static loading of CFRP samples with simultaneous application of positive or negative temperatures. The AE signals were located in real time, which allowed observation of the process of failure propagation upon increasing static load (Fig. 1).

For samples cooled down to −60 °C, active sources of the AE signals were observed near the opening, as well as in the lower part (Fig. 1, a), where the destruction occurred. For samples tested at −20 °C, the highest activity of the AE signals was observed near the opening (Fig. 1, b). A load increase up to P = 60 kN and beyond resulted in the destruction propagation into the upper part of the sample.

During sample loading at positive temperatures, it was found that the main source of the AE signal emission was the opening with a diameter of 12 mm at the center of the sample, where the onset of destruction occurred (Fig. 1, c–d). Raising temperature to +100 °C and increasing load resulted in delamination of the sample monolayers. This was caused by matrix transition into rubber-like state [8].

Location helped select for analysis of the main information bearing parameters (amplitude, dominant frequency and structural coefficient) only those AE signals, which were characteristic of the destruction process of the sample material. Load dependences of each of the parameters were considered in the range of 20 kN to 70 kN (Figs. 2, 3 and 4). This range corresponded to sample destruction from its onset to pre-destruction state or destruction.

It was demonstrated that at different temperatures the changes in the load dependences of the main information bearing parameters of the AE signals had their specific features. At the maximum and minimum temperatures, the AE signal amplitude did not exceed 700 mV. A load increase resulted in its change by $\Delta U = 350$ mV (Fig. 2, a).

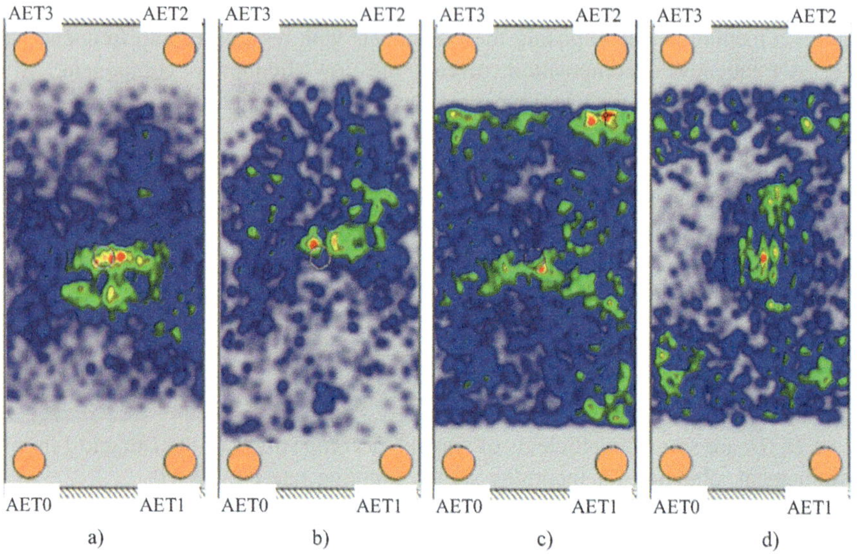

Fig. 1. Location of the AE signals upon sample loading at temperatures T_1 = −60 °C (a); T_2 = −20 °C (b); T_3 = +20 °C (c); T_4 = +100 °C (d).

At temperatures T_2 = −20 °C and T_3 = +20 °C, a more significant amplitude increase by 3,500 mV was observed (Fig. 2, b), while at loads of (50–60) kN, the maximum values preceding sample destruction were observed.

The analysis demonstrated a minor change in the dominant frequency by Δf = (20–25) kHz from load at maximum and minimum temperatures. The frequency changed in the range of (215–240) kHz (Fig. 3, a). The load dependence of the dominant frequency for samples tested at temperature T_3 = +20 °C also remained virtually constant. However, one sample at a load of P = 50 kN demonstrated a frequency drop down to f = 185 kHz. The greatest frequency change of Δf = 120 kHz was observed in the process of sample loading at temperature T_2 = −20 °C (Fig. 3, b). A load increase from 30 kN to 50 kN resulted in a frequency decrease from 210 kHz down to (140–160) kHz. Upon further loading, the dominant frequency changed by Δf = (10–20) kHz.

The highest sensitivity to changes in the shape of the AE signals was exhibited by the structural coefficient determined from Eq. (1). When processing the AE data, the coefficients of wavelet decomposition of the second and third levels of refinement were chosen for evaluation. The coefficient D2 corresponded to the frequency band of the AE signal of (250–500) kHz, and the coefficient D3 corresponded to (125–250) kHz. The selected frequency bands were within the bandwidth of AET. Thus, the equation to calculate the structural coefficient was as follows:

$$P_{D32} = \frac{\max D_3}{\max D_2}. \quad (2)$$

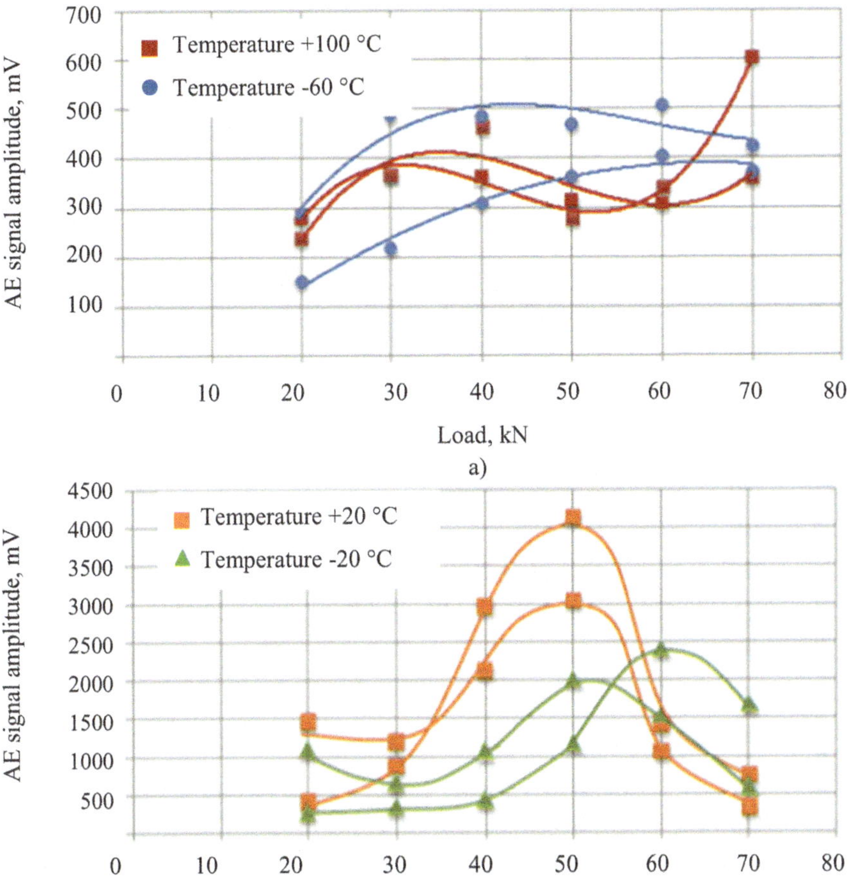

b)

Fig. 2. Load dependence of the AE signal amplitudes at temperatures $T_1 = -60$ °C and $T_4 = +100$ °C (a); $T_2 = -20$ °C and $T_3 = +20$ °C (b)

As the structural coefficient increased, the AE signal energy shifted to the frequency band of (125–250) kHz, and upon its decrease, it moved to the frequency band of (250–500) kHz. Such changes were characteristic of the destruction process of samples of matrix or fiber [8].

For samples tested at temperature $T_1 = -60$ °C, a decrease in the structural coefficient upon increasing load was observed (Fig. 4, a). In contrast, at temperature $T_4 = +100$ °C, the coefficient increased, which corresponded to a shift of the AE signal energy to the region of lower frequencies. A similar variation of the structural coefficient at temperature +100 °C was observed for samples from *T*700 CFRP [8]. At this temperature, the matrix was in elastic state, and thus the main destruction mechanism was fiber breakage, which corresponded to an increase in the structural coefficient.

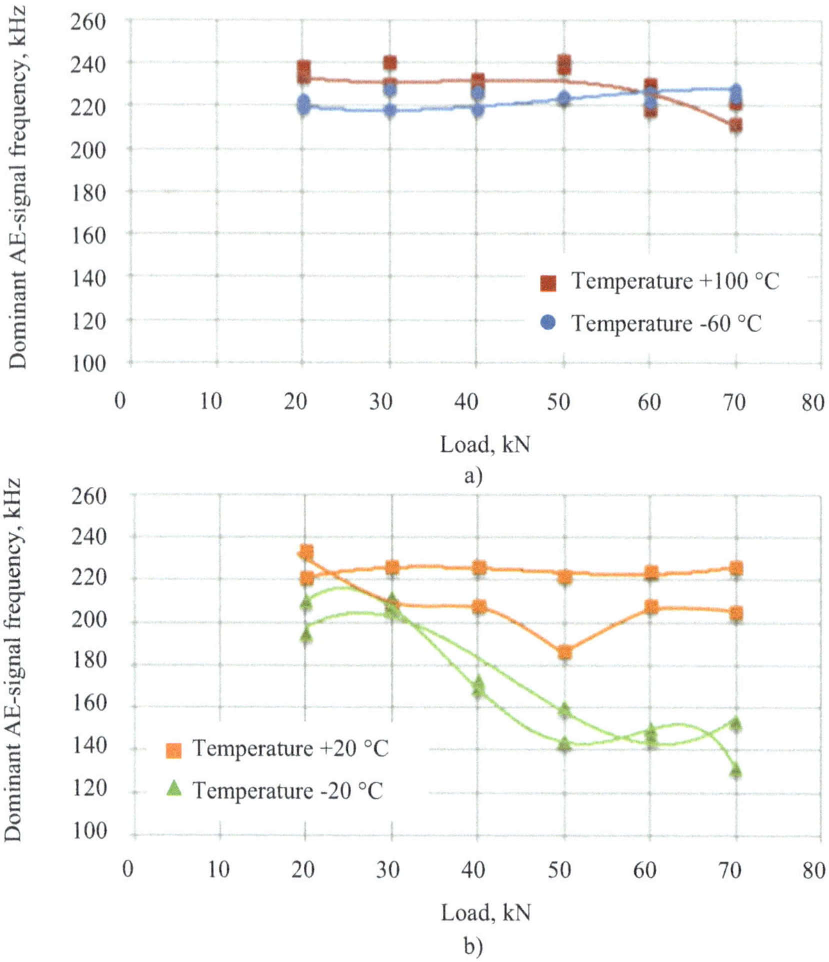

Fig. 3. Load dependence of the dominant frequency at temperatures $T_1 = -60$ °C and $T_4 = +100$ °C (a); $T_2 = -20$ °C and $T_3 = +20$ °C (b)

Sample loading at temperatures $T_2 = -20$ °C and $T_3 = +20$ °C was accompanied by an increase in the structural coefficient up to loads P = (40–50) kN (Fig. 4, b). Afterwards, this parameter virtually did not change until the end of testing. These load values corresponded to the onset of active sample failure. A local minimum was observed at load P = 30 kN.

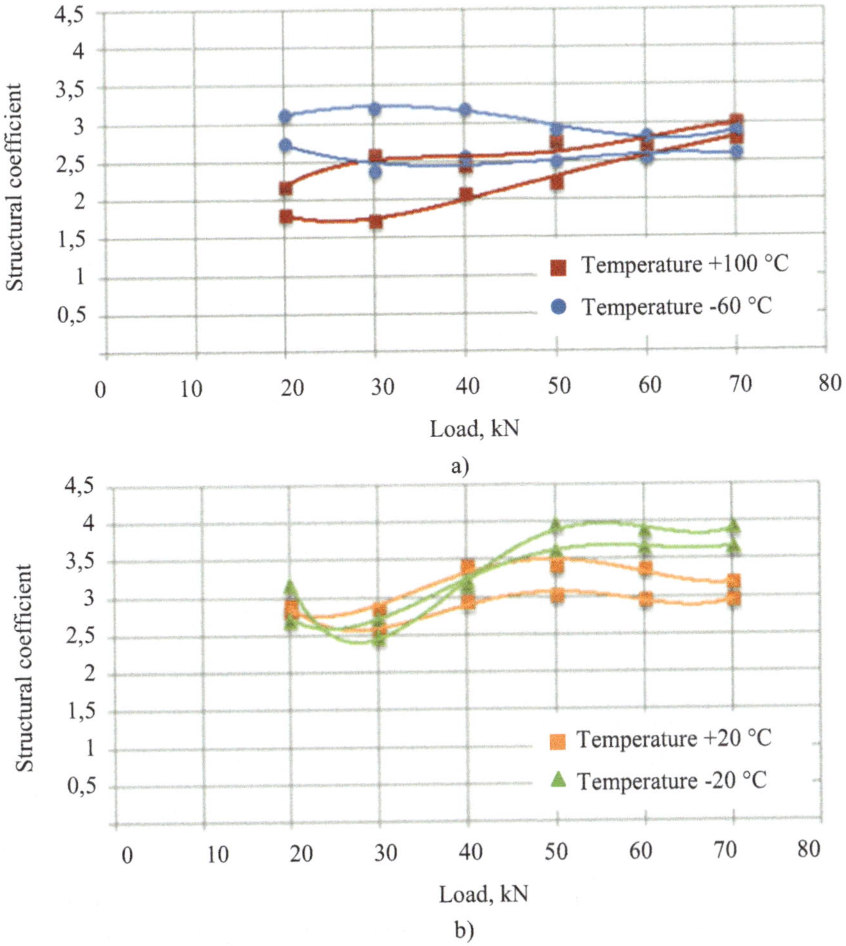

Fig. 4. Load dependence of the structural coefficient at temperatures $T_1 = -60$ °C and $T_4 = +100$ °C (a); $T_2 = -20$ °C and $T_3 = +20$ °C (b).

4 Discussion

To approach the problem of an early detection of failures in the CFRP structures, samples were tested under simultaneous static loading and an impact of positive or negative temperatures. Load dependences of the main information bearing parameters of the AE signals (amplitude, dominant frequency, and structural coefficient) were obtained.

The paper provides the results demonstrating that the considered parameters are the most informative for the tests performed in the temperature range of −20 °C to +20 °C. Analysis of the load dependences of the main information bearing parameters of the AE signals demonstrated that the onset of active sample destruction is accompanied by an increase in amplitude and structural coefficient and a decrease in frequency. The obtained data may be used to develop algorithms for processing AE data registered during tests of composite structural elements both at positive and negative temperatures.

References

1. Barsuk, V.E., Stepanova, L.N., Kabanov, S.I.: Acoustic emission monitoring of defects during static testing of a composite airplane structure. Control Diagn (4), 14–19 (2018). (In Russian)
2. Prosser, W.H., et al.: Structural Health Management for Future Aerospace Vehicles. NASA Technical Reports Server (2004)
3. Kanji, O., Antolino, G.: Research and applications of AE on advanced composites. Int. J. Psychophysiol. **30**, 180–229 (2012)
4. Aljets, D.: Acoustic emission source location in composite aircraft structures using modal analysis. Ph.D. thesis, University of Glamorgan, 163 p. (2011)
5. Cardoni, M., Giglio, M.: A low frequency lamb-waves based structural health monitoring of an aeronautical carbon fiber reinforced polymer composite. J. Acoust. Emiss. **32**, 1–20 (2014)
6. Sause Markus, G.R.: Acoustic emission signal propagation in damaged composite structures. J. Acoust. Emiss. **31**, 1–18 (2013)
7. Gorman, M.: Modal AE analysis of fracture and failure in composite materials, and quality and life of high composite pressure vessels. J. Acoust. Emiss. **29**, 1–28 (2011)
8. Stepanova, L.N., Chernova, V.V., Petrova, E.S., et al.: Acoustic-emission testing of failure in samples of CFRP exposed to static and heat loads. Russ. J. Nondestruct. Test. **54**(11), 748–756 (2018)
9. Stepanova, L.N., Ramazanov, I.S., Chernova, V.V.: A procedure for locating acoustic-emission signals during static testing of carbon composite samples. Russ. J. Nondestruct. Test. **51**(4), 227–235 (2015)
10. Oskouei, A.R., Ahmadi, M., Hajikhani, M.: Wavelet-based acoustic emission characterization of damage mechanism in composite materials under mode I delamination at different interfaces. Express Polym. Lett. **3**(12), 804–813 (2009)

Influence of Crack Propagation Parameters on Acoustic Emission Parameters During Low-Cycle Testing

Ludmila Stepanova, Aleksey Bobrov(✉), Sergey Bekher, and Maria Kuten

Siberian Transport University (STU), 191 Dusi Kovalchukst, 630049 Novosibirsk, Russia
{stepanova, beaver}@stu.ru

Abstract. This study is aimed at analyzing parameters of acoustic emission (AE) formed by propagating fatigue cracks during static tests of low-alloyed and low-carbon steels used for manufacture of numerous objects inspected by AE method. The purpose of the study is to make a comparative analysis of criteria that are used in AE inspection and characterize propagation of fatigue cracks.

Keywords: Acoustic emission · Crack propagation · Low-cycle testing

1 Introduction

Acoustic emission (AE) inspection method has an essential advantage over other nondestructive inspection methods which consists in recording defects that are just growing. This feature makes the method very important for detection of cracks and corrosion defects in technical objects, to which high requirements are imposed during their monitoring or service life extension. However, there is an evaluation criteria selection problem for rank-and-file specialists who make decision on technical condition of objects. Studies in this area [1–3] demonstrate that reliable evaluation of results on the basis of recorded AE data has certain probability and requires development of adequate algorithms to select objects condition evaluation criteria.

A fairly large number of papers are concerned with theoretical studies and experimental confirmation of connection between parameters of a propagating crack and AE parameters. The stress intensity factor (SIF) is most often selected as one of parameters characterizing the propagating crack, since this factor is universal with respect to the geometry of specimens under study and the type of loading. As shown by studies, connection between SIF and the total count of AE signals [4, 5], and between frequency characteristics and SIF [6, 7] is observed in certain cases. However, dynamic evaluation criteria [8] related to determining the increments of AE signals' parameters are often used in practice. Besides, frequency characteristics can be used when signals are received by broadband receiving transducers, though it is recommended to use narrow-banded receivers in the practical inspection of actual objects. Meanwhile, processing of signals will distort frequency parameters and, moreover, the latter will have influence on the amplitude and energetic parameters of recorded signals. Despite a

Z. Popovic et al. (Eds.): TransSiberia 2019, AISC 1115, pp. 885–893, 2020.
https://doi.org/10.1007/978-3-030-37916-2_87

rather large amount of research over the past years of AE method usage, searches for reliable and optimal criteria for identification of fatigue cracks still continue [11, 12].

2 Research Methods

The experiment was conducted using specimens whose shape is shown in Fig. 1. The specimens were made of St3ps and 20GFL steel blanks. Specimens had different thicknesses: 8 mm (tension test specimens), 9–12 mm (bending test specimens). Specimens from 20GFL cast steel, except an artificial stress concentrator, and also had real casting defects in the form of pits, pores, porosities, and segregations located randomly in the volume of each specimen. In total, 6 specimens were loaded in each group.

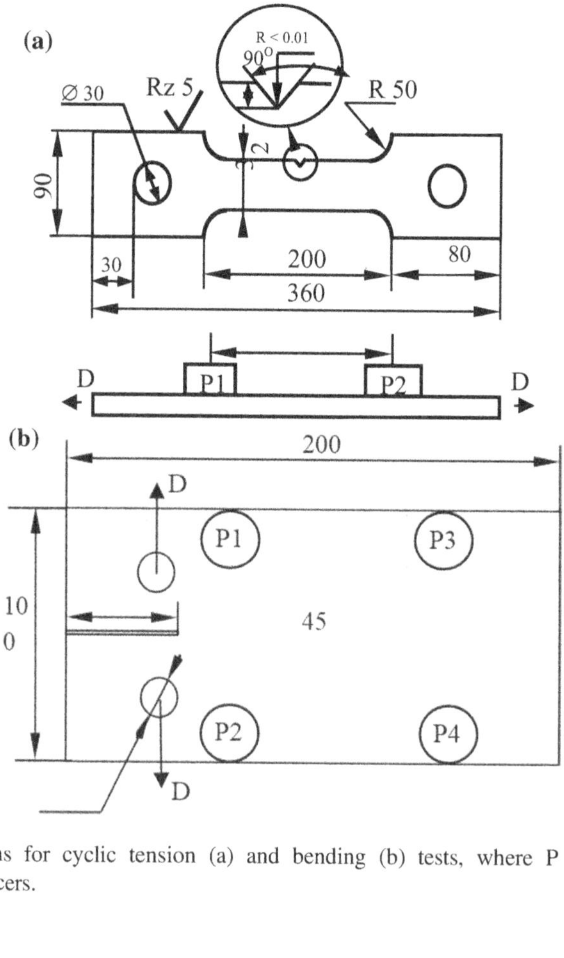

Fig. 1. Specimens for cyclic tension (a) and bending (b) tests, where P are piezoelectric receiving transducers.

The tests involved successively alternating cyclic and static loads applied to steel specimens being rolling stock molded pieces. The tests were conducted up to complete destruction of specimens, with recording of AE information at each stage of the tests. Some results are provided in papers [9]. A general photograph of one of the specimens installed in a testing machine with transducers is provided in Fig. 4.21.

Successively alternating cyclic and static loads were applied to specimens by means of tension. The cyclic load application frequency was 5 Hz. The number of cycles was 7,500…30,000 per test at the initial stages of cyclic loading, following which static loading was conducted at the rate of 0.5…1.0 kN/s. Amplitude value of tensile force during cyclic tests was 9.0–10.0 kN for the tension test specimens and 25 kN for the bending test specimens, respectively. Then, the tension test specimens were loaded to destruction. During bending tests, the maximum force of each static loading was 30 kN (i.e. complete simulation of load upon an object under inspection, when testing load is 1.25 times higher than dynamic load).

Static tests were conducted after a certain number of loading cycles that was varied from specimen to specimen and during loading of each specimen. Such pattern allows simulating a real loaded condition of a piece, when the distances between examination points are different, which results in an unequal number of loading cycles between the time intervals when railway car parts can be inspected. This method makes it possible to reveal the most generalized trends of AE parameters from propagating cracks at stress concentrators under simulated operational conditions.

When a crack was growing at a concentrator and the destruction point came nearer, the number of cycles between static tests was reduced to 2,000…8,000, so that the dynamics of variation of AE parameters in the developed crack when approaching the destruction point could be analyzed. AE signals were recorded using STAD 16.03 system at all testing stages, during both cyclic and static loading, by means of four receiving bandpass transducers located in such a way that signals unrelated to the crack propagation could be easily filtered out. Sensitivity of the channels was calibrated by means of an electronic simulator, where a signal with an amplitude of 1 V was alternatively supplied to each channel. Signals were recorded using receiving bandpass transducers operating at frequencies of 100–700 kHz that recorded signals with an amplitude of more than 20 μV, without regard to the gain factor. The value of the total gain factor was 1000.

In the course of the inspection, wave packets of AE events were recorded by each receiving transducer, following which the coordinates of sources were determined according to the procedure provided in [10]. Localized signals received from the deforming area at the tip of the propagating crack had an amplitude of at least 100 μV (without regard to the gain factor).

The stress intensity factor (SIF) was also calculated for the propagating crack in accordance with the procedure provided in the national standard.

3 Research Results

The experiment has demonstrated that equipment allows registering AE at the initial stage of a fatigue crack formation only at an elevated sensitivity level, when a large amount of noise is recorded. If this is the case, a cluster analysis should be used for identification of sources. Fatigue crack propagation gives a sustainable AE.

Throughout the testing, a fatigue crack formed and propagated at the concentrator of each specimen. The dynamics of increments in the length of cracks for three specimens of different thicknesses *h* is shown in Fig. 2.

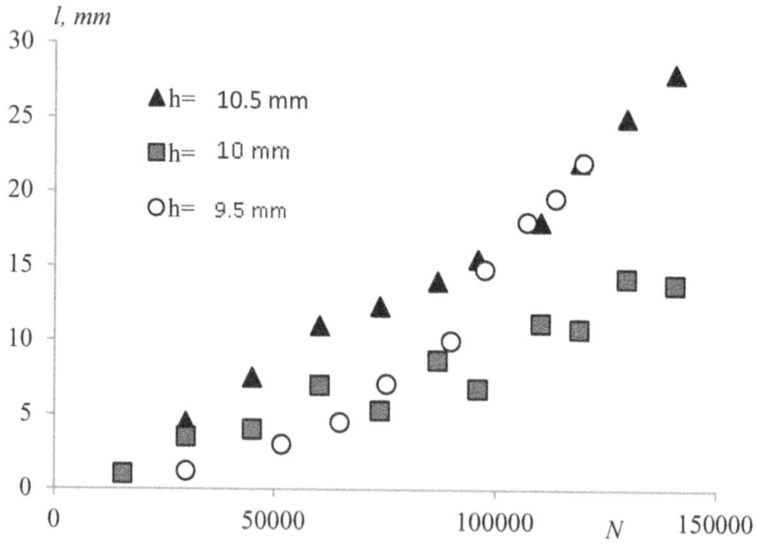

Fig. 2. Variation in the length of a fatigue crack being grown, upon increase in the number of loading cycles.

It should be noted that, having regard to the error for determination of coordinates of the crack development area as an AE source, the reference area of this source was continuously growing as the crack size increased. This is due to the fact that local sources forming individual AE events reside in a permanently expanding high-stress area around the crack tip as the length of this crack increases. An example of AE source area expansion would be results of location of events at the same specimen upon increase in the number of loading cycles, as shown in Fig. 3. With the fatigue crack length of 7.5 mm at a concentrator, the AE events coordinate area was approximately 300 mm^2 (Fig. 3a), and with the crack length of 22 mm, the AE events coordinate area was approximately 1,200 mm^2 (Fig. 3b).

a)

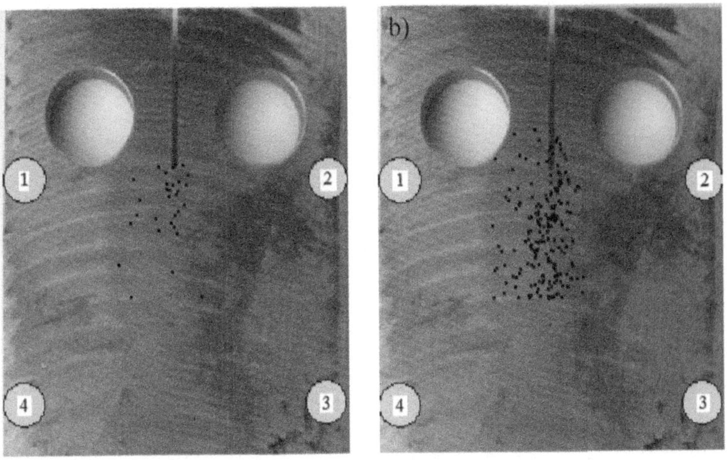

Fig. 3. Location of AE signals from a fatigue crack in specimen No. 2 after 45,000 and 119,000 loading cycles (destruction occurred after 140,500 cycles).

Growth of the crack length and SIF is accompanied by increase in the total number of recorded AE events and conditional total energy of these events for each subsequent static loading after a series of cyclic loads (Fig. 4a).

However, in some cases, certain elevations or reductions of these parameters were observed on molded pieces (Fig. 4b). Such variation in AE parameter is due to a strongly heterogeneous structure of cast steel, as was shown in paper [13]. Besides, indications of both the total events count and the total energy and, consequently, of all other derivative parameters for different specimens vary by more than an order of magnitude (as can be seen in Fig. 4), and this suggests an unacceptably low accuracy when this data is used to evaluate the condition of a growing defect.

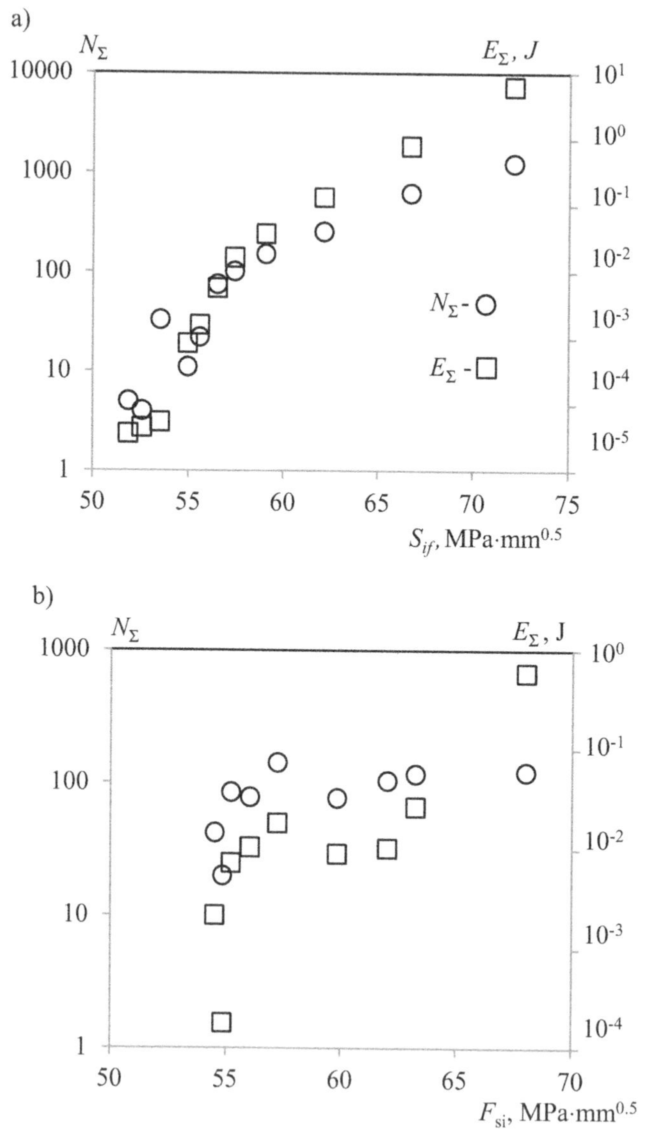

Fig. 4. Values of the total number of AE events and the total energy for cast steel specimens: (a) is specimen No. 2; (b) is specimen No. 3.

4 Results and Discussion

The obtained results (Fig. 4) demonstrate the possibility of inspecting specific objects with known record history by the main AE parameters, i.e. by the total number of AE events and the total energy of these signals during static tests. This allows not only detection of propagating cracks, but in conjunction with the results of previous studies

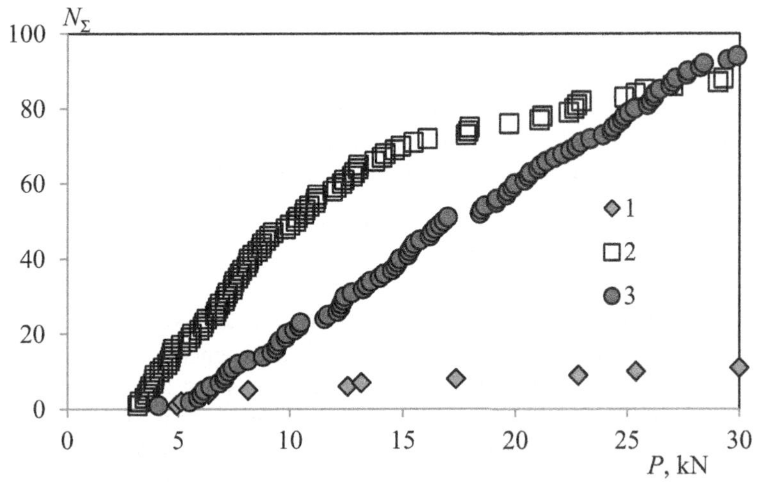

Fig. 5. Dynamics of growth of the total number of AE events upon increase in load: 1 – F_{si} = 52.6 MPa·mm$^{0.5}$; 2 – F_{si} = 57.4 MPa·mm$^{0.5}$; 3 – F_{si} = 59.1 MPa·mm$^{0.5}$.

determination of the crack size and prediction of its behavior. However, these results cannot be extended to a series of objects due to a wide scatter of data for each particular specimen with a growing defect.

Analysis of the signals arrival time shows that incoming signals are uniformly distributed during the entire loading, and there are no fundamental differences between the source activity at the initial stage and the one under increased loads (Fig. 5) and, moreover, the signal source activity can even decrease when approaching to the maximum load as compared to the initial stage.

In order to conduct a more detailed analysis of the AE parameter variation dynamics, the evaluation parameters of AE generated by a penetrating crack, that are traditionally used to assess the hazard of source, were studied.

Note that, from the viewpoint of inspection of real objects whose defect growth history before AE tests is, as a rule, unknown, derivative parameters can be of interest, first of all, for example, the activity that was determined from the formula:

$$\dot{N}_i = \frac{N_i}{t_i},$$

where N_i is the number of signals during time interval t_i. If the loading time interval is unstable, the value of the loading measurement unit can be used in the denominator. The obtained parameters can be considered differential activity of the source. Besides, average power of AE signals over a loading time interval or a normalized interval of load increment was determined:

$$W_i = \frac{E_{\Sigma i}}{F_i},$$

where $E_{\Sigma i}$ is the total energy of AE signals over load increment interval F_i. In addition, such parameters as the average amplitude of signals during a single test and source concentration factor C_k were evaluated from the formula:

$$C_k = \frac{N_{\Sigma k}}{S_k},$$

where $N_{\Sigma k}$ is the total number of AE events from the *k-th* source during testing; S_k is the AE source area including all coordinates of individual events forming the source.

Analysis of variations in the above parameters shows that a rigid correlation with the crack size and SIF for different specimens is only demonstrated by the total number of events N_{Σ}, the total number of events during a time interval of holding at the maximum load that were recorded from source N_{Σ}, and the source area. These parameters grow with increasing SIF and decreasing number of cycles to destruction (Figs. 3 and 6), except when the crack tip area falls on the casting discontinuity region. In these cases, both the number of signals and their total energy grow. As this takes place, the amplitude, integral and local dynamic criteria behave unstably, in general. However, some of said criteria have a significant coefficient of correlation with SIF for certain specimens.

References

1. Bykov, S.P., Yuraido, B.F., Ivanov, V.I.: Concerning reliability of acoustic emission control. Testing Diagn. **12**, 53–60 (2013). (in Russian)
2. Builo, S.I.: On the information value of the method of invariants of acoustic-emission signals in the diagnostics of pre-failure state in materials. Russ. J. Nondestr. Test. **54**(4), 237–242 (2018)
3. Vinogradov, A.Yu., Merson, D.L.: Nature of acoustic emission during deformation processes in metal materials. Low Temp. Phys. **44**(9), 1186–1195 (2018). (in Russian)
4. Holt, J., Goddard, D.J.: Acoustic emission during the elastic-plastic deformation of low alloy reactor pressure vessel steels II: deformation around a crack. Mater. Sci. Eng. **44**, 267–277 (1980)
5. Builo, S.I.: Association of propagating crack acoustic emission parameters with stress intensity factor and stress condition type. Defectoscopy **3**, 44–48 (2006). (in Russian)
6. Bigus, G.A., Travkin, A.A.: Definition of flaw-detective markers of fatigue crack detection by the acoustic emission method in specimens made of steel 20 and having a cast structure. Defectoscopy **1**, 40–47 (2015). (in Russian)
7. Seryeznov, A.N., Muravyev, V.V., Stepanova, L.N., et al.: Experimental determination of association between the AE signal spectrum and the fatigue crack length in steel specimens. Defectoscopy **2**, 73–78 (1999). (in Russian)
8. Ohtsu, M., Uchida, M., Okamato, T., Yuyama, S.: Damage assessment of reinforced concrete beams qualified by acoustic emission. ACI Struct. J. **99**(4), 411–417 (2002)

9. Stepanova, L.N., Bovrov, A.L., Kanifadin, K.V., Chernova, V.V.: Studying the main parameters of acoustic emission signals during static and cyclic testing of specimens made of steel 20GL. Deformation Fract. Mater. **6**, 41–45 (2014). (in Russian)
10. Stepanova, L.N., Lebedev, E.Yu., Kabanov, S.I.: Localization of acoustic emission signals in strength testing of engineered structures using a piezoelectric antenna of arbitrary shape. Russ. J. Nondestr. Test. **35**(9), 682–688 (1999)
11. Kachanov, V.K., Sokolov, I.V., Matyunin, V.M., Barat, V.A., Bardakov, V.V., Marchenkov, A.Y.: Evaluation of the fracture toughness of titanium nitride hardening coatings according to kinetic indentation and acoustic emission parameters. Measur. Tech. **60**(7), 706–710 (2017)
12. Chernov, D.V., Matyunin, V.M., Barat, V.A., Marchenkov, A.Y., Elizarov, S.V.: Studying mechanisms of acoustic emission during development of fatigue cracks in low-carbon steels. Defectoscopy **9**, 21–30 (2018). (in Russian)
13. Stepanova, L.N., Bekher, S.A., Boyarkin, E.V., Moreva, N.A.: A study of specimens from solebar material with boxlike cross sections using the acoustic-emission method. Russ. J. Nondestr. Test. **49**(4), 215–224 (2013)

Strength Characteristics Analysis of Structurally Inhomogeneous Steel Gears

Alexey Beskopylny[1(✉)], Besarion Meskhi[1], Nikolay Onishkov[1], and Victor Korotkin[2]

[1] Don State Technical University, Sq. Gagarin, 1, 344010 Rostov-on-Don, Russia
besk-an@yandex.ru
[2] Southern Federal University, Bolshaya Sadovaya Street 105/42, 344006 Rostov-on-Don, Russia

Abstract. The article is devoted to clarifying the methods for assessing the strength of thermally and thermochemical steel parts treatment, which are characterized by a structural heterogeneity of the surface layers with varying levels of resistance to normal and tangential stresses. The lack of knowledge of these factors leads to the need to introduce into the design regulations a significant number of correction factors obtained experimentally, as well as large safety factors. During operation, deep and surface destruction processes with fundamentally different fatigue life mechanisms occur simultaneously in any machine part. A complex of factors determines the leading type of destruction, the main of which are the level of tension and properties of the surface layers of the material. However, the near-surface layers and the core are zones fundamentally different in the energetic and structural state. This fact is especially clearly manifested in the surface hardening of the part when intercrystallite fracture mechanisms prevail in the surface layers, and transcrystalline mechanisms occur in the core. The article presents ways to increase the reliability of predicting the level of load capacity of thermally and thermochemical gears treatment under the conditions of the alternative forms of failure.

Keywords: Strength · Inhomogeneous · Steel gear · Thermochemical treatment · Carbonitriding

1 Introduction

The general trend of modern mechanical engineering is to increase reliability, energy intensity, and power-weight capacity of the mechanisms. This fact fully applies to gears. These are drive elements operating in the most general case of relative motion (rolling, sliding, spinning) under the action of bending stresses varying in the zero-to-tensile stress cycle, that could be compared with the time of the impact process even for medium speed gears. Thus, gears are extremely widely used in all areas of technology, determining in many cases, the most important technical, economic, and operational characteristics of the respective machines and devices. The global market for the production and consumption of transmissions reached about $100 billion by the

Z. Popovic et al. (Eds.): TransSiberia 2019, AISC 1115, pp. 894–903, 2020.
https://doi.org/10.1007/978-3-030-37916-2_88

beginning of the 21st century. The four most significant areas of the market for gears include general engineering (including machine tools, materials handling, and other industries), aerospace, shipbuilding, and especially the automotive industry. Industry in this sphere is constantly releasing new installations in which gears must withstand ever-increasing stresses, strains, and temperatures. In recent years, advanced methods of manufacturing, finishing and improving surface properties for solving these problems have appeared [1, 2].

Strength can be estimated by the ratio of the loads acting on the structural elements and the loads causing failure, i.e., a state in which a structure does not meet operational requirements. It can be expressed by the inequality $\Sigma Pi \leq F$, where ΣPi is the structure force of the sum of the design loads in the most unfavorable combination, and F is the bearing capacity of the structure, which is a function of the geometric dimensions of the structure and material mechanical properties. Moreover, it is possible to increase the bearing strength of the product either by lowering the effective stresses or by increasing the allowable ones. These are two separate directions. It is possible to reduce the effective stresses by changing the design of the product, increasing the accuracy of manufacturing its elements, etc. Increases allowed can be achieved by appropriate technological processing of the material. However, it should be taken in mind that the requirements for the properties of different zones of the material of the structural parts can be significantly different [3, 4].

The overwhelming part of the destruction of machine parts begins from the surface and in the surface layers of the part [5, 6]. This fact is typical since the surface layers of the parts are loaded more intensively than the internal volumes during operation [7], and they perceive all the actual loads with their surfaces. In addition, any part after manufacture has fundamentally different zones in the energy and structural state - the surface layer and the core. At the same time, the material of the surface layer (without additional hardening) has a lower strength relative to the core due to the appearance of various surface defects already at the manufacturing stage. However, in some cases, for example, if the element works under bending conditions, a solid surface must be combined with a viscous core, although less durable.

One of the priority tasks is the development of technologies for obtaining the mechanical properties of the surfaces of the teeth, providing the necessary strength and antifriction characteristics and methods for calculating gears with modified hardened surface layers. One of the main technological methods to achieve the desired effect for steel is thermal and chemical-thermal treatment.

2 Materials and Methods

The required performance properties are achieved by a combination of a complex of factors such as the mechanical characteristics of both the near-surface zone and the core, the structural and phase state of the hardened layer, and its thickness. Various types of mechanical processing and chemical heat treatment are used depending on the requirements of a particular product: the level of load, steel grade, operating conditions, the environment aggressiveness, etc. At present, the quality of thermally and chemically thermally hardened parts (in particular, gears) is estimated by the surface hardness

(H_0), core (H_C), and thickness of the hardened layer (h_t). Regarding the thickness of the layer, it should be noted that due to its "blurriness", and because of the wide permissible interval of H_C (29–43) HRC, the so-called the effective thickness (h_{te}) of the hardened layer is regulated. h_{te} is the distance along the normal from the surface to the depth with some predetermined level of effective hardness (H_e), carbon consistent or of a certain structure. H_e = 550 HV is most often used for cementation and carbonitriding. When nitriding H_e = 400 HV or $H_e = H_C + 50$ HV is appropriate. During surface hardening, the effective layer thickness is defined as the distance to the zone with a semi-martensitic structure, the hardness of which depends on the carbon content, i.e. from a particular steel grade.

The most important and most accessible to control the characteristic strength of steel is its hardness. Accordingly, the quality of the hardened layer is largely evaluated by the distribution of hardness over the thickness of this layer (Fig. 1). It is assumed that the effective thickness almost uniquely determines the strength of the layer.

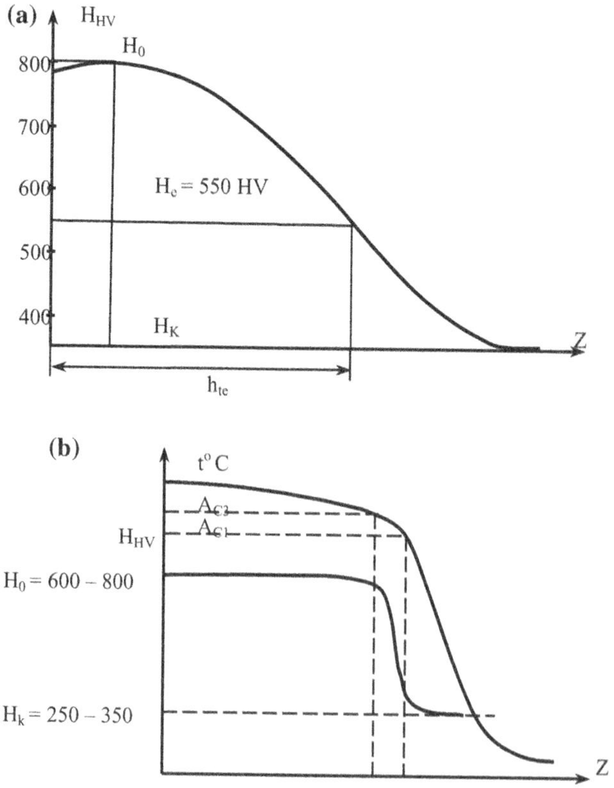

Fig. 1. The hardness distribution over the thickness of the hardened layer: (a) during cementation and carbonitriding; (b) in case of surface hardening by high frequency currents.

Therefore, such a wide range of HC values is allowed. This indicator is correct for traditional involute gears of general engineering, but in gears of on-board gearboxes of heavy tractors and mining machines, as well as in Novikov gears, destructions often occur with initial zone in the sublayer (Fig. 2).

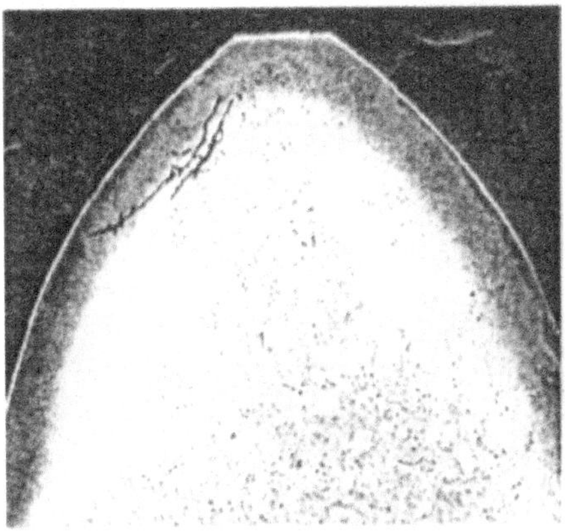

Fig. 2. Microsection of cemented Novikov gear with sublayer crack.

Even more dangerous is the occurrence of a primary focus of deep destruction during surface hardening, with a transition zone (region heated to temperatures of incomplete austenitic transformation), characterized by significant structural heterogeneity and a sharp decrease in hardness. The size of this zone is determined both by the chemical composition of the steel and by the mode of heat treatment. Figure 3 shows the distribution curves of the hardness of reduced-hardenability steel 55PP after quenching from deep induction heating, steel 50 after the "ordinary" quenching of high frequency and steel 18HGT (0.2% C, 1% Cr, 1% Mn) after cementation. All curves have the same reference point H = HV 550 at a depth of 1.9 mm. Curve 4 corresponds to the minimum values of the hardness of the transition zone in terms of the depth contact strength for Novikov transmission following GOST 30224-96, $m = 8$ mm, $\sigma_{Hmax} = 1700$ MPa, $h_{te} = 0.2375$ m. Obviously, in this case, the transmission capacity of the transfer by the criterion of the deep contact strength during quenching is not provided. Evaluation of the mechanical characteristics of inhomogeneous materials on the example of welded joints in [8], layered structures [7], bimodulus materials.

High hardness and wear resistance of the surface zone is achieved by nitriding. The temperature in the process of nitriding does not exceed 560… 600 °C, which does not cause phase recrystallization. Thermal deformations and surface roughness are such that they allow you to abandon the finishing operations. However, the limiting factor is the impossibility of obtaining diffusion layers with ht >0.6… 0.7 mm, which limits the area of its effective use by transfers of a small module, since load capacity will be determined by the strength of the core (especially in gears with theoretically point contact). However, overestimated the thickness of the hardened layer is no less dangerous. For traditional involute transmissions, the regulations of hardened (cemented,

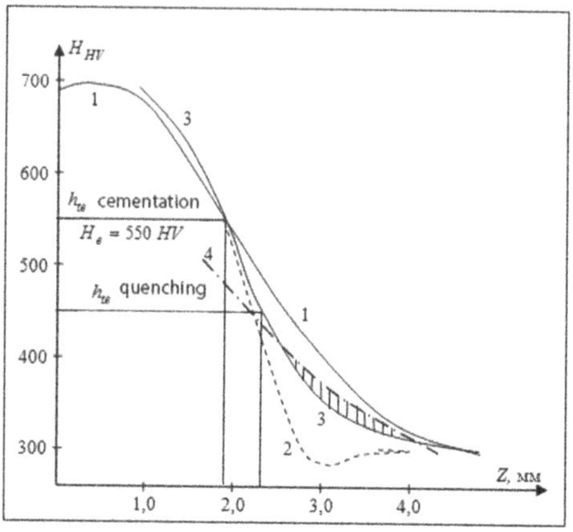

Fig. 3. The distribution of hardness over the depth of the hardened layer in various types of heat treatment and chemical-heat treatment: (1) cementation; (2) high-frequency currents quenching; (3) deep heating quenching; (4) minimum hardness of the layer according to the terms of deep contact strength.

carbonitriding) layers are based on extensive experimental data, even a slight deviation from which leads to catastrophic consequences (Fig. 4), which naturally requires increased attention to such transfers.

(a) (b)

Fig. 4. The destruction of thermochemical gears treatment: (a) pitting, breaking out of the teeth, face chipping due to the reduced thickness of the hardened layer: 0.7 mm instead of 1.4 mm; (b) chipping on the top of the teeth, fatigue crack at the base due to the increased layer thickness by 0.2 mm.

3 Results and Discussion

At present, piecewise linear dependences of the evaluation of the mechanical characteristics of steel parts on hardness are used in engineering practice. For example, the contact endurance limit of involute wheels: $\sigma_{H\,\lim} = 2.3\,\mathrm{HB}$ according to ISO 3663 or $\sigma_{H\,\lim} = 23\,\mathrm{HRC}$ according to GOST 21354-87. These dependencies, in general, correspond to the averaged values of the contact endurance limits. However, recommendations for cases of cementation and nitro-carbonizing are extended to the hardness interval (56… 65) HRC, while a number of standards (standards ASME, ISO) very high weight coefficients of hardness influence limit the level (58… 60), indicating that exceeding it gives contradictory results. This fact is consistent with the graphs of changes in the mechanical characteristics of hardened carbon steels (Fig. 5).

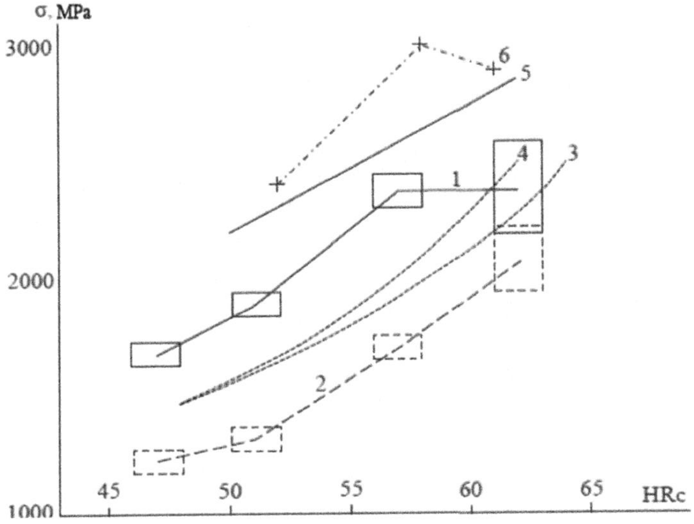

Fig. 5. Change of mechanical properties of hardened carbon steels depending on hardness: 1, 2 – σ_u and $\sigma_{0.5}$ steel ShH-15; 3 - σ_{Hlim}; 4 - $\sigma_{0.5}$ - with residual austenite up to 5%; 5 - σ_{cr} - allowable stresses on the crumbling of structural steels; 6 - σ_u- steel ShH-15.

Contradictions are obvious. So, with the number of loading cycles N = 10^7 and hardness 60 HRC, the limit of contact endurance $\sigma_{H\lim} = 2100\,\text{MPa}$, although even involute gears with ground teeth working in a favorable elastic-hydrodynamic lubrication mode are affected by progressive pitting at stresses of more than 1500… 1600 MPa already in the operation of the early stages, although alternative types of failure may be leading.

Table 1 shows the parameters, and Fig. 6 shows the test results of cemented rollers. Over 70 specimens of the C-30, C-60, C-120 series with diameters of 30, 60 and 120 mm, respectively, strengthened to different depths at H_e = 550 HV, were tested in the framework of the cited experiment.

Table 1. Specimens parameters.

Specimen	H_o (HV)	Number of specimens	h_{te}, mm	h_t, mm	Depth of destruction z_{max}, mm
CA-120	840	7	0.86	2.50	0.90–1.46
CB-120	860	8	1.60	3.00	0.32–0.97
CC-120	863	8	3.03	5.60	0.40–0.95
CD-120	910	8	5.16	7.50	0.40–0.84
CC-60	838	6	2.30	3.90	0.16-0.60

A sharp change in the exponent m of Wehler's lines draws attention in Fig. 6b: for the SS-120, CD-120 series, this indicator was at the level of m = 8… 10, typical for surface strength, whereas for CA-120, like all others, the indicator m = 18… 24, characteristic of fatigue life by the criterion of deep contact strength. The region of

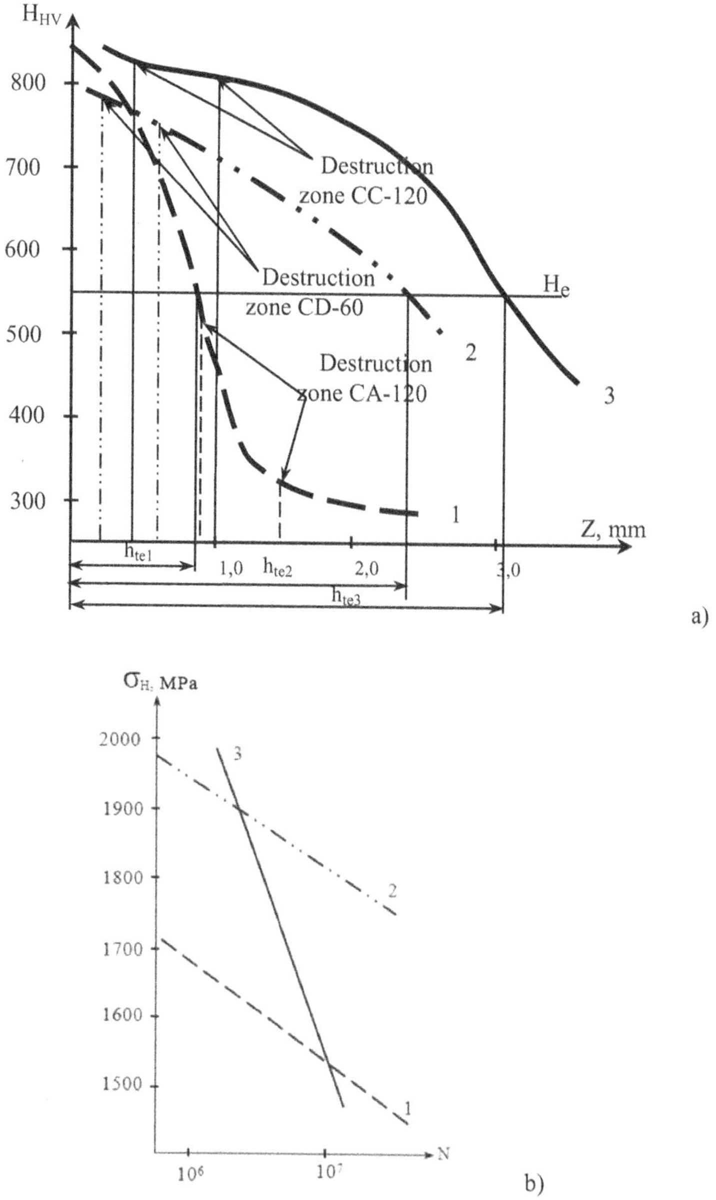

Fig. 6. Cemented Roller Testing: (1) CA-120; (2) CD-60; (3) CC-120. (a) the distribution of hardness over the thickness of the hardened layer and the area of destruction; (b) contact fatigue lifelines.

origin of the fatigue crack of samples CA-120 is a sublayer; samples CC-120, CD-120 is the effective zone; samples CB-120 is an initiation of cracks at two levels with advanced development (in 5 cases out of 7) in the effective zone. However, with calculated durability of N ≈ 107 cycles, the load capacity of samples SS-120, DM-120 is destruction in the effective zone with H > 800 HV is at the level of samples CA-120 with the crack initiation area in the sublayer with H ≈ 350 NV.

The authors carried out bench tests of nitro-carburizing gears with Novikov gearing and metallographic studies of destroyed specimens. With the calculated torques on the drive shaft, T1 = 1500… 2000 Nm, the failures had the character of chipping and breaking of the teeth.

Fractures of teeth are fatigue, destruction developed from several surface lesions. The zones of slow and accelerated crack development differ quite clearly. On sample No. 3 (Fig. 7a), there are folds on the side of the working surface - the result of the merging of microcracks. The zone of the outbreak is a smoothed area, which smoothly passes into the areas of selective and accelerated crack development. The presence of coarse scars characterizes the latter (located on the left). On the right, at the end, in the form of a step, there is a zone of dough of a fibrous structure.

However, if in the samples of gear 3 a pattern of predominantly viscous fracture is observed, then in the samples of gear 8 (Fig. 7b) the surface of the fracture is flatter. The zones characteristics of the fatigue fracture are mild; the site of rapid crack development is velvety with a thin stream structure, more typical of brittle fracture.

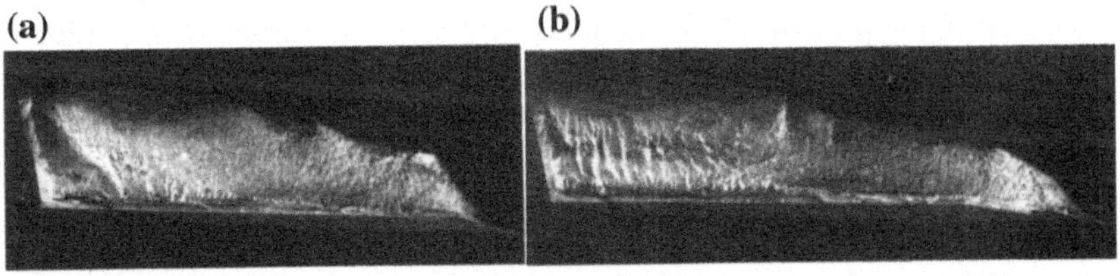

Fig. 7. Teeth fracture of nitro-carburizing gears: (a) specimen 3; (b) specimen 8.

The microstructures of the hardened layer and the core of all types do not have significant differences. Troostite-martensitic structure (defects, dining areas of the lung body), almost the same 0.10… 0.18 mm. Further differences are noted: in the gear layer from 8 to a depth of z ≈ 0.40 mm, the average needle martensite with a needle length of up to 8 μm (point 5) and residual austenite (point 4); In general, the microstructure with a hardened layer and cores with martensite, residual austenite, dark componen, and nitro-carbonizing corresponds to OST23.4.52-83. The microstructure of the transition zone is bainite-martensitic, the core is bainite; ferrite is missing.

The microstructures of the hardened layer and the core of all samples do not have significant differences. Troostite martensitic structure (defective, due to depletion of the solid solution of the surface zone by the alloying elements) is almost the same - 0.10… 0.18 mm. Further differences are noted: in the gear layer 8 to a depth of z ≈ 0.40 mm, the medium needle martensite with a needle length of up to 8 μm (point 5) and residual austenite (point 4); in the layer of gear 3 to the same depth - predominantly fine-needle martensite (needle length decreases to 4 μm, the amount of residual austenite is 3). In general, the microstructure of the hardened layer and core according to martensite, residual austenite, dark component, and nitro-carbonizing corresponds to OST 23.4.52-83. The microstructure of the transition zone is bainite-martensitic, the core is bainite; ferrite is absent.

When testing the unpolished transmissions of the DT-75 tractor: $a_w = 276.25$ mm, $m = 6.5$ mm, $z_1 = 13$, $z_2 = 65$, $\sigma_H = 1750$ MPa and $n_1 = 220$ min^{-1}, the cemented gears failed after 200 h due to the deep contact damage, extending to a depth of 1.4 mm. The same transmissions, but nitro-carburized, even at a lesser hardness at the indicated depths, failed due to pitting with an operating time of more than 700 h.

4 Conclusions

For viscous hypoeutectoid structures of the underlayer and the core, the fracture is typically transcrystalline, much more energy-intensive than the intercrystalline (intergranular), that is typical of the eutectoid and eutectoid structures of the properly hardened layer. The durability of rolling bearings after detecting a primary crack did not exceed 3% of the total resource, whereas the occurrence of primary defects in the sublayer of cemented rollers accounted for the first (10… 15)% of their life (until the crack reaches the surface). However, if the results of CA-120 and SS-120 rollers can be explained by different rates of propagation of cracks in areas with different plastic properties, then samples of SS-60 samples of SS-60, like SS-120, CD-120, failed according to the deep contact strength criterion with $m \approx 20$.

The dependences of the hardness distribution do not reflect the true distribution of a particular sample, being more often statistical. Despite the fact that the gears, which were discussed above, were made of steel of the same supply and chemical-heat treatment took place simultaneously (14 samples from 5 gears were investigated), the spread of hardness values reached $\Delta H \approx 150$ HV_1 within one tooth and $\Delta H \approx 250$ HV_1 for similar points of different samples. Except for CA-120 samples, the primary zones of deep contact damage were located at depths with a hardness of H > (725… 750) HV. The zones of delayed and accelerated development of cracks on the gear teeth also corresponded to the indicated interval with an insignificant increase in hardness in (50-70) HV in cases with a predominance of brittle fracture elements.

The exact ratio of the numbers of hardness according to Vickers and other scales does not exist, approximately (700… 860) HV ≈ (58… 64) HRC. A small amount of experimental data is insufficient for unambiguous conclusions, but it serves as a confirmation of the well-known, but not reflected in most calculation procedures, fact that to assess the load capacity of steel parts thermally or chemically thermally strengthened to high hardness, information about the magnitude of this hardness is completely insufficient.

Acknowledgements. The work was financially supported by Russian Foundation for Base Research (project 18-01-00715-a).

References

1. Gupta, K., Jain, N.K., Laubscher, R.: Advanced Gear Manufacturing and Finishing, p. 240. Academic Press, Cambridge (2017)
2. Beskopylny, A., Onishkov, N., Korotkin, V.: Adv. Tribol. (2018). https://doi.org/10.1155/2018/9607092
3. Chen, Z., Zhai, W., Wang, K.: Mech. Syst. Signal Process. **115**, 29–44 (2019). https://doi.org/10.1016/j.ymssp.2018.05.038
4. Hoffensona, S., Frischknecht, B., Papalambros, P.: Accid. Anal. Prev. **50**, 943–954 (2013). https://doi.org/10.1016/j.aap.2012.07.021
5. Beskopylny, A., Onishkov, N., Korotkin, V.: Adv. Intell. Syst. Comput. **692**, 184–191 (2017)
6. Beskopylny, A., Onishkov, N., Korotkin, V.: MATEC Web Conf. **224**, 0033 (2018)
7. Belen'kii, M., Beskopyl'nyi, A., Vernezi, N., Shamraev, L.: Welding Int. **11**(8), 642–645 (1997)
8. Afanasenko, V., Boev, E., Kulakov, P.: IOP Conf. Ser. Mater. Sci. Eng. **560**, 012003 (2019). https://doi.org/10.1088/1757-899x/560/1/012003

Limit-State Criteria and Their Use in Conditions of the Significant Structural Heterogeneity of the Gear Steel

Alexey Beskopylny[1(✉)], Besarion Meskhi[1], Nikolay Onishkov[1], and Victor Korotkin[2]

[1] Don State Technical University, sq. Gagarin1, Rostov-on-Don 344010, Russia
besk-an@yandex.ru
[2] Southern Federal University, Bolshaya Sadovaya Str. 105/42, Rostov-on-Don 344006, Russia

Abstract. The article discusses the problems of methods for assessing the strength of machine parts made of steel subjected to thermal and thermo-chemical processing, which are characterized by structural heterogeneity of surface layers with different levels of resistance to normal and tangential stresses. Insufficient knowledge of these processes leads to the need to use empirical calculations, a significant number of correction factors obtained experimentally. The processes of surface destruction and subsurface nucleation of cracks occur simultaneously in heavily loaded machine parts. Increasing the surface hardness often does not lead to the desired effect, since defects originate under the surface. Under these conditions, the load capacity of the material of the layer is investigated based on the generalized Pisarenko-Lebedev criterion of the limiting state of the structurally inhomogeneous material taking into account changes in its plastic properties as a result of surface hardening. The article presents ways to increase the carrying capacity of thermally and thermochemical machining of gear wheels in terms of alternative forms of destruction to increase the reliability of predicting their resource.

Keywords: Strength · Pisarenko-Lebedev criterion · Thermochemical processing · Steel gear · Subsurface crack · Inhomogeneous

1 Introduction

For most materials, the onset of the limit state is due to their ability to resist both normal and tangential stresses, and the relative effect, i.e., the "personal contribution" of each of these factors depends on many cases and is even different for the same material. So, for steels with hardness H < HB (450… 500), the ratio $\sigma_{u+} \approx 0.34$ HB is close to linear, and the fracture is viscous. However, a further increase in hardness (in order to increase the mechanical characteristics of the steel) leads to a violation of this ratio: $\sigma_{u+} \neq \sigma_{u-}; \sigma_{Y+} \approx \sigma_{u+}; \sigma_{Y-} \ll \sigma_{u-}$. The fracture will be brittle or mixed depending on the "rigidity" of loading (brittle under tension and torsion, mixed - under compression), which is typical of hardened tool and bearing steels. The danger of brittle fracture is not disputed, as in ISO 6336 [1], the upper limit of tooth hardness is limited

Z. Popovic et al. (Eds.): TransSiberia 2019, AISC 1115, pp. 904–912, 2020.
https://doi.org/10.1007/978-3-030-37916-2_89

to the level (62 … 63) of the H_{HRC}, however, the prediction of this type of failure is mostly qualitative.

In these circumstances, the hardness ceases to be a sufficient characteristic of strength, it is necessary to take into account other indicators, first of all, plastic, as well as possible structural defects, the probability of which increases, in particular, as a result of thermal or chemical-thermal treatment. This fact leads to a graduated microstructure and properties, which makes fracture prediction particularly difficult: a fragile outer layer, a plastic core material [2]. To assess the plastic properties of the material used so-called plasticity parameter χ, which takes into account the degree of shear deformation influence on the material microfracture. $\chi = \sigma_{e+}/\sigma_{e-}$, where σ_{e+}, σ_{e-} are the failure stresses of the material under uniaxial tension and compression, respectively.

Most of the methods used for the strength calculation of machine parts are based on extensive experimental material, and the criteria for using plasticity hypotheses (most often, maximum tangential and octahedral stresses). The most important characteristic of the strength of steel is its hardness [3]. This indicator is correct for normalized and improved steels, but not sufficient for hardened ones, especially if the crack originates in the root of the tooth [4]. With equal hardness, thermally or thermochemical treatment of steel alloyed with nickel shows higher strength and durability than its counterpart, which does not contain nickel. The reason is its increased plasticity. Thermal or chemical-thermal consolidation increases the hardness of steel while reducing its ductility. These two opposite effects act in parallel, and up to a certain level, due to both the chemical composition of the material and the nature of the stress-strain state (the so-called stiffness of loading), the positive influence of hardness prevails. For rolling bearings operating under three-dimensional compression, the optimum hardness of steel is 62-64HRC (steel ShH-15 C 1%, Si 0.2%, Mn 0.3%, Ni 0.3%, Cr 1.5%, Cu 0.25%), and for structural steel it is 48-50HRC (steel 40X C 0.4%, Si 0.2%, Mn 0.6%, Ni 0.3%, Cr 1.0%, Cu 0.25%). For gears (according to GOST 21354-87, the dependence $\sigma_{Hlimb} = 23H_{HRC}$ cementation, carbonitriding) is extended to a hardness range of 56 … 65 HRC. Meanwhile, several standards (ASME standards, ISO, recommendations) the very high weight coefficients of the effect of hardness limit the level to 58 … 60 HRC, indicating that with further increase in hardness, the experimental data are contradictory. The graphs of the change in the mechanical characteristics of hardened carbon steels can favour as confirmation.

2 Materials and Methods

The refinement of the quantitative estimates of strength is related to the plastic properties of the material and their change during thermal and thermochemical treatment. Under such conditions, the most promising is the use of generalized criteria for the limiting state, which takes into account the influence of both tangential and normal stresses. The first of these criteria was the Coulomb-Mohr criterion: $\sigma_e = \sigma_1 - \chi\sigma_3$, which satisfactorily determines the limiting state of sufficiently homogeneous materials with different resistance to stretching and compression. At present, several criteria have been developed for this direction: Yagna-Buzhinsky, Drucker-Prager, Johnson–Cook

[5] and others [6]. Of these, in our opinion, the most promising is the Lebedev – Pisarenko criterion for a structurally inhomogeneous material, used in [7–9], associated with predicting the occurrence of deep contact damage (DCD) of Novikov surface hardened gears, for which well-known recommendations for involute gear proved unacceptable:

$$\sigma_e = \chi\sigma_i + (1-\chi)\sigma_1 A^{1-(\sigma_1+\sigma_2+\sigma_3)/\sigma_i} \leq \sigma_{e+}, \quad (1)$$

where σ_{e+} is the limit equivalent stress at $\sigma_1 > 0$; $\chi = \sigma_{e+}/\sigma_{e-}$ is the parameter of plasticity of the material, taking into account the degree of influence on its micro destruction of shear deformations; σ_{e+}, σ_{e-} – stresses of failure (fracture) of the material under uniaxial tension and compression, respectively; A - the statistical parameter of defectiveness; for hardened steels A = 0.7 … 0.8.

For ductile materials, $\sigma_{e+} \approx \sigma_{e-}$, $\chi \to 1$, and Eq. (1) reflects the Huber-Mises-Henki criterion. For absolutely brittle materials, when $\chi \to 0$, the transition to the criterion of maximum normal stresses takes place. Since criterion (1) is resolved concerning tensile stresses, in the case of the stressed state with $\sigma 1 < 0$ (contact problems), the exponent with the defectiveness parameter should be taken in absolute value, and the limiting values of effective stresses can be represented

$$\sigma_{e+} = \chi\sigma_{HP}k_e \quad (2)$$

where σ_{HP} is the permissible normal contact stress, $k_e = \sigma_e/\sigma_{emax}$ is the equivalence factor. Thus, both the left and the right parts of (1) are in functional dependence on the parameter χ

If the plasticity criteria operate only with the components of the stress tensor, the use of the limit state criteria requires knowledge of the value of the plasticity parameter, as well as the nature of its change depending on the chemical composition of the steel and the type of its hardening. It is known that the value of χ varies from 0.9 … 1.0 for thermally improved structural steels to 0.5 … 0.7 for volumetrically medium- and high-carbon.

A simple example illustrates the expediency of using the limit state criterion. With a flat bend of a rectangular bar, the stresses are distributed by the Navier formula (Fig. 1a). The stress state at the boundaries of the beam: $\sigma_1 > 0$, $\sigma_2 = \sigma_3 = 0$ - on the stretched side and $\sigma_1 = \sigma_2 = 0$, $\sigma_3 < 0$ - on the compressed side. The equivalent stresses, according to the von Mises criterion, are the same at both boundaries: $\sigma_{i+} = \sigma_{i-} = \sigma_i$. But at $\chi = 0.7$ (carbon steel hardened to H $\approx$ 60 … 62HRC) by criterion (1): $\sigma_{e+} = \sigma_i$, $\sigma_{e-} = 0.7\sigma_i$ (Fig. 1b). Moreover, taking into account the different resistance of the material, the excess of the strength of the timber on the compressed side relative to the stretched one will increase even more, which, of course, requires an appropriate practical test.

In [7], an equivalent stress state was studied for the cases of linear and point contact. The distribution of relative maxima of equivalent stresses in planes parallel to the plane of contact, as well as the absolute maximum of these stresses $\sigma_{e\ max}$, was investigated depending on the contact geometry and the value of the plasticity parameter.

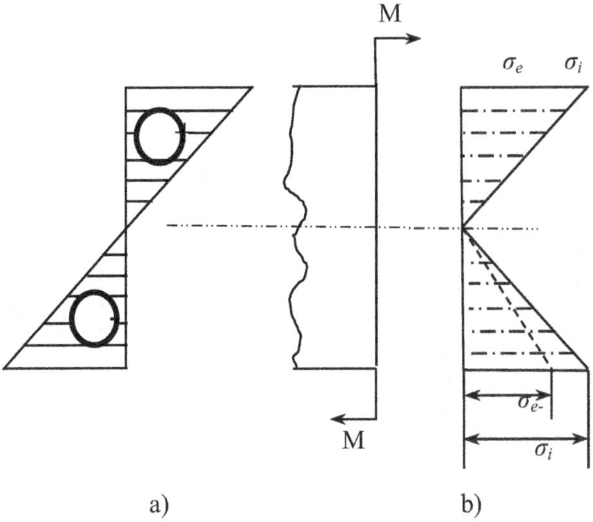

Fig. 1. Flat bend of a rectangular bar: (a) distribution of normal stresses; (b) equivalent stress distribution.

Figure 2 shows the effective stress fields for the case of linear contact (plane strain) according to the Huber-Mises-Genka (line 1) and Lebedev-Pisarenko criteria at $\chi = 0.6$ (line 2). This level of plasticity parameter is typical for steel type ShH-15, hardened to hardness 62 … 64 HRC. The reduction of ductility (other things being equal) leads to the expansion of the area of increased relative strength, expressed by the ratio of the current equivalent voltage to the maximum. So, the depth of the maximum equivalent stresses according to the Lebedev-Pisarenko criterion increased by 15% compared with the von Mises criterion. The zone $K_e > 0.95$ spread (along the central axis) in the interval (0.75…1.15)*b* (*b* is the half-width of the contact pad), whereas according to the Mises criterion it corresponds to the interval (0.50 … 0.95) b. At the same time, the total area of the potentially dangerous zone increased by 12 … 15%.

A comparison of the processes of cementation and carbonitriding within the limits of their alternative application shows that carbon cementation provides a higher level of bearing capacity. The difference in contact fatigue life is a multiple of [8]. However, in the practice of engineering calculations, this is not regulated in any way, and both processes are assumed to be equivalent. But with a hardness of the same and constant thickness of the sample (steel 12X2H4A) σu+ samples subjected to nitro-carburizing, (10 … 12)% higher than cemented, which, in the absence of data on the corresponding increase in σu− (Fig. 4), allowed, in the first approximation, to accept at the level of hardness 850HV values of the plasticity parameter $\chi = 0.6$ (cementation of nickel-free steels) and $\chi = 0.68 \ldots 0.7$ (cementation of nickel-containing steels and carbonitriding). In the absence of additional information, the change in the plasticity parameter in the range of 400 … 850HV was assumed linear.

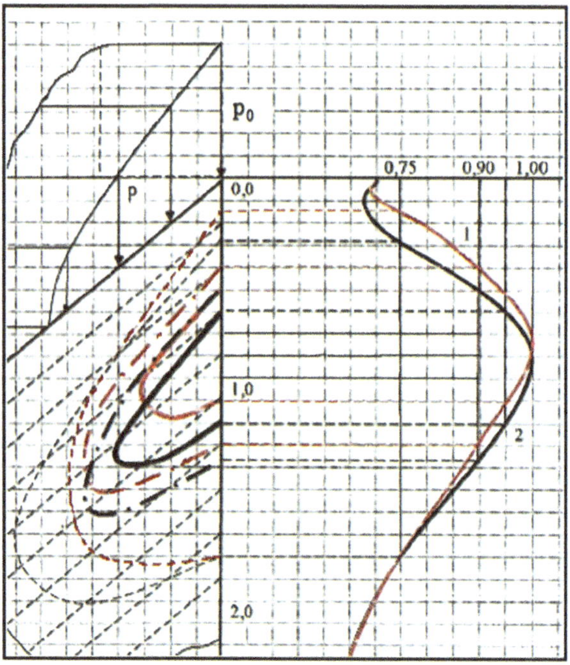

Fig. 2. The field of equivalent stresses. Linear contact: (1) $\chi = 1.0$; (2) $\chi = 0.6$. Z is the distance from the contact surface of bodies.

The correctness of the proposed model, as applied to contact calculations, was verified by its compliance with known experimental data and alternative methods of calculation. Total processed about 60 experiments performed on rollers and gears.

Concerning contact tasks, the estimated safety factor S_{KHe} takes the form:

$$S_{K\,He} = \zeta(\zeta - 0.11128)H_{HV}Z_{NK}K_1K_2K_3K_4K_5/\left[\chi\sigma_i + (1-\chi)\sigma_1 A^{1-(\sigma_1+\sigma_2+\sigma_3)/\sigma_i}\right] \quad (3)$$

where $Z_{NK} = (10^7/N_{EK})^{1/m}$ - the coefficient of durability; K_1, …, K_5 - coefficients that take into account the nature of the contact, the number of potentially dangerous zones, the effect of tangential load, the quality of the material and thermochemical treatment, the variation of material properties in the layer [10, 11].

3 Results and Discussion

Nowadays, in engineering practice, piecewise linear dependences of the contact endurance limit of steel parts on hardness are used: σ_{Hlim} = 2.3HB according to ISO 3663 or σ_{Hlim} = 23HRC according to GOST 21354-87. These dependencies, in general, correspond to the averaged values of the contact endurance limits. However, recommendations for thermochemical treatment for cases of cementation and carbonitriding are extended to the hardness range (56 … 65) HRC, although, for example, the number of loading cycles N = 10^7 and hardness 60 HRC, σ_{Hlim} = 2100 MPa, involute gears with ground teeth working in a favorable elastic-hydrodynamic lubrication regime, they are affected by progressive pitting at effective stresses σ_H > 1500 … 1600 MPa already in the early stages of operation, although alternative types of failure may be the leading ones. In [12] the quenching of carbon-nitrogen martensite resulted in a decrease in hardness to a lesser extent than that measured on carbon martensite with an equivalent embedding content.

Transmission electron microscopy analyses were performed to investigate this behavior. According to the described method, the contact strengths of the working surfaces of the teeth of the involute wheels were evaluated, the results of which are shown in the graphs in Fig. 3.

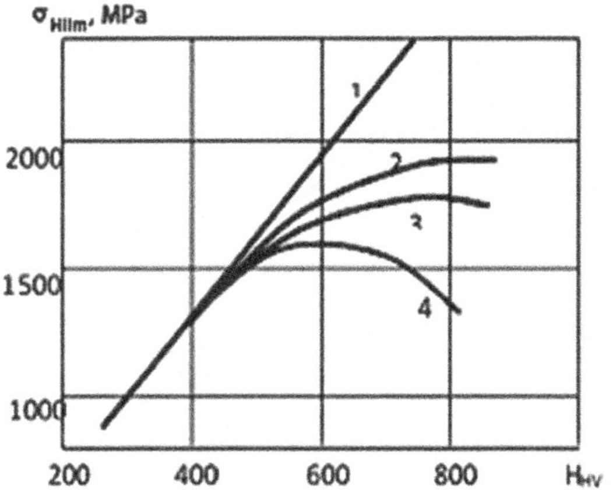

Fig. 3. The values of σ_{Hlim} based on 10^7 cycles with a calculated safety factor of 1.0: according to the von Mises criterion, graph 1; according to the Lebedev-Pisarenko criterion for carbonitriding, carburizing and quenching medium-carbon structural steels - charts 2, 3, 4, respectively.

For gears with a hardness of H $\geq$ 57 … 58 HRC the bending stresses determine, both traditional involute and known Novikov gear designs (W-N), the strength of the teeth. The evaluation of the flexural strength of involute gears is based on solving the problem of bending a cantilever clamped beam with a load distributed along the line of contact of the teeth. “Regular” fatigue failure is a fracture of a concave (into the wheel body) shape.

This model in one way or another modification is adopted in most of the design regulations (for example, DIN-3990 is a German standard or GOST 21354-87). However, for high-loaded evolvent transmissions, the results of calculations and experiments differ by order of magnitude, if not more, which is corrected by introducing a set of correction coefficients obtained experimentally. For example, ultrasound device for detection of inhomogeneities in two-layer sheet [13]. Also, in practice, the fracture is often found relatively flat, and sometimes even convex. Excluding a technological defect, then this type of destruction indicates not so much a bend as a cut.

Even greater difficulties arise in assessing the bending strength of Novikov gears (W-N) The region of localization of the maximum bending stresses in such gears is generally not defined. However, [1] during the run-in, the instantaneous contact area (ICA) extends (in height) to 80% of the active profile. The border of the ICA on the tooth stalk, if it does not coincide with the area of action of the maximum bending stress, is most likely close to it. In this case, along with other factors [14–16], the peculiarities of the stress-strain state in the contact area have a negative impact on the level of flexural strength.

We consider a possible effect on the example of the Hertz problem for the case of theoretically point contact – Fig. 4.

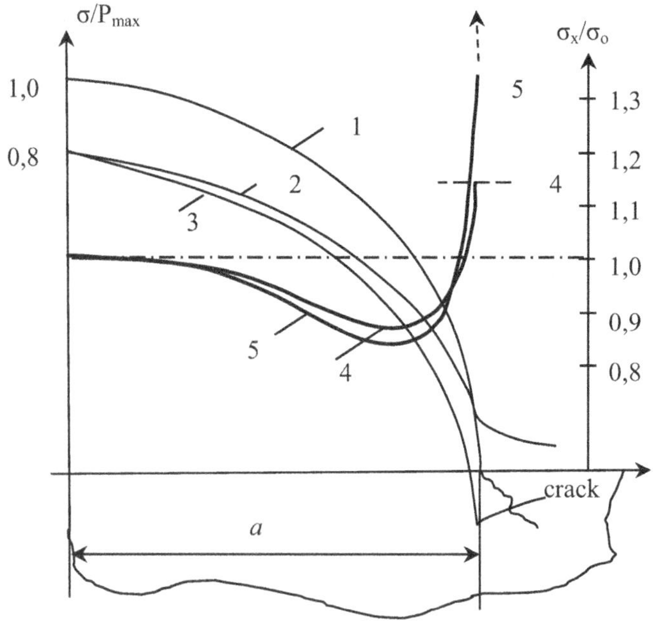

Fig. 4. Stresses on the half-space surface: 1. σ_n - normal; 2. σ_θ - circumferential; 3. σ_r - radial; 4, 5 - relative intensity - σ_x/σ_o, where σ_x - equivalent stresses at a distance "x" from the central axis, σ_o - equivalent stresses at the center of the ICA (4 - according to the Mises criterion, 5 - according to the Lebedev-Pisarenko criterion).

Stresses on the surface in the contact area are everywhere compressing, except for its boundary. Equivalent moduli stresses—radial (tensile) and circumferential (compressive)—appear at the ends of the principal axes of the conditional contact ellipse. Equivalent stresses according to the Mises criterion for a circular region are equal to σio = 0.2Pmax at the center of the ICA and σia = 0.225Pmax at the border of the ICA (Pmax is the maximum pressure). For hardened alloyed structural steels at H ≈ 60…62HRC and plasticity parameter $\chi \approx 0.7$ according to the Lebedev-Pisarenko criterion, the equivalent stresses will be σeo = 0.14 Pmax at the center of the contact patch and σea = 0.186Pmax at its boundary. According to the von Mises criterion: σia/σio = 1.125, i.e. congestion of the IPC boundary relative to the center - 12.5%). For hardened alloyed structural steels with H ≈ 60…62HRC and a plasticity parameter $\chi \approx 0.7$ according to the Lebedev-Pisarenko criterion, the equivalent stresses will be σeo = 0.14Pmax at the center of the contact patch and σea = 0.186Pmax: σea/σeo = 1.33. The relative tension of the border area increased by approximately another 15-18%. However, the matter is not limited to the growth of relative strength: in the center of the ICA, the level of permissible stresses is determined by the strength of the material in compression, while at the edge - in tension (in the first approximation - in $1/\chi$). A similar approach has been applied to the study of the occurrence of micropitting in the contacts of gear teeth [17].

4 Conclusions

The influence of the plastic properties of the material on the strength characteristics of gears is very significant. So even a slight increase in the plasticity parameter from 0.6 to 0.67… 0.7 at the same level of hardness determines an increase in σ_{Hlim} by 120… 150 MPa, which corresponds to the well-known difference in strength between cementation and carbonitriding.

For involute gears made of cemented structural steels, an increase in hardness above 650…700 HV practically does not lead to an increase in contact fatigue life, and with H > 800HV one can even expect a decrease. Recommendation GOST 21354-87: "Limits of contact endurance σ_{Hlim} for cementation and carbonitriding and subsequent hardening" should be limited to the upper hardness limit of surfaces 62HRC.

Under certain conditions, contact stresses can initiate flexural failure. At the border of the ICA, especially in materials with reduced plastic properties, there are radial microcracks that do not close when the load is removed, growing with repeated even minor stresses. The stress state outside the ICA is a net shift. The situation is exacerbated when the ICA is localized near areas with increased defectiveness of the surface layers. During cementation and carbonitriding, these are internal oxidation, decarburization of the solid solution, the dark component, etc. (depending on the type of thermo-chemical process), drastically reducing the strength characteristics of the material. For gears, the risk zone will be adjacent to the fillet of the active profile boundary. If the defective layer in the active areas can be neutralized by subsequent running-in, then for the transition zone the problem of its neutralization is difficult: there is no burn-in, mechanical removal, for example, grinding of fillets, is unacceptable because of the danger of burns, and electrolytic labor is intensive and expensive. For gears with theoretically point contact (such as W-N), the ICA-fillet boundary should be considered as potentially dangerous not only by contact but also by bending.

Accounting for the influence of the plastic properties of the material of the layer made it possible to increase the reliability of the prediction of the level of load capacity of the thermal process and chemically treatment of strengthened parts. Strength assessment should be carried out using the limiting state criteria and taking into account changes in the parameter χ.

The work was financially supported by the Russian Foundation for Base Research (project 18-01-00715-a).

References

1. Beskopylny, A., Onishkov, N., Korotkin, V.: Adv. Tribol. (2018), Article ID: 9607092, https://doi.org/10.1155/2018/9607092
2. Karolak, C., Montmitonnet, P., Parks, D., Delattre, G., Bouchard, P.: Proc. Eng. **207**, 2030–2035 (2017)
3. Beskopylny, A., Veremeenko, A., Kadomtseva, E., Beskopylnaia, N.: MATEC Web Conf. **129**, 02046 (2017)
4. Chen, Z., Zhai, W., Wang, K.: Mech. Syst. Sig. Proc. **115**, 29–44 (2019). https://doi.org/10.1016/j.ymssp.2018.05.038
5. Sobolev, A., Radchenko, M.: Nucl. Energy Technol. **2**, 272–276 (2016)
6. Mirone, G., Barbagallo, R., Corallo, D.: Proc. Struct. Integr. **2**, 3684–3696 (2016)
7. Beskopylny, A., Onishkov, N., Korotkin, V.: Advances in Intelligent Systems and Computing, vol. 692, pp. 184–191 (2017)
8. Beskopylny, A., Onishkov, N., Korotkin, V.: MATEC Web Conf. **224**, 0033 (2018)
9. Beskopylny, A., Kadomtseva, E., Strelnikov, G., Morgun, L., Berdnik, Y., Morgun, V.: IOP Conf. Ser. Mater. Sci. Eng. **365**, 032023 (2018). https://doi.org/10.1088/1757-899x/365/3/032023
10. Beskopylny, A., Liapin, A., Andreev, V.: MATEC Web Conf. **117**, 00018 (2017)
11. Belen'kii, M., Beskopyl'nyi, A., Vernezi, N., Shamraev, L.: Weld. Int. **11**(8), 642–645 (1997)
12. Silva, W., Dulcy, J., Ghanbaja, J., Redjaïmia, A., Michel, G., Thibault, S., Belmonte, T.: Mater. Sci. Eng., A **693**, 225–232 (2017)
13. Afanasenko, V., Boev, E., Kulakov, P.: IOP Conf. Ser. Mater. Sci. Eng. **560**, 012003 (2019). https://doi.org/10.1088/1757-899x/560/1/012003
14. Fukumasu, N., Machado, G., Souza, R., Machado, I.: Proc. CIRP **45**, 255–258 (2016)
15. Jolivet, S., Mezghani, S., Isselin, J., Mansori, M.: Tribol. Int. **102**, 436–443 (2016)
16. Beskopylny, A., Kadomtseva, E., Strelnikov, G., Berdnik, Y.: IOP Conf. Ser. Earth Environ. Sci. **90**(1), 012064 (2017)
17. Morales-Espejela, G., Rycerz, P., Kadiric, A.: Wear **398–399**, 99–115 (2018)

Parts Processing Technology for Transport Engineering

Mikhail Tamarkin(✉), Elina Tishchenko, Irina Chukarina, and Tamara Sosnitskaya

Don State Technical University, 1, Gagarin sq., Rostov-on-Don 344000, Russia
tehn_rostov@mail.ru

Abstract. The article presents the studies' results of processing multi-contact vibration impact tool - ball-rod hardener, which is designed for local processing methods of surface plastic deformation various parts, including engineering transport. The design of the hardener is described. Its scheme is presented. The acoustic characteristics of the process are analyzed. A linear source is used as a noise source model to simulate noise generation process for long parts. Sound pressure is determined. The formula for determination of natural oscillation frequencies for beam-type details is given. The speed of product oscillations at natural frequencies is determined. The real part of oscillation velocity is determined for calculations of octave-averaged sound pressure levels (noise spectra). The screen calculations are performed to ensure the acoustic safety of the worker to solve the problem of passive noise protection in the machine working area and to ensure safety of work. The maximum value of the acoustic screen height is determined. Technological possibilities of processing are revealed. The parameters of plastic print are determined. The dependences for determining surface roughness, hardened layer depth and the degree of deformation, the adequacy of which is confirmed by experimental studies, are given. The graphs of the dependence of surface roughness, hardened layer depth and the degree of deformation on the main technological parameters and the material hardness of the part obtained from the experiments are presented. Technological recommendations on the research results application are given. Elements of CAD TP system ball-rod hardener machining parts are developed.

Keywords: Ball-rod hardener processing · Surface roughness · Hardened layer depth · Degree of deformation · Acoustic safety

1 Introduction

Methods of processing of details by surface plastic deformation (SPD) are widely applied on final operations of their technological processes. These methods can significantly improve the performance properties of the processed parts, while the cost of processing is much lower than the introduction of many other finishing methods of processing. This is especially true for parts of transport engineering, the breakdown of which can lead to machine failure. In addition, many SPD methods can be applied locally, where stress concentrators (holes, fillets, threads, bevels, slots, samples, welds,

Z. Popovic et al. (Eds.): TransSiberia 2019, AISC 1115, pp. 913–922, 2020.
https://doi.org/10.1007/978-3-030-37916-2_90

grooves, etc.) are located, without processing the entire part, which also reduces the final cost of the resulting product. Local processing is also necessary on the unhardened parts of the surfaces, which have undergone general hardening treatment in vibration, bead-blasting and other installations (under clamps, in pockets, holes and other hard-to-reach for the processing environment areas), in those places of machined parts, where the hardened layer is removed during fitting work. These elements are characteristic for most of the processed engineering products, which leads to the widespread introduction of local methods of SPD processing in modern engineering production [1–3].

Details of a variety of sizes and shapes, made of different metals: steels, including stainless, non-ferrous alloys (aluminum, magnesium, etc.) can be subjected to local hardening processing by SPD methods.

Ball-rod hardener (BRH) is a tool invented by a group of researchers under the guidance of Professor Babichev. It is universal, allows processing not only flat surfaces, but also surfaces of complex configuration, with a small difference in height. The hardener is made on the basis of a pneumatic or electric hammer, which is a power drive with a striker 1 (Fig. 1). A block of round rods 2 in a special collet clamp 6, which are located in the body of the hardener 3, based on the surface of the work-piece 5, can copy the shape of its surface due to the presence of layers of steel balls 4. In this case, a layer of steel balls transmits the impact energy of the drive to the block of rods without significant losses. The rods have a spherical sharpening that allows you to leave plastic impressions on the surface of a workpiece. If we compare the process of BRH processing with other shock methods of SPD, we can distinguish a number of advantages that are typical for this method of processing: high intensity of processing, the formation of residual compression stresses, the possibility of local hardening, including long parts of a complex profile [1, 2].

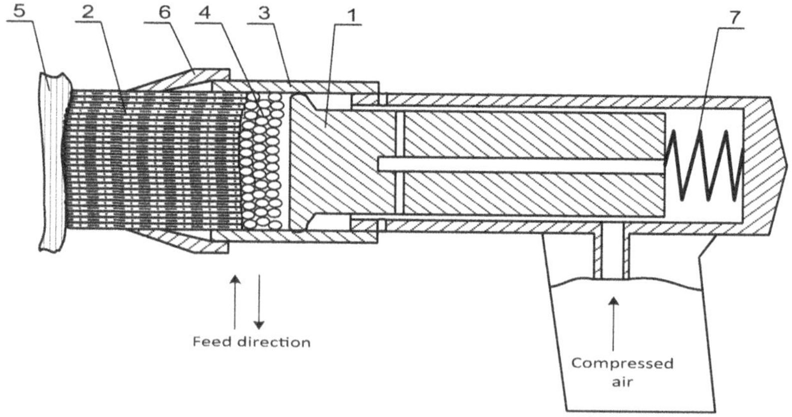

Fig. 1. Diagram of the multiple-contact vibro-impact tool BRH: 1 – power drive with the striker, 2 – block of round rods, 3 – case of the hardener, 4 – steel balls, 5 – the part, which is subjected to processing, 6 – collet clamp, 7 – elastic element.

2 Materials and Methods

Safe working conditions in modern production are of great importance and are stipulated by sanitary standards adopted at the level of state standards. Protection of workers from exposure to hazardous and harmful production factors is an urgent problem of labor protection. Noise radiation of high intensity, which is accompanied by many technological processes of parts processing, can lead to a decrease in productivity and deterioration in the health of workers. In this regard, the problem of noise reduction in processing is a very urgent task.

The process of hardening with the help of a device for BRH in most cases is realized on universal milling machines and as any shock treatment is accompanied by a significant noise impact on the worker. The hardener is attached to the spindle of the milling machine. The spindle does not rotate. Only the feed drive contributes to the formation of the sound field in the working area of the operator of the elements of the machine bearing system. Feeding box of a modern milling machine realizes table movement by a ball screw, the rotation of which is made from an adjustable motor. Such mechanisms are characterized by low noise impact due to the lack of gears. Therefore, it can be assumed that the noise spectra in the working area are created by the sound propagation from the hardened products and the ball-rod hardener itself.

To simulate the process of noise generation for long parts studied in this paper, a linear source is used as a noise source model, the sound pressure of which according to the work [3] can be calculated by the formula:

$$P = 9.5\frac{\upsilon_k}{r}(f_k F l)^{0.5} \tag{1}$$

where v_k - vibration speed, f_k - natural frequencies, F - the area of the sound emitting surface, l - the length of the part, r - the distance from the source to the design point.

If the shape of the workpiece allows, then it is possible to put a damping layer of the required thickness, which will absorb the natural frequencies of the part vibrations. For long parts such as beams, the natural oscillation frequencies can be determined by the following relationship:

$$f_k = \frac{1}{2\pi}\sqrt{\left(\frac{\pi\kappa}{l}\right)^4 \frac{EJ}{m_0} + \frac{j_{pr}}{m_0}} \tag{2}$$

where E – the modulus of elasticity, J – the moment of inertia of the part, m_O - the distributed mass, j_{pr} - the reduced stiffness of the technological system, k - the coefficient determining the natural frequencies of oscillations.

To determine the sound pressure levels, the following formula is obtained:

$$L = 20\lg \upsilon_k + 10\lg\left[\left(\frac{\pi\kappa}{l}\right)^4 \frac{EJ}{\rho} + \frac{j_{pr}}{\rho}\right] \times F + 20\lg\frac{l}{r} + 106, \tag{3}$$

where ρ - the density of the part material.

For engineering calculations of sound pressure levels it is necessary to determine the speed of vibrations of the product at natural frequencies, which is a complex value. To calculate the octave-averaged sound pressure levels (noise spectra), it will be enough to determine the real part of the oscillation rate by dependence:

$$\begin{aligned}\mathrm{Re}\{\upsilon_k\} = \frac{2\pi kP}{Ml}\sum \frac{k\left(\sin\frac{\pi k x_1}{l} + \sin\frac{\pi k x_2}{l} + \ldots + \sin\frac{\pi k x_K}{l}\right)}{\left\{\left[EJ\left(\frac{\pi k}{l}\right)^4 + \frac{j_{pr}}{m_0} - \left(\frac{\pi k \upsilon}{l}\right)^2\right]^2 + (\psi EJ)^2\left(\frac{\pi k}{l}\right)^3\right\}^{0.5}} \\ \times \sin arctg \frac{-EJ\psi\left(\frac{\pi k}{l}\right)^4}{EJ\left(\frac{\pi k}{l}\right)^4 + \frac{j_{pr}}{m_0} - \left(\frac{\pi k \upsilon}{l}\right)^2 + (\psi EJ)^2\left(\frac{\pi k}{l}\right)^3}\end{aligned} \quad (4)$$

where P - force action of the process, M - the mass of the part, x_k - the coordinate of the rod of the ball-rod hardener, ν - the speed of movement of the table with the hardened part, ψ - the effective coefficient of loss of the vibrational energy of the hardened part.

If the shape and configuration of the workpiece does not allow the use of vibration-absorbing layer or its use does not ensure the implementation of sanitary standards for noise, the use of passive noise protection in the working area of the machine will solve the safety problem. The installation of a flat screen made of polycarbonate glass, the acoustic efficiency of which should provide sanitary standards of sound pressure levels in the entire normalized range of 31.5–2000 Hz, will ensure the safety of the operator when using the device for BRH. The required acoustic efficiency is ensured by the height of the screen. Its length is determined by the size of the machine table. From the known dependence of acoustic efficiency [3, 4] we obtain:

$$\Delta \mathrm{L} = 10\lg\left(\mathrm{g}\pi^2 \cdot \frac{\mathrm{h}}{\lambda}\mathrm{tg}\frac{\varphi}{2}\right) \cdot 10\lg \mathrm{n} \quad (5)$$

where h - the height of the installed screen, λ - the wavelength in the air, ϕ - the angle of refraction of the sound wave, n - the number of edges of the screen through which the sound energy enters the design point.

The dependence of the height of the acoustic screen, provided that the sanitary standards for noise are followed is following:

$$\mathrm{h} = 10^{0.1\left(\mathrm{L_{pi}} - \mathrm{L_{ci}} + 1.1 - \lg \mathrm{f_i} - \lg \mathrm{tg}\frac{\varphi}{2}\right)} \quad (6)$$

where L_{pi} - the actual octave levels of sound pressure, L_{ci} - octave maximum permissible levels of sound pressure, fi - root-mean-square frequencies of octave bands.

The maximum value of the height of the acoustic screen is determined by the results of the calculation. The polycarbonate glass screen serves not only as an acoustic protection, but also protects the operator from injury in case of breakage of the tool or the workpiece. Figure 3 presents a graph of noise spectra in the working area of the operator performing processing using a ball-rod hardener. The graph shows that the implementation of sanitary standards for the noise level is achieved (Fig. 2).

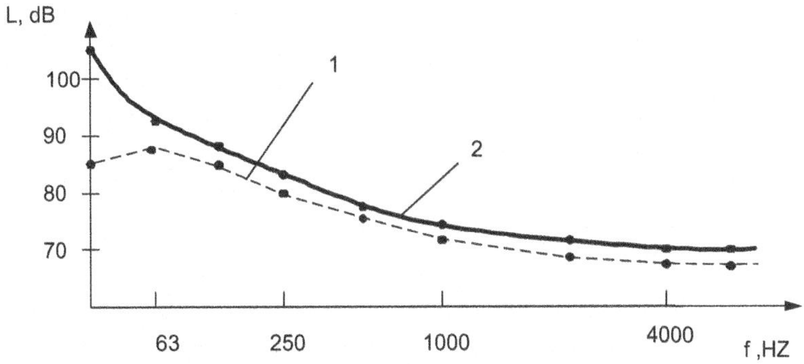

Fig. 2. Noise spectra: 1 – noise in processing with noise protection system, 2 – limitary spectrum.

The widespread introduction of BRH processing in machine-building production constrains the lack of theoretical studies of the impact of technological regimes on the quality of the treated surface. The main technological parameters of the process of BRH processing are the impact energy of the striker, the diameter and the number of rods, the radius of their spherical sharpening [5–7].

Calculation of the parameters of the plastic impression when introducing a spherical indenter into the deformable half-space; in this work the diameter and the depth of this impression are determined:

$$d = \sqrt[4]{\frac{D \cdot E_u}{0.1HD}} \tag{7}$$

$$h = \frac{d^2}{HD} \tag{8}$$

where D - the diameter of the sphere, E_u - the impact energy, HD - the dynamic hardness of the part material.

According to [1], the dynamic hardness can be determined from the ratio

$$HB = 0.2HD^{0.89}$$

where HB - the Brinell hardness of the part material.

Then

$$HD = 6.1HB^{1.12},$$

For BRH process, the formula (7) can be represented as follows:

$$d = \sqrt[4]{\frac{D \cdot E_u \cdot \eta}{N \cdot HB^{1.12}}} \tag{9}$$

where η - the efficiency of the device for the BRH, which depends on the tension created by pressing the device to the surface of the part, N - the number of rods in the block.

For the calculation of high-level parameters of roughness of the processed details surface, we can use P. P. Chebyshev method [1, 6] and write:

$$R_z = \frac{S^2}{8R} \tag{10}$$

where S – the device feed, R – the radius of the spherical sharpening of the indenter.

In this case, the feed value, which can vary within certain limits, should be calculated according to the following dependence: $s = 03d$.

This amount of feed will eliminate the appearance of untreated parts of the surface of the part.

Let us do the conversion and get the dependence to calculate the high-level roughness parameter processed by the device for BRH:

$$Rz = 0.03\sqrt{\frac{E_u \cdot \eta}{D \cdot N \cdot HB^{1.12}}} \tag{11}$$

The depth of the hardened layer and the degree of deformation of the workpiece treated with a ball-rod hardener affect its performance properties (fatigue strength, fatigue life, etc.). The depth of the hardened layer influences the surface area in which there is a residual deformation of grains and the increased density of crystal lattice dislocations formed as a result of the application of an external load. It is rather difficult to determine analytically the depth of the hardened layer and the degree of deformation depending on physical and mechanical properties of the part material and the process parameters [6]. In the works of researchers of SPD processing processes there are formulas that are obtained on the basis of the theory of formation of elastic and plastic deformation after the adoption of numerous simplifications and assumptions. The paper [5] shows the dependence for calculating the depth of the hardened layer during surface plastic deformation processing

$$h_n = 2.5K_o\sqrt{D \cdot h} \tag{12}$$

where K_O - the coefficient depending on the impression shape. For the BRH processing we can take $K_O = 1$.

Using this dependence, in the course of our own research the formula for calculating the depth of the hardened layer in the BRH processing is obtained:

$$h_n = \sqrt[8]{\left(\frac{E_u \cdot \eta \cdot D^2}{N \cdot HB^{1.12}}\right)^3} \tag{13}$$

For theoretical calculation of the obtained degree of deformation, to determine the degree of plastic deformation at SPD:

$$\varepsilon = \frac{d}{D} \tag{14}$$

We apply the dependence for the processing method under study and after the transformation we obtain the following formula to determine the degree of deformation:

$$\varepsilon = 1.13\sqrt[4]{\frac{E_u \cdot \eta}{D^3 \cdot N \cdot HB^{1.12}}} \tag{15}$$

3 Results

A number of experimental studies have been conducted to confirm the adequacy and application of theoretical dependencies (11), (13) and (15) in the design of technological processes of ball-rod hardener processing.

The ball-rod hardener was fixed on a miller or lathe. The air hammer of model KPM-14M was used. Samples were made of the following materials: aluminum alloys D16, B95, AVT, steels St3, 45, HVG. The choice of materials samples is due to the fact that they are often used for the manufacture of machine parts. The rods used for processing had different diameters of sharpening. Their number in the package varied. Dependency graphs were plotted using formulas (11), (13), and (15). In Figs. 4, 5, 6 and 7 they are shown in solid lines. The points indicate the results of experiments with indication of confidence intervals (with confidence probability of 95%). The analysis of the research results shows that the discrepancy between the theoretical and

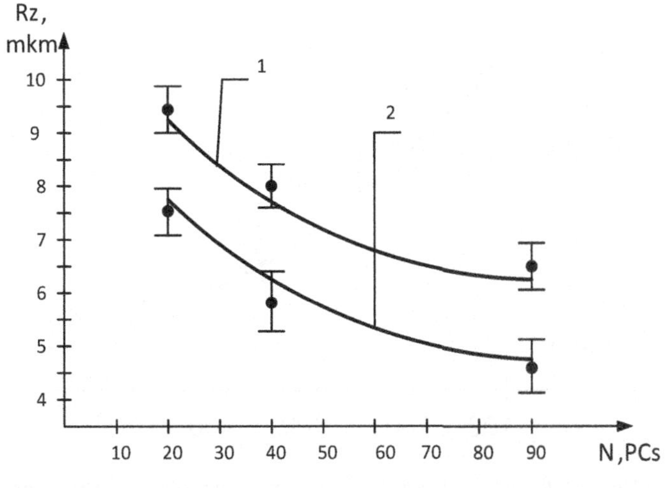

Fig. 3. Dependence of roughness of the processed surface on the number of rods in the nozzle (D = 5 mm, η = 0.7): 1 - part material – B95, 2 - part material – steel 45.

experimental values does not exceed 15%, so it can be argued that the proposed theoretical dependences are adequate and can be recommended for use in technological design.

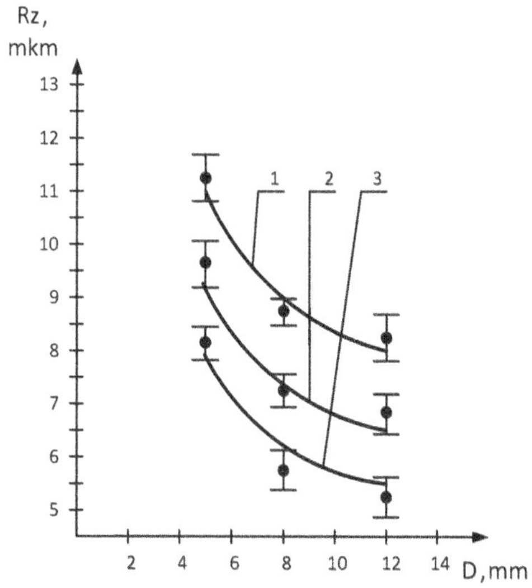

Fig. 4. Dependence of roughness of the processed surface on the diameter of spherical rods sharpening (N = 19, η = 0.7):1 - part material – D16, 2 - part material – B95, 3 - part material – steel 45.

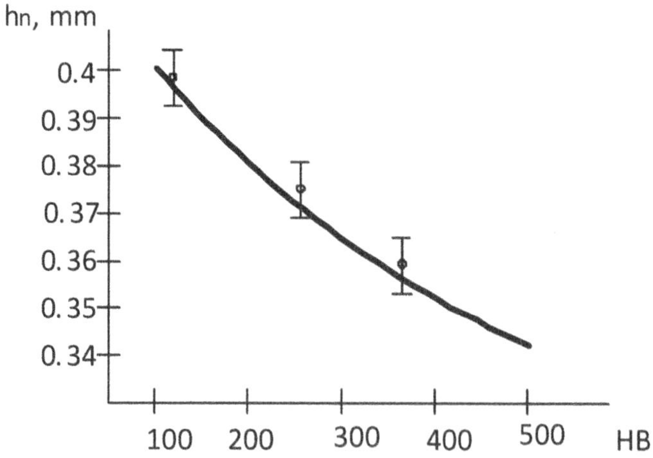

Fig. 5. Dependence of the depth of the hardened layer on Brinell hardness (number of rods N = 19, impact energy Eu = 2.5 j, η = 0.7, diameter of rods sharpening – D = 3 mm).

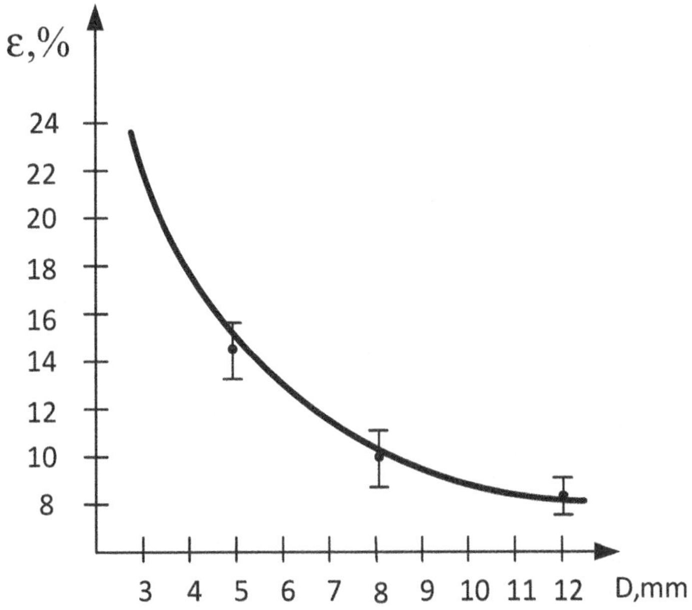

Fig. 6. Dependence of the degree of deformation of the surface layer of the diameter of rods sharpening (number of rods N = 40, impact energy Eu = 2.5 j, = 0.5, part material – steel HVG).

4 Discussion

The design of technological processes of local SPD processing includes a complex of works which is necessary for implementation of the technological process meeting the set requirements. At the same time, the multivariance of technological solutions and their considerable complexity necessitate the automation of design work. For this purpose, the existing systems of computer-aided design of technological processes are created or used. The use of CAD TP is possible with theoretical dependencies that describe the complex relationships studied by the technology, and determine the procedures that allow you to build design solutions. The theoretical models obtained above can be used as a basis for such a device for BRH processing. The results of the processing of the BRH process can be described by many criteria: the roughness of the treated surface, the depth of the hardened layer, the degree of deformation, processing performance, etc. It is necessary to ensure noise levels during processing within the normative values.

The design of a technological process should begin with a search of options for combinations of technological parameters available to the technologist (drive power, the number of rods in the block, the radius of rods sharpening, tension of the device). Of all the options such should be chosen that at the calculated dependences given above ensure the prescribed quality parameters of the surface layer. The processing time is between 10–20 s on the width of the block of rods in the flow direction. If several options of the technological process meet the specified conditions, the option with the highest processing capacity is chosen. Next, the sound pressure levels, noise spectra are determined and the acoustic screen is calculated, which must be installed in the processing area.

The developed CAD TP has been tested in production conditions and used in the implementation of the process of BRH parts processing at LLC "DonKuzlitMash" (Azov).

References

1. Tamarkin, M.A., Tishchenko, E.E., Fedorov, V.P.: IOP Conf. Ser. Mater. Sci. Eng. **124**(1), 012169 (2016)
2. Tamarkin, M.A., Tishchenko, E.E., Isaev, A.G., Razdorskiy S.A.: Taganrog: aspects of science, education and industrial modernization, p. 263 (2015)
3. Tamarkin, M.A., Tishchenko, E.E., Shvedova, A.S.: Russ. Eng. Res. **38**(9), 726–727 (2018)
4. Tamarkin, M.A., Tishchenko, E.E., Shvedova, A.S.: STIN, no. 3, pp. 26–28 (2018)
5. Artiukh, V., Raimberdiyev, T., Mazur, V.: Use of CAE-systems at evaluation of shock absorbers for metallurgical equipment. MATEC Web Conf. **53** (2016). http://dx.doi.org/10.1051/matecconf/20165301039, Article Number 01039
6. Artiukh, V., Mazur, V., Kargin, S., Zakharova, L.: Adapters for metallurgical equipment. MATEC Web Conf. **170** (2018). https://doi.org/10.1051/matecconf/201817003028, Article Number 03028
7. Gharaibeh, N.S., Matarneh, M.I., Artyukh, V.G.: Loading decrease in metallurgical machines. Res. J. Appl. Sci. Eng. Technol. **8**(12), 1461–1464 (2014)

Vacuum Thermal Magnetic-Pulse Welding of Cathode Assemblies

Evgeny Strizhakov(✉), Stanislav Nescoromniy, and Victor Vinogradov

Don State Technical University, 1, Gagarin Sq., Rostov-on-Don 344000, Russia
strizhakov@inbox.ru

Abstract. The paper presents studies on the vacuum thermal magnetic-pulse welding (VTMPW) of the cathode assemblies from dissimilar alloys. To activate the welded surfaces and remove contaminants, it is proposed to preheat the unit assembled for welding in vacuum, with subsequent magnetic-pulse radial action through the dielectric container on the joining surfaces heated to temperatures below the melting point. The VTMPW process parameters and the limitations related to the mechanical fracture of the welded assembly, burning-off or faulty fusion, affecting the fixed joint efficiency, are identified. The microstructure analysis did not detect the formation of common grains; hardness in the joining zone was increased, which is typical for the formation of the compound in the solid phase.

Keywords: Vacuum flame cleaning · Emitter · Secondary-emission cathodes · Pulse energy · Magnetic pressure · Inducer · Discharge frequency · Non-conducting gap · Dielectric container

1 Introduction

In the manufacture of microwave crossed-field tubes, emitter materials from metal alloys are increasingly used. They combine a fairly high level of thermal and secondary emissions, high electrical and thermal conductivity, and smooth surface, which ensure the required life cycle and reliability of products. High emissivity, its stability during the entire service life under the electron beam bombardment predetermine prospects and relevance of the development of a cost-effective technology for producing cathodes for electronic tubes based on such cathode materials as CuLi, AlBa, Pt, PtBa, PdBa, JrLa, ReLa, etc.

Metal-alloy secondary-emission cathodes are produced composite. The emitter in the form of foil or sleeves from emission material is fixed on the surface of cores made from copper, stainless steel, molybdenum, niobium or their alloys [1–3], Fig. 1.

The cathode assembly is the most heat-loaded component of the device. As a result of electron beam bombardment (in 107 W/m^2 pulse), the emitter is exposed to thermal shocks. For normal operation of the device, it is necessary to carry out an intensive heat removal from the cathode working surface providing a proper connection of the emitter and the core over the entire interface.

Z. Popovic et al. (Eds.): TransSiberia 2019, AISC 1115, pp. 923–929, 2020.
https://doi.org/10.1007/978-3-030-37916-2_91

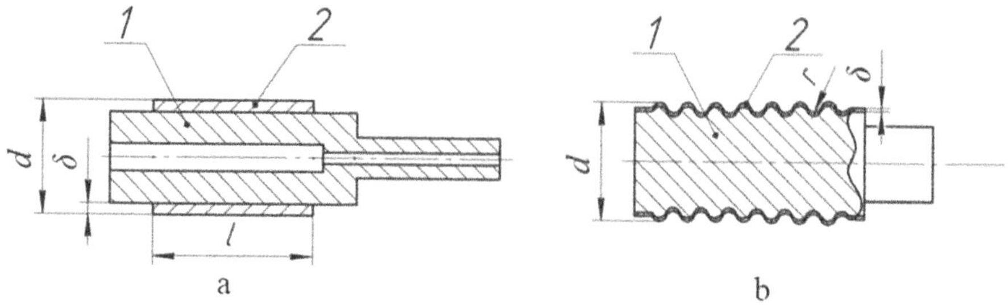

Fig. 1. Metal-alloy secondary-emission cathodes: *a* – on cylindrical; *b* – with developed surface. 1 is cathode base; 2 is emitter; *d* is diameter; *δ*, *l* are thickness and width of emitter; *r* is surface curvature.

The diffusion welding joining emitters and cathode bases provides quality welds, parameter stability, and device life [3]. To obtain lining emission coatings on cylindrical cores, devices are used that provide compressive pressure due to the difference in thermal linear expansion coefficients [4]. Residual deformation occurring in this case damages the device after receiving one cathode unit. The disadvantages of the diffusion welding also include the duration of high-temperature exposure, which leads to the appearance of brittle intermetallic phases in the weld junction, as well as the inability to get the emitter – core joints on the developed surface.

It is possible to enhance the process of diffusion welding of broadband (l/d > 1–3) lining joints using magnetic-pulse welding (MPW), which implements locking force ("crimping" scheme) [5]. As on the diffusion welding, the process must be carried out in vacuum to prevent oxidation of the joining surfaces (the oxygen hungry materials are used).

2 Research Procedure

In contrast to the conventional MPW, during vacuum-thermal magnetic-pulse welding (VTMPW), the loading is carried out along the normal to the joining part surfaces; there is no cumulative cleaning jet arising from glancing collision [2, 8]. Purification is performed through evaporation of oxide films upon heating in vacuum.

The VTMPW scheme is shown in Fig. 2.

The VTMPW process is as follows. The core (1) and emitter (2) are fixed in the collet clamp (7) of the vacuum chamber that consists of a ceramic container and two flanges. The welded assembly is introduced into the zone of the combined cooled inductor (3), which is connected to the high-frequency generator (HFG) (5) and the surge-current generator (SCG) (4). After the assemblage and installation, the chamber is evacuated to the operating pressure of $0.66 \cdot 10^{-2}$ Pa. the HFG is started. When the desired heating-up temperature is reached, the HFG is turned off. The HFG is charged. When the HFG is discharging, Ip discharge current (100–200 kA) is flowing through the coils of the working tool – inductor (3). A variable magnetic field (H), which induces Ii currents in the emitter, occurs in the operating space of the inductor. When Iи induced currents and the magnetic field of the inductor (H) interact, Pm magnetic pressure occurs [8]. Pm pressure acts on the emitter in radial axis; combined deformation of the joining juvenile surfaces and a solid state welding take place [9]. This device uses a unique feature of the magnetic-pulse processing, the impact remotability through a dielectric - ceramic container (6), which is a section of the vacuum chamber. The heated and welded assembly is in vacuum, and the combined inductor of HFG and VTMPW is in the atmosphere.

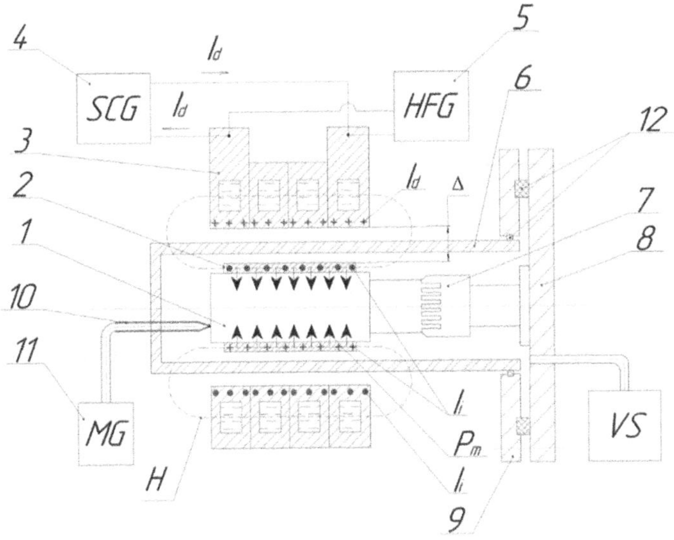

Fig. 2. VTMPW layout diagram. 1 is core; 2 is emitter; 3 is cooled inductor; 4 is surge-current generator (SCG); 5 is high-frequency generator (HFG); 6 is dielectric - ceramic container; 7 is collet clamp; 8 is flange-base; 9 is flange of container; 10 is thermal couple; 11 is measure gauge (MG); 12 is vacuum seal; Id is discharge current; H is magniflux; Ii is induced current; Pm is magnetic pressure; Δ is non-conducting gap; VS is vacuum system.

The staged experiments provided identifying the process variables affecting the welded joint quality: amount of vacuum in the working chamber, preheating temperature (T0), magnitude of the realizable magnetic pressure (Pm) dependent on pulse energy (W).

The studies of the effect of various parameters on the weld quality were carried out on the VTMPW installation assembled according to the scheme shown in Fig. 2. Operating frequency of the current pulse depends on the installation design, and it was a constant parameter (fd = 10 kHz) in a series of the staged experiments. This value is selected on the evaluation results to ensure that the first half-wave length of the magnetic pressure should be greater than the stress relaxation time of the joining materials in the interface. The non-conducting gap setting between the tool – inductor and the workpiece was determined from the design data analysis: $\Delta = (4 - 6) \cdot 10^{-3}$ m.

All experimental samples had base diameter of db = 18 · 10^{-3} m, lining coating width of l = 30 · 10^{-3} m, thickness of ts = 0.5 · 10^{-3} m, and the number of inductor windings of n = 4.

3 Research Results Discussion

The analysis of the welds through the shearing test under the microscope has shown that there are "craters" and partial traces of the lining on the base in the metal separation zone, Fig. 3.

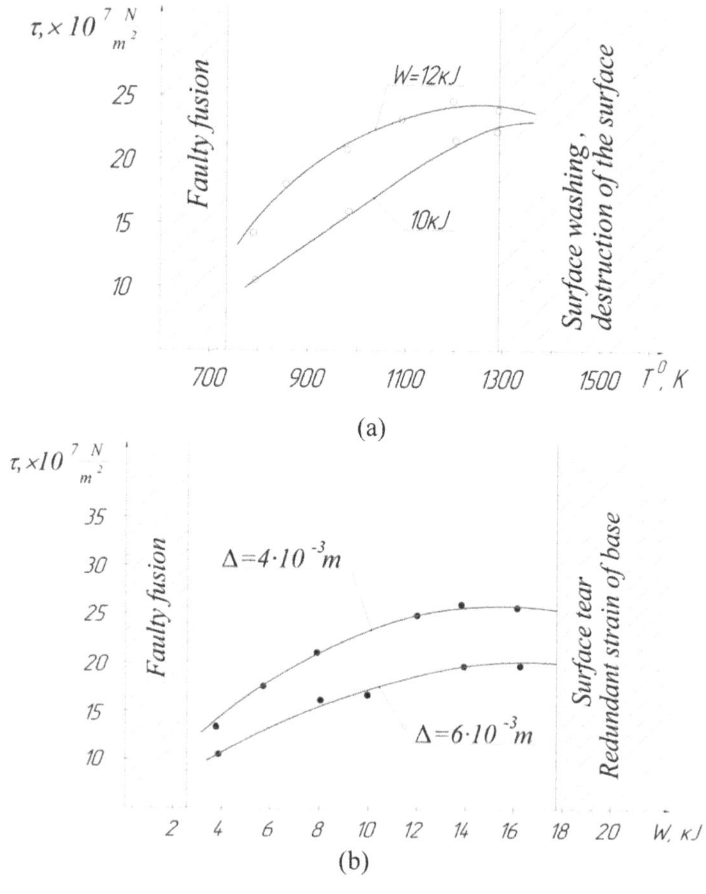

Fig. 3. Dependences of shear strength of welds from dissimilar alloys: *a* is Pt + MPVM (molybdenum) on preheating temperature; *b* is steel 12X18H10T + Pt, on pulse energy W, T^0 = 1200 K.

The upper limit of the temperature range of the VTMPW is related to the burning-off and destruction of the lining component under superheating, or with the appearance of heterogeneous structures in the joining zone, which degrades the welding quality and is unacceptable when producing emission coatings. The lower limit is related to insufficient activation of the surfaces; due to the lack of time for relaxation processes, it is not possible to obtain a quality weld (Fig. 3a).

Pressure under the VTMPW, as well as under other methods of welding in a solid state process, is the key factor in obtaining physical contact and activation of the joining surfaces. The welding pressure (Pm) behavior depends on the magnetic action energy (W) and the non-conducting gap (Δ) between the inductor and the "lining" under the VTMPW process. The dependence of the shear strength (τ) of the platinum – high-alloyed austenitic steel combination on the pulse energy (W) is shown in Fig. 3b. Low energy does not allow a physical contact and the required combined deformation. Energy restriction "from above" is related to the lining destruction and unacceptable deformation of the welded assembly.

As a result of the experimental studies, a range of definition and effective variation of the key process parameters was established. For experimental samples of the joining metal combinations, their values are as follows: preheating temperature is T0 = 850–1300 K; pulse energy is W = 5–17 kJ.

Heating the components to be joined in vacuum contributes to the activation of the surface layers, which is a prerequisite for setting, reduction of resistance to deformation, and acceleration of stress relaxation in the joining zone; and also, the contact surfaces are cleaned from adsorbed and oxide inclusions [6].

Heating the machined parts leads to a decrease in the mechanical strength of metals, which facilitates the deformation of components reducing the required treatment force. On the other hand, the electrical conductivity of components heated to the temperature of about 1300 K goes up about four times. This causes an increase in the equivalent resistance and inductance of the "inductor – workpiece" system, which contributes to a decrease in the magnetic pressure [8].

The VTMPW parameters, at which quality welds of the combinations (MN40A + Pt alloys, 12X18H10T steels + AlBa, MPVM (Molybdenum Pure Vacuum Melting) molybdenum + PtBa, HB7 alloy + PtBa) were obtained, are given in Table 1.

Table 1. VTMPW parameters of secondary-emission cathodes.

No.	Weld components	T°, K	W, kJ	P_m, 10^7 N/m^2
1	Core - MN40A alloy, d = 12 mm. Emitter is Pt, 0.4 mm thickness	970	6	10.2
2	Core - 12X18H10T steel, d = 12 mm. Emitter is AlBa, 0.5 mm thickness	970	8.67	12.0
3	Core - MPVM molybdenum, d = 17 mm. Emitter is PtBa, 0.4 mm thickness	1170	10.26	14.22
4	Core - HB7 alloy, d = 18 mm. Emitter is PtBa, 0.5 mm thickness	1270	13.23	18.33

The metallographic studies of the joining zone did not detect the formation of common grains on the welding of dissimilar materials. Figure 4 shows microhardness of the joining zone of the cathode combinations of MPVM molybdenum + PtBa, and HB7 alloy + PtBa.

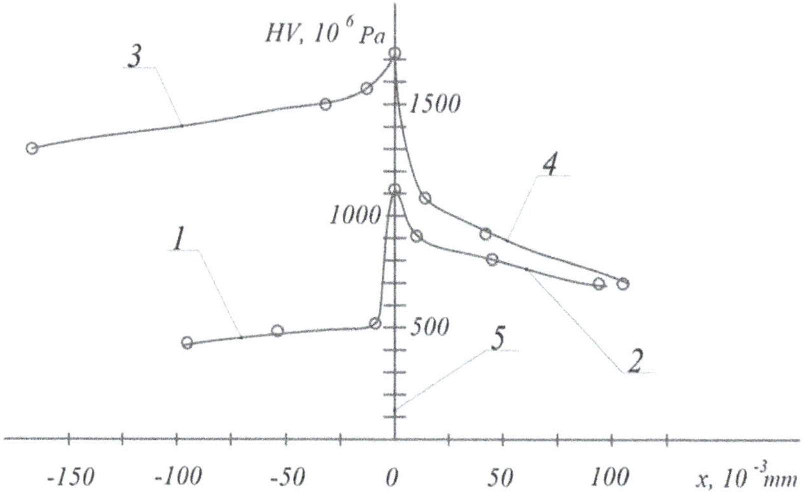

Fig. 4. Microhardness of weld junction: 1is MPVM molybdenum; 2 is PtBa; 3 is HB7 alloy; 4 is PtBa; 5 is weld junction; x is distance from weld junction.

Hardness in the joining zone (3) is increased, which is typical for a solid state welding [9].

Figure 5 shows welded assemblies of the secondary-emission cathodes.

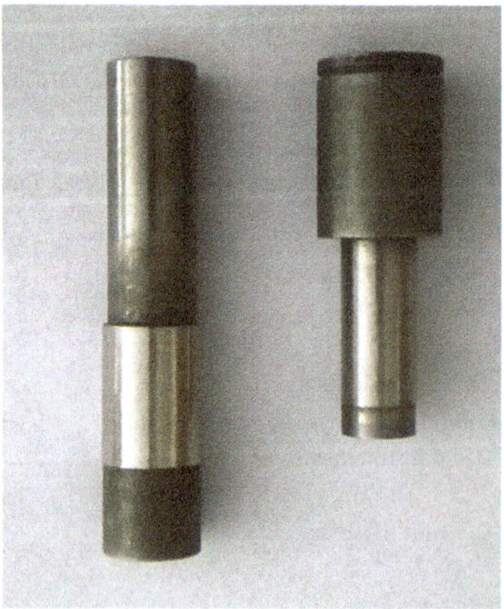

Fig. 5. Cathode welded assemblies obtained through VTMPW.

The VTMPW technology of secondary-emission cathodes of electronic tubes was developed at the "Microengineering" Research Institute in cooperation with DSTU.

4 Conclusions

The application of pulse magnetic pressure under the VTMPW to obtain broad-band lining coatings (l/d > 1–3) from dissimilar materials in the production of female joints instead of the diffusion welding, which enabled to develop the technology of obtaining secondary-emission cathode assemblies, is validated.

The VTMPW parameters and ranges of their variation are identified, which intensify obtaining quality welds.

The analysis of the welds obtained has shown that the joint is formed in a solid state process, which reduces the occurrence of intermetallic phases in the weld, there-by increasing the service life of electronic products.

References

1. Strizhakov, E.L., Neskoromny, S.V., Ageev, S.O.: Weld. Int. **29**(12), 988–990 (2015)
2. Sapanathan, T., Raoelison, R.N., Buiron, N., Rachik, M.: Industrial engineering and management. Joining Technologies, pp. 243–273 (2016)
3. Konyushkov, G.V., Musin R.A.: IPR Media, p. 632 (2009)
4. Baranov, Yu.V., Troitskiy, O.A., Avraamov, Yu.S.: MSIU Public Center, p. 843 (2001)
5. Weddeling, C., Hahn, M., Daehn, G.S., Tekkaya A.E.: International Conference on Manufacture of Lightweight Components, vol. 18, pp. 156–161 (2014)
6. Nescoromniy, S.V., Ageev, S.O., Strizhakov, E.L.: Key Engineering Materials, vol. 684, pp. 185–192 (2016)
7. Kem, A.Yu., Strizhakov, E.L., Sarychev, P.V.: Physics and Chemistry of Materials Treatment, vol. 3, pp. 53–56 (2006)
8. Kukhar, V., Yelistratova, N., Burko, V., Nizhelska, Y., Aksionova, O.: Estimation of occupation safety risks at energetic sector of iron and steel works. Int. J. Eng. Technol. (UAE) **7**(223), 216–220 (2018). https://doi.org/10.14419/ijet.v7i2.23.11922
9. Yakovlev, S.N., Mazurin, V.L.: Vibroisolating properties of polyurethane elastomeric materials, used in construction. Mag. Civ. Eng. **6**, 53–60 (2017). https://doi.org/10.18720/MCE.74.5
10. Strizhakov, E.L., Neskoromny, S.V., Ageev, S.O., Lemeshev, S.V.: Weld. Int. **30**(10), 813–816 (2016)

Expert System Software for Assessing the Technical Condition of Critical Lined Equipment

Vitaliy Yemelyanov[1(✉)], Alexey Nedelkin[2], and Nataliya Yemelyanova[1]

[1] Financial University Under the Government of the Russian Federation, 49 Leningradsky Prospekt, Moscow 125993, Russia
v.yemelyanov@gmail.com
[2] Plekhanov Russian University of Economics, 36, Stremyanny Lane, Moscow 117997, Russia

Abstract. The relevance of solving the problem of automating the process of assessing the condition of critical lined equipment has been described. The necessity of developing software to automate the process under consideration is substantiated. An object software model (including logical and physical models) for an expert system for assessing the condition of critical lined equipment has been developed and described. Based on the proposed object model, the expert system software was developed. This system allows generating control recommendations regarding the operating modes of critical lined equipment.

Keywords: Software · Expert system · Technical condition · Lined equipment

1 Introduction

Industrial production is characterized by a high concentration of hazardous and critical equipment. One of the types of critical equipment at metallurgical and machine-building enterprises is lined equipment, which includes stationary and torpedo ladle cars, steel-teeming ladles. The requirements for diagnostics, monitoring and evaluation of the technical condition of lined equipment are constantly increasing [1, 2], which necessitates the development of new systems and technologies and improvement of existing ones.

At present, the condition of critical lined equipment is assessed using means and systems whose operation under the industrial production conditions is characterized by an unacceptable level of measurement error [3–5]. In addition, the mode of operation of the lined equipment is selected by the technologist based on his personal experience. Also the process of weighing the cast iron in lined equipment is characterized by a sufficiently large measurement error introduced by the curvature of the railroad track and by other factors (subjective factor of a technologist), which causes the sampling of unreliable data on the weighing platform [6, 7]. The known methods and systems for controlling the amount of pig iron in the lined equipment, based on the use of strain gauges [8, 9], implement measurement methods characterized by low accuracy in determining the mass of cast iron in the lined equipment. However, they do not take into account the constant changes in the internal volume of the ladle car and the

Z. Popovic et al. (Eds.): TransSiberia 2019, AISC 1115, pp. 930–937, 2020.
https://doi.org/10.1007/978-3-030-37916-2_92

thickness of the slag layer on the metal surface. There is also a method for measuring the mass of cast iron, by the time of beginning and end of pouring of cast iron into the lined equipment. It is also characterized by an unacceptably large error, due to the different intensities of the fillings [1, 7, 8]. In addition, to implement this method, a technologist is involved, who introduces subjectivity into measurement and an additional error. Due to these facts, there are problematic situations associated with a low level of objectivity in decision-making. Therefore, the research aimed at developing systems and tools for automating the assessment of technical condition and decision-making support during the operation of critical lined equipment is relevant.

2 Development of an Object Model for an Expert System for Assessing the Condition of Critical Lined Equipment

For the purpose of the tool support of the proposed neural network approach for assessing the condition of critical lined equipment [10], the authors have designed and developed a client-server software of the expert system that implements the technologist's graphical interface providing input of the initial data, parameters for calculating and displaying the obtained results.

When designing the structure of the expert system software, an object-oriented approach was used. Unified Modeling Language was used to create an object model of the software being developed [11].

The specification of the developed software combines the following models:

1. The use model as a description of software functionality from the point of view of the technologist responsible for monitoring the condition of lined equipment.
2. The logical model that describes the main abstractions of the diagnostic processes and the assessment of the technical condition of lined equipment, which provide the required functionality and their interaction;
3. The implementation model that defines the real organization of software modules and files in the software developed.

The use model in the form of a diagram of use cases describing the functionality of the expert system. Expert system software is presented as a set of entities and actors interacting with the software using the use cases. The use case diagram is shown in Fig. 1. In order to make the logical presentation and analysis of the structural and functional relationships between the software components of the expert system, a conceptual model is proposed in the form of a set of several class diagrams. One of the class diagrams of the logical model is presented in Fig. 2.

According to Fig. 2, the "SSPrecedents" class includes methods and properties for implementing the method of finding solutions for similar diagnostic situations with lined equipment proposed in [12].

The "SSProductionModel" class implements the production method for solving diagnostic situations with the lined equipment.

The "SSNeuralDSS" class implements methods and properties for creating and training neural networks to support decision making in the diagnosis of the lined equipment. The "SSPredicating" class contains fields for implementing the neural network approach for predicting changes in the condition of the lined equipment proposed in [12].

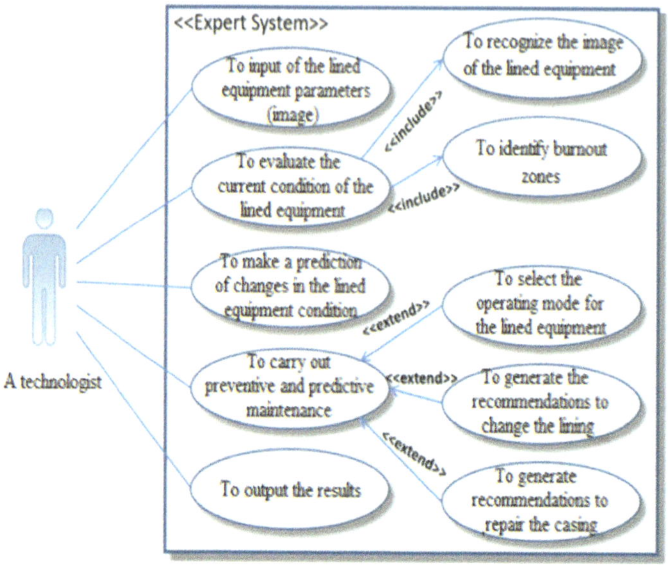

Fig. 1. The use cases diagram of the designed expert system software.

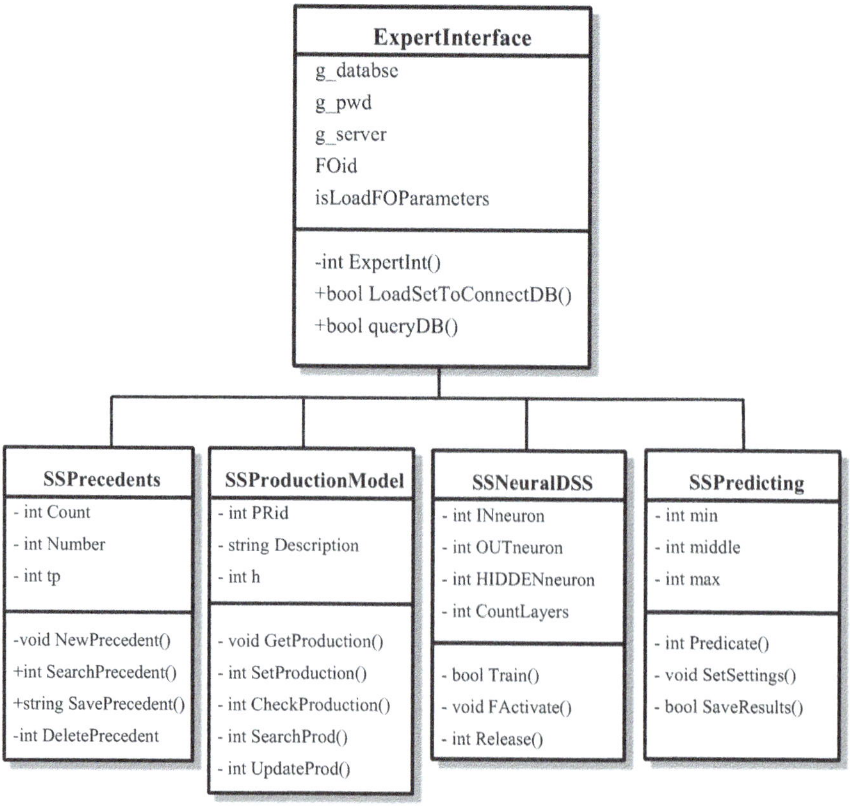

Fig. 2. The classes diagram of the designed expert system software.

To describe the real entities in the work, a physical design was performed and the implementation model presented in Fig. 3 was built. The model presents the features of the physical representation of the system, and establishes the dependencies between software components as well.

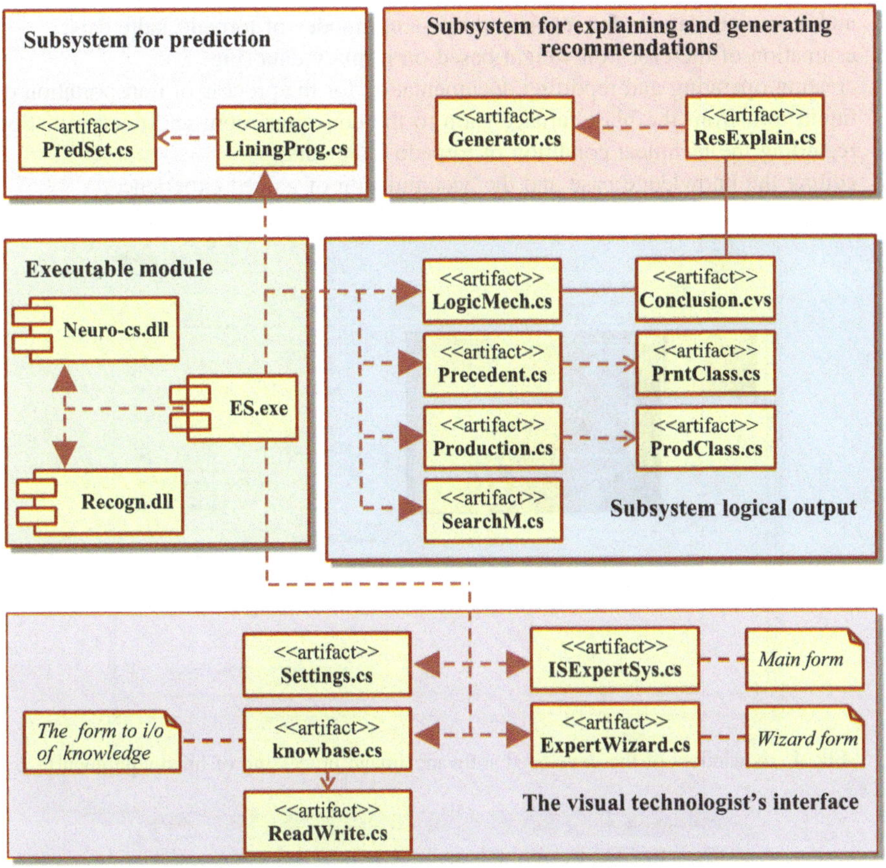

Fig. 3. The model of the implementation of the designed expert system software.

3 Development of an Expert System Software for Assessing the Condition of Critical Lined Equipment

Based on the proposed object model, an expert system software was developed to assess the condition of lined equipment (Fig. 4). The development was carried out in Visual Studio 2010 using the C # programming language.

The functions of the developed expert system software are as follows:

1. receiving and input the primary data on torpedo ladle cars required to evaluate the condition of the ladle car;
2. analysis and quantitative assessment of the lining condition of the ladle cars based (Fig. 4.) on the neural network approach proposed in [10];

3. quantitative assessment of the lining of the torpedo ladle car based on the weight of the empty torpedo ladle car before and after the transport of liquid iron (after each load) and the temperature of the body of the torpedo ladle car during transportation of liquid iron;
4. generating control recommendations regarding the technical condition of the lining and recommendations for repair and operating modes of torpedo ladle cars;
5. estimation of the cast iron weight based on primary data (Fig. 5.);
6. creating operating and reporting documentation for the process of transportation of liquid iron from the blast furnace shop to the converter room and documentation regarding the technical condition of torpedo ladle cars;
7. editing the knowledge base and the accumulation of gained experience.

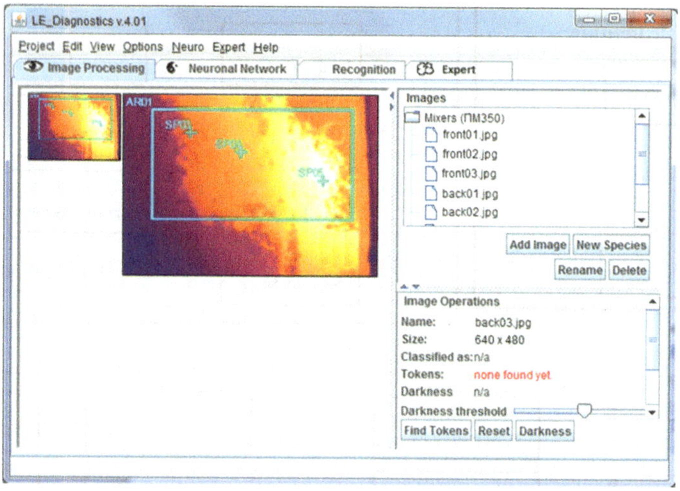

Fig. 4. A window of the developed software (image processing of lined equipment).

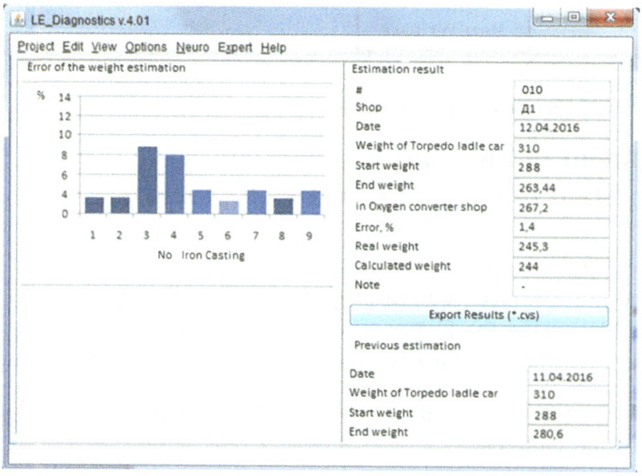

Fig. 5. A window of the developed software (estimation of the cast iron weight).

The developed software was tested in the conditions of the weighing facilities at the Alchevsk Iron and Steel Works, which uses torpedo ladle cars of the PM350 type to transport liquid iron from the blast furnace shop to the converter shop.

Table 1 demonstrates the comparative results of determination the operating modes of critical lined equipment. Analysis of the data from Table 1 provides information about increased operativeness of determination the operating modes of critical lined equipment.

Table 1. Results of determination the operating modes of critical lined equipment.

Lined equipment	Operativeness, average (min) (Basic diagnostic system)	Operativeness, average (min) (Diagnostic system with developed software)
Torpedo ladle cars (type PM350)	84	15
Steel ladle (50 tonnes)	55	10

To test the effectiveness of the developed system, an experiment was carried out to estimate the weight of liquid iron. In the experiment, the authors used data on 120 weighing operations and transportation of liquid iron in PM350 ladle cars at Alchevsk Iron and Steel Works. The results of evaluating the weight of the cast iron are shown in Table 2.

Table 2. Results of the cast iron weight estimation by developed software.

No.	The weight value calculated by a developed software	The weight value by the LS method	The cast iron weight obtained in the converter shop	Least Squares Error, δw	Errors with the developed software, δw
1	290.6	288.1	294.9	2.3%	1.5%
2	299.1	291	320.8	9.3%	6.3%
3	301.9	295.5	310.9	5%	2.8%
4	305.5	299.2	307.7	2.8%	0.7%
…	…	…	…	…	…
45	295.5	285.5	300.1	4.9%	1.5%
46	322.5	317.5	325.1	2.3%	0.8%
…	…	…	…	…	…
119	320.1	300.5	327.9	8.4%	2.4%
120	313.5	305.5	321.35	5%	2.5%
Minimum error		(Weighing operation No. 1, 46) 2.3%			0.6%
Maximum error		(Weighing operation No. 77) 12.5%			6.3%

Table 2 demonstrates the comparative results of estimating the weight of molten iron using the proposed neural network approach and the method of least squares (LS). Data analysis in Table 2 indicates a reduction in the maximum error to 6% while evaluating the weight of liquid iron due to the use of the developed system. The effectiveness of the neural network approach for evaluating the weight of liquid iron is confirmed by a reduction in the level of error in comparison with the method of least squares.

4 Conclusion

The following outputs were obtained:

1. An object model was proposed and described, reflecting the basic functions and entities of the expert system software for assessing the technical condition of critical lined equipment.
2. An expert system software was developed. It implements the functions of its individual subsystems, which allows generating control recommendations regarding the technical condition of the critical lined equipment and the modes of its operation. The developed software was tested under the conditions of metallurgical production in the process of diagnostics of torpedo ladle cars PM350 at Alchevsk Iron and Steel Works.

References

1. Chernyi, S.: Methods for ensuring fault tolerance of equipment based on reliability theory. In: IEEE Conference of Russian Young Researchers in Electrical and Electronic Engineering (EIConRus), pp. 24–48 (2018)
2. Veselkov, V., Vikhrov, N., Nyrkov, A., Chernyi, S., Titov, I.: Development of methods to identify risks to build up the automated diagnosis systems. In: IEEE Conference of Russian Young Researchers in Electrical and Electronic Engineering (EIConRus), pp. 598–601 (2017)
3. Chernopol'skii, A.D., Ageev, S.V., Dmitriev, A.A., Blinnikov, A.A., Glebova, E.S., Kartashev, R.A.: Evaluating the weight of pig iron tapped from a blast furnace. Metallurgist **61**(3–4), 278–282 (2017)
4. Boiko, V.S., Larionov, A.A., Sushchenko, A.V., Klimanchuk, V.V., Semenyuk, P.P.: The complex technology of converter melt at the OAO Mariupol. Metall. Complex Steel Transl. **37**(1), 40–43 (2007)
5. Stan, S., Chisamera, M., Riposan, I., Barstow, M.: Application of thermal analysis to monitor the quality of hypoeutectic cast irons during solidification in sand and metal moulds. J. Therm. Anal. Calorim. **110**(3), 1185–1192 (2012)
6. Petushkov, I.A., Chernopol'skii, A.D., Blinnikov, A.A.: Developing a method for measuring the level and assessing the weight of iron in a pouring ladle in the iron transfer area at the Severstal converter steelmaker. In: Proceedings of the 9th Congress of Steelmakers, pp. 202–205 (2007)
7. Dam, O., Siderúrgica, V.: Pig Iron. A Guide for transporting and handling at terminals, p. 21. International Iron Metallics Association (2014)

8. Anashkin, N.S., Polyakov, N.S., Usov, M.A., Polyakov, V.N.: Use of processed open-hearth dump slag for production of cast iron. Steel Transl. **37**(1), 5–6 (2007)
9. Zorya, V.N., Volynkina, E.P., Protopopov, E.V.: Metallurgical value of converter slag. Steel Transl. **43**(10), 620–625 (2013)
10. Yemelyanov, V.A.: Image processing method for thermal control of the lined objects. Naukovyi Visnyk Natsionalnoho Hirnychoho Universytetu **6**, 137–143 (2014)
11. Booch, G., Rumbaugh, J., Jacobson, I.: The Unified Modeling Language User Guide. Addison-Wesley, Boston (2005)
12. Yemelyanov, V.A.: Intelligent information technology of visual information processing for metals diagnostics. Naukovyi Visnyk Natsionalnoho Hirnychoho Universytetu **4**, 66–73 (2014)

Energy-Efficient Joints for Rolling Stands Screw-Down Mechanisms of Thick Strip Rolling Mills

Viktor Artiukh[1(✉)], Vladlen Mazur[2], Volodymyr Kukhar[3], and Dmitry Vorobev[4]

[1] Peter the Great St. Petersburg Polytechnic University, Polytechnicheskaya 29, 195251 St. Petersburg, Russia
artiukh@mail.ru

[2] LLC 'Saint-Petersburg Electrotechnical Company', Parkovaya Str. 56a Pushkin, 196603 St. Petersburg, Russia

[3] Pryazovskyi State Technical University, Str. Universytets'ka 7, Mariupol 87555, Ukraine

[4] Moscow State University of Technology and Management. K.G. Razumovsky, Zemlyanoy Val Street, 73, 109004 Moscow, Russia

Abstract. Work of hinge of screw-down mechanism of heavily loaded rolling stand of thick strip rolling mill is described and scrutinized. New technical solution in a form of elastic hinge with usage of polyurethane elastic element is proposed with acceptable characteristics. Done theoretical and experimental studies proves its theoretical efficiency.

Keywords: Energy-efficient joints · Rolling stands screw-down mechanisms · Thick strip rolling mills

1 Introduction. Problem Analysis

Screw-down mechanisms of heavily loaded stands of thick strip rolling mills are complex assemblies that consist of numerous details [1–5]. Many of these details work unsatisfactory and have low durability [6, 7]. E.g., breakdowns of screw-down mechanisms nuts are frequent [8] (refer to Fig. 1).

Particularly, above mentioned information applies to thrust bearing of screw-down mechanism [9]. 'Vítkovice' brand tapered roller thrust bearing as well as upgraded cylindrical roller thrust bearing do not meet requirements of efficient operation. Cylindrical rollers and separator of the thrust bearing are unevenly worn that is why it regularly (approximately once every two months) needs to re-equip these bearings discarding extremely deformed rollers. In addition, these rollers are destroyed (due to large contact stresses outer surface of the roller crumbles) while fragments of rollers create additional rolling resistance when screws are in operation and significantly overload drive motors [8].

Average service life of the bearings (including several re-equips) is 12…14 months that means purchase of four sets of the thrust bearings every year that costs around € 200 000. During operation of these bearings several attempts were made to improve

Z. Popovic et al. (Eds.): TransSiberia 2019, AISC 1115, pp. 938–949, 2020.
https://doi.org/10.1007/978-3-030-37916-2_93

Fig. 1. Crack in nut of screw-down mechanism screw of roughing stand of rolling mill 3000.

them including original solutions for bearing design changes, however, most of them were not introduced [10–12].

Analysis of operation of the bearings raises doubts about its necessity because rotation of the screws under load during rolling is absent and intended small torque is not used at all. In addition, this small torque is not so small. Friction torque between the screw and the nut is much higher than friction torque in the thrust (intact) bearing.

Under maximum load the bearing is stationary and its rollers are overloaded by resulting contact stresses that eventually leads to crumbling of surfaces material of the

Fig. 2. General view of crumbling of the rollers surfaces material.

rollers (refer to Fig. 2). These fragments damage previously intact surfaces of the rollers during operation of the screw and as a result they all fail.

Operation of device as bearing occurs only during pauses. In this case load acting on the screw and on the bearing is not 35 MN (as during rolling) and not more than 0.5 MN that is overbalancing force applied to the screw and the bearing. Under these conditions the rolling bearing can be eliminated and a step bearing (with a sliding element in the form of 'steel bearing runner – bronze bearing pad' friction pair which is constantly lubricated with a liquid lubricant) can be installed in its place.

Another detail that requires upgrade or replacement is a spherical hinge. Purpose of this hinge is to eliminate bending moment that can be transmitted to the screw, nut and housing due to distortions and displacements of chock (including due to displacements caused by gaps between chocks and housings). Operating experience of stands of thick strip rolling mill 3000 produced by 'Vítkovice' shows that real angles of rotation in the existing hinge are small (around a fraction of a degree). They are determined by the real gaps between the chocks of rolls and housings windows. On practice, such hinge does not meet operational requirements. Due to large values of force and diameter as well as due to small radius of curvature frictional torque arising in this hinge is so large that it can have a negative effect on strength of the screw, nut and housing [13]. In addition, this friction torque does not depend on angle of rotation in the hinge and occurs at any small angles of rotation.

Objectives of this paper is to scrutinize work of the hinge and propose new technical solutions with acceptable characteristics. New hinge can be created on the basis of elasticity and low compressibility in closed volume of modern polyurethane elastomers [14, 15]. This paper describes choice of material, dimensions, manufacturing errors, friction coefficients and friction torques in proposed hinge.

2 Formulation of Task

Development and study of new type hinge in which angle of rotation is proportional to applied torque and hinge taken out of its equilibrium position returns to its original position.

3 Materials and Methods

Since skew angles of the screws arising in the considering spherical hinge are very small it would be logical to suggest technical solution in which resulting bending moment is proportional to the skew angle. As a solution 'elastic hinge' (EH) can be offered. Principal design of EH is shown on Fig. 3.

EH consists of a durable steel base (housing) pos. 1 in which elastic element (EE) pos. 2 made from elastomer is placed without gaps and a plunger pos. 3 is supported on the EE. When there are no distortions force acting on the screw is evenly distributed on the EE which will be compressed.

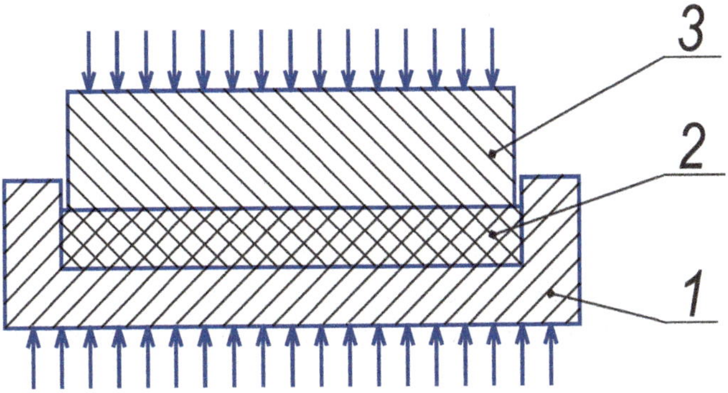

Fig. 3. Principal design of EH.

It is known [16] that elastomers are low-compressible materials. For these materials the bulk modulus of elasticity is approximately the same and equal to K = 3000… 3500 MPa for all class of elastomers starting from soft rubbers with compression modulus of elasticity E_c = 5…8 MPa till high hardness polyurethanes with E_c = 300… 450 MPa [18].

It is obvious that the bulk modulus of elasticity is a fundamental characteristic for elastomers. The constancy of the bulk modulus of elasticity for all elastomers suggests that rigidity of the designed EH cannot be adjusted by choosing type/brand of elastomer; if it is necessary to adjust its stiffness, first of all, diameter and thickness of the EE can be foreseen due to geometrical parameters of the EH.

For normal operation of rolling stand it is necessary to determine how its rigidity will be changed when the EH is installed into it. Initial data for calculation:

1. Overall dimensions of the EE: diameter D = 800 mm; thickness δ_{EE} = 30 mm.
2. Maximum compression force of the EE: P = 35 MN.

Compression of the EH under force P

$$\lambda = \frac{P \cdot \delta_{EE}}{K \cdot F} \tag{1}$$

where $F = \frac{\pi D^2}{4}$;

For described EH F = 5024·10–4 m2; K = 3500 MPa.

Then $\lambda = \frac{35 \cdot 10^6 \cdot 30 \cdot 10^{-3}}{3500 \cdot 10^6 \cdot 5024 \cdot 10^{-4}} = 0.597 \cdot 10^{-3}\,\mathrm{m} \approx 0.6\,\mathrm{m}$.

Thus, the EE compression (or additional roll gap) is 0.6 mm that can be neglected for roughing rolling stand or it can be taken into account during the first pass in finishing rolling stand. Maximum compressive stresses:

$$\sigma_{\max} = \frac{P}{F} = \frac{35 \cdot 10^6}{5024 \cdot 10^{-4}} = 69.7 \cdot 10^6\,\mathrm{N/m^2} \approx 70\ \mathrm{Mpa}.$$

Such stresses at cubic compression are not dangerous for polyurethane. There are devices in technics with polyurethane EEs operating under conditions of cubic compression with stress σmax = 160…240 MPa. At so high pressure the main problem is strength of the EH body and not strength of polyurethane EE that is almost always provided.

Another problem is with sealing. Polyurethane can reasonably be considered as sealing material, so in most cases additional seals are not required. Permissible radial clearance depends on two parameters: (a) working pressure p (MPa); (b) rigidity of polyurethane that is convenient to set by compression modulus of elasticity Ec.

Experiments with different materials (various grades polyurethanes) and pressures were carried out in laboratory of 'Hydraulics and Strength' department of 'Peter the Great St. Petersburg Polytechnic University' (SPbPU). EH model corresponded to one shown on Fig. 3. Inner diameter of the housing was D0 = 100 mm. Diameters of the plungers differed by δ = 0.1 mm. There were 10 plungers. The smallest plunger had dmin = 99 mm and the largest one had dmax = 99.9 mm. Thus, the total gap δ in diameters varied within 0.1 mm $\leq \delta \leq$ 1.0 mm. Pressure created in the EH were within 40 MPa $\leq p \leq$ 100 MPa.

Cases of the polyurethane EE entering the gap δ with its subsequent local destruction were recorded during the tests. Main results are presented in Table 1. For all loading cases maximum allowable gaps δ are shown ensuring safety of the EE and eliminating need for special seal. Friction coefficients were determined under conditions of friction without grease and friction with grease (refer to Tables 2, 3 and 4) where polyurethane grade SCU-PFL is CIS State brand.

Selection of the EE material for the EH was performed considering several parameters:

Table 1. Allowable gaps in the EH, mm.

Pressure, MPa	Test conditions							
	Without grease, hardness ShA				With grease, hardness ShA			
	80	90	95	97	80	90	95	97
40	0.8	>1.0	>1.0	>1.0	0.7	0.9	>1.0	>1.0
60	0.7	0.9	>1.0	>1.0	0.6	0.75	0.9	>1.0
80	0.6	0.8	>1.0	>1.0	0.5	0.65	0.8	0.9
100	0.5	0.7	0.9	>1.0	0.4	0.55	0.7	0.8

Table 2. Study of friction coefficients under conditions of friction without grease.

P, N	Measurements by dynamometer, kN							
	Steel	Bronze	Steel	Steel	Steel	Polyurethane (grade SCU-PFL)	Steel	Polyurethane (grade adiprene)
90	0.018		0.02		0.04		0.05	
180	0.036		0.04		0.08		0.1	
270	0.054		0.06		0.12		0.15	
360	0.072		0.08		0.16		0.2	
Friction coefficient	0.2		0.22		0.44		0.56	

Table 3. Study of friction coefficients under conditions of friction with liquid lubrication.

P, N	Measurements by dynamometer, kN							
	Steel	Bronze	Steel	Steel	Steel	Polyurethane (grade SCU-PFL)	Steel	Polyurethane (grade adiprene)
90	0.015		0.01		0.02		0.02	
180	0.03		0.02		0.04		0.04	
270	0.045		0.03		0.06		0.06	
360	0.06		0.04		0.08		0.08	
Friction coefficient	0.12		0.11		0.22		0.22	

Table 4. Study of friction coefficients of pair steel - polyurethane under conditions of friction with different lubrication.

P, N	Friction force, kN		
	Steel	Polyurethane (grade adiprene)	
	Grease (a lot)	Grease (little)	Water lubrication
90	0.004	0.02	0.055
180	0.008	0.03	0.11
270	0.012	0.04	0.165
360	0.016	0.05	0.22
Friction coefficient	0.04	0.16	0.61

Strength. Stress state of the EE of the proposed EH is close to the case of uniform cubic compression when equivalent stress is close to zero; any elastomer can withstand such loading.

Stiffness. Experiments on various elastomers (mainly different rubber grades and structural polyurethanes) show that during deformation in closed volume they have almost the same rigidity. For these materials the bulk modulus of elasticity is around K = 3200…3500 MPa.

High rigidity of used polyurethanes can significantly reduce requirements for precision manufacturing of the EH parts.

Nowadays, industry of the Russian Federation produces enough grades of molded structural polyurethanes of various rigidity. Materials brands are adiprene, SCU-PFL, vibratane, desmopan, elast, etc. Data given in Table 1 allows to select not only material of the EE but also to assign dimensions and tolerances for the main parts. It should be noted that the gap size of 0.4 mm (0.2 mm per side) in all cases ensures reliable operation of the EH. It does not require special seal even when grease is used. For the considered EH the presence of grease is almost impossible to avoid so it needs to be focused on the right side of Table 1. Any polyurethane with hardness of more than ShA 80 can be chosen. Compression modulus of elasticity of polyurethane depends on hardness of polyurethane. Table 1 covers values of the modules 20 MPa $\leq E_c \leq$ 100 MPa that provides wide selection.

Performance of the EH also should be considered with misalignment of the chocks and the screws due to gaps in their joints. In this case load component associated with a decrease of volume of the elastomer will not be taken into account. It means that objective of obtaining a working characteristic is to determine a bending moment M_b corresponding to this skew angle

$$\frac{1}{\rho} = \frac{M_b}{EI} \tag{2}$$

Skew angle

$$\varphi = \frac{\delta_{EE}}{\rho} \tag{3}$$

Then

$$M_b = \frac{\varphi \cdot EI}{\delta_{EE}} \tag{4}$$

Where ϕ is misalignment angle in the EH, rad;

EI is bending stiffness of the EE, N·m^2;

δ_{EE} is thickness of the EE, m.

Linear relation of applied moment M_b and misalignment angle ϕ can be found from formula (4). Graph of that relation is shown on Fig. 4.

With overall dimensions of the considered hinge and actually observed gaps the largest misalignment angle does not exceed 0.5°, i.e. $\phi \leq 0.01$ rad.

Bending moment is equal to

$$M_b^{\max} = 0.01 \cdot \frac{60 \cdot 10^6 \cdot 0.1 \cdot 0.8^4}{30 \cdot 10^{-3}} = 820 \cdot 10^3\,\mathrm{N \cdot m} = 820\,\mathrm{kN \cdot m}.$$

This moment is one order of magnitude less than the moment arising in the spherical hinge. In addition, the real misalignment angles are significantly less than $\phi_{max} = 0.01$. If necessary this bending moment can also be reduced by changing size of the EH and the EE material.

Replacing an initial cylindrical roller bearing (it is combined in terms of friction conditions because it has rolling of rollers inside bearing and their sliding on support plates) with a steel-bronze friction pair thrust sliding bearing will avoid: frequent stops, sorting of rolling elements, reloading the bearing and replacement of it when it fails due to increased contact stresses. Transition to a bronze bearing will significantly reduce contact stresses; at the same time friction torque during rotation of the screw may increase slightly. Below is strength and performance calculation of the thrust bronze bearing.

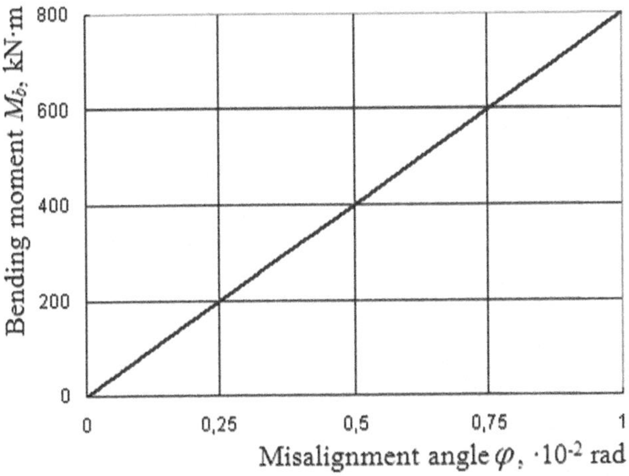

Fig. 4. Linear relation of applied moment M_b and misalignment angle φ.

Initial data for calculation:

1. Biggest load acting on the step bearing P = 35 MN;
2. Dimensions of the bearing: D = 860 mm is outer diameter of a bronze ring;

   d = 320 mm is inner diameter of the bronze ring.
3. Overbalancing force during rolling acting on one screw Q = 0.5 MN.

Maximal pressure acting on the bronze ring bearing pad

$$p = \frac{P}{F} = \frac{35 \cdot 10^6}{0.4} = 87.5 \cdot 10^6 N/m^2 = 87.5 \text{ MPa},$$

where $F = \varphi \frac{\pi}{4}(D^2 - d^2) = 0.8 \cdot \frac{3.14}{4}(0.86^2 - 0.32^2) = 0.4\,\text{m}^2$;

$\phi = 0.8$ is coefficient of utilization of support surface which depends on width and number of lubrication grooves.

Tensile strength of tin bronzes is 200…400 MPa that is substantially more than current stress.

This calculation shows that any tin bronze will be suitable for manufacture of the bearing pad. Overbalancing force is $Q = 0.5 \cdot 10^6$ $N = 0.5$ MN per screw at installation movement of the screw. This gives a specific pressure

$$p = \frac{Q}{F} = \frac{0.5 \cdot 10^6}{0.4} = 1.25 \cdot 10^6 \frac{N}{m^2} = 1.25 \text{ MPa}.$$

Allowable pressures are $15\ MPa \leq [p] \leq 200\ MPa$ that is significantly more than acting one. On the basis of the performed calculations it is possible to accept the bronze grade Br OCN-5-2-5 (CIS State grade) with a tensile strength of $\sigma_t = 350$ MPa.

In addition, verification is performed on the heating according to the criterion PV which is calculated by the formula

$$PV = \frac{Q \cdot n}{6000 \cdot b} \tag{5}$$

where n is speed of rotation of the screw, rev/min.

$b = \frac{D-d}{2} = \frac{86-32}{2} = 27$ cm is nominal radius (width) of the support.

Maximum speed of rotation is

$$n_{\max} = \frac{n_M}{U_G} = \frac{840}{13.25} = 63.4 \, \text{rev/min},$$

Where $n_M = 840$ rev/min is speed of rotation of drive motor;

$U_G = 13.25$ is gear ratio of drive gearbox.

Then value of PV is equal to

$$PV = \frac{0.5 \cdot 10^6 \cdot 63.4}{6000 \cdot 27} = 195 \frac{N \cdot m}{cm^2 \cdot \text{sec}}$$

Permissible values of [PV] = 1000…1200 for tin bronzes.

There will be no overheating of the thrust bearing because the calculated value PV is significantly less than the permissible ones [PV].

Experiments were conducted to determine friction coefficients for different friction pairs and friction conditions to test the drive motor on possibility of overcoming the friction torque. ‘Steel-polyurethane’, ‘steel-bronze’ and ‘steel-steel’ friction pairs were checked. In this case, three types of polyurethane were used, among them are adiprene, vibratane, SCU-PFL (refer to Tables 2, 3 and 4).

Frictional torque on the bronze ring bearing pad at installation pressure of the rolls

$$T_f = 0.5 \cdot Q_2 \cdot f \cdot d_m \tag{6}$$

where $Q_2 = 0.5\ MN = 500\ kN$ is axial load acting on the bearing when the screw rotates;

$f = 0.12$ is friction coefficient at ‘steel-bronze’ boundary friction;

$d_m = \frac{D+d}{2} = \frac{860+320}{2} = 590$ mm is middle diameter of the bronze ring bearing pad. It equals to

$$T_f = 0.5 \cdot 500 \cdot 0.12 \cdot 0.59 = 17.7\,\text{kN} \cdot \text{m}.$$

Friction torque in thread of the screw

$$T_{th} = Q_2 \cdot \frac{d_p}{2} \cdot tg(\beta + \rho) \tag{7}$$

where d_p is pitch diameter of the thread, $d_p = 0.68$ m;

$$tg(\beta + \rho) = tg(5°10' + 1°10') = tg6°20' = 0.111,$$

where $\beta = arctg\frac{t_{mid}}{\pi \cdot d_{mid}} = arctg\frac{40}{3.14 \cdot 680} = arctg0.002 = 1°10'$.
Where $t_{mid} = 40$ mm is middle pitch of the thread.
It equals to

$$T_{th} = 500 \cdot \frac{0.68}{2} \cdot 0.11 = 18.9\,\text{kN} \cdot \text{m}.$$

Total torque on the screw

$$T = T_f + T_{th} = 17.7 + 18.9 = 36.6\,\text{kN} \cdot \text{m}.$$

Torque on drive motor shaft

$$T_M = \frac{T}{i} = \frac{36.6}{13.25} = 2.76\,\text{kN} \cdot \text{m}.$$

Nominal torque of drive motor

$$T_{nom} = 3.35\,\text{kN} \cdot \text{m},$$

It means that $T_{nom} > T_M$ so there is no overheat of the drive motor.

4 Results

1. Proposed in this paper technical solution (EH) is theoretically efficient and needs experimental confirmation.
2. Investigated in this paper EH is ready for industrial test in roughing stand of thick strip rolling mill. It is proposed to use a polyurethane disk located in a confined space as the EH.
3. Industrial test of the EH allows to clarify results of its theoretical studies (for example, grade of polyurethane and the optimal size of radial gaps between the polyurethane disk and the housing).

5 Conclusions

1. Work of the hinge of screw-down mechanism of heavily loaded rolling stand of numerous thick strip rolling mills is described and scrutinized.
2. New technical solution in a form of the EH with usage of polyurethane EE is proposed with acceptable characteristics. Done theoretical and experimental studies proves its theoretical efficiency.
3. Industrial test of the EH needs to be done in roughing stand of thick strip rolling mill.

Acknowledgement. The reported study was funded by RFBR according to the research project № 19-08-01241a. The authors declare that there is no conflict of interest regarding the publication of this paper.

References

1. Mazur, V., Artyukh, V., Artyukh, G., Takadzhi, M.: Current views on the detailed design of heavily loaded components for rolling mills. Eng. Des. **37**(1), 26–29 (2012)
2. Artiukh, V.G.: Nagruzki i peregruzki v metallurgicheskih mashinah [Loads and overloads in metallurgical machines]. PSTU (Pryazovskyi State Technical Uni-versity), Mariupol (2008)
3. Baskov, S.N., et al.: Razrabotka matematicheskoy modeli vzaimosvyazannykh elektrotekhnicheskikh sistem kleti tolstolistovogo prokatnogo stana [Development of mathematical model of interconnected electrical systems of stand of thick plate rolling mill] news of higher educational institutions. Electromechanics **60**(6), 55–64 (2017)
4. Rudaev, Y.I., et al.: Sinergetika v mekhanike deformiruyemogo tela [Synergetics in mechanics of deformable body]. News of Kyrgyz-Russian Slavic Univ. **10**(10), 123–128 (2010)
5. Kukhar, V., Prysiazhnyi, A., Balalayeva, E., Anishchenko, O.: Designing of induction heaters for the edges of pre-rolled wide ultrafine sheets and strips correlated with the chilling end-effect. In: Modern Electrical and Energy System (MEES 2017), pp. 404–407. Kremenchuk Mykhailo Ostrohradskyi National University, Kremenchuk, Ukraine (2017). https://doi.org/10.1109/MEES.2017.8248945
6. Sorochan, E., Artiukh, V., Melnikov, B., Raimberdiyev, T.: Mathematical model of plates and strips rolling for calculation of energy power parameters and dynamic loads. In: MATEC Web of Conferences, vol. 73, p. 04009 (2016). http://dx.doi.org/10.1051/matecconf/20167304009
7. Artiukh, V., Mazur, V., Butyrin, A.: Analysis of stress conditions of rolling stand elements. In: Murgul, V., Popovic, Z. (eds.) International Scientific Conference Energy Management of Municipal Transportation Facilities and Transport EMMFT 2017. EMMFT 2017. Advances in Intelligent Systems and Computing, vol. 692, pp. 212–219. Springer, Cham (2018). https://doi.org/10.1007/978-3-319-70987-1_23
8. Artiukh, V.G.: Osnovi zashiti metallurgicheskih mashin ot polomok [Basics of protection of metallurgical machines from breakdowns]. PSTU (Pryazovskyi State Technical University), published group 'University', Mariupol (2015)
9. Adamenko, O.V., et al.: Diagnostika upornykh podshipnikov nazhimnykh vintov prokatnykh kletey [Diagnostics of thrust bearings of screws of screw-down mechanisms of rolling stands]. Metall. Process. Equipment **3**(37), 87–90 (2014)

10. Kitanov, A.A., et al.: Sovershenstvovaniye gidroborudovaniya kleti « kvarto » LPTS-1 OAO « Ural'skaya stal' » [Improve of hydraulic equipment of 4-hi rolling stand of PRS No.1 of OJSC 'Ural Steel']. Metall. Process. Equipment **3**, 59–63 (2014)
11. Gasiyarov, V.R., et al: Ogranicheniye dinamicheskikh nagruzok elektromekhanicheskikh sistem kleti prokatnogo stana za schet regulirovaniya zazora valkov [Reduction of dynamic loads of electromechanical systems of rolling mill stand by adjusting rolls gap]. News South Ural State Univ. **18**(2), 82–95 (2018)
12. Kitanov, A.A., et al.: Rekonstruktsiya ustroystva uravnoveshivaniya verkhnego opornogo valka kleti «kvarto» LPTS-1 OAO «Ural'skaya stal'» [Reconstruction of balancing device of top back-up roll of 4-hi rolling stand of PRS No.1 of OJSC 'Ural Steel']. Sci. Prod. Ural **11**, 121–124 (2015)
13. Gharaibeh, N.S., Matarneh, M.I., Artyukh, V.G.: Loading decrease in metallurgical machines. Res. J. Appl. Sci. Eng. Technol. **8**(12), 1461–1464 (2014)
14. Kukhar, V., Yelistratova, N., Burko, V., Nizhelska, Y., Aksionova, O.: Estimation of occupation safety risks at energetic sector of iron and steel works. Int. J. Eng. Technol. (UAE) **7**(2.23), 216–220 (2018). https://doi.org/10.14419/ijet.v7i2.23.11922
15. Yakovlev, S.N., Mazurin, V.L.: Vibroisolating properties of polyurethane elastomeric materials, used in construction. Mag. Civil Eng. **6**, 53–60 (2017). https://doi.org/10.18720/MCE.74.5
16. Al-Quran, F.M.F., Matarneh, M.E., Artukh, V.G.: Choice of elastomeric material for buffer devices of metallurgical equipment. Res. J. Appl. Sci. Eng. Technol. **4**(11), 1585–1589 (2012)
17. Artiukh, V.G., Karlushin, S.Y., Sorochan, E.N.: Peculiarities of mechanical characteristics of contemporary polyurethane elastomers. Procedia Eng. **117**, 938–944 (2015). https://doi.org/10.1016/j.proeng.2015.08.180
18. Kukhar, V., Artiukh, V., Prysiazhnyi, A., Pustovgar, A.: Experimental research and method for calculation of 'upsetting-with-buckling' load at the impression-free (dieless) preforming of workpiece. In: E3S Web of Conferences, vol. 33, p. 02031 (2018). https://doi.org/10.1051/e3sconf/20183302031

Protection of Metallurgical Machines from Breakdowns at Iron and Steel Works

Viktor Artiukh[1], Vladlen Mazur[2](✉), Yurii Sahirov[3], and Nadezhda Kapustina[4]

[1] Peter the Great St. Petersburg Polytechnic University, St. Petersburg 195251, Russia
[2] LLC "Saint-Petersburg Electrotechnical Company", St. Petersburg 196603, Russia
vladlenmazur@gmail.com
[3] Pryazovskyi State Technical University, Mariupol 87500, Ukraine
[4] Russian University of Transport, Obraztsova str., 9/9, Moscow 127994, Russia

Abstract. Problems of protection systems of metallurgical machines against breakdowns are described. Technical solutions to improve durability of metallurgical equipment are given. Engineering, scientific and educational activities of specialized laboratory 'Protection of metallurgical machines from breakdowns' with its program are described that allows reduction of expenses of equipment modernization by around ten times. Economic benefit of the laboratory is around €1.5 million per year due to absence of accidental breakdowns.

Keywords: Protection of metallurgical machines · Breakdowns

1 Introduction

Practically, Iron and Steel Works of the Commonwealth of Independent States (CIS) countries have full set of metallurgical production. For example, PJSC 'Ilyich Iron and Steel Works' (Mariupol city, Ukraine) has Sinter Factory, Blast Furnace Shop, Basic Oxygen Converter Shop with Steel Continuous Casters, Rolling production that consists of Reduction Shop 1150 Slabbing, Continuous Hot Wide Strip Rolling Mill (HRM 1700), Hot Plate Rolling Mill 3000, Cold Wide Strip Rolling Mill (CRM 1700) and Hot Pipe Rolling Shop with two Pilger stands 6 ÷ 12".

Mechanical equipment of these shops has been working during last 25…70 years. Better parts of it have become obsolete morally and demand reconstruction as well as equipment of other Iron and Steel Works [1–3]. Such modernizations are basically limited due to absence of funds because price of metallurgical equipment is very high.

That is why Head Mechanic Department (HMD) of PJSC 'Ilyich Iron and Steel Works' together with Pryazovskyi State Technical University (PSTU, Ukraine) and Peter the Great St. Petersburg Polytechnic University (SPbPU, Russian Federation) employees have developed program of metallurgical machines partial modernization by creation for them systems of protection against breakdowns (e.g. broken work rolls (WRs) of HRM are shown on Fig. 1). This program allows reduction of expenses for

Z. Popovic et al. (Eds.): TransSiberia 2019, AISC 1115, pp. 950–962, 2020.
https://doi.org/10.1007/978-3-030-37916-2_94

modernization by around ten times [4]. The program assumes to achieve the reduction or restriction of loads arising in machines during their work.

Fig. 1. Broken WRs of HRM.

Reductions of operating loads can probably be achieved for all metallurgical machines because in them there are very high levels of non-technological loads that do not take part in technological processes [5]. Non-technological loads depend on geometrical and constructive imperfections of machines, errors of parts installation, increased deterioration of details. It leads to impacts, fluctuations and, frequently, to accidental breakdowns of expensive details.

Attempts to use protection systems against accidental breakdowns were undertaken repeatedly however achieved results always were limited as there was no opportunity to solve 'as a complex' problems of protection against breakdowns [6–9].

Since 2001 these works have been concentrated and resolved in the HMD together with employees of 'Strength of Materials' department of PSTU and 'Hydraulics and Strength' department of SPbPU under the contract about scientific and technical cooperation. The main weak points have been revealed, among them are roller tables rollers, stand rollers of all rolling mills, big rolling bearings, etc. In 2001–2002 theoretical and laboratory researches and studies of various amortization systems of shock/impact loads were carried out. Thus, efficiency of energy-efficient elastic elements in a form of steel springs of various devices and springs from polymeric [10, 11] and elastomeric materials were analyzed. It allowed to choose type of elastic element with the biggest energy-efficiency. Such element has appeared to be a monoblock from constructional polyurethane [12–15] working with axial compression or shift. Similar shock-absorbers have been developed for roller table rollers of stand rollers of rolling mills of the Iron and Steel Works. They were produced in 2003–2004 and successfully passed industrial approbation in roller tables of stand rollers of 1150 Slabbing. Design and stress calculation of these shock-absorbers were done by means of FEM analysis where conditions of contact surfaces interactions, materials properties and geometry of the roller tables are actual according to scientific and practical suggestions [16–18] that helped to design mesh with great accuracy. Main task of FEM analysis was research of influence of friction coefficient on end surfaces of elastic element (refer to Fig. 2) working with axial compression on its rigidity and strength. Loading scheme of general design of the roller table roller shock-absorber based on FEM model (quarter of symmetrical assembly) is shown on Fig. 3.

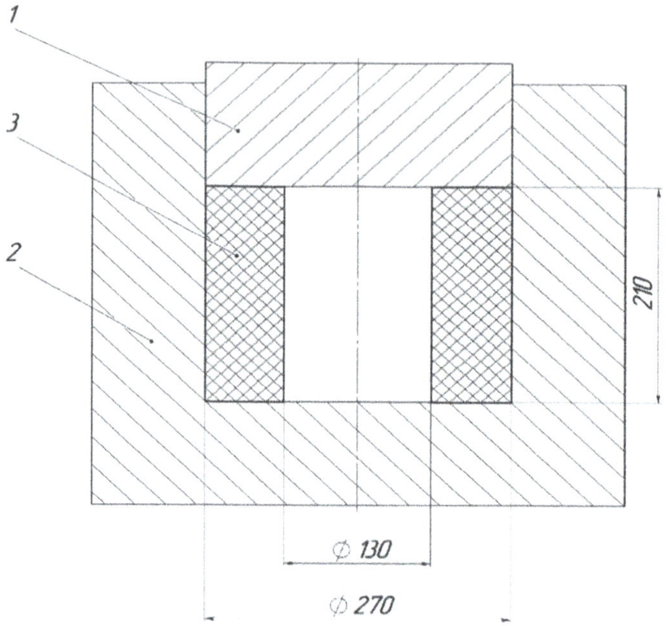

Fig. 2. Cross section of general design of the roller table roller shock-absorber/pos. 1 is movable top seat made from steel 080 M 46 according to B.S. 970; pos. 2 is fixed metal housing made from steel 080 M 46 according to B.S. 970; pos. 3 is monoblock made from polyurethane with technical characteristics: Young's modulus E = 37.9 MPa; Poisson's ratio $\mu = 0.498$; density $\rho = 1200$ kg/m^3.

Fig. 3. Loading scheme of FEM model by total load equal to 20 tons.

FEM calculations results of the roller table roller shock-absorber during loading by 20 tons when friction coefficient f = 0.3 are shown on Figs. 5, 6 and 7. FEM calculations results of the roller table roller shock-absorber during loading by load P = 5 ÷ 30 tons with friction coefficient f = 0.15 ÷ 0.30 are given in Table 1 (Fig. 4).

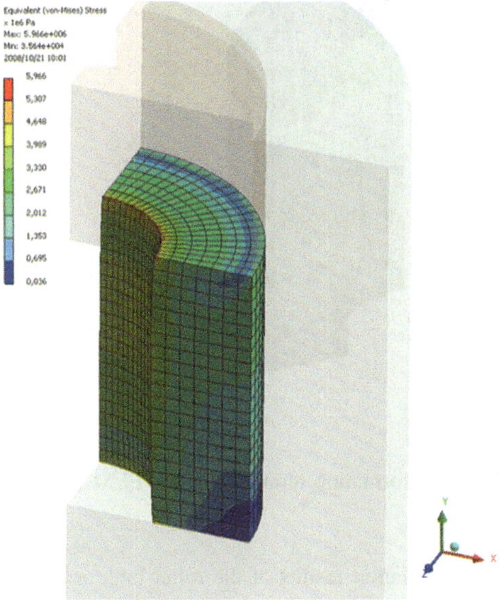

Fig. 4. Von-Mises Stress at FEM calculation.

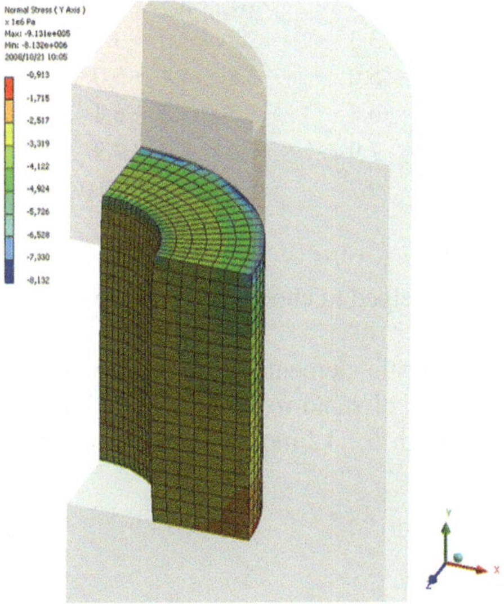

Fig. 5. Normal stress along Y axis at FEM calculation.

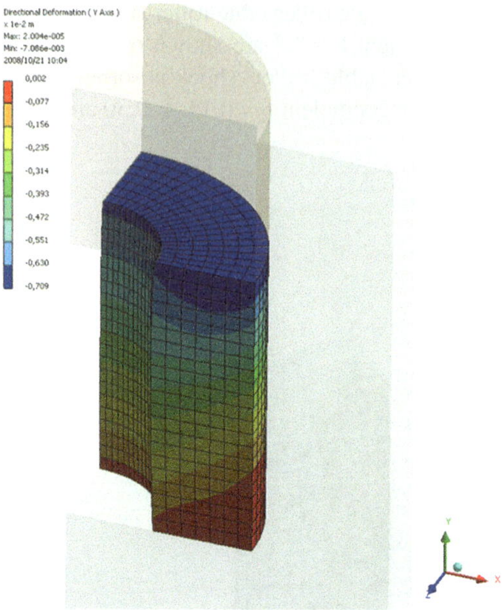

Fig. 6. Deformation along Y axis at FEM calculation.

Table 1. FEM calculation results of the roller table roller shock-absorber.

#	P, tons	Von-Mises stress, MPa			Normal stress along Y axis, MPa			Deformation along Y axis, mm		
		f = 0.15	f = 0.30	f = 0.45			f = 0.15	f = 0.30	f = 0.45	
1	5	1.8	2.0	1.6	−1.7	−2.3	−1.5	2.2	1.8	1.4
2	10	2.7	3.1	3.3	−2.7	−3.0	−3.3	4.4	3.6	2.9
3	15	4.0	4.3	5.0	−3.8	−4.4	−5.0	6.6	5.3	4.3
4	20	5.7	6.0	6.4	−5.3	−5.8	−6.4	8.8	7.1	5.7
5	25	6.7	7.2	7.9	−7.2	−7.0	−7.9	10.7	8.7	7.2
6	30	7.8	8.5	9.7	−8.6	−8.5	−10.0	13.1	10.6	8.6

FEM calculation results helped to choose proper type of polyurethane and design of elastic element.

Durability of installed elastic elements (general view of shock-absorbers is shown on Fig. 7 where pos. 1 is part of stand rollers unit, pos. 2 is elastic element) is around 24 months in comparison with 6…12 months of previous unit design.

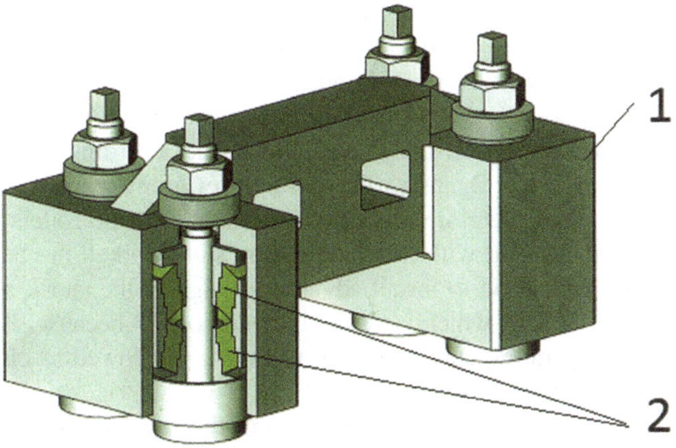

Fig. 7. General view of shock-absorbers installed in roller tables of stand rollers of 1150 Slabbing.

Practically, similar shock-absorbers installation does not require changes of roller tables frames because these shock-absorbers can be installed in initial slots intended for steel springs installation. Thus, shock-absorbers from constructional polyurethanes with initial overall dimensions are appropriate for available slots and provide better energy-efficiency (bigger by 2…4 times) that provides better protection of metallurgical equipment and greater durability of elastic elements [19].

As for design of elastic shock-absorbers for roller table rollers which reduce horizontal impacts [20] of equipment parts, these shock-absorbers can prevent rollers chocks and roller tables frames slots damages. On the basis of thermoplastic polyurethanes small-sized (thickness is equal to 15 ÷ 30 mm) shock-absorbers have been developed and tested in laboratory conditions. After good results of laboratory researches the shock-absorbers designed to reduce rollers chocks horizontal impacts [21–23] were produced for roller tables of Reduction Shop 1150 Slabbing. Provided design of the shock-absorbers consists of metal plate and ridged polyurethane plates glued to it (refer to Fig. 8). Industrial approbation of these shock-absorbers was in 2003–2004, proposed technical solution confirmed theoretical working capacity, however it was not possible to achieve full success because of low stability of the glutinous connection.

Fig. 8. General view of shock-absorbers designed to reduce rollers chocks horizontal impacts.

Durability of the shock-absorbers shown on Fig. 8 was from one to eight months. Such wide durability range is due to various operating conditions of the shock-absorbers on different rollers. The main reason is that slots of roller tables frames, repeatedly restored after damages, had significant deviations from initial/project sizes. Installed shock-absorbers in roller tables slots were made according to the project sizes that resulted in occurrences of gaps various for each of four experimental rollers (gap was around 3 ÷ 15 mm), therefore operating conditions of these rollers were various. The shock-absorbers installed with minimal gaps in slots worked the best.

A new decision was made to install advanced design of the shock-absorbers only on new roller tables sections which were not in operation yet because of impossibility to provide identical operating conditions for all rollers. It allowed to obtain objective assessment of similar shock-absorbers works.

The main characteristic of some roller tables of the Iron and Steel Works rolling mills is full absence of shock-absorbers e.g. on roller tables of HRM 1700. Thus, rolling bearings durability was several months. So, for the roller tables of HRM 1700 roughing stands group average durability of rolling bearings did not exceed 6 months, for conic roller tables of Hot Plate Rolling Mill 3000 durability of rolling bearings did not exceed 3 months.

Because of absence of shock-absorbers installation place small-sized polyurethane shock-absorbers in a form of rings installed on external rings of rolling bearings were developed and tested in laboratory conditions. These ring-shock-absorbers were installed on bearings of four roller tables rollers of HRM 1700 roughing stands group in 2004–2005 with industrial approbation within 20 months that showed their high efficiency. Installed technical solution provided increase by 3 times of durability of rolling bearings to around 18 months. In addition, breakdowns of chocks slots have completely stopped. Similar solutions after specifying of their optimum parameters can be installed on all equipment where place for shock-absorbers is not stipulated.

Due to mentioned scientific and technical cooperation some more objects representing safety devices were developed. One of them is a safety spindle (brehspindle) with advanced design intended for installation in main drive lines of two Pilger stands 6 ÷ 12" of Hot Pipe Rolling Shop.

In new design chosen form provides essential increase of fatigue durability of the brehspindle that is why charge/installation of the brehspindles with overload absence should be decreased by 4…5 times. Besides, constructive measures are taken for thrust effort exception after destruction of brehspindle (refer to Fig. 9). Similar brehspindle models were tested in laboratory conditions. Two new design brehspindles are passed industrial approbation in Hot Pipe Rolling Shop of the Iron and Steel Works.

One more designed safety device that is plastically deformable spigots installed between screwdowns and top back-up rolls chocks. They were made and passed industrial approbation in roughing stands group of HRM 1700 (refer to Fig. 10, where pos. 1 is plastically deformable spigot, pos. 2 is saddle body, pos. 3 is support, pos. 4 is saddle, pos. 5 is screwdown, pos. 6 is top back-up roll chock). The device has appeared effective but demanding some adaptation.

Another designed safety device was made to replace heavily loaded universal spindles which undergo dynamic loads and high contact stresses. It is called toothed elastic shaft (TES, refer to Fig. 10, where pos. 1 is half coupling, pos. 2 is polyurethane

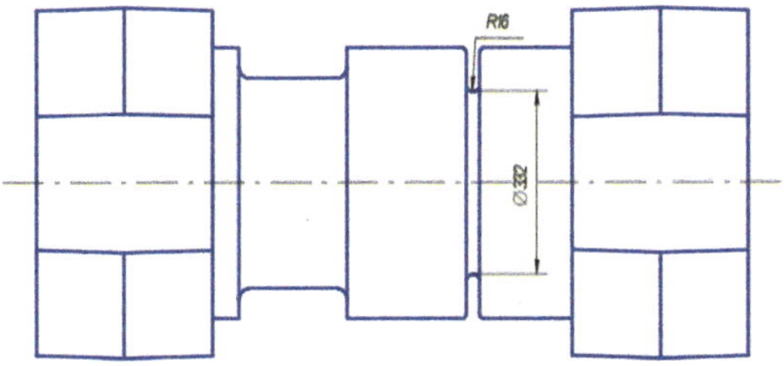

Fig. 9. General view of designed safety spindle (brehspindle).

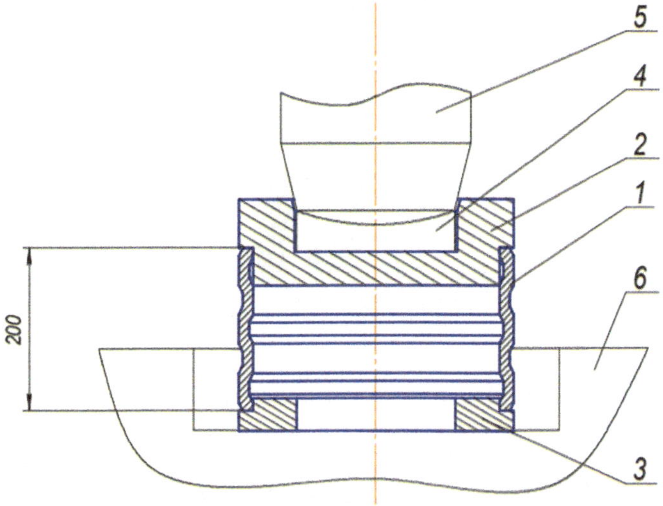

Fig. 10. General view of deformable spigot installed under screwdown.

element, pos. 3 is body, pos. 4 is shaft, pos. 5 is fixing ring). Polyurethane elements provide uniform distribution of contact stresses over surfaces of teeth. Its possible angle of angularity is up to 3°. TESs work in main drive lines of 4-stands CRM 1700 of PJSC 'Ilyich Iron and Steel Works'. There is no wear of steel teeth being in contact with polyurethane elements so far. Polyurethane elements are removable elements. One polyurethane element costs around 3% from TES cost (Fig. 11).

Another designed safety device was made to replace toothed shafts which are widely popular in cranes and roller tables drives. It is called universal elastic shaft. Its half coupling is similar to steel coupling with pins and polyurethane ring installed between half coupling (refer to Fig. 12, where pos. 1 is half coupling, pos. 2 is steel pin, pos. 3 is polyurethane element, pos. 4 is steel tubing). Such universal elastic shafts (refer to Fig. 13) are in use for 9 years in drives of roller tables of Rails Rolling Shop of PJSC 'Azovstal Iron and Steel Works' (Mariupol city, Ukraine). There is no need for lubrication. Durability of polyurethane elements is around 36 months.

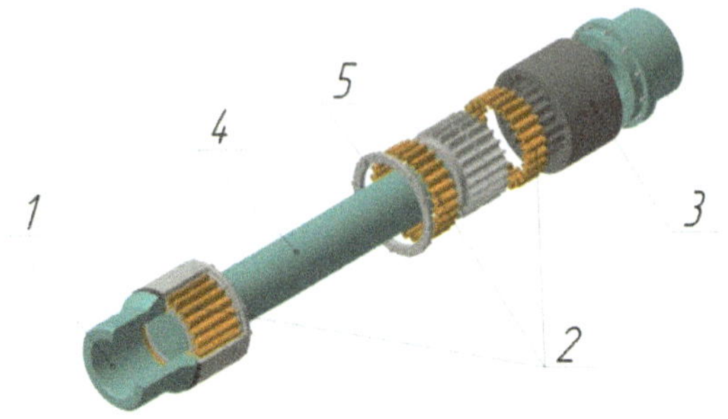

Fig. 11. General view of TES.

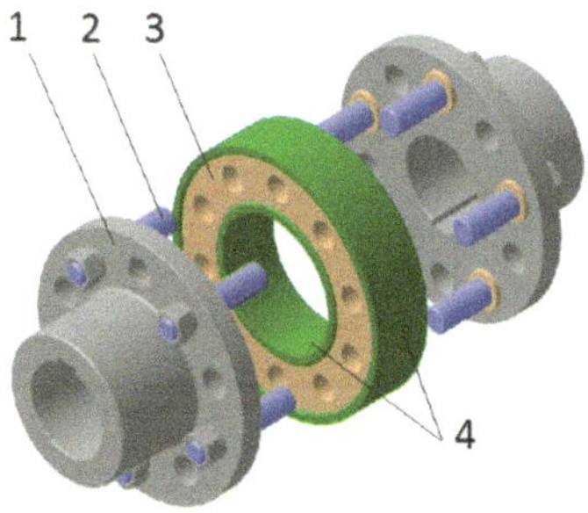

Fig. 12. General view of steel coupling with pins and polyurethane ring.

Fig. 13. General view of universal elastic shafts.

Described technical solutions are very important for the Iron and Steel Works and have good perspectives for widening of use. However, low introduction rates of new developments are characteristic of all variants of cooperation between universities and Iron and Steel Works. The most difficult stages are production and industrial approbation. That is why specialized laboratory 'Protection of metallurgical machines from breakdowns' based on HMD structure with employees of PSTU and SPbPU specialized in this sphere was established.

This laboratory should develop such safety and amortization devices that can be made using repair services of Iron and Steel Works. The laboratory was created in 2006. Work of laboratory is planned in four basic directions (refer to Table 2).

Table 2. Basic directions of the laboratory activity.

#	Basic directions of the laboratory works	Purpose
1	Fundamental research works	Research of reasons of overloads and devices development to increase durability of machines and units
2	Research and design development	Devices development for machines subassemblies and units modernization
3	Pilot study and development	New materials development for replacement of rolling bearings by sliding bearings and their installation technology
4	Carrying out of seminars and lecturing on problem of machines protection from breakdowns	Improvement of professional skills of specialists and revamp staff

Works described in Table 2 are offered according to below given schedule:

1. Fundamental research works. It is planned to execute a wide range of original works with obligatory patenting of developed designs, industrial introduction and essential economic benefit. These developments are important for metallurgical equipment of any Iron and Steel Works of CIS countries. Major works of given direction are:
   [1] Research of loads and overloads in stands and main drive lines of HRM 1700 and development of automatic safety devices for finishing stands group (lead time is around 24 months).
   [2] Development of elastic energy - accumulator shafts for roughing stands group drives of HRM 1700 (lead time is around 12 months).
   [3] Loading research of roughing stands group of HRM 1700 and development of safety devises for roughing stands (lead time is around 12 months).
   [4] Durability research of rolling stands screwdowns of Hot Plate Rolling Mill 3000 and development of designs to increase it (lead time is around 24 months).
2. Research and design development. These developments have big practical importance and are carried out operatively. It means that developments of designs which are carried out during 3…6 months are made and operatively installed by mechanical services of the Iron and Steel Works.
   [1] Modernization of roller tables of HRM 1700 for working with 20 tons slabs by installation of polyurethane shock-absorbers on rolling bearings (lead time is around 3 months).

   [2] Modernization and loads amortization of roller tables of Hot Plate Rolling Mill 3000 by installation of polyurethane shock-absorbers on rolling bearings (lead time is around 3 months).
   [3] Shock-absorbers modernization of roller tables of Reduction Shop 1150 Slabbing (lead time is around 3 months).
   [4] Development of universal spindles with elastic polyurethane elements for main drive lines of stands of CRM 1700 (lead time is around 6 months).
   [5] Development of elastic-compensating shafts for drives of auxiliary machines (lead time is around 3 months).
   [6] Units and details improvements of coilers of HRM 1700 in order to increase their durability (lead time is around 6 months).
3. Pilot study and development. These are theoretical researches and laboratory studies connected with search of new solutions and technologies referred to creation of new polymeric materials and new ways of classical materials surfaces machining. In addition, researches include working with glutinous compositions and sealants applied at installation and repairs. Performance of these works will give push to wide introduction of new technical solutions providing essential increase of durability of metallurgical machines. Such developments can be introduced by HMD services and repair shops of Iron and Steel Works.
   [1] Research of composite materials with antifrictional additives providing effect of self-greasing, for replacement of rolling bearings by autonomous sliding bearings e.g. due to composite material known as 'Romanit'.
   [2] Glutinous compositions research and selection of optimum technologies for elastomeric and steel details one-piece connections.
4. Carrying out of seminars and lecturing on problem of machines protection from breakdowns. Problems of weak points connected with strength and durability of machines are discussed on regular basis. Participants of seminars are workers of the laboratory, workers of main and repair shops, HMD management of Iron and Steel Works. Lecturing also includes problem of functional strength of machines and ways of its increase. Main purpose of the lecturing is improvements of professional skills of attendants.

2 Conclusions

1. Technical conditions of mechanical equipment of metallurgical shops that have been working during last 25…70 years are described.
2. Given description of loads arising in metallurgical machines helps to define loads in technological processes.
3. Technical solutions to improve durability of metallurgical equipment are given. Design and stress calculation of these technical solutions were done by means of FEM analysis where conditions of contact surfaces interactions, materials properties and geometry of equipment are actual.
4. Technical, scientific, engineering and educational activities of the specialized laboratory 'Protection of metallurgical machines from breakdowns' with program are

described that allows reduction of expenses for modernization in around ten times. Economic benefit of the laboratory is around €1.5 million per year due to absence of accidental breakdowns. It is reasonable to establish specialized laboratory 'Protection of metallurgical machines from breakdowns' on the basis of each Iron and Steel Works due to low introduction rates of new developments.

Acknowledgment. The reported study was funded by RFBR according to the research project №19-08-01252a. The authors declare that there is no conflict of interest regarding the publication of this paper.

References

1. Verenev, V.V., Bolshakov, V.I., Yunakov, A.M.: Fundamental and applied problems of ferrous metallurgy: works. Dnepropetrovsk. **19**, 346–358 (2009)
2. Kukhar, V., Yelistratova, N., Burko, V., Nizhelska, Y., Aksionova, O.: Estimation of occupation safety risks at energetic sector of iron and steel works. Int. J. Eng. Technol. (UAE) **7**(2.23), 216–220 (2018). https://doi.org/10.14419/ijet.v7i2.23.11922
3. Ishenko, A.A.: Issledovanie dinamicheskih nagruzok na stanini kleti stana 3000 v processe prokatki i razrabotka predlozeniy po optimizacii ego raboti. PSTU, Mariupol (2006)
4. Gharaibeh, N.S., Matarneh, M.I., Artyukh, V.G.: Loading decrease in metallurgical machines. Res. J. Appl. Sci. Eng. Technol. **8**(12), 1461–1464 (2014)
5. Sorochan, E., Artiukh, V., Melnikov, B., Raimberdiyev, T.: Mathematical model of plates and strips rolling for calculation of energy power parameters and dynamic loads. In: MATEC Web of Conferences, vol. 73, p. 04009 (2016). https://doi.org/10.1051/matecconf/20167304009
6. Kukhar, V., Prysiazhnyi, A., Balalayeva, E., Anishchenko, O.: Designing of induction heaters for the edges of pre-rolled wide ultrafine sheets and strips correlated with the chilling end-effect. In: Safarik, D., Tabunschikov, Y., Murgul, V. (eds.) Modern Electrical and Energy System (MEES 2017), pp. 404–407 (2017). https://doi.org/10.1109/MEES.2017.8248945
7. Kukhar, V., Artiukh, V., Prysiazhnyi, A., Pustovgar, A.: Experimental research and method for calculation of 'upsetting-with-buckling' load at the impression-free (Dieless) preforming of workpiece. In: E3S Web of Conference, vol. 33, p. 02031 (2018). https://doi.org/10.1051/e3sconf/20183302031
8. Kukhar, V., Balalayeva, E., Prysiazhnyi, A., Vasylevskyi, O., Marchenko, I.: Analysis of relation between edging ratio and deformation work done in pre-forming of workpiece by bulk buckling. In: MATEC Web of Conferences, vol. 178, pp. 02003 (6 p.) (2018). https://doi.org/10.1051/matecconf/201817802003
9. Kukhar, V.V., Grushko, A.V., Vishtak, I.V.: Shape indexes for dieless forming of elongated forgings with sharpened end by tensile drawing with rupture. In: Solid State Phenomena, vol. 284, pp. 408–415 (2018). https://doi.org/10.4028/www.scientific.net/SSP.284.408
10. Artiukh, V.G., Galikhanova, E.A., Mazur, V.M., Kargin, S.B.: Energy intensity of parts made from polyurethane elastomers. Mag. Civil Eng. **81**(5), 102–115 (2018). https://doi.org/10.18720/MCE.81.11
11. Alekseev, N.N., Kovalenko, I.A., Stolyarov, O.N., Melnikov, B.E.: Viscoelastic properties of geosynthetic materials. Constr. Unique Build. Struct. **5**(56), 17–32 (2017)

12. Al-Quran, F.M.F., Matarneh, M.E., Artukh, V.G.: Choice of elastomeric material for buffer devices of metallurgical equipment. Res. J. Appl. Sci. Eng. Technol. **4**(11), 1585–1589 (2012)
13. Artiukh, V.G., Karlushin, S.Y., Sorochan, E.N.: Peculiarities of mechanical characteristics of contemporary polyurethane elastomers. Procedia Eng. **117**, 938–944 (2015). https://doi.org/10.1016/j.proeng.2015.08.180
14. Yakovlev, S.N., Mazurin, V.L.: Vibroisolating properties of polyurethane elastomeric materials, used in construction. Mag. Civil Eng. **6**, 53–60 (2017). https://doi.org/10.18720/MCE.74.5
15. Yakovlev, S.N.: Proektirovanie i osnovy tekhnologii detaley mashin iz poliuretana. Renome, SPb (2013)
16. Artiukh, V., Raimberdiyev, T., Mazur, V.: Use of CAE-systems at evaluation of shock absorbers for metallurgical equipment. In: MATEC Web of Conferences, vol. 53, p. 01039 (2016). http://dx.doi.org/10.1051/matecconf/20165301039
17. Belevich, A.V., et al.: Issledovanie tehnologicheskih i konstruktivnih parametrov rabochih kletey vintovoy prokatki s pomoshyu sovremennih konechno - elementnih system [Study of technological and design parameters of helical rolling stands by means of modern FEM systems] Universities news: nonferrous metallurgy, vol. 9, pp. 32–35 (2002)
18. Artiukh, V., Mazur, V., Kargin, S., Zakharova, L.: Adapters for metallurgical equipment. In: MATEC Web of Conferences, vol. 170, p. 03028 (2018). https://doi.org/10.1051/matecconf/201817003028
19. Artiukh, V., Belyaev, M., Ignatovich, I., Miloradova, N.: Depreciation of bearing blocks of rollers of roller conveyers of rolling mills. In: IOP Conference Series: Earth and Environmental Science, vol. 90, p. 012228 (2017). https://doi.org/10.1088/1755-1315/90/1/012228
20. Mazur, V., Artiukh, V., Matarneh, M.I.: Horizontal force during rolling as indicator of rolling technology and technical conditions of main rolling equipment. Procedia Eng. **165**, 1722–1730 (2016). https://doi.org/10.1016/j.proeng.2016.11.915
21. Artiukh, V., Mazur, V., Shilova, L.: Device for making horizontal wedge thrust of rolling stand. In: MATEC Web of Conferences, vol. 106, p. 03002 (2017). https://doi.org/10.1051/matecconf/201710603002
22. Artiukh, V., Mazur, V., Adamtsevich, A.: Priority influence of horizontal forces at rolling on operation of main sheet rolling equipment. In: MATEC Web of Conferences, vol. 106, p. 04001 (2017). https://doi.org/10.1051/matecconf/201710604001
23. Artiukh, V., Mazur, V., Nidziy, E.: Evaluation of surface deformations of rolling mills stands elements resulting from horizontal forces acting at rolling. In: Advances in Intelligent Systems and Computing, vol. 692, pp. 1065–1073. Springer, Cham (2018). https://doi.org/10.1007/978-3-319-70987-1_115

Application of Nonlinear Dynamic Analysis in Calculation of Characteristics of Frictional Draft Gears

Andrew Nikitchenko[1(✉)], Viktor Artiukh[2], Denis Shevchenko[1], and Dmitriy Spitsov[3]

[1] All-Union Research and Development Centre for Transportation Technology, 23 Line Vasilievsky Island, 2, St. Petersburg 199106, Russia
anikita@i.ua

[2] Peter the Great St. Petersburg Polytechnic University, Polytechnicheskaya 29, St. Petersburg 195251, Russia

[3] Moscow Institute of Architecture, Rozhdestvenka Street 11, Moscow 107031, Russia

Abstract. The computing finite element model of the elastic-frictional draft gear was developed with application of the Siemens NX software package. The theoretical foundations of computational methods laid in the NX Nastran software system were analyzed. The optimal solver settings that make it possible to achieve the convergence of the solution without sacrificing acceptable accuracy were determined.

Keywords: Nonlinear dynamic analysis · Transient processes · Finite element method · NX Nastran · ADINA · Mechanics of impact · Ball collision

1 Introduction

The present article is concerned with an overview of possibilities of the nonlinear dynamic analysis implemented by the ADINA (Automatic Dynamic Incremental Nonlinear Analysis) toolset in the NX Nastran software package through the example of the draft gear calculation.

Traditionally, force characteristics calculations for elastic-frictional draft gears are carried out by means of forming systems of differential equations. In this case, equations are formed "manually" and can be quite cumbersome depending on the level of detail in the computational model, [1–5]. This approach causes quite high probability of errors occurring both at the stage of composing equations and at the stage of their typesetting in calculation programs.

It should be noted that works on the draft gear characteristics calculation using methods which do not require "manual" composition of differential equations have been carried out on a small scale. For example, the force characteristics calculation for the elastic-frictional draft gear was carried out using the NX Motion software system [6] that makes it possible to solve problems of dynamics of mechanical systems

Z. Popovic et al. (Eds.): TransSiberia 2019, AISC 1115, pp. 963–979, 2020.
https://doi.org/10.1007/978-3-030-37916-2_95

consisting of perfectly rigid bodies. Good convergence with previously developed analytical models represented in [7] and field test results was obtained.

In comparison with the above methods, the nonlinear dynamic analysis provides a number of advantages. In particular, the draft gear components have real stiffness, and at the same time, the appearance of plastic deformations in certain areas is taken into account. It provides not only more accurate modeling of contact interaction, but also obtaining a time-varying picture of the stress-strain state of the draft gear components.

2 Methods and Results

The prototype of the frictional draft gear being considered (Fig. 1) has the simplest design, and the version thereof is close to the draft gear of the model Sh-3-V (Ш-3-В) developed by Research and Production Corporation Uralvagonzavod JSC. It makes it possible to achieve energy intensity indicators corresponding to class T1. At first glance, it may seem that this is a rather simple mechanical design. It consists of a wedge system and a retaining block that creates contact interaction forces between component parts of the wedge system. Nevertheless, analytical estimation made by many researchers who deal with this issue is very different from experimental results.

This is due to the fact that the real stiffness of the draft gear components, their masses, the distribution of contact pressures and other factors are not taken into account.

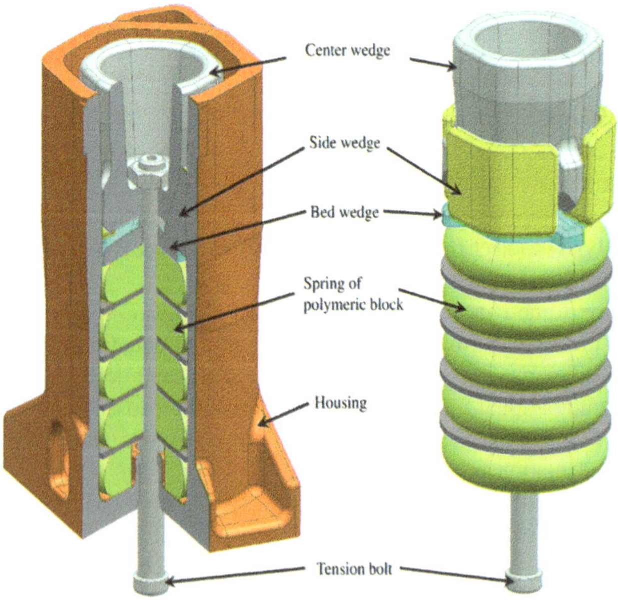

Fig. 1. Design of the draft gear.

To begin with, we carry out calculation on a simplified test model (Fig. 2) in order to determine solver settings satisfying the convergence and accuracy conditions. In this model, elastic interaction forces between elements are small and do not cause high-frequency elastic oscillations of large amplitude. In order to reduce the dimension of the problem, the symmetry property is used. This model can be calculated analytically, and it makes it possible to compare results obtained by different methods and evaluate effects of solver settings. Similar wedge systems are used not only in shock absorbers, but also in other mechanical systems, in particular, in wedge-type vibration dampers of freight car bogies.

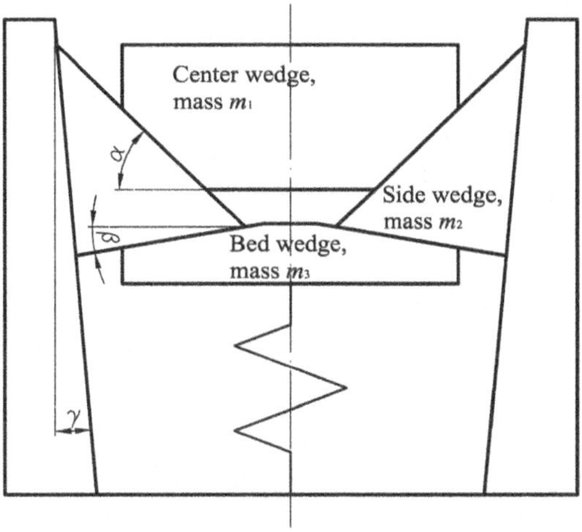

Fig. 2. Simplified test design model.

The most common method of forming differential equations of motion for a material system is the use of the Lagrange equations. In the case of ideal constraints, these equations do not include constraint reactions. At the same time, it is not practical to use the Lagrange equations in the case of friction forces depending on variable pressure [8]. In our case, the friction forces in the system depend on movements and accelerations; that is why we make an equilibrium equation for every element of the wedge system. The computational diagram for up and down movements of the system is shown in Fig. 3.

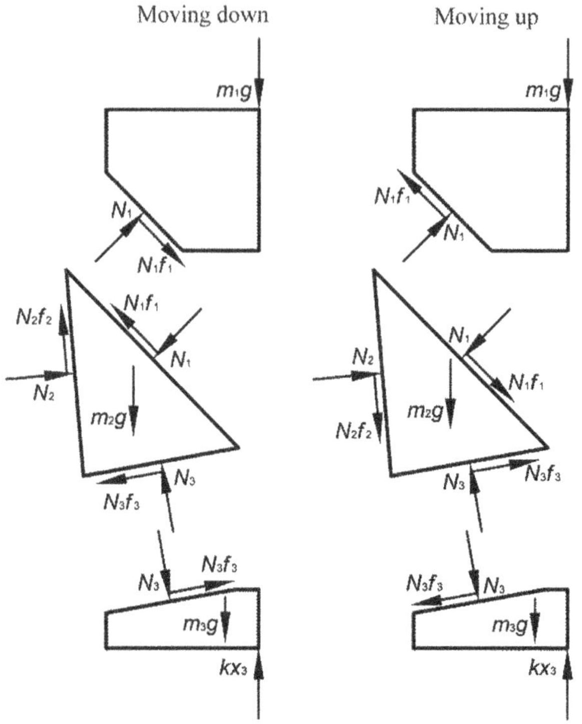

Fig. 3. Forces affecting the wedge system elements in the course of up and down movements.

The system of equations for the wedge system moving down

$$\begin{cases} -N_1(cos(\alpha) - f_1 sin(\alpha)) + m_1 g - m_1\ddot{x}_1 = 0 \\ N_1(cos(\alpha) - f_1 sin(\alpha)) - N_2(sin(\gamma) + f_2 cos(\gamma)) - N_3(cos(\beta) - f_3 sin(\beta)) + m_2 g - m_2\ddot{x}_2 cos(\gamma) = 0 \\ -N_1(sin(\alpha) + f_1 cos(\alpha)) + N_2(cos(\gamma) - f_2 sin(\gamma)) - N_3(sin(\beta) + f_3 cos(\beta)) - m_2\ddot{x}_2 sin(\gamma) = 0 \\ N_3(cos(\beta) - f_3 sin(\beta)) + m_3 g - m_3\ddot{x}_3 - kx_3 = 0 \end{cases} \quad (1)$$

For the wedge system moving up

$$\begin{cases} -N_1(cos(\alpha) + f_1 sin(\alpha)) + m_1 g - m_1\ddot{x}_1 = 0 \\ N_1(cos(\alpha) + f_1 sin(\alpha)) - N_2(sin(\gamma) - f_2 cos(\gamma)) - N_3(cos(\beta) + f_3 sin(\beta)) + m_2 g - m_2\ddot{x}_2 cos(\gamma) = 0 \\ -N_1(sin(\alpha) - f_1 cos(\alpha)) + N_2(cos(\gamma) \mp sin(\gamma)) - N_3(sin(\beta) - f_3 cos(\beta)) - m_2\ddot{x}_2 sin(\gamma) = 0 \\ N_3(cos(\beta) + f_3 sin(\beta)) + m_3 g - m_3\ddot{x}_3 - kx_3 = 0 \end{cases} \quad (2)$$

This system has one degree of freedom since the position of every part is uniquely determined by the position of the pressure cone, which we take as the generalized coordinate x. The wedge and the bed plate movements are expressed by the following kinematic dependencies

$$x_2 = i_2 \cdot x = x\frac{cos(\alpha)}{cos(\alpha+\gamma)}; \quad x_3 = i_p \cdot x = x\frac{cos(\alpha)cos(\beta-\gamma)}{\cos(\beta)cos(\alpha+\gamma)} \tag{3}$$

Having eliminated the internal forces of the system (1 and 2) and having plugged the dependencies (3), we obtain the equation of motion at the speed of the wedge system moving down

$$M_{a\downarrow} \cdot \ddot{x} + K_{\downarrow}k \cdot x = g \cdot M_{g\downarrow} \tag{4}$$

Consequently, when moving up

$$M_{a\uparrow} \cdot \ddot{x} + K_{\uparrow}k \cdot x = g \cdot M_{g\uparrow} \tag{5}$$

where

$$M_{a\downarrow} = m_1\left(\frac{cos(\gamma)-f_2sin(\gamma)}{sin(\gamma)+f_2cos(\gamma)} - \frac{sin(\alpha)+f_1cos(\alpha)}{cos(\alpha)-f_1sin(\alpha)}\right) + m_2i_2\frac{1}{sin(\gamma)+f_2cos(\gamma)} + m_3i_p\left(\frac{cos(\gamma)-f_2sin(\gamma)}{sin(\gamma)+f_2cos(\gamma)} + \frac{sin(\beta)+f_3cos(\beta)}{cos(\beta)-f_3sin(\beta)}\right) \tag{6}$$

$$M_{g\downarrow} = m_1\left(\frac{cos(\gamma)-f_2sin(\gamma)}{sin(\gamma)+f_2cos(\gamma)} - \frac{sin(\alpha)+f_1cos(\alpha)}{cos(\alpha)-f_1sin(\alpha)}\right) + m_2\frac{cos(\gamma)-f_2sin(\gamma)}{sin(\gamma)+f_2cos(\gamma)} + m_3\left(\frac{cos(\gamma)-f_2sin(\gamma)}{sin(\gamma)+f_2cos(\gamma)} + \frac{sin(\beta)+f_3cos(\beta)}{cos(\beta)-f_3sin(\beta)}\right) \tag{7}$$

$$K_{\downarrow} = i_p\left(\frac{cos(\gamma)-f_2sin(\gamma)}{sin(\gamma)+f_2cos(\gamma)} + \frac{sin(\beta)+f_3cos(\beta)}{cos(\beta)-f_3sin(\beta)}\right) \tag{8}$$

$$M_{a\uparrow} = m_1\left(\frac{cos(\gamma)+f_2sin(\gamma)}{sin(\gamma)-f_2cos(\gamma)} - \frac{sin(\alpha)-f_1cos(\alpha)}{cos(\alpha)+f_1sin(\alpha)}\right) + m_2i_2\frac{1}{sin(\gamma)-f_2cos(\gamma)} + m_3i_p\left(\frac{cos(\gamma)+f_2sin(\gamma)}{sin(\gamma)-f_2cos(\gamma)} + \frac{sin(\beta)-f_3cos(\beta)}{cos(\beta)+f_3sin(\beta)}\right) \tag{9}$$

$$M_{g\uparrow} = m_1\left(\frac{cos(\gamma)+f_2sin(\gamma)}{sin(\gamma)-f_2cos(\gamma)} - \frac{sin(\alpha)-f_1cos(\alpha)}{cos(\alpha)+sin(\alpha)}\right) + m_2\frac{cos(\gamma)+f_2sin(\gamma)}{sin(\gamma)-f_2cos(\gamma)} + m_3\left(\frac{cos(\gamma)+f_2sin(\gamma)}{sin(\gamma)-f_2cos(\gamma)} + \frac{sin(\beta)-f_3cos(\beta)}{cos(\beta)+f_3sin(\beta)}\right) \tag{10}$$

$$K_{\uparrow} = i_p\left(\frac{cos(\gamma)+f_2sin(\gamma)}{sin(\gamma)-f_2cos(\gamma)} + \frac{sin(\beta)-f_3cos(\beta)}{cos(\beta)+f_3sin(\beta)}\right) \tag{11}$$

The general form of the equation of motion for the system can be written as

$$\begin{aligned} \ddot{x} = \frac{g}{2} \cdot \left(\left(\frac{M_{g\downarrow}}{M_{a\downarrow}} + \frac{M_{g\uparrow}}{M_{a\uparrow}}\right) + \left(\frac{M_{g\downarrow}}{M_{a\downarrow}} - \frac{M_{g\uparrow}}{M_{a\uparrow}}\right)sign(\dot{x})\right) \\ - \frac{kx}{2}\left(\left(\frac{K_{\downarrow}}{M_{a\downarrow}} + \frac{K_{\uparrow}}{M_{a\uparrow}}\right) + \left(\frac{K_{\downarrow}}{M_{a\downarrow}} - \frac{K_{\uparrow}}{M_{a\uparrow}}\right)sign(\dot{x})\right) \end{aligned} \tag{12}$$

The resulting equation has an analytical solution which is too cumbersome. However, with no friction in the system, its solution looks rather concise.

$$x = x_0(1 - cos(\omega t)) \tag{13}$$

where x_0 is the amplitude of oscillation, i.e. the displacement from the initial position at which the spring deflection is equal to zero to the static equilibrium position; it is determined by the dependence $x_0 = \frac{g}{k} \cdot \frac{m_1 + m_2 i_2 cos(\gamma) + m_3 i_p}{i_p^2}$,

ω is the oscillation frequency, $\omega = \sqrt{\frac{k \cdot i_p^2}{m_1 + m_2 i_2^2 + m_3 i_p^2}}$

It should be noted that the Eq. (12) containing the function *sign* is nonlinear, and although its analytical solution can be derived, it has no practical meaning. The function *sign* is discontinuous; it sometimes leads to failure of solvers with step accuracy control. To avoid this, we replace the function *sign(x)* with the function *tanh(αx)*, where α is a large coefficient. It can be seen from the graphs represented in Fig. 4 that the curves $tanh(1000x)$ and $sign(x)$ almost coincide, and thus the approximation $sign(x) \rightarrow tanh(1000x)$ is quite successful.

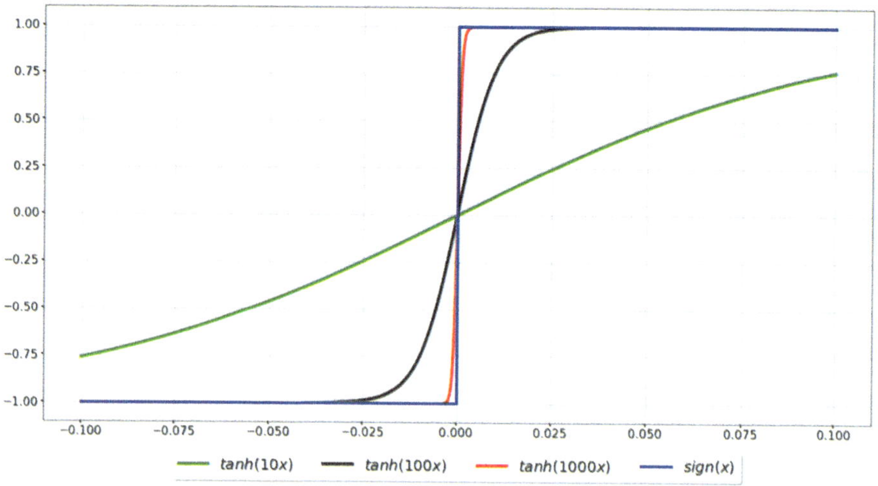

Fig. 4. Approximation of the discontinuous function *sign* by the functions *tanh*.

For the numerical solution of the Eq. (12), the function *odeint* of the python-*scipy* package is used.

The finite element implementation of the computational model being considered is represented in Fig. 5. To build the mesh, 8-node hexahedral finite elements CHEXA8 were used; the CELAS1 element was used as a spring with constant stiffness.

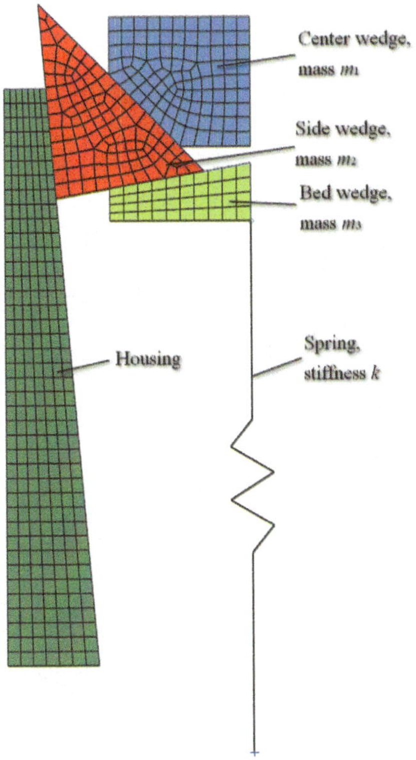

Fig. 5. Simplified finite element model.

We conduct the comparative calculation for the following initial data:

- masses of moving parts: $m_1 = 50$ kg, $m_2 = 20$ kg, $m_3 = 10$ kg;
- angles of the wedge system: $\alpha = 45°$, $\beta = 10°$, $\gamma = 5°$;
- free-fall acceleration $g = 10$ m/s^2;
- spring stiffness k = 8000 N/m;
- friction coefficients $f_1 = f_2 = f_3 = 0.05$.

Concurrently, in the settings of solvers, we disable all options that have a damping effect on the system.

The calculation results obtained by the explicit solver (Solver 701), the implicit solver (Solver 601) and analytically are shown in Fig. 6.

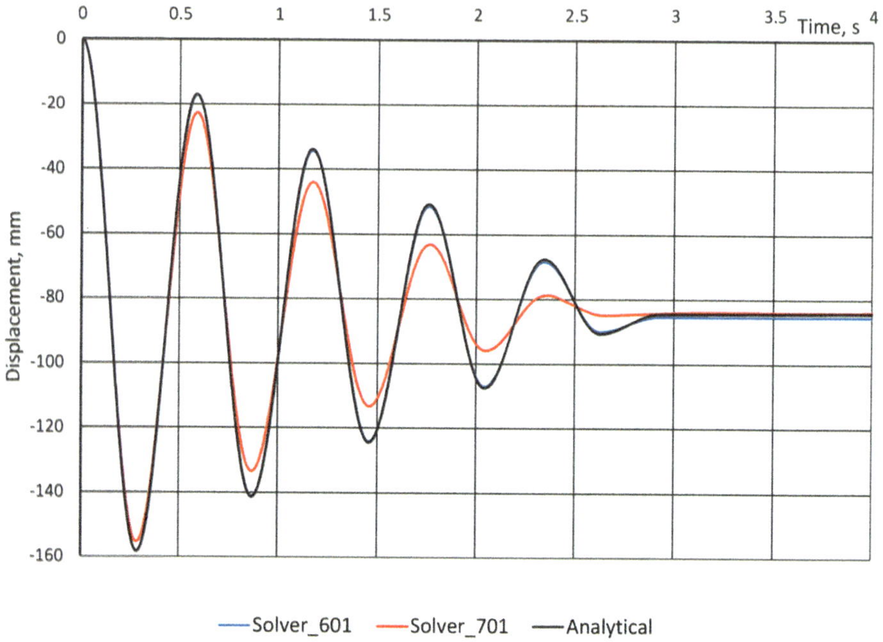

Fig. 6. Calculation results (the movement of the pressure cone).

It can be seen from the graphs in Fig. 6 that the results obtained using the implicit solver almost completely coincide with the analytical solution, however energy absorption occurs outside the friction surfaces when the explicit solver is used.

The time spent on the calculation by the implicit solver was 32 min, while the time spent on the calculation by the explicit solver was 4.8 h. This large difference is due to the fact that the integration step in an explicit solver is determined having regard to the speed of the sound-wave propagation, and it depends on the smallest element size in the model. In order to adequately model conditions of contact interaction between bodies, the model has to be divided into quite small elements. Consequently, the time step-out is also small. The implicit solver does not have this disadvantage, and it is more suitable for solving problems of low-speed dynamics, which include impact compression of the draft gear. In our further studies, we are planning to use only the implicit solver.

Taking into consideration that energy intensity is a key parameter for the draft gear models being developed, one should be very careful about solver settings that may affect it. As a rule, these settings are intended to improve the convergence of the solution, which is expressed in suppression of contact oscillations, stabilization of attenuation and damping, etc. Thus, an important task is to determine the solver settings which make it possible to obtain a convergent solution without significant distortion of the force characteristic of the draft gear.

Among the settings that clearly affect the change in the energy of the system, the following can be highlighted:

Table 1. Change in the energy of the system.

No. Sl.	Parameter name	Designation	Card name
1	Stiffness matrix stabilization	MSTAB	NXSTRAT
2	Low-speed dynamics option	ATSLOWS	NXSTRAT
3	Attenuation coefficient of low-speed dynamics	ATSDAMP	NXSTRAT
4	Attenuation stabilization	CTDAMP	NXSTRAT
5	Parameter of the frictional constraint function	EPST	BCTPARA
6	Rayleigh damping coefficients	ALPHA1, ALPHA2	PARAM

The given parameters are described quite well in [9] and [10], so we are not going to fix on the description thereof, but immediately give the results of the effect. For this purpose, we successively calculate the simplified model for various values of the options and coefficients presented in the table (Table 1).

The stiffness matrix stabilization is used to prevent occurrence of zero pivot. Since zero pivot is missing in the dynamic analysis due to the presence of the mass matrix, in the dynamic analysis, the inclusion of this option does not affect calculation results.

It is recommended to activate the low-speed dynamics option in order to overcome difficulties related to the convergence of the solution of the contact problem of the dynamic analysis. The equation of motion taking into account the low-speed dynamics option is given by the equation

$$\mathbf{M}\ddot{\mathbf{U}}_{t+\Delta t} + \mathbf{C}\dot{\mathbf{U}}_{t+\Delta t} + \mathbf{K}\Delta\mathbf{U}_{t+\Delta t} = \mathbf{R}_{t+\Delta t} - \mathbf{F}_{t+\Delta t} \tag{14}$$

where **M** is the generalized mass matrix,
K is the stiffness matrix,
C is the damping matrix,
R is the vector of external loads applied at the time moment t = Δt,
F is the vector of forces in nodes, which arise at the time moment t = Δt,
$\ddot{\mathbf{U}}$, $\dot{\mathbf{U}}$, **U** are accelerations, speeds and movements of nodes.

The damping matrix is calculated by the formula

$$\mathbf{C} = \beta\mathbf{K} \tag{15}$$

where β is the low-speed dynamics parameter ATSDAMP which by default is equal to $\beta = 10^{-4}$.

In addition to the above, it is recommended that the integration step shall be less than $10^5\beta$.

Figure 7 shows the calculation results for the simplified model with different values of the low-speed dynamics parameter ATSDAMP.

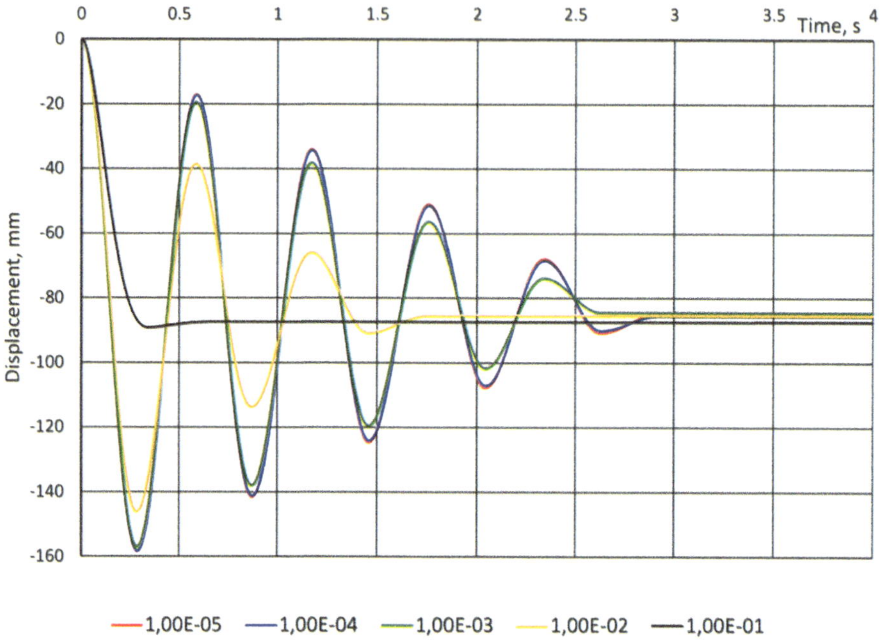

Fig. 7. Influence of the low-speed dynamics parameter (the movement of the pressure cone).

When the low-speed dynamics parameter is not activated, the solution does not converge. When $\beta = 10^{-4}$ and $\beta = 10^{-5}$, the results almost coincide, therefore, hereafter we use $\beta = 10^{-4}$ in the settings.

To reduce high-frequency oscillations arising from numerical integration, the Rayleigh damping matrix is added to the explicitly given damping matrix. The Rayleigh damping matrix is calculated using the mass matrix and the initial stiffness matrix by the formula

$$\mathbf{C}_R = \alpha\mathbf{M} + \beta\mathbf{K} \tag{16}$$

where **M** is the generalized mass matrix,
K is the stiffness matrix,
α and β are Rayleigh damping constants.

Generally, determining the coefficients α and β is not an easy task; that is why they should be used with caution and only in cases when there is no convergence.

Then, having determined the values of the coefficients affecting absorption of energy by the system and the solution convergence, we can proceed to the calculation of the real draft gear which is shown in Fig. 1. The computing FE model of the draft gear is shown in Fig. 8. In order to reduce the dimension of the problem, the symmetry property is used in this model. The model consists of 4-node elements TETRA4. In order to achieve the energy intensity required by regulatory documents, the retaining block of the draft gear has to be made of polymer materials, since steel springs with necessary parameters do not fit the internal space of the draft gear. As a result, the retaining block has a nonlinear concave characteristic shown in Fig. 9. In order to model a nonlinear spring, in the FE model, the element PBUSH1D1 is provided, for which the dependence of the force on the compression value shown in Fig. 9 is presented in tabular view. The lower end of the spring is fixed, and the upper one is connected with the nodes of the lower surface of the bed plate by means of RBE2 elements.

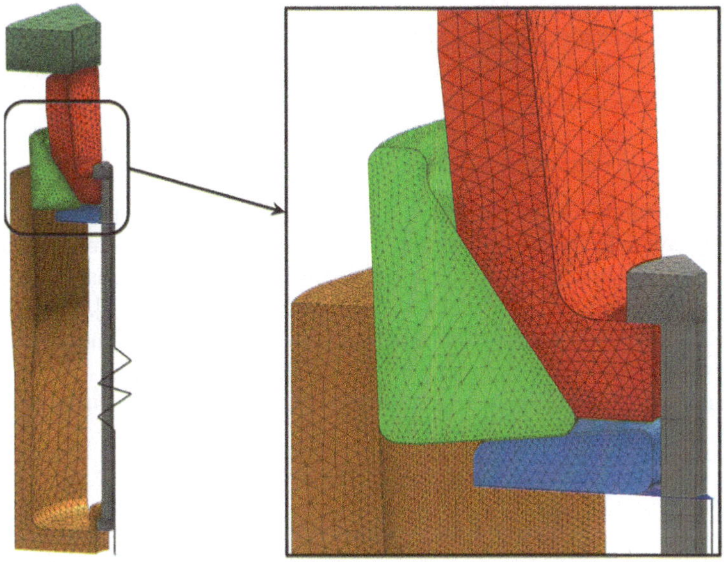

Fig. 8. Computing FE model of the draft gear.

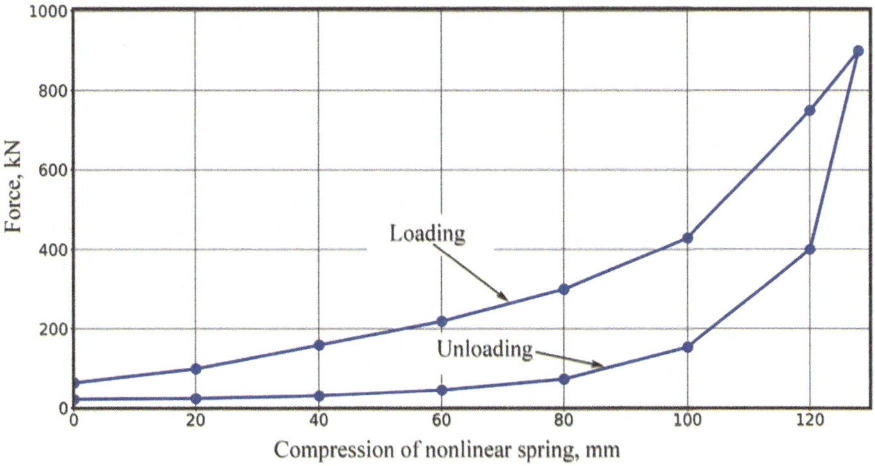

Fig. 9. Nonlinear characteristic of the retaining block used in the draft gear design.

We conduct the comparative calculation for the following initial data:

- the weight of the falling object: m_{dw} = 10000 kg;
- the angles of the wedge system: α = 57°, β = 7°, γ = 2°;
- the free-fall acceleration g = 9.81 m/s^2;
- the spring stiffness is represented by the upper branch of the graph in Fig. 9;
- the friction coefficients are constant and equal to f = 0.2;
- the initial velocity of the falling object V_0 = 4 m/s;
- the housing material is Steel 35NGML (35НГМЛ) with the following mechanical properties:
  - yield point σ_y = 589 MPa,
  - ultimate strength $\sigma_в$ = 763 MPa,
- the material of other component parts is Steel 38KhS (38ХС) with the following mechanical properties:
  - yield point σ_y = 540 MPa,
  - ultimate strength $\sigma_в$ = 685 MPa.

The applied method of the nonlinear dynamic analysis allows taking into account material behavior in the region of stresses exceeding the yield point. The analytical description of the material deformation diagram is made on the basis of mechanical characteristics according to the Bankin-Dzyuba-Hvatan method [11].

At each integration step, we obtained node-by-node values of movements, velocities, stresses, reactions in the nodes to which boundary conditions are applied and other parameters. According to these results, the energy intensity calculation was made, for which the trapezoid formula was applied.

$$E_{i+1} = E_i + \frac{P_i + P_{i+1}}{2} \cdot (x_{i+1} - x_i) \tag{17}$$

where E_i is the i-th step energy,
P_i is the total vertical reaction on fixed nodes of the draft gear housing and the spring,
x_i is the average movement of the pressure cone nodes.

For the analytical solution, the formula (12) which takes into account the nonlinear characteristic of the retaining block is used. In order to accomplish correct comparison with the results obtained by means of the nonlinear dynamic analysis, the upper branch of the diagram in Fig. 9 was applied in analytical calculations for both the loading and unloading stages of the draft gear. Figures 10, 11 and 12 show the comparison results obtained by means of analytical methods, using the nonlinear dynamic analysis.

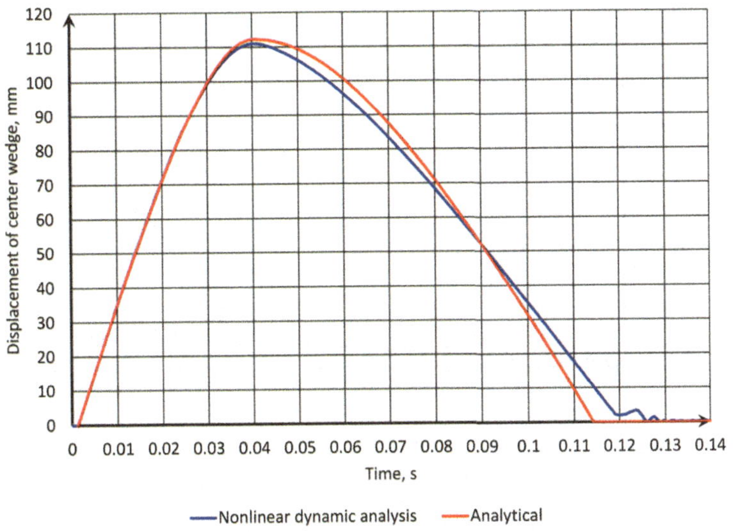

Fig. 10. Calculation results, the time history of the pressure cone movement

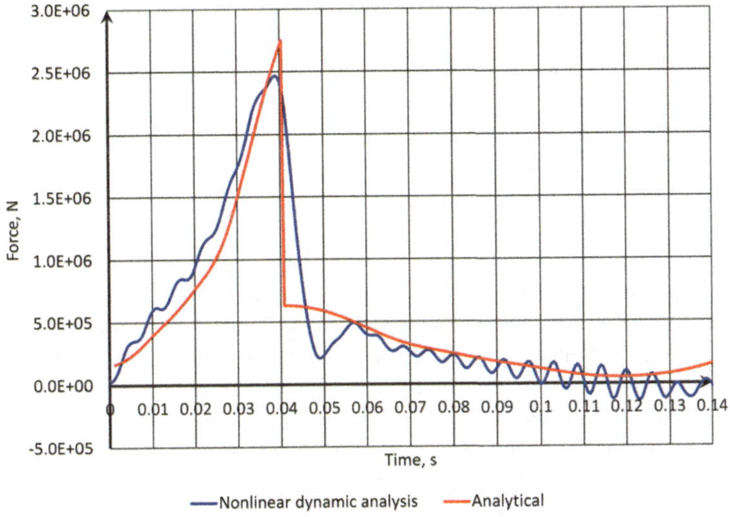

Fig. 11. Calculation results, the time history of the force

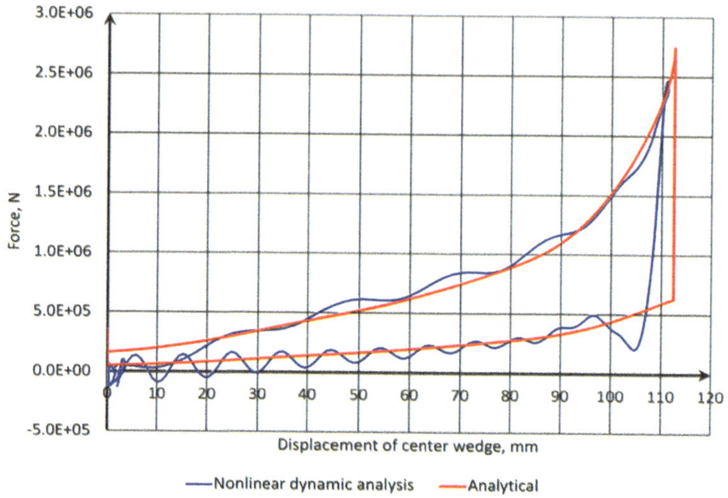

Fig. 12. Calculation results, the force characteristic of the draft gear.

From the "time-movement" diagram in Fig. 10, one can see the good convergence of the results, especially at the stage of compression. On the "time-force" and "movement-force" diagrams, the strong self-oscillating process that is conditioned upon the elastic properties of the draft gear housing and the presence of dry friction between its parts [12, 13] is clearly visible.

The obtained values of energy intensity of the draft gear, calculated by the formula (4), differ by 13% and are equal to:

- 84.1 kJ when obtained by means of the analytical method,
- 95.8 kJ when obtained by means of the nonlinear dynamic analysis.

Figure 13 shows the stress state of components at the moment of the maximum movement of the pressure cone when the stresses reach their maximum values. In some areas, the stresses significantly exceed the yield point, and it leads to appearance of residual deformations. From this, it follows that with each subsequent shock compression of the draft gear, the shape and dimensions of its components will change, and it may lead to the change in the force characteristics. Based on experience in operating frictional draft gears, it is known that cases of "inflation" of the housing are not uncommon, that is why the housing shall be made of steel with the minimum yield point of 700 MPa.

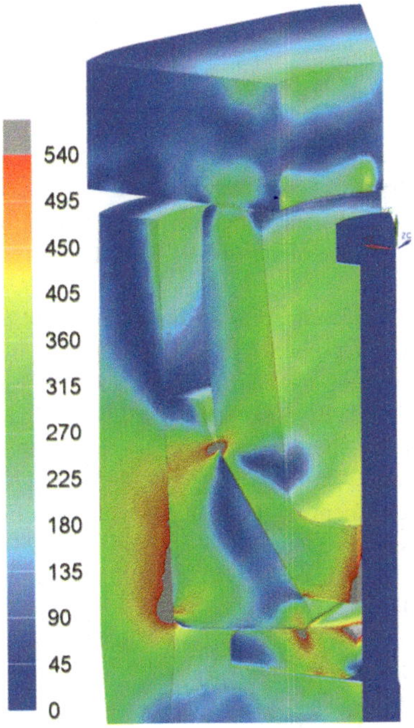

Fig. 13. Stress-strain state of components at the moment of maximum compression.

3 Conclusion

The shock compression of the draft gear was calculated using the nonlinear dynamic analysis. The satisfactory convergence of results was obtained, especially in such parameters as the maximum movement and the maximum force. At the same time, there is every reason to consider the results obtained by means of the nonlinear dynamic analysis to be more accurate, since the nonlinear dynamic analysis considers many factors that cannot be taken into account in the course of analytical calculation.

It should be noted that the stresses in some areas exceeded the material yield point; therefore, the field test prototype shall be made of materials with the minimum yield point of 700 MPa.

Even when powerful computer facilities are used, the calculation of one loading takes quite a long time, so the analytical methods of calculation will not lose their relevance in the immediate future. The most rational approach will involve the draft gear characteristics predicting by means of analytical methods, followed by improvement by means of the nonlinear dynamic analysis.

Acknowledgment. The reported study was funded by RFBR according to the research project №19-08-01241a. The authors declare that there is no conflict of interest regarding the publication of this paper.

References

1. Boldyrev, A.P.: Calculation and design of shock absorbers of rolling stock. Mashinostroyenie, Moscow (2004)
2. Vasiliev, A.S.: Development and research of the frictional shock absorber with the expansion block. Bull. Bryansk State Tech. Univ. **1**, 25–31 (2012)
3. Vasiliev, A.S.: Research of characteristics of the frictional elastomer shock absorber and calculation of stress-strain state of its nodes. Bull. Bryansk State Tech. Univ. **1**, 33–41 (2013)
4. Manashkin, L.A.: Oscillation Dampers and Shock Absorbers for Rail Vehicles (Mathematical Models). ART-Press, Dnepropetrovsk (2007)
5. Keglin, B.G.: Development of mathematical models and calculation of polymer shock absorbers characteristics. Bull. Bryansk State Tech. Univ. **4**, 18–26 (2013)
6. Nikitchenko, A., Artiukh, V., Shevchenko, D.: Modeling of operation of elastic-friction draft gear by NX Motion software. Procedia Eng. **187**, 790–796 (2017)
7. Nikitchenko, A.A.: Development of the mathematical model of a frictional-polymer draft gea. Bull. Dnepropetrovsk Nat. Univ. Railway Transp. **24**, 25–33 (2008)
8. Greiner, W.: Classical Mechanics. Systems of Particles and Hamiltonian Dynamics. Springer, Heidelberg (2010)
9. NX Nastran 10. Advanced Nonlinear Theory and Modeling Guide. Siemens Product Lifecycle Management Software Inc. (2014)
10. Bathe, K.J., Bucalem, M.L.: The Mechanics of Solids and Structures – Hierarchical Modeling and Finite Element Solution. Massachusetts Institute of Technology, Cambrige (2011)
11. Bankin, S., Dzyuba, A.S., Khvatan, A.M.: The method of the "σ-ε" deformation diagram plotting according to reference mechanical properties of the material. https://cae-club.ru/sites/default/files/users/files/6/dzyuba-hvatan_0.pdf
12. Shamberov, V.N.: Method for analytical study of the dry friction effect on the behavior of automatically regulated dynamic systems. Nauch. Priborostroyenie **13**, 77–83 (2003)
13. Shamberov, V.N.: Friction oscillations in servo electric drives. Design Technol. Electron. Means **3**, 56–62 (2005)
14. de Souza Neto, E.A., Peric, D., Owen, D.R.J.: Computational Methods for Plasticity: Theory and Applications. Wiley, Hoboken (2008)
15. Bathe, K.-J.: Finite Element Procedures. Prentice Hall, Pearson Education, Inc., Upper Saddle River (2016)
16. Cook, R.D., Malkus, D.S., Plesha, M.E.: Concepts and Applications of Finite Element Analysis. Wiley, Hoboken (2002)
17. Crisfield, M.A.: Non-linear Finite Element Analysis of Solids and Structure, vol. 1. Wiley, New York (2002)
18. Hughes, T.J.R.: The Finite Element Method, Linear Static and Dynamic Finite Element Analysis. Prentice-Hall, New York (2000)
19. Smith, I.M., Griffiths, D.V., Margetts, L.: Programming the Finite Element Method. Wiley, Hoboken (2013)
20. Khenanne, A.: Introdution to Finite Element Analysis Using MATHLAB and Abaqus. CRC Press, Boca Raton (2013)
21. Zienkiewicz, O.C., Taylor, R.L., Zhu, J.Z.: The Finite Element Method: Its Basis and Fundamental. Elsevier Science & Technology, Oxford (2013)
22. Franke, D.C.: Investigation of Mechanical Contact Problem with High-order Finite Element Method. Shaker Verlag GmbH, Aachen (2012)

23. Ota, N.S.N., Wilson, L., Gay Neto, A., Pellegrino, S., Pimenta, P.M.: Nonlinear dynamic analysis of creased shells. Finite Elem. Anal. Design. **121**, 64–74 (2016)
24. Tabiei, A., Tanov, R.: A nonlinear higher order shear deformation shell element for dynamic explicit analysis: Part II. Performance validation through standart tests. Finite Elem. Anal. Design. **72**, 39–50 (2004)
25. Landau, L.D., Lifshitz, E.M.: Theory of Elasticity. Pergamon, London (1986)
26. Stronge, W.J.: Impact Mechanics. Cambridge University Press, Cambridge (2004)
27. Grushko, A.V., Kukhar, V. V., Slobodyanyuk, Yu.O.: Phenomenological model of low-carbon steels hardening during multistage drawing. In: Solid State Phenomena, Materials Engineering and Technologies for Production and Processing III, vol. 265, pp. 114–123 (2017). https://doi.org/10.4028/www.scientific.net/SSP.265.114
28. Kukhar, V., Artiukh, V., Butyrin, A., Prysiazhnyi, A.: Stress-strain state and plasticity reserve depletion on the lateral surface of workpiece at various contact conditions during upsetting. In: Murgul, V., Popovic, Z. (eds.) International Scientific Conference Energy Management of Municipal Transportation Facilities and Transport EMMFT 2017, EMMFT 2017. Advances in Intelligent Systems and Computing, vol. 692, pp. 201–211. Springer, Cham (2018). https://doi.org/10.1007/978-3-319-70987-1_22
29. Balalayeva, E., Artiukh, V., Kukhar, V., Tuzenko, O., Glazko, V., Prysiazhnyi, A., Kankhva, V.: Researching of the stress-strain state of the open-type press frame using of elastic compensator of errors of "press-die" system. In: Murgul, V., Popovic, Z. (eds.) International Scientific Conference Energy Management of Municipal Transportation Facilities and Transport EMMFT 2017, EMMFT 2017, Advances in Intelligent Systems and Computing, vol. 692, pp. 220–235. Springer, Cham (2018). https://doi.org/10.1007/978-3-319-70987-1_24

Non-destructive Testing of Quality of Welded Joints of Titanium Plates of Superminiature Eddy-Current Probes

Sergey Dmitriev[1], Alexey Ishkov[2], Vladimir Malikov[1(✉)], and Anatoly Sagalakov[1]

[1] Altai State University, Lenina Avenue, 61, Altai Region, 656049 Barnaul, Russia
osysl1@gmail.com
[2] Altay State Agricultural University, Krasnoarmeyskiy Avenue, 98, Altai Region, 656049 Barnaul, Russia

Abstract. This article contains the main technical information on the measuring system and the eddy-current transducer used, and describes the measurement procedure to control defects in welds of titanium alloys, including the use of two subminiature eddy-current transducers, one of which is to be fixed above the weld, and the another is to be used directly for scanning. The experimental results obtained by means of the developed measuring system for samples of various titanium plates joined by welds are presented. Dependences of the eddy-current transducer signal that show the changes of the weak and defect-free weld signals are given. Defects in welds are defined by a dramatic drop in signal amplitude of the eddy-current transducer signal when scanning weld areas with defects. Experiments were conducted on welded BT1-0 titanium plates. The article contains the results of these measurements. The dependence data facilitates the assessment of the quality of welding seams and helps make an educated conclusion about welding quality.

Keywords: Eddy-current transducer · Titanium alloys · Electrical conductivity · Signal · Introduced voltage

1 Introduction

Titanium and its alloys occupy an important place among metal constructional materials due to their especially valuable physical and chemical properties [1–3] which include small specific gravity and big strength in normal and raised temperatures. These materials are highly resistant to atmospheric corrosion and various aggressive environments. Therefore, due to their properties, titanium and its alloys find multiple applications in construction of installations, pipelines, and chemical equipment [4–6], as, for example, a reactor made of an explosively clad steel titanium sheet [7, 8] or a welded technical titanium pipeline.

Military and civilian shipbuilding requirements for materials selection and technical design of seagoing vessels and their components are growing [9]. The tolerable load needs to be increased despite of the reduction in overall weight. Furthermore, the

Z. Popovic et al. (Eds.): TransSiberia 2019, AISC 1115, pp. 980–989, 2020.
https://doi.org/10.1007/978-3-030-37916-2_96

service life of the finished components needs to be prolonged, which goes hand in hand with a reduced susceptibility to corrosion. As an alternative to traditional materials like steel, titanium and its alloys are available, for example. Titanium has about half the density and similar strength compared to conventional steels [1, 2] Thanks to its exceptionally good corrosion resistance, titanium, although more costly, represents an alternative to the previously used steel. In offshore applications, titanium alloys are very widespread being no longer an exception [10]. The high technical requirements for measuring and working equipment make titanium an obvious choice for its application as an ideal construction material. Military shipbuilding has begun to use titanium alloys for seagoing vessels and submarines [9]. The plate thicknesses used here are in the double digit millimeter range. Joining of these thicknesses is a major challenge to the welding technique used.

Thereby, the increasingly higher use of titanium components to develop critical roles in the industry where high criteria of safety must be guaranteed requires an accurate NDT method. Penetrant testing (PT) methods are integrated as an NDT tool to locate surface flaws in non-porous materials from the manufacture up to the maintenance phase [11]. An hybrid PT method based on bacteria cells was successfully implemented as a NDT tool to inspect micro surface defects up to 700 μm diameter, in laser welds performed in titanium [12]. However its effectiveness is drastically affected by the surface conditions, requiring a complex procedure to avoid a false negative result [13]. On the other hand, digital radiography methods achieve a detection level of flaws with openings above 100 μm width in laser welded titanium specimens [14].

Under a non-complex sample preparation, eddy current testing is an high impact technology to detect hidden (low frequency regime) or surface defects (high frequency regime) in conductive materials upon the application of a time-dependent magnetic field [15]. The presence of a discontinuity acts as a resistive barrier that perturbs the eddy current flow changing the magnetic field generated by it. Furthermore, the resistive losses also promote a thermal distribution along the surface which can be captured by combining thermographic NDT techniques with ECT. In [16] defects with a length of 780 μm are the threshold value for eddy current induced thermography employed to fatigue cracks in titanium. From a sensing point of view, inductive coil sensors are a widespread ECT probe [17], however their poor spatial resolution and limited sensitivity at low frequency compromises the detection of deeply embedded flaws and subtle topographic variations. Superconducting quantum interference devices (SQUIDs) have the potential to surpass the inductive coil sensors by detecting deep buried defects [18], however their high field sensitivity compromises the spatial resolution and requires an apparatus which operates at cryogenic temperatures. Therefore, ECT tools based on magnetoresistive sensors offer advantages over inductive coil sensors and SQUIDs due to an enhanced spatial resolution, high sensitivity, large bandwidth and an operating point at room temperature [19], being very promising candidates for an universal integration in NDT tools [20–22] to overcome the specifications imposed by the industry and achieve a detection range of micrometric surface flaws on low conductive titanium alloys.

Despite a significant number of modern means and methods of nondestructive testing, the majority of the presented methods do not allow one to scan quickly. Most of flaw detection methods may be used only in the laboratory, using complex technical instruments. Portable material diagnostic methods are rather narrowly functional. Despite the possibility of finding defects, they are not designed to evaluate the degradation of the material, and they do not allow one to conclude the possible timing of its further use and the risk of breakdowns.

This is due to the inability to scan simultaneously at different depths, to search ultra-small defects and to analyze scans in real time. In this context, an urgent task is to develop eddy-current measurement systems based on miniature eddy-current transducers designed for finding defects in titanium alloys and predicting the reliability of technical systems.

The purpose of this work is the detection of defects in welded joints of titanium alloys with the application of the eddy-current transducer and the development of the eddy-current control procedure for such joints.

2 Methods

Subminiature ECT [23–25] is designed for experimentally local studies of titanium-alloy plates and weld seams. The developed subminiature ECT represents a core wrapped with the following windings: energizing, measuring and compensation. ECT consists of a core wrapped with the energizing, measuring and compensation windings. Both the windings and the core are impregnated with a compound. They are enclosed in a washer of corundum. This equates to increase the mechanical stability of the transducer.

To test different conductive materials, a developed transducer is used, which is connected to a personal computer via a sound card that is used as a generator and as a signal transducer. The signal thus is sent directly to the energizing winding.

The software is able to control the quantity of a signal applied to the energizing winding and also allows to read the voltage values from the measuring winding, which, taking into account the calibration, are converted into conductivity values. ECT winding coils consist of a copper wire with the thickness of 5 μm. The core is made of ferrite 2000NM3 with an initial magnetic permittivity value of 2000 and has a pyramidic shape. Characteristics of the developed transducer make it possible to achieve high localization of the control, namely, to localize the field within 2500 μm^2. The developed system provides a significant depth of penetration of the field into the prototype system up to values of ~5 mm (at frequencies of 500 Hz).

The software coded in C++ for Windows allows controlling the signal on the energizing winding and receiving the signal from the measuring winding. With the help of the software it is possible to effectively control the signal, which is applied directly to the energizing winding. Also with this software it is possible to receive a signal directly from the measuring winding. The impressed voltage can be controlled using a special mixer built into the Windows. With the help of this mixer, the frequency and amplitude parameters of the generator sinusoidal signal are set. In turn, the sound card makes it possible to extend the signal bandwidth, which is applied directly to the energizing winding.

To scan welds on titanium alloys, the eddy-current transducer was presented by two sensors with energizing and measuring windings. These sensors had the same electromagnetic characteristics and when scanning the weld were placed at a distance equal to or less than the width of the weld. This location of the sensors provided simultaneous consideration of signals corresponding both to the welded materials themselves and welds. The sensors were located at the control area in such a way as to excite the eddy currents at the boundaries:

- first welded material/weld,
- second welded material/weld,
- only in the area of the weld in a variety of combinations.

Due to the synchronous control of signal frequency on energizing winding and the frequency of the filtering system and selective amplification, it was possible to significantly reduce the signal interference level on the measuring winding, which served as an informative parameter of the weld under investigation. The signal amplitude on the energizing winding was also automatically generated, so that the complete subtraction of the informative signals of the two sensors in the absence of defects in the weld was made.

When scanning, the sensors were placed above the weld in accordance with Fig. 1. Material 1 and material 2 are joined by a weld 3. A sensor 4 and a sensor 5 were placed above the surface of a weld.

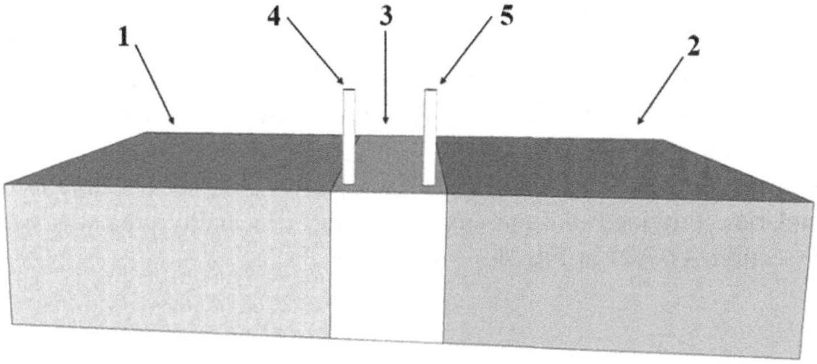

Fig. 1. Scheme of scanning the object under control

3 Experimental Results

3.1 Example 1. Inspection of a Weld of Type VT1-0/VT1-0

Analysis of the weld quality for titanium materials VT1-0/VT1-0. Signal frequency on energizing winding used for scanning was 1600 Hz. Signal amplitude on energizing winding was 1,5 V.

Sample No. 1—plates made of titanium and joined by welds. Plates' thickness—5 mm. Weld width—4–5 mm. Scanning was made both along and transvers to welds in various areas.

Experiment No. 1 with a sample No. 1 was carried out along the weld, two strong drops of the signal amplitude were recorded, this corresponded to the fault location (areas 1 and 3). The results of the experiment are shown in Fig. 2. The value of the voltage introduced to the measuring winding of the transducer in the weld area when scanning along the plate. A1-A2 – the boundaries of the first defect (area 1), B1-B2 – the boundaries of the second defect (area 3).

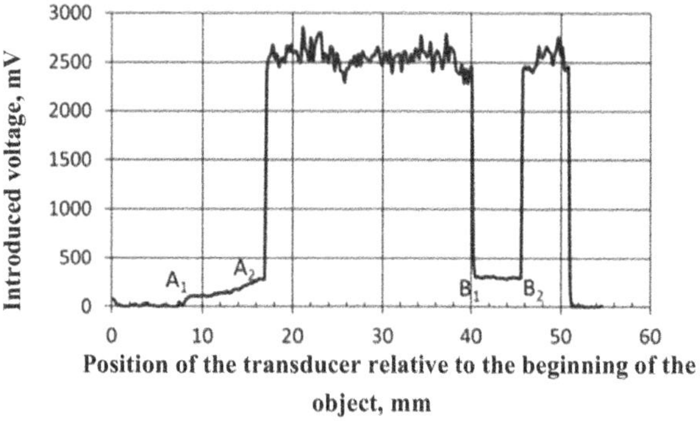

Fig. 2. Dependence of the introduced voltage on the scan coordinate. A_1-A_2—the boundaries of the first defect (area 1), B_1-B_2—the boundaries of the second defect (area 3).

Experiment No. 2 with sample No. 1 was carried out along the weld through areas 1 and 3 (defects) and area 2 in the middle of the weld and free of defects. Scanning of the area 1 had shown that the weld had basically no effect on the signal of eddy-current transducer. However, the defect (A1-A2) was significantly observed due to a significant drop in the signal amplitude (Fig. 3).

When studying area 3, which also had a defect, it was not possible to define the weld boundaries. Provided that the amplitude drop in a faulty area was significantly observed (section A1-A2) in Fig. 4.

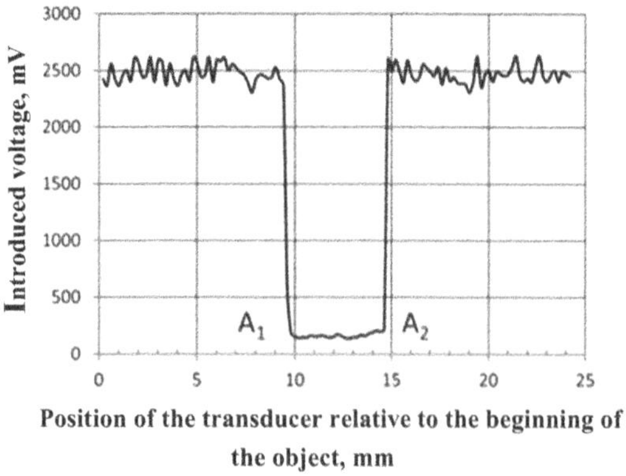

Fig. 3. Dependence of the introduced voltage on coordinates when scanning area 1 transvers to the plate. A_1-A_2—defect boundaries

While studying the area 2, which had any defects, the boundaries of the weld were still not observed. Since there was no defect in the area, the signal amplitude drops were not recorded (Fig. 5).

For comparison, the results of scanning of a sample area without a weld are presented. The results were almost identical (Fig. 6.).

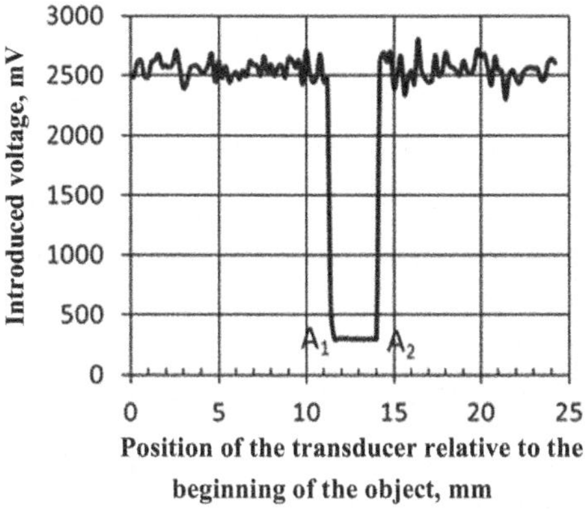

Fig. 4. Dependence of the introduced voltage on coordinates when scanning area 3 transvers to the plate. A_1-A_2—defect boundaries

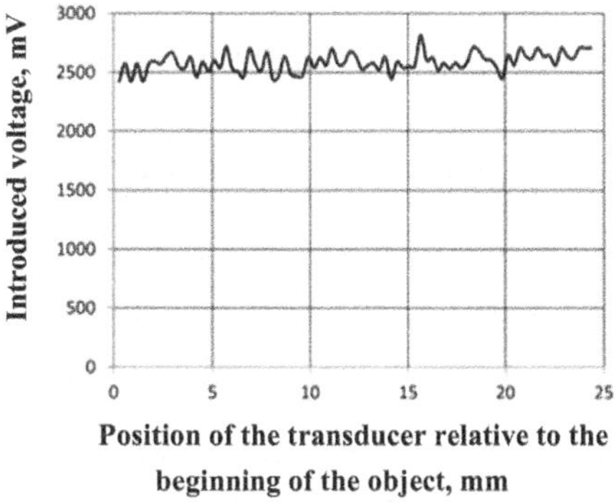

Fig. 5. Dependence of the voltage introduced on the coordinate when scanning across the plate through area 2(a) of the sample.

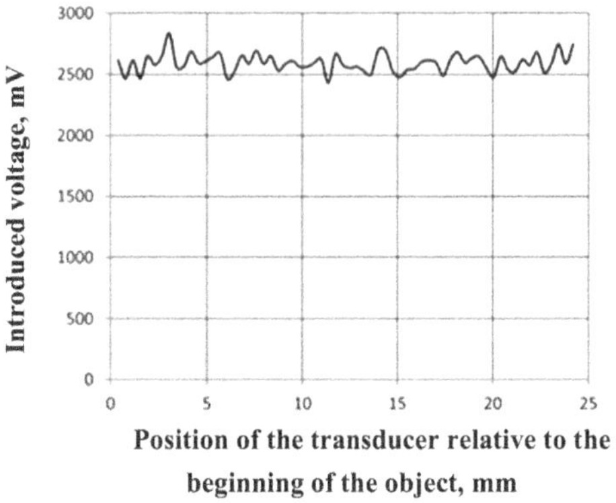

Fig. 6. Dependence of the voltage introduced on the coordinate when scanning across the plate through the defect-free part of the sample

3.2 Example 2. Inspection of a Weld of Type VT1-1/VT1-1

Sample No. 2: two titanium plates joined by a weld. The thickness of the plates was 5 mm. The width of the weld was 4–5 mm. Scanning was carried out along and transverse to the surface of the weld in different areas.

Scanning of Samples Along the Weld. To assess the uniformity of the weld, the samples were scanned along the surface of the weld. Significant fluctuations introduced into the measuring winding were not recorded (Fig. 7).

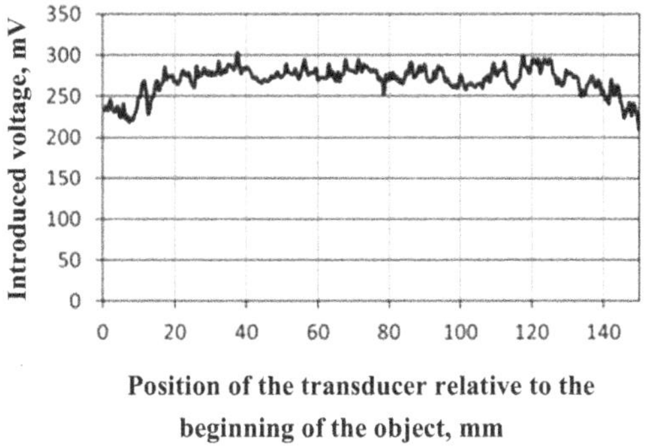

Fig. 7. The value of the voltage introduced to the measuring winding of the transducer when scanning along the weld

According to this experiment, the weld had a uniform texture. However, this study does not give information on the quality of the weld, since according to its results we can make a conclusion on the uniform distribution of defects in the weld, and the absence of any defective areas in it.

Investigation of Plates Transverse to the Weld. To compare the quality of the weld area with the area of the welded plates, a scanning was made transverse to the weld in such a way as to take both the weld area and the plate area. The results of the experiment are shown in Figs. 8 and 9.

A dramatic amplitude drop of the voltage introduced to the measuring winding of the transducer was observed in the scanning area corresponding to the boundaries of the weld. This drop was especially observed if compared with the signal level of corresponding areas of the plates under study.

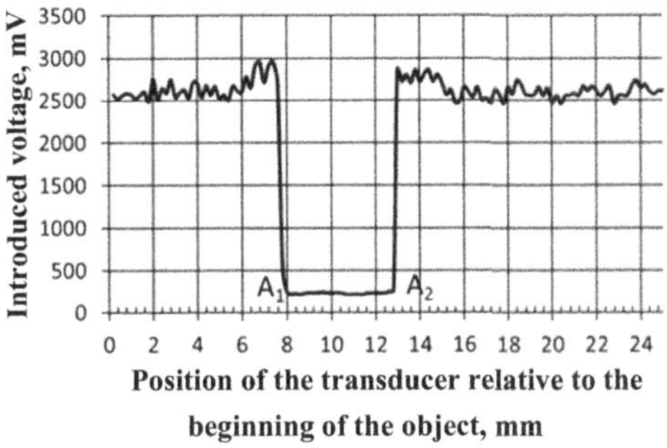

Fig. 8. Response rate when scanning the weld sample №1. The scanning frequency was 1,600 Hz. A1-A2 – weld boundaries

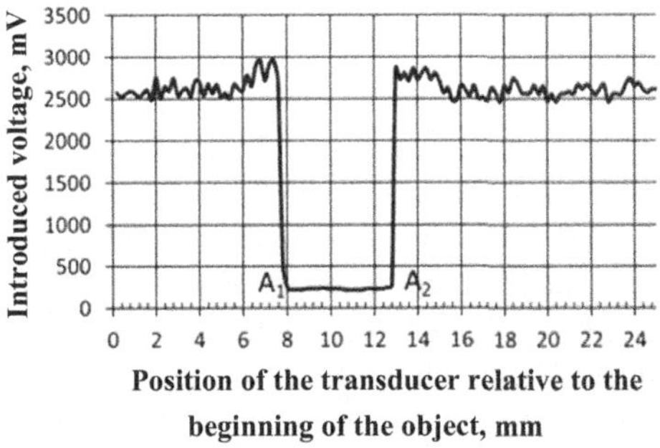

Fig. 9. Response rate when scanning the weld sample №2. The scanning frequency was 1,600 Hz. A_1-A_2—weld boundaries

Based on this experiment, we can make a conclusion on the poor quality of the weld that joined titanium plates of sample No. 1. The results of destructive testing confirmed the poor quality of the weld used in this sample. At the same time, the study of sample No. 2 did not reveal any significant fluctuations in the amplitude of the voltage introduced to the measuring winding in the weld area. High quality of the weld of sample No. 2 was confirmed by the results of destructive testing.

Attaching of Plates Without Welding. To explain the effect of the signal amplitude drop in the area of a poor quality weld, another experiment was carried out: two identical titanium plates were attached and then the interface was scanned with the use of a developed eddy-current transducer. The results of this experiment are shown in Fig. 10.

It is clearly seen that the dependence obtained in this experiment is qualitatively similar to the dependence obtained in Fig. 8 when scanning sample No. 1. In the same way, the amplitude of the voltage introduced to the measuring winding changed more than ten times in comparison with the amplitude of the signal from the plates themselves.

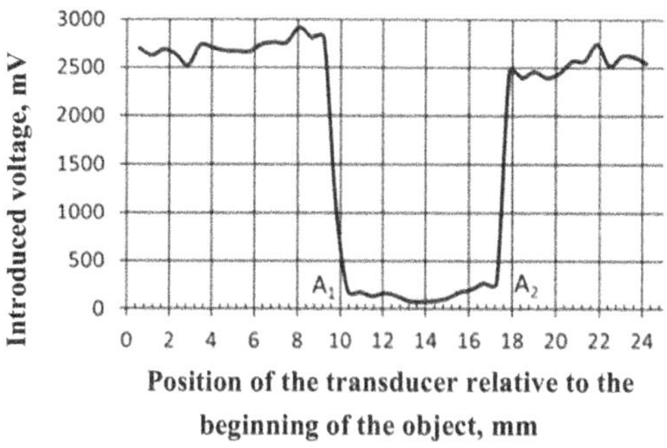

Fig. 10. Voltage value on the measuring winding of the transducer when scanning the area of two plates joint

4 Conclusions

The results of the experiments conducted showed a high efficiency of the developed method in the search for defects of the titanium slabs weld seams and in the evaluation of the welding quality. Thanks to the change of the eddy-current transducer signal amplitude in the area of the weld seam, we succeeded in defining the edges of the weld seam exactly, its low quality being obvious due to a steep drop in the signal amplitude. Likewise, the dependence, received as a result of sample No. 2 (Example 2), characterized by a high welding quality, scanning, showed the absence of any significant signal amplitude changes. So, the analysis of the eddy-current transducer response can be used for the evaluation of the performed welding quality.

References

1. Boyer, R., Collings, E.W., Welsch, G.: Materials Properties Handbook. Titanium Alloys. ASM International, Cleveland (1994)
2. Jenney, C.L., O'Brien, A.: Titanium and Titanium Alloys. Welding Handbook. Materials and Application, Part 2, vol. 4. American Welding Society, Miami (1998)

3. Deters, M., Leyens, C., Kumpfert, J.: Titan und Titanlegierungen. WILEY-WCH, Weinheim (1998)
4. Cebon, D., Ashby, M.F.: Information systems for material and process selection. Adv. Mater. Proc. **157**(6), 44–54 (2000)
5. Crauman, J.S., Willey, B.: Shedding new light on titanium in CPI construction. In: Proceedings of the Stainless Steel World Conference, vol. 1, pp. 487–494. KCI Publishing BV, The Hague (1999)
6. Madina, V., Azkarate, S.: Stress corrosion cracking studies of candidate materials for HLW disposal. Proc. Eur. Feder. Corr. **208**, 737–742 (1997)
7. Young, G.A.: Explosion clad works for reactors. Hydraul. Eng. **10**(3), 109–110 (2005)
8. Wronka, B.: Testing of explosive welding and welded joints. Wavy character of the process and joint quality. Int. J. Impact Eng **38**(5), 309–313 (2011)
9. Malletschek, A.: Einfluss von Titan auf den Entwurf von Unterwasserfahrzeugen. Ph.D. thesis, Hamburg, TUHH - Technische Universtat Hamburg (2011)
10. Caplan, I.L.: Ti-3Al-25 V for seawater piping applications. In: Industrial Applications of Titanium and Zirconium, vol. 1, pp. 43–53. ASTM International (1986)
11. Guirong, X., Xuesong, G.: Analysis and innovation for penetrant testing for airplane parts. Procedia Eng. **99**, 1438–1442 (2015)
12. Santos, T.G.: A new NDT technique based on bacterial cells to detect micro and nano surface defects. Soldag. Inspeção **20**(2), 253–259 (2015)
13. Santos, T.G., Miranda, R.M.: A new NDT technique based on bacterial cells to detect micro surface defects. NDT E Int. **63**, 43–49 (2014)
14. Alkhimov, Y.V., Gnyusov, S.F.: Investigation of laser welded titanium and stainless steel specimens using digital radiography methods. Russ. J. Nondestr. Test. **48**(4), 238–244 (2012)
15. García-Martín, J.: Nondestructive techniques based on eddy current testing. Sensors **11**(3), 2525–2565 (2011)
16. Weekes, B.: Eddy-current induced thermography - probability of detection study of small fatigue cracks in steel, titanium and nickel-based superalloy. NDT E Int. **49**, 47–56 (2012)
17. Theodoulidis, T.P., Kriezis, E.E.: Impedance evaluation of rectangular coils for eddy current testing of planar media. NDT E Int. **35**(6), 407–414 (2002)
18. Bonavolonta, C., Valentino, M.: Eddy current technique based on HTc-SQUID and GMR sensors for non-destructive evaluation of fiber/metal laminates. IEEE Trans. Appl. Supercond. **19**(3), 808–811 (2009)
19. Freitas, P.P., Cardoso, S., Ferreira, R.: Optimization and integration of magnetoresistive sensors. Spin **1**(1), 71–91 (2011)
20. He, D.F., Tachiki, M.: Highly sensitive anisotropic magnetoresistance magnetometer for Eddy-current nondestructive evaluation. Rev. Sci. Instrum. **80**(3), 1–5 (2009)
21. Yang, G., Tamburrino, A.: Pulsed eddy current based giant magnetoresistive system for the inspection of aircraft structures. IEEE Trans. Magn. **46**(3), 910–917 (2010)
22. Cardoso, F.A., Rosado, L.: Magnetic tunnel junction based eddy current testing probe for detection of surface defects. J. Appl. Phys. **115**(17), 17E516 (2014)
23. Ishkov, A., Dmitriev, S., Malikov, V., Sagalakov, A.: An alloy flaw measuring system using subminiature eddy-current transducers. In: AIP Conference Proceedings, vol. 1785, pp. 626–629. AIP Publishing Center, Melville (2016)
24. Dmitriev, S., Malikov, V., Sagalakov, A.: Investigation of steel to dielectric transition using microminiature eddy-current converter. MATEC Web Conf. **143**, 1–7 (2018)
25. Dmitriev, S., Malikov, V., Sagalakov, A.: Eddy current defectoscope for monitoring the duralumin and aluminum-magnesium alloys. IOP Conf. Ser. Mater. Sci. Eng. **189**, 1–5 (2018)

Methods for Assessing the Wear of Elements of Contact Pairs in High-Speed Motion

Victor Philippov(✉), Oleg Sidorov, and Elena Sidorova

Omsk State Transport University, Marx Av. 35, Omsk 644046, Russia
fvm-omgups@mail.ru

Abstract. The article deals with the assessment of frictional fatigue of materials, as well as wear-fatigue testing of contact pairs of current collection devices in high-speed motion. The test schemes, factors affecting the wear of contact pairs of current collection devices are presented. An analysis of methods for assessing wear, the types of integrated wear-fatigue damage to the elements of current collection systems are described.

Keywords: Contact pair · Conductors · Contact element · Wear-fatigue tests · Wear

1 Introduction

According to the concept of wear of solids during sliding, the fracture resistance under the action of alternating stresses is characterized by a frictional fatigue curve. This curve reflects the relationship between the amplitude of the current stress (or strain) and the number of cycles until the material of the surface layer is destroyed [1].

2 Methods

2.1 Evaluation of Friction Fatigue of Materials

Methods for assessing the characteristics of friction fatigue are divided direct and indirect. The direct methods are based on the testing of materials for frictional fatigue, in which the number of cycles before the material is destroyed is determined by the number of its frictional interactions with the indenter by the time the wear particle appears.

The basis of indirect methods for determining the parameters of friction fatigue are tests for wear in the laboratory. The essence of these methods lies in the experimental determination of the steady-state wear rate of samples of material under given conditions. The characteristics of friction fatigue of a material are determined using a mathematical apparatus. The most common test schemes used to assess wear resistance are shown in Fig. 1.

Z. Popovic et al. (Eds.): TransSiberia 2019, AISC 1115, pp. 990–999, 2020.
https://doi.org/10.1007/978-3-030-37916-2_97

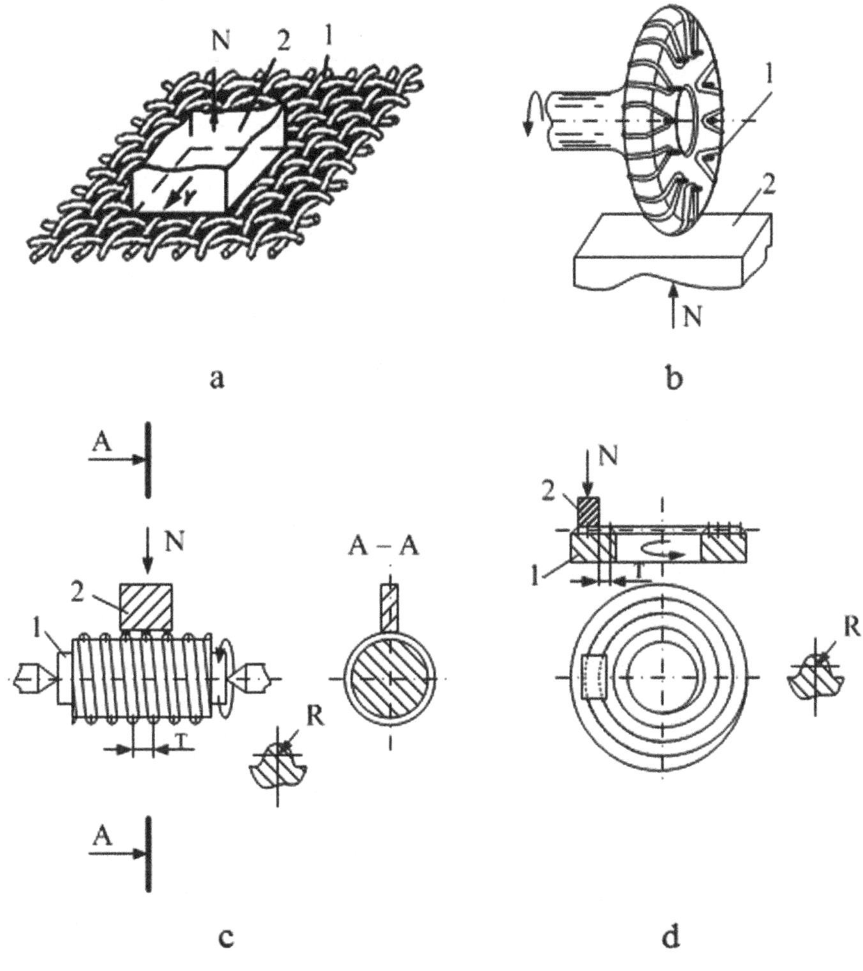

Fig. 1. Friction fatigue test circuits: a - on the grid; b - on the bracelet helix; c - on the screw; d - on the Archimedean spiral; 1 - counterbody; 2 - sample of the material under study

The main advantages of tests on the grid, screw and Archimedean spiral are high stability of the abrasive surface (intrinsic wear resistance), ability to eliminate the effect of abrasive particles, high stability of the results of wear measurements, good correlation with tests on smooth surfaces.

For the implementation of friction modeling methods and the introduction of a rational test cycle for tribotechnical materials, domestic and foreign universal friction machines and laboratory test complexes are used to implement the test schemes given in Table 1.

Table 1. Test schemes for a rational cycle

Unidirectional motion		Reciprocating motion	
mechanism	scheme	mechanism	scheme
Disc brake, clutch, thrust bearing	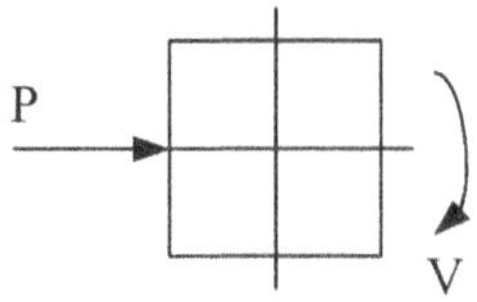	Hinge	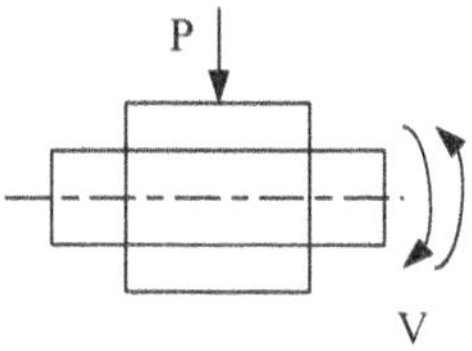
Sliding bearing	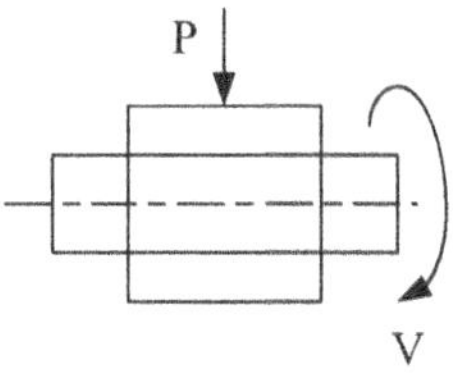	Heat exchanger tube bundle	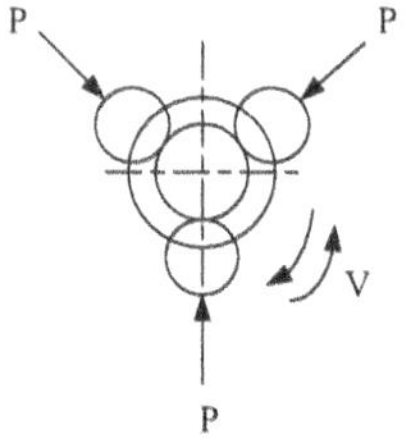
Car brake, crane brake	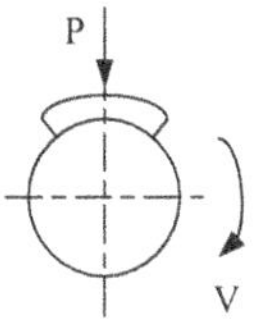	Slide supports and guides	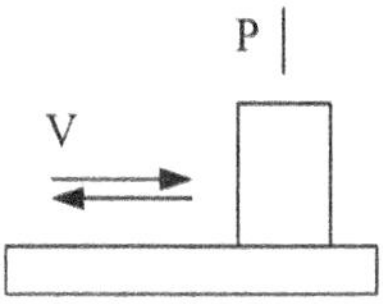
Friction at high speeds	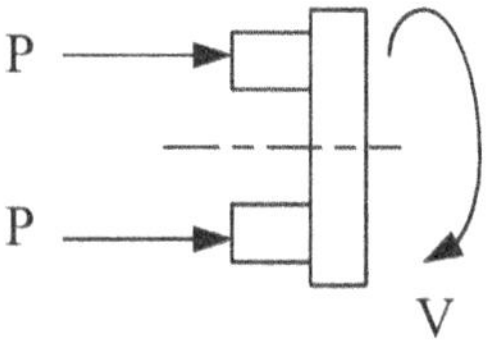	Any nodes at high contact pressures	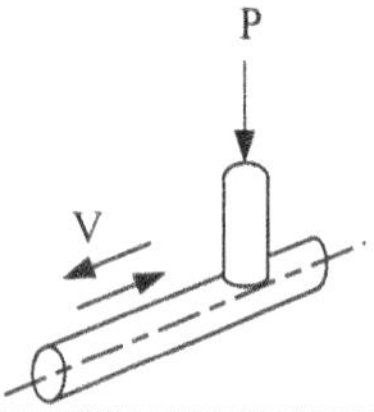

2.2 Wear-Fatigue Testing of Contact Pairs of Current Collection Devices in High-Speed Motion

Improving the level of reliability of power systems in conditions of multi-component exposure to external and internal factors is impossible without solving the problem of comprehensive assessment of their limiting state, which is characterized by various combinations of damaging phenomena. External and internal factors affecting the wear of the elements of the contact pairs of electrical transport current collection devices in high-speed motion are shown in Fig. 2.

In the general case, a comprehensive assessment of damage and the limiting state of power systems can be expressed on the basis of a new approach, including the theory of reliability of mechanical systems, tribology, mechanics of fatigue failure, mechanics of erosion damage, continuum mechanics, continuous and local damage mechanics, materials science, physics chemistry and thermodynamics.

Such an approach to the integrated assessment of the state of power systems, taking into account the criteria of the listed theories and sciences, has been called tribo-fatigue. The essence of tribo-fatigue corresponds to one of the main tendencies of the modern development of sciences: the tendencies of integration of individual sciences into complex sciences that study the totality of objects.

To implement complex experimental research, a new class of test equipment is being created - machines for wear-fatigue testing of materials and power systems under conditions as close as possible to operational.

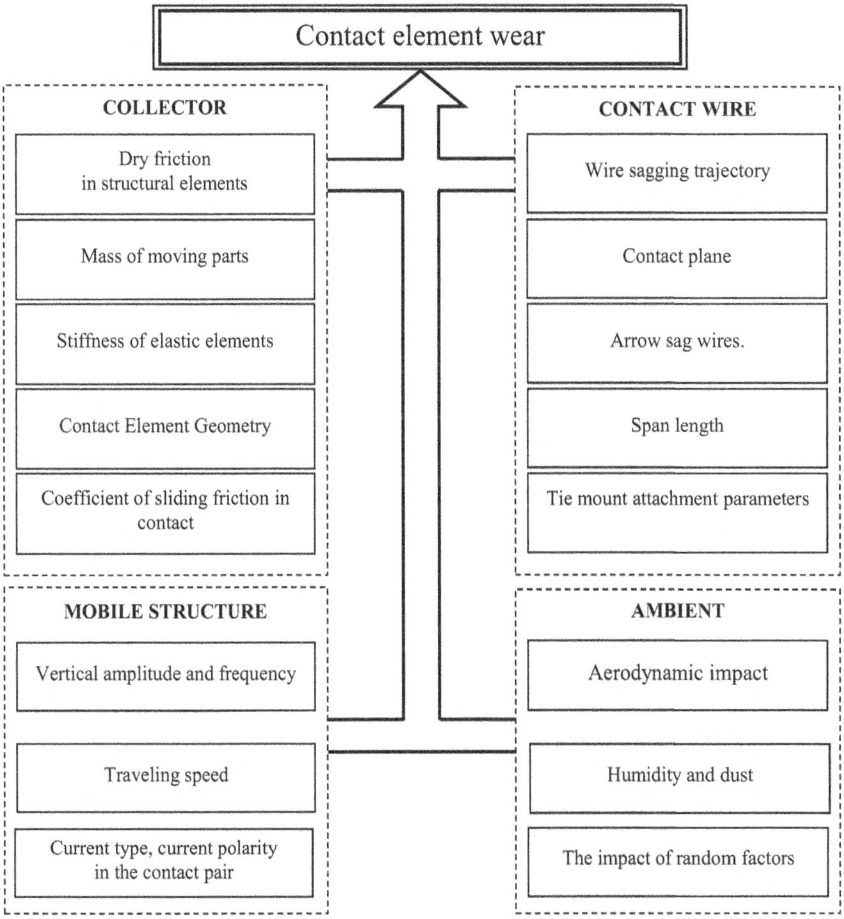

Fig. 2. External and internal factors affecting the wear of contact pairs of current collection devices during high-speed motion

In accordance with the general provisions of tribo-fatigue, only wear or fatigue damage cannot be considered as reasons for failure of any tribosystem. It is also necessary to take into account the complex nature of the processes of damage and destruction, representing wear-fatigue damage. Comprehensive wear-fatigue damage to the power system from the point of view of tribo-fatigue is caused by at least two specific sources: contact interaction of the system elements and repeated-variable volumetric deformation of the system elements.

One of the ways to develop methods for the integrated wear and fatigue testing of elements in high-speed motion is to combine the known methods of mechanical fatigue testing with the methods of friction and wear testing (Fig. 3).

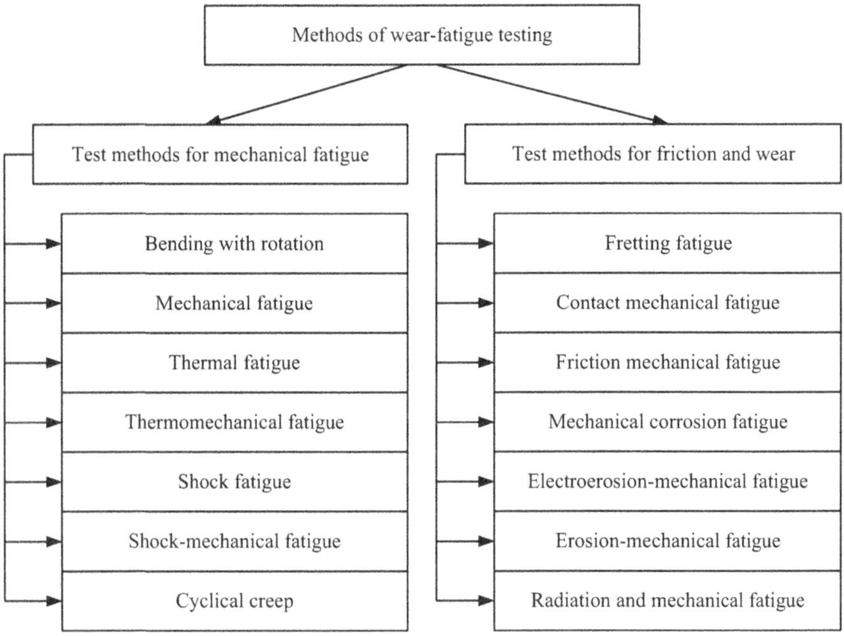

Fig. 3. Methods of wear-fatigue testing

As an example of a tribosystem study, consider the current collection system with a current collector and a contact wire (conductor). The traditional approach to the study of the functioning of the elements of the current collection system is mainly to study the characteristics of current collectors, the parameters of contact wires and to assess the nature of the processes of their interaction. This approach is based on many years of experience in operating current collection systems and basic research in the field of electrical sliding contact.

From the point of view of tribology, the study of the system «contact element – conductor» consists in considering the interaction process taking into account the parameters of the contact element and the surface of the conductor, which are a friction pair. The main damaging effects for this pair are contact fatigue and electrocorrosive damage to the working surfaces.

The limiting state of elements of a friction pair «contact element – conductor» is due to the following indicators: wear to critical values of elements of the tribosystem, critical density of contact surfaces, electrocorrosion of contact surfaces.

In the traditional assessment of the performance of a mechanical system according to individual criteria (wear resistance or fatigue resistance), the connection between the elements of this tribosystem is almost not taken into account. Studies and modeling of the tribosystem, taking into account the interrelations of its elements, should be carried out in accordance with the provisions of the tribo-fatigue.

According to Russian State Standard GOST 30638-99 "Tribo-fatigue. Terms and Definitions", a power system is a system that perceives and transmits a working alternating load and in which the process of friction in any of its manifestations occurs at the same time (during sliding, rolling, slipping, hitting, etc.). According to Russian State Standard GOST 30754-2001 "Tribo-fatigue. Methods of firmability testing. Tests for contact mechanical fatigue" a force system can also be defined as a friction pair, one of the elements of which is subject to volumetric cyclic deformation.

On the basis of modern concepts of the current collector as a power system, it is necessary to evaluate its performance and reliability using the basic principles of tribo-fatigue, that is, from the standpoint of complex wear-fatigue damage to the system [2].

Studies of the «current collector - current conductor» system show that the current conductor is subjected to an alternating load on the side of the contact element of the current collector, while being an element of a friction pair. Thus, this tribosystem should be considered as a power system that operates under conditions of complex wear-fatigue damage, which is a combination of sliding friction processes and mechanical fatigue phenomena.

However, the electric current rolling stock collector, in turn, is also a complex mechanical system in which the functional state of each element affects the quality of the current collection.

A detailed examination of the current collector as a power system can be divided into several subsystems operating under conditions of contact-mechanical and friction-mechanical fatigue. The state of the elements of the system also depends on aggressive environmental effects (corrosion-mechanical fatigue) and the flow of current in the contact (electro-corrosion-mechanical fatigue).

2.3 Types of Wear-Fatigue Damage to the Elements of Current Collection Systems

Studies of the dynamic processes of interaction between the elements of the "current collector – conductors" system during operation show that all types of damage listed in the Table 2 are present at the system nodes.

In addition, special attention should be paid to the inclusion of fretting-fatigue in the research methodology, since individual components of the current collector are exposed to its effects.

The criterion of the limiting state in assessing the durability of the elements of the tribosystem is fatigue fracture: volumetric resolution under alternating loading, surface fracture under friction.

Table 2. The main types of wear-fatigue damage to the system «current collector – conductors»

Type of wear-fatigue damage	The cause of wear-fatigue damage
Friction mechanical fatigue	Kinetic interaction of the phenomena of mechanical fatigue and sliding friction
Contact mechanical fatigue	Kinetic interaction of the phenomena of mechanical fatigue and sliding friction
Fretting fatigue	Kinetic interaction of mechanical fatigue and fretting phenomena
Mechanical corrosion fatigue	Exposure to alternating stress and corrosive environment
Corrosion and erosion fatigue	Kinetic interaction of the phenomena of mechanical fatigue, corrosion and erosion
Electrocorrosion-mechanical fatigue	Exposure to alternating stresses and electro-corrosive environment

When sliding friction, ultimate surface fatigue failure occurs when critical, or ultimate wear, of friction pair elements is reached.

Experimental and theoretical studies of the interaction of current collectors with a contact wire show that the main characteristics of friction and wear (ultimate contact pressure, wear rate, coefficient of friction) largely depend:

From the direct effect, that is, a significant change in the basic characteristics of fatigue resistance (endurance limit and fatigue durability) as a result of friction and wear processes;

From the opposite effect - the level of cyclic stresses arising from alternating loads (contact or non-contact).

Consideration of this problem also from the standpoint of micromechanics of diffuse damage makes it possible to note that the defects present in the material initially interact during operation with defects generated during the deformation process, which leads to a continuous process of accumulation of total damages by merging them and growing until the source appears. destruction.

Developed from the standpoint of tribo-fatigue approach to the study of the tribosystem is as follows:

1. For the power system «current collector – conductor» is characterized by complex wear-fatigue damage and destruction.
2. The operational performance of the system (reliability and durability) is determined by both direct and inverse effect.
3. Optimal durability of the system as a whole and its elements can be achieved by controlling the processes of complex wear-fatigue damage, taking into account the real interaction of irreversible damage (due to the contact interaction of the system elements and their volume re-variable deformation).

Friction Mechanical Fatigue. The wear-fatigue damage of this kind arises and interacts with each other from cyclic and frictional stresses, which cause various changes in the physicochemical and mechanical properties of the elements of the power tribosystem.

Friction stresses initiate surface damage to the elements of the power system, which manifests itself in the form of fatigue microcracks on the contact surface.

Cyclic stresses must be considered as a combination of stresses, the result of which manifestations are surface (microcracks) and bulk (macrocracks) fracture.

Analysis of the technical condition of high-speed electric vehicle current collector assemblies after their operation, as well as research under laboratory testing conditions, show that wear-fatigue damage of this type occurs in sliding bearings of the movable base and in a screw-nut pair of the transfer mechanism.

Contact Mechanical Fatigue. The criteria for limiting the state when testing for mechanical and contact-mechanical fatigue is the destruction of the tribosystem element and the critical density of chipping of the material on the friction surface.

In the first case, the determining parameter of the process is the amplitude of the cycle stresses, in the second - the maximum contact stresses.

This type of wear-fatigue damage occurs when the force interaction of the contact element and the conductors when moving the current collector of the rolling stock.

Mechanical Corrosion Fatigue, Corrosion and Erosion Fatigue and Electro-corrosion-Mechanical Fatigue. The limiting state of friction pair elements occurs as a result of additive and multiplicative effects of alternating voltages from the conductors and electric rolling stock, electro-corrosive environment during the flow of traction current in the contact zone and the corrosive effects from the environment. The contact element of the current collector is the least protected from this type of wear-fatigue damage.

Contact Mechanical Fatigue. In case of sliding friction, a critical and limiting condition occurs when critical or limiting wear is reached. The kinetic dependence of the change in wear over time is generally similar to the curve of the accumulation of residual deformations under cyclic loading. In comparison with other nodes, this type of wear-fatigue testing is more susceptible to the contact element of the current collector.

The most important load parameter affecting the wear process during sliding is the specific friction force, or frictional stress.

Fretting Fatigue. The damage to the tribosystem as a result of the flow of fretting processes is due to the kinetic interaction of variable loads in the elements and their mutual microdisplacements. The result of the interaction of elements is a violation of the surface layers, the emergence and development of microcracks and, as a result, fatigue failure.

Fretting phenomenon in current collector nodes manifests itself as a result of dynamic processes during the movement of high-speed electric trains (bolted connections, sliding bearings, etc.).

A methodology for conducting experimental studies of contact pairs of electric current collection devices has been developed and successfully tested at the Omsk State Transport University.

The methodology involves bench testing for each pair of contact materials "contact element - contact wire" in order to determine their optimal combination to reduce wear and increase resource and includes the following steps:

1. Determination of tribocompatibility, scoring resistance and wear resistance of materials (studies from mechanical load in contact without electric current flowing). For this, the following parameters vary: contact pressing, dustiness and humidity of the ambient air, aerodynamic effect of the air flow, speed of movement.
2. The study of the wear of a pair of "contact element - contact wire" depending on the magnitude of the flowing current. To determine the characteristics of wear resistance, it is necessary to change the following parameters: values of contact pressing, values of traction current (or current density), environmental parameters and contact suspension at critical values of pressing in contact and speed. When implementing the indicated stage of the research methodology, it is necessary to perform it with alternating and direct current.
3. Assessment of the amount of wear and prediction of the resource of the elements of the contact pair using mathematical models.

3 Results and Discussion

The frictional interaction of the elements of the "contact element - contact wire" tribosystem is a non-linear process, which is determined by a large number of both internal (physicochemical properties of the materials of the friction pair) and external (dynamic loads, the presence of electric current in the contact, environmental parameters, etc.) interrelated factors.

To ensure the reliability of the data array obtained as a result of the tests, an algorithm for the implementation and processing of experimental data is used (Fig. 4), developed taking into account the provisions of the theory of experimental design.

The results of experimental studies in high-speed motion depend on the parameters and characteristics of measuring instruments, the object of study, the quality of bench installations, as well as the human factor, the influence of which is manifested in the form of random errors and errors during the experiment [3, 4]. Processing the results of the experiment and predicting the wear and life of the element of the contact pair of the current collector devices are carried out using mathematical models.

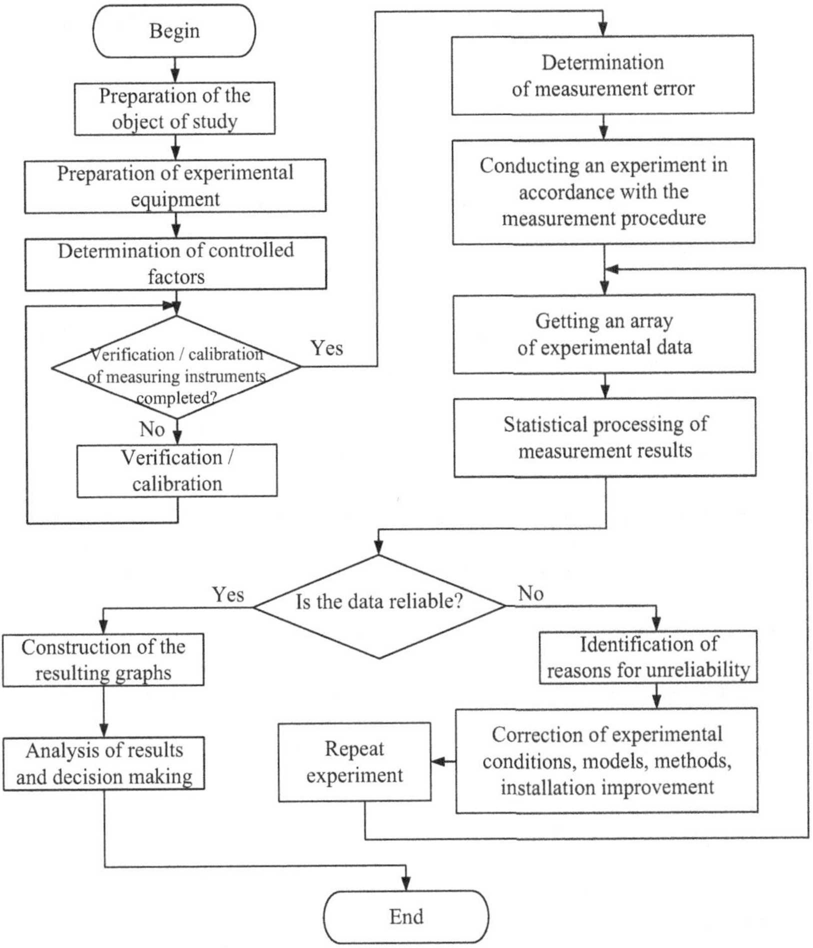

Fig. 4. Experiment planning algorithm

References

1. Chernyshev, D., Alpatov, V.: Research of the flow-over of the mechanical core of four pipes by a wind current, using control volume method. In: MATEC Web of Conferences, vol. 117, p. 00028 (2017). https://doi.org/10.1051/matecconf/201711700028
2. Shurinov, V.A., Koreshkov, V.N., Ozhigar, G.P., Sosnovskiy, L.A.: Modeling of power systems. Tribo-fatigue-96/97. Gomel, NPO TRIBOFATIKA (1999)
3. Voznesenski, V.A.: Statistical Methods of Experimental Design in Feasibility Studies, Moscow (1994)
4. Schenck, H.: Theories of Engineering Experimentation. Mcgraw Hill, New York (1997)

Study of the Load Distribution in Threaded Connection of Casings

Vladimir Papirovskiy, Sergei Konygin, Vera Zhivaeva, Nikolai Kats, and Pavel Bukin(✉)

Samara State Technical University, Samara 443100, Russia
mantoulenko@mail.ru

Abstract. The intention in this work is to identify the bearing capacity reserve of trapezoidal threaded connection of casings. A computer-based modelling of threaded connection screwing, with various values of preload and applied tensile force, was made. This modelling is performed using the method of finite elements, that allows to demonstrate the stress-strain state of the solid body when affected by external factors. The conducted experiments have shown that radial component of preload in the connection causes an increase in the diameter of a coupling and reduction in its longitudinal size with a simultaneous decrease in the coupling thread pitch. At the same time, a decrease in thread diameter of pipe nipple occurs, and its length and thread pitch increase. Although there are minor changes in coupling and pipe nipple thread pitches, they cause an abrupt redistribution of the components of longitudinal load, absorbed by the thread turns: the greater value preload has, the more uneven turn-wise distribution of load is.

Keywords: Axial preload · Radial preload · Threaded connection · Coupling · Pipe nipple

1 Introduction

To ensure strength and leak-tight integrity of threaded connection of oil-field pipes, it is important to decide on the value of axial preload. This work explores the problems of analyzing distribution of load across the turns of conical threaded connection of casings "Batress" with trapezoidal thread of 168 mm diameter and 8.9 mm wall thickness, D strength group (K55 as per the International classification) [1–8].

The important problem currently lies in the lack of well-proven methodologies for selecting the value of axial preload. It is commonly supposed that preload after manual screwing is created by screwing-in of pipe into coupling for ½ in. (12.7 mm). To do this requires 2.5 turns of pipe nipple, when thread pitch is 5.08 mm. Several computational experiments were carried out to analyze the effect of axial preload on the thread turns' load. In each of such experiments the threaded connection was screwed together applying a specified number of preload turns, and it was further loaded by tensile force.

Z. Popovic et al. (Eds.): TransSiberia 2019, AISC 1115, pp. 1000–1005, 2020.
https://doi.org/10.1007/978-3-030-37916-2_98

2 Methods

The thread stress-strain state was modelled using method of finite elements. To this effect, longitudinal load and rotation torque acted on the connection in question. It allowed to consider the problem axially symmetrical. The selected numbering of the thread turns is shown in Fig. 1. As per this numbering, turn 14 is in the basic plane of thread (the last turn with the full profile).

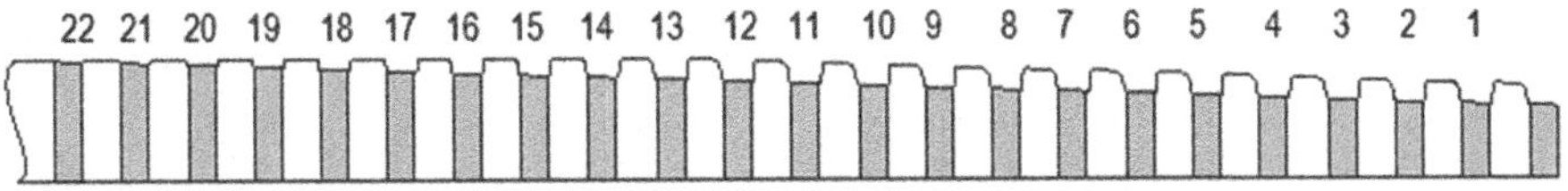

Fig. 1. Numbering of turns of the reviewed thread. Source: Authors

When creating the model of threaded connection (Fig. 2) it was supposed that surfaces 1 and 2 are in contact while screwing by hand. Surfaces 3 have a clearance defined by the production tolerance, and a clearance between surfaces 4 is within 0.03 mm. The model also considered potential reciprocal movement of the threaded connection contact surfaces. The value of the respective friction coefficient is taken as 0.08.

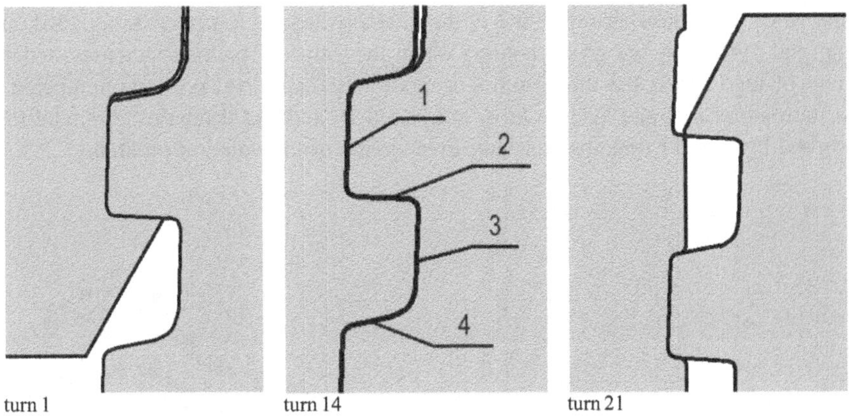

Fig. 2. State of turns after manual screwing. Source: Authors

The value of tensile load was specified to be 1345 kN that corresponds to reaching 80% of the yield stress for the selected type of pipes ($\sigma_T = 379$ MPa). The first stage of work involved a study of loading the threaded connection after its manual screwing. Although this alternate assembly is unacceptable, it is used to be compared with the other options of assembly. In the course of modelling the threaded connection operation, while applying tensile force it was identified that deformations, which appear, create substantial clearances between pipe and coupling (see Fig. 3). As a result, leak-tight integrity of the connection decreased, although its bearing capacity remained the same.

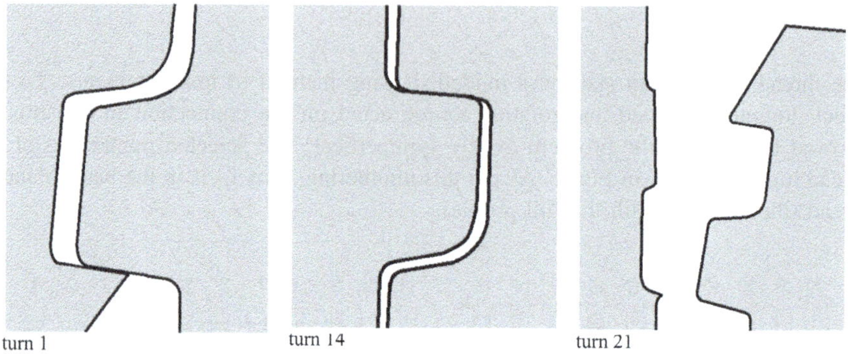

Fig. 3. Strained state of turns after manual screwing and applying tensile load (scale of movements increased by 30 times). Source: Authors

3 Results and Discussion

The results of calculations, made using the method of finite elements, are shown in Fig. 4. Examination of data in Fig. 4 shows that distribution of load across turns of the thread is very uneven, and this unevenness depends on the value of preload while screwing pipes. The most uniform distribution of load can be seen when screwing by hand, when most turns absorb 3–6% of the total tensile load. 1st (the most loaded one), 18th, and 19th turns are an exception. When the value of preload increases, redistribution of load across the thread turns occurs [9–12], and load on turns numbered 12 and more increases, and load on turns numbered 10 and less decreases. An axial force absorbed by turn 11 remains constant, irrespective of the value of preload.

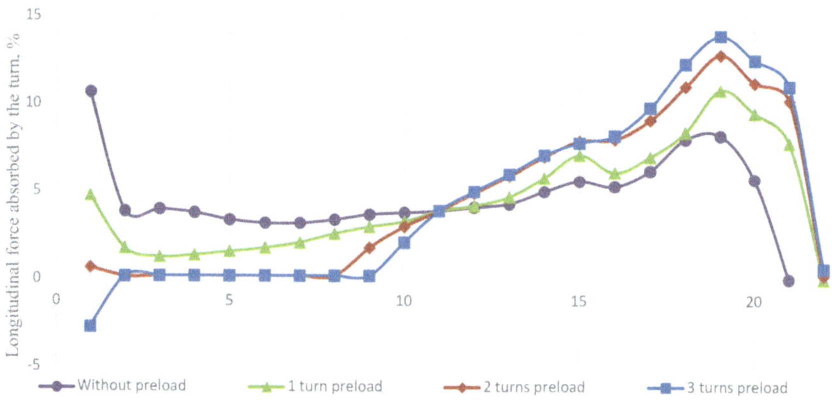

Fig. 4. Tensile force turn-wise distribution when there are various tightening forces of threaded connection. Source: Authors

The key point here is that when the value of preload increases, turns located at the end of casing (turns with small numbers) cease to absorb axial load. Thus, for example, with 2 turns preload value turns 2–8 do not work in tension, and with 3 turns preload turns 1–9 do not work. As noted above, axial force shall be absorbed by turns with high numbers, and turns 17–21 with incomplete profile absorb the most part of load. Turn 22 remains practically unloaded in all cases.

The above phenomenon occurs due to the following. When axial preload is created, radial sidewise preload appears. It amounts to 1/32 of the axial one. Here, a pipe nipple is compressed radially and lengthened axially. A coupling, conversely, elongates radially and shortens axially. Furthermore, since thickness of the pipe body is less as compared to thickness of the coupling body, the connection nipple has greater radial compression deformation.

Therefore, a clearance on bearing surface No. 2 is created even when preload is 1 turn, within the 1st turn a clearance is created on bearing surface No. 2, and a contact is formed on back surface No. 4 (see Fig. 5).

When preload is 3 turns, a significant axial force loading the remaining turns (see Fig. 4) appears on surface No. 4 of the first turn.

Thus, it can be noted that the value of preload affects distribution of axial load across the turns. Analyzing the results of modelling showed that the pipe nipple stress state also changes depending on the value of preload. Stresses distributed across the length of pipe nipple (starting from the pipe end), computed according to the maximum-strain-energy theory (by von Mises), are shown in Fig. 6.

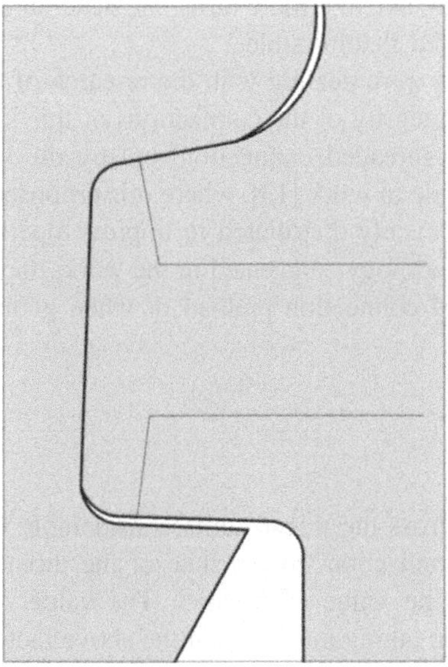

Fig. 5. Axial deformation of pipe nipple and coupling, when preload is 1 turn, which created a clearance on bearing surface of the 1st turn (this is a full-scale picture of movements). Source: Authors

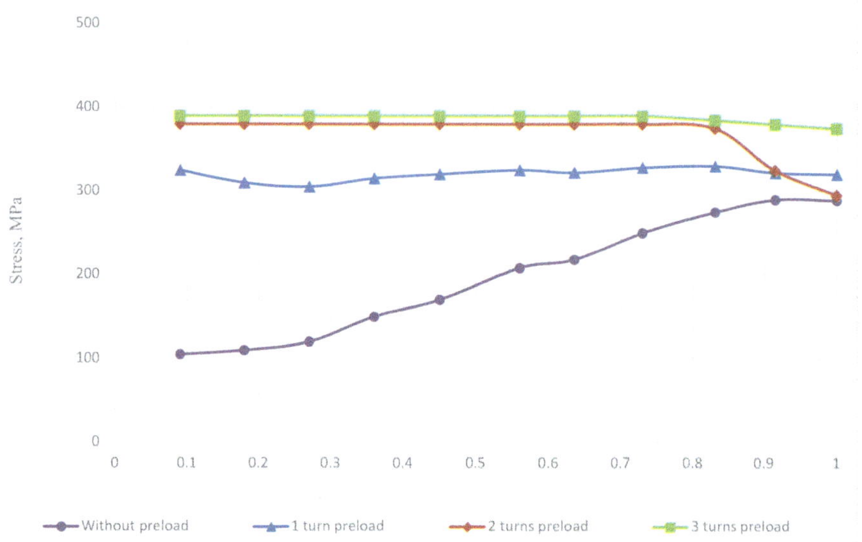

Fig. 6. Distribution of stresses across the pipe nipple length under various tightening conditions. Source: Authors

The value of yield stress for this strength group is 379 MPa. The data presented in Fig. 5 show that, when there is a 1 turn manual preload, the material of nipple is in the elastic range of stress. When there are two and more turns, the material goes into plastic strain range that is usually considered inadmissible.

Domestic and foreign specialists were dealing with the research of that kind. They proved that strength and leaktight integrity of the connections of this kind is specified not by the number of turns in the threaded connection, but by the value of tensile forces. Similar conclusions were made in work [13], where it is proposed to ensure that the load on threaded connection is evenly distributed to improve operability of aerial vehicles' units. Therefore, the methodology, suggested in the work, may be used when choosing the values of the threaded connection preload or when geometry of thread changes.

4 Conclusion

Distribution of axial tensile load across the thread turns considerably depends on the value of preload in the threaded connection. Stress state of the threaded connection nipple also changes according to the value of preload. The value of the threaded connection preload should be chosen taking into account the above factors. A principal method for improving uniformity of absorbing tensile load by the thread turns is to create a connection with various values of nipple and coupling thread pitch.

References

1. Izuppe, E.A.: Oilfield Tubes: Structure and Calculation. Samara Technical University, Samara (2015)
2. Faskhiev, K.A.: An increase in durability of threaded connection. Bull. Mach.-Build. **9**, 25–29 (2011)
3. Inzhener-Info: Distribution of load across thread turns. Engineer reference book. https://inzhener-info.ru. Accessed 6 Mar 2019
4. Requirements to thread, calibration, and control of threads of casing, oil-well, and line pipes. Standard of technical requirements. API 5 B. http://docs.cntd.ru/document/1200081853. Accessed 11 Mar 2019
5. Kantaria, S.N.: Enhancement of strength characteristics of conical connections of casings with trapezoidal thread. Synopsis of thesis for Ph.D. degree, Ufa (2014)
6. Ananchenko, V.N., Tsybriy, I.K., Morgunov, V.V.: Special aspects of fabricating and controlling thread on oil-well pipes. Bull. DTSU Spec. Issue, Tech. Sci. Part 1 **S1**, 48–54 (2009). https://cyberleninka.ru/article/n/osobennosti-izgotovleniya-i-kontrolya-rezby-na-trubah-neftyanogo-sortamenta
7. Saruev, A.A., Shadrina, A.V.: Analyzing operation of threaded connection of enclosed-type drilling pipes during percussive-rotary drilling. Mod. Probl. Sci. Educ. **1**(1) (2015). https://www.science-education.ru/ru/article/view?id=19422
8. Ivanov, A.S., Murkin, S.V.: Calculation and design of threaded connections of actuating devices. Innov. Mach.-Build. (2015). Moscow
9. Vochek, I.I., Kochetkov, D.V., Fadeev, N.A.: An impact of selective reinforcement on stress-strain state of pressure coupling parts. Electron. Res. Pract. J. Modern Sci. Res. Innov. **5**(73) (2017). http://web.snauka.ru/issues/2017/05/82761
10. Vochek, I.I., Kochetkov, D.V., Fadeev, N.A.: Complex design and technologic improvement of threaded connections' performance properties. Electron. Res. Pract. J. Modern Sci. Res. Innov. **5**(73) (2017). http://web.snauka.ru/issues/2017/05/82588
11. Vochek, I.I., Kochetkov, D.V., Vanteev, A.N., Kosova, M.A.: Assurance of corrosion resistance of threaded connections. Electron. Res. Pract. J. Modern Sci. Res. Innov. **6**(74) (2017). http://web.snauka.ru/issues/2017/06/83973
12. Sabanchiev, K.K., Galova, M.A.: Methods for calculating distribution of load across the turns of threaded connection **12**(18) (2015). https://cyberleninka.ru/article/n/metodika-rascheta-raspredeleniya-nagruzki-mezhdu-vitkami-rezbovogo-soedineniya. Scientific Collection of Articles of Kabardino-Bolgousskiy State University
13. Soloviev, V.L.: An increase in uniformity of tightening group threaded connections when assembling aerial vehicles' units. Electron. J. "Trudy MAI" **70** (2017). http://trudymai.ru/upload/iblock/0d1/0d13d4f00a135ec2ab1070e7148a2f65.pdf?lang=ru&issue=70

Earthquake Resistance of Buildings of Complex Macrostructure with Elastic-Plastic Bonds

Anton Smirnov(✉)

Saint-Petersburg University of Architecture and Engineering, Saint-Petersburg 190005, Russia
smirnovanton79@yandex.ru

Abstract. For high-rise buildings over 120 floors, it is proposed to use a new type of Buildings of Complex Macrostructure (BCM). The concept of a complex macrostructure is proposed by prof. V.I. Pletnev. The building of a complex macrostructure is a building consisting of two or more buildings united by lintels or connections. The building of a complex macrostructure differs high rigidity, since it is organized like a spatial truss structure. With low-frequency seismic shocks, the building becomes more vulnerable to large inertial internal forces arising in the rigid structure of the BCM. Eliminate this contradiction is proposed by introducing into the bonding elements between the towers of elastic-plastic bonds (EPD) which due to the correct selection of stiffness and the actuation force, go into the plastic phase of operation and turn off. Calculations show that such EPD can reduce horizontal acceleration by 30%.

Keywords: Seismic loads · Transport infrastructure · Complex macrostructure

1 Introduction

With an increase in the number of storeys of high-rise buildings of up to 120 floors or more, the one-volume layout exhausts its possibilities due to the cumbersome plan and insufficient rigidity. If the point building is capable of being rigid with the ratio of the height to the width of the smallest side in the plan equal to 7–8, then in this proportion the building is not able to grow indefinitely due to a number of limiting factors:

1. Dimensions of temperature blocks (60–70 m)
2. Depth of the shell by insolation (14 (60) m in Russia (USA))
3. Large costs for foundations for the lack of rolls and uneven sediment facades more than 60 m.

From the point of view of economic and constructive expediency, each of the traditional building schemes has both its own height range and the range of optimum number of storeys. Below are the boundaries of the optimal applicability of some traditional design schemes depending on the number of storeys (Table 1).

From the point of view of mechanics, an analogy can be made between a point building and a cantilevered composite shaft. With a certain outreach of the console, its

Z. Popovic et al. (Eds.): TransSiberia 2019, AISC 1115, pp. 1006–1018, 2020.
https://doi.org/10.1007/978-3-030-37916-2_99

Table 1. Applicability of some structural schemes depending on the number of storeys

Number of storeys building, quantity Floors	Height build-ing, m	Frame system	Wall system	Frame bond system with hardness kernel	Frame bond system with hardness kernel &bond stifness	Tube	Tube In Tube
10	30						
20	60						
30	100						
40	150						
50	200						
60	240						
70	280						
80	320						
90	360						
100	400						

solid cross section becomes impractical; the material in the neutral zones becomes ballast. The output is a truss console with a light grate, where the forces are perceived discretely. A similar output is possible in cases of macrostructural changes in the building. At the same time, a highly developed tower can be turned into a 2, 3-point building of a complex macrostructure united by a system of jumpers, diaphragms or links (hereinafter abbreviated to BCM). The building of a complex macrostructure is a building consisting of two or more buildings, or towers, connected by lintels or connections.

Work stiffness links BCM. As the calculations show [5], the greatest spatial effect of the work is due to the stiffness, which differs from the rigidity of individual bodies by more than an order of magnitude.

The stiffness ties between the bodies of the BCM can be regular in height. They can be hinged crossed or rigidly clamped and connected with the core of rigidity. This type of stiffeners creates the maximum effect of the compatibility of the bodies on bending not only in the plane, but also from the plane of the bonding. When working on horizontal loads, the system can be divided into three generalized cases according to the proportion of stiffnesses:

1. The stiffness of the links between the blocks is negligible-small - the systems coexist autonomously (example - Petronas Tower).
2. The rigidity of the connections is great, the systems work together, the neighboring building influences the forces and forms of the oscillations of this one.
3. The stiffness of the connections is comparable to the rigidities of the cases, the work of the BCM as a whole. (TVCC in Peking for example) TV-Center Peking).

The degree of compatibility of the work of two or more buildings as a whole depends on the configuration and stiffness of the connections with respect to free-standing buildings. Therefore, in terms of the degree of participation of bonds in the overall static and dynamic operation of the structure, several cases can be distinguished:

1. Links are fragmentary in height. Their rigidity has an articulated connection with the towers. The influence of these links on the compatibility of the work of the blocks is minimal.
2. The connections are regular in height, they have a hinged-mobile connection with the towers - they create the effect of joint operation of the blocks due to the work of connections as spacers to horizontal loads. The shear stiffnesses of the blocks are summarized.
3. Connections are fragmentary in height, but rigidly clamped and connected with a hard core or mega-strings, which creates a frame effect of the work.

2 Application of the EPD

At the same time, greater horizontal rigidity makes the BCM vulnerable to seismic influences. The building for optimal work on seismic must be low-frequency, i.e. own frequencies should be in the range of up to 0.5 Hz, and the period is more than 2 s. For buildings, it is important to simultaneously meet the sanitary and hygienic requirements for maximum horizontal acceleration ($a = 0.1$ m/s) and displacement (1/500 h). However, work on the wind load requires reverse, so the elastic-plastic bonds (hereinafter referred to as "EPD") allow creating a hybrid structure, equally well working both for wind and seismic. This contradiction can be resolved by using elastic-plastic bonds in the structure of the BCM. The actuation force of these elements should be greater than the forces arising from wind loads, and less than the forces from seismic action. In addition to the known types of elastic-plastic bonds, bonds can be used in the form of perforated steel strips used in the form of braces of the BCM of a farm species.

EPD fully retains all the positive properties of directional plastic dampers: stability of characteristics when external conditions change (temperature, humidity, etc.), long-term maintenance-free operation; simplicity and low cost of manufacture; reliability at operation. All these qualities are confirmed by the experience of using 3D EPD [3].

EPD went through a full cycle of design and development: design, development of mathematical models and software, calculations, manufacturing and experimental verification (Fig. 1).

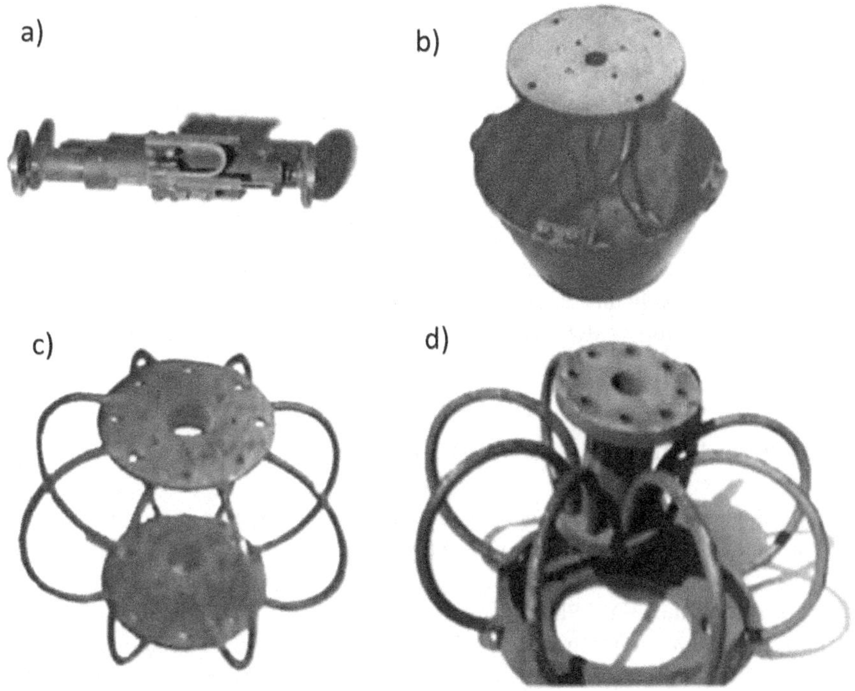

Fig. 1. Construction of plastic dampers Prof. Rutman [3]: a - with ribbon-kegelnymi elements, b, c, d - with curvilinear-rod elements

3 Calculation Program

At the center of this study, the work of the ZSM with the EPD on wind and seismic loads. The calculation was carried out in the software complex Ing +, developed in Moscow, Russia, in conjunction with Professor Rutman, which helped developers to block the module based on the method of all-fluidity [3, 4, 7], which allows you to take into account the EPD with a non-linear elastic characteristic and assign the stiffness of the EPD and the actuation force. In addition, when applied to the BCM elastic-plastic bonds, it is a question of physical nonlinearity, i.e. change in the elastic properties of the material along the course of the processes occurring in the material of the structure under the action and vibrations. The algorithm implemented to date on a computer does not allow us to calculate the effect of elastic-plastic bonds upon their operation on the shape of the structure's oscillations. This is a very difficult task. The algorithm of elastic-plastic bonds is realized so far only for the connections between the base and the structure. However, for BCM not only such seismic protection is applicable, but also the introduction of the EPD between separate buildings. Given this lack of software, in this study had to resort to a number of simplified calculation schemes.

The first attempt to apply elastic-plastic bonds in the design of the BCM imposes certain limitations on the methods of numerical studies of their work in the design of the BCM. Topical methods of calculation for seismic loads are the Spectral method of SP and the method of calculation by accelerograms. The spectral method is widely accepted in computational construction practice, however it extends only to systems with pre-predictable first forms, and therefore allows to register the parameters of the system only with respect to these modes of oscillation. The method of calculating accelerograms by means of a computer allows us to more finely combine the characteristics of the system by more eigenvalues of frequencies and shapes with the accelerogram of a particular earthquake, which makes this method suitable for elucidating the contribution of the work of physically-nonlinear elements to system accelerations in various modes of vibration.

4 Mathematical Model of the EPD

For calculations at the initial stage of 3D EPD design, the bilinear approximation of the force-displacement force diagram is usually used, with the linear-elastic discharge condition (Gerstner's law) being observed.

The force of the 3D EPD triggering along the selected direction - Pt can be determined on the assumption that the 3D EPD is a rigid-plastic structure and Pt is the ultimate force. The solution of the rigid-plastic problem in this case is supposed to be found by the pseudo-rigidity method. In this method, the rigid-plastic system is replaced by an equivalent (with respect to the distribution of internal forces) elastic system with specially constructed rigidity (pseudo-rigidities). To find the pseudo-rigidities and the maximum load, an iterative algorithm is developed. This algorithm is implemented in the PC "Stark" and "MicroFe" - the "limit analysis" option.

To more accurately solve the problem of the dynamics of objects located on 3D EPD, the use of detailed 3D EPD models is required. These models should take into account the incremental nature of the laws of plastic loading and the geometric nonlinearity of the deformation process of 3D EPD. Analysis of 3D EPD deformation processes using accurate models shows that for the design calculations, it is sufficient to assume that the power characteristics of the 3D EPD are mutually independent in different directions (that is, we can assume that the power characteristics of the 3D EPD are functions of one coordinate). A significantly more important task of the calculation is the need to take into account the influence of the nonlinearity factor of the 3D PSD characteristics on the parameters of elastic oscillations of a building standing on such supports. In other words, the linearization of the power characteristics of the 3D EPD is inadmissible.

To take into account the nonlinearity factor, Professor Rutman developed a generalized method of principal coordinates, implemented in the calculation software complexes "Stark" and "Ing +".

4.1 The Calculation Scheme

As a study, the calculation is considered of a two-tower BCM with a height of 160 m, 40 floors. The dimensions of the towers are 16 × 54 m, the distance between the towers is 24 m. In the design model 2, the BCM hulls are connected by a 6-cross link system with horizontal spacers. The estimated grossness of the district is 6 points. Calculation by the spectral theory gives a trigger force in the first level of 4000 KH. The cross section of each of the metal bonds is 200 cm^2.

To consider the effect of the work of elastic-plastic bonds, we shall calculate the effect of the work of the bond lattice as external bonds. To do this, we use the symmetry property of the system. In the skew-symmetric seismic horizontal action, the symmetric components are equal to zero, the effect of the work of the 6 belts of the cross bonds is replaced by the 6-moment couplings of equivalent stiffness along the axis of their symmetry. By simple transformations, we obtain that the rigidity of the moment coupling Rz with the slope of 45° is equal to

$$Rz = 8\,E\,F, \tag{1}$$

$$F = N_{cp}/\mathrm{R_y}, \tag{2}$$

where N_{cp} is the effort of the tripping of the brace, namely, such an effort in each particular diagonal, at which it passes into the plastic stage of its work, R_y – the yield strength of steel.

Then the moment of operation of the equivalent torque coupling is

$$M_{cp} = 8N_{cp} \tag{3}$$

Similarly, we can find the remaining equivalent rigidity of the moment constraints (Table 2).

Table 2. Stiffness and operating forces

Elements	Forces	Strength	Cross-sectional area	Modulus of elasticity	Stiffness	Stiffness	Operating force
Unit		kN/m^2	cm^2	kN/m^2	EA	Rz = 8EA	kN*m
Spacer	2355	240000	0,0098	206000000	2021375	16171000	18840
Diagonal coupling	13000	240000	0,0542	206000000	11158333		
Spacer	586	240000	0,0024	206000000	502983	4023867	4688
Diagonal coupling	15500	240000	0,0646	206000000	13304167		
Spacer	350	240000	0,0015	206000000	300417	2403333	2800

As is known, cross stiffeners and a regular bond lattice gives the building a high rigidity under wind action, and, as a consequence, increases its own frequencies and inertial forces during seismic action. As a more vivid manifestation, the behavior of a building with parameters similar to the previous ones with a connected triangular lattice, as a single cross stiffener, and a regular lattice in height is considered. As a design scheme, one tower is adopted, where in the places of attachment of diagonal stiffenings, boundary conditions are imbedded that simulate rigidities with variable rigidity. The horizontal spacer, typical for a triangular lattice, expressed in the calculation scheme by a horizontal superimposed stiffening connection, is excluded from the design scheme, because is a symmetric factor in the skew-symmetric action. Buildings complex macrostructure can improve the functional quality and improve the fire safety of high-rise buildings. But the main thing is that the comfort of high-rise buildings improves due to the decrease in the acceleration of wind oscillations. When calculating the BCM, as well as any high-rise buildings, the calculation of the static and dynamic load on the wind and the calculation of the seismic effect are decisive. The primary preparatory calculation of such buildings is the modal analysis of natural frequencies and forms of construction, as the basis for any dynamic calculation. With an approximate calculation, it is possible to use the method of successive loading and consideration of the problem with examples of simple static circuits (trusses, frames). A separate calculation is possible for horizontal (wind) and vertical loads as well as the use of symmetry.

4.2 Calculation of Wind Load

The structure of BCM as a subtype of high-rise buildings is largely dictated by high demands of rigidity and instability. Required requirements for the calculation of the average and pulsation wind load is the maximum deflection of the building's milestone equal to H/500–H/1000 and also the sanitary requirement of the acceleration limit equal to 0.1 ms^2. However, varying the stiffness of the BCS bonds and the formation of the cross lattice makes it possible to guarantee this rigidity. The reverse side of the high rigidity of the BCM is from vulnerability to seismic action.

4.3 Calculation of Seismic Loads

When calculating buildings of a complex macrostructure, we are dealing with a system whose oscillations are not characterized by the first forms, since this system is most often not brought to the cantilever rod. The building of a complex macrostructure consisting of two or more towers, united by different connections, has a very complex history of oscillations depending on the perturbation, which limits the application of the spectral method of SP (Russian Norm).

Consider the practical task of using EPD in a high-rise building of a complex macrostructure (BCM) to reduce seismic inertial forces by the example of introducing a linear EPD into the cross connection between the hulls of a two-point BCM with a height of 160 m.

In the 2006 version of the "MicroFE" software package, seismic calculation of structures using accelerograms is implemented, provided that the elastic-plastic bonds (EPD) are external. Therefore, taking into account the oblique symmetry of the problem, we turn to the consideration of a separate tower, where the bond cross lattice is replaced by external moment bonds, applied in the middle of the height of the panels (Fig. 3).

At the lateral jolt, the compressed connections of great flexibility are turned off, and the system actually operates as a truss with a one-sided ascending grid. This makes it possible to take into account the rigidity of only stretched braces.

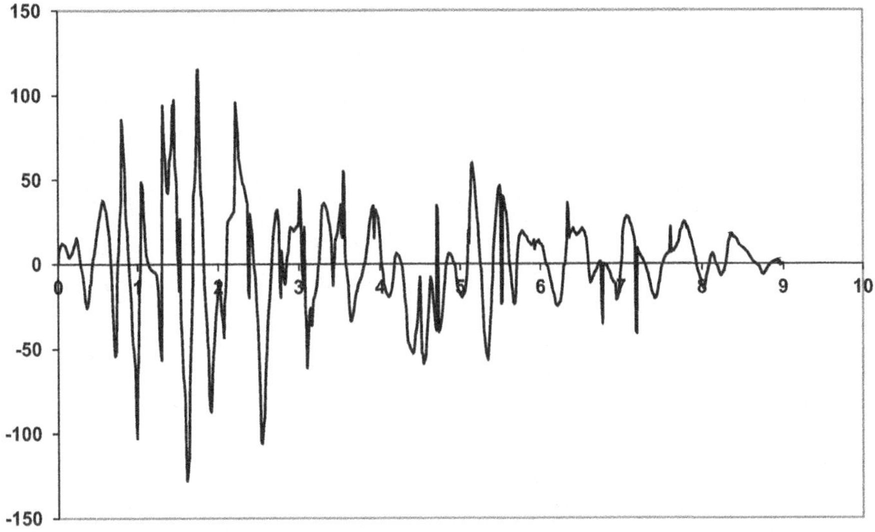

Fig. 2. Accelerogram of the earthquake “Holister” (USA) from 9.03.1949, making G-21, indicating the intensive section. Step of digitization 0,01 s/s^2

When the communication is working, a panel shift occurs at an angle ϕ (Fig. 5), then the absolute longitudinal elongation or shortening of the braces is equal to

$$e = \Delta \cdot \cos \alpha \tag{4}$$

where $\boldsymbol{e}$ is the absolute longitudinal elongation or shortening of the braces,

Δ is the magnitude of the absolute horizontal shift of the panel, α is the angle of inclination of the brace.

Axial relative deformation of the brace:

$$\varepsilon = \frac{e}{L} = \frac{\varphi \cdot h \cdot \cos \alpha}{h/\cos \alpha} = \varphi \cdot \cos^2 \alpha \tag{5}$$

where $\boldsymbol{e}$ is the absolute longitudinal elongation or shortening of the braces, $\boldsymbol{L}$ is the absolute length of the diagonal coupling, $\boldsymbol{\alpha}$ is the angle of inclination of the brace,

$\boldsymbol{h}$ is the height of the bond panel, $\boldsymbol{\varphi}$ is the angle of shear of the panel.

The force in the panel brace is:

$$N_p = \varepsilon \cdot E \cdot F_P = E \cdot F_P \cdot \varphi \cdot \cos^2 \alpha \tag{6}$$

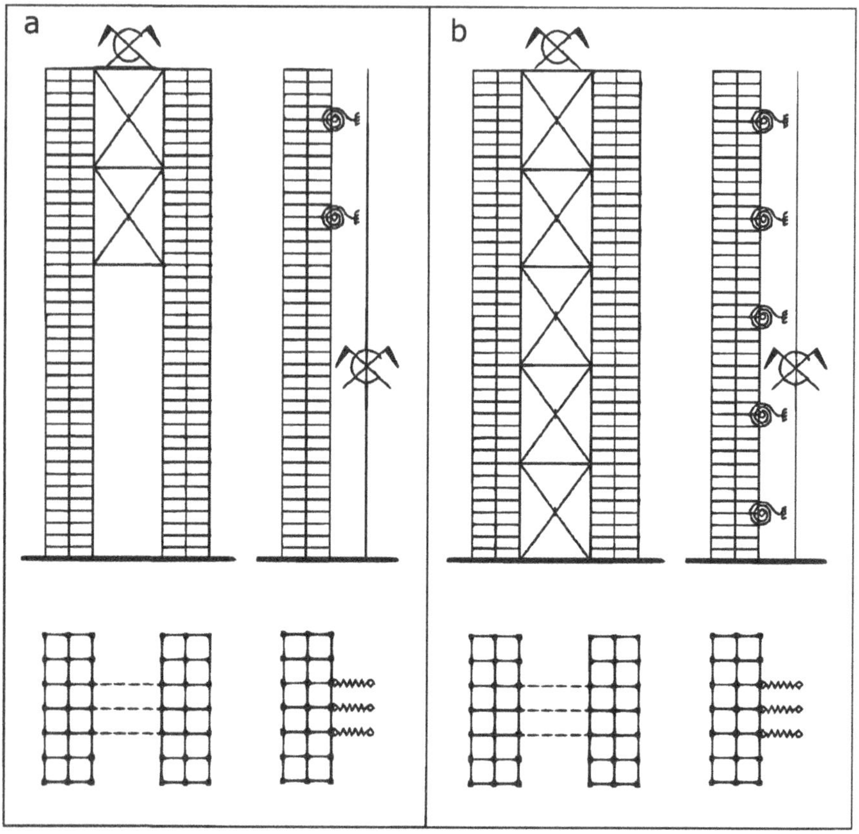

Fig. 3. A building: a - with a bridge; b - with connections

Then for a unit angle φ:

$$N_p = E \cdot F_p \cdot \cos^2 \alpha \quad (7)$$

A pair of forces gives a reactive moment on the shoulder h, then if the force in the belts:

$$N_{pr} = N_p \cdot \cos \alpha \quad (8)$$

The overall stiffness of the equivalent torque Rφ regard to both the tower is:

$$R_\varphi = E \cdot F_p \cdot \cos^3 \alpha \cdot h \quad (9)$$

However, taking into account the symmetry, the moment of operation of the equivalent torque link Msp for one building is:

$$M_{\text{cp}} = \frac{E \cdot F_{\text{p}} \cdot \cos^3 \alpha \cdot h}{2} = \frac{N_{\text{cp}} \cdot h \cdot \cos \alpha}{2} \tag{10}$$

For half of the building under consideration h = 24 m and the angle α = 45 the time response equivalent torque bracing is:

$$M_{\text{cp}} = 8 \cdot N_{\text{cp}} \tag{11}$$

By simple transformations we obtain that the total rigidity of the torque bracing Rz at an inclination of 45° connections and a height of 24 m panels for the two buildings is:

$$\text{R}_{\text{z}} = 8 \cdot E \cdot F \tag{12}$$

where ***E*** is the modulus of elasticity of steel, and ***F*** is the cross-sectional area of the diagonal stiffness bond.

In this case, the cross-sectional area is equal to the condition of the elastic stage of steel operation:

$$F = \frac{N_{\text{cp}}}{R_y} \tag{13}$$

where N_{cp} is the bias trigger force, that is, such force in each particular diagonal stiffness connection at which it passes into the plastic stage of its work, R_y is the yield strength of the steel.

Similarly, we can find the remaining equivalent rigidity of the moment constraints.

5 Results

In the framework of the BCM study, calculations were made for wind and seismic loads from the 6-point earthquake according to Holister's accelerogram (Fig. 2), a spatial model of a 40-storey building with a height of 160 m.

Two calculations have been made: a building with a double cross stiffness connection at the level of 2/3 of the height of the building (Fig. 3a) and a building with a tie grating, regular along the entire height of the building (Fig. 3b). The first scheme, when varying the rigidity of the EPD, showed the acceleration reduction zone to the values of a separately standing building, with a certain selection of the stiffness values and the force of switching on the EPD.

The selection of the stiffnesses showed the possibility of switching off the connections and the operation of the EPD as plastic at 500 kN • m (see Table 3). In this case, it is possible to reduce the acceleration from 0.41 m/s^2 to 0.27 m/s^2, equal to the acceleration of a single free tower.

The analysis showed that the double cross link established in the zone of 2/3 of the height of the building, with varying rigidity, showed the zone of reduction of accelerations to the values of a separately standing building with a certain choice of the stiffness and effort of inclusion of the EPD (see Table 3). The calculation showed that for a double cross bridge, and the rigidity is 4 times lower than the effort expected from the calculation of the transition from a double settlement scheme to a single building with instantaneous connections.

Table 3. Acceleration, speed and movement of 2-Tower-BCM with 2 EPD

Acceleration, speed and movement with 2 EPD					
Parameter of system/Type of building		1 tower	2 towers + EPD	2 towers + EPD	2 towers + EPD
Stiffness × 10^6, Operating force. × 10^2, kN		–	20,20	10.10	5.5
Acceleration	a, m/s^2	0.27	0.41	0.33	0.27
Speed	v, m/s	0.32	0.22	0.28	0.32
Movement	f, m	0.25	0.12	0.21	0.25

In the second scheme, the selection of the actuation and rigidity forces from the calculation by the spectral theory showed variable forces that decrease in a manner analogous to the values of the transverse forces in the composite rod.

The selection of the stiffnesses showed the possibility of deactivating the connections and the operation of the EPD as plastic at the actuation forces of 100–600 kN • m (see Table 4). In this case, it is possible to reduce the acceleration from 0.61 m/s^2 to 0.48 m/s^2, equal to the acceleration of the freestanding building (Fig. 4).

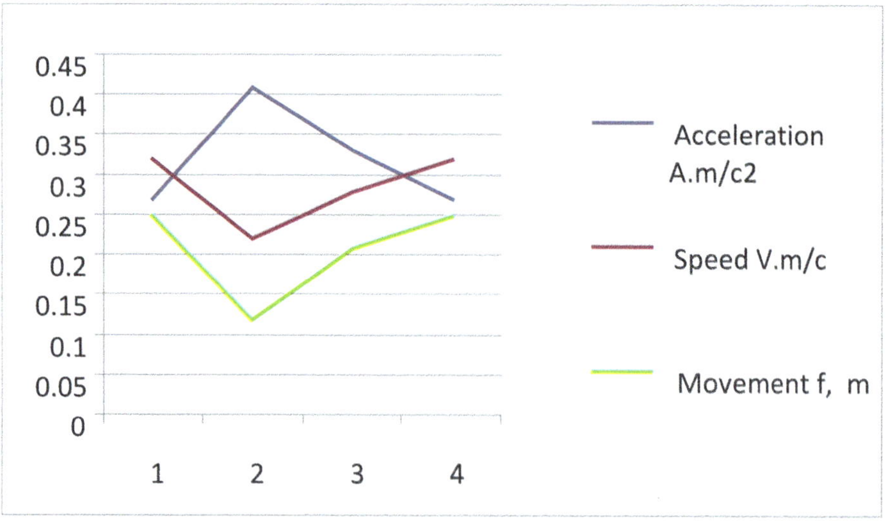

Fig. 4. Diagram of the dependence of acceleration, speed and movement from the rigidity of the EPD and the operating force

Table 4. Acceleration, speed and movement of 2-Tower-BCM with 5 EPD

Acceleration, speed and movement with 5 EPD					
Parameter of system/Type of building	Tower	2 towers EPD	2 towers EPD	2 towers EPD	2 towers EPD
Stiffness $\times$ 10^6, Operating Force. $\times$ 10^2, kN	–	$2–12\times10^7$	$1–6\times10^7$	$3–6\times10^6$	$1–18\times10^6$
Acceleration a, m/s^2	0,48	0.61	0.61	0.52	0.48
Speed v, m/s	0,27	0.21	0.17	0.32	0.27
Movement f, m	0,22	0.14	0.15	0.24	0.22

Calculations showed that the superposition of elastic-plastic bonds can reduce the acceleration by 30%, bringing it to a complete shutdown of the bonds to the values of a separately-standing building.

This study confirms that the imposition of elastic-plastic bonds can have a positive effect, but their application is only permissible when searching through various options for the relationship between the rigidity of the building and the connections based on a numerical calculation on the PC.

Regular links with the EPD did not show a positive effect, because when vibrations occur as a result of seismic activity, a number of connections involved in the entire height in the grid do not produce simultaneous shutdown of the simultaneous but act as alternately switched off, which at this stage of the research does not lead to a significant positive effect of the application of the EPD and a reduction in accelerations of more than 5–10%.

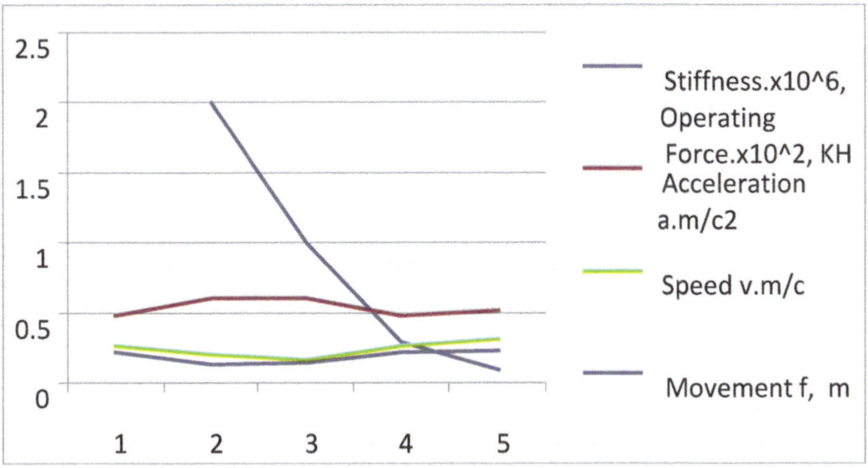

Fig. 5. Diagram of the dependence of acceleration, speed and movement from the rigidity of the EPD and the operating force

Recommendations for the rational design of BCM When building a building, give it as much as possible equal to the moments of inertia relative to the two mutually perpendicular axes of the BCM. With a strong difference, they are affected by torsional vibration. When assigning the section of the hulls and the distances between them, it is recommended to place individual bodies not more than 1–1.5 of the body diameter, other arrangement makes the jumper too flexible and the system is malleable [12].

With the prevailing wind effect, it is desirable to place the stiffeners in regular intervals in height. With the prevailing seismic action, it is recommended to use a single jumper with a moment of inertia of at least 20% of the moment of inertia of a separate hull, to create a minimum rigidity to guarantee deflection from the wind pulsating component of the wind load and then to investigate the building for seismic. If the jumper is insufficiently compliant, it is possible to use the EPD design Rutman by special calculation.

6 Conclusions

These results, of course, are still a scientific hypothesis and should be tested by a single calculating engineer before they can be accepted as a proven theory. During the confirmation of defense of the thesis in Moscow, one of the members of the Commission from CNIISK confirmed that such calculations were made for high-rise buildings in Moscow. Can this oscillation damping scheme compete with the vibration protection of buildings based on hydraulic cushions in the pillars of foundations [9]? Can it compete with pendulum dampers? Most likely yes, since the pendulum dampers are more high-frequency and are designed for wind loads, besides bulky.

These technologies have long been tested by projects and time. However, the theory of pseudo-rigidities and the suggestion of the designs of the EPD for the scientist, engineer, professor Rutman opens new possibilities for vibration protection of buildings and allows developing this method of damping accelerations and reducing inertial forces from vibro-taps, as promising.

References

1. Pletnev, V.I.: An effective method for calculating multi-storey buildings using discrete-continual models and continuum superelements. Ph.D. thesis, St. Petersburg (1995)
2. Rutman, Y.L.: Buildings of Complex Macrostructure. L.L.Academic Publishing, Saarbrücken (2011)
3. Rutman, Y.L.: The pseudo-rigidity method for solving problems on the limiting equilibrium of rigid-plastic constructions, St. Petersburg (1998)
4. Rutman, Y.L.: Aerodynamic instability of buildings and structures in the wind flow. Modern methods of calculation. Macro-structures. L.L.Academic Publishing, Saarbrücken (2012)

Author Index

Z. Popovic et al. (Eds.): TransSiberia 2019, AISC 1115, pp. 1019–1022, 2020.
https://doi.org/10.1007/978-3-030-37916-2

Y

Z

The manufacturer's authorised representative in the EU is Springer Nature Customer Service Centre GmbH, Europaplatz 3, 69115 Heidelberg, Germany. If you have any concerns regarding our products, please contact ProductSafety@springernature.com

Printed and bound by CPI Group (UK) Ltd, Croydon, CR0 4YY

30/09/2026

02225441-0004